D1236746

THE PROPERTIES OF GASES AND LIQUIDS

Bruce E. Poling
Professor of Chemical Engineering
University of Toledo

John M. Prausnitz
Professor of Chemical Engineering
University of California at Berkeley

John P. O'Connell
Professor of Chemical Engineering
University of Virginia

Fifth Edition

McGRAW-HILL
New York San Francisco Washington, D.C. Auckland Bogotá
Caracas Lisbon London Madrid Mexico City Milan
Montreal New Delhi San Juan Singapore
Sydney Tokyo Toronto

Library of Congress Cataloging-in-Publication Data

Poling, Bruce E.
The properties of gases and liquids / Bruce E. Poling, John M. Prausnitz, John P. O'Connell.—5th ed.
p. cm.
Includes bibliographical references and index.
ISBN 0-07-011682-2
1. Gases. 2. Liquids. I. Prausnitz, J. M. II. O'Connell, John P. (John Paul) III. Title.

TP242.P62 2000
660'.042—dc21 00-061622

McGraw-Hill

A Division of The ***McGraw·Hill*** *Companies*

1 2 3 4 5 6 7 8 9 0 DOC/DOC 0 6 5 4 3 2 1 0

ISBN 0-07-011682-2

The sponsoring editor for this book was Kenneth P. McCombs and the production supervisor was Sherri Souffrance. It was set in Times Roman by Pro-Image Corporation.

Printed and bound by R. R. Donnelley & Sons Company.

This book is printed on recycled, acid-free paper containing a minimum of 50% recycled de-inked fiber.

CONTENTS

PREFACE

Reliable values of the properties of materials are necessary for the design of industrial processes. An enormous amount of data has been collected and correlated over the years, but the rapid advance of technology into new fields seems always to maintain a significant gap between demand and availability. The engineer is still required to rely primarily on common sense, experience, and a variety of methods for estimating physical properties.

This book presents a critical review of various estimation procedures for a limited number of properties of gases and liquids: critical and other pure component properties; *PVT* and thermodynamic properties of pure components and mixtures; vapor pressures and phase-change enthalpies; standard enthalpies of formation; standard Gibbs energies of formation; heat capacities; surface tensions; viscosities; thermal conductivities; diffusion coefficients; and phase equilibria. For most cases, estimated properties are compared to experiment to indicate reliability. Most methods are illustrated by examples.

The procedures described are necessarily limited to those that appear to the authors to have the greatest validity and practical use. Wherever possible, we have included recommendations delineating the best methods for estimating each property and the most reliable techniques for extrapolating or interpolating available data.

Although the book is intended to serve primarily the practicing engineer, especially the process or chemical engineer, other engineers and scientists concerned with gases and liquids may find it useful.

The first edition of this book was published in 1958, the second in 1966, the third in 1977 and the fourth in 1987. In a sense, each edition is a new book because numerous estimation methods are proposed each year; over a (roughly) 10-year span, many earlier methods are modified or displaced by more accurate or more general techniques. While most estimation methods rely heavily on empiricism, the better ones—those that are most reliable—often have a theoretical basis. In some cases, the theory is outlined to provide the user with the foundation of the proposed estimation method.

There are some significant differences between the current edition and the preceding one:

1. Chapter 2 includes several extensive new group-contribution methods as well as discussion and comparisons of methods based on descriptors calculated with quantum-mechanical methods. Direct comparisons are given for more than 200 substances with data in Appendix A.
2. Chapter 3 includes several new methods as well as updated Benson-Method tables for ideal-gas properties of formation and heat capacities. Direct comparisons are given for more than 100 substances with data in Appendix A.
3. Chapter 4 includes presentation of current equations of state for pure components with complete formulae for many models, especially cubics. A new sec-

tion discusses issues associated with near-critical and very high pressure systems. The Lee–Kesler corresponding-states tables, readily available elsewhere, have been removed.

4. Chapter 5 includes presentation of current equations of state for mixtures with complete formulae for many models, especially cubics. A new section discusses current mixing and combining rules for equation-of-state parameters with attention to inconsistencies.
5. Chapter 6 includes a revised introduction to thermodynamic properties from equations of state with complete formulae for cubics. A new section discusses real-gas and liquid heat capacities. Because they are readily available elsewhere, the Lee–Kesler corresponding-states tables have been removed.
6. Chapter 7 gives attention to one form of the Wagner equation that appears to be particularly successful for representing vapor pressures, and to the useful tables of Majer and Svoboda for enthalpies of vaporization. Also included is a new discussion of the entropy of fusion.
7. Chapter 8 has been extended to include discussion of systems containing solids, a new correlation by Eckert et al. for activity coefficents at infinite dilution, and some new methods for high-pressure vapor-liquid equilibria, including those based on Wong–Sandler mixing rules.
8. In Chapters 9–12, most of the new methods for transport properties are based on thermodynamic data or molecular-thermodynamic models. The successful TRAPP method (from the National Institute of Science and Technology) is now explained in more detail.
9. The property data bank in Appendix A has been completely revised. Most of the properties are the same as in the last edition, but the format has been changed to identify the sources of values. The introduction to Appendix A describes the definitions and font usage of the data bank.

   We selected only those substances for which we could readily obtain an evaluated experimental critical temperature; the total number of compounds is fewer than in the last edition. All of the entries in Appendix A were taken from tabulations of the Thermodynamics Research Center (TRC), College Station, TX, USA, or from other reliable sources as listed in the Appendix. We also used experimentally-based results for other properties from the same sources whenever available. Some estimated values are also included.

   We tabulate the substances in alphabetical formula order. IUPAC names are used, with some common names added, and Chemical Abstracts Registry numbers are given for each compound. We indicate origins of the properties by using different fonts. We are grateful to TRC for permitting us to publish a significant portion of their values.
10. Appendix C presents complete tables of parameters for the multi-property group-contribution methods of Joback and of Constantinou and Gani.

The authors want to acknowledge with thanks significant contributions from colleagues who provided assistance in preparing the current edition; their help has been essential and we are grateful to them all: David Bush, Joe Downey, Charles Eckert, Michael Frenkel, Rafiqui Gani and students of the CAPEC Center at the Technical University of Denmark, Lucinda Garnes, Steven Garrison, Nathan Erb, K. R. Hall, Keith Harrison, Marcia Huber, Kevin Joback, Kim Knuth, Claude Leibovicci, Paul Mathias, Amy Nelson, Van Nguyen, Chorng Twu, Philippe Ungerer and Randolph Wilhoit.

For her patient and devoted service in performing numerous editorial tasks, we give special thanks to Nanci Poling. We are grateful to Nanci and also to Verna O'Connell and Susan Prausnitz for their encouragement and support during this project.

While we regret that the original author, Robert Reid, elected not to participate in the preparation of this edition, we nevertheless want to record here our gratitude to him for his pioneering leadership in establishing and collecting estimation methods for physical properties of fluids as required for chemical process and product design.

B. E. Poling
J. M. Prausnitz
J. P. O'Connell

CHAPTER ONE
THE ESTIMATION OF PHYSICAL PROPERTIES

1-1 INTRODUCTION

The structural engineer cannot design a bridge without knowing the properties of steel and concrete. Similarly, scientists and engineers often require the properties of gases and liquids. The chemical or process engineer, in particular, finds knowledge of physical properties of fluids essential to the design of many kinds of products, processes, and industrial equipment. Even the theoretical physicist must occasionally compare theory with measured properties.

The physical properties of every substance depend directly on the nature of the molecules of the substance. Therefore, the ultimate generalization of physical properties of fluids will require a complete understanding of molecular behavior, which we do not yet have. Though its origins are ancient, the molecular theory was not generally accepted until about the beginning of the nineteenth century, and even then there were setbacks until experimental evidence vindicated the theory early in the twentieth century. Many pieces of the puzzle of molecular behavior have now fallen into place and computer simulation can now describe more and more complex systems, but as yet it has not been possible to develop a complete generalization.

In the nineteenth century, the observations of Charles and Gay-Lussac were combined with Avogadro's hypothesis to form the gas "law," $PV = NRT$, which was perhaps the first important correlation of properties. Deviations from the ideal-gas law, though often small, were finally tied to the fundamental nature of the molecules. The equation of van der Waals, the virial equation, and other equations of state express these quantitatively. Such extensions of the ideal-gas law have not only facilitated progress in the development of a molecular theory but, more important for our purposes here, have provided a framework for correlating physical properties of fluids.

The original "hard-sphere" kinetic theory of gases was a significant contribution to progress in understanding the statistical behavior of a system containing a large number of molecules. Thermodynamic and transport properties were related quantitatively to molecular size and speed. Deviations from the hard-sphere kinetic theory led to studies of the interactions of molecules based on the realization that molecules attract at intermediate separations and repel when they come very close. The semiempirical potential functions of Lennard-Jones and others describe attraction and repulsion in approximately quantitative fashion. More recent potential functions allow for the shapes of molecules and for asymmetric charge distribution in polar molecules.

Although allowance for the forces of attraction and repulsion between molecules is primarily a development of the twentieth century, the concept is not new. In about 1750, Boscovich suggested that molecules (which he referred to as atoms) are "endowed with potential force, that any two atoms attract or repel each other with a force depending on their distance apart. At large distances the attraction varies as the inverse square of the distance. The ultimate force is a repulsion which increases without limit as the distance decreases without limit, so that the two atoms can never coincide" (Maxwell 1875).

From the viewpoint of mathematical physics, the development of a comprehensive molecular theory would appear to be complete. J. C. Slater (1955) observed that, while we are still seeking the laws of nuclear physics, "in the physics of atoms, molecules and solids, we have found the laws and are exploring the deductions from them." However, the suggestion that, in principle (the Schrödinger equation of quantum mechanics), everything is known about molecules is of little comfort to the engineer who needs to know the properties of some new chemical to design a commercial product or plant.

Paralleling the continuing refinement of the molecular theory has been the development of thermodynamics and its application to properties. The two are intimately related and interdependent. Carnot was an engineer interested in steam engines, but the second law of thermodynamics was shown by Clausius, Kelvin, Maxwell, and especially by Gibbs to have broad applications in all branches of science.

Thermodynamics by itself cannot provide physical properties; only molecular theory or experiment can do that. But thermodynamics reduces experimental or theoretical efforts by relating one physical property to another. For example, the Clausius-Clapeyron equation provides a useful method for obtaining enthalpies of vaporization from more easily measured vapor pressures.

The second law led to the concept of chemical potential which is basic to an understanding of chemical and phase equilibria, and the Maxwell relations provide ways to obtain important thermodynamic properties of a substance from $PVTx$ relations where x stands for composition. Since derivatives are often required, the $PVTx$ function must be known accurately.

The Information Age is providing a "shifting paradigm in the art and practice of physical properties data" (Dewan and Moore, 1999) where searching the World Wide Web can retrieve property information from sources and at rates unheard of a few years ago. Yet despite the many handbooks and journals devoted to compilation and critical review of physical-property data, it is inconceivable that all desired experimental data will ever be available for the thousands of compounds of interest in science and industry, let alone all their mixtures. Thus, in spite of impressive developments in molecular theory and information access, the engineer frequently finds a need for physical properties for which no experimental data are available and which cannot be calculated from existing theory.

While the need for accurate design data is increasing, the rate of accumulation of new data is not increasing fast enough. Data on multicomponent mixtures are particularly scarce. The process engineer who is frequently called upon to design a plant to produce a new chemical (or a well-known chemical in a new way) often finds that the required physical-property data are not available. It may be possible to obtain the desired properties from new experimental measurements, but that is often not practical because such measurements tend to be expensive and time-consuming. To meet budgetary and deadline requirements, the process engineer almost always must estimate at least some of the properties required for design.

1-2 ESTIMATION OF PROPERTIES

In the all-too-frequent situation where no experimental value of the needed property is at hand, the value must be estimated or predicted. "Estimation" and "prediction" are often used as if they were synonymous, although the former properly carries the frank implication that the result may be only approximate. Estimates may be based on theory, on correlations of experimental values, or on a combination of both. A theoretical relation, although not strictly valid, may nevertheless serve adequately in specific cases.

For example, to relate mass and volumetric flow rates of air through an air-conditioning unit, the engineer is justified in using $PV = NRT$. Similarly, he or she may properly use Dalton's law and the vapor pressure of water to calculate the mass fraction of water in saturated air. However, the engineer must be able to judge the operating pressure at which such simple calculations lead to unacceptable error.

Completely empirical correlations are often useful, but one must avoid the temptation to use them outside the narrow range of conditions on which they are based. In general, the stronger the theoretical basis, the more reliable the correlation.

Most of the better estimation methods use equations based on the form of an incomplete theory with empirical correlations of the parameters that are not provided by that theory. Introduction of empiricism into parts of a theoretical relation provides a powerful method for developing a reliable correlation. For example, the van der Waals equation of state is a modification of the simple $PV = NRT$; setting $N = 1$,

$$\left(P + \frac{a}{V^2}\right)(V - b) = RT \tag{1-2.1}$$

Equation (1-2.1) is based on the idea that the pressure on a container wall, exerted by the impinging molecules, is decreased because of the attraction by the mass of molecules in the bulk gas; that attraction rises with density. Further, the available space in which the molecules move is less than the total volume by the excluded volume b due to the size of the molecules themselves. Therefore, the "constants" (or parameters) a and b have some theoretical basis though the best descriptions require them to vary with conditions, that is, temperature and density. The correlation of a and b in terms of other properties of a substance is an example of the use of an empirically modified theoretical form.

Empirical extension of theory can often lead to a correlation useful for estimation purposes. For example, several methods for estimating diffusion coefficients in low-pressure binary gas systems are empirical modifications of the equation given by the simple kinetic theory for non-attracting spheres. Almost all the better estimation procedures are based on correlations developed in this way.

1-3 TYPES OF ESTIMATION

An ideal system for the estimation of a physical property would (1) provide reliable physical and thermodynamic properties for pure substances and for mixtures at any temperature, pressure, and composition, (2) indicate the phase (solid, liquid, or gas), (3) require a minimum of input data, (4) choose the least-error route (i.e., the best estimation method), (5) indicate the probable error, and (6) minimize computation time. Few of the available methods approach this ideal, but some serve remarkably well. Thanks to modern computers, computation time is usually of little concern.

In numerous practical cases, the most accurate method may not be the best for the purpose. Many engineering applications properly require only approximate estimates, and a simple estimation method requiring little or no input data is often preferred over a complex, possibly more accurate correlation. The simple gas law is useful at low to modest pressures, although more accurate correlations are available. Unfortunately, it is often not easy to provide guidance on when to reject the simpler in favor of the more complex (but more accurate) method; the decision often depends on the problem, not the system.

Although a variety of molecular theories may be useful for data correlation, there is one theory which is particularly helpful. This theory, called the law of corresponding states or the corresponding-states principle, was originally based on macroscopic arguments, but its modern form has a molecular basis.

The Law of Corresponding States

Proposed by van der Waals in 1873, the law of corresponding states expresses the generalization that equilibrium properties that depend on certain intermolecular forces are related to the critical properties in a universal way. Corresponding states provides the single most important basis for the development of correlations and estimation methods. In 1873, van der Waals showed it to be theoretically valid for all pure substances whose *PVT* properties could be expressed by a two-constant equation of state such as Eq. (1-2.1). As shown by Pitzer in 1939, it is similarly valid if the intermolecular potential function requires only two characteristic parameters. Corresponding states holds well for fluids containing simple molecules and, upon semiempirical extension with a single additional parameter, it also holds for "normal" fluids where molecular orientation is not important, i.e., for molecules that are not strongly polar or hydrogen-bonded.

The relation of pressure to volume at constant temperature is different for different substances; however, two-parameter corresponding states theory asserts that if pressure, volume, and temperature are divided by the corresponding critical properties, the function relating reduced pressure to reduced volume and reduced temperature becomes the same for all substances. The reduced property is commonly expressed as a fraction of the critical property: $P_r = P/P_c$; $V_r = V/V_c$; and $T_r = T/T_c$.

To illustrate corresponding states, Fig. 1-1 shows reduced *PVT* data for methane and nitrogen. In effect, the critical point is taken as the origin. The data for saturated liquid and saturated vapor coincide well for the two substances. The isotherms (constant T_r), of which only one is shown, agree equally well.

Successful application of the law of corresponding states for correlation of *PVT* data has encouraged similar correlations of other properties that depend primarily on intermolecular forces. Many of these have proved valuable to the practicing engineer. Modifications of the law are commonly made to improve accuracy or ease of use. Good correlations of high-pressure gas viscosity have been obtained by expressing η/η_c as a function of P_r and T_r. But since η_c is seldom known and not easily estimated, this quantity has been replaced in other correlations by other characteristics such as η_c°, η_T°, or the group $M^{1/2}P_c^{2/3}T_c^{1/6}$, where η_c° is the viscosity at T_c and low pressure, η_T° is the viscosity at the temperature of interest, again at

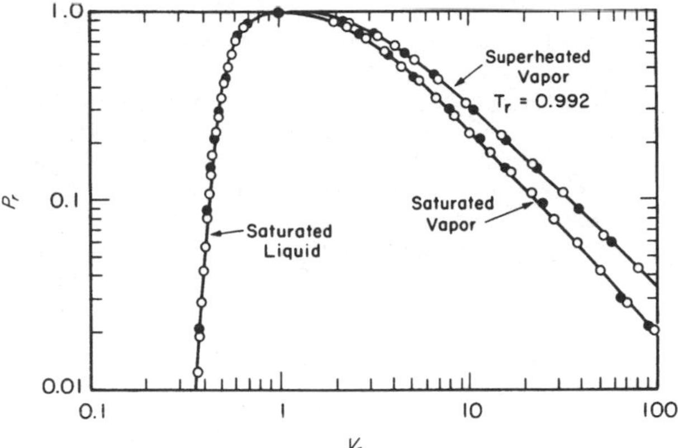

FIGURE 1-1 The law of corresponding states applied to the *PVT* properties of methane and nitrogen. Literature values (Din, 1961): ○ methane, ● nitrogen.

low pressure, and the group containing M, P_c, and T_c is suggested by dimensional analysis. Other alternatives to the use of η_c might be proposed, each modeled on the law of corresponding states but essentially empirical as applied to transport properties.

The two-parameter law of corresponding states can be derived from statistical mechanics when severe simplifications are introduced into the partition function. Sometimes other useful results can be obtained by introducing less severe simplifications into statistical mechanics to provide a more general framework for the development of estimation methods. Fundamental equations describing various properties (including transport properties) can sometimes be derived, provided that an expression is available for the potential-energy function for molecular interactions. This function may be, at least in part, empirical; but the fundamental equations for properties are often insensitive to details in the potential function from which they stem, and two-constant potential functions frequently serve remarkably well. Statistical mechanics is not commonly linked to engineering practice, but there is good reason to believe it will become increasingly useful, especially when combined with computer simulations and with calculations of intermolecular forces by computational chemistry. Indeed, anticipated advances in atomic and molecular physics, coupled with ever-increasing computing power, are likely to augment significantly our supply of useful physical-property information.

Nonpolar and Polar Molecules

Small, spherically-symmetric molecules (for example, CH_4) are well fitted by a two-constant law of corresponding states. However, nonspherical and weakly polar molecules do not fit as well; deviations are often great enough to encourage development of correlations using a third parameter, e.g., the acentric factor, ω. The acentric factor is obtained from the deviation of the experimental vapor pressure–temperature function from that which might be expected for a similar substance consisting of small spherically-symmetric molecules. Typical corresponding-states correlations express a desired dimensionless property as a function of P_r, T_r, and the chosen third parameter.

Unfortunately, the properties of strongly polar molecules are often not satisfactorily represented by the two- or three-constant correlations which do so well for nonpolar molecules. An additional parameter based on the dipole moment has often been suggested but with limited success, since polar molecules are not easily characterized by using only the dipole moment and critical constants. As a result, although good correlations exist for properties of nonpolar fluids, similar correlations for polar fluids are often not available or else show restricted reliability.

Structure and Bonding

All macroscopic properties are related to molecular structure and the bonds between atoms, which determine the magnitude and predominant type of the intermolecular forces. For example, structure and bonding determine the energy storage capacity of a molecule and thus the molecule's heat capacity.

This concept suggests that a macroscopic property can be calculated from group contributions. The relevant characteristics of structure are related to the atoms, atomic groups, bond type, etc.; to them we assign weighting factors and then determine the property, usually by an algebraic operation that sums the contributions from the molecule's parts. Sometimes the calculated sum of the contributions is not for the property itself but instead is for a correction to the property as calculated by some simplified theory or empirical rule. For example, the methods of Lydersen and of others for estimating T_c start with the loose rule that the ratio of the normal boiling temperature to the critical temperature is about 2:3. Additive structural increments based on bond types are then used to obtain empirical corrections to that ratio.

Some of the better correlations of ideal-gas heat capacities employ theoretical values of C_p° (which are intimately related to structure) to obtain a polynomial expressing C_p° as a function of temperature; the constants in the polynomial are determined by contributions from the constituent atoms, atomic groups, and types of bonds.

1-4 ORGANIZATION OF THE BOOK

Reliable experimental data are always to be preferred over results obtained by estimation methods. A variety of tabulated data banks is now available although many of these banks are proprietary. A good example of a readily accessible data bank is provided by DIPPR, published by the American Institute of Chemical Engineers. A limited data bank is given at the end of this book. But all too often reliable data are not available.

The property data bank in Appendix A contains only substances with an evaluated experimental critical temperature. The contents of Appendix A were taken either from the tabulations of the Thermodynamics Research Center (TRC), College Station, TX, USA, or from other reliable sources as listed in Appendix A. Substances are tabulated in alphabetical-formula order. IUPAC names are listed, with some common names added, and Chemical Abstracts Registry numbers are indicated.

In this book, the various estimation methods are correlations of experimental data. The best are based on theory, with empirical corrections for the theory's defects. Others, including those stemming from the law of corresponding states, are based on generalizations that are partly empirical but nevertheless have application to a remarkably wide range of properties. Totally empirical correlations are useful only when applied to situations very similar to those used to establish the correlations.

The text includes many numerical examples to illustrate the estimation methods, especially those that are recommended. Almost all of them are designed to explain the calculation procedure for a single property. However, most engineering design problems require estimation of several properties; the error in each contributes to the overall result, but some individual errors are more important that others. Fortunately, the result is often adequate for engineering purposes, in spite of the large measure of empiricism incorporated in so many of the estimation procedures and in spite of the potential for inconsistencies when different models are used for different properties.

As an example, consider the case of a chemist who has synthesized a new compound (chemical formula CCl_2F_2) that boils at −20.5°C at atmospheric pressure. Using only this information, is it possible to obtain a useful prediction of whether or not the substance has the thermodynamic properties that might make it a practical refrigerant?

Figure 1-2 shows portions of a Mollier diagram developed by prediction methods described in later chapters. The dashed curves and points are obtained from estimates of liquid and vapor heat capacities, critical properties, vapor pressure, en-

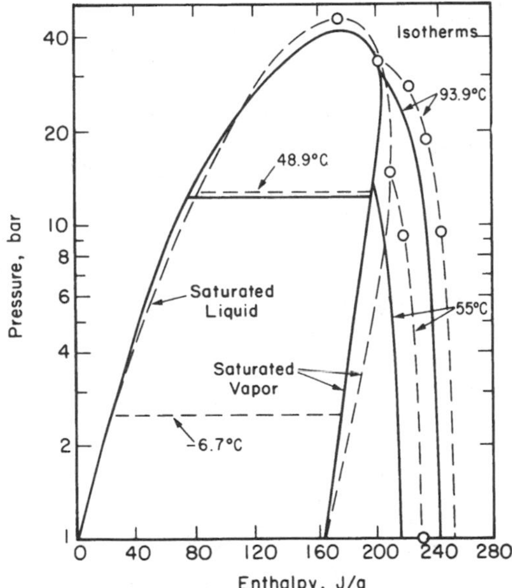

FIGURE 1-2 Mollier diagram for dichlorodifluoromethane. The solid lines represent measured data. Dashed lines and points represent results obtained by estimation methods when only the chemical formula and the normal boiling temperature are known.

thalpy of vaporization, and pressure corrections to ideal-gas enthalpies and entropies. The substance is, of course, a well-known refrigerant, and its known properties are shown by the solid curves. While environmental concerns no longer permit use of CCl_2F_2, it nevertheless serves as a good example of building a full description from very little information.

For a standard refrigeration cycle operating between 48.9 and −6.7°C, the evaporator and condenser pressures are estimated to be 2.4 and 12.4 bar, vs. the known values 2.4 and 11.9 bar. The estimate of the heat absorption in the evaporator checks closely, and the estimated volumetric vapor rate to the compressor also shows good agreement: 2.39 versus 2.45 m^3/hr per kW of refrigeration. (This number indicates the size of the compressor.) Constant-entropy lines are not shown in Fig. 1-2, but it is found that the constant-entropy line through the point for the low-pressure vapor essentially coincides with the saturated vapor curve. The estimated coefficient of performance (ratio of refrigeration rate to isentropic compression power) is estimated to be 3.8; the value obtained from the data is 3.5. This is not a very good check, but it is nevertheless remarkable because the only data used for the estimate were the normal boiling point and the chemical formula.

Most estimation methods require parameters that are characteristic of single pure components or of constituents of a mixture of interest. The more important of these are considered in Chap. 2.

The thermodynamic properties of ideal gases, such as enthalpies and Gibbs energies of formation and heat capacities, are covered in Chap. 3. Chapter 4 describes the *PVT* properties of pure fluids with the corresponding-states principle, equations of state, and methods restricted to liquids. Chapter 5 extends the methods of Chap. 4 to mixtures with the introduction of mixing and combining rules as well as the special effects of interactions between different components. Chapter 6 covers other thermodynamic properties such as enthalpy, entropy, free energies and heat capacities of real fluids from equations of state and correlations for liquids. It also introduces partial properties and discusses the estimation of true vapor-liquid critical points.

Chapter 7 discusses vapor pressures and enthalpies of vaporization of pure substances. Chapter 8 presents techniques for estimation and correlation of phase equilibria in mixtures. Chapters 9 to 11 describe estimation methods for viscosity, thermal conductivity, and diffusion coefficients. Surface tension is considered briefly in Chap. 12.

The literature searched was voluminous, and the lists of references following each chapter represent but a fraction of the material examined. Of the many estimation methods available, in most cases only a few were selected for detailed discussion. These were selected on the basis of their generality, accuracy, and availability of required input data. Tests of all methods were often more extensive than those suggested by the abbreviated tables comparing experimental with estimated values. However, no comparison is adequate to indicate expected errors for new compounds. The average errors given in the comparison tables represent but a crude overall evaluation; the inapplicability of a method for a few compounds may so increase the average error as to distort judgment of the method's merit, although efforts have been made to minimize such distortion.

Many estimation methods are of such complexity that a computer is required. This is less of a handicap than it once was, since computers and efficient computer programs have become widely available. Electronic desk computers, which have become so popular in recent years, have made the more complex correlations practical. However, accuracy is not necessarily enhanced by greater complexity.

The scope of the book is inevitably limited. The properties discussed were selected arbitrarily because they are believed to be of wide interest, especially to chemical engineers. Electrical properties are not included, nor are the properties of salts, metals, or alloys or chemical properties other than some thermodynamically derived properties such as enthalpy and the Gibbs energy of formation.

This book is intended to provide estimation methods for a limited number of physical properties of fluids. Hopefully, the need for such estimates, and for a book of this kind, may diminish as more experimental values become available and as the continually developing molecular theory advances beyond its present incomplete state. In the meantime, estimation methods are essential for most process-design calculations and for many other purposes in engineering and applied science.

REFERENCES

Dewan, A. K., and M. A. Moore: "Physical Property Data Resources for the Practicing Engineer/Scientist in Today's Information Age," Paper 89C, AIChE 1999 Spring National Mtg., Houston, TX, March, 1999. Copyright Equilon Enterprise LLC.

Din, F., (ed.): *Thermodynamic Functions of Gases,* Vol. 3, Butterworth, London, 1961.

Maxwell, James Clerk: "Atoms," *Encyclopaedia Britannica,* 9th ed., A. & C. Black, Edinburgh, 1875–1888.

Slater, J. C.: *Modern Physics,* McGraw-Hill, New York, 1955.

CHAPTER TWO
PURE COMPONENT CONSTANTS

2-1 SCOPE

Though chemical engineers normally deal with mixtures, pure component properties underlie much of the observed behavior. For example, property models intended for the whole range of composition must give pure component properties at the pure component limits. In addition, pure component property constants are often used as the basis for models such as corresponding states correlations for *PVT* equations of state (Chap. 4). They are often used in composition-dependent mixing rules for the parameters to describe mixtures (Chap. 5).

As a result, we first study methods for obtaining *pure component constants* of the more commonly used properties and show how they can be estimated if no experimental data are available. These include the vapor-liquid critical properties, atmospheric boiling and freezing temperatures and dipole moments. Others such as the liquid molar volume and heat capacities are discussed in later chapters. Values for these properties for many substances are tabulated in Appendix A; we compare as many of them as possible to the results from estimation methods. Though the origins of current group contribution methods are over 50 years old, previous editions show that the number of techniques were limited until recently when computational capability allowed more methods to appear. We examine most of the current techniques and refer readers to earlier editions for the older methods.

In Secs. 2-2 (critical properties), 2-3 (acentric factor) and 2-4 (melting and boiling points), we illustrate several methods and compare each with the data tabulated in Appendix A and with each other. All of the calculations have been done with spreadsheets to maximize accuracy and consistency among the methods. It was found that setting up the template and comparing calculations with as many substances as possible in Appendix A demonstrated the level of complexity of the methods. Finally, because many of the methods are for multiple properties and recent developments are using alternative approaches to traditional group contributions, Sec. 2-5 is a general discussion about choosing the best approach for pure component constants. Finally, dipole moments are treated in Sec. 2-6.

Most of the estimation methods presented in this chapter are of the *group, bond,* or *atom contribution* type. That is, the properties of a molecule are usually established from contributions from its elements. The conceptual basis is that the intermolecular forces that determine the constants of interest depend mostly on the bonds between the atoms of the molecules. The elemental contributions are principally determined by the nature of the atoms involved (*atom contributions*), the bonds between pairs of atoms (*bond contributions* or equivalently *group interaction contributions*), or the bonds within and among small groups of atoms (*group contributions*). They all assume that the elements can be treated independently of their arrangements or their neighbors. If this is not accurate enough, corrections for specific multigroup, conformational or resonance effects can be included. Thus, there can be *levels* of contributions. The identity of the elements to be considered (*group*, *bond*, or *atom*) are normally assumed in advance and their contributions obtained by fitting to data. Usually applications to wide varieties of species start with saturated hydrocarbons and grow by sequentially adding different types of bonds, rings, heteroatoms and resonance. The formulations for pure component constants are quite similar to those of the ideal gas formation properties and heat capacities of Chap. 3; several of the group formulations described in Appendix C have been applied to both types of properties.

Alternatives to *group/bond/atom* contribution methods have recently appeared. Most are based on adding weighted contributions of measured properties such as molecular weight and normal boiling point, etc. (*factor analysis*) or from "quantitative structure-property relationships" (QSPR) based on contributions from molecular properties such as electron or local charge densities, molecular surface area, etc. (*molecular descriptors*). Grigoras (1990), Horvath (1992), Katritzky, et al. (1995; 1999), Jurs [Egolf, et al., 1994], Turner, et al. (1998), and St. Cholakov, et al. (1999) all describe the concepts and procedures. The descriptor values are computed from molecular mechanics or quantum mechanical descriptions of the substance of interest and then property values are calculated as a sum of contributions from the descriptors. The significant descriptors and their weighting factors are found by sophisticated regression techniques. This means, however, that there are no tabulations of molecular descriptor properties for substances. Rather, a molecular structure is posed, the descriptors for it are computed and these are combined in the correlation. We have not been able to do any computations for these methods ourselves. However, in addition to quoting the results from the literature, since some tabulate their estimated pure component constants, we compare them with the values in Appendix A.

The methods given here are not suitable for *pseudocomponent properties* such as for the poorly characterized mixtures often encountered with petroleum, coal and natural products. These are usually based on measured properties such as average molecular weight, boiling point, and the specific gravity (at 20°C) rather than molecular structure. We do not treat such systems here, but the reader is referred to the work of Tsonopoulos, et al. (1986), Twu (1984, Twu and Coon, 1996), and Jianzhong, et al. (1998) for example. Older methods include those of Lin and Chao (1984) and Brule, et al. (1982), Riazi and Daubert (1980) and Wilson, et al. (1981).

2-2 VAPOR-LIQUID CRITICAL PROPERTIES

Vapor-liquid critical temperature, T_c, pressure, P_c, and volume, V_c, are the pure-component constants of greatest interest. They are used in many corresponding states correlations for volumetric (Chap. 4), thermodynamic (Chaps. 5–8), and transport (Chaps. 9 to 11) properties of gases and liquids. Experimental determination of their values can be challenging [Ambrose and Young, 1995], especially for larger components that can chemically degrade at their very high critical temperatures [Teja and Anselme, 1990]. Appendix A contains a data base of properties for all the substances for which there is an evaluated critical temperature tabulated by the Thermodynamics Research Center at Texas A&M University [TRC, 1999] plus some evaluated values by Ambrose and colleagues and by Steele and colleagues under the sponsorship of the Design Institute for Physical Properties Research (DIPPR) of the American Institute of Chemical Engineers (AIChE) in New York and NIST (see Appendix A for references). There are fewer evaluated P_c and V_c than T_c. We use only evaluated results to compare with the various estimation methods.

Estimation Techniques

One of the first successful *group contribution* methods to estimate critical properties was developed by Lydersen (1955). Since that time, more experimental values have been reported and efficient statistical techniques have been developed that allow determination of alternative group contributions and optimized parameters. We examine in detail the methods of Joback (1984; 1987), Constantinou and Gani (1994), Wilson and Jasperson (1996), and Marrero and Pardillo (1999). After each is described and its accuracy discussed, comparisons are made among the methods, including descriptor approaches, and recommendations are made. Earlier methods such as those of Lyderson (1955), Ambrose (1978; 1979; 1980), and Fedors (1982) are described in previous editions; they do not appear to be as accurate as those evaluated here.

Method of Joback. Joback (1984; 1987) reevaluated Lydersen's *group contribution* scheme, added several new functional groups, and determined new contribution values. His relations for the critical properties are

$$T_c(\text{K}) = T_b\left[0.584 + 0.965\left\{\sum_k N_k(tck)\right\} - \left\{\sum_k N_k(tck)\right\}^2\right]^{-1} \tag{2-2.1}$$

$$P_c\ (\text{bar}) = \left[0.113 + 0.0032N_{\text{atoms}} - \sum_k N_k(pck)\right]^{-2} \tag{2-2.2}$$

$$V_c\ (\text{cm}^3\text{mol}^{-1}) = 17.5 + \sum_k N_k(vck) \tag{2-2.3}$$

where the contributions are indicated as *tck*, *pck* and *vck*. The group identities and Joback's values for contributions to the critical properties are in Table C-1. For T_c, a value of the normal boiling point, T_b, is needed. This may be from experiment or by estimation from methods given in Sec. 2-4; we compare the results for both. An example of the use of Joback's groups is Example 2-1; previous editions give other examples, as do Devotta and Pendyala (1992).

Example 2-1 Estimate T_c, P_c, and V_c for 2-ethylphenol by using Joback's group method.

solution 2-ethylphenol contains one —CH_3, one —CH_2—, four =CH(ds), one ACOH (phenol) and two =C(ds). Note that the group ACOH is only for the OH and does not include the aromatic carbon. From Appendix Table C-1

Group k	N_k	$N_k\ (tck)$	$N_k\ (pck)$	$N_k\ (vck)$
$—CH_3$	1	0.0141	−0.0012	65
$—CH_2—$	1	0.0189	0	56
=CH(ds)	4	0.0328	0.0044	164
=C(ds)	2	0.0286	0.0016	64
—ACOH (phenol)	1	0.0240	0.0184	−25
$\sum_{k=1}^{5} N_k F_k$		0.1184	0.0232	324

The value of $N_{atoms} = 19$, while $T_b = 477.67$ K. The Joback estimation method (Sec. 2-4) gives $T_b = 489.74$ K.

$$T_c = T_b[0.584 + 0.965(0.1184) - (0.1184)^2]^{-1}$$

$$= 698.1 \text{ K (with exp. } T_b), = 715.7 \text{ K (with est. } T_b)$$

$$P_c = [0.113 + 0.0032(19) - 0.0232]^{-2} = 44.09 \text{ bar}$$

$$V_c = 17.5 + 324 = 341.5 \text{ cm}^3 \text{ mol}^{-1}$$

Appendix A values for the critical temperature and pressure are 703 K and 43.00 bar. An experimental V_c is not available. Thus the differences are

$$T_c \text{ Difference (Exp. } T_b) = 703 - 698.1 = 4.9 \text{ K or } 0.7\%$$

$$T_c \text{ Difference (Est. } T_b) = 703 - 715.7 = -12.7 \text{ K or } -1.8\%$$

$$P_c \text{ Difference} = 43.00 - 44.09 = -1.09 \text{ bar or } -2.5\%.$$

A summary of the comparisons between estimations from the Joback method and experimental Appendix A values for T_c, P_c, and V_c is shown in Table 2-1. The results indicate that the Joback method for critical properties is quite reliable for T_c of all substances regardless of size if the experimental T_b is used. When estimated values of T_b are used, there is a significant increase in error, though it is less for compounds with 3 or more carbons (2.4% average increase for entries indicated by [b] in the table, compared to 3.8% for the whole database indicated by [a]).

For P_c, the reliability is less, especially for smaller substances (note the difference between the [a] and [b] entries). The largest errors are for the largest molecules, especially fluorinated species, some ring compounds, and organic acids. Estimates can be either too high or too low; there is no obvious pattern to the errors. For V_c, the average error is several percent; for larger substances the estimated values are usually too small while estimated values for halogenated substances are often too large. There are no obvious simple improvements to the method. Abildskov (1994) did a limited examination of Joback predictions (less than 100 substances) and found similar absolute percent errors to those of Table 2-1.

A discussion comparing the Joback technique with other methods for critical properties is presented below and a more general discussion of group contribution methods is in Sec. 2-5.

Method of Constantinou and Gani (CG). Constantinou and Gani (1994) developed an advanced *group contribution* method based on the UNIFAC groups (see Chap. 8) but they allow for more sophisticated functions of the desired properties

TABLE 2-1 Summary of Comparisons of Joback Method with Appendix A Database

Property	# Substances	AAE[c]	A%E[c]	# Err > 10%[d]	# Err < 5%[e]
T_c (Exp. T_b)[f], K	352[a]	6.65	1.15	0	345
	289[b]	6.68	1.10	0	286
T_c (Est. T_b)[g], K	352[a]	25.01	4.97	46	248
	290[b]	20.19	3.49	18	229
P_c, bar	328[a]	2.19	5.94	59	196
	266[b]	1.39	4.59	30	180
V_c, cm^3 mol^{-1}	236[a]	12.53	3.37	13	189
	185[b]	13.98	3.11	9	148

[a] The number of substances in Appendix A with data that could be tested with the method.

[b] The number of substances in Appendix A having 3 or more carbon atoms with data that could be tested with the method.

[c] AAE is average absolute error in the property; A%E is average absolute percent error.

[d] The number of substances for which the absolute percent error was greater than 10%.

[e] The number of substances for which the absolute percent error was less than 5%. The number of substances with errors between 5% and 10% can be determined from the table information.

[f] The experimental value of T_b in Appendix A was used.

[g] The value of T_b used was estimated by Joback's method (see Sec. 2-4).

and also for contributions at a "Second Order" level. The functions give more flexibility to the correlation while the Second Order partially overcomes the limitation of UNIFAC which cannot distinguish special configurations such as isomers, multiple groups located close together, resonance structures, etc., at the "First Order." The general CG formulation of a function $f[F]$ of a property F is

$$F = f\left[\sum_k N_k(F_{1i}) + W \sum_j M_j(F_{2j})\right] \tag{2-2.4}$$

where f can be a linear or nonlinear function (see Eqs. 2-2.5 to 2-2.7), N_k is the number of First-Order groups of type k in the molecule; F_{1k} is the contribution for the First-Order group labeled $1k$ to the specified property, F; N_j is the number of Second-Order groups of type j in the molecule; and F_{2j} is the contribution for the Second-Order group labeled $2j$ to the specified property, F. The value of W is set to zero for First-Order calculations and set to unity for Second-order calculations.

For the critical properties, the CG formulations are

$$T_c(K) = 181.128 \ln\left[\sum_k N_k(tc1k) + W \sum_j M_j(tc2j)\right] \tag{2-2.5}$$

$$P_c(bar) = \left[\sum_k N_k(pc1k) + W \sum_j M_j(pc2j) + 0.10022\right]^{-2} + 1.3705 \tag{2-2.6}$$

$$V_c(cm^3\ mol^{-1}) = -0.00435 + \left[\sum_k N_k(vc1k) + W \sum_j M_j(vc2j)\right] \tag{2-2.7}$$

Note that T_c does not require a value for T_b. The group values for Eqs. (2-2.5) to (2-2.7) are given in Appendix Tables C-2 and C-3 with sample assignments shown in Table C-4.

Example 2-2 Estimate T_c, P_c, and V_c for 2-ethylphenol by using Constantinou and Gani's group method.

solution The First-Order groups for 2-ethylphenol are one CH_3, four ACH, one ACCH2, and one ACOH. There are no Second-Order groups (even though the ortho proximity effect might suggest it) so the First Order and Second Order calculations are the same. From Appendix Tables C-2 and C-3

Group k	N_k	$N_k(tc1k)$	$N_k(pc1k)$	$N_k(vc1k)$
CH_3	1	1.6781	0.019904	0.07504
ACH	4	14.9348	0.030168	0.16860
ACCH2	1	10.3239	0.012200	0.10099
ACOH	1	25.9145	−0.007444	0.03162
$\sum_{k=1}^{5} N_k F_k$		52.8513	0.054828	0.37625

$$T_c = 181.128 \ln[52.8513 + W(0)] = 718.6 \text{ K}$$

$$P_c = [0.054828 + W(0) + 0.10022]^{-2} + 1.3705 = 42.97 \text{ bar}$$

$$V_c = (-0.00435 + [0.37625 + W(0)])1000 = 371.9 \text{ cm}^3 \text{ mol}^{-1}$$

The Appendix A values for the critical temperature and pressure are 703.0 K and 43.0 bar. An experimental V_c is not available. Thus the differences are

$$T_c \quad \text{Difference} = 703.0 - 718.6 = -15.6 \text{ K or } -2.2\%$$

$$P_c \quad \text{Difference} = 43.0 - 42.97 = 0.03 \text{ kJ mol}^{-1} \text{ or } 0.1\%.$$

Example 2-3 Estimate T_c, P_c, and V_c for the four butanols using Constantinou and Gani's group method

solution The First- and Second-Order groups for the butanols are:

Groups/Butanol	1-butanol	2-methyl-1-propanol	2-methyl-2-propanol	2-butanol
# First-Order groups, N_k	—	—	—	—
CH_3	1	2	3	2
CH_2	3	1	0	1
CH	0	1	0	1
C	0	0	1	0
OH	1	1	1	1
Second-Order groups, M_j	—	—	—	—
$(CH_3)_2CH$	0	1	0	0
$(CH_3)_3C$	0	0	1	0
CHOH	0	1	0	1
COH	0	0	1	0

Since 1-butanol has no Second-Order group, its calculated results are the same for both orders. Using values of group contributions from Appendix Tables C-2 and C-3 and experimental values from Appendix A, the results are:

Property/Butanol	1-butanol	2-methyl-1-propanol	2-methyl-2-propanol	2-butanol
T_c, K				
Experimental	563.05	547.78	506.21	536.05
Calculated (First Order)	558.91	548.06	539.37	548.06
Abs. percent Err. (First Order)	0.74	0.05	6.55	2.24
Calculated (Second Order)	558.91	543.31	497.46	521.57
Abs. percent Err. (Second Order)	0.74	0.82	1.73	2.70
P_c, bar				
Experimental	44.23	43.00	39.73	41.79
Calculated (First Order)	41.97	41.91	43.17	41.91
Abs. percent Err. (First Order)	5.11	2.52	8.65	0.30
Calculated (Second Order)	41.97	41.66	42.32	44.28
Abs. percent Err. (Second Order)	5.11	3.11	6.53	5.96
V_c, cm^3 mol^{-1}				
Experimental	275.0	273.0	275.0	269.0
Calculated (First Order)	276.9	272.0	259.4	272.0
Abs. percent Err. (First Order)	0.71	0.37	5.67	1.11
Calculated (Second Order)	276.9	276.0	280.2	264.2
Abs. percent Err. (Second Order)	0.71	1.10	1.90	1.78

The First Order results are generally good except for 2-methyl-2-propanol (*t*-butanol). The steric effects of its crowded methyl groups make its experimental value quite different from the others; most of this is taken into account by the First-Order groups, but the Second Order contribution is significant. Notice that the Second Order contributions for the other species are small and may change the results in the wrong direction so that the Second Order estimate can be slightly worse than the First Order estimate. This problem occurs often, but its effect is normally small; including Second Order effects usually helps and rarely hurts much.

A summary of the comparisons between estimations from the Constantinou and Gani method and experimental values from Appendix A for T_c, P_c, and V_c is shown in Table 2-2.

The information in Table 2-2 indicates that the Constantinou/Gani method can be quite reliable for all critical properties, though there can be significant errors for some smaller substances as indicated by the lower errors in Table 2-2B compared to Table 2-2A for T_c and P_c but not for V_c. This occurs because group additivity is not so accurate for small molecules even though it may be possible to form them from available groups. In general, the largest errors of the CG method are for the very smallest and for the very largest molecules, especially fluorinated and larger ring compounds. Estimates can be either too high or too low; there is no obvious pattern to the errors.

Constantinou and Gani's original article (1994) described tests for 250 to 300 substances. Their average absolute errors were significantly less than those of Table 2-2. For example, for T_c they report an average absolute error of 9.8 K for First Order and 4.8 K for Second Order estimations compared to 18.5K and 17.7 K here for 335 compounds. Differences for P_c and V_c were also much less than given here. Abildskov (1994) made a limited study of the Constantinou/Gani method (less than 100 substances) and found absolute and percent errors very similar to those of Table 2-2. Such differences typically arise from different selections of the substances and data base values. In most cases, including Second Order contributions improved the

TABLE 2-2 Summary of Constantinou/Gani Method Compared to Appendix A Data Base

A. All substances in Appendix A with data that could be tested with the method

Property	T_c, K	P_c, bar	V_c, cm^3 mol^{-1}
# Substances (1st)[a]	335	316	220
AAE (1st)[b]	18.48	2.88	15.99
A%E (1st)[b]	3.74	7.37	4.38
# Err > 10% (1st)[c]	28	52	18
# Err < 5% (1st)[d]	273	182	160
# Substances (2nd)[e]	108	99	76
AAE (2nd)[b]	17.69	2.88	16.68
A%E (2nd)[b]	13.61	7.33	4.57
# Err > 10% (2nd)[c]	29	56	22
# Err < 5% (2nd)[d]	274	187	159
# Better (2nd)[f]	70	58	35
Ave. Δ% 1st to 2nd[g]	0.1	0.2	−0.4

B. All substances in Appendix A having 3 or more carbon atoms with data that could be tested with the method

Property	T_c, K	P_c, bar	V_c, cm^3 mol^{-1}
# Substances (1st)[a]	286	263	180
AAE (1st)[b]	13.34	1.8	16.5
A%E (1st)[b]	2.25	5.50	3.49
# Err > 10% (1st)[c]	4	32	10
# Err < 5% (1st)[d]	254	156	136
# Substances (2nd)[e]	104	96	72
AAE (2nd)[b]	12.49	1.8	17.4
A%E (2nd)[b]	2.12	5.50	3.70
# Err > 10% (2nd)[c]	6	36	15
# Err < 5% (2nd)[d]	254	160	134
# Better (2nd)[f]	67	57	32
Ave. Δ% 1st to 2nd[g]	0.3	0.1	−0.5

[a] The number of substances in Appendix A with data that could be tested with the method.

[b] AAE is average absolute error in the property; A%E is average absolute percent error.

[c] The number of substances for which the absolute percent error was greater than 10%.

[d] The number of substances for which the absolute percent error was less than 5%. The number of substances with errors between 5% and 10% can be determined from the table information.

[e] The number of substances for which Second-Order groups are defined for the property.

[f] The number of substances for which the Second Order result is more accurate than First Order.

[g] The average improvement of Second Order compared to First Order. A negative value indicates that overall the Second Order was less accurate.

results 1 to 3 times as often as it degraded them, but except for ring compounds and olefins, the changes were rarely more than 1 to 2%. Thus, Second Order contributions make marginal improvements overall and it may be worthwhile to include the extra complexity only for some individual substances. In practice, examining the magnitude of the Second Order values for the groups involved should provide a user with the basis for including them or not.

A discussion comparing the Constantinou/Gani technique with other methods for critical properties is presented below and a more general discussion is found in Sec. 2-5.

Method of Wilson and Jasperson. Wilson and Jasperson (1996) reported three methods for T_c and P_c that apply to both organic and inorganic species. The Zero-Order method uses *factor analysis* with boiling point, liquid density and molecular weight as the descriptors. At the First Order, the method uses *atomic contributions* along with boiling point and number of rings, while the Second Order method also includes *group contributions*. The Zero-Order has not been tested here; it is iterative and the authors report that it is less accurate by as much as a factor of two or three than the others, especially for P_c. The First Order and Second Order methods use the following equations:

$$T_c = T_b \Big/ \left[(0.048271 - 0.019846N_r + \sum_k N_k(\Delta tck) + \sum_j M_j(\Delta(tcj)\right]^{0.2} \tag{2-2.8}$$

$$P_c = 0.0186233T_c/[-0.96601 + \exp(Y)] \tag{2-2.9a}$$

$$Y = -0.00922295 - 0.0290403N_r + 0.041\left(\sum_k N_k(\Delta pck) + \sum_j M_j(\Delta pcj)\right) \tag{2-2.9b}$$

where N_r is the number of rings in the compound, N_k is the number of atoms of type k with First Order atomic contributions Δtck and Δpck while M_j is the number of groups of type j with Second-Order group contributions Δtcj and Δpcj. Values of the contributions are given in Table 2-3 both for the First Order Atomic Contributions and for the Second-Order Group Contributions. Note that T_c requires T_b. Application of the Wilson and Jasperson method is shown in Example 2-4.

Example 2-4 Estimate T_c and P_c for 2-ethylphenol by using Wilson and Jasperson's method.

solution The atoms of 2-ethylphenol are 8 −C, 10 −H, 1 −O and there is 1 ring. For groups, there is 1 −OH for "C5 or more." The value of T_b from Appendix A is 477.67 K; the value estimated by the Second Order method of Constantinou and Gani (Eq. 2-4.4) is 489.24 K. From Table 2-3A

Atom k	N_k	$N_k(\Delta tck)$	$N_k(\Delta pck)$
C	8	0.06826	5.83864
H	10	0.02793	1.26600
O	1	0.02034	0.43360
$\sum_{k=1}^{3} N_k F_k$	—	0.11653	7.53824

TABLE 2-3A Wilson-Jasperson (1996) Atomic Contributions for Eqs. (2-2.8) and (2-2.9)

Atom	Δtck	Δpck
H	0.002793	0.12660
D	0.002793	0.12660
T	0.002793	0.12660
He	0.320000	0.43400
B	0.019000	0.91000
C	0.008532	0.72983
N	0.019181	0.44805
O	0.020341	0.43360
F	0.008810	0.32868
Ne	0.036400	0.12600
Al	0.088000	6.05000
Si	0.020000	1.34000
P	0.012000	1.22000
S	0.007271	1.04713
Cl	0.011151	0.97711
Ar	0.016800	0.79600
Ti	0.014000	1.19000
V	0.018600	*****
Ga	0.059000	*****
Ge	0.031000	1.42000
As	0.007000	2.68000
Se	0.010300	1.20000
Br	0.012447	0.97151
Kr	0.013300	1.11000
Rb	−0.027000	*****
Zr	0.175000	1.11000
Nb	0.017600	2.71000
Mo	0.007000	1.69000
Sn	0.020000	1.95000
Sb	0.010000	*****
Te	0.000000	0.43000
I	0.005900	1.315930
Xe	0.017000	1.66000
Cs	−0.027500	6.33000
Hf	0.219000	1.07000
Ta	0.013000	*****
W	0.011000	1.08000
Re	0.014000	*****
Os	−0.050000	*****
Hg	0.000000	−0.08000
Bi	0.000000	0.69000
Rn	0.007000	2.05000
U	0.015000	2.04000

TABLE 2-3B Wilson-Jasperson (1996) Group Contributions for Eqs. (2-2.8) and (2-2.9)

Group	Δtcj	Δpcj
—OH, C_4 or less	0.0350	0.00
—OH, C_5 or more	0.0100	0.00
—O—	−0.0075	0.00
—NH_2, >NH, >N—	−0.0040	0.00
—CHO	0.0000	0.50
>CO	−0.0550	0.00
—COOH	0.0170	0.50
—COO—	−0.0150	0.00
—CN	0.0170	1.50
—NO_2	−0.0200	1.00
Organic Halides (once/molecule)	0.0020	0.00
—SH, —S—, —SS—	0.0000	0.00
Siloxane bond	−0.0250	−0.50

Thus the First Order estimates are

$$T_c = 477.67/[0.048271 - 0.019846 + 0.11653]^{0.2} = 702.9 \text{ K}$$

$$P_c = 0.0186233(704.1)/[-0.96601 + \exp(Y)] = 37.94 \text{ bar}$$

$$Y = -0.0092229 - 0.0290403 + 0.3090678 = 0.2708046$$

From Table 2-3B there is the "−OH, C5 or more" contribution of $N_k \Delta tck = 0.01$ though for P_c there is no contribution. Thus only the Second Order estimate for T_c is changed to

$$T_c = 477.67/[0.048271 - 0.019846 + 0.11653 + 0.01]^{0.2} = 693.6 \text{ K}$$

If the estimated value of T_b is used, the result is 710.9 K. The Appendix A values for the critical properties are 703.0 K and 43.0 bar, respectively. Thus the differences are

First Order	T_c (Exp. T_b)	Difference = 703.0 − 702.9 = 0.1 K or 0.0%
	T_c (Est. T_b)	Difference = 703.0 − 719.9 = −16.9 K or −2.4%
	P_c	Difference = 43.0 − 37.9 = 5.1 bar or 11.9%.
Second Order	T_c (Exp. T_b)	Difference = 703.0 − 693.6 = 9.4 K or 1.3%
	T_c (Est. T_b)	Difference = 703.0 − 710.9 = −7.9 K or −1.1%
	P_c (= First Order)	Difference = 43.0 − 37.9 = 5.1 bar or 11.9%.

The First Order estimate for T_c is more accurate than the Second Order estimate which occasionally occurs.

A summary of the comparisons between estimations from the Wilson and Jasperson method and experimental values from Appendix A for T_c and P_c are shown in Table 2-4. Unlike the Joback and Constantinou/Gani method, there was no dis-

TABLE 2-4 Summary of Wilson/Jasperson Method Compared to Appendix A Data Base

Property	T_c, K (Exp. T_b)*	T_c, K (Est T_b)+	P_c, bar (Exp T_c)#	P_c, bar (Est T_c)@
# Substances[a]	353	—	348	348
AAE (First Order)[b]	8.34	—	2.08	2.28
A%E (First Order)[b]	1.50	—	5.31	5.91
# Err > 10% (First Order)[c]	0	—	54	66
# Err < 5% (First Order)[d]	220	—	234	220
# Substances[e]	180	289	23	23
AAE (Second Order)[b]	6.88	16.71	1.82	2.04
A%E (Second Ordcr)[b]	1.22	2.95	4.74	5.39
# Err > 10% (Second Order)[c]	0	15	46	57
# Err < 5% (Second Order)[d]	348	249	245	226
# Better (Second Order)[f]	120	77	19	18
Ave. Δ% First to Second Order[g]	0.5	−1.8	8.6	7.9

* Eq. (2-2.8) with experimental T_b.
+ Eq. (2-2.8) with T_b estimated from Second Order Method of Constantinou and Gani (1994).
# Eq. (2-2.9) with experimental T_c.
@ Eq. (2-2.9) with T_c estimated using Eq. (2-2.8) and experimental T_b.
[a] The number of substances in Appendix A with data that could be tested with the method.
[b] AAE is average absolute error in the property; A%E is average absolute percent error.
[c] The number of substances for which the absolute percent error was greater than 10%.
[d] The number of substances for which the absolute percent error was less than 5%. The number of substances with errors between 5% and 10% can be determined from the table information.
[e] The number of substances for which Second-Order groups are defined for the property.
[f] The number of substances for which the Second Order result is more accurate than First Order.
[g] The average improvement of Second Order compared to First Order. A negative value indicates that overall the Second Order was less accurate.

cernible difference in errors between small and large molecules for either property so only the overall is given.

The information in Table 2-4 indicates that the Wilson/Jasperson method is very accurate for both T_c and P_c. When present, the Second Order group contributions normally make significant improvements over estimates from the First Order atom contributions. The accuracy for P_c deteriorates only slightly with an estimated value of T_c if the experimental T_b is used. The accuracy of T_c is somewhat less when the required T_b is estimated with the Second Order method of Constantinou and Gani (1994) (Eq. 2-4.4). Thus the method is remarkable in its accuracy even though it is the simplest of those considered here and applies to all sizes of substances equally.

Wilson and Jasperson compared their method with results for 700 compounds of all kinds including 172 inorganic gases, liquids and solids, silanes and siloxanes. Their reported average percent errors for organic substance were close to those found here while they were somewhat larger for the nonorganics. The errors for organic acids and nitriles are about twice those for the rest of the substances. Nielsen (1998) studied the method and found similar results.

Discussion comparing the Wilson/Jasperson technique with other methods for critical properties is presented below and a more general discussion is in Sec. 2-5.

Method of Marrero and Pardillo. Marrero-Marejón and Pardillo-Fontdevila (1999) describe a method for T_c, P_c, and V_c that they call a group interaction contribution technique or what is effectively a *bond contribution* method. They give

equations that use values from pairs of atoms alone, such as >C< & —N<, or with hydrogen attached, such as CH_3— & —NH_2. Their basic equations are

$$T_c = T_b / \left[0.5851 - 0.9286 \left(\sum_k N_k tcbk \right) - \left(\sum_k N_k tcbk \right)^2 \right] \tag{2-2.10}$$

$$P_c = \left[0.1285 - 0.0059 N_{\text{atoms}} - \sum_k N_k pcbk \right]^{-2} \tag{2-2.11}$$

$$V_c = 25.1 + \sum_k N_k vcbk \tag{2-2.12}$$

where N_{atoms} is the number of atoms in the compound, N_k is the number of atoms of type k with contributions *tcbk*, *pcbk*, and *vcbk*. Note that T_c requires T_b, but Marrero and Pardillo provide estimation methods for T_b (Eq. 2-4.5).

Values of contributions for the 167 pairs of groups (bonds) are given in Table 2-5. These were obtained directly from Dr. Marrero and correct some misprints in the original article (1999). The notation of the table is such that when an atom is bonded to an element other than hydrogen, — means a single bond, > or < means 2 single bonds, = means a double bond and ≡ means a triple bond, [r] means that the group is in a ring such as in aromatics and naphthenics, and [rr] means the pair connects 2 rings as in biphenyl or terphenyl. Thus, the pair >C< & F— means that the C is bonded to 4 atoms/groups that are not hydrogen and one of the bonds is to F, while =C< & F— means that the C atom is doubly bonded to another atom and has 2 single bonds with 1 of the bonds being to F. Bonding by multiple bonds is denoted by both members of the pair having [=] or [≡]; if they both have a = or a ≡ without the brackets [], they will also have at least 1 — and the bonding of the pair is via a single bond. Therefore, the substance CHF=CFCF_3 would have 1 pair of [=]CH— & [=]C<, 1 pair of =CH— & F—, 1 pair of =C< & —F, 1 pair of =C< and >C<, and 3 pairs of >C< & —F. The location of bonding in esters is distinguished by the use of [] as in pairs 20, 21, 67, 100 and 101. For example, in the pair 20, the notation CH_3— & —COO[—] means that CH_3— is bonded to an O to form an ester group, CH_3—O—CO—, whereas in the pair 21, the notation CH_3— & [—]COO— means that CH_3— is bonded to the C to form CH_3—CO—O—. Of special note is the treatment of aromatic rings; it differs from other methods considered in this section because it places single and double bonds in the rings at specific locations, affecting the choice of contributions. This method of treating chemical structure is the same as used in traditional Handbooks of Chemistry such as Lange's (1999). We illustrate the placement of side groups and bonds with 1-methylnaphthalene in Example 2-5. The locations of the double bonds for pairs 130, 131, and 139 must be those illustrated as are the single bonds for pairs 133, 134 and 141. The positions of side groups must also be carefully done; the methyl group with bond pair 10 must be placed at the "top" of the diagram since it must be connected to the 131 and 141 pairs. If the location of it or of the double bond were changed, the contributions would change.

Example 2-5 List the pairs of groups (bonds) of the Marrero/Pardillo (1999) method for 1-methylnaphthalene.

solution The molecular structure and pair numbers associated with the bonds from Table 2-5 are shown in the diagram.

TABLE 2-5 Marrero-Pardillo (1999) Contributions for Eqs. (2-2.10) to (2-2.12) and (2-4.5)

Pair #	Atom/Group Pairs	tcbk	pcbk	vcbk	tbbk
1	CH_3— & CH_3—	−0.0213	−0.0618	123.2	113.12
2	CH_3— & —CH_2—	−0.0227	−0.0430	88.6	194.25
3	CH_3— & >CH—	−0.0223	−0.0376	78.4	194.27
4	CH_3— & >C<	−0.0189	−0.0354	69.8	186.41
5	CH_3— & =CH—	0.8526	0.0654	81.5	137.18
6	CH_3— & =C<	0.1792	0.0851	57.7	182.20
7	CH_3— & ≡C—	0.3818	−0.2320	65.8	194.40
8	CH_3— & >CH— [r]	−0.0214	−0.0396	58.3	176.16
9	CH_3— & >C< [r]	0.1117	−0.0597	49.0	180.60
10	CH_3— & =C< [r]	0.0987	−0.0746	71.7	143.58
11	CH_3— & F—	−0.0370	−0.0345	88.1	160.83
12	CH_3— & Cl—	−0.9141	−0.0231	113.8	453.70
13	CH_3— & Br—	−0.9166	−0.0239	*****	758.44
14	CH_3— & I—	−0.9146	−0.0241	*****	1181.44
15	CH_3— & —OH	−0.0876	−0.0180	92.9	736.93
16	CH_3— & —O—	−0.0205	−0.0321	66.0	228.01
17	CH_3— & >CO	−0.0362	−0.0363	88.9	445.61
18	CH_3— & —CHO	−0.0606	−0.0466	128.9	636.49
19	CH_3— & —COOH	−0.0890	−0.0499	145.9	1228.84
20	CH_3— & —COO[—]	0.0267	0.1462	93.3	456.92
21	CH_3— & [—]COO—	−0.0974	−0.2290	108.2	510.65
22	CH_3— & —NH_2	−0.0397	−0.0288	*****	443.76
23	CH_3— & —NH—	−0.0313	−0.0317	*****	293.86
24	CH_3— & >N—	−0.0199	−0.0348	76.3	207.75
25	CH_3— & —CN	−0.0766	−0.0507	147.9	891.15
26	CH_3— & —NO_2	−0.0591	−0.0385	148.1	1148.58
27	CH_3— & —SH	−0.9192	−0.0244	119.7	588.31
28	CH_3— & —S—	−0.0181	−0.0305	87.9	409.85
29	—CH_2— & —CH_2—	−0.0206	−0.0272	56.6	244.88
30	—CH_2— & >CH—	−0.0134	−0.0219	40.2	244.14
31	—CH_2— & >C<	−0.0098	−0.0162	32.0	273.26
32	—CH_2— & =CH—	0.8636	0.0818	50.7	201.80
33	—CH_2— & =C<	0.1874	0.1010	24.0	242.47
34	—CH_2— & ≡C—	0.4160	−0.2199	33.9	207.49
35	—CH_2— & >CH— [r]	−0.0149	−0.0265	31.9	238.81
36	—CH_2— & >C< [r]	0.1193	−0.0423	*****	260.00
37	—CH_2— & =C< [r]	0.1012	−0.0626	52.1	167.85
38	—CH_2— & F—	−0.0255	−0.0161	49.3	166.59
39	—CH_2— & Cl—	−0.0162	−0.0150	80.8	517.62
40	—CH_2— & Br—	−0.0205	−0.0140	101.3	875.85
41	—CH_2— & I—	−0.0210	−0.0214	*****	1262.80
42	—CH_2— & —OH	−0.0786	−0.0119	45.2	673.24
43	—CH_2— & —O—	−0.0205	−0.0184	34.5	243.37
44	—CH_2— & >CO	−0.0256	−0.0204	62.3	451.27
45	—CH_2— & —CHO	−0.0267	−0.0210	106.1	648.70
46	—CH_2— & —COOH	−0.0932	−0.0253	114.0	1280.39
47	—CH_2— & —COO[—]	0.0276	0.1561	69.9	475.65
48	—CH_2— & [—]COO—	−0.0993	−0.2150	79.1	541.29
49	—CH_2— & —NH_2	−0.0301	−0.0214	63.3	452.30
50	—CH_2— & —NH—	−0.0248	−0.0203	49.4	314.71

TABLE 2-5 Marrero-Pardillo (1999) Contributions for Eqs. (2-2.10) to (2-2.12) and (2-4.5) (*Continued*)

Pair #	Atom/Group Pairs	tcbk	pcbk	vcbk	tbbk
51	—CH$_2$— & >N—	−0.0161	−0.0170	32.7	240.08
52	—CH$_2$— & —CN	−0.0654	−0.0329	113.5	869.18
53	—CH$_2$— & —SH	−0.0137	−0.0163	93.3	612.31
54	—CH$_2$— & —S—	−0.0192	−0.0173	57.9	451.03
55	>CH— & CH—	−0.0039	−0.0137	18.3	291.41
56	>CH— & >C<	0.0025	−0.0085	8.6	344.06
57	>CH— & =CH—	0.8547	0.0816	48.9	179.96
58	>CH— & =C<	0.1969	0.1080	4.3	249.10
59	>CH— & >CH— [r]	0.0025	−0.0168	*****	295.33
60	>CH— & =C< [r]	0.1187	−0.0556	*****	132.66
61	>CH— & F—	−0.0200	−0.0147	37.7	68.80
62	>CH— & Cl—	−0.0142	−0.0131	68.6	438.47
63	>CH— & —OH	−0.0757	−0.0093	45.6	585.99
64	>CH— & —O—	−0.0162	−0.0155	23.7	215.94
65	>CH— & >CO	−0.0194	−0.0112	39.3	434.45
66	>CH— & —CHO	−0.0406	−0.0280	92.2	630.07
67	>CH— & [—]COO—	−0.0918	−0.2098	72.3	497.58
68	>CH— & —COOH	−0.1054	−0.0358	110.2	1270.16
69	>CH— & —NH$_2$	−0.0286	−0.0212	39.2	388.44
70	>CH— & —NH—	−0.0158	−0.0162	*****	260.32
71	>C< & >C<	0.0084	0.0002	22.7	411.56
72	>C< & =CH—	0.8767	0.0953	23.4	286.30
73	>C< & =C<	0.2061	0.1109	8.8	286.42
74	>C< & >C< [r]	0.0207	0.0213	*****	456.90
75	>C< & >CH— [r]	0.0049	−0.0111	*****	340.00
76	>C< & =C< [r]	0.1249	−0.0510	*****	188.99
77	>C< & F—	−0.0176	−0.0161	30.0	−16.64
78	>C< & Cl—	−0.0133	−0.0129	63.7	360.79
79	>C< & Br—	−0.0084	−0.0121	85.7	610.26
80	>C< & —OH	−0.0780	−0.0094	40.6	540.38
81	>C< & —O—	−0.0156	−0.0103	40.8	267.26
82	>C< & >CO	−0.0114	−0.0085	62.1	373.71
83	>C< & —COOH	−0.1008	−0.0455	89.0	1336.54
84	[=]CH$_2$ & [=]CH$_2$	−0.9129	−0.0476	105.3	51.13
85	[=]CH$_2$ & —CH[=]	−0.8933	−0.1378	77.4	205.73
86	[=]CH$_2$ & >C[=]	−0.4158	−0.2709	99.2	245.27
87	[=]CH$_2$ & =C[=]	−0.0123	−0.0239	68.4	183.55
88	—CH[=] & —CH[=]	−1.7660	−0.2291	47.8	334.64
89	—CH[=] & >C[=]	−1.2909	−0.3613	73.6	354.41
90	—CH[=] & =C[=]	−0.8945	−0.1202	43.6	316.46
91	=CH— & =CH—	1.7377	0.1944	42.1	174.18
92	=CH— & =C<	1.0731	0.2146	16.6	228.38
93	=CH— & ≡C—	1.2865	−0.1087	*****	174.39
94	=CH— & =C< [r]	0.9929	0.0533	*****	184.20
95	=CH— & F—	0.8623	0.0929	41.4	5.57
96	=CH— & Cl—	0.8613	0.0919	68.7	370.60
97	=CH— & —O—	0.8565	0.0947	36.4	204.81
98	=CH— & —CHO	0.8246	0.0801	*****	658.53
99	=CH— & —COOH	0.7862	0.0806	107.4	1245.86
100	=CH— & —COO[—]	0.8818	0.2743	55.2	423.86

TABLE 2-5 Marrero-Pardillo (1999) Contributions for Eqs. (2-2.10) to (2-2.12) and (2-4.5) (*Continued*)

Pair #	Atom/Group Pairs	tcbk	pcbk	vcbk	tbbk
101	=CH— & [—]COO—	0.7780	−0.1007	64.1	525.35
102	=CH— & —CN	0.8122	0.0771	107.4	761.36
103	>C[=] & >C[=]	−0.8155	−0.4920	93.7	399.58
104	>C[=] & =C[=]	−0.4009	−0.2502	58.1	321.02
105	=C< & =C< [r]	0.3043	0.0705	*****	250.88
106	=C< & F—	0.1868	0.1064	14.6	−37.99
107	=C< & Cl—	0.1886	0.1102	43.3	367.05
108	=C[=] & O[=]	−0.0159	−0.0010	51.4	160.42
109	CH[≡] & CH[≡]	−0.0288	−0.0226	87.6	120.85
110	CH[≡] & —C[≡]	−0.4222	0.1860	73.1	222.40
111	—C[≡] & —C[≡]	−0.7958	0.3933	64.3	333.26
112	—CH_2— [r] & —CH_2— [r]	−0.0098	−0.0221	47.2	201.89
113	—CH_2— [r] & >CH— [r]	−0.0093	−0.0181	47.5	209.40
114	—CH_2— [r] & >C< [r]	−0.1386	0.0081	49.9	182.74
115	—CH_2— [r] & =CH— [r]	0.0976	−0.1034	42.5	218.07
116	—CH_2— [r] & =C< [r]	0.1089	−0.0527	*****	106.21
117	—CH_2— [r] & —O— [r]	−0.0092	−0.0119	29.2	225.52
118	—CH_2— [r] & >CO [r]	−0.0148	−0.0177	50.7	451.74
119	—CH_2— [r] & —NH— [r]	−0.0139	−0.0127	38.8	283.55
120	—CH_2— [r] & —S— [r]	−0.0071	*****	*****	424.13
121	>CH— [r] & >CH— [r]	−0.0055	−0.0088	33.9	210.66
122	>CH— [r] & >C< [r]	−0.1341	0.0162	*****	220.24
123	>CH— [r] & >CH— [rr]	*****	*****	*****	254.50
124	>CH— [r] & >C[=] [rr]	*****	*****	*****	184.36
125	>CH— [r] & —O— [r]	−0.0218	−0.0091	19.2	169.17
126	>CH— [r] & —OH	−0.0737	−0.0220		597.82
127	>C< [r] & >C< [r]	0.0329	−0.0071	36.2	348.23
128	>C< [r] & =C< [r]	*****	*****	*****	111.51
129	>C< [r] & F—	−0.0314	−0.0119	18.4	−41.35
130	—CH[=] [r] & —CH[=] [r]	−0.2246	0.1542	36.5	112.00
131	—CH[=] [r] & >C[=] [r]	−0.3586	0.1490	34.4	291.15
132	—CH[=] [r] & —N[=] [r]	0.3913	0.1356	8.3	221.55
133	=CH— [r] & =CH— [r]	0.2089	−0.1822	39.3	285.07
134	=CH— [r] & =C< [r]	0.2190	−0.1324	29.8	237.22
135	=CH— [r] & —O— [r]	0.1000	−0.0900	40.3	171.59
136	=CH— [r] & —NH— [r]	0.0947	*****	*****	420.54
137	=CH— [r] & =N— [r]	−0.4067	−0.1491	65.9	321.44
138	=CH— [r] & —S— [r]	0.1027	−0.0916	40.8	348.00
139	>C[=] [r] & >C[=] [r]	−0.4848	0.1432	37.8	477.77
140	>C[=] [r] & —N[=] [r]	0.2541	*****	*****	334.09
141	=C< [r] & =C< [r]	0.2318	−0.0809	20.6	180.07
142	=C< [r] & =C< [rr]	0.2424	−0.0792	51.7	123.05
143	=C< [r] & —O— [r]	0.1104	−0.0374	−0.3	134.23
144	=C< [r] & =N— [r]	−0.3972	−0.0971	35.6	174.31
145	=C< [r] & F—	0.1069	−0.0504	23.7	−48.79
146	=C< [r] & Cl—	0.1028	−0.0512	60.3	347.33
147	=C< [r] & Br—	0.1060	−0.0548	83.2	716.23
148	=C< [r] & I—	0.1075	−0.0514	110.2	1294.98
149	=C< [r] & —OH	0.0931	−0.0388	8.5	456.25
150	=C< [r] & —O—	0.0997	−0.0523	*****	199.70

TABLE 2-5 Marrero-Pardillo (1999) Contributions for Eqs. (2-2.10) to (2-2.12) and (2-4.5) (*Continued*)

Pair #	Atom/Group Pairs	tcbk	pcbk	vcbk	tbbk
151	=C< [r] & >CO	0.1112	−0.0528	46.3	437.51
152	=C< [r] & —CHO	0.0919	−0.0597	*****	700.06
153	=C< [r] & —COOH	0.0313	−0.0684	100.2	1232.55
154	=C< [r] & [—]COO—	0.0241	−0.2573	55.2	437.78
155	=C< [r] & —NH$_2$	0.0830	−0.0579	33.2	517.75
156	=C< [r] & —NH—	0.0978	−0.0471	*****	411.29
157	=C< [r] & >N—	0.0938	−0.0462	*****	422.51
158	=C< [r] & —CN	0.0768	−0.0625	*****	682.19
159	Cl— & >CO	−0.0191	−0.0125	84.0	532.24
160	[—]COO— & [—]COO—	−0.1926	−0.0878	*****	1012.51
161	—O— [r] & =N— [r]	−0.5728	*****	*****	382.25
162	>CO & —O—	−0.3553	−0.0176	*****	385.36
163	—H & —CHO	−0.0422	−0.0123	*****	387.17
164	—H & —COOH	−0.0690	*****	*****	1022.45
165	—H & [—]COO—	−0.0781	−0.1878	51.2	298.12
166	—NH— & —NH$_2$	−0.0301	*****	*****	673.59
167	—S— & —S—	−0.0124	*****	*****	597.59

Pair #	Atom/Group Pair	N_k
10	CH3— & =C< [r]	1
130	—CH[=] [r] & —CH[=] [r]	3
131	—CH[=] [r] & >C[=] [r]	1
133	=CH— [r] & =CH— [r]	2
134	=CH— [r] & =C< [r]	3
139	>C[=] [r] & >C[=] [r]	1
141	=C< [r] & =C< [r]	1

Other applications of the Marrero and Pardillo method are shown in Examples 2-6 and 2-7. There are also several informative examples in the original paper (1999).

Example 2-6 Estimate T_c, P_c, and V_c for 2-ethylphenol by using Marrero and Pardillo's method.

solution The chemical structure to be used is shown. The locations of the various bond pairs are indicated on the structure shown. The value of N_{atoms} is 19.

Pair #	Atom/Group Pair	N_k	$N_k tck$	$N_k pck$	$N_k vck$
2	CH$_3$— & —CH$_2$—	1	−0.0227	−0.0430	88.6
37	—CH$_2$— & =C< [r]	1	0.1012	−0.0626	52.1
130	—CH[=] [r] & —CH[=] [r]	1	−0.2246	0.1542	36.5
131	—CH[=] [r] & >C[=] [r]	2	−0.7172	0.2980	68.8
133		2	0.4178	−0.3644	78.6
141	=CH— [r] & =CH— [r]	1	0.2318	−0.0809	20.6
149	=C< [r] & —OH	1	0.0931	−0.0388	8.5
—	$\sum_{k=1}^{7} N_k F_k$	—	−0.1206	−0.1375	353.7

The estimates from Eqs. (2-2.10) to (2-2.12) are

$$T_c = 477.67/[0.5851 + 0.1120 - 0.0145] = 699.8 \text{ K}$$

$$P_c = [0.1285 - 0.1121 + 0.1375]^{-2} = 42.2 \text{ bar}$$

$$V_c = 25.1 + 353.7 = 378.8 \text{ cm}^3 \text{ mol}^{-1}$$

The Appendix A values for the critical temperature and pressure are 703.0 K and 43.0 bar. An experimental V_c is not available. Thus the differences are

$$T_c \quad \text{Difference} = 703.0 - 699.8 = 3.2 \text{ K or } 0.5\%$$

$$P_c \quad \text{Difference} = 43.0 - 42.2 = 0.8 \text{ bar or } 1.8\%$$

If Marrero and Pardillo's recommended method for estimating T_b is used (Eq. 2-4.5), the result for T_c is 700.6, an error of 0.3% which is more accurate than with the experimental T_b.

Example 2-7 Estimate T_c, P_c, and V_c for the four butanols using Marrero and Pardillo's method.

solution The Atom/Group Pairs for the butanols are:

Pair #	Atom/Group Pair	1-butanol	2-methyl-1-propanol	2-methyl-2-propanol	2-butanol
2	CH_3— & —CH_2—	1	0	0	1
3	CH_3— & >CH—	0	2	0	1
4	CH_3— & >C<	0	0	3	0
29	—CH_2— & —CH_2—	2	0	0	0
30	—CH_2— & >CH—	0	1	0	1
42	—CH_2— & —OH	1	1	0	0
63	>CH— & —OH	0	0	0	1
80	>C< & —OH	0	0	1	0

Using values of group contributions from Table 2-5 and experimental values from Appendix A, the results are:

Property/Butanol	1-butanol	2-methyl-1-propanol	2-methyl-2-propanol	2-butanol
T_c, K				
Experimental	563.05	547.78	506.21	536.05
Calculated (Exp T_b)[a]	560.64	549.33	513.80	538.87
Abs. percent Err. (Exp T_b)[a]	0.43	0.28	1.50	0.53
Calculated (Est T_b)[b]	560.40	558.52	504.56	533.93
Abs. percent Err. (Est T_b)[b]	0.47	1.96	0.33	0.40
P_c, bar				
Experimental	44.23	43.00	39.73	41.79
Calculated	44.85	45.07	41.38	43.40
Abs. percent Err.	1.41	4.81	4.14	3.86
V_c, cm^3 mol^{-1}				
Experimental	275.0	273.0	275.0	269.0
Calculated	272.1	267.2	275.0	277.8
Abs. percent Err.	1.07	2.14	0.01	3.26

[a] Calculated with Eq [2-2.10] using T_b from Appendix A.
[b] Calculated with Eq [2-2.10] using T_b estimated with Marrero/Pardillo method Eq. (2-4.5).

The results are typical of the method. Notice that sometimes the value with an estimated T_b is more accurate than with the experimental value. As shown in Table 2-4, this occurs about ¼ of the time through coincidence.

A summary of the comparisons between estimations from the Marrero and Pardillo method and experimental values from Appendix A for critical properties is shown in Table 2-6. It shows that there is some difference in errors between small and large molecules.

The information in Table 2-6 indicates that the Marrero/Pardillo is accurate for the critical properties, especially T_c. The substances with larger errors in P_c and V_c are organic acids and some esters, long chain substances, especially alcohols, and those with proximity effects such as multiple halogens (including perfluorinated species) and stressed rings.

A discussion comparing the Marrero and Pardillo technique with other methods for the properties of this chapter is presented in Sec. 2-5.

Other methods for Critical Properties. There are a large number of other *group/bond/atom* methods for estimating critical properties. Examination of them indicates that they either are restricted to only certain types of substances such as paraffins, perfluorinated species, alcohols, etc., or they are of lower accuracy than those shown here. Examples include those of Tu (1995) with 40 groups to obtain

TABLE 2-6 Summary of Marrero/Pardillo (1999) Method Compared to Appendix A Data Base

A. All substances in Appendix A with data that could be tested with the method

Property	T_c*, K	T_c#, K	P_c, bar	V_c, cm³ mol⁻¹
# Substances[a]	343	344	338	296
AAE[b]	6.15	15.87	1.79	13.25
A%E[b]	0.89	2.93	5.29	3.24
# Err > 10%[c]	1	22	47	18
# Err < 5%[d]	336	288	228	241
# Better Est T_b[e]		83		

B. All substances in Appendix A having 3 or more carbon atoms with data that could be tested with the method

Property	T_c*, K	T_c#, K	P_c, bar	V_c, cm³ mol⁻¹
# Substances[a]	285	286	280	243
AAE[b]	5.78	15.53	1.68	14.72
A%E[b]	0.94	2.62	5.38	3.28
# Err > 10%[c]	1	14	39	15
# Err < 5%[d]	282	248	188	200
# Better with Est T_b[e]		68		

*Calculated with Eq [2-2.10] using T_b from Appendix A.
#Calculated with Eq [2-2.10] using T_b estimated with Marrero/Pardillo method Eq. (2-4.5).
[a] The number of substances in Appendix A with data that could be tested with the method.
[b] AAE is average absolute error in the property; A%E is average absolute percent error.
[c] The number of substances for which the absolute percent error was greater than 10%.
[d] The number of substances for which the absolute percent error was less than 5%. The number of substances with errors between 5% and 10% can be determined from the table information.
[e] The number of substances for which the T_c result is more accurate when an estimated T_b is used than with an experimental value.

T_c for all types of organics with somewhat better accuracy than Joback's method; Sastri, et al. (1997) treating only V_c and obtaining somewhat better accuracy than Joback's method; Tobler (1996) correlating V_c with a substance's temperature and density at the normal boiling point with improved accuracy over Joback's method, but also a number of substances for which all methods fail; and Daubert [Jalowka and Daubert, 1986; Daubert and Bartakovits, 1989] using Benson groups (see Sec. 3.3) and obtaining about the same accuracy as Lydersen (1955) and Ambrose (1979) for all properties. Within limited classes of systems and properties, these methods may be more accurate as well as easier to implement than those analyzed here.

As mentioned in Sec. 2.1, there is also a great variety of other estimation methods for critical properties besides the above *group/bond/atom* approaches. The techniques generally fall into two classes. The first is based on *factor analysis* that builds correlation equations from data of other measurable, macroscopic properties such as densities, molecular weight, boiling temperature, etc. Such methods include those of Klincewicz and Reid (1984) and of Vetere (1995) for many types of substances. Somayajulu (1991) treats only alkanes but also suggests ways to approach other homologous series. However, the results of these methods are either reduced accuracy or extra complexity. The way the parameters depend upon the type of substance and their need for other input information does not yield a direct or universal computational method so, for example, the use of spreadsheets would be much more complicated. We have not given any results for these methods.

The other techniques of estimating critical and other properties are based on molecular properties, *molecular descriptors*, which are not normally measurable. These "Quantitative Structure-Property Relationships" (QSPR) are usually obtained from on-line computation of the structure of the whole molecule using molecular mechanics or quantum mechanical methods. Thus, no tabulation of descriptor contributions is available in the literature even though the weighting factors for the descriptors are given. Estimates require access to the appropriate computer software to obtain the molecular structure and properties and then the macroscopic properties are estimated with the QSPR relations. It is common that different methods use different computer programs. We have not done such calculations, but do compare with the data of Appendix A the results reported by two recent methods. We comment below and in Sec. 2.5 on how they compare with the *group/bond/atom* methods. The method of Gregoras is given mainly for illustrative purposes; that of Jurs shows the current status of *molecular descriptor* methods.

Method of Grigoras. An early molecular structural approach to physical properties of pure organic substances was proposed by Grigoras (1990). The concept was to relate several properties to the molecular surface areas and electrostatics as generated by combining quantum mechanical results with data to determine the proper form of the correlation. For example, Grigoras related the critical properties to molecular properties via relations such as

$$V_c = 2.217A - 93.0 \tag{2-2.13}$$

$$T_c = 0.633A - 1.562A_- + 0.427A_+ + 9.914A_{HB} + 263.4 \tag{2-2.14}$$

where A is the molecular surface area, A_- and A_+ are the amounts of negatively and positively charged surface area on the molecule and A_{HB} is the amount of charged surface area involved in hydrogen bonding. Examples of values of the surface area quantities are given in the original reference and comparisons are made for several properties of 137 compounds covering many different types. This is the only example where a tabulation of descriptors is available.

These relationships can be used to obtain other properties such as P_c by correlations such as

$$P_c = 2.9 + 20.2(T_c/V_c) \tag{2-2.15}$$

Similar equations are available for liquid molar volume and T_b. Table 2-7 gives comparisons we computed from Eqs. (2-2.14) and (2-2.15) using information given by Grigoras (1990). It can be seen that the accuracy is quite poor. Since the only comparisons given in the original were statistical quality of fits, detailed agreement with the author's results cannot be verified.

Method of Jurs. Jurs and coworkers have produced a series of papers describing extensions and enhancements of molecular descriptor concepts (see, e.g., Egolf, et al., 1994; Turner, et al., 1998). Compared to the early work of Grigoras, the quantum mechanical calculations are more reliable and the fitting techniques more refined so that the correlations should be much better. In particular, the descriptors ultimately used for property estimation are now sought in a sophisticated manner rather than fixing on surface area, etc. as Grigoras did. For example, in the case of T_c, the descriptors are dipole moment, μ, area A_+, a connectivity index, number of oxygens, number of secondary carbon bonds of the sp^3 type, gravitation index, a function of acceptor atom charge, and average positive charge on carbons. A completely different descriptor set was used for P_c. Since the descriptor values must be obtained from a set of calculations consistent with the original fitting, and everything is contained in a single computer program, the particular choice of descriptors is of little importance to the user.

Turner, et al. (1998) list results for T_c and P_c which are compared in Table 2-7 with data of Appendix A. It can be seen that these new results are very good and are generally comparable with the *group/bond/atom* methods. The principal difficulty is that individual access to the computational program is restricted.

TABLE 2-7 Summary of Grigoras and Jurs Methods Compared to Appendix A Data Base

All substances in Appendix A with data that could be tested with the method

Method	Grigoras		Jurs	
Property	T_c, K	P_c, bar	T_c, K	P_c, bar
Equation	(2-2.14)	(2-2.15)	—	—
# Substances[a]	83	83	130	127
AAE[b]	58.50	43.60	6.53	1.45
A%E[b]	10.90	7.23	1.20	3.92
# Err > 10%[c]	39	17	0	11
# Err < 5%[d]	31	37	129	94

[a] The number of substances in Appendix A with data that could be tested with the method.

[b] AAE is average absolute error in the property; A%E is average absolute percent error.

[c] The number of substances for which the absolute percent error was greater than 10%.

[d] The number of substances for which the absolute percent error was less than 5%. The number of substances with errors between 5% and 10% can be determined from the table information.

A discussion comparing the QSPR techniques with other methods for the properties of this chapter is presented below and in Sec. 2-5.

Discussion and Recommendations for Critical Properties. The methods of Joback (1984; 1987), Constantinou and Gani (1994), Wilson and Jasperson (1996) and Marrero and Pardillo (1999) were evaluated. Summaries of comparisons with data from Appendix A are given in Tables 2-1, 2-2, 2-4, and 2-6. A few results from QSPR methods are given in Table 2-7. Overall, the methods are all comparable in accuracy.

A useful method for determining consistency among T_c, P_c, and V_c is to use Eq. (2-3.2) relating the critical compressibility factor, $Z_c = P_cV_c/RT_c$, to the acentric factor (Sec. 2-3). The theoretical basis of the acentric factor suggests that except for substances with $T_c < 100$ K, Z_c must be less than 0.291. When Eq. (2-3.2) was tested on the 142 substances of Appendix A for which reliable values of Z_c and ω are available and for which the dipole moment was less than 1.0 debye, the average absolute percent error in Z_c was 2% with only 9 substances having errors greater than 5%. When applied to 301 compounds of all types in Appendix A, the average percent error was 5% with 32 errors being larger than 10%. Some of these errors may be from data instead of correlation inadequacy. In general, data rather than estimation methods should be used for substances with one or two carbon atoms.

Critical Temperature, T_c. The methods all are broadly applicable, describing nearly all the substances of Appendix A; the average percent of error is around 1% with few, if any substances being off by more than 10%. If an experimental T_b is available, the method of Marrero and Pardillo has higher accuracy than does that of Wilson and Jasperson. On the other hand, for simplicity and breadth of substances, the Wilson/Jasperson method is best since it has the fewest groups to tabulate, is based mostly on atom contributions, and treats inorganic substances as well as organics. Finally, Joback's method covers the broadest range of compounds even though it is somewhat less accurate and more complex.

However, if there is no measured T_b available and estimated values must be used, the errors in these methods increase considerably. Then, if the substance has fewer than 3 carbons, either the Wilson/Jasperson or Marrero/Pardillo method is most reliable; if the substance is larger, the Constantinou/Gani approach generally gives better results with Second Order calculations being marginally better than First Order. The Joback method is somewhat less accurate than these.

The *molecular descriptor* method of Jurs is as accurate as the *group/bond/atom* methods, at least for the substances compared here, though the earlier method of Grigoras is not. While the method is not as accessible, current applications show that once a user has established the capability of computing descriptors, they can be used for many properties.

Critical Pressure, P_c. The methods all are broadly applicable, describing nearly all the substances of Appendix A. All methods give average errors of about 5% with about the same fraction of substances (20%) having errors greater than 10%. The Wilson/Jasperson method has the lowest errors when an experimental value of T_c is used; when T_c is estimated the errors in P_c are larger than the other methods. All show better results for substances with 3 or more carbons, except for a few species. The Constantinou/Gani Second Order contributions do not significantly improve agreement though the Second Order contributions to the Wilson/Jasperson method are quite important. Thus, there is little to choose among the methods. The decision can be based less on accuracy and reliability than on breadth of applicability and ease of use.

The *molecular descriptor* method of Jurs is as accurate as the *group/bond/atom* methods, at least for the substances compared here, though the earlier method of Grigoras is not.

Critical Volume, V_c. The methods of Joback, Constantinou/Gani and Marrero/Pardillo are all broadly applicable, describing nearly all the substances of Appendix A. The Joback method has the lowest error, around 3% with better results for larger substances (3 or more carbons). The Constantinou/Gani results averaged the highest error at about 4.5% for all compounds and 4% for larger ones. For V_c estimation with the CG method, second-order contributions often yield higher error than using only first-order contributions. About 5% of the estimates were in excess of 10% for all methods. There is a little better basis to choose among the methods here, but still a decision based on breadth of applicability and ease of use can be justified.

There have been no *molecular descriptor* methods applied to V_c.

2-3 ACENTRIC FACTOR

Along with the critical properties of Sec. 2-2, a commonly used pure component constant for property estimation is the acentric factor which was originally defined by Pitzer, et al. (1955) as

$$\omega \equiv -\log_{10}\left[\lim_{(T/T_c)=0.7} (P_{vp}/P_c)\right] - 1.0 \tag{2-3.1}$$

The particular definition of Eq. (2-3.1) arose because the monatomic gases (Ar, Kr, Xe) have $\omega \sim 0$ and except for quantum gases (H_2, He, Ne) and a few others (*e.g.*, Rn), all other species have positive values up to 1.5. To obtain values of ω from its definition, one must know the constants T_c, P_c, and the property P_{vp} at the reduced temperature, $T/T_c = 0.7$. Typically, as in Appendix A, a value of ω is obtained by using an accurate equation for $P_{vp}(T)$ along with the required critical properties.

Pitzer's principal application of the acentric factor was to the thermodynamic properties of "normal" fluids (1955). We describe in detail the use and limitations of this correlation in Chaps. 4 and 6, but an example of interest here is for Z_c

$$Z_c = 0.291 - 0.080\omega \tag{2-3.2}$$

There are two useful procedures to estimate an unknown acentric factor. The common and most accurate technique is to obtain (or estimate) the critical constants T_c and P_c and use one or more experimental P_{vp} such as T_b. Then ω can be found from equations given in Chap. 7. We have found the most reliable to be Eq. (7-4.1) with $T_r = T_{br} = T_b/T_c$.

$$\omega = -\frac{\ln(P_c/1.01325) + f^{(0)}(T_{br})}{f^{(1)}(T_{br})} \tag{2-3.3}$$

where P_c is in bars while T_b and T_c are both absolute temperatures. The functions $f^{(0)}$ and $f^{(1)}$ are given in Eqs. (7-4.2) and (7-4.3), respectively. Equation (2-3.3) results from ignoring the term in ω^2 in Eq. (7-4.1) and solving for ω. For 330 compounds of Appendix A, the average absolute deviation in ω is 0.0065 or 2.4%

with only 19 substances having an error greater than 0.02. Retaining the ω^2 term makes almost no difference in the error because $f^{(2)}$ of Eq. (7-4.4) is close to zero for most values of T_{br}.

Example 2-8 Estimate ω for benzene using Eq. (2-3.3).

solution Properties for benzene from Appendix A are T_b = 353.24 K, T_c = 562.05 K and P_c = 48.95 bar.

Then T_{br} = 353.24/562.05 = 0.6285 and $\tau = 1 - T_{br} = 0.3715$.

$$f^{(0)} = \frac{-5.97616(0.3715) + 1.29874(0.3715)^{1.5} - 0.60394(0.3715)^{2.5} - 1.06841(0.3715)^{5}}{0.6285}$$

$$= -3.1575$$

$$f^{(1)} = \frac{-5.03365(0.3715) + 1.11505(0.3715)^{1.5} - 5.41217(0.3715)^{2.5} - 7.46628(0.3715)^{5}}{0.6285}$$

$$= -3.3823$$

Eq. (2-3.3) gives $\omega = -[-3.1575 + \ln(48.95/1.01325)]/(-3.3823) = 0.213$

The value of ω for benzene from Appendix A is 0.210. Error = 0.213 − 0.210 = 0.003 or 1.4%

Using Eq. (2-3.3) is preferable to using empirical vapor pressure equations as described by Chen, et al. (1993) who have used the Antoine Equation (Eq. 7-3.1) to predict ω with an average accuracy of 3.7% for almost 500 compounds. They state that the Antoine Equation shows significant improvement over the Clausius-Clapeyron expression (Eq. 7-2.3) used by Edmister (1958).

It is also possible to directly estimate ω via *group/bond/atom* contribution methods. Only the Constantinou and Gani method (Constantinou, et al., 1995) attempts to do this for a wide range of substances from *group contributions* only. The basic relation of the form of Eq. (2-2.4) is

$$\omega = 0.4085\left\{\ln\left[\sum_k N_k(w1k) + W\sum_j M_j(w2j) + 1.1507\right]\right\}^{(1/0.5050)} \quad (2\text{-}3.4)$$

where the contributions $w1k$ and $w2j$ are given in Appendix Table C-2 and C-3 and the application is made in the same way as described in Sec. (2-2) and (3-4). Example 2-9 shows the method and Table 2-8 gives a summary of the results for ω compared to Appendix A.

Example 2-9 Estimate ω for 2,3,3 trimethylpentane by using Constantinou and Gani's group method.

solution The First-Order groups for are 5 -CH_3, 1 -CH_2, 1 -CH, and 1 -C. The Second-Order group is 1 -CH(CH3)C(CH3)2. From Appendix Tables C-2 and C-3

Group k	N_k	$N_k(w1k)$	Group j	M_j	$N_j\,(w2j)$
CH_3	5	1.4801	$CH(CH_3)C(CH_3)_2$	1	−0.0288
CH_2	1	0.1469			
CH	1	−0.0706			
C	1	−0.3513			
$\sum_{k=1}^{4} N_k w1k$		1.2051	$\sum_{j=1}^{1} M_j w2j$		−0.0288

$$\omega = 0.4085\{\ln[1.2051 + W(-0.0288) + 1.1507]\}^{(1/0.5050)}$$
$$= 0.301 \text{ (First Order, } W = 0)$$
$$= 0.292 \text{ (Second Order, } W = 1)$$

The Appendix A value for ω is 0.291. The differences are 3.4% and 0.3% for the First and Second Order estimates. While these are much lower than given in Table 2-8 for all substances, this is not atypical of estimates for normal fluids.

Table 2-8 shows that the errors can be significant though it covers all kinds of substances, not just "normal" ones.

Discussion and Recommendations for Acentric Factor. The acentric factor defined in Eq. (2-3.1) was originally intended for corresponding states applications of "normal" fluids as defined by Eq. (4.3-2). With care it can be used for predicting properties of more strongly polar and associating substances, though even if the "best" value is used in equations such as Eq. (2-3.2) or those in Sec. (4-3), there is no guarantee of accuracy in the desired property.

TABLE 2-8 Summary of Constantinou/Gani Method for ω Compared to Appendix A Data Base

First Order	All*	$N_c > 2^+$	Second Order	All*	$N_c > 2^+$
# Substances[a]	239	208	# Substances[e]	80	78
AAE[b]	0.050	0.047	AAE[b]	0.048	0.045
A%E[b]	12.73	10.17	A%E[b]	11.98	9.53
# Err > 10%[c]	84	61	# Err > 10%[c]	77	56
# Err < 5%[d]	123	116	# Err < 5%[d]	123	114
			# Better (2nd)[f]	48	44
			Ave. Δ% 1st to 2nd[g]	1.60	0.80

* All Substances in Appendix A with data that could be tested with the method.

\+ All Substances in Appendix A having 3 or more carbon atoms with data that could be tested with the method.

[a] The number of substances in Appendix A with data that could be tested with the method.

[b] AAE is average absolute error in the property; A%E is average absolute percent error.

[c] The number of substances for which the absolute percent error was greater than 10%.

[d] The number of substances for which the absolute percent error was less than 5%. The number of substances with errors between 5% and 10% can be determined from the table information.

[e] The number of substances for which Second-Order groups are defined for the property.

[f] The number of substances for which the Second Order result is more accurate than First Order.

[g] The average improvement of Second Order compared to First Order.

As shown by Liu and Chen (1996), the sensitivity of ω to errors of input information is very great. The recommended procedure for obtaining an unknown value of ω is to use a very accurate correlation for P_{vp} such as Eqs. (7-3.2), (7-3.3) or (7-3.7) directly in Eq. (2-3.1). The next most reliable approach is to use accurate experimental values of T_c, P_c, T_b in Eq. (2-3.3). Finally, the method of Constantinou and Gani with Eq. (2-3.4) can be used with some confidence.

Estimated property values will not yield accurate acentric factors. For example, approximate correlations for P_{vp} such as the Clausius-Clapeyron Equation (7-2.3) as used by Edmister (1958) or the Antoine Equation (7-3.1) as used by Chen, et al. (1993) are about as good as (2-3.4). Further, we tried using this chapter's best estimated values of T_c, P_c, T_b, or estimated T_c and P_c with experimental T_b, or other combinations of estimated and experimental data for nearly 300 substances in Appendix A. The results generally gave large errors, even for "normal" substances. Earlier methods for ω described in the 4th Edition are not accurate or have limited applications.

Along these lines, Chappelear (1982) has observed that "accepted" values of the acentric factor can change due to the appearance of new vapor pressure or critical constant data, changing predicted properties. In addition, using revised acentric factors in a correlation developed from earlier ω values can lead to unnecessary errors. Chappelear's example is carbon dioxide. In Appendix A, we show ω = 0.225; others have quoted a value of 0.267 (Nat. Gas Proc. Assoc., 1981). The differences result from the extrapolation technique used to extend the liquid region past the freezing point to T_r = 0.7. Also, Eq. (2-3-2) yields ω = 0.213. Yet, in the attractive parameter in the Peng-Robinson equation of state (1976) (see Chapter 4), the value should be 0.225, since that was what was used to develop the equation of state relations. One should always choose the value used for the original correlation of the desired property.

2-4 BOILING AND FREEZING POINTS

Boiling and freezing points are commonly assumed to be the phase transition when the pressure is 1 atm. A more exact terminology for these temperatures might be the "normal" boiling and "normal" freezing points. In Appendix A, values for T_{fp} and T_b are given for many substances. Note that estimation methods of Sec. 2-2 may use T_b as input information for T_c. The comparisons done there include testing for errors introduced by using T_b from methods of this section; they can be large.

A number of methods to estimate the normal boiling point have been proposed. Some were reviewed in the previous editions. Several of *group/bond/atom* methods described in Sec. 2-2 have been applied to T_{fp} and T_b, as have some of the *molecular descriptor* techniques of Sec. 2-2. We describe the application of these in a similar manner to that used above for critical properties.

Method of Joback for T_{fp} and T_b. Joback (1984; 1987) reevaluated Lydersen's *group contribution* scheme, added several new functional groups, and determined new contribution values. His relations for T_{fp} and T_b are

$$T_{fp} = 122 + \sum_k N_k(tfpk) \tag{2-4.1}$$

$$T_b = 198 + \sum_k N_k(tbk) \tag{2-4.2}$$

where the contributions are indicated as *tfpk* and *tbk*. The group identities and

Joback's values for contributions to the critical properties are in Table C-1. Example 2-10 shows the method.

Example 2-10 Estimate T_{fp} and T_b for 2,4-dimethylphenol by using Joback's group method.

solution 2,4-dimethylphenol contains two —CH_3, three =CH(ds), three =C(ds), and one —OH (phenol). From Appendix Table C-1

Group k	N_k	$N_k(tfpk)$	$N_k(tbk)$
—CH_3	2	−10.20	47.16
=CH(ds)	3	24.39	80.19
=C(ds)	3	111.06	93.03
—ACOH (phenol)	1	82.83	76.34
$\sum_{k=1}^{5} N_k F_k$		208.08	296.72

The estimates are:

$$T_{fp} = 122 + \sum_k N_k(tfpk) = 330.08 \text{ K}$$

$$T_b = 198 + \sum_k N_k(tbk) = 494.72 \text{ K}$$

Appendix A values for these properties are T_{fp} = 297.68K and T_b = 484.09 K Thus the differences are

$$T_{fp} \text{ Difference} = 297.68 - 330.08 = -32.40 \text{ K or } -10.9\%$$

$$T_b \text{ Difference} = 484.09 - 494.72 = -10.63 \text{ K or } -2.2\%$$

Devotta and Pendyala (1992) modified the Joback method to more accurately treat T_b of halogenated compounds. They report that the average percent deviations for refrigerants and other substances was 12% in the original method; this is consistent with our comparison and is much larger than the overall average given below. Devotta and Pendyala did not change Joback's basic group contribution values; they only changed the groups and values for halogen systems. Their results showed an average percent deviation of 6.4% in T_b.

A summary of the comparisons between estimations from the Joback method and experimental Appendix A values for T_{fp} and T_b are shown along with those from other methods in Tables 2-9 and 2-10 below.

Method of Constantinou and Gani (CG) for T_{fp} and T_b. Constantinou and Gani (1994, 1995) developed an advanced *group contribution* method based on the UNIFAC groups (see Chap. 8) but enhanced by allowing for more sophisticated functions of the desired properties and by providing contributions at a "Second Order" level (see Secs. 2-2 and 3-3 for details).

For T_{fp} and T_b, the CG equations are

$$T_{fp} = 102.425 \ln \left[\sum_k N_k(tfp1k) + W \sum_j M_j (tfp2j) \right] \tag{2-4.3}$$

$$T_b = 204.359 \ln \left[\sum_k N_k(tb1k) + W \sum_j M_j (tb2j) \right] \tag{2-4.4}$$

The group values $tfp1k$, $tfp2j$, $tb1k$, and $tb2j$ for Eqs. (2-4.3) and (2-4.4) are given in Appendix Tables C-2 and C-3 with sample assignments shown in Table C-4. Examples 2-11 and 2-12 illustrate this method.

Example 2-11 Estimate T_{fp} and T_b for 2,4-dimethylphenol using Constantinou and Gani's group method.

solution The First-Order groups for 2,4-dimethylphenol are three ACH, two $ACCH_3$, and one ACOH. There are no Second-Order groups so the First Order and Second Order calculations are the same. From Appendix Tables C2 and C3

Group k	N_k	$N_k tfp1k$	$N_{ik} tb1k$
ACH	3	4.4007	2.7891
$ACCH_3$	2	3.7270	3.9338
ACOH	1	13.7349	4.4014
$\sum_{k=1}^{5} N_k F_k$		21.8626	11.1243

$$T_{fp} = 102.425 \ln (21.8626) = 315.96 \text{ K}$$

$$T_b = 204.359 \ln (11.1243) = 492.33 \text{ K}$$

Appendix A values for these properties are $T_{fp} = 297.68$K and $T_b = 484.09$ K. Thus the differences are

$$T_{fp} \quad \text{Difference} = 297.68 - 315.96 = -18.28 \text{ K or } -6.1\%$$

$$T_b \quad \text{Difference} = 484.09 - 492.33 = -8.24 \text{ K or } -1.7\%$$

Example 2-12 Estimate T_{fp}, and T_b for five cycloalkanes with formula C_7H_{14} using Constantinou and Gani's group method.

solution The First- and Second-Order groups for the cycloalkanes are:

	cycloheptane	methyl cyclohexane	ethyl cyclopentane	cis-1,3-dimethyl cyclopentane	trans-1,3-dimethyl cyclopentane
First-Order groups, N_k					
CH_3	0	1	1	2	2
CH_2	7	5	5	3	3
CH	0	1	1	2	2
Second-Order groups, M_j					
7-ring	1	0	0	0	0
6-ring	0	1	0	0	0
5-ring	0	0	1	1	1
Alicyclic Side Chain	0	1	1	0	0

All of the substances have one or more Second-Order groups. Using values of group contributions from Appendix Tables C-2 and C-3 and experimental values from Appendix A, the results are

Property	cycloheptane	methyl cyclohexane	ethyl cyclopentane	cis-1,3-dimethyl cyclopentane	trans-1,3-dimethyl cyclopentane
T_{fp}, K					
Experimental	265.15	146.56	134.70	139.45	139.18
Calculated (First Order)	191.28	173.54	173.54	152.06	152.06
Abs. percent Err. (First Order)	27.85	18.41	21.41	9.04	9.25
Calculated (Second Order)	266.15	146.46	122.14	166.79	166.79
Abs. percent Err. (Second Order)	0.38	0.07	9.32	19.64	19.84
T_b, K					
Experimental	391.95	374.09	376.59	364.71	363.90
Calculated (First Order)	381.18	369.71	369.71	357.57	357.57
Abs. percent Err. (First Order)	2.75	1.17	1.82	1.96	1.77
Calculated (Second Order)	391.93	377.81	377.69	367.76	367.76
Abs. percent Err. (Second Order)	0.00	0.99	0.29	0.84	1.06

The First Order results are generally good for boiling but not melting. The Second Order contributions are significant in all cases and improve the agreement for all but T_{fp} for the dimethylpentanes, where the correction goes in the wrong direction. The errors shown are about average for the method.

A summary of the comparisons between estimations from the Second Order Constantinou and Gani method and experimental Appendix A values for T_{fp} and T_b are shown along with those from other estimation methods in Tables 2-9 and 2-10.

Method of Marrero and Pardillo for T_b. Marrero-Marejón and Pardillo-Fontdevila (1999) give two equations for estimating T_b. They call their preferred method a group interaction contribution technique; it can also be considered as a method of *bond contributions*. They tabulate contributions from 167 pairs of atoms alone, such as >C< & —N<, or with hydrogen attached, such as CH_3— & —NH_2 (see Table 2-5 and the discussion of section 2-2). For T_b their basic equation is

$$T_b = M^{-0.404} \sum_k N_k(tbbk) + 156.00 \tag{2-4.5}$$

where M is the molecular weight and N_k is the number of atoms of type k with contributions *tbbk*. The method is shown in Examples 2-13 and 2-14.

Example 2-13 Estimate T_b for 2,4-dimethylphenol by using Marrero and Pardillo's method.

solution The chemical structure and the required locations of the various bond pairs are indicated on the structure shown (see the discussion in Sec. 2-2 and Examples 2-5 and 2-6 about this important aspect of the model). The molecular weight is 122.167.

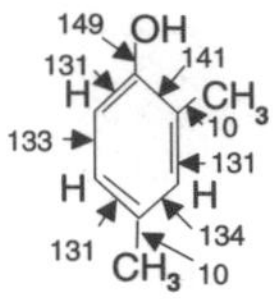

Pair #	Atom/Group Pair	N_k	$N_k tbbk$
10	—CH3 & =C< [r]	2	291.12
131	—CH[=] [r] & >C[=] [r]	3	873.45
133	=CH— [r] & =CH— [r]	1	285.07
134	=CH— [r] & =C< [r]	1	237.22
141	=C< [r] & =C< [r]	1	180.07
149	=C< [r] & —OH	1	456.25
—	$\sum_{k=1}^{7} N_k F_k$	—	2323.18

The estimate using Eq. (2-4.5) is

$$T_b = 122.167^{-0.404}(2323.18) + 156.00 = 489.49 \text{ K}$$

The Appendix A value is T_b = 484.09 K. Thus the difference is

$$T_b \text{ Difference} = 484.09 - 489.49 = -5.40 \text{ K or } -1.1\%$$

Example 2-14 Estimate T_b for five cycloalkanes with formula C_7H_{14} using Marrero and Pardillo's method.

solution The group pairs for the cycloalkanes are:

Pair #	Atom/Group Pair	cyclo-heptane	methyl cyclo-hexane	ethyl cyclo-pentane	cis-1,3-dimethyl cyclo-pentane	trans-1,3-dimethyl cyclo-pentane
2	CH_3— & —CH_2—	0	0	1	0	0
8	CH_3— & >CH— [r]	0	1	0	2	2
35	—CH_2— & >CH— [r]	0	0	1	0	0
112	—CH_2— [r] & —CH_2— [r]	7	4	3	1	1
113	—CH_2— [r] & >CH— [r]	0	2	2	4	4

Using values of bond contributions from Table 2-5 and experimental values from Appendix A, the results are:

T_b, K	cycloheptane	methyl cyclohexane	ethyl cyclopentane	cis-1,3-dimethyl cyclopentane	trans-1,3-dimethyl cyclopentane
Experimental	391.95	374.09	376.59	364.71	363.90
Calculated	377.52	375.84	384.47	374.16	374.16
Abs. percent Err.	3.68	0.47	2.09	2.59	2.54

These errors are a little above this method's average.

A summary of the comparisons between estimations from the Marrero and Pardillo method and experimental Appendix A values for T_b are shown along with those from other estimation methods in Table 2-10.

Other Methods for Normal Boiling and Normal Freezing Points. There are several other *group/bond/atom* and *molecular descriptor* methods that have been applied for estimating T_{fp} and T_b. Most of the former are restricted to individual classes of substances such as hydrocarbons or complex substances such as triglycerides (Zeberg-Mikkelsen and Stenby, 1999). The *molecular descriptor* techniques are based on properties of the molecules of interest which are not normally measurable. As described in Sec. 2-2, these "Quantitative structure-property relationships (QSPR)" are usually obtained from on-line computation from quantum mechanical methods. Thus, in most of these methods, there is no tabulation of descriptor values. Katritzky, et al. (1998) summarize the literature for such methods applied to T_b. Egolf, et al. (1994), Turner, et al. (1998), and St. Cholokov, et al. (1999) also give useful descriptions of the procedures involved.

We present here comparisons with T_{fp} and T_b data of Appendix A for some recent methods. The method of Yalkowsky (Yalkowsky, et al., 1994; Krzyzaniak, et al., 1995; Zhao and Yalkowsky, 1999) for both properties is a hybrid of group contributions and molecular descriptors; direct comparisons are possible for T_{fp}. We also examine the more extensive results for T_b published for full molecular descriptor methods of Katrizky, et al. (1996) and Wessel and Jurs (1995). The early method of Grigoras (1990) described above is not as bad for T_b as for the critical properties, but is not as good as the others shown here. Finally, in Sec. 2-5, we comment on how these techniques compare with the *group/bond/atom* methods.

Method of Yalkowsky for T_{fp} and T_b. Yalkowsky and coworkers (Yalkowsky, et al., 1994; Krzyzaniak, et al., 1995; Zhao and Yalkowsky, 1999) have explored connections between T_{fp} and T_b as well as have proposed correlations for T_{fp}. This is part of an extensive program to correlate many pure component and mixture properties of complex substances (Yalkowsky, et al., 1994). The method consists of both *group contributions* which are additive and *molecular descriptors* which are not additive. The main properties of the latter are the symmetry number, σ, and the flexibility number, Φ. The former is similar to that used for ideal gas properties (see Sec. 3-5 and Wei, 1999) and the latter is strictly defined as the inverse of the probability that a molecule will be in the conformation of the solid crystalline phase of interest. It is argued that flexibility affects both melting and boiling while symmetry affects only melting. The methodology has been to determine easily accessible molecular properties to estimate the flexibility contribution while values of σ are to be obtained directly from molecular structure such as described in Sec. 3-4 for the Benson group method for ideal gas properties. Thus, Krzyzanaik, et al. (1995) and Zhao and Yalkowsky (1999) use

$$T_b = \sum_k N_k(b_k)/(86 + 0.4\tau) \tag{2-4.6}$$

$$T_{fp} = \sum_k N_k(m_k)/(56.5 - 19.2\log_{10}\sigma + 9.2\tau) \tag{2-4.7}$$

where b_k and m_k are selected from among 61 group contribution terms and 9 molecular correction terms that they tabulate. The value of τ is estimated by

$$\tau = \text{SP3} + 0.5\text{SP2} + 0.5\text{RING} - 1 \tag{2-4.8}$$

Here, SP3 is the number of "non-ring, nonterminal sp^3 atoms," SP2 is the number of "non-ring, nonterminal sp^2 atoms," and RING is the number of "monocyclic fused ring systems in the molecule." Examples of the appropriate assignments are given in the original papers.

Though no detailed comparisons have been made with this method for T_b, the authors report their average deviations were 14.5 K compared to an average of 21.0 for Joback's method, Eq. (2-4.2). Table 2-9 below shows about 17 K as our average for Joback's method, which is consistent given the probable difference of data bases. Thus, Yalkowsky's method represents some improvement though the T_b errors with the more recent *group/bond/atom* methods are much less. For example, we find that Marrero and Pardillo's technique has an average error of 7.5 K. For T_{fp}, Zhao and Yalkowsky provide a table of 1040 substances with an average error of about 34 K. Direct comparison of their predictions with our data base show a similar average error of 26.7 K.

A summary of the comparisons between estimations from the Zhao and Yalkowsky method and experimental Appendix A values for T_{fp} are shown along with those from other estimation methods in Table 2-9.

Method of Jurs for T_b. Jurs and coworkers have produced a series of papers describing extensions and enhancements of *molecular descriptor* concepts, especially to T_b. The significant descriptors are sought in a sophisticated manner; for T_b of complex organics, these include the partial positive and negative surface areas, the relative positive charge, the number of ring atoms, the molecular weight, the surface area of donatable hydrogens, the number of fluorine atoms, a "ketone indicator," the number of sulfide groups, and the fraction of atoms that are sulfur. This set is different from those used for critical properties (see Sec. 2-2) and is also different from that used for hydrocarbons. However, since descriptor values are obtained from a set of calculations consistent with the original fitting, and everything is contained in a single computer program, the particular choice is of little importance to the user.

While we have not been able to do calculations with the model due to not having values of the molecular descriptors, Wessel and Jurs (1995) list results for 633 substances, many of which can be compared with the data of Appendix A in Table 2-6 as shown in Table 2-10.

Method of Katritzky for T_b. Katritzky and coworkers (1996, 1998, 1999) have developed another *molecular descriptor* approach and applied it to T_b for diverse organic compounds. Their 8 descriptors include the "gravitation index," a "hydrogen bonding descriptor," surface area of hydrogen acceptors, fraction of atoms that are fluorine, number of nitriles, a "topographic electronic index" and the charged surface area of the hydrogens and of the chlorines. They have tabulated results for almost 900 compounds, many of which can be compared with the data of Appendix A. Table 2-10 compares the published results for T_b with 175 substances of Appendix A.

Method of St. Cholakov, et al. for T_b. St. Cholakov, et al. (1999) have also developed a *molecular descriptor* method applied initially for T_b of hydrocarbons. They use 8 descriptors. The group/atom descriptors are: number of carbons, number of CH_2 groups, and number of >C< groups, number of carbons in —CH=HC— groups. The molecular mechanics descriptors are: total energy, bond energy, "van der Waals energy," and unsaturated van der Waals surface area. They say their estimates of T_b are essentially at the accuracy of the data.

Summary for T_{fp} and T_b. Table 2-9 summarizes our results for T_{fp} estimations. Results from the Second Order estimations for the Constantinou/Gani (CG) method are listed. All of the methods are similar and none are very reliable. As many as one-half of the estimates are in error by more than 10%. There is no general pattern to the errors, though Yalkowsky's method consistently under predicts T_{fp} of long chain substances.

TABLE 2-9 Estimation Methods for T_{fp} Compared to Appendix A Data Base

Method	Joback	CG	Yalkowsky
Equation	(2-4.1)	(2-4.3)	(2-4.7)
# Substances[a]	307	273	146
AAE[b]	28.8	25.8	26.70
A%E[b]	14.4	13.2	15.10
# Err > 10%[c]	154	116	80
# Err < 5%[d]	80	80	35

[a] The number of substances in Appendix A with data that could be tested with the method.

[b] AAE is average absolute error in the property; A%E is average absolute percent error.

[c] The number of substances for which the absolute percent error was greater than 10%.

[d] The number of substances for which the absolute percent error was less than 5%. The number of substances with errors between 5% and 10% can be determined from the table information.

TABLE 2-10 Summary of Estimation Methods for T_b Compared to Appendix A Data Base

Method	Joback	Constantinou/ Gani	Marrero/ Pardillo	Katritzky	Jurs
Equation	(2-4.2)	(2-4.4)	(2-4.5)	—	—
# Substances[a]	353	341	347	175	242
AAE[b]	16.8	13.4	7.5	9.20	5.30
A%E[b]	5.0	4.0	2.0	2.70	1.50
# Err > 10%[c]	42	39	10	6	3
# Err < 5%[d]	242	279	318	154	231

[a] The number of substances in Appendix A with data that could be tested with the method.

[b] AAE is average absolute error in the property; A%E is average absolute percent error.

[c] The number of substances for which the absolute percent error was greater than 10%.

[d] The number of substances for which the absolute percent error was less than 5%. The number of substances with errors between 5% and 10% can be determined from the table information.

Table 2-10 summarizes the results of the methods for T_b estimations. It can be seen that the Jurs *molecular descriptor* method is the most accurate with the Marrero/Pardillo *bond contribution* technique also quite reliable. All of the others yield larger average errors and they often describe individual systems poorly. As mentioned above, the method of Yalkowsky is not as accurate as the best methods here.

2-5 DISCUSSION OF ESTIMATION METHODS FOR PURE COMPONENT CONSTANTS

This chapter has described a variety of methods for predicting critical properties, acentric factor and normal boiling and freezing temperatures. Unlike previous editions where most of the methods were of the *group/bond/atom* type and were limited in the classes of properties or types of substances treated, recent work in these techniques has both improved their accuracy and broadened their range. Thus, there are now two methods (Joback and Constantinou/Gani) that provide all pure component constants and heat capacities and properties of formation of ideal gases (see Chap. 3) with a single group formulation.

All of the *group/bond/atom* methods examined here were set up on spreadsheets since their application was the same regardless of the property and component. Some methods required larger data bases than others, but implementation and execution for new substances and properties would be straightforward. It is also possible to obtain a complete suite of estimation methods in the program Cranium (1998). The Constantinou/Gani method for the broadest set of properties is available directly from CAPEC (1999). It is expected that methods of currently limited application, such as the Marrero/Pardillo approach, will be expanded to include other properties. There are still enough errors and limitations in the methods that new research will continue with this approach. It is likely that an individual user with a typical individual computer will be able to use both current and future versions of these methods.

This edition has also introduced the *molecular descriptor* and QSPR relations which add another dimension to the methodology since they can be applied not only to pure component constants but to a variety of solution systems (Mitchell and Jurs, 1998; Katritzsky, et al., 1998). This presents users with opportunities to obtain more reliable values, but also may require greater expertise and investment in the selection of computer software for estimations. As mentioned above, there is no tabulation of contributions for these methods since the molecular structure and descriptors of each new substance are computed from molecular and quantum mechanical programs. While complex, the estimation methods are established by a generally agreed upon process of fitting limited data (Mitchell and Jurs, 1996; St. Cholakov, et al., 1999) to establish the weights of the significant descriptors from a large set of possibilities. As described above, the results can be very good and it is likely that further improvements in computational techniques will add even greater reliability and applicability. However, the computational power required is extensive and care must be exercised to use the same computational programs as the developer in order to insure that the values for the descriptors will be consistent with those fitted. This is likely to require expertise and computers of a large organization and beyond that of an individual. At this time, these methods have not been implemented in process simulation software, but that would certainly be possible in the future.

It will be important that users follow the developments in this area so that the most prudent decisions about investment and commitment can be made.

2-6 DIPOLE MOMENTS

Dipole moments of molecules are often required in property correlations for polar materials such as for virial coefficients (Chap. 4) and viscosities (Chap. 9). The best sources of this constant are the compilations by McClellan (1963] and Nelson, et al. [1967). There also is a large number of values in the compilation of Lide (1996). For the rare occasions when an estimated value is needed, vector group contribution methods such as summarized by Minkin, et al. (1970) can be used. In addition, all of the programs used for *molecular descriptors* yield molecular dipole moments as a part of the analysis.

Dipole moments for many materials are listed in Appendix A. They do not vary with temperature and we have ignored the difference between gas and liquid phase values. Such differences are not large enough to affect the estimation result.

It should be noted that the dipole is only the lowest of a series of electrostatic effects on intermolecular forces; higher order terms such as quadrupoles can also be important such as for CO_2. It is often of interest to determine whether electrostatic contributions are significant compared to van der Waals attraction (dispersion). The theory of intermolecular forces (Prausnitz, et al., 1999) shows that the importance of the dipolar forces depends on the ratio of electrostatic to van der Waals energies which can be estimated in dimensionless fashion by

$$\mu^* = N^2_A \mu^2 / RT_c V_c = 4300\mu^2 / T_c V_c \tag{2-6.1}$$

where μ is the dipole moment and N_A is Avogadro's number. The factor in Eq. (2-6.1) is for T_c in K, V_c in cm^3 mol^{-1} (as in Appendix A) and μ in debye units, a debye being 3.162×10^{-25} (J-m^3)$^{1/2}$. It is estimated that if μ^* of Eq. (2-6.1) is less than 0.03, dipolar effects can be neglected. Another estimate can be made using the surface tension test of Pitzer (Eq. 4-3.2); if the substance is "normal," polar forces are not important.

2-7 AVAILABILITY OF DATA AND COMPUTER SOFTWARE

There are several readily available commercial products for obtaining pure component constants. These include data and correlation-based tabulations and computer-based group contribution methods. Those which were used in developing this chapter are referenced below or in Appendix C including Web sites as of the date of publication. The data for Appendix A were obtained from the Thermodynamics Research Center at Texas A&M University (TRC, 1999); there is a similar tabulation available from DIPPR (1999). Joback has established a program (Cranium, 1998) for computing many properties by various group contribution methods though the current version only includes the Joback version for ideal gas properties. Gani and coworkers at the Center for Computer-Aided Process Engineering (CAPEC) at the Danish Technical University also have a program available (ProPred, 1999) for many properties including the Joback, Constantinou/Gani and Wilson/Jasperson methods. The *molecular descriptor* methods can be obtained by contacting the developers directly (St. Cholakov, et al. 1999; Turner, et al., 1998; Katritzky, et al., 1999; Zhao and Yalkowsky, 1999).

NOTATION

A, A_+, A_-, A_{HB}	molecular areas in Grigoras method, Eqs. (2-2.13) and (2-2.14)
b_k, m_k	group contribution and molecular correction terms in the Yalkowsky method, Eqs. (2-4.6) and (2-4.7)

F, F_{1k}, F_{2j}	properties in the method of Constantinou and Gani, Eq. (2-2.4)
$f^{(0)}$, $f^{(1)}$, $f^{(2)}$	functions in Pitzer's correlation for P_{vp}, Eqs. (2-3.3) and (7-4.2) to (7-4.4)
M	molecular weight of substance
N_A	Avogadro's Number, 6.022142×10^{23} molecules mol^{-1}
N_{atoms}	number of atoms in substance
N_k, M_j	number of groups of type k in a molecule; N_k for First-Order groups in all methods and M_j for Second-Order groups in Constantinou/Gani, Eq. (2-2.4), and Wilson /Jasperson methods, Eqs. (2-2.8) and (2-2.9b)
N_r	number of rings in substance, Eqs. (2-2.8) and (2-2.9b)
P_c	critical pressure, bar
pc1k, tb1k, tc1k, tfp1k, vc1k, w1k	First-Order group Contributions for Constantinou/Gani method, Table C-2.
pc2j, tb2j, tc2j, tfp2j, vc2j, w2j	Second-Order group Contributions for Constantinou/Gani method, Table C-3.
pck, tbk, tck, tfpk, vck	Group Contributions for Joback method, Table C-1.
Δpck, Δtck	First-Order Group Contributions for Wilson/ Jasperson method, Table 2-3A.
Δpcj, Δtcj	Second-Order Group Contributions for Wilson /Jasperson method, Table 2-3B.
pcbk, tbbk, tcbk, vcbk	Group Contributions for Marrero/Pardillo method, Table 2-5.
R	gas constant, 8.31447 J mol^{-1} K^{-1}
SP3, SP2, RING	number of various bond types in Yalkowsky method, Eq. (2-4.8)
T	absolute temperature, K
T_b	atmospheric boiling temperature, K
T_c	vapor-liquid critical temperature, K
T_{fp}	atmospheric freezing/melting temperature, K
V_c	critical volume, cm^3 mol^{-1}
W	weight for Second-Order groups in Constantinou/Gani method; = 0 for First Order only, = 1 for full estimation
Y	function in Wilson-Jasperson method for critical pressure, Eq. (2-2.9)

Greek

μ	dipole moment
μ^*	reduced dipole moment, Eq. (2-6.1)
τ	sum of bond types in Yalkowsky method, Eqs. (2-4.6) and (2-4.7)
σ	symmetry number in Yalkowsky method, Eq. (2-4.7)

REFERENCES

Abildskov, J.: "Development of a New Group Contribution Method," M.S. Thesis, Inst. for Kemiteknik, Danish Tech. Univ., 1994.

Ambrose, D.: "Correlation and Estimation of Vapour-Liquid Critical Properties. I. Critical Temperatures of Organic Compounds," National Physical Laboratory, Teddington, *NPL Rep. Chem. 92.*, 1978, 1980.

Ambrose, D.: "Correlation and Estimation of Vapour-Liquid Critical Properties. I. Critical Temperatures of Organic Compounds," National Physical Laboratory, Teddington, *NPL Rep. Chem. 98,* 1979.

Ambrose, D., and C. L. Young: *J. Chem. Eng. Data,* **40:** 345 (1995).

Brulé, M. R., C. T. Lin, L. L. Lee, and K. E. Starling: *AIChE J.*, **28:** 616 (1982).

Chappelear, P. S.: *Fluid Phase Equil.,* **9:** 319 (1982).

Chen, D. H., M. V. Dinivahi, and C.-Y. Jeng:, *IEC Res.,* **32:** 241 (1993).

Constantinou, L., and R. Gani: *AIChE J.,* **40:** 1697 (1994).

Constantinou, L., R. Gani, and J. P. O'Connell: *Fluid Phase Equil.,* **104:** 11 (1995).

Cranium Version 1.0: Molecular Knowledge Systems, Inc., PO Box 10755, Bedford, NH 03110-0755 USA; http://www.molknow.com, 1998.

Daubert, T. E., and R. Bartakovits: *IEC Res.,* **28:** 638 (1989).

Design Institute for Physical Properties Research (DIPPR): American Institute of Chemical Engineers, http://www.aiche.org/dippr, 1999.

Devotta, S., and V. R. Pendyala: *IEC Res.,* **31:** 2046 (1992).

Edmister, W. C.: *Pet. Refin.,* **37:** 173 (1958).

Egolf, L. M., M. D. Wessel, and P. C. Jurs: *J. Chem. Inf. Comput. Sci.,* **34:** 947 (1994).

Fedors, R. F.: *Chem. Eng. Comm.:* **16:** 149 (1982).

Grigoras, S.: *J. Comp. Chem.,* **11:** 493 (1990).

Horvath, A. L.: "Molecular Design: Chemical Structure Generation from the Properties of Pure Organic Compounds," Elsevier, 1992.

Jalowka, J. W., and T. E. Daubert: *Ind Eng. Chem. Proc. Des. Dev.,* **25:** 139 (1986).

Jianzhong, Z., Z. Biao, Z. Suoqi, W. Renan, and Y. Guanghua: *IEC Res.,* **37:** 2059 (1998).

Joback, K. G.: "A Unified Approach to Physical Property Estimation Using Multivariate Statistical Techniques," S.M. Thesis, Department of Chemical Engineering, Massachusetts Institute of Technology, Cambridge, MA, 1984.

Joback, K. G., and R. C. Reid: *Chem. Eng. Comm.,* **57:** 233 (1987).

Katritzky, A. R., V. S. Lobanov, and M. Karelson: *Chem Soc. Rev.,* **24:** 279 (1995).

Katritsky, A. R., V. S. Lobanov, and M. Karelson: *J. Chem. Inf. Comput. Sci,* **38:** 28 (1998).

Katritzky, A. R., L. Mu, V. S. Lobanov, and M. Karelson: *J. Phys. Chem.,* **100:** 10400 (1996).

Katritzky, A. R., T. Tamm, Y. Wang, S. Sild, and M. Karelson: *J. Chem. Inf. Comput. Sci.,* **39:** 684 (1999).

Klincewicz, K. M., and R. C. Reid: *AIChE J.,* **30:** 137 (1984).

Krzyzaniak, J. F., P. B. Myrdal, P. Simamora, and S. H. Yalkowsky: *IEC Res.,* **34:** 2530 (1995).

Lange's Handbook of Chemistry, 15th ed., New York, McGraw-Hill, 1999.

Lin, H.-M., and K. C. Chao: *AIChE J.,* **30:** 981 (1984).

Liu, L., and S. Chen: *IEC Res.,* **35:** 2484 (1996).

Lydersen, A. L.: "Estimation of Critical Properties of Organic Compounds," *Univ. Wisconsin Coll. Eng., Eng. Exp. Stn. rept. 3,* Madison, WI, April, 1955.

Marrero-Marejón, J., and E. Pardillo-Fontdevila: *AIChE J.,* **45:** 615 (1999).

Mitchell, B. E., and P. C. Jurs: *Chem. Inf. Comp. Sci.,* **38:** 489 (1998).

Nat. Gas Proc. Assoc. (NGPA): *Engineering Data Book,* 1981.

Nielsen, T. L.: "Molecular Structure Based Property Prediction," 15-point Project Department of Chemical Engineering, Technical University of Denmark., Lyngby, DK-2800, 1998.

Peng, D.-Y. and D. B. Robinson: *Ind. Eng. Chem. Fund.,* **15:** 59 (1976).

Pitzer, K. S.: *J. Am. Chem. Soc.,* **77:** 3427 (1955).

Pitzer, K. S., D. Z. Lippmann, R. F. Curl, C. M. Huggins, and D. E. Petersen: *J. Am. Chem. Soc.,* **77:** 3433 (1955).

ProPred, Computer Aided Process Engineering Center, Danish Technical University, DK-2800 Lyngby Denmark; http://www.capec.kt.dtu.dk, 1999.

Riazi, M. R., and T. E. Daubert: *Hydrocarbon Process. Petrol. Refiner,* **59**(3): 115 (1980).

St. Cholokov, G., W. A. Wakeham, and R. P. Stateva: *Fluid Phase Equil.,* **163:** 21 (1999).

Sastri, S. R. S., S. Mohanty, and K. K. Rao: *Fluid Phase Equil.,* **129:** 49 (1997).

Somayajulu, G. R.: *Int. J. Thermophys.,* **12:** 1039 (1991).

Teja, A. S., and M. J. Anselme: *AIChE Symp. Ser.,* **279:** 115, 122 (1990).

Tobler, F. C.: *IEC Res.,* **35:** 811 (1996).

Thermodynamics Research Center (TRC), Texas A&M University, http://trcweb.tamu.edu, 1999.

Tsonopoulos, C., J. L. Heidman, and S. C. Hwang: *Thermodynamic and Transport Properties of Coal Liquids*, New York, Wiley, 1986.

Tu, C. H.: *Chem. Eng. Sci.,* **50:** 3515 (1995).

Turner. B. E., C. L. Costello, and P. C. Jurs: *J. Chem. Inf. Comput. Sci.,* **38:** 639 (1998).

Twu, C.: *Fluid Phase Equil.,* **16:** 137 (1984).

Twu, C. H. and J. E. Coon: *Fluid Phase Equil.,* **117:** 233 (1996).

Vetere, A.: *Fluid Phase Equil.,* **109:** 17 (1995); **119:** 231 (1996).

Wei, J.: *IEC Res.,* **38:** 5019 (1999).

Wessel, M. D., and P. C. Jurs: *J. Chem. Inf. Comput. Sci.,* **35:** 68, 841 (1995).

Wilson, G. M., R. H. Johnston, S. C. Hwang, and C. Tsonopoulos: *Ind. Eng. Chem. Proc. Des. Dev.,* **20:** 94 (1981).

Wilson. G. M., and L. V. Jasperson: "Critical Constants T_c, P_c, Estimation Based on Zero, First and Second Order Methods," AIChE Spring Meeting, New Orleans, LA, 1996.

Yalkowsky, S. H., R.-M. Dannenfelser, P. Myrdal, and P. Simamora: *Chemosphere,* **28:** 1657 (1994).

Zeberg-Mikkelsen, C. K., and E. H. Stenby: *Fluid Phase Equil.,* **162:**7 (1999).

Zhao, L., and S. H. Yalkowsky: *IEC Res.,* **38:** 3581 (1999).

CHAPTER THREE
THERMODYNAMIC PROPERTIES OF IDEAL GASES

3-1 SCOPE AND DEFINITIONS

Methods are described to estimate the standard-state enthalpy and Gibbs energy of formation, $\Delta H_f^\circ(T)$ and $\Delta G_f^\circ(T)$ respectively, and the entropy for organic compounds in the ideal gas standard state, $S^\circ(T)$. The *reference* temperature is 298.15 K, and the reference pressure is one *atmosphere* (1.01325×10^5 Pa). In addition, techniques are given for estimating the ideal-gas heat capacity, $C_p^\circ(T)$, as a function of temperature.

The enthalpy of formation is defined as the enthalpy change to form a species from chemical elements in their *standard* states by an isothermal reaction. In such a reaction scheme, the elements are assumed initially to be at the reaction temperature, at 1 atm., and in their most stable phase, e.g., diatomic oxygen as an ideal gas, carbon as a solid in the form of β-graphite, bromine as a pure saturated liquid, etc. Numerical values of properties of the constituent elements are not of concern, since, when the standard enthalpy of a reaction with several species is calculated, all the enthalpies of formation of the elements cancel. For a reaction at other than *standard* conditions, corrections must be made such as for fluid nonidealities.

Any reaction can be written in mathematical notation as

$$\sum_{i=\text{species}} \nu_i(A_i) = 0 \tag{3-1.1}$$

where the species (reactants and products) are identified by the subscript i and are named A_i. The stoichiometric coefficients ν_i are positive for products and negative for reactants. An example of this notation is steam oxidation of propane which is usually written as

$$-1(C_3H_8) - 3(H_2O) + 3(CO) + 7(H_2) = 0$$

where the names are in parentheses and the stoichiometric coefficients for propane (C_3H_8), water (H_2O), carbon monoxide (CO), and hydrogen (H_2) are -1, -3, 3 and 7 respectively. In more familiar form, this would be

$$(C_3H_8) + 3(H_2O) = 3(CO) + 7(H_2)$$

The purpose of the notation of Eq. (3-1.1) is to express more compactly the properties associated with the reaction. Thus, the enthalpy change when stoichiometric amounts of reactants are reacted to completion in the standard state (ideal gases at T, 1 atm.) is obtained from the enthalpies of formation of the species at the same temperature, $\Delta H^\circ_{fi}(T)$

$$\Delta H^\circ(T) = \sum_i \nu_i \Delta H^\circ_{fi}(T) \tag{3-1.2}$$

where

$$\Delta H^\circ_{fi}(T) = H^\circ_i(T) - \sum_e \nu_{ei} H^\circ_e(T) \tag{3-1.3}$$

where ν_{ei} is the number of atoms of an element of type e that are found in species i. Note that some elements are diatomic. Thus, for propane, $\nu_{eC} = 3$, $\nu_{eH_2} = 4$ while for carbon monoxide, $\nu_{eC} = 1$, $\nu_{eO_2} = ½$. Since all of the values of $H^\circ_e(T)$ cancel out in Eq. (3-1.2), their values are never obtained explicitly. For our steam oxidation example, Eq. (3.1.2) becomes

$$\Delta H^\circ(T) = 3\Delta H^\circ_{fCO}(T) + 7\Delta H^\circ_{fH_2}(T) - \Delta H^\circ_{fC_3H_8}(T) - 3\Delta H^\circ_{fH_2O}(T) \tag{3-1.4}$$

Enthalpies of formation, ΔH°_f(298.15 K), are normally tabulated only for the *reference* state of 298.15 K, 1 atm. with enthalpy values for all elements in the *reference* state effectively set to zero. At other temperatures, we use

$$\Delta H^\circ_{fi}(T) = \Delta H^\circ_{fi}(298.15 \text{ K}) + \int_{298.15 \text{ K}}^{T} \Delta C^\circ_{pi}(T)dT + \Delta H_{ti} \tag{3-1.5}$$

where the temperature effects on the elements e in species i are taken into account by

$$\Delta C^\circ_{pi}(T) = C^\circ_{pi}(T) - \sum_e \nu_{ei} C^\circ_{pe}(T) \tag{3-1.6}$$

where $C^\circ_{pe}(T)$ is the heat capacity of element e at T.

Also, ΔH_{ti} is the sum of contributions made by enthalpy effects of phase and structural changes, such as melting and crystal habit, that the elements undergo in the temperature range from 298.15 K to T. If transitions are present, the C°_{pi} value of the integral must be consistent with the physical state of the species and will be different in different T ranges. In our example, to find $\Delta H^\circ_{fC_3H_8}(T)$ for C_3H_8 from carbon and hydrogen, $C^\circ_{pC}(T)$ would be for carbon (β-graphite) and $C^\circ_{pH_2}(T)$ would be for diatomic hydrogen ideal gas and there are no transitions. If, however, the elements change phase between 298.15 K and T, the enthalpy change for this process must be included. Consider obtaining ΔH°_f for bromobenzene at T = 350 K. The elements have $\nu_{eC} = 6$, $\nu_{eH_2} = 5/2$, and $\nu_{eBr_2} = ½$. Since the standard-state pressure of Br_2 is 1 atmosphere and T is greater than bromine's normal boiling temperature of T_b = 332 K, one must use the liquid $C^\circ_{pBr_2(1)}(T)$ up to T_b, subtract ½ of the enthalpy of vaporization of Br_2 at T_b, and then use the vapor $C^\circ_{pBr_2(g)}(T)$ between T_b and 350 K. That is,

$$\begin{aligned}\Delta H^\circ_{fC_6H_5Br}(350\ K) &= \Delta H^\circ_{fC_6H_5Br}(298.15\ K) \\ &\quad + \int_{298.15 \text{ K}}^{332 \text{ K}} \Delta C^\circ_{pC_6H_5Br}(T)dT - ½\Delta H_{vBr_2} \\ &\quad + \int_{332 \text{ K}}^{350} \Delta C'^\circ_{pC_6H_5Br}(T)dT\end{aligned} \tag{3-1.7}$$

where $C^\circ_{pBr_2(1)}(T)$ is used in Eqs. (3-1.6) and (3-1.7) for $\Delta C^\circ_{pC_6H_5Br}$ and $C^\circ_{pBr_2(g)}(T)$ is used in Eq. (3-1.6) for $\Delta C'^\circ_{pC_6H_5Br}$.

A similar analysis can be done for standard-state entropies and there are equivalent relations to Eqs. (3-1.2) and (3-1.5). Thus, for our example reaction,

$$\Delta S°(T) = 3S°_{CO}(T) + 7S°_{H_2}(T) - S°_{C_3H_8} - 3S°_{H_2O}(T) \tag{3-1.8}$$

However, there is one apparent difference for entropy when obtaining and tabulating values in practice. Unlike energy and enthalpy, there can be found the *absolute entropy*, $S°(T)$, which has a zero value when the species and the elements are at $T = 0$ in a perfectly ordered solid state. This means that the entropy of formation is not normally used explicitly; the expression is a combination of Eqs. (3-1.2) and (3-1.3)

$$\Delta S°(T) = \sum_i \nu_i \left[S_i°(T) - \sum_e \nu_{ei} S_e°(T) \right] = \sum_i \nu_i [S_i°(T) - S_{ei}°(T)] \tag{3-1.9}$$

where the $S_i°$ and $S_e°$ values are *absolute*. Though all of the $S_e°$ values cancel out of Eq. (3-1.9), they are tabulated separately (see the end of Table 3-4, for example), because they, like the values for all species, can be found experimentally from

$$S_i°(T) = \int_0^T \frac{C_{pi}°(T)}{T} dT + \sum_t \frac{\Delta H_{ti}}{T_{ti}} \tag{3-1.10}$$

where T_i is the temperature of a transition. The same procedure for the integral in (3-1.10) must be used as in Eq. (3-1.5).

The Gibbs energy change of reaction, $\Delta G°(T)$, is defined analogously to $\Delta H°(T)$ and $\Delta S°(T)$. It is especially useful because it is related to chemical equilibrium constants by

$$\ln K = -\frac{\Delta G°(T)}{RT} \tag{3-1.11}$$

There are a variety of routes to determine $\Delta G°(T)$. The first is to compute enthalpy and entropy changes individually from Eqs. (3-1.2) and (3-1.9) and then use

$$\Delta G°(T) = \Delta H°(T) - T\Delta S°(T) \tag{3-1.12}$$

Another way to obtain $\Delta G°(T)$ is to use tabulated values of $\Delta G_f°(298.15 \text{ K})$ in a manner similar to Eq. (3-1.5).

$$\frac{\Delta G°(T)}{T} = \frac{\sum_i \nu_i G_{fi}°(298.15\ K)}{298.15} + \Delta H°(298.15) \left(\frac{1}{T} - \frac{1}{298.15} \right)$$
$$+ \frac{1}{T} \int_{298.15 \text{ K}}^T \sum_i \nu_i \Delta C_{pi}°(T) dT - \int_{298.15 \text{ K}}^T \sum_i \nu_i \frac{\Delta C_{pi}°(T)}{T} dT \tag{3-1.13}$$

In this case, there are no explicit terms for transitions since $\Delta G_{ti} = 0$. However, appropriate ranges of T and values of $C_p°$ must still be used.

If tabulated property values are all consistent, results from the different treatments will be equal. When estimation methods for different properties are employed or errors occur in doing the calculations, inconsistencies can occur and it is best to check important values by using different routes.

Instead of using properties of formation for pure component ideal gas properties, it is common with multiproperty equations of state (see Section 4-7) to select a zero-value reference state for the substance's $H°$ and $S°$, such as at 298.15 K and 1 atm. With an equation for $C_p^\circ(T)$, an expression for the ideal gas Helmholtz energy, $A°/RT$, can be obtained for all V and T. An example is given by Setzmann and Wagner (1991) for methane.

$$\frac{A°}{RT} = \ln\left(\frac{V_c}{V}\right) + \sum_{i=1}^{8} a_i f_i \left(\frac{\Theta_i}{T}\right) \tag{3-1.14}$$

where the a_i and Θ_i are fitted parameters and the functions f_i are either simple or of the form $\ln[1 - \exp(\Theta_i/T)]$. Then all other properties relative to the chosen reference states can be obtained by differentiation of Eq. (3-1.14).

In the case of reaction equilibrium constants, the exponential character of Eq. (3-1.11) for K amplifies small errors in $\Delta G°(T)$ since the percentage error in K is exponentially related to the error in the value of $\Delta G°(T)/RT$. Thus, percen-

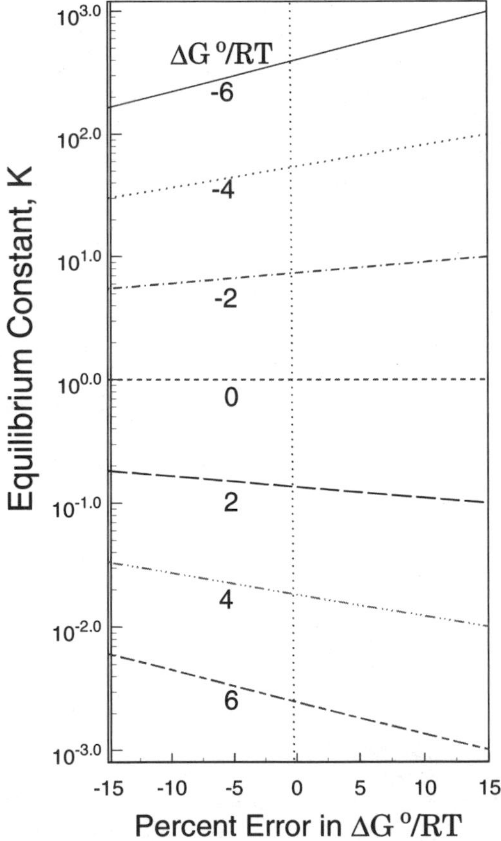

FIGURE 3-1 Effect of errors in $\Delta G°(T)/RT$ on the equilibrium constant, K.

tage errors in $\Delta G°(T)/RT$ are not indicative. We illustrate this in Fig. 3-1 where values of K are plotted versus percentage error in $\Delta G°(T)/RT$ for different values of $\Delta G°(T)/RT$. The correct values for K are the intersections of the dotted vertical line with the lines for the computed values. If $\Delta G°(T)/RT$ is 6 and too small by 15%, the computed value of K is 0.0061 rather than 0.0025; it is too large by a factor of almost 2.5!

The measurement of properties of formation is difficult because of many problems. Impurities and instrument errors can give results that are in error by as much as a few kJ mol^{-1}. Evidence of this can be found by examining values from different sources. The result of these uncertainties is that estimation methods may be more accurate than experimental data and it is now becoming common to compute properties of formation from quantum mechanical methods (see, for example, CHETAH, 1998; O'Connell and Neurock, 1999). Not only are values obtained much more rapidly, better reliability is often found from modern chemistry and powerful computers.

3-2 ESTIMATION METHODS

Since the properties of most of the species treated in this chapter are for ideal gases, intermolecular forces play no role in their estimation and, as a result, the law of corresponding states is inapplicable. Rather, $\Delta H_f^\circ(T)$, $S°(T)$, $\Delta G_f^\circ(T)$, and $C_p^\circ(T)$ are estimated from schemes related to the molecular structure of the compound. Benson (1968) and Benson and Buss (1969) have pointed out a hierarchy of such methods. The most simple methods would use contributions based on the atoms present in the molecule. Although exact for molecular weights and occasionally reasonable for, *e.g.*, the liquid molar volume at the normal boiling point, such methods are completely inadequate for the properties discussed in this chapter.

The next level consists of methods which assign contributions to various chemical bonds and are often not much more complicated. Such techniques are easy to use, but usually have unacceptable errors and are often unreliable. A more successful method assigns contributions to common molecular groupings, *e.g.*, —CH_3, —NH_2—, —COOH, which are simply added. The Joback method (1984) discussed in Sec. 3-3 employs this approach.

Proceeding to more complicated, and usually more accurate, methods, atoms or molecular groups are chosen, and then allowance is made for interactions with next-nearest neighbors to the atom or group. The methods of both Constantinou and Gani (1994) and Benson (1968), discussed in Secs. 3-4 and 3-5, illustrate this type. This is generally the limit for estimation methods because allowance for atoms or groups that are two or more atoms removed from the one of interest treats very small effects while making the technique quite cumbersome. Further, insufficient data exist to develop a reliable table of contributions of second next-nearest neighbor effects.

In this chapter, we present details of the commonly used estimation methods for ideal gas properties *i.e.*, those of Joback (Sec. 3-3), Constantinou and Gani (Sec. 3-4), and Benson, et al. (Sec. 3-5). They all provide estimates of ΔH_f°(298.15 K) and $C_p^\circ(T)$. The Joback and Constantinou/Gani procedures also provide schemes for estimating ΔG_f°(298.15 K), whereas Benson's method yields $S°$(298.15 K) which is used with ΔH_f°(298.15 K) in Eq. (3-1.12) to obtain ΔG_f°(298.15 K). Earlier editions of this book describe alternative and limited methods, such as those of

Yoneda (1979), Thinh, et al. (1971), Thinh and Trong (1976) and Cardozo (1983; 1986), for these properties but they are not repeated here. All methods of this chapter are evaluated and discussed in Sec. 3-6 while their application to computing heats of combustion is described in Sec. 3-7.

Comparisons with data in Appendix A are made where experimental property values are available and all of the group contribution values have been determined. The substances which have been compared are generally organic in nature, but the details depend on the method and the limitations of the data base. Where available in the literature, discussion from other comparisons is included. In general, the results here are typical of what others have found.

3-3 METHOD OF JOBACK

Choosing the same atomic and molecular groups as for the properties in Chap. 2, Joback (1984; 1987) used the values given in Stull, et al. (1956; 1969) to obtain group contributions for $\Delta G_f^\circ(298.15 \text{ K})$, $\Delta H_f^\circ(298.15 \text{ K})$ and polynomial coefficients for $C_p^\circ(T)$. His group values are shown in Appendix Table C-1, and they are to be used in Eqs. (3-3.1) to (3-3.3). Distinctions are made among nonring, nonaromatic ring and aromatic ring groups as well as different atoms and bonding.

$$\Delta G_f^\circ(298.15 \text{ K}) = 53.88 + \sum_k N_k \Delta gfk \qquad (\text{kJ mol}^{-1}) \qquad (3\text{-}3.1)$$

$$\Delta H_f^\circ(298.15 \text{ K}) = 68.29 + \sum_k N_k \Delta hfk \qquad (\text{kJ mol}^{-1}) \qquad (3\text{-}3.2)$$

$$C_p^\circ(T) = \left\{\sum_k N_k CpAk - 37.93\right\} + \left\{\sum_k N_k CpBk + 0.210\right\} T + \left\{\sum_k N_k CpCk - 3.91E-04\right\} T^2 + \left\{\sum_k N_k CpDk + 2.06E-07\right\} T^3 \qquad (\text{J mol}^{-1}\text{ K}^{-1}) \qquad (3\text{-}3.3)$$

where N_k is the number of groups of type k in the molecule, F_K is the contribution for the group labeled k to the specified property, F, and T is the temperature in kelvins.

Example 3-1 Estimate $\Delta H_f^\circ(298.15 \text{ K})$, $\Delta G_f^\circ(298.15 \text{ K})$, and $C_p^\circ(700 \text{ K})$ for 2-ethylphenol by using Joback's group method.

solution 2-ethylphenol contains one $—CH_3$, one $—CH_2—$, four $=CH(ds)$, two $=C(ds)$, and one —ACOH (phenol). From Appendix Table C-1

Group k	N_k	$N_i\Delta hfk$	$N_i\Delta gfk$	N_iCpAk	N_iCpBk	N_iCpCk	N_iCpDk
$—CH_3$	1	−76.45	−43.96	19.500	−0.00808	1.53E-04	−9.70E-08
$—CH_2—$	1	−20.64	8.42	−0.909	0.09500	−0.54E-04	1.19E-08
=CH(ds)	4	8.36	45.20	−8.560	0.22960	−0.06E-04	−6.36E-08
=C(ds)	2	92.86	108.10	−16.500	0.20200	−2.80E-04	13.56E-08
—ACOH(phenol)	1	−221.65	−197.37	−2.810	0.11100	−1.16E-04	4.94E-08
$\sum_{k=1}^{5} N_kF_k$		−217.52	−79.61	−9.279	0.62952	−3.07E-04	3.59E-08

$$\Delta H_f^\circ(298.15 \text{ K}) = 68.29 + \sum_{k=1}^{5} N_k\Delta H_{fk} = -149.23 \text{ kJ mol}^{-1}$$

$$\Delta G_f^\circ(298.15 \text{ K}) = 53.88 + \sum_{k=1}^{5} N_k\Delta G_{fk} = -25.73 \text{ kJ mol}^{-1}$$

$$\begin{aligned} C_p^\circ(700) &= \left\{\sum_k N_kC_{pAk} - 37.93\right\} + \left\{\sum_k N_kC_{pBk} + 0.210\right\} T \\ &\quad + \left\{\sum_k N_kC_{pCk} - 3.91E - 04\right\} T^2 \\ &\quad + \left\{\sum_k N_kC_{pDk} + 2.06E - 07\right\} T^3 \\ &= \{-9.279 - 37.93\} + \{0.62952 + .210\}(700) \\ &\quad + \{-3.074 - 3.91\}(700/100)^2 + \{0.0359 + 0.206\}(700/100)^3 \\ &= 281.21 \text{ J mol}^{-1} \text{ K}^{-1} \end{aligned}$$

The Appendix A values for the formation properties are −145.23 and −23.15 kJ mol^{-1}, respectively, while the heat capacity calculated from the coefficients of Appendix A is 283.14 J mol^{-1} K^{-1}. Thus the differences, are

$$\Delta H_f^\circ(298.15 \text{ K}) \text{ Difference} = -145.23 - (-149.23) = 4.00 \text{ kJ mol}^{-1} \text{ or } 2.75\%$$

$$\Delta G_f^\circ(298.15 \text{ K}) \text{ Difference} = -23.15 - (-25.73) = 2.58 \text{ kJ mol}^{-1} \text{ or } 11.14\%.$$

However, since $\Delta G_f^\circ/RT$ is small (0.934), the error in the equilibrium constant is only 10.97%.

$$C_p^\circ(700 \text{ K}) \text{ Difference} = 283.14 - 281.21 = 1.93 \text{ J mol}^{-1} \text{ K}^{-1} \text{ or } 0.68\%$$

A summary of the comparisons between estimations from the Joback method and experimental Appendix A values for ΔH_f°(298.15 K), ΔG_f°(298.15 K), and C_p° at various temperatures are shown in Table 3-1.

The information in Table 3-1 indicates that the Joback method is marginally accurate for the formation properties of all substances regardless of size and good for ideal gas heat capacities for temperatures at ambient and above. The substances with major errors are halogenated species, suggesting that the group contributions for —F, —Cl, —Br, and —I might be revised in light of the greater abundance of data available now than when the correlation was developed. The terms in the heat capacity correlation are usually of opposite sign so there is no consistent error at

TABLE 3-1 Summary of Comparisons of Joback Method with Appendix A Data Base

Property	# Substances	AAE[c]	A%E[c]	# Err > 10%[d]	# Err < 5%[e]
ΔG_f°(298.15 K)	291[a]	11.9	13.3	86	141
kJ mol^{-1}	234[b]	9.9	13.4	69	113
ΔH_f°(298.15 K)	307[a]	17.7	11.3	59	200
kJ mol^{-1}	246[b]	10.2	9.2	42	171
C_p°(100 K)	121[a]	20.2	43.4	111	2
J mol^{-1} K^{-1}	78[b]	25.7	53.4	74	1
C_p°(298 K)	248[a]	4.0	3.2	10	195
J mol^{-1} K^{-1}	193[b]	4.4	3.0	7	152
C_p°(700 K)	248[a]	5.9	2.3	4	225
J mol^{-1} K^{-1}	193[b]	6.5	2.0	1	179
C_p°(1000 K)	248[a]	9.7	3.3	18	201
J mol^{-1} K^{-1}	193[b]	10.6	2.7	10	168

[a] The number of substances in Appendix A with data that could be tested with the method.

[b] The number of substances in Appendix A having 3 or more carbon atoms with data that could be tested with the method.

[c] AAE is average absolute error in the property; A%E is average absolute percent error. For ΔG_f°(298.15 K), the 21 substances with absolute values less than 10 kJ mol^{-1} were not counted in the A%E. Note the discussion of Fig. 3-1 about errors in ΔG_f°(298.15 K) and the reaction equilibrium constant. Thus, the average absolute percent errors in K were more than 50%, mainly due to the species with errors greater than 25 kJ mol^{-1}.

[d] The number of substances for which the absolute percent error was greater than 10%.

[e] The number of substances for which the absolute percent error was less than 5%. The number of substances with errors between 5% and 10% can be determined from the table information.

low and high temperatures that might be easily corrected. Abildskov (1994) studied Joback results for properties of formation while Nielsen (1998) studied the method for formation properties and for heat capacities. Both did a much more limited examination than for the properties of Chap. 2 but found absolute percent errors that were similar to those of Table 3-1.

Discussion comparing the Joback technique with other methods for the properties of this chapter is presented in Sec. 3-6.

3-4 METHOD OF CONSTANTINOU AND GANI (CG)

Choosing the same first and second order atomic and molecular groups as for the properties in Chap. 2, Constantinou and Gani (1994) obtained group contributions for ΔG_f°(298.15 K) and ΔH_f°(298.15 K). Following this approach and the initial developments of Coniglio and Daridon (1997) for hydrocarbons, Nielsen (1998) developed correlations for polynomial coefficients to obtain $C_p^\circ(T)$ for all classes of organic substances. The group values are shown in Appendix Tables C-2 and C-3 with sample assignments in Table C-4. These values are to be used in Eqs. (3-4.1) to (3-4.3).

$$\Delta G_f^\circ = -14.83 + \left[\sum_k N_k(gf1k) + W\sum_j M_j(gf2j)\right] \qquad (\text{kJ mol}^{-}) \qquad (3\text{-}4.1)$$

$$\Delta H_f^\circ = 10.835 + \left[\sum_k N_k(hf1k) + W\sum_j M_j(hf2j)\right] \qquad (\text{kJ mol}^{-1}) \qquad (3\text{-}4.2)$$

$$\begin{aligned} C_p^\circ = & \left[\sum_k N_k(C_{pA1k}) + W\sum_j M_j(C_{pA2j}) - 19.7779\right] \\ & + \left[\sum_k N_k(C_{pB1k}) + W\sum_j M_j(C_{pB2j}) + 22.5981\right]\theta \qquad (\text{J mol}^{-1}\ \text{K}^{-1}) \\ & + \left[\sum_k N_k(C_{pC1k}) + W\sum_j M_j(C_{pC2j}) - 10.7983\right]\theta^2 \end{aligned}$$

$$\theta = (T - 298)/700 \qquad (3\text{-}4.3)$$

where N_k is the number of First-Order groups of type k in the molecule, F_{1k} is the contribution for the First-Order group labeled $1k$, and N_j is the number of Second-Order groups of type j in the molecule, F_{2j} is the contribution for the Second-Order group labeled $2j$ to the specified property, F, and T is the temperature in kelvins. The value of W is set to zero for first-order calculations and unity for second-order calculations.

Example 3-2 Estimate ΔH_f°(298.15 K), ΔG_f°(298.15 K), and C_p°(700 K) for 2-ethylphenol by using Constantinou and Gani's group method.

solution The First-Order groups for 2-ethylphenol are one CH_3, four ACH, one $ACCH_2$, and one ACOH. There are no Second-Order groups even though there is the ortho proximity effect in this case so the First Order and Second Order calculations are the same. From Appendix Tables C-2 and C-3

Group k	N_k	$N_k hf\ 1k$	$N_{ik} gf\ 1k$	$N_{ik} C_{pA1k}$	$N_k C_{pB1k}$	$N_{ki} C_{pC1k}$
CH_3	1	−45.9470	−8.0300	35.1152	39.5923	−9.9232
ACH	4	44.7560	90.1320	65.5176	130.9732	−52.6768
$ACCH_2$	1	9.4040	41.2280	32.8206	70.4153	−28.9361
ACOH	1	−164.6090	−132.0970	39.7712	35.5676	−15.5875
$\sum_{k=1}^{5} N_k F_k$		−156.3960	−8.7670	173.2246	276.5484	−107.1236

$$\Delta H_f^\circ(298.15\ \text{K}) = 10.835 + \sum_{k=1}^{5} N_k(hfk) = -145.561\ \text{kJ mol}^{-1}$$

$$\Delta G_f^\circ(298.15\ K) = -14.83 + \sum_{k=1}^{5} N_k(gfk) = -23.597\ \text{kJ mol}^{-1}$$

$$C_p^\circ(700) = \left\{\sum_k N_k C_{pAk} - 19.7779\right] + \left[\sum_k N_k C_{pBk} + 22.5981\right]\theta$$
$$+ \left[\sum_k N_k C_{pCk} - 10.7983\right]\theta^2$$
$$= \{173.2256 - 19.7779\} + \{276.5484 + 22.5981\}\{(0.5743)$$
$$+ \{-107.1236 - 10.7983\}(0.5743)^2$$
$$= 286.35 \text{ J mol}^{-1} \text{ K}^{-1}$$

The Appendix A values for the formation properties are −145.23 and −23.15 kJ mol^{-1}, respectively, while the heat capacity calculated from the coefficients of Appendix A is 283.14 J mol^{-1} K^{-1}. Thus the differences are

ΔH_f°(298.15 K) Difference = −145.23 − (−145.56) = 0.33 kJ mol^{-1} or 0.23%

ΔG_f°(298.15 K) Difference = −23.15 − (−23.60) = 0.45 kJ mol^{-1} or 1.94%.

The error in the equilibrium constant is 1.83%.

C_p°(700 K) Difference = 283.14 − 286.35 = 3.21 J mol^{-1} K^{-1} or 1.13%

Example 3-3 Estimate ΔH_j°(298.15 K), ΔG_f°(298.15 K), and C_p°(298 K) for the four butanols using Constantinou and Gani's group method as was done in Example 2-3.

solution The First- and Second-Order groups for the butanols are given in Example 2-3. Since 1-butanol has no Second-Order group, its calculated results are the same for both orders. Using values of group contributions from Appendix Tables C-2 and C-3 and experimental values from Appendix A, the results are:

Property	1-butanol	2-methyl-1-propanol	2-methyl-2-propanol	2-butanol
ΔH_f°(298.15 K), kJ mol^{-1}				
Experimental	−274.60	−282.90	−325.81	−292.75
Calculated (First Order)	−278.82	−287.01	−291.31	−287.01
Abs. % Err. (First Order)	1.54	1.45	10.59	1.96
Calculated (2nd-Order)	−278.82	−287.87	−316.77	−290.90
Abs. % Err. (2nd-Order)	1.54	1.76	2.77	0.63
ΔG_f°(298.15 K), kJ mol^{-1}				
Experimental	−150.17	−167.40	−191.20	−167.71
Calculated (First Order)	−156.75	−161.40	−159.53	−161.40
Abs. % Err. (First Order)	4.38	3.59	16.57	3.76
Calculated (2nd-Order)	−156.75	−161.10	−180.70	−168.17
Abs. % Err. (2nd-Order)	4.38	3.76	5.49	0.27

Property	1-butanol	2-methyl-1-propanol	2-methyl-2-propanol	2-butanol
C_p°(298 K), J mol^{-1} K^{-1}				
Experimental	108.40	—	114.00	113.10
Calculated (First Order)	110.50	109.30	113.20	109.30
Abs. % Err. (First Order)	1.90	—	0.70	3.40
Calculated (2nd-Order)	110.50	109.90	112.00	111.70
Abs. % Err. (2nd-Order)	1.90	—	1.80	1.20

As was seen in Example 2-3, the First Order results are generally good except for 2-methyl-2-propanol (t-butanol). The steric effects of its crowded methyl groups make its experimental value quite different from the others; most of this is taken into account by the First-Order groups, but the Second Order contribution is significant. Notice that the Second Order effects for the other species are small and may change the results in the wrong direction so that the Second Order estimate is slightly worse than the First Order estimate. However, when it occurs, this effect is not large.

A summary of the comparisons between estimations from the CG method and experimental values from Appendix A for ΔH_f°(298.15 K), ΔG_f°(298.15 K), and C_p° at various temperatures is shown in Table 3-2.

The information in Table 3-2 indicates that the CG method can be quite reliable for the formation properties, especially for species with three or more carbon atoms and when Second Order contributions are included such as for smaller species with rings and multiple bonds. There are instances with large errors, mainly for perfluorinated substances and for small molecules that probably should be treated as single groups, though they are not done this way in other methods. The differences between Table 3-2A and 3-2B illustrate the importance of these effects. First Order heat capacity results are quite good for ambient temperatures and above, while significant improvement is found with Second Order contributions only for special cases.

Constantinou and Gani's original article (1994) described tests for about 350 substances with average absolute errors of 4.8 kJ mol^{-1} in ΔG_f°(298.15 K) and 5.4 kJ mol^{-1} in ΔH_f°(298.15 K) which is somewhat less than reported in Table 3-2. Abildskov (1996) studied the CG results for properties of formation while Nielsen (1998) studied the method for formation properties and for heat capacities. Both did a much more limited examination (44 substances) but found absolute percent errors that were slightly larger than the original but still less than those found here. These differences are due to selection of the substances for comparison and different data bases. In most cases, including Second Order contributions improved the results 1 to 3 times as often as it degraded them, but except for ring compounds and olefins, the changes were rarely more than a few kJ mol^{-1}. Thus, the overall improvement from Second Order is about 1–2 kJ mol^{-1} and so the extra complexity may not be worthwhile. In practice, an examination of the magnitude of possible Second Order values for a given case should provide the basis for including them or not.

TABLE 3-2 Summary of CG Method Compared to Appendix A Data Base

A. All substances in Appendix A with data that could be tested with the method

Property	ΔG_f° kJ mol^{-1}	ΔH_f° kJ mol^{-1}	C_f°(100 K) J mol^{-1} K^{-1}	C_p°(298 K) J mol^{-1} K^{-1}	C_p°(700 K) J mol^{-1} K^{-1}	C_p°(1000 K) J mol^{-1} K^{-1}
# Substances (1st)[a]	266	279	95	217	217	215
AAE (1st)[b]	13	13	11.7	3.7	3.8	5.6
A%E (1st)[b]	12	8	25.2	3.2	2	2.1
# Err > 10% (1st)[c]	75	42	73	11	4	3
# Err < 5% (1st)[d]	135	187	8	183	207	205
# Substances (2nd)[e]	74	93	24	67	68	66
AAE (2nd)[b]	11.4	10.5	13.3	4	4.3	6.90
A%E (2nd)[b]	10	4.7	27.9	3	1.5	1.6
# Err > 10% (2nd)[c]	62	28	56	9	3	2
# Err < 5% (2nd)[d]	144	209	6	152	203	206
# Better (2nd)[f]	41	66	9	40	39	27
Ave. Δ% 1st to 2nd[g]	6.1	8.8	−5.4	0.8	0	−0.2

B. All substances in Appendix A having 3 or more carbon atoms with data that could be tested with the method

Property	ΔG_f° kJ mol^{-1}	ΔH_f° kJ mol^{-1}	C_f°(100 K) J mol^{-1} K^{-1}	C_p°(298 K) J mol^{-1} K^{-1}	C_p°(700 K) J mol^{-1} K^{-1}	C_p°(1000 K) J mol^{-1} K^{-1}
# Substances (1st)[a]	224	279	69	182	182	180
AAE (1st)[b]	7.4	12.7	13.3	4.0	3.9	6.4
A%E (1st)[b]	11.3	7.9	27.9	3.0	1.4	1.7
# Err > 10% (1st)[c]	64	42	56	9	3	2
# Err < 5% (1st)[d]	110	187	6	152	175	172
# Substances (2nd)[e]	75	92	24	65	64	64
AAE (2nd)[b]	5.8	5.4	13.8	3.6	4.0	8.4
A%E (2nd)[b]	8.9	4.2	29.3	2.7	1.5	1.7
# Err > 10% (2nd)[c]	52	23	55	7	4	3
# Err < 5% (2nd)[d]	121	184	7	158	170	173
# Better (2nd)[f]	42	65	9	40	35	25
Ave. Δ% 1st to 2nd[g]	6.6	8.7	−5.4	0.8	−0.1	−0.2

[a] The number of substances in Appendix A with data that could be tested with the method.

[b] AAE is average absolute error in the property; A%E is average absolute percent error. The 16 substances for which ΔG_f°(298.15 K) and the 9 substances for ΔH_f°(298.15 K) that have absolute values less than 10 kJ mol^{-1} were not counted in the A%E. Note the discussion of Figure 3-1 about errors in ΔG_f°(298.15 K) and the reaction equilibrium constant. Thus, the average absolute percent errors in K were more than 25%, mainly due to the species with errors greater than 15 kJ mol^{-1}.

[c] The number of substances for which the absolute percent error was greater than 10%.

[d] The number of substances for which the absolute percent error was less than 5%. The number of substances with errors between 5% and 10% can be determined from the table information.

[e] The number of substances for which Second-Order groups are defined for the property.

[f] The number of substances for which the Second Order result is more accurate than First Order.

[g] The average improvement of Second Order compared to First Order. A negative value indicates that overall the Second Order was less accurate.

Discussion comparing the CG technique with other methods for the properties of this chapter is presented in Sec. 3-6.

3-5 METHOD OF BENSON [1968; 1969]

Benson and coworkers have developed extensive techniques for estimating ΔH_f°(298.15 K), S°(298.15), and C_p° which then allow one to obtain ΔG_f°(298.15 K), energy release information, heats of combustion (see Sec. 3-7) and lower flammability limits. There are several references to Benson's work (1968; Benson and Buss, 1969; Benson et al., 1969; O'Neal and Benson, 1970; Eigenmann et al., 1973; Stein et al., 1977) and the CHETAH program (1998). Here, we adopt the notation of the CHETAH (version 7.2) program from ASTM, distributed by NIST as Special Database 16. This differs from Benson's original and also from that of previous editions of this book because it makes clearer the distinction between the *structural* groups and the *neighbor* groups.

It should also be mentioned that other versions of Benson's method exist. One correlation of significance is that of Domalski and coworkers (see especially Domalski and Hearing, (1993)) which includes condensed phases as well as ideal gases.

There are contributions from all of the bonding arrangements ("type") that the chosen groups can have with every other type of group or atom (except hydrogen). Thus the method involves next-nearest neighbor interactions. Table 3-3 shows some of the many distinct groups of the elements C, N, O, and S that bond to more than one neighbor. The column "valence" contains the number of single-bonded groups, such as H or a halogen, that can be attached to the group. Thus, for C, 4 single-bonded groups can be attached, for Ct, only 1 can be and for =C=, no single-bonded groups (only double-bonded groups) can be attached. There is also a word description of the group. In addition to the above elements the method can treat

TABLE 3-3 Some Multivalent Groups in Benson's Method for Ideal Gas Properties

Group	Valence	Definition
C	4	tetravalent carbon (alkanes)
=C	2	double bonded carbon (alkenes), note that Cd represents cadmium
Cb	1	benzene-type carbon (aromatic)
Cp	3	aromatic carbon at ring junction (polyaromatics)
Ct	1	triple bonded carbon (alkynes)
=C=	0	allene carbon
—Cim	2	carbon double bonded to nitrogen (C in >C=N—)
CO	2	carbonyl group (aldehydes, ketones, esters, carboxylic acids)
O	2	oxygen (non-carbonyl oxygen atom in ethers, esters, acids, alcohols)
N	3	trivalent nitrogen (amines)
=Nim	1	imino nitrogen (N in >C=N—)
=Naz	1	azo, nitrogen (N in —N=N—)
Nb	0	aromatic nitrogen (pyridine, pyrazine and pyrimidine, but not pyridazine)
CS	2	thiocarbonyl
S	2	divalent sulfur (sulfides)
SO_2	2	sulfoxide group
SO	2	sulfone group

TABLE 3-4 Group Contributions for Benson Method

Group	ΔH_f° 298K kJ mol^{-1}	S° 298K J mol^{-1}K^{-1}	C_p° 298K J mol^{-1}K^{-1}	C_p° 400K J mol^{-1}K^{-1}	C_p° 500K J mol^{-1}K^{-1}	C_p° 600K J mol^{-1}K^{-1}	C_p° 800K J mol^{-1}K^{-1}	C_p° 1000K J mol^{-1}K^{-1}	C_p° 1500K J mol^{-1}K^{-1}
CH_3 Groups									
CH_3—(Al)	−42.19								
CH_3—(BO_3)	−42.19								
CH_3—(B)	−42.19								
CH_3—(Cb)	−42.19	127.29	25.91	32.82	39.35	45.17	54.5	61.83	73.59
CH_3—(Cd)	−42.19								
CH_3—(CO)	−42.19	127.29	25.91	32.82	39.35	45.17	54.5	61.83	73.59
CH_3—(Ct)	−42.19	127.29	25.91	32.82	39.35	45.17	54.5	61.83	73.59
CH_3—(C)	−42.19	127.29	25.91	32.82	39.35	45.17	54.5	61.83	73.59
CH_3—(Ge)	−42.19								
CH_3—(Hg)	−42.19								
CH_3—(N)	−42.19	127.25	25.95	32.65	39.35	45.21	54.42	61.95	73.67
CH_3—(O)	−42.19	127.29	25.91	32.82	39.35	45.17	54.54	61.83	73.59
CH_3—(Pb)	−42.19								
CH_3—(PO)	−42.19	127.25	25.91	32.82	39.35	45.17	54.54	61.83	73.59
CH_3—(P)	−42.19	127.25	25.91	32.82	39.35	45.17	54.54	61.83	73.59
CH_3—(P=N)	−42.19	127.25	25.91	32.82	39.35	45.17	54.54	61.83	73.59
CH_3—(Si)	−42.19	127.29	25.91	32.82	39.35	45.17	54.5	61.83	73.59
CH_3—(Sn)	−42.19								
CH_3—(SO_2)	−42.19	127.29	25.91	32.82	39.35	45.17	54.5	61.83	
CH_3—(SO_3)	−42.19	127.29							
CH_3—(SO_4)	−42.19	127.29							
CH_3—(SO)	−42.19	127.29	25.91	32.82	39.35	45.17	54.5	61.83	
CH_3—(S)	−42.19	127.29	25.91	32.82	39.35	45.17	54.5	61.83	
CH_3—(Zn)	−42.19								
CH_3—(=C)	−42.19	127.29	25.91	32.82	39.35	45.17	54.5	61.83	73.59

TABLE 3-4 Group Contributions for Benson Method (*Continued*)

Group	ΔH_f° 298K kJ mol^{-1}	S° 298K J $mol^{-1}K^{-1}$	C_p° 298K J $mol^{-1}K^{-1}$	C_p° 400K J $mol^{-1}K^{-1}$	C_p° 500K J $mol^{-1}K^{-1}$	C_p° 600K J $mol^{-1}K^{-1}$	C_p° 800K J $mol^{-1}K^{-1}$	C_p° 1000K J $mol^{-1}K^{-1}$	C_p° 1500K J $mol^{-1}K^{-1}$
CH_3—(=Naz)	−42.19								
CH_3—(=Nim)	−42.28								
Ct Groups									
Ct—(Cb)	122.23	26.92	10.76	14.82	14.65	20.59	22.35	23.02	24.28
Ct—(Ct)	123.78	24.57	14.82	16.99	18.42	19.42	20.93	21.89	23.32
Ct—(C)	115.32	26.58	13.1	14.57	15.95	17.12	19.25	20.59	26.58
Ct—(=C)	122.23	26.92	10.76	14.82	14.65	20.59	22.35	23.02	24.28
CtBr	98.79	151.11	34.74	36.42	37.67	38.51	39.77	40.6	
CtCl	74.51	139.81	33.07	35.16	36.42	37.67	39.35	40.18	
CtF	10.46	122.02	28.55	31.65	33.99	35.79	38.3	39.85	41.77
CtH	112.72	103.39	22.06	25.07	27.17	28.76	31.27	33.32	37.04
CtI	141.48	158.64	35.16	36.84	38.09	38.93	40.18	41.02	
Ct(CN)	267.06	148.18	43.11	47.3	50.65	53.16	56.93	59.86	64.04
CH_2 Groups									
CH_2—(2Cb)	−27.21								
CH_2—(2CO)	−31.81	47.3	16.03	26.66	32.15	37.8	45.46	51.74	
CH_2—(2C)	−20.64	39.43	23.02	29.09	34.53	39.14	46.34	51.65	59.65
CH_2—(2O)	−67.39	32.65	11.85	21.18	31.48	38.17	43.2	47.26	
CH_2—(2=C)	−17.96	42.7	19.67	28.46	35.16	40.18	47.3	52.74	60.28
CH_2—(Cb,CO)	−22.6	40.18							
CH_2—(Cb,Ge)	−18.63								
CH_2—(Cb,N)	−24.4								
CH_2—(Cb,O)	−33.91	40.6	15.53	26.26	34.66	40.98	49.35	55.25	
CH_2—(Cb,Sn)	−32.52								
CH_2—(Cb,SO_2)	−29.8	40.18	15.53	27.5	34.66	40.98	49.77	55.25	

CH_2—(Cb,S)	−19.8	20.51	38.09	49.02	57.43	63.71	72.58	78.82	
CH_2—(Cb,=C)	−17.96	42.7	19.67	28.46	35.16	40.13	47.3	52.74	60.28
CH_2—(CO,N)	−22.27								
CH_2—(CO,O)	−28.46								
CH_2—(Ct,CO)	−22.6	44.37							
CH_2—(Ct,O)	−27.21								
CH_2—(C,Al)	2.93								
CH_2—(C,BO_3)	−9.21								
CH_2—(C,B)	−8.66								
CH_2—(C,Cb)	−20.34	39.1	24.45	31.85	37.59	41.9	48.1	52.49	57.6
CH_2—(C,Cd)	−1.26								
CH_2—(C,CO)	−21.77	40.18	25.95	32.23	36.42	39.77	46.46	51.07	
CH_2—(C,Ct)	−19.8	43.11	20.72	27.46	33.19	38.01	45.46	51.03	59.44
CH_2—(C,Ge)	−18.33								
CH_2—(C,Hg)	−11.22								
CH_2—(C,N)	−27.63	41.02	21.77	28.88	34.74	39.35	46.46	51.49	
CH_2—(C,N=P)	81.21								
CH_2—(C,O)	−33.91	41.02	20.89	28.67	34.74	39.47	46.5	51.61	61.11
CH_2—(C,Pb)	−7.12								
CH_2—(C,PO)	−14.23								
CH_2—(C,P)	−10.34								
CH_2—(C,Si)	−31.94								
CH_2—(C,Sn)	−9.13								
CH_2—(C,SO_2)	−32.11	39.35	17.12	24.99	31.56	36.84	44.58	49.94	
CH_2—(C,SO_3)	−35.58	41.02							
CH_2—(C,SO_4)	−36.42	41.02							
CH_2—(C,SO)	−29.18	39.35	19.05	26.87	33.28	38.34	45.84	51.15	
CH_2—(C,S)	−23.65	41.36	22.52	29.64	36	41.73	51.32	59.23	
CH_2—(C,Zn)	−7.45								
CH_2—(C,=C)	−19.92	41.02	21.43	28.71	34.83	39.72	46.97	52.24	60.11
CH_2—(C,=Naz)	−25.12								
CH_2—(=C,CO)	−15.91								

TABLE 3-4 Group Contributions for Benson Method (*Continued*)

Group	ΔH_f° 298K	S° 298K	C_p° 298K	C_p° 400K	C_p° 500K	C_p° 600K	C_p° 800K	C_p° 1000K	C_p° 1500K
	kJ mol^{-1}	J mol^{-1}K^{-1}	J mol^{-1}K^{-1}	J mol^{-1}K^{-1}	J mol^{-1}K^{-1}	J mol^{-1}K^{-1}	J mol^{-1}K^{-1}	J mol^{-1}K^{-1}	J mol^{-1}K^{-1}
CH_2—(=C,O)	−27.21	37.25	19.51	29.18	36.21	41.36	48.3	53.29	
CH_2—(=C,SO_2)	−29.51	43.95	20.34	28.51	34.95	40.1	47.17	52.49	
CH_2—(=C,SO)	−27.58	42.28	18.42	26.62	29.05	38.72	45.92	51.28	
CH_2—(=C,S)	−27	45.63	22.23	28.59	34.45	40.85	50.98	59.48	
CH Groups									
CH—(2C,Al)	−31.48								
CH—(2C,B)	4.6								
CH—(2C,Cb)	−4.1	−50.86	20.43	27.88	33.07	36.63	40.73	42.9	44.7
CH—(2C,CO)	−7.12	−50.23	18.96	25.87	30.89	35.12	41.11	43.99	
CH—(2C,Ct)	−7.2	−46.84	16.7	23.48	28.67	32.57	38.09	41.44	46.55
CH—(2C,Hg)	15.15								
CH—(2C,N)	−21.77	−48.97	19.67	26.37	31.81	35.16	40.18	42.7	
CH—(2C,O)	−30.14	−46.04	20.09	27.79	33.91	36.54	41.06	43.53	
CH—(2C,Sn)	14.15								
CH—(2C,SO_2)	−18.75	−50.23	18.5	26.16	31.65	35.5	40.35	43.11	
CH—(2C,SO_3)	−14.65	−48.97							
CH—(2C,SO_4)	−25.12	−48.97							
CH—(2C,SO)	−20.93	−48.97							
CH—(2C,S)	−11.05	−47.38	20.3	27.25	32.57	36.38	41.44	44.24	
CH—(2C,=C)	−6.2	−48.93	17.41	24.74	30.72	34.28	39.6	42.65	47.22
CH—(2C,=Naz)	−14.15								
CH—(3Cb)	−5.06								
CH—(3CO)	−51.36								
CH—(3C)	−7.95	−50.52	19	25.12	30.01	33.7	38.97	42.07	46.76
CH—(C,2CO)	−22.6	−42.7							
CH—(C,2O)	−68.23	−48.56	22.02	23.06	27.67	31.77	35.41	38.97	

C Groups									
C—(2C,2O)	−77.86	−149.85	19.25	19.25	23.02	25.53	27.63	28.46	
C—(3C,Cb)	11.76	−147.26	19.72	28.42	33.86	36.75	38.47	37.51	31.94
C—(3C,CO)	5.86	−138.13	9.71	18.33	23.86	27.17	30.43	31.69	
C—(3C,Ct)	1.93	−146.5	0.33	7.33	14.36	19.97	25.2	26.71	
C—(3C,N)	−13.39	−142.74	18.42	25.95	30.56	33.07	35.58	35.58	
C—(3C,O)	−27.63	−140.48	18.12	25.91	30.35	32.23	34.32	34.49	
C—(3C,Sn)	34.16								
C—(3C,SO_2)	2.09	−144.41	9.71	18.33	23.86	27.17	30.43	31.23	
C—(3C,SO_3)	−6.28	−143.57							
C—(3C,SO_4)	−16.74	−143.57							
C—(3C,SO)	−9.29	−144.41	12.81	19.17	20.26	27.63	31.56	33.32	
C—(3C,S)	−2.3	−144.04	19.13	26.25	31.18	34.11	36.5	33.91	
C—(3C,=C)	7.03	−145.33	16.7	25.28	31.1	34.58	37.34	37.51	34.45
C—(3C,=Naz)	−12.56								
C—(4Cb)	29.3								
C—(4C)	2.09	−146.92	18.29	25.66	30.81	33.99	36.71	36.67	33.99
C—(4N)	129.89								
Aromatic (Cb and Cp Groups)									
Cb-(B)	46.04								
Cb—(Cb)	20.76	−36.17	13.94	17.66	20.47	22.06	24.11	24.91	25.32
Cb—(CO)	15.49	−32.23	11.18	13.14	15.4	17.37	20.76	22.77	
Cb—(Ct)	23.78	−32.65	15.03	16.62	18.33	19.76	22.1	23.48	24.07
Cb—(C)	23.06	−32.19	11.18	13.14	15.4	17.37	20.76	22.77	25.03
Cb—(Ge)	20.76								
Cb—(Hg)	−7.53								
Cb—(N)	−2.09	40.56	16.53	21.81	24.86	26.45	27.33	27.46	
Cb—(O)	−3.77	−42.7	16.32	22.19	25.95	27.63	28.88	28.88	
Cb—(Pb)	23.06								
Cb—(PO)	9.63								
Cb—(P)	−7.53								
Cb—(P=N)	9.63								
Cb—(Si)	23.06	−32.19	11.18	13.14	15.4	17.37	20.76	22.77	25.03
Cb—(Sn)	23.06								

TABLE 3-4 Group Contributions for Benson Method (*Continued*)

Group	ΔH_f° 298K	S° 298K	C_p° 298K	C_p° 400K	C_p° 500K	C_p° 600K	C_p° 800K	C_p° 1000K	C_p° 1500K
	kJ mol^{-1}	J $mol^{-1}K^{-1}$	J $mol^{-1}K^{-1}$	J $mol^{-1}K^{-1}$	J $mol^{-1}K^{-1}$	J $mol^{-1}K^{-1}$	J $mol^{-1}K^{-1}$	J $mol^{-1}K^{-1}$	J $mol^{-1}K^{-1}$
Cb—(SO_2)	9.63	36	11.18	13.14	15.4	17.37	20.76	22.77	
Cb—(SO)	9.63	43.53	11.18	13.14	15.4	17.37	20.76	22.77	
Cb—(S)	−7.53	42.7	16.32	22.19	25.95	27.63	28.88	28.88	
Cb—(=C)	23.78	−32.65	15.03	16.62	18.33	19.76	22.1	23.48	24.07
Cb—(=Nim)	−2.09	−40.56	16.53	21.81	24.86	26.45	27.33	27.46	
CbBr	44.79	90.41	32.65	36.42	39.35	41.44	43.11	43.95	
CbCl	−15.91	79.11	30.98	35.16	38.51	40.6	42.7	43.53	
CbF	−183.34	67.39	26.37	31.81	35.58	38.09	41.02	42.7	
CbH	13.81	48.26	13.56	18.59	22.85	26.37	31.56	35.2	40.73
CbI	94.18	99.2	33.49	37.25	40.18	41.44	43.11	43.95	
Cb(CHN_2)	215.57	167.43	47.3						
Cb(CN)	149.85	85.81	41.86	48.14	52.74	55.67	59.86	62.79	
Cb(N_3)	317.29	116.37	34.74						
Cb(NCO)	−29.3	120.13	55.25	64.04	70.32	74.51	79.95	82.88	85.81
Cb(NCS)	190.87	123.06	32.23						
Cb(NO_2)	−1.67	129.76	38.93	50.23	59.44	66.56	76.18	80.37	
Cb(NO)	22.6								
Cb(SO_2N_3)	312.26								
Cb(SO_2OH)	−547.93	123.48	65.42	79.49	84.51	97.61	109.25	113.31	
Cp—(2Cb,Cp)	20.09	−20.93	12.56	15.49	17.58	19.25	21.77	23.02	
Cp—(3Cp)	6.28	5.86	8.37	12.14	14.65	16.74	19.67	21.35	
Cp—(Cb,2Cp)	15.49	−20.93	12.56	15.49	17.58	19.25	21.77	23.02	
=C=, =C—, =CH—, =Cim Groups									
=C—(2Cb)	33.49	−53.16							
=C—(2C)	43.28	−53.16	17.16	19.3	20.89	22.02	24.28	25.45	26.62
=C—(2C)	194.22								

=C—(2=C)	19.25	−36.84							
=C—(CO,O)	48.56	−52.74	23.4	29.3	31.31	32.44	33.57	34.03	
=C—(C,Cb)	36.17	−61.11	18.42	22.48	24.82	25.87	27.21	27.71	28.13
=C—(C,CO)	31.39	−49.39	22.94	29.22	31.02	31.98	33.53	34.32	
=C—(C,N)	−53.96								
=C—(C,O)	43.11	−53.16	17.16	19.3	20.89	22.02	24.28	25.45	
=C—(C,SO_2)	60.69	−40.18	15.49	26.04	33.32	38.51	44.62	47.47	
=C—(C,S)	45.75	−51.95	14.65	14.94	16.03	17.12	18.46	20.93	
=C—(C,=C)	37.17	−61.11	18.42	22.48	24.82	25.87	27.21	27.71	28.13
=CC—(=C,O)	37.25	−61.11	18.42	22.9	24.82	26.29	27.21	27.71	
=CH—(B)	65.3								
=CH—(Cb)	28.38	26.71	18.67	24.24	28.25	31.06	34.95	37.63	41.77
=CH—(CO)	20.93	33.36	31.73	37.04	38.8	40.31	43.45	46.21	
=CH—(Ct)	28.38	26.71	18.67	24.24	28.25	31.06	34.95	37.63	41.77
=CH—(C)	35.96	33.36	17.41	21.05	24.32	27.21	32.02	35.37	40.27
=CH—(O)	36	33.49	17.41	21.05	24.32	27.21	32.02	35.37	40.27
=CH—(Sn)	36.71								
=CH—(SO_2)	52.32	49.81	12.72	19.55	24.82	28.63	32.94	36.29	
=CH—(S)	35.83	33.49	17.41	21.05	24.32	27.21	32.02	35.37	
=CH—(=C)	28.38	26.71	18.67	24.24	28.25	31.06	34.95	37.63	41.77
=CH—(=Nim)	28.38								
=CH_2	26.2	115.57	21.35	26.62	31.44	35.58	42.15	47.17	55.21
=Cim—(2C)	43.11								
=Cim—(Cb,N)	−5.86								
=Cim—(Cb,O)	−12.98								
=Cim—(C,N)	−59.86								
=Cim—(C,O)	−66.97								
=CimH—(C)	36								
=CimH—(N)	−47.72								
=CimH—(O)	−54.83								
=$CimH_2$	26.37								
=C=	143.16	25.12	16.32	18.42	19.67	20.93	22.19	23.02	23.86

TABLE 3-4 Group Contributions for Benson Method (*Continued*)

Group	ΔH_f° 298K kJ mol^{-1}	S° 298K J mol^{-1}K^{-1}	C_p° 298K J mol^{-1}K^{-1}	C_p° 400K J mol^{-1}K^{-1}	C_p° 500K J mol^{-1}K^{-1}	C_p° 600K J mol^{-1}K^{-1}	C_p° 800K J mol^{-1}K^{-1}	C_p° 1000K J mol^{-1}K^{-1}	C_p° 1500K J mol^{-1}K^{-1}
Oxygen Groups									
O—(2C)	−97.11	36.33	14.23	15.49	15.49	15.91	18.42	19.25	
O—(2O)	79.53	39.35	15.49	15.49	15.49	15.49	17.58	17.58	20.09
O—(2PO)	−228.13								
O—(2SO$_2$)	−16.74								
O—(2=C)	−138.13	42.28	14.02	16.32	17.58	18.84	21.35	22.6	
O—(Cb,CO)	−153.62	42.7	8.62	11.3	13.02	14.32	16.24	17.5	
O—(CO,O)	−79.53	34.32	1.51	6.28	9.63	11.89	15.28	17.33	
O—(C,B)	−290.62								
O—(C,Cb)	−92.27	49.81	2.6	3.01	4.94	7.45	11.89	14.99	
O—(C,CO)	−180.41	35.12	11.64	15.86	18.33	19.8	20.55	21.05	
O—(C,Cr)	−98.37								
O—(C,O)	−18.84	39.35	15.49	15.49	15.49	15.49	17.58	17.58	20.009
O—(C,PO)	−170.36								
O—(C,P)	−98.37								
O—(C,P=N)	−170.36								
O—(C,Ti)	−98.37								
O—(C,V)	−98.37								
O—C,=C)	−127.67	40.6	12.72	13.9	14.65	15.49	17.54	18.96	
O—(O,SO$_2$)	12.56								
O—(=C,CO)	−189.2	15.91	6.03	12.47	16.66	18.79	20.8	21.77	
OH—(B)	−483.47								
OH—(Cb)	−158.64	121.81	18	18.84	20.09	21.77	25.12	27.63	
OH—(CO)	−243.2	102.64	15.95	20.85	24.28	26.54	30.01	32.44	37.34
OH—(Ct)	−158.64	146.21							
OH—(C)	−158.56	121.68	18.12	18.63	20.18	21.89	25.2	27.67	33.65

OH—(O)	−68.1	116.58	21.64	24.24	26.29	27.88	29.93	31.44	34.2
OH—(PO)	−272.08								
OH—(P)	−245.71								
OH—(SO_2)	−159.06								
OH—(SO)	−159.06								
OH—(S)	−159.06								
OH—(=C)	−158.64	146.21							
O(CN)—(Cb)	29.3	122.23	34.74						
O(CN)—(C)	8.37	165.34	41.86						
O(CN)—(=C)	31.39	180.41	54.42						
O(NO_2)—(C)	−81.21	203.01	39.93	48.3	55.5	65.3	68.61	72.75	
O(NO)—(C)	−24.7	175.39	38.09	43.11	46.88	50.23	55.67	58.18	60.69
(CO)Br—(Cb)	−158.23								
(CO)Cl—(Cb)	−218.92	167.43							
(CO)Cl—(C)	−200.92	176.64	42.28	46.04	49.39	51.9	55.67	57.76	
(CO)H—(Cb)	−121.81	148.18	33.53	44.2	48.77	59.48	68.56	74.01	
(CO)H—(CO)	−105.9	89.16	28.13	32.78	37.25	41.4	47.84	50.73	
(CO)H—(Ct)	−121.81	148.18							
(CO)H—(C)	−121.81	146.21	29.43	32.94	36.92	40.52	46.71	51.07	
(CO)H—(N)	−123.9	146.21	29.43	32.94	36.92	40.52	46.71	51.07	
(CO)H—(O)	−134.37	146.21	29.43	32.94	36.92	40.52	46.71	51.07	
(CO)H—(=C)	−121.81	148.18	24.32	30.22	39.77	48.77	63.12	74.68	
(CO)I—(Cb)	−99.2								
Halide Groups									
CBr—(3C)	−1.67	−8.37	39.35	47.72	52.74	55.25	56.93	56.09	
CBr_3—(C)	37.67	245.29	72.12	78.65	82.92	85.64	88.66	89.66	
$CBrF_2$—(C)	−395.56								
CCl—(2C,O)	−54.58								
CCl—(3C)	−53.58	−22.6	36.96	43.87	47.72	49.52	52.07	53.12	
CCl_2—(2C)	−92.93	93.76	51.07	62.29	66.76	68.98	70.99	71.24	
CCl_2F—(C)	−266.22								
CCl_3—(CO)	−43.11								

TABLE 3-4 Group Contributions for Benson Method (*Continued*)

Group	ΔH_f° 298K kJ mol^{-1}	S° 298K J mol^{-1}K^{-1}	C_p° 298K J mol^{-1}K^{-1}	C_p° 400K J mol^{-1}K^{-1}	C_p° 500K J mol^{-1}K^{-1}	C_p° 600K J mol^{-1}K^{-1}	C_p° 800K J mol^{-1}K^{-1}	C_p° 1000K J mol^{-1}K^{-1}	C_p° 1500K J mol^{-1}K^{-1}
CCl_3—(C)	−104.23	210.97	68.23	75.35	79.95	82.88	86.23	87.9	
CClF—(2C)	−225.2								
$CClF_2$—(C)	−444.96	185.06	57.35	67.39	73.25	77.86	82.88	85.39	
CF—(3C)	−216.83	−32.23	28.46	37.09	42.7	46.71	52.03	53.24	
CF—(3N)	−103.14								
CF_2—(2C)	−414.4	74.51	39.01	46.97	53.24	57.85	63.46	65.84	
CF_2—(2N)	−391.08								
CF_2—(C,CO)	−396.36								
CF_2—(C,O)	−466.72								
CF_3—(Cb)	−691.5	179.15	52.32	64.04	72	77.44	84.14	87.9	
CF_3—(CO)	−641.02								
CF_3—(C)	−702.97	177.9	53.16	62.79	68.65	74.93	80.79	83.72	
CF_3—(N)	−674.76								
CF_3—(S)	−629.13	162.83	41.36	54.46	62.08	68.52	76.06	79.99	
$CF(NO_2)_2$—(Cb)	−277.1								
$CF(NO_2)_2$—(C)	−195.9								
CH_2Br—(Cb)	−16.53	176.64	30.51	46.46	52.2	57.3	65.26	69.95	
CH_2Br—(C)	−22.6	170.78	38.09	46.04	52.74	57.35	64.88	70.32	
CH_2Br—(=C)	−16.53	171.62	40.6	47.72	54.42	59.86	67.81	73.67	
CH_2Cl—(Cb)	−73.42								
CH_2Cl—(CO)	−44.79								
CH_2Cl—(C)	−69.07	158.23	37.25	44.79	51.49	56.09	64.04	69.9	
CH_2Cl—(=C)	−68.65								
CH_2F—(C)	−223.11	148.18	33.91	41.86	50.23	54.42	63.62	69.49	
CH_2I—(Cb)	35.16	186.27	33.91	45.17	53.7	59.9	68.15	73.8	
CH_2I—(CO)	43.07								

CH_2I—(C)	33.49	179.99	38.51	46.04	54	58.18	66.14	72
CH_2I—(O)	15.91	170.36	34.41	43.91	51.19	56.72	64.25	69.36
CH_2I—(=C)	33.28							
CHBr—(2C)	−14.23	79.95	37.38	44.62	50.06	53.75	58.81	61.62
CHBrCl—(C)	−37.67	191.29	51.9	58.6	63.3	68.23	74.93	79.53
CHBrF—(C)	−228.13							
CHCl—(2C)	−61.95	69.78	35.45	42.7	48.89	53.41	59.82	64.38
CHCl—(C,CO)	−94.22							
CHCl—(C,O)	−90.41	66.56	37.67	41.44	43.95	46.88		
$CHCl_2$—(CO)	48.93							
$CHCl_2$—(C)	−87.9	182.92	50.65	58.6	64.46	69.07	74.93	78.28
CHClF—(C)	−256.59							
CHF—(2C)	−205.11	58.6	30.56	37.84	43.83	48.39	54.83	58.64
CHF_2—(C)	−455	163.67	41.44	50.23	57.35	63.21	69.9	74.51
CHI—(2C)	43.95	89.16	38.64	45.67	50.9	54.42	59.31	61.95
CHI_2—(C)	108.83	228.55	56.93	63.42	69.61	74.17	79.7	81.58
CI—(3C)	54.42		41.15	49.18	54.08	56.3	57.72	56.93
$=CBr_2$	31.39	199.25	51.49	55.25	58.18	59.86	62.37	63.62
=CBrCl	27.21	188.78	50.65	53.16	56.51	59.02	61.53	62.79
=CBrF	−131.02	177.9	45.21	50.23	53.58	56.51	59.86	61.53
=CCl—(C)	−8.79	62.79						
$=CCl_2$	−7.53	176.22	47.72	52.32	55.67	58.13	61.11	62.79
=CClF	−180.83	166.6	43.11	48.97	52.74	55.67	59.44	61.53
=CF—(=C)	−144.83							
$=CF_2$	−324.4	156.13	40.6	46.04	50.23	53.15	57.76	60.69
=CHBr	46.04	160.32	33.91	39.77	44.37	47.72	51.9	55.25
=CHCl	−5.02	148.18	33.07	38.51	43.11	46.88	51.49	54.83
=CHF	−157.39	137.3	28.46	35.16	39.77	43.95	49.39	53.16
=CHI	102.55	169.53	36.84	41.86	45.63	48.56	52.74	55.67
=CimBr—(Cb)	29.3							
=CimBr—(C)	−24.7							
=CimCl—(Cb)	14.65							

TABLE 3-4 Group Contributions for Benson Method (*Continued*)

Group	$\Delta H°_f$ 298K kJ mol^{-1}	$S°$ 298K J mol^{-1}K^{-1}	$C°_p$ 298K J mol^{-1}K^{-1}	$C°_p$ 400K J mol^{-1}K^{-1}	$C°_p$ 500K J mol^{-1}K^{-1}	$C°_p$ 600K J mol^{-1}K^{-1}	$C°_p$ 800K J mol^{-1}K^{-1}	$C°_p$ 1000K J mol^{-1}K^{-1}	$C°_p$ 1500K J mol^{-1}K^{-1}
=CimCl—(C)	−39.35								
=CimF—(Cb)	−184.6								
=CimF—(C)	−238.59								
=CimHBr	−12.56								
=CimHCl	−27.21								
=CimHF	−226.45								
=CimHI	92.09								
=CimI—(Cb)	133.95								
=CimI—(C)	79.95								
Nitrogen Groups									
$CH_2(N_3)$—(C)	267.89	195.48	64.46						
=$CH(N_3)$	340.73	182.08	54.42						
N—(2C,B)	−41.57								
N—(2C,Cb)	109.67	−64.88	2.6	8.46	13.69	17.29	21.89	23.4	
N—(2C,CO)	25.53	−70.74	13.02	19.17	23.52	26.16	28.42	28.76	
N—(2C,N)	122.23	−57.76							
N—(2C,PO)	74.51								
N—(2C,P)	134.78								
N—(2C,SO_2)	−85.39		25.2	26.58	31.56	34.45	37.8	38.47	
N—(2C,SO)	66.97		17.58	24.61	25.62	27.33	28.59	34.91	
N—(2C,S)	125.16		15.99	21.64	25.99	29.05	30.93	38.68	
N—(2C,Ti)	163.67								
N—(2C,=C)	102.13								
N—(2C,=Nim)	122.23								
N—(3Cb)	125.99								
N—(3C)	102.13	−56.34	14.57	19.09	22.73	24.99	27.46	27.92	27.21

N—(Cb,2CO)	−2.09	−69.9	4.1	12.81	17.71	20.3	22.1	22.14	
N—(C,2CO)	−24.7	63.62	4.48	12.99	18.04	20.93	22.94	27.08	
N—(C,=C,N)	124.74								
Nb pyrid—N	70.74	47.38	10.88	13.48	15.95	17.66	20.05	21.43	
NbO	18.84								
NF_2—(C)	−32.65		26.5	34.58	40.9	45.63	50.9	53.54	
NH—(2Cb)	68.23	18	9.04	13.06	17.29	21.35	28.3	32.98	
NH—(2CO)	−77.44	31.81	15.03	23.19	28.05	30.93	33.28	34.28	
NH—(2C)	64.46	37.42	17.58	21.81	25.66	28.59	33.07	36.21	39.97
NH—(Cb,CO)	1.67	−12.14	2.39	6.32	9.96	13.94	16.91	18.21	
NH—(Cb,N)	92.51	47.72							
NH—(C,Cb)	62.37	28.46	15.99	20.47	23.9	26.29	30.1	32.36	
NH—(C,CO)	−18.42	16.32	2.76	6.49	10.3	14.57	17.75	18.96	
NH—(C,N)	87.48	40.18	20.09	24.28	27.21	29.3	32.65	34.74	37.67
NH—(C,=C)	64.46								
NH—(C,=Nim)	87.9								
NH—(=C,N)	90								
NH_2—(Cb)	20.09	124.36	23.94	27.25	30.64	33.78	39.39	43.83	51.4
NH_2—(CO)	−62.37	103.35	17.04	24.03	29.85	34.7	41.69	46.97	
NH_2—(C)	20.09	124.36	23.94	27.25	30.64	33.78	39.39	43.83	51.4
NH_2—(N)	47.72	121.81	25.53	30.98	35.16	38.93	43.95	48.14	55.25
NH_2—(=C)	20.09								
NH_2—(=Nim)	47.72								
$N(NO_2)$—(2C)	50.23								
=Naz—(Cb)	132.69								
=Naz—(C)	113.02	35.58	11.3	17.16	20.59	22.35	23.82	23.9	
=Naz—(N)	96.27	35.58	8.87	17.5	23.06	28.34	28.71	29.51	
=NazH	105.06	112.18	18.33	20.47	22.77	24.86	28.34	31.06	35.33
=Nim—(Cb)	65.3	25.12	12.56						
=Nim—(C)	89.16	24.7	10.38	13.98	16.53	17.96	19.21	19.25	
=Nim—(N)	104.23								
=NimH	50.23	51.49	12.35	19.17	27	32.27	38.22	41.52	

TABLE 3-4 Group Contributions for Benson Method (*Continued*)

Group	ΔH_f° 298K kJ mol^{-1}	S° 298K J mol^{-1}K^{-1}	C_p° 298K J mol^{-1}K^{-1}	C_p° 400K J mol^{-1}K^{-1}	C_p° 500K J mol^{-1}K^{-1}	C_p° 600K J mol^{-1}K^{-1}	C_p° 800K J mol^{-1}K^{-1}	C_p° 1000K J mol^{-1}K^{-1}	C_p° 1500K J mol^{-1}K^{-1}
Sulfur Groups									
S—(2Cb)	108.41	−113.02	8.37	8.41	9.38	11.47	15.91	19.72	
S—(2C)	48.18	55.04	20.89	20.76	21.01	21.22	22.65	23.98	
S—(2O)	37.67								
S—(2S)	13.39	56.09	19.67	20.93	21.35	21.77	22.19	22.6	
S—(2=C)	−19	68.98	20.05	23.36	23.15	26.33	33.24	40.73	
S—(Cb,B)	−32.65								
S—(Cb,S)	60.69	−33.49	12.1	14.19	15.57	17.37	20.01	21.35	
S—(C,B)	−60.69								
S—(C,Cb)	80.2	−32.65	12.64	14.19	15.53	16.91	19.34	20.93	
S—(C,S)	29.51	51.78	21.89	22.69	23.06	23.06	22.52	21.43	
S—(C,=C)	41.73	55.25	17.66	21.26	23.27	24.15	24.57	24.57	
S—(N,S)	−20.51								
SH—(Cb)	50.06	52.99	21.43	22.02	23.32	25.24	29.26	32.82	
SH—(CO)	−5.9	130.6	31.94	33.86	33.99	34.2	35.58	34.49	
SH—(C)	19.34	137	24.53	25.95	27.25	28.38	30.56	32.27	
SH—(=C)	25.53								
SO—(2Cb)	−66.97	−99.2	23.94	38.05	40.6	47.93	47.97	47.09	
SO—(2C)	−66.97	75.76	37.17	41.98	43.95	45.17	45.96	46.76	
SO—(2N)	−132.11								
SO—(2O)	−213.48								
SO—(C,Cb)	−72.04	−12.56							
SO_2—(2Cb)	−296.44	−72.42	34.99	46.17	56.72	62.54	66.39	66.81	
SO_2—(2C)	−288.82	87.48	48.22	50.1	55.88	59.77	64.38	66.47	
SO_2—(2N)	−132.11								
SO_2—(2O)	−417.5								

SO_2—(2=C)	−308.08	56.51	48.22	50.1	55.88	59.77	64.38	66.47
SO_2—(Cb,SO_2)	−325.32	−13.39	41.06	48.14	56.59	61.66	65.76	67.1
SO_2—(Cb,=C)	−296.44	−26.37	41.4	48.14	55.88	61.16	65.8	66.64
SO_2—(C,Cb)	−289.24	5.86	41.61	48.14	56.3	60.74	65.38	66.64
SO_2—(C,=C)	−316.95	75.76						
SO_3—(2C)	−396.82	126.83						
SO_4—(2C)	−602.34	138.55						
S(CN)—(Cb)	196.74	138.55	39.77					
S(CN)—(C)	175.81	181.67	46.88					
S(CN)—(=C)	198.83	196.74	59.44					
(SO_2)Cl—(O)	−406.03							
(SO_2)F—(O)	−594.39							
Phosphorus Groups								
P—(3Cb)	118.46							
P—(3C)	29.3							
P—(3N)	−279.61							
P—(3O)	−279.61							
PCl_2—(C)	−209.71							
PO—(3Cb)	−221.43							
PO—(3C)	−304.73							
PO—(3N)	−437.84							
PO—(3O)	−437.84							
PO—(C,2O)	−416.49							
P=N—(2Cb,N=P,P=N)	−95.86							
P=N—(2C,N=P,P=N)	−64.88							
P=N—(2O,N=P,P=N)	−181.67							
P=N—(C,3Cb)	−107.58							
P=N—(C,3C)	2.09							
(PO)Cl—(C,O)	−471.33							
(PO)Cl_2—(C)	−514.86	221.72						
(PO)F—(2O)	−701.97							
(P=N)Cl_2—(N=P,P=N)	−243.62							

TABLE 3-4 Group Contributions for Benson Method (*Continued*)

Group	ΔH_f° 298K kJ mol^{-1}	S° 298K J mol^{-1}K^{-1}	C_p° 298K J mol^{-1}K^{-1}	C_p° 400K J mol^{-1}K^{-1}	C_p° 500K J mol^{-1}K^{-1}	C_p° 600K J mol^{-1}K^{-1}	C_p° 800K J mol^{-1}K^{-1}	C_p° 1000K J mol^{-1}K^{-1}	C_p° 1500K J mol^{-1}K^{-1}
Boron and Silicon Groups									
B—(2C,O)	122.65								
B—(3C)	3.73								
B—(3N)	102.13								
B—(3O)	102.13								
B—(3S)	102.13								
B—(B,2O)	−376.56								
B—(C,N,O)	−43.95								
BBr—(2Cb)	−239.64								
BBr—(2C)	−112.6								
BBr_2—(Cb)	−244.45								
BCl—(2C)	−178.74								
BCl—(2N)	−83.05								
BCl—(2O)	−82.46								
BCl_2—(Cb)	−381.25								
BCl_2—(N)	−284.22								
BCl_2—(O)	−256.17								
BF_2—(C)	−786.52								
B—(2C,O)	122.65								
B—(3C)	3.73								
B—(3N)	102.13								
B—(3O)	102.13								
B—(3S)	102.13								
B—(B,2O)	−376.56								
B—(C,N,O)	−43.95								
BBr—(2Cb)	−239.64								

BBr—(2C)	−112.6								
BBr_2—(Cb)	−244.45								
BCl—(2C)	−178.74								
BCl—(2N)	−83.05								
BCl—(2O)	−82.46								
BCl_2—(Cb)	−381.25								
BCl_2—(N)	−284.22								
BCl_2—(O)	−256.17								
BF_2—(C)	−786.52								
BF_2—(=C)	−807.45								
BH—(2O)	83.3								
BI—(2C)	−37.25								
BO_3—(3C)	−873.59								
Si—(2C,2Si)	−19.88								
Si—(3C,Si)	−55.04								
Si—(4Cb)	−608.2								
Si—(4C)	−64.46	184.51	113.23	134.95	154.5	171.2	198.62	219.72	252.91
SiBr—(3C)	−251.15								
SiCl—(3C)	−227.38								
$SiCl_2$—(2C)	−364.13								
$SiCl_3$—(C)	−529.85								
SiH—(3C)	−59.36								
SiH_2—(2C)	−88.15								
SiH_3—(C)	−8.37	129.13	−39.64						
SiHCl—(2C)	208.33								
$SiHCl_2$—(C)	−359.98								
Metal Groups									
Al—(3C)	38.51								
AlH—(2C)	−2.76								
Cd—(2C)	194.22								
Cr—(4O)	−267.89								
Ge—(3Cb,Ge)	124.24								

TABLE 3-4 Group Contributions for Benson Method (*Continued*)

Group	ΔH_f° 298K kJ mol^{-1}	S° 298K J $mol^{-1}K^{-1}$	C_p° 298K J $mol^{-1}K^{-1}$	C_p° 400K J $mol^{-1}K^{-1}$	C_p° 500K J $mol^{-1}K^{-1}$	C_p° 600K J $mol^{-1}K^{-1}$	C_p° 800K J $mol^{-1}K^{-1}$	C_p° 1000K J $mol^{-1}K^{-1}$	C_p° 1500K J $mol^{-1}K^{-1}$
Ge—(3C,Ge)	26.41								
Ge—(4Cb)	86.4								
Ge—(4C)	98.03								
Hg—(2Cb)	269.57								
Hg—(2C)	177.9								
HgBr—(Cb)	75.76								
HgBr—(C)	20.43								
HgCl—(Cb)	41.44								
HgCl—(C)	−11.8								
HgI—(Cb)	116.79								
HgI—(C)	66.05								
Pb—(4Cb)	341.57								
Pb—(4C)	305.15								
Sn—(3Cb,Sn)	147.3								
Sn—(3C,Cb)	146.21								
Sn—(3C,Sn)	110.51								
Sn—(3C,=C)	157.39								
Sn—(4Cb)	109.92								
Sn—(4C)	151.53								
Sn—(4=C)	151.53								
SnBr—(3C)	−7.53								
SnCl—(3C)	−41.02								
SnCl—(3=C)	−34.32								
$SnCl_2$—(2C)	−205.94								
$SnCl_2$—(2=C)	−212.22								
$SnCl_3$—(C)	−374.63								

$SnCl_3$—(=C)	−344.08								
SnH—(3C)	145.67								
SnI—(3C)	41.44								
Ti—(4N)	−514.86								
Ti—(4O)	−657.18								
V—(4O)	−364.17								
Zn—(2C)	139.39								
Monovalent Ligands									
$CH_2(CN)$—(Ct)	108.41								
$CH_2(CN)$—(C)	94.18	168.27	47.72	56.93	64.04	70.74	80.79	85.81	
$CH_2(CN)$—(=C)	95.86								
$CH_2(NCS)$—(C)	120.97	213.48	61.95						
$CH_2(NO_2)$—(C)	−60.28	202.6	52.7	66.22	77.52	86.48	99.58	108.41	
$CH_2(NO)$—(C)	74.09								
CH(CN)—(2C)	107.99	82.88	45.21	54	60.69	66.14	72	79.11	
$CH(NO_2)$—(2C)	−56.93	112.6	50.19	63.67	74.17	82.08	92.84	99.2	
$CH(NO_2)_2$—(Cb)	−57.35								
$CH(NO_2)_2$—(C)	−38.09	276.68	80.79	101.3	117.2	129.76	146.09	156.13	
CH(NO)—(2C)	82.04								
C(CN)—(3C)	123.9	−12.14	36.21	46.71	53.96	58.81	64.92	67.77	
$C(CN)_2$—(2C)	293.43	118.46	61.62	74.47	83.72	90.46	99.54	104.48	
$C(CN)_3$—(C)	479.28								
$C(NO_2)$—(3C)	−50.65	16.32	41.4	55.84	66.39	73.75	82.92	87.32	
$C(NO_2)_2$—(2C)	−34.32								
$C(NO_2)_3$—(C)	−6.07								
C(NO)—(3C)	86.23								
=$CH(CHN_2)$	251.15	193.8	72.42						
=CH(CN)	155.71	156.13	43.11	50.23	56.09	61.11	68.65	73.67	
=CH(NCS)	178.74	187.11	51.9						
=$CH(NO_2)$	29.72	185.85	51.49	63.21	72.83	80.37	90.41	97.11	105.9
=$C(CN)_2$	339.89	66.56	56.93	69.28	78.19	84.75	93.51	98.74	
=$C(NO_2)$—(C)	18.42								

TABLE 3-4 Group Contributions for Benson Method (*Continued*)

Group	ΔH_f° 298K kJ mol^{-1}	S° 298K J mol^{-1}K^{-1}	C_p° 298K J mol^{-1}K^{-1}	C_p° 400K J mol^{-1}K^{-1}	C_p° 500K J mol^{-1}K^{-1}	C_p° 600K J mol^{-1}K^{-1}	C_p° 800K J mol^{-1}K^{-1}	C_p° 1000K J mol^{-1}K^{-1}	C_p° 1500K J mol^{-1}K^{-1}
3,4 Member Ring Corrections									
azetidine ring	109.67	122.65							
beta-propiolactone ring	100.04	116.79							
cyclobutane ring	109.67	124.74	−19.3	−16.28	−13.14	−11.05	−7.87	−5.78	−2.8
cyclobutanone ring	94.6	−116.79							
cyclobutene ring	124.74	121.39	−10.59	−9.17	−7.91	−7.03	−6.2	−5.57	−5.11
cyclopropane ring	115.53	134.37	−12.77	−10.59	−8.79	−7.95	−7.41	−6.78	−6.36
cyclopropene ring	224.78	140.64							
diketene ring	92.09								
dimethylsila-cyclobutane ring	−53.62								
ethylene oxide ring	112.18	127.67	−8.37	−11.72	−12.56	−10.88	−9.63	−8.63	
ethylene sulfide ring	74.09	123.36	−11.93	−10.84	−11.13	−12.64	18.09	24.36	
ethyleneimine ring	115.95	132.27							
malonic anhydride ring	92.09	116.79							
methylenecyclobutane ring	109.67								
methylenecyclopropane ring	171.2								
thietane ring	81.08	113.77	−19.21	−17.5	−16.37	−16.37	−19.25	−23.86	
trimethylene oxide ring	107.58	115.95	−19.25	−20.93	−17.58	−14.65	−10.88	0.84	
5,6 Member Ring Corrections									
1,2dihydrothiophene 1, 1 dioxide	24.03	85.81							
1,3-cyclohexadiene ring	20.09	100.46							
1,3-dioxane ring	0.84	66.14							
1,3-dioxolane ring	25.12	92.09							
1,3,5-trioxane ring	27.63	53.58							
1,4-cyclohexadiene ring	2.09	106.32							

1,4-dioxane ring	13.81	69.28	−19.21	−20.8	−15.91	−10.97	−6.4	−1.8	
2-Thiolene ring	20.93	106.32							
3-Thiolene ring	20.93	106.32							
cyclohexane ring		78.69	−24.28	−17.16	−12.14	−5.44	4.6	9.21	13.81
cyclohexanone ring	9.21	66.56							
cyclohexene ring	5.86	82.88	−17.92	−12.72	−8.29	−5.99	−1.21	0.33	3.39
cyclopentadiene ring	25.12	117.2	−14.44	−11.85	−8.96	−6.91	−5.36	−4.35	
cyclopentane ring	26.37	114.27	−27.21	−23.02	−18.84	−15.91	−11.72	−8.08	−1.55
cyclopentanone ring	21.77	102.97							
cyclopentene ring	24.7	107.99	−25.03	−22.39	−20.47	−17.33	−12.26	−9.46	−4.52
dihydrofuran ring	19.67	92.09							
dihydropyran ring	5.02	84.55							
dimethylsila-cyclopentane ring	11.59								
furan ring	37.25	110.51	−20.51	−18	−15.07	−12.56	−10.88	−10.05	
glutaric anhydride ring	3.35	84.14							
maleic anhydride ring	15.07	114.69							
piperidine ring	5.86	77.86	−24.7	−19.67	−12.14	−3.77	9.21	17.58	
pyrrolidine ring	28.46	111.76	−25.83	−23.36	−20.09	−16.74	−12.01	−9.08	
succinic anhydride ring	18.84	126.41							
succinimide ring	33.49								
tetrahydrofuran ring	24.7	105.9	−25.12	−24.28	−20.09	−15.91	−11.3	−7.53	
tetrahydropyran ring	2.09	78.69							
thiacyclohexane ring		73.08	−26.04	−17.83	−9.38	−2.89	3.6	5.4	
thiolane ring	7.24	98.62	−20.51	−19.55	−15.4	−15.32	−18.46	−23.32	
thiophene ring	7.12	98.62	−20.51	−19.55	−15.4	−15.32	−18.46	−23.32	
7-17 Member Ring Corrections									
1,3-cycloheptadiene ring	27.63	81.21							
1,3,5-cycloheptatriene ring	19.67	99.2							
1,3,5-cyclooctatriene ring	37.25	88.32							
cis-cyclononene ring	41.44	46.88							
cis-cyclooctene ring	25.12	50.23							
cyclodecane ring	52.74								
cyclodecanone ring	15.07	49.81							
cyclododecane ring	18.42								
cyclododecanone ring	12.56	28.05							

TABLE 3-4 Group Contributions for Benson Method (*Continued*)

Group	ΔH_f° 298K kJ mol^{-1}	S° 298K J mol^{-1}K^{-1}	C_p° 298K J mol^{-1}K^{-1}	C_p° 400K J mol^{-1}K^{-1}	C_p° 500K J mol^{-1}K^{-1}	C_p° 600K J mol^{-1}K^{-1}	C_p° 800K J mol^{-1}K^{-1}	C_p° 1000K J mol^{-1}K^{-1}	C_p° 1500K J mol^{-1}K^{-1}
cycloheptadecanone ring	4.6	−10.05							
cycloheptane ring	26.79	66.56	−38.01						
cycloheptanone ring	9.63	72							
cycloheptene ring	0	65.3							
cyclononane ring	53.58	51.07							
cyclononanone ring	19.67	58.18							
cyclooctane ring	41.44	51.49	−44.16						
cyclooctanone ring	6.28	64.46							
cyclooctatetriene ring	71.58	115.53							
cyclopentadecanone ring	8.79	7.95							
cycloundecanone ring	18.42	39.77							
thiacycloheptane ring	16.28	72.42							
trans-cyclononene ring	53.58	46.88							
trans-cyclooctene ring	44.79	62.79							
Polycyclic Ring Corrections									
1,3-benzodioxole ring	69.49								
1,4-benzodioxole ring	8.37	66.14							
1,4-diazabicyclo(2.2.2)octane	14.23								
benzenetetracarboxylic anhydr ring	88.32	226.04							
bicyclo-(1.1.)-butane ring	238.59	289.66							
bicyclo-(2.1.)-pentane ring	231.48	270.82							
bicyclo-(2.2.1)-heptane ring	67.48								
bicyclo-(3.1.)-hexane ring	136.88	254.92							
bicyclo-(4.1.)-heptane ring	121.01	232.31							
bicyclo-(5.1.)-octane ring	123.9	211.8							
bicyclo-(6.1.)-nonane ring	130.18	205.94							

bicyclo(2.2.1)hepta-2,5-diene ring	132.27	
biphenylene ring	246.13	
cis-bicyclo-(3.3.)-2-one ring	22.6	226.04
cis-decahydronapthalene-2-one ring	64.04	
cis-octahydro-1H-indene ring	34.32	
dibenzofuran ring	17.58	117.2
dodecahydrodibenzofuran ring	47.72	
phthalic anhydride ring	43.11	114.69
spiropentane ring	265.8	282.96
trans-bicyclo-(3.3.)-2-one ring	46.04	226.04
trans-decahydronapthalene-2-one ring	87.48	
trans-octahydro-1H-indene ring	30.14	
xanthene ring	5.02	92.09
Gauche and 1,5 Repulsion Corrections		
1,5 H-repulsion (crowded methyls)	6.28	
di-tertiary ether structure	32.65	
gauche— across C—B bond	3.35	
gauche— across ether oxygen	2.09	
gauche— group attached to ether O	1.26	
gauche—(alkane/alkane)	3.35	
gauche—(alkene/non-halogen)	2.09	
gauche—(alkyl/CN group)	0.42	
gauche—(alkyl/NO_2)		
gauche—(alkyl/ONO)	3.35	
gauche—(halogen/halogen)	10.46	
gauche—(NO_2/NO_2)	27.63	
gauche—(vinyl/CN group)	−15.49	

TABLE 3-4 Group Contributions for Benson Method (*Continued*)

Group	ΔH_f° 298K	S° 298K	C_p° 298K	C_p° 400K	C_p° 500K	C_p° 600K	C_p° 800K	C_p° 1000K	C_p° 1500K
	kJ mol^{-1}	J $mol^{-1}K^{-1}$	J $mol^{-1}K^{-1}$	J $mol^{-1}K^{-1}$	J $mol^{-1}K^{-1}$	J $mol^{-1}K^{-1}$	J $mol^{-1}K^{-1}$	J $mol^{-1}K^{-1}$	J $mol^{-1}K^{-1}$
Cis, Ortho/Para Interactions									
2nd cis— across 1 double bond	8.37								
but-2-ene structure C—C=C—C		5.02	−5.61	−4.56	−3.39	−2.55	−1.63	−1.09	
but-3-ene structure C—C—C=C		−2.51	−5.61	−4.56	−3.39	−2.55	−1.63	−1.09	
cis- between 2 t-butyl groups	41.86		−5.61	−4.56	−3.39	−2.55	−1.63	−1.09	
cis-halogen/(alkane,alkene)	−3.35								
cis- involving 1 t-butyl group	16.74		−5.61	−4.56	−3.39	−2.55	−1.63	−1.09	
cis-(alkyl/CN group)	−14.65	−11.72							
cis-(CN group/CN group)	20.93								
cis-(halogen/halogen)	−1.26								
cis-(not with t-butyl group)	4.19		−5.61	−4.56	−3.39	−2.55	−1.63	−1.09	
number of =Naz to Nb (resonance)	−25.12								
ortho— between Cl atoms	9.21	−9.63	−2.09	5.02	2.09	−2.51	−1.26		
ortho— between F atoms	18	−5.86		−0.84	−0.42	1.26	2.93		
ortho— between NH_2 & NO_2	−5.02								
ortho— on pyridine ring	−6.28								
ortho— (alkane,alkene)/NO_2	18.84								
ortho— (alkane,alkene)/(Br,Cl,I)	2.51								
other ortho—(nonpolar—nonpolar)	3.14	−6.74	4.69	5.65	5.44	4.9	3.68	2.76	−0.21
other ortho—(nonpolar—polar)	1.42								
other ortho—(polar—polar)	10.05								
para— on pyridine ring	−6.28								
Elements									
Al	0	28	24	26	27	28	31	32*	32
B	0	6	11	15	19	21	23	25	28
Br_2	0	152	76	37*	37	37	38	38	38

C	0	6	9	12	15	17	20	22	24
Cd	0	52	26	27	28	30*	30	30	21*
Cl_2	0	223	34	35	36	37	37	37	38
F_2	0	203	31	33	34	35	36	37	38
Ge	0	31	23	24	25	25	26	27	28*
H_2	0	130	29	29	29	29	30	30	32
Hg	0	76	28	27	27	27	21*	21	21
I_2	0	116	54	81*	37*	38	38	38	38
N_2	0	191	29	29	30	30	31	33	35
O_2	0	205	29	30	31	32	34	35	37
Pb	0	65	27	27	28	29	30*	29	29
S	0	32	23	32*	38	34	18*	19*	19
Si	0	19	20	22	23	24	26	27	28
Sn	0	51	27	29	31	29*	28	28	28
Ti	0	31	25	26	27	28	30	32	33*
V	0	29	25	26	27	27	29	31	36
Zn	0	42	25	26	27	28	31*	31	21*

*Means that a transition (melting, vaporization, crystal habit) occurs between T and the next lower temperature.

many other atoms and groups containing F, Cl, Br, I, P, B, Si, and 10 different metals. For each type, the notation gives the key group or atom followed in parentheses by the groups or atoms it is bonded to. Thus, the repeating CH_2 group in polyethylene is CH_2—(2C) since each CH_2 is bonded to two C atoms. The methylene group attached to the oxygen in methylethylether is CH_2—(C,O). The carbon which is bonded to the ring and to the 2 methyl groups of the side group in 1-(1-methylethyl)-4-methylbenzene (cumene) is CH—(2C,Cb).

Table 3-4 lists the contributions from the 623 groups that are distinguished in 16 categories along with the 98 different ring configurations, 13 gauche and 1,5 repulsion types, and 22 cis and ortho/para interactions. In addition to having new values since the last edition of this book, the table contains corrections for erroneous contributions that were given there. The CHETAH program allows one to compute all of the relevant properties for species made of these groups and includes a sizable database of values for molecules as obtained from the literature.

The notation described above is also used in Table 3-4 of group contributions. When adding the contributions, there should be terms from both the group before a — and from the group in the parenthesis following the —. Thus, the group Cb—(C) would need to be accompanied by a group such as CH_3—(Cb) to complete the side group contributions for toluene. If there is no — even though there may be a parenthetical group, there is only the contribution from the group listed. Thus, the groups CbBr and Cb(CN) are for both the aromatic carbons and their side groups for bromobenzene and cyanobenzene, respectively. Finally, if multiple bonds are indicated for a group, it must be the parenthetical group to another group. For example, the group =CHI must be accompanied by a group such as =CHF which would complete the species 1-fluoro-2-iodoethene or by a group such as =CCl-(C), which when additionally accompanied by CH_3—(=C) would give the species 1-iodo-2-chloropropene. Example 3-4 shows some other examples of Benson groups to construct species.

Example 3-4 Examples of Benson Groups (CHETAH, 1998)

Name	Formula	Group	# Groups	Name	Formula	Group	# Groups
4-hydroxy-2-heptanone	$C_7H_{14}O$	CH_3—(CO)	1	propene	C_3H_6	CH_3—(=C)	1
		CO—(2C)	1			=CH—(C)	1
		CH_2—(C,CO)	1			$=CH_2$	1
		CH—(2C,O)	1	3-chloropropanoic acid	$C_3H_5O_2Cl$	CH_2Cl—(C)	1
		OH—(C)	1			CH_2—(C,CO)	1
		CH_2—(2C)	2			CO—(C,O)	1
		CH_3—(C)	1			OH—(CO)	1
benzylideneaniline	$C_{13}H_{11}N$	CbH	10	anthracene	$C_{14}H_{10}$	CbH	10
		Nim—(Cb)	1			Cp—(2Cb,Cp)	4
		=CimH—(Cb)	1	phenanthrene	$C_{14}H_{10}$	CbH	10
		Cb—(=C)	1			Cp—(2Cb,Cp)	2
		Cb—(N)	1			Cp—(Cb,2Cp)	2
1-butanol	$C_4H_{10}O$	CH_3—(C)	1	2-butanol	$C_4H_{10}O$	CH_3—(C)	2
		CH_2—(2C)	2			CH_2—(2C)	1
		CH_2—(C,O)	1			CH_2—(C,O)	1
		OH—(C)	1			OH—(C)	1
2-methyl-1-propanol	$C_4H_{10}O$	CH_3—(C)	2	2-methyl-2-propanol	$C_4H_{10}O$	CH_3—(C)	3
		CH—(3C)	1			C—(3C,O)	1
		CH_2—(C,O)	1			OH—(C)	1
		OH—(C)	1				

Values from the Benson groups can be summed directly to obtain $\Delta H_f^\circ(298.15\text{ K})$ and $C_p^\circ(T)$ values. However, obtaining $S^\circ(298.15)$ also requires taking molecular symmetry into account. Finally, obtaining $\Delta G_f^\circ(298.15)$ requires subtracting the entropy of the elements. The relations are

$$\Delta H_f^\circ(298.15\text{ K}) = \sum_k N_k(\Delta H_{fk}^\circ) \tag{3-5.1}$$

$$S^\circ(298.15\text{ K}) = \sum_k N_k(S_k^\circ) + S_S^\circ \tag{3-5.2}$$

$$S_{el}^\circ(298.15\text{ K}) = \sum_e \nu_e(S_e^\circ) \tag{3-5.3}$$

$$\Delta G_f^\circ(298.15\text{ K}) = \Delta H_f^\circ(298.15\text{K}) - 298.15\,[S^\circ(298.15\text{K}) - S_{el}^\circ(298.15\text{K})] \tag{3-5.4}$$

$$C_p^\circ(T) = \sum_k N_k C_{pk}^\circ(T) \tag{3-5.5}$$

where the group contribution values are in Table 3-4 and the symmetry entropy, S_S°, is given below in Eq. (3-5.6). Though Table 3-4 gives C_p° at only a few temperatures, the CHETAH program can provide values at any specified T. Though apparently complicated, the rules for these adjustments are straightforward and the CHETAH (1998) program performs all of the necessary calculations.

A stepwise procedure for obtaining symmetry numbers is described in the CHETAH manual as adapted from internal memoranda of the Dow Chemical Company. Statistical mechanics shows that entropy varies as $R \ln W$, where W is the number of distinguishable configurations of a compound. If, by rotating a molecule either totally as if it were rigid or along bonds between atoms, one can find indistinguishable configurations, the result will be an overcounting and W must be reduced. There are also cases where structural isomers of the substance exist; this could cause not enough configurations to be counted. These subtle and often complicated aspects of estimation require great care to implement correctly. General rules and examples based on the CHETAH program manual are given here; the reader is referred to Benson, et al. (1969) and CHETAH (1998) for more complete treatments.

The symmetry entropy, S_s°, is independent of T and given by

$$S_s^\circ = R \ln(N_{oi}) - R \ln(N_{ts}) \tag{3-5.6}$$

where N_{oi} is the number of structural isomers of the molecule and N_{ts} is the total symmetry number. Normally $N_{oi} = 1$ so it makes no contribution in Eq. (3-5.6), but two cases will lead to nonunit values. The first is when there is a plane of symmetry where the atoms can form mirror image arrangements (optical isomers) so that the atom in the plane has asymmetric substitutions. For example, the four atoms (H, F, Cl, I) bonded to the carbon in CHFClI can be arranged in two distinct ways, so its $N_{oi} = 2$. The second way for N_{oi} to be different from unity is if an otherwise symmetrical molecule is frozen by steric effects into an asymmetrical conformation. For example, 2,2′,6,6′-tetramethylbiphenyl cannot rotate about the bond between the two benzene rings due to its 2,2′ steric effects. Therefore, the plane of the rings can have two distinct arrangements ($N_{oi} = 2$) which must be included in the entropy calculation. If the desired species is the racemic mixture (equal amounts of the isomers), each asymmetric center contributes two to N_{oi}, but if the species is a pure isomer, $N_{oi} = 1$.

To obtain N_{ts}, one multiplies the two distinct types of indistinguishability that can occur: "internal," designated N_{is}, and "external," designated N_{es}. The value of N_{is} can be found by rotating terminal groups about their bonds to interior groups. An example is methyl ($—CH_3$) which has three indistinguishable conformations ($N_{is} = 3$) and phenyl which has $N_{is} = 2$. Other examples are given in Table 3-5. (Note that in the 2,2′,6,6′-tetramethylbiphenyl example above, the expected indistinguishability rotation cannot occur so $N_{is} = 1$.) The value of N_{es} comes from indistinguishability when the whole molecule is rotated as if it were rigid. Thus, diatomics have $N_{es} = 2$ from rotation about their bond axis, benzene has $N_{es} = 6$ from rotation about its ring center, etc.

Finally N_{ts} is found from

$$N_{ts} = N_{es} \cdot \prod_{k=term} (N_{is})_k \tag{3-5.7}$$

Table 3-5 shows examples of the analysis.

The method is illustrated in Examples 3-5 and 3-6, and estimated values of $\Delta H_f^\circ(298.15\text{ K})$, $S^\circ(298.15\text{ K})$, and $C_p^\circ(T)$ are compared with literature values in Tables 3-6 and 3-7. A summary of the Benson method for properties of formation and heat capacities is discussed along with other methods in Sec. 3-6.

Example 3-5 Estimate $\Delta H_f^\circ(298.15\text{ K})$, $\Delta G_f^\circ(298.15\text{ K})$, and $C_p^\circ(700\text{ K})$ for 2-ethylphenol by using Benson's group method.

solution The Benson groups for 2-ethylphenol are one CH_3—(C), one CH_2—(C,Cb), four Cb—H, one Cb—(C), one Cb—(O), one OH—(Cb), and there is an ortho-(nonpolar-polar) ring effect. The methyl group makes $N_{is} = 3$ while there are 2 optical isomers about the CH_2 group. The elements are 8 Carbon, 5 H_2 and ½ O_2. Using the CHETAH Program for Eqs. (3-5.1) to (3-5.5) with the values in Table 3-4, the results are

TABLE 3-5 Examples of Benson Group Indistinguishabilities (CHETAH, 1998)

Molecule	Formula	N_{is}	N_{es}	N_{ts}	N_{oi}
Methane	CH_4	1	$4 \times 3 = 12$	12	1
Benzene	C_6H_6	1	$6 \times 2 = 12$	12	1
Phosphorus Pentafluoride	PF_5	1	$3 \times 2 = 6$	6	1
1,1-dichloroethene	$C_2H_2Cl_2$	1	$1 \times 2 = 2$	2	1
Hydrogen Peroxide	H_2O_2	1	$1 \times 2 = 2$	2	1
N-methylmethanamine	C_2H_7N	$3^2 = 9$	1	9	1
2,2-dimethylpropane	C_5H_{12}	$3^4 = 81$	$4 \times 3 = 12$	972	1
1,4-di-(2-methyl-2-propyl)benzene	$C_{14}H_{22}$	$3^6 \times 3^2 = 6561$	$2 \times 2 = 4$	26244	1
2,2-dimethyl-4-nitro-3-(4-nitrophenyl)	$C_{12}H_{16}N_2O_4$	$3^3 \times 2^2 \times 3^1 \times 2^1 = 648$	1	648	2
2-(3,5-di-(3-trichloromethylphenyl)-phenyl)-butane	$C_{24}H_{20}Cl_6$	$3^2 \times 3^2 \times 2^1 = 162$	1	162	2

TABLE 3-6 Comparisons of Estimated and Literature Values for Properties of Formation at 298.15 K

Substance	ΔG_f°(298.15K) kJ mol^{-1}	% Error Joback	% Error C/G	% Error Benson	ΔH_f°(298.15K) kJ mol^{-1}	% Error Joback	% Error C/G	% Error Benson
propane	−24.29	5.48	−6.71	0.00	−104.68	0.54	−2.73	0.29
heptane	8.20	−1.71	25.24	0.12	−187.80	0.01	−1.56	−0.17
2,2,3-trimethylbutane	5.00	69.20	6.20	6.40	−204.40	−1.25	−2.96	−0.52
trans-2-butene	63.34	11.53	10.99	−5.46	−11.00	−21.18	−80.36	13.36
3,3-dimethyl-1-butene	81.84[a]	10.36	2.14	4.03	−59.62[a]	−15.31	−3.54	−7.30
2-methyl-1,3-butadiene	145.77[b]	8.63	4.36	0.18	75.73[b]	24.84	11.83	0.00
2-pentyne	190.99[c]	1.59	1.40	0.23	128.95[c]	−2.47	−2.64	−1.98
1-methyl-4-ethylbenzene	130.28	−2.00	0.56	−4.34	−2.05	96.59	−90.24	77.56
2-methylnaphthalene	215.00	16.74	5.29	1.01	114.90	26.86	7.88	2.44
cis-1,3-dimethyl-cyclopentane	39.23	13.71	0.00	−0.92	−135.90	−6.31	0.88	−0.10
4-methylphenol	−31.55	8.24	8.68	4.98	−125.35	2.58	2.32	0.70
di-(methylethyl)ether	−122.07[b]	−9.69	−2.83	7.52	−318.82[b]	−2.78	−3.42	0.94
1,4-dioxane	−180.20	−12.72	16.35	0.91	−314.70	0.17	8.87	0.38
butanone	−146.50	−0.26	−0.40	−0.60	−238.60	−0.05	−0.57	−0.47
ethylethanoate (ethyl acetate)	−328.00	−0.13	1.60	1.66	−444.50	−0.30	0.25	0.21
N,N-dimethylmethanamine (trimethylamine)	99.30	−14.24	−1.60	−1.13	−23.60	59.83	10.72	3.56
propanenitrile	95.97[c]	12.08	2.83	1.60	50.66[c]	17.71	4.99	2.59
2-nitrobutane	−5.54[c]	−431.23	−308.84	−19.31	−162.70[c]	−16.01	−12.24	−2.56
3-methylpyridine	184.62	2.43	−4.07	−0.08	106.36	−3.96	−6.97	0.43
1,1-difluoroethane	−443.30	−3.88	−3.35	−0.89	−500.80	−3.73	−3.02	−0.76
octafluorocyclobutane	−1394.60[c]	12.73	5.56	1.79	−1529.00[c]	6.49	5.19	1.19
bromobenzene	138.51[b]	−8.76	1.62	6.04	105.02[b]	−8.88	2.76	8.37
trichloroethene	19.72[c]	−90.67	−128.04	−38.84	−5.86[c]	316.38	154.78	135.67
1-thiahexane (methylbutylsulfide)	26.27[c]	−6.20	22.04	5.63	−102.24[c]	2.37	−2.69	−1.12
2-methyl-2-butanethiol	9.03[c]	159.69	166.00	41.53	−127.11[c]	−8.11	9.28	−2.90
4,5-dithiaoctane (propyldisulfide)	36.58[c]	80.10	70.23	9.49	−117.27[c]	−28.86	−26.20	−2.88
3-methylthiophene	121.75[c]	3.19	—	1.05	82.86[c]	0.94	—	0.28

Literature source Appendix A except
[a] TRC
[b] CHETAH (1998)
[c] 4th Edition, Chapter 6

TABLE 3-7 Comparisons of Estimated and Literature Values for Ideal Gas Heat Capacities at 298.15 and 700K

Substance	C_p°(298.15K) kJ mol^{-1}	% Error Joback	% Error C/G	% Error Benson	C_p°(700K) kJ mol^{-1}	% Error Joback	% Error C/G	% Error Benson
propane	73.76	1.08	−0.87	0.33	143.11	−0.47	−0.67	−0.08
heptane	165.80	0.29	−1.27	0.12	318.38	−0.97	−0.89	−1.06
2,2,3-trimethylbutane	163.39	1.10	1.08	1.60	326.73	−3.02	−0.78	−1.14
trans-2-butene	80.31	3.42	8.65	0.86	160.54	0.02	1.99	−1.58
3,3-dimethyl-1-butene	126.40[a]	4.86	7.24	0.47	245.71[a]	3.66	5.14	3.37
2-methyl-1,3-butadiene	102.64[a]	−3.34	−1.47	2.30	187.53[a]	−1.96	−2.48	0.25
2-pentyne	98.70[a]	2.21	1.18	−0.71	178.86[a]	0.15	0.22	0.08
1-methyl-4-ethylbenzene	148.60	4.05	5.38	1.62	303.30	0.14	0.53	−1.75
2-methylnaphthalene	159.79	−1.12	−0.70	−2.37	319.32	−0.61	0.44	−1.35
cis-1,3-dimethyl-cyclopentane	134.60	0.03	−3.14	−2.67	292.64	1.63	1.19	−1.24
4-methylphenol	124.86	−0.30	2.85	0.11	240.91	−1.21	0.31	−1.21
di-(methylethyl)ether	158.27[a]	−0.84	−1.11	−0.80	288.65[a]	−1.47	−2.09	2.20
1,4-dioxane	92.35	1.68	9.07	−0.38	200.12	0.87	0.62	−0.56
butanone	103.40	−5.36	−2.21	−2.32	177.66	0.83	0.93	0.19
ethylethanoate (ethyl acetate)	113.58	−0.14	−7.68	−4.03	200.00	−0.33	−0.52	−0.50
N,N-dimethylmethanamine (trimethylamine)	91.77	−0.02	0.14	0.25	177.08	−0.69	0.14	−0.05
propanenitrile	73.92[a]	8.67	−0.37	−1.24	126.62[a]	12.82	−0.60	−0.49
2-nitrobutane	123.59[a]	−3.88	0.69	0.33	232.82[a]	−8.98	−0.75	−0.78
3-methylpyridine	99.88	3.33	0.27	2.12	206.07	0.16	0.41	−0.52
1,1-difluoroethane	68.49	−1.12	0.32	−2.18	117.53	−1.22	1.97	−0.45
octafluorocyclobutane	156.08[a]	−13.75	3.06	−12.87	236.99[a]	0.70	13.00	−0.84
bromobenzene	100.71[a]	−2.70	−2.37	−0.70	190.04[a]	−0.98	−0.89	−1.07
trichloroethene	80.25[a]	−4.04	−1.53	0.93	109.30[a]	−4.71	0.00	−0.27
1-thiahexane (methylbutylsulfide)	140.84[b]	−0.27	0.16	0.11	254.60[b]	0.27	−0.15	0.16
2-methyl-2-butanethiol	143.50[a]	−5.48	16.28	0.35	259.49[a]	−5.16	18.02	−0.57
4,5-dithiaoctane (propyldisulfide)	187.09[a]	−1.96	−0.66	−0.58	322.66[a]	0.71	1.19	0.73
3-methylthiophene	95.71[a]	−18.56	—	1.35	181.04[a]	−13.78	—	1.08

Literature source values computed from constants in Appendix A except
[a] TRC
[b] 4th Edition, Chapter 6

$$\Delta H_f^\circ(298.15\text{ K}) = \sum_{k=1}^{7} N_k(\Delta H_{fk}^\circ) = -145.14\text{ kJ mol}^{-1}$$

$$S^\circ(298.15\text{ K}) = \sum_{k=1}^{7} N_k(S_k^\circ) = 0.388\text{ kJ mol}^{-1}\text{ K}^{-1}$$

$$S_{el}^\circ(298.15\text{ K}) = \sum_{e=1}^{3} N_e(S_e^\circ) = 0.801\text{ kJ mol}^{-1}\text{ K}^{-1}$$

$$\Delta G_f^\circ(298.15\text{ K}) = \Delta H_f^\circ(298.15\text{K}) - 298.15[S^\circ(298.15\text{K}) - S_{el}^\circ(298.15\text{K})]$$
$$= -22.01\text{ kJ mol}^{-1}$$

$$C_p^\circ(700\text{ K}) = \sum_{k=1}^{7} N_k(C_{pk}^\circ) = 283\text{ J mol}^{-1}\text{ K}^{-1}$$

The Appendix A values for the formation properties are -145.23 and -23.15 kJ mol^{-1}, respectively, while the heat capacity calculated from the coefficients of Appendix A is 283.14 J mol^{-1} K^{-1}. Thus the differences are

$$\Delta H_f^\circ(298.15\text{ K})\text{ Difference} = -145.23 - (-145.14) = 0.09\text{ kJ mol}^{-1}\text{ or }0.06\%$$

$$\Delta G_f^\circ(298.15\text{ K})\text{ Difference} = -23.15 - (-22.01) = 1.14\text{ kJ mol}^{-1}\text{ or }4.92\%.$$

The error in the equilibrium constant is 4.3%.

$$C_p^\circ(700\text{ K})\text{ Difference} = 283.14 - 283 = 0.14\text{ J mol}^{-1}\text{ K}^{-1}\text{ or }0.05\%$$

Example 3-6 Estimate $\Delta H_f^\circ(298.15\text{ K})$, $\Delta G_f^\circ(298.15\text{ K})$, and $C_p^\circ(298\text{K})$ for the four butanols using Benson's group method.

solution The groups are given in Example 3-4 and their values are from Table 3-4. The symmetry numbers and optical isomer numbers are listed in the table below.

Property	1-butanol	2-methyl-1-propanol	2-methyl-2-propanol	2-butanol
N_{is}	3	9	27	9
N_{es}	1	1	1	1
N_{ts}	3	9	27	9
N_{oi}	1	1	1	2
$\Delta H_f^\circ(298.15\text{ K})$, kJ mol^{-1}				
Experimental	−274.60	−282.90	−325.81	−292.75
Calculated	−275.81	−284.68	−312.63	−293.59
Abs. % Err.	0.44	0.63	4.05	0.28
$S^\circ(298.15\text{ K})$, J mol^{-1} K^{-1}	360	348	336	357
$S_{el}^\circ(298.15\text{ K})$, J mol^{-1} K^{-1}	777	777	777	777
$\Delta G_f^\circ(298.15\text{ K})$, kJ mol^{-1}				

Property	1-butanol	2-methyl-1-propanol	2-methyl-2-propanol	2-butanol
Experimental	−150.17	−167.40	−191.20	−167.71
Calculated	−151.13	−156.65	−180.78	−168.15
Abs. % Err.	0.64	6.42	5.45	0.26
C_p°(298.15 K), J mol^{-1} K^{-1}				
Experimental	108.40	—	114.00	113.10
Calculated	110	109	113	112
Abs. % Err.	1.5	—	0.9	1.0

Because of the very large number of groups in the Benson method, it was not possible to compare estimates from it to the many values of Appendix A. Section 3.6 shows how the method compares with both data and the other methods of this chapter as was done in the 4th Edition. In general the Benson method is the most accurate and reliable technique available, but it is not completely reliable.

Bures, et al. (1981) have attempted to fit the tabular values of $C_p^\circ(T)$ presented by Benson to various equation forms. In many tests of Bures et al.'s equations, it was found that the calculated group values did agree well, but in other cases, however, serious errors were found. As a result, the method cannot be considered reliable enough and the equations are not given.

3-6 DISCUSSION AND RECOMMENDATIONS

A. Standard State Enthalpy of Formation, ΔH_f°(298.15 K) and Gibbs Energy of Formation, ΔG_f°(298.15 K)

The methods of Joback (1984), Constantinou and Gani (1994) and Benson (as encoded in CHETAH, 1998) were evaluated. While the Benson method computes entropies from group contributions, ΔG_f°(298.15 K) is of greater interest and is the property tabulated in the literature. However, these properties are related by Eq. (3-1-12) so the method can be directly tested. In addition to the comparisons of Tables 3-1 and 3-2, estimations from the methods for both properties of a variety of substances are compared with literature values in Table 3-6. The method of Benson (Sec. 6-6) yields the smallest errors with only a few large percentage deviations, usually when the absolute value of the property is less than 10 kJ mol^{-1}. However, the method has too many groups to be easily done on a spreadsheet as can the methods of Joback and CG. The Thinh, et al. (1971) method, described in the 4th Edition and limited to hydrocarbons, has errors normally less than 1 kJ mol^{-1} which for these substances is comparable to those given here. Joback's technique is the simplest broadly applicable method and can be of adequate accuracy. However, it can yield large errors for some compounds. These errors can usually be avoided by the CG method at the cost of somewhat increased complexity. For highest accuracy, the Benson method should be selected. Contributions to ΔH_f° for groups not currently tabulated are being determined by Steele and coworkers (1997) and references therein. Methods for other systems include that of Ratkey and Harrison (1992) and Mavrovouniotis and Constantinou (1994) for ionic compounds and by Forsythe, et al. (1997) for equilibrium constants including biotransformations.

B. Ideal Gas Heat Capacity, $C_p^\circ(T)$

The methods of Joback (1984), Constantinou and Gani (Nielsen, 1998) and Benson (as encoded in CHETAH, 1998) were evaluated. In addition to the comparisons of Tables 3-1 and 3-2, estimations from the methods of a variety of substances are compared with literature values in Table 3-7. All techniques are similar in accuracy, and except for quite unusual structures, errors are less than 1 to 2%. Joback's (1984) method is the easiest to use and has wide applicability. The CG method of Nielsen (1998) is more reliable but has many more groups. However, both can be set up on a spreadsheet. The Benson method (CHETAH, 1998) requires extensive programming, but it gives the highest accuracy. In case contributions are not available for groups of interest, Harrison and Seaton (1988) describe a simple and reliable method for creating new values that is based on the atoms involved. Other authors have tabulated polynomial constants for $C_p^\circ(T)$ [Seres, 1977; 1981], and equations to express $C_p^\circ(T)$ as a function of temperature have been suggested [Aly and Lee, 1981; Fakeeha, et al., 1983; Harmens, 1979; Thompson, 1977]. Of importance in some of these equations is that, rather than a polynomial, the mathematical form has some theoretical origin and it extrapolates in a more reliable fashion. One strategy in practice is to estimate $C_p^\circ(T)$ values in the range where the prediction is reliable and fit the parameters of these equations to the estimates if properties are needed outside the range of the group method.

C. Availability of Data and Computer Software

There are several readily available commercial products for obtaining ideal-gas properties. These include data and correlation-based tabulations and computer-based group contribution methods. Those which were used in developing this chapter are referenced below or in Appendix C including web sites as of the date of publication. The data for Appendix A were obtained from the book by Frenkel, et al. (1994); there is a similar tabulation available from DIPPR (1999). Joback has established a program (Cranium, 1998) for computing many properties by various group contribution methods though the current version only includes the Joback version for ideal gas properties. Gani and coworkers at the Center for Computer-Aided Process Engineering (CAPEC) at the Danish Technical University also have a program available (Propred, 1999) for many properties including both the Joback and Constantinou and Gani methods for ideal gas properties. The most complete program for the Benson method is from ASTM (CHETAH, 1998).

3-7 HEAT OF COMBUSTION

The standard-state heat of combustion is defined as the difference in enthalpy of a compound and that of its products of combustion in the gaseous state, all at 298 K and 1 atm. The products of combustion are assumed to be $H_2O(g)$, $CO_2(g)$, $SO_2(g)$, $N_2(g)$, and $HX(g)$, where X is a halogen atom. Since product water is in the gaseous state, this enthalpy of combustion would be termed the lower enthalpy of combustion. For the case where the water product is liquid, the energy released by condensation would be added to the lower enthalpy to give the higher enthalpy of combustion.

There is a direct relation between the standard-state enthalpy of combustion and the standard-state enthalpy of formation. The general equation for reactants and products in their standard states at $T = 298.15$ K is

$$\Delta H_f^\circ(298.15\ \text{K}) = -393.78N_\text{C} - 121.00(N_\text{H} - N_\text{X}) - 271.81N_\text{F} - 92.37N_\text{Cl} - 36.26N_\text{Br} + 24.81N_\text{I} - 297.26N_\text{S} - \Delta H_\text{C}^\circ(298.15\ \text{K}) \tag{3-7.1}$$

where N_C, N_H, N_F, N_Cl N_Br, N_I, and N_S are the numbers of atoms of carbon, hydrogen, fluorine, chlorine, bromine, iodine, and sulfur in the compound and N_X is the total number of halogen (F, Cl, Br, I) atoms. Each numerical coefficient of Eq. (3-7.1) is the value of $\Delta H_\text{C}^\circ(298.15\ \text{K})$ for the element product indicated. Measuring the heat of combustion is relatively easy, so this is a common way to obtain values of $\Delta H_f^\circ(298.15\ \text{K})$. There is an estimation method for $\Delta H_\text{C}^\circ(298.15\ \text{K})$ due to Cardozo (1983, 1986) described in the 4th Edition. It was recommended for complex organic substances found in the liquid or solid phase when pure at the temperature of interest. The accuracy was similar to that of the Joback method.

NOTATION

$C_p^\circ(T)$	ideal-gas heat capacity at constant pressure at T, J mol^{-1} K^{-1}
$\Delta G^\circ(T)$	standard Gibbs energy change in a reaction at T, kJ mol^{-1}; Eqs. (3-1.12) and (3-1.13)
$\Delta G_{fi}^\circ(T)$	standard Gibbs energy of formation at T and 1.01325 bar(1 atm.), kJ mol^{-1}; Eq. (3-1.13)
$\Delta H^\circ(T)$	standard enthalpy of reaction at T, kJ mol^{-1}; Eqs. (3-1.2) and (3-1.4)
$\Delta H_\text{C}^\circ(298.15\ \text{K})$	standard (lower) enthalpy of combustion at 298.15 K, kJ mol^{-1}; Eq. (3-7.1)
$\Delta H_{fi}^\circ(T)$	standard enthalpy of formation of species i at T, kJ mol^{-1}; Eq. (3-1.3)
ΔH_t, T_t	transition enthalpy change from a change of phase or solid structure at temperature T_t; Eqs. (3-1.5) and (3-1.10)
K	reaction equilibrium constant; Eq. (3-1.11)
N_C	number of carbon atoms in a compound; similarly for N_F, N_{Cl}, N_{Br}, N_I, N_S, N_X for fluorine, chlorine, bromine, iodine, sulfur, and total halogen atoms; Eq. (3-7.1)
N_k, N_j	number of groups of type k in a molecule; k for First-Order groups and j for Second-Order groups in Constantinou and Gani method, Sec 3-4
N_{is}, N_{es}, N_{ts}	internal, external and total symmetry numbers, respectively, for indistinguishability of a molecule's conformations, Sec. 3-5; Eq. (3-5.7)
N_{oi}	number of optical isomers of a molecule, Sec. 3-5
R	gas constant, 8.31447 J mol^{-1} K^{-1}
$S^\circ(T)$	absolute entropy at T and 1.01326 bar (1 atm.), kJ mol^{-1} K^{-1}; Eq. (3-1.8), Sec. 3-5
S_S°	entropy contribution from symmetry and optical isomers; Eq. (3-5.6)

$S_e^\circ(T)$ absolute entropy of an element used to make a species at T and 1.01325 bar (1 atm.), kJ mol^{-1} K^{-1}; Eq. (3-1.9)
$S_{el}^\circ(T)$ absolute entropy of the elements used to make a species at T and 1.01325 bar (1 atm.), kJ mol^{-1} K^{-1}; Eq. (3-1.9)
$\Delta S^\circ(T)$ standard entropy change of reaction at T, kJ mol^{-1} K^{-1}; Eq. (3-1.8)
T absolute temperature, K
W weight for Second-Order groups in Constantinou and Gani method; = 0 for First Order only, = 1 for full estimation

Greek

ν_i stoichiometric coefficient for species i in a reaction, > 0 for products, < 0 for reactants
ν_{ei} stoichiometric coefficient for element e in species i, > 0; Eq. (3-1.3)

REFERENCES

Abildskov, J: "Development of a New Group Contribution Method," M.S. Thesis, Inst. for Kemiteknik, Danish Technical University, 1994.

Aly, F. A., and L. L. Lee: *Fluid Phase Equil.,* 6:169 (1981).

Benson, S. W.: *Thermochemical Kinetics,* Wiley, New York, 1968, chap. 2.

Benson, S. W., and J. H. Buss: *J. Chem. Phys.,* **29:** 279 (1969).

Benson, S. W., F. R. Cruickshank, D. M. Golden, G. R. Haugen, H. E. O'Neal, A.S. Rodgers, R. Shaw and R. Walsh: *Chem. Rev.,* **69:** 279 (1969).

Bures, M., V. Maier, and M. Zabransky: *Chem. Eng. Sci.,* **36:** 529 (1981).

Cardozo, R. L.: private communication, Akzo Zout Chemie Nederland bv, January 1983; *AIChE J.,* **32:** 844 (1986).

CHETAH Version 7.2: *The ASTM Computer Program for Chemical Thermodynamic and Energy Release Evaluation* (NIST Special Database 16), 4th ed 1998. The values here were reprinted with permission from the ASTM Computer Program for Chemical Thermodynamic and Energy Release Evaluation (CHETAH™), copyright American Society for Testing and Materials (ASTM). Copies of the complete program are available from ASTM, 100 Barr Harbor Drive, West Conshohocken, PA 19428-2959 USA, phone: 1-610-832-9585, fax: 1-610-832-9555, e-mail service@astm.org, URL http://www.astm.org.

Coniglio, L. and J. L. Daridon: *Fluid Phase Equil.,* **139:** 15 (1997).

Constantinou, L., and R. Gani: *AIChE J.,* **40:** 1697–1710, 1994.

Cranium Version 1.0: Molecular Knowledge Systems, Inc., PO Box 10755, Bedford, NH 03110-0755 USA, 1998.

Design Institute for Physical Properties Research (DIPPR): American Institute of Chemical Engineers, http://www.aiche.org/dippr, 1999.

Domalski, E. S., and E. D. Hearing: *J. Phys. Chem. Ref. Data,* **22:** 805 (1993); **23:** 157 (1994).

Eigenmann, H. K., D. M. Golden, and S. W. Benson: *J. Phys. Chem.,* **77:** 1687 (1973).

Fakeeha, A., A. Kache, Z. U. Rehman, Y. Shoup, and L. L. Lee: *Fluid Phase Equil.,* **11:** 225 (1983).

Forsythe, Jr., R. G., P. D. Karp and M. L. Mavrovouniotis: *Computer Apps. in Biosciences,* **13:** 537 (1997).

Frenkel, M., K. N. Marsh, G. J. Kabo, R. C. Wilhoit, and G. N. Roganov: *Thermodynamics of Organic Compounds in the Gas State,* Vol. I., Thermodynamics Research Center, College Station, TX, (1994).

Harmens, A.: *Correlation of Thermodynamic Data of Fluids and Fluid Mixtures: Their Estimation, Correlation, Use,* Proc. NPL Cong., 1978 (pub. 1979) p. 112–20.

Harrison, B. K., and W. H. Seaton: *Ind. Eng. Chem. Res.,* **27:** 1536 (1988).

Joback, K. G.: *A Unified Approach to Physical Property Estimation Using Multivariate Statistical Techniques,* S.M. Thesis, Department of Chemical Engineering, Massachusetts Institute of Technology, Cambridge, MA, 1984.

Joback, K. G., and R. C. Reid: *Chem. Eng. Comm.,* **57:** 233–243, (1987).

Mavrovouniotis, M. L., and L. Constantinou: *J. Phys. Chem.,* **98,** 404 (1994).

Nielsen, T. L.: *Molecular Structure Based Property Prediction,* 15-point Project Department of Chemical Engineering, Technical University of Denmark., Lyngby, DK-2800, 1998.

O'Connell, J. P. and M. Neurock: Foundations of Computer-Aided Process Design Conference, July, 1999. Proceedings to be published by CACHE Corporation, Austin, TX.

O'Neal, H. E., and S. W. Benson: *J. Chem. Eng. Data,* **15:** 266 (1970).

ProPred: CAPEC, Department of Chemical Engineering, Building 229, Danish Technical Univeristy, DK-2800 Lyngby Denmark.

Ratkey, C. D, and B. K. Harrison: *Ind. Eng. Chem. Res.,* **31:** 2362 (1992).

Seres, L., L. Zalotal, and F. Marta: *Acta. Phys. Chem.,* **23**(4)**:** 433 (1977).

Seres, L.: *Acta. Phys. Chem.,* **27**(1–4): 31 (1981).

Setzmann, U., and W. Wagner: *J. Phys. Chem. Ref. Data,* **20:** 1061 (1991).

Steele, W. V., R. D. Chirico, A. B. Crowell, S. E. Knipmeyer, and A. Nguyen: *J. Chem. Eng. Data,* **42:** 1053 (1997).

Stein, S. E., D. M. Golden, and S. W. Benson: *J. Phys. Chem.,* **81:** 314 (1977).

Stull, D. R., and G. C. Sinke: "Thermodynamic Properties of the Elements," *Advan. Chem. Ser.,* **8:** (1956).

Stull, D. R., E. F. Westrum and G. C. Sinke: *The Chemical Thermodynamics of Organic Compounds,* Wiley, New York, 1969.

Thinh, T.-P., J.-L. Duran, and R. S. Ramalho: *Ind. Eng. Chem. Process Design Develop.,* **10:** 576 (1971).

Thinh, T.-P., and T. K. Trong: *Can. J. Chem. Eng.,* **54:** 344 (1976).

Thompson, P. A.: *J. Chem. Eng. Data,* **22:** 431 (1977).

Yoneda, Y.: *Bull. Chem. Soc. Japan,* **52:** 1297 (1979).

CHAPTER FOUR

PRESSURE-VOLUME-TEMPERATURE RELATIONSHIPS OF PURE GASES AND LIQUIDS

4-1 Scope

Methods are presented in this chapter for estimating the volumetric behavior of pure gases and liquids as a function of temperature and pressure. Section 4-2 introduces the framework of *PVT* relations. Their generalized basis in the corresponding states principle is discussed in section 4-3. Techniques for estimation and calculation from equations of state (EoS) are given in Secs. 4-4 to 4-7. Typically these are used in computer-based systems. The models include the virial series, analytical density relations (EoS which can be solved analytically to find the volume of a fluid at a specified T and P, mainly cubic EoS), and more complex relations requiring numerical root-solving algorithms for the volume. Section 4-8 summarizes our evaluation of EoS methods. Sections 4-9 to 4-12 give estimation methods for saturated and compressed liquid densities that are not based on equations of state.

Extension of this chapter's methods to mixtures is given in Chap. 5. Chapter 6 describes the application of the models of these chapters to thermodynamic properties of pure and mixed nonideal fluids.

4-2 INTRODUCTION TO VOLUMETRIC PROPERTIES

The volumetric properties of a pure fluid in a given state are commonly expressed with the compressibility factor Z, which can be written as a function of T and P or of T and V

$$Z \equiv \frac{PV}{RT} = f_P(T, P) \tag{4-2.1a}$$

$$= f_V(T, V) \tag{4-2.1b}$$

where V is the molar volume, P is the absolute pressure, T is the absolute temperature, and R is called the universal gas constant. The value of R depends upon the units of the variables used. Common values are shown in Table 4-1 (NIST, 1998, with unit conversions from links). In this book, unless otherwise noted, P is in bars, V in cm^3 mol^{-1}, and the term "mol" refers to gram-moles. Note that 1 bar = 10^5 Pa = 10^5 N m^{-1} and 1 atm = 1.01325 bar.

The choice of independent variables, T and P in Eq. (4-2.1a) and T and V Eq. (4-2.1b), depends upon the application. Commonly, a user specifies the state with T and P since these are most easily measured and so Eq. (4-2.1a) is considered most convenient. However, if one seeks an equation which can describe both gaseous and liquid phases with the same parameters, the needed multiplicity of volume or density roots demands a function of the form (4-2.1b). Thus, the well-known cubic equations of state are in the form of Eq. (4-2.1b).

For an ideal gas, as in Chap. 3, $Z^{ig} = 1.0$. For real gases, Z is somewhat less than one except at high reduced temperatures and pressures. For liquids that are saturated between the triple or melting point and the boiling point or under low applied pressure, Z is normally quite small. Near the vapor-liquid critical point, Z is usually between 0.15 and 0.4, but it varies rapidly with changes in T and P (see below).

Since the compressibility factor is dimensionless, it is often represented by a function of dimensionless (reduced) temperature, $T_r = T/T^*$, and dimensionless (reduced) pressure, $P_r = P/P^*$, where T^*, and P^* are characteristic properties for the substance, such as the component's vapor-liquid criticals, T_c, and P_c. It could also be given as a function of T_r and reduced volume, $V_r = V/V^*$, where V^* could be chosen as V_c or RT_c/P_c or another quantity with units of volume. Then Z is considered a function of dimensionless variables

$$Z = f_{P_r}(T_r, P_r) \tag{4-2.2a}$$

$$= f_{V_r}(T_r, V_r) \tag{4-2.2b}$$

This scaling allows many substances to be represented graphically in generalized form. For example, $f_{P_r}(T_r, P_r)$ was obtained by Nelson and Obert (1954) for several substances from experimental PVT data and they constructed the graphs of Figs. 4-1 to 4-3. Except as noted below, using these figures to obtain Z at a given T/T_c and P/P_c should lead to errors of less than 4 to 6% except near the saturation curve or near the critical point. Appendix A lists values of T_c and P_c for many substances and methods to estimate them are described in Sec. 2-2.

Figures 4-1 to 4-3 should not be used for strongly polar fluids, nor are they recommended for helium, hydrogen, or neon unless special, modified critical constants are used. These aspects are considered in section 4-3 below. For very high pressures and temperatures, Breedveld and Prausnitz (1973) have generated more accurate extensions of these graphs.

TABLE 4-1 Values of the Gas Constant, R

Value of R	Units* on R
83.145	bar cm^3 mol^{-1} K^{-1}
8.3145	J mol^{-1} K^{-1}
10.740	psia ft^3 $lb\text{-}mol^{-1}R^{-1}$
1.986	btu $lb\text{-}mol^{-1}$ R^{-1}
82.058	atm cm^3 mol^{-1} K^{-1}

*The unit mol refers to gram moles.

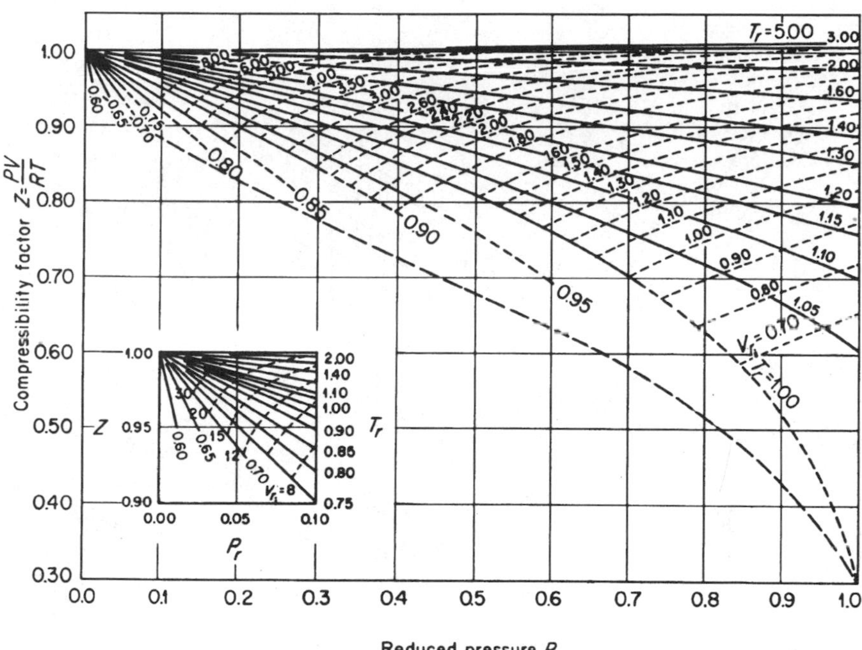

FIGURE 4.1 Generalized compressibility chart at low P_r. $V_{r_i} = V/(RT_c/P_c)$. *(Nelson and Obert, 1954)*

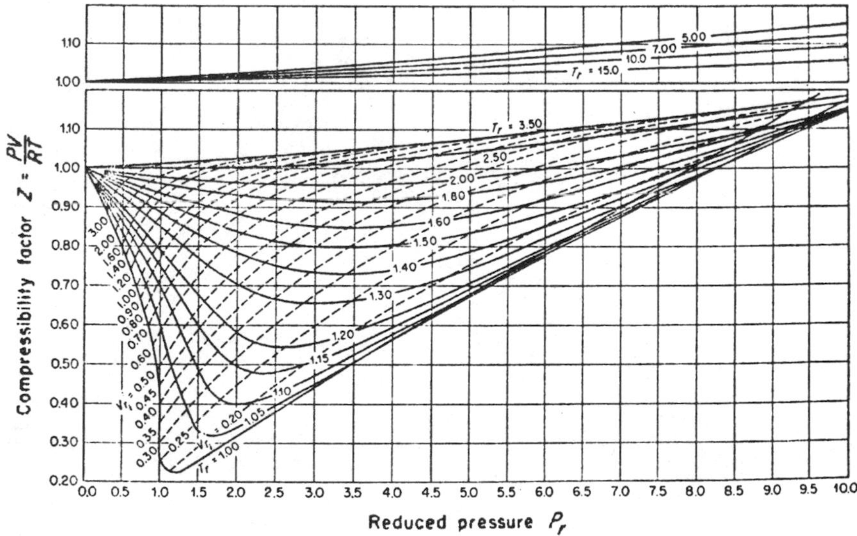

FIGURE 4.2 Generalized compressibility chart at moderate P_r. $V_{r_i} = V/(RT_c/P_c)$. *(Nelson and Obert, 1954)*.

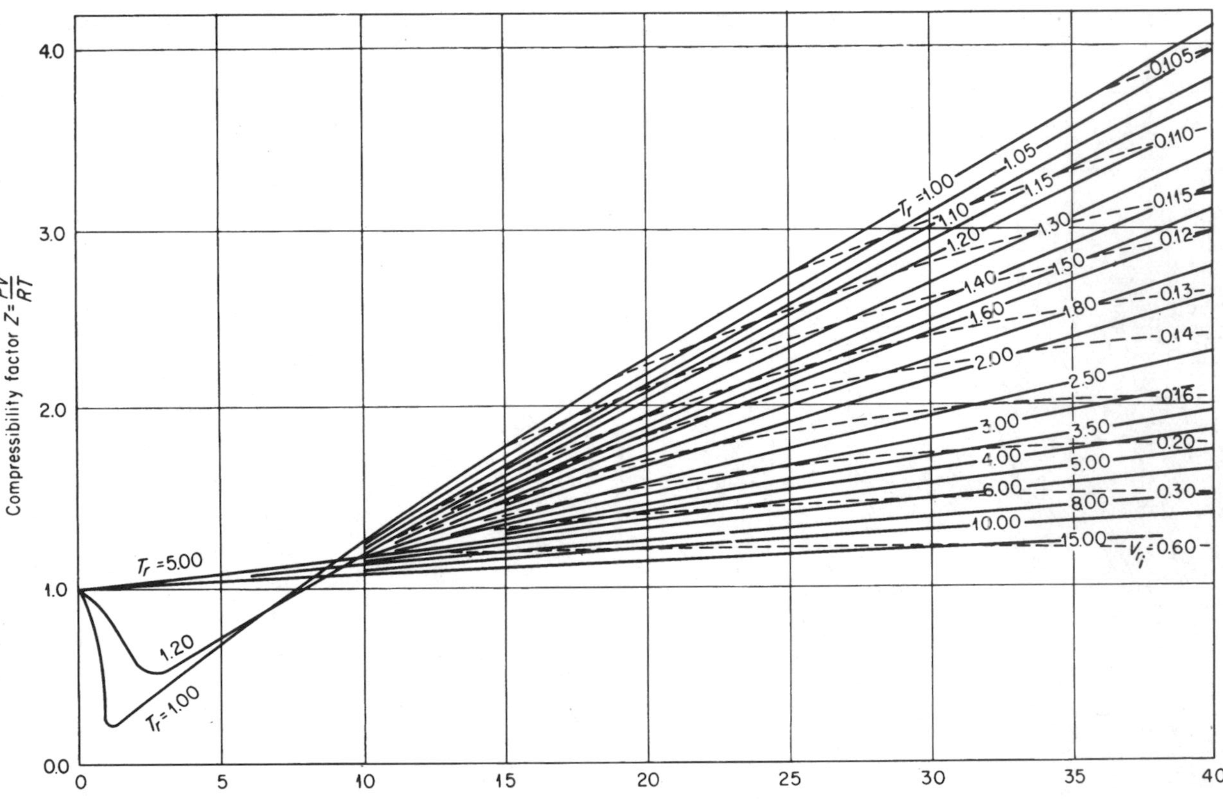

FIGURE 4.3 Generalized compressibility chart for all P_r. $V_{r_i} = V/(RT_c/P_c)$. (*Nelson and Obert 1954.*)

Many versions of Figs. 4-1 to 4-3 have been published. All differ somewhat, as each reflects the choice of experimental data and how they are smoothed. Those shown are as accurate as any two-parameter plots published, and they have the added advantage that V can be found directly from the lines of $V_{r_i} = V/(RT_c/P_c)$.

Equations of state (EoS) are mathematical representations of graphical information such as shown in Figs. 4-1 to 4-3. Modern computers obviate the need to manually obtain volumetric behavior such as from graphs and also allow more accurate results by using equations with component-specific parameters. There have been an enormous number of EoS functions generated, especially in the last few years (see, *e.g.*, Anderko, 1990 and Sandler, et al., 1994 and especially Sengers, et al., 2000, who give comprehensive reviews and references to the state of the art). It is not possible to evaluate all of such models here, but the discussion of this chapter is intended to provide guidance about the variations of accuracy, reliability and computational difficulties encountered when doing pure component volumetric analysis with literature models. Mixtures are treated in Chap. 5 and other thermodynamic properties are considered in Chap. 6 and 8.

4-3 CORRESPONDING STATES PRINCIPLE

Equations (4-2.2*a*) and (4-2.2*b*), Figs. 4-1 to 4-3, and Equations of State are all formulations of a general principle of dimensionless functions and dimensionless variables called the corresponding states principle (CSP) or sometimes the "Law of Corresponding States." It asserts that suitably dimensionless properties of all substances will follow universal variations of suitably dimensionless variables of state and other dimensionless quantities. Its general and specific forms can be derived from molecular theory (Hakala, 1967). Useful practical guidelines are given by Leland and Chappelar (1968). The number of parameters characteristic of the substance determines the level of CSP.

Two-Parameter CSP

Pitzer (1939) and Guggenheim (1945) describe the molecular conditions for this lowest level CSP when only two characteristic properties, such as T_c and P_c, are used to make the state conditions dimensionless, and the dimensionless function may be Z. Thus, Figs. 4-1 to 4-3 and Eqs. (4-2.2) are examples of this two-parameter form of CSP where the characteristics are T_c and P_c or T_c and V_c. Only the monatomic substances Ar, Kr, Xe, or "simple fluids" (Guggenheim, 1945), follow this behavior accurately; all others show some deviation. For example, two-parameter CSP requires that all substances have the same critical compressibility factor, $Z_c = P_cV_c/RT_c$. Simple transformations of EoS models show that all those with only two substance-specific constant parameters such as the original van der Waals EoS (1890), also predict the same Z_c for all compounds. Appendix A shows that the experimental values for monatomic substances give $Z_c = 0.291$. However, for most organic compounds Z_c ranges from 0.29 down to 0.15. Analysis shows that popular two-parameter cubic EoS models yield values greater than 0.3. This behavior suggests that more than two dimensionless characteristics must be used both in concept and in modeling.

Three-Parameter CSP

In general, successful EoS have included one or more dimensionless characteristic parameters into the function expressed by Eqs. (4-2.2), especially to obtain good agreement for liquid properties. The first step in accomplishing this is to introduce a third parameter; usually it is related to the vapor pressure, P_{vp}, or to a volumetric property at or near the critical point. This improves accuracy for many substances, though not all. In fact, its success might not have been expected. Molecular theory (Hakala, 1967) suggests that effects of nonpolar nonsphericity and of "globularity" (the range and sharpness of nonpolar repulsive forces) should require separate CSP parameters, but in practice, only a single characteristic accounts for both.

Historically, several different third parameters were introduced at about the same time but the most popular have been Z_c (Lydersen, et al., 1955) and the acentric factor, ω, (Pitzer and Curl, 1955, 1957ab). Lydersen, et al. (1955) and a later revision by Hougan, et al. (1959) tabulated Z (and reduced thermodynamic properties, see Chap. 6) at increments of T/T_c and P/P_c for different values of Z_c. In practice, this correlation has been used only occasionally, such as by Edwards and Thodos (1974) for estimating saturated vapor densities of nonpolar compounds.

The much more commonly used third parameter is the Pitzer acentric factor ω defined in Eq. (2-3.1) (Pitzer and Curl, 1955; 1957ab; Pitzer, et al., 1955).

Instead of different tables for incremental values, Pitzer's assumption was that ω would describe deviations from the monatomic gases in a linear (Taylor's series) fashion, implying the corrections would be small. Otherwise nonlinear terms as in the quadratic Eq. (7-4.1) for P_{vp} or interpolation techniques (see Sec. 4-7) would have to be employed. For example, the compressibility factor was given as

$$Z = Z^{(0)}(T/T_c, P/P_c) + \omega Z^{(1)}(T/T_c, P/P_c) \tag{4-3.1}$$

where $Z^{(0)}$ and $Z^{(1)}$ are generalized functions of reduced temperature and pressure with $Z^{(0)}$ obtained from the monatomic species and $Z^{(1)}$ by averaging $(Z - Z^{(0)})/\omega$ for different substances. The same formulation was used for thermodynamic properties (see Chap. 6).

Equation (2-3.2) shows that $Z^{(0)}$ (1,1) = 0.291 and $Z^{(1)}$ (1, 1) = −0.080. Accessibility from measured or correlated data and its equation form made the acentric factor the third parameter of choice from the time of its development and this preference still continues.

It was expected that ω would describe only "normal fluids," not strongly polar or associating species such as via hydrogen-bonding, and limited to smaller values of ω (Pitzer, 1995). In a series of discussions on the subject of when the acentric factor could be expected to describe a compound, Pitzer (1995) focused on the surface tension as being the most sensitive property to indicate when the molecular forces were more complex than those for "normal" substances. These ideas can be coalesced into a single equation for surface tension, σ, made dimensionless with critical properties

$$\frac{\sigma}{T_c}\left(\frac{RT_c}{P_c}\right)^{(2/3)} = \left(1 - \frac{T}{T_c}\right)^{(11/9)} (1.86 + 1.18\omega)\left[\frac{3.74 + 0.91\omega}{0.291 - 0.080\omega}\right]^{(2/3)} \tag{4-3.2}$$

Use of this equation for estimating σ is described in Chap. 12. For CSP usage, Pitzer (1995) states that if a substance deviates more than 5% from Eq. (4-3.2) it "appears to indicate significant abnormality." Otherwise, the substance is expected to be "normal" and three-parameter CSP should be reliable.

The issue of whether a linear variation with ω is adequate has also been considered and some correlations such as for P_{vp} are best done with a quadratic function (Pitzer and Curl, 1955). However, many useful correlations for equations of state parameters and other properties based on only linear variations of ω have been developed for all varieties of substances. See, for example a discussion in Sec. 4-7 about this technique for EoS. In cases where the compound of interest is not "normal" or ω is greater than 0.6, the user should use caution about applying the acentric factor concept without verification of its validity in a model for a property or for the substance's class of components.

Several revisions of the original tables and graphs (Edmister, 1958) as well as extensions to wider ranges of T_r and P_r have been published; the 4th Edition lists several references for these and gives the tabulations for $Z^{(0)}$ and $Z^{(1)}$ prepared by Lee and Kesler (1975) calculated from an equation of state as described in Sec. 4-7. The tables for $Z^{(0)}$ and $Z^{(1)}$ are also given in Smith, et al. (1996). Extensive testing (Tarakad and Daubert, 1974; Lee and Kesler, 1975) indicates that this formulation is more accurate than the original.

Example 4-1 Estimate the specific volume of 1,1,1,2-tetrafluoroethane at 24.35 bars and 355.55 K using CSP. A literature value is 800 $cm^3\ mol^{-1}$ (Lemmon, et al., 1998).

solution From Appendix A, T_c = 374.26 K, P_c = 40.59 bars. Then $T_r = 0.95$, $P_r = 0.60$. From Fig. 4-1, $Z = 0.695$ and thus $V = ZRT/P = 844\ cm^3\ mol^{-1}$. The error is 44 $cm^3\ mol^{-1}$ or 5.5%.

If the Pitzer-Curl method were to be used with $\omega = 0.326$ from Appendix A, the 4th Edition Tables 3-2 and 3-3 give $Z^{(0)} = 0.697$ and $Z^{(1)} = -0.111$. From Eq. (4-3.1), $Z = 0.697 + 0.326\ (-0.111) = 0.661$ so $V = 802.5\ cm^3\ mol^{-1}$. The error is 2.5 $cm^3\ mol^{-1}$ or 0.3%.

Higher Parameter and Alternative CSP Approaches

One method to extend CSP to substances more complex than normal fluids is to use more terms in Eq. (4-3.1) with new characteristic parameters to add in the effects of polarity and association on the properties. Though none has been widely adopted, the 4th Edition lists several references suggesting ways this was tried. Most of these correlations were not very successful though the approach of Wilding (Wilding and Rowley, 1986 and Wilding, et al., 1987) showed reasonable accuracy. This lack of reliability may be due to the polar and associating forces affecting not only the volumetric behavior as the model tries to treat, but also the critical properties. The result is that to use them for dimensionless scaling is inconsistent with the molecular theory basis for CSP (Hakala, 1967).

Alternative expansions to Eq. (4-3.1) have also appeared. Rather than use simple fluids as a single reference, multiple reference fluids can be used. For example, Lee and Kesler (1975) and Teja, et al. (1981) developed CSP treatments for normal fluids that give excellent accuracy. The concept is to write Z in two terms with two reference substances having acentric factors $\omega^{(R1)}$ and $\omega^{(R2)}$, $Z^{(R1)}(T_r, P_r, \omega^{(R1)})$ and $Z^{(R2)}(T_r, P_r, \omega^{(R2)})$. The expression is

$$Z(T_r, P_r, \omega) = Z^{(R1)}(T_r, P_r, \omega^{(R1)}) + \frac{\omega - \omega^{(R1)}}{\omega^{(R2)} - \omega^{(R1)}} [Z^{(R2)}(T_r, P_r, \omega^{(R2)}) - Z^{(R1)}(T_r, P_r, \omega^{(R1)})] \tag{4-3.3}$$

Typical reference fluids to provide properties are simple fluids as $R1$ and a larger hydrocarbon such as n-octane for $R2$. Also, rather than use tables, the functions $Z^{(R1)}$ and $Z^{(R2)}$ can be computed from EoS (see Secs. 4-6 and 4-7). Further, though this approach is strictly applicable only to normal fluids, it has also been used for polar substances. In that case, it can be as reliable as for normal fluids if $R2$ is from the same class of the substance as the one of interest such as alcohols, alkyl halides, etc.

It is also possible to use more reference fluids to extend CSP to more complex substances. For example, Golobic and Gaspersic (1994) use three reference fluids (simple fluids, n-octane and water) with a modification of the second term in Eq. (4-3.3) and an additional term, both of which include another characteristic parameter. They provide property tables rather than analytic equations. Golobic and Gaspersic compare 20 different models for saturated vapor and liquid volumes with their method. Their comparisons with eight strongly polar alcohols and others were the most reliable, giving errors that were rarely more than any of the others and with maximum errors less than twice the average. Platzer and Maurer (1989) also used a three-reference fluid approach but with an equation for the EoS. For the 24 normal and 18 polar and associating fluids for which they obtained T_c, P_c, ω, and a polar factor, the correlation was as good as other methods available.

Finally, for mixtures, see Eqs. (5-6.1) to (5-6.6). For "quantum fluids," H_2, He, Ne, Gunn, et al. (1966) showed how to obtain useful values for CSP over wide ranges of conditions. The equations are

$$T^* = T_c = \frac{T_c^{\text{cl}}}{1 + 21.8/MT} \tag{4-3.4a}$$

$$P^* = P_c = \frac{P_c^{\text{cl}}}{1 + 44.2/MT} \tag{4-3.4b}$$

$$V^* = V_c = \frac{V_c^{\text{cl}}}{1 - 9.91/MT} \tag{4-3.4c}$$

where M is the molecular weight, T_c^{cl}, P_c^{cl} and V_c^{cl} are "classical" critical constants found empirically and some are given in Table 4-2. The "classical" acentric factor for these substances is defined as zero. Values for other cryogenic isotopes can be found in Prausnitz, et al. (1999).

4-4 EQUATIONS OF STATE

An equation of state (EoS) is an algebraic relation between P, V, and T. This section discusses what behavior of Nature must be described by EoS models. Then, one

TABLE 4-2 Classical Critical Constants for Quantum Fluids in CSP

Quantum Substance	T_c^{cl}, K	P_c^{cl}, bar	V_c^{cl}, cm^3 mol^{-1}
Helium	10.47	6.76	37.5
Hydrogen	43.6	20.5	51.5
Neon	45.5	27.3	40.3

general class of EoS is presented in each of the next three sections. First, the virial equation, which can be derived from molecular theory, but is limited in its range of applicability, is discussed in Sec. 4-5. It is a polynomial in P or $1/V$ (or density) which, when truncated at the Second or Third Order term, can represent modest deviations from ideal gas behavior, but not liquid properties. Next, in Sec. 4-6, semitheoretical EoS which are cubic or quartic in volume, and therefore whose volumes can be found analytically from specified P and T, are discussed. These equations can represent both liquid and vapor behavior over limited ranges of temperature and pressure for many but not all substances. Finally, Sec. 4-7 describes several empirical EoS in which volume cannot be found analytically. Nonanalytic equations are applicable over much broader ranges of P and T than are the analytic equations, but they usually require many parameters that require fitting to large amounts of data of several properties. These models include empirical forms of original and modified Benedict-Webb-Rubin (MBWR) as well as Wagner models, semitheoretical models such as perturbation models that include higher order polynomials in density, chemical theory equations for strongly associating species such as carboxylic acids and hydrogen fluoride, and crossover relations for a more rigorous treatment of the critical region.

This discussion is not comprehensive, but does illustrate the immense amount of work that has been done in this area. Readers are referred to the papers of Deiters (1999; Deiters and de Reuck, 1997) for full descriptions of how EoS models should be developed and communicated. Following Deiters' recommendations, generators of new models will have a greater opportunity to be considered more thoroughly while users of new models will understand better their possibilities and limitations.

Challenges to EoS Models: The Critical and High Pressure Regions

Fluid properties in states near a pure component's vapor-liquid critical point are the most difficult to obtain from both experiments and from models such as EoS (see the collected articles in Kiran and Levelt Sengers, 1994). The principal experimental difficulty is that the density of a near-critical fluid is so extremely sensitive to variations in P, T, that maintaining homogeneous and stable conditions takes extreme care (Wagner, et al., 1992; Kurzeja, et al., 1999). Even gravity influences the measurements.

The principal model difficulty is that near-critical property variations do not follow the same mathematics as at conditions well-removed from the critical. For example, the difference of the saturation volumes, V_s from V_c near the critical point varies as

$$\lim_{T \to T_c} (V_s - V_c) \sim (T - T_c)^{\beta_c} \tag{4-4.1}$$

Careful experiments have shown that $\beta_c = 0.32 \pm 0.01$. This is close to the results from theories that account for the molecular density fluctuations that cause critical opalescence. However, typical EoS models give a smaller β_c value. Thus, for example, all cubics have $\beta_c = 0.25$. Also, the variation of P with V along the critical isotherm is found to be

$$\lim_{V \to V_c} (P - P_c) \sim (V - V_c)^{\delta_c} \quad \text{for } T = T_c \tag{4-4.2}$$

Careful experiments have shown that $\delta_c = 4.8 \pm 0.2$. Again, this is close to theoretical results, but EoS models give a smaller exponent. All cubics have $\delta_c = 3.0$.

Differences also occur in the variation of C_V in the near-critical region where quite large values persist over fairly wide conditions, but cubics and other models do not show this (Gregorowicz, et al., 1996; Konttorp, 1998).

Only complex EoS expressions of the form of Eq. (4-2.1*b*) can capture these strong variations, but even they are not rigorous very close to the critical point. To overcome this deficiency, a variety of EoS models that attempt to include both "classical" and "nonclassical" behavior of models have been developed (see Sec. 4-7).

The other region where EoS are often inaccurate is at very high pressures both above and below the critical temperature. The form of the *PV* isotherms of EoS functions often do not correspond to those which best correlate data as described in Sec. 4-12, unless careful modifications are made (see Sec. 4-6 and 4-8).

To illustrate the difficulties of these two regions, Table 4-3 (de Hemptinne and Ungerer, 1995) tabulates the maximum relative deviation in density for several equations of state applied to light hydrocarbons in the near-critical region and the high *P*, high *T* region. Figure 4-4 shows results from a classical EoS that shows minimum deviations over all regions except the highest pressures. Similar plots of

TABLE 4-3 Estimates of the Maximum Relative Percent Deviation in Density for Several EoS Applied to Methane and Butane See Table 4-6 for EoS Functions. (From de Hemptinne and Ungerer, 1995)

Substance		Methane		*n*-Butane	
EoS	Region	Critical	High P^+	Critical	High P^+
Peng-Robinson (1976)		8	15	10	8
Peng-Robinson with Translation*		10	1.5	12	4
Behar, et al. (1985, 1986)		7	5	4	3
Lee and Kesler (1975)		3	2	1.5	1.5

*The Peng-Robinson (1976) EoS of Table 4-6 with $\delta = 2b + 4c$, $\varepsilon = 2c^2 - b^2$.
$^+$For $1000 < P < 2000$ bar, $400 < T$ 500 K.

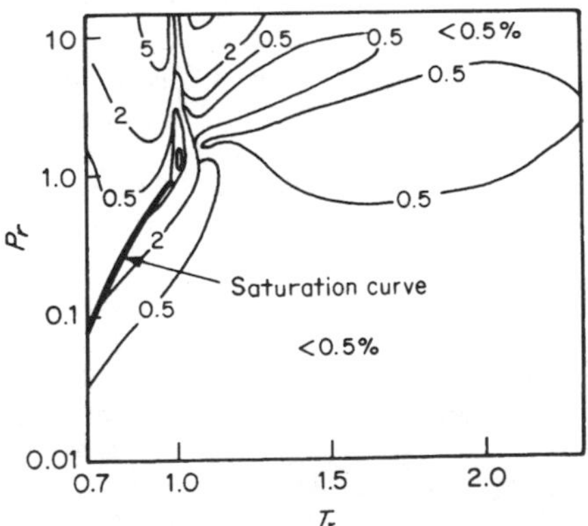

FIGURE 4.4 Contours of percent error in molar volume of CO_2 calculated from the Redlich-Kwong (1949) EoS with parameters from Morris and Turek (1986).

de Hemptinne and Ungerer (1995) and de Sant'Ana, et al. (1999) suggest that the latter errors can increase as T decreases and that similar errors are found for larger hydrocarbons.

The effects of these errors in the PVT relation are carried through to all thermodynamic property variations because they involve derivatives. Major errors for the heat capacities, isothermal compressibility, and sound speed have been shown by Gregorowycz, et al. (1996). See the discussion in Sec. 4-7 and in Chap. 6.

4.5 VIRIAL EQUATION OF STATE

The virial equation of state is a polynomial series in pressure or in inverse volume whose coefficients are functions only of T for a pure fluid. The consistent forms for the initial terms are

$$Z = 1 + B\left(\frac{P}{RT}\right) + (C - B^2)\left(\frac{P}{RT}\right)^2 + \cdots \tag{4-5.1a}$$

$$= 1 + \frac{B}{V} + \frac{C}{V^2} + \cdots \tag{4-5.1b}$$

where the coefficients $B, C, \ldots$ are called the second, third, . . . virial coefficients. Except at high temperatures, B is negative and, except at very low T where they are of little importance, C and higher coefficients are positive. This can be inferred from the behavior of the isotherms in Fig. 4-1. Formulae relating B to molecular pair interactions, C to molecular trio interactions, etc., can be derived from statistical mechanics. Much has been written about the virial EoS; see especially Mason and Spurling (1968) and Dymond and Smith (1980).

Because 1) the virial expansion is not rigorous at higher pressures, 2) higher-order molecular force relations are intractable, and 3) alternative EoS forms are more accurate for dense fluids and liquids, the virial equation is usually truncated at the second or third term and applied only to single-phase gas systems. The general ranges of state for applying Eqs. (4-5.1) and (4-5.2) are given in Table 4-4; they were obtained by comparing very accurately correlated Z values of Setzmann and Wagner (1991) with those computed with their highly accurate virial coefficients over the entire range of conditions that methane is described. When only B is used, the Eqs. (4-5.1*a*) and (4-5.1*b*) are equivalent at the lowest densities. Equation (4-5.1*b*) in density is more accurate to somewhat higher densities but if it is used at higher pressures, it can yield negative Z values. Thus, it is common to use Eq. (4-5.1*a*) in pressure if only the second virial, B, is known. If the term in C is included, Eq. (4-5.1*b*) in density is much more accurate than Eq. (4-5.1*a*). Application ranges for virial equations have also been discussed elsewhere (Chueh and Prausnitz, 1967; Van Ness and Abbott, 1982). Another indication of the range covered by the second virial form of Eq. (4-5.1*a*) is the initial relative linearity of isotherms in Fig. 4-1.

Uncertainties in virial coefficients can affect user results. However, because errors affect $Z - 1$, which is often smaller than Z, tolerances in B may be large. Absolute errors of $0.05V_c$ will generally cause the same level of error as truncating at B (Table 4-4) rather than using C. Thus, for methane with $V_c \sim 100$ cm^3 mol^{-1}, an error of 5 cm^3 mol^{-1} in B causes an error in Z of about 1% for both the saturated vapor at $T = 160$ K where $B = -160$ cm^3 mol^{-1} and $Z = 0.76$ as well as at $T =$

TABLE 4-4 Ranges of Conditions for Accurate Z Values from Virial Equations Using Methane Expressions from Setzmann and Wagner (1991)

Equation	<1% Error*	<1% Error+	<5% Error*	<5% Error+
$Z = 1 + B/V$	$\rho V_c < 0.18$	$T/T_c < 0.82$	$\rho V_c < 0.35$	$T/T_c < 0.9$
$Z = 1 + BP/RT$	$\rho V_c < 0.1$	$T/T_c < 0.7$	$\rho V_c < 0.2$	$T/T_c < 0.8$
$Z = 1 + B/V + C/V^2$	$\rho V_c < 0.8$	$T/T_c < 0.95$	$\rho V_c < 1.5$	$T/T_c < 0.99$
$Z = 1 + BP/RT + (C - B^2)(P/RT)^2$	$\rho V_c < 0.15$	$T/T_c < 0.8$	$\rho V_c < 0.35$	$T/T_c < 0.9$

* Stated density conditions generally accurate when $T/T_c > 1.05$ or when $P/P_c < 5$.
\+ For saturated vapor.

220 K and $P = 30$ bar where $B = -86$ cm^3 mol^{-1} and $Z = 0.85$. An error in B of $0.25V_c$ is acceptable for estimating Z within 5% at these conditions. Since the ideal gas law would be wrong by 25% and 15% respectively, some estimate of nonideality based on B is likely to be better than assuming ideal gas behavior.

The most extensive compilations of second virial coefficients are those of Dymond and Smith (1980) and Cholinski, et al. (1986). Newer values for alkanes, linear 1-alkanols and alkyl ethers are given by Tsonopoulos and Dymond (1997) and measurements using indirect thermodynamic methods have been reported recently by McElroy and coworkers (see, *e.g.*, McElroy and Moser, 1995) and Wormald and coworkers (see, *e.g.*, Massucci and Wormald, 1998). Some third virial coefficient values are also given by Dymond and Smith (1980).

Estimation of Second Virial Coefficients

Though it is possible to derive correlations from molecular theory, such expressions are usually much more complicated than those cited, even for simple substances, and so they have not been considered here. Also, some specialized correlations have not been evaluated. Rather, we list references for a number of practical techniques for estimating values for most types of pure substances (Tsonopoulos, 1974; Hayden and O'Connell, 1975; Tarakad and Danner, 1977; McCann and Danner, 1984; Orbey, 1988, and Kis and Orbey, 1989; Abusleme and Vera, 1989; Olf, et al., 1989; Lee and Chen, 1998; Vetere, 1999) and cite some recent discussions about these methods at the end of this subsection. Section 5-4 treats cross coefficients for mixtures from these methods.

Unlike for empirical EoS, there is direct theoretical justification for extending simple CSP for B to complex substances by merely adding terms to those for simple substances. Thus, essentially all of the methods referenced above can be written in the form

$$\frac{B(T)}{V^*} = \sum_i a_i f^{(i)}(T/T^*) \tag{4-5.2}$$

where V^* is a characteristic volume, such as V_c or P_c/RT_c, the a_i are strength parameters for various intermolecular forces, and the $f^{(i)}$ are sets of universal functions of reduced temperature, T/T^*, with T^* typically being T_c. Then, $f^{(0)}$ is for simple substances with a_0 being unity, $f^{(1)}$ corrects for nonspherical shape and globularity of normal substances with a_1 commonly being, ω, $f^{(2)}$ takes account of polarity with a_2 being a function of the dipole moment, μ (see Sec. 2-6), and $f^{(3)}$ takes account of association with a_3 an empirical parameter. In methods such as those of Tsonopoulos (1974) and Tarakad and Danner (1977), the terms in the various $f^{(i)}(T)$ are obvious; in the Hayden-O'Connell (1975) and Abusleme-Vera (1989) methods, the derivation and final expressions might disguise this simple division, but the correlations can be expressed this way.

The principal distinctions of the Hayden-O'Connell method are that 1) it attempts to remove polar and associating effects from the critical properties and uses an effective acentric factor for shape and globularity so that more appropriate CSP characteristics are used for the nonpolar contributions, and 2) it applies directly to carboxylic acids by relating the equilibrium constant for dimerization to the estimated B according to the "chemical theory" of nonideal gas behavior. Olf, et al. (1989) also use chemical theory expressions, but none of the other methods can treat strongly dimerizing substances.

Detailed discussion about the methods is given below. Because there is no single technique that is significantly better than the others, we illustrate the expressions and use in detail only the Tsonopoulos correlation since it is one of the most popular and reliable.

The Tsonopoulos (1974) correlation uses $V^* = RT_c/P_c$ and $T^* = T_c$. The substance dependent strength coefficients are $a_1 = \omega$, $a_2 = a$, and $a_3 = b$ which can be constant parameters or variable functions of the dipole moment, μ, (see Sec. 2-6) that may depend upon the "family" of the substance or the substance itself (see Table 4-5). The full form is

$$\frac{BP}{RT_c} = f^{(0)} + \omega f^{(1)} + af^{(2)} + bf^{(3)} \tag{4-5.3}$$

where
$$f^{(0)} = 0.1445 - 0.330/T_r - 0.1385/T_r^2 - 0.0121/T_r^3 - 0.000607/T_r^8 \tag{4-5.4a}$$

$$f^{(1)} = 0.0637 + 0.331/T_r^2 - 0.423/T_r^3 - 0.008/T_r^8 \tag{4-5.4b}$$

$$f^{(2)} = 1/T_r^6 \tag{4-5.4c}$$

$$f^{(3)} = -1/T_r^8 \tag{4-5.4d}$$

and $T_r = T/T_c$. Equations (4-5.4*ab*) are modifications of the early correlation of Pitzer and Curl (1955). There is considerable sensitivity to the values of a and b in this model because of the large powers on T_r in $f^{(2)}$ and $f^{(3)}$. As a result, for highest accuracy, fitting data to one of the parameters, probably b, should be considered.

Several revisions and extensions have appeared for the Tsonopoulos model (Tsonopoulos, et al., 1975, 1978, 1979, 1989, 1990, 1997) mainly treating new data for alkanes and alcohols with revised parameters and making comparisons with other models. Table 4-5 summarizes current recommendations for a and b in Eq. (4-5.3).

For normal fluids, simpler equations for $f^{(0)}$ and $f^{(1)}$ were obtained by Van Ness and Abbott (1982)

$$f^{(0)} = 0.083 - 0.422/T_r^{1.6} \tag{4-5.5a}$$

$$f^{(1)} = 0.139 - 0.172/T_r^{4.2} \tag{4-5.5b}$$

Equations (4-5.5*a*,*b*) agree with Eqs. (4-5.4*a*,*b*) to within 0.01 for T_r above 0.6 and ω less than 0.4, but the difference rapidly grows for lower T_r.

Example 4-2 Estimate the second virial coefficient of ethanol at 400 K using the Tsonopoulos method.

solution From Appendix A, T_c = 513.92 K, P_c = 61.48 bar, V_c = 167 ω = 0.649, and μ = 1.7 debyes. From Table 4-5, a = 0.0878 and with μ_r = 66.4, b = 0.0553. With T_r = 400/512.64 = 0.778, Eqs. (4-5.4) give $f^{(0)}$ = − 0.538, $f^{(1)}$ = − 0.346, $f^{(2)}$ = 4.498, $f^{(3)}$ = −7.4248. Then BP_c/RT_c = −0.7786, giving B = −541 cm^3 mol^{-1}. The recommended experimental value is − 535 cm^3 mol^{-1} (Tsonopoulos, et al., 1989). The agreement is within the experimental uncertainty of ±40 cm^3 mol^{-1} and is within the uncertainty limit for 1% agreement in Z as described above since here $0.05V_c$ = 8 cm^3 mol^{-1}. The first row of Table 4-4 suggests that at this temperature $Z = 1 + B/V$

TABLE 4-5 Estimation Methods for Tsonopoulos Parameters for Polar and Associating Species (Tsonopoulos and Heidman, 1990; Tsonopoulos and Dymond, 1997)

Species Class	a	b
Simple, Normal	0	0
Ketones, Aldehydes, Alkyl Nitriles, Ethers, Carboxylic Acid Esters	$-2.14 \times 10^{-4}\ \mu_r - 4.308 \times 10^{-21}(\mu_r)^8$	0
Alkyl Halides, Mercaptans, Sulfides, Disulfides	$-2.188 \times 10^{-4}\ (\mu_r)^4 - 7.831 \times 10^{-21}\ (\mu_r)^8$	0
1-Alkanols (except Methanol)	0.0878	$0.00908 + 0.0006957\ \mu_r$
Methanol	0.0878	0.0525
Water	−0.0109	0

$\mu_r = 10^5 \mu^2 P_c / T_c^2$ where μ is in debye, P_c is in atm (1.01325 bar) and T_c is in K.

would be within 1% up to the density for the saturated vapor, but that $Z = 1 + BP/RT$ would give somewhat more than 1% error. Changing b to 0.0541 would yield the recommended value exactly.

Lee and Chen (1998) have revised the Tsonopoulos expression for $f^{(1)}$, claiming some improvement for n-alkanes over that of Dymond, et al. (1986). For polar substances, Lee and Chen use a "nonpolar" acentric factor based on the radius of gyration, a new expression relating a to μ_r and combine $f^{(2)}$ and $f^{(3)}$ into a single term of a/T_r^n where $n = 6$ for nonassociating substances. They correlate a with μ_r for associating species. These refinements provide some improvement in accuracy, though errors of up to 100 cm^3 mol^{-1} remain in some cases.

There have been some updates to the original methods cited above. The model of Hayden and O'Connell (1975) has been discussed by several authors. Stein and Miller (1980) noted sensitivities and nonuniqueness with the association/solvation parameter. Prausnitz, et al. (1980) provided Fortran programs for the method. O'Connell (1984a) discussed sources of association/solvation parameters and also noted (1984b) that C-H. Twu had found small errors in the Prausnitz, et al. (1980) programs (Eq. A-16 has the conversion from atm to bar included for a second time and the factor in Eq. A-21 should be 1.7491 instead of 1.7941). Upon request, O'Connell will provide access to a Fortran program for the Hayden method with parameters and the original data base.

Many methods are limited because they often require at least one fitted parameter to yield accurate results and data are often not available (*e.g.*, Tarakad and Danner, 1977; Orbey, 1988). The Tsonopoulos and Hayden-O'Connell methods use empirical parameters for different classes of compounds so some predictability is possible, but the highest accuracy is obtained for complex substances when one parameter is fitted to data. Several group parameterization methods have been developed (McCann and Danner, 1984; Abusleme and Vera, 1989; Olf, et al., 1989); these attempt to use only generalized molecular structure parameters and not those which are component-specific and thus require data. Of these, however, only the Abusleme method is not restricted to pure components. The most recent version of Vetere's method (1999) uses other properties such as T_b, V_{liq} and ΔH_b with class-dependent expressions; his limited comparisons show errors similar to those from the Tsonopoulos method.

Literature discussion and our own comparisons show that none of the correlations referenced above is significantly more accurate or reliable than the others. All of the methods show average deviations in the range of 10 to 100 cm^3 mol^{-1} depending upon the class of compound. Some methods are better for some classes of substances and worse for others; no consistency is apparent. Often all methods are poor, suggesting that the data may be incorrect.

In any case, for the range of conditions that the second virial coefficient should be applied to obtain fluid properties, all models are likely to be adequate.

Estimation of Third Virial Coefficients As with second virial coefficients, it is possible to derive third virial coefficient correlations from molecular theory, but these are not very successful. The principal theoretical problem is that the trio intermolecular potential includes significant contributions that cannot be determined from the pair potentials that describe second virial coefficients. Thus, CSP is also used for C, though the range of substances considered has been much more limited. This means that users often must choose to use a complete equation of state such as described in Secs. 4-6 and 4-7 rather than try to estimate B and C to use in Eq. (4-5.1a,b). However, there are cases where it is worthwhile, especially in supercritical mixtures (see Sec. 5-6) and a few estimation methods have been developed for normal fluids.

The principal techniques for C are the CSP methods of Chueh and Prausnitz (1967), De Santis and Grande (1979) and Orbey and Vera (1983). All use $T^* = T_c$ in the equation

$$\frac{C(T)}{V^{*2}} = \sum_i a_i g^{(i)}(T/T^*) \tag{4-5.6}$$

but they differ in the choice of V^* and of the third parameter. Chueh and Prausnitz select $V^* = V_c$ and use a special third parameter that must be found from C data. De Santis and Grande use $V^* = V_c$ while reformulating Chueh and Prausnitz' expressions for the $g^{(i)}$ (T/T^*), and choose to correlate their special third parameter with ω and other molecular properties. The correlation of Orbey and Vera uses the more accessible $V^* = RT_c/P_c$ and ω directly. They take two terms in the series of Eq. (4-5.6) with $a_0 = 1$ and $a_1 = \omega$ and

$$g^{(0)} = 0.01407 + 0.02432/T_r^{2.8} - 0.00313/T_r^{10.5} \tag{4-5.7a}$$

$$g^{(1)} = -0.02676 + 0.01770/T_r^{2.8} + 0.04/T_r^3 - 0.003/T_r^6 - 0.00228/T_r^{10.5} \tag{4-5.7b}$$

The correlation is the best available and its estimates should be adequate for simple and normal substances over the range of conditions that Table 4-4 indicates that Eq. (4-5.1b) should be used. There is no estimation method for third virial coefficients of polar and associating substances.

4-6 ANALYTICAL EQUATIONS OF STATE

As pointed out above, an EoS used to describe both gases and liquids requires the form of Eq. (4-2.1b) and it must be at least cubic in V. The term "analytical equation of state" implies that the function f_V (T, V) has powers of V no higher than quartic. Then, when T and P are specified, V can be found analytically rather than only numerically. We focus here on cubic EoS because of their widespread use and simple form. One quartic equation, that of Shah, et al. (1996), has been developed for pure components only.

This section introduces a generalized way to consider cubic equations and then addresses their use to describe the gaseous and liquid volumetric behavior of pure components. In particular, the issues of what parameterizations are valid, methods of model selection and techniques for obtaining parameter values are addressed. Similar aspects of mixtures are treated in Chap. 5 and the use of EoS for other thermodynamic properties is examined in Chap. 6 and 8.

Formulations of Cubic EoS

It is possible to formulate all possible cubic EoS in a single general form with a total of five parameters (Abbott, 1979). If one incorporates the incompressibility of liquids to have P go to infinity as V approaches a particular parameter b, the general cubic form for P is

$$P = \frac{RT}{V - b} - \frac{\Theta(V - \eta)}{(V - b)(V^2 + \delta V + \varepsilon)} \tag{4-6.1}$$

where, depending upon the model, the parameters Θ, b η, δ, and ε may be constants, including zero, or they may vary with T and/or composition. Thus, the distinctions among cubic EoS models for pure components are which of the parameters in Eq. (4-6.1) are nonzero and how they are made to vary with T. A common notation for recent EoS is to use $\Theta(T) = aa(T)$, where $\alpha(T_c) = 1$. Composition dependence is considered by combining and mixing rules as discussed in Sec. 5.6.

Table 4-6 gives relations among the Eq. (4-6.1) parameters for several common cubic EoS. Note that in all cases, b is a positive constant and $\eta = b$. Also given in Table 4-6 are the total number of substance-specific parameters of each model. Table 4-7 gives the expressions for $\alpha(T)$. As discussed in Sec. 4-3, CSP would suggest that two-parameter models would apply to simple substances, three-parameter models to normal substances and four-parameter models to polar and perhaps associating substances. Table 4-6 indicates if the CSP equations described below have been used for a model; if so, the resulting relations are given below in Table 4-8. If CSP is not used, some or all of the parameter values must be found by regression of data. Strategies for obtaining parameter values are discussed below.

Equation (4-6.1) in the form Eq. (4-2.1*b*) is

$$Z = \frac{V}{V - b} - \frac{(\Theta/RT)V(V - \eta)}{(V - b)(V^2 + \delta V + \varepsilon)} \tag{4-6.2}$$

When it is rewritten as the form to be solved when T and P are specified and Z is to be found analytically, it is

$$Z^3 + (\delta' - B' - 1)Z^2 + [\Theta' + \varepsilon' - \delta'(B' + 1)Z - [\varepsilon'(B' + 1) + \Theta'\eta'] = 0 \tag{4-6.3}$$

where the dimensionless parameters are defined as

$$B' \equiv \frac{bP}{RT} \qquad \delta' \equiv \frac{\delta P}{RT} \qquad \Theta' \equiv \frac{\Theta P}{(RT)^2} \qquad \varepsilon' \equiv \frac{\varepsilon P}{RT} \qquad \eta' \equiv \frac{\eta P}{RT} \tag{4-6.4}$$

When a value of Z is found by solving Eq. (4-6.3) from given T, P and parameter values, V is found from $V = ZRT/P$. V must always be greater than b.

Parameterizations

The expressions in Table 4-6 show explicitly how models have been developed to adjust density dependence through different choices of δ and ε. Temperature dependence is mainly included in $\alpha(T)$, though b, c, d, etc. may be varied with T. The decisions about how parameters are included focuses on what properties are to be described. The principal methods attempt to adjust the formulations to obtain the most reliable liquid densities and good vapor pressures, though connections to virial coefficients can also be made (Abbott, 1979; Tsonopoulos and Heidman, 1990). We will not show all the possible variations, but do describe the main themes.

As discussed in great detail by Martin (1979), experimental density variations from the ideal gas to compressed liquid with the saturation and critical conditions

TABLE 4-6 Equation (4-6.1) Parameters for Popular Cubic EoS*

EoS/Eq. (4-6.1) Parameter	δ	ε	Θ	# Parameters[+]	Generalized?[#]
van der Waals (1890)	0	0	a	2:a, b	Y(T_c, P_c)
Redlich and Kwong (1949)	0	0	$a/T_r^{0.50}$	2:a, b	Y(T_c, P_c)
Wilson (1964)	b	0	$a\alpha(T_r)$	3:a, b, $\alpha(1)$	Y(T_c, P_c, ω)
Soave (1972)	b	0	$a\alpha(T_r)$	3:a, b, $\alpha(1)$	Y(T_c, P_c, ω)
Fuller (1976)	bc	0	$a\alpha(T_r)$	4:a, b, c, $\alpha(1)$	Y(T_c, P_c, Z_c, ω)
Peng and Robinson (1976)	$2b$	$-b^2$	$a\alpha(T_r)$	3:a, b, $\alpha(1)$	Y(T_c, P_c, ω)
Martin (1979)	$0.25 - 2b$	$(0.125 - b)^2$	a/T_r^n	3:a, b, n	N(2)
Soave (1979)	b	0	$a\alpha(T_r)$	4:a, b, $\alpha(2)$	N(2)
Patel and Teja (1982)	$b + c$	$-bc$	$a\alpha(T_r)$	4:a, b, c, $\alpha(1)$	Y(T_c, P_c, ω), N(1)
Peneloux, et al. (1982)	$b + 3c$	$2c^2$	$a\alpha(T_r)$	4:a, b, c, $\alpha(1)$	N(1)
Adachi, et al. (1983)	$b_3 - b_2$	$-b_2b_3$	$a\alpha(T_r)$	5:a, b, b_2, b_3, $\alpha(1)$	Y(T_c, P_c, ω)
Mathias (1983)	b	0	$a\alpha(T_r)$	4:a, b, $\alpha(2)$	N(1)
Mathias and Copeman (1983)	$2b$	$-b^2$	$a\alpha(T_r)$	5:a, b, $\alpha(3)$	N(3)
Soave (1984)	$2c$	c^2	$a\alpha(T_r)$	4–5: a, b, c, $\alpha(1–2)$	Y(T_c, P_c, ω), N(2)
Adachi, et al. (1985)	$2c$	$-c^2$	$a\alpha(T_r)$	4:a, b, c, $\alpha(1)$	Y(T_c, P_c, ω)
Stryjek and Vera (1986)	$2b$	$-b^2$	$a\alpha(T_r)$	4:a, b, $\alpha(2)$	N(2)
Trebble and Bishnoi (1987)	$b + c$	$-bc - d^2$	$a\alpha(T_r)$	6:a, $b(2)$, c, d, $\alpha(1)$	N(2)
Mathias, et al. (1989)	$2b + 3c$	$2c^2 - b^2$	$a\alpha(T_r)$	6:a, b, c, $\alpha(3)$	N(4)
Rogalski, et al. (1990)[&]	$\gamma(b - c)$	$c[\gamma(b - c) + c]$	$a\alpha(T_r)$	5:a, b, c, $\alpha(2)$	N(3)
Twu, et al. (1992)	$4b + c$	bc	$a\alpha(T_r)$	6:a, b, c, $\alpha(3)$	N(3)
Soave (1993)	b	0	$a\alpha(T_r)$	3–4:a, b, $\alpha(1–2)$	N(1–2)
Twu, et al. (1995)	$2b$	$-b^2$	$a\alpha(T_r)$	3:a, b, $\alpha(1)$	Y(T_c, P_c, ω)
Stamateris and Olivera-Fuentes (1995)	0	0	$a\alpha(T_r)$	4:a, b, $\alpha(2)$	N(2)
Patel (1996)	$b + c$	$-bc$	$a\alpha(T_r)$	6:a, b, c, $\alpha(3)$	N(4)
Zabaloy and Vera (1996, 1998)	$2b$	$-b^2$	$a\alpha(T_r)$	6–8:a, b, $\alpha(4–6)$	N(3–6)

* Single letters such as a, b, c, etc., are substance-specific parameters that are usually constants or may be simple funcitons of T. Expressions such as $\alpha(T)$ are multiterm functions of T containing from 1 to 3 parameters and are shown in Table 4-7. In all cases here, b of Eq. (4-6.1) is retained as a positive parameter and $\eta = b$.

[+] The total number of substance-specific constant parameters including a, b, c, d, etc. and those explicit in the expression for $\alpha(T)$ given in Table 4-7. Additional parameters may be included in any T dependence of b, c, d, etc.

[#] Y means that CSP relations exist to connect all of the parameters a, b, c, d, etc., to T_c, P_c, Z_c, ω, etc.; see Table 4-8. In some cases, this reduces the number of substance-specific parameters; compare the number of parameters listed in the last two columns. N means that at least some of the parameter values are found by data regression of liquid densities and/or vapor pressures while others are critical properties or ω. The number of such fitted parameters is in parentheses. Parameters obtained by matching with a correlation such as the Peneloux, et al. c value, from the Rackett Model, Eq. (4-11.1), or Rogalski, et al. with T_b are considered fitted.

[&] $\gamma = 4.82843$.

TABLE 4-7 Expressions for $a(T_r)$ in common EoS for $T_r = T/T_c \leq 1$

EoS	$a(T_r)$*	# Parameters*
van der Waals (1890)	1	0
Redlich Kwong (1949)	$1/T_r^{1/2}$	0
Wilson (1964)	$[1 + (1.57 + 1.62\omega)(1/T_r - 1)]T_r$	$1(\omega)$
Soave (1972); Fuller (1976)	$[1 + (0.48 + 1.574\omega - 0.176\omega^2)(1 - T_r^{1/2})]^2$	$1(\omega)$
Peng and Robinson (1976)	$[1 + (0.37464 + 1.54226\omega - 0.2699\omega^2)(1 - T_r^{1/2})]^2$	$1(\omega)$
Martin (1979)	$1/T_r^n$	$1(n)$
Soave (1979)	$[1 + (1 - T_r)(m + n/T_r]$	$2(m, n)$
Patel and Teja (1982)	$\{1 + F[1 - (T_r)^{1/2}]\}^2$	$1(F)$
Patel and Teja (1982)	$[1 + (0.452413 + 1.38092\omega - 0.295937\omega^2)(1 - T_r^{1/2})]^2$	$1(\omega)$
Peneloux, et al. (1982)	$[1 + (0.48 + 1.574\omega - 0.176\omega^2)(1 - T_r^{1/2})]^2$	$1(\omega)$
Adachie, et al. (1983)	$[1 + (0.407 + 1.3787\omega - 0.2933\omega^2)(1 - T_r^{1/2})]^2$	$1(\omega)$
Mathias (1983)	$[1 + (0.48508 + 1.55191\omega - 0.15613\omega^2)(1 - T_r^{1/2}) - p(1 - T_r)(0.7 - T_r)]^2$	$2(\omega, p)$
Mathias and Copeman (1983); Mathias, et al. (1989)	$[1 + c_1(1 - T_r^{1/2}) + c_2(1 - T_2^{1/2})^2 + c_3(1 - T_r^{1/2})^3]^2$	$3(c_1, c_2, c_3)$
Soave (1984)	$[1 + (1 - T_r)(m + n/T_r]$	$2(m, n)$
Soave (1984)	$[1 + (0.4998 + 1.5928\omega - 0.19563\omega^2 + 0.025\omega^3)(1 - T_r^{1/2})]^2$	$1(\omega)$
Adachie et al. (1985)	$[1 + (0.26332 + 1.7379\omega - 1.2990\omega^2 + 1.5199\omega^3)(1 - T_r^{1/2})]^2$	$1(\omega)$
Stryjek and Vera (1986)	$[1 + (0.378893 + 1.4897153\omega - 0.17131848\omega^2 + 0.0196554\omega^3)(1 - T_2^{1/2}) + \kappa_1(1 - T_r)(0.7 - T_r)]^2$	$2(\omega, \kappa_1)$
Trebble and Bishnoi (1987)	$\exp[q_1(1 - T_r)]$	$1(q_1)$
Rogalski, et al. (1990)	$a(T_r, T_b, m)^+$	$2(T_b, m)$
Twu, et al. (1992)	$T_r^{N(M-1)} \exp[L(1 - T_r^{NM})]$	$3\ (L, M, N)$
Soave (1993)	$1 + n(1 - T_r^{1/2})^2 + m(1 - T_r)$	$2(m, n)$
Soave (1993)	$1 + (2.756m - 0.7)(1 - T_r^{1/2})^2 + m(1 - T_r)$; $m = 0.484 + 1.515\omega - 0.44\omega^2$	$1(\omega)$
Twu, et al. (1995)	$T_r^{-0.171813} \exp[0.125283(1 - T_r^{1.77634})] + \omega\{T_r^{-0.607352} \exp[0.511614(1 - T_r^{2.20517})] - T_r^{-0.171813} \exp[0.125283(1 - T_r^{1.77634})]\}$	$1(\omega)$
Stamateria Olivera-Fuentes (1995)	$1 + \dfrac{m}{n - 1}(1 - T_r^{1-n})$	$2(m, n)$
Patel (1996)	$1 + c_1(T_r - 1) + c_2(T_r^{1/2} - 1) + c_3(T_r^N - 1)$	$4(c_1, c_2, c_3, N)$
Zabaloy and Vera (1998)	$1 + C_1 T_r \ln T_r + C_2(T_r - 1) + C_3(T_r^2 - 1)$	$3\ (C_1, C_2, C_3)$

*The substance-specific parameters, n, counted in Table 4-6 as $\alpha(n)$. In many cases, the parameter is the acentric factor, ω, which can be obtained from independent measurements. If it is the only parameter, the model is fully generalized, but it is then limited to normal substances.

+The function of Rogalski, et al. (1990) has different expressions for different temperature ranges covering the triple to the critical temperatures of hydrocarbons.

between them suggest that the simple forms adopted by van der Waals (1890) and Redlich and Kwong (1949) are inadequate. One idea was that of a "volume translation" where V computed from an original EoS is shifted so that the translated volume matches some experimental value(s) or values from an estimation method. Thus, Peneloux, et al. (1982) used the Rackett Equation, Eq. (4-11.1), for their estimation. The translation is small and does not materially change the gas or vapor phase densities. It is common to express the shift by substituting $V - c$ for V in Eqs. (4-6.1) and (4-6.2). It can also be accomplished by reformulation as done here. For example, in Table 4-6 the Soave (1984) EoS is the van der Waals (1890) EoS with a translation of c by merely using new expressions for δ and ε. Note that doing this will make inappropriate any prediction of the b parameter of Eq. (4-6.1), such as with a CSP relation (see below).

Most forms of volume translation have been chosen to avoid changing the EoS vapor pressure; Zabaloy and Brignole (1997) point out that care must be taken in the expression to insure this and give an example where the vapor pressure was affected by translation.

In addition, it is possible to make the translation parameter dependent upon T as, for example, Mathias, et al. (1989) and de Sant' Ana, et al. (1999) have. Pfohl (1999) warns of the dangers of this; Ungerer and de Sant' Ana (1999) agree with his comments. Another aspect of temperature-dependent parameters was noted by Trebble and Bishnoi (1986) for some cases where b was made dependent on T; they found that negative C_P values can occur at high pressures.

Rather than explicitly express a volume translation, it can be incorporated directly into the expression. Twu, et al. (1992) studied the results for 21 equations of state and concluded that their own relations for δ and ε were best to fit saturation densities. Trebble and Bishnoi (1986) found that among 10 untranslated cubic EoS models, the Fuller (1976) model was best. de Sant'Ana, et al. (1999) have extensively studied the *PVT* behavior of petroleum substances and made recommendations for EoS forms.

A major advance in the use of EoS came (Wilson, 1964; Barner, et al., 1966; Soave, 1972) with the generalization of the description of the vapor-liquid boundary by solving the isofugacity condition, *i.e.*, vapor-liquid equilibrium at the vapor pressure (see Chap. 6 and 8). This made the models much more useful for phase equilibria than when only the liquid and/or vapor density was described. This approach required a new temperature dependence of $\alpha(T)$ even though it was not intended to change the number of CSP parameters. Thus, for simple and normal substances functions of T/T_c and ω were often utilized. Otherwise, one or more additional parameters were fitted to pure component P_{vp} data. Table 4-7 lists functions for $\alpha(T)$ in several common cubic EoS. Forms generalized with T_r and ω are given where possible, and the total number of substance-specific constants is also listed.

Most of the expressions in Table 4-7 for $\alpha(T)$ were developed considering only $T/T_c \leq 1$. As a result, there has been some uncertainty about whether the expressions should also be used at high temperatures. Examination of the generalized forms of Table 4-7 shows that all are similar up to $T_r \sim 2$ and less than unity. Above this condition, the Soave (1972) and Soave (1993) models give values that are somewhat higher than those of Twu, et al. (1995) and Mathias (1983) as well as those of the Trebble-Bishnoi (1987) model. Those without an exponential form will give minima in $\alpha(T)$ when T_r is 3 to 5, while others will give negative values. For normal fluids, except at such high temperatures, all functions will give similar *PVT* behavior. However, for polar and associating substances, accuracy will probably be best with fitted, not generalized, functions like those indicated as N in Table 4-6. This will be especially true for high temperature systems.

Obtaining Cubic EoS Parameter Values

There are several approaches which have been used to set the values of the parameters listed in Tables 4-6 and 4-7, ranging from completely empirical to completely CSP based on critical property characteristics and acentric factor.

Corresponding States Principle. Critical constants have been used so commonly for cubic EoS that nearly all models contain them, though it is not necessary. It is convenient to force the EoS to obey the critical conditions along the true critical isotherm because then three relations exist to set parameter values. For the typical form shown here where $\eta = b$, these can be expressed as

$$Z_c = \frac{1}{1 - b/V_c} - \frac{(\Theta/BRT_c)(b/V_c)}{(1 + \delta/b)(b/V_c) + (\varepsilon/b^2)(b/V_c)^2} \tag{4-6.5a}$$

$$\lim_{V \to Vc} \left(\frac{\partial P}{\partial V}\right)_{T=T_c} = 0 \tag{4-6.5b}$$

$$\lim_{V \to V_c} \left(\frac{\partial^2 P}{\partial V^2}\right)_{T=T_c} = 0 \tag{4-6.5c}$$

An alternative to solving all of Eqs. (4-6.5) is to realize that Eq. (4-6.5*a*) has three roots of b/V_c. As implied above, forcing a cubic equation to meet the true critical conditions will cause errors in the results because of the different nature of the critical region from the rest. Abbott (1979) comments that this is seen in the low density pressures being not too bad, but at supercritical densities, the pressures are much too low and volume translation does not eliminate the errors on the critical isotherm. However, if the critical region is not so important, Eqs. (4-6.5) can be used (Abbott, 1973, 1979), to obtain the following parameter relations

$$\frac{bP_c}{RT_c} = Z_c + \Omega - 1; \qquad \frac{\delta P_c}{RT_c} = \Omega - 2Z_e; \quad \varepsilon\left(\frac{P_c}{RT_c}\right)^2 = Z_c^2 - \Omega Z_c - \Omega^2(\Omega - 1) \tag{4-6.6}$$

where $\Omega = aP_c/R^2T_c^2$.

Depending upon the model, not all the parameters may be set independently; Sec. 4-3 notes that if there are only two constants in the EoS, only two of the three equations can be used. Typically these are Eqs. (4-6.5*b*) and (4-6.5*c*). Sometimes the true Z_c (or equivalently, V_c) is not used even if there are three parameters; accuracy at the critical point is sacrificed for accuracy in other regions by choosing a Z_c that is not the true value. Table 4-8 shows the dimensionless values for some of the models of Tables 4-6 and 4-7.

For volume translations, the CSP relations must be changed. A direct relation may be retained for the translated parameters. For example, the Soave (1984) relation is a translated van der Waals (1890) equation; the expressions in Table 4-8 show the difference. On the other hand, the relationships may become complex. Thus, in the methods of Fuller (1976), Patel and Teja (1982), Trebble and Bishnoi (1987) and Twu (1992), the value bP_c/RT_c is found by solving a cubic equation.

Regression of Data for EoS Parameter Values. In cases where CSP has not been employed, the parameter values must be obtained by fitting data. The most common data are P_{vp} and saturated liquid densities, either experimental or correlated, such

TABLE 4-8 Generalized Parameters for cubic EoS Models of Table 4-6 Using Critical Properties

EoS\Parameter	Z_c	$\frac{bP_c}{RT_c}$	$\frac{\delta P_c}{RT_c}$	$\varepsilon\left(\frac{P_c}{RT_c}\right)^2$	$\frac{aP_c}{(RT_c)^2}$
van der Waals (1890)	0.3750	0.125	0	0	0.42188
Redlich and Kwong (1949)	0.3333	0.08664	0.08664	0	0.42748
Wilson (1964)	0.3333	0.08664	0.08664	0	0.42748
Soave (1972)	0.3333	0.08664	0.08664	0	0.42748
Fuller (1976)	Z_c	$f_{F1}(Z_c)$*	$f_{F2}(Z_c)$*	0	$f_{F3}(Z_c)$*
Peng and Robinson (1976)	0.3070	0.0778	0.15559	−0.006053	0.45724
Patel and Teja (1982)	$f_{PT1}(\omega)$*	$f_{PT2}(\omega)$*	$f_{PT3}(\omega)$*	$f_{PT4}(\omega)$*	$f_{PT5}(\omega)$*
Adachie, et al. (1983)	$0.3242 - 0.0576\omega$	$f_{A1}(\omega)$*	$f_{A2}(\omega)$*	$f_{A3}(\omega)$*	$f_{A4}(\omega)$*
Soave (1984)	0.3333	0.08333	0.08333	0.001736	0.42188
Adachi, et al. (1985)	$f_{A5}(\omega)$*	$f_{A6}(\omega)$*	$f_{A7}(\omega)$*	$f_{A8}(\omega)$*	$f_{A9}(\omega)$*
Twu et al. (1995)	0.03070	0.0778	0.15559	−0.006052	0.457236

* f_{Fn} (Z_c), f_{PTn} (ω) and f_{An} (ω) for various n values are functions given in the original articles.

as with the methods of Sec. 4-9. We mention here those methods with three or more parameters which include at least one not related to critical properties.

Many workers use a combination of CSP and fitting. Table 4-6 shows the number of data fitted parameters of each method that are not critical properties. Thus, of Martin's (1979) four parameters, three are T_c, P_c, and Z_c, while of the four parameters used by Mathias (1983), three are T_c, P_c, and ω. Peneloux, et al. (1982) use the CSP formulation of Soave (1972) but match the volume translation to results from the Rackett Equation, Eq. (4-9.11), which gives accurate saturation densities but misses the critical point. For α (T) parameters, P_{vp} values are the typical data. This may consist of regression of data over the entire range of the liquid or force matching at a particular state such as at $T_r = 0.7$ to obtain ω, the triple and boiling points and the critical point. Zabaloy and Vera (1996, 1997, 1998) present detailed discussion of such strategies. They also describe in depth the matching of saturation volumes to obtain EoS model parameters.

Martin (1979) discusses ranges of data to choose for fitting parameters. Morris and Turek (1986) used P_{vp} and volumetric data over a range of pressures (at fixed temperatures) to determine optimal values of a and b in the Redlich and Kwong (1949) EoS for eight substances. Soave, et al. (1994) show that most CSP formulations require only T_c/P_c, which can be estimated, one liquid density data point and one to two vapor pressures including T_b rather than elaborate data sets. An example of fitting to obtain parameters for many substances is the work of Sandarusi, et al. (1986) who tabulate parameters for 286 organic systems from P_{vp} data to be used in the Soave (1979) EoS.

Example 4-3 Find the molar volumes of saturated liquid and vapor propane at $T =$ 300 K and $P_{vp} = 9.9742$ bar and at $P_c = 42.477$ bar for models in Table 4-6 for which parameters have been listed.

solution Table 4-9 shows the results including the percent errors for V^V, V^L, and ΔV for the compression from P_{vp} to P_c at T = 300 K. Parameters for propane were not readily available for the EoS models of Martin (1979), Rogalski, et al. (1990), Patel (1996), Soave (1993), Mathias (1983). It was not possible to obtain reasonable values with the equations and parameters given in the original article of Adachi, et al. (1983).

Analytic EoS Model Selection for Volumes

The issue of which model with the same number of parameters is best does not have a universal answer. Martin's (1979) extremely detailed analysis of the volumetric descriptions of two-parameter cubic models compared to precise data for argon concludes that "no one equation stands clearly above the others." It also appears that there is no obvious choice among the three and four-parameter models.

All analytic models show deviations in the critical region even when the parameters are chosen to give the correct Z_c. As Table 4-9 shows, most methods do not attempt to do this; the decision is to accept larger errors very near the critical rather than compromise results further away. Some do have adjustments to the volume translation to improve agreement in the near-critical region, but they appear not to be as effective as the crossover methods described in the next section.

As suggested by Gregorowicz, et al. (1996) and verified in Table 4-9, even if analytic equations are accurate for saturation volumes, they do not give very reliable

TABLE 4-9 Saturated Vapor and Liquid and Compressed Liquid Molar Volumes of Propane at T = 300K from EoS Models listed in Table 4-6. Experimental values from Lemmon, et al. (1998) are for P_{vp} = 9.9742 bar, V^V = 2036.5 cm^3 mol^{-1}, V^L = 90.077 cm^3 mol^{-1} and for P_c = 42.477 bar, V = 88.334 cm^3 mol^{-1}, ΔV = −1.743 cm^3 mol^{-1}

EoS\Calculated Volumes, cm^3 mol^{-1}	V^V	Percent Err*	V^L	Percent Err*	V	ΔV	Percent Err* ΔV
van der Waals (1890)	2177	6.9	145.4	61.5	135.5	9.9	467
Redlich and Kwong (1949)	2085	2.4	101.4	12.5	97.3	4.1	134
Wilson (1964)	2061	1.2	98.0	8.8	94.8	3.2	84
Soave (1972)	2065	1.4	98.5	9.3	95.1	3.4	90
Fuller (1976)	2127	4.5	78.1	−13.3	75.6	2.5	42
Peng and Robinson (1976)	2038	0.1	86.8	−3.7	84.1	2.7	51
Patel and Teja (1982)	2048	0.6	91.0	1.0	88.1	2.9	65
Patel and Teja (1982)	2049	0.6	91.4	1.5	88.5	2.9	67
Peneloux, et al. (1982)	2061	1.2	94.2	4.6	90.6	3.6	107
Soave (1984)	2066	1.4	97.1	7.8	93.7	3.4	99
Adachi et al. (1985)	2051	0.7	92.9	3.1	90.0	2.9	66
Stryek and Vera (1986)	2039	0.1	86.9	−3.6	84.2	2.7	53
Trebble and Bishnoi (1987)	2025	−0.6	89.4	−0.7	88.3	1.1	−37
Twu, et al. (1992)	2026	−0.5	87.8	−2.5	85.7	2.1	21
Twu, et al. (1995)	2017	−0.9	84.6	−6.1	82.4	2.2	25
Stamateris and Olivera-Fuentes (1995)	1928	−5.3	114.8	27.5	113.6	1.2	−29

*Defined as 100 $(V_{calc} - V_{exp})/V_{exp}$.

changes of volume upon compression. Most of the predicted isothermal compressibilities are too large.

4-7 NONANALYTIC EQUATIONS OF STATE

The complexity of property behavior cannot be described with high accuracy with the cubic or quartic EoS that can be solved analytically for the volume, given T and P. Though the search for better models began well before computers, the ability to rapidly calculate results or do parameter regression with complicated expressions has introduced increasing levels of complexity and numbers of fitted parameters. This section describes five approaches that are available for pure components. Two are strictly empirical: BWR/MBWR models and Wagner formulations. Two are semiempirical formulations based on theory: perturbation methods and chemical association models. The last method attempts to account for the fundamentally different behavior of the near-critical region by using "Crossover" expressions.

BWR and MBWR Models

The BWR expressions are based on the pioneering work of Benedict, Webb and Rubin (1940, 1942) who combined polynomials in temperature with power series and exponentials of density into an eight-parameter form. Additional terms and parameters were later introduced by others to formulate modified Benedict-Webb-Rubin (MBWR) EoS.

The general form of BWR/MBWR correlations is

$$Z = 1 + f_1(T)/V + f_2(T)/V^2 + f_3(T)/V^n + f_4(T)[(\alpha + \gamma/V^2)/V^m] \exp(-\gamma/V^2) \tag{4-7.1}$$

the T-dependent functions $f_i(T)$ can contain more than 30 parameters in addition to m, n, α, and γ. Until very recently, this equation form was standard for IUPAC and NIST compilations of pure component fluid volumetric and thermodynamic properties. Kedge and Trebble (1999) have investigated an expression similar to Eq. (4-7.1) with 16 parameters that provides high accuracy (within 0.3% of validated data for volumetric properties and P_{vp}).

However, other formulations described below have become more prevalent in use. As a result, we refer readers to previous editions of this book which describe this approach, especially in corresponding states form using generalized parameters for normal fluids (see Sec. 4-2).

Wagner Models

Setzmann and Wagner (1989, 1991) describe a computer-intensive optimization strategy for establishing highly accurate EoS models by a formulation for the residual Helmholtz energy,

$$A^r/RT = [A(T, V) - A^\circ(T, V)]/RT \tag{4-7.2}$$

where $A^\circ(T, V)$ is the ideal gas Helmholtz energy at T and V. (See Eq. (3-1.14)). The model expressions contain large numbers of parameters whose values are obtained by regression on data for many properties over wide ranges of conditions. Recently, the trend is that empirical EoS for pure components use based on this highly accurate methodology, provided sufficient data exist. Most of these EoS have been published in the Journal of Physical and Chemical Reference Data and the International Journal of Thermophysics.

The technique first establishes a "bank of terms" that are functions of temperature and density in the forms $\Phi_{ij}(T/T_c, V/V_c) = n_{ij}\delta^{d_i}\tau^{t_j}$ and $\Phi_{ijk}(T/T_c, V/V_c) = n_{ijk}\delta^{d_i}\tau^{t_j}\exp(-\delta^{c_k})$, where n_{ijk} is a fitted coefficient, the reduced density is, $\delta = V_c/V$, the reduced temperature is $\tau = T_c/T$, the d_i are integers from 1 to 10, the t_j are integers from 0 to 22 and half-integers from $-\frac{1}{2}$ to $\frac{9}{2}$ and the c_k are integers ranging from 1 to 6. There can also be as many as 27 additional terms designed to make significant contributions only near the critical point. Thus, up to 393 total terms and associated parameters may be used (Setzmann and Wagner, 1991).

Their "optimization strategy" is to regress all available and rigorously validated volumetric, calorimetric (see Chap. 6) and speed of sound data by finding optimal linear parameters, n_{ijk}, as different numbers of terms are included in the model. Ultimately only those terms which significantly improve the fit are included in the model. In the case of methane, the optimum was for 40 terms and parameters plus values of T_c and V_c. This number varies with different substances and ranges of data conditions. For methane they also fitted eight parameters to the ideal gas heat capacities (see Chap. 3) to obtain the accurate temperature dependence of $A^{\circ}(T, V)/RT$. The compressibility factor of Eq. (4-2.1) is found using a thermodynamic partial derivative

$$Z = 1 - V\frac{\partial(A^r/RT)}{\partial V}\bigg|_{\tau} = 1 + \delta\left\{\sum_i\sum_j n_{ij}d_i\delta^{d_i-1}\,\tau^{t_j}\right. \left. + \sum_i\sum_j\sum_k n_{ijk}\delta^{d_i-1}\,\tau^{t_j}[\exp(-\delta^{c_k})][d_i - c_k\delta^{c_k}]\right\} \qquad (4\text{-}7.3)$$

There are actually a few additional terms in the expression of Setzmann and Wagner (1991).

Equations in this form can describe all measured properties of a pure substance with an accuracy that probably exceeds that of the measurements. It gives excellent agreement with the second virial coefficient (see Sec. 4-4); where B/V_c is all terms in the sums of Eq. (4-7.3) when $\delta = 0$. It can predict the properties of fluids at hyperpressures and hypertemperatures (accessible only at explosive conditions, Span and Wagner, 1997). All other thermodynamic properties are straightforward derivatives of the terms in Eq. (4-7.2).

Thus, if the analysis and regression have been done for a substance (nearly 20 have been completed at the time of this writing), readers who wish benchmark descriptions of a common substance can use equations of this form with confidence. Generally, saturation properties (vapor pressures and liquid and vapor volumes) are fitted to separate parameterized equations by the workers in this area. These expressions, also known as Wagner Equations, are described in Chap. 7 and in Sec. 4-9.

Perturbation Models

The technique of perturbation modeling uses reference values for systems that are similar enough to the system of interest that good estimates of desired values can be made with small corrections to the reference values. For EoS models, this means that the residual Helmholtz energy of Eq. (4-7.3) is written as

$$A^r/RT = [A^r_R(T, V)/RT] + \Sigma[A^r_{Att}(T, V)/RT]^{(i)} \tag{4-7.4}$$

where the form of the perturbation terms $[A^r_{Att}(T, V)/RT]^{(i)}$ can be obtained from a rigorous or approximate theory, from a Taylor's expansion or from intuition. The result is that there are very many models obtained in this manner and expressed in this form. For example, the virial EoS of Sec. 4-4 is a Taylor's series in density with the ideal gas as the reference so its first term in Eq. (4-7.4) is zero and the terms in the summation have increasing powers of density. Alternatively, like most models that seek to describe liquids, the cubic EoS of Sec. 4-6 use the hard sphere fluid as the reference. Choosing the particular expression of van der Waals (vdW)

$$[A^r_R(T, V)/RT]^{(vdW)} = -\ln[(V - b)/V] \tag{4-7.5a}$$

leads to the compressibility factor of

$$Z^{(vdW)}_R = V/(V - b) \tag{4-7.5b}$$

Then there are one or more terms in the summation that match the forms of Eq. (4-6.1) or (4-6.2) such as those listed in Table 4-6 and 4-7. A useful discussion of this approach is given by Abbott and Prausnitz (1987).

Our purpose here is to mention the possible options and give a few references to specific models which have become popular, especially for phase equilibria (see Sec. 8.12). Much more complete reviews are given by Anderko (1990), Sandler, et al. (1994) and Prausnitz, et al. (1999). A very important point is that models of the form of Eq. (4-7.4) inevitably have a positive reference term and negative perturbation terms. This is necessary to be able to describe both vapors and liquids, but the consequence is that the perturbation terms at high density are typically about the same magnitude as the reference term is. This can cause difficulties in evaluation and errors in estimation. Further, the isotherms in the two-phase region can be quite complex or even unrealistic, especially at low temperatures (see, for example, Koak, et al. 1999).

Reference Fluid Expressions. The first adaptation is to choose a different form for the reference expression. Equation (4-7.5*b*) is not very accurate (Henderson, 1979). It was retained because, except for a few possibilities, any more complicated function of V would make the EoS of higher power than cubic when used with even the simplest appropriate perturbation expression. This limitation has been overcome by increased computational ability, so it is now common to use noncubic expressions known to be more accurate for hard spheres even though the resulting EoS is noncubic. The most common reference now is that of Carnahan and Starling (1969) (*CS*) which is typically written in terms of the compressibility factor

$$Z^{CS}_R = \frac{1 - \eta^3}{(1 - \eta)^4} \tag{4-7.6}$$

where the covolume, $\eta = \pi N_A \sigma^3/(6V) = V^*/V$ with σ being the diameter of the hard sphere and V^* being a characteristic volume for the species. With the simplest form of perturbation term, the model EoS is fifth order in volume so it is nonanalytic.

This idea has been expanded to deal with nonspherical and flexible molecules in three principal ways. The first is to assume that the rigid bodies are not spheres, but have different shapes so there are several different terms to replace those in η in Eq. (4-7.6). The expressions have been reviewed by Boublik (1981). The second approach is in the Perturbed Hard Chain Theory (PHCT) which multiplies a hard-sphere compressibility factor (as in Eq. (4-7.6), for example) by a factor, c, which is a substance specific parameter (for a review, see, e.g., Cotterman, et al., 1986).

$$Z_{\mathrm{R}}^{\mathrm{PHCT}} = cZ_{\mathrm{R}}^{\mathrm{HS}} \tag{4-7.7}$$

where Z^{HS} can be any appropriate hard sphere model such as given in Eqs. (4-7.5) and (4-7.6). This idea has been used by many workers with success (see Anderko, 1990, and Prausnitz, et al., 1999 for details). Alternatives include those of Siddiqi and Lucas (1989) and of Chiew (1990) who derived terms to be added to Eq. (4-7.5) for chains of hard spheres (*HC*). This can be called a perturbed hard-sphere-chain theory (PHSC). Chiew's form is

$$Z_{\mathrm{R}}^{\mathrm{PHSC}} = Z_{\mathrm{R}}^{\mathrm{HS}} - \frac{r-1}{r}\frac{1+\eta/2}{(1-\eta)^2} \tag{4-7.8}$$

where r is the number of segments in the substance of interest. This has been adopted for example, by Fermeglia, et al. (1997) for alternative refrigerants, by Song, et al. (1994) and Hino and Prausnitz (1997) for all substances, and by Feng and Wang (1999) for polymers. A final alternative reference expression from similar origins to Eq. (4-7.5) is that of the Statistical Associating Fluid Theory (SAFT) derived by Chapman and coworkers (Chapman, et al., 1989, 1990; Prausnitz, et al., 1999)

$$Z_{\mathrm{R}}^{\mathrm{SAFT}} = 1 + r\left[\frac{4\eta - 2\eta^2}{(1-\eta)^3}\right] + (1-r)\left[\frac{5\eta - 2\eta^2}{(1-\eta)(2-\eta)}\right] \tag{4-7.9}$$

Perturbation (Attraction) Expressions. The perturbation terms, or those which take into account the attraction between the molecules, have ranged from the very simple to extremely complex. For example, the simplest form is that of van der Waals (1890) which in terms of the Helmholtz energy is

$$[A^r(T, V)/RT]_{\mathrm{Att}}^{(vdW)} = -a/RTV \tag{4-7.10}$$

and which leads to an attractive contribution to the compressibility factor of

$$Z_{\mathrm{Att}}^{(vdW)} = -a/(RTV) \tag{4-7.11}$$

This form would be appropriate for simple fluids though it has also been used with a variety of reference expressions such as with the CS form of Eq. (4-7.6) by Aly and Ashour (1994) for a great variety of substances including organic acids, and the PHSC form of Eq. (4-7.8) in the model of Song, et al. (1994) for polymers. Other terms such as those found in Tables 4-6 and 4-7 can be used for normal fluids. The most complex expressions for normal substances are those used in the BACK (Chen and Kreglewski, 1977), PHCT, and SAFT EoS models. In this case there are many terms in the summation of Eq. (4-7.3). Their general form is

$$Z_{\mathrm{Att}}^{\mathrm{BACK}} = r\sum_{i=1}^{n}\sum_{j=1}^{m} jD_{ij}\left[\frac{u}{kT}\right]^i\left[\frac{\eta}{\tau}\right]^j \tag{4-7.12}$$

where the number of terms may vary, but generally $n \sim 4\text{-}7$ and $m \sim 10$, the D_{ij}

coefficients and τ are universal, and u and η are substance-dependent and may also be temperature-dependent as in the SAFT model.

Statistical mechanical perturbation theory gives first and second order terms which depend upon the intermolecular potential. A convenient potential model is the square-well which allows analytic expressions. These have been used by Fleming and Brugman (1987) for aqueous systems, and by Hino and Prausnitz (1997) to simplify and increase the accuracy of previous models (e.g., Song, et al., 1994) for small substances and polymers. Additional terms can be put in Z_{Att} to account for polarity such as by Muller, et al. (1996). Thus, there are many possible expressions and they can be very complicated. However, the total number of pure component parameters ranges from only three to five with the rest of the quantities being universal constants so the input information is mostly the same in all models. In general, the results are also similar.

Chemical Theory EoS

In many practical systems, the interactions between the molecules are quite strong due to charge-transfer and hydrogen bonding (see Prausnitz, et al., 1999 for a description of the origin and characteristics of these interactions). This occurs in pure components such as alcohols, carboxylic acids, water and HF and leads to quite different behavior of vapors of these substances. For example, Twu, et al. (1993) show that Z for the saturated vapor of acetic acid increases with temperature up to more than 450 K as increased numbers of molecules appear due to a shift in the dimerization equilibrium. However, the liquid Z behaves like most other polar substances. Also, the apparent second virial coefficients of such components species are much more negative than suggested by corresponding states and other correlations based on intermolecular forces and the temperature dependence is much stronger.

Instead of using parameters of a model from only nonpolar and polar forces, one approach has been to consider the interactions so strong that new "chemical species" are formed. Then the thermodynamic treatment assumes that the properties deviate from an ideal gas mainly due to the "speciation" (the actual number of molecules in the system is not the number put in) plus some physical effects. It is assumed that all of the species are in reaction equilibrium. Thus, their concentrations can be determined from equilibrium constants having parameters such as enthalpies and entropies of reaction in addition to the usual parameters for their physical interactions.

An example is the formation of dimers (D) from two monomers (M)

$$2\text{M} = \text{D} \tag{4-7.13}$$

The equilibrium constant for this reaction can be exactly related to the second virial coefficient of Eq. (4-5.1)

$$K_{\text{D}} = y_{\text{D}}/y_{\text{M}}^2 P = -B/RT \tag{4-7.14}$$

The model of Hayden and O'Connell (1975) described in Sec. 4-5 explicitly includes such contributions so that it can also predict the properties of strongly interacting substances.

Anderko (1990) notes that there are two general methods for analyzing systems with speciation. The first, exemplified by the work of Gmehling (Gmehling, et al., 1979; Grensheuzer and Gmehling, 1986), is to postulate the species to be found, such as dimers, and then obtain characteristic parameters of each species such as critical properties from those of the monomers along with the enthalpy and Gibbs energy of each reaction. The alternative approach was first developed by Heidemann and Prausnitz (1976) and extended by Ikonomu and Donohue (1986), Anderko (1989, 1991), Twu, et al. (1993) and Visco and Kofke, (1999). This approach builds the species from linear polymers whose characteristics and reaction equilibrium constants can be predicted for all degrees of association from very few parameters. By proper coupling of the contributions of the physical and chemical effects, the result is a closed form equation. A similar formulation is made with the SAFT equation (Chapman, et al., 1989) where molecular level association is taken into account by a reaction term that is added to the free energy terms from reference, dispersion, polarity, etc.

Though this form of the EoS may appear to be not very different from those considered earlier in this chapter, the computational aspects are somewhat more complex because obtaining the numbers of moles of the species is an additional nonlinear equation to be solved. However, there is no other practical way to deal with the large and unique nonidealities of such systems.

Economou and Donohue (1992) and Anderko, et al. (1993) show that care must be exercised in treating the chemical and physical contributions to equations of state since some methods introduce thermodynamic inconsistencies.

EoS Models for the Near-Critical Region

Conditions near the vapor-liquid critical point of a substance show significantly different behavior from simple extrapolations of the EoS models described so far in this chapter. The shape of the critical isotherm, the variations of C_V, the isothermal compressibility, and other properties and the vapor-liquid coexistence curve are all different than that given by most EoS models. This is because the molecular correlations are much longer ranged and fluctuate differently in this region. The result is that, unlike in the "classical" region where Taylor's series expansions can be taken of properties about a given state, such a mathematical treatment breaks down near the "nonclassical" critical point. Research into this effect shows that certain universalities appear in the properties, but substance-specific quantities also are involved.

There are a variety of ways to define the "critical region." Anisimov, et al. (1992) define a criterion of $0.96 < T/T_c < 1.04$ along the critical isochore with effects on derivative properties felt at densities as far as 50 to 200% from ρ_c.

Considerable work has been done to develop EoS models that will suitably bridge the two regimes. There are several different approaches taken. The first is to use a "switching function" that decreases the contribution to the pressure of the classical EoS and increases that from a nonclassical term (e.g., Chapela and Rowlinson, 1974). The advantage of this method is that no iterative calculations are needed. Another approach is to "renormalize" T_c and ρ_c from the erroneous values that a suitable EoS for the classical region gives to the correct ones. Examples of this method include Fox (1983), Pitzer and Schreiber (1988), Chou and Prausnitz (1989), Vine and Wormald (1993), Solimando, et al. (1995), Lue and Prausnitz (1998) and Fornasiero, et al. (1999). These have different levels of rigor, but all involve approximations and iterative calculations. The technique of Fornasiero, et al. (1999) was applied to the corresponding states forms of the van der Waals (1890), Soave (1972) and Peng-Robinson (1976) cubic EoS models described in Sec. 4-6 and used Z_c as an additional piece of data. Comparisons of saturated liquid densities with data for 17 normal fluids and 16 polar and associating substances showed RMS deviations of 1 to 5% which appears to be comparable with the direct methods described above and the liquid density correlations described below.

The final approach to including nonclassical behavior has been the more rigorous approach via crossover functions of Sengers and coworkers (e.g., Tang and Sengers, 1991; Tang, et al., 1991; Kiselev, 1998; Anisimov, et al., 1999; Kiselev and Ely, 1999; Kostrowicka Wyczalkowska, et al., 1999). The original method was to develop an EoS model that was accurate from the critical point to well into the classical region, but did not cover all conditions. Anisimov, et al. (1992) and Tang, et al. (1991) show results for several substances. Recent efforts with this method have led to EoS models applicable to all ranges. Though not applied extensively yet, indications are that it should be broadly applicable with accuracies similar to the scaling methods. In addition, theoretical analyses of this group (e.g., Anisimov, et al. 1999) have considered the differences among approaches to the critical point of different kinds of systems such as electrolytes, micelles and other aggregating substances, and polymers where the range of the nonclassical region is smaller than molecular fluids and the transition from classical to nonclassical can be sharper and even nonuniversal.

4-8 DISCUSSION OF EQUATIONS OF STATE

In this section, we discuss the use of the EoS described above. In the low density limit, all reduce to the ideal-gas law. In the critical region, only those equations that give nonclassical behavior can be satisfactory. The primary differences among the myriad of forms are computational complexity and quality of the results at high pressures, for liquids, for polar and associating substances and for polymers. While equations of state were previously limited to vapor phase properties, they now are commonly applied to the liquid phase as well. Thus, the most desirable expressions give the *PVT* behavior of both vapor and liquid phases and also all other pure component properties with extensions to mixtures while remaining as simple as possible for computation. Of course, since not all of these constraints can be satisfied simultaneously, which model to use requires judgment and optimization among the possibilities.

The truncated virial equation, Eq. (4-5.1) is simple but it can be used only for the vapor phase. Temperatures and pressures for which this equation applies are given in Table 4-4 and generally in the regions of Figs. 4-1 to 4-3 for which V_r is greater than about 0.5.

Cubic EoS have often been chosen as the optimal forms because the accuracy is adequate and the analytic solution for the phase densities is not too demanding. The most comprehensive comparisons of different cubic EoS models for 75 fluids of all types have been performed by Trebble and Bishnoi (1986). Because the translated forms were not widely applied at the time, their liquid volume comparisons showed most widely used models were not as accurate as some less popular ones (for example, Fuller, 1976). However, the improvement with shifting reported by de Sant'Ana, et al. (1999) and others suggests that translated forms can be quite good with average errors of the order of less than 2% in liquid volumes for simple and normal substances. This is consistent with the results shown in Table 4-9. However, higher accuracy can normally be obtained from experiment, from nonanalytic EoS in Sec. 4-7, and from methods given in Secs. 4-10 to 4-12.

When selecting a cubic EoS for *PVT* properties, users should first evaluate what errors they will accept for the substances and conditions of interest, as well as the effort it would take to obtain parameter values if they are not available in the literature. Sometimes this takes as much effort as implementing a more complex, but accurate model such as a nonanalytic form.

Except for MBWR and Wagner EoS models, nearly all methods have been developed to give good results for mixtures (Sec. 5.5) and for phase equilibria of mixtures (Sec. 8.12). This is especially true of perturbation methods and chemical theory treatments for complex substances like carboxylic acids and polymers.

No EoS models should be extrapolated outside the temperature and pressure range for which it has been tested. Within their ranges however, they can be both accurate and used for many properties. Unlike what was presented in the 4th Edition, there are now both cubic and other EoS models that can be used to predict with confidence the *PVT* behavior of polar molecules. Complex substances require more than three parameters, but when these are obtained from critical properties and measured liquid volumes and vapor pressures, good agreement can now be obtained.

Recommendations

To characterize small deviations from ideal gas behavior use the truncated virial equation with either the second alone or the second and third coefficients, B and C, Eq. (4-5.1). Do not use the virial equation for liquid phases.

For normal fluids, use a generalized cubic EoS with volume translation. The results shown in Table 4-9 are representative of what can be expected. All models give equivalent and reliable results for saturated vapors except for the dimerizing substances given above.

For polar and associating substances, use a method based on four or more parameters. Cubic equations with volume translation can be quite satisfactory for small molecules, though perturbation expressions are usually needed for polymers and chemical models for carboxylic acid vapors.

If one wishes to calculate only saturated or compressed liquid volumes, one of the correlations in the following sections may be the best choice.

4-9 PVT PROPERTIES OF LIQUIDS—GENERAL CONSIDERATIONS

Liquid specific volumes are relatively easy to measure and for most common organic liquids, at least one experimental value is available. Values at a single temperature for many compounds may be found in Dean (1999), Perry and Green (1997), and Lide (1999). Daubert, et al. (1997) list over 11,000 references to physical property data for over 1200 substances. This compilation includes references to original density data for many of these compounds. The highest quality data have been used to determine constants in a four-parameter equation with the same temperature dependence as Eq. (4-11.7). These constants can be used to calculate saturated-liquid volumes at any temperature. Other summaries of literature density data may be found in Spencer and Adler (1978), Hales (1980), and Tekac, et al. (1985). In Appendix A, single-liquid volumes are tabulated for many compounds at a given temperature. Section 4-10 describes methods for the molar volume at the normal boiling point while section 4-11 gives methods for saturated liquid molar volumes over a range of temperature up to the critical point. Section 4-12 describes correlations for compressed liquids.

4-10 ESTIMATION OF THE LIQUID MOLAR VOLUME AT THE NORMAL BOILING POINT

Three methods are presented to estimate the liquid volume at the normal boiling point temperature. In addition, methods presented later that give the volume as a function of temperature may also be used for obtaining V_b at T_b. Equations of state may also be used for estimating volumes as described in Secs. 4-6 to 4-8.

Additive Methods

Schroeder (Partington, 1949) suggested a simple additive method for estimating molar volumes at the normal boiling point. His rule is to count the number of atoms of carbon, hydrogen, oxygen, and nitrogen, add one for each double bond, two for each triple bond and multiply the sum by seven. Schroeder's original rule has been expanded to include halogens, sulfur, and triple bonds. This gives the volume in cubic centimeters per gram mole. This rule is surprisingly accurate, giving results within 3 to 4% except for highly associated liquids. Table 4-10 gives the contributions to be used. The values in the table may be expressed in equation form as

$$\mathrm{V}_b = 7(N_\mathrm{C} + N_\mathrm{H} + N_\mathrm{O} + N_\mathrm{N} + N_\mathrm{DB} + 2\,N_\mathrm{TB}) + 31.5\,N_\mathrm{Br} + 24.5\,N_\mathrm{Cl} + 10.5\,N_\mathrm{F} + 38.5\,N_\mathrm{I} + 21\,N_\mathrm{S} - 7^{*} \tag{4-10.1}$$

where subscripts DB and TB stand for double and triple bonds and the last value * is counted once if the compound has one or more rings. V_b for benzene, for example, is $7(6 + 6 + 3) - 7 = 98$ cm^3 mol^{-1} compared to the experimental value of 95.8 or 2.3% error. The accuracy of this method is shown in the third column of Table 4-11. The average error for the compounds tested is 3.9% with 5 strongly polar and associating substances having errors greater than 10%.

The additive volume method of Le Bas (1915) is an alternative to Schroeder's rule. Volume increments from Le Bas are shown in Table 4-10, and calculated values of V_b are compared with experimental values in the fourth column of Table 4-11. The average error for the compounds tested is 3.9% with 5 substances having errors greater than 10%.

Tyn and Calus Method (1975)

In this method, V_b is related to the critical volume by

$$V_b = 0.285\,V_c^{1.048} \tag{4-10.2}$$

where both V_b and V_c are expressed in cubic centimeters per gram mole. Comparisons with the substances of Table 4-11 shows that this method is somewhat more accurate and has greater reliability since only 1 substance has an error of more than

TABLE 4-10 Group/Atom Contributions for Schroeder (Eq. 4-10.1) and Le Bas Methods

	Increment, cm^3/mol	
	Schroeder	Le Bas
Carbon	7.0	14.8
Hydrogen	7.0	3.7
Oxygen	7.0	7.4
In methyl esters and ethers		9.1
In ethyl esters and ethers		9.9
In higher esters and ethers		11.0
In acids		12.0
Joined to S, P, and N		8.3
Nitrogen	7.0	
Doubly bonded		15.6
In primary amines		10.5
In secondary amines		12.0
Bromine	31.5	27.0
Chlorine	24.5	24.6
Flourine	10.5	8.7
Iodine	38.5	37.0
Sulfur	21.0	25.6
Ring, three-membered	−7.0	−6.0
Four-membered	−7.0	−8.5
Five-membered	−7.0	−11.5
Six-membered	−7.0	−15.0
Naphthalene	−7.0	−30.0
Anthracene	−7.0	−47.5
Double bond	7.0	
Triple bond	14.0	

10%. The table results are representative of the method where errors exceed 3% only for the low-boiling permanent gases (He, H_2, Ne, Ar, Kr) and some polar nitrogen and phosphorus compounds (HCN, PH_3, BF_3).

Example 4-4 Estimate the liquid molar volume of acetone (C_3H_6O) at its normal boiling point using the methods of Schroeder, Le Bas and Tyn and Calus. The critical volume is 209 $cm^3\ mol^{-1}$ (Appendix A). The accepted value of 77.6 $cm^3\ mol^{-1}$ is from Daubert, et al. (1997).

solution Schroeder Method. From Eq. (4-10.1) with $N_C = 3$, $N_H = 6$, $N_O = 1$, and $N_{DB} = 1$,

$$V_b = (7)(3 + 6 + 1 + 1) = 77\ cm^3\ mol^{-1}$$

$$\text{Error} = (77 - 77.6)/77.6 = -0.008 \text{ or } -0.8\%$$

Le Bas Method. From Table 4-10, C = 14.8, H = 3.7, and O = 7.4. Therefore,

$$V_b = (3)(14.8) + (6)(3.7) + 7.4 = 74\ cm^3\ mol^{-1}$$

$$\text{Error} = (77 - 77.6)/77.6 = -0.046 \text{ or } -4.6\%$$

Tyn and Calus Method. With Eq. (4-10.2),

$$V_b = 0.285\ V_c^{1.048} = 0.285(209)^{1.048} = 77.0\ \text{cm}^3\ \text{mol}^{-1}$$

$$\text{Error} = (77 - 77.6)/77.6 = -0.008 \text{ or } -0.8\%$$

Results for 35 substances with the method of Tyn and Calus are shown in the fifth column of Table 4-11. The average deviation for the three methods of this section is 3 to 4%. Examination of Table 4-11 shows that there is no pattern of error in the Schroeder and Le Bas methods that would suggest either is to be preferred but in any case, the Tyn and Calus method is more reliable than both. The simplicity of the methods makes them attractive but none of them are as accurate as those described in the next section (the last three columns of Table 4-11).

4-11 SATURATED LIQUID DENSITIES AS A FUNCTION OF TEMPERATURE

A number of techniques are available to estimate pure saturated-liquid molar or specific volumes or densities as a function of temperature. Here, one group contribution technique and several corresponding states methods are presented to estimate saturated-liquid densities.

Rackett Equation

Rackett (1970) proposed that saturated liquid volumes be calculated by

$$V_s = V_c Z_c^{(1-T/T_c)^{2/7}} \tag{4-11.1}$$

where V_s = saturated liquid volume, V_c = critical volume, Z_c = critical compressibility factor, T_c = critical temperature. Eq. (4-11.1) is often written in the equivalent form

$$V_s = \frac{RT_c}{P_c} Z_c^{[1+(1-T/T_c)^{2/7}]} \tag{4-11.2}$$

While Eq. (4-11.1) is remarkably accurate for many substances, it underpredicts V_s when $Z_c < 0.22$.

Yamada and Gunn (1973) proposed that Z_c in Eq. (4-11.1) be correlated with the acentric factor:

$$V_s = V_c(0.29056 - 0.08775\omega)^{(1-T/T_c)^{2/7}} \tag{4-11.3}$$

If one experimental density, V_s^R, is available at a reference temperature, T^R, Eqs. (4-11.1) and (4-11.3) can be modified to give

$$V_s = V_s^R(0.29056 - 0.08775\omega)^{\phi} \tag{4-11.4a}$$

$$V_s = V_s^R Z_c^{\phi} \tag{4-11.4b}$$

where

TABLE 4-11 Comparisons of Estimations of Liquid Molar Volumes at the Boiling Temperature

Substance	V_b^* $cm^3\ mol^{-1}$	Percent error[#] when calculated by method of Schroeder	LeBas	Tyn & Calus	Elbro	Eq. (4-11.3)[+]	Eq. (4-11.4a)[+]
Methane	37.9	−7.8	−22.0	−7.7		−0.8	0.2
Propane	75.7	1.7	−2.2	−2.9	−1.6	0.2	−1.7
Heptane	163.0	−1.2	−0.1	0.1	−2.8	−0.1	0.5
n-octadecane	452.8	−13.4	−10.1	5.1	−8.6	−0.6	1.0
Cyclohexane	116.8	1.9	1.2	−1.1		−0.3	0.0
Ethylene	49.2	−0.5	−9.8	−4.1	5.4	0.8	0.1
Benzene	95.8	2.3	0.2	−0.6	1.6	0.2	−0.1
Fluorobenzene	101.5	0.0	−0.5				
Bromobenzene	120.1	2.0	−0.7				
Chlorobenzene	114.5	0.9	2.1	0.9	−1.4	0.8	0.1
Iodobenzene	130.2	−0.5	−0.7				
Methanol	42.7	−1.7	−13.4	−1.1		−3.1	1.5
n-propanol	82.1	2.3	−0.9	−1.6	1.9	−5.2	2.4
Dimethyl ether	63.0	−0.1	−3.4	−1.7	3.7	1.0	−0.2
Ethyl propyl ether	129.5	−2.7	−0.9	−1.3	−2.3	−1.1	−0.4
Acetone	77.6	0.8	−4.7	−0.8	1.1	0.1	−0.2
Acetic acid	66.0	−4.5	3.7	−5.5		−2.9	1.3
Isobutyric acid	108.7	−3.4	3.8	−0.2		−4.0	1.9
Methyl formate	62.7	0.4	−0.2	0.0		1.5	0.1
Ethyl acetate	106.3	−1.2	2.1	0.6	1.3	0.5	0.1
Propyl butanoate	174.7	−7.8	1.5	−1.8	−1.1	−2.8	0.1
Diethyl amine	109.3	2.5	2.4	3.2			
Acetonitrile	55.5	0.8	1.4				
Methyl chloride	50.1	4.7	0.7	3.2		6.2	0.7
Carbon tetrachloride	103.6	1.3	9.3	−0.9		0.1	−0.1
Dichlorodiflouromethane	81.4	−5.4	0.0	−1.6		1.0	1.4

Ethyl mercaptan	75.7	1.7	2.3	0.7		2.2	−1.1
Diethyl sulfide	118.0	0.8	3.2	0.2		0.1	0.3
Phosgene	70.5	−0.8	1.2				
Ammonia	25.0	12.0	−13.6	1.4		4.5	0.1
Chlorine	45.4	8.0	8.4	−1.9		1.9	0.0
Water	18.8	11.7	−21.3	2.8		3.8	3.6
Hydrogen chloride	30.6	3.1	−7.4	−6.7		−2.0	2.3
Sulfur dioxide	43.8	11.9	−3.6	0.0		2.0	0.5
Bromine triflouride	55.0	14.6	−3.4	−25.3			
Average Error		3.9	3.9	2.8	2.8	1.8	0.8

*Calculated from correlation of Daubert, et al. (1997), Eq. 4-11.7.
#Percent Error = [(calc. − exp.)/exp.] × 100.
+Z_{RA} from Eq. (4-11.4).

$$\phi = (1 - T/T_c)^{2/7} - (1 - T^R/T_c)^{2/7} \tag{4-11.5}$$

Eq. (4-11.4a) is obtained from Eq. (4-11.3) by using the known reference volume to eliminate V_c. The same approach is used to obtain Eq. (4-11.4b) from Eq. (4-11.1). It is also possible to eliminate Z_c from Eq. (4-11.1), but then V_c appears in the final equation and it is generally known less accurately than the quantities that appear in Eq. (4-11.4).

An often-used variation of Eq. (4-11.3) is

$$V_s = \frac{RT_c}{P_c}(0.29056 - 0.08775\omega)^{[1+(1-T/T_c)^{2/7}]} \tag{4-11.6}$$

However, this form does not predict V_c correctly unless the actual $Z_c = 0.29056 - 0.08775\omega$, in which case it is identical to Eq. (4-11.2).

Equation (4-11.1) has been used as the starting point to develop a variety of equations for correlating liquid densities. For example, Spencer and Danner (1972) replaced Z_c with an adjustable parameter, Z_{RA}, values of which are tabulated in Spencer and Danner (1972), Spencer and Adler (1978), and the 4th Edition of this book. Daubert, et al. (1997) changed the physical quantities and constants of Eq. (4-11.1) into four adjustable parameters to give

$$V_s = B^{[1+(1-T/C)^D]}/A \tag{4-11.7}$$

Values of the four constants A through D, are tabulated in Daubert, et al. (1997) for approximately 1200 compounds. The value of C is generally equal to T_c while A, B, and D are generally close to the values used in Eq. (4-11.3).

Of all the forms of the Rackett Equation shown above and including versions of the 4th Edition, we recommend Eq. (4-11.4a). This form uses a known reference value, V_s^R, does not require V_c or P_c, and is more accurate when Z_c is low. Errors associated with the assumption that a correlation in ω applies to all substances is mitigated by use of the reference value.

When various forms of the Rackett equation based on critical properties were used to predict the liquid volumes tabulated in Appendix A, Eq. (4-11.3) performed better than did either Eq. (4-11.1) or (4-11.6). Results of these calculations are shown in Table 4-12. For $\omega < 0.4$, Eq. (4-11.3) gave an average deviation of 2.6% for 225 substances. For $\omega > 0.4$, the average deviation was 6.1% for 65 substances. It is likely that this conclusion would be valid at other conditions as well because comparisons of reduced volumes at other reduced temperatures and acentric factors all gave essentially the same results.

Another liquid volume correlation was proposed by Hankinson and Thomson (1979) and further developed in Thomson, et al. (1982). This correlation, herein referred to as the HBT correlation, is

TABLE 4-12 Average Absolute Percent Deviations for Predictions of Liquid Molar Volumes Tabulated in Appendix A

	Eq. (4-11.1)	Eq. (4-11.3)	Eq. (4-11.6)	HBT*
225 substances, $\omega < 0.4$	4.09	2.56	4.88	2.56
65 substances, $\omega > 0.4$	7.17	6.07	10.1	5.66

* Equation (4-11.8) to (4-11.10) with $\omega_{SRK} = \omega$ and $V^* = V_c$.

$$V_s = V^* V^{(0)}[1 - \omega_{\mathrm{SRK}} V^{(\delta)}] \tag{4-11.8}$$

$$V^{(0)} = 1 + a(1 - T_r)^{1/3} + b(1 - T_r)^{2/3} + c(1 - T_r) + d(1 - T_r)^{4/3} \tag{4-11.9}$$

$$V^{(\delta)} = \frac{e + fT_r + gT_r^2 + hT_r^3}{T_r - 1.00001} \tag{4-11.10}$$

Equation (4-11.9) may be used in the range $0.25 < T_r < 0.95$ and Eq. (4-11.10) may be used when $0.25 < T_r < 1.0$. Constants a through h are given by

a	−1.52816	b	1.43907
c	−0.81446	d	0.190454
e	−0.296123	f	0.386914
g	−0.0427258	h	−0.0480645

In Eqs. (4-11.8) to (4-11.10), ω_{SRK} is that value of the acentric factor that causes the Soave equation of state to give the best fit to pure component vapor pressures, and V^* is a parameter whose value is close to the critical volume. Values of ω_{SRK} and V^* are tabulated for a number of compounds in Hankinson and Thomson (1979) and in the 4th Edition of this book. We have found that ω_{SRK} and V^* can be replaced with ω and V_c with little loss in accuracy. Thus, we have used ω and V_c for the comparisons shown in Table 4-12.

The dependence on temperature and acentric factor expressed by the HBT correlation is nearly identical to that described by Eq. (4-11.3). In fact for $T_r < 0.96$ and $\omega < 0.4$, the difference in these two sets of equations is always less than 1% when $V_c = V^*$ and $\omega = \omega_{\mathrm{SRK}}$. Thus, it can be expected that any improvement seen in the HBT correlation by using the empirical parameters ω_{SRK} and V^* could be reproduced with the same values used in place of ω and V_c in Eq. (4-11.3). The errors shown in Table 4-12 suggest that the HBT method is marginally better than Eq. (4-11.3) when $\omega > 0.4$. The HBT correlation continues to be used with success (Aalto, 1997; Aalto et al., 1996; Nasrifar and Moshfeghian, 1999).

Example 4-5 Use various forms of the Rackett equation to calculate the saturated liquid volume of 1,1,1-trifluoroethane (R143a) at 300 K. The literature value for the liquid volume at this temperature is 91.013 cm^3 mol^{-1} (Defibaugh and Moldover, 1997).

solution From Appendix A for 1,1,1-trifluoroethane, T_c = 346.30 K, V_c = 193.60 cm^3 mol^{-1}, P_c = 37.92 bar, Z_c = 0.255, and ω = 0.259. Also from Appendix A, V_{liq} = 75.38 cm^3 mol^{-1} at T_{liq} = 245 K. T_r = 300/346.3 = 0.8663, so $1 - T_r$ = 0.1337. With Eq. (4-11.1)

$$V_s = (193.60)(0.255)^{(0.1337)^{0.2857}} = 89.726$$

$$\text{Error} = (89.726 - 91.013)/91.013 = 0.0141 \text{ or } 1.41\%$$

With Eqs. (4-11.4a) and (4-11.5)

$$\phi = (0.1337)^{0.2857} - \left(1 - \frac{245}{346.30}\right)^{0.2857} = -0.1395$$

$$V_s = (75.38)[0.29056 - 0.08775(0.259)]^{-0.1395} = 90.59$$

$$\text{Error} = (90.59 - 91.01)/91.01 = 0.005 = 0.5\%$$

Figure 4-5 shows the percent deviation in liquid volume of 1,1-difluoroethane

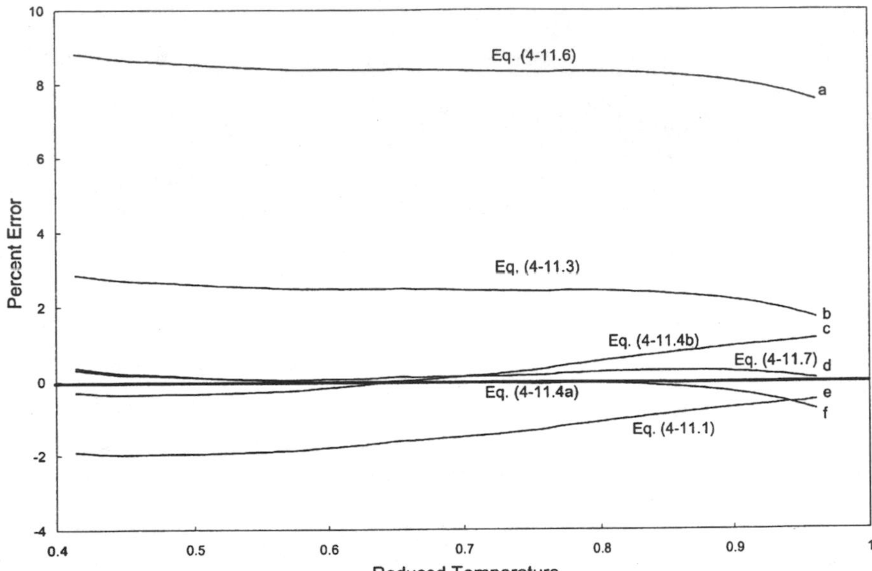

FIGURE 4.5 Percent deviation of liquid volumes of 1,1 difluoroethane (R152a) calculated by various equations. Literature values from Blanke and Weiss (1992) and Defibaugh and Morrison (1996). Lines: a—Eq. (4-11.6); b—Eq. (4-11.3); c—Eq. (4-11.4b); d—Eq. (4-11.7); e—Eq. (4-11.1); f—Eq. (4-11.4a).

(R152a) when calculated by the different equations. The experimental data are best reproduced by line d which is Eq. (4-11.7) with parameters *A-D* from Daubert, et al. (1997). However, using V_{liq} from Appendix A as V_s^R in Eq. (4-11.4a) is nearly as accurate (line f). Line c is Eq. (4-11.4b); with the quantity $0.29056 - 0.08775\omega = 0.266$ replaced by $Z_c = 0.255$, there is some loss of accuracy. Line b (Eq. 4-11.3) and line e (Eq. 4-11.1) are less accurate than Eqs. (4-11.7) and (4-11.4ab) but more accurate than Eq. (4-11.6) which is line a. This comparison among equations is consistent with the results shown in Table 4-12 and suggests that among the simpler models, Eq. (4-11.4a) is the most accurate.

Method of Elbro, et al. (1991)

Elbro, et al. (1991) have presented a group contribution method for the prediction of liquid densities as a function of temperature from the triple point to the normal boiling point. In addition to being applicable to simple organic compounds, the method can also be used for amorphous polymers from the glass transition temperature to the degradation temperature. The method should not be used for cycloalkanes. To use the method, the volume is calculated by

$$V = \Sigma n_i \Delta v_i \tag{4-11.11}$$

where n_i is the number of group i in the substance and Δv_i is a temperature dependent group molar volume given by

$$\Delta v_i = A_i + B_i T + C_i T^2 \tag{4-11.12}$$

Values for the group volume temperature constants are given in Table 4-13. To calculate the density of a polymer, only groups present in the repeat unit need be considered. The technique first obtains the molar volume of the repeat unit and then divides this into the repeat unit molecular weight to obtain the polymer density. The method is illustrated in Examples 4-6 and 4-7.

Example 4-6 Estimate V_s of hexadecane at 298.15 K with the method of Elbro, et al. (1991). From Appendix A, $V_{\text{liq}} = 294.11$, cm^3 mol^{-1} at $T_{\text{liq}} = 298.15$ K.

solution Using values from Table 4-13 in Eq. (4-11.12):

TABLE 4-13 Elbro Group Contributions for Saturated Liquid Volume

No.	Group	A, cm^3/mol	10^3B, cm^3/(mol K)	10^6C, cm^3/(mol K^2)
1	CH_3	18.960	45.58	0
2	CH_2	12.520	12.94	0
3	CH	6.297	−21.92	0
4	C	1.296	−59.66	0
5	ACH	10.090	17.37	0
6	$ACCH_3$	23.580	24.43	0
7	$ACCH_2$	18.160	−8.589	0
8	ACCH	8.925	−31.86	0
9	ACC	7.369	−83.60	0
10	CH_2=	20.630	31.43	0
11	CH=	6.761	23.97	0
12	C=	−0.3971	−14.10	0
13	CH_2OH	39.460	−110.60	23.31
14	CHOH	40.920	−193.20	32.21
15	ACOH	41.20	−164.20	22.78
16	CH_3CO	42.180	−67.17	22.58
17	CH_2CO	48.560	−170.40	32.15
18	CHCO	25.170	−185.60	28.59
19	CHOH	12.090	45.25	0
20	CH_3COO	42.820	−20.50	16.42
21	CH_2COO	49.730	−154.10	33.19
22	CHCOO	43.280	−168.70	33.25
23	COO	14.230	11.93	0
24	ACCOO	43.060	−147.20	20.93
25	CH_3O	16.660	74.31	0
26	CH_2O	14.410	28.54	0
27	CHOH	35.070	−199.70	40.93
28	COO	30.120	−247.30	40.69
29	CH_2Cl	25.29	49.11	0
30	CHCl	17.40	27.24	0
31	CCl	37.62	−179.1	32.47
32	$CHCl_2$	36.45	54.31	0
33	CCl_3	48.74	65.53	0
34	ACCl	23.51	9.303	0
35	Si	86.71	−555.5	97.90
36	SiO	17.41	−22.18	0

group	number	A	10^3B	C	Δv_i
$—CH_3$	2	18.960	45.58	0	32.550
$—CH_2$	14	12.520	12.94	0	16.378

With Eq. (4-11.11)

$$V_s = 2 \times 32.550 + 14 \times 16.378 = 294.39 \text{ cm}^3 \text{ mol}^{-1}$$

$$\text{Error} = (294.39 - 294.11)/294.11 = 0.001 \text{ or } 0.1\%$$

Example 4-7 Estimate the density of poly (methyl acrylate) at 298.15 K with the method of Elbro, et al. (1991). The value given in van Krevelen and Hoftyzer (1972) is 1.220 g cm^{-3}.

solution For poly (methyl acrylate), the repeat unit is

$$—(—H_2C—CH—)_n— \text{ with } M = 86.09$$
$$| $$
$$COOCH_3$$

Using values from Table 4-13 in Eq. (4-11.12):

group i	number	A_i	10^3B_i	10^5C_i	Δv_i
$—CH_3$	1	18.960	45.58	0	32.550
$—CH_2$	1	12.520	12.94	0	16.378
—CHCOO	1	43.280	−168.70	33.25	22.539

Here,

$$V_s = 32.550 + 16.378 + 22.539 = 71.4 \text{ cm}^3\text{mol}^{-1}$$

$$\rho_s = \frac{M}{V_s} = \frac{86.09}{71.48} = 1.205 \text{ g cm}^3$$

$$\text{Error} = (1.205 - 1.220)/1.220 = 0.012 = 1.2\%$$

Discussion and Recommendations

For the saturated liquid volume at any temperature, including the normal boiling point temperature, if constants for the compound are available from Daubert, et al. (1997), these should be used. If these constants are not available, but T_c, ω, and one liquid density value are, then Eq. (4-11.4*a*) should be used. If only critical properties and ω are available, Eq. (4-11.3) should be used. If critical properties are not available, the Elbro method, Eq. (4-11.11) may be used at temperatures below the normal boiling point when group contribution values are available in Table 4-13. At the normal boiling point, the simple methods of Schroeder or Le Bas can be used with errors generally less than 5%. Above the normal boiling point, it is possible to use estimated values of V_b from these methods as V_s^R values in Eq. (4-11.4*a*).

4-12 Compressed Liquid Densities

Critical evaluations of the literature have been recently published by Cibulka and coworkers (see Cibulka and Takagi, 1999 and references to earlier work). Sun, et al. (1990) describe many different relations that have been used for the effect of pressure on liquid volumes and densities. They are generally of the forms

$$P - P_0 = f(T, \rho, \rho_0) \tag{4-12.1a}$$

$$\rho - \rho_0 = f(T, P, P_0) \tag{4-12.1b}$$

where the reference density, ρ_0, is the density at T and P_0, often picked as atmospheric. Since the desired quantity is ρ at T and P, Eq. (4-12.1*b*) can require iterative calculations, but often the expressions are simpler than for Eq. (4-12.1*a*). Sun, et al. (1990) compared limited data with eight different equations and concluded that Eq. (4-12.1*b*) was the more effective form; a cubic in density gave errors that were as low as 1/10 that from similar forms of Eq. (4-12.1*a*).

One of the most often used equations which is similar in form to Eq. (4-12.1*b*) is the Tait Equation (Dymond and Malhotra, 1988), which can be written as

$$V - V_0 = -C \ln \frac{B + P}{B + P_0} \tag{4-12.2}$$

where V is the compressed volume, V_0 is a reference volume, and P_0 is the reference pressure. Often, the reference state is the saturation state at T, so $V_0 = V_s$ and $P_0 = P_{vp}$. Also, B and C are substance specific parameters. Though C may be a constant for small, nonpolar species, it is common to make both B and C functions of T. Thomson et al. (1982) present generalized correlations for B and C in terms of T_c, P_c, and ω for $T_r < 0.95$. Dymond and Malhotra (1988) show other correlations for the parameters B and C.

An alternative form of Eq. (4-14-1*b*) is the equation of Chang and Zhao (1990) to $T_r = 1$

$$V = V_s \frac{AP_c + C^{(D-T_r)^B}(P - P_{vp})}{AP_c + C(P - P_{vp})} \tag{4-12.3}$$

where V_s is the saturated liquid molar volume, P_c is the critical pressure, and P_{vp} is the vapor pressure while A and B were polynomials in T/T_c and ω, respectively. Aalto, et al. (1996) modified the Chang-Zhao model by substituting ω_{SRK}, the Soave-Redlich-Kwong EoS acentric factor used in the HBT method for saturated liquid volumes (Thomson, et al., 1982) as described in Sec. 4-11. They also changed the formulation of A and B:

$$A = a_0 + a_1 T_r + a_2 T_r^3 + a_3 T_r^6 + a_4 / T_r \tag{4-12.4a}$$

$$B = b_0 + \omega_{SRK} b_1 \tag{4-12.4b}$$

The constants of Aalto et al. (1996) are shown in Table 4-14. In addition to the ω_{SRK} values tabulated in the 4th Edition for a number of compounds, Aalto (1997) gives values for refrigerants. Of course, it is possible to use the true value of ω as given in Appendix A; we find little difference in the final results with this substitution. Examples 4-8 and 4-9 show results from the correlation.

TABLE 4-14 Constants for Aalto, et al. (1996) Correlation for Compressed Liquids of Eqs. (4-12.3) and (4-12.4)

a_0	a_1	a_2	a_3	a_4	b_0	b_1	C	D
−170.335	−28.578	124.809	−55.5393	130.01	0.164813	−0.0914427	exp(1)	1.00588

Huang and O'Connell (1987) extended the generalized method of Brelvi and O'Connell (1972) for obtaining the change in liquid density with pressure in the form of Eq. (4-12.1*b*) which can be applied to within 2 to 4% of the critical temperature. The correlation is of the form

$$\frac{(P - P_o)}{RT} = (1 - C^*b_1)(\rho - \rho_o) - C^* \left[b_2 V^* \frac{(\rho^2 - \rho_o^2)}{2} + b_3 V^{*2} \frac{(\rho^3 - \rho_o^3)}{3} + b_4 V^{*3} \frac{(\rho^4 - \rho_o^4)}{4} \right] \tag{4-12.5}$$

where the coefficients b_i are calculated at reduced temperatures, $\tau = T/T^* < 1$

$$b_i = a_{1i} + a_{2i}\tau + a_{3i}\tau^2 \tag{4-12.6}$$

The universal coefficients a_{ji} are given in Table 4-15. Values of the parameters C^*, T^*, and V^* for some representative substances fitted to compression data are given in Table 4-16. Huang (1986) gives parameter values for over 300 substances including simple fluids, nonpolar and polar organic and inorganic substances, fused salts, polymers and metals. The pressure range was up to 10^4 bar and the temperature range to $T/T_c = 0.95$. The average error in the change in pressure for a given change of density when fitted parameters are used in Eq. (4-12.5) is 1.5% or less for essentially all substances; this means that when a density change is calculated from a specified pressure change, the error in $\rho - \rho_0$ is normally within experimental error over the entire pressure range of the data.

Since the correlation is not very sensitive to T^*, Huang and O'Connell (1987) estimate $T^* = 0.96\ T_c$ and give group contributions for estimating C^* and V^*. However, they strongly recommend that the value of V^* be obtained from fitting at least one isothermal compressibility or speed of sound measurement in order to maintain high accuracy. Examples 4-8 and 4-9 show results for the Huang-O'Connell correlation.

Example 4-8 Obtain the compressed liquid volume of ammonia at 400 bar and 300 K using the methods of Aalto, et al. (1996) and Huang and O'Connell (1987). Repeat for 400 K. Haar and Gallagher (1978) give the volumes at these conditions to be 1.5831 and 2.0313 cm^3 g^{-1}, respectively.

TABLE 4-15 Constants of the Huang-O'Connell (1987) Correlation for Compressed Liquids for Eqs. (4-12.5) and (4-12.6)

a_{11}	a_{21}	a_{31}	a_{21}	a_{22}	a_{23}
9.8642	−10.191	−1.5356	−28.465	30.864	6.0294
a_{13}	a_{23}	a_{33}	a_{14}	a_{24}	a_{34}
27.542	−32.898	−8.7130	−8.2606	12.737	4.0170

TABLE 4-16 Parameters of the Huang-O'Connell (1987) Correlation for Compressed Liquids for Eqs. (4-12.5) and (4-12.6)

Formula	Name	$-C^*$	V^*, cm^3 mol^{-1}	T^*, K
CH_4	methane	14.7	38.72	191
n-C_9H_{20}	n-nonane	70.8	177.20	537
neo-C_9H_{20}	tetraethylmethane	68.3	171.64	626
CH_4O	methanol	16.5	39.59	492
$C_2H_6O_2$	ethylene glycol	34	59.61	603
CCl_4	tetrachloromethane	44.4	95.01	508
$C_{30}H_{62}$	squalane	223.5	541.49	594
NH_3	ammonia	12.2	26.01	381
$BC_{20}H_{48}N$	tetraethylammonium tetrapropylborate	143.7	295.68	490
[—C_8H_9—]	polystyrene	273.2	1.05$^+$	955

$^+$Units of cm^3 g^{-1}.

solution From Appendix A, for ammonia, M = 17.031, T_c = 405.4 K, P_c = 113.53 bar, and ω = 0.256. With vapor pressures from the constants in Appendix A, V_s from Haar and Gallagher (1978) and using Eqs. (4-12.3) and (4-12.4) with the constants of Table 4-14 for the Aalto, et al. correlation, the following results are obtained.

T, K	P_{vp}, bar	V_s, cm^3 mol^{-1}	T_r	A	V_{calc}, cm^3 mol^{-1}	V_{lit}, cm^3 mol^{-1}	Error in V, %	Error in $V - V_s$, %
300	10.61	28.38	0.7398	25.66	27.19	26.96	0.8	−16.2
400	102.97	49.15	0.9864	1.875	35.60	34.59	2.9	−6.9

Using ω_{SRK} = 0.262 in place of ω changes the error in $V - V_s$ by less than 0.3%. Using Eqs. (4-12.5) and (4-12.6) and the parameters of Tables 4-15 and 4-16 for the Huang-O'Connell correlation, the following results are obtained at 300K. The temperature of 400 K is greater than T^*, so the correlation should not be used.

T, K	P_{vp}, bar	V_s, cm^3 mol^{-1}	T/T^*	V_{calc}, cm^3 mol^{-1}	V_{lit}, cm^3 mol^{-1}	Error in V,%	Error in $V - V_s$, %
300	10.61	28.38	0.7874	27.37	26.96	1.5	−28.9

Example 4-9 Predict the compressed liquid volume of 3-methylphenol (m-cresol) at 3000 bar and 503.15 K using the methods of Aalto, et al. (1996) and Huang and O'Connell (1987). Randzio, et al. (1995) give V = 105.42 cm^3 mol^{-1}.

solution From Appendix A, for 3-methylphenol, M = 108.14, T_c = 705.7 K, P_c = 45.6 bar, and ω = 0.452. The value of P_0 is assumed to be 1 bar and Randzio, et al. (1995) give V_0 = 127.31 cm^3 mol^{-1}. We note that the equations of Cibulka, et al. (1997) do not give these volumes.

Using Eqs. (4-12.3) and (4-12.4) for the Aalto, et al. correlation, the following results are obtained

T, K	P_0, bar	V_0, cm^3 mol^{-1}	T_r	A	V_{calc}, cm^3 mol^{-1}	V_{lit}, cm^3 mol^{-1}	Error in V, %	Error in $V - V_0$, %
503.15	1.0	127.31	0.7130	29.576	112.77	105.42	7.0	−33.6

An estimated value of $\omega_{SRK} \doteq 0.43$ in place of ω decreases the error in $V - V_0$ by less than 1%.

Using Eqs. (4-12.5) and (4-12.6) and estimating parameters with the group contributions of Huang (1996), $-C^* = 6*6.8 + 9.7 - 6 = 44.5$, $V^* = 6*14.77 + 26.02 + 23 = 137.64$ cm^3 mol^{-1}, and $T^* = 0.96T_c = 677.47$ K, the following results are obtained

T, K	P_0, bar	V_0, cm^3 mol^{-1}	T/T^*	V_{calc}, cm^3 mol^{-1}	V_{lit}, cm^3 mol^{-1}	Error in V, %	Error in $V - V_s$, %
503.15	1.0	127.31	0.7427	104.99	105.42	−0.1	1.9

In this case, the Huang-O'Connell method describes the large change in density much better than the method of Aalto, et al., which has been applied only at much lower pressures (<250 bar).

NOTATION

In many equations in this chapter, special constants or parameters are defined and usually denoted $a, b, \ldots, A, B, \ldots$. They are not defined in this section because they apply only to the specific equation and are not used anywhere else in the chapter.

- A — molar Helmholtz energy, J mol^{-1}
- b — cubic EoS variable, Eq. (4-6.1), Table 4-6
- B — second virial coefficient in Eqs. (4-5.1), cm^3 mol^{-1}
- C — third virial coefficient in Eqs. (4-5.1), cm^6 mol^{-2}
- C^* — parameter in Huang-O'Connell correlation for compressions, Eq. (4-12.5)
- C_V — heat capacity at constant volume, J mol^{-1} K^{-1}
- C_P — heat capacity at constant pressure, J mol^{-1} K^{-1}
- ΔH_b — enthalpy change of boiling, kJ mol^{-1}
- K — reaction equilibrium constant
- M — molecular weight, g mol^{-1}
- n — number of groups in a molecule, Eq. (4-11.11)
- N — number of atoms in molecule, Eq. (4-10.1)
- N_A — Avogadro's number, $= 6.022\,142 \times 10^{23}$ mol^{-1}
- P — pressure, bar
- P_{vp} — vapor pressure, bar
- r — number of segments in a chain, Eq. (4-7.8)
- R — universal gas constant, Table 4-1
- T — temperature, K
- V — molar volume, cm^3 mol^{-1}

V_{ri} dimensionless volume = $V/(RT_c/P_c)$, Figs. 4-1 to 4-3
y mole fraction, Eq. (4-7.12)
Z compressibility factor, = PV/RT Eqs. (4-2.1)

Greek

α cubic EoS variable, Table 4-7
β_c critical index for liquid-vapor density difference, Eq. (4-4-1)
δ cubic EoS variable, Eq. (4-6.1), Table 4-6
δ_c critical index for critical isotherm PV relation, Eq. (4-4-2)
ε cubic EoS variable, Eq. (4-6.1), Table 4-6
η cubic EoS variable, Eq. (4-6.1), Table 4-6
Θ cubic EoS variable, Eq. (4-6.1), Table 4-6
μ dipole moment, Sec. 2-6
Δv group molar volume, Eq. (4-11.12)
ρ molecular density, cm^{-3}
σ hard sphere diameter, nm
ϕ exponent, Eqs. (4-11.5) and (4-11.6)
ω acentric factor, Eq. (2-3.1)

Superscripts

$'$ dimensionless cubic EoS variable, Eqs. (4-6.3) to (4-6.4)
$*$ characteristic property, Sec. 4-2
$\circ$ ideal gas
L liquid phase
v vapor phase
r residual property, Eq. (4-7.2)
$(R1),(R2)$ functions for multiple reference corresponding states relations, Eq. (4-3.3)
cl "classical" critical point property, Table 4-2
(0),(1) functions for corresponding states relations, Eq. (4-3.1), Eq. (4-5.2), Eq. (4-5.3)

Subscripts

0 reference state for compressions, Eqs. (4-12.1)
Att attractive perturbation, Sec. 4-7
c critical property
liq liquid property
b boiling state
calc, exp calculated value, experimental value
D dimer in chemical theory, Section 4-7
r reduced property, Sec. 4-2
R repulsive, Sec. 4-7
M monomer in chemical theory, Sec. 4-7

REFERENCES

Aalto, M.: *Fluid Phase Equil.,* **141:** 1 (1997).
Aalto, M., K. I. Keskinen, J. Aittamaa, and S. Liukkonen: *Fluid Phase Equil.,* **114:** 1 (1996).
Abbott, M. M.: *AIChE J.,* **19:** 596 (1973).
Abbott, M. M.: *Adv. in Chem. Ser.,* **182**: 47 (1979).
Abbott, M. M., and J. M. Prausnitz: *Fluid Phase Equil.,* **37:** 29 (1987).
Abusleme, J. A., and J. H. Vera: *AIChE J.,* **35:** 481 (1989).

Adachi, Y., B. C.-Y. Lu and H. Sugie: *Fluid Phase Equil.,* **11:** 29 (1983).
Adachi, Y., H. Sugie and B. C.-Y. Lu: *J. Chem. Eng. Japan,* **18:** 20 (1985).
Aly, G. and I. Ashour: *Fluid Phase Equil.,* **101:** 137 (1994).
Anderko, A: *Fluid Phase Equil.,* **45:** 39 (1989).
Anderko, A: *Fluid Phase Equil.,* **61:** 145 (1990).
Anderko, A: *Fluid Phase Equil.,* **65:** 89 (1991).
Anderko, A, I. G. Economou, and M. D. Donohue: *IEC Res.,* **32:** 245 (1993).
Anisimov, M. A., S. B. Kiselev, J. V. Sengers, and S. Tang: *Physica* **A188:** 487 (1992).
Anisimov, M. A., A. A. Povodyrev, and J. V. Sengers: *Fluid Phase Equil.,* **158-60:** 537 (1999).
Barner, H. E., R. L. Pigford, and W. C. Schreiner: Paper Presented at 31st Midyear API Meeting, Houston, TX (1966).
Benedict, M., G. B. Webb, and L. C. Rubin: *J. Chem. Phys.,* **8:**: 334 (1940).
Benedict, M., G. B. Webb, and L. C. Rubin: *J. Chem. Phys.,* **10:** 747 (1942).
Blanke, W., and R. Weiss: *Fluid Phase Equil.,* **80:** 179 (1992).
Boublik, T.: *Mol. Phys.,* **42:** 209 (1981).
Behar, E., R. Simonet, and E. Rauzy: *Fluid Phase Equil.,* **21:** 237 (1985); **31:** 319 (1986).
Breedveld, G. J. F., and J. M. Prausnitz: *AIChE J.,* **19:** 783 (1973).
Carnahan, N., and K. E. Starling: *J. Chem. Phys.,* **51:** 635 (1969).
Chapela, G., and J. S. Rowlinson: *J. Chem. Soc. Faraday Trans. I,* **70:** 584 (1974).
Chapman, W. G., K. E. Gubbins, G. Jackson, and M. Radosz: *Fluid Phase Equil.,* **52:**, 31 (1989).
Chapman, W. G., K. E. Gubbins, G. Jackson, and M. Radosz: *IEC Res.,* **29:** 1709 (1990).
Chang, C. H., and X. Zhao: *Fluid Phase Equil.,* **58:** 231 (1990).
Chen, S. S., and A. Kreglewski: *Ber. Bunsenges. Phys. Chem.,* **81:** 1048 (1977).
Chiew, Y.: *Mol. Phys.,* **70:** 129 (1990).
Cholinski, J., A. Szafranski, and D. Wyrzykowska-Stankiewicz: *Second Virial Coefficients for Organic Compounds,* Thermodynamic Data for Technology, Institute of Physical Chemistry, Polish Academy of Sciences and Insitute of Industrial Chemistry, Warsaw, 1986.
Chou, G. F., and J. M. Prausnitz: *AIChE J.,* **35:** 1487 (1989).
Chueh, P. L., and J. M. Prausnitz: *AIChE J.,* **13:** 896 (1967).
Cibulka, I., L. Hnedkovsky, and T. Takagi: *J. Chem. Eng. Data,* **42:** 415 (1997).
Cibulka, I., and T. Takgi: *J. Chem. Eng. Data,* **44:** 411 (1999).
Cotterman, R. L., B. J. Schwarz, and J. M. Prausnitz: *AIChE J.,* **32:** 1787 (1986).
Daubert, T. E., R. P. Danner, H. M. Sibel, and C. C. Stebbins: *Physical and Thermodynamic Properties of Pure Chemicals: Data Compilation,* Taylor & Francis, Washington, D.C., 1997.
Dean, J. A.: *Lange's Handbook of Chemistry,* 15th ed., McGraw-Hill, New York, 1999.
Defibaugh, D. R., and G. Morrison: *J. Chem. Eng. Data,* **41:** 376 (1996).
Defibaugh, D. R., and M. R. Moldover: *J. Chem. Eng. Data,* **42:** 160 (1997).
de Hemptinne, J.-C., and P. Ungerer: *Fluid Phase Equil.,* **106:** 81 (1995).
Deiters, U.: *Fluid Phase Equil.,* **161:** 205 (1999).
Deiters, U., and K. M. de Reuck: *Pure and Appl. Chem.,* **69:** 1237 (1997).
de Sant'Ana, H. B., P. Ungerer, and J. C. de Hemptinne: *Fluid Phase Equil.,* **154:** 193 (1999).
DeSantis, R., and B. Grande: *AIChE J.,* **25:** 931 (1979).
Dymond, J. H., J. A. Cholinski, A. Szafranski, and D. Wyrzykowska-Stankiewicz: *Fluid Phase Equil.,* **27:** 1 (1986).
Dymond, J. H., and R. Malhotra: *Int. J. Thermophys.,* **9:** 941 (1988).
Dymond, J. H., and E. B. Smith: *The Virial Coefficients of Pure Gases and Mixtures; A Critical Compilation,* Clarendon Press, Oxford, 1980.

Economou, I. G., and M. D. Donohue: *IEC Res.,* **31:** 1203 (1992).
Edmister, W. C.: *Petrol. Refiner,* **37**(4): 173 (1958).
Edwards, M. N. B., and G. Thodos: *J. Chem. Eng. Data,* **19:** 14 (1974).
Elbro, H. S., A. Fredenslund, and P. Rasmussen: *Ind. Eng. Chem. Res.,* **30:** 2576 (1991).
Feng, W, and W. Wang: *Ind. Eng. Chem.,* **38:** 4966 (1999).
Fermeglia, M., A. Bertucco, and D. Patrizio: *Chem. Eng. Sci.,* **52:** 1517 (1997).
Fleming, P. D., and R. J. Brugman: *AIChE J.* **33:** 729 (1987).
Fornasiero, F., L. Lue, and A. Bertucco: *AiChE J.,* **45:** 906 (1999).
Fox, J.: *Fluid Phase Equil.,* **14:** 45 (1983).
Fuller, G. G.: *Ind. Eng. Chem. Fundam.,* **15:** 254 (1976).
Gmehling, J., D. D. Liu, and J. M. Prausnitz: *Chem Eng. Sci.,* **34:** 951 (1979).
Golobic, I., and B. Gaspersic: *Chem. Eng. Comm.,* **130:** 105 (1994).
Gregorowicz, J., J. P. O'Connell, and C. J. Peters: *Fluid Phase Equil.,* **116:** 94 (1996).
Grensheuzer, P., and J. Gmehling: *Fluid Phase Equil.,* **25:** 1 (1986).
Guggenheim, E. A.: *J. Chem. Phys.,* **13:** 253 (1945).
Guggenheim, E. A., and C. J. Wormald: *J. Chem. Phys.,* **42:** 3775 (1965).
Gunn, R. D., P. L. Chueh, and J. M. Prausnitz: *AIChE J.,* **12:** 937 (1966).
Haar, L., and J. S. Gallagher: *J. Phys. Chem. Ref. Data,* **7:** 635 (1978).
Hakala, R.: *J. Phys. Chem.,* **71:** 1880 (1967).
Hales, J. L.: "Bibliography of Fluid Density," *Nat. Phys. Lab. Rep. Chem.,* **106,** February 1980.
Hankinson, R. W., and G. H. Thomson: *AIChE J.,* **25:** 653 (1979).
Hayden, J. G., and J. P. O'Connell: *Ind. Eng. Chem. Process Des. Dev.,* **14:** 209 (1975).
Heidemann, R. A., and J. M. Prausnitz: *Proc. Nat. Acad. Sci.,* **73:** 1773 (1976).
Henderson, D.: *Adv. in Chem. Ser.,* **182:** 1 (1979).
Hino, T., and J. M. Prausnitz: *Fluid Phase Equil.,* **138:** 105 (1997).
Hougen, O., K. M. Watson, and R. A. Ragatz: *Chemical Process Principles,* pt. II, Wiley, New York, 1959.
Huang, Y.-H.: "Thermodynamic Properties of Compressed Liquids and Liquid Mixtures from Fluctuation Solution Theory; PhD Dissertation, University of Florida, Gainesville, FL, 1986.
Huang, Y.-H., and J. P. O'Connell: *Fluid Phase Equil.,* **37:** 75 (1987).
Ikonomou, G. D., and M. D. Donohue: *AIChE J.* **32:** 1716 (1986).
Kedge, C. J., and M. A. Trebble: *Fluid Phase Equil.,* **158–60:** 219 (1999).
Kiran, E., and J. M. H. Levelt Sengers, eds: *Supercritical Fluids: Fundamentals for Applications,* NATO ASI Series E: **273:** Klewer Academic Publishers, Dordrecht, Holland, 1994.
Kiselev, S. B.: *Fluid Phase Equil.,* **147:** 7 (1998).
Kiselev, S. B., and J. F. Ely: *IEC Res.,* **38:** 4993 (1999).
Kis, K. and H. Orbey: *Chem Eng. J.,* **41:** 149 (1989).
Koak, N., T. W. de Loos, and R. A. Heidemann: *IEC Res.,* **38:** 1718 (1999).
Konttorp, M.: *MS Thesis,* T. U. Delft, Netherlands, 1998.
Kostowicka Wyczalkowska, A., M. A. Anisimov, and J. V. Sengers: *Fluid Phase Equil.,* **158–60:** 523 (1999).
Kurzeja, N., T. Tielkes, and W. Wagner: *Int. J. Thermophys.,* **20:** 531 (1999).
Le Bas, G.: *The Molecular Volumes of Liquid Chemical Compounds,* Longmans, Green, New York, 1915.
Lee, B. I., and M. G. Kesler: *AIChE J.,* **21:** 510 (1975).
Lee, M.-J., and J.-T. Chen: *J. Chem. Eng. Japan,* **31:** 518 (1998).
Leland, T. L., Jr., and P. S. Chappelear: *Ind. Eng. Chem.,* **60**(7): 15 (1968).

Lemmon, E. W., M. O. McLinden and D. G. Friend: "Thermophysical Properties of Fluid Systems" in *NIST Chemistry WebBook,* NIST Standard Reference Database Number 69, eds. W. G. Mallard and P. J. Linstrom, November 1998, Nat. Inst. Stand. Tech., Gaithersburg, MD 20899 (http://webbook.nist.gov).

Lide, D. R.: *CRC Handbook of Chemistry and Physics,* 80th ed., CRC Press, Boca Raton, FL, 1999.

Lue, L., and J. M. Prausnitz: *AIChE J.,* **44:** 1455 (1998).

Lydersen, A. L., R. A. Greenkorn, and O. A. Hougen: *Generalized Thermodynamic Properties of Pure Fluids, Univ. Wisconsin, Coll. Eng., Eng. Exp. Stn. Rep. 4,* Madison, Wis., October 1955.

Martin, J. J.: *Ind. Eng. Chem. Fundam.,* **18:** 81 (1979).

Mason, E. A., and T. H. Spurling: *The Virial Equation of State,* Pergamon, New York, 1968.

Massucci, M., and C. J. Wormald: *J. Chem. Thermodynamics,* **30:** 919 (1998).

Mathias, P. M.: *Ind. Eng. Chem. Proc. Des. Dev.,* **27:** 385 (1983).

Mathias, P. M., and T. W. Copeman: *Fluid Phase Equil.,* **13:** 91 (1983).

Mathias, P. M., T. Naheiri, and E. M. Oh: *Fluid Phase Equil.,* **47:** 77 (1989).

McCann, D. W., and R. P. Danner: *Ind. Eng. Chem. Process Des. Dev.,* **23:** 529 (1984).

McElroy, P. J., and J. Moser: *J. Chem. Thermo.,* **27:** 267 (1995).

Morris, R. W., and E. A. Turek: *ACS Symp. Ser.* **300:** 389 (1986).

Muller, A., J. Winkelmann, and J. Fischer: *AIChE J.,* **42:** 1116 (1996).

Nasrifar, K., and M. Moshfeghian: *Fluid Phase Equil.,* **158–160:** 437 (1999).

Nelson, L. C., and E. F. Obert: *Trans. ASME,* **76:** 1057 (1954).

NIST: http://physics.nist.gov/cuu/Constants/index.html, 1998.

O'Connell, J. P.: *AIChE J.,* **30:** 1037 (1984a).

O'Connell, J. P.: University of Florida, Personal Communication, (1984b).

Olf, G., J. Spiske, and J. Gaube: *Fluid Phase Equil.,* **51:** 209 (1989).

Orbey, H., and J. H. Vera: *AIChE J.,* **29:** 107 (1983).

Orbey, H.: *Chem. Eng. Comm.,* **65:** 1 (1988).

Partington, J.: *An Advanced Treatise on Physical Chemistry,* Vol. I, *Fundamental Principles: The Properties of Gases,* Longmans, Green, New York, 1949.

Patel, N. C., and A. S. Teja: *Chem. Eng. Sci.,* **37:** 463 (1982).

Patel, N. C.: *Int. J. Thermophys.,* **17:** 673 (1996).

Peneloux, A., E. Rauzy and R. Freze: *Fluid Phase Equil.,* **8:** 7 (1982).

Peng, D. Y., and D. B. Robinson: *Ind. Eng. Chem. Fundam.,* **15:** 59 (1976).

Perry, R. H., and D. W. Green: *Perry's Chemical Engineers' Handbook,* 7th ed., McGraw-Hill, New York, 1997.

Pfohl, O.: *Fluid Phase Equil.,* **163:** 157 (1999).

Pitzer, K. S.: J. *Chem. Phys.,* **7:** 583 (1939).

Pitzer, K. S.: Thermodynamics, 3rd ed., McGraw-Hill, New York, 1995, p. 521.

Pitzer, K. S., and R. F. Curl: *J. Am. Chem. Soc.,* **77:** 3427 (1955).

Pitzer, K. S., and R. F. Curl: *J. Am. Chem. Soc.,* **79:** 2369 (1957a).

Pitzer, K. S., and R. F. Curl: *The Thermodynamic Properties of Fluids,* Inst. Mech. Eng., London, 1957b.

Pitzer, K. S., D. Z. Lippmann, R. F. Curl, C. M. Huggins, and D. E. Petersen: *J. Am. Chem. Soc.,* **77:** 3433 (1955).

Pizer, K. S., and D. R. Scheiber: *Fluid Phase Equil.,* **41:** 1 (1988).

Platzer, B., and G. Maurer: *Fluid Phase Equil.,* **51:** 223 (1989).

Prausnitz, J. M., T. F. Anderson, E. A. Grens, C. A. Eckert, R. Hsieh, and J. P. O'Connell: *Computer Calculations for Multicomponent Vapor-Liquid and Liquid-Liquid Equilibria,* Prentice-Hall, Englewood Cliffs, N.J., 1980.

Prausnitz, J. M., R. N. Lichtenthaler, and E. G. de Azevedo: *Molecular Thermodynamics of Fluid-Phase Equilibria,* 3rd ed.," Upper Saddle River, N.J., Prentice-Hall, 1999.

Rackett, H. G.: *J. Chem. Eng. Data,* **15:** 514 (1970).

Redlich, O., and J. N. S. Kwong: *Chem. Rev.,* **44:** 233 (1949).

Rodgers, P. A.: *J. Appl. Polym. Sci.,* **48:** 1061 (1993).

Rogalski, M., B. Carrier, R. Solimando, and A. Peneloux: *IEC Res.,* **29:** 659 (1990).

Sandarusi, J. A., A. J. Kidnay, and V. F. Yesavage: *Ind. Eng. Chem. Proc. Des. Dev.,* **25:** 957 (1986).

Sandler, S.: in Kiran, E., and J. M. H. Levelt Sengers, eds: *Supercritical Fluids: Fundamentals for Applications,* NATO ASI Series E: **273:** 147. Klewer Academic Publishers, Dordrecht, Holland, 1994.

Sandler, S. I., H. Orbey, and B.-I. Lee: S. I. Sandler, ed., in *Models for Thermodynamic and Phase Equilibria Calculations,* Marcel Dekker, New York, 1994.

Sengers, J. V., R. F. Kayser, C. J. Peters, and H. J. White, Jr. (eds.): *Equations of State for Fluids and Fluid Mixtures,* Elsevier, Amsterdam, 2000.

Setzmann, U., and W. Wagner: *Int. J. Thermophys.,* **10:** 1103 (1989).

Setzmann, U., and W. Wagner: *J. Phys. Chem. Ref. Data,* **20:** 1061 (1991).

Shah, V. M., Y.-L. Lin, and H. D. Cochran: *Fluid Phase Equil.,* **116:** 87 (1996).

Siddiqi, M. A., and K. Lucas: *Fluid Phase Equil.,* **51:** 237 (1989).

Smith, J. M., H. C. Van Ness, and M. M. Abbott: *Introduction to Chemical Engineering Thermodynamics,* 5th ed., New York, McGraw-Hill, 1996.

Soave, G.: *Chem. Eng. Sci.,* **27:** 1197 (1972).

Soave, G.: *Inst. Chem. Eng. Symp. Ser.,* **56**(1.2)**:** 1 (1979).

Soave, G.: *Chem. Eng. Sci.,* **39:** 357 (1984).

Soave, G.: *Fluid Phase Equil.,* **82:** 345 (1993).

Soave, G., A. Bertucco, and M. Sponchiado: 14th European Seminar on Applied Thermodynamics, Athens, June, 1994.

Solimando, R., M. Rogalski, E. Neau, A. Peneloux: *Fluid Phase Equil.,* **106:** 59 (1995).

Song, Y., S. M. Lambert, and J. M. Prausnitz: *IEC Res.,* **33:** 1047 (1994).

Span, R., and W. Wagner: *Int. J. Thermophys.,* **18:** 1415 (1997).

Spencer, C. F., and S. B. Adler: *J. Chem. Eng. Data,* **23:** 82 (1978).

Spencer, C. F., and R. P. Danner: *J. Chem. Eng. Data,* **17:** 236 (1972).

Stamateris, B., and C. Olivera-Fuentes: Personal Communication, 1995.

Stein, F. P., and E. J. Miller: *Ind. Eng. Chem. Process Des. Dev.* **19:** 123 (1980).

Stryjek, R., and J. H. Vera: *Can. J. Chem. Eng.,* **64:** 323 (1986).

Sun, T. F., J. A. Schouten, P. J. Kortbeek, and S. N. Biswas: *Phys Chem. Liq.,* **21:** 231 (1990).

Tang, S., and J. V. Sengers: *J. Supercrit. Fluids,* **4:** 209 (1991).

Tang, S., J. V. Sengers and Z. Y. Chen: *Physica,* **A179:** 344 (1991).

Tarakad, R. R., and R. P. Danner: *AIChE J.,* **23:** 685 (1977).

Tarakad, R. R., and T. E. Daubert, Pennsylvania State University, University Park: Personal Communication, 1974.

Teja, A., S. I. Sandler, and N. C. Patel: *Chem. Eng. J.,* **21:** 21 (1981).

Tekac, V., I. Cibulka, and R. Holub: *Fluid Phase Equil.* **19:** 33 (1985).

Thomson, G. H., K. R. Brobst, and R. W. Hankinson: *AIChE J.,* **28:** 671 (1982).

Trebble, M. A., and P. R. Bishnoi: *Fluid Phase Equil.,* **29:** 465 (1986).

Trebble, M. A., and P. R. Bishnoi: *Fluid Phase Equil.,* **35:** 1 (1987).

Tsonopoulos, C.: *AIChE J.,* **20:** 263 (1974).

Tsonopoulos, C.: *AIChE J.,* **21:** 827 (1975).

Tsonopoulos, C.: *AIChE J.,* **24:** 1112 (1978).

Tsonopoulos, C.: *Adv. in Chem. Ser.,* **182:** 143 (1979).
Tsonopoulos, C., J. H. Dymond, and A. M. Szafranski: *Pure. Appl. Chem.* **61:** 1387 (1989).
Tsonopoulos, C., and J. L. Heidman: *Fluid Phase Equil.,* **57:** 261 (1990).
Tsonopoulos, C., and J. H. Dymond: *Fluid Phase Equil.,* **133:** 11 (1997).
Twu, C. H., J. E. Coon, and J. R. Cunningham: *Fluid Phase Equil.,* **75:** 65 (1992).
Twu, C. H., J. E. Coon, and J. R. Cunningham: *Fluid Phase Equil.,* **82:** 379 (1993).
Twu, C. H., J. E. Coon, and J. R. Cunningham: *Fluid Phase Equil.,* **105:** 49 (1995).
Tyn, M. T., and W. F. Calus: *Processing,* **21(4):** 16 (1975).
Ungerer, P., and H. B. de Sant'Ana: *Fluid Phase Equil.,* **163:** 161 (1999).
van der Waals, J. H.: *Z. Phys. Chem.,* **5:** 133 (1890).
van Krevelen, D. W., and P. J. Hoftyzer: *Properties of Polymers, Correlations with Chemical Structure,* Elsevier, Amsterdam, 1972.
Van Ness, H. C. and M. M. Abbott: *Classical Thermodynamics of Non Electrolyte Solutions,* McGraw-Hill, New York, 1982.
Vetere, A.: *Fluid Phase Equil.,* **164:** 49 (1999).
Visco, D. P., Jr., and D. A. Kofke: *IEC Res.,* **38:** 4125 (1999).
Vilmachand, P., and M. D. Donohue: *J. Phys. Chem.,* **93:** 4355 (1989).
Vine, M. D., and C. J. Wormald: *J. Chem. Soc. Faraday Trans.,* **89:** 69 (1993).
Wagner, W., N. Kurzeja, and B. Pieperbeck: *Fluid Phase Equil.,* **79:** 151 (1992).
Wilding, W. V., and R. L. Rowley: *Int. J. Thermophys.,* **7:** 525 (1986).
Wilding, W. V., J. K. Johanson, and R. L. Rowley: *Int. J. Thermophys.,* **8:** 717 (1987).
Wilson, G. M.: *Adv. Cryogenic Eng.,* **9:** 168 (1964).
Yamada, T., and R. D. Gunn: *J. Chem. Eng. Data,* **18:** 234 (1973).
Zabaloy, M. S., and E. A. Brignole: *Fluid Phase Equil.,* **140:** 87 (1997).
Zabaloy, M. S., and J. H. Vera: *IEC Res.,* **35:** 829 (1996).
Zabaloy, M. S., and J. H. Vera: *Can. J. Chem. Eng.,* **75:** 214 (1997).
Zabaloy, M. S., and J. H. Vera: *IEC Res.,* **37:** 1591 (1998).

CHAPTER FIVE

PRESSURE-VOLUME-TEMPERATURE RELATIONSHIPS OF MIXTURES

5-1 SCOPE

Methods are presented in this chapter for estimating the volumetric behavior of mixtures of gases and liquids as a function of temperature, pressure, and composition as expressed in mole, mass, or volume fractions. Section 5-2 introduces the extension of pure component models to mixtures, especially the concepts of *mixing rules* and *combining rules*. Their use in models based on the corresponding states principle (CSP) is discussed in Sec. 5-3. Techniques for estimation and calculation from equations of state (EoS) models, are given in Secs. 5-4 to 5-7. The models include the virial equation (Sec. 5-4), analytical density relations (EoS, mainly cubics, which can be solved analytically to find the volume of a fluid at a specified T, P and composition, Sec. 5-5), and more complex relations requiring numerical root-solving algorithms for the volume (Sec. 5-6). Section 5-7 summarizes our evaluation of mixing rules for volumetric properties. Sections 5-8 and 5-9 give estimation methods for saturated and compressed liquid densities that are not based on equations of state.

As in Chap. 4 for pure components, the discussion here is not comprehensive, and focuses on formulations of models for mixtures and on their application to volumetric properties. However, the treatments of composition here have implications for all other thermodynamic properties, since mathematical manipulations done on volumetric expressions with composition held constant provide *calorimetric* properties (such as internal energy, enthalpy and heat capacities) and *free energy* properties (such as Helmholtz and Gibbs energies). Chapter 6 describes application of EoS methods to calorimetric, free energy and partial properties which are required for the phase equilibria modeling described in Chap. 8. Thus, Sec. 8-12 has illustrations of the use of fugacities from EoS models derived in Chapter 6. Mixing rules are also used in the estimation of transport properties and the surface tension of mixtures as in Chaps. 9 through 12. Thus, the mixing rules developed here appear in much of the rest of the book.

Readers are referred to the papers of Deiters (1999; Deiters and de Reuck, 1997) for full descriptions of how EoS models should be developed and communicated. Following Deiters' recommendations, generators of new models will have a greater opportunity to be considered more thoroughly while users of new models will understand better their possibilities and limitations. Excellent article reviews of EoS models, especially for mixtures, are given by Vidal (1983), Anderko (1990), Sandler (1994), Sandler, et al. (1994), Orbey and Sandler (1998) and Prausnitz, et al. (1999). A nice historical perspective on the evolution of a very popular EoS model is given by Soave (1993). See Sengers, et al. (2000) for a comprehensive treatment.

5-2 MIXTURE PROPERTIES—GENERAL DISCUSSION

Typically, a model for a pure component physical property contains parameters that are constant or temperature-dependent and found either by fitting to data or by CSP. Thus, the EoS models of Secs. 4-4 to 4-8 express the relationship among the variables P, V, and T. To describe mixture properties, it is necessary to include composition dependence which adds considerable richness to the behavior, and thus complicates modeling. Therefore, a mixture equation of state (EoS) is an algebraic relation between P, V, T, and $\{y\}$, where $\{y\}$ is the set of $n - 1$ independent mole fractions of the mixture's n components.

Challenges to EoS Models: Composition Dependence of Liquid Partial Properties, Multiphase Equilibria, the Critical Region and High Pressures

The composition dependence of the properties of liquid mixtures is fundamentally different from that of a vapor or gas. The strongest effect on gaseous fluids is caused by changes in system density from changes in pressure; composition effects are usually of secondary importance, especially when mixing is at constant volume. Except at high pressures, vapors are not dissimilar to ideal gases and deviations from ideal mixing (Van Ness and Abbott, 1982) are small. However, changes in pressure on liquids make little difference to the properties, and volumetric, calorimetric and phase variations at constant T and P are composition-dominated. The extreme example is at a composition near infinite dilution where the solute environment is both highly dense and far from pure-component. These phenomena mean that comprehensive property models such as EoS must show different composition/pressure connections at low and high densities.

This distinction between low and high density phases is most obviously seen in the liquid-liquid and vapor-liquid-liquid systems discussed in Chap. 8, especially at low concentrations of one or more species. The standard state for EoS models is the ideal gas where no phase separation can occur. As a result, when a model must quantitatively predict deviations from ideal liquid solution behavior from subtle differences between like and unlike interactions, complex relationships among the parameters are usually required. A number of issues in these formulations, such as inconsistencies and invariance in multicomponent systems, are discussed in Sec. 5-7. However, the last few years have seen tremendous advances that have firmly established useful expressions and computational tools for EoS to yield reliable results for many complex systems.

Fluid properties in states near a mixture vapor-liquid critical point are less difficult to obtain from experiment than near pure component critical points, since the fluid compressibility is no longer divergent. However, there are composition fluctuations that lead to both universalities and complex near-critical phase behavior (see the collected articles in Kiran and Levelt Sengers, 1994, reviews by Rainwater such as in 1991, and work by Sengers and coworkers such as Jin, et al., 1993 and Anisimov, et al., 1999). Describing the crossover from nonclassical to classical behavior is even more difficult than for pure components because of the additional degrees of freedom from composition variations (see, for example, Kiselev and Friend, 1999). A brief discussion of such treatments is made in Sec. 5-6.

As in pure components, mixture EoS expressions are often inaccurate at very high pressures, both above and below the critical temperature. The forms of EoS *PV* isotherms at constant composition often do not correspond to those which best correlate data such as described in Sec. 5-9 unless careful modifications are made (see Secs. 5-5 and 5-6).

The effects of errors in *PVTy* relations are carried through to all thermodynamic property variations because they involve derivatives, including those with respect to composition. See the discussions in Sec. 5-7 and Chaps. 6 and 8.

Composition Variations

Typically composition is specified by some fractional weighting property such as the mole fraction, y_i, the mass fraction, w_i, or the superficial volume fraction, ϕ_i.

$$y_i = \frac{N_i}{N_m} \tag{5-2.1a}$$

$$w_i = \frac{y_i M_i}{M_m} \tag{5-2.1b}$$

$$\phi_i = \frac{y_i V_i^\circ}{V_m^\circ} \tag{5-2.1c}$$

where N_i is the number of moles of component *i*, M_i is the molecular weight of component *i*, and V_i° is the pure component molar volume of component *i*. The denominators in Eqs. (5-2.1) perform the normalization function by summing the numerators over all components. Thus, $N_m = \sum_{i=1}^{n} N_i$; $M_m = \sum_{i=1}^{n} y_i M_i$, $V_m^\circ = \sum_{i=1}^{n} y_i V_i^\circ$. Often, representation of the properties of mixtures are via plots versus the mole fraction of one of the components as expressed by theories. However, experimental data are often reported in mass fractions. Sometimes, asymmetries in these plots can be removed if the composition variable is the volume fraction, which allows simpler correlations.

There are two principal ways to extend the methods of Chap. 4 to include composition variations. One is based on molecular theory which adds contributions from terms that are associated with interactions or correlations of properties among pairs, trios, etc., of the components. The virial equation of state described in Secs. 4-4 and 5-4 is an example of this approach; the mixture expression contains pure-component and "cross" virial coefficients in a quadratic, cubic, or higher-order summation of mole fractions. The other approach to mixtures, which is more convenient, uses the same equation formulation for a mixture as for pure components, and composition dependence is included by making the parameters vary with composition. This leads to *mixing rules* for the Corresponding States Principle (CSP) of Secs. 4-3 and 5-3, and for the parameters of equations of state (EoS) discussed in Secs. 4-6, 4-7, 5-5, and 5-6. Essentially all of the models for pure components discussed in Chap. 4 have been extended to mixtures, often within the original articles; we will not cite them again in this chapter. Generally, there is no rigorous basis for either the mixing rule composition dependence or the parameter *combining rules* that bring together pairwise or higher-order contributions from the interactions of different components. The empiricism of this situation yields many possibilities which must be evaluated individually for accuracy and reliability.

Mixing and Combining Rules. The concept of a *one-fluid* mixture is that, for fixed composition, the mixture properties and their variations with T and P are the same as some pure component with appropriate parameter values. To describe all pure components as well as mixtures, the mixture parameters must vary with composition so that if the composition is actually for a pure component, the model describes that substance. Though other variations are possible, a common *mixing rule* for a parameter Q is to have a quadratic dependence on mole fractions of the components in the phase, y_i

$$Q_m = \sum_{i=1}^{n} \sum_{j=1}^{n} y_i y_j Q_{ij} \tag{5-2.2}$$

In Eq. (5-2.2), the parameter value of pure component i would be Q_{ii}.

Depending upon how the "interaction" parameter, Q_{ij} for $i \neq j$ is obtained from a *combining rule,* the resulting expression can be simple or complicated. For example, linear mixing rules arise from arithmetic and geometric combining rules.

$$\text{For } Q_{ij}^{(a)} = \frac{Q_{ii} + Q_{jj}}{2} \qquad Q_m = \sum_{i=1}^{n} y_i Q_{ii} \tag{5-2.3a}$$

$$\text{For } Q_{ij}^{(g)} = (Q_{ii} Q_{jj})^{1/2} \qquad Q_m = \left(\sum_{i=1}^{n} y_i Q_{ii}^{1/2} \right)^2 \tag{5-2.3b}$$

There is also the harmonic mean combining rule $Q_{ij}^{(h)} = 2/[(1/Q_{ii}) + (1/Q_{jj})]$, but no linear relationship arises with it. The order of values for positive Q_{ii} and Q_{jj}, is $Q_{ij}^{(h)} < Q_{ij}^{(g)} < Q_{ij}^{(a)}$.

However, as will be shown in Sec. 5-5, these relationships are not adequate to describe most composition variations, especially those in liquids. Thus, it is common to use parameters that only apply to mixtures and whose values are obtained by fitting mixture data or from some correlation that involves several properties of the components involved. Examples include *binary interaction parameters,* which modify the combining rules at the left of Eqs. (5-2.3). These parameters can appear in many different forms. They may be called simply binary parameters or interaction parameters, and they are often given symbols such as k_{ij} and l_{ij}.

The reader is cautioned to know precisely the definition of binary interaction parameters in a model of interest, since the same symbols may be used in other models, but with different definitions. Further, values may be listed for a specific formulation, but are likely to be inappropriate for another model even though the expressions are superficially the same. For instance, consider Eqs. (5-2.4) below. It is expected that in Eqs. (5-2.4*a*) and (5-2.4*c*) the values of k_{ij} and l_{ij} for $i \neq j$ would be close to unity, while in Eqs. (5-2.4*b*) and (5-2.4*d*) the values would be close to zero. Significant errors would be encountered if a value for the wrong parameter were used.

$$Q_{ij} = (Q_{ii}Q_{jj})^{1/2}k_{ij} \qquad k_{ii} = 1 \tag{5-2.4a}$$

$$Q_{ij} = (Q_{ii}Q_{jj})^{1/2}(1 - k_{ij}) \qquad k_{ii} = 0 \tag{5-2.4b}$$

$$Q_{ij} = \frac{Q_{ii} + Q_{jj}}{2} l_{ij} \qquad l_{ii} = 1 \tag{5-2.4c}$$

$$Q_{ij} = \frac{Q_{ii} + Q_{jj}}{2} (1 - l_{ij}) \qquad l_{ii} = 0 \tag{5-2.4d}$$

These binary interaction parameters may be constants, functions of T, or even functions of the mixture density, $\rho = 1/V$; model formulations have been made with many different types.

The sensitivity of solution properties to binary interaction parameters can be very high or be negligible, depending upon the substances in the system and the property of interest. For example, mixture volumes from EoS change very little with k_{ij} of Eq. (5-2.4*b*) if it is used for a Θ parameter of Eq. (4-6.1), but l_{ij} of Eq. (5-2.4*d*) can be quite important for b parameters of Eq. (4-6.1) when the substances are very different in size or at high pressures (Arnaud, et al., 1996). On the other hand, partial properties, such as fugacities, are very sensitive to k_{ij} in the Θ parameter and change little with l_{ij} in the b parameter.

In addition to one-fluid mixing rules for EoS, recent research has generated many different ways to connect EoS mixture parameters to liquid properties such as excess Gibbs energies (see Chap. 6). These are described in detail in Sec. 5-5.

There are theories of mixing and combining rules which suggest practical expressions; these are discussed elsewhere (Gunn, 1972; Leland and Chappelear, 1968; Reid and Leland, 1965; Prausnitz, et al., 1999) and we will not consider them here.

5-3 CORRESPONDING STATES PRINCIPLE (CSP): THE PSEUDOCRITICAL METHOD

The direct application of mixing rules to the CSP correlations in Secs. 4-2 and 4-3 to describe mixtures assumes that the behavior of a mixture in a reduced state is the same as some pure component in the same reduced state. When the reducing parameters are critical properties and these are made functions of composition, they are called *pseudocritical properties* because the values are not generally expected to be the same as the true mixture critical properties. Thus the assumption in applying corresponding states to mixtures is that the *PVT* behavior will be the same as that of a pure component whose T_c and P_c are equal to the pseudocritical temperature, T_{cm}, and pseudocritical pressure of the mixture, P_{cm}, and other CSP parameters such as acentric factor can also be made composition dependent adequately for reliable estimation purposes.

Two-Parameter and Three-Parameter CSP

The assumptions about intermolecular forces that allow CSP use for mixtures are the same as for pure components (Sec. 4-3). However, here it is necessary to deal with the effects of interactions between unlike species as well as between like species. As described above, this is commonly done with mixing and combining rules.

Thus, for the pseudocritical temperature, T_{cm}, the simplest mixing rule is a mole-fraction average method. This rule, often called one of Kay's rules (Kay, 1936), can be satisfactory.

$$T_{cm} = \sum_{i=1}^{n} y_i T_{ci} \tag{5-3.1}$$

Comparison of T_{cm} from Eq. (5-3.1) with values determined from other, more complicated rules considered below shows that the differences in T_{cm} are usually less than 2% if, for all components the pure component critical properties are not extremely different. Thus, Kay's rule for T_{cm} is probably adequate for $0.5 < T_{ci}/T_{ij} < 2$ and $0.5 < P_{ci}/P_{cj} < 2$ (Reid and Leland, 1965).

For the pseudocritical pressure, P_{cm}, a mole-fraction average of pure-component critical pressures is normally unsatisfactory. This is because the critical pressure for most systems goes through a maximum or minimum with composition. The only exceptions are if all components of the mixture have quite similar critical pressures and/or critical volumes. The simplest rule which can give acceptable P_{cm} values for two-parameter or three-parameter CSP is the modified rule of Prausnitz and Gunn (1958)

$$P_{cm} = \frac{Z_{cm} R T_{cm}}{V_m} = \frac{\left(\sum_{i=1}^{n} y_i Z_{ci}\right) R \left(\sum_{i=1}^{n} y_i T_{ci}\right)}{\left(\sum_{i=1}^{n} y_i V_{ci}\right)} \tag{5-3.2}$$

where all of the mixture pseudocriticals Z_{cm}, T_{cm}, and V_{cm} are given by mole-fraction averages (Kay's rule) and R is the universal gas constant of Table 4-1.

For three-parameter CSP, the mixture pseudo acentric factor is commonly given by a mole fraction average (Joffe, 1971)

$$\omega_{cm} = \sum_{i=1}^{n} y_i \omega_{ci} \tag{5-3.3}$$

though others have been used (see, *e.g.*, Brule, et al., 1982). While no empirical binary (or higher order) interaction parameters are included in Eqs. (5-3.1) to (5-3.3), good results may be obtained when these simple pseudomixture parameters are used in corresponding-states calculations for determining mixture properties.

Example 5-1 Estimate the molar volume of an equimolar mixture of methane (1) and propane (2) at $T = 310.92$ K, $P = 206.84$ bar and mixtures of 22.1 and 75.3 mol percent methane at $T = 153.15$ K, $P = 34.37$ bar using CSP. Literature values are 79.34, 48.06 and 60.35 cm^3 mol^{-1} respectively (Huang, et al., 1967).

solution The characteristic properties of methane and propane from Appendix A are listed in the table below. Also, the computed pseudoproperties from Eqs. (5-3.1) to (5-3.3) for the three cases are given.

Pure Component/Property	T_c, K	P_c, bar	V_c, cm^3 mol^{-1}	Z_c	ω
Methane	190.56	45.99	98.60	0.286	0.011
Propane	369.83	42.48	200.00	0.276	0.152
Mixture Pseudoproperty	T^*_{cm}, K	P^+_{cm}, bar	V^*_{cm}, cm^3 mol^{-1}	Z^*_{cm}	$\omega^\#_{cm}$
$y_1 = 0.5$	280.20	44.24	149.30	0.281	0.082
$y_1 = 0.221$	330.21	43.26	177.59	0.278	0.121
$y_1 = 0.753$	234.84	45.12	123.65	0.284	0.046

*Mole fraction average as in Eq. (5-3.1)
+Eq. (5-3.2)
#Eq. (5-3.3)

The value of Z can be found from Fig. 4-1 only for the first case, but Tables 3-2 and 3-3 of the 4th Edition give values of $Z^{(0)}$ and $Z^{(1)}$ for the Pitzer-Curl method to use in Eq. (4-3.1), $Z = Z^{(0)} + \omega Z^{(1)}$.

T K	P bar	y_1	T_r	P_r	Fig. 4-1 Z	Fig. 4-1 V, cm^3 mol^{-1}	Error %	$Z^{(0)}$	$Z^{(1)}$	Z	V cm^3 mol^{-1}	Error %
310.92	206.84	0.500	1.110	4.676	0.64	79.99	0.8	0.655	−0.092	0.647	80.92	2.0
153.15	34.37	0.221	0.652	0.762	—	—	—	0.137	−0.058	0.134	49.62	3.2
153.15	34.37	0.753	0.464	0.797	—	—	—	0.173	−0.073	0.164	60.80	0.7

The errors for these compressed fluid mixtures with components having significantly different T_c's and V_c's are typical for CSP. In general, accuracy for normal fluid mixtures is slightly less than for pure components unless one or more binary interaction parameters are used.

As discussed in Sec. 4-3, CSP descriptions are less reliable for substances with strong dipoles or showing molecular complexation (association). The same limitations apply to mixtures of such compounds. Mixtures can also bring in one additional dimension; there can be mixtures involving normal substances with complex substances. Though the interactions between a nonpolar species and a polar or associating species involve only nonpolar forces (Prausnitz, et al., 1999), because the critical or other characteristic properties of the polar species involve more than just the nonpolar forces, combining rules such as Eqs. (5-3.1) to (5-3.3) are usually in error. The common approach to treating polar/nonpolar systems, and also mixtures of normal compounds where the sizes are significantly different, is to use binary interaction parameters as described in Sec. 5-2 and in Eqs. (5-2.4). For example, Barner and Quinlan (1969) found optimal values for T_{cij} for Eq. (5-2.4c) [the notation here is that the property Q of Eq. 5-2.4c is T_c and Barner and Quinlan used k^*_{ij} for their binary interaction parameter]. Tabulated values were given in Table 4-1 of the 4th Edition, as was a plot of the k^*_{ij} values versus the ratio of larger to smaller V_c values. The V_{cii}/V_{cjj} values ranged from unity to nearly 5. For normal fluids, the range of k^*_{ij} was from 0.98 to 1.3 with the largest values being for the greatest V_c ratio. For normal fluids with CO_2, H_2S, HCl, and C_2H_2, the k^*_{ij} values were 0.92 ± 0.04 unless the size ratio became very large. Further, with these polar and quadrupolar substances, the correlation was much less sensitive to size differences.

Regardless of which combining rule (5-2.4) is used, it is common for the fitted binary interaction parameter of polar/nonpolar systems to reduce the value of T_{cij} to less than that of the geometric or arithmetic mean of the pure component values. The need for a binary interaction parameter may be overcome by a combination of the geometric and harmonic means as in the second virial coefficient of Hayden and O'Connell (1975).

It should be recognized that there is no rigorous basis for implementing binary interaction parameters in this manner. For example, it can be argued that for T_{cij}, the geometric mean of Eqs. (5-2.4*ab*) may be more appropriate than the arithmetic mean of Eqs. (5-2.4*cd*). Also, it is likely that binary interaction parameters may be inadequate for multicomponent mixtures, requiring ternary and higher interaction parameters. Finally, it would be surprising if a constant value of k_{ij} would be adequate to describe properties over wide ranges of conditions. There is much experience supporting a single k_{ij} value for many purposes, especially for binary volumetric behavior. However, for more exacting requirements and for multicomponent systems, more complex formulations and additional empirical parameters may be needed.

There seem to be no extensive applications of the higher order CSP methods described in Sec. 4-3. However, the quantum corrections of Eqs. (4-3.4) have been successfully used.

5-4 VIRIAL EQUATIONS OF STATE FOR MIXTURES

As described in Sec. 4-5, the virial equation of state is a polynomial series in pressure or in inverse volume, but for mixtures the coefficients are functions of both T and $\{y\}$. The consistent forms for the initial terms are

$$Z = 1 + B\left(\frac{P}{RT}\right) + (C - B^2)\left(\frac{P}{RT}\right)^2 + \ldots \tag{5-4.1a}$$

$$= 1 + \frac{B}{V} + \frac{C}{V^2} + \ldots \tag{5-4.1b}$$

where the coefficients $B, C, \ldots$ are called the second, third, ... virial coefficients. Except at high temperatures, B is negative and, except at very low T where they are of little importance, C and higher coefficients are positive. Mixture isotherms at constant composition are similar to those of Fig. 4-1. Formulae relating B to molecular pair interactions, C to molecular trio interactions, etc., can be derived from statistical mechanics. In particular, their composition dependence is rigorous.

$$B(T, \{y\}) = \sum_{i=1}^{n}\sum_{j=1}^{n} y_i y_j B_{ij}(T) \tag{5-4.2a}$$

$$C(T, \{y\}) = \sum_{i=1}^{n}\sum_{j=1}^{n}\sum_{k=1}^{n} y_i y_j y_k C_{ijk}(T) \tag{5-4.2b}$$

where the virial coefficients for pure component i would be B_{ii} and C_{iii} with the pairs and trios being the same substance. When $i \neq j, k$, the pairs and trios are unlike and the coefficients are called *cross coefficients.* There is symmetry with the subscripts so that $B_{ij} = B_{ji}$ and $C_{iij} = C_{jii} = C_{iji}$. In the case of a three-component system

$$B(T, \{x\}) = y_1^2 B_{11}(T) + 2y_1 y_2 B_{12}(T) + y_2^2 B_{22}(T) + 2y_1 y_3 B_{13}(T) + 2y_2 y_3 B_{23}(T) + y_3^2 B_{33}(T) \tag{5-4.3a}$$

while in the case of a two-component system

$$C(T, \{x\}) = y_1^3 C_{111}(T) + 3y_1^2 y_2 C_{112}(T) + 3y_1 y_2^2 C_{122}(T) + y_2^3 C_{222}(T) \tag{5-4.3b}$$

Much has been written about the virial EoS; see especially Mason and Spurling (1968) and Dymond and Smith (1980).

The general ranges of state for applying Eqs. (5-4.1) to mixtures are the same as described in Sec. 4-5 for pure fluids; the virial equation should be truncated at the second or third term and applied only to single-phase gas systems.

The most extensive compilations of second cross virial coefficients are those of Dymond and Smith (1980) and Cholinski, et al. (1986). Newer values for B_{ij} of pairs among alkanes, linear 1-alkanols and alkyl ethers are given by Tsonopoulos and Dymond (1997) and Tsonopoulos, et al. (1989) and measurements of cross coefficients using indirect thermodynamic methods have been reported recently by McElroy and coworkers (see, *e.g.* McElroy and Moser, 1995) and Wormald and coworkers (see, *e.g.*, Massucci and Wormald, 1998). Tsonopoulos and Heidman (1990) review water-*n*-alkane systems. Some third cross virial coefficient values are also given by Dymond and Smith (1980). Iglesias-Silva, et al. (1999) discuss methods to obtain cross coefficients from density measurements.

Estimation of Second Cross Virial Coefficients

Our treatment of cross virial coefficients is the same as for pure coefficients in Sec. 4-5. All of the methods there can be used here if the parameters are suitably adjusted. As before, the formulation is in CSP for all pairs of components in the mixture, i and j.

$$\frac{B_{ij}(T)}{V_{ij}^*} = \sum_m a_{mij} f^{(m)}(T/T_{ij}^*) \tag{5-4.4}$$

where V_{ij}^* is a characteristic volume for the pair, the a_{mij} are strength parameters for various pair intermolecular forces described in Sec. 4-5, and the $f^{(m)}$ are sets of universal functions of reduced temperature, T/T_{ij}^*, with T_{ij}^* a characteristic temperature for the pair. Then, $f^{(0)}$ is for simple substances with a_0 being unity, $f^{(1)}$ corrects for nonspherical shape and globularity of normal substances with a_1 commonly being, ω_{ij}. If one or both of the components are dipolar, $f^{(2)}$ takes account of polarity with a_2 being a function of the dipole moments (see Sec. 2-6), μ_i and μ_j, when both are dipolar, or, if only one species is dipolar, another function of the dipole of the polar species and the polarizability of the other component. Finally, $f^{(3)}$ takes account of association among like molecules or solvation among unlike molecules with a_3 an empirical parameter. The value of a_3 may be the same for cross coefficients as for pure coefficients among substances of the same class such as alcohols. On the other hand, an a_3 value can be required even if none exists for pure interactions, should the unlikes solvate as do $CHCl_3$ and $(CH_3)_2C{=}O$.

When treating cross coefficients, most of the methods of Sec. 4-5 use combining rules for T^*_{ij} of the form of Eq. (5-2.4*b*) with a constant binary interaction parameter, k_{ij}. Often, there are methods for estimating k_{ij} such as the same value for all pairs of particular classes of components. These methods commonly omit the polar/associating contribution for polar/nonpolar pairs and also may use an empirical parameter for solvation. There is normally considerable sensitivity to the values of the parameters; Stein and Miller (1980) discuss this issue with the Hayden-O'Connell (1975) model and provide useful guidance about obtaining solvation parameters. Example 5-2 below illustrates the sensitivity for the Tsonopoulos (1974) correlation.

Detailed discussion of second virial coefficient correlations is given in Sec. 4-5. Just as for pure components, no single technique is significantly better than the others for cross coefficients, except for systems involving very strongly associating/solvating species such as carboxylic acids where the correlation of Hayden and O'Connell (1975) is the only one that applies. We illustrate the expressions and use in detail only the Tsonopoulos (1974, 1975, 1978, 1979) correlation (Eqs. 4-5.3 to 4-5.4), since it is one of the most popular and reliable.

For second cross coefficients, the Tsonopoulos correlation uses $V^*_{ij} = RT_{cij}/P_{cij}$ and $T^*_{ij} = T_{cij}$. The substance-dependent strength coefficients are $a_{1ij} = \omega_{ij}$, $a_2 = a_{ij}$, and $a_3 = b_{ij}$. Table 4-5 summarizes current recommendations for a_{ii} and b_{ii} in Eq. (4-5.3). The following combining rules were established (these expressions are rearrangements of the original expressions):

$$T^*_{ij} = T_{cij} = (T_{cii}T_{cjj})^{1/2}(1 - k_{ij}) \tag{5-4.5a}$$

$$V^*_{ij} = \frac{(V_{cii}^{1/3} + V_{cjj}^{1/3})^3}{4(Z_{cii} + Z_{cjj})} = \frac{RT_{cij}}{P_{cij}} = \frac{R(V_{cii}^{1/3} + V_{cjj}^{1/3})^3}{4(P_{cii}V_{cii}/T_{cii} + P_{cjj}V_{cjj}/T_{cjj})} \tag{5-4.5b}$$

$$a_{1ij} = \omega_{ij} = (\omega_{ii} + \omega_{jj})/2 \tag{5-4.5c}$$

where a binary interaction parameter, k_{ij}, has been included. For either i or j or both without a significant dipole moment

$$a_{ij} = 0 = b_{ij} \qquad \mu_i \sim 0 \text{ and/or } \mu_j \sim 0 \tag{5-4.5d}$$

For both i and j having a significant dipole moment

$$a_{ij} = (a_{ii} + a_{jj})/2 \qquad \mu_i \neq 0 \neq \mu_j \tag{5-4.5e}$$

$$b_{ij} = (b_{ii} + b_{jj})/2 \tag{5-4.5f}$$

Values of the binary interaction parameter, k_{ij}, are given in the references cited in Sec. 4-5. Estimations of k_{ij} for nonpolar pairs usually involve critical volumes. For example, Tsonopoulos, et al. (1989) reconfirm the relationship of Chueh and Prausnitz (1967b) for nonpolar pairs which is apparently reliable to within ± 0.02:

$$k_{ij} = 1 - \left[\frac{2(V_{cii}V_{cjj})^{1/6}}{(V_{cii}^{1/3} + V_{cjj}^{1/3})}\right]^3 \qquad \mu_i \sim 0 \sim \mu_j \tag{5-4.6}$$

For polar/nonpolar pairs, constant values of k_{ij} are used. Thus, for 1-alkanols with ethers, Tsonopoulos and Dymond (1997) recommend $k_{ij} = 0.10$.

Example 5-2 Estimate the second virial coefficient of a 40 mole percent mixture of ethanol (1) with benzene (2) at 403.2 K and 523.2 K using the Tsonopoulos method and compare the results with the data of Wormald and Snowden (1997).

solution From Appendix A, the critical properties and acentric factors of the substances are given in the table below. In addition, from Table 4-5, $a_{11} = 0.0878$ and with $\mu_r = 66.4$, $b_{11} = 0.0553$ while these quantities are zero for both B_{12} and B_{22}. The recommendation (Tsonopoulos, 1974) for the binary interaction constant is $k_{12} = 0.20$; however this seems too large and an estimate of $k_{12} = 0.10$ is more consistent with later analyses such as by Tsonopoulos, et al. (1989). The values of T^*_{ij}, V^*_{ij}, and ω_{ij} are given for each pair followed by all computed quantities for the B_{ij} values.

Quantity	For 1-1 pair	For 1-2 pair	For 2-2 pair
T_{cij}, K	513.92	483.7	562.05
P_{cij}, bar	61.48	—	48.95
V_{cij}, cm³ mol⁻¹	167.00	208.34	256.00
Z_{cij}	0.240	0.254	0.268
V^*_{ij}, cm³ mol⁻¹	695.8	820.2	955.2
ω_{ij}	0.649	0.429	0.209
T/T_{cij}	0.785	0.834	0.717
$f^{(0)}_{ij}$	−0.530	−0.474	−0.626
$f^{(1)}_{ij}$	−0.330	−0.225	−0.553
$f^{(2)}_{ij}$	4.288	2.981	7.337
$f^{(3)}_{ij}$	−6.966	−4.290	−14.257
B_{ij}/V^*_{ij}	−0.753	−0.571	−0.742
B_{ij}, cm³ mol⁻¹	−524	−468	−708
B_{ij} (exp), cm³ mol⁻¹	−529	−428	−717
Error, cm³ mol⁻¹	−5	40	−9

The computed mixture value is B_{mix} (calc) = −591 cm³ mol⁻¹ while the experimental value, B_{mix} (exp) = −548 cm³ mol⁻¹. The difference of 43 cm³ mol⁻¹ is almost within the experimental uncertainty of ±40 cm³ mol⁻¹. Most of the error is from B_{12}. If the original value of $k_{12} = 0.2$ is used, $B_{12} = -342$ cm³ mol⁻¹. If Eq. (5-4.6) were used, $k_{12} = 0.006$, and $B_{12} = -613$ cm³ mol⁻¹. To reproduce B_{12} (exp) precisely, $k_{12} = 0.13$.

If the same procedure is used at $T = 523.2$, the results are:

Quantity	For 1-1 pair	For 1-2 pair	For 2-2 pair
T/T_{cij}	1.018	1.082	0.931
$f^{(0)}_{ij}$	−0.325	−0.289	−0.386
$f^{(1)}_{ij}$	−0.025	0.008	−0.093
$f^{(2)}_{ij}$	0.898	0.624	1.537
$f^{(3)}_{ij}$	−0.867	−0.534	−1.774
B_{ij}/V^*_{ij}	−0.310	−0.285	−0.405
B_{ij}, cm³ mol⁻¹	−216	−234	−387
B_{ij} (exp), cm³ mol⁻¹	−204	−180	−447
Error, cm³ mol⁻¹	12	54	−60

The agreement for the individual coefficients is not as good as at $T = 403.2$ K, but the errors compensate and the calculated mixture value, B_{mix} (calc) = −278 cm³ mol⁻¹ is very close to the experimental value, B_{mix} (exp) = −280 cm³ mol⁻¹, though this would not occur when y_2 is near 1. To reproduce B_{12} (exp) precisely, $k_{12} = 0.18$. This example illustrates the sensitivity of the calculations to the value of the binary interaction parameter and how results may appear accurate at certain conditions but not at others.

See the discussion in Sec. 4-5 for updates to the methods for second virial coefficients. Lee and Chen (1998) have revised the Tsonopoulos expression. The model of Hayden and O'Connell (1975) has been discussed by several authors as noted in Sec. 4-5. In particular, Stein and Miller (1980) made improvements in this model to treat solvating systems such as amines with methanol.

Literature discussion and our own comparisons show that none of the correlations referenced above is significantly more accurate or reliable than the others except for systems with carboxylic acids where the Hayden-O'Connell method is best. Thus, for the range of conditions that the second virial coefficient should be applied to obtain fluid properties, all models are likely to be adequate.

Estimation of Third Cross Virial Coefficients

The limitations on predicting third cross virial coefficients are the same as described in Sec. 4-5 for pure third virial coefficients. In particular, no comprehensive models have been developed for systems with polar or associating substances. Further, there are very few data available for the C_{ijk} or for C_{mix} in Eqs. (5-4.2*b*) and (5-4.3*b*).

For third cross coefficients of nonpolar substances, the CSP models of Chueh and Prausnitz (1967a), De Santis and Grande (1979) and Orbey and Vera (1983) can be used. In all cases, the approach is

$$C_{ijk} = (C_{ij}C_{jk}C_{ik})^{1/3} \tag{5-4.7}$$

where the pairwise C_{ij} are computed from the pure component formula with characteristic parameters obtained from pairwise combining rules such as Eqs. (5-4.5) including binary interaction parameters. The importance of accurate values of C_{112} for describing solid-fluid equilibria of a dilute solute (2) in a supercritical solvent (1) is nicely illustrated in Chueh and Prausnitz (1967a).

5-5 ANALYTICAL EQUATIONS OF STATE FOR MIXTURES

As discussed in Sec. 4-6, analytical EoS models allow the solution for the density of a fluid from a specified T, P, and for mixtures, $\{y\}$, to be noniterative. Thus, the formulation must be cubic or quartic in V. As discussed in Sec. 5-2, the common way to include the effects of composition in the parameters is with *mixing rules* and *combining rules* as in the CSP treatments of Sec. 5-3. Commonly, the *one-fluid* approach is used where the mixture is assumed to behave as a pure component with appropriate parameters. There have been *two-fluid* methods (see, for example, Prausnitz, et al., 1999), but these have been used mainly for modeling liquid excess properties.

Because of the different responses of gases and liquids to changes in pressure and composition as discussed in Sec. 5-2, there are three kinds of mixing rules. The first, often called *van der Waals* or *conformal mixing rules,* use relationships such as Eqs. (5-2.2) and (5-2.4) with direct estimates of the binary interaction parameters k_{ij} and l_{ij}. However, these expressions have very limited flexibility for composition variation and are inadequate for complex liquid mixtures. The second, often called *density-dependent mixing rules,* add functions of density and composition to the conformal expressions (Gupte and Daubert, 1990). The third alternative, called *excess free energy rules*, has recently been implemented to a great extent, especially for phase equilibria (see Chap. 8). This last method first recognizes that excess properties (such as the excess Gibbs energy, G^E, (see Sec. 8-5) are crafted to give suitable composition effects in liquid solutions, even though pressure dependence is rarely included. Second, it utilizes the relationship between the fugacity coefficient from an EoS (see Sec. 6-8) and the activity coefficient from a G^E model (see Secs. 8-2 and 8-5) to connect G^E models to EoS models.

Here we develop the general relationships of these alternatives and discuss the issues associated with them. In general, volumetric and caloric properties of gases and many liquids can be adequately described by van der Waals rules. They are also adequate for mixtures of simple and normal fluids. While density-dependent mixing rules can be quite adequate for many cases, for solutions of different classes of species, *e.g.*, nonpolar with polar or associating components, excess free energy models are now recommended.

van der Waals and Density-Dependent Mixing Rules

As described above, the basic approach of van der Waals mixing rules is in Eqs. (5-2.2) and (5-2.4), while density-dependent mixing rules add functions to the van der Waals expressions.

The simplest justification of the van der Waals concept is to match the composition dependence of the second virial coefficient from an EoS model to the rigorous relationship (5-4.2*a*). For the general cubic equation of Eq. (4-6.1), and therefore all of the models of Tables 4-6 and 4-7, this is (Abbott, 1979)

$$B = b - \frac{\Theta}{RT} \tag{5-5.1}$$

which, with Eq. (5-4.2*a*) gives

$$b_m = \sum_{i=1}^{n} \sum_{j=1}^{n} y_i y_j b_{ij} \tag{5-5.2a}$$

$$\Theta_m = \sum_{i=1}^{n} \sum_{j=1}^{n} y_i y_j \Theta_{ij} \tag{5-5.2b}$$

where, when $i \neq j$, a combining rule such as (5-2.4) is used. This development assumes symmetry in the parameters, e.g., $\Theta_{ij} = \Theta_{ji}$. However this is not necessary; Patel, et al. (1998) develop asymmetric mixing rules from perturbation theory (see Sec. 4-7).

For b_{ij}, Eq. (5-2.3*a*) is often used, so that Eq. (5-5.2*a*) becomes

$$b_m = \sum_{i=1}^{n} y_i b_i \tag{5-5.3}$$

However, Eq. (5-2.4*d*) has also been recommended with a positive binary interaction parameter, l_{ij}, especially if the values of b_i are very different (Arnaud, et al., 1996). For the parameters c and d in Table 4-6, it is common to use linear mole fraction averages in the form of Eq. (5-5.3). See, for example, Peneloux, et al. (1982).

It is to the parameter Θ that more complex expressions are applied. The most general way to express these is with the quadratic rule of Eq. (5-5.2*b*) with an added "nonconformal" function, f_{NC}, such as from the density-dependent mixing rules

$$\Theta_m = \sum_{i=1}^{n}\sum_{j=1}^{n} y_i y_j \Theta_{ij} + f_{NC}(V, \{y\}; \{k\}) \qquad (5\text{-}5.4)$$

where $\{k\}$ is a set of binary parameters. The simplest approach is to set $f_{NC} = 0$ and obtain Θ_{ij} from a combining rule such as Eqs. (5-4.2*ab*) with a single binary interaction parameter. The simplest two-parameter mixing rule is

$$\Theta_{ij} = (\Theta_{ii}\Theta_{jj})^{1/2}; \qquad f_{NC} = \sum_{i=1}^{n}\sum_{j=1}^{n} y_i y_j \Theta_{ij}(k_{ij}x_i + k_{ji}x_j) \qquad (5\text{-}5.5)$$

where $k_{ij} \neq k_{ji}$. Specific functions of f_{NC} for density-dependent mixing rules are given by Gupte and Daubert (1990), Mathias (1983), Mathias and Copeman (1983), Mollerup (1981, 1983, 1985, 1986) and Patel, et al. (1998). Care should be taken in adopting mixing rules with f_{NC} functions, especially those with asymmetries in composition dependence; there have been formulations that give inconsistencies in the properties of multicomponent mixtures (see Sec. 5-7).

Because there are so many EoS models plus options for mixing and combining rules, there are few comprehensive tabulations of binary interaction parameters. Knapp, et al. (1982) give values of k_{12} of Eq. (5-4.2*b*) for the Soave (1972) and Peng-Robinson (1976) EoS models for many systems. For example, aqueous systems at extreme conditions have been treated by Akinfiev (1997), and Carroll and Mather (1995) discuss Peng-Robinson parameters for paraffins with H_2S. It is common for developers of an EoS model to list some values obtained with the models, but usually these tabulations are not extensive. Users are expected to fit data for their own systems.

When components of mixtures are above their critical temperature, such as H_2, CH_4, etc., there still can be multiphase equilibria. However, the α functions of Table 4-7 may not have been studied for $T/T_c > 1$. Mathias (1983), Soave (1993) and Twu, et al. (1996) have discussed this issue and suggested modifications of the common models.

Excess Free Energy (G^E) Mixing Rules

As discussed above, to treat more complex solutions, mixing rules based on expressions more appropriate for the composition dependence of liquids have been developed, especially for phase equilibrium calculations. The first widely recognized analysis of this approach was by Huron and Vidal (1979). Since that time, a very large literature has arisen with many different expressions; reviews have been given, for example, by Orbey (1994), Orbey and Sandler (1995), Kalospiros, et al. (1995), and Heidemann (1996). This method is also discussed in the more general reviews by Anderko (1990), Sandler (1994), Sandler, et al. (1994), Orbey and Sandler (1998) and Prausnitz, et al. (1999).

The concept is that an excess property such as G^E is normally directly correlated by models such as the forms shown in Table 8-3 or directly predicted by group contribution or other methods such as described in Sec. 8-10. However, thermodynamics also allows it to be computed from EoS expressions. Thus, an EoS mixing rule for the liquid phase can reflect the composition variations of a desirable G^E model if the different expressions are matched.

Precise matching cannot be made over all conditions because direct methods do not include density dependence as EoS models do. However, there have been developed many strategies for making the connection. We will briefly outline the procedure; full details must be obtained from the literature. Since there are many subtle consequences of these analyses (see, e.g., Heidemann, 1993; Michelsen, 1996; Michelsen and Heidemann, 1996), care in implementation should be exercised.

The fundamental EoS expressions for the excess Gibbs and Helmholtz functions are

$$\frac{G^E_{EoS}}{RT} = \left[Z_m - \ln Z_m - \int_\infty^{V_m} \frac{Z_m - 1}{V_m}\bigg|_{T,\{y\}} dV\right] - \sum_{i=1}^{n} y_i \left[Z_i - \ln Z_i - \int_\infty^{V_i} \frac{Z_i - 1}{V_i}\bigg|_{T} dV\right] \quad (5\text{-}5.6a)$$

$$\frac{A^E_{EoS}}{RT} = \left[-\ln Z_m - \int_\infty^{V_m} \frac{Z_m - 1}{V_m}\bigg|_{T,\{y\}} dV\right] - \sum_{i=1}^{n} y_i \left[-\ln Z_i - \int_\infty^{V_i} \frac{Z_i - 1}{V_i}\bigg|_{T} dV\right] \quad (5\text{-}5.6b)$$

where $Z_m = PV_m/RT$, the mixture volume, V_m, is evaluated at the mixture conditions $(T, P, \{y\})$, $Z_i = PV_i/RT$, and the pure component volume, V_i, is evaluated at the pure-component conditions of T, P. Typically, the goal is to obtain an EoS mixing rule that gives the first terms on the right-hand-side of Eq. (5-5.6*b*) the same composition dependence as the G^E model used to obtain the left-hand-side by setting A^E_{EoS} equal to a well-defined function of a direct G^E. To accomplish this, the excess property and the P, V_m and V_i at the matching condition must be chosen.

There have been many strategies developed to select the optimal conditions for matching. Fischer and Gmehling (1996) show that the general process is to select a pressure and the G^E function (which we denote G^{E0} or $G^{E\infty}$, depending upon the pressure chosen) on the lhs. Then values of the inverse packing fraction, $u_m = V_m/b_m$ and $u_i = V_i/b_i$, are selected for use on the rhs so that the parameter mixing rule that gives Z_m and V_m can be found. Twu and Coon (1996) also discuss matching with constraints.

Here we show some of the more popular expressions, but we will not be able to describe all. We focus on generalized van der Waals cubic EoS models (such as those in Table 4-6), since essentially all EoS-G^E treatments have been done with them and it is possible to write out analytical equations for these cases. Twu, et al. (1999) give an interesting perspective on the issue of matching.

For cubic equations shown in Table 4-6, the EoS expressions for the rhs of Eq. (5-5.6*b*) are

$$\frac{A^E_{EoS}}{RT} = \sum_{i=1}^{n} y_i \left\{-\ln \frac{Z_m}{Z_i} - \ln \left(\frac{1 - \dfrac{b_m}{V_m}}{1 - \dfrac{b_i}{V_i}}\right) + \left[\frac{\Theta_m}{b_m RT} C(u_m) - \frac{\Theta_i}{b_i RT} C(u_i)\right]\right\} \quad (5\text{-}5.7)$$

Here an EoS dependent dimensionless function appears for both the mixture, $C(u_m)$, and the pure component, $C(u_i)$,

$$C(u) = \frac{b}{\sqrt{\delta^2 - 4\varepsilon}} \ln \left[\frac{1 + \dfrac{(\delta - \sqrt{\delta^2 - 4\varepsilon})}{2bu}}{1 + \dfrac{(\delta + \sqrt{\delta^2 - 4\varepsilon})}{2bu}} \right] \tag{5-5.8}$$

where u is the inverse packing fraction, $u = V/b$ and δ and ε are the EoS parameters in Eq. (4-6.1). Table 5-1 shows the different matching conditions and relations that were tabulated by Fischer and Gmehling (1996). For illustration we also tabulate the results of Eq. (5-5.8) for the Soave (1972) (SRK) model [$\delta_{SRK} = b$ and $\varepsilon_{SRK} = 0$ gives $C_{SRK}(u) = -\ln(1 + u)$].

Some specific results from Table 5-1 are:

Huron-Vidal

$$b_m = \sum_{i=1}^{n} y_i b_i \tag{5-5.9a}$$

with $A^{E\infty}_{EoS} = G^{E\infty}$, and

$$\Theta_m^\infty = b_m \left[\sum_{i=1}^{n} \frac{y_i \Theta_i}{b_i} + \frac{A^{E\infty}_{EoS}}{C(V)} \right] = b_m \left[\sum_{i=1}^{n} \frac{y_i \Theta_i}{b_i} + \frac{G^{E\infty}}{C(V)} \right] \tag{5-5.9b}$$

Equations (5-5.9) do not obey the second virial composition dependence of Eq. (5-4.2*a*) and parameters obtained from data are not necessarily appropriate for high pressures.

Wong-Sandler

$$b_m - \frac{\Theta_m}{RT} = \sum_{i=1}^{n} \sum_{j=1}^{n} y_i y_j \left(b - \frac{\Theta}{RT} \right)_{ij} \tag{5-5.10a}$$

with $A^{E\infty}_{EoS} = G^{E0}$, and

$$\Theta_m = b_m \left[\sum_{i=1}^{n} \frac{y_i \Theta_{ii}}{b_{ii}} + \frac{A^{E\infty}_{EoS}}{C(V)} \right] = b_m \left[\sum_{i=1}^{n} \frac{y_i \Theta_{ii}}{b_{ii}} + \frac{G^{E0}}{C(V)} \right] \tag{5-5.10b}$$

After selecting a combining rule for $(b - \Theta/RT)_{ij}$ and substituting in Eq. (5-5.10*b*) to eliminate Θ_m from Eq. (5-5.10*a*), b_m is found. Then Θ_m is found from Eq. (5-5.10*b*). Unlike for Eqs. (5-5.9), the combining rules for b_{ij} and Θ_{ij} can be chosen independently so the second virial relation is preserved. Note that all $(b - \Theta/RT)$ values must be positive to avoid b_m becoming zero or negative (Orbey and Sandler, 1995, 1998).

Recognizing that there will not be an exact match of computed and the input experimental or correlated G^{E0}, Wong and Sandler (1992) and Wong, et al. (1992) suggest that the combining rule be of the form

$$1 - k_{ij} = \frac{2(b - a/RT)_{ij}}{(b_{ii} - a_{ii}/RT) + (b_{jj} - a_{jj}/RT)} \tag{5-5.11}$$

where the optimal value of k_{ij} is obtained by minimizing the difference between calculated and input G^E over the whole composition range. Another choice is to

TABLE 5-1 EoS-G^E Matching Conditions (after Fischer and Gmehling, 1996).

Mixing rule	P match	u match	G^E match*	b_m	$C_{SRK}(u)$
Huron/Vidal (1979)	∞	1.000	$G^{E\infty}/RT$	$\Sigma y_i b_{ii}$	−0.693
MHV1 (Michelsen, 1990)	0	1.235	$G^{E0}/RT + \sum_{i=1}^{n} y_i \ln \frac{b_m}{b_i}$	$\Sigma y_i b_{ii}$	−0.593
PSRK (Holderbaum and Gmehling, 1991)	1 atm	1.100	$G^{E0}/RT + \sum_{i=1}^{n} y_i \ln \frac{b_m}{b_i}$	$\Sigma y_i b_{ii}$	−0.647
MHV2 (Dahl and Michelsen, 1990)	0	1.632	$\left(G^{E0}/RT + \sum_{i=1}^{n} y_i \ln \frac{b_m}{b_i}\right)^a$	$\Sigma y_i b_{ii}$	0.478
LCVM (Boukouvalas, et al., 1994)	0	—	$\left(\frac{G^{E0}}{RT} + \frac{1-\lambda}{C(1.235)} \sum_{i=1}^{n} y_i \ln \frac{b_m}{b_i}\right)^b$	$\Sigma y_i b_{ii}$	−0.553
Wong and Sandler (1992)	∞	1.000	$(G^{E0}/RT)^c$	+	−0.693
Tochigi, et al. (1994)	0	1.235	$G^{E0}/RT + \sum_{i=1}^{n} y_i \ln \frac{b_m}{b_i}$	+	−0.593
Orbey and Sandler (1995)	∞	1.000	$G^{E0}/RT + \sum_{i=1}^{n} y_i \ln \frac{b_m}{b_i}$	+	−0.693

* Expression to set equal to $\frac{A^E_{EoS}}{RT}$. For those not indicated [abc], $\frac{A^E_{EoS}}{RT} = C(u)\left[\frac{\Theta_m}{b_m RT} - \sum_{i=1}^{n} y_i \frac{\Theta_i}{b_i RT}\right]$

[a] $\frac{A^E_{EoS}}{RT} = C(u)\left[\frac{\Theta_m}{b_m RT} - \sum_{i=1}^{n} y_i \frac{\Theta_i}{b_i RT}\right] + A_2\left[\left(\frac{\Theta_m}{b_m RT}\right)^2 - \sum_{i=1}^{n} y_i \left(\frac{\Theta_i}{b_i RT}\right)^2\right]$; $(A_2)_{SRK} = -0.0047$

[b] $\frac{A^E_{EoS}}{RT} = F(\lambda)\left[\frac{\Theta_m}{b_m RT} - \sum_{i=1}^{n} y_i \frac{\Theta_i}{b_i RT}\right]$; $F(\lambda) = 1\Big/\left[\frac{\lambda}{C(1)} + \frac{1-\lambda}{C(1.235)}\right]$; $\lambda_{SRK} = 0.36$

[c] $\frac{A^E_{EoS}}{RT} = C(u)\left\{\frac{\Theta_m}{b_m RT} - \frac{1}{2b_m}\sum_{i=1}^{n}\sum_{j=1}^{n} y_i y_j \left[\left(b - \frac{\Theta}{RT}\right)_{ii} + \left(b - \frac{\Theta}{RT}\right)_{jj}\right][1 - k_{ik}]\right\}$

[+] $b_m = \sum_{i=1}^{n}\sum_{j=1}^{n} y_i y_j \left(b - \frac{\Theta}{RT}\right)_{ij} \Big/ \left[1 - \left[\sum_{i=1}^{n} \frac{y_i \Theta_{ii}}{b_{ii}} + \frac{G^{E0}}{C(1)}\right]\right]$

match EoS second cross virial coefficients and pure virial coefficients to experiment with (Eubank, et al., 1995)

$$\frac{2(b - a/RT)_{ij}}{(b_{ii} - a_{ii}/RT) + (b_{jj} - a_{jj}/RT)} = \frac{2B_{ij}}{B_{ii} + B_{jj}} \tag{5-5.12}$$

Kolar and Kojima (1994) match infinite dilution activity coefficients from the input experiment or correlation to those computed from the EoS. There have been revisions of the original Wong-Sandler rule (Orbey and Sandler, 1995; Satyro and Trebble, 1996, 1998).

MHV1 and MHV2

A linear or quadratic function q is used to cover the variation of $\alpha = \Theta/(bRT)$ over all possible values

$$q = q_0 + q_1\alpha + q_2\alpha^2 \tag{5-5.13}$$

for both ideal solution, $\alpha_m^I = \sum_{i=1}^{n} y_i\Theta_i/(b_iRT)$ and real solution, $\alpha_m = \Theta_m/(b_mRT)$. The value of q_1 is $C(1.235)$. The "MHV1" ($q_2 = 0$) and "MHV2" (q_2 optimized) mixing rules that result are (Michelsen, 1990; Dahl and Michelsen, 1990)

$$q_m = q_m^I + \frac{G^{e0}}{RT} + \sum_{i=1}^{n} y_i \ln\left(\frac{b_m}{b_i}\right) \tag{5-5.14}$$

The LCVM model is a linear combination of the Huron-Vidal and MHV1 mixing rules with the coefficients optimized for application. Other models with a G^E basis are those of Heidemann and Kokal (1990) which involve an iterative calculation to obtain the parameters and Kolar and Kojima (1993) where the u matching involves parameters.

5-6 NONANALYTIC EQUATIONS OF STATE FOR MIXTURES

As described in Sec. 4-7, the complexity of property behavior cannot generally be described with high accuracy by cubic or quartic EoS that can be solved analytically for the volume when given T, P, and y. Though the search for better models began well before computers, the ability to rapidly calculate results or do parameter regression with complicated expressions has introduced increasing levels of complexity and numbers of fitted parameters. This section briefly covers the mixture forms of the nonanalytic EoS models of Sec. 4-7. There are MBWR forms but no Wagner formulations. Perturbation methods and chemical association models have been developed and mixture "crossover" expressions for the near-critical region exist.

MBWR Model

The most extensively explored MBWR EoS is the "LKP" model of Ploecker, et al. (1978). The expressions, including tables of binary interaction parameters and recommendations for light-gas systems were described fully in the 4th Edition. No other recent treatments of this method seem to have been developed. Phase equilibrium results for this method are shown in Tables 8-36 and 8-37.

Perturbation Models

The technique of perturbation modeling uses reference values for systems that are similar enough to the system of interest that good estimates of desired values can be made with small corrections to the reference values. For EoS models, this means that the residual Helmholtz energy of Eq. (4-7.3) is written as

$$A^r/RT = [A_R^r(T, V, \{y\})/RT] + \Sigma[A_{Att}^r(T, V, \{y\})/RT]^{(i)} \tag{5-6.1}$$

where the form of the perturbation terms $[A_{Att}^r(T, V, \{y\})/RT]^{(i)}$ can be obtained from a rigorous or approximate theory, from a Taylor's expansion, or from intuition. The result is that there are very many models obtained in this manner and expressed in this form; the general ideas for pure components have often been applied to mixtures. A useful discussion of this approach is given by Abbott and Prausnitz (1987).

Our purpose here is to mention the possible options and give a few references to specific models which have become popular, especially for phase equilibria (see Sec. 8-12). Much more complete reviews are given by Anderko (1990), Sandler, et al. (1994) and Prausnitz, et al. (1999). A very important point is that models of the form of Eq. (4-7.4) inevitably have a positive reference term and negative perturbation terms. This is necessary to be able to describe both vapors and liquids, but the consequence is that the perturbation terms at high density are typically about the same magnitude as the reference term. This can cause difficulties in evaluation and errors in estimation. Further, the isotherms in the two-phase region can be quite complex, especially at low temperatures (see, for example, Koak, et al. 1999).

Reference Fluid Expressions. The most common mixture reference is that of Mansoori, et al. (MCSL) (1971) which is typically written in terms of the compressibility factor

$$Z_{\mathrm{R}}^{\mathrm{MCSL}} = \frac{6}{\pi}\left[\frac{\xi_0}{(1-\xi_3)} + \frac{3\xi_1\xi_2}{(1-\xi_3)^2} + \frac{3\xi_2^3}{(1-\xi_3)^3}\right] \tag{5-6.2a}$$

where the covolumes are

$$\xi_j = \frac{\pi}{6}\sum_{i=1}^{n} y_i \frac{N_A\sigma_i^j}{V} \qquad j = 0, 1, 2, 3 \tag{5-6.2b}$$

with σ_i being the diameter of the hard sphere of species i. With the simplest form of perturbation term, the model EoS is fifth order in volume so it is nonanalytic.

As with pure components, this idea has been expanded to deal with nonspherical and flexible molecules in three principal ways. The first is to assume that the rigid bodies are not spheres, but have different shapes so there are several different terms to replace those in ξ_j in Eq. (5-6.2). The expressions have been reviewed by Boublik (1981). The second approach is in the Perturbed Hard Chain Theory (PHCT) which multiplies the right-hand-side of Eq. (4-7.6) by a factor, c, which is a substance specific parameter (for a review, see, e.g., Cotterman, et al., 1986 and Cotterman and Prausnitz, 1996).

$$Z_{\mathrm{R}}^{\mathrm{PHCT}} = cZ_{\mathrm{R}}^{\mathrm{HS}} \tag{5-6.3}$$

where Z^{HS} can be any appropriate hard sphere model such as given in Eqs. (4-7.5) and (5-6.2) and the mixture c is found by a simple mixing rule such as Eq (5-2.3a). This idea has been used by many workers with success (see Anderko, 1990, and Prausnitz, et al., 1999 for details). Alternatives include those of Siddiqi and Lucas (1989) and of Chiew (1990) who derived terms to be added to Eq. (5-6.2) for chains of hard spheres (HC). This can be called a perturbed hard-sphere-chain theory (PHSC). Chiew's mixture form is (see, for example, Feng and Wang, 1999)

$$Z_{\mathrm{R}}^{\mathrm{PHSC}} = Z_{\mathrm{R}}^{\mathrm{MCSL}} - \frac{1}{2(1-\xi_3)}\sum_{i=i}^{n} y_i \frac{N_A(r_i-1)}{V}\left(2 + 3\sigma_i\frac{\xi_2}{(1-\xi_3)}\right) \tag{5-6.4}$$

where r_i is the number of segments in species i. This has been adopted by, for example, Donaggio, et al. (1996) for high-pressure systems, Fermeglia, et al. (1997) for alternative refrigerants, by Song, et al. (1994) and Hino and Prausnitz (1997) for all substances, and by Feng and Wang (1999) for polymers. A final alternative reference expression from similar origins to Eq. (4-7.6) is that of the Statistical Associating Fluid Theory (SAFT) derived by Chapman and coworkers (Chapman, et al. 1989, 1990; Prausnitz, et al., 1999). The reference fluid expressions for this model are complex.

Perturbation (Attraction) Expressions. The perturbation terms, or those which take into account the attraction between the molecules, have ranged from very simple to extremely complex. For example, the simplest form is that of van der Waals (1890) which in terms of the Helmholtz energy is

$$[A^r(T, V, \{y\})/RT]_{\text{Att}}^{(vdW)} = -a_m/(RTV) \tag{5-6.5}$$

and which leads to an attractive contribution to the compressibility factor of

$$Z_{\text{Att}}^{(vdW)} = -a_m/(RTV) \tag{5-6.6}$$

Here, the parameter a_m is usually a function of T and $\{y\}$ and obtained with mixing and combining rules such as given in Sec. 5-4. This form would be appropriate for simple fluids, though it has also been used with a variety of reference expressions. The MCSL form of Eq. (5-6.2) has been used by Campanella, et al. (1987) for solution densities and solubilities of gases in many different solvents, by Aly and Ashour (1994) for a great range of mixtures including organic esters, and by Song, et al. (1994) in the PHSC model of Eq. (5-6.3) for polymer solutions. Other terms such as those found in Tables 4-6 and 4-7, but suitably modified for mixtures, can be used for normal fluids. The most complex expressions for normal substances are those used in the BACK, PHCT, and SAFT EoS models. In this case, there are many terms in the summation of Eq. (4-7.3) which have composition dependence as well. Their general form remains

$$Z_{\text{Att}}^{\text{BACK}} = r \sum_{i=1}^{n} \sum_{j=1}^{m} jD_{ij} \left[\frac{u}{kT}\right]^i \left[\frac{\eta}{\tau}\right]^j \tag{5-6.7}$$

where now u and η are also composition dependent. Typically, 1-fluid rules of the form of Eq. (5-2.4*b*) for u and Eq. (5-2.4*d*) for η are used.

Most of the references cited in Sec. 4-7 discuss mixtures as well as pure components using statistical mechanical perturbation theory. These include Fleming and Brugman (1987) for aqueous systems, and Hino and Prausnitz (1997) to simplify and increase the accuracy of previous models (e.g., Song, et al., 1994) for small substances and polymers. Additional terms can be put in Z_{Att} to account for polarity such as by Muller, et al. (1996). Again, there are many possible expressions and they can be very complicated. The important point is that claims are made that no binary interaction parameters are needed. When true, the model is predictive.

Chemical Theory EoS

In many practical systems, the interactions between the molecules are quite strong due to charge-transfer and hydrogen bonding (see Sec. 4-7 for a brief introduction and Prausnitz, et al., 1999 for a fuller description of the origin and characteristics of these interactions). This occurs in mixtures if alcohols, carboxylic acids, water, HF, etc. are present. It can lead to quite complex behavior of vapors of these substances though not so much different in the liquids. One approach has been to consider the interactions so strong that new "chemical species" are formed. Then the thermodynamic treatment assumes that the properties deviate from an ideal gas mainly due to the "speciation" (the actual number of molecules in the system is not the number put in) plus some physical effects. It is assumed that all of the species are in reaction equilibrium. Thus, their concentrations can be determined from equilibrium constants having parameters such as enthalpies and entropies of reaction as well as parameters for species physical interactions.

In mixtures, an example is the formation of unlike dimers (D) from two different monomers (M_1 and M_2)

$$M_1 + M_2 = D \tag{5-6.8}$$

The equilibrium constant for this reaction can be exactly related to the second cross virial coefficient of Eq. (5-4.2*a*)

$$K_D = \frac{y_D}{y_{M_1} y_{M_2} P} = -2 \frac{B_{12}}{RT} \tag{5-6.9}$$

where the factor of 2 arises here and not in Eq. (4-7.14) because of the distinct monomers of Eq. (5-6.8). The model of Hayden and O'Connell (1975) described in Secs. 4-5 and 5-4 explicitly includes such contributions so that it can also predict the properties of strongly interacting unlike substances.

Anderko's (1991) treatment of systems with speciation also has been extended to mixtures as has that of Gmehling, et al. (1979) described in Sec. 8-12. The procedures are quite similar to those for pure components. Though this form of the EoS may appear to be not very different from those considered earlier in this chapter, the computational aspects are somewhat more complex because obtaining the numbers of moles of the species is an additional nonlinear equation to be solved. However, there is no other practical way to deal with the large and unique non-idealities of such systems.

Economou and Donohue (1992) and Anderko, et al. (1993) show that care must be exercised in treating the chemical and physical contributions to equations of state since some methods introduce thermodynamic inconsistencies.

EoS Models for the Near-Critical Region

Conditions near a mixture vapor-liquid critical point show significantly different behavior from simple extrapolations of the EoS models described so far in this chapter. The molecular correlations mentioned in Sec. 4-7 are long ranged and concentration fluctuations dominate. The formulation must be in terms of chemical potentials as the independent variables, not the composition variables of Eqs. (5-2.1). Research into this effect shows that certain universalities appear in the properties, but substance-specific quantities also are involved.

Kiselev (1998) has published a general procedure for adapting equations of state to describe both classical and near critical regions of mixtures. This has been applied to cubic EoS by Kiselev and Friend (1999). Their model predicts two-phase behavior and also excess properties (see Chap. 6) using parameters that are fitted only to volumetric behavior in the one-phase region.

It has also been found that complex solutions such as ionic solutions and polymers do not have the same universalities as simpler fluids (Anisimov, et al., 1999). This is because the long-range forces among such species also affect long-range correlations. Thus, crossovers in these systems commonly appear at conditions closer to the critical than for small and nonionic systems. Therefore, classical models should apply over larger ranges of conditions, but the crossovers can cause very sharp property changes.

5-7 DISCUSSION OF MIXTURE EQUATIONS OF STATE

Mixture EoS models must reflect the complexity of real mixtures. As a result, the expressions can be very complicated. Fortunately, the formulation of mixing rules such as for cubic EoS, does not depend upon which model is used; the ideas of Secs. 5-5 and 5-6 can work for all. For example, Twu, et al. (1998) show that their mixing rule gives reliable results for both the Soave (1972) and Peng-Robinson (1976) EoS models.

Though few very extensive comparisons have been made among the many models available, there have been a few that do treat more than one model for a range of systems and data. For example, Knudsen, et al. (1993) compared five versions of the Soave (1972) model with one-fluid, density-dependent and EoS-G^E mixing rules for hydrocarbon and polar-containing systems. They concluded that one-fluid models fail for binaries and multicomponent systems, especially when water or methanol is present. However, these systems can be correlated with density-dependent and EoS-G^E mixing rules if three or four parameters are fitted. They also caution about trying to fit with five or more parameters because "overcorrelation" occurs. Huang and Sandler (1993) compared two EoS-G^E mixing rules in two different EoS models and concluded that the Wong-Sandler (1992) approach was better than MHV2 (Dahl and Michelsen, 1990), especially for wide ranges of conditions.

However, the accuracy obtained, and the interaction parameter values to be used for best results, do depend upon the model. Further, the form of mixing rule to be used depends upon the substances involved and the desired property. For example, normal fluids can usually be described with simple rules like Eqs. (5-2.3) and (5-2.4), often without any binary interaction parameter. However, if some, but not all, of the components in a solution are polar or associating such as halogenated species, alcohols, and especially water, usually multiple terms and interaction parameters are needed.

A subject of importance in mixing rules is that of *invariance*. Thermodynamics requires that some properties should not change when another property or parameters vary. An example is the volume translation of the pure component equation of state that should not change the vapor pressure. Another is that if one of the components of a mixture is divided into two distinct subcomponents with the same characteristic properties, the mixture parameters and partial molar properties (see Chap. 6) should not change. Finally, if a mixing rule has multiples of more than two mole fractions in a double summation that are not normalized, as the number of components increases, the importance of the term decreases—the so-called "dilution" effect. Michelsen and Kistenmacher (1990) first noted the mixture issues by pointing out that the mixing rule of Schwarzentruber and Renon (1989) does not meet these requirements. Neither do the rules of Panagiotopoulos and Reid (1986). There are several articles that discuss the subject of invariance in detail (Mathias, et al., 1991; Leibovici, 1993; Brandani and Brandani, 1996; Zabaloy and

Vera, 1996) and list mixing rules that do not meet thermodynamic constraints. In addition to mathematical errors, programming errors can also be made. Mollerup and Michelsen (1992) describe useful relationships to numerically check code for errors. Brandani and Brandani (1996) also point out an inconsistency that arises for infinite pressure matching.

All the mixing rules presented thus far in this chapter are for a discrete number of components. For mixtures with a very large number of components, such as crude oil, the computations required to evaluate the summations can be prohibitively expensive. As described in the 4th Edition, both pseudocomponent and continuous distribution functions have been used for such systems.

5-8 DENSITIES OF LIQUID MIXTURES AT THEIR BUBBLE POINT

In order to extend equations such as Eq. (4-11.3) to mixtures, mixing rules are required. Li (1971) and Spencer and Danner (1973) recommended

$$V_m = R\left(\sum_i \frac{x_i T_{ci}}{P_{ci}}\right) Z_{RAm}^{[1+(1-T_r)^{0.2857}]} \qquad (5\text{-}8.1a)$$

$$Z_{RAm} = \sum x_i Z_{RAi} \qquad (5\text{-}8.1b)$$

with the relation of Yamada and Gunn (1973)

$$Z_{RAi} = 0.29056 - 0.08775\omega_i \qquad (5\text{-}8.1c)$$

where $T_r = T/T_{cm}$. Spencer and Danner (1973) recommend the mixing rules of Chueh and Prausnitz (1967b).

$$T_{cm} = \sum_i \sum_j \phi_i \phi_j T_{cij} \qquad (5\text{-}8.2a)$$

$$\phi_i = \frac{x_i V_{ci}}{\sum_j x_j V_{cj}} \qquad (5\text{-}8.2b)$$

$$1 - k_{ij} = \frac{8(V_{ci}V_{cj})^{1/2}}{(V_{ci}^{1/3} + V_{cj}^{1/3})^3} \qquad (5.8.2c)$$

$$T_{cij} = (1 - k_{ij})(T_{ci}T_{cj})^{1/2} \qquad (5\text{-}8.2d)$$

Li's method sets $k_{ij} = 0$ for Eq. (5-8.2*d*). The HBT method of Eq. (4-11.8) to (4-11.10) has been extended to mixtures by Hankinson and Thomson (1979) with

$$T_{cm} = \frac{\left[\sum_i x_i (V_{ci}T_{ci})^{1/2}\right]^2}{V_{cm}} \qquad (5\text{-}8.3a)$$

$$V_{cm} = 0.25\left[\sum_i x_i V_i^* + 3\left(\sum_i x_i V_i^{*2/3}\right)\left(\sum_i x_i V_i^{*1/3}\right)\right] \qquad (5\text{-}8.3b)$$

$$\omega_m = \sum_{i=1}^{n} x_i \omega_{SRKi} \tag{5-8.3c}$$

$$V_m = V^* V^{(0)}(1 - \omega_m V^{(\delta)}) \tag{5-8.3d}$$

where $V^{(0)}$ and $V^{(\delta)}$ are from Eqs. (4-11.9) and (4-11.10) using T_{cm} from Eq. (5-8.3*a*) to obtain T_r. As in Eq. (4-11.8), V^* of Eq. (5-8.3*b*) is a parameter fit to experimental data and is nearly identical to the pure component V_c, while ω_{SRKi} is that value of ω that causes the Soave EoS to most closely match experimental vapor pressure behavior and is nearly equal to the true value of ω_i. Values of the pure component parameters V^* and ω_{SRKi} are tabulated in the 4th Edition of this book and in Hankinson and Thomson (1979). However, results are only marginally affected if V^* and ω_{SRK} are replaced with the true values of V_c and ω. Unlike the pure component case where the HBT method and Eq. (4-11.3) gave nearly the same temperature dependence for V_s, the averaging in Eqs. (5-8.3*b*) to (5-8.3*d*) changes the HBT results for V_m mainly through computing a much smaller T_{cm}.

Figure 5-1 shows experimental liquid volumes as a function of temperature up to 511 K and the results from the three methods up to $T/T_{cm} = 1$. At low temperatures, all three methods give reasonably accurate results, although for the example shown, the Li method is most accurate. At temperatures above 479 K, only the Li method can be used since for the other methods, $T_r > 1$ because of their low values of T_{cm}. Thus, the Li method can be used for calculations over the largest liquid range and also shows the greatest accuracy.

Spencer and Danner (1973) reported the best results with the Chueh-Prausnitz rules, but they did not compare the Li method with the same data set. Thus, their recommendation of the Chueh-Prausnitz rule should perhaps not be so strong. In light of the fact that the Li approach more closely approximates the true liquid

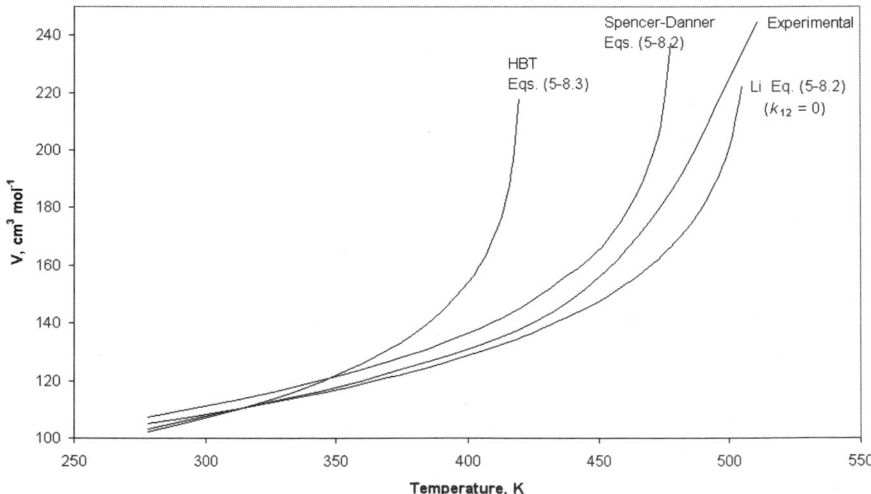

FIGURE 5-1 Bubble point volumes of a mixture of 70 mol % ethane and 30 mol % decane. (*Experimental data from Reamer and Sage (1962).*)

range, the Li may be the best overall. Example 5-3 illustrates the use of the three methods.

Example 5-3 Calculate the volume of a saturated liquid mixture of 70 mol % ethane (1) and 30 mol % *n*-decane (2) at 344.26 K (160°F). The experimental value (Reamer and Sage, 1962) is 116.43 cm³ mol⁻¹ (1.865 ft³ lb-mol⁻¹) at a bubble point pressure of 53.99 bar (783 psia).

solution From Appendix A

	T_{ci}, K	P_{ci}, bar	V_{ci}, cm³ mol⁻¹	ω	x_i
ethane	305.32	48.72	145.5	0.099	0.7
n-decane	617.70	21.10	624	0.491	0.3

From Eq. (5-8.2*b*)

$$\phi_1 = \frac{0.7 \times 145.5}{0.7 \times 145.5 + 0.3 \times 624} = 0.3524$$

$$\phi_2 = 1 - \phi_1 = 0.6476$$

Chueh-Prausnitz Method (1967b)

Equations (5-8.2*c*) and (5-8.2*d*) give $1 - k_{ij} = 0.9163$ and $T_{cij} = 397.92$ K. Then, Eq. (5-8.2*a*) gives $T_{cm} = (0.3524)^2(305.32) + (0.3524)(0.6476)(397.92) + (0.6476)^2(617.70) = 478.6$ K

$T_r = 344.26/478.6 = 0.7193$. Eq. (5-8.1*c*) gives $Z_{\text{RA1}} = 0.282$ and $Z_{\text{RA2}} = 0.247$. Equation (5-8.1*b*) then gives $Z_{RAm} = 0.7 \times 0.282 + 0.3 \times 0.247 = 0.2715$

With Eq. (5-8.1*a*) $V_m = 83.14 \left(0.7\,\dfrac{305.32}{48.72} + 0.3\,\dfrac{617.70}{21.10}\right)(0.2715)^{[1+(1-0.7193)^{0.2857}]}$ $= 120.1$ cm³/mol

$$\text{Error} = \frac{120.1 - 116.43}{116.43} \times 100 = 3.1\%$$

Li Method (1971)

The procedure for this method is identical to the Chueh-Prausnitz method except that now, $k_{ij} = 0$. This leads to $T_{cm} = 507.6$ K, $T_r = 0.6782$, and $V_m = 115.8$ cm³ mol⁻¹ for an error of −0.5%.

HBT (Hankinson and Thomson, 1979; Thomson, et al., 1982)

In this approach, T_{cm}, and V_{cm} are calculated with Eqs. (5-8.3). Using V_c values for V^* in Eq. (5-8.3*b*) leads to $V_{cm} = 265.6$ cm³ mol⁻¹ and Eq. (5-8.3*a*) gives $T_{cm} = 419.5$ K.

Thus, $T_r = 344.26/478.6 = 0.8206$ and $1 - T_r = 0.1794$. Application of Eqs. (4-11.8) through (4-11.10) and Eq. (5-8.3*c*) gives $V^{(0)} = 0.4690$, $V^{(\delta)} = 0.1892$, $\omega_m = 0.2166$, and $V_m = 119.45$ cm³ mol⁻¹ for an error of 2.66%

If values of V^* and ω_{SRK} from Hankinson and Thomson (1979) are used, the calculated volume is 118.2 rather than 119.45 cm³ mol⁻¹. The above example illustrates the method used to calculate the results shown in Fig. 5-1.

5-9 DENSITIES OF COMPRESSED LIQUID MIXTURES

Aalto, et al. (1996, 1997) have extended Eq. (4-12.2) to mixtures by using Eqs. (5-8.3*a*) and (5-8.3*b*) for T_{cm} and V_{cm}. They calculate P_{cm} by

$$P_{cm} = \frac{(0.291 - 0.080\omega_{SRKm})RT_{cm}}{V_{cm}} \tag{5-9.1}$$

$$\omega_{SRKm} = \left(\sum_i x_i \omega_{SRKi}^{1/2}\right)^2 \tag{5-9.2}$$

Aalto, et al. (1996) recommend that the vapor pressure used to calculate P_{vpr} in Eq. (4-12.2) be obtained from a generalized Riedel vapor pressure equation which, with some algebraic manipulation, can be put in the form

$$\ln P_{vpr} = P_{vpr}^0 + \omega P_{vpr}^1 \tag{5-9.3a}$$

$$P_{vpr}^0 = 6.13144 - \frac{6.30662}{T_{rm}} - 1.55663 \ln T_{rm} + 0.17518 T_{rm}^6 \tag{5-9.3b}$$

$$P_{vpr}^1 = 2.99938 - \frac{3.08508}{T_{rm}} + 1.26573 \ln T_{rm} + 0.08560 T_{rm}^6 \tag{5-9.3c}$$

As previously mentioned, Eqs. (5-8.3*a*) and (5-8.3*b*) do not represent estimates of the true T_{cm} and V_{cm}. Thus, Eq. (5-9.1) is not an estimate of the true P_{cm}, but is rather a pseudocritical pressure, and in the same way, the P_{vp} value predicted by Eqs. (5-9.3) is not an estimate of the true bubble point pressure, but is rather a "pseudo" vapor pressure associated with the pseudocritical values of T_{cm} and P_{cm}. As a result, the Aalto method does not predict the correct volume at the bubble point unless the correct bubble point pressure is used in Eq. (4-12.2) instead of P_{vpr} from Eqs. (5-9.3). The Aalto method is illustrated with Example 5-4.

Example 5-4 Calculate the volume of a compressed liquid mixture of 70 mol % ethane (1) and 30 mol % *n*-decane (2) at 344.26 K (160°F) and 689.47 bar (10,000 psia). The experimental value (Reamer and Sage, 1962) is 100.8 cm^3 mol^{-1} (1.614 ft^3 lb-mol^{-1}). The experimental value for the volume at the bubble point pressure of 53.99 bar (783 psia) is 116.43 cm^3 mol^{-1} (1.865 ft^3 lb-mol^{-1}).

solution Using Eqs. (5-9.1) and (5-9.2) along with values from Example 5-3

$$\omega_{SRKm} = (0.7 \times 0.099^{1/2} + 0.3 \times 0.491^{1/2})^2 = 0.185$$

$$P_{cm} = \frac{(0.291 - 0.080 \times 0.185)(83.14)(419.5)}{265.6} = 36.27 \text{ bar}$$

With Eq. (5-9.3*b*) and with $T_r = 0.8206$ from Example 5-3,

$$P_{vpr}^0 = 6.13144 - \frac{6.30662}{0.8206} - 1.55663 \ln(0.8206) + 0.17518(0.8206)^6 = -1.1927$$

Similarly, with Eq. (5-9.3*c*), $P_{vpr}^1 = -0.9842$. Then Eq. (5-9.3*a*) gives

$$P_{vpr} = \exp(-1.1927 - 0.185 \times 0.9842) = 0.2529$$

A value for P_{vpr} of 0.2529 corresponds to a pressure of 9.2 bar, a value considerably less than the true bubble point pressure of 53.99 bar. Equations (4-12.2) to (4-12.4) give $A = 16.643$, $B = 0.1479$, $P_r = 689.47/36.27 = 19.01$

$$V = 116.43 \frac{16.643 + 2.7183^{(1.00588-0.8206)^{0.1479}}(19.01 - 0.2529)}{16.643 + 2.7183(19.01 - .2529)} = 99.05$$

$$\text{Error} = \frac{99.05 - 100.8}{100.8} \times 100 = -1.7\%$$

Because Aalto's approach uses the Hankinson equations for T_{cm}, there is a range of temperatures between T_{cm} and the true T_c for which the Aalto approach cannot be used. For the mixture in Example 5-4, this temperature range is 419.5 K to 510.7 K. In this range, equation of state methods described earlier in this chapter should be used.

The method of Huang and O'Connell (1987) for obtaining volume changes with increased pressure, as described in Sec. 4-12, has been applied to a few mixtures but none at temperatures approaching the critical point.

NOTATION

In many equations in this chapter, special constants or parameters are defined and usually denoted $a, b, \ldots, A, B, \ldots$. They are not defined in this section because they apply only to the specific equation and are not used anywhere else in the chapter.

A^E	excess Helmholtz energy, J mol^{-1}
a_{mij}	coefficient for terms in second virial coefficient correlations, Eq. (5-4.4)
b	cubic EoS variable, Eq. (4-6.1), Table 4-6, Eqs. (5-5.1) to (5-5.3)
B_{ij}	second virial coefficient in Eqs. (5-4.1*a*) to (5-4.3*a*), cm^3 mol^{-1}
C_{ijk}	third virial coefficient in Eqs. (5-4.1*b*) to (5-4.3*b*), cm^6 mol^{-2}
$C(u)$	EoS matching variable for G^E mixing rules, Eq. (5-5.8)
G^E	excess Gibbs energy, J mol^{-1}
k_{ij}, l_{ij}	binary interaction parameters, Eq. (5-2.4)
n	number of components in a mixture
P	pressure, bar
P_{vp}	vapor pressure, bar
Q	generalized property, Eqs. (5-2.2) to (5-2.4)
q	quantity in MHV1 and MHV2 mixing rules, Table 5-1
R	gas constant, Table 4-1
r	number of segments in a chain, Eqs. (4-7.8), (5-6.4)
T	temperature, K
u	inverse packing fraction $= V/b$, Table 5-1
V	volume, cm^3 mol^{-1}
V^*	parameter in HBT correlation, Eq. (5-8.3*d*)
w	weight fraction, Eq. (5-2.1*b*)
x, y	mole fraction, Eq. (5.2-1*a*)
Z	compressibility factor $= PV/RT$
Z_{RA}	Rackett compressibility factor, Eq. 5-8.1

Greek

α	quantity in MHV1 and MHV2 mixing rules $= \Theta/bRT$, Eq. (5-5.13)
δ, ε	EoS variables, Table 4-6
ϕ	volume fraction, Eqs. (5-2.1*c*) and (5-9.3)

Θ	EoS variable, Table 4-6, Eqs. (5-5.2*b*), (5-5.4) and (5-5.5)
μ	dipole moment, Sec. 2-6
ξ	mixture hard-sphere packing fraction, Eq. (5-6.2*b*)
σ	hard sphere diameter, Eq. (5-6.2*b*)
ω	acentric factor, Eq. (2-3.1)

Superscript

0	zero pressure
∞	infinite pressure
BACK	BACK EoS, Eq. (5-6.7)
I	ideal solution, Eq. (5-5.14)
L	liquid
MCSL	Mansoori, et al., hard-sphere EoS, Eq. (5-6.2*a*)
PHCT	Perturbed Hard-Chain EoS, Eq. (5-6.3)
o	pure component property
r	residual property, Eqs. (4-7.2) and (5-6.1)
V	vapor
*	characteristic property

Subscripts

Att	attractive forces, Eqs. (5-6.5) to (5-6.7)
c	critical
cm	mixture pseudocritical
EOS	equation of state result, Eq. (5-5.6)
i	component i
ij	component pair i and j
m	mixture
NC	nonconformal, Eq. (5-5.4)
R	Repulsive, Eq. (5-6.1)
vp$r$	pseudo reduced vapor pressure, Eq. (5-9.3)

REFERENCES

Aalto, M.: *Fluid Phase Equil.,* **141:** 1 (1997).

Aalto, M., K. I. Keskinen, J. Aittamaa, and S. Liukkonen: *Fluid Phase Equil.,* **114:** 1 (1996).

Abbott, M. M.: *Adv. in Chem. Ser.,* **182:** 47 (1979).

Abbott, M. M., and J. M. Prausnitz: *Fluid Phase Equil.,* **37:** 29 (1987).

Akinfiev, N. N.: *Geochem. Internat.,* **35:** 188 (1997).

Aly, G. and I. Ashour: *Fluid Phase Equil.,* **101:** 137 (1994).

Anderko, A: *Fluid Phase Equil.,* **61:** 145 (1990).

Anderko, A: *Fluid Phase Equil.,* **65:** 89 (1991).

Anderko, A, I. G. Economou, and M. D. Donohue: *IEC Res.,* **32:** 245 (1993).

Anisimov, M. A., A. A. Povodyrev, and J. V. Sengers: *Fluid Phase Equil.,* **158-160:** 537 (1999).

Arnaud, J. F., P. Ungerer, E. Behar, B. Moracchini, and J. Sanchez: *Fluid Phase Equil.,* **124:** 177 (1996).

Barner, H. E., and C. W. Quinlan: *IEC Process Des. Dev.,* **9:** 407 (1969).

Boublik, T.: *Mol. Phys.,* **42:** 209 (1981).

Boukouvalas, C., N. Spiliotis, P. Coutsikos, and N. Tzouvaras: *Fluid Phase Equil.*, **92:** 75 (1994).

Brandani, S. and V. Brandani: *Fluid Phase Equil.*, **121:** 179 (1996).

Brulé, M. R., C. T. Lin, L. L. Lee, and K. E. Starling: *AIChE J.*, **28:** 616 (1982).

Campanella, E. A., P. M. Mathias, and J. P. O'Connell: *AIChE J.*, **33:** 2057 (1987).

Carroll, J. J., and A. E. Mather: *Fluid Phase Equil.*, **105:** 221 (1995).

Chapman, W. G., K. E. Gubbins, G. Jackson, and M. Radosz: *Fluid Phase Equil.*, **52:** 31 (1989).

Chapman, W. G., K. E. Gubbins, G. Jackson, and M. Radosz: *IEC Res.*, **29:** 1709 (1990).

Chiew, Y.: *Mol. Phys.*, **70:** 129 (1990).

Cholinski, J., A. Szafranski, and D. Wyrzykowska-Stankiewicz: *Second Virial Coefficients for Organic Compounds*, Thermodynamic Data for Technology, Institute of Physical Chemistry, Polish Academy of Sciences and Insitute of Industrial Chemistry, Warsaw, 1986.

Chueh, P. L., and J. M. Prausnitz: *AIChE J.*, **13:** 896 (1967a).

Chueh, P. L., and J. M. Prausnitz: *Ind. Eng. Chem. Fundam.*, **6:** 492 (1967b).

Cotterman, R. L., and J. M. Prausnitz: *AIChE J.*, **32:** 1799 (1986).

Cotterman, R. L., B. J. Schwarz, and J. M. Prausnitz: *AIChE J.*, **32:** 1787 (1986).

Dahl, S. and M. L. Michelseon: *AIChE. J.*, **36:** 1829 (1990).

Deiters, U.: *Fluid Phase Equil.*, **161:** 205 (1999).

Deiters, U., and K. M. de Reuck: *Pure and Appl. Chem.*, **69:** 1237 (1997).

De Santis, R., and B. Grande: *AIChE J.*, **25:** 931 (1979).

Donaggio, F., M. Fermeglia, and V. deLeeuw: *J. Supercrit. Fluids*, **9:** 207 (1996).

Dymond, J. H., and E. B. Smith: *The Virial Coefficients of Pure Gases and Mixtures; A Critical Compilation,* Clarendon Press, Oxford 1980.

Economou, I. G., and M. D. Donohue: *IEC Res.*, **31:** 1203 (1992).

Eubank, P. T., G-S. Shyu, and N. S. M. Hanif: *IEC Res.*, **34:** 314 (1995).

Feng, W., and W. Wang: *Ind. Eng. Chem.*, **38:** 4966 (1999).

Fermeglia, M., A. Bertucco, and D. Patrizio: *Chem. Eng. Sci.*, **52:** 1517 (1997).

Fischer, K., and J. Gmehling: *Fluid Phase Equil.*, **121:** 185 (1996).

Fleming, P. D., and R. J. Brugman: *AIChE J.*, **33:** 729 (1987).

Gmehling, J., D. D. Liu, and J. M. Prausnitz: *Chem. Eng. Sci.*, **34:** 951 (1979).

Gunn, R. D.: *AIChE J.*, **18:** 183 (1972).

Gupte, P., and T. E. Daubert: *Fluid Phase Equil.*, **59:** 171 (1990).

Hankinson, R. W., and G. H. Thomson: *AIChE J.*, **25:** 653 (1979).

Hayden, J. G., and J. P. O'Connell: *Ind. Eng. Chem. Process Des. Dev.*, **14:** 209 (1975).

Heidemann, R. A.: *Fluid Phase Equil.*, **116:** 454 (1996).

Heidemann, R. A., and S. L. Kokal: *Fluid Phase Equil.*, **56:** 17 (1990).

Hino, T., and J. M. Prausnitz: *Fluid Phase Equil.*, **138:** 105 (1997).

Huang, E. T. S., G. W. Swift, and F. Kurata: *AIChE J.*, **13:** 846 (1967).

Huang, H., and S. I. Sandler: *IEC Res.*, **32:** 1498 (1993).

Huang, Y.-H., and J. P. O'Connell: *Fluid Phase Equil.*, **37:** 75 (1987).

Huron, M.-J., and J. Vidal: *Fluid Phase Equil.*, **3:** 255 (1979).

Iglesias-Silva, G. A., M. S. Mannan, F. Y. Shaikh, and K. R. Hall: *Fluid Phase Equil.*, **161:** 33 (1999).

Jin, G. X., S. Tang, and J. V. Sengers: *Phys. Rev. A*, **47:** 388 (1993).

Joffe, J.: *Ind. Eng. Chem. Fundam.*, **10:** 532 (1971).

Kalospiros, N. S., N. Tzouvaras, P. Coutsikos, and D. P. Tassios: *AIChE J.*, **41:** 928 (1995).

Kay, W. B.: *Ind. Eng. Chem.*, **28:**1014 (1936).

Kiran, E., and J. M. H. Levelt Sengers (eds): *Supercritical Fluids: Fundamentals for Applications,* NATO ASI Series E: **273:** Klewer Academic Publishers, Dordrecht, Holland, 1994.

Kiselev, S. B.: *Fluid Phase Equil.,* **147:** 7 (1998).

Kiselev, S. B. and D. G. Friend: *Fluid Phase Equil.,* **162:** 51 (1999).

Knapp, H., R. Doring, L. Oellrich, U. Plocker, and J. M. Prausnitz: *Vapor-Liquid Equilibria for Mixtures of Low Boiling Substances,* Chem. Data. Ser., Vol. VI, (1982), DECHEMA.

Knudsen, K., E. H. Stenby, and A. Fredenslund: *Fluid Phase Equil.,* **82:** 361 (1993).

Kolar, P. and K. Kojima: *J. Chem. Eng. Japan,* **26:** 166 (1993).

Kolar, P. and K. Kojima: *J. Chem. Eng. Japan,* **27:** 460 (1994).

Koak, N., T. W. de Loos, and R. A. Heidemann: *IEC Res.,* **38:** 1718 (1999).

Lee, M.-J., and J.-T. Chen: *J. Chem. Eng. Japan,* **31:** 518 (1998).

Leibovici, C. F.: *Fluid Phase Equil.,* **84:** 1 (1993).

Leland, T. W., Jr., and P. S. Chappelear: *Ind. Eng. Chem.,* **60**(7): 15 (1968).

Li, C. C.: *Can. J. Chem. Eng.,* **19:** 709 (1971).

Mansoori, G. A., N. F. Carnahan, K. E. Starling, and T. W. Leland, Jr.: *J. Chem. Phys.,* **54:** 1523 (1971).

Mason, E. A., and T. H. Spurling: *The Virial Equation of State,* Pergamon, New York, 1968.

Massucci, M., and C. J. Wormald: *J. Chem. Thermodynamics,* **30:** 919 (1998).

Mathias, P. M.: *Ind. Eng. Chem. Proc. Des. Dev.,* **27:** 385 (1983).

Mathias, P. M., and T. W. Copeman: *Fluid Phase Equil.,* **13:** 91 (1983).

Mathias, P. M., H. C. Klotz, and J. M. Prausnitz: *Fluid Phase Equil.,* **67:** 31 (1991).

McElroy, P. J., and J. Moser: *J. Chem. Thermodynamics,* **27:** 267 (1995).

Michelsen, M. L.: *Fluid Phase Equil.,* **60:** 47, 213 (1990).

Michelsen, M. L., and H. Kistenmacher: *Fluid Phase Equil.,* **58:** 229 (1990).

Michelsen, M. L.: *Fluid Phase Equil.,* **121:** 15 (1996).

Michelsen, M. L., and R. A. Heidemann: *IEC Res.,* **35:** 278 (1996).

Mollerup, J.: *Fluid Phase Equil.,* **7:** 121 (1981).

Mollerup, J.: *Fluid Phase Equil.,* **15:** 189 (1983).

Mollerup, J.: *Fluid Phase Equil.,* **22:** 139 (1985).

Mollerup, J.: *Fluid Phase Equil.,* **25:** 323 (1986).

Mollerup, J. and M. L. Michelsen: *Fluid Phase Equil.,* **74:** 1 (1992).

Muller, A., J. Wenkelmann and J. Fischer: *AIChE J.,* **42:** 1116 (1996).

Orbey, H.: in Kiran, E., and J. M. H. Levelt Sengers, eds: *Supercritical Fluids: Fundamentals for Applications,* NATO ASI Series E, **273:** 177, Klewer Academic Publishers, Dordrecht, Holland, 1994.

Orbey, H., and S. I. Sandler: *Fluid Phase Equil.,* **111:** 53 (1995).

Orbey, H., and S. I. Sandler: *Modeling Vapor-Liquid Equilibria.* Cambridge U. Press, Cambridge, 1998.

Orbey, H., and J. H. Vera: *AIChE J.,* **29:** 107 (1983).

Panagiotopoulos, A. Z., and R. C. Reid: *Adv. in Chem. Ser.,* **300:** 571 (1986).

Patel, N. C., V. Abousky and S. Watanasiri: *Fluid Phase Equil.,* **152:** 219 (1998).

Peng, D. Y., and D. B. Robinson: *Ind. Eng. Chem. Fundam.,* **15:** 59 (1976).

Peneloux, A., E. Rauzy, and R. Freze: *Fluid Phase Equil.,* **8:** 7 (1982).

Ploecker, U., H. Knapp, and J. M. Prausnitz: *IEC Process Des. Dev.,* **17:** 324 (1978).

Prausnitz, J. M., and R. D. Gunn: *AIChE J.,* **4:** 430, 494 (1958).

Prausnitz, J. M., R. N. Lichtenthaler, and E. G. de Azevedo: *Molecular Thermodynamics of Fluid-Phase Equilibria,* 3d ed.," Prentice-Hall, Upper Saddle River, NJ, 1999.

Rainwater, J. C.: in Ely, J. F., and T. J. Bruno (eds): *Supercritical Fluid Technology,* CRC Press, Boca Raton, FL, 1991, p. 57.

Reamer, H. H., and B. H. Sage: *J. Chem. Eng. Data,* **7:** 161 (1962).

Reid, R. C., and T. W. Leland, Jr.: *AIChE J.,* **11:** 228 (1965), **12:** 1227 (1966).

Sandler, S. I.: in Kiran, E., and J. M. H. Levelt Sengers, eds: *Supercritical Fluids: Fundamentals for Applications,* NATO ASI Series E, **273:** 147, Klewer Academic Publishers, Dordrecht, Holland, 1994.

Sandler, S. I., H. Orbey, and B.-I. Lee: in *Models for Thermodynamic and Phase Equilibria Calculations,* S. I. Sandler (ed.), Marcel Dekker, New York, 1994.

Satyro, M. A., and M. A. Trebble: *Fluid Phase Equil.,* **115:** 135 (1996).

Satyro, M. A., and M. A. Trebble: *Fluid Phase Equil.,* **143:** 89 (1998).

Schwartzentruber, J. and H. Renon: *IEC Res.,* **28:** 1049 (1989).

Sengers, J. V., R. F. Kayser, C. J. Peters, and H. J. White, Jr. (eds.): *Equations of State for Fluids and Fluid Mixtures,* Elsevier, Amsterdam, 2000.

Siddiqi, M. A., and K. Lucas: *Fluid Phase Equil.,* **51:** 237 (1989).

Soave, G.: *Chem. Eng. Sci.,* **27:** 1197 (1972).

Soave, G.: *Fluid Phase Equil.,* **82:** 345 (1993).

Song, Y., S. M. Lambert, and J. M. Prausnitz: *Ind. Eng. Chem. Res.,* **33:** 1047 (1994).

Spencer, C. F., and R. P. Danner: *J. Chem. Eng. Data,* **18:** 230 (1973).

Stein, F. P., and E. J. Miller: *Ind. Eng. Chem. Process Des. Dev.,* **19:** 123 (1980).

Thomson, G. H., K. R. Brobst, and R. W. Hankinson: *AIChE J.,* **28:** 671 (1982).

Tochigi, K., P. Kolar, T. Izumi, and K. Kojima: *Fluid Phase Equil.,* **96:** 53 (1994).

Tsonopoulos, C.: *AIChE J.,* **20:** 263 (1974).

Tsonopoulos, C.: *AIChE J.,* **21:** 827 (1975).

Tsonopoulos, C.: *AIChE J.,* **24:** 1112 (1978).

Tsonopoulos, C.: *Adv. in Chem. Ser.,* **182:** 143 (1979).

Tsonopoulos, C., J. H. Dymond, and A. M. Szafranski: *Pure. Appl. Chem.,* **61:** 1387 (1989).

Tsonopoulos, C., and J. H. Dymond: *Fluid Phase Equil.,* **133:** 11 (1997).

Tsonopoulos, C., and J. L. Heidman: *Fluid Phase Equil.,* **57:** 261 (1990).

Twu, C. H., J. E. Coon, A. H. Harvey, and J. R. Cunningham: *IEC Res.,* **35:** 905 (1996).

Twu, C. H., and J. E. Coon: *AIChE J.,* **42:** 32312 (1996).

Twu, C. H., J. E. Coon, and D. Bluck: *IEC Res.,* **37:** 1580 (1998).

Twu, C. H., J. E. Coon, D. Bluck, and B. Tilton: *Fluid Phase Equil.,* **158-160:** 271 (1999).

van der Waals, J. H.: *J. Phys. Chem.,* **5:** 133 (1890).

Van Ness, H. C. and M. M. Abbott: *Classical Thermodynamics of Non Electrolyte Solutions,* McGraw-Hill, New York, 1982.

Vidal, J.: *Fluid Phase Equil.,* **13:** 15 (1983).

Wormald, C. J., and C. J. Snowden: *J. Chem. Thermodynamics,* **29:** 1223 (1997).

Wong, D. S. H., and S. I. Sandler: *AIChE J.,* **38:** 671 (1992).

Wong, D. S. H., H. Orbey, and S. I. Sandler: *IEC Res.,* **31:** 2033 (1992).

Yamada, T., and R. D. Gunn: *J. Chem Eng. Data,* **18:** 234 (1973).

Zabaloy, M. S., and J. H. Vera: *Fluid Phase Equil.,* **119:** 27 (1996).

CHAPTER SIX
THERMODYNAMIC PROPERTIES OF PURE COMPONENTS AND MIXTURES

6-1 SCOPE

In Secs. 6-2 and 6-3, we develop rigorous relations for the enthalpies, entropies, Helmholtz and Gibbs energies, and fugacities that are used with equations of state (EoS) and ideal gas heat capacities to obtain caloric and vapor-liquid equilibrium properties of pure components. In Sec. 6-4, these relations are analyzed for EoS models of Chap. 4 to obtain estimation techniques for such properties. In Sec. 6-5, methods are presented for determining the heat capacities of real gases from ideal gas heat capacities and EoS models, while heat capacities of liquids are treated in Sec. 6-6. Expressions for partial properties and fugacity coefficients of components in mixtures are considered in Sec. 6-7. The true fluid-phase critical properties of mixtures are discussed in Sec. 6-8.

6-2 FUNDAMENTAL THERMODYNAMIC RELATIONSHIPS FOR PURE COMPONENTS

Thermodynamics provides relationships among many useful properties. These include the *PVT* relationship of the EoS for volumetric behavior; the enthalpy (H), internal energy (U), and heat capacity (C_p) used in evaluating energy effects of processes via energy balances; the entropy (S) used in evaluating the properties of reversible processes and in evaluating the consequences of irreversibilities in real processes; and the fugacity (f) used for obtaining vapor-liquid equilibrium conditions. Except for the EoS and the ideal-gas heat capacity (C_p°), the above properties are not directly measurable; they may be called *conceptuals* (O'Connell and Haile, 2000). Their changes can be obtained from experiment by using thermodynamic relations among *measurables* and they can be estimated from models for the EoS and for C_p°.

This section gives the general relations for these properties and shows how they are usually put into the most convenient form for calculations by using *departure functions* based on EoS. Section 6-3 describes general relationships among the departure functions, while Sec. 6-4 shows how to obtain departure function expressions with EoS models such as described in Secs. 4-6 and 4-7.

In addition to not being measurable, it is not possible to determine absolute values of the above conceptual properties. Only differences can be established. Their value is that they are *state properties.* This means that, unlike heat and work effects, changes in their values depend only on the initial and final states. For example, in evaluating the heat for steadily changing the temperature or phase of a pure component, the enthalpy change between the inlet and outlet states depends only on those state conditions, not on the details of the heating or cooling. The consequence of this also allows us to establish calculational techniques for obtaining the changes from a minimum of information and with the use of the most readily accessible models.

The other advantage of these state properties is that many mathematical operations can be done to both interrelate them and to evaluate them. In particular, partial derivatives and integrals are extensively used. Familiarity with such mathematics is useful, but not necessary, in order to fully understand the developments and applications. Since such procedures have been used for so long, the final formulae for the most interesting cases have been well established and can be directly used. However, subtle errors can arise as described in Sec. 5-7 and reliable use of new models requires careful computer programming.

To illustrate path independence and the use of properties in establishing expressions for changes in conceptuals, consider the molar enthalpy change of a pure component. The properties of H allow us to directly integrate the total differential of the enthalpy

$$H_2 - H_1 = \int_{H_1}^{H_2} dH \tag{6-2.1}$$

However, the way we characterize the two different states is by the variables T_1, P_1 and T_2, P_2. This implies that enthalpy is a function only of T and P, $H(T, P)$. This particular choice of variables is for convenience and essentially any two others such as V and C_P could be chosen. Then we use mathematics to obtain dH in terms of changes of T and P.

$$dH = \left(\frac{\partial H}{\partial T}\right)_P dT + \left(\frac{\partial H}{\partial P}\right)_T dP \tag{6-2.2}$$

So

$$H(T_2, P_2) - H(T_1, P_1) = \int_{T_1,P_1}^{T_2,P_2} \left[\left(\frac{\partial H}{\partial T}\right)_P dT + \left(\frac{\partial H}{\partial P}\right)_T dP\right] \tag{6-2.3}$$

The integration can be done in many different ways but all must yield the same expression and, when calculated, the same numerical answer. To illustrate, we choose apparently convenient paths along isobars and isotherms. As Fig. 6-1 shows, there are two possibilities, path ABC or path ADC

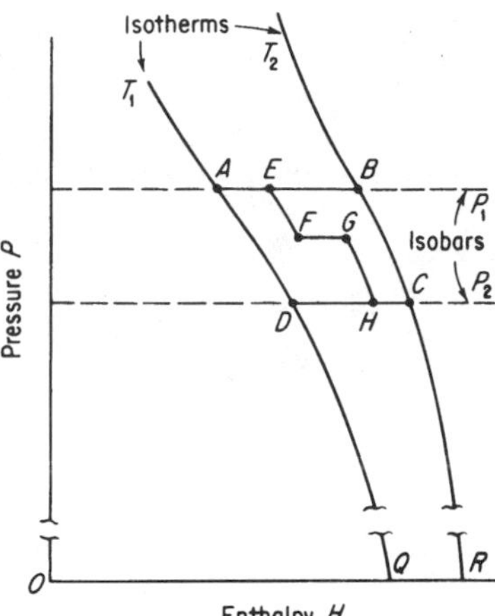

FIGURE 6-1 Schematic diagram of paths to evaluate changes in enthalpy, H, with changes in pressure, P, and temperature, T.

$$
\begin{aligned}
[H(T_2, P_2) - H(T_1, P_1)]_{ABC} &= H(T_2, P_2) - H(T_2, P_1) \\
&\quad + H(T_2, P_1) - H(T_1, P_1) \\
&= \int_{P_1}^{P_2} \left(\frac{\partial H}{\partial P}\right)_{T=T_2} dP + \int_{T_1}^{T_2} \left(\frac{\partial H}{\partial T}\right)_{P=P_1} dT \\
[H(T_2, P_2) - H(T_1, P_1)]_{ADC} &= H(T_2, P_2) - H(T_1, P_2) \\
&\quad + H(T_1, P_2) - H(T_1, P_1) \\
&= \int_{T_1}^{T_2} \left(\frac{\partial H}{\partial T}\right)_{P=P_2} dT + \int_{P_1}^{P_2} \left(\frac{\partial H}{\partial P}\right)_{T=T_1} dP
\end{aligned}
\tag{6-2.4}
$$

Though the individual integrals in Eq. (6-2.4) are not the same, they must yield the same sum. Further, it is possible to use more apparently complicated paths such as *AEFGHC* in Fig. 6-1 and still obtain the desired answer given the variations of the partial derivatives $(\partial H/\partial T)_P$ and $(\partial H/\partial P)_T$ over the states of the chosen path.

It turns out that the most convenient path is an artificial one, chosen because it only requires an EoS that relates P, V, and T as described in Chaps. 4 and 5, and the ideal gas heat capacity, C_p°, described in Chap. 3. This process changes the fluid from interacting molecules at T_1, P_1 to an ideal gas (noninteracting molecules) at T_1, P_1. Then, the ideal gas is changed from T_1, P_1 to T_2, P_2. Finally, the ideal gas is returned to its real fluid state at T_2, P_2 by restoring the intermolecular forces). We choose this path because we know how to evaluate the property changes of ideal gases between any two states and this requires knowledge of only C_p° if $T_1 \neq T_2$. For H but not other conceptuals, this path is equivalent to *ADQRC* in Fig. 6-1.

Thermodynamic manipulations yield the property changes for the molecular transformations in terms of *departure functions.* We define the departure function of a conceptual property, F, as $F^d \equiv F^{ig}(T, P) - F(T, P)$. The distinction between departure functions as defined here and residual functions, F^r, defined in Section 4-7 is that $F^r = F(T, V) - F^{ig}(T, V)$. There are subtle differences between F^d and F^r that will be discussed further in the next section. Here, the enthalpy change of the process *AC* is found from the sum of 1) the departure function, $H^d(T_2, P_2)$, 2) the ideal gas enthalpy change which, because the enthalpy of an ideal gas does not depend upon pressure, is an integral only over temperature, and 3) the departure function, $H^d(T_1, P_1)$

$$\begin{aligned} H(T_2, P_2) - H(T_1, P_1) &= -[H^{ig}(T_2, P_2) - H(T_2, P_2)] \\ &\quad + H^{ig}(T_2, P_2) - H^{ig}(T_1, P_1) \\ &\quad + [H^{ig}(T_1, P_1) - H(T_1, P_1)] \\ &= -\int_0^{P_2} \left(V - T\left(\frac{\partial V}{\partial T}\right)_P \right)_{T=T_2} dP \\ &\quad + \int_{T_1}^{T_2} C_p^\circ \, dT \\ &\quad + \int_0^{P_1} \left(V - T\left(\frac{\partial V}{\partial T}\right)_P \right)_{T=T_1} dP \end{aligned} \tag{6-2.5a}$$

For the entropy, the relation is

$$\begin{aligned} S(T_2, P_2) - S(T_1, P_1) &= -[S^{ig}(T_2, P_2) - S(T_2, P_2)] \\ &\quad + S^{ig}(T_2, P_2) - S^{ig}(T_1, P_1) \\ &\quad + [S^{ig}(T_1, P_1) - S(T_1, P_1)] \\ &= -\int_0^{P_2} \left(\frac{R}{P} - \left(\frac{\partial V}{\partial T}\right)_P \right)_{T=T_2} dP \\ &\quad + \int_{T_1}^{T_2} \frac{C_p^\circ}{T} \, dT - R \ln \frac{P_2}{P_1} \\ &\quad + \int_0^{P_1} \left(\frac{R}{P} - \left(\frac{\partial V}{\partial T}\right)_P \right)_{T=T_1} dP \end{aligned} \tag{6-2.5b}$$

The departure functions need only the EoS for evaluation while the ideal gas change needs only C_p°. Chapter 3 gives methods for estimating C_p°; its temperature dependence may be significant.

The next section describes departure functions for all of the properties of interest, while Section 6-4 gives results for EoS models of Chapters 4 and 5.

6-3 DEPARTURE FUNCTIONS FOR THERMODYNAMIC PROPERTIES

The departure functions of Section 6-2 are widely used for the evaluation of changes in *conceptual* thermodynamic properties such as *H* and *S*. This section gives general expressions for the departure functions for properties of interest in applications based on the EoS models of Chapters 4 and 5. Prausnitz, et al. (1999), O'Connell and Haile (2000), and most textbooks in chemical engineering thermodynamics, derive the expressions given below. They can also be derived from molecular theory, from which model ideas for complex systems often arise.

The principal issue in expressing departure functions is the choice of the variable, P or V, at which the ideal gas property value is to be compared to the real fluid property value. As implied in Chapter 4, EoS models are based on T and V as the independent variables because that is the only way the multiple values of V at phase equilibrium can be obtained. Thus, except for the virial equation, all of the model expressions of Sections 4-6 and 4-7 are of a form where the compressibility factor is $Z = PV/RT = f(T, V, \{y\})$ where $\{y\}$ is the set of composition variables such as mole fractions.

The major consequence of this form is that changes in certain "natural" properties are determined from departure function integrals and then the others are found from algebraic relations. The most important departure function relations are collected in Table 6-1 with the effect of the variables taken into account. Our thermodynamic relations are based on the following partial derivatives:

$$Z = -V\left[\frac{\partial(A/RT)}{\partial V}\right]_{T,\{y\}} \qquad \frac{U}{RT} = -T\left[\frac{\partial(A/RT)}{\partial T}\right]_{V,\{y\}} \tag{6-3.1}$$

Then

$$\begin{aligned} 1 - Z &= V\left[\frac{\partial(A^r/RT)}{\partial V}\right]_{T,\{y\}} \\ \frac{U^r}{RT} &= -T\left[\frac{\partial(A^r/RT)}{\partial T}\right]_{V,\{y\}} \end{aligned} \tag{6-3.2}$$

Also,

$$\frac{A^r}{RT} = -\frac{A^d}{RT} + \ln Z \tag{6-3.3}$$

TABLE 6-1 General Expressions Used to Obtain Departure Functions for Pure Components and Mixtures at Fixed Composition, $F^d = F^{ig}(T, P, \{y\}) - F(T, P, \{y\})$ from EoS Models of the Form $Z = PV/RT = f(T, V, \{y\})$ with $\lim_{V\to\infty} Z = 1$

Property	$F^d = F^{ig}(T, P) - F(T, P)$	Integral* (if used)	Relation (if used)
Compressibility Factor, Z	$1 - Z$	—	$1 - Z$
Internal Energy, U	$\frac{U^{ig} - U}{RT}$	$\int_V^\infty \left[T\left(\frac{\partial Z}{\partial T}\right)_V\right]\frac{dV}{V}$	—
Enthalpy, H	$\frac{H^{ig} - H}{RT}$	—	$\frac{U^{ig} - U}{RT} + 1 - Z$
Entropy, S	$\frac{S^{ig} - S}{R}$	—	$\frac{U^{ig} - U}{RT} - \frac{A^{ig} - A}{RT}$
Helmholtz Energy, A	$\frac{A^{ig} - A}{RT}$	$\int_V^\infty [1 - Z]\frac{dV}{V} + \ln Z$	—
Gibbs Energy, G	$\frac{G^{ig} - G}{RT}$	—	$\frac{A^{ig} - A}{RT} + 1 - Z$
Fugacity, f	$\ln\left(\frac{f}{P}\right)$	—	$-\frac{G^{ig} - G}{RT}$

*T and $\{y\}$ are held constant in the integrations.

Note the relationship implied by the table entries:

$$\ln\left(\frac{f}{P}\right) = -\frac{G^{ig} - G}{RT} = -\frac{A^{ig} - A}{RT} - (1 - Z) \qquad (6\text{-}3.4)$$

Explicit departure function expressions for important EoS models described in Chapter 4 will be given in the next section.

Corresponding States Principle (CSP)

Just as volumetric properties can be determined from CSP as described in Section 4-3, so also can departure functions. It has been common for authors of CSP correlations to give charts, tables and/or equations for Z, $H^d/RT^* = (H^{ig} - H)/RT^*$ (or $(H^{ig} - H)/T^*$ with dimensions), $S^d/R = (S^{ig} - S)/R$ (or $S^{ig} - S$ with dimensions), and $\ln(f/P)$ in terms of reduced temperature, T/T^*, and pressure, P/P^* and other characteristics. All of these can be conveniently used when CSP is applicable but care should be exercised (see Section 4-3). The 4th Edition provided extensive tables based on the Lee-Kesler (1975) formulation where $T^* = T_c$, $P^* = P_c$, and the acentric factor, ω, (Section 2-3) is used. These tables are also reproduced in Smith, et al. (1996). It should be noted (Cuthbert and Downey, 1999) that there are errors in some equations but not in the table values for these properties in these references. A typical formulation equivalent to Eq. (4-3.1) is

$$\frac{H^d}{RT_c} = \frac{H^{ig} - H}{RT_c} = \left(\frac{H^{ig} - H}{RT_c}\right)^{(0)} + \omega\left(\frac{H^{ig} - H}{RT_c}\right)^{(1)} \qquad (6\text{-}3.5)$$

where tables or graphs of the values of $((H^{ig} - H)/RT_c)^{(0)}$ and $((H^{ig} - H)/RT_c)^{(1)}$ are given for specified values of $T_r = T/T_c$ and $P_r = P/P_c$. The equivalent of the two-reference approach of Eq. (4-4.3) is

$$\left(\frac{H^d}{RT_c}\right)(T_r, P_r, \omega) = \left(\frac{H^{ig} - H}{RT_c}\right)^{(R1)}(T_r, P_r, \omega^{(R1)}) + \frac{\omega - \omega^{(R1)}}{\omega^{(R2)} - \omega^{(R1)}} \left[\left(\frac{H^{ig} - H}{RT_c}\right)^{(R2)}(T_r, P_r, \omega^{(R2)}) - \left(\frac{H^{ig} - H}{RT_c}\right)^{(R1)}(T_r, P_r, \omega^{(R1)})\right] \qquad (6\text{-}3.6)$$

All of the opportunities and limitations of CSP expressed in Section 4-3 for volumetric properties apply to obtaining departure functions. Also, all of the mixing rules for CSP described in Section 5-3 apply to mixture departure functions. A calculation of CSP for departure functions is given in Example 6-1.

6-4 EVALUATION OF DEPARTURE FUNCTIONS FOR EQUATIONS OF STATE

The departure functions of Table 6-1 can be evaluated from data or equations that express Z as a function of T and V at fixed composition. In this section, the expressions are given for the equation of state models of Sections 4-5 to 4-7. Virial and analytic EoS models are commonly expressed as $Z(T, V)$ so the integrals of Table 6-1 are used. Most nonanalytic models are expressed as $A^r(T, V)/RT$, so the derivatives of the form of Eq. (6-3.2) are used rather than Table 6-1.

Properties from the Virial EoS

The form of Z for the virial EoS is given in Eqs. (4-5.1).

$$Z = 1 + B\left(\frac{P}{RT}\right) + (C - B^2)\left(\frac{P}{RT}\right)^2 + \ldots \tag{4-5.1a}$$

$$= 1 + \frac{B}{V} + \frac{C}{V^2} + \ldots \tag{4-5.1b}$$

This is the only case of EoS models for gases in which it can be useful to choose the independent variables T and P as in Eq. (4-5.1a) rather than T and V as in Table 6-1. Thus, Table 6-2 tabulates the departure functions for both forms of the EoS of Eqs. (4-5.1).

For departure functions, the limitations of applying virial coefficient models are the same as given in Table 4-4. A calculation of departure functions from the virial EoS is given in Example 6-1 using the correlation of Tsonopoulos (1974).

Analytic EoS Models

The general form of Z for cubic EoS models is written in Eq. (4-6.2)

$$Z = \frac{V}{V - b} - \frac{(\Theta/RT)V(V - \eta)}{(V - b)(V^2 + \delta V + \varepsilon)} \tag{4-6.2}$$

The models given in Table 4-6 use $\eta = b$, so the general form which can be directly integrated is

$$Z = \frac{V}{V - b} - \frac{(\Theta/RT)V}{(V^2 + \delta V + \varepsilon)} \tag{6-4.1}$$

Table 6-3 shows results for the cubic forms that have T dependence of only $\Theta(T) = a\alpha(T)$. The expressions for $T(d\alpha/dT)$ from the models given in Table 4-7 are given in Table 6-4. If the other parameters b, δ, and ε depend on T, additional temperature derivative terms arise. Caution should be exercised when this is done because extrapolation of such variations to high temperatures can lead to negative heat capacities for both pure components (Salim and Trebble, 1991) and mixtures (Satyro and Trebble, 1996).

Excess Properties and EoS Mixing Rules

For mixtures of gases and liquids, if possible, departure functions are calculated directly from EoS models. As described in Sec. 5-2, obtaining reliable results for liquid properties requires that the mixing rules for the EoS parameters describe their strong composition dependence. As indicated in Sec. 5-5, this is often accomplished by matching EoS results to G^E correlations. The same idea can be used to match other excess properties to mixture departure functions over the composition range. For example, Orbey and Sandler (1996) compared various cubic EoS mixing rules for describing G^E and H^E simultaneously. The success was mixed; no parameter sets yielded optimal descriptions of both properties and extrapolation over ranges of conditions were generally unsuccessful even when the model parameters were allowed to be temperature-dependent. Also, Satyro and Trebble (1996) showed

TABLE 6-2 Departure Functions F^d for Pure Components and Mixtures at Fixed Composition from Virial Equations of State Eqs. (4-5.1)

Property	Departure function	$Z(T, P)$, Eq. (4-5.1a)
Compressibility Factor, Z	$1 - Z$	$-B\left(\frac{P}{RT}\right) - (C - B^2)\left(\frac{P}{RT}\right)^2 - \ldots$
Internal Energy, U	$\frac{U^{ig} - U}{RT}$	$\left(\frac{P}{RT}\right)\left(T\frac{dB}{dT}\right) + \left(\frac{P}{RT}\right)^2\left(\frac{T}{2}\frac{dC}{dT} - B\frac{dB}{dT}\right) + \ldots$
Enthalpy, H	$\frac{H^{ig} - H}{RT}$	$\left(\frac{P}{RT}\right)\left(B - T\frac{dB}{dT}\right) - \left(\frac{P}{RT}\right)^2\left(C - \frac{T}{2}\frac{dC}{dT} - B^2 + B\frac{dD}{dT}\right) + \ldots$
Entropy, S	$\frac{S^{ig} - S}{R}$	$\frac{dB}{dT}\left(\frac{P}{R}\right) - \frac{1}{2}\left[C - T\frac{dC}{dT} - \left(B^2 - 2BT\frac{dB}{dT}\right)\right]\left(\frac{P}{RT}\right)^2 + \ldots$
Helmholtz Energy, A	$\frac{A^{ig} - A}{RT}$	$\frac{(C - B^2)}{2}\left(\frac{P}{RT}\right)^2 + \ldots$
Gibbs Energy, G	$\frac{G^{ig} - G}{RT}$	$-B\left(\frac{P}{RT}\right) - \frac{(C - B^2)}{2}\left(\frac{P}{RT}\right)^2 - \ldots$
Fugacity, f	$\ln\left(\frac{f}{P}\right)$	$B\left(\frac{P}{RT}\right) + \frac{(C - B^2)}{2}\left(\frac{P}{RT}\right)^2 + \ldots$
Property	**Departure function**	**$Z(T, V)$, Eq. (4-5.1b)**
Compressibility Factor, Z	$1 - Z$	$-\left(\frac{B}{V}\right) - \frac{C}{V^2} - \ldots$
Internal Energy, U	$\frac{U^{ig} - U}{RT}$	$\frac{T}{V}\frac{dB}{dT} + \frac{T}{V^2}\frac{dC}{dT} + \ldots$
Enthalpy, H	$\frac{H^{ig} - H}{RT}$	$-\frac{\left(B - T\frac{dB}{dT}\right)}{V} - \frac{\left(2C - T\frac{dC}{dT}\right)}{2V^2} - \ldots$
Entropy, S	$\frac{S^{ig} - S}{R}$	$\frac{\left(B + T\frac{dB}{dT}\right)}{V} + \frac{\left(C - T\frac{dC}{dT}\right)}{2V^2} + \ldots -\ln Z$
Helmholtz Energy, A	$\frac{A^{ig} - A}{RT}$	$-\frac{B}{V} - \frac{1}{2}\frac{C}{V^2} - \ldots + \ln Z$
Gibbs Energy, G	$\frac{G^{ig} - G}{RT}$	$-2\frac{B}{V} - \frac{3}{2}\frac{C}{V^2} - \ldots + \ln Z$
Fugacity, f	$\ln\left(\frac{f}{P}\right)$	$2\frac{B}{V} + \frac{3}{2}\frac{C}{V^2} + \ldots -\ln Z$

that the Wong-Sandler (1992) mixing rules can lead to anomalous behavior of excess properties if only the k_{ij} parameter of Eq. (5-5.11) is fitted to established G^E values. There have been many publications by Djordjevic and coworkers (see Djordjevic, et al., 1999; Kijevcanin, et al., 1998 and earlier papers) treating excess properties with cubic EoS. In this work, several properties were fitted to obtain parameters for mixing and combining rules (see Sec. 5-5) of both the van der Waals and G^E model forms. They conclude, for example, (Djordjevic, et al., 1999) that vapor-liquid equilibria, H^E, and C_p^E of extremely nonideal systems can be correlated

TABLE 6-3 Some Constant Pressure Departure Function Expressions for Cubic EoS Models. See Table 6-1 and Eqs. (6-3.1) to (6-3.3) for Other Departure Functions

Property	Departure Function	$Z(T, V) = \dfrac{V}{V-b} - \dfrac{(\Theta/RT)V}{(V^2 + \delta V + \varepsilon)}$, Eq. (6-4.1)*
Compressibility Factor, Z	$1 - Z$	$-\dfrac{b}{V-b} + \dfrac{(\Theta/RT)V}{V^2 + \delta V + e}$
Enthalpy, H	$\dfrac{H^{ig} - H}{RT}$	$-\dfrac{(\Theta + Td\Theta/dT))}{RT(\delta^2 - 4\varepsilon)^{1/2}} \ln \left[\dfrac{2V + \delta - (\delta^2 - 4\varepsilon)^{1/2}}{2V + \delta + (\delta^2 - 4\varepsilon)^{1/2}}\right] + 1 - Z$
Entropy, S	$\dfrac{S^{ig} - S}{R}$	$\dfrac{d\Theta/dT}{R(\delta^2 - 4\varepsilon)^{1/2}} \ln \left\{\dfrac{2V + \delta - (\delta^2 - 4\varepsilon)^{1/2}}{2V + \delta + (\delta^2 - 4\varepsilon)^{1/2}}\right\} - \ln[Z(1 - b/V)]$
Helmholtz Energy, A	$\dfrac{A^{ig} - A}{RT}$	$-\dfrac{\Theta}{RT(\delta^2 - 4\varepsilon)^{1/2}} \ln \left\{\dfrac{2V + \delta - (\delta^2 - 4\varepsilon)^{1/2}}{2V + \delta + (\delta^2 - 4\varepsilon)^{1/2}}\right\} + \ln[Z(1 - b/V)]$
Fugacity, f	$\ln\left(\dfrac{f}{P}\right)$	$\dfrac{\Theta}{RT(\delta^2 - 4\varepsilon)^{1/2}} \ln \left\{\dfrac{2V + \delta - (\delta^2 - 4\varepsilon)^{1/2}}{2V + \delta + (\delta^2 - 4\varepsilon)^{1/2}}\right\} - \ln[Z(1 - b/V)] - (1 - Z)$

*If $\delta^2 - 4\varepsilon = 0$, terms with Θ and $Td\Theta/dT$ have $-\dfrac{1}{V + \delta/2}$ instead of $\dfrac{1}{(\delta^2 - 4\varepsilon)^{1/2}} \ln \left\{\dfrac{2V + \delta - (\delta^2 - 4\varepsilon)^{1/2}}{2V + \delta + (\delta^2 - 4\varepsilon)^{1/2}}\right\}$.

satisfactorily if the Wong-Sandler (1992) mixing rules are used with fitted parameters from the NRTL model for G^E (Table 8-3) that are linear in T (a total of six constant parameters). However they did not attempt to predict any multicomponent systems, nor correlate any behavior near critical points.

Example 6-1 Compute the differences between properties for propane between a very low pressure and the conditions of Example 4-3 given by the virial equation, CSP, and all cubic equations for which parameters are known. Compare with the following values from the NIST Webbook:

Property/state	1. Saturated vapor	2. Saturated liquid	3. Critical fluid
T, K	300	300	369.85
P, bar	9.9742	9.9742	42.477
Z	0.8143	0.0360	0.2763
$\dfrac{\Delta H}{RT} = \dfrac{H_i - H(T, P = 0.001)}{RT}$	−0.6046	−6.4736	−2.9568
H^d/RT	0.6046	6.4736	2.9568
$\dfrac{\Delta S}{R} = \dfrac{S_i - S(T, P = 0.001)}{R}$	−9.6441	−15.5131	−13.1997
S^d/RT	0.4364	6.3053	2.5429
$\dfrac{\Delta A}{RT} = \dfrac{A_i - A(T, P = 0.001)}{RT}$	9.2251	10.0035	10.9666
A^d/RT	−0.0174	−0.7958	−0.3099
$\ln\left(\dfrac{f}{P}\right)$	−0.1682	−0.1682	−0.4138

TABLE 6-4 Temperature Derivatives of $\alpha = \Theta/a$ of Some Cubic EoS*

EoS	$T(d\alpha/dT)$
van der Waals (1890)	0
Redlich & Kwong (1949)	$-1/(2T_r^{1/2})$
Wilson (1964)	$[1 - f_\omega]T_r \quad f_\omega = 1.57 + 1.62\omega$
SP-R* (1972–85)	$-[1 + f_\omega(1 - T_r^{1/2})]f_\omega T_r^{1/2}$
Martin (1979)	$-n/T_r^n$
Soave (1979)	$-(mT_r + n/T_r)$
Mathias (1983)	$-[1 + f_\omega(1 - T_r^{1/2})][f_\omega T_r^{1/2} + p(1.4T_r - 4T_r^2)]$ $f_\omega = 0.48508 + 1.55191\omega - 0.15613\omega^2$
Mathias and Copeman (1983); Mathias, et al. (1989)	$-[1 + c_1(1 - T_r^{1/2}) + c_2(1 - T_r^{1/2})^2 + c_3(1 - T_r^{1/2})^3]$ $\times [c_1 + 2c_2(1 - T_r^{1/2}) + 3c_3(1 - T_r^{1/2})^2]T_r^{1/2}$
Soave (1984)	$-(mT_r + n/T_r)$
Stryjek and Vera (1986)	$-[1 + f_\omega(1 - T_r^{1/2})[f_\omega T_r^{1/2} + \kappa_1(1.4T_r - 4T_r^2)]$ $f_\omega = 0.378893 + 1.4897153\omega - 0.17131848\omega^2 + 0.0196554\omega^3$
Trebble and Bishnoi (1987)	$-q_1T_r \exp[q_1(1 - T_r)]$
Twu, et al. (1992)	$[N(M - 1)T_r^{-N} - LNM]T_r^{MN} \exp[L(1 - T_r^{NM})]$
Soave (1993)	$-n(1 - T_r^{1/2})T_r^{1/2} - mT_r$
Soave (1993)	$-(2.756f_\omega - 0.7)(1 - T_r^{1/2})T_r^{1/2} - f_\omega T_r$ $f_\omega = 0.484 + 1.515\omega - 0.44\omega^2$
Twu, et al. (1995)	$-[0.171813T_r^{-0.171813} + 0.222545T_r^{1.77634}] \exp[0.125283(1 - T_r^{1.77634})]$ $-\omega\{[0.607352T_r^{-0.607352} + 1.12820T_r^{2.20517}] \exp[0.511614(1 - T_r^{2.20517})]$ $-[0.171813T_r^{-0.171813} + 0.222545T_r^{1.77634}] \exp[0.125283(1 - T_r^{1.77634})]\}$
Stamateris and Olivera-Fuentes (1995)	mT_r^{1-n}
Patel (1996)	$c_1T_r + 0.5c_2T_r^{1/2} + Nc_3T_r^N$
Zabaloy and Vera (1998)	$C_1(T_r \ln T_r + 1) + C_2T_r + 2C_3T_r^2$

*The general form of the Soave/Peng-Robinson (SP-R) attractive parameter is $\Theta = a[1 + f_\omega(1 - T_r^{1/2})]^2$ where f_ω can be found by comparison with the expressions in Table 4-7. For these models, the final expression is $T(d\alpha/dT) = -[1 + f_\omega(1 - T_r^{1/2})]f_\omega T_r^{1/2}$. This form is used in the models of Soave (1972), Fuller (1976), Peng and Robinson (1976), Patel and Teja (1982), Peneloux, et al. (1982), Adachie, et al. (1983), Soave (1984), Adachie, et al. (1985).

Example 6-1 (cont.) The low pressure of 0.001 bar was chosen so that the departure function for State 1 in Eq. (6-2.5) is zero and then, *e.g.*, $H^d/RT = -(\Delta H/RT)$ and $A^d/RT = -(\Delta A/RT) + \ln(P_2/0.001)$.

solution The expressions for the second virial coefficient from the Tsonopoulos (1974) model can be obtained by the operations suggested in Table 6-2. This model is only applied to the saturated vapor since it cannot be used for liquids or at high pressures. The CSP results use Tables 5-3 to 5-7 of the 4th Edition. The calculations for the cubic EoS models use the expressions of Tables 6-3 and 6-4.

Since most of the newer models have been optimized for vapor pressure and perhaps liquid densities, the results show the best agreement with ln (f/P) and Z. For the enthalpy and entropy departure functions, many have unacceptable deviations, especially for the saturated vapor. This is consistent with the work of Kumar, et al. (1999) who showed that gas compressor efficiencies for the same inlet and outlet conditions can vary several percent depending upon the EoS model used.

STATE 1 Saturated Vapor at $T = 300$ K*

EoS Model	Z	% Err	$\frac{H^d}{RT}$	% Err	$\frac{S^d}{R}$	% Err	$\frac{A^d}{RT}$	% Err$^+$	$\ln\left(\frac{f}{P}\right)$	Abs. Err. ×100
Virial, Eq. (4-5.1a)$^+$	0.8438	3.6	0.4340	−28.2	0.2777	−36.4	0.0000	−100.0	−0.1562	7.1
Virial, Eq. (4-5.1b)$^+$	0.8100	−1.0	0.5383	−11.0	0.3661	−16.1	−0.0216	24.2	−0.1722	−2.4
van der Waals (1890)	0.8704	−6.9	0.3026	50.0	0.1813	58.5	−0.0083	52.3	−0.1213	−4.7
Redlich and Kwong (1949)	0.8338	−2.4	0.5666	6.3	0.4125	5.5	−0.0122	30.0	−0.1541	−1.4
Wilson (1964)	0.8240	−1.2	0.5074	16.1	0.3454	20.9	−0.0140	19.5	−0.1620	−0.6
Soave (1972)	0.8256	−1.4	0.5691	5.9	0.4084	6.4	−0.0137	21.3	−0.1608	−0.8
Fuller (1976)	0.8507	−4.5	0.4279	29.2	0.2879	34.0	−0.0093	46.3	−0.1400	−2.8
Peng and Robinson (1976)	0.8151	−0.1	0.5161	14.6	0.3447	21.0	−0.0135	22.5	−0.1714	0.3
Patel and Teja (1982)	0.8188	−0.6	0.5146	14.9	0.3470	20.5	−0.0136	21.8	−0.1676	−0.1
Patel and Teja (1982)	0.8195	−0.6	0.5124	15.3	0.3455	20.8	−0.0135	22.1	−0.1670	−0.1
Peneloux, et al. (1982)	0.8241	−1.2	0.5098	15.7	0.3472	20.4	−0.0133	23.3	−0.1626	−0.6
Soave (1984)	0.8261	−1.4	0.5090	15.8	0.3488	20.1	−0.0137	21.0	−0.1602	−0.8
Adachi, et al. (1985)	0.8198	−0.7	0.5155	14.8	0.3489	20.0	−0.0137	21.4	−0.1665	−0.2
Stryjek and Vera (1986)	0.8155	−0.1	0.6523	−7.9	0.4812	−10.3	−0.0134	22.9	−0.1711	0.3
Trebble and Bishnoi (1987)	0.8097	0.6	0.5100	15.6	0.3328	23.7	−0.0131	24.7	−0.1773	0.9
Twu, et al. (1992)	0.8102	0.5	0.5170	14.5	0.3405	22.0	−0.0132	23.8	−0.1765	0.8
Twu, et al. (1995)	0.8066	0.9	0.5253	13.1	0.3471	20.5	−0.0151	12.8	−0.1782	1.0
Stamateris and Olivera-Fuentes (1995)	0.7711	5.3	0.9559	−58.1	0.7569	−73.4	−0.0299	−71.9	−0.1990	3.1

*Calculated with Expressions in Tables 4-6 to 4-8 and 6-2 to 6-4.

$^+$Note that the virial expressions are at least as accurate as most of the EoS models with Eq. (4-5.1b) being better.

STATE 2 Saturated Liquid at T = 300 K*

EoS Model	Z	% Err	$\frac{H^d}{RT}$	% Err	$\frac{S^d}{R}$	% Err	$\frac{A^d}{RT}$	% Err	$\ln\left(\frac{f}{P}\right)$	Abs. Err. ×100
CSP, Eq. (6-6.3)	0.0366	1.6	6.4910	0.3	6.2698	−0.6	−0.8092	1.7	−0.1543	8.3
van der Waals (1890)	0.0582	−61.5	3.5309	45.5	3.8182	39.4	−1.2291	−54.5	0.2873	−45.5
Redlich and Kwong (1949)	0.0405	−12.6	7.4642	−15.3	7.4215	−17.7	−0.9168	−15.2	−0.0427	−12.6
Wilson (1964)	0.0392	−8.8	6.4299	0.7	6.2567	0.8	−0.7876	1.0	−0.1732	0.5
Soave (1972)	0.0394	−9.3	7.4606	−15.2	7.3083	−15.9	−0.8083	−1.6	−0.1523	−1.6
Fuller (1976)	0.0312	13.3	6.4686	0.1	6.4568	−2.4	−0.9570	−20.3	−0.0118	−15.6
Peng and Robinson (1976)	0.0347	3.7	6.4336	0.6	6.2620	0.7	−0.7936	0.3	−0.1717	0.3
Patel and Teja (1982)	0.0364	−1.0	6.4634	0.2	6.2968	0.1	−0.7970	−0.2	−0.1666	−0.2
Patel and Teja (1982)	0.0366	−1.5	6.4352	0.6	6.2735	0.5	−0.8017	−0.7	−0.1617	−0.6
Peneloux, et al. (1982)	0.0377	−4.6	6.3703	1.6	6.2522	0.8	−0.8442	−6.1	−0.1181	−5.0
Soave (1984)	0.0388	−7.8	6.4790	−0.1	6.3290	−0.4	−0.8111	−1.9	−0.1501	−1.8
Adachi, et al. (1985)	0.0372	−3.1	6.4972	−0.4	6.3279	−0.4	−0.7936	0.3	−0.1692	0.1
Stryjek and Vera (1986)	0.0347	3.6	8.6836	−34.1	8.5174	−35.1	−0.7990	−0.4	−0.1662	−0.2
Trebble and Bishnoi (1987)	0.0358	0.7	6.3041	2.6	6.1143	3.0	−0.7744	2.7	−0.1898	2.2
Twu, et al. (1992)	0.0351	2.5	6.3889	1.3	6.2042	1.6	−0.7802	2.0	−0.1846	1.6
Twu, et al. (1995)	0.0338	6.1	6.4986	−0.4	6.2139	1.4	−0.6815	14.4	−0.2847	11.6
Stamateris and Olivera-Fuentes (1995)	0.0459	−27.5	13.1612	−103.3	12.1688	−93.0	0.0383	104.8	−0.9924	82.4

*Calculated with Expressions in Tables 4-6 to 4-8, 6-3, 6-4 and in Tables 5-2 to 5-7 of the 4th Edition.

STATE 3 Critical Point Fluid at T = 369.85 K, P = 42.477 bar*

EoS Model	Z	% Err	$\frac{H^d}{RT}$	% Err$^+$	$\frac{S^d}{R}$	% Err	$\frac{A^d}{RT}$	% Err$^+$	$\ln\left(\frac{f}{P}\right)$	Abs. Err. ×100
CSP, Eq. (6-3.3)	0.2767	0.10	2.9605	0.1	2.5436	0	−0.32	−2.6	0.4053	2.1
van der Waals (1890)	0.3644	−31.9	2.0630	30.2	1.4297	43.8	−0.0023	99.2	−0.6333	21.9
Redlich and Kwong (1949)	0.3335	−20.7	2.9457	0.4	2.5386	0.4	−0.2595	16.3	−0.4071	0.7
Wilson (1964)	0.3335	−20.7	2.7370	7.4	2.3299	21.3	−0.2595	16.3	−0.4071	0.7
Soave (1972)	0.3335	−20.7	2.9747	−0.6	2.5676	−2.5	−0.2595	16.3	−0.4071	0.7
Fuller (1976)	0.2763	0.0	2.7236	7.9	2.2506	29.2	−0.2507	19.1	−0.4731	5.9
Peng and Robinson (1976)	0.3142	−13.7	2.6077	11.8	2.1655	37.7	−0.2436	21.4	−0.4422	2.8
Patel and Teja (1982)	0.3170	−14.8	2.6442	10.6	2.2152	32.8	−0.2539	18.1	−0.4290	1.5
Patel and Teja (1982)	0.3179	−15.1	2.6354	10.9	2.2075	33.5	−0.2542	18.0	−0.4279	1.4
Peneloux, et al. (1982)	0.3135	−13.5	2.6959	8.8	2.2905	25.2	−0.2812	9.3	−0.4054	0.8
Soave (1984)	0.3302	−19.5	3.1010	−4.9	2.4314	11.2	−0.0003	99.9	−0.6696	25.6
Adachi, et al. (1985)	0.2989	−8.2	2.8483	3.7	2.4236	11.9	−0.2764	10.8	−0.4246	1.1
Stryjek and Vera (1986)	0.3112	−12.6	3.4453	−16.5	3.0031	−46.0	−0.2467	20.4	−0.4422	2.8
Trebble and Bishnoi (1987)	0.2997	−8.5	2.6543	10.2	2.1920	35.1	−0.2379	23.2	−0.4624	4.9
Twu, et al. (1992)	0.2999	−8.6	2.6430	10.6	2.1843	35.9	−0.2414	22.1	−0.4587	4.5
Twu, et al. (1995)	0.3198	−15.8	2.5371	14.2	2.0950	44.8	−0.2380	23.2	−0.4422	2.8
Stamateris and Olivera-Fuentes (1995)	0.3644	−31.9	4.6017	−55.6	3.9684	−142.5	−0.0023	99.2	−0.6333	21.9

*Calculated with Expressions in Tables 4-6 to 4-8, 6-3, 6-4 and in Tables 5-2 to 5-7 of the 4th Edition.
$^+$Note that there can be much larger errors in H^d/RT and S^d/RT for the EoS models than in $\ln(f/P)$.

Nonanalytic EoS Models

The thermodynamic analysis used for the above equations is also used to obtain departure functions for more complex EoS models such as from MBWR and perturbation theories (see Secs. 4-7 and 5-6). Typically, these expressions are developed in terms of the Helmholtz departure function, A^d/RT or the residual Helmholtz function, A^r/RT, rather than the compressibility factor, Z, as, for example, Eqs. (4-7.2) and (4-7.4). Z is then obtained from Eq. (6-3.2).

$$Z = 1 - V\left(\frac{\partial(A^r/RT)}{\partial V}\right)_{T,\{y\}} \tag{6-4.2}$$

where, if the model is for a mixture, the $_{\{y\}}$ indicates that composition is also to be held constant when taking the derivative. The advantage of expressing the model in this way is that from the expressions derived, comparisons with many different data can be done and parameters of the model can be fitted to the data directly. As pointed out by Gregorowicz, et al. (1996), ignoring the consequences of departure functions when establishing EoS models can lead to errors if the models are extended with thermodynamic manipulations without recognizing that taking derivatives can exacerbate model limitations.

Often, such as in the case of Wagner models (Setzmann and Wagner, 1989; 1991), the ideal gas Helmholtz function, A^{ig}/RT, is also developed to give expressions for all ideal gas properties including C_p° (see Sec. 3-1). There is a very useful table in Setzmann and Wagner (1991) that lists all of the general temperature and density derivatives of A^r/RT and A^{ig}/RT to obtain thermodynamic properties of interest, including Z and C_p°. These can be used for any EoS model formulated in terms of A^r/RT and any ideal gas correlation expressed as A^{ig}/RT.

Chemical theory models have also been used with some success in systems of very strong interactions such as carboxylic acids. For example, Nagy, et al. (1987) successfully correlated enthalpies for *p*-dioxane systems with solvating substances with the EoS model of Gmehling, et al. (1979).

Discussion and Recommendations

Here we summarize our recommendations about thermodynamic properties of fluids from EoS models. In general the possibilities and limitations are similar to those described in Secs. 4-8 and 5-7.

Departure Functions for Gases and Gas Mixtures. In the zero-pressure or infinite-volume limit, all models must give zero for the departure functions. In the critical region, only those equations that give nonclassical behavior can be satisfactory. The primary differences among the myriad of forms are computational complexity and quality of the results at high pressures, for liquids, for polar and associating substances and for polymers. While equations of state were previously limited to vapor phase properties, they now are commonly applied to the liquid phase as well. The most desirable EoS expressions give the *PVT* behavior and all other property values for vapors and liquids of pure components and mixtures while being as simple as possible computationally. Of course, since not all of these constraints can be satisfied simultaneously, deciding which model to use requires judgment and optimization among the possibilities.

For hydrocarbons and hydrocarbon gas mixtures (including light gases such as N_2, CO_2, and H_2S) calculate $(H^{ig} - H)$ from the Soave, Peng-Robinson, or Lee-Kesler equation. Errors should be less than 4 J g^{-1} (Dillard, et al., 1968; Peng and Robinson, 1976; Tarakad and Daubert, 1974; West and Erbar, 1973).

The truncated virial equations, Eq. (4-5.1) are simple but can be used only for the vapor phase. Temperatures and pressures for which this condition applies are given in Table 4-4 and generally in the regions of Figs. 4-1 to 4-3 for which V_{r_i} is greater than about 0.5.

Cubic EoS have often been chosen as the optimal forms because the accuracy is adequate and the analytic solution for the phase densities is not too demanding. For example, the EoS model of Mathias and Copeman (1983) has been used to accurately describe the thermal (Mathias and Stein, 1984) and phase (Mathias, et al., 1984) behavior of synthetic coal fluids from the SRC-I process. Marruffo and Stein (1991) used the Soave EoS (1972) with temperature independent binary parameters to describe the thermal properties of CF_4—CHF_3 and N_2—CF_4—CHF_3 mixtures. However, higher accuracy can normally be obtained only from correlating experimental data directly, from nonanalytic EoS as in Secs. 4-7 and 5-6, and, for liquids, from methods given in Secs. 4-10 to 4-12 and 5-8 and 5-9. At very high pressures, the correlation of Breedveld and Prausnitz (1973) can be used.

When selecting a cubic EoS for *PVT* properties, users should first evaluate what errors they will accept for the substances and conditions of interest, as well as the effort it would take to obtain parameter values if they are not available in the literature. Sometimes this takes as much effort as implementing a more complex, but accurate, model such as a nonanalytic form.

No EoS models should be extrapolated outside the temperature and pressure range for which it has been tested. Within their ranges however, they can be both accurate and used for many properties. Unlike what was presented in the 4th Edition, there are now both cubic and other EoS models that can be used to predict with confidence the *PVT* behavior of polar molecules. Complex substances require more than three parameters, but when these are obtained from critical properties and measured liquid volumes and vapor pressures, good agreement can now be obtained.

To characterize small deviations from ideal gas behavior, use the truncated virial equation with either the second alone or the second and third coefficients, B and C, Eq. (4-5.1). Do not use the virial equation for liquid phases.

For normal fluids, use either CSP tables such as those from Lee and Kesler (1975) or a generalized cubic EoS with volume translation. The results shown in Example 6-1 are representative of what can be expected. All models give equivalent and reliable results for saturated vapors except for dimerizing substances.

For polar and associating substances, use a method based on four or more parameters. Cubic equations with volume translation can be quite satisfactory for small molecules, though perturbation expressions are usually needed for polymers and chemical models for carboxylic acid vapors.

If one wishes to calculate property changes for liquids at low reduced temperatures, the best choice may be to obtain C_{pL} from a method in the following section and integrating

$$H_2 - H_1 = \int_{T_1}^{T_2} C_{pL}\, dT \qquad S_2 - S_1 = \int_{T_1}^{T_2} \frac{C_{pL}}{T}\, dT \tag{6-4.3}$$

If there is condensation from vapor to liquid in the process of interest, a useful alternative is to choose a path that has the following steps: 1) Change real vapor to ideal gas at (T_1, P_1); 2) change ideal gas to saturated vapor state (T_1, P_{vp1}); 3) condense saturated vapor to saturated liquid at (T_1); and 4) change liquid to final state (T_2, P_2). In this case, ignoring the effect of pressure on the liquid properties, the enthalpy and entropy changes would be

$$H(T_2, P_2) - H(T_1, P_1) = \int_{T_1}^{T_2} C_{pL}\, dT - \Delta H_v(T_1) - H^d(T_1, P_{\text{vp}1}) + H^d(T_1, P_1) \tag{6 4.4a}$$

$$S(T_2, P_2) - S(T_1, P_1) = \int_{T_1}^{T_2} \frac{C_{pL}}{T}\, dT - \frac{\Delta H_v(T_1)}{T_1} - S^d(T_1, P_{\text{vp}1}) - R \ln \frac{P_{\text{vp}1}}{P_1} + S^d(T_1, P_1) \tag{6-4.4b}$$

Obviously, since the differences are independent of the path, other processes could be devised that would allow the use of property change values from other, and more convenient sources.

6-5 HEAT CAPACITIES OF REAL GASES

The heat capacities of ideal gases have been discussed in Chap. 3. In this section we discuss the behavior of real gases, including methods for estimating their behavior using equations of state. The focus is on the constant pressure heat capacity, $C_p = (\partial H/\partial T)_p$, since it is normally obtained from calorimetric measurements and tabulated. The constant volume heat capacity, $C_v = (\partial U/\partial T)_V$, is also of interest and the two are related by the thermodynamic equation

$$C_p = C_v - T\left(\frac{\partial V}{\partial T}\right)_P^2 \Big/ \left(\frac{\partial V}{\partial P}\right)_T = C_v - T\left(\frac{\partial P}{\partial T}\right)_V^2 \Big/ \left(\frac{\partial P}{\partial V}\right)_T \tag{6-5.1}$$

Equation (6-5.1) shows that the difference between the heat capacities can be found from an EoS; the two forms written here make explicit the different independent variables, P or V. For ideal gases, $C_p^\circ = C_v^\circ + R$.

As will be discussed in the next section, the effects of pressure and temperature for liquids are not great. However, both C_p and C_v diverge at the critical point of a pure fluid. In the neighborhood of the critical, $(\partial P/\partial V)_T$ approaches zero, so C_p increases much faster than C_v. Figure 6-2 shows slightly supercritical isotherms for C_v and C_p of propane as calculated from the EoS of Younglove, et al. (1987) by Konttorp (1998). At both high and low densities, the differences are small, but for T_r near unity, they increase rapidly as the critical density is approached. At fixed density in this region, C_p actually decreases as T increases.

As with the other thermodynamic properties treated in this chapter, there are departure and residual functions for the heat capacities related to the properties in Table 6-2 when the appropriate derivatives of U and H are used. The departure function for C_p is obtained from the residual function for C_v and Eq. (6-5.1) by integrating the partial derivative relation

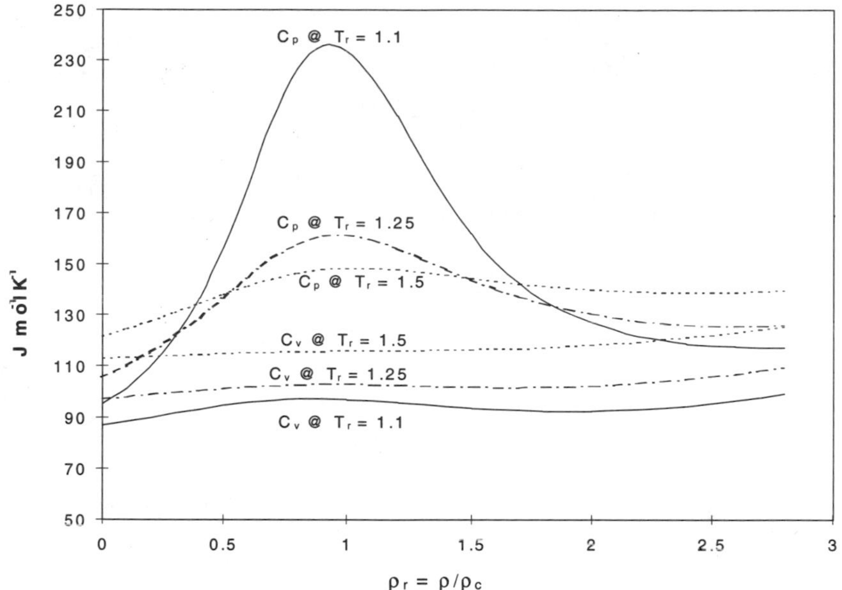

FIGURE 6-2 Reduced density variations at constant temperature of C_v and C_p for propane from the EoS of Younglove, et al. (1987) as calculated by Konttorp (1998).

$$\left(\frac{\partial C_v}{\partial V}\right)_T = \left(\frac{\partial^2 P}{\partial T^2}\right)_V \tag{6-5.2}$$

$$\begin{aligned}\frac{C_p}{R} &= \frac{C_p^\circ}{R} - \frac{C_p^\circ - C_p}{R} = \frac{C_p^\circ}{R} + \frac{C_p^r}{R} \\ &= \frac{C_p^\circ}{R} - 1 - \frac{C_v^\circ - C_v}{R} + \frac{C_p - C_v}{R} \\ &= \frac{C_p^\circ}{R} - 1 - T\int_V^\infty \left[\left(\frac{\partial^2 P}{\partial T^2}\right)_V\right]_T dV - T\left(\frac{\partial P}{\partial T}\right)_V^2 \Big/ \left(\frac{\partial P}{\partial V}\right)_T\end{aligned} \tag{6-5.3}$$

CSP tables are given in the 4th Edition and in Smith, et al. (1996) for $(C_p^r/R)^{(0)}$ and $(C_p^r/R)^{(1)}$ from the Lee-Kesler (1975) correlation at values of $T_r = T/T_c$ and $P_r = P/P_c$. Then

$$\frac{C_p}{R} = \frac{C_p^\circ}{R} + \left(\frac{C_p^r}{R}\right)^{(0)} + \omega\left(\frac{C_p^r}{R}\right)^{(1)} \tag{6-5.4}$$

A two-reference formulation such as Eq. (6-3.4) can also be used for C_p.

6-6 HEAT CAPACITIES OF LIQUIDS

There are three liquid heat capacities in common use: C_{pL}, $C_{\sigma L}$, and $C_{\text{sat}L}$. The first represents the change in enthalpy with temperature at constant pressure; the second shows the variation in enthalpy of a saturated liquid with temperature; the third indicates the energy required to affect a temperature change while maintaining the liquid in a saturated state. The three heat capacities are related as follows:

$$C_{\sigma L} = \frac{dH_{\sigma L}}{dT} = C_{pL} + \left[V_{\sigma L} - T\left(\frac{\partial V}{\partial T}\right)_p\right]\left(\frac{dP}{dT}\right)_{\sigma L} = C_{satL} + V_{\sigma L}\left(\frac{dP}{dT}\right)_{\sigma L} \quad (6\text{-}6.1)$$

where $C_{pL} = (\partial H/\partial T)_p$ and $(dP/dT)_{\sigma L}$ is the change of P_{vp} with T. Except at high reduced temperatures, all three forms of the liquid heat capacity are in close numerical agreement. Most estimation techniques yield either C_{pL} or $C_{\sigma L}$, although C_{satL} is often the quantity measured experimentally.

Liquid heat capacities are not strong functions of temperature except above $T_r = 0.7$ to 0.8. In fact, a shallow minimum is often reported at temperatures slightly below the normal boiling point. At high reduced temperatures, C_p values are large and strong functions of temperature, approaching infinity at the critical point. The general trend is illustrated in Fig. 6-3 for propylene.

Near the normal boiling point, most liquid organic compounds have heat capacities between 1.2 and 2 J g^{-1} K^{-1}. In this temperature range, there is essentially no effect of pressure (Gambill, 1957).

Experimentally reported liquid heat capacities for over 1600 substances have been compiled and evaluated by Zábranský, et al. (1996) and values at 298.15 K for over 2500 compounds are given by Domalski and Hearing (1996). Constants for equations that may be used to calculate liquid heat capacities are presented by Daubert, et al. (1997). Most of the heat capacity data are for temperatures below the normal boiling point temperature, data for higher temperatures are far less plentiful.

Usually group contribution or corresponding states methods are used for estimating liquid heat capacities. Examples of these two approaches are described below, and recommendations are presented at the end of the section.

Group Contribution Methods for Liquid C_p

A number of different group contribution methods have been proposed to estimate liquid heat capacities. In some of these, the assumption is made that various groups

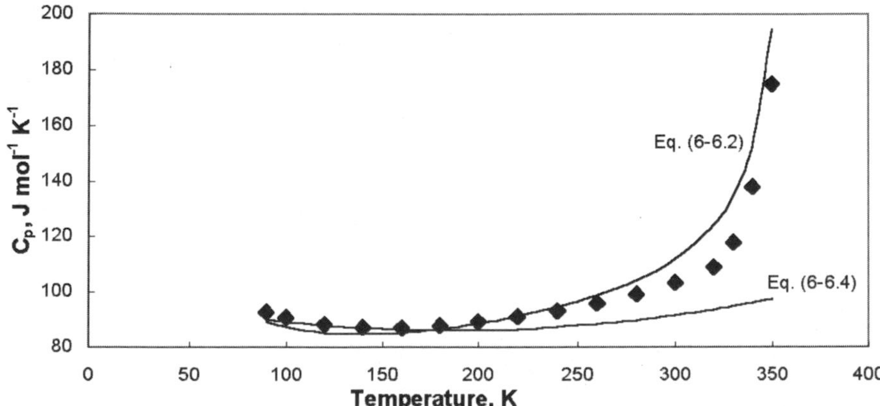

FIGURE 6-3 Constant pressure heat capacities of liquid propene. Symbols are from the tabulation of Zábranský, et al. (1996). Lines from Eqs. (6-6.2) and (6-6.4).

in a molecule contribute a definite value to the total molar heat capacity that is independent of other groups present. Such methods include those of Chueh and Swanson (1973ab) and Missenard (1965) which are described in the 4th Edition of this book and the Chueh and Swanson method is also described in Perry's Handbook (Perry and Green, 1997).

More recently, methods have been presented that account for differing contributions depending on what a particular atom is bonded to in the same way as Benson's method, which was described in Chap. 3. Domalski and Hearing (1993) present such a method to estimate liquid heat capacities at 298.15 K. The method employs over 600 groups and energy corrections and covers 1512 compounds. These groups are used by DIPPR (1999). Another Benson-type method, that of Rùzicka and Domalski (1993) is described below.

Method of Rùzicka and Domalski (1993). This is a group contribution method that can be used to develop heat capacity expressions that can be used to calculate liquid heat capacities from the melting point to the boiling point. The method can be expressed by:

$$C_{pL} = R\left[A + B\frac{T}{100} + D\left(\frac{T}{100}\right)^2\right] \tag{6-6.2}$$

where R is the gas constant (see Table 4-1) and T is temperature in K. Parameters A, B, and D are obtained from

$$A = \sum_{i=1}^{k} n_i a_i \qquad B = \sum_{i=1}^{k} n_i b_i \qquad D = \sum_{i=1}^{k} n_i d_i \tag{6-6.3}$$

where n_i is the number of groups of type i, k is the total number of different kinds of groups, and the parameters a_i, b_i, and d_i are listed in Table 6-5 for 114 different groups and Table 6-6 for 36 different ring strain corrections (rsc). Twenty-one more groups can be accommodated by the method with the group equivalency table, Table 6-7. See Chap. 3 (or Rùzicka and Domalski, 1993) for discussion and examples of group assignments in a molecule. The nomenclature in Rùzicka and Domalski has been modified here to match that used in Sec. 3-5 and Table 3-4.

Example 6-2 Estimate the liquid heat capacity of 1,3-cyclohexadiene at 300 K by using the Rùzicka-Domalski group contribution method. The recommended value given by Zábranský, et al. (1996) is 142 J mol^{-1} K^{-1}.

solution Since 300 K is less than the boiling point temperature of 353.49 K (Daubert, et al., 1997), the Rùzicka-Domalski method can be used. The six groups of 1,3-cyclohexadiene are two each of: =C—(H,C), =C—(H,=C), and C—(2H,C,=C). There is also a cyclohexadiene rsc. Using values from Tables 6-5 and 6-6 along with Eq. (6-6.3) gives

$$A = 2(4.0749) + 2(3.6968) + 2(2.0268) - 8.9683 = 10.6287$$

Similarly, $B = 1.54424$ and $D = 0.23398$. When a value of 300 is used for T in Eq. (6-6.2), the result is $C_p = 144.4$ J mol^{-1} K^{-1} for an error of 1.7%.

Corresponding States Methods (CSP) for Liquid C_p. Several CSP methods for liquid C_p estimation have been developed using the residual C_p^r of Eq. (6-5.3). One such equation is

TABLE 6-5 Group Contribution Parameters for the Rùzicka-Domalski Method, Eq. (6-6.3)

	a_i	b_i	d_i	T range, K
Hydrocarbon Groups				
C—(3H,C)	3.8452	−0.33997	0.19489	80–490
C—(2H,2C)	2.7972	−0.054967	0.10679	80–490
C—(H,3C)	−0.42867	0.93805	0.0029498	85–385
C—(4C)	−2.9353	1.4255	−0.085271	145–395
=C—(2H)	4.1763	−0.47392	0.099928	90–355
=C—(H,C)	4.0749	−1.0735	0.21413	90–355
=C—(2C)	1.9570	−0.31938	0.11911	140–315
=C—(H,=C)	3.6968	−1.6037	0.55022	130–305
=C—(C,=C)	1.0679	−0.50952	0.33607	130–305
C—(2H,C,=C)	2.0268	0.20137	0.11624	90–355
C—(H,2C,=C)	−0.87558	0.82109	0.18415	110–300
C—(3C,=C)	−4.8006	2.6004	−0.040688	165–295
C—(2H,2=C)	1.4973	−0.46017	0.52861	130–300
Ct—(H)	9.1633	−4.6695	1.1400	150–275
Ct—(C)	1.4822	1.0770	−0.19489	150–285
=C=	3.0880	−0.62917	0.25779	140–315
Ct—(Cb)	12.377	−7.5742	1.3760	230–550
Cb—(H)	2.2609	−0.25000	0.12592	180–670
Cb—(C)	1.5070	−0.13366	0.011799	180–670
Cb—(=C)	−5.7020	5.8271	−1.2013	230–550
Cb—(Cb)	5.8685	−0.86054	−0.063611	295–670
C—(2H,C,Cb)	1.4142	0.56919	0.0053465	180–470
C—(H,2C,Cb)	−0.10495	1.0141	−0.071918	180–670
C—(3C,Cb)	1.2367	−1.3997	0.41385	220–295
C—(2H,2Cb)	−18.583	11.344	−1.4108	300–420
C—(H,3Cb)	−46.611	24.987	−3.0249	375–595
Cp—(Cp,2Cb)	−3.5572	2.8308	−0.39125	250–510
Cp—(2Cp,Cb)	−11.635	6.4068	−0.78182	370–510
Cp—(3Cp)	26.164	−11.353	1.2756	385–480
Halogen Groups				
C—(C,3F)	15.42300	−9.24640	2.86470	125–345
C—(2C,2F)	−8.95270	10.55000	−1.99860	125–345
C—(C,3Cl)	8.54300	2.69660	−0.42564	245–310
C—(H,C,2Cl)	10.88000	−0.35391	0.08488	180–355
C—(2H,C,Cl)	9.66630	−1.86010	0.41360	140–360
C—(H,2C,Cl)	−2.06000	5.32810	−0.82721	275–360
C—(2H,C,Br)	6.39440	−0.10298	0.19403	168–360
C—(H,2C,Br)	10.78400	−2.47540	0.33288	190–420
C—(2H,C,I)	0.03762	5.62040	−0.92054	245–340
C—(C,2Cl,F)	13.53200	−3.27940	0.80145	240–420
C—(C,Cl,2F)	7.22950	0.41759	0.15892	180–420
C—(C,Br,2F)	8.79560	−0.19165	0.24596	165–415
=C—(H,Cl)	7.15640	−0.84442	0.27199	120–300
=C—(2F)	7.66460	−2.07500	0.82003	120–240
=C—(2Cl)	9.32490	−1.24780	0.44241	155–300
=C—(Cl,F)	7.82040	−0.69005	0.19165	120–240
Cb—(F)	3.07940	0.46959	−0.00557	210–365
Cb—(Cl)	4.54790	0.22250	−0.00979	230–460
Cb—(Br)	2.28570	2.25730	−0.40942	245–370

TABLE 6-5 Group Contribution Parameters for the Rùzicka-Domalski Method, Eq. (6-6.3) (*Continued*)

	a_i	b_i	d_i	T range, K
Cb—(I)	2.90330	2.97630	−0.62960	250–320
C—(Cb,3F)	7.44770	0.92230	0.39346	210–365
C—(2H,Cb,Cl)	16.75200	−6.79380	1.25200	245–345
Nitrogen Groups				
C—(2H,C,N)	2.45550	1.04310	−0.24054	190–375
C—(H,2C,N)	2.63220	−2.01350	0.45109	240–370
C—(3C,N)	1.96300	−1.72350	0.31086	255–375
N—(2H,C)	8.27580	−0.18365	0.03527	185–455
N—(H,2C)	−0.10987	0.73024	0.89323	170–400
N—(3C)	4.39420	−2.21340	0.55316	160–360
N—(H,C,Cb)	0.49631	3.46170	−0.57161	240–380
N—(2C,Cb)	−0.23640	16.26000	−2.52580	285–390
Cb—(N)	−0.78169	1.50590	−0.25287	240–455
N—(2H,N)	6.80500	−0.72563	0.15634	215–465
N—(H,C,N)	1.14110	3.59810	−0.69350	205–300
N—(2C,N)	−1.05700	4.00380	−0.71494	205–300
N—(H,Cb,N)	−0.74531	3.62580	−0.53306	295–385
C—(2H,C,CN)	11.97600	−2.48860	0.52358	185–345
C—(3C,CN)	2.57740	3.52180	−0.58466	295–345
=C—(H,CN)	9.07890	−0.86929	0.32986	195–345
Cb—(CN)	1.93890	3.02690	−0.47276	265–480
C—(2H,C,NO2)	18.52000	−5.45680	1.05080	190–300
O—(C,NO2)	−2.01810	10.50500	−1.83980	180–350
Cb—(NO2)	15.27700	−4.40490	0.71161	280–415
N—(H,2Cb) (pyrrole)	−7.36620	6.36220	−0.68137	255–450
Nb—(2Cb)	0.84237	1.25560	−0.20336	210–395
Oxygen Groups				
O—(H,C)	12.95200	−10.14500	2.62610	155–505
O—(H,C) (diol)	5.23020	−1.51240	0.54075	195–475
O—(H,Cb)	−7.97680	8.10450	−0.87263	285–400
C—(2H,C,O)	1.45960	1.46570	−0.27140	135–505
C—(2H,Cb,O)	−35.12700	28.40900	−4.95930	260–460
C—(H,2C,O) (alcohol)	2.22090	−1.43500	0.69508	185–460
C—(H,2C,O) (ether,ester)	0.98790	0.39403	−0.01612	130–170
C—(3C,O) (alcohol)	−44.69000	31.76900	−4.87910	200–355
C—(3C,O) (ether, ester)	−3.31820	2.63170	−0.44354	170–310
O—(2C)	5.03120	−1.57180	0.37860	130–350
O—(C,Cb)	−22.52400	13.11500	−1.44210	320–350
O—(2Cb)	−4.57880	0.94150	0.31655	300–535
C—(2H,2O)	1.08520	1.54020	−0.31693	170–310
C—(2C,2O)	−12.95500	9.10270	−1.53670	275–335
Cb—(O)	−1.06860	3.52210	−0.79259	285–530
C—(2H,C,CO)	6.67820	−2.44730	0.47121	180–465
C—(H,2C,CO)	3.92380	−2.12100	0.49646	185–375
C—(3C,CO)	−2.26810	1.75580	−0.25674	225–360
CO—(H,C)	−3.82680	7.67190	−1.27110	180–430
CO—(H,=C)	−8.00240	3.63790	−0.15377	220–430
CO—(2C)	5.43750	0.72091	−0.18312	185–380
CO—(C,=C)	41.50700	−32.63200	6.03260	275–355

TABLE 6-5 Group Contribution Parameters for the Rùzicka-Domalski Method, Eq. (6-6.3) (*Continued*)

	a_i	b_i	d_i	T range, K
CO—(C,Cb)	−47.21100	24.36800	−2.82740	300–465
CO—(H,O)	13.11800	16.12000	−5.12730	280–340
CO—(C,O)	29.24600	3.42610	−2.89620	180–445
CO—(=C,O)	41.61500	−12.78900	0.53631	195–350
CO—(O,CO)	23.99000	6.25730	−3.24270	320–345
O—(C,CO)	−21.43400	−4.01640	3.05310	175–440
O—(H,CO)	−27.58700	−0.16485	2.74830	230–500
=C—(H,CO)	−9.01080	15.14800	−3.04360	195–355
=C—(C,CO)	−12.81800	15.99700	−3.05670	195–430
Cb—(CO)	12.15100	−1.67050	−0.12758	175–500
CO—(Cb,O)	16.58600	5.44910	−2.68490	175–500
Sulfur Groups				
C—(2H,C,S)	1.54560	0.88228	−0.08349	130–390
C—(H,2C,S)	−1.64300	2.30700	−0.31234	150–390
C—(3C,S)	−5.38250	4.50230	−0.72356	190–365
Cb—(S)	−4.45070	4.43240	−0.75674	260–375
S—(H,C)	10.99400	−3.21130	0.47368	130–380
S—(2C)	9.23060	−3.00870	0.45625	165–390
S—(C,S)	6.65900	−1.35570	0.17938	170–350
S—(2Cb) (thiophene)	3.84610	0.36718	−0.06131	205–345

$$\frac{C_p^r}{R} = \frac{C_p - C_p^\circ}{R}$$

$$= 1.586 + \frac{0.49}{1 - T_r} + \omega\left[4.2775 + \frac{6.3(1 - T_r)^{1/3}}{T_r} + \frac{0.4355}{1 - T_r}\right] \quad (6\text{-}6.4)$$

Equation (6-6.4) is similar to one given by Bondi (1968) but we have refitted the first two constants to more accurately describe liquid argon behavior than Bondi's form. Of the substances in Appendix A, there are 212 that have values of C_{pL} at 298 K along with T_c and ω for use in Eq. (6-6.4). The deviation in C_{pL} calculated with Eq. (6-6.4) was greater than 10% for 18 of the 212 substances. These 18 substances included the C_1 to C_4 alcohols and acids, water, D_2O, bromoethane, hydrazine, HF, SO_3, N_2O_4, 1,2-oxazole, C_6F_{14}, and isobutyl amine. Most of these 18 substances associate by forming hydrogen bonds or dimers. For the other 194 substances, the average absolute percent deviation in C_{pL} from Eq. (6-6.4) was 2.5%.

If the substance follows CSP behavior, C_{pL}, $C_{\sigma L}$, and C_{satL} can also be related to each other by CSP relations or the EoS quantities of Eq. (6-6.1)

$$\frac{C_p - C_\sigma}{R} = \exp(20.1T_r - 17.9) \quad (6\text{-}6.5)$$

$$\frac{C_\sigma - C_{sat}}{R} = \exp(8.655T_r - 8.385) \quad (6\text{-}6.6)$$

Equations (6-6.5) and (6-6.6) are valid for $T_r < 0.99$. Below $T_r \sim 0.8$, C_{pL}, $C_{\sigma L}$, and C_{satL} may be considered to have the same value.

TABLE 6-6 Ring Strain Contributions (rsc) for the Rùzicka-Domalski Method, Eq. (6-6.3)

	a_i	b_i	d_i	T Range, K
Hydrocarbons				
cyclopropane	4.4297	−4.3392	1.0222	155–240
cyclobutane	1.2313	−2.8988	0.75099	140–300
cyclopentane (unsub)	−0.33642	−2.8663	0.70123	180–300
cyclopentane (sub)	0.21983	−1.5118	0.23172	135–365
cyclohexane	−2.0097	−0.72656	0.14758	145–485
cycloheptane	−11.460	4.9507	−0.74754	270–300
cyclooctane	−4.1696	0.52991	−0.018423	295–320
spiropentane	5.9700	−3.7965	0.74612	175–310
cyclopentene	0.21433	−2.5214	0.63136	140–300
cyclohexene	−1.2086	−1.5041	0.42863	160–320
cycloheptene	−5.6817	1.5073	−0.19810	220–300
cyclooctene	−14.885	7.4878	−1.0879	260–330
cyclohexadiene	−8.9683	6.4959	−1.5272	170–300
cyclooctadiene	−7.2890	3.1119	−0.43040	205–320
cycloheptatriene	−8.7885	8.2530	−2.4573	200–310
cyclooctatetraene	−12.914	13.583	−4.0230	275–330
indan	−6.1414	3.5709	−0.48620	170–395
1H-indene	−3.6501	2.4707	−0.60531	280–375
tetrahydronaphthalene	−6.3861	2.6257	−0.19578	250–320
decahydronaphthalene	−6.8984	0.66846	−0.070012	235–485
hexahydroindan	−3.9271	−0.29239	0.048561	210–425
dodecahydrofluorene	−19.687	8.8265	−1.4031	315–485
tetradecahydrophenanthrene	−0.67632	−1.4753	−0.13087	315–485
hexadecahydropyrene	61.213	−30.927	3.2269	310–485
Nitrogen Compounds				
ethyleneimine	15.281	−2.3360	−0.13720	195–330
pyrrolidine	12.703	1.3109	−1.18130	170–400
piperidine	25.681	−7.0966	0.14304	265–370
Oxygen Compounds				
ethylene oxide	6.8459	−5.8759	1.2408	135–325
trimethylene oxide	−7.0148	7.3764	−2.1901	185–300
1,3-dioxolane	−2.3985	−0.48585	0.10253	175–300
furan	9.6704	−2.8138	0.11376	190–305
tetrahydrofuran	3.2842	−5.8260	1.2681	160–320
tetrahydropyran	−13.017	3.7416	−0.15622	295–325
Sulfur Compounds				
thiacyclobutane	−0.73127	−1.3426	0.40114	200–320
thiacyclopentane	−3.2899	0.38399	0.089358	170–390
thiacyclohexane	−12.766	5.2886	−0.59558	295–340

Example 6-3 Estimate the liquid heat capacity of *cis*-2-butene at 350 K by using Eq. (6-6.4). The recommended value given by Zábranský et al. (1996) is 151 J mol^{-1} K^{-1}.

solution From Appendix A, $T_c = 435.5$ K and $\omega = 0.203$. The ideal gas heat capacity constants from Appendix A give $C_p^\circ = 91.21$ J mol^{-1} K^{-1}. The reduced temperature, $T_r = 350/435.5 = 0.804$. Equation (6-6.4) gives

TABLE 6-7 List of Equivalent Groups for the Rùzicka-Domalski Method for C_{pL} (≡ placed between each equivalent group)

C—(3H,C)≡C—(3H,=C)≡C—(3H,Ct)≡C—(3H,Cb)
C—(2H,C,Ct)≡C—(2H,C,=C)
Cb—(Ct)≡Cb—(=C)
=C—(H,Cb)≡=C—(H,=C)
=C—(C,Cb)≡=C—(C,=C)
C—(3H,C)≡C—(3H,N)≡C—(3H,O)≡C—(3H,CO)≡C—(3H,S)
N—(2H,Cb)≡N—(2H,C)
S—(H,Cb)≡S—(H,C)
O—(H,Cb) (diol)≡O—(H,C) (diol)
CO—(H,Cb)≡CO—(H,=C)
C—(2H,=C,Cl)≡C—(2H,C,Cl)
C—(2H,Cb,N)≡C—(2H,C,N)
N—(C,2Cb)≡N—(3C)
C—(2H,=C,O)≡C—(2H,Cb,O)
S—(Cb,S)≡S—(2C)
S—(2Cb)≡S—(2C)

$$\frac{C_{pL} - C_p^\circ}{R} = 1.586 + \frac{0.49}{1 - 0.804} + 0.203 \left[4.2775 + \frac{6.3(1 - 0.804)^{1/3}}{0.804} + \frac{0.4355}{1 - 0.804}\right] = 6.027$$

$$C_{pL} = 91.21 + (8.3145)(6.027) = 141.3 \text{ J mol}^{-1} \text{ K}^{-1}$$

$$\text{Error} = \frac{141.3 - 151}{151} \times 100 = -6.4\%$$

Discussion

Two methods for estimating liquid heat capacities have been described. The Rùzicka–Domalski method, Eq. (6-6.2), is a group contribution method that is applicable between the melting and boiling points. At higher temperatures, it generally underpredicts the heat capacity. Figure 6-3 illustrates this for propylene. Equation (6-6.4) is a CSP correlation that works well for all compounds except those that associate. Equation (6-6.4) requires T_c, ω, and C_p°. Table 6-8 illustrates the capabilities and limitations of both estimation methods. When the reduced temperature is above 0.9, the Rùzicka–Domalski method gives large negative deviations, while Eq. (6-6.4) gives large deviations for ethanol, acetic acid, and ethyl bromide. These latter deviations become proportionally smaller at higher reduced temperatures.

Recommendations

Use the Rùzicka–Domalski method for temperatures below the boiling point. However, at higher temperatures, use Eq. (6-6.4).

TABLE 6-8 Comparisons of Experimental and Estimated C_{pL} Values

Substance	T, K	C_{pL}, lit* J mol^{-1} K^{-1}	T_r	% error Eq. (6-6.4)	% err Eq. (6-6.2)
argon	90	44.96	0.60	−2.0	—
	148	250.7	0.98	−0.7	—
methane	100	54.09	0.52	4.1	—
	140	61.45	0.73	1.2	—
	180	108.6	0.94	11	—
propane	100	85.2	0.27	7.9	0.0
	200	93.5	0.54	1.8	−2.2
	230	98.88	0.62	1.5	3.9
	300	122	0.81	−0.9	−13
	360	260.1	0.97	0.7	−54
n-pentane	150	141.6	0.32	3.0	−3.6
	250	153.6	0.53	2.4	−0.4
	350	186.4	0.75	0.3	−2.7
	390	204.4	0.83	0.5	−4.0
isobutane	115	99.26	0.28	8.0	−1.2
	200	114.9	0.49	0.4	−3.8
	300	143.4	0.74	−0.8	−6.4
	400	371.1	0.98	−2.8	−55
heptane	190	201.82	0.35	−0.4	−4.4
	300	225.58	0.56	1.2	−0.0
	400	270.87	0.74	−0.7	0.2
	480	324.3	0.89	−1.9	−1.6
decane	250	297.5	0.40	0.2	−2.1
	460	413.3	0.74	−2.3	3.1
cyclohexane	280	149.3	0.51	−2.1	0.2
	400	202.8	0.72	−3.7	−5.0
	500	271	0.90	−4.3	−10
cis-2-butene	150	112.4	0.34	−3.2	−0.6
	250	116.3	0.57	−0.5	−0.7
	350	151	0.80	−6.4	−12.1
	370	165	0.85	−8.5	−16
benzene	290	134.3	0.52	−5.9	−3.6
	400	161.6	0.71	−2.1	1.1
	490	204.5	0.87	−3.6	−1.0
chlorobenzene	230	140	0.36	−1.8	−0.4
	300	151	0.47	0.8	1.0
	360	163	0.57	2.3	3.0
ethanol	160	87.746	0.31	114	−1.9
	300	113	0.58	40	4.1
	380	156.6	0.74	3.9	10
acetone	180	117.1	0.35	9.3	−1.5
	300	126.6	0.59	3.3	−0.8
	330	133.1	0.65	1.2	−3.1
diethyl ether	160	147.6	0.34	3.3	−2.5
	200	155.3	0.43	−0.0	−4.2
	300	173.1	0.64	−0.3	−5.3
	440	306	0.94	−16	−38

TABLE 6-8 Comparisons of Experimental and Estimated C_{pL} Values (*Continued*)

Substance	T, K	C_{pL}, lit* J mol^{-1} K^{-1}	T_r	% error Eq. (6-6.4)	% err Eq. (6-6.2)
ethyl mercaptan	130	114.3	0.26	−4.4	1.1
	250	113.4	0.50	−2.4	−2.0
	320	120.8	0.64	−0.9	−4.8
bromine	300	75.63	0.51	−4.9	
chlorine	230	66.01	0.55	−6.8	
ethyl chloride	140	96.91	0.30	−0.0	−0.3
	250	98.73	0.54	1.5	−0.5
	320	115	0.70	−4.1	−8.2
ethyl bromide	170	89.1	0.34	24	−1.0
	300	100	0.60	15	3.2
acetic acid	290	121.2	0.49	17	−1.4
	350	135.8	0.59	5.6	−0.2
	400	148.7	0.67	−0.4	0.3
Ave. Abs. Error	—	—	—	6.4	5.9

* Values from Younglove (1982), Younglove and Ely (1987), and Zabransky, et al. (1996).

6-7 PARTIAL PROPERTIES AND FUGACITIES OF COMPONENTS IN MIXTURES

The thermodynamic properties of mixtures can be obtained by adding contributions from the partial properties of the components (Van Ness and Abbott, 1982). In addition, the fundamental engineering equation used for phase equilibrium is that the fugacity of a component must have the same value in all phases. As described in Sec. 8-2, these equations are used to compute the values of the dependent state conditions from independent values. We describe here a brief basic analysis for obtaining partial properties and fugacities from equations of state as considered in Chaps. 4 and 5.

The basic relation for what is commonly called the partial molar property, $\overline{F}_i(T, P, \{y\})$, of a general property, $F(T, P, \{y\})$, is

$$\overline{F}_i(T, P, \{y\}) = \left(\frac{\partial(\underline{F})}{\partial N_i}\right)_{T,P,N_{j\neq i}} \tag{6-7.1}$$

The notation is that the total property, $\underline{F}$, is found from algebraically multiplying the molar property, F, by the total number of moles in the system, N, so that $\underline{F} = NF$ is expressed only in terms of T, P, and $\{N\}$. The partial derivative of that expression is taken with respect to the number of moles of the component of interest, N_i, while holding constant T, P, and all the other numbers of moles for all other components j different from i, ($j \neq i$).

The consequences of this definition (Smith, et al., 1996) are three important equations. The first is the Gibbs-Duhem equation

$$\sum_{i=1}^{n} y_i d\overline{F}_i + \left(\frac{\partial F}{\partial T}\right)_{P,\{y\}} dT + \left(\frac{\partial F}{\partial P}\right)_{T,\{y\}} dP = 0 \tag{6-7.2}$$

The second is the additive rule for mixture properties

$$F(T, P, \{y\}) = \sum_{i=1}^{n} y_i \overline{F}_i(T, P, \{y\}) \tag{6-7.3}$$

The third relation is the pure component limit

$$\lim_{y_i \to 1} \overline{F}_i = F_i^\circ(T, P) \tag{6-7.4}$$

where $F_i^\circ(T, P)$ is the molar property of pure component i. It should be apparent from Eq. (6-7.3) that it is valuable to know or be able to compute partial properties. From EoS models, partial properties are obtained by taking the derivative of Eq. (6-7.1) on the departure functions of Table 6-3 and adding the ideal gas term. For $\underline{U}_m^{ig}/RT$ and $\underline{H}_m^{ig}/RT$, the ideal gas terms are $\underline{U}_i^{ig}/RT$ and $\underline{H}_i^{ig}/RT$, while for $\underline{S}_m^{ig}/R$, $\underline{A}_m^{ig}/RT$ and $\underline{G}_m^{ig}/RT$, the ideal gas terms are $\underline{S}_i^{ig}/R - \ln y_i$, $\underline{A}_i^{ig}/RT - \ln y_i$ and $\underline{G}_i^{ig}/RT - \ln y_i$.

Equation (6-7.2) is useful for checking the consistency when models for the partial properties of different components have been derived or modeled separately, and the consistency of phase equilibrium data if more properties have been measured than are truly independent (Prausnitz, et al., 1999). Numerical results from a computer program with Eq. (6-7.2) can be examined to see if correct derivations and program code have been established (Mollerup and Michelsen, 1992).

The chemical potential of a component, μ_i, is the partial molar Gibbs energy, $\overline{G}_i$, but it is also a partial derivative of other properties

$$\begin{aligned}\mu_i(T, P, \{y\}) = \overline{G}_i &= \left(\frac{\partial(G)}{\partial N_i}\right)_{T,P,N_{j\neq i}} = \left(\frac{\partial(A)}{\partial N_i}\right)_{T,\underline{V},N_{j\neq i}} \\ &= \left(\frac{\partial(\underline{U})}{\partial N_i}\right)_{\underline{S},\underline{V},N_{j\neq i}} = \left(\frac{\partial(\underline{H})}{\partial N_i}\right)_{\underline{S},P,n_{j\neq i}}\end{aligned} \tag{6-7.5}$$

where the fugacity of component i, $\hat{f}_i$, is related to its departure function chemical potential (see Sec. 6-3) as

$$\ln \frac{\hat{f}_i(T, P, \{y\})}{y_i P} = \frac{\mu_i(T, P, \{y\}) - \mu_i^{ig}(T, P)}{RT} = \frac{\mu_i^r(T, P, \{y\})}{RT} \tag{6-7.6}$$

The combination of one of the forms of Eqs. (6-7.5) with (6-7.6) yields an expression for $\hat{f}_i$ from an EoS with independent T, V, $\{y\}$ as

$$\ln \frac{\hat{f}_i(T, P, \{y\})}{y_i P} = \ln \phi_i = \int_{\underline{V}}^{\infty} \left[\left(\frac{\partial(P\underline{V}/RT)}{\partial N_i}\right)_{T,\underline{V},N_{j\neq i}} - 1\right] \frac{d\underline{V}}{\underline{V}} - \ln Z \tag{6-7.7}$$

where the quantity $P\underline{V}/RT = N[Z(T, V, \{y\})]$ which is a function of T, V, and $\{N\}$ so the derivative of Eq. (6-7.7) can be taken before doing the integration. The integral is equivalent to the derivative of the residual Helmholtz energy

$$\ln \phi_i = \left(\frac{\partial(\underline{A}^r/RT)}{\partial N_i}\right)_{T,\underline{V},N_{j\neq i}} - \ln Z \tag{6-7.8}$$

Olivera-Fuentes (1991) gives alternative derivations of ϕ_i.

Evaluation of Fugacity Coefficients

Table 6-9 shows Eq. (6-7.9) for the fugacity coefficient from the virial EoS models, Eqs. (5-4.1) and (5-4.2). The third virial coefficient term in Eq. (6-7.9*a*) from Eq. (5-4.1*a*) has been omitted because if the third virial coefficients C_{ijk} are known, Eq. (5-4.1*b*) and (6-7.9*b*) should be used.

Equation (6-7.10) shows the general expression for the EoS of Eq. (4-6) when all of the parameters can depend on composition through mixing rules as described in Sec. 5-5.

$$\ln \phi_i = \left\{ \frac{1}{(\delta_m^2 - 4\varepsilon_m)^{1/2}} \frac{1}{N} \left(\frac{\partial [N^2 \Theta_m / RT]}{\partial N_i} \right)_{T, N_{j \neq i}} - \frac{(\Theta_m / RT) \Delta N_i}{2(\delta_m^2 - 4\varepsilon_m)^{3/2}} \right\}$$

$$\ln \left\{ \frac{2V + \delta_m - (\delta_m^2 - 4\varepsilon_m)^{1/2}}{2V + \delta_m + (\delta_m^2 - 4\varepsilon_m)^{1/2}} \right\} + \frac{\Theta_m / RT}{(\delta_m^2 - 4\varepsilon_m)^{1/2}}$$

$$\left\{ \frac{\left(\frac{\partial [N\delta_m]}{\partial N_i} \right)_{T, N_{j \neq i}} - \frac{\Delta N_i}{2(\delta_m^2 - 4\varepsilon_m)^{1/2}}}{2V + \delta_m - (\delta_m^2 - 4\varepsilon_m)^{1/2}} - \frac{\left(\frac{\partial [N\delta_m]}{\partial N_i} \right)_{T, N_{j \neq i}} + \frac{\Delta N_i}{2(\delta_m^2 - 4\varepsilon_m)^{1/2}}}{2V + \delta_m + (\delta_m^2 - 4\varepsilon_m)^{1/2}} \right\}$$

$$+ \frac{\left(\frac{\partial [N b_m]}{\partial N_i} \right)_{T, N_{j \neq i}}}{V - b_m} - \ln \left[\frac{V - b_m}{V} \right] - \ln Z \qquad (6\text{-}7.10)$$

where

$$\Delta_{N_i} = \frac{1}{N} \left[\left(\frac{\partial [N^2 \delta^2]}{\partial N_i} \right)_{T, N_{j \neq i}} - 4 \left(\frac{\partial [N^2 \varepsilon]}{\partial N_i} \right)_{T, N_{j \neq i}} \right] \qquad (6\text{-}7.11)$$

As in Table 6-3, if $\delta_m^2 - 4\varepsilon_m = 0$ in a model, terms with Θ are simpler since there is no complex term with ln{ }. Particular EoS model results are obtained by substituting in the formulae of Tables 4-6 and 4-7 for the EoS model and then substituting in the mixing and combining rules selected.

Special manipulations must be used for the G^E mixing rules of Table 5-1. Equations (8-12.37) to (8-12.39) show the set of expressions that result for the case of the Wong-Sandler (1992) mixing rules in the PRSV EoS (Stryjek and Vera, 1986).

Example 6-4 shows the use of Eqs. (6-7.7) and (6-7.8) for a cubic equation of state with van der Waals mixing rules, Eqs. (5-5.2). Equations (8-12.28) show the expression for $\ln\phi_i$ from the Peng-Robinson (1976) EoS model with van der Waals mixing rules.

TABLE 6-9 Expressions for the Fugacity Coefficient from Virial Equations (5-4.1) and (5-4.2)

Virial Equation	$\ln \phi_i$	
(5-4.1*a*) & (5-4.2*a*)	$\left[2 \sum_{j=1}^{n} y_j B_{ij}(T) - B(T, \{y\}) \right] \frac{P}{RT}$	(6-7.9*a*)
(5-4.1*b*) & (5-4.2*a,b*)	$\frac{2}{V} \sum_{j=1}^{n} y_j B_{ij} + \frac{3}{2V^2} \sum_{j=1}^{n} \sum_{k=1}^{n} y_j y_k C_{ijk} - \ln Z$	(6-7.9*b*)

Example 6-4 Obtain the expression for the fugacity coefficient, ϕ_i, from Eq. (6-7.7) and from Eq. (6-7.8) for the van der Waals EoS (1890)

$$Z = \frac{V}{V - b_m} - \frac{a_m}{RTV}$$

with van der Waals mixing rules $a_m = \sum_{i=1}^{n}\sum_{j=1}^{n} y_i y_j a_{ij}$; $b_m = \sum_{i=1}^{n}\sum_{j=1}^{n} y_i y_j b_{ij}$ and the combining rules,

$$a_{ij} = (a_{ii}a_{jj})^{1/2}(1 - k_{ij}); \quad b_{ij} = \frac{b_{ii} + b_{jj}}{2}(1 - l_{ij})$$

solution In the notation of the integrand of Eq. (6-7.7),

$$NZ = \frac{N\underline{V}}{\underline{V} - (Nb)} - \frac{(N^2 a_m)}{RT\underline{V}}$$

so the integrand is

$$\frac{\left[\left(\frac{\partial(NZ)}{\partial N_i}\right)_{T,\underline{V},N_{j\neq i}} - 1\right]}{\underline{V}} = \frac{1}{\underline{V} - (Nb)} - \frac{1}{\underline{V}} + \frac{N[\partial(Nb)/\partial N_i]_{N_{j\neq i}}}{(\underline{V} - Nb)^2} - \frac{[\partial N^2 a_m)/\partial N_i]_{N_{j\neq i}}}{RT\underline{V}^2}$$

Then

$$\ln \phi_i = -\ln\left[\frac{Z(V - b_m)}{V}\right] + \frac{[\partial(Nb_m)/\partial N_i]_{T,N_{j\neq i}}}{(V - b_m)} - \frac{[\partial(N^2 a_m)/\partial N_i]_{T,N_{j\neq i}}}{RT\underline{V}}$$

With the mixing and combining rules given,

$$\frac{1}{V - b_m}\left[\frac{\partial(Nb_m)}{\partial N_i}\right]_{T,N_{j\neq i}} = \frac{\left[\sum_{j=1}^{n} 2y_j b_{ij} - b_m\right]}{(V - b_m)} = \frac{\sum_{j=1}^{n} 2y_j b_{ij} - \sum_{j=1}^{n}\sum_{k=1}^{n} y_j y_k b_{jk}}{V - b_m}$$

$$\frac{1}{RT\underline{V}}\left[\frac{\partial(N^2 a_m)}{\partial N_i}\right]_{T,N_{j\neq i}} = 2\sum_{j=1}^{n} y_j \frac{a_{ij}}{RTV} = \frac{2a_{ii}^{1/2}}{RTV}\sum_{j=1}^{n} y_j a_{jj}^{1/2}(1 - k_{ij})$$

Some simplification can be made of the expression for the term containing the derivative of (Nb_m); when all $l_{ij} = 0$, the term becomes $b_{ii}/(V - b_m)$.

In Eq. (6-7.8), we use Eq. (6-3.3) to obtain the residual Helmholtz energy departure function for the van der Waals EoS.

$$\frac{A^r}{RT} = N\left[-\frac{A^d}{RT} + \ln Z\right] = -N\ln\left[\frac{\underline{V} - Nb_m}{\underline{V}}\right] - \frac{(N^2 a_m)}{RT\underline{V}}$$

Taking the derivative and substituting into Eq. (6-7.8) gives

$$\ln \phi_i = -\ln\left[\frac{V - b_m}{V}\right] + \frac{[\partial(Nb_m)/\partial N_i]_{T,N_{j\neq i}}}{(V - b_m)} - \frac{[\partial(N^2 a_m)/\partial N_i]_{T,N_{j\neq i}}}{RTV} - \ln Z$$

which is the same answer as above. When all $l_{ij} = 0$, the final result is

$$\ln \phi_i = -\ln\left[\frac{V - b_m}{V}\right] + \frac{b_i}{V - b_m} - \frac{2a_i^{1/2}}{RTV}\sum_{j=1}^{n} y_j a_{ij}^{1/2}(1 - k_{ij}) - \ln Z$$

6-8 TRUE CRITICAL POINTS OF MIXTURES

In Chap. 5, emphasis was placed upon the estimation of pseudocritical constants for mixtures. Such constants are necessary if one is to use most corresponding-states correlations to estimate mixture $PVT\{y\}$ or derived properties. However, these pseudocritical constants often differ considerably from the true critical points for mixtures. Estimation techniques for the latter can be evaluated by comparison with experimental data. A summary of experimental values is given by Hicks and Young (1975); there seems not to have been an evaluated review since then though many systems are listed by Sadus (1992).

In this section, we briefly discuss methods of estimating the true critical properties of mixtures. The 4th Edition described several methods in detail; since that time there has been limited activity in estimation methods based on groups because EoS methods are now more accessible.

Group Contribution Methods

The methods of Chueh and Prausnitz (1967) and Li (1971) were described in detail in the 4th Edition and are mentioned in Sec. 5-8. Li and Kiran (1990) developed another method for binary systems based on group contributions to the factors in the simple method of Klincewicz and Reid (1984). Li and Kiran developed the correlation from 41 systems representing a variety of hydrocarbon, hydrocarbon-polar and hydrocarbon-CO_2 binaries. Then they used it to predict values for 15 more systems; the predicted results were about as good as the correlations. For the true critical temperature, T_{cT}, the standard deviations ranged from 1 to 20 K or up to 4% which was comparable to the method of Li (1971). For V_{cT}, the errors were 2 to 7% which is somewhat better than reported by Spencer, et al. (1973) in their extensive testing of hydrocarbon systems. For P_{cT}, the standard deviations were 6 to 17 bar or as high as 15%, which is somewhat higher than reported by Spencer, et al. (1973).

Liu (1998) relates P_{cT} to estimated values for T_{cT} and V_{cT} along with the CSP method of Chen (1965) for ΔH_v. His results for dilute hydrocarbon and CO_2-containing hydrocarbon mixtures are as good as the EoS method of Anselme and Teja (1990) and can avoid the occasional large errors found in some systems by other methods.

Rigorous Methods

Thermodynamics provides mathematical criteria for phase stability and critical points of pure components and mixtures. Though the pure component critical point is determined from an equation related to the isothermal compressibility, Eqs. (4-6.5), for mixtures the criterion involves matrices of second and third partial derivatives of energy functions with respect to numbers of moles of the components. For example, with the EoS models of Table 4-6 and 4-7, the quantities that must be obtained are the $(\partial^2(\underline{A})/\partial N_i \partial N_j)_{T,\underline{V},N_k}$. The review of Heidemann (1994) and the monograph of Sadus (1992) describe the concepts and equations as do the discussions of cubic equations, such as the Soave (1972) model, in Michelsen and Heidemann (1981), Michelsen (1982), and in Heidemann and Khalil (1980). Abu-Eishah, et al. (1998) show an implementation for all critical properties with the

PRSV EoS (Stryjek and Vera, 1986), though an adjustment with a translation volume (see Sec. 4-6) has been implemented to improve agreement for V_{cT}.

Recommendations

For the most reliable estimates of the true critical temperature and pressure of a mixture, use the rigorous methods based on an EoS with the methods suggested by Heidemann (1994). The computer method of Michelsen (1982) is probably the most efficient. The implementation of Abu-Eishah (1999) with the PRSV EoS (Stryjek and Vera, 1986) is considerably better than any of the group contribution estimates, especially for V_{cT}.

For more rapid estimates, the recommended methods from the 4th Edition should still be used. The method of Li (1971) should be used for hydrocarbons but for others, the Chueh-Prausnitz (1967) correlation is preferred, especially when an interaction parameter can be estimated. For rapid estimates of the true critical pressure of a mixture, the Chueh-Prausnitz (1967) or Liu (1998) methods may be used. Neither are very accurate for systems containing methane. Percentage errors are usually larger for P_{cT} estimations than for T_{cT}.

To quickly estimate the true critical volume of a mixture, use the method of Schick and Prausnitz (1968), especially if reliable estimates of the binary parameter can be obtained. However, such values may be in error for complex systems.

NOTATION

A	Helmholtz energy, J mol^{-1}
a, b	cubic EoS variables, Table 4-6
B_{ij}	second virial coefficient in Eqs. (5-4.1*a*) to (5-4.3*a*), cm^3 mol^{-1}
C_{ijk}	third virial coefficient in Eqs. (5-4.1*b*) to (5-4.3*b*), cm^6 mol^{-2}
C_v, C_p	heat capacity at constant volume, constant pressure, J mol^{-1} K^{-1}
C_p°	ideal gas heat capacity, J mol^{-1} K^{-1}
C_{pL}, $C_{\sigma L}$, C_{satL}	heat capacities of liquids, Eq. (6-6.1), J mol^1 K^{-1}
F	general molar (mol^{-1}) thermodynamic property such as U, H, S, A, G, V
$\underline{F}$	general total thermodynamic property such as $\underline{U}$, $\underline{H}$, $\underline{S}$, $\underline{A}$, $\underline{G}$, $\underline{V}$
$\overline{F}_i$	general partial thermodynamic property of component i such as $\overline{U}_i$, $\overline{H}_i$, $\overline{S}_i$, $\overline{A}_i$, $\overline{G}_i$, $\overline{V}_i$, Eq. (6-7.1)
$\hat{f}_i$	fugacity of component i, Eq. (6-7.6)
G	Gibbs energy, J mol^{-1}
H	enthalpy, J mol^{-1}
ΔH_v	enthalpy of vaporization, J mol^{-1}
k_{ij}, l_{ij}	binary interaction parameters, Eq. (5-2.4)
n	number of components in a mixture
P	pressure, bar
P_{vp}	vapor pressure, bar
R	gas constant, Table 4-1
T	temperature, K
U	internal energy, J mol^{-1}
V	volume, cm^3 mol^{-1}

y	mole fraction, Eq. (5.2-1a)
Z	compressibility factor = PV/RT

Greek

α	quantity in cubic EoS models, Θ/a
δ, ε, η	EoS variables, Table 4-6, Eq. (4-6.2)
ϕ_i	fugacity coefficient, Eq. (6-7.7)
Θ	EoS variable, Table 4-6, Eq. (4-6.2)
ω	acentric factor, Eq. (2-3.1)

Superscript

(0), (1)	corresponding states (CSP) functions, Eq. (6-3.5)
d	departure function, Table 6-1
ig	ideal gas
o	pure component property
r	residual property, Eq. (4-7.2)
(R1), (R2)	CSP reference functions, Eq. (6-3.6)
*	characteristic property

Subscripts

c	critical
cm	mixture pseudocritical
cT	true critical of mixture
i	component i
m	mixture
vp	vapor pressure
r	reduced, as in $T_r = T/T_c$
v	change on vaporization

REFERENCES

Abu-Eishah, S. I.: *Fluid Phase Equil.,* **157:** 1 (1999).

Abu-Eishah, S. I., N. A. Darwish, and I. H. Aljundi: *Int. J. Thermophys.,* **19:** 239 (1998).

Adachi, Y., B. C.-Y. Lu and H. Sugie: *Fluid Phase Equil.,* **11:** 29 (1983).

Adachi, Y., H. Sugie and B. C.-Y. Lu: *J. Chem. Eng. Japan,* **18:** 20 (1985).

Anselme, M. J., and A. S. Teja: *AIChE J.,* **36:** 897 (1990).

Bondi, A.: "Physical Properties of Molecular Crystals, Liquids and Glasses," Wiley, New York, 1968.

Breedveld, G. J. F., and J. M. Prausnitz: *AIChE J.,* **19:** 783 (1973).

Chen, N. H.: *J. Chem. Eng. Data,* **10:** 207 (1965).

Chueh, P. L., and J. M. Prausnitz: *AIChE J.,* **13:** 1099 (1967).

Chueh, C. F., and A. C. Swanson: *Chem Eng. Progr.,* **69**(7): 83 (1973a).

Chueh, C. F., and A. C. Swanson: *Can. J. Chem. Eng.,* **51:** 596 (1973b).

Cuthbert, J., and J. R. Downey: Dow Chemical Co., Personal communication, 1999.

Daubert, T. E., R. P. Danner, H. M. Sibel, and C. C. Stebbins: *Physical and Thermodynamic Properties of Pure Chemicals: Data Compilation,* Taylor & Francis, Washington, D. C., 1997.

Dillard, D. D., W. C. Edmister, J. H. Erbar, and R. L. Robinson, Jr.: *AIChE J.,* **14:** 923 (1968).

DIPPR, Design Institute for Physical Property Research: *Amer. Inst. Chem. Eng.,* http://www.aiche.org/dippr, 1999.

Domalski, E. S., and E. D. Hearing: *J. Phys. Chem. Ref. Data,* **22:** 805 (1993).

Domalski, E. S., and E. D. Hearing: *J. Phys. Chem. Ref. Data,* **25:** 1 (1996).

Djordjevic, B. D., M. LJ. Kijevcanin, S. P. Serbanovic: *Fluid Phase Equil.,* **155:** 205 (1999).

Fuller, G. G.: *Ind. Eng. Chem. Fundam.,* **15:** 254 (1976).

Gambill, W. R.: *Chem. Eng.,* **64**(5): 263, **64**(6): 243, **64**(7): 263; **64**(8): 257 (1957).

Gmehling, J., D. D. Liu, and J. M. Prausnitz: *Chem. Eng. Sci.,* **34:** 951 (1979).

Gregorowicz, J., J. P. O'Connell, and C. J. Peters: *Fluid Phase Equil.,* **116:** 94 (1996).

Heidemann, R. A., and A. M. Khalil: *AIChE J.,* **26:** 769 (1980).

Heidemann, R. A.: in Kiran, E., and J. M. H. Levelt Sengers (eds): "Supercritical Fluids: Fundamentals for Applications," NATO ASI Series E: **273:** 39. Klewer Academic Publishers, Dordrecht, Holland, 1994.

Hicks, C. P., and C. L. Young: *Chem. Rev.,* **75:** 119 (1975).

Kijevcanin, M. LJ., B. D. Djordjevic, P. S. Veselinovic, and S. P. Serbanovic: *J. Serb. Chem. Soc.,* **63:** 237, 251 (1998).

Klincewicz, K. M., and R. C. Reid: *AIChE J.,* **30:** 137 (1984).

Konttorp, M.: M. S. Thesis, Tech. Univ. Delft, 1998.

Kumar, S., R. Kurz, and J. P. O'Connell: "Equations of State for Gas Compressor Design and Testing," ASME/IGTI TurboExpo, Indianapolis, June, 1999. Proceedings to be published by ASME.

Lee, B. I., and M. G. Kesler: *AIChE J.,* **21:** 510 (1975).

Li, C. C.: *Can. J. Chem. Eng.,* **19:** 709 (1971).

Li, L., and E. Kiran: *Chem. Eng. Comm.,* **94:** 131 (1990).

Liu, Z.-Y.: *AIChE. J.,* **44:** 1709 (1998).

Marruffo, F., and F. P. Stein: *Fluid Phase Equil.,* **69:** 235 (1991).

Martin, J. J.: *Ind. Eng. Chem. Fundam.,* **18:** 81 (1979).

Mathias, P. M.: *Ind. Eng. Chem. Proc. Des. Dev.,* **27:** 385 (1983).

Mathias, P. M., and T. W. Copeman: *Fluid Phase Equil.,* **13:** 91 (1983).

Mathias, P. M., and F. P. Stein: Report DOE/OR/03054-110, Energy Res. Abstr., **9:** #39255 (1984).

Mathias, P. M., J. B. Gonsalves, and F. P. Stein: Report DOE/OR/03054-111, Energy Res. Abstr., **9:** #39256 (1984).

Mathias, P. M., T. Naheiri, and E. M. Oh: *Fluid Phase Equil.,* **47:** 77 (1989).

Michelsen, M. L: *Fluid Phase Equil.,* **9:** 21 (1982).

Michelsen, M. L., and R. A. Heidemann: *AIChE J.,* **27:** 521 (1981).

Missenard, F.-A.: *Compte Rend.,* **260:** 5521 (1965).

Mollerup, J. and M. L. Michelsen: *Fluid Phase Equil.,* **74:** 1 (1992).

Nagy, P. E., G. L. Bertrand, and B. E. Poling: *J. Chem. Eng. Data,* **32:** 439 (1987).

O'Connell, J. P., and J. M. Haile: "Multicomponent Thermodynamics: Fundamentals for Applications," Cambridge U. Press, Under Contract (2000).

Olivera-Fuentes, C.: *Chem. Eng. Sci.,* **46:** 2019 (1991).

Orbey, H. and S. I. Sandler: *Fluid Phase Equil.,* **121:** 67 (1996).

Patel, N. C., and A. S. Teja: *Chem. Eng. Sci.,* **37:** 463 (1982).

Patel, N. C.: *Int. J. Thermophys.,* **17:** 673 (1996).

Peneloux, A., E. Rauzy, and R. Freze: *Fluid Phase Equil.,* **8:** 7 (1982).

Peng, D. Y., and D. B. Robinson: *Ind. Eng. Chem. Fundam.,* **15:** 59 (1976).

Perry, R. H., and D. W. Green: "Perry's Chemical Engineers' Handbook," 7th ed., McGraw-Hill, New York, 1997.

Prausnitz, J. M., R. N. Lichtenthaler, and E. G. de Azevedo: "Molecular Thermodynamics of Fluid-Phase Equilibria," 3rd ed., Upper Saddle River, NJ, Prentice-Hall, 1999.

Redlich, O., and J. N. S. Kwong: *Chem. Rev.,* **44:** 233 (1949).

Rùzicka, V., and E. S. Domalski: *J. Phys. Chem. Ref. Data,* **22:** 597, 619 (1993).

Sadus, R. J.: "High Pressure Phase Behavior of Multicomponent Fluid Mixtures," Elsevier Science, New York, 1992.

Salim, P. H., and M. A. Trebble: *Fluid Phase Equil.,* **65:** 59 (1991).

Satyro, M. A., and M. A. Trebble: *Fluid Phase Equil.,* **115:** 135 (1996).

Schick, L. M., and J. M. Prausnitz: *AIChE J.,* **14:** 673 (1968).

Setzmann, U., and W. Wagner: *Int. J. Thermophys.,* **10:** 1103 (1989).

Setzmann, U., and W. Wagner: *J. Phys. Chem. Ref. Data,* **20:** 1061 (1991).

Smith, J. M., II. C. Van Ness, and M. M. Abbott: "Introduction to Chemical Engineering Thermodynamics," 5th ed., New York, McGraw-Hill, 1996.

Soave, G.: *Chem. Eng. Sci.,* **27:** 1197 (1972).

Soave, G.: *Inst. Chem. Eng. Symp. Ser.,* **56**(1.2): 1 (1979).

Soave, G.: *Chem. Eng. Sci.,* **39:** 357 (1984).

Soave, G.: *Fluid Phase Equil.,* **82:** 345 (1993).

Spencer, C. F., T. E. Daubert, and R. P. Danner: *AIChE J.,* **19:** 522 (1973).

Stamateris, B., and C. Olivera-Fuentes: Personal Communication, 1995.

Stryjek, R., and J. H. Vera: *Can. J. Chem. Eng.,* **64:** 323 (1986).

Tarakad, R., and T. E. Daubert: API-5-74, Penn. State Univ., University Park, Sept. 23, 1974.

Trebble, M. A., and P. R. Bishnoi: *Fluid Phase Equil.,* **35:** 1 (1987).

Tsonopoulos, C.: *AIChE J.,* **20:** 263 (1974).

Twu, C. H., J. E. Coon, and J. R. Cunningham: *Fluid Phase Equil.,* **75:** 65 (1992).

Twu, C. H., J. E. Coon, and J. R. Cunningham: *Fluid Phase Equil.,* **105:** 49 (1995).

van der Waals, J. H.: *J. Phys. Chem.,* **5:** 133 (1890).

Van Ness, H. C. and M. M. Abbott: "Classical Thermodynamics of Non Electrolyte Solutions," McGraw-Hill, New York, 1982.

West, E. W., and J. H. Erbar: An Evaluation of Four Methods of Predicting Properties of Light Hydrocarbon Systems, paper presented at NGPA 52nd Ann. Meeting, Dallas, TX, March 1973.

Wilson, G. M.: *Adv. Cryogenic Eng.,* **9:** 168 (1964).

Wong, D. S. H., and S. I. Sandler: *AIChE J.,* **38:** 671 (1992).

Younglove, B. A: *J. Phys. Chem. Ref. Data,* **11:** Supp. 1 (1982).

Younglove, B. A., and J. F. Ely: *J. Phys. Chem. Ref. Data,* **16:** 577 (1987).

Zabaloy, M. S., and J. H. Vera: *Ind. Eng. Chem. Res.,* **37:** 1591 (1998).

Zábranský, M., V. Rùzicka, V. Majer, and E. S. Domalski: *Heat Capacity of Liquids: Critical Review and Recommended Values,* Amer. Chem. Soc., and Amer. Inst. Phys. For NIST, Washington D. C., 1996.

CHAPTER SEVEN

VAPOR PRESSURES AND ENTHALPIES OF VAPORIZATION OF PURE FLUIDS

7-1 SCOPE

This chapter covers methods for estimating and correlating vapor pressures of pure liquids. Since enthalpies of vaporization are often derived from vapor pressure-temperature data, the estimation of this property is also included.

7-2 THEORY

When the vapor phase of a pure fluid is in equilibrium with its liquid phase, the equality of chemical potential, temperature, and pressure in both phases leads to the Clapeyron equation (Smith, et al., 1996)

$$\frac{dP_{\text{vp}}}{dT} = \frac{\Delta H_v}{T\Delta V_v} = \frac{\Delta H_v}{(RT^2/P_{\text{vp}})\Delta Z_v} \tag{7-2.1}$$

$$\frac{d \ln P_{\text{vp}}}{d(1/T)} = -\frac{\Delta H_v}{R\Delta Z_v} \tag{7-2.2}$$

In this equation, ΔH_v and ΔZ_v refer to differences in the enthalpies and compressibility factors of saturated vapor and saturated liquid.

Most vapor-pressure estimation and correlation equations stem from an integration of Eq. (7-2.2). To integrate, an assumption must be made regarding the dependence of the group $\Delta H_v/\Delta Z_v$ on temperature. Also a constant of integration is obtained which must be evaluated using one vapor pressure-temperature point.

The simplest approach is to assume that the group $\Delta H_v/\Delta Z_v$ is constant and independent of temperature. Then, with the constant of integration denoted as A, integration of Eq. (7-2.2) leads to

$$\ln P_{\text{vp}} = A - \frac{B}{T} \tag{7-2.3}$$

where $B = \Delta H_v / R\Delta Z_v$. Equation (7-2.3) is sometimes called the Clausius-Clapeyron equation. Surprisingly, it is a fairly good relation for approximating vapor pressure over small temperature intervals. Except near the critical point, ΔH_v and ΔZ_v are both weak functions of temperature; since both decrease with rising temperature, they provide a compensatory effect. However, over large temperature ranges, especially when extrapolated below the normal boiling point, Eq. (7-2.3) normally represents vapor pressure data poorly as shown in Fig. 7-1. The ordinate in Fig. 7-1 is the ratio $[P_{exp} - P_{calc}]/P_{exp}$ and the abscissa $T_r = T/T_c$. P_{calc} is obtained from Eq. (7-2.3) where constants A and B are set by the value of P_{vp} at $T = 0.7T_c$ and $P_{vp} = P_c$ at T_c. Thus, P_{calc} is obtained from

$$\ln (P_{calc}/P_c) = -\beta \left(1 - \frac{1}{T_r}\right) \tag{7-2.4a}$$

$$\text{where } \beta = \frac{7}{3} \ln (\mathrm{P}_{\mathrm{vp}_r} \text{ at } T_r = 0.7) \tag{7-2.4b}$$

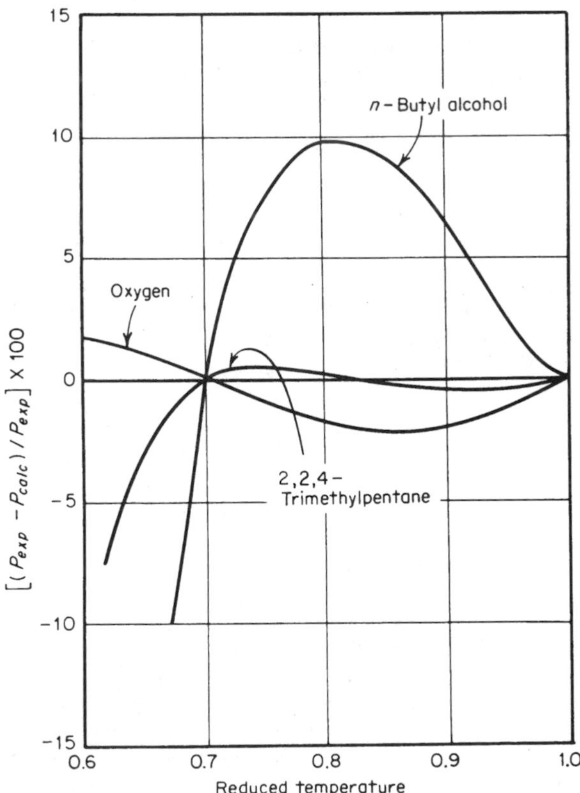

FIGURE 7-1 Comparison of the simple Clapeyron equation with experimental vapor pressure data. (*Ambrose, 1972.*)

Figure 7-1 is a plot of the deviation of the true vapor pressure from that described by Eq. (7-2.4). The figure shows that at high reduced temperatures, the fit of Eq. (7-2.4) is reasonably good for oxygen and a typical hydrocarbon, 2,2,4-trimethylpentane, but for the associating liquid, *n*-butanol, errors as high as 10% can result for reduced temperatures between 0.7 and 1.0. When Eq. (7-2.4) is used to extrapolate to lower temperatures, much larger errors result. For ethanol, for example, Eq. (7-2.4) predicts a vapor pressure at the melting temperature that is 24 times too high.

Extending our consideration of Eq. (7-2.3) one step further, a common practice is to use both the normal boiling point (rather than the vapor pressure at $T_r = 0.7$ as in Eq. 7-2.4) and the critical point to obtain generalized constants. Expressing pressure in bars and temperature on the absolute scale (kelvins or degrees Rankine), with $P_{vp} = P_c$ at $T = T_c$ and $P_{vp} = 1.01325$ at $T = T_b$, the normal boiling temperature at 1 atm = 1.01325 bar, Eq. (7-2.3) becomes

$$\ln P_{vp_r} = h\left(1 - \frac{1}{T_r}\right) \tag{7-2.5}$$

$$h = T_{br}\frac{\ln (P_c/1.01325)}{1 - T_{br}} \tag{7-2.6}$$

The behavior of Eq. (7-2.5) is similar to that of Eq. (7-2.4), i.e., the equation is satisfactory for describing vapor-pressure behavior over small temperature ranges but over large temperature ranges, or when used to extrapolate data, can lead to unacceptably large errors.

7-3 CORRELATION AND EXTRAPOLATION OF VAPOR-PRESSURE DATA

Vapor pressures have been measured for many substances. When reliable measurements are available, they are preferred over results from the estimation methods presented later in this chapter. Boublik (1984) presents tabulations of experimental data that have been judged to be of high quality for approximately 1000 substances. Numerous additional tabulations of "experimental" vapor pressure exist. However, sometimes these vapor pressures are calculated rather than original data and therefore the possibility exists that errors have been introduced in fitting, interpolation, or extrapolation of these data. Literature references to experimental vapor pressure data can be found in Dykyj and Repá (1979), Dykyj, et al. (1984), Lide (1999), Majer, et al. (1989), Ohe (1976), and Perry and Green (1997). Data for environmentally significant solids and liquids including polycyclic aromatics, polychlorinated biphenyls, dioxins, furans, and selected pesticides are compiled in Dellesite (1997).

Many different equations have been presented to correlate vapor pressures as a function of temperature. Two of these, the Antoine and Wagner equations are discussed below.

Antoine Equation

Antoine (1888) proposed a simple modification of Eq. (7-2.3) which has been widely used over limited temperature ranges.

$$\log_{10} P_{vp} = A - \frac{B}{T + C - 273.15} \tag{7-3.1}$$

where T is in kelvins. When $C = 0$, Eq. (7-3.1) reverts to the Clapeyron equation (7-2.3). Simple rules have been proposed (Fishtine, 1963; Thompson, 1959) to relate C to the normal boiling point for certain classes of materials, but these rules are not reliable and the only reliable way to obtain values of the constants A, B, and C is to regress experimental data.

Values of A, B, and C are tabulated for a number of materials in Appendix A with P_{vp} in bars and T in K. Additional tabulations of Antoine constants may be found in Boublik, et al. (1984), Dean (1999), and Yaws (1992). The applicable temperature range is not large and in most instances corresponds to a pressure interval of about 0.01 to 2 bars. The Antoine equation should never be used outside the stated temperature limits. Extrapolation beyond these limits may lead to absurd results. The constants A, B, and C form a set. Never use one constant from one tabulation and the other constants from a different tabulation.

Cox (1923) suggested a graphical correlation in which the ordinate, representing P_{vp} is a log scale, and a straight line (with a positive slope) is drawn. The sloping line is taken to represent the vapor pressure of water (or some other reference substance). Since the vapor pressure of water is accurately known as a function of temperature, the abscissa scale can be marked in temperature units. When the vapor pressure and temperature scales are prepared in this way, vapor pressures for other compounds are often found to be nearly straight lines, especially for homologous series. Calingaert and Davis (1925) have shown that the temperature scale on this Cox chart is nearly equivalent to the function $(T + C)^{-1}$, where C is approximately -43 K for many materials boiling between 273 and 373 K. Thus the Cox chart closely resembles a plot of the Antoine vapor pressure equation. Also, for homologous series, a useful phenomenon is often noted on Cox charts. The straight lines for different members of the homologous series often converge to a single point when extrapolated. This point, called the infinite point, is useful for providing one value of vapor pressure for a new member of the series. Dreisbach (1952) presents a tabulation of these infinite points for several homologous series.

Example 7-1 Calculate the vapor pressure of furan at 309.429 K by using the Antoine equation. The literature value (Boublik, et al., 1984) is 1.20798 bar.

solution From Appendix A constants for Eq. (7-3.1) are $A = 4.11990$, $B = 1070.2$, and $C = 228.83$. With Eq. (7-3.1),

$$\log_{10} P_{vp} = 4.11990 - \frac{1070.2}{309.429 + 228.83 - 273.15}$$

$$P_{vp} = 1.2108 \text{ bar}$$

$$\text{Error} = \frac{1.2108 - 1.20798}{1.20798} \times 100 = 0.2\%$$

Wagner Equation

Wagner (1973, 1977) used an elaborate statistical method to develop an equation for representing the vapor pressure behavior of nitrogen and argon over the entire temperature range for which experimental data were available. In this method, the actual terms as well as their coefficients were variables; i.e., a superfluity of terms was available and the most significant ones were chosen according to statistical criteria. The resulting equation is

$$\ln P_{vp_r} = (a\tau + b\tau^{1.5} + c\tau^3 + d\tau^6)/T_r \qquad (7\text{-}3.2)$$

P_{vp_r} is the reduced vapor pressure, T_r is the reduced temperature, and τ is $1 - T_r$. However, since Eq. (7-3.1) was first presented, the following form has come to be preferred (Ambrose, 1986; Ambrose and Ghiassee, 1987):

$$\ln P_{vp_r} = (a\tau + b\tau^{1.5} + c\tau^{2.5} + d\tau^5)/T_r \qquad (7\text{-}3.3)$$

Both Eqs. (7-3.2) and (7-3.3) can represent the vapor pressure behavior of most substances over the entire liquid range. Various forms of the Wagner equation that employ a fifth term have been presented; for some substances, e.g. water (Wagner, 1973a, 1977), oxygen (Wagner, et al., 1976), and some alcohols (Poling, 1996), this fifth term can be justified. Ambrose (1986) however points out that except in such cases, a fifth term cannot be justified and is not necessary. The constants in Eq. (7-3.2) have been given by McGarry (1983) for 250 fluids. More recently, constants for Eq. (7-3.3) have been given by Ambrose and Ghiassee (1987, 1987a, 1987b, 1988, 1988a, 1990), Ambrose, et al. (1988, 1990), and Ambrose and Walton (1989) for 92 fluids. Table 7-1 presents some of these values while values of the constants a, b, c, d as well as the values of T_c and P_c to be used in Eq. (7-3.3) for all 92 fluids are listed in Appendix A.

Often it is desired to extrapolate a set of vapor pressure data to either lower or higher temperatures. Extrapolation of the Antoine equation or Eq. (7-2.3) is not reliable. One procedure that has been recommended (Ambrose, 1980; Ambrose, et al., 1978; Ambrose and Ghiassee, 1987, and McGarry, 1983) is to use either Eq. (7-3.2) or (7-3.3) where the constants are determined by a constrained fit to the data. Three constraints are commonly used to reproduce features of the vapor-pressure curve that are believed to be valid for all substances. The first of these features is a minimum in the $\Delta H_v/\Delta Z_v$ vs. T_r curve at some reduced temperature between 0.8 and 1.0. This minimum was first observed by Waring (1954) for water. Ambrose and Ghiassee (1987) point out that this constraint causes b and c in Eqs. 7-3.2 and 7-3.3 to have different signs. The second characteristic feature first identified by Thodos (1950) requires that there be an inflection point in the ln P_{vp}

TABLE 7-1 Wagner Constants* for Eq. (7-3.3)

Substance	T_c, K	P_c, bar	ω	a	b	c	d
Propane	369.85	42.47	0.152	−6.76368	1.55481	−1.5872	−2.024
Octane	568.95	24.90	0.399	−8.04937	2.03865	−3.3120	−3.648
Benzene	562.16	48.98	0.21	−7.01433	1.55256	−1.8479	−3.713
Pentaflourotoluene	566.52	31.24	0.415	−8.08717	1.76131	−2.72838	−4.138

*Literature references for constants given in text.

vs. $1/T$ curve. The presence of this inflection point is insured by requiring that the quantity $\ln(P_{vp}/P_{calc})$ at a T_r of 0.95 take on a value that falls within some specified range, where P_{calc} is determined from Eq. (7-2.4). For example, Ambrose, et al. (1978) impose this constraint by requiring that the selected constants generate a P_{vp} value at $T_r = 0.95$ such that $-0.010 < \ln(P_{vp}/P_{calc}) < -0.002$ for non-associated compounds. The third constraint employs the Watson equation, Eq. (7-11.1), to insure that the low-temperature behavior of the vapor-pressure equation matches the temperature dependence of the enthalpy of vaporization predicted by Eq. (7-11.1). To do this, Ambrose, et al. (1978) calculate the quantity $g = \Delta H_v/(1 - T_r)^{0.375}$ at several reduced temperatures between 0.5 and 0.6, where it is supposed to be approximately constant. (ΔH_v is calculated as described in Sec. 7-8.) The constraint is satisfied if the standard deviation of g from its mean value g', over this range, is less than 5%.

Because the Watson equation is not exact, low-temperature behavior is best established by combining vapor-pressure information with thermal data. Introduction of the temperature dependence of ΔH_v into Eq. (7-2.2) leads to the following equation that relates heat capacities to vapor pressure (King and Al-Najjar, 1974)

$$\frac{d}{dT}\left(T^2 \frac{d \ln P_{vp}}{dT}\right) = \frac{C_p^o - C_p^L - \Gamma}{R} \tag{7-3.4}$$

C_p^o is the ideal-gas heat capacity, C_p^L the saturated-liquid heat capacity, and R the gas constant. Γ represents deviations of ΔZ_v from unity and when the truncated virial equation is used for the vapor phase and the liquid volume is assumed independent of pressure, Γ is given by (King and Al-Najjar, 1974)

$$\Gamma = T\left(P_{vp}\left(\frac{d^2B}{dT^2}\right) + 2\left(\frac{dP_{vp}}{dT}\right)\left(\frac{dB}{dT} - \frac{dV^L}{dT}\right) + \left(\frac{d^2P_{vp}}{dT^2}\right)(B - V^L)\right) \tag{7-3.5}$$

where B is the second virial coefficient. When Eq. (7-3.3) is used for vapor pressures, Eq. (7-3.4) leads to

$$\frac{C_p^o - C_p^L - \Gamma}{R} = T_r\left(\frac{3b}{4\sqrt{\tau}} + 3.75c\sqrt{\tau} + 20d\tau^3\right) \tag{7-3.6}$$

By simultaneously fitting heat capacities at low temperatures with Eq. (7-3.6) and vapor pressures at higher temperatures with Eq. (7-3.3), sets of constants can be generated that accurately reproduce vapor pressures down to the melting point. Vapor pressure and thermal data have been simultaneously used by several authors (Ambrose and Davies, 1980; King and Al-Najjar, 1974; Majer, et al., 1989; Moelwyn-Hughes, 1961; Poling, 1996; Růžička and Majer, 1994) to generate more reliable low-temperature vapor-pressure equations. When heat-capacity information is available for a compound, this procedure is preferred to the use of the Watson equation to establish low-temperature vapor pressure behavior. Myrdal and Yalkowsky (1997) have developed an equation in which estimated values are used for ΔC_p and ΔH_v to calculate vapor pressures below one atmosphere.

Example 7-2 Estimate the vapor pressure of ethylbenzene at 347.25 and 460 K with Eq. 7-3.3. Experimental values are 0.13332 bar (Chaiyavech and van Winkle, 1959) and 3.325 bar (Ambrose, et al., 1967), respectively.

solution From Appendix A as well as Ambrose and Ghiassee (1987a), constants for Eq. (7-3.3) are $a = -7.53139$, $b = 1.75439$, $c = -2.42012$, $d = -3.57146$, $T_c = 617.2$ K and $P_c = 36.00$ bar. Eq. (7-3.3) leads to the following results:

				P_{vp}, bar		$\frac{P_{calc} - P_{exp}}{P_{exp}} \times 100$
T, K	T_r	τ	$\ln P_{vpr}$	calc.	exp.	
347.25	0.5626	0.4374	−5.5987	0.13330	0.13332	0.00
460	0.7453	0.2547	−2.3826	3.323	3.325	−0.06

Extended Antoine Equation

The Thermodynamics Research Center at Texas A&M has used the following equation to extend the description of vapor pressure behavior to high temperatures:

$$\log_{10} P_{vp} = A - \frac{B}{T + C - 273.15} + 0.43429x^n + Ex^8 + Fx^{12} \quad (7\text{-}3.7)$$

where P_{vp} is in bar, T is in K, and $x = (T - t_o - 273.15)/T_c$. Values of constants A, B, C, n, E, F and t_o as well as the value of T_c to be used with Eq. (7-3.7) are listed in Appendix A for a number of fluids. Values of T_{min} and T_{max} are also listed in Appendix A. Eq. (7-3.7) is not meant to be extrapolated outside the range of T_{min} and T_{max} and merely provides a best fit of existing vapor-pressure data. At low temperatures, when x becomes negative the last three terms in Eq. (7-3.7) are not used and Eq. (7-3.7) reverts to Eq. (7-3.1).

7-4 AMBROSE-WALTON CORRESPONDING-STATES METHOD

Equation (7-2.4) is a two-parameter corresponding-states equation for vapor pressure. To improve accuracy, several investigators have proposed three-parameter forms. The Pitzer expansion is one of the more successful:

$$\ln P_{vp_r} = f^{(0)}(T_r) + \omega f^{(1)}(T_r) + \omega^2 f^{(2)}(T_r) \quad (7\text{-}4.1)$$

Although a number of analytical expressions have been suggested for $f^{(0)}$, $f^{(1)}$, and $f^{(2)}$ (Brandani, 1993; Schreiber and Pitzer, 1989; Twu, et al., 1994) we recommend the following developed by Ambrose and Walton (1989).

$$f^{(0)} = \frac{-5.97616\tau + 1.29874\tau^{1.5} - 0.60394\tau^{2.5} - 1.06841\tau^5}{T_r} \quad (7\text{-}4.2)$$

$$f^{(1)} = \frac{-5.03365\tau + 1.11505\tau^{1.5} - 5.41217\tau^{2.5} - 7.46628\tau^5}{T_r} \quad (7\text{-}4.3)$$

$$f^{(2)} = \frac{-0.64771\tau + 2.41539\tau^{1.5} - 4.26979\tau^{2.5} + 3.25259\tau^5}{T_r} \quad (7\text{-}4.4)$$

In Eqs. (7-4.2) to (7-4.4), $\tau = (1 - T_r)$. This set of equations was fit to the vapor pressure behavior of the n-alkanes and more accurately describes this behavior than the earlier equations of Lee and Kesler (1975). The quantity, $f^{(2)}$, is important only for fluids with large acentric factors and at low reduced temperatures. In fact, it is zero at $T_r = 0.7$.

Equation (7-4.1) relies on a fluid's properties being similar to those on the n-alkanes. Several authors (Armstrong, 1981; Teja, et al., 1981) have suggested that reference fluids more similar to the unknown fluid be used according to

$$\ln P_{vp_r} - \ln P_{vp_r}^{(R1)} + (\ln P_{vp_r}^{(R2)} - \ln P_{vp_r}^{(R1)}) \frac{\omega - \omega^{(R1)}}{\omega^{(R2)} - \omega^{(R1)}} \tag{7-4.5}$$

The superscripts, R1 and R2, refer to the two reference substances. Ambrose and Patel (1984) used either propane and octane, or benzene and pentafluorotoluene as the reference fluids. However, it is permissible to use any two substances chemically similar to the unknown fluid whose vapor-pressure behavior is well established. The vapor pressure-behavior of the above four fluids can be calculated with Eq. (7-3.3) and the constants in Table 7-1. In Eq. (7-4.5), all vapor pressures are calculated at the reduced temperature of the substance whose vapor pressure is to be predicted. Equation (7-4.5) is written so as to estimate vapor pressures. However, if two or more vapor pressures are known in addition to T_c and ω, Eq. (7-4.5) can be used to estimate the critical pressure. Ambrose and Patel (1984) have examined this application and report average errors in P_c of about 2% for 65 fluids. This can be as good as the methods in Chap. 2. When at least three vapor pressures are known, it is mathematically possible to estimate T_c also, but Ambrose and Patel (1984) indicate that this procedure does not yield accurate results.

When using Eq. (7-4.5), more reliable estimates are obtained when

$$\omega^{(R1)} < \omega < \omega^{(R2)} \tag{7-4.6}$$

Equation (7-4.6) represents an interpolation in the acentric factor rather than an extrapolation. Use of Eq. (7-4.5) to estimate vapor pressures is illustrated by Example 7-3 and Fig. 7-2. Within the accuracy of the graph, the dashed line in Fig. 7-2 coincides with the literature data for ethylbenzene in (Ambrose, et al., 1967; Willingham, et al., 1945).

Example 7-3 Repeat Example 7-2 using Eq. (7-4.5) and benzene and pentafluorotoluene as reference fluids.

solution For ethylbenzene, from Appendix A, $T_c = 617.15$ K, $\omega = 0.304$, and $P_c = 36.09$ bar. Using benzene as R1 and pentafluorotoluene as R2, Eqs. (7-3.3) and (7-4.5) along with the values in Table 7-1 leads to results that are nearly as accurate as the correlation results in Example 7-2.

T, K	T_r	$\ln P_{vp_r}^{(R1)}$	$\ln P_{vp_r}^{(R2)}$	$\ln P_{vp_r}$	P_{vpcalc} bar	P_{vpexp} bar	$\frac{P_{calc} - P_{exp}}{P_{exp}} \times 100$
347.25	0.5627	−5.175	−6.111	−5.604	0.1329	0.1333	0.32
460	0.7454	−2.215	−2.585	−2.385	3.325	3.325	−0.01

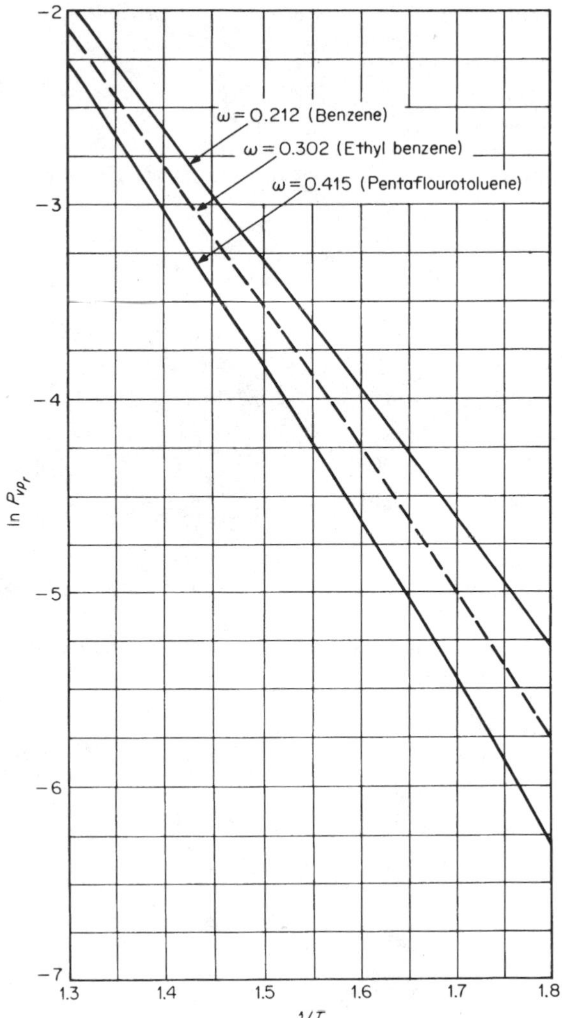

FIGURE 7-2 Vapor pressure prediction for ethylbenzene by the two-reference fluid method; — from Eq. (7-3.2); -- from Eq. (7-4.5).

7-5 RIEDEL CORRESPONDING-STATES METHOD

Riedel (1954) proposed a vapor pressure equation of the form

$$\ln P_{vp} = A + \frac{B}{T} + C \ln T + DT^6 \tag{7-5.1}$$

The T^6 term allows description of the inflection point of the vapor pressure curve in the high-pressure region. To determine the constants in Eq. (7-5.1), Riedel defined a parameter α

$$\alpha \equiv \frac{d \ln P_{vp_r}}{d \ln T_r} \tag{7-5.2}$$

From a study of experimental vapor pressure data, Plank and Riedel (1948) showed that

$$\frac{d\alpha}{dT_r} = 0 \qquad \text{at } T_r = 1 \tag{7-5.3}$$

Using Eq. (7-5.3) as a constraint on (7-5.1), Riedel found that

$$\ln P_{vp_r} = A^+ - \frac{B^+}{T_r} + C^+ \ln T_r + D^+ T_r^6 \tag{7-5.4}$$

where $\quad A^+ = -35Q \qquad B^+ = -36\,Q \qquad C^+ = 42\,Q + \alpha_c$

$$D^+ = -Q \qquad Q = K(3.758 - \alpha_c) \tag{7-5.5}$$

where α_c is α at the critical point. Riedel originally chose K to be 0.0838, but Vetere (1991) has found that for alcohols and acids, improved predictions result if the expressions shown in Table 7-2 are used. The correlating parameter in these expressions, h, is defined by Eq. (7-2.6).

Since it is not easy (or desirable) to determine α_c by its defining equation at the critical point, α_c is usually found from Eqs. (7-5.4) and (7-5.5) by inserting $P =$ 1.01325 bar at $T = T_b$ and calculating α_c. The equations that result from this process are:

$$\alpha_c = \frac{3.758K\psi_b + \ln(P_c/1.01325)}{K\psi_b - \ln T_{br}} \tag{7-5.6}$$

$$\psi_b = -35 + \frac{36}{T_{br}} + 42 \ln T_{br} - T_{br}^6 \tag{7-5.7}$$

Example 7-4 Repeat Example 7-2 using the Riedel Correlation.

solution For ethylbenzene, $T_b = 409.36$ K, $T_c = 617.15$ K, and $P_c = 36.09$ bar. Thus, $T_{br} = 409.36/617.15 = 0.663$. From Eq. (7-5.7)

$$\psi_b = -35 + \frac{36}{0.663} + 42 \ln 0.663 - (0.663)^6 = 1.9525$$

Then, with Eq. (7-5.6) and $K = 0.0838$,

TABLE 7-2 Vetere Rules for Riedel Constant K for Acids and Alcohols

acids	$K = -0.120 + 0.025h$
alcohols	$K = 0.373 - 0.030h$

$$\alpha_c = \frac{(3.758)(0.0838)(1.9525) + \ln(36.09/1.01325)}{(0.0838)(1.9525) - \ln 0.663} = 7.2881$$

The constants in Eq. (7-5.5) become

$$Q = (0.838)(3.758 - 7.2881) = -0.2958$$

$$A+ = -35Q = 10.354$$

$$B+ = -36Q = 10.650$$

$$C+ = 42Q + \alpha_c = -5.136$$

$$D+ = -Q - 0.2958$$

and Eq. (7-5.4) becomes

$$\ln P_{vp_r} = 10.354 - \frac{10.650}{T_r} - 5.136 \ln T_r + 0.2958 T_r^6$$

At 347.25 and 460 K,

T, K	T_r	P_{vpcalc} bar	P_{vpexp} bar	$\dfrac{P_{\text{calc}} - P_{\text{exp}}}{P_{\text{exp}}} \times 100$
347.25	0.5627	0.1321	0.1333	−0.66
460	0.7454	3.361	3.325	1.08

These errors are marginally higher than those obtained in Examples 7-2 and 7-3.

7-6 DISCUSSION AND RECOMMENDATIONS FOR VAPOR-PRESSURE ESTIMATION AND CORRELATION

Starting from the Clausius-Clapeyron equation, Eq. (7-2.2), we have shown only a few of the many vapor-pressure equations which have been published. We have emphasized those which appear to be most accurate and general for correlation (section 7-3) and estimation (sections 7-4 and 7-5). Properties required for the different estimation equations are T_b, T_c, and P_c for Vetere's modification of the Riedel method, and ω, T_c, and P_c for the Ambrose-Walton corresponding-states methods. For typical fluids, these techniques accurately predict vapor pressures over wide ranges of temperature with little input. We show in Table 7-3 a detailed comparison between calculated and experimental vapor pressures for acetone, 1-octanol, and tetradecane for the three estimation techniques described in this chapter. For acetone, the temperature range covers the melting point to the critical point, 47 bars. For n-tetradecane, the lowest temperature is the triple point. The least accurate correlation is, as expected, the Clapeyron equation, especially at lower temperatures.

The Antoine equation should not be used outside the range of the experimental data to which the constants have been correlated. In the range for which the fitted

TABLE 7-3 Comparison Between Calculated and Experimental Vapor Pressures

					Percent error				
	T, K	P, exp bar	Ref.*	T_r	Clapeyron Eq. (7-2.5)	Antoine Eq. (7-3.1)	Wagner Eq. (7-3.3)	Ambrose Eq. (7-4.1)	Riedel Eq. (7-5.4)
acetone	178.2	2.31E-05	1	0.351	328	−21	−3.9	−23	−12
	209.55	0.000944	7	0.412	113	−4.9	−1	−13	−11
	237.04	0.009965	7	0.467	47	−2.4	−1.8	−9.4	−10
	259.175	0.04267	4	0.510	25	−0.1	−0.1	−4.9	−6.5
	285.623	0.1748	4	0.562	9.9	0	0	−2.3	−3.6
	320.47	0.74449	4	0.631	1.1	0	−0.1	−0.3	−0.5
	390.32	5.655	4	0.768	−1.3	0.1	0.3	1.3	2.1
	446.37	17.682	4	0.879	0.1	−2	−0.1	0.5	0.9
	470.61	26.628	4	0.926	0.6	−3.5	−0.2	0.1	0.4
	499.78	41.667	4	0.984	0.3	−6.4	−0.2	−0.2	0
	508.1	47	4	1	0	−7.4	0	0	0
1-octanol	328.03	0.00142	2	0.503	353	−17	−3	95	6.2
	395.676	0.0781	3	0.606	48	0	0	17	−2.6
	457.45	0.7451	3	0.701	2.7	−0.1	−0.1	−0.3	−0.7
	521.1	3.511	5	0.799	−4.3	0.2	0	−2.6	3
	554	6.499	5	0.849	−2.8	0.1	0	−1.5	4.2
n-tetradecane	279	2.65E-06	8	0.403	1274	−50	0	1.3	27
	350	1.14E-03	8	0.505	164	−7.5	−0.2	0.5	3.7
	400	1.71E-02	8	0.577	54	−0.9	0	0.3	0.3
	464.384	0.19417	6	0.670	11	0.1	−0.1	0	−0.2
	509.163	0.67061	6	0.735	1.6	0	−0.2	−0.1	−0.1
	527.315	1.0285	6	0.761	−0.1	0	−0.2	−0.2	−0.1

Percent error = [(calc.-exp.)/exp.] × 100. Wagner constants from Ambrose and Ghiassee (1987a) and Ambrose and Walton (1989). Antoine constants from pages 179, 688, and 833 of Boublik (1984).

*Data refs.: 1, Ambrose and Davies (1980); 2, Ambrose et al. (1974); 3, Ambrose and Sprake (1970); 4, Ambrose et al., (1974a); 5, Ambrose et al. (1975); 6, Camin and Rossini (1955); 7, Felsing (1926); 8, King and Al-Najjar (1974).

parameters are applicable, it is very accurate. Of all the methods shown in Table 7-3, the Wagner equation has the largest number of correlated parameters and is the most accurate. Both the Ambrose-Walton and Riedel estimation methods perform well at higher temperatures. At low temperatures the Ambrose-Walton method does better for the hydrocarbon, *n*-tetradecane, while the Riedel method does better for the alcohol, 1-octanol.

The estimation methods presented in this chapter require the critical properties of a compound. When the critical properties are not known, a predictive method recently presented by Li, et al. (1994) may be used. This is a combination group-contribution, corresponding-states method that requires only the normal boiling point and the chemical structure of the compound. The authors claim more accurate results than those obtained with the Riedel method. If the normal boiling point is not known, the predictive group-contribution method reported by Tu (1994) may be used. Tu reports average errors of 5% with his method. Both of these latter two methods are applicable to polar as well as nonpolar fluids. When no vapor-pressure data are available, when one is uncertain of the molecular structure of a compound, or if a petroleum fraction is being considered, the SWAP equation (Edwards, et al., 1981; Macknick, et al., 1978; Smith, et al. 1976) may be used. Other correlations that have been published may be found in Bloomer, 1990; Campanella, 1995; Ledanois, et al., 1997 and Xiang and Tan, 1994.

Recommendations. If constants are available in Appendix A or other reference for a particular fluid, use these along with the appropriate equation. The Antoine equation should not be used for temperatures outside the range listed in Appendix A. The Wagner equation may be extrapolated to higher temperatures with confidence. The Wagner equation may be used down to a reduced temperature of 0.5 or to the value of T_{min} listed in Appendix A. At reduced temperatures below 0.5, it is most desirable to use correlations that have incorporated thermal information such as those in Ambrose and Davies (1980), King and Al-Najjar (1974), Majer, et al. (1989), Moelwyn-Hughes (1961), Poling (1996), and Růžička and Majer (1994). Poling (1996) presents an example that shows the magnitude of errors one can expect by different extrapolation methods, and thus suggests the minimum information required to achieve a desired accuracy. If constants based on thermal information are not available, the Wagner equation constrained to fit the Watson equation is recommended for low temperature predictions. The Ambrose-Walton method and Riedel methods are recommended over the Clapeyron or Antoine equation at low temperatures. For polar compounds at reduced temperatures between 0.5 and 1.0, the two-reference-fluid or Riedel method is recommended. If no data are available for a compound, and its normal boiling point is unknown, one of the group-contribution methods mentioned above may be used, but in that event, calculated results may not be highly accurate.

7-7 ENTHALPY OF VAPORIZATION OF PURE COMPOUNDS

The enthalpy of vaporization ΔH_v is sometimes referred to as the latent heat of vaporization. It is the difference between the enthalpy of the saturated vapor and that of the saturated liquid at the same temperature.

Because molecules in the vapor do not have the energy of attraction that those in the liquid have, energy must be supplied for vaporization to occur. This is the internal energy of vaporization ΔU_v. Work is done on the vapor phase as vaporization proceeds, since the vapor volume increases if the pressure is maintained constant at P_{vp}. This work is $P_{vp}(V^G - V^L)$. Thus

$$\Delta H_v = \Delta U_v + P_{vp}(V^G - V^L) = \Delta U_v + RT(Z^G - Z^L)$$
$$= \Delta U_v + RT\Delta Z_v \tag{7-7.1}$$

Many "experimental" values of ΔH_v have been calculated from Eq. (7-2.2), where it was shown that ΔH_v is related to the slope of the vapor pressure-temperature curve. More recently, experimental techniques have developed to the point that many experimentally determined values are available; these are often more accurate than those calculated with Eq. (7-2.2). Majer and Svoboda (1985) is a comprehensive and critical compilation of experimental values of ΔH_v measured since 1932 for approximately 600 organic compounds. They give recommended values for ΔH_v at both the normal boiling point and 298 K, as well as values for the three constants, α, β, and A, for many of the 600 compounds that can be used in the following equation to correlate ΔH_v with reduced temperature:

$$\Delta H_v = A(1 - T_r)^{\beta} \exp(-\alpha T_r) \tag{7-7.2}$$

Additional compilations of heat-of-vaporization information can be found in Tamir, et al. (1983). In spite of the increased availability of experimental values, it is usually necessary to supplement data with results calculated or extrapolated by some method. Majer, et al. (1989) presents a comprehensive description of the various methods that have been used to determine ΔH_v. Three of the methods are reviewed in Secs. 7-8 to 7-10.

7-8 ESTIMATION OF ΔH_v FROM VAPOR-PRESSURE EQUATIONS

The vapor-pressure relations presented in Secs. 7-2 to 7-5 can be used to estimate enthalpies of vaporization. From Eq. (7-2.2), we can define a dimensionless group ψ

$$\psi \equiv \frac{\Delta H_v}{RT_c\Delta Z_v} = \frac{-d \ln P_{vp_r}}{d(1/T_r)} \tag{7-8.1}$$

Differentiating the vapor-pressure equations discussed earlier, we can obtain various expressions for ψ. These are shown in Table 7-4. To use these expressions, one must refer to the vapor-pressure equation given earlier in this chapter for the definition of the various parameters.

In Fig. 7-3, we show experimental values of ψ for propane. These were calculated from smoothed values tabulated in Das and Eubank (1973) and Yarbrough and Tsai (1978). Note the pronounced minimum in the curve around $T_r = 0.8$. Since

$$\psi = \frac{-d \ln P_{vp_r}}{d(1/T_r)} \tag{7-8.2}$$

we have

TABLE 7-4 Expressions for ψ for Various Vapor-Pressure Equations

Vapor-pressure equation	Expression for ψ	
Clapeyron, Eq. (7-2.5)	h, defined in Eq. (7-2.6)	(T7.4*a*)
Antoine, Eq. (7-3.1)	$\frac{2.303B}{T_c}\left(\frac{T_r}{T_r + (C - 273.15)/T_c}\right)^2$	(T7.4*b*)
Wagner, Eq. (7-3.2)	$-a + b\tau^{0.5}(0.5\tau - 1.5) + c\tau^2(2\tau - 3) + d\tau^5(5\tau - 6)$	(T7.4*c*)
Wagner, Eq. (7-3.3)	$-a + b\tau^{0.5}(0.5\tau - 1.5) + c\tau^{1.5}(1.5\tau - 2.5) + d\tau^4(4\tau - 5)$	(T7.4*d*)
Extended Antoine, Eq. (7-3.7)	$\frac{2.303B}{T_c}\left(\frac{T_r}{T_r + (C - 273.15)/T_c}\right)^2 + T_r^2(nx^{n-1} + 18.421Ex^7 + 27.631Fx^{11}$	(T7.4*e*)
Ambrose-Walton Eq. (7-4.1)	$5.97616 + 1.29874\tau^{0.5}(0.5\tau - 1.5) - 0.60394\tau^{1.5}(1.5\tau - 2.5) - 1.06841\tau^4(4\tau - 5) + \omega[5.03365 + 1.11505\tau^{0.5}(0.5\tau - 1.5) - 5.41217\tau^{1.5}(1.5\tau - 2.5) - 7.46628\tau^4(4\tau - 5)] + \omega^2[0.64771 + 2.41539\tau^{0.5}(0.5\tau - 1.5) - 4.26979\tau^{1.5}(1.5\tau - 2.5) + 3.25259\tau^4(4\tau - 5)]$	(T7.4*f*)
Riedel, Eq. (7-5.4)	$B^+ + C^+T_r + 6D^+T_r^7$	(T7.4*g*)

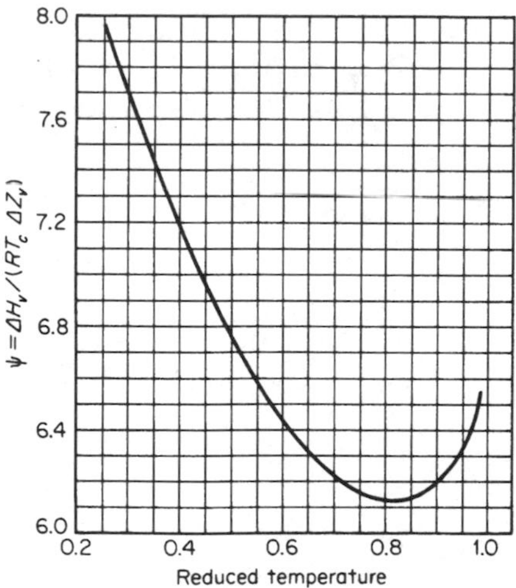

FIGURE 7-3 Literature values of ψ for propane.

$$\frac{d\psi}{dT_r} = \frac{1}{T_r^2}\frac{d^2 \ln P_{vp_r}}{d(1/T_r)^2} \tag{7-8.3}$$

At low values of T_r, $d\psi/dT_r < 0$ so that $(d^2 \ln P_{vp_r})/d(1/T_r)^2$ is also < 0. At high values of T_r, the signs reverse. When $d\psi/dT_r = 0$, there is an inflection point in the ln P_{vp} vs. $1/T$ curve. Thus the general (though exaggerated) shape of a log (Antoine constants from App. A were used) vapor-pressure-inverse-temperature curve is that shown in Fig. 7-4.

Figure 7-5 illustrates how well the Riedel and Antoine vapor-pressure equations (Antoine constants from Appendix A were used) are able to predict the shape of Fig. 7-3. The Antoine equation does not predict the $\psi - T_r$ minimum, and deviates from the true behavior outside the range over which the constants were fit, which was $0.46 < T_r < 0.67$. The Riedel equation reproduces the true behavior at all but low temperatures. The Clapeyron equation predicts a constant value of ψ of 6.22. The curves generated with the Wagner constants in Appendix A along with Eq. (T7.4*d*) or with the Ambrose-Walton equation, Eq. (T7.4*f*) are not shown in Fig. 7-5 because they agree with the literature values over the T_r range shown in the figure to within 0.3%. Thus, except for the Clapeyron and Antoine equations as discussed above, we may recommend any of the vapor-pressure correlations in Table 7-4 to predict ψ, and thus ΔH_v. However, accurate values of ΔZ_v must be available. ΔZ_v is determined best as a difference in the Z values of saturated vapor and saturated liquid. These Z values may be determined by methods in Chap. 4.

7-9 ESTIMATION OF ΔH_v FROM THE LAW OF CORRESPONDING STATES

Equation (7-8.1) can be rearranged to

$$\frac{\Delta H_v}{RT_c} = -\Delta Z_v \frac{d \ln P_{vp_r}}{d(1/T_r)} \tag{7-9.1}$$

The reduced enthalpy of vaporization $-\Delta H_v/RT_c$ is a function of $(d \ln P_{vp_r})/$

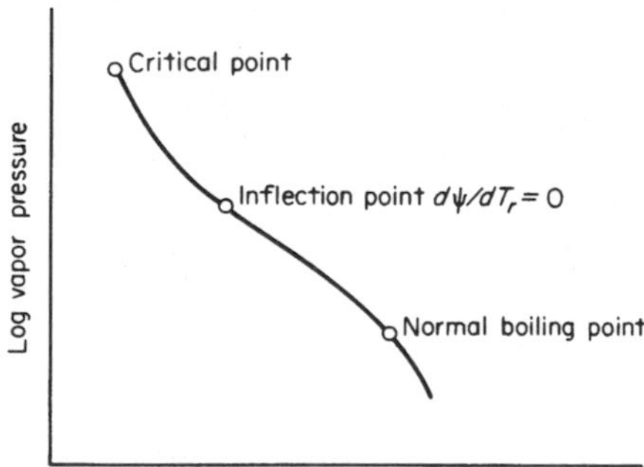

FIGURE 7-4 Schematic vapor pressure plot.

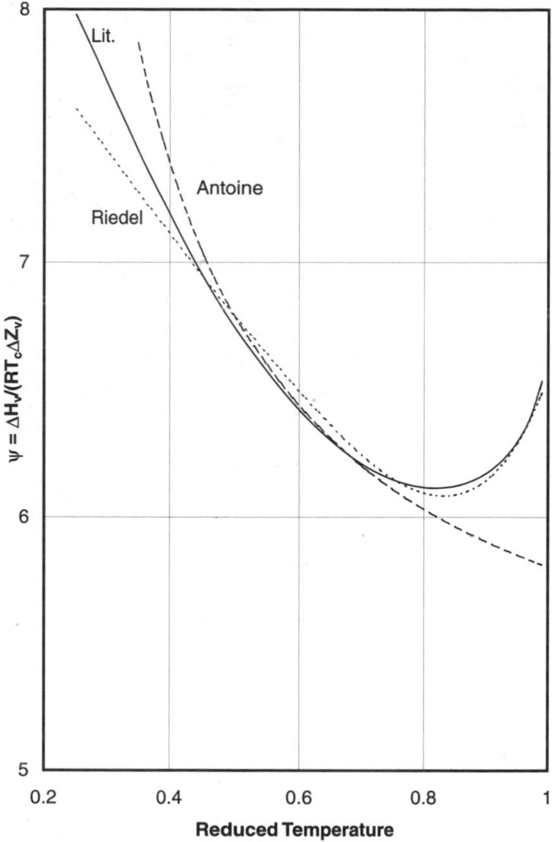

FIGURE 7-5 Comparison between calculated and experimental values of $\Delta H_v/(RT_c\Delta Z_v)$ for propane.

$d(1/T_r)$ and ΔZ_v. Both these properties are commonly assumed to be functions of T_r or P_{vp_r} and some third parameter such as ω or Z_c.

Pitzer, et al. have shown that ΔH_v can be related to T, T_r and ω by an expansion similar to that used to estimate compressibility factors Eq. (4-3.1), i.e.,

$$\frac{\Delta H_v}{T} = \Delta S_v^{(0)} + \omega \Delta S_v^{(1)} \tag{7-9.2}$$

where $\Delta S_v^{(0)}$ and $\Delta S_v^{(1)}$ are expressed in entropy units, for example, J/(mol·K), and are functions only of T_r. Multiplying Eq. (7-9.2) by T_r/R gives

$$\frac{\Delta H_v}{RT_c} = \frac{T_r}{R} (\Delta S_v^{(0)} + \omega \Delta S_v^{(1)}) \tag{7-9.3}$$

Thus $\Delta H_v/RT_c$ is a function only of ω and T_r. From the tabulated $\Delta S_v^{(0)}$ and $\Delta S_v^{(1)}$

functions given by Pitzer, et al., Fig. 7-6 was constructed. For a close approximation, an analytical representation of this correlation for $0.6 < T_r < 1.0$ is

$$\frac{\Delta H_v}{RT_c} = 7.08(1 - T_r)^{0.354} + 10.95\,\omega(1 - T_r)^{0.456} \tag{7-9.4}$$

Example 7-5 Using the Pitzer, et al. corresponding-states correlation, estimate the enthalpy of vaporization of propionaldehyde at 321.1 K. The literature value is 28,310 J/mol (Majer and Svoboda, 1985).

solution For propionaldehyde (Daubert, et al., 1997), $T_c = 504.4$ K and $\omega = 0.313$. $T_r = 321.1/504.4 = 0.637$, and from Eq. (7-9.4)

$$(\Delta H_v/RT_c) = 7.08(1 - 0.637)^{0.354} + (10.95)(0.313)(1 - 0.637)^{0.456} = 7.11$$

$$\Delta H_v = (7.11)(8.314)(504.4) = 29{,}816 \text{ J/mol}$$

$$\text{Error} = \frac{29{,}816 - 28{,}310}{28{,}310} \times 100 = 5.3\%$$

This error is not unexpected since propionaldehyde is not a "normal" fluid (see Section 4-3).

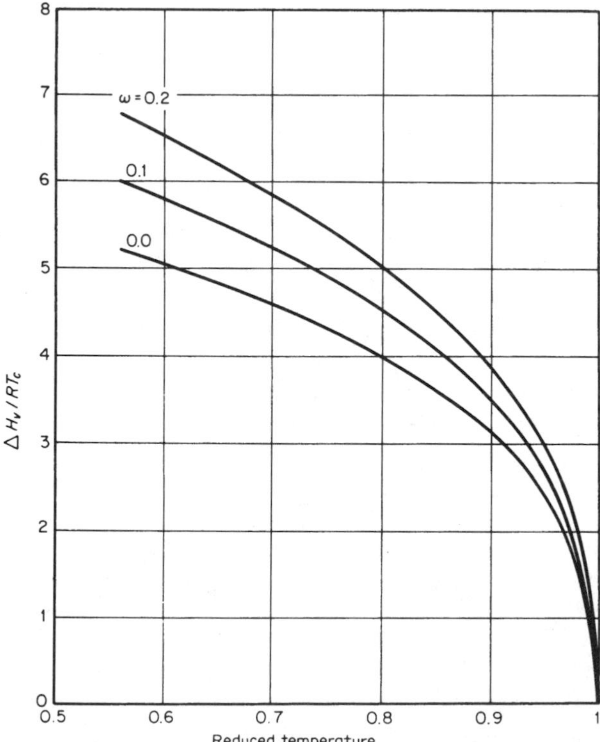

FIGURE 7-6 Plot of Pitzer, et al. correlation for enthalpies of vaporization.

7-10 ΔH_v AT THE NORMAL BOILING POINT

A pure-component constant that is occasionally used in property correlations is the enthalpy of vaporization at the normal boiling point ΔH_{vb}. Any one of the correlations discussed in Sec. 7-8 or 7-9 can be used for this state where $T = T_b$, $P =$ 1.01325 bar. Some additional techniques are discussed below. Several special estimation methods are also suggested.

ΔH_{vb} From Vapor-Pressure Relations

In Table 7-4, we show equations for $\psi = \Delta H_v/(RT_c\Delta Z_v)$ as determined from a few of the more accurate vapor-pressure equations. Each can be used to determine ψ (T_b). With $\psi(T_b)$ and $\Delta Z_v(T_b)$, ΔH_{vb} can be estimated.

When the Clapeyron equation is used to calculate ψ [see Eq. (T7.4*a*) in Table 7-4], ψ is equal to h regardless of T_r, that is,

$$\psi(T_r) = \psi(T_b) = T_{br}\frac{\ln(P_c/1.013)}{1 - T_{br}} \tag{7-10.1}$$

and

$$\Delta H_{vb} = RT_c\Delta Z_{vb}\, T_{br}\frac{\ln(P_c/1.013)}{1 - T_{br}} \tag{7-10.2}$$

Equation (7-10.2) has been widely used to make rapid estimates of ΔH_{vb}; usually, in such cases, ΔZ_{vb} is set equal to unity. In this form, it has been called the Giacalone Equation (Giacalone, 1951). Extensive testing of this simplified form indicates that it normally overpredicts ΔH_{vb} by a few percent.

The Kistiakowsky rule (see p. 165 of Majer, et al., 1989) is another simple equation that can be used to estimate ΔH_{vb}:

$$\Delta H_{vb} = (36.1 + R\ln T_b)T_b \tag{7-10.3}$$

Correction terms have been suggested (Fishtine, 1963 and Klein, 1949) to improve the accuracy of the Giacalone and Kistiakowsky equations, but better results are obtained with other relations, noted below.

Riedel Method

Riedel (1954) modified Eq. (7-10.2) slightly and proposed that

$$\Delta H_{vb} = 1.093\, RT_c\, T_{br}\frac{\ln P_c - 1.013}{0.930 - T_{br}} \tag{7-10.4}$$

Chen Method

Chen (1965) used Eq. (7-9.3) and a similar expression proposed by Pitzer, et al. to correlate vapor pressures so that the acentric factor is eliminated. He obtained a relation between ΔH_v, P_{vp_r}, and T_r. When applied to the normal boiling point,

$$\Delta H_{vb} = RT_c T_{br} \frac{3.978T_{br} - 3.958 + 1.555 \ln P_c}{1.07 - T_{br}} \tag{7-10.5}$$

Vetere Methods

Vetere (1979, 1995) proposed a relation similar to the one suggested by Chen. When applied to the normal boiling point:

$$\Delta H_{vb} = RT_b \frac{(1 - T_{br})^{0.38}(\ln P_c - 0.513 + 0.5066/(P_c T_{br}^2))}{1 - T_{br} + F(1 - (1 - T_{br})^{0.38}) \ln T_{br}} \tag{7-10.6}$$

F is 1.05 for C_2+ alcohols and dimerizing compounds such as SO_3, NO, and NO_2. For all other compounds investigated by Vetere, F is 1.0.

When T_c and P_c are not available, Vetere proposed

$$\Delta H_{vb} = RT_b \left(A + B \ln T_b + \frac{CT_b^{1.72}}{M'}\right) \tag{7-10.7}$$

Constants A, B, and C are given in Table 7-5 for a few classes of compounds. M' is a fictitious molecular weight that is equal to the true molecular weight for most compounds. But for fluids that contain halogens or phosphorus, the molecular-weight contributions for these atoms are those shown in Table 7-6.

Table 7-7 compares calculated and experimental values of ΔH_{vb} using the estimation methods described in this section. The Riedel, Chen, and the Vetere equation (7-10.6) are generally accurate to 2%. For these three methods, T_b, T_c, and P_c must be known or estimated. The Vetere equation (7-10.7) performs as well; it has the advantage that T_c and P_c are not required.

TABLE 7-5 Constants for Eq. (7-10.7)

	A	B	C
hydrocarbons and CCl_4	3.298	1.015	0.00352
alcohols	−13.173	4.359	0.00151
esters	4.814	0.890	0.00374
other polar compounds	4.542	0.840	0.00352

TABLE 7-6 Contributions to Fictitious Molecular Weight, M'

Atom	Contribution
F	1
Cl	19.6
Br	60
I	60
P	24

TABLE 7-7 Comparison Between Calculated and Literature Values of ΔH_{vb}

						Percent error*				
	M'	T_c, K	T_b, K	ΔH_{vb} kJ/mol	P_c bar	Giacalone Eq. (7-10.2)	Riedel Eq. (7-10.4)	Chen Eq. (7-10.5)	Vetere Eq. (7-10.6)	Vetere Eq. (7-10.7)
pentane	72.15	469.6	309.2	25.79	33.7	2.3	0.5	0.1	−0.1	0.2
octane	114.231	568.8	398.8	34.41	24.9	3.2	1.3	0.1	−0.7	−0.8
3-methyl pentane	86.177	504.4	336.4	28.06	31.2	2.6	0.5	0.0	−0.1	0.8
cyclohexane	84.161	553.4	353.9	29.97	40.7	0.6	−0.5	−0.8	−1.0	0.8
1-pentene	70.134	464.7	303.1	25.2	35.3	2.1	0.4	0.0	0.0	0.3
1-octene	112.215	566.6	394.4	34.07	26.2	3.0	1.3	0.1	−0.7	−1.1
benzene	78.114	562.1	353.3	30.72	48.9	−0.2	−0.3	−0.6	−1.1	−1.1
ethylbenzene	106.167	617.1	409.3	35.57	36	1.4	0.8	0.0	−0.7	−0.2
hexafluorobenzene	78.066	516.7	353.4	31.66	33	2.3	2.4	1.0	−0.4	−2.0
1,2-dichloroethane	67.254	561.2	356.6	31.98	53.7	1.0	2.2	1.6	0.3	−0.2
$C_2Br_2ClF_3$	206.43	560.7	366	31.17	36.1	0.5	−1.0	−1.4	−1.6	−3.0
propylamine	59.111	497	321.7	29.55	48.1	−0.9	0.1	−0.7	−2.0	−3.9
pyridine	79.101	620	388.4	35.09	56.3	−1.0	0.0	−0.4	−1.5	−0.5
ethyl propyl ether	88.15	500.2	336.3	28.94	33.7	3.3	2.6	1.7	0.8	−0.3
methyl phenyl ether	108.14	644.1	426.8	38.97	42.5	0.8	1.9	0.8	−0.8	−2.4
ethanol	46.069	513.9	351.4	38.56	61.4	−1.7	4.4	1.3	1.4	−0.3
1-pentanol	88.15	588.2	411.1	44.36	39.1	−6.5	−3.3	−6.1	−4.9	4.8
propanal	58.08	496.2	321.1	28.31	63.3	0.8	1.3	0.8	−0.2	0.3
acetone	58.08	508.2	329.3	29.10	47	2.6	3.5	2.7	1.4	0.7
3-methyl-2-butanone	86.134	553.4	367.4	32.35	38.5	2.2	2.3	1.4	0.2	−0.3
acetic acid	60.053	592.7	391.1	37.48	57.9	3.2	6.9	5.3	2.5	−2.5
ethyl acetate	88.106	523.2	350.3	31.94	38.3	0.2	0.7	−0.4	−1.8	0.7
tetrahydrofuran	72.107	540.2	339.1	29.81	51.9	0.0	0.4	0.1	−0.7	−0.4
carbon disulphide	76.143	552	319.4	26.74	79	2.7	3.7	3.8	3.3	2.5
thiophene	84.142	579.4	357.3	31.48	56.9	−0.8	−0.3	−0.5	−1.2	−0.8
ethyl mercaptan	62.136	499	308.2	26.79	54.9	−0.1	0.2	0.0	−0.6	−0.2
nitromethane	61.04	588	374.4	33.99	63.1	4.2	6.9	6.1	3.9	1.3
$C_3H_3Cl_2F_3O$	97.256	559.3	384.9	35.67						−5.3
$C_3H_2Cl_2F_4O$	97.25	518.5	360.5	32.69						−4.7

*Percent error = [(calc. − exp.)/exp.] × 100.
Lit. value for acetic acid from Majer, et al. (1989), others from Majer and Svoboda (1985).

Example 7-6 Use Vetere's two methods, Eqs. (7-10.6) and (7-10.7) to estimate the enthalpy of vaporization of propionaldehyde at the normal boiling point. The experimental value is 28310 J/g-mol (Majer and Svoboda, 1985).

solution For propionaldehyde Daubert, et al. (1997) give T_b = 321.1 K, T_c = 504.4 K, and P_c = 49.2 bar. Thus T_{br} = 0.6366.

Vetere Method with Eq. (7-10.5). With F = 1.0, Eq. (7-10.5) becomes

$$\Delta H_{vb} = (8.314)(321.1) \frac{(1 - 0.6366)^{0.38}(\ln 49.2 - 0.513 + 0.5066/[(49.2)(0.6366)^2]}{1 - 0.6366 + (1 - (1 - 0.6366)^{0.38}) \ln 0.6366}$$

$$= 28{,}260 \text{ J/g-mol}$$

$$\text{Error} = \frac{28{,}260 - 28{,}310}{28{,}310} \times 100 = -0.2\%$$

Vetere Method With Eq. (7-10.6). Since propionaldehyde contains none of the atoms listed in Table 7-6, M' is the same as M and is 58.08. Equation (7-10.6) becomes

$$\Delta H_{vb} = (8.314)(321.1)\left(4.542 + 0.840 \ln 321.1 + \frac{(0.00352)(321.1)^{1.72}}{58.08}\right)$$

$$= 28{,}383 \text{ J/g-mol}$$

$$\text{Error} = \frac{28{,}380 - 28{,}310}{28{,}310} \times 100 = 0.3\%$$

Various group contribution methods have been proposed (Constantinou and Gani, 1994; Fedors, 1974; Guthrie and Taylor, 1983; Hoshino, et al., 1983; Lawson, 1980; Ma and Zhao, 1993; McCurdy and Laidler, 1963; Tu and Liu, 1996) to estimate ΔH_{vb}, Majer, et al. (1989) summarize these methods and constants for the Joback (1984, 1987) and Constantinou-Gani (1994) methods are in Appendix C. These methods do not require a value of T_b. The Constantinou-Gani (1994) method for ΔH_v at 298 K, ΔH_{v298}, uses the equation

$$\Delta H_{v298} = 6.829 + \sum_i N_i(hv1i) + W \sum_j M_j(hv2j) \tag{7-10.8}$$

where N_i and M_j and the number of occurrences of First-Order group i and Second-Order group j respectively. The group-contribution values $hv1$ and $hv2$ are in Appendix C. The value of W is set to zero for First-Order calculations and to unity for Second-Order calculations. This method is illustrated with Example 7-7.

Example 7-7 Use Eq. (7-10.8) to estimate ΔH_{v298} for n-butanol and 2-butanol. Literature values are 52.35 and 49.72 kJ/mol respectively.

solution n-butanol contains groups CH_3, CH_2, and OH and no Second-Order groups. Thus, with Eq. (7-10.8) and group contributions from Table C-2.

$$\Delta H_{v298} = 6.829 + 4.116 + 3 \times 4.650 + 24.529 = 49.42 \text{ kJ/mol}$$

$$\text{Error} = \frac{49.42 - 52.35}{52.35} \times 100 = -5.6\%$$

2-butanol contains First-Order groups CH_3, CH_2, CH and OH and also contains the Second-Order group, CHOH, with a contribution of −1.398. Again, with Eq. (7-10.8) and group contributions from Appendix C

$$\Delta H_{v298} = 6.829 + 2 \times 4.116 + 4.650 + 2.771$$
$$+ 24.529 - 1.398 = 45.61 \text{ kJ/mol}$$

$$\text{Error} = \frac{45.61 - 49.72}{49.72} \times 100 = -8.3\%$$

Note that including the Second-Order term for 2-butanol in the previous example actually makes the prediction worse, rather than better. Including the Second-Order terms improves predictions about two-thirds of the time and makes it worse in the other one-third. For further discussion of the Constantinou-Gani method, see Chaps. 2 and 3.

7-11 VARIATION OF ΔH_v WITH TEMPERATURE

The latent heat of vaporization decreases steadily with temperature and is zero at the critical point. Typical data are shown in Fig. 7-7. The shapes of these curves agree with most other enthalpy-of-vaporization data. The variation of ΔH_v with temperature could be determined from any of the ψ relations shown in Table 7-4, although the variation of ΔZ_v with temperature would also have to be specified.

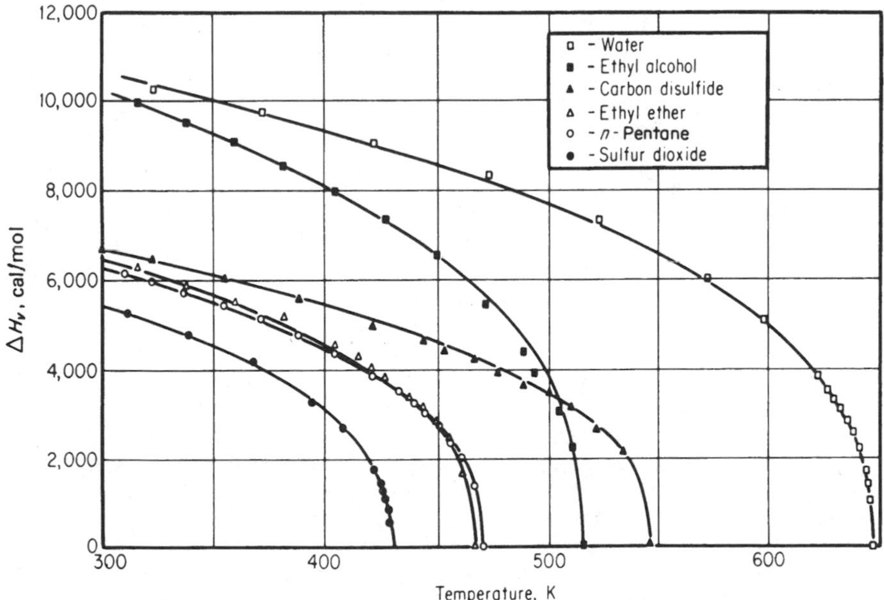

FIGURE 7-7 Enthalpies of vaporization.

A widely used correlation between ΔH_v and T_r is the Watson relation (Thek and Stiel, 1967)

$$\Delta H_{v2} = \Delta H_{v1} \left(\frac{1 - T_{r2}}{1 - T_{r1}} \right)^n \tag{7-11.1}$$

where subscripts 1 and 2 refer to reduced temperatures T_{r1} and T_{r2}. A common choice for n is 0.375 or 0.38 (Thodos, 1950).

Table 7-8 shows a comparison between experimental results and those calculated by Eqs. (7-9.4) and (7-11.1). In Table 7-8 the value of n in Eq. (7-11.1) was 0.38. The value of ΔH_v calculated by Eq. (7-11.1) necessarily agrees with that measured at the normal boiling point. For propane, both methods give good agreement over nearly the entire liquid range; for 1-pentanol, agreement is not so good with errors as high as 13% and 8% for the two methods, respectively.

7-12 DISCUSSION AND RECOMMENDATIONS FOR ENTHALPY OF VAPORIZATION

If experimental values, or constants that correlate experimental values for a particular fluid are available in Majer, et al., (1989) for example, use those. Otherwise, values of ΔH_v may be estimated by one of the three techniques described above. The first is based on Eq. (7-2.1) and requires finding dP_{vp}/dT from a vapor-pressure-temperature correlation (Sec. 7-8). A separate estimate of ΔZ_v must be made before ΔH_v can be obtained. This procedure can give results as accurate as those measured experimentally, especially if the vapor-pressure equation is accurate and if ΔZ_v is obtained from reliable P-V-T correlations discussed in Chap. 4. Any number of combinations can be used. At low temperatures, the method is most accurate if thermal information has been used to establish the low temperature vapor-pressure behavior.

In the second category are the techniques from the principle of corresponding states. The Pitzer et al. form is one of the most accurate and convenient. In an analytical form, this equation for ΔH_v is approximated by Eq. (7-9.4). Thompson and Braun (1964) also recommended the Pitzer, et al. form for hydrocarbons. The critical temperature and acentric factor are required.

In the third method, first estimate ΔH_{vb} as recommended in Sec. 7-10. Then scale with temperature with the Watson equation discussed in Sec. 7-11. All three of these techniques are satisfactory and yield approximately the same error when averaged over many types of fluids and over large temperature ranges.

Finally, for most correlations discussed here, T_c and P_c are required either directly or indirectly. Although these constants are available for many fluids—and can be estimated for most others—there are occasions when one would prefer not to use critical properties. (For example, for some high-molecular-weight materials or for polyhydroxylated compounds, it is difficult to assign reliable values to T_c and P_c.) For such cases, Vetere's equation (7-10.6) may be used first to estimate the enthalpy of vaporization at the normal boiling point. Then scale with temperature with the Watson relation (described in Sec. 7-11) with an estimated T_c.

TABLE 7-8 Comparison Between Calculated and Literature Values of ΔH_v

T, K	$\Delta H_{v,\text{lit}}$** kJ/mol	T_r	Percent error* Eq. (7-9.4)	Percent error* Eq. (7-11.1)
Propane, T_c = 369.85, ω = 0.153, T_b = 231.1				
90	24.56	0.243	−1.2	1.2
100	24.12	0.270	−0.8	1.6
150	22.06	0.406	0.5	2.8
200	20.10	0.541	0.2	2.3
231.1	19.04	0.625	−1.9	0.0
250	17.80	0.676	−0.6	1.2
277.6	16.28	0.751	−1.4	0.2
299.9	14.67	0.811	−1.3	0.1
316.5	13.22	0.856	−0.9	0.2
338.8	10.77	0.916	−0.5	0.1
349.9	9.11	0.946	−0.1	0.0
361	6.73	0.976	0.3	−0.6
1-pentanol, T_c = 588.2, ω = 0.579, T_b = 411.1				
298.2	56.94	0.507	−13.2	−6.0
328.2	54.43	0.558	−13.1	−5.7
358.2	51.22	0.609	−12.1	−4.3
374.4	49.16	0.637	−11.0	−3.0
411.2	44.37	0.699	−8.6	0.0
431.1	41.32	0.733	−6.4	2.6
479.4	34.62	0.815	−3.5	6.5
499.7	31.61	0.850	−2.6	7.9

* Percent error = [(calc − exp)/exp] × 100
For Eq (7-11.1), n = 0.38.
** $\Delta H_{v,\text{lit}}$ from Das and Eubank (1973), Majer and Svoboda (1985), and Yarbrough and Tsai (1978).

7-13 ENTHALPY OF FUSION

The enthalpy change on fusion or melting is commonly referred to as the latent heat of fusion. References for literature values are listed in Tamir, et al. (1983). Domalski and Hearing (1996) is an extensive tabulation of experimental values. The enthalpy of fusion is related to the entropy change on fusion and the melting point temperature by

$$\Delta H_m = T_{fp} \Delta S_m \tag{7-13.1}$$

Reliable methods to estimate the enthalpy or entropy of fusion at the melting point have not been developed. However, as first suggested by Bondi (1963), methods have been developed to estimate the total change in entropy, ΔS_{tot}, due to phase changes when a substance goes from the solid state at 0 K to a liquid at its melting point. For substances that do not have solid-solid transitions, ΔS_{tot} and ΔS_m are the same. For these substances, the method to estimate ΔS_{tot} along with Eq. (7-13.1) can be used to estimate ΔH_m. But for substances that do demonstrate solid-solid transitions, ΔS_{tot} can be much greater than ΔS_m. This can be seen in Table 7-9 which lists ΔS_{tot} and ΔS_m for 44 hydrocarbons. For 14 of the 44 hydrocarbons listed, ΔS_{tot} and ΔS_m are different due to solid-solid transitions below the melting point. This difference, ΔS_{sol}, is tabulated in the third column of the table. When there is a solid-solid transition, ΔS_m can be much less than ΔS_{tot}. For example, 2,2-dimethylbutane has two solid-solid transitions, one at 127 K for which $\Delta S = 42.66$ J/(mol·K) and one at 141 K for which $\Delta S = 2.03$ J/(mol·K). For this compound, $\Delta S_{tot} = 47.9$ J/(mol·K) while $\Delta S_m = 3.3$ J/(mol·K).

For substances that do have solid-solid transitions, no reliable method exists for the estimation of ΔS_m because there is no way to predict whether solid-solid transitions occur. Still the methods to estimate ΔS_{tot} represent a significant development. One of these methods is described below.

Dannenfelser-Yalkowski Method (Dannenfelser and Yalkowsky, 1996)

In this method, ΔS_{tot} is calculated by

$$\Delta S_{tot} = 50 - R \ln \sigma + 1.047\, R\tau \qquad (7\text{-}13.2)$$

where R is the gas constant, 8.314 J/(mol·K), ΔS_{tot} is in J/(mol·K), σ is a symmetry number, and τ is the number of torsional angles. The symmetry number is the number of ways a molecule can be *rigidly* rotated so that atoms are in a different position but the molecule occupies the same space. Thus, σ for benzene and carbon tetraflouride is 12, and for flourobenzene it is 2. In the assignment of a value to σ, the structure is hydrogen suppressed and the following groups are assumed to be radially symmetrical and/or freely rotating: halogens, methyl, hydroxyl, mercapto, amine, and cyano. This is different than in Benson's method for ideal gas properties (see Chap. 3). Thus, for propane, $\sigma = 2$; for all higher n-alkanes, $\sigma = 1$. For spherical molecules, such as methane or neon, $\sigma = 100$. Molecules that are conical (e.g., hydrogen cyanide and chloromethane) or cylindrical (e.g., carbon dioxide and ethane) have one infinite rotational axis and are empirically assigned σ values 10 and 20 respectively. Dannenfelser and Yalkowsky (1996) gives additional examples and the value of σ for 949 compounds is given in the supplementary material of Dannenfelser and Yalkowsky (1996) which may be found at http://pubs.acs.org. The quantities σ and τ are also used for estimating T_{fp} (see Section 2-4).

The number of torsional angles, τ, is calculated by

$$\tau = \text{SP3} + 0.5\text{SP2} + 0.5\text{RING} - 1 \qquad (7\text{-}13.3)$$

SP3 is the number of sp^3 chain atoms, SP2 is the number of sp^2 chain atoms, and RING is the number of fused-ring systems. τ cannot be less than zero. Note that the radially symmetrical end groups mentioned above, as well as carbonyl oxygen and *tert*-butyl groups are not included in the number of chain atoms. This method is illustrated in Example 7-8, results for 44 hydrocarbons are listed in Table 7-9, and results for 949 compounds (as well as the value of τ) are listed at the above web address. For the 44 compounds in Table 7-9, the average absolute difference (AAD) between the literature values and those calculated with Eq. (7-13.2) was 9.9 J/(mol·K). For the 43 compounds when tetratricontane is excluded, this represents an everage error in ΔH_m of 1700 J/mol. For the 29 compounds in Table 7-9 without solid-solid phase transitions, this represents an average error of 18% in ΔH_m.

Chickos, et al. (1990, 1991, 1998) have developed a group contribution method to estimate ΔS_{tot}. Their method is more accurate than Eq. (7-13.2) (AAD of 7.4 as

TABLE 7-9 Entropies and Enthalpies of Fusion

	ΔS_m at T_{fp}	ΔS_{tot}	ΔS_{sol}	ΔS_{tot} Eq. (7-13.2)	diff	T_{fp}, K	ΔH_m
Methane	10.4	14.9	4.5	11.7	−3.2	90.7	943
Ethane	6.5	31.9	25.4	25.1	−6.8	89.9	584
Propane	41.2	41.2		44.2	3	85.5	3523
n-Butane	34.6	53.8	19.2	58.7	4.9	134.8	4664
Isobutane	39.9	39.9		40.9	1	113.6	4533
n-Pentane	58.6	58.6		67.4	8.8	143.4	8403
Isopentane	45.4	45.4		58.7	13.3	113.3	5144
Neopentane	12.4	30.9	18.5	29.3	−1.6	256.6	3182
n-Hexane	71	71		76.1	5.1	177.8	12624
2-Methylpentane	52.4	52.4		67.4	15	119.5	6262
2,2-Dimethylbutane	3.3	47.9	44.6	50	2.1	173.3	572
2,3-Dimethylbutane	5.5	53	47.5	58.7	5.7	144.6	795
n-Heptane	78	78		84.8	6.8	182.6	14243
2-Methylhexane	59.3	59.3		76.1	16.8	154.9	9186
3-Ethylpentane	61.8	61.8		76.1	14.3	154.6	9554
2,2-Dimethylpentane	39	39		58.7	19.7	149.4	5827
2,4-Dimethylpentane	44.5	44.5		67.4	22.9	154	6853
3,3-Dimethylpentane	49.3	55.3	6	67.4	12.1	138.7	6838
2,2,3-Trimethylbutane	8.9	28.5	19.6	50	21.5	248.3	2210
n-Octane	95.8	95.8		93.5	−2.3	216.4	20731
2-Methylheptane	72.6	72.6		84.8	12.2	164.2	11921
n-Nonane	70.4	99.3	28.9	102.2	2.9	219.7	15467
n-Decane	117.9	117.9		111	−6.9	243.5	28709
n-Dodecane	139.7	139.7		128.4	−11.3	263.6	36825
n-Octadecane	203	203		180.6	−22.4	301.3	61164
n-Nonadecane	153	199	46	189.3	−9.7	305	46665
n-tetratricontane	231.2	371	139.8	319.9	−51.1	346	79995
Benzene	35.4	35.4		29.3	−6.1	278.7	9866
Toluene	37.2	37.2		44.2	7	178	6622
Ethylbenzene	51.4	51.4		54.4	3	178.2	9159
o-Xylene	54.9	54.9		44.2	−10.7	248	13615
m-Xylene	51.4	51.4		44.2	−7.2	225.3	11580
p-Xylene	59.8	59.8		38.5	−21.3	286.4	17127
n-Propylbenzene	53.4	53.4		63.1	9.7	173.7	9276
Isopropylbenzene	41.4	41.4		54.4	13	177.1	7332
1,2,3-Trimethylbenzene	33	41.8	8.8	44.2	2.4	247.1	8154
1,2,4-Trimethylbenzene	57.5	57.5		50	−7.5	227	13053
1,3,5-Trimethylbenzene	41.7	41.7		35.1	−6.6	228.4	9524
Cyclohexane	9.5	45.5	36	35.1	−10.4	279.6	2656
Methylcyclohexane	46.1	46.1		50	3.9	146.6	6758
Ethylcyclohexane	51.5	51.5		54.4	2.9	161.8	8333
1,1-Dimethylcyclohexane	8.4	47.4	39	50	2.6	239.7	2013
1,cis-2-Dimethylcyclohexane	7.4	55.2	47.8	50	−5.2	223.1	1651
1,trans-2-Dimethylcyclohexane	56.7	56.7		50	−6.7	185	10490

ΔS in J/mol K, ΔH in J/mol, lit. values of ΔS from Domalski and Hearing (1996). T_{fp} values from Dreisbach (1995, 1959).

diff = calc. − lit.

compared to 9.9 J/(mol·K) for the 44 compounds in Table 7-9), but it is not as easy to use as Eq. (7-13.2).

Example 7-8 Calculate ΔS_{tot} and ΔH_m for isobutane with Eqs. (7-13.1) and (7-13.2). Literature values (Domalski and Hearing, 1996) are 39.92 J/(mol·K) and 4540 J/mol respectively. The melting point temperature is 113.2 K.

solution For isobutane, SP3 = 1 so $\tau = 0$ and $\sigma = 3$. With Eqs. (7-13.2), then (7-13.1)

$$\Delta S_{tot} = 50 - 8.314 \ln 3 = 40.9 \text{ J/(mol·K)}$$

$$\Delta H_m = 113.2 \times 40.9 = 4630 \text{ J/mol}$$

$$\text{Error} = \frac{4630 - 4540}{4540} \times 100 = 2.0\%$$

7-14 ENTHALPY OF SUBLIMATION; VAPOR PRESSURES OF SOLIDS

Solids vaporize without melting (sublime) at temperatures below the triple-point temperature. Sublimation is accompanied by an enthalpy increase, or latent heat of sublimation. This may be considered to be the sum of a latent heat of fusion and a hypothetical latent heat of vaporization, even though liquid cannot exist at the pressure and temperature in question.

The latent heat of sublimation ΔH_s is best obtained from solid vapor-pressure data. For this purpose, the Clausius-Clapeyron equation, Eq. (7-2.1) is applicable. In only a very few cases is the sublimation pressure at the melting point known with accuracy. At the melting point, the sublimation pressure is the same as the vapor pressure of the liquid which can be determined by using thermal data as discussed in Sec. 7-3. However, even if P_{vp} at T_m is known, at least one other value of the vapor pressure of the solid is necessary to calculate ΔH_s from the integrated form of the Clausius-Clapeyron equation. Vapor-pressure data for solids may be found in Dellesite (1997), Oja and Suuberg (1998, 1999), and Pouillot, et al. (1996).

In some cases, it is possible to obtain ΔH_s from thermochemical data by subtracting known values of the enthalpies of formation of solid and vapor. This is hardly a basis for estimation of an unknown ΔH_s, however, since the enthalpies of formation tabulated in the standard references are often based in part on measured values of ΔH_s. If the enthalpies of dissociation of both solid and gas phases are known, it is possible to formulate a cycle including the sublimation of the solid, the dissociation of the vapor, and the recombination of the elements to form the solid compound.

Finally, as a rough engineering rule, one might estimate ΔH_v and ΔH_m separately and obtain ΔH_s as the sum. The latent heat of fusion is usually less than one-quarter of the sum, therefore the estimate may be fair even though that for ΔH_m is crude.

If enthalpy-of-fusion information is available, Eq. (7-2.1) may be used to estimate vapor pressures of solids in addition to liquid vapor pressures. The technique, along with its limitations, is illustrated with Example 7-9 (Prausnitz, et al., 1999).

Example 7-9 Use information at the triple point and Eq. (7-2.1) to estimate the vapor pressure of ice at 263 K.

solution Equation (7-2.1) may be written for the solid in equilibrium with vapor

$$\frac{dP^S_{vp}}{dT} = \frac{\Delta H_s}{T\Delta V_s} \tag{i}$$

and for hypothetical subcooled liquid in equilibrium with vapor

$$\frac{dP^L_{vp}}{dT} = \frac{\Delta H_v}{T\Delta V_v} \tag{ii}$$

In this example ΔV in both (i) and (ii) may be taken as

$$\Delta V = V^G - V^L \text{ (or } V^S) \cong V^G = \frac{RT}{P}$$

Subtracting Eq. (i) from (ii), and using $\Delta H_v - \Delta H_s = -\Delta H_m$, we obtain

$$\ln\left(\frac{P_{263}}{P_{273}}\right)^L - \ln\left(\frac{P_{263}}{P_{273}}\right)^S = -\int_{273}^{263} \frac{\Delta H_m}{RT^2}\, dT \tag{iii}$$

ΔH_m is given as a function of temperature by

$$\Delta H_m = (\Delta H_m \text{ at } T_1) + \int_{T_1}^{T} (C_p^L - C_p^S)\, dT \tag{iv}$$

For H_2O, ΔH_m at 273 K is 6008 J/mol; for liquid, $C_p^L = 75.3$ J/mol K, and for ice, $C_p^S \cong 37.7$ J/(mol K).

$$\begin{aligned}\Delta H_m &= 6008 + (37.6)(T - 273) \\ &= 37.6\, T - 4284\end{aligned} \tag{v}$$

Substitution into (iii) and integration gives

$$\ln \frac{P^S_{263}}{P^S_{273}} - \ln \frac{P^L_{263}}{P^L_{273}} = \frac{37.6}{8.314} \ln \frac{263}{273} + \frac{4257}{8.314}\left(\frac{1}{263} - \frac{1}{273}\right) \tag{vi}$$

$P^S_{273} = P^L_{273}$ and P^L_{263} is the vapor pressure of subcooled liquid at 263 K. An extrapolation based on the assumption that $\ln P^L_{vp}$ vs $1/T$ is linear gives $P^L_{263} = 0.00288$ bar. Solving Eq. (vi) gives $P^S_{263} = 0.00261$ bars, which is the same as the literature value.

The technique used in Example 7-9 requires care. Discontinuities can occur in ΔH_m because of the solid-solid transition as discussed in Sec. 7-14. Unless these are accounted for, the integration as shown above in (iii) will not be correct. Also, the extrapolation of vapor pressures for hypothetical sub-cooled liquid over large temperature ranges is uncertain (see Sec. 7-3).

NOTATION

a, A, b, B, c, C, d, D	empirical coefficients
A^+, B^+, C^+, D^+	parameters in Riedel equation
B	second virial coefficient used in Eq. (7-3.5)
C_p°	ideal gas heat capacity

C_p^L	heat capacity of saturated liquid; C_p^s, heat capacity of solid
$f^{(0)}$, $f^{(1)}$, $f^{(2)}$	functions of reduced temperature in Eq. (7-4.1) and defined in Eqs. (7-4.2) and (7-4.4)
h	parameter defined in Eq. (7-2.6)
ΔH_m	enthalpy change on melting, J/mol
ΔH_v	enthalpy of vaporization, J/mol, ΔH_{vb}, at T_b; ΔH_v°, at 298 K
ΔH_s	enthalpy of sublimation, J/mol
K	Vetere constant, see Table 7-2
M	molecular weight, M' is fictitious molecular weight, see Table 7-6
n	exponent in Eq. (7-12.1), usually chosen as 0.38
P	pressure, bar unless stated otherwise, $P_r = P/P_c$, P_c is critical pressure, P_{vp} is vapor pressure, $P_{vp_r} = P_{vp}/P_c$
Q	parameter in Riedel equation
R	gas constant, 8.3145 J/mol K
ΔS_m	entropy change on melting, J/mol K; ΔS_{tot}, total phase change entropy between 0 K and T_m
ΔS_v	entropy of vaporization, J/mol K; $\Delta S^{(0)}$ and $\Delta S^{(1)}$, Pitzer parameters in Eq. (7-9.2)
T	temperature, K; $T_r = T/T_c$; T_b, normal boiling point; $T_{br} = T_b/T_c$, T_{fp}, melting point, T_c, critical temperature, K
ΔU_v	internal energy of vaporization, J/mol
V	volume, cm^3/mol, V^G, saturated vapor, V^L, saturated liqud
V_c	critical volume, cm^3/mol
ΔV_v	volume change on vaporization, cm^3/mol
ΔV_S	volume change on sublimation, cm^3/mol
Z	compressibility factor, PV/RT, Z^G, saturated vapor, Z^L, saturated liquid
ΔZ_v	$Z^G - Z^L$, ΔZ_{vb}, at normal boiling point
Z_c	Z at the critical point

Greek

α, β	empirical constants in Eq. (7-7.2)
α	Riedel factor, see Eq. (7-5.2), α_c, at critical point
Γ	defined in Eq. (7-3.5)
τ	$1 - T_r$
ψ	$\Delta H_v/(RT_c\Delta Z_v)$
ψ_b	Riedel parameter, defined in Eq. (7-5.7)
ω	acentric factor

Superscripts

(0)	simple fluid property
(1)	deviation function
(R1)	property for reference fluid 1
(R2)	property for reference fluid 2

REFERENCES

Ambrose, D.: "Vapor-Pressure Equations," *Natl. Phys. Lab. Rep. Chem.*, **19:** November 1972.

Ambrose, D.: "Vapor-Pressure Equations," *Natl. Phys. Lab. Rep. Chem.*, **114:** August 1980.

Ambrose, D.: *J. Chem. Thermodynamics,* **18:** 45 (1986).

Ambrose, D., and R. H. Davies: *J. Chem. Thermodynamics,* **12:** 871 (1980).

Ambrose, D., and N. B. Ghiassee: *J. Chem. Thermodynamics,* **19:** 505 (1987).

Ambrose, D., and N. B. Ghiassee: *J. Chem. Thermodynamics,* **19:** 903 (1987a).

Ambrose, D., and N. B. Ghiassee: *J. Chem. Thermodynamics,* **19:** 911 (1987b).

Ambrose, D., and N. B. Ghiassee: *J. Chem. Thermodynamics,* **20:** 765 (1988).

Ambrose, D., and N. B. Ghiassee: *J. Chem. Thermodynamics,* **20:** 1231 (1988a).

Ambrose, D., and N. B. Ghiassee: *J. Chem. Thermodynamics,* **22:** 307 (1990).

Ambrose, D., and N. C. Patel: *J. Chem. Thermodynamics,* **16:** 459 (1984).

Ambrose, D., and C. H. S. Sprake: *J. Chem. Thermodynamics,* **2:** 631 (1970).

Ambrose, D., and J. Walton: *Pure & Appl. Chem.*, **61:** 1395 (1989).

Ambrose, D., B. E. Broderick, and R. Townsend: *J. Chem. Soc.,* **1967A:** 633.

Ambrose, D., J. H. Ellender, and C. H. S. Sprake: *J. Chem. Thermodynamics,* **6:** 609 (1974).

Ambrose, D., C. H. S. Sprake, and R. Townsend: *J. Chem. Thermodynamics,* **6:** 693 (1974a).

Ambrose, D., C. H. S. Sprake, and R. Townsend: *J. Chem. Thermodynamics,* **7:** 185 (1975).

Ambrose, D., J. F. Counsell, and C. P. Hicks: *J. Chem. Thermodynamics,* **10:** 771 (1978).

Ambrose, D., N. B. Ghiassee, and R. Tuckerman: *J. Chem. Thermodynamics,* **20:** 767 (1988).

Ambrose, D., M. B. Ewing, N. B. Ghiassee, and J. C. Sanchez Ochoa: *J. Chem. Thermodynamics,* **22:** 589 (1990).

Antoine, C.: *C.R.,* **107:** 681, 836 (1888).

Armstrong, B.: *J. Chem. Eng. Data,* **26:** 168 (1981).

Bloomer, O. T.: *Ind. Eng. Chem. Res.,* **29:** 128 (1990).

Bondi, A.: *J. Chem. Eng. Data,* **8:** 371 (1963).

Boublik, T., V. Fried, and E. Hala: *The Vapor Pressures of Pure Substances,* 2d rev. ed., Elsevier, New York, 1984.

Brandani, S.: *Ind. Eng. Chem. Res.,* **32:** 756 (1993).

Calingaert, G., and D. S. Davis: *Ind. Eng. Chem.,* **17:** 1287 (1925).

Camin, D. L., and F. D. Rossini: *J. Phys. Chem.,* **59:** 1173 (1955).

Campanella, E. A.: *J. Chem. Eng. Japan,* **28:** 234 (1995).

Chaiyavech, P., and M. van Winkle: *J. Chem. Eng. Data,* **4:** 53 (1959).

Chen, N. H.: *J. Chem. Eng. Data,* **10:** 207 (1965).

Chickos, J. S. and D. G. Hesse: *J. Org. Chem.,* **55:** 3833 (1990).

Chickos, J. S., C. M. Braton, D. G. Hesse, and J. F. Liebman: *J. Org. Chem.,* **56:** 927 (1991).

Chickos, J. S., W. E. Acree Jr., and J. F. Liebman: *Estimating Phase Change Enthalpies and Entropies, in Computational Thermochemistry, Prediction and Estimation of Molecular Thermodynamics.* D. Frurip and K. Irikura, (eds.), ACS Symp. Ser. 677, p. 63, ACS, Washington, D. C., 1998.

Constantinou, L., and R. Gani: *AIChE J.,* **40:** 1697 (1994).

Cox, E. R.: *Ind. Eng. Chem.,* **15:** 592 (1923).

Dannenfelser, R.-M., and S. H. Yalkowsky: *Ind. Eng. Chem. Res.,* **35:** 1483 (1996).

Das, T. R. and P. T. Eubank: *Adv. Cryog. Eng.,* **18:** 208 (1973).

Daubert, T. E., R. P. Danner, H. M. Sibel, and C. C. Stebbins: *Physical and Thermodynamic Properties of Pure Chemicals: Data Compilation,* Taylor & Francis, Washington, D. C., 1997.

Dean, J. A.: *Lange's Handbook of Chemistry,* 15th ed., McGraw-Hill, New York, 1999.

Dellesite, A.: *J. Phys. Chem. Ref. Data,* **26:** 157 (1997).

Domalski, E. S., and E. D. Hearing: *J. Phys. Chem. Ref. Data,* **25:** 1 (1996).

Dreisbach, R. R.: *Pressure-Volume-Temperature Relationships of Organic Compounds,* 3d ed., McGraw-Hill, New York, 1952.

Dreisbach, R. R.: Physical Properties of Chemical Compounds, *Advan. Chem. Ser., ACS Monogr. 15 and 22,* Washington, D.C., 1955, 1959.

Dykyj, J., and M. Repá: *The Vapour Pressures of Organic Compounds* (in Slovak), Veda, Bratislava, 1979.

Dykyj, J., M. Repá and J. Svoboda: *The Vapour Pressures of Organic Compounds* (in Slovak), Veda, Bratislava, 1984.

Edwards, D., Van de Rostyne, C. G., Winnick, J., and J. M. Prausnitz: *Ind. Eng. Chem. Process Des. Dev.,* **20:** 138 (1981).

Fedors, R. F.: *Polym. Eng. Sci.,* **14:** 147 (1974).

Felsing, W.A.: *J. Am Chem. Soc.,* **26:** 2885 (1926).

Fishtine, S. H.: *Ind. Eng. Chem.,* **55**(4): 20, **55**(5): 49, **55**(6): 47(1963); *Hydrocarbon Process. Pet. Refiner,* **42**(10): 143(1963).

Giacalone, A.: *Gazz. Chim. Ital.,* **81:** 180 (1951).

Guthrie, J. P., and K. F. Taylor: *Can. J. Chem.,* **61:** 602 (1983).

Hoshino, D., K. Nagahama, and M. Hirata: *Ind. Eng. Chem. Fundam.,* **22:** 430 (1983).

Joback, K. G.: "A Unified Approach to Physical Property Estimation Using Multivariate Statistical Techniques," S. M. Thesis, Department of Chemical Engineering, Massachusetts Institute of Technology, Cambridge, MA, 1984.

Joback, K. G., and R. C. Reid: *Chem. Eng. Comm.,* **57:** 233 (1987).

King, M. G., and H. Al-Najjar: *Chem. Eng. Sci.,* **29:** 1003 (1974).

Klein, V. A.: *Chem. Eng. Prog.,* **45:** 675 (1949).

Lawson, D. D.: *App. Energy,* **6:** 241 (1980).

Ledanois, J.-M., C. M. Colina, J. W. Santos, D. González-Mendizabal, and C. Olivera-Fuentes: *Ind. Eng. Chem. Res.,* **36:** 2505 (1997).

Lee, B. I., and M. G. Kesler: *AIChE J.,* **21:** 510 (1975).

Li, P., P.-S. Ma, S.-Z. Yi, Z.-G. Zhao, and L.-Z. Cong: *Fluid Phase Equil,* **101:** 101 (1994).

Lide, D. R.: *CRC Handbook of Chemistry and Physics,* 80th ed., CRC Press, Boca Raton, FL, 1999.

Ma, P., and X. Zhao: *Ind. Eng. Chem. Res.,* **32:** 3180 (1993).

Macknick, A. B., J. Winnick, and J. M. Prausnitz: *AIChE J.,* **24:** 731 (1978).

Majer, V., and V. Svoboda: Enthalpies of Vaporization of Organic Compounds, A Critical Review and Data Compilation, *IUPAC Chem. Data Ser No.* 32, Blackwell Sci. Pub., Oxford, 1985.

Majer, V., V. Svoboda, and J. Pick: Heats of Vaporization of Fluids, *Studies in Modern Thermodynamics* **9,** Elsevier, Amsterdam, 1989.

McCurdy, K. G., and K. J. Laidler: *Can. J. Chem.,* **41:** 1867 (1963).

McGarry, J.: *Ind. Eng. Chem. Process Des. Dev.,* **22:** 313 (1983).

Moelwyn-Hughes, E. A.: *Physical Chemistry,* 2d ed., pp 699–701, Pergamon Press, New York, 1961.

Myrdal, P. B., and S. H. Yalkowsky: *Ind. Eng. Chem. Res.,* **36:** 2494 (1997).

Ohe, S.: *Computer Aided Data Book of Vapor Pressure, Data Book,* Tokyo, 1976.

Oja, V., and E. M. Suuberg: *J. Chem Eng. Data,* **43:** 486 (1998).

Oja, V., and E. M. Suuberg: *J. Chem Eng. Data,* **44:** 26 (1999).

Perry, R. H., and D. W. Green: *Perry's Chemical Engineers' Handbook,* 7th ed., McGraw-Hill, New York, 1997.

Plank, R., and L. Riedel: *Ing. Arch.,* **16:** 255 (1948).

Plank, R., and L. Riedel: *Tex. J. Sci.,* **1:** 86 (1949).

Poling, B. E.: *Fluid Phase Equil.,* **116:** 102 (1996).

Pouillot, F. L. L., D. Chandler, and C. A. Eckert: *Ind. Eng. Chem. Res.,* **35:** 2408 (1996).

Prausnitz, J. M., R. N. Lichtenthaler, and E. G. de Azevedo: *Molecular Thermodynamics of Fluid-Phase Equilibria,* 3d ed. Prentice Hall, Englewood Cliffs, New Jersey, 1999, pp. 638–641.

Riedel, L.: *Chem. Ing. Tech.,* **26:** 679 (1954).

Růžička, K., and V. Majer: *J. Phys. Chem. Ref. Data,* **23:** 1 (1994).

Schreiber, D. R., and K. S. Pitzer: *Fluid Phase Equil.,* **46:** 113 (1989).

Smith, G., J. Winnick, D. S. Abrams, and J. M. Prausnitz: *Can. J. Chem. Eng.,* **54:** 337 (1976).

Smith, J. M., H. C. Van Ness, and M. M. Abbott: *Introduction to Chemical Engineering Thermodynamics,* 5th ed., McGraw-Hill, New York, 1996.

Tamir, A., E. Tamir, and K. Stephan: *Heats of Phase Change of Pure Components and Mixtures,* Elsevier, Amsterdam, 1983.

Teja, A. S., S. I. Sandler, and N. C. Patel: *Chem. Eng. J.* (Lausanne), **21:** 21 (1981).

Thek, R. E., and L. I. Stiel: *AIChE J.,* **12:** 599 (1966), **13:** 626 (1967).

Thodos, G.: *Ind. Eng. Chem.* **42:** 1514 (1950).

Thompson, W. H., and W. G. Braun: *29th Midyear Meet., Am. Pet. Inst., Div. Refining,* St. Louis, Mo., May 11, 1964, prepr. 06-64.

Thomson, G. W.: *Techniques of Organic Chemistry,* A. Weissberger (ed.), 3d. ed., vol. I, pt. I, p. 473, Interscience, New York, 1959.

Tu, C.-H.: *Fluid Phase Equil.,* **99:** 105 (1994).

Tu, C.-H, and C.-P. Liu: *Fluid Phase Equil,* **121:** 45 (1996).

Twu, C.-H., J. E. Coon, and J. R. Cunningham: *Fluid Phase Equil,* **19:** 96 (1994).

Vetere, A.: *Chem. Eng. J.,* **17:** 157 (1979).

Vetere, A.: *Ind. Eng. Chem Res.,* **30:** 2487 (1991).

Vetere, A.: *Fluid Phase Equil.,* **106:** 1 (1995).

Wagner, W.: *Cyrogenics,* **13:** 470 (1973).

Wagner, W.: *Bull. Inst. Froid. Annexe,* no. 4, p. 65 (1973a).

Wagner, W.: *A New Correlation Method for Thermodynamic Data Applied to the Vapor-Pressure Curve of Argon, Nitrogen, and Water*, J. T. R. Watson (trans. and ed.). IUPAC Thermodynamic Tables Project Centre, London, 1977.

Wagner, W., J. Evers, and W. Pentermann: *J. Chem. Thermodynamics,* **8:** 1049 (1976).

Waring, W.: *Ind. Eng. Chem.,* **46:** 762 (1954).

Willingham, C. B., W. J. Taylor, J. M. Pignorco, and F. D. Rossini: *J. Res. Bur. Stand.* **35:** 219 (1945).

Xiang, H. W., and L. C. Tan: *Int. J. Thermophys.,* **15:** 711 (1994).

Yarbrough, D. W., and C.-H. Tsai: *Advan. Cryog. Eng.,* **23:** 602 (1978).

Yaws, C. L.: *Thermodynamic and Physical Property Data,* Gulf Pub. Co., Houston, 1992.

CHAPTER EIGHT
FLUID PHASE EQUILIBRIA IN MULTICOMPONENT SYSTEMS

8-1 SCOPE

In the chemical process industries, fluid mixtures are often separated into their components by diffusional operations such as distillation, absorption, and extraction. Design of such separation operations requires quantitative estimates of the partial equilibrium properties of fluid mixtures. Whenever possible, such estimates should be based on reliable experimental data for the particular mixture at conditions of temperature, pressure, and composition corresponding to those of interest. Unfortunately, such data are often not available. In typical cases, only fragmentary data are at hand and it is necessary to reduce and correlate the limited data to make the best possible interpolations and extrapolations. This chapter discusses some techniques that are useful toward that end. Although primary attention is given to nonelectrolytes, a few paragraphs are devoted to aqueous solutions of electrolytes, Emphasis is given to the calculation of fugacities in liquid solutions; fugacities in gaseous mixtures are discussed in Section 6-8.

The scientific literature on fluid phase equilibria goes back well over 150 years and has reached monumental proportions, including thousands of articles and hundreds of books and monographs. Tables 8-1*a* and 8-1*b* give the authors and titles of some books useful for obtaining data and for more detailed discussions. The lists are not exhaustive; they are restricted to publications likely to be useful to the practicing engineer in the chemical process industries.

There is an important difference between calculating phase equilibrium compositions and calculating typical volumetric, energetic, or transport properties of fluids of known composition. In the latter case we are interested in the property of the mixture as a whole, whereas in the former we are interested in the *partial* properties of the individual components which constitute the mixture. For example, to find the pressure drop of a liquid mixture flowing through a pipe, we need the viscosity and the density of that liquid mixture at the particular composition of interest. But if we ask for the composition of the vapor which is in equilibrium with the liquid mixture, it is no longer sufficient to know the properties of the liquid mixture at that particular composition; we must now know, in addition, how certain of its properties (in particular the Gibbs energy) *depend on composition.* In phase equilibrium calculations, we must know *partial* properties, and to find them, we typically differentiate data with respect to composition. Whenever experimental data are differentiated, there is a loss of accuracy, often a serious loss. Since partial,

TABLE 8-1a Some Useful Books on Fluid-Phase Equilibria

Book	Remarks
Balzhiser, R. E., M. R. Samuels, and J. D. Eliassen: *Chemical Engineering Thermodynamics: The Study of Energy, Entropy, and Equilibrium,* Prentice Hall, Englewood Cliffs, 1972.	An introductory text with numerous examples.
Chao, K. C., and R. A. Greenkorn: *Thermodynamics of Fluids,* Dekker, New York, 1975.	Introductory survey including an introduction to statistical thermodynamics of fluids; also gives a summary of surface thermodynamics.
Danner, R., and T. Daubert: *Manual for Predicting Chemical Process Design Data,* AIChE Publications, 1987.	Presents various techniques for estimating physical properties of pure and mixed fluids for process-design applications.
Danner, R., and M. High: *Handbook of Polymer Solution Thermodynamics,* AIChE Publications, 1993.	Provides densities and phase equilibria for polymer systems.
Elliott, J. R., and C. T. Lira: *Introductory Chemical Engineering Thermodynamics,* Prentice Hall, Upper Saddle River, N.J., 1999 (see http://www.egr.msu.edu/~lira/thermtxt.htm)	An introductory text with numerous examples.
Francis, A. W.: *Liquid-Liquid Equilibriums,* Wiley-Interscience, New York, 1963.	Phenomenological discussion of liquid-liquid equilibria with extensive data bibliography.
Gess, M., R. Danner, and M. Nagvekar: *Thermodynamic Analysis of Vapor-Liquid Equilibria: Recommended Models and a Standard Data Base,* AIChE Publications, 1991.	Presents a critical discussion of models for predicting and correlating vapor-liquid equilibria.
Hala, E., et al.: *Vapour-Liquid Equilibrium,* 2d English ed., trans. By George Standart, Pergamon, Oxford, 1967.	Comprehensive survey, including a discussion of experimental methods.
Hildebrand, J. H., and R. L. Scott: *Solubility of Nonelectrolytes,* 3d ed., Reinhold, New York, 1950. (Reprinted by Dover, New York, 1964.)	A classic in its field, it gives a survey of solution chemistry from a chemist's point of view. Although out of date, it nevertheless provides physical insight into how molecules "behave" in mixtures.
Hildebrand, J. H., and R. L. Scott: *Regular Solutions,* Prentice Hall, Englewood Cliffs, N.J., 1962.	Updates some of the material in Hildebrand's 1950 book.
Hildebrand, J. H., J. M. Prausnitz, and R. L. Scott: *Regular and Related Solutions,* Van Nostrand Reinhold, New York, 1970.	Further updates some of the material in Hildebrand's earlier books.
Kiran, E., P. G. Debenedetti, and C. J. Peters, (eds.): "Supercritical Fluids: Fundamentals and Applications," NATO Science Series E: Applied Sciences, **366:** Kluwer Academic Publishers, Dordrecht, 2000.	Leaders of the field describe all aspects of properties and processing systems that involve fluids in near-critical states. From a NATO Advanced Study Institute. See also the earlier NATO Science Series E: Applied Sciences, **273:** edited by E. Kiran and J. M. H. Levelt Sengers, 1994.
Kyle, B., *Chemical and Process Thermodynamics,* 2d ed., Prentice Hall, Englewood Cliffs, N.J., Chapters 9, 10, 11, 1992.	An introductory text with numerous examples.
MacKay, D., W. Shiu, and K. Ma: *Illustrated Handbook of Physical-* [illegible]	Provides data and examples relevant to water-pollution problems.

Reference	Description
Majer, V., V. Svoboda, and J. Pick: Heats of Vaporization of Fluids, *Studies in Modern Thermodynamics* 9, Elsevier, Amsterdam, 1989.	In depth treatment of thermodynamics of vapor pressure and enthalpy of vaporization.
Murrell, J. M., and E. A. Boucher: *Properties of Liquids and Solutions,* Wiley, New York, 1982.	A short introduction to the physics and chemistry of the liquid state.
Null, H. R.: *Phase Equilibrium in Process Design,* Wiley-Interscience, New York, 1970.	An engineering-oriented monograph with a variety of numerical examples.
Palmer, D.: *Handbook of Applied Thermodynamics,* CRC Press, Boca Raton, Fla. 1987.	A practical guide from an industrial point of view.
Pitzer, K.: *Activity Coefficients in Electrolyte Solutions,* 2d ed., CRC Press, 1991.	A semiadvanced discussion of models (especially Pitzer's model) for electrolyte solutions.
Prausnitz, J. M., R. N. Lichtenthaler, and E. G. Azevedo: *Molecular Thermodynamics of Fluid-Phase Equilibria,* 3d ed., Prentice Hall, Englewood Cliffs, N.J., 1999.	A text which attempts to use molecular-thermodynamic concepts useful for engineering. Written from a chemical engineering point of view.
Prausnitz, J. M., T. F. Anderson, E. A. Grens, C. A. Eckert, R. Hsieh, and J. P. O'Connell: *Computer Calculations for Vapor-Liquid and Liquid Equilibria,* Prentice Hall, Englewood Cliffs, N.J., 1980.	A monograph with detailed computer programs and a (limited) data bank.
Prigogine, I., and R. Defay, *Chemical Thermodynamics,* trans. and rev. by D. H. Everett, Longmans, Green, London, 1954.	A semiadvanced text from a European chemist's point of view. It offers many examples and discusses molecular principles. Although out of date, it contains much useful information not easily available in standard American texts.
Rowlinson, J. S., and F. L. Swinton: *Liquids and Liquid Mixtures,* 3d ed., Butterworth, London, 1982.	Presents a thorough treatment of the physics of fluids and gives some statistical mechanical theories of the equilibrium properties of simple pure liquids and liquid mixtures; contains data bibliography. Primarily for research-oriented readers.
Sandler, S. (ed.): *Models for Thermodynamic and Phase Equilibria Calculations,* Dekker, New York, 1995.	Presents summaries of various models that are useful for correlating phase equilibria in a variety of systems, including polymers and electrolytes.
Sandler, S.: *Chemical and Engineering Thermodynamics,* 3d ed., Wiley, Chapters 6, 7 and 8, 1999.	An introductory text with numerous examples.
Sengers, J. V., R. F. Kayser, C. J. Peters, and H. J. White, Jr. (eds.): *Equations of State for Fluids and Fluid Mixtures,* Elsevier, Amsterdam, 2000.	A very comprehensive treatment of all aspects of equation of State fundamentals and applications including phase equilibria.

TABLE 8-1*a* Some Useful Books on Fluid-Phase Equilibria

Book	Remarks
Smith, J., H. Van Ness, and M. Abbott: *Introduction to Chemical Engineering Thermodynamics,* 5th ed., McGraw-Hill, New York, Chapters 11–14, 1996.	An introductory text with numerous examples.
Tester, J., and M. Modell: *Thermodynamics and Its Applications:* 3d ed., Prentice Hall, Englewood Cliffs, N.J., 1997.	This semiadvanced text emphasizes the solution of practical problems through application of fundamental concepts of chemical engineering thermodynamics and discusses surface thermodynamics and systems in potential fields.
Van Ness, H. C., and M. M. Abbott: *Classical Thermodynamics of Nonelectrolyte Solutions,* McGraw Hill, New York, 1982.	Systematic, comprehensive, and clear exposition of the principles of classical thermodynamics applied to solutions of nonelectrolytes. Discusses phase equilibria in fluid systems with numerous examples.
Walas, S.: *Phase Equilibria in Chemical Engineering,* Butterworth, 1985.	A practical "how-to-do-it" manual with examples.
Winnick, J.: *Chemical Engineering Thermodynamics,* Chapters 11–14, Wiley, 1997.	An introductory text with numerous examples.

TABLE 8-1*b* Some Useful Books on Fluid-Phase Equilibria Data Sources

Book	Remarks
API Research Project 42: *Properties of Hydrocarbons of High Molecular Weight,* American Petroleum Institute, New York, 1966.	A compilation of physical properties (vapor pressure, liquid density, transport properties) for 321 hydrocarbons with carbon number 11 or more.
API Research Project 44: *Handbook of Vapor Pressures and Heats of Vaporization of Hydrocarbons and Related Compounds,* Thermodynamics Research Center, College Station, Texas, 1971.	A thorough compilation of the vapor pressures and enthalpies of vaporization of alkanes (up to C_{100}), aromatics, and naphthenes (including some with heteroatoms). Other API-44 publications include data on a variety of thermodynamic properties of hydrocarbons and related compounds.
Barton, A. F. M.: *Handbook of Polymer-Liquid Interaction Parameters,* CRC Press, Boca Raton, Fla. 1990.	Contains an extensive compilation of data for solubility parameters, cohesive energies, and molar volumes for a variety of substances including polymers. Some correlations (group contributions) are also presented.
Behrens, D., and R. Eckermann: *Chemistry Data Series,* DECHEMA, Frankfurt a.M., Vol I, (subdivided into nineteen separate volumes) *VLE Data Collection,* by J. Gmehling, U. Onken, W. Arlt, P. Grenzheuser, U. Weidlich, and B. Kolbe, 1980–1996; Vol II, *Critical Data,* by K. H. Simmrock, 1986; Vol. III, (divided into four volumes) *Heats of Mixing Data Collection,* by C. Christensen, J. Gmehling, P. Rasmussen, and U. Weidlich; Vol. V, (divided into four volumes) *LLE-Data Collection,* by J. M. Sorensen and W. Arlt, 1979–1987; Vol. VI, (divided into four volumes) *VLE for Mixtures of Low-Boiling Substances,* by H. Knapp, R. Döring, L. Oellrich, U. Plöcker, J. M. Prausnitz, R. Langhorst and S. Zeck, 1982–1987; Vol. VIII, *Solid-Liquid Equilibrium Data Collection,* by H. Knapp, R. Langhorst and M. Teller, 1987; Vol IX, (divided into four volumes) *Activity Coefficients of Infinite Dilution,* by D. Tiegs, J. Gmehling, A. Medina, M. Soares, J. Bastos, P. Alessi, I. Kikic, 1986–1994; Vol. XII (divided into nine volumes) Electrolyte Data Collection, by J. Barthel, R. Neueder, R. Meier et al, 1992–1997.	An extensive compilation (in six volumes, some of them consisting of several parts) of thermodynamic property data for pure compounds and mixtures, *PVT* data, heat capacity, enthalpy, and entropy data, phase equilibrium data, and transport and interfacial tension data for a variety of inorganic and organic compounds including aqueous mixtures.
Boublik, T., V. Fried, and E. Hala: *The Vapour Pressures of Pure Substances,* 2d ed., Elsevier, Amsterdam, 1984.	Experimental and smoothed data are given for the vapor pressures of pure substances in the normal and low-pressure region; Antoine constants are reported.
Brandrup, J., and E. H. Immergut (eds.): *Polymer Handbook,* 4th ed., Wiley-Interscience, New York, 1999.	A thorough compilation of polymerization reactions and of solution and physical properties of polymers and their associated oligomers and monomers.

TABLE 8-1*b* Some Useful Books on Fluid-Phase Equilibria (*Continued*)

Book	Remarks
Broul, M., J. Nyult, and O. Söhnel: *Solubility in Inorganic Two-Component Systems,* Elsevier, Amsterdam, 1981.	An extensive compilation of data on solubility of inorganic compounds in water.
Caruthers, J. M., K. C. Chao, V. Venkatasubramanian, R. Sy-Siong-Kiao, C. Novenario, and A. Sundaram: *Handbook of Diffusion and Thermal Properties of Polymers and Polymer Solutions,* (*DIPPR*) AIChE Publications, 1999.	Evaluated physical and thermodynamic data for polymeric systems.
Constants of Binary Systems, Interscience, New York, 1959.	A four-volume compilation of vapor-liquid and liquid-liquid equilibria, densities of the coexisting phases, transport properties, and enthalpy data for binary concentrated solutions.
Christensen, J. J., L. D. Hansen, and R. M. Isatt: *Handbook of Heats of Mixing,* Wiley, New York, 1982.	Experimental heat-of-mixing data for a variety of binary mixtures.
Cunningham, J., and D. Jones, *Experimental Results for DIPPR,* 1990–1991, *Projects on Phase Equilibria and Pure Component Properties,* DIPPR Data Series No. 2, DIPPR, AIChE Publications, 1994.	A summary of selected experimental data obtained by DIPPR.
Danner, R. P., and T. E. Daubert (eds.): *Technical Data Book-Petroleum Refining,* American Petroleum Institute, Washington, D.C., 1983.	A two-volume data compilation of the physical, transport, and thermodynamic properties of petroleum fractions and related model compound mixtures of interest to the petroleum-refining industry.
Daubert, T. E., R. P. Danner, H. M. Sibel, and C. C. Stebbins, *Physical and Thermodynamic Properties of Pure Chemicals: Data Compilation,* Taylor & Francis, Washington, D.C., 1997.	A five-volume compilation of pure-component properties.
DIPPR's Manual for Predicting Chemical Process Design Data, Documentation Reports, Chapter 6: Phase Equilibria, (DIPPR) AIChE Publications, 1987.	Provides useful background information.
DIPPR, Results from DIPPR Experimental Projects, 1991–1994, J. Chem. & Eng. Data **41:** 1211 (1996); **42:** 1007 (1997).	Presents experimental data for vapor pressure, critical properties, enthalpies of formation and phase equilibria.
Dymond, J. H., and E. B. Smith: *The Virial Coefficients of Pure Gases and Mixtures,* Oxford, Clarendon Press, 1980.	A critical compilation of data for virial coefficients of pure gases and binary mixtures published through 1979.
Gmehling, J. et al: *Azeotropic Data,* 2 vols., Wiley-VCH, Weinheim and New York, 1994.	A comprehensive compilation of azeotropic compositions, temperatures and pressures.

Hao, W., H. Elbro, and P. Alessi: *Polymer Solution Data Collection, 3* vols., DECHEMA, Frankfurt, Germany, 1992.	Presents phase-equilibria for many polymer solutions, mostly binaries and mostly vapor-liquid equilibria.
Hicks, C. P., K. N. Marsh, A. G. Williamson, I. A. McLure, and C. L. Young: *Bibliography of Thermodynamic Studies,* Chemical Society, London, 1975.	Literature references for vapor-liquid equilibria, enthalpies of mixing, and volume changes of mixing of selected binary systems.
Hirata, M., S. Ohe, and K. Nagahama: *Computer-Aided Data Book of Vapor-Liquid Equilibria,* Elsevier, Amsterdam, 1975.	A compilation of binary experimental data reduced with the Wilson equation and, for high pressures, with a modified Redlich-Kwong equation.
Hiza, M. J., A. J. Kidnay, and R. C. Miller: *Equilibrium Properties of Fluid Mixtures,* 2 vols., IFL/Plenum, New York, 1975, 1982.	Volume 1 contains references for experimental phase equilibria and thermodynamic properties of fluid mixtures of cryogenic interest. Volume 2 updates to January 1980 the references given in Vol. 1. Includes mixtures containing pentane and some aqueous mixtures.
IUPAC: *Solubility Data Series,* Pergamon, Oxford, 1974.	A multivolume compilation of the solubilities of inorganic gases in pure liquids, liquid mixtures, aqueous solutions, and miscellaneous fluids and fluid mixtures.
Kehiaian, H. V., (ed.-in-chief), and B. J. Zwolinski (exec. officer), *International Data Series: Selected Data on Mixtures,* Thermodynamics Research Center, Chemistry Dept., Texas A&M University, College Station, TX 77843, continuing since 1973.	Presents a variety of measured thermodynamic properties of binary mixtures. These properties are often represented by empirical equations.
V. Lobo and J. Quaresma: *Handbook of Electrolyte Solutions,* Elsevier 1989.	Presents vapor-liquid and solid-liquid equilibria for numerous aqueous electrolyte solutions.
Majer, V. and V. Svoboda: *Enthalpies of Vaporization of Organic Compounds, A Critical Review and Data Compilation: IUPAC Chem. Data Ser No.* 32, Blackwell Scientific Pub., Oxford, 1985.	An extensive tabulation of enthalpies of vaporization.
Maczynski, A.: *Verified Vapor-Liquid Equilibrium Data,* Polish Scientific Publishers, Warszawa, 1976. Thermodynamics Data Center, Warszawa, 1997.	A four-volume compilation of binary vapor-liquid equilibrium data for mixtures of hydrocarbons with a variety of organic compounds; includes many data from the East European literature.
Marsh, K., Q. Dong, and A. Dewan: *Transport Properties and Related Thermodynamic Data of Binary Mixtures,* (DIPPR) AIChE Publications, 5 vol. 1993–1998.	Evaluated thermodynamic and transport properties for a large number of mostly binary mixtures.
Ohe, S.: *Computer-Aided Data Book of Vapor Pressure,* Data Book Publishing Company, Tokyo, 1976.	Literature references for vapor pressure data for about 2,000 substances are given. The data are presented in graphical form, and Antoine constants also are given.
Ohe, S.: *Vapor-Liquid Equilibrium Data-Salt Effect,* Kodanshah-Elsevier, 1991.	Summarizes data for the effect of a salt on fluid-phase equilibria.

TABLE 8-1*b* Some Useful Books on Fluid-Phase Equilibria (*Continued*)

Book	Remarks
Pytkowicz, R. M. (ed.): "Activity Coefficients in Electrolyte Solutions, I & II," Boca Raton, Fla. CRC Press, 1979.	Tabulation of activity coefficients in electrolyte solutions.
Seidell, A.: *Solubilities of Inorganic and Organic Compounds,* 3d ed., Van Nostrand, New York, 1940 (1941, 1952). Linke, W. L.: *Solubilities of Inorganic and Organic Compounds,* 4th ed., Van Nostrand, Princeton, N.J., 1958 and 1965, vols. 1 and 2.	A two-volume (plus supplement) data compilation. The first volume concerns the solubilities of inorganic and metal organic compounds in single compounds and in mixtures; the second concerns organics; and the supplement updates the solubility references to 1949.
Silcock, H., (ed.): *Solubilities of Inorganic and Organic Compounds,* translated from Russian, Pergamon, Oxford, 1979.	A systematic compilation of data to 1965 on solubilities of ternary and multi-component systems of inorganic compounds.
Stephen, H. and T. Stephen (eds): *Solubilities of Inorganic and Organic Compounds,* translated from Russian, Pergamon, Oxford, 1979.	A five-volume data compilation of solubilities for inorganic, metal-organic and organic compounds in binary, ternary and multicomponent systems.
Tamir, A., E. Tamir, and K. Stephan: *Heats of Phase Change of Pure Components and Mixtures,* Elsevier, Amsterdam, 1983.	An extensive compilation of data published to 1981 on the enthalpy of phase change for pure compounds and mixtures.
Timmermans, J.: *Physico-chemical constants of Pure Organic Compounds,* Elsevier, Amsterdam, 1950 and 1965, vols. 1 and 2.	A compilation of data on vapor pressure, density, melting and boiling point, heat-capacity constants, and transport properties of organic compounds (2 vols.)
Van Krevelen, D. W.: *Properties of Polymers,* 3rd ed., Elsevier, Amsterdam, 1990.	Presents methods to correlate and predict thermodynamic, transport, and chemical properties of polymers as a function of chemical structure.
Vargaftik, N. B.: *Handbook of Physical Properties of Liquids and Gases: Pure Substances and Mixtures,* 2d ed., Hemisphere, Washington, 1981.	A compilation of thermal, caloric, and transport properties of pure compounds (including organic compounds, SO_2, and halogens) and mainly transport properties of binary gas mixtures, liquid fuels, and oils.
Wisniak, J.: *Phase Diagrams,* Elsevier, Amsterdam, 1981.	A literature-source book for published data to 1980 on phase diagrams for a variety of inorganic and organic compounds (2 vols.).
Wisniak, J., and A. Tamir: *Liquid-Liquid Equilibrium and Extraction,* Elsevier, Amsterdam, 1980.	A two-volume literature source book for the equilibrium distribution between two immiscible liquids for data published to 1980.
Wisniak, J., and A. Tamir: *Mixing and Excess Thermodynamic Properties,* Elsevier, Amsterdam, 1978 (supplement 1982).	An extensive bibliographic compilation of data references on mixing and excess properties published between 1900 and 1982.
Yalkowsky, S.: *Aquasol Database,* 5th ed., College of Pharmacy, Univ. of Arizona, Tucson, Ariz. 85721, 1997.	Provides solubility data for numerous organics (mostly solids) in water.
Zaytsev, I., and G. Aseyev, *Properties of Aqueous Solutions of Electrolytes,* CRC Press, Boca Raton, Fla., 1992.	Presents extensive data collection. However, many "data" are not from direct measurements but from data smoothing.
Zematis, J., D. Clark, M. Raful, and N. Scrivner: *Handbook of Aqueous Electrolyte Thermodynamics,* AIChE Publications, 1986.	Provides discussion of models and gives data for numerous aqueous systems. A useful guide.

rather than total, properties are needed in phase equilibria, it is not surprising that phase equilibrium calculations are often more difficult and less accurate than those for other properties encountered in chemical process design.

In one chapter it is not possible to present a complete review of a large subject. Also, since this subject is so wide in its range, it is not possible to recommend to the reader simple, unambiguous rules for obtaining quantitative answers to a particular phase equilibrium problem. Since the variety of mixtures is extensive, and since mixture conditions (temperature, pressure, and composition) cover many possibilities, and, finally, since there are large variations in the availability, quantity, and quality of experimental data, the reader cannot escape responsibility for using judgment, which, ultimately, is obtained only by experience.

This chapter, therefore, is qualitatively different from the others in this book. It does not give specific advice on how to calculate specific quantities. It provides only an introduction to some (by no means all) of the tools and techniques which may be useful for an efficient strategy toward calculating particular phase equilibria for a particular process design.

8-2 THERMODYNAMICS OF VAPOR-LIQUID EQUILIBRIA

We are concerned with a liquid mixture that, at temperature T and pressure P, is in equilibrium with a vapor mixture at the same temperature and pressure. The quantities of interest are the temperature, the pressure, and the compositions of both phases. Given some of these quantities, our task is to calculate the others.

For every component i in the mixture, the condition of thermodynamic equilibrium is given by

$$f_i^V = f_i^L \tag{8-2.1}$$

where f = fugacity
V = vapor
L = liquid

The fundamental problem is to relate these fugacities to mixture composition. In the subsequent discussion, we neglect effects due to surface forces, gravitation, electric or magnetic fields, semipermeable membranes, or any other special conditions.

The fugacity of a component in a mixture depends on the temperature, pressure, and composition of that mixture. In principle any measure of composition can be used. For the vapor phase, the composition is nearly always expressed by the mole fraction y. To relate f_i^V to temperature, pressure, and mole fraction, it is useful to introduce the vapor-phase fugacity coefficient ϕ_i^V

$$\phi_i^V = \frac{f_i^V}{y_i P} \tag{8-2.2a}$$

which can be calculated from vapor phase $PVTy$ data, usually given by an equation of state as discussed in Sec. 6-7. For ideal gases $\phi_i^V = 1$.

The fugacity coefficient ϕ_i^V depends on temperature and pressure and, in a multicomponent mixture, on *all* mole fractions in the vapor phase, not just y_i. The fugacity coefficient is, by definition, normalized such that as $P \rightarrow 0$, $\phi_i^V \rightarrow 1$ for all i. At low pressures, therefore, it is usually a good assumption to set $\phi_i^V = 1$. But just what "low" means depends on the composition and temperature of the mixture. For typical mixtures of nonpolar (or slightly polar) fluids at a temperature near or above the normal boiling point of the least volatile component, "low" pressure means a pressure less than a few bars. However, for mixtures containing a strongly associating carboxylic acid, e.g., acetic acid-water at 25°C, fugacity coefficients may differ appreciably from unity at pressures much less than 1 bar.† For mixtures containing one component of very low volatility and another of high volatility, e.g., decane-methane at 25°C, the fugacity coefficient of the light component may be close to unity for pressures up to 10 or 20 bar while at the same pressure the fugacity coefficient of the heavy component is typically much less than unity. A detailed discussion is given in chap. 5 of Prausnitz, et al., (1999).

The fugacity of component i in the liquid phase is generally calculated by one of two approaches: the equation of state approach or the activity coefficient approach. In the former, the liquid-phase fugacity coefficient, ϕ_i^L, is introduced

$$\phi_i^L = \frac{f_i^L}{x_i P} \tag{8-2.2b}$$

where x_i is the liquid phase mole fraction. Certain equations of state are capable of representing liquid-phase, as well as vapor-phase behavior. The use of such equations of state to calculate phase equilibria is discussed in Sec. 8-12.

In the activity coefficient approach, the fugacity of component i in the liquid phase is related to the composition of that phase through the activity coefficient γ_i. In principle, any composition scale may be used; the choice is strictly a matter of convenience. For some aqueous solutions, frequently used scales are molality (moles of solute per 1000 g of water) and molarity (moles of solute per liter of solution); for polymer solutions, a useful scale is the volume fraction, discussed briefly in Sec. 8-15. However, for typical solutions containing nonelectrolytes of normal molecular weight (including water), the most useful measure of concentration is mole fraction x. Activity coefficient γ_i is related to x_i and to standard-state fugacity f_i^o by

$$\gamma_i \equiv \frac{a_i}{x_i} = \frac{f_i^L}{x_i f_i^o} \tag{8-2.3}$$

where a_i is the activity of component i. The standard-state fugacity f_i^o is the fugacity of component i at the temperature of the system, i.e., the mixture, and at some arbitrarily chosen pressure and composition. The choice of standard-state pressure and composition is dictated only by convenience, but it is important to bear in mind that the numerical values of γ_i and a_i have no meaning unless f_i^o is clearly specified.

While there are some important exceptions, activity coefficients for most typical solutions of nonelectrolytes are based on a standard state where, for every component i, f_i^o is the fugacity of *pure* liquid i at system temperature and pressure; i.e., the arbitrarily chosen pressure is the total pressure P, and the arbitrarily chosen composition is $x_i = 1$. Frequently, this standard-state fugacity refers to a hypothetical state, since it may happen that component i cannot physically exist as a pure liquid at system temperature and pressure. Fortunately, for many common mixtures it is possible to calculate this standard-state fugacity by modest extrapolations with respect to pressure; and since liquid-phase properties remote from the critical region

† For moderate pressures, fugacity coefficients can often be estimated with good accuracy as discussed for example, by Prausnitz, et al., (1980).

are not sensitive to pressure (except at high pressures), such extrapolation introduces little uncertainty. In some mixtures, however, namely, those that contain supercritical components, extrapolations with respect to temperature are required, and these, when carried out over an appreciable temperature region, may lead to large uncertainties.

Whenever the standard-state fugacity is that of the pure liquid at system temperature and pressure, we obtain the limiting relation that $\gamma_i \rightarrow 1$ as $x_i \rightarrow 1$.

8-3 FUGACITY OF A PURE LIQUID

To calculate the fugacity of a pure liquid at a specified temperature and pressure, we may use an equation of state capable of representing the liquid phase and first calculate ϕ_i^L (See Chap. 6) and then use Eq. (8-2.2*b*). Alternatively, we may use the two primary thermodynamic properties: the saturation (vapor) pressure, which depends only on temperature, and the liquid density, which depends primarily on temperature and to a lesser extent on pressure. Unless the pressure is very large, it is the vapor pressure which is by far the more important of these two quantities. In addition, we require volumetric data (equation of state) for pure vapor *i* at system temperature, but unless the vapor pressure is high or unless there is strong dimmerization in the vapor phase, this requirement is of minor, often negligible, importance.

The fugacity of pure liquid *i* at temperature *T* and pressure *P* is given by

$$f_i^L(T, P, x_i = 1) = P_{\text{vp}i}(T)\phi_i^s(T) \exp \int_{P_{\text{vp}i}}^{P} \frac{V_i^L(T, P)}{RT}\, dP \tag{8-3.1}$$

where P_{vp} is the vapor pressure (see Chap. 7) and superscript *s* stands for saturation. The fugacity coefficient ϕ_i^s is calculated from vapor-phase volumetric data, as discussed in Sec. 6-7; for typical nonassociated fluids at temperatures well below the critical, ϕ_i^s is close to unity.

The molar liquid volume V_i^L is the ratio of the molecular weight to the density, where the latter is expressed in units of mass per unit volume.† At a temperature well below the critical, a liquid is nearly incompressible. In that case the effect of pressure on liquid-phase fugacity is not large unless the pressure is very high or the temperature is very low. The exponential term in Eq. (8-3.1) is called the *Poynting factor.*

To illustrate Eq. (8-3.1), the ratio of the fugacity of pure liquid water to the vapor pressure (equal to the product of ϕ^s and the Poynting factor) is shown in Table 8-2 at four temperatures and three pressures, the vapor pressure, 40 bar, and 350 bar. Since ϕ^s for a pure liquid is always less than unity, the ratio is always less than one at saturation. However, at pressures well above the vapor pressure, the product of ϕ^s and the Poynting factor may easily exceed unity, and then the fugacity is larger than the vapor pressure.

Sometimes it is necessary to calculate a liquid fugacity for conditions when the substance does not exist as a liquid. At 300°C, for example, the vapor pressure exceeds 40 bar, and therefore pure liquid water cannot exist at this temperature and 40 bar. Nevertheless, the value 0.790 shown in Table 8-2 at these conditions can

† For volumetric properties of liquids, see Chap. 4.

TABLE 8-2 Fugacity of Liquid Water, bar

Temp., °C	P_{vp}, bar	f/P_{vp} Saturation	40 bar	350 bar
75	0.3855	0.992	1.020	1.243
150	4.760	0.960	0.985	1.171
250	29.78	0.858	0.873	1.026
300	85.93	0.784	0.790*	0.931

*Hypothetical because $P < P_{vp}$

be calculated by a mild extrapolation; in the Poynting factor we neglect the effect of pressure on molar liquid volume.

Table 8-2 indicates that the vapor pressure is the primary quantity in Eq. (8-3.1). When data are not available, the vapor pressure can be estimated, as discussed in Chapter. 7. Further, for nonpolar (or weakly polar) liquids, the ratio of fugacity to pressure can be estimated from generalized (corresponding states) methods.

8-4 SIMPLIFICATIONS IN THE VAPOR-LIQUID EQUILIBRIUM RELATION

Equation (8-2.1) gives the rigorous, fundamental relation for vapor-liquid equilibrium. Equations (8-2.2), (8-2.3), and (8-3.1) are also rigorous, without any simplifications beyond those indicated in the paragraph following Eq. (8-2.1). Substitution of Eqs. (8-2.2), (8-2.3), and (8-3.1) into Eq. (8-2.1) gives

$$y_i P = x_i \gamma_i P_{vpi} \mathcal{F}_i \tag{8-4.1}$$

where

$$\mathcal{F}_i = \frac{\phi_i^s}{\phi_i^V} \exp \int_{P_{vpi}}^{P} \frac{V_i^L \, dP}{RT} \tag{8-4.2}$$

For subcritical components, the correction factor $\mathcal{F}_i$ is often near unity when the total pressure P is sufficiently low. However, even at moderate pressures, we are nevertheless justified in setting $\mathcal{F}_i = 1$ if only approximate results are required and, as happens so often, if experimental information is sketchy, giving large uncertainties in γ.

If, in addition to setting $\mathcal{F}_i = 1$, we assume that $\gamma_i = 1$, Eq. (8-4.1) reduces to the familiar relation known as *Raoult's law.*

In Eq. (8-4.1), both ϕ_i^V and γ_i depend on temperature, composition, and pressure. However, remote from critical conditions, and unless the pressure is large, the effect of pressure on γ_i is usually small.

8-5 ACTIVITY COEFFICIENTS; GIBBS-DUHEM EQUATION AND EXCESS GIBBS ENERGY

In typical mixtures, Raoult's law provides no more than a rough approximation; only when the components in the liquid mixture are similar, e.g., a mixture of *n*-

butane and isobutane, can we assume that γ_i is essentially unity for all components at all compositions. The activity coefficient, therefore, plays a key role in the calculation of vapor-liquid equilibria.

Classical thermodynamics has little to tell us about the activity coefficient; as always, thermodynamics does not give us the experimental quantity we desire but only relates it to other experimental quantities. Thus thermodynamics relates the effect of pressure on the activity coefficient to the partial molar volume, and it relates the effect of temperature on the activity coefficient to the partial molar enthalpy, as discussed in any thermodynamics text. These relations are of limited use because good data for the partial molar volume and for the partial molar enthalpy are rare.

However, there is one thermodynamic relation that provides a useful tool for correlating and extending limited experimental data: the Gibbs-Duhem equation. This equation is not a panacea, but, given some experimental results, it enables us to use these results efficiently. In essence, the Gibbs-Duhem equation says that, in a mixture, the activity coefficients of the individual components are not independent of one another but are related by a differential equation. In a binary mixture the Gibbs-Duhem relation is

$$x_1 \left(\frac{\partial \ln \gamma_i}{\partial x_1} \right)_{T,P} = x_2 \left(\frac{\partial \ln \gamma_2}{\partial x_2} \right)_{T,P} \qquad (8\text{-}5.1)\dagger$$

Equation (8-5.1) has several important applications.

1. If we have experimental data for γ_1 as a function of x_1, we can integrate Eq. (8-5.1) and calculate γ_2 as a function of x_2. That is, in a binary mixture, activity coefficient data for one component can be used to predict the activity coefficient of the other component.

2. If we have extensive experimental data for *both* γ_1 and γ_2 as a function of composition, we can test the data for thermodynamic consistency by determining whether or not the data obey Eq. (8-5.1). If the data show serious inconsistencies with Eq. (8-5.1), we may conclude that they are unreliable.

3. If we have limited data for γ_1 and γ_2, we can use an integral form of the Gibbs-Duhem equation; the integrated form provides us with thermodynamically consistent equations that relate γ_1 and γ_2 to x. These equations contain a few adjustable parameters that can be determined from the limited data. It is this application of the Gibbs-Duhem equation which is of particular use to chemical engineers. However, there is no *unique* integrated form of the Gibbs-Duhem equation; many forms are possible. To obtain a particular relation between γ and x, we must assume some model consistent with the Gibbs-Duhem equation.

For practical work, the utility of the Gibbs-Duhem equation is best realized through the concept of excess Gibbs energy, i.e., the observed Gibbs energy of a mixture above and beyond what it would be for an ideal solution at the same temperature, pressure, and composition. By definition, an ideal solution is one

† Note that the derivatives are taken at constant temperature T and constant pressure P. In a binary, two-phase system, however, it is not possible to vary x while holding *both* T and P constant. At ordinary pressures γ is a very weak function of P, and therefore it is often possible to apply Eq. (8-5.1) to isothermal data while neglecting the effect of changing pressure. This subject has been amply discussed in the literature; see, for example, Appendix D of Prausnitz, et al., (1999).

where all $\gamma_i = 1$. The *total* excess Gibbs energy G^E for a binary solution, containing n_1 moles of component 1 and n_2 moles of component 2, is defined by

$$G^E = RT(n_1 \ln \gamma_1 + n_2 \ln \gamma_2) \tag{8-5.2}$$

Equation (8-5.2) gives G^E as a function of *both* γ_1 and γ_2. Upon applying the Gibbs-Duhem equation, we can relate the *individual* activity coefficients γ_1 or γ_2 to G^E by differentiation

$$RT \ln \gamma_1 = \left(\frac{\partial G^E}{\partial n_1}\right)_{T,P,n_2} \tag{8-5.3}$$

$$RT \ln \gamma_2 = \left(\frac{\partial G^E}{\partial n_2}\right)_{T,P,n_1} \tag{8-5.4}$$

Equations (8-5.2) to (8-5.4) are useful because they enable us to interpolate and extrapolate limited data with respect to composition. To do so, we must first adopt some mathematical expression for G^E as a function of composition. Second, we fix the numerical values of the constants in that expression from the limited data; these constants are independent of x, but they usually depend on temperature. Third, we calculate activity coefficients at any desired composition by differentiation, as indicated by Eqs. (8-5.3) and (8-5.4).

To illustrate, consider a simple binary mixture. Suppose that we need activity coefficients for a binary mixture over the entire composition range at a fixed temperature T. However, we have experimental data for only one composition, say $x_1 = x_2 = ½$. From that one datum we calculate $\gamma_1(x_1 = ½)$ and $\gamma_2(x_2 = ½)$; for simplicity, let us assume symmetrical behavior, that is, $\gamma_1(x_1 = ½) = \gamma_2(x_2 = ½)$.

We must adopt an expression relating G^E to the composition subject to the conditions that at fixed composition G^E is proportional to $n_1 + n_2$ and that $G^E = 0$ when $x_1 = 0$ or $x_2 = 0$. The simplest expression we can construct is

$$G^E = (n_1 + n_2)g^E = (n_1 + n_2)Ax_1x_2 \tag{8-5.5}$$

where g^E is the excess Gibbs energy per mole of mixture and A is a constant depending on temperature. The mole fraction x is simply related to mole number n by

$$x_1 = \frac{n_1}{n_1 + n_2} \tag{8-5.6}$$

$$x_2 = \frac{n_2}{n_1 + n_2} \tag{8-5.7}$$

The constant A is found from substituting Eq. (8-5.5) into Eq. (8-5.2) and using the experimentally determined γ_1 and γ_2 at the composition midpoint:

$$A = \frac{RT}{(½)(½)}[½ \ln \gamma_1(x_1 = ½) + ½ \ln \gamma_2(x_2 = ½)] \tag{8-5.8}$$

Upon differentiating Eq. (8-5.5) as indicated by Eqs. (8-5.3) and (8-5.4), we find

$$RT \ln \gamma_1 = Ax_2^2 \tag{8-5.9}$$

$$RT \ln \gamma_2 = Ax_1^2 \tag{8-5.10}$$

With these relations we can now calculate activity coefficients γ_1 and γ_2 at any desired x even though experimental data were obtained only at one point, namely, $x_1 = x_2 = \frac{1}{2}$.

This simplified example illustrates how the concept of excess function, coupled with the Gibbs-Duhem equation, can be used to interpolate or extrapolate experimental data with respect to composition. Unfortunately, the Gibbs-Duhem equation tells nothing about interpolating or extrapolating such data with respect to temperature or pressure.

Equations (8-5.2) to (8-5.4) indicate the intimate relation between activity coefficients and excess Gibbs energy G^E. Many expressions relating g^E (per mole of mixture) to composition have been proposed, and a few are given in Table 8-3. All these expressions contain adjustable parameters which, at least in principle, depend on temperature. That dependence may in some cases be neglected, especially if the temperature interval is not large. In practice, the number of adjustable constants per binary is typically two or three; the larger the number of constants, the better the representation of the data but, at the same time, the larger the number of reliable experimental data points required to determine the constants. Extensive and highly accurate experimental data are required to justify more than three empirical constants for a binary mixture at a fixed temperature.†

For many moderately nonideal binary mixtures, all equations for g^E containing two (or more) binary parameters give good results; there is little reason to choose one over another except that the older ones (Margules, van Laar) are mathematically easier to handle than the newer ones (Wilson, NRTL, UNIQUAC). The two-suffix (one-parameter) Margules equation is applicable only to simple mixtures where the components are similar in chemical nature and in molecular size.

For strongly nonideal binary mixtures, e.g., solutions of alcohols with hydrocarbons, the equation of Wilson is probably the most useful because, unlike the NRTL equation, it contains only two adjustable parameters and it is mathematically simpler than the UNIQUAC equation. For such mixtures, the three-suffix Margules equation and the van Laar equation are likely to represent the data with significantly less success, especially in the region dilute with respect to alcohol, where the Wilson equation is particularly suitable.

With rare exceptions, the four-suffix (three-parameter) Margules equation has no significant advantages over the three-parameter NRTL equation.

Numerous articles in the literature use the Redlich-Kister expansion [see Eq. (8-9.20)] for g^E. This expansion is mathematically identical to the Margules equation.

The Wilson equation is not applicable to a mixture which exhibits a miscibility gap; it is inherently unable, even qualitatively, to account for phase splitting. Nevertheless, Wilson's equation may be useful even for those mixtures where miscibility is incomplete provided attention is confined to the one-phase region.

Unlike Wilson's equation, the NRTL and UNIQUAC equations are applicable to *both* vapor-liquid and liquid-liquid equilibria.‡ Therefore, mutual solubility data [See Sec. 8-10] can be used to determine NRTL or UNIQUAC parameters but not Wilson parameters. While UNIQUAC is mathematically more complex than NRTL,

† The models shown in Table 8-3 are not applicable to solutions of electrolytes; such solutions are not considered here. However, brief attention is given to aqueous solutions of volatile weak electrolytes in a later section of this chapter. An introduction to the thermodynamics of electrolyte solutions is given in Chap. 9 of Prausnitz, et al., (1999). See also Table 8-1*b*.

‡ Wilson (1964) has given a three-parameter form of his equation that is applicable also to liquid–liquid equilibria. The three-parameter Wilson equation has not received much attention, primarily because it is not readily extended to multicomponent systems.

TABLE 8-3 Some Models for the Excess Gibbs Energy and Subsequent Activity Coefficients for Binary Systems[a]

Name	g^E	Binary parameters	$\ln \gamma_1$ and $\ln \gamma_2$
Two-suffix[b] Margules	$g^E = Ax_1x_2$	A	$RT \ln \gamma_1 = Ax_2^2$
			$RT \ln \gamma_2 = Ax_1^2$
Three-suffix[b] Margules	$g^E = x_1x_2[A + B(x_1 - x_2)]$	A, B	$RT \ln \gamma_1 = (A + 3B)x_2^2 - 4Bx_2^3$
			$RT \ln \gamma_2 = (A - 3B)x_1^2 + 4Bx_1^3$
van Laar	$g^E = \dfrac{Ax_1x_2}{x_1(A/B) + x_2}$	A, B	$RT \ln \gamma_1 = A\left(1 + \dfrac{A}{B}\dfrac{x_1}{x_2}\right)^{-2}$
			$RT \ln \gamma_2 = B\left(1 + \dfrac{B}{A}\dfrac{x_2}{x_1}\right)^{-2}$
Wilson	$\dfrac{g^E}{RT} = -x_1 \ln(x_1 + \Lambda_{12}x_2) - x_2 \ln(x_2 + \Lambda_{21}x_1)$	$\Lambda_{12}, \Lambda_{21}$	$\ln \gamma_1 = -\ln(x_1 + \Lambda_{12}x_2) + x_2\left(\dfrac{\Lambda_{12}}{x_1 + \Lambda_{12}x_2} - \dfrac{\Lambda_{21}}{\Lambda_{21}x_1 + x_2}\right)$
			$\ln \gamma_2 = -\ln(x_2 + \Lambda_{21}x_1) - x_1\left(\dfrac{\Lambda_{12}}{x_1 + \Lambda_{12}x_2} - \dfrac{\Lambda_{21}}{\Lambda_{21}x_1 + x_2}\right)$
Four-suffix[b] Margules	$g^E = x_1x_2[A + B(x_1 - x_2) + C(x_1 - x_2)^2]$	A, B, C	$RT \ln \gamma_1 = (A + 3B + 5C)x_2^2 - 4(B + 4C)x_2^3 + 12Cx_2^4$
			$RT \ln \gamma_2 = (A - 3B + 5C)x_1^2 + 4(B - 4C)x_1^3 + 12Cx_1^4$
NRTL[c]	$\dfrac{g^E}{RT} = x_1x_2\left(\dfrac{\tau_{21}G_{21}}{x_1 + x_2G_{21}} + \dfrac{\tau_{12}G_{12}}{x_2 + x_1G_{12}}\right)$		$\ln \gamma_1 = x_2^2\left[\tau_{21}\left(\dfrac{G_{21}}{x_1 + x_2G_{21}}\right)^2 + \dfrac{\tau_{12}G_{12}}{(x_2 + x_1G_{12})^2}\right]$
	where $\tau_{12} = \dfrac{\Delta g_{12}}{RT}$ $\tau_{21} = \dfrac{\Delta g_{21}}{RT}$; $\ln G_{12} = -\alpha_{12}\tau_{12}$ $\ln G_{21} = -\alpha_{12}\tau_{21}$	$\Delta g_{12}, \Delta g_{21}, \alpha_{12}$[d]	$\ln \gamma_2 = x_1^2\left[\tau_{12}\left(\dfrac{G_{12}}{x_2 + x_1G_{12}}\right)^2 + \dfrac{\tau_{21}G_{21}}{(x_1 + x_2(G_{21})^2}\right]$

UNIQUAC[e]	$g^E = g^E \text{ (combinatorial)} + g^E \text{ (residual)}$ $\dfrac{g^E \text{ (combinatorial)}}{RT} = x_1 \ln \dfrac{\Phi_1}{x_1} + x_2 \ln \dfrac{\Phi_2}{x_2} + \dfrac{z}{2}\left(q_1 x_1 \ln \dfrac{\theta_1}{\Phi_1} + q_2 x_2 \ln \dfrac{\theta_2}{\Phi_2}\right)$ $\dfrac{g^E \text{ (residual)}}{RT} = -q_1 x_1 \ln[\theta_1 + \theta_2 \tau_{21}] - q_2 x_2 \ln[\theta_2 + \theta_1 \tau_{12}]$ $\Phi_1 = \dfrac{x_1 r_1}{x_1 r_1 + x_2 r_2}$ $\theta_1 = \dfrac{x_1 q_1}{x_1 q_1 + x_2 q_2}$ $\ln \tau_{21} = -\dfrac{\Delta u_{21}}{RT} \quad \ln \tau_{12} = -\dfrac{\Delta u_{12}}{RT}$ r and q are pure-component parameters and coordination number $z = 10$	Δu_{12} and Δu_{21}[f]	$\ln \gamma_i = \ln \dfrac{\Phi_i}{x_i} + \dfrac{z}{2} q_i \ln \dfrac{\theta_i}{\Phi_i} + \Phi_j \left(l_i - \dfrac{r_i}{r_j} l_j\right) - q_i \ln(\theta_i + \theta_j \tau_{ji}) + \theta_j q_i \left(\dfrac{\tau_{ji}}{\theta_i + \theta_j \tau_{ji}} - \dfrac{\tau_{ij}}{\theta_j + \theta_i \tau_{ij}}\right)$ where $i = 1 \quad j = 2$ or $i = 2 \quad j = 1$ $l_i = \dfrac{z}{2}(r_i - q_i) - (r_i - 1)$ $l_j = \dfrac{z}{2}(r_j - q_j) - (r_j - 1)$

[a] Prausnitz, et al. (1999) discuss the Margules, van Laar, Wilson, UNIQUAC, and NRTL equations.

[b] Two-suffix signifies that the expansion for g^E is quadratic in mole fraction. Three-suffix signifies a third-order, and four-suffix signifies a fourth-order equation.

[c] NRTL = Non Random Two Liquid.

[d] $\Delta g_{12} = g_{12} - g_{22}$ $\Delta g_{21} = g_{21} - g_{11}$.

[e] UNIQUAC = *Uni*versal *Qua*si *C*hemical. Parameters q and r can be calculated from Eq. (8-10.62).

[f] $\Delta u_{12} = u_{12} - u_{22}$; $\Delta u_{21} = u_{21} - u_{11}$.

it has three advantages: (1) it has only two (rather than three) adjustable parameters, (2) UNIQUAC's parameters often have a smaller dependence on temperature, and (3) because the primary concentration variable is a surface fraction (rather than mole fraction), UNIQUAC is applicable to solutions containing small or large molecules, including polymers.

Simplifications: One-parameter Equations

It frequently happens that experimental data for a given binary mixture are so fragmentary that it is not possible to determine two (or three) *meaningful* binary parameters; limited data can often yield only one significant binary parameter. In that event, it is tempting to use the two-suffix (one-parameter) Margules equation, but this is usually an unsatisfactory procedure because activity coefficients in a real binary mixture are rarely symmetric with respect to mole fraction. In most cases better results are obtained by choosing the van Laar, Wilson, NRTL, or UNIQUAC equation and reducing the number of adjustable parameters through reasonable physical approximations.

To reduce the van Laar equation to a one-parameter form, for mixtures of nonpolar fluids, the ratio A/B can often be replaced by the ratio of molar liquid volumes: $A/B = V_1^L/V_2^L$. This simplification, however, is not reliable for binary mixtures containing one (or two) polar components.

To simplify the Wilson equation, we first note that

$$\Lambda_{ij} = \frac{V_j^L}{V_i^L} \exp\left(-\frac{\lambda_{ij} - \lambda_{ii}}{RT}\right) \tag{8-5.11}$$

where V_i^L is the molar volume of pure liquid i and λ_{ij} is an energy parameter characterizing the interaction of molecule i with molecule j.

The Wilson equation can be reduced to a one-parameter form by assuming that $\lambda_{ij} = \lambda_{ji}$† and that

$$\lambda_{ii} = -\beta(\Delta H_{vi} - RT) \tag{8-5.12}$$

where β is a proportionality factor and ΔH_{vi} is the enthalpy of vaporization of pure component i at T. A similar equation is written for λ_{jj}. When β is fixed, the only adjustable binary parameter is λ_{ij}.

Theoretical considerations suggest that $\beta = 2/z$, where z is the coordination number (typically, $z = 10$). This assumption, used by Wong and Eckert (1971) and Schreiber and Eckert (1971), gives good estimates for a variety of binary mixtures.

Ladurelli et al. (1975) have suggested $\beta = 2/z$ for component 2, having the smaller molar volume, while for component 1, having the larger molar volume, $\beta = (2/z)(V_2^L/V_1^L)$. This suggestion follows from the notion that a larger molecule has a larger area of interaction; parameters λ_{ii}, λ_{jj}, and λ_{ij} are considered as interaction energies per segment rather than per molecule. In this particular case the unit segment is that corresponding to one molecule of component 2.

Using similar arguments, Bruin and Prausnitz (1971) have shown that it is possible to reduce the number of adjustable binary parameters in the NRTL equation

† The simplifying assumption that cross-parameter $\lambda_{ij} = \lambda_{ji}$ (or, similarly, $g_{ij} = g_{ji}$ or $u_{ij} = u_{ji}$) is often useful but is not required by theory unless severe simplifying assumptions are made concerning liquid structure.

by making a reasonable assumption for α_{12} and by substituting NRTL parameter G_{ii} for Wilson parameter Λ_{ii} in Eq. (8-5.12). Bruin gives some correlations for g_{ij}, especially for aqueous systems.

The UNIQUAC equation can be simplified by assuming that

$$u_{11} = \frac{-\Delta U_1}{q_1} \quad \text{and} \quad u_{22} = \frac{-\Delta U_2}{q_2} \tag{8-5.13}$$

and that

$$u_{12} = u_{21} = (u_{11}u_{22})^{1/2}(1 - c_{12}) \tag{8-5.14}$$

where, remote from the critical temperature, energy ΔU_i is given very nearly by $\Delta U_i \approx \Delta H_{vi} - RT$. The only adjustable binary parameter is c_{12}, which, for mixtures of nonpolar liquids, is positive and small compared with unity. For some mixtures containing polar components, however, c_{12} is of the order of 0.5; and when the unlike molecules in a mixture are attracted more strongly than like molecules, c_{12} may be negative, e.g., in acetone-chloroform.

For mixtures of nonpolar liquids, a one-parameter form (van Laar, Wilson, NRTL, UNIQUAC) often gives results nearly as good as those obtained by using two, or even three, parameters. However, if one or both components are polar, significantly better results are usually obtained by using two parameters, provided that the experimental data used to determine the parameters are of sufficient quantity and quality.

8-6 CALCULATION OF LOW-PRESSURE BINARY VAPOR-LIQUID EQUILIBRIA WITH ACTIVITY COEFFICIENTS

First consider the isothermal case. At some constant temperature T, we wish to construct two diagrams; y vs. x and P vs. x. We assume that, since the pressure is low, we can use Eq. (8-4.1) with $\mathcal{F}_i = 1$. The steps toward that end are:

1. Find the pure liquid vapor pressures P_{vp1} and P_{vp2} at T.

2. Suppose a few experimental points for the mixture are available at temperature T. Arbitrarily, to fix ideas, suppose there are five points; i.e., for five values of x there are five corresponding experimental equilibrium values of y and P. For each of these points calculate γ_1 and γ_2 according to

$$\gamma_1 = \frac{y_1 P}{x_1 P_{vp1}} \tag{8-6.1}$$

$$\gamma_2 = \frac{y_2 P}{x_2 P_{vp2}} \tag{8-6.2}$$

3. For each of the five points, calculate the molar excess Gibbs energy g^E:

$$g^E = RT(x_1 \ln \gamma_1 + x_2 \ln \gamma_2) \tag{8-6.3}$$

4. Choose one of the equations for g^E given in Table 8.3. Adjust the constants in that equation to minimize the deviation between g^E calculated from the equation and g^E found from the experimental data as in step 3.

5. Using the equations for γ that go with the chosen g^E model, find γ_1 and γ_2 at arbitrarily selected values of x_1 from $x_1 = 0$ to $x_1 = 1$.

6. For each selected x_1 find the corresponding y_1 and P by solving Eqs. (8-6.1) and (8-6.2) coupled with the mass balance relations $x_2 = 1 - x_1$ and $y_2 = 1 - y_1$. The results obtained give the desired y-vs.-x and P-vs.-x diagrams.

The simple steps outlined above provide a rational, thermodynamically consistent procedure for interpolation and extrapolation with respect to composition. The crucial step is 4. Judgment is required to obtain the best, i.e., the most representative, constants in the expression chosen for g^E. To do so, it is necessary to decide on how to weight the five individual experimental data; some may be more reliable than others. For determining the constants, the experimental points that give the most information are those at the ends of the composition scale, that is, y_1 when x_1 is small and y_2 when x_2 is small. Unfortunately, however, these experimental data are often the most difficult to measure. Thus it frequently happens that the data that are potentially most valuable are also the ones that are likely to be least accurate.

Now consider the more complicated isobaric case. At some constant pressure P, we wish to construct two diagrams: y vs. x and T vs. x. Assuming that the pressure is low, we again use Eq. (8-4.1) with $\mathcal{F}_i = 1$. The steps toward construction of these diagrams are:

1. Find pure-component vapor pressures P_{vp1} and P_{vp2}. Prepare plots (or obtain analytical representation) of P_{vp1} and P_{vp2} vs. temperature in the region where $P_{vp1} \approx P$ and $P_{vp2} \approx P$. (See Chap. 7.)

2. Suppose there are available a few experimental data points for the mixture at pressure P or at some other pressure not far removed from P or, perhaps, at some constant temperature such that the total pressure is in the general vicinity of P. As in the previous case, to fix ideas, we arbitrarily set the number of such experimental points at five. By experimental point we mean, as before, that for some value of x_1 we have the corresponding experimental equilibrium values of y_1, T, and total pressure.

For each of the five points, calculate activity coefficients γ_1 and γ_2 according to Eqs. (8-6.1) and (8-6.2). For each point the vapor pressures P_{vp1} and P_{vp2} are evaluated at the experimentally determined temperature for that point. In these equations, the experimentally determined total pressure is used for P; the total pressure measured is not necessarily the same as the pressure for which we wish to construct the equilibrium diagrams.

3. For each of the five points, calculate the molar excess Gibbs energy according to Eq. (8-6.3).

4. Choose one of the equations for g^E given in Table 8-3. As in step 4 of the previous (isothermal) case, find the constants in that equation which give the smallest deviation between calculated values of g^E and those found in step 3. When the experimental data used in Eq. (8-6.3) are isobaric rather than isothermal, it may be advantageous to choose an expression for g^E that contains the temperature as one of the explicit variables. Such a choice, however, complicates the calculations in step 6.

5. Find γ_1 and γ_2 as functions of x by differentiation according to Eqs. (8-5.3) and (8-5.4).†

6. Select a set of arbitrary values for x_1 for the range $x_1 = 0$ to $x_1 = 1$. For each x_1, by iteration, solve simultaneously the two equations of phase equilibrium [Eqs. (8-6.1) and (8-6.2)] for the two unknowns, y_1 and T. In these equations the total pressure P is now the one for which the equilibrium diagrams are desired.

Simultaneous solution of Eqs. (8-6.1) and (8-6.2) requires trial and error because, at a given x, both y and T are unknown and both P_{vp1} and P_{vp2} are strong, nonlinear functions of T. In addition, γ_1 and γ_2 may also vary with T (as well as x), depending on which expression for g^E has been chosen in step 4. For simultaneous solution of the two equilibrium equations, the best procedure is to assume a reasonable temperature for each selected value of x_1. Using this assumed temperature, calculate y_1 and y_2 from Eqs. (8-6.1) and (8-6.2). Then check if $y_1 + y_2 = 1$. If not, assume a different temperature and repeat the calculation. In this way, for fixed P and for each selected value of x, find corresponding equilibrium values y and T.

Calculation of isothermal or isobaric vapor-liquid equilibria can be efficiently performed with a computer. Further, it is possible in such calculations to include the correction factor $\mathcal{F}_i$ [Eq. 8-4.1)] when necessary. In that event, the calculations are more complex in detail but not in principle.

When the procedures outlined above are followed, the accuracy of any vapor-liquid equilibrium calculation depends primarily on the extent to which the expression for g^E accurately represents the behavior of the mixture at the particular conditions (temperature, pressure, composition) for which the calculation is made. This accuracy of representation often depends not so much on the algebraic form of g^E as on the reliability of the constants appearing in that expression. This reliability, in turn, depends on the quality and quantity of the experimental data used to determine the constants.

Some of the expressions for g^E shown in Table 8-3 have a better theoretical foundation than others, but all have a strong empirical flavor. Experience has indicated that the relatively more recent equations for g^E (Wilson, NRTL, and UNIQUAC) are more consistently reliable than the older equations in the sense that they can usually reproduce accurately even highly nonideal behavior by using only two or three adjustable parameters.

The oldest equation of g^E, that of Margules, is a power series in mole fraction. With a power series it is always possible to increase accuracy of representation by including higher terms, where each term is multiplied by an empirically determined coefficient. (The van Laar equation, as shown by Wohl (1946) is also a power series in effective volume fraction, but in practice this series is almost always truncated after the quadratic term.) However, inclusion of higher-order terms in g^E is dangerous because subsequent differentiation to find γ_1 and γ_2 can then lead to spurious maxima or minima. Also, inclusion of higher-order terms in binary data reduction often leads to serious difficulties when binary data are used to estimate multicomponent phase equilibria.

It is desirable to use an equation for g^E that is based on a relatively simple model and which contains only two (or at most three) adjustable binary parameters. Ex-

† Some error is introduced here because Eqs. (8-5.3) and (8-5.4) are based on the isobaric and isothermal Gibbs-Duhem equation. For most practical calculations this error is not serious. See, for example, Appendix D of Prausuitz, et al., (1999).

perimental data are then used to find the "best" binary parameters. Since experimental data are always of limited accuracy, it often happens that several sets of binary parameters may equally well represent the data within experimental uncertainty. Only in rare cases, when experimental data are both plentiful and highly accurate, is there any justification for using more than three adjustable binary parameters.

8-7 EFFECT OF TEMPERATURE ON LOW-PRESSURE VAPOR-LIQUID EQUILIBRIA

A particularly troublesome question is the effect of temperature on the molar excess Gibbs energy g^E. This question is directly related to s^E, the molar excess entropy of mixing about which little is known.† In practice, either one of two approximations is frequently used.

(a) Athermal Solution. This approximation sets $g^E = -Ts^E$, which assumes that the components mix at constant temperature without change of enthalpy ($h^E = 0$). This assumption leads to the conclusion that, at constant composition, $\ln \gamma_i$ is independent of T or, its equivalent, that g^E/RT is independent of temperature.

(b) Regular Solution. This approximation sets $g^E = h^E$, which is the same as assuming that $s^E = 0$. This assumption leads to the conclusion that, at constant composition, $\ln \gamma_i$ varies as $1/T$ or, its equivalent, that g^E is independent of temperature.

Neither one of these extreme approximations is valid, although the second one is often better than the first. Good experimental data for the effect of temperature on activity coefficients are rare, but when such data are available, they suggest that, for a moderate temperature range, they can be expressed by an empirical equation of the form

$$(\ln \gamma_i)_{\substack{\text{constant}\\ \text{composition}}} = c + \frac{d}{T} \tag{8-7.1}$$

where c and d are empirical constants that depend on composition. In most cases constant d is positive. It is evident that, when $d = 0$, Eq. (8-7.1) reduces to assumption (a) and, when $c = 0$, it reduces to assumption (b). Unfortunately, in typical cases c and d/T are of comparable magnitude.

Thermodynamics relates the effect of temperature on γ_i to the partial molar enthalpy $\bar{h}_i$

$$\left[\frac{\partial \ln \gamma_i}{\partial(1/T)}\right]_{x,P} = \frac{\bar{h}_i - h_i^\circ}{R} \tag{8-7.2}$$

where h_i° is the enthalpy of liquid i in the standard state, usually taken as pure liquid i at the system temperature and pressure. Sometimes (but rarely) experimental

† From thermodynamics, $s^E = -(\partial g^E/\partial T)_{P,x}$ and $g^E = h^E - Ts^E$.

data for $\bar{h}_i - h_i^\circ$ may be available; if so, they can be used to provide information on how the activity coefficient changes with temperature. However, even if such data are at hand, Eq. (8-7.2) must be used with caution because $\bar{h}_i - h_i^\circ$ depends on temperature and often strongly so.

Some of the expressions for g^E shown in Table 8-3 contain T as an explicit variable. However, one should not therefore conclude that the parameters appearing in those expressions are independent of temperature. The explicit temperature dependence indicated provides only an approximation. This approximation is usually, but not always, better than approximation (*a*) or (*b*), but, in any case, it is not exact.

Fortunately, the primary effect of temperature on vapor-liquid equilibria is contained in the pure-component vapor pressures or, more precisely, in the pure-component liquid fugacities [Eq. (8-3.1)]. While activity coefficients depend on temperature as well as composition, the temperature dependence of the activity coefficient is usually small when compared with the temperature dependence of the pure-liquid vapor pressures. In a typical mixture, a rise of 10°C increases the vapor pressures of the pure liquids by a factor of 1.5 or 2, but the change in activity coefficient is likely to be only a few percent, often less than the experimental uncertainty. Therefore, unless there is a large change in temperature, it is frequently satisfactory to neglect the effect of temperature on g^E when calculating vapor-liquid equilibria. However, in calculating liquid-liquid equilibria, vapor pressures play no role at all, and therefore the effect of temperature on g^E, although small, may seriously affect liquid-liquid equilibria. Even small changes in activity coefficients can have a large effect on multicomponent liquid-liquid equilibria, as briefly discussed in Sec. 8-14.

8-8 BINARY VAPOR-LIQUID EQUILIBRIA: LOW-PRESSURE EXAMPLES

To introduce the general ideas, we present first two particularly simple methods for reduction of vapor-liquid equilibria. These are followed by a brief introduction to more accurate, but also mathematically more complex, procedures.

Example 8-1 Given five experimental vapor-liquid equilibrium data for the binary system methanol (1)–1,2-dichloroethane (2) at 50°C, calculate the *P*-*y*-*x* diagram at 50°C and predict the *P*-*y*-*x* diagram at 60°C.

Experimental Data at 50°C [Udovenko and Frid, 1948]

100 x_1	100 y_1	P, bar
30	59.1	0.6450
40	60.2	0.6575
50	61.2	0.6665
70	65.7	0.6685
90	81.4	0.6262

solution To interpolate in a thermodynamically consistent manner, we must choose an algebraic expression for the molar excess Gibbs energy. For simplicity, we choose the van Laar equation (See Table 8-3). To evaluate the van Laar constants A' and B', we rearrange the van Laar equation in a linear form†

$$\frac{x_1 x_2}{g^g/RT} = D + C(2x_1 - 1) \qquad \text{where} \qquad \begin{aligned} A' &= (D - C)^{-1} \\ B' &= (D + C)^{-1} \end{aligned} \tag{8-8.1}$$

Constants D and C are found from a plot of $x_1 x_2 (g^E/RT)^{-1}$ vs. x_1. The intercept at $x_1 = 0$ gives $D - C$, and the intercept at $x_1 = 1$ gives $D + C$. The molar excess Gibbs energy is calculated from the definition

$$\frac{g^g}{RT} = x_1 \ln \gamma_1 + x_2 \ln \gamma_2 \tag{8-8.2}$$

For the five available experimental points, activity coefficients γ_1 and γ_2 are calculated from Eq. (8-4.1) with $\mathcal{F}_i = 1$ and from pure-component vapor-pressure data.

Table 8-4 gives $x_1 x_2 (g^E/RT)^{-1}$ as needed to obtain van Laar constants. Figure 8-1 shows the linearized van Laar equation. The results shown are obtained with Antoine constants given in Appendix A.

From Fig. 8-1 we obtain the van Laar constants

$$A' = 1.93 \qquad B' = 1.62 \qquad \frac{A'}{B'} = 1.19$$

We can now calculate γ_1 and γ_2 at any mole fraction:

$$\ln \gamma_1 = 1.93 \left(1 + 1.19 \frac{x_1}{x_2}\right)^{-2} \tag{8-8.3}$$

$$\ln \gamma_2 = 1.62 \left(1 + \frac{x_2}{1.19\, x_1}\right)^{-2} \tag{8-8.4}$$

By using Eqs. (9-8.3) and (8-8.4) and the pure-component vapor pressures, we can now find y_1, y_2, and total pressure P. There are two unknowns: y_1 (or y_2) and P. To find them, we must solve the two equations of vapor-liquid equilibrium

TABLE 8-4 Experimental Activity Coefficients for Linearized van Laar Plot, Methanol (1)–1,2-Dichloroethane (2) at 50°C

x_1	γ_1	γ_2	$\frac{x_1 x_2}{g^E/RT}$
0.3	2.29	1.21	0.551
0.4	1.78	1.40	0.555
0.5	1.47	1.66	0.562
0.7	1.13	2.45	0.593
0.9	1.02	3.74	0.604

† From Table 8-3, $A' = A/RT$ and $B' = B/RT$.

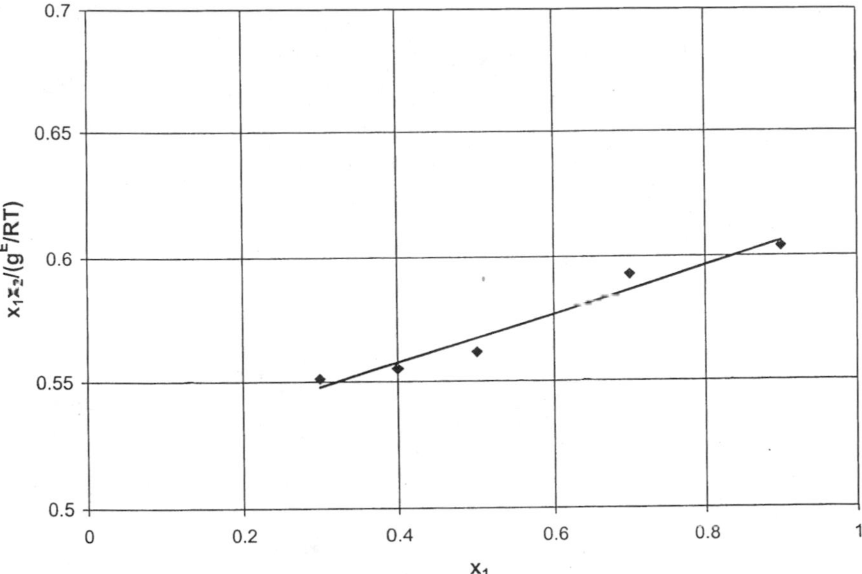

FIGURE 8-1 Determination of van Laar constants for the system methanol (1)–1,2-dichloroethane (2) at 50°C.

$$P = x_1\gamma_1 P_{vp1} + x_2\gamma_2 P_{vp2} \tag{8-8.5}$$

$$y_1 = \frac{x_1\gamma_1 P_{vp1}}{P} \tag{8-8.6}$$

Calculated results at 50°C are shown in Table 8-5. To predict vapor-liquid equilibria at 60°C, we assume that the effect of temperature on activity coefficients is given by the regular-solution approximation (see Sec. 8-7):

$$\frac{\ln \gamma_i(60°\text{C})}{\ln \gamma_i(50°\text{C})} = \frac{273 + 50}{273 + 60} \tag{8-8.7}$$

Pure-component vapor pressures at 60°C are found from the Antoine relations. The two equations of equilibrium [Eqs. (8-8.5) and (8-8.6)] are again solved simultaneously to obtain y and P as a function of x. Calculated results at 60°C are shown in Table 8-5 and in Fig. 8-2.

Predicted y's are in good agreement with experiment [Udovenko and Frid, 1948], but predicted pressures are too high. This suggests that Eq. (8-8.7) is not a good approximation for this system.

Equation (8-8.7) corresponds to approximation (*b*) in Sec. 8-7. If approximation (*a*) had been used, the predicted pressure would have been even higher.

Example 8-2 Given five experimental vapor-liquid equilibrium data for the binary system propanol (1)-water (2) at 1.01 bar, predict the *T-y-x* diagram for the same system at 1.33 bar.

TABLE 8-5 Calculated Vapor-Liquid Equilibria in the System Methanol (1)–1,2-Dichloroethane (2) at 50 and 60°C

	γ_1		γ_2		100 y_1		P, bar		100 Δy		$10^3\ \Delta P$, bar	
100 x_1	50°C	60°C	50°C	60°C	50°C	60°C	50°C	60°C	50°C	60°C	50°C	60°C
5	5.52	5.25	1.01	1.01	34.0	33.7	0.4513	0.6574				
10	4.51	4.31	1.02	1.02	46.6	46.5	0.5373	0.7836	−1.2	0.1	−1.6	2.0
20	3.15	3.04	1.09	1.09	56.3	56.5	0.6215	0.9106	0.1	0.9	11.4	20.9
40	1.82	1.79	1.38	1.36	61.1	61.9	0.6625	0.9782	0.9	2.0	4.9	28.3
60	1.28	1.27	1.95	1.91	63.7	64.9	0.6710	0.9946	1.2	1.7	−0.1	26.1
80	1.06	1.06	3.03	2.93	71.4	72.8	0.6601	0.9830	0.3	0.9	3.1	36.2
90	1.01	1.01	3.88	3.73	80.7	81.9	0.6282	0.9413	−0.7	−0.3	2.0	34.7

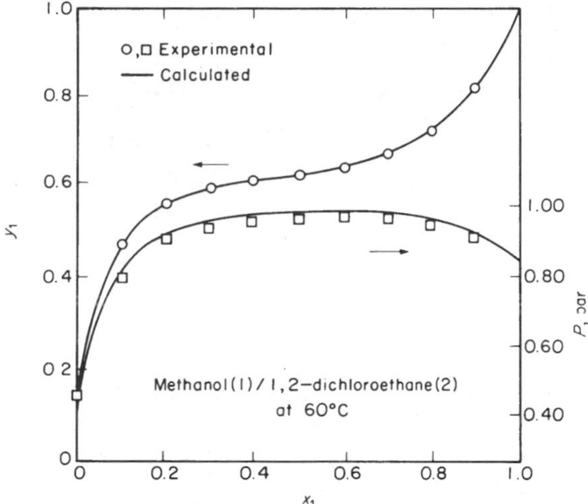

FIGURE 8-2 Calculated and experimental vapor compositions and total pressures of the system methanol (1)–1,2-dichloroethane (2) at 60°C.

Experimental Data at 1.01 bar [Murte and Van Winkle, 1978]

100 x_1	100 y_1	T, °C
7.5	37.5	89.05
17.9	38.8	87.95
48.2	43.8	87.80
71.2	56.0	89.20
85.0	68.5	91.70

solution To represent the experimental data, we choose the van Laar equation, as in Example 8-1. Since the temperature range is small, we neglect the effect of temperature on the van Laar constants.

As in Example 8-1, we linearize the van Laar equation as shown in Eq. (8-8.1). To obtain the van Laar constants A' and B', we need, in addition to the data shown above, vapor pressure data for the pure components.

Activity coefficients γ_1 and γ_2 are calculated from Eq. (8-4.1) with $\mathcal{F}_i = 1$, and g^E/RT is calculated from Eq. (8-8.2). Antoine constants are from Appendix A. Results are given in Table 8-6. The linearized van Laar plot is shown in Fig. 8-3. From the intercepts in Fig. 8-3 we obtain

$$A' = 2.67 \qquad B' = 1.13 \qquad \frac{A'}{B'} = 2.37 \tag{8-8.8}$$

Activity coefficients γ_1 and γ_2 are now given by the van Laar equations

TABLE 8-6 Experimental Activity Coefficients for Linearized van Laar Plot, *n*-Propanol (1)–water (2) at 1.01 Bar

$100x_1$	T, °C	γ_1	γ_2	$\frac{x_1 x_2}{g^E/RT}$
7.5	89.05	6.85	1.01	0.440
17.9	87.95	3.11	1.17	0.445
48.2	87.80	1.31	1.71	0.612
71.2	89.20	1.07	2.28	0.715
85.0	91.70	0.99	2.85	0.837

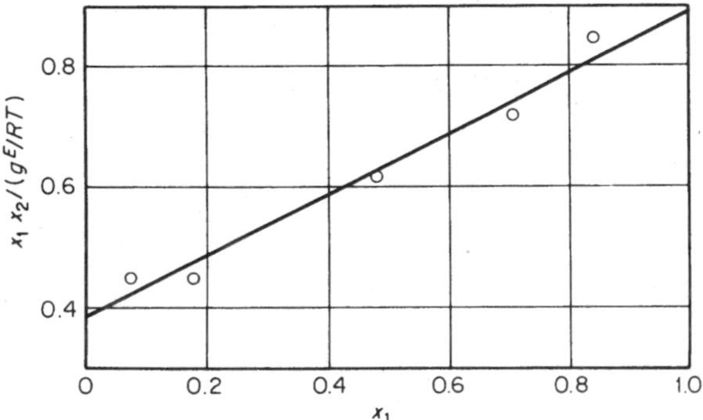

FIGURE 8-3 Determination of van Laar constants for the system *n*-propanol (1)–water at 1.01 bar.

$$\ln \gamma_1 = 2.67 \left(1 + 2.37 \frac{x_1}{x_2}\right)^{-2} \tag{8-8.9}$$

$$\ln \gamma_2 = 1.13 \left(1 + \frac{x_2}{2.37\, x_1}\right)^{-2} \tag{8-8.10}$$

To obtain the vapor-liquid equilibrium diagram at 1.33 bar, we must solve simultaneously the two equations of equilibrium

$$y_1 = \frac{x_1 \gamma_1 P_{vp1}(T)}{1.33} \tag{8-8.11}$$

$$1 - y_1 = y_2 = \frac{x_2 \gamma_2 P_{vp2}(T)}{1.33} \tag{8-8.12}$$

In this calculation we assume that γ_1 and γ_2 depend only on x (as given by the van Laar equations) and not on temperature. However, P_{vp1} and P_{vp2} are strong functions of temperature.

The two unknowns in Eqs. (8-8.11) and (8-8.12) are y_1 and T. To solve for these unknowns, it is also necessary to use the Antoine relations for the two pure components. The required calculations contain the temperature as an implicit variable; solution of the equations of equilibrium must be by iteration.

While iterative calculations are best performed with a computer, in this example it is possible to obtain results rapidly by hand calculations. Dividing one of the equations of equilibrium by the other, we obtain

$$y_1 = \left(1 + \frac{x_2 \gamma_2}{x_1 \gamma_1} \frac{P_{vp2}}{P_{vp1}}\right)^{-1} \tag{8-8.13}$$

Although P_{vp2} and P_{vp1} are strong functions of temperature, the ratio P_{vp2}/P_{vp1} is a much weaker function of temperature.

For a given x_1, and γ_2/γ_1 from the van Laar equations. Choose a reasonable temperature and find P_{vp2}/P_{vp1} from the Antoine relations. Equation (8-8.13) then gives a first estimate for y_1. Using this estimate, find P_{vp1} from

$$P_{vp1} = \frac{1.33\ y_1}{x_1 \gamma_1} \tag{8-8.14}$$

The Antoine relation for component 1 then gives a first estimate for T. By using this T, find the ratio P_{vp2}/P_{vp1} and, again using Eq. (9-8.13), find the second estimate for y_1. This second estimate for y_1 is then used with the Antoine relation to find the second estimate for T. Repeat until there is negligible change in the estimate for T.

It is clear that Eq. (8-8.14) for component 1 could be replaced with the analogous equation for component 2. Which one should be used? In principle, either one may be used, but for components of comparable volatility, convergence is likely to be more rapid if Eq. (8-8.14) is used for $x_1 > ½$ and the analogous equation for component 2 is used when $x_1 < ½$. However, if one component is much more volatile than the other, the equation for that component is likely to be more useful. Table 8-7 presents calculated results at 1.33 bar. Unfortunately, no experimental results at this pressure are available for comparison.

The two simple examples above illustrate the essential steps for calculating vapor-liquid equilibria from limited experimental data. Because of their illustrative nature, these examples are intentionally simplified. For more accurate results it is desirable to replace some of the details by more sophisticated techniques. For example, it may be worthwhile to include corrections for vapor phase nonideality and perhaps the Poynting factor, i.e., to relax the simplifying assumption $\mathcal{F}_i = 1$ in Eq. (8-4.1). At the modest pressures encountered here, however, such modifications are likely to have a small effect. A more important change would be to replace the van Laar equation with a better equation for the activity coefficients, e.g., the Wilson equation or the UNIQUAC equation. If this is done, the calculational procedure is

TABLE 8-7 Calculated Vapor-Liquid Equilibria for *n*-Propanol (1)–water (2) at 1.33 Bar

$100x_1$	γ_1	γ_2	T, °C	$100y_1$
5	8.25	1.01	97.8	32.2
10	5.33	1.05	95.7	38.3
20	2.87	1.17	95.1	40.3
40	1.49	1.53	95.0	41.8
50	1.27	1.75	95.0	44.4
60	1.14	1.99	95.4	48.6
80	1.02	2.52	98.0	64.3
90	1.01	2.80	100.6	78.2

the same but the details of computation are more complex. Because of algebraic simplicity, the van Laar equations can easily be linearized, and therefore a convenient graphical procedure can be used to find the van Laar constants.† An equation like UNIQUAC or that of Wilson cannot easily be linearized, and therefore, for practical application, it is necessary to use a computer for data reduction to find the binary constants in the equation.

In Examples 8-1 and 8-2 we have not only made simplifications in the thermodynamic relations but have also neglected to take into quantitative consideration the effect of experimental error.

It is beyond the scope of this chapter to discuss in detail the highly sophisticated statistical methods now available for optimum reduction of vapor-liquid equilibrium data. Nevertheless, a very short discussion may be useful as an introduction for readers who want to obtain the highest possible accuracy from the available data.

A particularly effective data reduction method is described by Anderson, Abrams, and Grens (1978) who base their analysis on the principle of maximum likelihood while taking into account probable experimental errors in all experimentally determined quantities.

To illustrate the general ideas, we define a calculated pressure (constraining function) by

$$P^c = \exp\left[x_1 \ln\left(\frac{x_1 \gamma_1 P_{\text{vp1}}}{y_1} \mathcal{F}_1\right) + x_2 \ln\left(\frac{x_2 \gamma_2 P_{\text{vp2}}}{y_2} \mathcal{F}_2\right)\right] \tag{8-8.15}$$

where $\mathcal{F}_i$ is given in Eq. (8-4.2). The most probable values of the parameters (for the function chosen for g^E) are those which minimize the function I:

$$I = \sum_{\text{data}} \left[\frac{(x_i^\circ - x_i^M)^2}{\sigma_{x_i}^2} + \frac{(y_i^\circ - y_i^M)^2}{\sigma_{y_i}^2} + \frac{(P_i^\circ - P_i^M)^2}{\sigma_{P_i}^2} + \frac{(T_i^\circ - T_i^M)^2}{\sigma_{T_i}^2}\right] \tag{8-8.16}$$

In Eq. (8-8.16), the superscript ^{M} means a measured value of the variable and $^\circ$ means a statistical estimate of the true value of the variable which is used to calculate all of the properties in Eq. (8-8.15). The σ's are estimates of the variances of the variable values, *i.e.*, an indication of the experimental uncertainty. These may or may not be varied from data point to data point.

By using experimental *P-T-x-y* data and the UNIQUAC equation with estimated parameters $u_{12} - u_{22}$ and $u_{21} - u_{11}$, we obtain estimates of x_i°, y_i°, T_i°, and P_i°. The last of these is found from Eq. (8-8.15) with true values, x_i°, y_i°, and T_i°. We then evaluate I, having previously set variances σ_x^2, σ_y^2, σ_P^2, and σ_T^2 from a critical inspection of the data's quality. Upon changing the estimate of UNIQUAC parameters, we calculate a new I; with a suitable computer program, we search for the parameters that minimize I. Convergence is achieved when, from one iteration to the next, the relative change in I is less than 10^{-5}. After the last iteration, the variance of fit σ_F^2 is given by

$$\sigma_F^2 = \frac{I}{D - L} \tag{8-8.17}$$

where D is the number of data points and L is the number of adjustable parameters.

Since all experimental data have some experimental uncertainty, and since any equation for g^E can provide only an approximation to the experimental results, it

† The three-suffix Margules equation is also easily linearized, as shown by H. C. Van Ness, "Classical Thermodynamics of Nonelectrolyte Solutions," p. 129, Pergamon, New York, 1964.

follows that the parameters obtained from data reduction are not unique; there are many sets of parameters which can equally well represent the experimental data within experimental uncertainty. To illustrate this lack of uniqueness, Fig. 8-4 shows results of data reduction for the binary mixture ethanol (1)–water (2) at 70°C. Experimental data reported by Mertl (1972) were reduced using the UNIQUAC equation with the variances

$$\sigma_x = 10^{-3} \qquad \sigma_y = 10^{-2} \qquad \sigma_P = 6.7 \times 10^{-4} \text{ bar} \qquad \sigma_T = 0.1 \text{ K}$$

For this binary system, the fit is very good; $\sigma_F^2 = 5 \times 10^{-4}$.

The ellipse in Fig. 8-4 clearly shows that, although parameter $u_{21} - u_{11}$ is strongly correlated with parameter $u_{12} - u_{22}$, there are many sets of these parameters that can equally well represent the data. The experimental data used in data reduction are not sufficient to fix a unique set of "best" parameters. Realistic data reduction can determine only a region of parameters.†

While Fig. 8-4 pertains to the UNIQUAC equation, similar results are obtained when other equations for g^E are used; only a region of acceptable parameters can

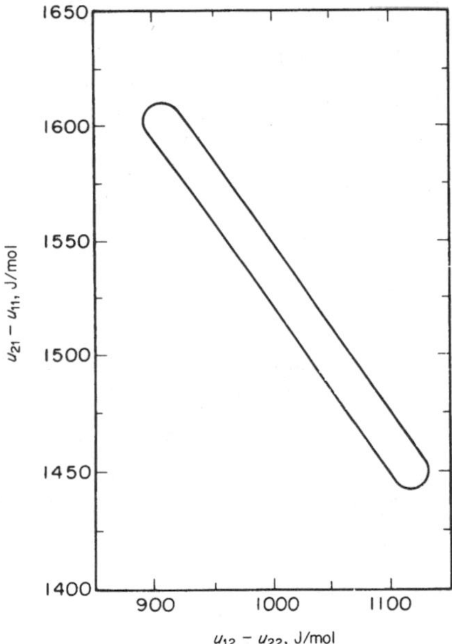

FIGURE 8-4 The 99 percent confidence ellipse for UNIQUAC parameters in the system ethanol (1)–water (2) at 70°C.

† Instead of the constraint given by Eq. (8-8.15), it is sometimes preferable to use instead two constraints; first, Eq. (8-8.18), and second,

$$y_1 = \frac{x_1 \gamma_1 P_{vp1} \mathfrak{F}_1}{x_1 \gamma_1 P_{vp1} \mathfrak{F}_1 + x_2 \gamma_2 P_{vp2} \mathfrak{F}_2}$$

or the corresponding equation for y_2.

be obtained from a given set of *P-T-y-x* data. For a two-parameter equation, this region is represented by an area; for a three-parameter equation, it is represented by a volume. If the equation for g^E is suitable for the mixture, the region of acceptable parameters shrinks as the quality and quantity of the experimental data increase. However, considering the limits of both theory and experiment, it is unreasonable to expect this region to shrink to a single point.

As indicated by numerous authors, notably Abbott and Van Ness (1975), experimental errors in vapor composition y are usually larger than those in experimental pressure P, temperature T, and liquid phase composition x. Therefore, a relatively simple fitting procedure is provided by reducing only *P-x-T* data; y data, even if available, are not used.† The essential point is to minimize the deviation between calculated and observed pressures.

The pressure is calculated according to

$$P_{\text{calc}} = y_1 P + y_2 P = x_1 \gamma_1 P_{\text{vp}1} \mathcal{F}_1 + x_2 \gamma_2 P_{\text{vp}2} \mathcal{F}_2 \tag{8-8.18}$$

where $\mathcal{F}_i$ is given by Eq. (8-4.2).

Thermodynamically consistent equations are now chosen to represent γ_1 and γ_2 as functions of x (and perhaps T); some are suggested in Table 8-3. These equations contain a number of adjustable binary parameters. With a computer, these parameters can be found by minimizing the deviation between calculated and measured pressures.

At low pressures we can assume that $\mathcal{F}_1 = \mathcal{F}_2 = 1$. However, at higher pressures, correction factors $\mathcal{F}_1$ and $\mathcal{F}_2$ are functions of pressure, temperature, and vapor compositions y_1 and y_2; these compositions are calculated from

$$y_1 = \frac{x_1 \gamma_1 P_{\text{vp}1} \mathcal{F}_1(P, T, y)}{P} \quad \text{and} \quad y_2 = \frac{x_2 \gamma_2 P_{\text{vp}2} \mathcal{F}_2(P, T, y)}{P} \tag{8-8.19‡}$$

The data reduction scheme, then, is iterative; to get started, it is necessary first to assume an estimated y for each x. After the first iteration, a new set of estimated y's is found from Eq. (8-8.19). Convergence is achieved when, following a given iteration, the y's calculated differ negligibly from those calculated after the previous iteration and when the pressure deviation is minimized.

8-9 MULTICOMPONENT VAPOR-LIQUID EQUILIBRIA AT LOW PRESSURE

The equations required to calculate vapor-liquid equilibria in multicomponent systems are, in principle, the same as those required for binary systems. In a system containing N components, we must solve N equations simultaneously: Eq. (8-4.1) for each of the N components. We require the saturation (vapor) pressure of each component, as a pure liquid, at the temperature of interest. If all pure-component vapor pressures are low, the total pressure also is low. In that event, the factor $\mathcal{F}_i$ [Eq. (8-4.2)] can often be set equal to unity.

† This technique is commonly referred to as *Barker's method*.

‡ If the Lewis fugacity rule is used to calculate vapor-phase fugacity coefficients, $\mathcal{F}_1$ and $\mathcal{F}_2$ depend on pressure and temperature but are independent of y. The Lewis rule provides mathematical simplification, but, unfortunately, it can be a poor rule. If a computer is available, there is no need to use it.

Activity coefficients γ_i are found from an expression for the excess Gibbs energy, as discussed in Sec. (8-5). For a mixture of N components, the total excess Gibbs energy G^E is defined by

$$G^E = RT \sum_{i=1}^{N} n_i \ln \gamma_i \tag{8-9.1}$$

where n_i is the number of moles of component i. The molar excess Gibbs energy g^E is simply related to G^E by

$$g^E = \frac{G^E}{n_T} \tag{8-9.2}$$

where n_T, the total number of moles, is equal to $\sum_{i=1}^{N} n_i$.

Individual activity coefficients can be obtained from G^E upon introducing the Gibbs-Duhem equation for a multicomponent system at constant temperature and pressure. That equation is

$$\sum_{i=1}^{N} n_i \, d \ln \gamma_i = 0 \tag{8-9.3}$$

The activity coefficient γ_i is found by a generalization of Eq. (8-5.3):

$$RT \ln \gamma_i = \left(\frac{\partial G^E}{\partial n_i} \right)_{T,P,n_{j \neq i}} \tag{8-9.4}$$

where $n_{j \neq i}$ indicates that all mole numbers (except n_i) are held constant in the differentiation.

The key problem in calculating multicomponent vapor-liquid equilibria is to find an expression for g^E that provides a good approximation for the properties of the mixture. Toward that end, the expressions for g^E for binary systems, shown in Table 8-3, can be extended to multicomponent systems. A few of these are shown in Table 8-8.

The excess Gibbs energy concept is particularly useful for multicomponent mixtures because in many cases, to a good approximation, extension from binary to multicomponent systems can be made in such a way that only binary parameters appear in the final expression for g^E. When that is the case, a large saving in experimental effort is achieved, since experimental data are then required only for the mixture's constituent binaries, not for the multicomponent mixture itself. For example, activity coefficients in a ternary mixture (components 1, 2, and 3) can often be calculated with good accuracy by using only experimental data for the three binary mixtures: components 1 and 2, components 1 and 3, and components 2 and 3.

Many physical models for g^E for a binary system consider only two-body intermolecular interactions, i.e., interactions between two (but not more) molecules. Because of the short range of molecular interaction between nonelectrolytes, it is often permissible to consider only interactions between molecules that are first neighbors and then to sum all the two-body, first-neighbor interactions. A useful consequence of these simplifying assumptions is that extension to ternary (and higher) systems requires only binary, i.e., two-body, information; no ternary (or higher) constants appear. However, not all physical models use this simplifying assumption, and those which do not often require additional simplifying assump-

TABLE 8-8 Three Expressions for the Molar Excess Gibbs Energy and Activity Coefficients of Multicomponent Systems Using Only Pure-Component and Binary Parameters. Symbols defined in Table 8-3; the number of components is N

Name	Molar excess Gibbs energy	Activity coefficient for component i
Wilson	$\frac{g^E}{RT} = -\sum_i^N x_i \ln\left(\sum_j^N x_j\Lambda_{ij}\right)$	$\ln\gamma_i = -\ln\left(\sum_j^N x_j\Lambda_{ij}\right) + 1 - \sum_k^N \frac{x_k\Lambda_{ki}}{\sum_j^N x_j\Lambda_{kj}}$
NRTL	$\frac{g^E}{RT} = \sum_i^N x_i \frac{\sum_j^N \tau_{ji}G_{ji}x_j}{\sum_k^N G_{ki}x_k}$	$\ln\gamma_i = \frac{\sum_j^N \tau_{ji}G_{ji}x_j}{\sum_k^N G_{ki}x_k} - \sum_j^N \frac{x_jG_{ij}}{\sum_k^N G_{kj}x_k}\left(\tau_{ij} - \frac{\sum_k^N x_k\tau_{kj}G_{kj}}{\sum_k^N G_{kj}x_k}\right)$
UNIQUAC†	$\frac{g^E}{RT} = \sum_i^N x_i \ln\frac{\Phi_i}{x_i} + \frac{z}{2}\sum_i^N q_ix_i \ln\frac{\theta_i}{\Phi_i}$ $- \sum_i^N q_ix_i \ln\left(\sum_j^N \theta_j\tau_{ji}\right)$	$\ln\gamma_i = \ln\frac{\Phi_i}{x_i} + \frac{z}{2}q_i \ln\frac{\theta_i}{\Phi_i} + l_i$ $- \frac{\Phi_i}{x_i}\sum_j^N x_jl_j - q_i \ln\left(\sum_j^N \theta_j\tau_{ji}\right) + q_i$ $- q_i\sum_j^N \frac{\theta_j\tau_{ij}}{\sum_k^N \theta_k\tau_{kj}}$ where $\Phi_i = \frac{r_ix_i}{\sum_k^N r_kx_k}$ and $\theta_i = \frac{q_ix_i}{\sum_k^N q_kx_k}$

† Parameters q and r can be calculated from Eq. (8-10-62).

tions if the final expression for g^E is to contain only constants derived from binary data.

To illustrate with the simplest case, consider the two-suffix Margules relation for g^E (Table 8-3). For a binary mixture, this relation is given by Eq. (8-5.5), leading to activity coefficients given by Eqs. (8-5.9) and (8-5.10). The generalization to a system containing N components is

$$g^E = \frac{1}{2} \sum_{i=1}^{N} \sum_{j=1}^{N} A_{ij} x_i x_j \tag{8-9.5}$$

where the factor ½ is needed to avoid counting molecular pairs twice. The coefficient A_{ij} is obtained from data for the ij binary. [In the summation indicated in Eq. (8-9.5), $A_{ii} = A_{jj} = 0$ and $A_{ij} = A_{ji}$.] For a ternary system Eq. (8-9.5) becomes

$$g^E = A_{12} x_1 x_2 + A_{13} x_1 x_3 + A_{23} x_2 x_3 \tag{8-9.6}$$

Activity coefficients are obtained by differentiating Eq. (8-9.6) according to Eq. (8-9.4), remembering that $x_i = n_i / n_T$, where n_T is the total number of moles. Upon performing this differentiation, we obtain for component k

$$RT \ln \gamma_k = \sum_{i=1}^{N} \sum_{j=1}^{N} (A_{ik} - \tfrac{1}{2} A_{ij}) x_i x_j \tag{8-9.7}$$

For a ternary system, Eq. (8-9.7) becomes

$$RT \ln \gamma_1 = A_{12} x_2^2 + A_{13} x_3^2 + (A_{12} + A_{13} - A_{23}) x_2 x_3 \tag{8-9.8}$$

$$RT \ln \gamma_2 = A_{12} x_1^2 + A_{23} x_3^2 + (A_{12} + A_{23} - A_{13}) x_1 x_3 \tag{8-9.9}$$

$$RT \ln \gamma_3 = A_{13} x_1^2 + A_{23} x_2^2 + (A_{13} + A_{23} - A_{12}) x_1 x_2 \tag{8-9.10}$$

All constants appearing in these equations can be obtained from binary data; no ternary data are required.

Equations (8-9.8) to (8-9.10) follow from the simplest model for g^E. This model is adequate only for nearly ideal mixtures, where the molecules of the constituent components are similar in size and chemical nature, e.g., benzene-cyclohexane-toluene. For most mixtures encountered in the chemical process industries, more elaborate models for g^E are required.

First, it is necessary to choose a model for g^E. Depending on the model chosen, some (or possibly all) of the constants in the model may be obtained from binary data. Second, individual activity coefficients are found by differentiation, as indicated in Eq. (8-9.4).

Once we have an expression for the activity coefficients as functions of liquid phase composition and temperature, we can then obtain vapor-liquid equilibria by solving simultaneously *all* the equations of equilibrium. For every component i in the mixture,

$$y_i P = \gamma_i x_i P_{vpi} \mathcal{F}_i \tag{8-9.11}$$

where $\mathcal{F}_i$ is given by Eq. (8-4.2).

Since the equations of equilibrium are highly nonlinear, simultaneous solution is almost always achieved only by iteration. Such iterations can be efficiently performed with a computer.

Example 8-3 A simple example illustrating how binary data can be used to predict ternary equilibria is provided by Steele, Poling, and Manley (1974) who studied the system 1-butene (1)-isobutane (2)–1,3-butadiene (3) in the range 4.4 to 71°C.

solution Steele et al. measured isothermal total pressures of the three binary systems as functions of liquid composition. For the three pure components, the pressures are given as functions of temperature by the Antoine equation

$$\ln P_{vp} = a + b(c + t)^{-1} \tag{8-9.12}$$

where P_{vp} is in bars and t is in degrees Celsius. Pure component constants a, b, and c are shown in Table 8-9.

For each binary system the total pressure P is given by

$$P = \sum_{i=1}^{2} y_i P = \sum_{i=1}^{2} x_i \gamma_i P_{vpi} \exp \frac{(V_i^L - B_{ii})(P - P_{vpi})}{RT} \tag{8-9.13}$$

where γ_i is the activity coefficient of component i in the liquid mixture, V_i^L is the molar volume of pure liquid i, and B_{ii} is the second virial coefficient of pure vapor i, all at system temperature T. Equation (8-9.13) assumes that vapor phase imperfections are described by the (volume explicit) virial equation truncated after the second term (See Sec. 3-5). Also, since the components are chemically similar, and since there is little difference in the molecular size, Steele, et al. used the Lewis fugacity rule $B_{ij} = (\frac{1}{2})(B_{ii} + B_{jj})$. For each pure component, the quantity $(V_i^L - B_{ii})/RT$ is shown in Table 8-10.

For the molar excess Gibbs energy of the binary liquid phase, a one-parameter (two-suffix) Margules equation was assumed:

TABLE 8-9 Antoine Constants for 1-Butene (1)–Isobutene (2)–1,3-Butadiene (3) at 4.4 to 71°C [Eq. (8-9.12)] (Steele et al., 1974)

Component	a	$-b$	c
(1)	9.37579	2259.58	247.658
(2)	9.47209	2316.92	256.961
(3)	9.43739	2292.47	247.799

TABLE 8-10 Pure Component Parameters for 1–Butene (1)–Isobutane (2)–1,3-Butadiene (3) (Steele, et al., 1974)

Temperature, °C	$10^3(V_i^L - B_{ii})/RT$, bar^{-1}		
	(1)	(2)	(3)
4.4	35.13	38.62	33.92
21	33.04	33.04	31.85
38	28.30	28.82	27.60
54	24.35	25.02	23.32
71	21.25	22.12	20.33

$$\frac{g_{ij}^E}{RT} = A'_{ij}x_ix_j \tag{8-9.14}$$

From Eq. (8-9.14) we have

$$\ln \gamma_i = A'_{ij}x_j^2 \qquad \text{and} \qquad \ln \gamma_j = A'_{ij}x_i^2 \tag{8-9.15}$$

Equation (8-9.15) is used at each temperature to reduce the binary, total-pressure data yielding Margules constant A'_{ij}. For the three binaries studied, Margules constants are shown in Table 8-11.

To predict ternary phase equilibria, Steele, et al. assume that the molar excess Gibbs energy is given by

$$\frac{g^E}{RT} = A'_{12}x_1x_2 + A'_{13}x_1x_3 + A'_{23}x_2x_3 \tag{8-9.16}$$

Activity coefficients γ_1, γ_2, and γ_3 are then found by differentiation. [See Eqs. (8-9.8) to (8-9.10), noting that $A'_{ij} = A_{ij}/RT$.]

Vapor-liquid equilibria are found by writing for each component

$$y_iP = x_i\gamma_iP_{vpi}\mathfrak{F}_i \tag{8-9.17}$$

where, consistent with earlier assumptions,

$$\mathfrak{F}_i = \exp\frac{(V_i^L - B_{ii})(P - P_{vpi})}{RT} \tag{8-9.18}$$

Steele and coworkers find that predicted ternary vapor-liquid equilibria are in excellent agreement with their ternary data.

Example 8-4 A simple procedure for calculating multicomponent vapor-liquid equilibria from binary data is to assume that for the multicomponent mixture, regardless of the model for g^E,

$$g^E = \sum_{\text{all binary pairs}} g_{ij}^E \tag{8-9.19}$$

solution To illustrate Eq. (8-9.19), we consider the ternary mixture acetonitrile-benzene–carbon tetrachloride studied by Clarke and Missen (1974) at 45°C.

The three sets of binary data were correlated by the Redlich-Kister expansion, which is equivalent to the Margules equation

TABLE 8-11 Margules Constants A'_{ij} for Three Binary Mixtures formed by 1–Butene (1), Isobutane (2), and 1,3-Butadiene (3) (Steele, et al., 1974)

Temp. °C	$10^3\ A'_{12}$	$10^3\ A'_{13}$	$10^3\ A'_{23}$
4.4	73.6	77.2	281
21	60.6	64.4	237
38	52.1	54.8	201
54	45.5	47.6	172
71	40.7	42.4	147

$$g_{ij}^E = x_i x_j [A + B(x_i - x_j) + C(x_i - x_j)^2 + D(x_i - x_j)^3] \tag{8-9.20}$$

The constants are given in Table 8-12.

When Eq. (8-9.20) for each binary is substituted into Eq. (8-9.19), the excess Gibbs energy of the ternary is obtained. Clarke and Missen compared excess Gibbs energies calculated in this way with those obtained from experimental data for the ternary system according to the definition

$$g^E = RT(x_1 \ln \gamma_1 + x_2 \ln \gamma_2 + x_3 \ln \gamma_3) \tag{8-9.21}$$

Calculated and experimental excess Gibbs energies were in good agreement, as illustrated by a few results sown in Table 8-13. Comparison between calculated and experimental results for more than 60 compositions showed that the average deviation (without regard to sign) was only 16 J/mol. Since the uncertainty due to experimental error is about 13 J/mol, Clarke and Missen concluded that Eq. (8-9.19) provides an excellent approximation for this ternary system.

Since accurate experimental studies on ternary systems are not plentiful, it is difficult to say to what extent the positive conclusion of Clarke and Missen can be applied to other systems. It appears that, for mixtures of typical organic fluids, Eq. (8-9.19) usually gives reliable results, although some deviations have been observed, especially for systems with appreciable hydrogen bonding. In many cases

TABLE 8-12 Redlich-Kister Constants for the Three Binaries Formed by Acetonitrile (1), Benzene (2), and Carbon Tetrachloride (3) at 45°C (see Eq. (8-9.20)) (Clark and Missen, 1974)

Binary system		J/mol			
i	j	A	B	C	D
1	2	2691.6	−33.9	293	0
2	3	317.6	−3.6	0	0
3	1	4745.9	497.5	678.6	416.3

TABLE 8-13 Calculated and Observed Molar Excess Gibbs Energies for Acetonitrile (1)–Benzene (2)–Carbon Tetrachloride (3) at 45°C (Clarke and Missen, 1974)

Calculations from Eq. (8-9.19)

Composition		g^E, J/mol	
x_1	x_2	Calc.	Obs.
0.156	0.767	414	431
0.422	0.128	1067	1063
0.553	0.328	808	774
0.673	0.244	711	686
0.169	0.179	690	724
0.289	0.506	711	707

the uncertainties introduced by assuming Eq. (8-9.19) are of the same magnitude as the uncertainties due to experimental error in the binary data.

Example 8-5 Although the additivity assumption [Eq. (8-9.19)] often provides a good approximation for strongly nonideal mixtures, there may be noticeable deviations between experimental and calculated multicomponent equilibria. Such deviations, however, are significant only if they exceed experimental uncertainty. To detect significant deviations, data of high accuracy are required, and such data are rare for ternary systems; they are extremely rare for quaternary (and higher) systems. To illustrate, we consider the ternary system chloroform-ethanol-heptane at 50°C studied by Abbott, et al., 1975. Highly accurate data were first obtained for the three binary systems. The data were reduced using Barker's method, as explained by Abbott and Van Ness (1975) and by Prausnitz, et al. (1999). The essential feature of this method is that it uses only *P-x* data (at constant temperature); it does not use data for vapor composition *y*.

solution To represent the binary data, Abbott et al. considered a five-suffix Margules equation and a modified Margules equation

$$\frac{g^E}{RT} = x_1x_2[A'_{21}x_1 + A'_{12}x_2 - (\lambda_{21}x_1 + \lambda_{12}x_2)x_1x_2] \tag{8-9.22}†$$

$$\frac{g^E}{RT} = x_1x_2\left(A'_{21}x_1 + A'_{12}x_2 - \frac{\alpha_{12}\alpha_{21}x_1x_2}{\alpha_{12}x_1 + \alpha_{21}x_2 + \eta x_1x_2}\right) \tag{8-9.23}†$$

If in Eq. (8-9.22), $\lambda_{21} = \lambda_{12} = D$, and if in Eq. (8-9.23) $\alpha_{12} = \alpha_{21} = D$ and $\eta = 0$, both equations reduce to

$$\frac{g^E}{RT} = x_1x_2(A'_{21}x_1 + A'_{12}x_2 - Dx_1x_2) \tag{8-9.24}$$

which is equivalent to the four-suffix Margules equation shown in Table 8-3. If, in addition, $D = 0$, Eqs. (8-9.22) and (8-9.23) reduce to the three-suffix Margules equations.

For the two binaries chloroform-heptane and chloroform-ethanol, experimental data were reduced using Eq. (8-9.22); however, for the binary ethanol-heptane, Eq. (8-9.23) was used. Parameters reported by Abbott et al. are shown in Table 8-14. With these

TABLE 8-14 Binary Parameters in Eq. (8-9.22) or (8-9.23) and rms Deviation in Total Pressures for the Systems Chloroform–Ethanol–*n*-Heptane at 50°C (Abbott, et al., 1975)

	Chloroform (1), ethanol (2)	Chloroform (1), heptane (2)	Ethanol (1) heptane (2)
A'_{12}	0.4713	0.3507	3.4301
A'_{21}	1.6043	0.5262	2.4440
α_{12}		0.1505	11.1950
α_{21}		0.1505	2.3806
η		0	9.1369
λ_{12}	−0.3651		
λ_{21}	0.5855		
rms ΔP, bar	0.00075	0.00072	0.00045

† The α's and λ's are not to be confused with those used in the NRTL and Wilson equations.

parameters, calculated total pressures for each binary are in excellent agreement with those measured.

For the ternary, Abbott and coworkers expressed the excess Gibbs energy by

$$\frac{g^E_{123}}{RT} = \frac{g^E_{12}}{RT} + \frac{g^E_{13}}{RT} + \frac{g^E_{23}}{RT} + (C_0 - C_1x_1 - C_2x_2 - C_3x_3)x_1x_2x_3 \qquad (8\text{-}9.25)$$

where C_0, C_1, C_2, and C_3 are ternary constants and g^E_{ij} is given by Eq. (8-9.22) or (8-9.23) for the ij binary. Equation (8-9.25) successfully reproduced the ternary data within experimental error (rms $\Delta P = 0.0012$ bar).

Abbott et al. considered two simplifications:

Simplification a: $\quad C_0 = C_1 = C_2 = C_3 = 0$

Simplification b: $\quad C_1 = C_2 = C_3 = 0 \qquad C_0 = \frac{1}{2}\sum\sum_{i \neq j} A'_{ij}$

where the A'_{ij}'s are the binary parameters shown in Table 8-14.

Simplification b was first proposed by Wohl (1953) on semitheoretical grounds. When calculated total pressures for the ternary system were compared with experimental results, the deviations exceeded the experimental uncertainty.

Simplification	rms ΔP, bar
a	0.0517
b	0.0044

These results suggest that Wohl's approximation (simplification b) provides significant improvement over the additivity assumption for g^E (simplification a). However, one cannot generalize from results for one system. Abbott et al. made similar studies for another ternary (acetone-chloroform-methanol) and found that for this system simplification a gave significantly better results than simplification b, although both simplifications produced errors in total pressure beyond the experimental uncertainty.

Although the results of Abbott and coworkers illustrate the limits of predicting ternary (or higher) vapor-liquid equilibria for nonelectrolyte mixtures from binary data only, these limitations are rarely serious for engineering work unless the system contains an azeotrope. As a practical matter, it is common that experimental uncertainties in binary data are as large as the errors that result when multicomponent equilibria are calculated with some model for g^E using only parameters obtained from binary data.

Although Eq. (8-9.19) provides a particularly simple approximation, the UNIQUAC, NRTL, and Wilson equations can be generalized to multicomponent mixtures without using that approximation but also without requiring ternary (or higher) parameters. Experience has shown that multicomponent vapor-liquid equilibria can usually be calculated with satisfactory engineering accuracy using the Wilson equation, the NRTL equation, or the UNIQUAC equation provided that care is exercised in obtaining binary parameters.

Example 8-6 A liquid mixture at 1.013 bar contains 4.7 mole % ethanol (1), 10.7 mole % benzene (2), and 84.5 mole % methylcyclopentane (3). Find the bubble-point temperature and the composition of the equilibrium vapor.

solution There are three unknowns: the bubble-point temperature and two vapor-phase mole fractions. To find them we use three equations of equilibrium:

$$y_i \phi_i P = x_i \gamma_i f_i^{oL} \qquad i = 1, 2, 3 \tag{8-9.26}$$

where y is the vapor-phase mole fraction and x is the liquid-phase mole fraction. Fugacity coefficient ϕ_i is given by the truncated virial equation of state

$$\ln \phi_i = \left(2 \sum_{j=1}^{3} y_j B_{ij} - B_M\right) \frac{P}{RT} \tag{8-9.27}$$

where subscript M stands for mixture:

$$B_M = y_1^2 B_{11} + y_2^2 B_{22} + y_3^2 B_{33} + 2y_1 y_2 B_{12} + 2y_1 y_3 B_{13} + 2y_2 y_3 B_{33} \tag{8-9.28}$$

All second virial coefficients B_{ij} are found from the correlation of Hayden and O'Connell (1975).

The standard-state fugacity f_i^{oL} is the fugacity of pure liquid i at system temperature and system pressure P.

$$f_i^{oL} = P_{vpi} \phi_i^s \exp \frac{V_i^L (P - P_{vpi})}{RT} \tag{8-9.29}$$

where P_{vpi} is the saturation pressure (i.e., the vapor pressure) of pure liquid i, ϕ_i^s is the fugacity coefficient of pure saturated vapor i, and V_i^L is the liquid molar volume of pure i, all at system temperature T.

Activity coefficients are given by the UNIQUAC equation with the following parameters:

Pure-Component Parameters

Component	r	q	q'
1	2.11	1.97	0.92
2	3.19	2.40	2.40
3	3.97	3.01	3.01

Binary Parameters

$$\tau_{ij} = \exp\left(-\frac{a_{ij}}{T}\right) \quad \text{and} \quad \tau_{ji} = \exp\left(-\frac{a_{ji}}{T}\right)$$

i	j	a_{ij}, K	a_{ji}, K
1	2	−128.9	997.4
1	3	−118.3	1384
2	3	−6.47	56.47

For a bubble-point calculation, a useful objective function $F(1/T)$ is

$$F\left(\frac{1}{T}\right) = \ln\left[\sum_{i=1}^{3} K_i x_i\right] \rightarrow \text{zero}$$

where $K_i = y_i/x_i$. In this calculation, the important unknown is T (rather than y) because P_{vpi} is a strong function of temperature, whereas ϕ_i is only a weak function of y.

A suitable program for these iterative calculations uses the Newton-Raphson method, as discussed, for example, by Prausnitz, et al., (1980). This program requires initial estimates of T and y.

The calculated bubble-point temperature is 335.99 K. At this temperature, the second virial coefficients (cm^3/mole) and liquid molar-volumes (cm^3/mole) are:

$$\begin{array}{llll} B_{11} & = -1155 & & \\ B_{12} = B_{21} & = -587 & V_1^L = 61.1 \\ B_{22} & = -1086 & V_2^L = 93.7 \\ B_{23} = B_{32} & = -1134 & V_3^L = 118 \\ B_{33} & = -1186 & \\ B_{31} = B_{13} & = -618 & \\ B_M & = -957.3 & \end{array}$$

The detailed results at 335.99 K are:

				100 y_i	
Component	γ_i	f_i^{OL} (bar)	ϕ_i	Calculated	Observed
1	10.58	0.521	0.980	26.1	25.8
2	1.28	0.564	0.964	7.9	8.4
3	1.03	0.739	0.961	66.0	65.7

The experimental bubble-point temperature is 336.15 K. Experimental results are from Sinor and Weber (1960).

In this particular case, there is very good agreement between calculated and experimental results. Such agreement is not unusual, but it is, unfortunately, not guaranteed. For many mixtures of nonelectrolyte liquids (including water), agreement between calculated and observed VLE is somewhat less satisfactory than that shown in this example. However, if there is serious disagreement between calculated and observed VLE, do not give up. There may be some error in the calculation, or there may be some error in the data, or both.

8-10 DETERMINATION OF ACTIVITY COEFFICIENTS

As discussed in Secs. 8-5 and 8-6, activity coefficients in binary liquid mixtures can often be estimated from a few experimental vapor-liquid equilibrium data for the mixtures by using some empirical (or semiempirical) excess function, as shown in Table 8-3. The excess functions provide a thermodynamically consistent method for interpolating and extrapolating limited binary experimental mixture data and for extending binary data to multicomponent mixtures. Frequently, however, few or no mixture data are at hand, and it is necessary to estimate activity coefficients from some suitable prediction method. Unfortunately, few truly reliable prediction methods have been established. Theoretical understanding of liquid mixtures is limited.

Therefore, the few available prediction methods are essentially empirical. This means that estimates of activity coefficients can be made only for systems similar to those used to establish the empirical prediction method. Even with this restriction, with few exceptions, the accuracy of prediction is not likely to be high whenever predictions for a binary system do not utilize at least some reliable binary data for that system or for another that is closely related. In the following sections we summarize a few of the activity-coefficient prediction methods useful for chemical engineering applications.

Activity Coefficient from Regular Solution Theory

Following ideas first introduced by van der Waals and van Laar, Hildebrand and Scatchard working independently (Hildebrand and Scott, 1962), showed that for binary mixtures of nonpolar molecules, activity coefficients γ_1 and γ_2 can be expressed by

$$RT \ln \gamma_1 = V_1^L \Phi_2^2 (c_{11} + c_{22} - 2c_{12}) \tag{8-10.1}$$

$$RT \ln \gamma_2 = V_2^L \Phi_1^2 (c_{11} + c_{22} - 2c_{12}) \tag{8-10.2}$$

where V_i^L is the liquid molar volume of pure liquid i at temperature T, R is the gas constant, and volume fraction Φ_1 and Φ_2 are defined by

$$\Phi_1 = \frac{x_1 V_1^L}{x_1 V_1^L + x_2 V_2^L} \tag{8-10.3}$$

$$\Phi_2 = \frac{x_2 V_2^L}{x_1 V_1^L + x_2 V_2^L} \tag{8-10.4}$$

with x denoting mole fraction. Note that the above equations are obtained from the van Laar equations if the van Laar constants A and B are set equal to $V_1^L(c_{11} + c_{22} - 2c_{12})$ and $V_2^L(c_{11} + c_{22} - 2c_{12})$, respectively.

For pure liquid i, the cohesive energy density c_{ii} is defined by

$$c_{ii} = \frac{\Delta U_i}{V_i^L} \tag{8-10.5}$$

where ΔU_i is the energy required to isothermally evaporate liquid i from the saturated liquid to the ideal gas. At temperatures well below the critical,

$$\Delta U_i \approx \Delta H_{vi} - RT \tag{8-10.6}$$

where ΔH_{vi} is the molar enthalpy of vaporization of pure liquid i at temperature T.

Cohesive energy density c_{12} reflects intermolecular forces between molecules of component 1 and 2; this is the key quantity in Eqs. (8-10.1) and (8-10.2). Formally, c_{12} can be related to c_{11} and c_{22} by

$$c_{12} = (c_{11}c_{22})^{1/2}(1 - l_{12}) \tag{8-10.7}$$

where l_{12} is a binary parameter, positive or negative, but usually small compared with unity. Upon substitution, Eqs. (8-10.1) and (8-10.2) can be rewritten

$$RT \ln \gamma_1 = V_1^L \Phi_2^2[(\delta_1 - \delta_2)^2 + 2\, l_{12}\, \delta_1\, \delta_2] \tag{8-10.8}$$

$$RT \ln \gamma_2 = V_2^L \Phi_1^2[(\delta_1 - \delta_2)^2 + 2\, l_{12}\, \delta_1\, \delta_2] \tag{8-10.9}$$

where solubility parameter δ_i is defined by

$$\delta_i = (c_{ii})^{1/2} = \left(\frac{\Delta U_i}{V_i^L}\right)^{1/2} \tag{8-10.10}$$

For a first approximation, Hildebrand and Scatchard assume that $l_{12} = 0$. In that event, Eqs. (8-10.8) and (8-10.9) contain no binary parameters, and activity coefficients γ_1 and γ_2 can be predicted using only pure-component data.

Although δ_1 and δ_2 depend on temperature, the theory of regular solutions assumes that the excess entropy is zero. It then follows that, at constant composition,

$$RT \ln \gamma_i = \text{const} \tag{8-10.11}$$

Therefore, the right-hand sides of Eqs. (8-10.8) and (8-10.9) may be evaluated at any convenient temperature provided that all quantities are calculated at the same temperature. For many applications the customary convenient temperature is 25°C. A few typical solubility parameters and molar liquid volumes are shown in Table 8-15, and some calculated vapor-liquid equilibria (assuming $l_{12} = 0$) are shown in Fig. 8-5 to 8-7 and are compared to Raoult's Law. For typical nonpolar mixtures, calculated results are often in reasonable agreement with experiment.

The regular solution equations are readily generalized to multicomponent mixtures similar to Eq. (8-9.5). For component k

$$RT \ln \gamma_k = V_k^L \sum_i \sum_j (A_{ik} - \tfrac{1}{2}A_{ij})\Phi_i\Phi_j \tag{8-10.12}$$

where

$$A_{ij} = (\delta_i - \delta_j)^2 + 2l_{ij}\delta_i\delta_j \tag{8-10.13}$$

If all binary parameters l_{ij} are assumed equal to zero, Eq. (8-10.12) simplifies to

$$RT \ln \gamma_k = V_k^L(\delta_k - \bar{\delta})^2 \tag{8-10.14}$$

where

$$\bar{\delta} = \sum_i \phi_i\delta_i \tag{8-10.15}$$

where the summation refers to all components, including component k.

The simplicity of Eq. (8-10.14) is striking. It says that, in a multicomponent mixture, activity coefficients for all components can be calculated at any composition and temperature by using only solubility parameters and molar liquid volumes for the pure components. For mixtures of hydrocarbons, Eq. (8-10.14) often provides a good approximation.

Although binary parameter l_{12} is generally small compared with unity in nonpolar mixtures, its importance may be significant, especially if the difference between δ_1 and δ_2 is small. To illustrate, suppose $T = 300$ K, $V_1^L = 100$ cm^3/mol, $\delta_1 = 14.3$, and $\delta_2 = 15.3$ (J/cm^3)$^{1/2}$. At infinite dilution ($\Phi_2 = 1$) we find from Eq. (8-10.8) that $\gamma_1^\infty = 1.04$ when $l_{12} = 0$. However, if $l_{12} = 0.01$, we obtain $\gamma_1^\infty = 1.24$, and if $l_{12} = 0.03$, $\gamma_1^\infty = 1.77$. These illustrative results indicate that calculated activity coefficients are often sensitive to small values of l_{12} and that much im-

TABLE 8-15 Molar Liquid Volumes and Solubility Parameters of Some Nonpolar Liquids

	V^L, cm^3 mol^{-1}	δ, (J cm^{-3})$^{1/2}$
Liquified gases at 90 K:		
Nitrogen	38.1	10.8
Carbon monoxide	37.1	11.7
Argon	29.0	13.9
Oxygen	28.0	14.7
Methane	35.3	15.1
Carbon tetrafluoride	46.0	17.0
Ethane	45.7	19.4
Liquid solvents at 25°C:		
Perfluoro-*n*-heptane	226	12.3
Neopentane	122	12.7
Isopentane	117	13.9
n-Pentane	116	14.5
n-Hexane	132	14.9
1-Hexene	126	14.9
n-Octane	164	15.3
n-Hexadecane	294	16.3
Cyclohexane	109	16.8
Carbon tetrachloride	97	17.6
Ethyl benzene	123	18.0
Toluene	107	18.2
Benzene	89	18.8
Styrene	116	19.0
Tetrachloroethylene	103	19.0
Carbon disulfide	61	20.5
Bromine	51	23.5

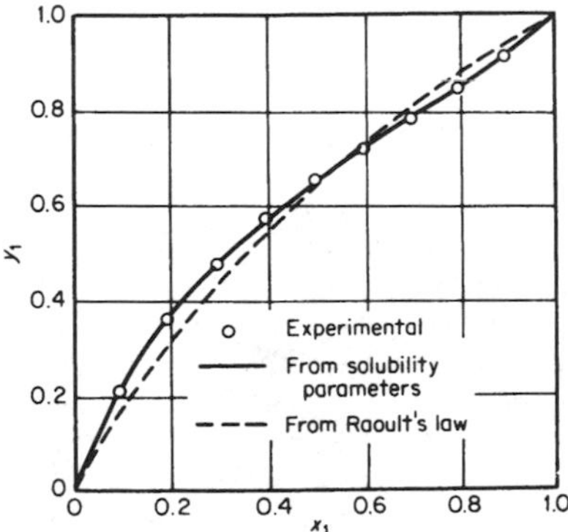

FIGURE 8-5 Vapor-liquid equilibria for benzene (1)–normal heptane (2) at 70°C. (*Prausnitz, et al., 1999*)

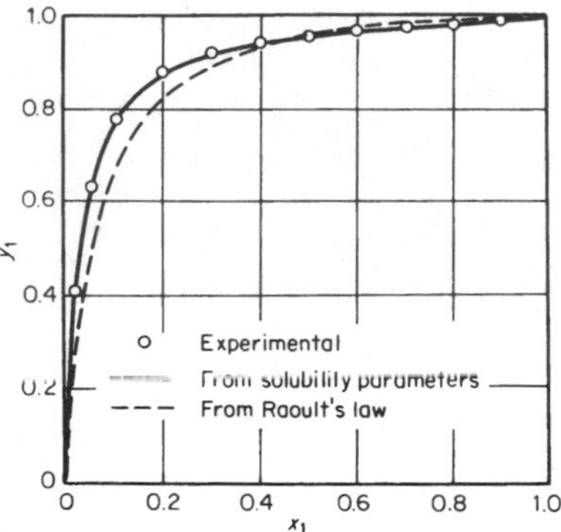

FIGURE 8-6 Vapor-liquid equilibria for carbon monoxide (1)–methane (2) at 90.7 K. (*Prausnitz, et al., 1999*)

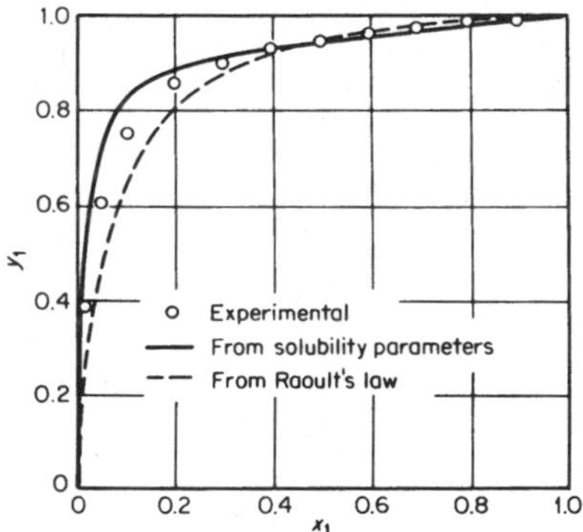

FIGURE 8-7 Vapor-liquid equilibria for neopentane (1)–carbon tetrachloride (2) at 0°C. (*Prausnitz, et al., 1999*)

provement in predicted results can often be achieved when just one binary datum is available for evaluating l_{12}.

Efforts to correlate l_{12} have met with little success. In their study of binary cryogenic mixtures, Bazúa and Prausnitz (1971) found no satisfactory variation of l_{12} with pure component properties, although some rough trends were found by Cheung and Zander (1968) and by Preston and Prausnitz (1970). In many typical cases l_{12} is positive and becomes larger as the differences in molecular size and chemical nature of the components increase. For example, for carbon dioxide–paraffin mixtures at low temperatures, Preston and Prausnitz (1970) found that $l_{12} = -0.02$ (methane); $+ 0.08$ (ethane); $+ 0.08$ (propane); $+ 0.09$ (butane).

Since l_{12} is an essentially empirical parameter it depends on temperature. However, for typical nonpolar mixtures over a modest range of temperature, that dependence is usually small.

For mixtures of aromatic and saturated hydrocarbons, Funk and Prausnitz (1970) found a systematic variation of l_{12} with the structure of the saturated component, as shown in Fig. 8-8. In that case a good correlation could be established because experimental data are relatively plentiful and because the correlation is severely restricted with respect to the chemical nature of the components. Figure 8-9 shows the effect of l_{12} on calculating relative volatility in a typical binary system considered by Funk and Prausnitz.

Our inability to correlate l_{12} for a wide variety of mixtures follows from our lack of understanding of intermolecular forces, especially between molecules at short separations.

Several authors have tried to extend regular solution theory to mixtures containing polar components; but unless the classes of components considered are restricted, such extension has only semiquantitative significance. In establishing the extensions, the cohesive energy density is divided into separate contributions from nonpolar (dispersion) forces and from polar forces:

$$\left(\frac{\Delta U}{V^L}\right)_{\text{total}} = \left(\frac{\Delta U}{V^L}\right)_{\text{nonpolar}} + \left(\frac{\Delta U}{V^L}\right)_{\text{polar}} \tag{8-10.16}$$

Equations (8-10.1) and (8-10.2) are used with the substitutions

$$c_{11} = \tau_1^2 + \lambda_1^2 \tag{8-10.17}$$

$$c_{22} = \tau_2^2 + \lambda_2^2 \tag{8-10.18}$$

$$c_{12} = \lambda_1\lambda_2 + \tau_1\tau_2 + \psi_{12} \tag{8-10.19}$$

where λ_i is the nonpolar solubility parameter $[\lambda_1^2 = (\Delta U_i/V_i^L)_{\text{nonpolar}}]$ and τ_i is the

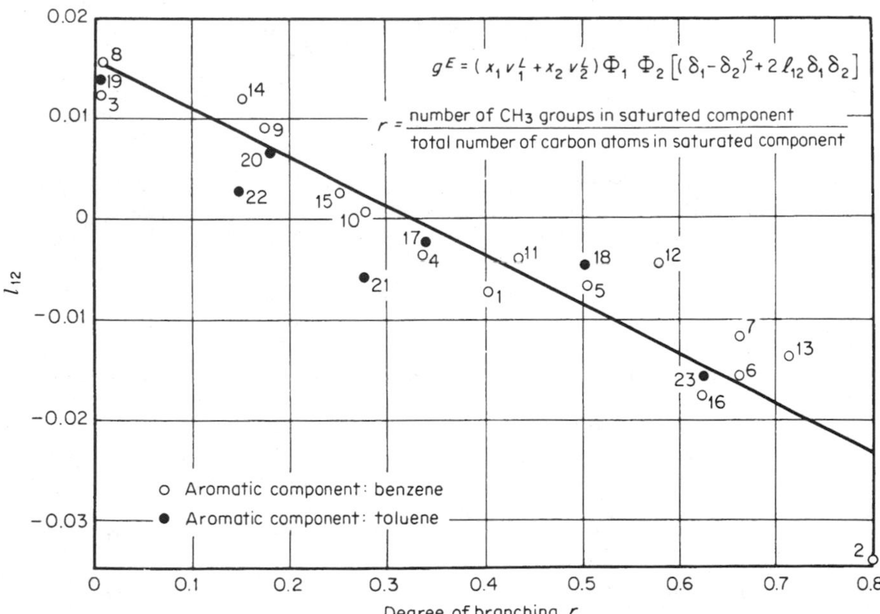

FIGURE 8-8 Correlation of the excess Gibbs energy for aromatic-saturated hydrocarbon mixtures at 50°C. Numbers relate to list of binary systems in Funk and Prausnitz (1970).

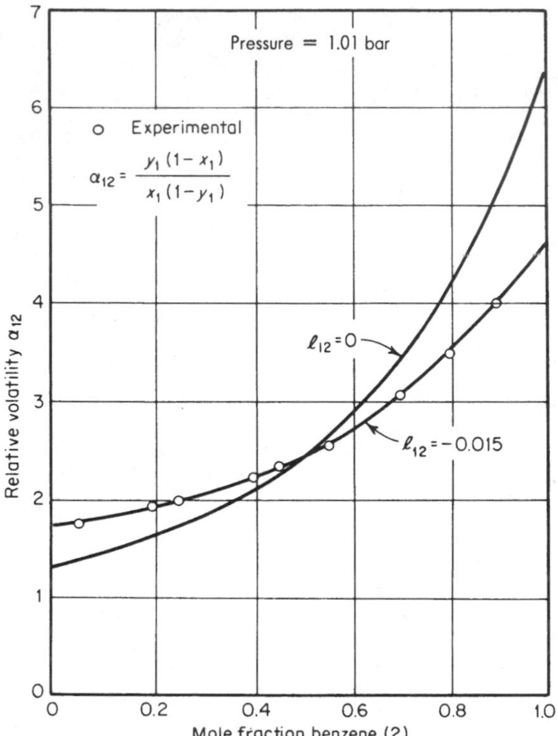

FIGURE 8-9 Comparison of experimental volatilities with volatilities calculated by Scatchard-Hildebrand theory for 2,2-dimethylbutane (1)–benzene (2). (*Funk and Prausnitz, 1970*)

polar solubility parameter $[\tau_i^2 = (\Delta U_i/V_i^L)_{\text{polar}}]$. The binary parameter ψ_{12} is not negligible, as shown by Weimer and Prausnitz (1965) in their correlation of activity coefficients at infinite dilution for hydrocarbons in polar non-hydrogen-bonding solvents.

Further extension of the Scatchard-Hildebrand equation to include hydrogen-bonded components makes little sense theoretically, because the assumptions of regular solution theory are seriously in error for mixtures containing such components. Nevertheless, some semiquantitative success has been achieved by Hansen, et al., (1967, 1971) and others (Burrell, 1968) interested in establishing criteria for formulating solvents for paints and other surface coatings. Also, Null and Palmer (1969) have used extended solubility parameters for establishing an empirical correlation of activity coefficients. A compendium of solubility parameters and pertinent discussion is given in a monograph by Barton (1990).

Activity Coefficients at Infinite Dilution

Experimental activity coefficients at infinite dilution are particularly useful for calculating the parameters needed in an expression for the excess Gibbs energy (Table

8-3). In a binary mixture, suppose experimental data are available for infinite-dilution activity coefficients γ_1^∞ and γ_2^∞. These can be used to evaluate two adjustable constants in any desired expression for g^E. For example, consider the van Laar equation

$$g^E = Ax_1x_2 \left(x_1 \frac{A}{B} + x_2\right)^{-1} \tag{8-10.20}$$

As indicated in Sec. 8-5, this gives

$$RT \ln \gamma_1 = A \left(1 + \frac{A}{B}\frac{x_1}{x_2}\right)^{-2} \tag{8-10.21}$$

and

$$RT \ln \gamma_2 = B \left(1 + \frac{B}{A}\frac{x_2}{x_1}\right)^{-2} \tag{8-10.22}$$

In the limit, as $x_1 \rightarrow 0$ or as $x_2 \rightarrow 0$, Eqs. (8-10.21) and (8-10.22) become

$$RT \ln \gamma_1^\infty = A \tag{8-10.23}$$

and

$$RT \ln \gamma_2^\infty = B \tag{8-10.24}$$

Calculation of parameters from γ^∞ data is particularly simple for the van Laar equation, but in principle, similar calculations can be made by using any two-parameter equation for the excess Gibbs energy. If a three-parameter equation, e.g., NRTL, is used, an independent method must be chosen to determine the third parameter α_{12}.

Relatively simple experimental methods have been developed for rapid determination of activity coefficients at infinite dilution. These are based on gas-liquid chromatography and on ebulliometry.

Schreiber and Eckert (1971) have shown that if reliable values of γ_1^∞ and γ_2^∞ are available, either from direct experiment or from a correlation, it is possible to predict vapor-liquid equilibria over the entire range of composition. For completely miscible mixtures the Wilson equation is particularly useful. Parameters Λ_{12} and Λ_{21} are found from simultaneous solution of the relations

$$\ln \gamma_1^\infty = -\ln \Lambda_{12} - \Lambda_{21} + 1 \tag{8-10.25}$$

$$\ln \gamma_2^\infty = -\ln \Lambda_{21} - \Lambda_{12} + 1 \tag{8-10.26}$$

Table 8-16 shows some typical results obtained by Schreiber and Eckert. The average error in vapor composition using Λ_{ij} from γ^∞ data alone is only slightly larger than that obtained when γ data are used over the entire composition range. Schreiber and Eckert also show that reasonable results are often obtained when γ_1^∞ or γ_2^∞ (but not both) are used. When only one γ^∞ is available, it is necessary to use the one-parameter Wilson equation, as discussed earlier. [See Eq. (8-5.12).]

Activity coefficients at infinite dilution are tabulated by Tiogs, et al., (1986). An extensive correlation for γ^∞ data in binary systems has been presented by Pierotti, et al., (1959). This correlation can be used to predict γ^∞ for water, hydrocarbons, and typical organic components, e.g., esters, aldehydes, alcohols, ketones, nitriles, in the temperature region 25 to 100°C. The pertinent equations and tables are summarized by Treybal (1963) and, with slight changes, are reproduced in Tables 8-17 and 8-18. The accuracy of the correlation varies considerably from one system to

TABLE 8-16 Fit of Binary Data Using Limiting Activity Coefficients in the Wilson Equation (Schreiber and Eckert, 1971)

System and γ^∞	Temp., °C	Average absolute error in calc. $y \times 10^3$	
		All points	γ_1^∞ and γ_2^∞ only
Acetone (1.65)–benzene (1.52)	45	2	4
Carbon tetrachloride (5.66)–acetonitrile (9.30)	45	7	11
Ethanol (18.1)–*n*-hexane (9.05)	69–79	10	12
Chloroform (2.00)–methanol (9.40)	50	10	28
Acetone (8.75)–water (3.60)	100	10	15

another; provided that λ^∞ is not one or more orders of magnitude removed from unity, the average deviation in γ^∞ is about 8 percent.

To illustrate use of Table 8-17, an example, closely resembling one given by Treybal, follows.

Example 8-7 Estimate infinite-dilution activity coefficients for the ethanol-water binary system at 100°C.

solution First we find γ^∞ for ethanol. Subscript 1 stands for ethanol, and subscript 2 stands for water. From Table 8-17, $\alpha = -0.420$, $\epsilon = 0.517$, $\zeta = 0.230$, $\theta = 0$, and $N_1 = 2$. Using Eq. (*a*) at the end of Table 8-17, we have

$$\log \gamma^\infty = -0.420 + (0.517)(2) + \frac{0.230}{2} = 0.729$$

$$\gamma^\infty \text{ (ethanol)} = 5.875$$

Next, for water, we again use Table 8-17. Now subscript 1 stands for water and subscript 2 stands for ethanol.

$$\alpha = 0.617 \quad \epsilon = \zeta = 0 \quad \theta = -0.280 \quad N_2 = 2$$

$$\log \gamma^\infty = 0.617 - \frac{0.280}{2} = 0.477$$

$$\gamma^\infty \text{ (water)} = 3.0$$

These calculated results are in good agreement with experimental data of Jones, et. al., (1943)

A predictive method for estimating γ^∞ is provided by the solvatochromic correlation of Bush and Eckert (2000) through the SPACE equation. SPACE stands for Solvatochromic Parameters for Activity-Coefficient Estimation.[1]

One of the most serious limitations of the majority of the g^E expressions such as Wilson, NRTL, and UNIQUAC, and thus of the various versions of UNIFAC

[1] The authors of this book are grateful to D. M. Bush and C. A. Eckert [Georgia Institute of Technology, Atlanta, Georgia] for providing this discussion prior to publication in the literature.

for γ^∞ estimation is the absence of any accounting for strong interactions, such as hydrogen bonds. Since many separation processes seek specifically to take advantage of these interactions, such as extraction or extractive distillation, it is useful to have available methods for property estimation that account for strong, specific interactions. Perhaps the most useful of these, and most widely used currently, is the method of solvatochromism. (Kamlet, et al., 1983)

The basic Kamlet-Taft multiparameter approach gives any configurational property XYZ in terms of the sum of an intercept XYZ_o, a cavity-formation term related to the energy required to make a cavity in the solvent large enough to accommodate a solute molecule, and a term summing the solvent-solute intermolecular interactions.

$$XYZ = XYZ_o + \text{cavity formation term} + \Sigma(\text{solvent-solute interactions}) \qquad (8\text{-}10.27)$$

Most often this is expressed in terms of parameters π^* (polarity/polarizability), α (hydrogen-bond donor strength), and β (hydrogen-bond acceptor strength). In its simplest form it is,

$$XYZ = (XYZ)_0 + s\pi^* + a\alpha + b\beta \qquad (8\text{-}10.28)$$

XYZ a solvent-dependent physico-chemical property such as $\ln \gamma^\infty$
$(XYZ)_o$ the property in the gas phase or in an inert solvent
s, a, b relative susceptibilities of the property XYZ to the solvent parameters
π^* dipolarity/polarizability scale (dispersive, inductive and electrostatic forces)
α hydrogen-bond donor (HBD)/electron-pair acceptor (EPA) scale
β hydrogen-bond acceptor (HBA)/electron-pair donor (EPD) scale

Equation (8-10.28) is an example of a linear solvation energy relationship (LSER) and is successful for describing a wide variety of medium-related processes, including γ^∞ in ambient water over more than six orders of magnitude variation, (Sherman et al., 1996) and extending to such diverse applications as predictions of dipole moments, fluorescence lifetimes, reaction rates, NMR shifts, solubilities in blood, and biological toxicities (Kamlet, et al., 1988; Carr, 1993; Taft, et al., 1985). Tables are available in the literature for the parameters for solvents. Solute parameters sometimes vary somewhat from those for the substance as a solvent, as the solute parameter represents the forces for a single molecule, and the solvent parameters for the aggregate. For example, the acidity of a hydrogen-bonded alcohol is different from that for an unbonded alcohol. Table 8-19 gives the best current values of these parameters. In Table 8-19, π^{*KT}, α^{KT}, and β^{KT} represent values when a substance is in the solvent state while π^{*H}, α^H, β^H are values in the solute state.

Often additional parameters are used for various applications; a typical example is the prediction of the Henry's law constant $H_{2,1}$ for a solute (2) in ambient water (1) at 25°C (Sherman, et al., 1996):

$$\ln H_{2,1} = -0.536 \log L_2^{16} - 5.508\, \pi_2^{*H} - 8.251\, \alpha_2^H - 10.54\, \beta_2^H - 1.598 \left[\ln \left(\frac{V_2}{V_1}\right)^{0.75} + 1 - \left(\frac{V_2}{V_1}\right)^{0.75} \right] + 16.10 \qquad (8\text{-}10.29)$$

where L^{16} is the hexadecane-air partition coefficient, usually measured by gas chro-

TABLE 8-17 Correlating Constants for Activity Coefficients at Infinite Dilution; Homologous Series of Solutes and Solvents (Pierotti, et al., 1959)

Solute (1)	Solvent (2)	Temp., °C	α	ϵ	ζ	η	θ	Eq.
n-Acids	Water	25	−1.00	0.622	0.490	...	0	(*a*)
		50	−0.80	0.590	0.290	...	0	(*a*)
		100	−0.620	0.517	0.140	...	0	(*a*)
n-Primary alcohols	Water	25	−0.995	0.622	0.558	...	0	(*a*)
		60	−0.755	0.583	0.460	...	0	(*a*)
		100	−0.420	0.517	0.230	...	0	(*a*)
n-Secondary alcohols	Water	25	−1.220	0.622	0.170	0	...	(*b*)
		60	−1.023	0.583	0.252	0	...	(*b*)
		100	−0.870	0.517	0.400	0	...	(*b*)
n-Tertiary alcohols	Water	25	−1.740	0.622	0.170	...	...	(*c*)
		60	−1.477	0.583	0.252	...	...	(*c*)
		100	−1.291	0.517	0.400	...	...	(*c*)
Alcohols, general	Water	25	−0.525	0.622	0.475	0	...	(*d*)
		60	−0.33	0.583	0.39	0	...	(*d*)
		100	−0.15	0.517	0.34	0	...	(*d*)
n-Allyl alcohols	Water	25	−1.180	0.622	0.558	...	0	(*a*)
		60	−0.929	0.583	0.460	...	0	(*a*)
		100	−0.650	0.517	0.230	...	0	(*a*)
n-Aldehydes	Water	25	−0.780	0.622	0.320	...	0	(*a*)
		60	−0.400	0.583	0.210	...	0	(*a*)
		100	−0.03	0.517	0	...	0	(*a*)
n-Alkene aldehydes	Water	25	−0.720	0.622	0.320	...	0	(*a*)
		60	−0.540	0.583	0.210	...	0	(*a*)
		100	−0.298	0.517	0	...	0	(*a*)
n-Ketones	Water	25	−1.475	0.622	0.500	0	...	(*b*)
		60	−1.040	0.583	0.330	0	...	(*b*)
		100	−0.621	0.517	0.200	0	...	(*b*)

n-Acetals	Water	25	−2.556	0.622	0.486	...	...	(*e*)
		60	−2.184	0.583	0.451	...	...	(*e*)
		100	−1.780	0.517	0.426	...	...	(*e*)
n-Ethers	Water	20	−0.770	0.640	0.195	0	...	(*b*)
n-Nitriles	Water	25	−0.587	0.622	0.760	...	0	(*a*)
		60	−0.368	0.583	0.415	...	0	(*a*)
		100	−0.095	0.517	0	...	0	(*a*)
n-Alkene nitriles	Water	25	−0.520	0.622	0.760	...	0	(*a*)
		60	−0.323	0.583	0.413	...	0	(*a*)
		100	−0.074	0.517	0	...	0	(*a*)
n-Esters	Water	20	−0.930	0.640	0.260	0	...	(*b*)
n-Formates	Water	20	−0.585	0.640	0.260	...	0	(*a*)
n-Monoalkyl chlorides	Water	20	1.265	0.640	0.073	...	0	(*a*)
n-Paraffins	Water	16	0.688	0.642	0	...	0	(*a*)
n-Alkyl benzenes	Water	25	3.554	0.622	−0.466	...	...	(*f*)
n-Alcohols	Paraffins	25	1.960	0	0.475	−0.00049	...	(*d*)
		60	1.460	0	0.390	−0.00057	...	(*d*)
		100	1.070	0	0.340	−0.00061	...	(*d*)
n-Ketones	Paraffins	25	0.0877	0	0.757	−0.00049	...	(*b*)
		60	0.016	0	0.680	−0.00057	...	(*b*)
		100	−0.067	0	0.605	−0.00061	...	(*b*)
Water	*n*-Alcohols	25	0.760	0	0	...	−0.630	(*a*)
		60	0.680	0	0	...	−0.440	(*a*)
		100	0.617	0	0	...	−0.280	(*a*)
Water	*sec*-Alcohols	80	1.208	0	0	...	−0.690	(*c*)
Water	*n*-Ketones	25	1.857	0	0	...	−1.019	(*c*)
		60	1.493	0	0	...	−0.73	(*c*)
		100	1.231	0	0	...	−0.557	(*c*)

TABLE 8-17 Correlating Constants for Activity Coefficients at Infinite Dilution; Homologous Series of Solutes and Solvents (Pierotti, et al., 1959) (*Continued*)

Solute (1)	Solvent (2)	Temp., °C	α	ϵ	ζ	η	θ	Eq.
Ketones	*n*-Alcohols	25	−0.088	0.176	0.50	−0.00049	−0.630	(*g*)
		60	−0.035	0.138	0.33	−0.00057	−0.440	(*g*)
		100	−0.035	0.112	0.20	−0.00061	−0.280	(*g*)
Aldehydes	*n*-Alcohols	25	−0.701	0.176	0.320	−0.00049	−0.630	(*h*)
		60	−0.239	0.138	0.210	−0.00057	−0.440	(*h*)
Esters	*n*-Alcohols	25	0.212	0.176	0.260	−0.00049	−0.630	(*g*)
		60	0.055	0.138	0.240	−0.00057	−0.440	(*g*)
		100	0	0.112	0.220	−0.00061	−0.280	(*g*)
Acetals	*n*-Alcohols	60	−1.10	0.138	0.451	−0.00057	−0.440	(*i*)
Paraffins	Ketones	25	...	0.1821	...	−0.00049	0.402	(*j*)
		60	...	0.1145	...	−0.00057	0.402	(*j*)
		90	...	0.0746	...	−0.00061	0.402	(*j*)

Equations

(*a*) $$\log \gamma_1^\infty = \alpha + \epsilon N_1 + \frac{\zeta}{N_1} + \frac{\theta}{N_2}$$

(*b*) $$\log \gamma_1^\infty = \alpha + \epsilon N_1 + \zeta\left(\frac{1}{N_1'} + \frac{1}{N_1''}\right) + \eta(N_1 - N_2)^2$$

(c) $$\log \gamma_1^\infty = \alpha + \epsilon N_1 + \zeta\left(\frac{1}{N_1'} + \frac{1}{N_1''} + \frac{1}{N_1'''}\right) + \theta\left(\frac{1}{N_2'} + \frac{1}{N_2''}\right)$$

(*d*) $$\log \gamma_1^\infty = \alpha + \epsilon N_1 + \zeta\left(\frac{1}{N_1'} + \frac{1}{N_1''} + \frac{1}{N_1'''} - 3\right) + \eta(N_1 - N_2)^2$$

(*e*) $\log \gamma_1^\infty = \alpha + \epsilon N_1 + \zeta \left(\frac{1}{N_1'} + \frac{1}{N_1''} + \frac{2}{N_1'''} \right)$

(*f*) $\log \gamma_1^\infty = \alpha + \epsilon N_1 + \zeta \left(\frac{1}{N_1} - 4 \right)$

(*g*) $\log \gamma_1^\infty = \alpha + \epsilon \frac{n_1}{N_2} + \zeta \left(\frac{1}{N_1'} + \frac{1}{N_1''} \right) + \eta (N_1 - N_2)^2 + \frac{\theta}{N_2}$

(*h*) $\log \gamma_1^\infty = \alpha + \epsilon \frac{N_1}{N_2} + \frac{\zeta}{N_1} + \eta (N_1 - N_2)^2 + \frac{\theta}{N_2}$

(*i*) $\log \gamma_1^\infty = \alpha + \epsilon \frac{N_1}{N_2} + \zeta \left(\frac{1}{N_1'} + \frac{1}{N_1''} + \frac{2}{N_1'''} \right) + \eta (N_1 - N_2)^2 + \frac{\theta}{N_2}$

(*j*) $\log \gamma_1^\infty = \epsilon \frac{N_1}{N_2} + \eta (N_1 - N_2)^2 + \theta \left(\frac{1}{N_2'} + \frac{1}{N_2''} \right)$

N_1, N_2 = total number of carbon atoms in molecules 1 and 2, respectively

N', N'', N''' = number of carbon atoms in respective branches of branched compounds, counted from the polar grouping; thus, for *t*-butanol, $N' = N'' = N''' = 2$, for 2-butanol, $N' = 3$, $N'' = 2$, $N''' = 0$

TABLE 8-18 Correlating Constants for Activity Coefficients at Infinite Dilution; Homologous Series of Hydrocarbons in Specific Solvents (Pierotti, et al. 1959)

Temperature, °C				Solvent: Heptane	Methyl ethyl ketone	Furfural	Phenol	Ethanol	Triethylene glycol	Diethylene glycol	Ethylene glycol
			η	Value of ϵ							
25			−0.00049	0	0.0455	0.0937	0.0625	0.088	...	0.191	0.275
50			−0.00055	0	0.033	0.0878	0.0590	0.073	0.161	0.179	0.249
70			−0.00058	0	0.025	0.0810	0.0586	0.065	...	0.173	0.236
90			−0.00061	0	0.019	0.0686	0.0581	0.059	0.134	0.158	0.226
				Value of θ							
25				0.2105	0.1435	0.1152	0.1421	0.2125	0.181	0.2022	0.275
70				0.1668	0.1142	0.0836	0.1054	0.1575	0.129	0.1472	0.2195
130				0.1212	0.0875	0.0531	0.0734	0.1035	0.0767	0.0996	0.1492
				Value of κ							
25				0.1874	0.2079	0.2178	0.2406	0.2425	0.3124	0.3180	0.4147
70				0.1478	0.1754	0.1675	0.1810	0.1753	0.2406	0.2545	0.3516
130				0.1051	0.1427	0.1185	0.1480	0.1169	0.1569	0.1919	0.2772
	Solute (1)	Eq.	ζ	Value of α							
25	Paraffins	(*a*)	0	0	0.335	0.916	0.870	0.580	...	0.875	...
50			0	0	0.332	0.756	0.755	0.570	0.72	0.815	1.208
70			0	0	0.331	0.737	0.690	0.590	...	0.725	1.154
90			0	0	0.330	0.771	0.620	0.610	0.68	0.72	1.089
25	Alkyl cyclohexanes	(*a*)	−0.260	0.18	0.70	1.26	1.20	1.06	...	1.675	...
50			−0.220	...	0.650	1.120	1.040	1.01	1.46	1.61	2.36
70			−0.195	0.131	0.581	1.020	0.935	0.972	...	1.550	2.22
90			−0.180	0.09	0.480	0.930	0.843	0.925	1.25	1.505	2.08

25	Alkyl benzenes	(*a*)	−0.466	0.328	0.277	0.67	0.694	1.011	...	1.08	
50			−0.390	0.243	...	0.55	0.580	0.938	0.80	1.00	1.595
70			−0.362	0.225	0.240	0.45	0.500	0.900	...	0.96	1.51
90			−0.350	0.202	0.239	0.44	0.420	0.862	0.74	0.935	1.43
25	Alkyl naphthalenes	(*a*)	−0.10	0.53	0.169	0.46	0.595	1.06	...	1.00	
50			−0.14	0.53	0.141	0.40	0.54	1.03	0.75	1.00	1.92
70			−0.173	0.53	0.215	0.39	0.497	1.02	...	0.991	1.82
90			−0.204	0.53	0.232	...	0.445	...	0.83	1.01	1.765
25	Alkyl tetralins	(*a*)	0.28	0.244	0.179	0.652	0.378	...	...	1.43	
50			0.24	...	...	0.528	0.364	...	1.00	1.38	
70			0.21	0.220	0.217	0.447	0.371	...	...	1.33	
90			0.19	...	...	0.373	0.348	...	0.893	1.28	
25	Alkyl decalins	(*a*)	−0.43	...	0.871	1.54	1.411	...	...	2.46	
50			−0.368	...	...	1.367	1.285	...	1.906	2.25	
70			−0.355	0.356	0.80	1.253	1.161	...	...	2.07	
90			−0.320	...	...	1.166	1.078	...	1.68	2.06	
25	Unalkylated aromatics, naphthenes, naphthene aromatics	(*b*)	1.176† 1.845‡	−1.072	−0.7305	−0.230	−0.383	−0.485	−0.406	−0.377	−0.154
70			0.846† 1.362‡	−0.886	−0.625	−0.080	−0.226	−0.212	−0.186	−0.0775	−0.0174
130			0.544† 0.846‡	−0.6305	−0.504	0.020	−0.197	0.47	0.095	0.181	0.229

† Condensed, naphthalene-like.
‡ Tandem, diphenyl-like.

TABLE 8-18 Correlating Constants for Activity Coefficients at Infinite Dilution; Homologous Series of Hydrocarbons in Specific Solvents (Pierotti, et al. 1959) (*Continued*)

Equations

(*a*) $\log \gamma_1^\infty = \alpha + \epsilon N_p + \dfrac{\zeta}{N_p + 2} + \eta(N_1 - N_2)^2$

(*b*) $\log \gamma_1^\infty = \alpha + \theta N_a + \kappa N_n + \zeta\left(\dfrac{1}{r} - 1\right)$

where N_1, N_2 = total number of carbon atoms in molecules 1 and 2, respectively
N_p = number of paraffinic carbon atoms in solute

N_a = number of aromatic carbon atoms, including $=\overset{|}{\mathrm{C}}-$, $=$CH—, ring-juncture naphthenic carbons$-\underset{|}{\overset{|}{\mathrm{C}}}-$H, and naphthenic carbons in the α position to an aromatic nucleus
N_n = number of naphthenic carbon atoms not counted in N_a
r = number of rings

Examples:

Butyl decalin:	$N_p = 4$	$N_a = 2$	$N_n = 8$	$N_1 = 14$	$r = 2$
Butyl tetralin:	$N_p = 4$	$N_a = 8$	$N_n = 2$	$N_1 = 14$	$r = 2$

TABLE 8-19 Solvatochromic Parameters

Name	RI	V	π^{*KT}	π^{*H}	α^{KT}	α^{H}	β^{KT}	β^{H}	$\log L^{16}$
n-butane	1.32594	96.5	−0.11	0	0	0	0	0	1.615
2-methylpropane	1.3175	105.5	−0.11	0	0	0	0	0	1.409
n-pentane	1.35472	115.1	−0.08	0	0	0	0	0	2.162
2-methylbutane	1.35088	117.5	−0.08	0	0	0	0.01	0	2.013
n-hexane	1.37226	131.6	−0.04	0	0	0	0	0	2.668
2,2-dimethylbutane	1.36595	133.7	−0.1	0	0	0	0	0	2.352
2,3-dimethylbutane	1.37231	131.2	−0.08	0	0	0	0	0	2.495
2-methylpentane	1.36873	132.9	−0.02	0	0	0	0	0	2.503
3-methylpentane	1.37386	130.6	−0.04	0	0	0	0	0	2.581
n-heptane	1.38511	147.5	−0.01	0	0	0	0	0	3.173
2-methylhexane	1.38227	148.6	0	0	0	0	0	0	3.001
3-ethylpentane	1.3934	143.5	−0.05	0	0	0	0	0	
3-methylhexane	1.38609	146.7	−0.03	0	0	0	0	0	3.044
2,4-dimethylpentane	1.37882	149.9	−0.07	0	0	0	0	0	2.809
2,2-dimethylpentane	1.3822	148.7	−0.08	0	0	0	0	0	2.796
n-octane	1.39505	163.5	0.01	0	0	0	0	0	3.677
2,3,4-trimethylpentane	1.4042	158.9	−0.06	0	0	0	0	0	3.481
2,2,4-trimethylpentane	1.38898	166.1	−0.04	0	0	0	0	0	3.106
n-nonane	1.40311	179.7	0.02	0	0	0	0	0	4.182
2,5-dimethylheptane	1.4033	178.2	−0.05	0	0	0	0	0	
1-pentene	1.36835	112.0	0.08	0.08	0	0	0.1	0.07	2.047
2-methyl-2-butene	1.3874	105.9	0.08	0.08	0	0	0.07	0	2.226
2-methyl-1,3-butadiene	1.3869	107.3	0.12	0.23	0	0	0.1	0.1	2.101
1,3-cyclopentadiene	1.444	82.4	0.13	0.1	0	0	0.14	0.07	
cyclohexene	1.44377	101.9	0.1	0.2	0	0	0.07	0.1	3.021
1-hexene	1.38502	125.9	0.1	0.08	0	0	0.07	0.07	2.572
3-methyl-1-pentene	1.3841	126.1	0.1	0.08	0	0	0.07	0.07	
1-heptene	1.39713	141.7	0.12	0.08	0	0	0.07	0.07	
1-octene	1.4062	157.9	0.13	0.08	0	0	0.07	0.07	3.568
1-nonene	1.41333	174.1	0.13	0.08	0	0	0.07	0.07	4.073

TABLE 8-19 Solvatochromic Parameters (*Continued*)

Name	RI	V	π^{*KT}	π^{*H}	α^{KT}	α^{H}	β^{KT}	β^{H}	$\log L^{16}$
diethylamine	1.3825	104.3	0.24	0.3	0.03	0.08	0.7	0.68	2.395
dipropylamine	1.4018	138.1	0.25	0.3	0	0.08	0.7	0.68	3.351
triethylamine	1.398	140.0	0.14	0.15	0	0	0.71	0.79	3.04
benzene	1.49792	89.4	0.59	0.52	0	0	0.1	0.14	2.786
toluene	1.49413	106.9	0.54	0.52	0	0	0.11	0.14	3.325
1,2-dimethylbenzene	1.50295	121.2	0.51	0.56	0	0	0.12	0.16	3.939
1,3-dimethylbenzene	1.49464	123.4	0.47	0.52	0	0	0.12	0.16	3.839
1,4-dimethylbenzene	1.49325	123.9	0.43	0.52	0	0	0.12	0.16	3.839
ethylbenzene	1.4932	123.1	0.53	0.51	0	0	0.12	0.15	3.778
propylbenzene	1.492	139.4	0.51	0.5	0	0	0.12	0.15	4.23
isopropylbenzene	1.4889	140.2	0.51	0.49	0	0	0.12	0.16	4.084
1,3,5-trimethylbenzene	1.49684	139.6	0.41	0.52	0	0	0.13	0.19	4.344
phenol	1.5509	87.8	0.72	0.89	1.65	0.6	0.3	0.31	3.766
m-cresol	1.5396	105.0	0.68	0.88	1.13	0.57	0.34	0.34	4.31
anisole	1.5143	109.3	0.73	0.74	0	0	0.32	0.29	3.89
acetophenone	1.53423	117.4	0.9	1.01	0.04	0	0.49	0.49	4.501
carbon disulfide	1.62409	60.6	0.61	0.21	0	0	0.07	0.07	2.353
cyclopentane	1.40363	94.7	−0.02	0.1	0	0	0	0	2.477
cyclohexane	1.42354	108.8	0	0.1	0	0	0	0	2.964
methylcyclopentane	1.407	113.2	0.01	0.1	0	0	0	0	2.816
methylcyclohexane	1.42058	128.3	0.01	0.1	0	0	0	0	3.323
tetrahydrofuran	1.40496	81.9	0.58	0.52	0	0	0.55	0.48	2.636
diethyl ether	1.34954	104.7	0.27	0.25	0	0	0.47	0.45	2.015
methyl acetate	1.3589	79.8	0.6	0.64	0	0	0.42	0.45	1.911
ethyl acetate	1.36978	98.5	0.55	0.62	0	0	0.45	0.45	2.314
propyl acetate	1.3828	115.7	0.53	0.6	0	0	0.4	0.45	2.819
butyl acetate	1.3918	132.6	0.46	0.6	0	0	0.45	0.45	3.353
dichloromethane	1.42115	64.5	0.82	0.57	0.13	0.1	0.1	0.05	2.019
chloroform	1.44293	80.7	0.58	0.49	0.2	0.15	0.1	0.02	2.48
1-chloropropane	1.3851	89.0	0.39	0.4	0	0	0.1	0.1	2.202
1-chlorobutane	1.39996	105.1	0.39	0.4	0	0	0	0.1	2.722
bromoethane	1.4212	75.1	0.47	0.4	0	0	0.05	0.12	2.12

iodoethane	1.5101	81.1	0.5	0.4	0	0	0.03	0.15	2.573
chlorobenzene	1.52185	102.3	0.71	0.65	0	0	0.07	0.07	3.657
bromobenzene	1.55709	105.5	0.79	0.73	0	0	0.06	0.09	4.041
acetone	1.35596	74.1	0.71	0.7	0.08	0.04	0.43	0.51	1.696
2-butanone	1.37685	90.2	0.67	0.7	0.06	0	0.48	0.51	2.287
2-pentanone	1.38849	107.5	0.65	0.68	0.05	0	0.5	0.51	2.755
cyclohexanone	1.4505	104.4	0.76	0.86	0	0	0.53	0.56	3.792
propionaldehyde	1.3593	73.4	0.65	0.65	0	0	0.41	0.45	1.815
butyraldehyde	1.3766	90.5	0.63	0.65	0	0	0.41	0.45	2.27
acetonitrile	1.34163	52.9	0.75	0.9	0.19	0.04	0.4	0.33	1.739
propionitrile	1.3636	70.9	0.71	0.9	0	0.02	0.39	0.36	2.082
butyronitrile	1.382	87.9	0.71	0.9	0	0	0.4	0.36	2.548
nitromethane	1.37964	54.0	0.85	0.95	0.22	0.06	0.06	0.32	1.892
nitroethane	1.38973	71.9	0.8	0.95	0.22	0.02	0.25	0.33	2.414
2-nitropropane	1.39235	90.6	0.75	0.92	0.22	0	0.27	0.32	2.55
ethanol	1.35941	58.7	0.54	0.42	0.86	0.37	0.75	0.48	1.485
1-propanol	1.3837	75.2	0.52	0.42	0.84	0.37	0.9	0.48	2.031
1-butanol	1.39741	92.0	0.47	0.42	0.84	0.37	0.84	0.48	2.601
2-methyl-1-propanol	1.39389	92.9	0.4	0.39	0.79	0.37	0.84	0.48	2.413
1-pentanol	1.408	108.7	0.4	0.42	0.84	0.37	0.86	0.48	3.106
3-methyl-1-butanol	1.4052	109.2	0.4	0.39	0.84	0.37	0.86	0.48	3.011
1-hexanol	1.4157	125.3	0.4	0.42	0.8	0.37	0.84	0.48	3.61
1-octanol	1.4276	158.5	0.4	0.42	0.77	0.37	0.81	0.48	4.619
2-propanol	1.3752	76.9	0.48	0.36	0.76	0.33	0.84	0.56	1.764
2-methyl-2-propanol	1.3852	94.9	0.41	0.3	0.42	0.31	0.93	0.6	1.963
methanol	1.32652	40.8	0.6	0.44	0.98	0.43	0.66	0.47	0.97
carbon tetrachloride	1.45739	97.1	0.28	0.38	0	0	0.1	0	2.823
ethylcyclohexane	1.43073	143.2	0.01	0.1	0	0	0	0	3.877
1,4-dioxane	1.42025	85.7	0.55	0.75	0	0	0.37	0.64	2.892

matography retention on a hexadecane column, (Table 8-19) and V is the molal volume (Table 8-19). Hexadecane is a convenient solvent for characterizing solutes in that it is easy to run as a stationary phase in a gas chromatography, and it has no polar or hydrogen-bonding interactions. L^{16} gives a good measure of the cavity term plus the dispersive interactions. Then values for γ^∞ may be found from (See Sec. 8-11)

$$\ln \gamma_2^\infty = \ln H_{2,1} - \ln f_2^o \tag{8-10.30}$$

where f_2^o is the reference-state fugacity as in Eq. (8-9.29).

Example 8-8 To illustrate application of this technique, we calculate γ_2^∞ for benzene (2) in water (1) at 25°C.

solution For water, $V_1 = 18$ and from Table 8-19, for benzene

$$L^{16} = 2.786;\ \pi_2^{*H} = 0.52;\ \alpha_2^H = 0;\ \beta_2^H = 0.14;\ V_2 = 89.4$$

Then from Eq. (8-10.29), $H_{2,1} = 174 \times 10^3$ torr. Using the vapor pressure at 25°C, 95.14 torr, for f_2^o in Eq. (8-10.30), we get $\gamma_2^\infty = 1830$. The experimental value is 2495 (Li, et al., 1993).

The solvatochromic technique has been coupled with modifications of the Hildebrand solubility parameter to give estimation techniques for nonionic liquids (other than water) at 25°C. These methods differ substantially from the UNIFAC methods as they are not made up of group contributions but rather reflect measurements or estimates of molecular properties. In other words, these methods sum contributions to the cohesive energy density by different types of contributions, and the most recent and most successful of these, SPACE, uses the methods described above to include specific chemical interactions into the estimation technique (Hait, et al., 1993).

The SPACE formulation for γ_2^∞ in solvent 1 is

$$\ln \gamma_2^\infty = \frac{V_2}{RT}[(\lambda_1 - \lambda_2)^2 + (\tau_1 - \tau_{2\text{eff}})^2 + (\alpha_1 - \alpha_{2\text{eff}})(\beta_1 - \beta_{2\text{eff}})] + \ln\left(\frac{V_2}{V_1}\right)^{0.936} + 1 - \left(\frac{V_2}{V_1}\right)^{0.936} \tag{8-10.31}$$

The dispersion terms are calculated as functions of the molar refractivity n_D

$$\lambda = k\left(\frac{n_D^2 - 1}{n_D^2 + 2}\right) \tag{8-10.32}$$

where constant k is 15.418 for aliphatic compounds, 15.314 for aromatics, and 17.478 for halogen compounds. In Eq. (8-10.31), R is 1.987, T is in kelvins, and V is in cm^3/mol.

The polarity and hydrogen-bond parameters for the solvent are

$$\tau_1 = \left|\frac{A_1\pi_1^{*KT} + B}{\sqrt{V_1}}\right| \tag{8-10.33}$$

$$\alpha_1 = \left|\frac{C_1\alpha_1^{KT} + D_1}{\sqrt{V_1}}\right| \tag{8-10.34}$$

$$\beta_1 = \left|\frac{E_1\beta_1^{KT} + F_1}{\sqrt{V_1}}\right| \tag{8-10.35}$$

Parameters $\tau_{2\text{eff}}$, $\alpha_{2\text{eff}}$, and $\beta_{2\text{eff}}$ are for the solute. Subscript eff means they are normalized such that limiting activity coefficients for a solute in itself (solvent) will be unity. Calculation of these quantities requires both solvent and solute parameters for the solute.

$$\tau_{2\text{eff}} = \tau_2^o + (\tau_2 - \tau_2^o)\frac{|\pi_1^{*KT} - \pi_2^{*KT}|}{1.33} \tag{8-10.36}$$

$$\alpha_{2\text{eff}} = \alpha_2^o + (\alpha_2 - \alpha_2^o)\frac{|\alpha_1^{KT} - \alpha_2^{KT}|}{1.20} \tag{8-10.37}$$

$$\beta_{2\text{eff}} = \beta_2^o + (\beta_2 - \beta_2^o)\frac{|\beta_1^{KT} - \beta_2^{KT}|}{0.95} \tag{8-10.38}$$

where

$$\tau_2 = \left|\frac{A_2\pi_2^{*H} + B}{\sqrt{V_2}}\right| \tag{8-10.39}$$

$$\alpha_2 = \left|\frac{C_2\alpha_2^H + D_2}{\sqrt{V_2}}\right| \tag{8-10.40}$$

$$\beta_2 = \left|\frac{E_2\beta_2^H + F_2}{\sqrt{V_2}}\right| \tag{8-10.41}$$

and

$$\tau_2^o = \left|\frac{A_1\pi_2^{*KT} + B}{\sqrt{V_2}}\right| \tag{8-10.42}$$

$$\alpha_2^o = \left|\frac{C_1\alpha_2^{KT} + D_1}{\sqrt{V_2}}\right| \tag{8-10.43}$$

$$\beta_2^o = \left|\frac{E_1\beta_2^{KT} + F_1}{\sqrt{V_2}}\right| \tag{8-10.44}$$

Parameters π^{*KT}, α^{KT}, and β^{KT} for the solvent-like state and parameters π^H, α^H, β^H for the solute state are given in Table 8-19. Superscript ^{o} means that we are calculating properties for the solute in its solvent-like state.

There are 19 families or classes of compounds based on functional group as well as methanol, carbon tetrachloride, and THF. Each constitutes an independent family. Parameters A_1, A_2, B, D_1, D_2, F_1, and F_2 are class-dependent. Parameters C_1, C_2, E_1, and E_2 are class-independent. Of these, A_1, D_1, F_1, C_1, and E_1 are for the solvent state while A_2, D_2, F_2, C_2, and E_2 are for the solute state. B is the same in both solvent and solute states. These parameters are given in Table 8-20 and Example 8-9 illustrates the use of SPACE.

SPACE is similar to the earlier MOSCED model, (Thomas, et al., 1984; Howell, et al., 1989) discussed in a previous edition of this book. SPACE is similar to MOSCED, but reduces the three adjustable parameters for each compound to 0 and adds 7 adjustable parameters per functionality of compound. Thus, for a database

TABLE 8-20 SPACE Equation Parameters, Eqs. (8-10.33) to (8-10.44)

	A_1	A_2	B	D_1	D_2	F_1	F_2
Alkanes	18.522	12.802	−4.055	0	0	0	0
Alkenes	65.738	−85.606	−0.245	0	0	−4.095	−0.15
Amides	19.929	—	17.142	0	—	14.944	—
Amines	−66.112	−173.878	19.903	0	0	11.346	−22.023
Aromatics	30.949	31.114	1.081	0	0	−0.273	−0.306
Aromatics with N	−4.569	—	−23.106	−25.226	—	−0.685	—
Aromatics with O	17.295	—	11.889	−10.671	—	0.877	—
Naphthenes	−113.528	56.428	−2.632	0	0	0	0
1,4-Dioxane	−10.866	−22.769	−19.37	0	0	15.902	−1.375
Ethers/Esters	32.795	42.132	8.889	0	0	5.147	−0.941
Halog. Aliphatics	−12.186	−18.479	−12.359	−9.141	−1.737	−2.756	−0.103
Halog. Arom.	12.576	—	11.083	0	—	−9.124	—
Heterocycles	−50.622	—	16.334	0	—	−48.531	—
Ketones	8.067	27.477	22.021	0.336	5.3	10.617	−1.095
Nitriles	11.419	38.912	22.092	−2.269	−50.619	−25.227	−65.782
Nitroalkanes	−14.439	−31.992	−20.876	17.172	−162.981	−10.864	9.39
Methanol	−30.494	59.387	29.223	2.659	−5.007	9.377	40.347
Alcohols	10.132	−83.652	−16.27	−4.851	73.597	−45.278	−31.23
Sec. Alcohols	−10.079	—	−12.643	−2.86	—	0.054	—
Sulfoxides	18.402	—	21.287	0	—	0	—
Carbon Tet.	−12.227	−76.165	13.971	0	0	−1.63	7.374
THF	15.102	−12.603	7.983	0	0	28.447	61.373

$C_1 = 26.92$
$C_2 = -132.494$
$C_2(\text{alcohols}) = 3.702$
$E_1 = 27.561$
$E_2 = 2.147$

containing 100 different solvents, MOSCED will have 300 parameters (3 per solvent) and SPACE about 100 parameters (the values in Table 8-20). The main advantage of SPACE over MOSCED is the prediction of activity coefficients of compounds that were not in the original database provided they have the same functionality as others in the database as well as the required solvent and solute parameters. There are minor differences between the parameters given here and those used in the original 1993 formulation of SPACE (Hait, et al., 1993) reflecting new measurements.

In some cases, it is necessary to estimate γ^{∞} at temperatures other than 25°C. Few good methods exist for the temperature dependence of γ^{∞}, which is a function of the partial molal excess enthalpy of mixing at infinite dilution, $\bar{h}_2^{E\infty}$.

$$\frac{\partial(\ln \gamma_2^{\infty})}{\partial(1/T)} = \frac{\bar{h}_2^{E\infty}}{R} \tag{8-10.45}$$

Often one may simply assume that ln γ^{∞} is linear in $1/T$ with good results. A more accurate method also uses the solvatochromic techniques (Sherman, et al., 1995). In this work, partial molal heats of transfer $\bar{h}_{TR}$ are correlated, and the partial molal excess enthalpy of mixing at infinite dilution is given in an expression that includes the enthalpy of vaporization of the solute, ΔH_v:

$$\bar{h}_2^{E\infty} = \Delta H_v - l \log L_2^{16} - s\, \pi_2^{*H} - d\delta_2 - a\, \alpha_2^H - b\beta_2^H - \text{Intercept} \tag{8-10.46}$$

where parameter δ_2 is a commonly used correction term, equal to zero for aliphatic compounds, 0.5 for polychlorinated aliphatics, and unity for aromatics. Table 8-21

TABLE 8-21 Parameters for Estimation of Partial Molal Excess Enthalpy at Infinite Dilution

Solvent	Intercept	l	s	d	a	b
Cyclohexane	1.71	2.131	−0.75	−0.59		
Heptane	1.22	2.32	−0.94	−0.13		
Dibutyl Ether	1.79	2.219	1.12	−0.50	7.30	
Diethyl Ether	1.44	2.294	3.55	−0.29	9.84	
Ethyl Acetate	2.50	1.707	4.47	−0.52	6.07	
Carbon Tetrachloride	2.35	2.020	2.07	−0.83		
Dichloromethane	1.61	1.852	5.35	−0.34		2.13
Benzene	2.13	1.901	4.28	−1.14	−0.94	
Mesitylene	1.73	2.207	2.23	−0.77	1.04	
Toluene	2.14	2.023	3.01	−0.61	0.51	
Methanol	2.49	1.657	1.41	−0.21	9.49	3.35
1-Butanol	0.39	2.15	−1.78	−0.42	11.2	4.62
1-Octanol	0.48	2.375	−2.70	0.55	11.1	3.83
Acetonitrile	1.51	1.628	6.45	−0.63	7.77	−0.05
Dimethylsulfoxide	2.97	1.05	7.3	−0.84	8.3	
N,N-Dimethylformamide	2.26	1.65	5.59	−0.69	7.49	
Nitromethane	0.66	1.711	7.36	−0.61	6.7	−0.57
Triethylamine	2.32	2.150	1.73	−1.50	12.61	

gives parameters *l*, *s*, *d*, *a*, *b*, and the intercepts for a number of common solvents.

Example 8-9 Use SPACE to calculate γ^∞ at 25°C for methanol (2) in 2-nitropropane (1).

solution Use Eqs. (8-10.31) to (8-10.44) to determine γ^∞ at 25°C:

$$\lambda_1 = 15.418 \cdot \frac{1.39235^2 - 1}{1.39235^2 + 2} = 3.674$$

$$\lambda_2 = 15.418 \cdot \frac{1.32652^2 - 1}{1.32652^2 + 2} = 3.115$$

$$\tau_1 = \left| \frac{-14.439 \cdot 0.75 - 20.876}{\sqrt{90.6}} \right| = 3.33$$

$$\tau_{2\text{eff}} = \left| \frac{-30.494 \cdot 0.6 + 29.223}{\sqrt{40.8}} \right| + \left(\left| \frac{59.387 \cdot 0.44 + 29.223}{\sqrt{40.8}} \right| - \left| \frac{-30.494 \cdot 0.6 + 29.223}{\sqrt{40.8}} \right| \right) \frac{|0.75 - 0.6|}{1.33} = 2.495$$

$$\alpha_1 = \left| \frac{-26.92 \cdot 0.22 + 17.172}{\sqrt{90.6}} \right| = 1.182$$

$$\alpha_{2\text{eff}} = \left| \frac{-26.92 \cdot 0.98 + 2.659}{\sqrt{40.8}} \right| + \left(\left| \frac{-132.494 \cdot 0.43 - 5.007}{\sqrt{40.8}} \right| - \left| \frac{-26.92 \cdot 0.98 + 2.659}{\sqrt{40.8}} \right| \right) \frac{|0.22 - 0.98|}{0.95} = 7.507$$

$$\beta_1 = \left| \frac{27.561 \cdot 0.27 - 10.864}{\sqrt{90.64}} \right| = 0.359$$

$$\beta_{2\text{eff}} = \left| \frac{27.561 \cdot 0.66 + 9.377}{\sqrt{40.8}} \right| + \left(\left| \frac{2.147 \cdot 0.47 + 40.347}{\sqrt{40.8}} \right| - \left| \frac{27.561 \cdot 0.66 + 9.377}{\sqrt{40.8}} \right| \right) \frac{|0.27 - 0.66|}{0.95} = 5.202$$

$$\ln \gamma_2^\infty = \frac{40.8}{1.987 \cdot 298.15} [(3.674 - 3.115)^2 + (3.33 - 2.495)^2 + (1.182 - 7.507)(0.359 - 5.202)] + \ln \left(\frac{40.8}{90.6} \right)^{0.936} + 1 - \left(\frac{40.8}{90.6} \right)^{0.936} = 1.96$$

$$\gamma_2^\infty = 7.10$$

Thomas, et al. (1982) report an experimental value at 20°C of 8.35. Equation (8-10.46) cannot be used to correct the above value to 20°C because although solute properties are available in Table 8-19 for methanol, solvent properties for

2-nitropropane are not available in Table 8-21. Nevertheless, the correction for a temperature difference of 5°C is likely to be small so the error is 10 to 15%.

The SPACE method is probably the best general method now available for estimating activity coefficients at infinite dilution. An alternate method, restricted to aqueous systems and easier to use is described below. For design of water-pollution abatement processes, it is often necessary to estimate the activity coefficient of a pollutant dilute in aqueous solution. A useful correlation for such estimates was presented by Hwang, et al., (1992) as illustrated in the following example.

Example 8-10 Estimate some infinite-dilution activity coefficients for organic pollutants in aqueous solution.

Hwang, et al., (1992) collected vapor-liquid distribution coefficients at infinite dilution (K_1^∞) for 404 common organic pollutants in aqueous solution at 100°C and proposed an empirical correlation based on the molecular structure of the organic pollutants:

$$\begin{aligned}\log K_1^\infty(100°\text{C}) = {} & 3.097 + 0.386n_{\text{satC}} + 0.323n_{=\text{C}} + 0.097n_{\equiv\text{C}} + 0.145n_{\text{aroC}} \\ & - 0.013n_{\text{C}}^2 + 0.366n_{\text{F}} - 0.096n_{\text{Cl}} - 0.496n_{\text{BrI}} \\ & - 1.954n_{-\text{O}-} - 2.528n_{=\text{O}} - 3.464n_{\text{OH}} + 0.331n_{\text{O}}^2 \\ & - 2.674n_{\text{N}} - 2.364n_{\equiv\text{N}} - 1.947n_{\text{NO}_2} - 1.010n_{\text{S}} \end{aligned} \tag{8-10.47}$$

where subscript 1 denotes the organic solute; the distribution coefficient K_1^∞ (100°C) at infinite dilution is defined as the ratio of the mole fraction of the solute in the vapor phase to that of the solute in the liquid phase at 100°C; n denotes the number of atoms or groups specified in the subscript. Atoms or groups in the subscripts represent the following categories:

Subscript	Atom/group identity
satC	Saturated carbon atoms or those bonded to carbonyl oxygens or nitrile nitrogens
=C	Double-bonded carbon atoms
≡C	Triple-bonded carbon atoms
aroC	Aromatic carbon atoms
C	All carbon atoms, including the four above
F	Fluorine atoms
Cl	Chlorine atoms
BrI	Bromide or Iodine atoms
—O—	Ether oxygen atoms
=O	Carbonyl oxygen atoms
OH	Hydroxyl groups
O	All oxygen atoms, including the three above
N	Amine or amide nitrogen atoms
≡N	Nitrile nitrogen atoms
NO_2	Nitro groups
S	Sulfur atoms

In this example, we use Hwang's correlation to calculate the infinite-dilution activity coefficients (γ_1^∞) at 100°C for the following six solutes: benzene, toluene, chlorobenzene, phenol, aniline and nitrobenzene.

solution At a low pressure, if we assume unity for fugacity coefficients and Poynting factors, Eq. (8-4.1) for the solute (1) is

$$y_1 P = x_1 \gamma_1 P_{vp1}$$

where y_1 and x_1 are, respectively, vapor-phase and liquid-phase mole fractions of the solute; P is the total pressure; P_{vp1} is the vapor pressure of the solute at 100°C.

The relation between γ and K is

$$K_1 \equiv \frac{y_1}{x_1} = \frac{\gamma_1 P_{vp1}}{P}$$

As $x_1 \rightarrow 0$, $K_1 \rightarrow K_1^\infty$, $\gamma_1 \rightarrow \gamma_1^\infty$, and $P \rightarrow P_{vp2}$.
Because $P_{vp2} = 1$ atm at 100°C,

$$K_1^\infty = \gamma_1^\infty P_{vp1}$$

where P_{vp1} is in atm.
Combining the above equation with Eq. (8-10.42), we obtain

$$\begin{aligned}\log \gamma_1^\infty(100°\text{C}) = 3.097 &+ 0.386n_{\text{satC}} + 0.323n_{=\text{C}} + 0.097n_{\equiv\text{C}} \\ &+ 0.145n_{\text{aroC}} - 0.013n_{\text{C}}^2 + 0.366n_{\text{F}} - 0.096n_{\text{Cl}} \\ &- 0.496n_{\text{BrI}} - 1.954n_{-\text{O}-} - 2.528n_{=\text{O}} - 3.46n_{\text{OH}} \\ &+ 0.331n_{\text{O}}^2 - 2.67n_{\text{N}} - 2.364n_{=\text{N}} - 1.947n_{\text{NO}_2} \\ &- 1.010n_{\text{S}} - \log P_{vp1}\end{aligned}$$

To use this equation, we must (a) calculate P_{vp1} (b) make an inventory of the constitutive atoms and groups in the solute molecule.

(a) calculating P_{vp1}

The vapor pressure of nitrobenzene was obtained from Daubert, et al., (1997). All others were obtained from Appendix A. The vapor pressures at 100°C are tabulated below.

Solute	P_{vp1} (bar)
benzene	1.80
toluene	0.742
chlorobenzene	0.395
phenol	0.0547
aniline	0.0595
nitrobenzene	0.0280

(b) Counting atoms and groups

Molecular structures of the six solutes are

	Number of occurrences					
Atom/Group	benzene	toluene	chlorobenzene	phenol	aniline	nitrobenzene
satC	0	1	0	0	0	0
=C	0	0	0	0	0	0
≡C	0	0	0	0	0	0
aroC	6	6	6	6	6	6
C	6	7	6	6	6	6
F	0	0	0	0	0	0
Cl	0	0	1	0	0	0
Br/I	0	0	0	0	0	0
—O—	0	0	0	0	0	0
=O	0	0	0	0	0	0
OH	0	0	0	1	0	0
O	0	0	0	1	0	0
N	0	0	0	0	1	0
=N	0	0	0	0	0	0
NO_2	0	0	0	0	0	1
S	0	0	0	0	0	0

Substitution of P_{vp1} and various n's into Eq. (8-10.47) leads to the six calculated γ_1^∞ (100°C) below. Experimental results at the same temperature (100°C) are from Hwang, et al., (1992).

Solute	Calculated	Measured	% error
benzene	1.78E3	1.30E3	36.9
toluene	7.10E3	3.40E3	109
chlorobenzene	6.69E3	3.60E3	85.8
phenol	4.33E1	4.39E1	1.40
aniline	1.12E2	2.82E2	60.3
nitrobenzene	1.27E3	1.70E3	25.3

where

$$\% \text{ error} = \frac{|\text{Calculated} - \text{Observed}|}{\text{Observed}} \times 100$$

These results suggest that the correlation of Hwang, et al. can predict infinite-dilution activity coefficients in aqueous solutions within a factor of about 2 or less. However, the accuracy of the experimental data is often not significantly better.

Azeotropic Data

Many binary systems exhibit azeotropy, i.e., a condition in which the composition of a liquid mixture is equal to that of its equilibrium vapor. When the azerotropic conditions (temperature, pressure, composition) are known, activity coefficients γ_1 and γ_2 at that condition are readily found. These activity coefficients can then be used to calculate two parameters in some arbitrarily chosen expresison for the excess Gibbs energy (Table 8-3). Extensive compilations of azeotropic data are available (Horsley, 1952, 1962, 1973; Gmehling, et al., 1994)

For a binary azeotrope, $x_1 = y_1$ and $x_2 = y_2$; therefore, Eq. (8-4.1), with $\mathcal{F}_i = 1$, becomes

$$\gamma_1 = \frac{P}{P_{vp1}} \quad \text{and} \quad \gamma_2 = \frac{P}{P_{vp2}} \tag{8-10.48}$$

Knowing total pressure P and pure-component vapor pressures P_{vp1} and P_{vp2}, we determine γ_1 and γ_2. With these activity coefficients and the azeotropic composition x_1 and x_2 it is now possible to find two parameters A and B by simultaneous solution of two equations of the form

$$RT \ln \gamma_1 = f_1(x_2, A, B) \tag{8-10.49a}$$

$$RT \ln \gamma_2 = f_2(x_1, A, B) \tag{8-10.49b}$$

where, necessarily, $x_1 = 1 - x_2$ and where functions f_1 and f_2 represent thermodynamically consistent equations derived from the choice of an expression for the excess Gibbs energy. Simultaneous solution of Eqs. (8-10.49*a*) and (8-10.49*b*) is simple in principle, although the necessary algebra may be tedious if f_1 and f_2 are complex.

Example 8-11 To illustrate, consider an example similar to one given by Treybal (1963) for the system ethyl acetate (1)—ethanol (2). This system forms an azeotrope at 1.01 bar, 71.8°C, and $x_2 = 0.462$.

solution At 1.01 bar and 71.8°C, we use Eq. (8-10.48):

$$\gamma_1 = \frac{1.01}{0.839} = 1.204 \qquad \gamma_2 = \frac{1.01}{0.772} = 1.308$$

where 0.839 and 0.772 bar are the pure component vapor pressures at 71.8°C. For functions f_1 and f_2 we choose the van Laar equations shown in Table 8-3. Upon algebraic rearrangement, we obtain explicit solutions for A and B.

$$\frac{A}{RT} = \ln 1.204 \left(1 + \frac{0.462 \ln 1.308}{0.538 \ln 1.204}\right)^2 = 0.93$$

$$\frac{B}{RT} = \ln 1.308 \left(1 + \frac{0.538 \ln 1.204}{0.462 \ln 1.308}\right)^2 = 0.87$$

and $A/B = 1.07$.

At 71.8°C, the activity coefficients are given by

$$\ln \gamma_1 = \frac{0.93}{(1 + 1.07x_1/x_2)^2}$$

$$\ln \gamma_2 = \frac{0.87}{(1 + x_2/1.07x_1)^2}$$

Figure 8-10 shows a plot of the calculated activity coefficients. Also shown are exper-

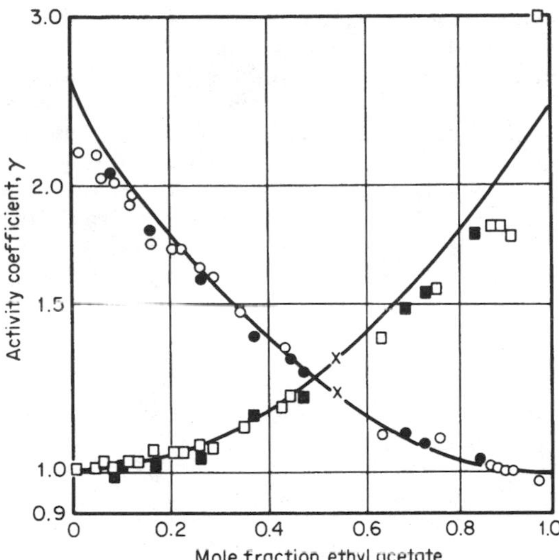

FIGURE 8-10 Activity coefficients in the system ethylacetate-ethanol. Calculated lines from azeotropic data (indicated by x) at 1.01 bar. Points are experimental (*Treybal, 1963*)

imental results at 1.01 bar by Furnas and Leighton (1937) and by Griswold, et al. (1949). Since the experimental results are isobaric, the temperature is not constant. However, in this example, the calculated activity coefficients are assumed to be independent of temperature.

Figure 8-10 shows good overall agreement between experimental and calculated activity coefficients. Generally, fair agreement is found if the azeotropic data are accurate, if the binary system is not highly complex, and, most important, if the azeotropic composition is in the midrange $0.25 < x_1$ (or x_2) < 0.75. If the azeotropic composition is at either dilute end, azeotropic data are of much less value for estimating activity coefficients over the entire composition range. This negative conclusion follows from the limiting relation $\gamma_1 \rightarrow 1$ as $x_1 \rightarrow 1$. Thus, if we have an azeotropic mixture where $x_2 << 1$, the experimental value of γ_1 gives us very little information, since γ_1 is necessarily close to unity. For such a mixture, only γ_2 supplies significant information, and therefore we cannot expect to calculate two meaningful adjustable parameters when we have only one significant datum. However, if the azeotropic composition is close to unity, we may, nevertheless, use the azeotropic data to find one activity coefficient, namely, γ_2 (where $x_2 << 1$), and then use that γ_2 to determine the single adjustable parameter in any of the one-parameter equations for the molar excess Gibbs energy, as discussed in Sec. 8-5.

Activity Coefficient Parameters from Mutual Solubilities of Liquids

When two liquids are only partially miscible, experimental data for the two mutual solubilities can be used to estimate activity coefficients over the entire range of composition in the homogeneous regions. Suppose the solubility (mole fraction) of component 1 in compound 2 is $x_1^{s'}$ and that of component 2 in component 1 is $x_2^{s''}$, where superscript s denotes saturation and the primes designate the two liquid phases. If $x_1^{s'}$ and $x_2^{s''}$ are known at some temperature T, it is possible to estimate activity coefficients for both components in the homogeneous regions $0 \leq x_1' \leq x_1^{s'}$ and $0 \leq x_2'' \leq x_2^{s''}$.

To estimate the activity coefficients, it is necessary to choose some thermodynamically consistent analytical expression which relates activity coefficients γ_1 and γ_2 to mole fraction x. (See Sec. 8-5.) Such an expression contains one or more parameters characteristic of the binary system; these parameters are generally temperature-dependent, although the effect of temperature is often not large. From the equations of liquid-liquid equilibrium, it is possible to determine two of these parameters. The equations of equilibrium are

$$(\gamma_1 x_1)^{s'} = (\gamma_1 x_1)^{s''} \quad \text{and} \quad (\gamma_2 x_2)^{s'} = (\gamma_2 x_2)^{s''} \tag{8-10.50}$$

Suppose we choose a two-constant expression for the molar excess Gibbs energy g^E. Then, as discussed in Sec. 8-5,

$$RT \ln \gamma_1 = f_1(x_2, A, B) \quad \text{and} \quad RT \ln \gamma_2 = f_2(x_1, A, B) \tag{8-10.51}$$

where f_1 and f_2 are known functions and the two (unknown) constants are designated A and B. These constants can be found by simultaneous solution of Eqs. (8-10.50) and (8-10.51) coupled with experimental values for $x_1^{s'}$ and $x_2^{s''}$ and the material balances

$$x_2^{s'} = 1 - x_1^{s'} \quad \text{and} \quad x_1^{s''} = 1 - x_2^{s''} \tag{8-10.52}$$

In principle, the calculation is simple although the algebra may be tedious, depending on the complexity of the functions f_1 and f_2.

To illustrate, Table 8-22 presents results obtained by Brian (1965) for five binary aqueous systems, where subscript 2 refers to water. Calculations are based on both the van Laar equation and the three-suffix (two-parameter) Margules equation (see Table 8-3). Table 8-22 shows the calculated activity coefficients at infinite dilution, which are easily related to the constants A and B. [See Eqs. (8-10.23) and (8-10.24).]

Brian's calculations indicate that results are sensitive to the expression chosen for the molar excess Gibbs energy. Brian found that, compared with experimental

TABLE 8-22 Limiting Activity Coefficients as Calculated from Mutual Solubilities in Five Binary Aqueous Systems (Brian, 1965)

		Solubility limits		log γ_1^∞		log γ_2^∞	
Component (1)	Temp., °C	$x_1^{s'}$	$x_2^{s''}$	van Laar	Margules	van Laar	Margules
Aniline	100	0.01475	0.372	1.8337	1.5996	0.6076	−0.4514
Isobutyl alcohol	90	0.0213	0.5975	1.6531	0.6193	0.4020	−3.0478
1-Butanol	90	0.0207	0.636	1.6477	0.2446	0.3672	−4.1104
Phenol	43.4	0.02105	0.7325	1.6028	−0.1408	0.2872	−8.2901
Propylene oxide	36.3	0.166	0.375	1.1103	1.0743	0.7763	0.7046

vapor-liquid equilibrium data for the homogeneous regions, the Margules equations gave poor results and the van Laar equation gave fair, but not highly accurate, results.

Calculations of this sort can also be made by using a three-parameter equation for g^E, but in that event, the third parameter must be estimated independently. A nomogram for such calculations, using the NRTL equation, has been given by Renon and Prausnitz (1969).

Estimation of Activity Coefficients from Group-Contribution Methods

For correlating thermodynamic properties, it is often convenient to regard a molecule as an aggregate of functional groups; as a result, some thermodynamic properties of pure fluids, e.g., heat capacity (Chaps. 3 and 6) and critical volume (Chap. 2), can be calculated by summing group contributions. Extension of this concept to mixtures was suggested long ago by Langmuir, and several attempts have been made to establish group-contribution methods for heats of mixing and for activity coefficients. Here we mention only two methods, both for activity coefficients, which appear to be particularly useful for making reasonable estimates for those nonideal mixtures for which data are sparse or totally absent. The two methods, called ASOG and UNIFAC, are similar in principle but differ in detail.

In any group-contribution method, the basic idea is that whereas there are thousands of chemical compounds of interest in chemical technology, the number of functional groups that constitute these compounds is much smaller. Therefore, if we assume that a physical property of a fluid is the sum of contributions made by the molecule's functional groups, we obtain a possible technique for correlating the properties of a very large number of fluids in terms of a much smaller number of parameters that characterize the contributions of individual groups.

Any group-contribution method is necessarily approximate because the contribution of a given group in one molecule is not necessarily the same as that in another molecule. The fundamental assumption of a group-contribution method is additivity: the contribution made by one group within a molecule is assumed to be independent of that made by any other group in that molecule. This assumption is valid only when the influence of any one group in a molecule is not affected by the nature of other groups within that molecule.

For example, we would not expect the contribution of a carbonyl group in a ketone (say, acetone) to be the same as that of a carbonyl group in an organic acid (say, acetic acid). On the other hand, experience suggests that the contribution of a carbonyl group in, for example, acetone, is close to (although not identical with) the contribution of a carbonyl group in another ketone, say 2-butanone.

Accuracy of correlation improves with increasing distinction of groups; in considering, for example, aliphatic alcohols, in a first approximation no distinction is made between the position (primary or secondary) of a hydroxyl group, but in a second approximation such a distinction is desirable. In the limit as more and more distinctions are made, we recover the ultimate group, namely, the molecule itself. In that event, the advantage of the group-contribution method is lost. For practical utility, a compromise must be attained. The number of distinct groups must remain small but not so small as to neglect significant effects of molecular structure on physical properties.

Extension of the group-contribution idea to mixtures is attractive because, although the number of pure fluids in chemical technology is already very large, the number of different mixtures is larger by many orders of magnitude. Thousands, perhaps millions, of multicomponent liquid mixtures of interest in the chemical industry can be constituted from perhaps 30, 50, or at most 100 functional groups.

ASOG Method. The analytical solution of groups (ASOG) method was developed by Wilson and Deal (1962) and Wilson (1964) following earlier work by Redlich, Derr, Pierotti, and Papadopoulos (1959). An introduction to ASOG was presented by Palmer (1975).

For component i in a mixture, activity coefficient γ_i consists of a configurational (entropic) contribution due to differences in molecular size and a group-interaction contribution due primarily to differences in intermolecular forces:

$$\ln \gamma_i = \ln \gamma_i^S + \ln \gamma_i^G \tag{8-10.53}$$

where superscript S designates size and superscript G designates group interaction.

Activity coefficient γ_i^S depends only on the number of size groups, e.g., CH_2, CO, OH, in the various molecules that constitute the mixture. From the Flory-Huggins theory for athermal mixtures of unequal-size molecules:

$$\ln \gamma_i^S = 1 - R_i + \ln R_i \tag{8-10.54}$$

where

$$R_i = \frac{s_i}{\sum_j s_j x_j} \tag{8-10.55}$$

where x_j = jmole fraction of component j in mixture
s_j = number of size groups in molecule j

Parameter s_j is independent of temperature. The summation extends over all components, including component i.

To calculate γ_i^G, we need to know the *group* mole fractions X_k, where subscript k stands for a particular group in molecule j

$$X_k = \frac{\sum_j x_j \nu_{kj}}{\sum_j x_j \sum_k \nu_{kj}} \tag{8-10.56}$$

where ν_{kj} is the number of interaction groups k in molecule j. Activity coefficient γ_i^G is given by

$$\ln \gamma_i^G = \sum_k \nu_{ki} \ln \Gamma_k - \sum_k \nu_{ki} \ln \Gamma_k^* \tag{8-10.57}$$

where Γ_k = activity coefficient of group k in the mixture of groups
Γ_k^* = activity coefficient of group k in the standard state

This standard state depends on molecule i.
Activity coefficient Γ_k is given by Wilson's equation

$$\ln \Gamma_k = -\ln \sum_i X_i A_{ki} + \left(1 - \sum_i \frac{X_i A_{ik}}{\sum_m x_m A_{im}}\right) \tag{8-10.58}$$

where the summations extend over all groups present in the mixture.

Equation (8-10.58) is also used to find Γ_k^* for component i, but in that case it is applied to a "mixture" of groups as found in pure component i. For example, if i is water, hexane,† or benzene, there is only one kind of group and $\ln \Gamma_k^*$ is zero. However, if i is methanol, $\ln \Gamma_k^*$ has a finite value for both hydroxyl and methyl groups.

Parameters A_{kt} and A_{tk} ($A_{kt} \neq A_{tk}$) are group-interaction parameters that depend on temperature. These parameters are obtained from reduction of vapor-liquid equilibria, and a substantial number of such parameters have been reported by Derr and Deal (1969) and by Kojima and Tochigi (1979). The important point here is that, at fixed temperature, these parameters depend only on the nature of the groups and, by assumption, are independent of the nature of the molecule. Therefore, group parameters obtained from available experimental data for some mixtures can be used to predict activity coefficients in other mixtures that contain not the same molecules, but the same groups. For example, suppose we wish to predict activity coefficients in the binary system dibutyl ketone-nitrobenzene. To do so, we require group interaction parameters for characterizing interactions between methyl, phenyl, keto, and nitrile groups. These parameters can be obtained from other binary mixtures that contain these groups, e.g., acetone-benzene, nitropropane-toluene, and methyl ethyl ketone-nitroethane.

UNIFAC Method. The fundamental idea of a solution-of-groups model is to utilize existing phase equilibrium data for predicting phase equilibria of systems for which no experimental data are available. In concept, the UNIFAC method follows the ASOG method, wherein activity coefficients in mixtures are related to interactions between structural groups. The essential features are:

1. Suitable reduction of experimentally obtained activity-coefficient data to yield parameters characterizing interactions between pairs of structural groups in nonelectrolyte systems.
2. Use of those parameters to predict activity coefficients for other systems that have not been studied experimentally but that contain the same functional groups.

The molecular activity coefficient is separated into two parts: one part provides the contribution due to differences in molecular size and shape, and the other provides the contribution due to molecular interactions. In ASOG, the first part is arbitrarily estimated by using the athermal Flory-Huggins equation; the Wilson equation, applied to functional groups, is chosen to estimate the second part. Some of this arbitrariness is removed by combining the solution-of-groups concept with the UNIQUAC equation (see Table 8-3); first, the UNIQUAC model per se contains a combinatorial part, essentially due to differences in size and shape of the molecules in the mixture, and a residual part, essentially due to energy interactions, and second, functional group sizes and interaction surface areas are introduced from independently obtained, pure-component molecular structure data.

The UNIQUAC equation often gives good representation of vapor-liquid and liquid-liquid equilibria for binary and multicomponent mixtures containing a variety of nonelectrolytes such as hydrocarbons, ketones, esters, water, amines, alcohols, nitriles, etc. In a multicomponent mixture, the UNIQUAC equation for the activity coefficient of (molecular) component i is

† It is assumed here that with respect to group interactions, no distinction is made between groups CH_2 and CH_3.

$$\ln \gamma_i = \underset{\text{combinatorial}}{\ln \gamma_i^c} + \underset{\text{residual}}{\ln \gamma_i^R} \tag{8-10.59}$$

where
$$\ln \gamma_i^c = \ln \frac{\Phi_i}{x_i} + \frac{z}{2} q_i \ln \frac{\theta_i}{\Phi_i} + l_i - \frac{\Phi_i}{x_i} \sum_j x_j l_j \tag{8-10.60}$$

and
$$\ln \gamma_i^R = q_i \left[1 - \ln \left(\sum_j \theta_j \tau_{ji} \right) - \sum_j \frac{\theta_j \tau_{ij}}{\sum_k \theta_k \tau_{kj}} \right]$$

$$l_i = \frac{z}{2}(r_i - q_i) - (r_i - 1) \quad z = 10 \tag{8-10.61}$$

$$\theta_i = \frac{q_i x_i}{\sum_j q_j x_j} \quad \Phi_i = \frac{r_i x_i}{\sum_j r_j x_j} \quad \tau_{ji} = \exp\left(- \frac{u_{ji} - u_{ii}}{RT} \right)$$

In these equations x_i is the mole fraction of component i and the summations in Eqs. (8-10.60) and (8-10.61) are over all components, including component i; θ_i is the area fraction, and Φ_i is the segment fraction, which is similar to the volume fraction. Pure-component parameters r_i and q_i are, respectively, measures of molecular van der Waals volumes and molecular surface areas.

In UNIQUAC, the two adjustable binary parameters τ_{ij} and τ_{ji} appearing in Eq. (8-10.61) must be evaluated from experimental phase equilibrium data. No ternary (or higher) parameters are required for systems containing three or more components.

In the UNIFAC method (Fredenslund, et al., 1975, 1977), the combinatorial part of the UNIQUAC activity coefficients, Eq. (8-10.60), is used directly. Only pure component properties enter into this equation. Parameters r_i and q_i are calculated as the sum of the group volume and area parameters R_k and Q_k, given in Table 8-23:

$$r_i = \sum_k \nu_k^{(i)} R_k \quad \text{and} \quad q_i = \sum_k \nu_k^{(i)} Q_k \tag{8-10.62}$$

where $\nu_k^{(i)}$, always an integer, is the number of groups of type k in molecule i. Group parameters R_k and Q_k are obtained from the van der Waals group volume and surface areas V_{wk} and A_{wk}, given by Bondi (1968):

$$R_k = \frac{V_{wk}}{15.17} \quad \text{and} \quad Q_k = \frac{A_{wk}}{2.5 \times 10^9} \tag{8-10.63}$$

The normalization factors 15.17 and 2.5×10^9 are determined by the volume and external surface area of a CH_2 unit in polyethylene.

The residual part of the activity coefficient, Eq. (8-10.61), is replaced by the solution-of-groups concept. Instead of Eq. (8-10.61), we write

$$\ln \gamma_i^R = \sum_{\substack{k \\ \text{all groups}}} \nu_k^{(i)} (\ln \Gamma_k - \ln \Gamma_k^{(i)} \tag{8-10.64}$$

where Γ_k is the group residual activity coefficient and $\Gamma_k^{(i)}$ is the residual activity coefficient of group k in a reference solution containing only molecules of type i. (In UNIFAC, $\Gamma^{(i)}$ is similar to ASOG's Γ^* of Eq. (8-10.57).) In Eq. (8-10.64) the term $\ln \Gamma_k^{(i)}$ is necessary to attain the normalization that activity coefficient γ_i becomes unity as $x_i \rightarrow 1$. The activity coefficient for group k in molecule i depends on the molecule i in which k is situated. For example, $\Gamma_k^{(i)}$ for the COH group† in ethanol refers to a "solution" containing 50 group percent COH and 50 group percent CH_3 at the temperature of the mixture, whereas $\Gamma_k^{(i)}$ for the COH group in n-butanol refers to a "solution" containing 25 group percent COH, 50 group percent CH_2, and 25 group percent CH_3.

The group activity coefficient Γ_k is found from an expression similar to Eq. (8-10.61):

$$\ln \Gamma_k = Q_k \left[1 - \ln \left(\sum_m \theta_m \Psi_{mk} \right) - \sum_m \frac{\theta_m \Psi_{km}}{\sum_n \theta_n \Psi_{nm}} \right] \tag{8-10.65}$$

Equation (8-10.65) also holds for $\ln \Gamma_k^{(i)}$. In Eq. (8-10.65), θ_m is the area fraction of group m, and the sums are over all different groups. θ_m is calculated in a manner similar to that for θ_i:

$$\theta_m = \frac{Q_m X_m}{\sum_n Q_n X_n} \tag{8-10.66}$$

where X_m is the mole fraction of group m in the mixture. The group-interaction parameter Ψ_{mn} is given by

$$\Psi_{mn} = \exp \left(- \frac{U_{mn} - U_{nn}}{RT} \right) = \exp \left(- \frac{a_{mn}}{T} \right) \tag{8-10.67}$$

where U_{mn} is a measure of the energy of interaction between groups m and n. The group interaction parameters a_{mn} must be evaluated from experimental phase equilibrium data. Note that a_{mn} has units of kelvins and $a_{mn} \neq a_{nm}$. Parameters a_{mn} and a_{nm} are obtained from a database using a wide range of experimental results. Some of these are shown in Table 8-24. Efforts toward updating and extending Table 8-24 are in progress in several university laboratories. (See Gmehling, et al., 1993, for example).

The combinatorial contribution to the activity coefficient [Eq. (8-10.60)] depends only on the sizes and shapes of the molecules present. As the coordination number z increases, for large-chain molecules $q_i/r_i \rightarrow 1$ and in that limit, Eq. (8-10.60) reduces to the Flory-Huggins equation used in the ASOG method.

The residual contribution to the activity coefficient [Eqs. (8-10.64) and (8-10.65)] depends on group areas and group interaction. When all group areas are equal, Eqs. (8-10.64) and (8-10.65) are similar to those used in the ASOG method.

The functional groups considered in this work are those given in Table 8-23. Whereas each group listed has its own values of R and Q, the subgroups within the same main group, e.g., subgroups 1, 2, and 3 are assumed to have identical energy interaction parameters. We present one example that illustrates (1) the nomenclature and use of Table 8-23 and (2) the UNIFAC method for calculating activity coefficients.

Example 8-12 Obtain activity coefficients for the acetone (1) n-pentane (2) system at 307 K and $x_1 = 0.047$.

† COH is shortened notation for CH_2OH.

TABLE 8-23 UNIFAC Group Specifications and Sample Group Assignments (Hansen, et al., 1991)

Group numbers: Main	Group numbers: Secondary	Name	Volume R	Surface Area Q	Sample Assignments = (Number of Occurrences) × (Secondary Group Number)
1	1	CH_3	0.9011	0.848	Hexane = (2)(1) + (4)(2)
	2	CH_2	0.6744	0.540	2-Methylpropane = (3)(1) + (1)(3)
	3	CH	0.4469	0.228	Neopentane = (4)(1) + (1)(4)
	4	C	0.2195	0.000	2,2,4-Trimethylpentane = (5)(1) + (1)(2) + (1)(3) + (1)(4)
2	5	$CH_2{=}CH$	1.3454	1.176	3-Methyl-1-hexene = (2)(1) + (2)(2) + (1)(3) + (1)(5)
	6	$CH{=}CH$	1.1167	0.867	Hexene-2 = (2)(1) + (2)(2) + (1)(6)
	7	$CH_2{=}C$	1.1173	0.988	2-Methyl-1-butene = (2)(1) + (1)(2) + (1)(7)
	8	$CH{=}C$	0.8886	0.676	2-Methyl-2-butene = (3)(1) + (1)(8)
	70	$C{=}C$	0.6605	0.485	2,3-Dimethylbutene = (4)(1) + (1)(70)
3	9	ACH	0.5313	0.400	Benzene = (6)(9)
	10	AC	0.3652	0.120	Styrene = (1)(5) + (5)(9) + (1)(10)
4	11	$ACCH_3$	1.2663	0.968	Toluene = (5)(9) + (1)(11)
	12	$ACCH_2$	1.0396	0.660	Ethylbenzene = (1)(1) + (5)(9) + (1)(12)
	13	ACCH	0.8121	0.348	Cumene = (2)(1) + (5)(9) + (1)(13)
5	14	OH	1.0000	1.200	Ethanol = (1)(1) + (1)(2) + (1)(14)
6	15	CH_3OH	1.4311	1.432	Methanol = (1)(15)
7	16	H_2O	0.9200	1.400	Water = (1)(16)
8	17	ACOH	0.8952	0.680	Phenol = (5)(9) + (1)(17)
9	18	CH_3CO	1.6724	1.488	Methylethylketone = (1)(1) + (1)(2) + (1)(18)
	19	CH_2CO	1.4457	1.180	Ethylphenylketone = (1)(1) + (1)(19) + (5)(9) + (1)(10)
10	20	CHO	0.9980	0.948	Hexanal = (1)(1) + (4)(2) + (1)(20)
11	21	CH_3COO	1.9031	1.728	Butyl acetate = (1)(1) + (3)(2) + (1)(21)
	22	CH_2COO	1.6764	1.420	Methyl propionate = (2)(1) + (1)(22)
12	23	HCOO	1.2420	1.188	Ethyl formate = (1)(1) + (1)(2) + (1)(23)

13	24	CH_3O	1.1450	1.088	Dimethyl ether = (1)(1) + (1)(24)
	25	CH_2O	0.9183	0.780	Diethyl ether = (2)(1) + (1)(2) + (1)(25)
	26	CHO	0.6908	0.468	Diisopropyl ether = (4)(1) + (1)(3) + (1)(26)
	27	THF	0.9183	1.100	Tetrahydrofuran = (3)(2) + (1)(27)
14	28	CH_3NH_2	1.5959	1.544	Methylamine = (1)(28)
	29	CH_2NH_2	1.3692	1.236	Ethylamine = (1)(1) + (1)(29)
	30	$CHNH_2$	1.1417	0.924	Isopropylamine = (2)(1) + (1)(30)
15	31	CH_3NH	1.4337	1.244	Dimethyl amine = (1)(1) + (1)(31)
	32	CH_2NH	1.2070	0.936	Diethylamine = (2)(1) + (1)(2) + (1)(32)
	33	CHNH	0.9795	0.624	Diisopropylamine = (4)(1) + (1)(3) + (1)(33)
16	34	CH_3N	1.1865	0.940	Trimethylamine = (2)(1) + (1)(34)
	35	CH_2N	0.9597	0.632	Triethylamine = (3)(1) + (2)(2) + (1)(35)
17	36	$ACNH_2$	1.0600	0.816	Aniline = (5)(9) + (1)(36)
18	37	C_5H_5N	2.9993	2.113	Pyridine = (1)(37)
	38	C_5H_4N	2.8332	1.833	2-Methylpyridine = (1)(1) + (1)(38)
	39	C_5H_3N	2.6670	1.553	2,3-Dimethylpyridine = (2)(1) + (1)(39)
19	40	CH_3CN	1.8701	1.724	Acetonitrile = (1)(40)
	41	CH_2CN	1.6434	1.416	Proprionitrile = (1)(1) + (1)(41)
20	42	COOH	1.3013	1.224	Acetic Acid = (1)(1) + (1)(42)
	43	HCOOH	1.5280	1.532	Formic Acid = (1)(43)
21	44	CH_2Cl	1.4654	1.264	1-Chlorobutane = (1)(1) + (2)(2) + (1)(44)
	45	CHCl	1.2380	0.952	2-Chloropropane = (2)(1) + (1)(45)
	46	CCl	1.0060	0.724	2-Chloro-2-methylpropane = (3)(1) + (1)(46)
22	47	CH_2Cl_2	2.2564	1.988	Dichloromethane = (1)(47)
	48	$CHCl_2$	2.0606	1.684	1,1-Dichloroethane = (1)(1) + (1)(48)
	49	CCl_2	1.8016	1.448	2,2-Dichloropropane = (2)(1) + (1)(49)
23	50	$CHCl_3$	2.8700	2.410	Chloroform = (1)(50)
	51	CCl_3	2.6401	2.184	1,1,1-Trichloroethane = (1)(1) + (1)(51)

TABLE 8-23 UNIFAC Group Specifications and Sample Group Assignments (Hansen, et al., 1991) (*Continued*)

Group numbers: Main	Group numbers: Secondary	Name	Volume R	Surface Area Q	Sample Assignments = (Number of Occurrences) × (Secondary Group Number)
24	52	CCl_4	3.3900	2.910	Tetrachloromethane = (1)(52)
25	53	ACCl	1.1562	0.844	Chlorobenzene = (5)(9) + (1)(53)
26	54	CH_3NO_2	2.0086	1.868	Nitromethane = (1)(54)
	55	CH_2NO_2	1.7818	1.560	Nitroethane = (1)(1) + (1)(55)
	56	$CHNO_2$	1.5544	1.248	2-Nitropropane = (2)(1) + (1)(56)
27	57	$ACNO_2$	1.4199	1.104	Nitrobenzene = (5)(9) + (1)(57)
28	58	CS_2	2.5070	1.650	Carbon disulfide = (1)(58)
29	59	CH_3SH	1.8770	1.676	Methanethiol = (1)(59)
	60	CH_2SH	1.6510	1.368	Ethanethiol = (1)(1) + (1)(60)
30	61	Furfural	3.1680	2.481	Furfural = (1)(61)
31	62	DOH	2.4088	2.248	1,2-Ethanediol = (1)(62)
32	63	I	1.2640	0.992	Iodoethane = (1)(1) + (1)(2) + (1)(63)
33	64	Br	0.9492	0.832	Bromoethane = (1)(1) + (1)(2) + (1)(64)
34	65	CH≡C	1.2929	1.088	1-Hexyne = (1)(1) + (3)(2) + (1)(65)
	66	C≡C	1.0613	0.784	2-Hexyne = (2)(1) + (2)(2) + (1)(66)
35	67	DMSO	2.8266	2.472	Dimethylsulfoxide = (1)(67)
36	68	Acrylonitrile	2.3144	2.052	Acrylonitrile = (1)(68)
37	69	Cl—(C=C)	0.7910	0.724	Trichloroethylene = (1)(8) + (3)(69)
38	71	ACF	0.6948	0.524	Fluorobenzene = (5)(9) + (1)(71)
39	72	DMF	3.0856	2.736	*N,N*-Dimethylformamide = (1)(72)
	73	$HCON(CH_2)_2$	2.6322	2.120	*N,N*-Diethylformamide = (2)(1) + (1)(73)
40	74	CF_3	1.4060	1.380	Perfluoroethane = (2)(74)
	75	CF_2	1.0105	0.920	
	76	CF	0.6150	0.460	Perfluoromethylcyclohexane = (1)(74) + (5)(75) + (1)(76)
41	77	COO	1.3800	1.200	Butylacetate = (2)(1) + (3)(2) + (1)(77)

42	78	SiH_3	1.6035	1.263	Methylsilane = (1)(1) + (1)(78)
	79	SiH_2	1.4443	1.006	Diethylsilane = (2)(1) + (2)(2) + (1)(79)
	80	SiH	1.2851	0.749	Trimethylsilane = (3)(1) + (1)(80)
	81	Si	1.0470	0.410	Tetramethylsilane = (4)(1) + (1)(81)
43	82	SiH_2O	1.4838	1.062	
	83	SiHO	1.3030	0.764	
	84	SiO	1.1044	0.466	Hexamethyldisiloxane = (6)(1) + (1)(81) + (1)(84)
44	85	NMP	3.9810	3.200	*N*-Methylpyrrolidone = (1)(85)
45	86	CCl_3F	3.0356	2.644	Trichlorofluoromethane = (1)(86)
	87	CCl_2F	2.2287	1.916	Tetrachloro-1,2-difluoroethane = (2)(87)
	88	$HCCl_2F$	2.4060	2.116	Dichlorofluoromethane = (1)(88)
	89	HCClF	1.6493	1.416	2-Chloro-2-fluoroethane = (1)(1) + (1)(89)
	90	$CClF_2$	1.8174	1.648	2-Chloro-2,2-difluoroethane = (1)(1) + (1)(90)
	91	$HCClF_2$	1.9670	1.828	Chlorodifluoromethane = (1)(91)
	92	$CClF_3$	2.1721	2.100	Chlorotrifluoromethane = (1)(92)
	93	CCl_2F_2	2.6243	2.376	Dichlorodifluoromethane = (1)(93)
46	94	$CONH_2$	1.4515	1.248	Acetamide = (1)(1) + (1)(94)
	95	$CONHCH_3$	2.1905	1.796	*N*-Methylacetamide = (1)(1) + (1)(95)
	96	$CONHCH_2$	1.9637	1.488	*N*-Ethylacetamide = (2)(1) + (1)(96)
	97	$CON(CH_3)_2$	2.8589	2.428	*N,N*-Dimethylacetamide = (1)(1) + (1)(97)
	98	$CONCH_3CH_2$	2.6322	2.120	*N,N*-Methylethylacetamide = (2)(1) + (1)(98)
	99	$CON(CH_2)_2$	2.4054	1.812	*N,N*-Diethylacetamide = (3)(1) + (1)(99)
47	100	$C_2H_5O_2$	2.1226	1.904	2-Ethoxyethanol = (1)(1) + (1)(2) + (1)(100)
	101	$C_2H_4O_2$	1.8952	1.592	2-Ethoxy-1-propanol = (2)(1) + (1)(2) + (1)(101)
48	102	CH_3S	1.6130	1.368	Dimethylsulfide = (1)(1) + (1)(102)
	103	CH_2S	1.3863	1.060	Diethylsulfide = (2)(1) + (1)(2) + (1)(103)
	104	CHS	1.1589	0.748	Diisopropylsulfide = (4)(1) + (1)(3) + (1)(104)
49	105	MORPH	3.4740	2.796	Morpholine = (1)(105)
50	106	C_4H_4S	2.8569	2.140	Thiophene = (1)(106)
	107	C_4H_3S	2.6908	1.860	2-Methylthiophene = (1)(1) + (1)(107)
	108	C_4H_2S	2.5247	1.580	2,3-Dimethylthiophene = (2)(1) + (1)(108)

TABLE 8-24 UNIFAC Group-Group Interaction Parameters, a_{mn}, in Kelvins

Main group	n = 1	2	3	4	5	6	7	8	9
m = 1	0.0	86.02	61.13	76.50	986.5	697.2	1318	1333	476.4
2	−35.36	0.0	38.81	74.15	524.1	787.6	270.6	526.1	182.6
3	−11.12	3.446	0.0	167.0	636.1	637.4	903.8	1329	25.77
4	−69.70	−113.6	−146.8	0.0	803.2	603.3	5695	884.9	−52.10
5	156.4	457.0	89.60	25.82	0.0	−137.1	353.5	−259.7	84.00
6	16.51	−12.52	−50.00	−44.50	249.1	0.0	−181.0	−101.7	23.39
7	300.0	496.1	362.3	377.6	−229.1	289.6	0.0	324.5	−195.4
8	275.8	217.5	24.34	244.2	−451.6	−265.2	−601.8	0.0	−356.1
9	26.76	42.92	140.1	365.8	164.5	108.7	472.5	−133.1	0.0
10	505.7	56.30	23.39	106.0	529.0	−340.2	480.8	−155.6	128.0
11	114.8	132.1	85.84	−170.0	245.4	249.6	200.8	−36.72	372.2
12	329.3	110.4	18.12	428.0	139.4	227.8	NA	NA	385.4
13	83.36	26.51	52.13	65.69	237.7	238.4	−314.7	−178.5	191.1
14	−30.48	1.163	−44.85	296.4	−242.8	−481.7	−330.4	NA	NA
15	65.33	−28.70	−22.31	223.0	−150.0	−370.3	−448.2	NA	394.6
16	−83.98	−25.38	−223.9	109.9	28.60	−406.8	−598.8	NA	225.3
17	1139	2000	247.5	762.8	−17.40	−118.1	−341.6	−253.1	−450.3
18	−101.6	−47.63	31.87	49.80	−132.3	−378.2	−332.9	−341.6	29.10
19	24.82	−40.62	−22.97	−138.4	185.4	162.6	242.8	NA	−287.5
20	315.3	1264	62.32	89.86	−151.0	339.8	−66.17	−11.00	−297.8
21	91.46	40.25	4.680	122.9	562.2	529.0	698.2	NA	286.3
22	34.01	−23.50	121.3	140.8	527.6	669.9	708.7	NA	82.86
23	36.70	51.06	288.5	69.90	742.1	649.1	826.8	NA	552.1
24	−78.45	160.9	−4.700	134.7	856.3	709.6	1201	10000	372.0
25	106.8	70.32	−97.27	402.5	325.7	612.8	−274.5	622.3	518.4
26	−32.69	−1.996	10.38	−97.05	261.6	252.6	417.9	NA	−142.6
27	5541	NA	1824	−127.8	561.6	NA	360.7	NA	−101.5
28	−52.65	16.62	21.50	40.68	609.8	914.2	1081	1421	303.7
29	−7.481	NA	28.41	19.56	461.6	448.6	NA	NA	160.6
30	−25.31	82.64	157.3	128.8	521.6	NA	23.48	NA	317.5

Main group	n = 1	2	3	4	5	6	7	8	9
m = 31	139.9	NA	221.4	150.6	267.6	240.8	−137.4	838.4	135.4
32	128.0	NA	58.68	26.41	501.3	431.3	NA	NA	138.0
33	−31.52	174.6	−154.2	1112	524.9	494.7	NA	NA	−142.6
34	−72.88	41.38	NA	NA	68.95	NA	NA	NA	443.6
35	50.49	64.07	−2.504	−143.2	−25.87	695.0	−240.0	NA	110.4
36	−165.9	573.0	−123.6	397.4	389.3	218.8	386.6	NA	NA
37	47.41	124.2	395.8	419.1	738.9	528.0	NA	NA	−40.90
38	−5.132	−131.7	−237.2	−157.3	649.7	645.9	NA	NA	NA
39	−31.95	249.0	−133.9	−240.2	64.16	172.2	−287.1	NA	97.04
40	147.3	62.40	140.6	NA	NA	NA	NA	NA	NA
41	529.0	1397	317.6	615.8	88.63	171.0	284.4	−167.3	123.4
42	−34.26	NA	787.9	NA	1913	NA	180.2	NA	992.4
43	110.2	NA	234.4	NA	NA	NA	NA	NA	NA
44	13.89	−16.11	−23.88	6.214	796.9	NA	832.2	−234.7	NA
45	30.74	NA	167.9	NA	794.4	762.7	NA	NA	NA
46	27.97	9.755	NA	NA	394.8	NA	−509.3	NA	NA
47	−11.92	132.4	−86.88	−19.45	517.5	NA	−205.7	NA	156.4
48	39.93	543.6	NA	NA	NA	420.0	NA	NA	NA
49	−23.61	161.1	142.9	274.1	−61.20	−89.24	−384.3	NA	NA
50	−8.479	NA	23.93	2.845	682.5	597.8	NA	810.5	278.8

TABLE 8-24 UNIFAC Group-Group Interaction Parameters, a_{mn}, in Kelvins (*Continued*)

Main group	n = 10	11	12	13	14	15	16	17	18
m = 1	677.0	232.1	507.0	251.5	391.5	255.7	206.6	920.7	287.8
2	448.8	37.85	333.5	214.5	240.9	163.9	61.11	749.3	280.5
3	347.3	5.994	287.1	32.14	161.7	122.8	90.49	648.2	−4.449
4	586.8	5688	197.8	213.1	19.02	−49.29	23.50	664.2	52.80
5	−203.6	101.1	267.8	28.06	8.642	42.70	−323.0	−52.39	170.0
6	306.4	−10.72	179.7	−128.6	359.3	−20.98	53.90	489.7	580.5
7	−116.0	72.87	NA	540.5	48.89	168.0	304.0	243.2	459.0
8	−271.1	−449.4	NA	−162.9	NA	NA	NA	119.9	−305.5
9	−37.36	−213.7	−190.4	−103.6	NA	−174.2	−169.0	6201	7.341
10	0.0	−110.3	766.0	304.1	NA	NA	NA	NA	NA
11	185.1	0.0	−241.8	−235.7	NA	−73.50	−196.7	475.5	NA
12	−236.5	1167	0.0	−234.0	NA	NA	NA	NA	−233.4
13	−7.838	461.3	457.3	0.0	−78.36	251.5	5422	NA	213.2
14	NA	NA	NA	222.1	0.0	−107.2	−41.11	−200.7	NA
15	NA	136.0	NA	−56.08	127.4	0.0	−189.2	NA	NA
16	NA	2889	NA	−194.1	38.89	865.9	0.0	NA	NA
17	NA	−294.8	NA	NA	−15.07	NA	NA	0.0	89.70
18	NA	NA	554.4	−156.1	NA	NA	NA	117.4	0.0
19	NA	−266.6	99.37	38.81	−157.3	−108.5	NA	777.4	134.3
20	−165.5	−256.3	193.9	−338.5	NA	NA	NA	493.8	−313.5
21	−47.51	35.38	NA	225.4	131.2	NA	NA	429.7	NA
22	190.6	−133.0	NA	−197.7	NA	NA	−141.4	140.8	587.3
23	242.8	176.5	235.6	−20.93	NA	NA	−293.7	NA	18.98
24	NA	129.5	351.9	113.9	261.1	91.13	316.9	898.2	368.5
25	NA	−171.1	383.3	−25.15	108.5	102.2	2951	334.9	NA
26	NA	129.3	NA	−94.49	NA	NA	NA	NA	NA
27	NA	NA	NA	NA	NA	NA	NA	134.9	2475
28	NA	243.8	NA	112.4	NA	NA	NA	NA	NA
29	NA	NA	201.5	63.71	106.7	NA	NA	NA	NA
30	NA	−146.3	NA	−87.31	NA	NA	NA	NA	NA

Main group	n = 10	11	12	13	14	15	16	17	18
m = 31	NA	152.0	NA	9.207	NA	NA	NA	192.3	NA
32	245.9	21.92	NA	476.6	NA	NA	NA	NA	NA
33	NA	24.37	NA	736.4	NA	NA	NA	NA	−42.71
34	NA	NA	NA	NA	NA	NA	NA	NA	NA
35	NA	41.57	NA	−93.51	NA	NA	−257.2	NA	NA
36	354.0	175.5	NA	NA	NA	NA	NA	NA	NA
37	183.8	611.3	134.5	−217.9	NA	NA	NA	NA	281.6
38	NA	NA	NA	167.3	NA	−198.8	116.5	NA	159.8
39	13.89	−82.12	−116.7	−155.2	49.70	NA	−185.2	343.7	NA
40	NA	NA	NA	NA	NA	NA	NA	NA	NA
41	577.5	−234.9	145.4	−247.8	NA	284.5	NA	−22.10	NA
42	NA	NA	NA	448.5	961.8	1464	NA	NA	NA
43	NA	NA	NA	NA	−125.2	1604	NA	NA	NA
44	NA	NA	NA	NA	NA	NA	NA	NA	NA
45	NA	NA	NA	NA	NA	NA	NA	NA	NA
46	NA	NA	NA	NA	NA	NA	NA	NA	NA
47	NA	−3.444	NA	NA	NA	NA	NA	NA	NA
48	NA	NA	NA	NA	NA	NA	NA	NA	NA
49	NA	NA	NA	NA	NA	NA	NA	NA	NA
50	NA	NA	NA	NA	NA	NA	NA	NA	221.4

TABLE 8-24 UNIFAC Group-Group Interaction Parameters, a_{mn}, in Kelvins (*Continued*)

Main group	n = 19	20	21	22	23	24	25	26	27
m = 1	597.0	663.5	35.93	53.76	24.90	104.3	11.44	661.5	543.0
2	336.9	318.9	−36.87	58.55	−13.99	−109.7	100.1	357.5	NA
3	212.5	537.4	−18.81	−144.4	−231.9	3.000	187.0	168.0	194.9
4	6096	872.3	−114.1	−111.0	−80.25	−141.3	−211.0	3629	4448
5	6.712	199.0	75.62	65.28	−98.12	143.1	123.5	256.5	157.1
6	53.28	−202.0	−38.32	−102.5	−139.4	−44.76	−28.25	75.14	NA
7	112.6	−14.09	325.4	370.4	353.7	497.5	133.9	220.6	399.5
8	NA	408.9	NA	NA	NA	1827	6915	NA	NA
9	481.7	669.4	−191.7	−130.3	−354.6	−39.20	−119.8	137.5	548.5
10	NA	497.5	751.9	67.52	−483.7	NA	NA	NA	NA
11	494.6	660.2	−34.74	108.9	−209.7	54.57	442.4	−81.13	NA
12	−47.25	−268.1	NA	NA	−126.2	179.7	24.28	NA	NA
13	−18.51	664.6	301.1	137.8	−154.3	47.67	134.8	95.18	NA
14	358.9	NA	−82.92	NA	NA	−99.81	30.05	NA	NA
15	147.1	NA	NA	NA	NA	71.23	−18.93	NA	NA
16	NA	NA	NA	−73.85	−352.9	−262.0	−181.9	NA	NA
17	−281.6	−396.0	287.0	−111.0	NA	882.0	617.5	NA	−139.3
18	−169.7	−153.7	NA	−351.6	−114.7	−205.3	NA	NA	2845
19	0.0	NA	4.933	−152.7	−15.62	−54.86	−4.624	−0.5150	NA
20	NA	0.0	13.41	−44.70	39.63	183.4	−79.08	NA	NA
21	54.32	519.1	0.0	108.3	249.6	62.42	153.0	32.73	86.20
22	258.6	543.3	−84.53	0.0	0.0	56.33	223.1	108.9	NA
23	74.04	504.2	−157.1	0.0	0.0	−30.10	192.1	NA	NA
24	492.0	631	11.80	17.97	51.90	0.0	−75.97	490.9	534.7
25	363.5	993.4	−129.7	−8.309	−0.2266	248.4	0.0	132.7	2213
26	0.2830	NA	113.0	−9.639	NA	−34.68	132.9	0.0	533.2
27	NA	NA	1971	NA	NA	514.6	−123.1	−85.12	0.0
28	335.7	NA	−73.09	NA	−26.06	−60.71	NA	277.8	NA
29	161.0	NA	−27.94	NA	NA	NA	NA	NA	NA
30	NA	570.6	NA	NA	48.48	−133.2	NA	NA	NA

Main group	n = 19	20	21	22	23	24	25	26	27
m = 31	169.6	NA	NA	NA	NA	NA	NA	481.3	NA
32	NA	616.6	NA	−40.82	21.76	48.49	NA	64.28	2448
33	136.9	5256	−262.3	−174.5	NA	77.55	−185.3	125.3	4288
34	329.1	NA	NA	NA	NA	NA	NA	174.4	NA
35	NA	−180.2	NA	−215.0	−343.6	−58.43	NA	NA	NA
36	−42.31	NA	NA	NA	NA	−85.15	NA	NA	NA
37	335.2	898.2	383.2	301.9	−149.8	−134.2	NA	379.4	NA
38	NA	NA	NA	NA	NA	−124.6	NA	NA	NA
39	150.6	−97.77	NA	NA	NA	−186.7	NA	223.6	NA
40	NA	NA	NA	NA	NA	NA	NA	NA	NA
41	−61.6	1179	182.2	305.4	−193.0	335.7	955.1	NA	NA
42	NA	NA	NA	NA	NA	NA	NA	NA	NA
43	NA	NA	NA	NA	NA	70.81	NA	NA	NA
44	NA	NA	NA	NA	−196.2	NA	161.5	NA	NA
45	NA	NA	NA	NA	NA	NA	NA	844	NA
46	NA	−70.25	NA	NA	NA	NA	NA	NA	NA
47	119.2	NA	NA	−194.7	NA	3.163	7.082	NA	NA
48	NA	NA	NA	NA	−363.1	−11.30	NA	NA	NA
49	NA	NA	NA	NA	NA	NA	NA	NA	NA
50	NA	NA	NA	NA	NA	−79.34	NA	176.3	NA

TABLE 8-24 UNIFAC Group-Group Interaction Parameters, a_{mn}, in Kelvins (*Continued*)

Main group	n = 28	29	30	31	32	33	34	35	36
m = 1	153.6	184.4	354.6	3025	335.8	479.5	298.9	526.5	689.0
2	76.30	NA	262.9	NA	NA	183.8	31.14	179.0	−52.87
3	52.07	−10.43	−64.69	210.7	113.3	261.3	NA	169.9	383.9
4	−9.451	393.6	48.49	4975	259.0	210.0	NA	4284	−119.2
5	488.9	147.5	−120.5	−318.9	313.5	202.1	727.8	−202.1	74.27
6	−31.09	17.50	NA	−119.2	212.1	106.3	NA	−399.3	−5.224
7	887.1	NA	188.0	12.72	NA	NA	NA	−139.0	160.8
8	8484	NA	NA	−687.1	NA	NA	NA	NA	NA
9	216.1	−46.28	−163.7	71.46	53.59	245.2	−246.6	−44.58	NA
10	NA	NA	NA	NA	117.0	NA	NA	NA	−339.2
11	183.0	NA	202.3	−101.7	148.3	18.88	NA	52.08	−28.61
12	NA	103.9	NA	NA	NA	NA	NA	NA	NA
13	140.9	−8.538	170.1	−20.11	−149.5	−202.3	NA	128.8	NA
14	NA	−70.14	NA	NA	NA	NA	NA	NA	NA
15	NA	NA	NA	NA	NA	NA	NA	NA	NA
16	NA	NA	NA	NA	NA	NA	NA	243.1	NA
17	NA	NA	NA	0.1004	NA	NA	NA	NA	NA
18	NA	NA	NA	NA	NA	−60.78	NA	NA	NA
19	230.9	0.4604	NA	177.5	NA	−62.17	−203.0	NA	81.57
20	NA	NA	−208.9	NA	228.4	−95.00	NA	−463.6	NA
21	450.1	59.02	NA	NA	NA	344.4	NA	NA	NA
22	NA	NA	NA	NA	177.6	315.9	NA	215.0	NA
23	116.6	NA	−64.38	NA	86.40	NA	NA	363.7	NA
24	132.2	NA	546.7	NA	247.8	146.6	NA	337.7	369.5
25	NA	NA	NA	NA	NA	593.4	NA	NA	NA
26	320.2	NA	NA	139.8	304.3	10.17	−27.70	NA	NA
27	NA	NA	NA	NA	2990	−124.0	NA	NA	NA
28	0.0	NA	NA	NA	292.7	NA	NA	NA	NA
29	NA	0.0	NA	NA	NA	NA	NA	31.66	NA
30	NA	NA	0.0	NA	NA	NA	NA	NA	NA

Main group	n = 28	29	30	31	32	33	34	35	36
m = 31	NA	NA	NA	0.0	NA	NA	NA	−417.2	NA
32	−27.45	NA	NA	NA	0.0	NA	NA	NA	NA
33	NA	NA	NA	NA	NA	0.0	NA	32.90	NA
34	NA	NA	NA	NA	NA	NA	0.0	NA	NA
35	NA	85.70	NA	535.8	NA	−111.2	NA	0.0	NA
36	NA	NA	NA	NA	NA	NA	NA	NA	0.0
37	167.9	NA	NA	NA	NA	NA	631.5	NA	837.2
38	NA	NA	NA	NA	NA	NA	NA	NA	NA
39	NA	−71.00	NA	−191.7	NA	NA	6.699	136.6	5.150
40	NA	NA	NA	NA	NA	NA	NA	NA	NA
41	885.5	NA	−64.28	−264.3	288.1	627.7	NA	−29.34	−53.91
42	NA	NA	NA	NA	NA	NA	NA	NA	NA
43	NA	NA	NA	NA	NA	NA	NA	NA	NA
44	NA	−274.1	NA	262.0	NA	NA	NA	NA	NA
45	NA	NA	NA	NA	NA	NA	NA	NA	NA
46	NA	NA	NA	NA	NA	NA	NA	NA	NA
47	NA	NA	NA	515.8	NA	NA	NA	NA	NA
48	NA	6.971	NA	NA	NA	NA	NA	NA	NA
49	NA	NA	NA	NA	NA	NA	NA	NA	NA
50	NA	NA	NA	NA	NA	NA	NA	NA	NA

TABLE 8-24 UNIFAC Group-Group Interaction Parameters, a_{mn}, in Kelvins (*Continued*)

Main group	n = 37	38	39	40	41	42	43	44	45
m = 1	−4.189	125.8	485.3	−2.859	387.1	−450.4	252.7	220.3	−5.869
2	−66.46	359.3	−70.45	449.4	48.33	NA	NA	86.46	NA
3	−259.1	389.3	245.6	22.67	103.5	−432.3	238.9	30.04	−88.11
4	−282.5	101.4	5629	NA	69.26	NA	NA	46.38	NA
5	225.8	44.78	−143.9	NA	190.3	−817.7	NA	−504.2	72.96
6	33.47	−48.25	−172.4	NA	165.7	NA	NA	NA	−52.1
7	NA	NA	319.0	NA	−197.5	−363.8	NA	−452.2	NA
8	NA	NA	NA	NA	−494.2	NA	NA	−659.0	NA
9	−34.57	NA	−61.70	NA	−18.80	−588.9	NA	NA	NA
10	172.4	NA	−268.8	NA	−275.5	NA	NA	NA	NA
11	−275.2	NA	85.33	NA	560.2	NA	NA	NA	NA
12	11.40	NA	308.9	NA	−122.3	NA	NA	NA	NA
13	240.20	−274.0	254.8	NA	417.0	1338	NA	NA	NA
14	NA	NA	−164.0	NA	NA	−664.4	275.9	NA	NA
15	NA	570.9	NA	NA	−38.77	448.1	−1327	NA	NA
16	NA	−196.3	22.05	NA	NA	NA	NA	NA	NA
17	NA	NA	−334.4	NA	−89.42	NA	NA	NA	NA
18	160.7	−158.8	NA	NA	NA	NA	NA	NA	NA
19	−55.77	NA	−151.5	NA	120.3	NA	NA	NA	NA
20	−11.16	NA	−228.0	NA	−337.0	NA	NA	NA	NA
21	−168.2	NA	NA	NA	63.67	NA	NA	NA	NA
22	−91.80	NA	NA	NA	−96.87	NA	NA	NA	NA
23	111.2	NA	NA	NA	255.8	NA	NA	−35.68	NA
24	1187.1	215.2	498.6	NA	256.5	NA	233.1	NA	NA
25	NA	NA	NA	NA	−71.18	NA	NA	−209.7	NA
26	10.76	NA	−223.1	NA	248.4	NA	NA	NA	−218.9
27	NA	NA	NA	NA	NA	NA	NA	NA	NA
28	−47.37	NA	NA	NA	469.8	NA	NA	NA	NA
29	NA	NA	78.92	NA	NA	NA	NA	1004	NA
30	NA	NA	NA	NA	43.37	NA	NA	NA	NA

Main group	n = 37	38	39	40	41	42	43	44	45
m = 31	NA	NA	302.2	NA	347.8	NA	NA	−262	NA
32	NA	NA	NA	NA	68.55	NA	NA	NA	NA
33	NA	NA	NA	NA	−195.1	NA	NA	NA	NA
34	2073	NA	−119.8	NA	NA	NA	NA	NA	NA
35	NA	NA	−97.71	NA	153.7	NA	NA	NA	NA
36	−208.8	NA	−8.804	NA	423.4	NA	NA	NA	NA
37	0.0	NA	255.0	NA	730.8	NA	NA	26.35	NA
38	NA	0.0	NA	−117.2	NA	NA	NA	NA	NA
39	137.7	NA	0.0	−5.579	72.31	NA	NA	NA	NA
40	NA	185.6	55.80	0.0	NA	NA	NA	NA	111.8
41	−198.0	NA	−28.65	NA	0.0	NA	NA	NA	NA
42	NA	NA	NA	NA	NA	0.0	−2166	NA	NA
43	NA	NA	NA	NA	NA	745.3	0.0	NA	NA
44	−66.31	NA	NA	NA	NA	NA	NA	0.0	NA
45	NA	NA	NA	−32.17	NA	NA	NA	NA	0.0
46	NA	NA	NA	NA	NA	NA	NA	NA	NA
47	NA	NA	NA	NA	101.2	NA	NA	NA	NA
48	148.9	NA	NA	NA	NA	NA	NA	NA	NA
49	NA	NA	NA	NA	NA	NA	NA	NA	NA
50	NA	NA	NA	NA	NA	NA	NA	NA	NA

TABLE 8-24 UNIFAC Group-Group Interaction Parameters, a_{mn}, in Kelvins (*Continued*)

Main group	n = 46	47	48	49	50
m = 1	390.9	553.3	187.0	216.1	92.99
2	200.2	268.1	−617.0	62.56	NA
3	NA	333.3	NA	−59.58	−39.16
4	NA	421.9	NA	−203.6	184.9
5	−382.7	−248.3	NA	104.7	57.65
6	NA	NA	37.63	−59.40	−46.01
7	835.6	139.6	NA	407.9	NA
8	NA	NA	NA	NA	1005
9	NA	37.54	NA	NA	−162.6
10	NA	NA	NA	NA	NA
11	NA	151.8	NA	NA	NA
12	NA	NA	NA	NA	NA
13	NA	NA	NA	NA	NA
14	NA	NA	NA	NA	NA
15	NA	NA	NA	NA	NA
16	NA	NA	NA	NA	NA
17	NA	NA	NA	NA	NA
18	NA	NA	NA	NA	−136.6
19	NA	16.23	NA	NA	NA
20	−322.3	NA	NA	NA	NA
21	NA	NA	NA	NA	NA
22	NA	361.1	NA	NA	NA
23	NA	NA	565.9	NA	NA
24	NA	423.1	63.95	NA	108.5
25	NA	434.1	NA	NA	NA
26	NA	NA	NA	NA	−4.565
27	NA	NA	NA	NA	NA
28	NA	NA	NA	NA	NA
29	NA	NA	−18.27	NA	NA
30	NA	NA	NA	NA	NA
					NA

Main group	n = 46	47	48	49	50
m = 31	NA	−353.5	NA	NA	NA
32	NA	NA	NA	NA	NA
33	NA	NA	NA	NA	NA
34	NA	NA	NA	NA	NA
35	NA	NA	NA	NA	NA
36	NA	NA	NA	NA	NA
37	NA	NA	2429	NA	NA
38	NA	NA	NA	NA	NA
39	NA	NA	NA	NA	NA
40	NA	122.4	NA	NA	NA
41	NA	NA	NA	NA	NA
42	NA	NA	NA	NA	NA
43	NA	NA	NA	NA	NA
44	NA	NA	NA	NA	NA
45	NA	NA	NA	NA	NA
46	0.0	0.0	NA	NA	NA
47	NA	NA	0.0	NA	NA
48	NA	NA	NA	NA	NA
49	NA	NA	NA	0.0	NA
50	NA	NA	NA	NA	0.0

solution Acetone has one ($\nu_1 = 1$) CH_3 group (main group 1, secondary group 1) and one ($\nu_9 = 1$) CH_3CO (main group 9, secondary group 18). *n*-Pentane has two ($\nu_1 = 2$) CH_3 groups (main group 1, secondary group 1), and three ($\nu_1 = 3$) CH_2 groups (main group 1, secondary group 2).

Based on the information in Table 8-23, we can construct the following table:

Molecule (i)	Group Identification					
	Name	Main No.	Sec. No.	$\nu_j^{(i)}$	R_j	Q_j
Acetone (1)	CH_3	1	1	1	0.9011	0.848
	CH_3CO	9	18	1	1.6724	1.488
n-Pentane (2)	CH_3	1	1	2	0.9011	0.848
	CH_2	1	2	3	0.6744	0.540

We can now write:

$$r_1 = (1)(0.9011) + (1)(1.6724) = 2.5735$$

$$q_1 = (1)(0.848) + (1)(1.488) = 2.336$$

$$\Phi_1 = \frac{(2.5735)(0.047)}{(2.5735)(0.047) + (3.8254)(0.953)} = 0.0321$$

$$\theta_1 = \frac{(2.336)(0.047)}{(2.336)(0.047) + (3.316)(0.953)} = 0.0336$$

$$l_1 = (5)(2.5735 - 2.336) - 1.5735 = -0.3860$$

or in tabular form:

Molecule (i)	r_i	q_i	100 Φ_i	100 θ_i	l_i
Acetone (1)	2.5735	2.336	3.21	3.36	−0.3860
n-Pentane (2)	3.8254	3.316	96.79	96.64	−0.2784

We can now calculate the combinatorial contribution to the activity coefficients:

$$\ln \gamma_1^C = \ln \frac{0.0321}{0.047} + (5)(2.336) \ln \frac{0.0336}{0.0321} - 0.3860$$
$$+ \frac{0.0321}{0.047} [(0.047)(0.3860) + (0.953)(0.2784)] = -0.0403$$

$$\ln \gamma_2^C = -0.0007$$

Next, we calculate the residual contributions to the activity coefficients. Since only two main groups are represented in this mixture, the calculation is relatively simple. The group interaction parameters, a_{mn}, are obtained from Table 8-24.

$$a_{1,9} = 476.40$$

$$\Psi_{1,9} = \exp\left(\frac{-476.40}{307}\right) = 0.2119$$

$$a_{9,1} = 26.760$$

$$\Psi_{9,1} = \exp\left(\frac{-26.760}{307}\right) = 0.9165$$

Note that $\Psi_{1,1} = \Psi_{9,9} = 1.0$, since $a_{1,1} = a_{9,9} = 0$. Let 1 = CH_3, 2 = CH_2, and 18 = CH_3CO.

Next, we compute $\Gamma_k^{(i)}$, the residual activity coefficient of group k in a reference solution containing only molecules of type i. For pure acetone (1), the mole fraction of group m, X_m, is

$$X_1^{(1)} = \frac{\nu_1^{(1)}}{\nu_1^{(1)} + \nu_{18}^{(1)}} = \frac{1}{1+1} = \frac{1}{2} \qquad X_{18}^{(1)} = \frac{1}{2}$$

Hence

$$\theta_1^{(1)} = \frac{\frac{1}{2}(0.848)}{\frac{1}{2}(0.848) + \frac{1}{2}(1.488)} = 0.363 \qquad \theta_{18}^{(1)} = 0.637$$

$$\ln \Gamma_1^{(1)} = 0.848\left\{1 - \ln[0.363 + (0.637)(0.9165)] - \left[\frac{0.363}{0.363 + (0.637)(0.9165)} + \frac{(0.637)(0.2119)}{(0.363)(0.2119) + 0.637}\right]\right\} = 0.409$$

$$\ln \Gamma_{18}^{(1)} = 1.488\left\{1 - \ln[(0.363)(0.2119) + 0.637] - \left[\frac{(0.363)(0.9165)}{0.363 + (0.637)(0.9165)} + \frac{0.637}{(0.363)(0.2119) + 0.637}\right]\right\} = 0.139$$

For pure n-pentane (2), the mole fraction of group m, X_m is

$$X_1^{(2)} = \frac{\nu_1^{(2)}}{\nu_1^{(2)} + \nu_2^{(2)}} = \frac{2}{2+3} = \frac{2}{5} \qquad X_2^{(2)} = \frac{3}{5}$$

Since only one mian group is in n-pentane (2),

$$\ln \Gamma_1^{(2)} = \ln \Gamma_2^{(2)} = 0.0$$

The group residual activity coefficients can now be calculated for $x_1 = 0.047$:

$$X_1 = \frac{(0.047)(1) + (0.953)(2)}{(0.047)(2) + (0.953)(5)} = 0.4019$$

$$X_2 = 0.5884 \qquad X_{18} = 0.0097$$

$$\theta_1 = \frac{(0.848)(0.4019)}{(0.848)(0.4019) + (0.540)(0.5884) + (1.488)(0.0097)} = 0.5064$$

$$\theta_2 = 0.4721 \qquad \theta_{18} = 0.0214$$

$$\ln \Gamma_1 = 0.848 \left\{ 1 - \ln[0.5064 + 0.4721 + (0.0214)(0.9165)] - \left[\frac{0.5064 + 0.4721}{0.5064 + 0.4721 + (0.0214)(0.9165)} + \frac{(0.0214)(0.2119)}{(0.5064 + 0.4721)(0.2119) + 0.0214} \right] \right\} = 1.45 \times 10^{-3}$$

$$\ln \Gamma_2 = 0.540 \left\{ 1 - \ln[0.5064 + 0.4721 + (0.0214)(0.9165)] - \left[\frac{0.5064 + 0.4721}{0.5064 + 0.4721 + (0.0214)(0.9165)} + \frac{(0.0214)(0.2119)}{(0.5064 + 0.4721)(0.2119) + 0.0214} \right] \right\} = 9.26 \times 10^{-4}$$

$$\ln \Gamma_{18} = 1.488 \left\{ 1 - \ln[(0.5064 + 0.4721)(0.2119) + 0.0214] - \left[\frac{(0.5064 + 0.4721)(0.9165)}{0.5064 + 0.4721 + (0.0214)(0.9165)} + \frac{0.0214}{(0.5064 + 0.4721)(0.2119) + 0.0214} \right] \right\} = 2.21$$

The residual contributions to the activity coefficients follow

$$\ln \gamma_1^R = (1)(1.45 \times 10^{-3} - 0.409) + (1)(2.21 - 0.139) = 1.66$$

$$\ln \gamma_2^R = (2)(1.45 \times 10^{-3} - 0.0) + (3)(9.26 x 10^{-4} - 0.0) = 5.68 \times 10^{-3}$$

Finally, we calculate the activity coefficients:

$$\ln \gamma_1 = \ln \gamma_1^C + \ln \gamma_1^R = -0.0403 + 1.66 = 1.62$$

$$\ln \gamma_2 = \ln \gamma_2^C + \ln \gamma_2^R = -0.0007 + 5.68 \times 10^{-3} = 4.98 \times 10^{-3}$$

Hence,

$$\gamma_1 = 5.07 \qquad \gamma_2 = 1.01$$

Retaining more significant figures in the values for ϕ and θ, as would be the case if calculations were done on a computer, leads to slightly different answers. In this case, $\ln \gamma_1^c = -0.0527$, $\ln \gamma_2^c = -0.0001$, $\gamma_1 = 4.99$, and $\gamma_2 = 1.005$.

The corresponding experimental values of Lo, et al. (1962) are:

$$\gamma_1 = 4.41 \qquad\qquad \gamma_2 = 1.11$$

Although agreement with experiment is not as good as we might wish, it is not bad and it is representative of what UNIFAC can do. The main advantage of UNIFAC is its wide range of application for vapor-liquid equilibria of nonelectrolyte mixtures.

Because UNIFAC is so popular in the chemical (and related) industries, and because it is used so widely to estimate vapor-liquid equilibria for a large variety of mixtures, five illustrative examples are presented to indicate both the power and the limitations of the UNIFAC method.

Example 8-13 Using UNIFAC for liquid-phase activity coefficients, calculate vapor-liquid equilibria *Pxy* for the binary methanol (1)–water (2) at 50 and 100°C.

solution

a. $t = 50°\text{C}$

The two governing equations at equilibrium are

$$y_i \varphi_i P = x_i\, \gamma_i\, \phi_i^s\, P_{\text{vp}i}(PC)_i \qquad (i = 1, 2)$$

where subscript vp*i* stands for vapor pressure of component *i*; y_i and x_i are, respectively, vapor-phase and liquid-phase mole fractions of component *i*; *P* is the equilibrium total pressure, φ_i and φ_i^s are fugacity coefficient in the mixture and pure-component fugacity coefficient at saturation, respectively; γ_i is the activity coefficient. $(PC)_i$ is the Poynting factor, given by:

$$(PC)_i = \exp\left[\frac{1}{R(t + 273.15)}\, V_i^L(P - P_{\text{vp}i})\right] \qquad (i = 1, 2)$$

where *R* is the universal gas constant; $t = 50°\text{C}$; and V_i^L is the molar volume of pure liquid *i* at 50°C.

The pure-component vapor pressure $P_{\text{vp}i}$ is from McGlashan and Williamson (1976), the vapor pressures are 417.4 mmHg for methanol and 92.5 mmHg for water.

Because both vapor pressures are much less than atmospheric (760 mmHg), the vapor phase can be assumed to be ideal. Hence, all fugacity coefficients are set equal to unity. Further, both Poynting factors are assumed to be unity. The equilibrium relations reduce to:

$$y_i P = x_i P_{\text{vp}i} \gamma_i \qquad (i = 1, 2)$$

where activity coefficient γ_i is calculated from UNIFAC at 50°C.

To obtain phase equilibria at 50°C:

Step 1. Assign liquid-phase mole fraction of methanol (x_1) from 0 to 1 with intervals of 0.1. At each chosen x_1, we have two equations with two unknowns: *P* and y_1; x_2 and y_2 are not independent variables because they follow from material balances $x_1 + x_2 = 1$ and $y_1 + y_2 = 1$.

Step 2. At each x_1, use UNIFAC to calculate the activity coefficients for both components by the procedure described in Example 8-12.

In this particular example, the two components are not broken into groups because, in UNIFAC, methanol and water are themselves distinct groups.

Group-volume (R_k) and surface-area (Q_k) parameters are

Group	R_k	Q_k
CH_3OH	1.4311	1.432
H_2O	0.9200	1.400

Group-group interaction parameters used in the calculation of the residual activity coefficient (K) are

Group	CH_3OH	H_2O
CH_3OH	0	−181.0
H_2O	289.6	0

Step 3. Calculate the total pressure from:

$$P = (y_1 + y_2)P = x_1\gamma_1 P_{\text{vp1}} + x_2\gamma_2 P_{\text{vp2}}$$

with γ_i from Step 2.

Step 4. Evaluate vapor-phase mole fraction y_1 by:

$$y_1 = \frac{x_1\gamma_1 P_{\text{vp1}}}{P}$$

with γ_1 from Step 2 and P from Step 3.

Step 5. Return to Step 2 with the next value of x_1.

Following these steps leads to the results shown in Table 8-25. Also shown are experimental data at the same temperature from McGlashan and Williamson (1976). Figure 8-11 compares calculated and observed results.

In this example, UNIFAC calculations are in excellent agreement with experiment.

b. t = 100°C

The experimental pure-component vapor pressures are 2650.9 mmHg for methanol and 760 mmHg for water (Griswold and Wong, 1952). Following the same procedure illustrated in part a. but with t = 100°C, leads to the results in Table 8-26. Experimental data are from Griswold and Wong (1952). Figure 8-12 shows VLE data for the mixture at 100°C.

At 100°C, Fig. 8-12 shows good agreement between calculated and experimental results. But agreement at 100°C is not as good as that at 50°C.

This first UNIFAC example is particularly simple because the molecules, methanol and water, are themselves groups. In a sense, therefore, this example does not provide a real test for UNIFAC whose main idea is to substitute an assembly of groups for a particular molecule. Nevertheless, this UNIFAC example serves to introduce the general procedure. The next examples, where the molecules are subdivided into groups, provide a more realistic test for the UNIFAC method.

TABLE 8-25 Vapor-liquid Equilibria for Methanol (1)–Water (2) at 50°C

Calculated			Experimental		
x_1	y_1	P(mmHg)	x_1	y_1	P(mmHg)
0.0	0.00	92.5	0.0000	0.0000	92.50
0.1	0.47	158.4	0.0453	0.2661	122.73
0.2	0.62	201.8	0.0863	0.4057	146.74
0.3	0.70	235.1	0.1387	0.5227	174.21
0.4	0.76	263.6	0.1854	0.5898	194.62
0.5	0.81	289.8	0.3137	0.7087	239.97
0.6	0.85	315.0	0.4177	0.7684	266.99
0.7	0.89	339.9	0.5411	0.8212	298.44
0.8	0.93	365.0	0.6166	0.8520	316.58
0.9	0.96	390.5	0.7598	0.9090	352.21
1.0	1.00	417.4	0.8525	0.9455	376.44
			0.9514	0.9817	403.33
			1.0000	1.0000	417.40

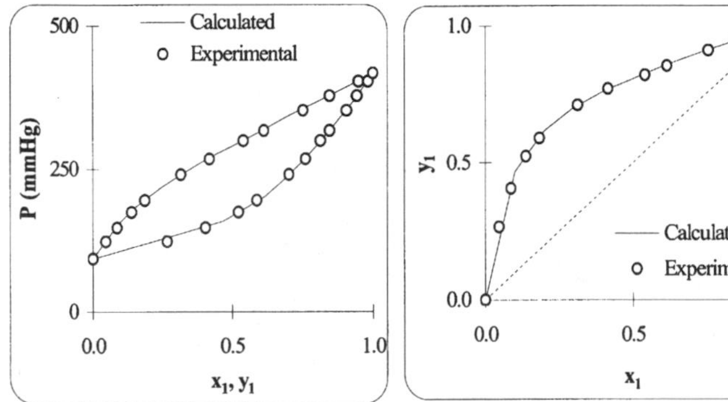

FIGURE 8-11 *Pxy* and *yx* diagrams for methanol (1)–water (2) at 50°C.

Example 8-14 Using UNIFAC for liquid-phase activity coefficients, calculate vapor-liquid equilibria *txy* for the binary 2,3-dimethylbutane (1)–chloroform (2) at 760 mmHg.

solution The procedure for solving this problem is similar to that discussed in Example 8-13. However, we are now given the equilibrium total pressure and need to calculate the temperature. At 760 mmHg, the gas phase is assumed to be ideal. Further, both Poynting factors are set to unity. The two equations of equilibrium are

$$y_i P = x_i \gamma_i P_{\text{vp}i} \qquad (i = 1, 2)$$

where all symbols are defined in Example 8-13.

Because the pure-component vapor pressures, as well as the liquid-phase activity coefficients, depend on temperature, solution for this problem requires iteration.

The activity coefficient γ_i is calculated from UNIFAC. The temperature dependence of vapor pressure $P_{\text{vp}i}$ is expressed by Antoine's equation:

TABLE 26 Vapor-Liquid Equilibrium for Methanol (1)–Water (2) at 100°C

Calculated			Experimental		
x_1	y_1	P(mmHg)	x_1	y_1	P(mmHg)
0.0	0.000	760.0	0.000	0.000	760.0
0.1	0.410	1173.7	0.011	0.086	827.6
0.2	0.561	1443.6	0.035	0.191	931.0
0.3	0.648	1645.6	0.053	0.245	1003.2
0.4	0.712	1813.6	0.121	0.434	1235.8
0.5	0.765	1964.3	0.281	0.619	1536.0
0.6	0.813	2106.1	0.352	0.662	1624.1
0.7	0.860	2243.7	0.522	0.750	1882.5
0.8	0.906	2379.6	0.667	0.824	2115.1
0.9	0.952	2515.1	0.826	0.911	2337.8
1.0	1.000	2650.9	0.932	0.969	2508.0
			1.000	1.000	2650.9

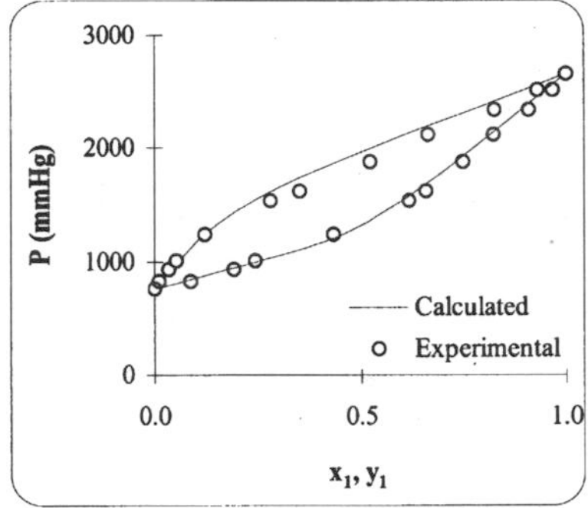

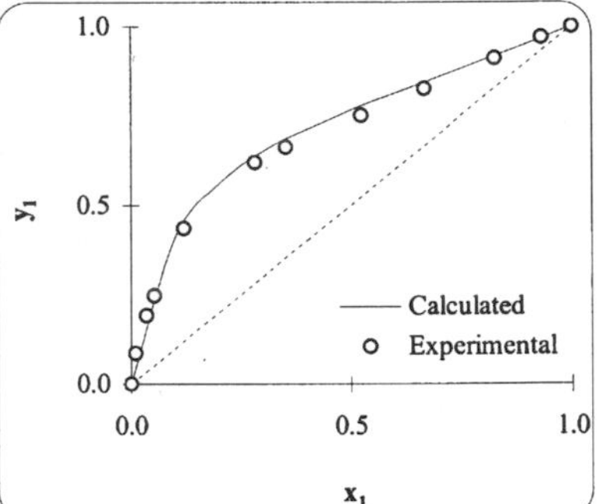

FIGURE 8-12 *Pxy* and *yx* diagrams for methanol (1)–water (2) at 100°C.

$$\log_{10} P_{vpi} = A_i - \frac{B_i}{t + C_i} \qquad (i = 1, 2)$$

where P_{vpi} is in mmHg and t is in °C.

Coefficients A, B, and C are given by Willock and van Winkle (1970):

Component	A	B	C
1	6.8161	1130.7	229.32
2	7.0828	1233.1	232.20

The iteration procedure is as follows:

Step 1. Assign liquid-phase mole fraction x_1 from 0 to 1 with intervals of 0.1. At each chosen x_1, we have two equations with two unknowns: t and y_1. Again, x_2 and y_2 are not independent variables because they follow from material balances $x_1 + x_2 = 1$ and $y_1 + y_2 = 1$.

Step 2. At each x_1, make a crude initial guess of temperature t (°C):

$$t = x_1 t_1 + x_2 t_2$$

where t_1 and t_2 are, respectively, calculated from the Antoine equation at 760 mmHg.

Step 3. Use initial-guess t from Step 2 to calculate the corresponding pure-component vapor pressures with the Antoine equation. We denote these as P^0_{vp1} and P^0_{vp2} for components 1 and 2, respectively (superscript 0 denotes the initial guess from Step 3).

Step 4. Compute both activity coefficients from UNIFAC (see Example 8-12), using t from Step 2. For this example, the two components are broken into groups as follows:

Component	Constitutive groups
1	$4CH_3 + 2CH$
2	$CHCl_3$

While the molecules of component 1 are broken into six groups, component 2 is itself a group. Group-volume (R_k) and surface-area (Q_k) parameters are

Group	R_k	Q_k
CH_3	0.9011	0.848
CH	0.4469	0.228
$CHCl_3$	2.8700	2.410

Group-group interaction parameters (K) are

Group	CH_3	CH	$CHCl_3$
CH_3	0	0	24.90
CH	0	0	24.90
$CHCl_3$	36.70	36.70	0

Step 5. Recalculate the vapor pressure for component 2 (because component 2 is more volatile) from:

$$P_{\text{vp2}} = \frac{P^0_{\text{vp2}} P}{\sum_i x_i \gamma_i P^0_{\text{vp}i}}$$

where $P = 760$ mmHg; P^0_{vp1} and P^0_{vp2} are from Step 3; γ_1 and γ_2 are from Step 4.

Step 6. Recalculate the equilibrium temperature from:

$$t = \frac{B_2}{A_2 - \log_{10} P_{vp2}} - C_2$$

P_{vp2} is from Step 5; coefficients A_2, B_2, and C_2 are given above.

Step 7. Compare this new t with the earlier one. If the absolute value of the difference is less than or equal to a small value ε (here we choose $\varepsilon = 0.1°C$), go to Step 8. Otherwise, go back to Step 3 using the new t in place of the t used previously. Repeat until the difference in the temperatures is less than or equal to ε.

Step 8. Calculate P_{vp1} with the Antoine equation using the final t from Step 7. Finally, the vapor-phase mole fraction y_1 is given by:

$$y_1 = \frac{x_1 \gamma_1 P_{vp1}}{P}$$

Step 9. Return to Step 2 with the next value of x_1.

Calculated and experimental results at 760 mmHg are shown in Table 8-27. Experimental data are from Willock and van Winkle (1970). Figure 8-13 compares calculated and observed results.

TABLE 8-27 Vapor-Liquid Equilibrium for 2,3-dimethylbutane (1)–chloroform (2) at 760 mmHg

Calculated			Experimental		
x_1	y_1	t(°C)	x_1	y_1	t(°C)
0.0	0.000	61.3	0.087	0.130	59.2
0.1	0.163	58.7	0.176	0.230	28.1
0.2	0.268	57.5	0.275	0.326	57.0
0.3	0.352	56.7	0.367	0.406	56.5
0.4	0.430	56.3	0.509	0.525	56.0
0.5	0.508	56.1	0.588	0.588	56.0
0.6	0.590	56.1	0.688	0.671	56.1
0.7	0.678	56.3	0.785	0.760	56.5
0.8	0.774	56.7	0.894	0.872	57.0
0.9	0.881	57.3			
1.0	1.000	58.0			

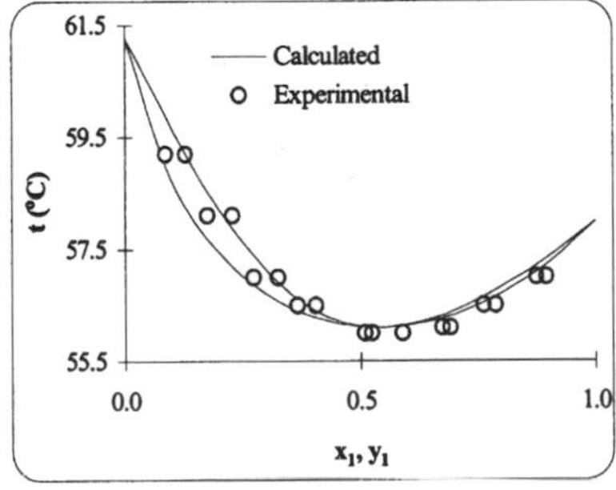

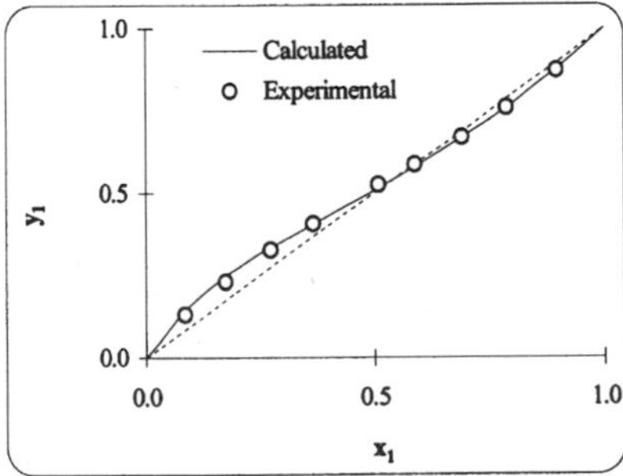

FIGURE 8-13 *txy* and *yx* plots for 2,3-dimethylbutane (1)–chloroform (2) at 760 mmHg.

Agreement of calculated and observed vapor-liquid equilibria is fair. The azeotropic temperature and composition, however, are in good agreement. It is particularly difficult to fit this system with high accuracy because, although the system shows an azeotrope, nonideality (as indicated by how much infinite-dilution activity coefficients deviate from unity) is small.

Example 8-15 Using UNIFAC, calculate vapor-liquid equilibria for the ternary acetone (1)–methanol (2)–cyclohexane (3) at 45°C.

solution The three governing equations at equilibrium are

$$y_i P = x_i \gamma_i P_{\text{vp}i} \qquad (i = 1, 2, 3)$$

where all symbols are defined in Example 8-13. Because the total pressure is low, the Poynting factors as well as corrections for vapor-phase nonideality have been neglected.

To obtain phase equilibria at 45°C:

Step 1. Assign values of x_1 and x_2 from 0 to 1. To facilitate comparison with experiment, we choose sets (x_1, x_2) identical to those used by Marinichev and Susarev (1965). At each set (x_1, x_2), we have three equations with three unknowns: P, y_1, and y_2. Mole fractions x_3 and y_3 are not independent variables because they are constrained by material balances $x_1 + x_2 + x_3 = 1$ and $y_1 + y_2 + y_3 = 1$.

Step 2. At each set (x_1, x_2), use UNIFAC to calculate the activity coefficients for the three components at 45°C.

The three molecules are broken into groups as follows:

Component	Constitutive groups
1	$CH_3 + CH_3CO$
2	CH_3OH
3	$6CH_2$

Group-volume (R_k) and surface-area (Q_k) parameters are

Group	R_k	Q_k
CH_3	0.9011	0.848
CH_2	0.6744	0.540
CH_3CO	1.6724	1.448
CH_3OH	1.4311	1.432

Group-group interaction parameters (in K) are

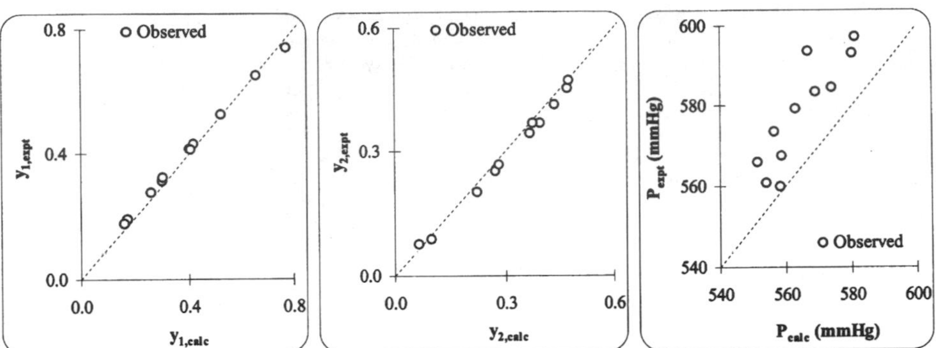

FIGURE 8-14 Vapor-liquid-equilibrium data for acetone (1)–methanol (2)–cyclohexane (3) at 45°C.

Group	CH_3	CH_2	CH_3CO	CH_3OH
CH_3	0	0	476.4	697.2
CH_2	0	0	476.4	679.2
CH_3CO	26.76	26.76	0	108.7
CH_3OH	16.51	16.51	23.39	0

Step 3. Calculate the equilibrium total pressure from:

$$P = (y_1 + y_2 + y_3)P = x_1\gamma_1P_{vp1} + x_2\gamma_2P_{vp2} + x_3\gamma_3P_{vp3}$$

with γ_i from Step 2.

Step 4. Evaluate mole fractions y_1 and y_2 from:

$$y_1 = \frac{x_1\gamma_1P_{vp1}}{P}$$

$$y_2 = \frac{x_2\gamma_2P_{vp2}}{P}$$

with γ_i from Step 2 and P from Step 3.

Step 5. Return to Step 2 with the next set (x_1, x_2).

Following these steps, Table 8-28 gives calculated results. Also shown are experimental data at the same temperature from Marinichev and Susarev (1965). Figure 8-14 compares calculated and observed results.

While UNIFAC gives a good representation of vapor-phase mole fractions, there is appreciable error in the total pressure.

Example 8-16 Using UNIFAC, calculate vapor-liquid equilibria *txy* for the ternary acetone (1)–2-butanone (2)–ethyl acetate (3) at 760 mmHg.

TABLE 8-28 Vapor-Liquid Equilibrium Data for Acetone (1)–Methanol (2)–Cyclohexane (3) at 45°C

		Calculated			Experimental			(Calculated-Experimental)		
x_1	x_2	y_1	y_2	P (mmHg)	y_1	y_2	P (mmHg)	Δy_1	Δy_2	ΔP (mmHg)
0.117	0.127	0.254	0.395	558	0.276	0.367	560	−0.022	0.028	−1.6
0.118	0.379	0.169	0.470	559	0.191	0.452	568	−0.022	0.018	−8.9
0.123	0.631	0.157	0.474	554	0.176	0.471	561	−0.019	0.003	−6.8
0.249	0.120	0.400	0.272	567	0.415	0.252	594	−0.016	0.020	−27.0
0.255	0.369	0.296	0.375	574	0.312	0.367	585	−0.016	0.008	−10.8
0.250	0.626	0.299	0.435	551	0.325	0.412	566	−0.026	0.023	−14.6
0.382	0.239	0.418	0.281	580	0.433	0.267	593	−0.015	0.014	−13.0
0.379	0.497	0.405	0.367	563	0.414	0.343	579	−0.009	0.024	−16.3
0.537	0.214	0.521	0.222	581	0.526	0.202	597	−0.005	0.020	−16.5
0.669	0.076	0.656	0.098	569	0.654	0.088	584	0.002	0.010	−14.7
0.822	0.054	0.772	0.064	557	0.743	0.076	574	0.029	−0.012	−17.2
							Stdv.	**0.015**	**0.011**	**6.5**

solution The three equations of equilibrium are

$$y_i P = x_i \gamma_i P_{\text{vp}i} \qquad (i = 1, 2, 3)$$

where all symbols are defined in Example 8-13. Simplifying assumptions are the same as those in Example 8-15.

From Gmehling, et al. (1979a), coefficients in Antoine's equation (see Example 8-14) are

Component	A	B	C
1	7.117	1210.6	229.7
2	7.064	1261.3	222.0
3	7.102	1245.0	217.9

To obtain phase equilibria at 760 mmHg, we use an iteration procedure (Step 1 to Step 9) similar to that in Example 8-12.

Step 1. Assign values of x_1 and x_2 from 0 to 1. Similar to Example 8-15, we choose sets (x_1, x_2) identical to those used by Babich, et al. (1969). At each set (x_1, x_2), we have three equations with three unknowns: t, y_1, and y_2.

At each set (x_1, x_2), the iteration is as follows:

Step 2. Make an initial guess of the temperature t (°C):

$$t = x_1 t_1 + x_2 t_2 + x_3 t_3$$

where t_1, t_2, and t_3 are calculated from Antoine's equation using $P_{\text{vp}i} = 760$ mmHg.

$$t_i = \frac{B_i}{A_i - \log_{10} P_{\text{vp}i}} - C_i \qquad (i = 1, 2, 3)$$

Step 3. Calculate the three pure-component vapor pressures (we denote these as $P^0_{\text{vp}i}$; superscript 0 refers the initial guess from Step 3) from Antoine's equation using t from Step 2.

Step 4. Compute the liquid-phase activity coefficients from UNIFAC (see Example 8-12) at $T = t + 273.15$.

Molecules of the three components are broken into groups as follows:

Component	Constitutive groups
1	$CH_3 + CH_3CO$
2	$CH_3 + CH_2 + CH_3CO$
3	$CH_3 + CH_2 + CH_3COO$

Group-volume (R_k) and surface-area (Q_k) parameters are

Group	R_k	Q_k
CH_3	0.9011	0.848
CH_2	0.6744	0.540
CH_3CO	1.6724	1.448
CH_3COO	1.9031	1.728

Group-group interaction parameters (in K) are

Group	CH_3	CH_2	CH_3CO	CH_3COO
CH_3	0	0	476.4	232.1
CH_2	0	0	476.4	232.1
CH_3CO	26.76	26.76	0	−213.7
CH_3COO	114.8	114.8	372.2	0

Step 5. Because component (1) is the most volatile, recalculate P_{vp1} from:

$$P_{vp1} = \frac{P(P^0_{vp1})}{\sum_i^3 x_i \gamma_i P^0_{vpi}}$$

where P = 760 mmHg. P^0_{vpi} is from Step 3, and γ_i is from Step 4.

Step 6. Compute the new value of t based on P_{vp1} from Step 5:

$$t = \frac{B_1}{A_1 - \log_{10} P_{vp1}} - C_1$$

Step 7. Compare this new t with the previous one. If the absolute value of the difference is less than or equal to a small number ε (here we choose ε = 0.1°C), go to Step 8. Otherwise, go back to Step 3 using the new t in place of the one used previously. Repeat until the difference in the temperatures is less than or equal to ε.

Step 8. Calculate the vapor pressure of component 2 from Antoine's equation using the new t (°C) from Step 7:

$$\log_{10} P_{vp2} = A_2 - \frac{B_2}{t + C_2}$$

Compute the vapor-phase mole fractions:

$$y_i = \frac{x_i \gamma_i P_{vpi}}{P} \qquad (i = 1, 2)$$

Step 9. Return to Step 2 with a new set (x_1, x_2).

Table 8-29 gives calculated and experimental results at 760 mmHg. Experimental data are from Babich, et al. (1969).

TABLE 8-29 Vapor-Liquid Equilibria for Acetone (1)–2-Butanone (2)–Ethyl acetate (3) at 760 mmHg

		Calculated			Experimental			(Calculated-Experimental)		
x_1	x_2	y_1	y_2	t (°C)	y_1	y_2	t (°C)	Δy_1	Δy_2	Δt (°C)
0.200	0.640	0.341	0.508	72.5	0.290	0.556	72.6	0.051	−0.048	−0.1
0.400	0.480	0.583	0.322	67.5	0.525	0.370	67.6	0.058	−0.048	−0.1
0.600	0.320	0.761	0.185	63.2	0.720	0.215	63.5	0.041	−0.031	−0.3
0.800	0.160	0.896	0.081	59.4	0.873	0.095	59.8	0.023	−0.014	−0.4
0.200	0.480	0.336	0.374	71.8	0.295	0.420	71.8	0.041	−0.046	0.0
0.400	0.360	0.576	0.238	67.1	0.535	0.285	67.3	0.041	−0.047	−0.3
0.600	0.240	0.755	0.137	63.0	0.725	0.170	63.3	0.030	−0.033	−0.3
0.800	0.120	0.893	0.060	59.3	0.880	0.075	59.6	0.013	−0.015	−0.3
0.200	0.320	0.334	0.248	71.2	0.302	0.276	71.3	0.032	−0.029	−0.1
0.400	0.240	0.571	0.157	66.6	0.540	0.180	67.0	0.031	−0.023	−0.4
0.600	0.160	0.750	0.091	62.7	0.729	0.105	62.9	0.021	−0.014	−0.2
0.200	0.160	0.339	0.125	70.7	0.295	0.145	71.1	0.044	−0.020	−0.4
0.400	0.120	0.570	0.078	66.2	0.530	0.095	66.5	0.040	−0.017	−0.3
0.600	0.080	0.746	0.045	62.5	0.720	0.095	62.7	0.026	−0.050	−0.2
0.800	0.040	0.887	0.020	59.1	0.873	0.024	59.6	0.014	−0.004	−0.4
							Stdv.	**0.013**	**0.015**	**0.1**

Figure 8-15 compares calculated and observed results. Agreement is good. However, we must not conclude from this example that agreement will necessarily be equally good for other systems.

Example 8-17 Peres and Macedo (1997) studied mixtures containing D-glucose, D-fructose, sucrose, and water. From these studies, they proposed new groups for the modified UNIFAC method (Larsen, et al., 1987) to calculate vapor-liquid equilibria (VLE) and solid-liquid equilibria (SLE) for these mixtures.

In this example, the modified UNIFAC method is used to calculate vapor-liquid equilibria for the ternary D-glucose (1)–sucrose (2)–water (3) at 760 mmHg.

solution Because the vapor phase contains only water, and because the pressure is low, the vapor phase can be assumed to be ideal. At equilibrium,

$$P = x_3 \gamma_3 P_{vp3}$$

where P and P_{vp3} are total pressure and water vapor pressure, respectively; x is liquid-phase mole fraction; γ is the liquid-phase activity coefficient.

The temperature dependence of P_{vp3} is expressed by Antoine's equation:

$$\log P_{vp3} - A_3 - \frac{B_3}{t + C_3}$$

where P_{vp3} is in mmHg and t is in °C.

From Gmehling, et al. (1981), Antoine parameters are

	A_3	B_3	C_3
water	8.071	1730.6	233.4

In the modified UNIFAC, the combinatorial part of the activity coefficient is calculated from Larsen, et al. (1987).

$$\ln \gamma_i^C = \ln \left(\frac{\varphi_i}{x_i} \right) + 1 - \frac{\varphi_i}{x_i}$$

where x_i is mole fraction of component i.

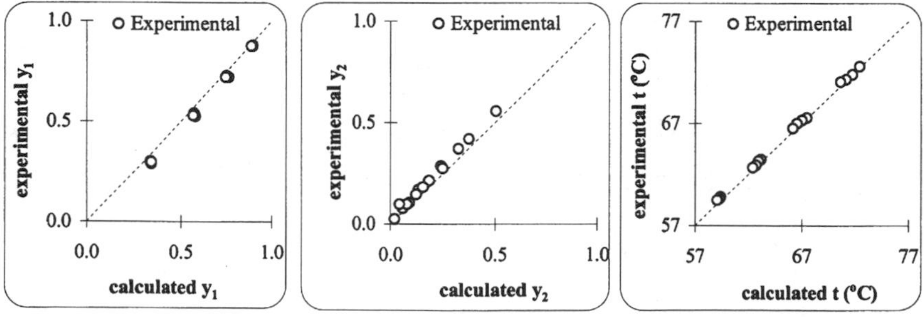

FIGURE 8-15 VLE for acetone (1)–2-butanone (2)–ethyl acetate (3) at 760 mm Hg.

Volume fraction φ_i of component i is defined as:

$$\varphi_i = \frac{x_i r_i^{2/3}}{\sum_{j=1}^{3} x_j r_j^{2/3}}$$

The volume parameter r_i is calculated from:

$$r_i = \sum_k R_k \nu_k^{(i)}$$

where R_k is the volume parameter of group k; $\nu_k^{(i)}$ is the number of times that group k appears in molecule i.

The residual activity coefficient is calculated as in the original UNIFAC (see Example 8-12).

To break the two sugars into groups, Peres and Macedo (1997) used the three groups proposed by Catte, et al. (1995): pyranose ring (PYR),[1] furanose ring (FUR),[1] and osidic bond (—O—), that is the ether bond connecting the PYR and FUR rings in sucrose. Further, because there are many OH groups in D-glucose and sucrose and because these OH groups are close to one another, their interactions with other groups are different from those for the usual alcohol group. Peres and Macedo (1997) proposed a new OH_{ring} group.

The two sugars and water are broken into groups as follows:

Component	Constitutive groups
D-glucose	CH_2 + PYR + 5OH_{ring}
sucrose	3CH_2 + PYR + FUR + (—O—) + 8OH_{ring}
water	H_2O

Group-volume and surface area parameters are

Group	R_k	Q_k
CH_2	0.6744	0.5400
PYR	2.4784	1.5620
FUR	2.0315	1.334
(—O—)	1.0000	1.200
OH_{ring}	0.2439	0.442
H_2O	0.9200	1.400

1

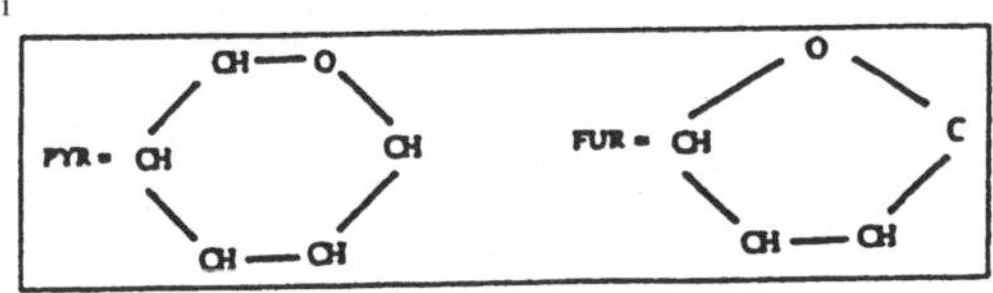

TABLE 8-30 Vapor-Liquid Equilibria for D-glucose (1)–Sucrose (2)–Water (3) at 760 mmHg

x_1	x_2	Expt. T (°C)	Calc. T (°C)
0.0014	0.0147	101.0	100.2
0.0023	0.0242	102.0	100.6
0.0036	0.0390	103.0	101.2
0.0054	0.0579	104.0	102.2
0.0068	0.0714	105.0	103.0
0.0075	0.0803	105.5	103.5
0.0098	0.1051	107.0	105.2
0.0150	0.1576	108.5	109.4
0.0167	0.1756	107.0	111.0

Group-group interaction parameters (in K) suggested by Peres and Macedo (1997)

	CH_2	PYR	FUR	(—O—)	OH_{ring}	H_2O
CH_2	0	0	0	0	0	0
PYR	0	0	0	0	0	−43.27
FUR	0	0	0	0	0	−169.23
(—O—)	0	0	0	0	0	0
OH_{ring}	0	0	0	0	0	591.93
H_2O	0	−599.04	−866.91	0	−102.54	0

For any given set (x_1, x_2), we first substitute γ_3 into the equilibrium equation and solve for P_{vp3}. We then use this P_{vp3} in the Antoine equation and solve for T. Mole fraction x_3 is obtained by mass balance $x_1 + x_2 + x_3 = 1$.

Table 8-30 shows calculated boiling temperatures. Experimental data are from Abderafi and Bounahmidi (1994).

Figure 8-16 compares calculated and experimental results.

8-11 PHASE EQUILIBRIUM WITH HENRY'S LAW

Although the compositions of liquid mixtures may span the entire composition range from dilute up to the pure component, many multiphase systems contain compounds only in the dilute range ($x_i < 0.1$). This is especially true for components where the system T is above their critical T_c (gases) or where their pure-component vapor pressure, P_{vp}, is well above the system pressure. Liquid-liquid systems also often do not span the entire composition range. In such cases, the thermodynamic description of Secs. 8-2 to 8-9 using the pure-component standard state may not be most convenient. This section describes methods based on the Henry's law standard state. Details are given by Prausnitz, et al. (1999).

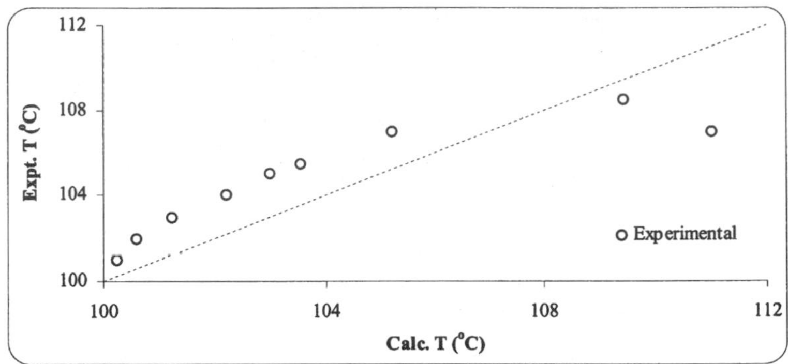

FIGURE 8-16 Vapor-liquid equilibria for D-glucose (1)–sucrose (2)–water (3) at 760 mm Hg.

Solubility of Gases

At modest pressures, most gases are only sparingly soluble in typical liquids. For example, at 25°C and a partial pressure of 1.01 bar, the (mole fraction) solubility of nitrogen in cyclohexane is $x = 7.6 \times 10^{-4}$ and that in water is $x = 0.18 \times 10^{-4}$. Although there are some exceptions (notably, hydrogen), the solubility of a gas in typical solvents usually falls with rising temperature. However, at higher temperatures, approaching the critical temperature of the solvent, the solubility of a gas usually rises with temperature, as illustrated in Fig. 8-17.

Experimentally determined solubilities have been reported in the chemical literature for over 100 years, but many of the data are of poor quality. Although no truly comprehensive and critical compilation of the available data exists, Table 8-31 gives some useful data sources.

Unfortunately, a variety of units has been employed in reporting gas solubilities. The most common of these are two dimensionless coefficients: *Bunsen coefficient,* defined as the volume (corrected to 0°C and 1 atm) of gas dissolved per unit volume of solvent at system temperature T when the partial pressure (mole fraction times total pressure, yP) of the solute is 1 atm; *Ostwald coefficient,* defined as the volume of gas at system temperature T and partial pressure p dissolved per unit volume of solvent. If the solubility is small and the gas phase is ideal, the Ostwald and Bunsen coefficients are simply related by

$$\text{Ostwald coefficient} = \frac{T}{273} \text{(Bunsen coefficient)}$$

where T is in kelvins. Adler (1983), Battino (1971, 1974 and 1984), Carroll (1999), and Friend and Adler (1957) have discussed these and other coefficients for expressing solubilities as well as some of their applications for engineering calculations.

These coefficients are often found in older articles. In recent years it has become more common to report solubilities in units of mole fraction when the solute partial pressure is 1 atm or as Henry's constants. Gas solubility is a case of phase equilibrium where Eq. (8-2.1) holds. We use Eq. (8-2.2) for the gas phase, mostly dominated by the normally supercritical solute (2), but for the liquid dominated by one or more subcritical solvents (1, 3, . . .) since x_2 is small, Eq. (8-2.3) is not

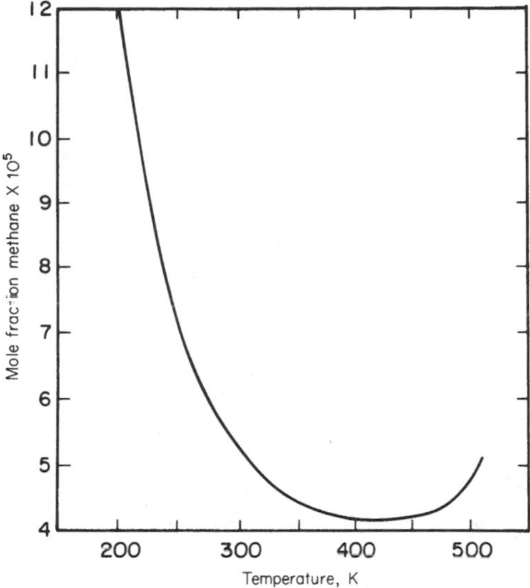

FIGURE 8-17 Solubility of methane in *n*-heptane when vapor phase fugacity of methane is 0.0101 bar (*Prausnitz, et al., 1999*)

convenient. As a result, instead of using an ideal solution model based on Raoult's law with standard-state fugacity at the pure-component saturation condition, we use the Henry's law ideal solution with a standard-state fugacity based on the infinitely dilute solution. Henry's law for a binary system need not assume an ideal gas phase; we write it as

$$y_2 \phi_2^V P = x_2 H_{2,1} \tag{8-11.1}$$

Subscript 2,1 indicates that Henry's constant H is for solute 2 in solvent 1. Then, Henry's constant in a single solvent is rigorously defined by a limiting process

$$H_{2,1} = \lim_{x_2 \to 0} \left(\frac{y_2 \phi_2^V P}{x_2} \right) \tag{8-11.2}$$

where in the limit, $P = P_{vp1}$. When $\phi_2^V = 1$, and when the total pressure is low, as in Fig. 8-18, Henry's constant is proportional to the inverse of the solubility. $H_{2,1}$ depends only on T, but often strongly, as the figure shows (see also Prausnitz, et al., 1999).

If total pressure is not low, some changes are needed in Eq. (8-11.2). High pressure is common in gas-liquid systems. The effects of pressure in the vapor are accounted for by ϕ_i^V, while for the liquid the effect of pressure is in the Poynting factor as in Eq. (8-4.2) which contains the partial molar volume. For typical dilute solutions, this is close to the infinite dilution value, $\overline{V}_{2,1}^{\infty}$. However, in addition, an effect of liquid nonideality can often occur because as P increases at constant T, so must x_2. This nonideality is taken into account by an activity coefficient γ_2^* which is usually less than unity and has the limit

TABLE 8-31 Some Sources for Solubilities of Gases in Liquids

Washburn, E. W. (ed.): *International Critical Tables,* McGraw-Hill, New York, 1926.

Markam, A. E., and K. A. Kobe: *Chem. Rev.,* **28:** 519 (1941).

Seidell, A.: *Solubilities of Inorganic and Metal-Organic Compounds,* Van Nostrand, New York, 1958, and *Solubilities of Inorganic and Organic Compounds, ibid.,* 1952.

Linke, W. L.: *Solubilities of Inorganic and Metal-Organic Compounds,* 4th ed., Van Nostrand, Princeton, N.J., 1958 and 1965, Vols. 1 and 2. (A revision and continuation of the compilation originated by A. Seidell.)

Stephen, H., and T. Stephen: *Solubilities of Inorganic and Organic Compounds,* Vols. 1 and 2, Pergamon Press, Oxford, and Macmillan, New York, 1963 and 1964.

Battino, R., and H. L. Clever: *Chem. Rev.,* **66:** 395 (1966).

Wilhelm, E., and R. Battino: *Chem. Rev.,* **73:** 1 (1973).

Clever, H. L., and R. Battino: "The Solubility of Gases in Liquids," in M. R. J. Dack (ed.), *Solutions and Solubilities,* Vol. 8, Part 1, Wiley, New York, 1975, pp. 379–441.

Kertes, A. S., O. Levy, and G. Y. Markovits: "Solubility," in B. Vodar (ed.), *Experimental Thermodynamics of Nonpolar Fluids,* Vol. II, Butterworth, London, 1975, pp. 725–748.

Gerrard, W.: *Solubility of Gases and Liquids,* Plenum, New York, 1976.

Landolt-Börnstein: 2. Teil, Bandteil b, *Lösungsgleichgewichte I,* Springer, Berlin, 1962; IV. Band, Technik, 4. Teil, Wärmetechnik; Bandteil c, *Gleichgewicht der Absorption von Gasen in Flüssigkeiten, ibid.,* 1976.

Wilhelm, E., R. Battino, and R. J. Wilcock: *Chem. Rev.,* **77:** 219 (1977).

Gerrard, W.: *Gas Solubilities, Widespread Applications,* Pergamon, Oxford, 1980.

Battino, R., T. R. Rettich, and T. Tominaga: *J. Phys. Chem. Ref. Data,* **12:** 163 (1963).

Wilhelm, E.: *Pure Appl. Chem.,* **57**(2)**:** 303–322 (1985).

Wilhelm, E.: *CRC Crit. Rev. Anal. Chem.,* **16**(2)**:** 129–175 (1985).

IUPAC: *Solubility Data Series,* A. S. Kertes, editor-in-chief, Pergamon, Oxford (1979–1996).

Tominaga, T., R. Battino, and H. K. Gorowara: *J. Chem & Eng. Data,* **31:** 175–180 (1986).

Chang, A. Y., K. Fitzner, and M. Zhang: "The Solubility of Gases in Liquid Metals and Alloys," *Progress in Materials Science,* Vol. 32. No. 2-3, Oxford, New York, Pergamon Press (1988).

Hayduk, W., H. Asatani, and Y. Miyano: *Can. J. Chem. Eng.,* **66**(3)**:** 466–473 (1988).

Sciamanna, S. F., and S. Lynn: *Ind. Eng. Chem. Res.,* **27**(3)**:** 492–499 (1988).

Luhring, P., and A. Schumpe: *J. Chem. & Eng. Data,* **34:** 250–252 (1989).

Tomkins, R., P. T. Bansal, and P. Narottam (eds.) "Gases in Molten Salts," *Solubility Data Series,* Vol. 45/46, Pergamon, Oxford, UK (1991).

Fogg, P. G. T., and W. Gerrard: *Solubility of Gases in Liquids: A Critical Evaluation of Gas/Liquid Systems in Theory and Practice,* Chichester, New York, J. Wiley (1991).

Japas, M. L., C. P. Chai-Akao, and M. E. Paulaitis: *J. Chem. & Eng. Data,* **37:** 423–426 (1992).

Srivastan, S., N. A. Darwish, and K. A. M. Gasem: *J. Chem. & Eng. Data,* **37:** 516–520 (1992).

Xu, Y., R. P. Schutte, and L. G. Hepler: *Can. J. Chem. Eng.,* **70**(3)**:** 569–573 (1992).

Bo, S., R. Battino, and E. Wilhelm: *J. Chem. & Eng. Data,* **38:** 611–616 (1993).

Bremen, B., A. A. C. M. Beenackers, and E. W. J. Rietjens: *J. Chem. & Eng. Data,* **39:** 647–666 (1994).

TABLE 8-31 Some Sources for Solubilities of Gases in Liquids (*Continued*)

Hesse, P. J., R. Battino, P. Scharlin: *J. Chem. & Eng. Data,* **41:** 195–201 (1996).

Darwish, N. A., K. A. M. Gasem, and R. L. Robinson, Jr.: *J. Chem. & Eng. Data,* **43**(2): 238–240 (1998).

Dhima, A., J. C. de Hemptinne, and G. Moracchini: *Fluid Phase Equil.,* **145**(1): 129–150 (1998).

Abraham, M. H., G. S. Whiting, P. W. Carr, and Ouyang, H.: *J. Chem. Soc.,* Perkin Trans. **2**(6): 1385–1390 (1998).

Gao, W., K. A. M. Gasem, and R. L. Robinson, Jr.: *J. Chem. & Eng. Data,* **44**(2): 185–190 (1999).

Pardo, J., A. M. Mainar, M. C. Lopez, F. Royo, and J. S. Urieta: *Fluid Phase Equil.,* **155**(1): 127–137 (1999).

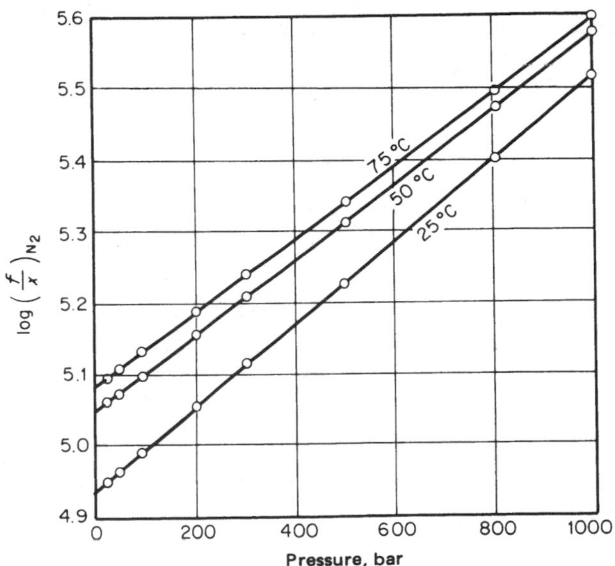

FIGURE 8-18 Solubility of nitrogen in water at high pressures. (*Prausnitz, et al., 1999*)

$$\lim_{x_2 \to 0} \gamma_2^* = 1 \tag{8-11.3}$$

At high pressures, Eq. (8-11.1) becomes

$$y_2 \phi_2^V P = x_2 \gamma_2^* H_{2,1} \exp \left[\int_{P_{\text{vp}1}}^{P} \frac{\overline{V}_{2,1}^{\infty}}{RT} \, dP \right] \tag{8-11.4}$$

A convenient form of Eq. (8-11.4) is

$$\ln \frac{y_2 \phi_2^V P}{x_2} = \ln H_{2,1} + \ln \gamma_2^* + \frac{\overline{V}_{2,1}^{\infty}(P - P_{vp1})}{RT} \tag{8-11.5}$$

where it has been assumed that $\overline{V}_{2,1}^{\infty}$ is independent of pressure. The issues associated with using Eq. (8-11.5) are described in detail by Prausnitz, et al. (1999). Briefly, for $T_{c2} << T << T_{c1}$, the last term in Eq. (8-11.5) makes the solubility less than expected from Eq. (8-11.1). Figure 8-18 shows this effect for nitrogen in water near ambient temperatures. It is possible to obtain useful values of $H_{2,1}$ from such plots and under the right conditions, values of $\overline{V}_{2,1}^{\infty}$. Alternatively, the volume change of a solution upon dissolution of a gas, dilatometry, also gives experimental data for $\overline{V}_{2,1}^{\infty}$.

Table 8-32 shows typical values of $\overline{V}_{2,1}^{\infty}$ for gases in liquids at 25°C; they are similar to the pure liquid volume of the solute at its normal boiling point, T_{b2}. Correlations for $\overline{V}_{2,1}^{\infty}$ have been developed by Brelvi and O'Connell (1972, 1975), Campanella, et al. (1987), Lyckman, et al. (1965), and by Tiepel and Gubbins (1972, 1973).

As T approaches T_{c1}, $\overline{V}_{2,1}^{\infty}$ diverges to infinity as the isothermal compressibility of the solvent, $\kappa_1 = -1/V(\partial V/\partial P)_T$ also diverges (Levelt Sengers, 1994; O'Connell, 1994). Under these conditions, the integral of Eq. (8-11.4) is not a simple function of pressure such as in Fig. 8-18. Sharygin, et al. (1996) show that an excellent correlation of $\overline{V}_{2,1}^{\infty}$ for nonelectrolyte gases in aqueous solution from ambient conditions to well above the critical point of pure water can be obtained from correlating $\overline{V}_{2,1}^{\infty}/\kappa_1 RT$ with the density of water. This has been extended to other properties such as fugacity coefficients, enthalpies and heat capacities by Sedlbauer, et al. (2000) and by Plyasunov, et al. (2000*a*, 2000*b*).

As discussed by Orentlicher and Prausnitz (1964), Campanella, et al. (1987), Mathias and O'Connell (1979, 1981), and Van Ness and Abbott (1982), when x_2 increases because yP is large, or because the system nears T_{c1} or due to solvation effects as with CO_2 in water, the middle term of Eq. (8-11.5) can become important and partially cancel the last term. Then, though the variation of $\ln(f_2/x_2)$ may be linear with P, the slope will not be $\overline{V}_{2,1}^{\infty}/RT$. Using only a volumetrically determined $\overline{V}_{2,1}^{\infty}$ will underestimate the solubility at elevated pressures and solute compositions.

Many attempts have been made to correlate gas solubilities, but success has been severely limited because, on the one hand, a satisfactory theory for gas-liquid solutions has not been established and, on the other, reliable experimental data are

TABLE 8-32 Partial Molal Volumes $\overline{V}^{\infty}$ of Gases in Liquid Solution at 25°C, cm^3/mol†

	H_2	N_2	CO	O_2	CH_4	C_2H_2	C_2H_4	C_2H_4	CO_2	SO_2
Ethyl ether	50	66	62	56	58					
Acetone	38	55	53	48	55	49	58	64	···	68
Methyl acetate	38	54	53	48	53	49	62	69	···	47
Carbon tetrachloride	38	53	53	45	52	54	61	67	···	54
Benzene	36	53	52	46	52	51	61	67	···	48
Methanol	35	52	51	45	52	···	···	···	43	
Chlorobenzene	34	50	46	43	49	50	58	64	···	48
Water	26	40	36	31	37	···	···	···	33	
Molar volume of pure solute at its normal boiling point	28	35	35	28	39	42	50	55	40	45

†J. H. Hildebrand and R. L. Scott (1950).

not plentiful, especially at temperatures remote from 25°C. Among others, Battino and Wilhelm (1971, 1974) have obtained some success in correlating solubilities in nonpolar systems near 25°C by using concepts from perturbed-hard-sphere theory, but, as yet, these are of limited use for engineering work. A more useful graphical correlation, including polar systems, was prepared by Hayduk, et al. (1970, 1971), and a correlation based on regular solution theory for nonpolar systems was established by Prausnitz and Shair (1961) and, in similar form, by Yen and McKetta (1962). The regular solution correlation is limited to nonpolar (or weakly polar) systems, and although its accuracy is not high, it has two advantages: it applies over a wide temperature range, and it requires no mixture data. Correlations for nonpolar systems, near 25°C, are given by Hildebrand and Scott (1962).

A method for predicting Henry's constant in a different solvent (3), $H_{2,3}$, from $H_{2,1}$ at the same T is given by Campanella, et al. (1987).

A crude estimate of solubility can be obtained rapidly by extrapolating the vapor pressure of the gaseous solute on a linear plot of log P_{vp} vs. $1/T$. The so-called *ideal solubility* is given by

$$x_2 = \frac{y_2 P}{P_{vp2}} \tag{8-11.6}$$

where P_{vp2} is the (extrapolated) vapor pressure of the solute at system temperature T. The ideal solubility is a function of temperature, but it is independent of the solvent. Table 8-33 shows that for many typical cases, Eq. (8-11.6) provides an order-of-magnitude estimate.

Gas solubility in mixed solvents and, therefore, Henry's constant, varies with solvent composition. The simplest approximation for this is (Prausnitz, et al., 1999)

$$\ln H_{2,\text{mix}} = \sum_{i=\text{solvents}} x_i \ln H_{2,i} \tag{8-11.7}$$

The next-order estimate (Prausnitz, et al., 1999), which usually gives the proper correction to Eq. (8-11.7) for typical systems except aqueous alcohols (Campanella, et al., 1987), is

$$\ln H_{2,\text{mix}} = \sum_{i=\text{solvents}} x_i H_{2,i} - \frac{g^E_{\text{solvents}}}{RT} \tag{8-11.8}$$

where g^E_{solvents}/RT is the excess Gibbs energy for the solvent mixture found from models such as in Sec. 8-9. The method of Campanella, et al. (1987) can also be applied to mixed-solvent systems including aqueous alcohols and ternary solvents.

Henry's law can give reliable results for many systems. When nonidealities do arise, it is common to use Equation-of-State methods as described in Sec. 8-12.

TABLE 8-33 Solubilities of Gases in Several Liquid Solvents at 25°C and 1.01 bar Partial Pressure. Mole Fraction × 10^4

	Ideal†	n-C_7F_{16}	n-C_7H_{16}	CCl_4	CS_2	$(CH_3)_2CO$
H_2	8	14.01	6.88	3.19	1.49	2.31
N_2	10	38.7	. . .	6.29	2.22	5.92
CH_4	35	82.6	. . .	28.4	13.12	22.3
CO_2	160	208.8	121	107	32.8	

† Eq. (8-11.6).

Prausnitz, et al. (1999) describe techniques for such cases. Also, the method of Campanella, et al. (1987) was developed to provide good estimates of the properties of liquids containing supercritical components. These estimates can be used either directly or to obtain reliable EoS parameters.

Other Dilute-Solution Methods

In some practical situations, especially those concerned with water-pollution abatement, it is necessary to estimate the solubilities of volatile organic solutes in water. While such estimates could be obtained with the methods of Secs. 8-3 to 8-9, the Henry's law approach is preferable. The basis is that Henry's constant is related to the fugacity of the pure component by the activity coefficient at infinite dilution as suggested in Sec. 8-10 and Example 8-8.

$$H_{2,1} = \gamma_{2,1}^{\infty} f_2^o \tag{8-11.9}$$

Typically, f_2^o is the saturation fugacity,

$$f_2^o = \phi_2^s P_{vp2} \tag{8-11.10}$$

A common method for obtaining $H_{2,1}$ is to estimate $\gamma_{2,1}^{\infty}$ from experimental measurements and then use Eq. (8-11.9). Alternatively, it is possible to use a direct estimation method. For example, Sherman, et al. (1996) have presented a useful correlation for estimating aqueous solubility based on the linear solvation energy relationship (LSER) of Abraham (1993). They have tested it extensively with a data base of 326 organic solutes in dilute aqueous solution. Both methods are illustrated in Example 8-18.

Example 8-18 Estimate Henry's constants for some organic solutes (2) in water (1). First use solubility data with corrections, if necessary, for finite-composition effects and then use the correlation of Sherman, et al. (1996).

solution We consider, *n*-butane, 1-butanol and diethyl ether at 25°C.

Using Experimental Solubilities. We assume that $f_2^o = P_{vp2}$ and estimate $\gamma_{2,1}^{\infty}$ from solubility data. Riddick, et al. (1986) give P_{vp2}.

Solute	P_{vp2} (mmHg)
n-butane	1850
1-butanol	6.83
diethyl ether	537

n-butane:

At 25°C, liquid *n*-butane is only sparingly soluble in water. Therefore, its infinite-dilution activity coefficient is essentially equal to its activity coefficient γ_2 at its aqueous solubility x_2 ($x_2 << 1$).

If we neglect the very small solubility of water (1) in n-butane (2), the activity a_1 of pure n-butane (2) is equal to unity:

$$a_2 = x_2\gamma_2 \approx x_2\gamma_2^\infty = 1$$

Hence,

$$\gamma_2^\infty = \frac{1}{x_2} \tag{8-11.11}$$

At 25°C, Hwang, et al. (1992) reported solubility $x_2 = 4.21\text{E}{-5}$ for n-butane in water. Substituting into Eq. (8-11.11), we have $\gamma_2^\infty = 2.37\text{E}4$.

1-butanol and diethyl ether:

To obtain the infinite-dilution activity coefficients for these two solutes from experimental vapor-liquid-equilibrium data at 25°C, we use the van Laar model for the liquid-phase molar excess Gibbs energy. For the van Laar model

$$\frac{g^E}{RT} = \frac{Ax_1x_2}{\left(\dfrac{A}{B}\right)x_2 + x_1}$$

Using experimental data at 25°C reported by Butler, et al. (1933) for the 1-butanol (2)–water (1) system and those reported by Villamanan, et al. (1984) for the diethyl ether (1)–water (2) system, Gmehling and Onken (1977) and Gmehling, et al. (1981) give parameters A and B:

Parameter	1–butanol (2)–water (1)	diethyl ether (2)–water (1)
A	4.11	1.63
B	1.47	4.46

From the van Laar model, ln γ_2^∞ = A. Hence, γ_2^∞ (25°C) for 1-butanol and diethyl ether are

Solute	γ_2^∞ (25°C)
1-butanol	61.1
diethyl ether	5.10

Substituting P_{vp2} and γ_2^∞ into Eq. (8-11.9), Henry's constants for the three solutes are shown in Table 8-34.

(b) Using the modified LSER method:

The modified LSER by Sherman, et al. (1996) is given by Eq. (8-10.29)

TABLE 8-34 Henry's Constants for Solutes in Water from Solubilities

Solute	P_{vp2} (mmHg)	γ_2^∞	H_2 (mmHg)
n-butane	1.85E3	2.37E4	4.38E7
1-butanol	6.83E0	6.11E1	4.17E2
diethyl ether	5.37E2	5.10	2.74E3

TABLE 8-35 Henry's Constants (mmHg) at 25°C of Solute (2) in Water (1)

Solute	Experimental*	From solubility†	% error	Eq. (8-10.29)	% error
n-butane	3.50E7	4.38E7	25	3.11E7	−11
1-butanol	3.27E2	4.17E2	28	4.73E2	45
diethyl ether	5.87E4	2.74E3	−95	7.15E4	21

* Sherman, et al. (1996).
† Table 8-34.

$$\ln H_{2,1} = -0.536 \log L_2^{16} - 5.508\, \pi_2^{*H} - 8.251\alpha_2^H - 10.54\beta_2^H - 1.598\left[\ln\left(\frac{V_2}{V_1}\right)^{0.75} + 1 - \left(\frac{V_2}{V_1}\right)^{0.75}\right] + 16.10$$

where V_1, the molar volume of water, is 18 cm³/mol. Solute parameters from Table 8-19 and results from Eq. (8-10.29) are shown below:

Solute	Log L^{16}	π^{*H}	α^H	β^H	V_2	$H_{2,1}$ (mmHg)
n-butane	1.615	−0.11	0	0	96.5	3.11E7
1-butanol	2.601	0.47	0.37	0.48	92.0	4.73E2
diethyl ether	2.015	0.27	0	0.45	104.7	7.15E4

Table 8-35 compares experimental with calculated Henry's constants. Experimental results are from Sherman, et al. (1996).
where

$$\%\text{error} = \frac{[\text{Calculated} - \text{Observed}]}{\text{Observed}} \times 100$$

Results shown in Table 8-35 suggest that calculated and experimental Henry's constants agree within a factor of about 2. However, the experimental uncertainty is probably also a factor of 2. It is difficult to measure Henry's constants with high accuracy.

8-12 VAPOR-LIQUID EQUILIBRIA WITH EQUATIONS OF STATE

Thermodynamics provides the basis for using Equations of State (EoS) not only for the calculation of the *PVT* relations of Chaps. 4 and 5 and the caloric property relations of Chap. 6, but, as Sec. 8-2 shows, EoS can also be used for computing phase equilibria among fluid phases. We consider this alternative in detail because there are several disadvantages to the methods of Secs. 8-2 to 8-9, especially for high-pressure systems.

1. While liquid properties are generally insensitive to pressure changes, pressure effects become significant in high-pressure systems. Although the correction term of Eq. (8-4.2) can be effective to several atmospheres, it may become inaccurate when significant composition variations or compressibility of the liquid appear. This is especially true near the solution critical point (see Chap. 5) or if supercritical components are present (Sec. 8-11).

2. Common low-pressure models for the vapor fugacity, such as the ideal gas ($\phi_i^V = 1$) and second virial model (Secs. 4-5, 5-4, 8-8 and 8-9) become inaccurate and must be replaced by models valid at higher pressures.

3. The presence of components above their T_c prevents us from directly obtaining the commonly-used pure-component standard-state fugacity that is determined primarily by the vapor pressure. The supercritical standard state can be defined by a Henry's constant with the unsymmetric convention for activity coefficients (Prausnitz, et al., 1999) and some correlations for engineering use have been established on that basis (Sec. 8-11). However, because conceptual complexities arise in ternary and higher-order systems and because computational disadvantages can occur, this approach has not been popular.

4. The use of different formulae for computing fugacities in the vapor and liquid phases, such as Eq. (8-4.1), leads to a discontinuity as the mixture critical point is approached. This can cause considerable difficulty in computational convergence as well as large inaccuracies, especially for liquid-phase properties.

As a result of these disadvantages, vapor-liquid equilibrium calculations using the same EoS model for both phases have become popular with an enormous number of articles describing models, methods and results. Many of the books cited in Table 8-1*a* discuss EoS methods for the calculation of phase equilibria.

The basis is Eq. (8-2.1) with vapor and liquid fugacity coefficients, ϕ_i^V and ϕ_i^L, as given in Sec. 6-7:

$$f_i^V = y_i \phi_i^V P = x_i \phi_i^L P = f_i^L \tag{8-12.1}$$

The K-factor commonly used in calculations for process simulators is then simply related to the fugacity coefficients of Eq. (8-12.1).

$$K_i = \frac{y_i}{x_i} = \frac{\phi_i^L}{\phi_i^V} \tag{8-12.2}$$

As shown in Sec. 6-7, to obtain ϕ_i^V, we need the vapor composition, y, and volume, V^V, while for the liquid phase, ϕ_i^L is found using the liquid composition, x, and volume, V^L. Since state conditions are usually specified by T and P, the volumes must be found by solving the PVT relationship of the EoS.

$$P = P(T, V^V, y) = P(T, V^L, x) \tag{8-12.3}$$

In principle, Eqs. (8-12.1) to (8-12.3) with (Sec. 6-8) are sufficient to find all K factors in a multicomponent system of two or more phases. This kind of calculation is not restricted to high-pressure systems. A great attraction of the EoS method is that descriptions developed from low-pressure data can often be used for high-temperature, high-pressure situations with little adjustment. One difficulty is that EoS relations are highly nonlinear and thus can require sophisticated numerical initialization and convergence methods to obtain final solutions.

In principle, Eqs. (8-12.2) are sufficient to find all K factors in a multicomponent system containing two (or more) fluid phases.

To fix ideas, consider a two-phase (vapor-liquid) system containing m components at a fixed total pressure P. The mole fractions in the liquid phase are x_1, x_2, . . . , x_{m-1}. We want to find the bubble-point temperature T and the vapor phase mole fractions y_1, y_2, . . . , y_{m-1}. The total number of unknowns, therefore, is m. However, to obtain ϕ_i^V and ϕ_i^L, we also must know the molar volumes V^L and V^V. Therefore, the total number of unknowns is $m + 2$.

To find $m + 2$ unknowns, we require $m + 2$ independent equations. These are:

Equation (8-12.2) for each component i:	m equations
Equation (8-12.3), once for the vapor phase and once for the liquid phase:	2 equations
Total number of independent equations:	$m + 2$

This case, in which P and x are given and T and y are to be found, is called a bubble-point T problem. Other common cases are:

Given variables	Variables to be found	Name
P, y	T, x	Dew-point T
T, x	P, y	Bubble-point P
T, y	P, x	Dew-point P

However, the most common way to calculate phase equilibria in process design and simulation is to solve the "flash" problem. In this case, we are given P, T, and the mole fractions, z, of a feed to be split into fractions α of vapor and $(1 - \alpha)$ of liquid. We cannot go into details about the procedure here; numerous articles have discussed computational procedures for solving flash problems with EoS (see, *e.g.*, Heidemann, 1983; Michelsen, 1982; Topliss, 1985).

Representative Results for Vapor-Liquid Equilibria from Equations of State

Knapp, et al. (1982) have presented a comprehensive monograph on EoS calculations of vapor-liquid equilibria. It contains an exhaustive literature survey (1900 to 1980) for binary mixtures encountered in natural-gas and petroleum technology: hydrocarbons, common gases, chlorofluorocarbons and a few oxygenated hydrocarbons. The survey has been extended by Dohrn, et al. (1995).

Knapp, et al. (1982) considered in detail four EoS models applicable to both vapor and liquid phases of the above substances. Because two of the expressions were cubic in volume, an analytic solution is possible while two others could not be solved analytically for the volume:

1. The Redlich-Kwong-Soave (Soave, 1972) cubic EoS in Tables 4-6 and 4-7 with generalized CSP parameters of Table 4-8 and the mixing rules of Eq. (5-5.2).

2. The Peng-Robinson (Peng and Robinson, 1976) cubic EoS in Tables 4-6 and 4-7 with generalized CSP parameters of Table 4-8 and the mixing rules of Eq. (5-5.2).

3. The LKP nonanalytic EoS model which, for pure components, is the Lee-Kesler (1975) expression and for mixtures is the extension of Plöcker, et al. (1978).

4. The BWRS nonanalytic EoS model of Starling (1973).

All four of these EoS require essentially the same input parameters: for each pure fluid, critical properties T_c and P_c, and acentric factor, ω, and for each pair of components, one binary parameter, designated here and in Chaps. 5 and 6 as k_{ij} whose value is usually close to zero.†

To determine binary parameters, Knapp, et al. (1984) fit calculated vapor-liquid equilibria to experimental ones. The optimum binary parameter is the one which minimizes DP/P defined by

$$\frac{DP}{P} = \frac{100}{N} \sum_{n=1}^{N} \frac{|P_n^e - P_n^c|}{P_n^e} \tag{8-12.4}$$

where P_n^e is the experimental total pressure of point n and P_n^c is the corresponding calculated total pressure, given temperature T and liquid phase mole fraction x. The total number of experimental points is N.

Similar definitions hold for Dy_1/y_1 and for DK_1/K_1. Here y_1 is the vapor phase mole fraction and K_1 is the K factor ($K_1 = y_1/x_1$) for the more volatile component. In addition, Knapp, et al. calculated Df/f by

$$\frac{Df}{f} = \frac{100}{N} \sum_{n=1}^{N} \frac{|f_{1n}^V - f_{1n}^L|}{f_{1n}^V} \tag{8-12.5}$$

where f_1^V is the calculated fugacity of the more volatile component in the vapor phase and f_1^L is that in the liquid phase.

When the binary parameter is obtained by minimizing DP/P, the other deviation functions are usually close to their minima. However, for a given set of data, it is unavoidable that the optimum binary parameter depends somewhat on the choice of objective function for minimization. Minimizing DP/P is preferred because that objective function gives the sharpest minimum and pressures are usually measured with better accuracy than compositions (Knapp, et al., 1982).

Tables 8-36 and 8-37 show some results reported by Knapp, et al. (1982). Table 8-36, for propylene-propane, concerns a simple system in which the components are similar; in that case, excellent results are obtained by all four equations of state with only very small values of k_{12}.

However, calculated results are not nearly as good for the system nitrogen-isopentane. Somewhat larger k_{12} values are needed but, even with such corrections, calculated and observed K factors for nitrogen disagree by about 6% for LKP, RKS, and PR and by nearly 12% for BWRS.

These two examples illustrate the range of results obtained by Knapp, et al. for binary mixtures containing nonpolar components. (Disagreement between calcu-

† In the original publication of Plöcker, et al. (1978), the symbol for the binary parameter was $K_{ij} = 1 - k_{ij}$. In the monograph by Knapp, et al. (1982), k_{ij}^* was used instead of K_{ij}. In Chap. 5 of this book, several notations for binary parameters are used. Unfortunately, the plethora of symbols and quantities related to binary parameters can be very confusing for a user; see Chap. 5.

TABLE 8-36 Comparison of Calculated and Observed Vapor-Liquid Equilibria for the System Propylene (1)–Propane (2)† (From Knapp, et al., 1982)
Temperature range: 310 to 344 K; Pressure range: 13 to 31 bar

Equation of state	Binary constant, k_{12}‡	Percent Deviation in Properties			
		DP/P	Dy_1/y_1	DK_1/K_1	Df_1/f_1
LKP	0.0081	0.31	0.10	0.38	0.46
BWRS	0.0025	0.55	0.06	0.27	0.46
RKS	0	0.56	0.23	0.61	0.90
PR	0.0063	0.31	0.08	0.40	0.29

† Experimental data (77 points) from Laurence and Swift (1972).
‡ Binary constants obtained by minimizing DP/P.

TABLE 8-37 Comparison of Calculated and Observed Vapor-Liquid Equilibria for the System Nitrogen (1)–Isopentane (2)† (From Knapp, et al., 1982)
Temperature range: 277 to 377 K; Pressure range: 1.8 to 207 bar

Equation of state	Binary constant, k_{12}‡	Percent Deviation in Properties			
		DP/P	Dy_1/y_1	DK_1/K_1	Df_1/f_1
LKP	0.347	5.14	0.87	6.12	4.99
BWRS	0.1367	12.27	3.73	11.62	10.70
RKS	0.0867	4.29	1.58	6.66	5.78
PR	0.0922	3.93	1.61	5.98	5.26

† Experimental data (47 points) from Krishnan, et al. (1977).
‡ Binary constants obtained by minimizing DP/P.

lated and observed vapor-liquid equilibria is often larger when polar components are present.) For most nonpolar binary mixtures, the accuracy of calculated results falls between the limits indicated by Tables 8-36 and 8-37.

While Knapp, et al. found overall that the BWRS equation did not perform as well as the others, it is not possible to conclude that, of the four equations used, one particular equation is distinctly superior to the others. Further, it is necessary to keep in mind that the quality of experimental data varies appreciably from one set of data to another. Therefore, if calculated results disagree significantly with experimental ones, one must not immediately conclude that the disagreement is due to a poor equation of state.

Knapp's monograph is limited to binary mixtures. If pure-component equation-of-state constants are known and if the mixing rules for these constants are simple, requiring only characteristic binary parameters, then it is possible to calculate vapor-liquid equilibria for ternary (and higher) mixtures using only pure-component and binary data. Although few systematic studies have been made, it appears that this "scale-up" procedure usually provides good results for vapor-liquid equilibria, especially in nonpolar systems. (However, as defined in Sec. 8-14, this scale-up procedure is usually not successful for ternary liquid-liquid equilibria, unless special precautions are observed.)

Regardless of what equation of state is used, it is usually worthwhile to make an effort to obtain the best possible equation-of-state constants for the fluids that comprise the mixture. Such constants can be estimated from critical data, but it is usually better to obtain them from vapor-pressure and density data as discussed in

Sec. 4-6. Knapp, et al. (1982) describe results for binary systems with several EoS models using generalized pure-component corresponding-states parameters and a single binary parameter. For greater accuracy, especially in multicomponent systems, modification of this procedure can be useful. We cite an example from Turek, et al. (1984) who used a modified Redlich-Kwong EoS to correlate their extensive phase-behavior data for CO_2-hydrocarbon systems encountered in miscible enhanced oil recovery.

The Redlich-Kwong model is that shown in Table 4-6 with corresponding-states parameters from Table 4-8 except that the values for bP_c/RT_c and $aP_c/(RT_c)^2$ have a special reduced temperature dependence for CO_2 and are generalized functions of T/T_c and ω for the hydrocarbons (Yarborough, 1979). The combining rules are those of Eqs. (5-2.4*b*) and (5-2.4*d*) while the mixing rules are Eq. (5-5.2*b*) with fitted binary parameter k_{ij}, and Eq. (5-5.2*a*) with fitted binary parameter l_{ij}. The values of Yarborough (1979) were used for the hydrocarbon pairs; the binary parameters are nonzero only for substances of greatly different carbon number. For the CO_2-hydrocarbon pairs, the values depend on temperature as well as on hydrocarbon acentric factor as shown in Fig. 8-19. Finally, comparisons were made with new measurements of CO_2 with a synthetic oil whose composition is shown in Table 8-38. Typical results are shown in Fig. 8-20 for the *K*-factors of all components at 322 K as a function of pressure. All of the many results shown are quite good, especially for the heavy components which are commonly very challenging to describe.

Mollerup (1975, 1980) has shown similar success in describing properties of natural-gas mixtures when careful analysis is used to obtain the EoS parameters.

Liquid and Vapor Volumes in EoS Calculations of Phase Equilibria

In a typical EoS, the pressure, P, is given as a function of T, V and composition, z as in Eq. (8-12.3). If P, T, and z are specified, it is necessary to find V, a task that may not be simple, especially in phase equilibrium calculations.

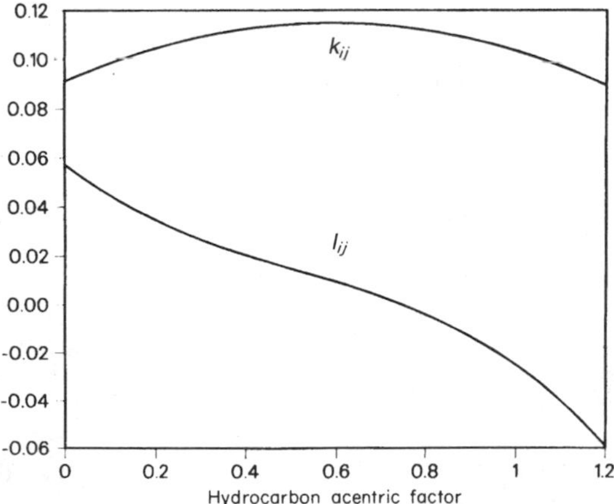

FIGURE 8-19 Carbon dioxide-hydrocarbon binary interaction parameters. (*Turek, et al., 1984*)

TABLE 8-38 Composition of Synthetic Oil Used by Turek, et al., (1984) for Experimental Studies of Vapor-Liquid Equilibria with Carbon Dioxide

Component	Mole percent	Component	Mole percent
Methane	34.67	*n*-Hexane	3.06
Ethane	3.13	*n*-Heptane	4.95
Propane	3.96	*n*-Octane	4.97
n-Butane	5.95	*n*-Decane	30.21
n-Pentane	4.06	*n*-Tetradecane	5.04

Density at 322.0 K and 15.48 MPa is 637.0 kg/m^3.
Density at 338.7 K and 14.13 MPa is 613.5 kg/m^3.

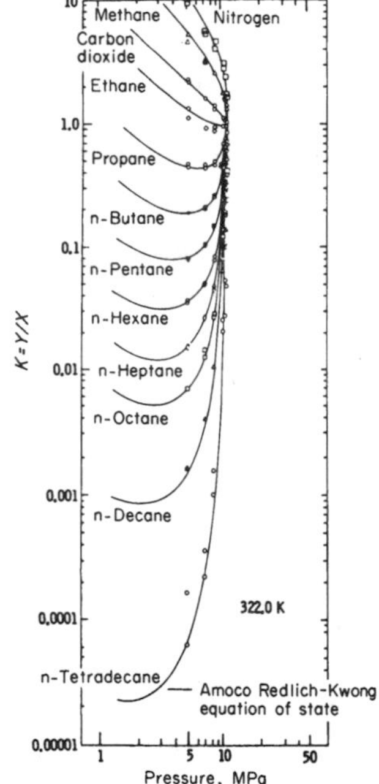

FIGURE 8-20 *K* factors for carbon dioxide-synthetic oil. Turek et al., (1984)

In particular, as the algorithm attempts to converge, there can be situations in which a guessed condition is specified for one phase, but the properties are computed for another phase because the identification of the volume is incorrect. For example, consider the system ethane-heptane at fixed composition computed with the Soave EoS (Table 4-6 to 4-8) as shown in Fig. 8-21. The phase envelope consists of the higher line for the bubble-point P and T relation for the liquid and the lower line for the dew-point P and T relation for the vapor. The two phases meet at the critical point which is normally neither the maximum pressure nor the maximum temperature.

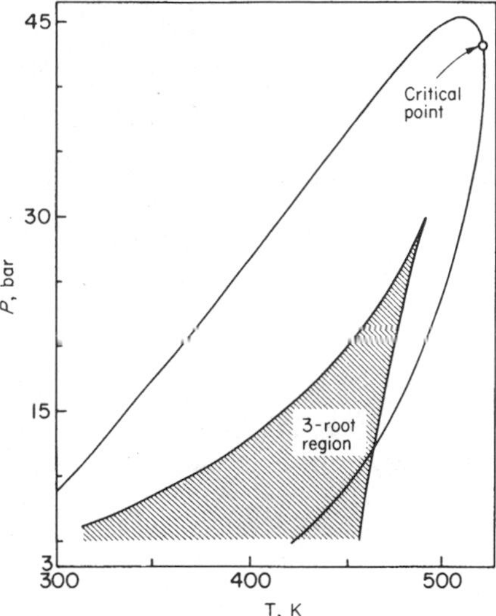

FIGURE 8-21 Phase behavior calculated by the Soave (1972) EoS for 26.54% ethane with 73.46% *n*-heptane. In the 3-root region, three different real volumes satisfy Eq. (8-12.3) but only one will correspond to the phase of interest.

Since this is a cubic EoS, there are always three volume roots, but there are regions where two are complex conjugates and therefore not real. The region of three real roots is identified in the figure. Outside of this region there is only one real root. At certain conditions, a calculation arrives at what is called the trivial-root problem (Coward, et al., 1978; Gunderson, 1982; Mathias, et al., 1984; Poling, et al., 1981). One case is when the phases attempting to become separated converge to identical compositions and therefore are actually the same phase, not different as a vapor and liquid normally would be. Another is when the volume of the phase with the current vapor composition is actually that of a liquid of the same composition or vice versa. This can happen outside of the 3-root region of Fig. 8-21. The result is incorrect fugacities for the desired phase leading to a lack of convergence to the desired multiphase system. This phase misidentification can best be avoided by using accurate initial guesses for the iterative solution. Example 8-19 shows how this might be accomplished.

Example 8-19 Estimate the *K* factor for a 26.54% ethane (1)–73.46% *n*-heptane (2) mixture at 400 K and 15 bar as the first step in a bubble-point *P* calculation. Use the generalized Soave EoS (Tables 4-6 to 4-8).

solution The vapor composition must also be guessed to obtain the vapor fugacity coefficient. If the specified liquid composition is chosen, only one volume root is found and it corresponds to the liquid. Thus, the calculated ϕ_i will be that of the liquid, not of a vapor, giving a trivial root. This would not be the case if this guess were in the 3-root region (*e.g.*, at 440 K and 15 bar or 400 K and 7 bar) because then the smallest volume root would be for the liquid and the largest volume root would correspond to a phase with vapor properties. Since it would have a ϕ different from that of the liquid, it could be used as the initial guess for phase-equilibrium convergence.

However, if one guesses a different vapor composition such as $y_1 = 0.9$, the EoS provides a volume root more appropriate for a vapor. The table below summarizes the results that have been calculated with the following pure-component parameters:

fluid	T_c, K	P_c, bar	ω
Ethane	305.4	48.8	0.099
Heptane	540.3	27.4	0.349

Note two important subtleties: The proposed vapor composition is not on Fig. 8-21 which is only for 26.54% ethane whether liquid or vapor. Also the numerical results are not for converged phase equilibrium. In that case the K factors must follow $x_1 K_1 + (1 - x_1)K_2 = 1$. This is not the case here because we used guessed, not converged, P and y_1 values. In this case, the pressure is probably too high.

Phase	x_1, y_1	$Z = PV/RT$	ϕ_1	ϕ_2
Liquid	0.2654	0.0795	5.3780	0.1484
Vapor	0.9000	0.9327	0.9607	0.7377

$$K_1 = 5.378/0.9607 = 5.598 \qquad K_2 = 0.1484/0.7377 = 0.2012$$

$$x_1K_1 + (1 - x_1)K_2 = 1.486 + 0.148 = 1.634$$

Phase Envelope Construction—Dew and Bubble-Point Calculations

Phase-equilibrium calculations at low pressures are normally easy, but high-pressure calculations can be complicated by both trivial-root and convergence difficulties. Trivial-root problems can be avoided by starting computations at a low pressure and marching toward the critical point in small increments of temperature or pressure. When the initial guess of each calculation is the result of a previous calculation, trivial roots are avoided. Convergence difficulties are avoided if one does dew or bubble-point *pressure* calculations when the phase envelope is flat and dew or bubble-point *temperature* calculations when the phase envelope is steep. These conditions have been stated by Ziervogel and Poling (1983) as follows:

$$\text{If } \left|\frac{d \ln P}{d \ln T}\right| < 2, \text{ calculate dew or bubble-point pressures}$$

$$\text{If } \left|\frac{d \ln P}{d \ln T}\right| > 20, \text{ calculate dew or bubble-point temperatures.}$$

This technique allows convergence on a single variable. Several multivariable Newton-Raphson techniques have been presented in the literature (Michelsen, 1980). The advantage of these multivariable techniques is that fewer iterations are required for calculations near the critical point (within several degrees or bars). Calculation of dew and bubble points by any method near the critical point is tedious, and although the above methods are often successful, the most efficient approach is to calculate the critical point by the direct method outlined in Sec. 6-8.

To calculate dew or bubble points, partial derivatives of K_i are used to converge to the equilibrium T and P. These derivatives may be determined numerically or analytically; in both cases, they require certain partial derivatives of ϕ_i^V and ϕ_i^L.

$$\left.\frac{\partial K_i}{\partial \theta}\right|_{V,z} = K_i \left(\left.\frac{\partial \ln \phi_i^L}{\partial \theta}\right|_{V,x} - \left.\frac{\partial \ln \phi_i^V}{\partial \theta}\right|_{V,y} \right) \tag{8-12.6}$$

where θ can be either T or P. Analytical expressions are most easily written for ϕ_i from a cubic EoS in the form given in Sec. 6-7.

$$\begin{aligned}\frac{\partial \ln \phi_i}{\partial T} &= \frac{b_i}{b}\frac{\partial Z}{\partial T} + \frac{\partial Z/\partial T + B^*/T}{B^* - Z} + \frac{A^*}{Z + B^*}\left(\delta_i - \frac{b_i}{b}\right) \\ &\times \left(\frac{1}{Z}\frac{\partial Z}{\partial T} + \frac{1}{T}\right) + \left(\frac{b_i}{b} - \delta_i\right)\left(\frac{\partial A^*}{\partial T} + \frac{A^*}{T}\right) \\ &\times \frac{\ln(1 + B^*/Z)}{B^*} - \frac{A^*}{B^*}\left[\left(\frac{1}{a_i^{1/2}}\frac{\partial a_i^{1/2}}{\partial T} - \frac{1}{a}\frac{\partial a}{\partial T}\right)\delta_i \right. \\ &\left. + \frac{2a_i^{1/2}}{a}\sum_j x_j(1 - k_{ij})\frac{\partial a_j^{1/2}}{\partial T}\right]\ln\left(1 + \frac{B^*}{Z}\right)\end{aligned} \tag{8-12.7}$$

$$\frac{\partial Z}{\partial T} = \frac{(\partial A^*/\partial T)(B^* - Z) - B^*(A^* + Z + 2B^*Z)T}{3Z^2 - 2Z + (A^* - B^* - B^{*2})} \tag{8-12.8}$$

$$\frac{\partial A^*}{\partial T} = A^*\left(\frac{1}{a}\frac{\partial a}{\partial T} - \frac{2}{T}\right) \tag{8-12.9}$$

$$\frac{\partial a_i^{1/2}}{\partial T} = -R\frac{f\omega_i}{2}\left(\frac{0.42748 T_{ci}}{T P_{ci}}\right)^{1/2} \tag{8-12.10}$$

For $\theta = P$,

$$\begin{aligned}\frac{\partial \ln \phi_i}{\partial P} &= \frac{b_i}{b}\frac{\partial Z}{\partial P} + \frac{(\partial Z/\partial P) - (B^*/P)}{B^* - Z} \\ &+ \frac{A^*}{Z + B^*}\left(\delta_i - \frac{b_i}{b}\right)\left(\frac{1}{Z}\frac{\partial Z}{\partial P} - \frac{1}{P}\right)\end{aligned} \tag{8-12.11}$$

$$\frac{\partial Z}{\partial P} = \frac{B^*(2A^* + 2B^*Z + Z) - A^*Z}{P(3Z^2 - 2Z + A^* - B^* - B^{*2})} \tag{8-12.12}$$

where a_i and b_i are given in Tables 4-7 and 4-8.

$$A^* = \frac{aP}{R^2T^2} \tag{8-12.13}$$

$$B^* = \frac{bP}{RT} \tag{8.12.14}$$

$$f\omega = 0.48 + 1.574\ w - 0.176\ s^2 \tag{8-12.15}$$

We next describe dew-point and flash calculations.

Dew-Point Example

Example 8-20 Use the Soave (1972) equation to calculate the dew-point temperature at 40 bar for a 26.54% ethane-73.46% heptane mixture.

solution By using the pure-component properties in Example 8-19, dew points may be calculated at 5-bar intervals from 5 to 35 bar and then at 1-bar intervals from 35 to 39 bar. The dew point at 39 bar is 521.07 K and $x_1 = 0.1818$. This is used as the initial guess for the dew-point calculation at 40 bar. When T is adjusted according to Newton's method and a convergence criterion that $\Sigma y_i/K_i < 1 \times 10^{-8}$ is used, 20 iterations are required. The answer is

$$T = 521.54 \text{ K} \quad x_1 = 0.1944$$

The point where $\partial \ln P/\partial \ln T$ changes sign (cricondentherm) occurs at 41.4 bar and 521.81 K.

Flash Example

Example 8-21 For a mixture of 26.54 mole% ethane and 73.46 mole % *n*-heptane at 10 bar and 430 K, calculate the fraction of liquid and the compositions of the vapor and liquid phases. Use the Soave (1972) equation of state.

solution Use the pure-component properties listed in Example 8-19. The following procedure leads to the solution:

1. Guess $L = 0.5$ and $x_i = y_i = z_i$ ($z_i \equiv$ overall mole fraction of i, L = fraction liquid).
2. Solve the EoS for Z^L and Z^V.
3. Calculate ϕ_i^L and ϕ_i^V with Eq. (6-7.10).
4. Calculate K_i with $K_i = \phi_i^L/\phi_i^V$.
5. See if $\Sigma_i (x_i - y_i) = 0$, where $x_i = z_i/[K_i + L(1 - K_i)]$ and $y_i = K_i x_i$.
6. If $\Sigma_i (x_i - y_i)$ is not close enough to zero, adjust L according to King (1980)

$$L_{\text{new}} = L_{\text{old}} - \frac{\sum_i \{[z_i(K_i - 1)]/[K_i + (1 - K_i)L]\}}{\sum_i \{[z_i(K_j - 1)^2]/[K_i + (1 - K_i)L]^2\}}$$

7. Go back to step 2 and keep going until $\Sigma_i (x_i - y_i)$ is close enough to zero.

This procedure leads to the following:

$L = 0.545$	x_i	y_i	K_i
Ethane	0.0550	0.518	9.410
Heptane	0.9450	0.482	0.5106

For the above procedure, 10 iterations were required to obtain the condition that

$$\left|\sum_i (y_i - x_i)\right| < 10^{-8}$$

Although well-known equations of state (e.g., Soave (1972) and Peng-Robinson, 1976) are suitable for calculating vapor-liquid equilibria for nonpolar mixtures, these equations of state, using conventional mixing rules, are not satisfactory for mixtures containing strongly polar and hydrogen-bonded fluids in addition to the common gases and hydrocarbons. For those mixtures, the assumption of simple (random) mixing is poor because strong polarity and hydrogen bonding can produce significant segregation or ordering of molecules in mixtures. For example, at ordinary temperatures, water and benzene form a strongly nonrandom mixture; the mixture is so far from random that water and benzene are only partially miscible at ordinary temperatures because preferential forces of attraction between water molecules tend to keep these molecules together and prevent their mixing with benzene molecules.

It is possible to describe deviations from simple mixing by using complex (essentially empirical) mixing rules, as shown, for example, by Vidal (1978, 1983). For thermodynamic consistency, however, these mixing rules must be density-dependent because at low densities, the equation of state must give the second virial coefficient which is quadratic in mole fraction (Sec. 5-4). While many mixing rules do satisfy that requirement (Secs. 5-6 to 5-8), the Vidal mixing rules do not because they are independent of density. On the other hand, the mixing rule of Wong and Sandler (1992), described in Sec. 5-6 and discussed below, avoids explicit density dependence but does allow for quadratic composition dependence of the second virial coefficient including a binary parameter.

Chemical Theory

A useful technique for describing systems with strong attractions such as hydrogen bonds among the components is provided by the chemical hypothesis which postulates the existence of chemical species formed by virtual reactions among the components. This is not only useful for *PVT* properties as described in Chaps. 4 and 5, but also for phase equilibrium in such systems.

Consider, for example, a mixture of components A and B. The chemical theory assumes that the mixture contains not only monomers A and B but, in addition, dimers, trimers, etc., of A and of B and, further, complexes of A and B with the general formula A_nB_m, where n and m are positive integers. Concentrations of the various chemical species are found from chemical equilibrium constants coupled with material balances.

The chemical hypothesis was used many years ago to calculate activity coefficients in liquid mixtures and also to calculate second virial coefficients of pure and mixed gases. However, the early work was restricted to liquids or to gases at moderate densities, and most of that early work assumed that the "true" chemical species form ideal mixtures. It was not until 1976 that Heidemann and Prausnitz (1976) combined the chemical hypothesis with an equation of state valid for all fluid densities. Unfortunately, Heidemann's work is limited to pure fluids; for extension to mixtures additional assumptions are required as discussed by Hu, et al., (1984). However, the chemical hypothesis, coupled with an equation of state, becomes tractable for mixtures provided that association is limited to dimers as shown in 1979 by Gmehling, et al. Since then, several other authors have presented similar ideas. Particularly noteworthy is the work of Anderko, (1990) where attention is given to mixtures containing hydrogen fluoride (Lencka and Anderko, 1993) and to aqueous mixtures (Anderko, 1991).

Gmehling, et al., (1979) used an equation of state of the van der Waals form (in particular, the perturbed-hard-chain equation of state) coupled with a dimerization hypothesis. A binary mixture of nominal components A and B is considered to be a five-species mixture containing two types of monomer (A_1 and B_1) and three types of dimer (A_2, B_2, AB).

There are three chemical equilibrium constants:

$$K_{A_2} = \frac{z_{A_2}}{z_{A_1}^2} \frac{\phi_{A_2}}{\phi_{A_1}^2} \frac{1}{P} \tag{8-12.16a}$$

$$K_{B_2} = \frac{z_{B_2}}{z_{B_1}^2} \frac{\phi_{B_2}}{\phi_{B_1}^2} \frac{1}{P} \tag{8-12.16b}$$

$$K_{AB} = \frac{z_{AB}}{z_{A_1} z_{B_1}} \frac{\phi_{AB}}{\phi_{A_1} \phi_{B_1}} \frac{1}{P} \tag{8-12.16c}$$

where z_{A_1} is the mole fraction of A_1 (etc.) and ϕ_{A_1} is the fugacity coefficient of A_1 (etc.). The fugacity coefficient is found from the equation of state by using physical interaction parameters to characterize monomer-monomer, monomer-dimer, and dimer-dimer interactions.

Mole fractions z are related to nominal mole fractions x_A and x_B through chemical equilibrium constants and material balances.

To reduce the number of adjustable parameters, Gmehling established physically reasonable relations between parameters for monomers and those for dimers.

The temperature dependence of equilibrium constant K_{A_2} is given by

$$\ln K_{A_2} = -\frac{\Delta H^\circ_{A_2}}{RT} + \frac{\Delta S^\circ_{A_2}}{R} \tag{8-12.17}$$

where $\Delta H^\circ_{A_2}$ is the enthalpy and $\Delta S^\circ_{A_2}$ is the entropy of formation of dimer A_2 in the standard state. Similar equations hold for K_{B_2} and K_{AB}.

All pure-component parameters (including K_{A_2} and K_{B_2}) are obtained from experimental density and vapor-pressure data.

A reasonable estimate for ΔH°_{AB} is provided by

$$\Delta H^\circ_{AB} = \tfrac{1}{2}(\Delta H^\circ_{A_2} + \Delta H^\circ_{B_2}) \tag{8-12.18}$$

but a similar relation of ΔS°_{AB} does not hold. For a binary mixture of A and B, ΔS°_{AB} must be found from binary data.

The equations for vapor-liquid equilibrium are

$$f^V_A = f^L_A \quad \text{and} \quad f^V_B = f^L_B \tag{8-12.19}$$

where f stands for fugacity and superscripts V and L stand for vapor and liquid, respectively. As shown by Prigogine and Defay (1954), Eq. (8-12.19) can be replaced without loss of generality by

$$f^V_{A_1} = f^L_{A_1} \quad \text{and} \quad f^V_{B_1} = f^L_{B_1} \tag{8-12.20}$$

Figure 8-22 shows calculated and observed vapor-liquid equilibria for methanol-

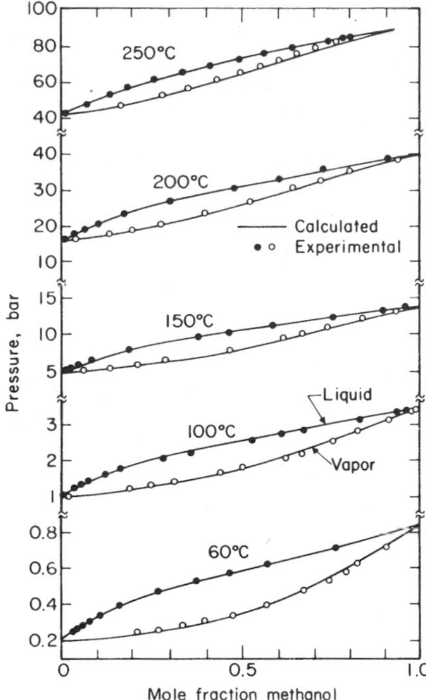

FIGURE 8-22 Vapor-liquid equilibria for methanol-water. Calculations based on chemical theory.

water at modest and advanced pressures. Calculations are based on Gmehling's equation as outlined above. For this mixture, the calculations require only two adjustable binary parameters that are independent of temperature over the indicated temperature range. One of those is ΔS°_{AB}, and the other is $k_{A_1-B_1}$, a physical parameter to characterize $A_1 - B_1$ interactions.

Gmehling's equation of state, coupled with a (chemical) dimerization hypothesis, is particularly useful for calculating vapor-liquid equilibria at high pressures for fluid mixtures containing polar and nonpolar components, some subcritical and some supercritical. By using an equation of state valid for both phases, the equations of phase equilibrium avoid the awkward problem of defining a liquid-phase standard state for a supercritical component. By superimposing dimerization equilibria onto a "normal" equation of state, Gmehling achieves good representation of thermodynamic properties for both gaseous and liquid mixtures containing polar or hydrogen-bonded fluids in addition to "normal" fluids (such as common gases and hydrocarbons) by using the same characteristic parameters for both phases.

Buck (1984) tested Gmehling's method by comparing calculated and observed vapor-liquid equilibria for several ternary systems containing polar and hydrogen-bonded fluids. Encouraged by favorable comparisons, Buck then described an application of Gmehling's method to an isothermal flash calculation at 200°C and 100 bar. Table 8-39 shows specified feed compositions and calculated compositions for the vapor and for the liquid at equilibrium. All required parameters were obtained from pure-component and binary experimental data.

TABLE 8-39 Isothermal Flash Calculation Using Gmehling's Equation of State at 200°C and 100 bar (Buck, 1984)

Component	Mole percent of		
	Feed	Vapor	Liquid
Hydrogen	6.0	33.86	2.03
Carbon monoxide	5.5	24.63	2.77
Methane	0.3	1.08	0.19
Methyl acetate	27.2	13.34	29.18
Ethanol	39.9	19.35	42.83
Water	3.3	1.81	3.51
1,4-Dioxane	17.8	5.93	19.49
Total moles	100.00	12.49	87.51

To implement Gmehling's method for multicomponent fluid mixtures, it is necessary to construct a far-from-trivial computer program requiring a variety of iterations. The calculations summarized in Table 8-39 are for seven components, but the number of (assumed) chemical species is much larger. For H_2, CO, and CH_4 it is reasonable to assume that no dimers are formed; further, it is reasonable to assume that these components do not form cross-dimers with each other or with the other components in the mixture. However, the four polar components form dimers with themselves and with each other. In Gmehling's method, therefore, this 7-component mixture is considered to be a mixture of 17 chemical species.

Example 8-22 Use Gmehling's method to calculate the bubble-point pressure and vapor-phase composition for a mixture of 4.46 mole % methanol (1) and 95.54 mole % water (2) at 60°C.

solution In the Gmehling model, a water-methanol mixture contains five species: methanol and water monomers, methanol dimers, water dimers, and a methanol-water cross-dimer. The mole fractions of these are, respectively, z_{M1}, z_{M2}, z_{D1}, z_{D2}, and z_{D12}. There are 13 unknowns and 13 equations. The unknowns include five liquid z values, five vapor z values, the pressure, and molar volumes of the liquid and vapor phases. The 13 equations are the equation of state for both the liquid and vapor phases, the three reaction-equilibrium equations (8-12.16*a*) through (8-12.16*c*), five fugacity equalities (that is, $f_i^L = f_i^V$), $\Sigma z_i = 1$ in both the liquid and vapor, and a material balance accounting for the overall mixture composition. Pure-component parameters from Gmehling, et al., (1979) are as follows:

Component	T^*, K	V^*, cm³/mol	$\Delta S°/R$	$\Delta H°/R$, K
Methanol (1)	348.09	26.224	−16.47	−5272
Water (2)	466.73	12.227	−14.505	−4313

From Eq. (8-12.17),

$$K_1 = \exp\left(-16.47 + \frac{5272}{333.15}\right) = 0.525$$

Similarly, $K_2 = 0.210$
From Gmehling, et al., (1979), $\Delta S^\circ_{12}/R = -15.228$ and $K_{12} = 0.0371$ so that $K_{12} = \exp(-15.228 + (5272 + 4313)(0.5)/333.15) = 0.431$
The problem may be solved by the following procedure:

1. Guess P
2. Guess all $\phi_i = 1$
3. Guess $y_i = x_i$
4. Solve the reaction-equilibria problem for values of z_i in each phase [Eqs. (8-12.16*a*) to (8-12.16*c*)].
5. Calculate mixture parameters with mixing rules from Gmehling, et al. (1979)
6. Solve the equation of state for V^L and V^V.
7. Calculate ϕ_i for each of the five species in both phases.
8. Go back to step 4 and recalculate z_i values. When z_i values no longer change, reaction equilibria are satisfied, but phase equilibria are not.
9. Calculate K_i by $K_i = \phi_i^L/\phi_i^V$, where $K_i \equiv z_i^V/z_i^L$.
10. See if $\sum_i K_i z_i^L = 1$; if not, adjust P according to $P_{\text{new}} = P_{\text{old}} (\sum_i K_i z_i^L)$ and go back to step 4.

This procedure converges to the following values:

	z_{M1}	z_{M2}	z_{D1}	z_{D2}	z_{D12}	V, L/mol
Liquid	0.001680	0.05252	0.003510	0.8642	0.07808	0.03706
Vapor	0.2641	0.6794	0.00980	0.02603	0.0207	103.3

	ϕ_{M1}	ϕ_{M2}	ϕ_{D1}	ϕ_{D2}	ϕ_{D12}
Liquid	156.6	12.91	2.775	0.02996	0.2636
Vapor	0.9976	0.9983	0.9937	0.9947	0.9942

$P = 0.2675$ bar

The above numbers satisfy the material balance and reaction-equilibria equations:

$$x_1 = \frac{z^L_{M1} + 2z^L_{D1} + z^L_{D12}}{z^L_{M1} + z^L_{M2} + 2(z^L_{D1} + z^L_{D12} + z^L_{D2})}$$

$$= \frac{0.00168 + (2)(0.00351) + 0.0781}{0.00168 + 0.0525 + (2)(0.00351 + 0.864 + 0.0781)} = 0.0446$$

$$K_1 = \frac{z_{D1}\phi_{D1}}{z^2_{M1}\phi^2_{M1}}\frac{1}{P} = \frac{0.00351}{(0.00168)^2}\frac{2.775}{(156.6)^2}\frac{1}{0.2675} = 0.526$$

$$K_{12} = \frac{z_{D12}}{z_{M1}z_{M2}}\frac{\phi_{D12}}{\phi_{M1}\phi_{M2}}\frac{1}{P} = \frac{0.07808}{(0.00168)(0.05252)}\frac{(0.2336)}{(156.6)(12.91)}\frac{1}{0.2675} = 0.431$$

$$K_2 = \frac{z_{D2}}{z^2_{M2}}\frac{\phi_{D2}}{\phi^2_{M2}}\frac{1}{P} = \frac{0.8642}{(0.05252)^2}\frac{0.02996}{(12.91)^2}\frac{1}{0.2675} = 0.210$$

The reaction expressions are verified above for liquid-phase values. They are also satisfied for vapor-phase values, since $f^L_i = f^V_i$ is satisfied for each of the five-components (as can easily be verified). Mixture parameters to be used can be obtained from mixing rules (Guehling, et al., 1979). For example, for the liquid phase,

$$\begin{aligned}
\langle cT^*V^*\rangle &= z_{M1}c_{M1}T^*_{M1}\,[z_{M1}V^*_{M1} + z_{M2}V^*_{M2}(1 - k_{12}) + z_{D1}\,V^*_{D1} + z_{D2}V^*_{D2}(1 - k_{12}) \\
&\quad + z_{D12}V^*_{D12}(1 - k_{12})^{1/2} + z_{M2}c_{M2}T^*_{M2}(z_{M1}V^*_{M1}(1 - k_{12}) + z_{M2}V^*_{M2} \\
&\quad + z_{D1}\,V^*_{D1}(1 - k_{12}) \\
&\quad + z_{D2}V^*_{D2} + z_{D12}V^*_{D12}(1 - k_{12})^{1/2}] + z_{D1}c_{D1}T^*_{D1}[z_{M1}V^*_{M1} \\
&\quad + z_{M2}V^*_{M2}(1 - k_{12}) + z_{D1}V^*_{D1} + z_{D2}V^*_{D2}(1 - k_{12}) \\
&\quad + z_{D12}V^*_{D12}(1 - k_{12})^{1/2}] + z_{D2}c_{D2}T^*_{D2}[z_{M1}V^*_{M1}(1 - k_{12}) \\
&\quad + z_{M2}V^*_{M2} + z_{D1}V^*_{D1}(1 - k_{12}) + z_{D2}V^*_{D2} \\
&\quad + z_{D12}V^*_{D12}(1 - k_{12})^{1/2}] + z_{D12}c_{D12}T^*_{D12}[z_{M1}V^*_{M1}(1 - k_{12})^{1/2} \\
&\quad + z_{M2}V^*_{M2}(1 - k_{12})^{1/2} + z_{D1}V^*_{D1}(1 - k_{12})^{1/2} \\
&\quad + z_{D2}V^*_{D2}(1 - k_{12})^{1/2} + z_{D12}V^*_{D12}(1 - k_{12})^{1/2}] \\
&= 12 + 537 + 45 + 15465 + 1200
\end{aligned}$$

$$\langle cT^*V^*\rangle = 17{,}259\ (\mathrm{K}\cdot\mathrm{cm}^3)/\mathrm{mol}$$

Values of all parameters are as follows:

	$\langle c\rangle$	$\langle V^*\rangle$	$\langle cT^*V^*\rangle$	$\langle T^*\rangle^{(2)}$
Vapor	1.017	16.93	7503	444.92
Liquid	1.284	21.97	17259	614.72

As the last step, values of y_i and V in liters per mol of original monomer may be calculated:

$$y_1 = \frac{z^V_{M1} + 2z^V_{D1} + z^V_{D12}}{z^V_{M1} + z^V_{M2} + 2(z^V_{D1} + z^V_{D2} + z^V_{D12})} = 0.2881$$

$$y_2 = 0.7119$$

$$V^V = \frac{103.3}{z^V_{M1} + z^V_{M2} + 2(z^V_{D1} + z^V_{D2} + z^V_{D12})} = 97.8 \text{ L/mol}$$

$$V^L = \frac{0.03706}{z^L_{M1} + z^L_{M2} + 2(z^L_{D1} + z^L_{D2} + z^L_{D12})} = 0.01905 \text{ L/mol}$$

Experimental values are $P = 0.2625$ bar and $y_1 = 0.2699$.

Grenzheuser and Gmehling (1986) have presented a revised version of this EoS but the essential ideas and procedures remain as before.

Calculations Based on 1-vdW Mixing Rules. For direct use of EoS methods for computing vapor-liquid equilibria, the traditional mixing rules can be attributed to van der Waals (See Sec. 5-5). The *PVT* expression is identical to the pure-component equation and the composition dependence is put into the equation-of-state constants in a simple fashion. We illustrate this formulation in the next two examples using a Peng-Robinson EoS as modified by Stryek and Vera (1986) (See Tables 4-6 and 4-7).

Example 8-23 Use the one-parameter van der Waals (1−vdW) mixing rules Eq. (5-5.2) and the Peng-Robinson-Stryek-Vera EoS (Tables 4-6 and 4-7) to calculate vapor-liquid equilibria for the acetone (1)—benzene (2) binary at 50°C. To estimate the binary interaction parameters, k_{12} in the 1-vdW combining rule, fit pressures from the PRSV EoS to the experimental data reported by Kraus and Linek (1971).

solution There are two equations of the form (8-12.1), one for each component

$$y_i \phi_i^V = x_i \phi_i^L \qquad (i = 1,2) \tag{8-12.21}$$

where y_i and x_i are the vapor and liquid mole fractions of component i, respectively: ϕ_i is the fugacity coefficient with the superscript^V for vapor and ^{L} for liquid.

Equation (6-7.10) is used to obtain the expression for ϕ_i^V and ϕ_i^L. The PRSV EoS is, from Table 4-6,

$$P = \frac{RT}{(V - b)} - \frac{a}{V^2 + 2bV - b^2} \tag{8-12.22}$$

The 1-vdW mixing rules for the parameters in the liquid phase, a^L and b^L, and in the vapor phase, a^V and b^V, are

$$a^L = \sum_{i=1}^{N}\sum_{j=1}^{N} x_i x_j a_{ij} \qquad a^V = \sum_{i=1}^{N}\sum_{j=1}^{N} y_1 y_j a_{ij} \tag{8-12.23a}$$

$$b^L = \sum_{i=1}^{N} x_i b_i \qquad b^V = \sum_{i=1}^{N} y_i b_i \tag{8-12.23b}$$

Where N is the number of components in the mixture (here N = 2). The pure-component parameters for the PRSV EoS from Tables 4-7 and 4-8 are

$$a_i = \left(4.57235 \frac{R^2 T_{c,i}^2}{P_{c,i}}\right) \alpha_i \left(\frac{T}{T_{c,i}}\right) \tag{8-12.24a}$$

$$b_i = 0.077796 \frac{RT_{c,i}}{P_{c,i}} \tag{8-12.24b}$$

with

$$\alpha_i \left(\frac{T}{T_{c,i}}\right) = \left[1 + \kappa_i \left(1 - \frac{T}{T_{c,i}}\right)^{1/2}\right]^2 \tag{8-12.25}$$

and the parameter κ_i is found from

$$\kappa_i = 0.378893 + 1.4897\omega_i - 0.17138\omega_i^2 + 0.0196554\omega_i^3$$
$$+ \kappa_i^{(1)} \left[1 + \left(\frac{T}{T_{c,i}}\right)^{1/2}\right]\left[0.7 - \frac{T}{T_{c,i}}\right] \tag{8-12.26a}$$

where ω_i is the component's acentric factor (see Sec. 2-3). The parameter $\kappa_i^{(1)}$ is found by fitting experimental P_{vp} data over some temperature range. In this case, $\kappa_1^{(1)} = -0.0089$ and $\kappa_2^{(1)} = 0.0702$.

Finally, the common 1-vdW combining rule [Eq. (5-2.4*b*) with *a* in place of *Q*] is used for a_{ij}.

$$a_{ij} = (a_i a_j)^{1/2}(1 - k_{ij}) \tag{8-12.26b}$$

The binary interaction parameter, k_{ij}, is found by minimizing the objective function on the total pressure:

$$F = \sum_{\text{data}} |P_{\text{EoS}} - P_{\text{Exp}}| \tag{8-12.27}$$

where EoS stands for calculations from the above EoS and Exp stands for experimental data, here at 50°C. Using the 12 data prints from Kraus and Linek (1971), the optimal value for this binary is $k_{12} = 0.032$.

Substituting the PRSV EoS into Eq. (6-7.10) for acetone (1) in both the liquid and vapor phases yields

$$\ln \phi_1^L = \frac{b_1}{b^L}(Z^L - 1) - \ln(Z^L - B^L) - \frac{A^L}{2\sqrt{2}\, B^L}$$
$$\left(\frac{2(x_1 a_{11} + x_2 a_{12})}{a^L} - \frac{b_1}{b^L}\right) \ln \left[\frac{Z^L + (1 + \sqrt{2})B^L}{Z^L + (1 - \sqrt{2})B^L}\right] \tag{8-12.28a}$$

$$\ln \phi_1^V = \frac{b_1}{v^V}(Z^V - 1) - \ln(Z^V - B^V) - \frac{A^V}{2\sqrt{2}B^V}$$
$$\left(\frac{2(y_1 a_{11} + y_2 a_{12})}{a^V} - \frac{b_1}{b^V}\right) \ln \left[\frac{Z^V + (1 + \sqrt{2})B^V}{Z^V + (1 - \sqrt{2})B^V}\right] \tag{8-12.28b}$$

Here $Z = PV/RT$ is the compressibility factor and *A* and *B* are dimensionless quantities whose values depend upon the phase of interest:

$$A^L = \frac{a^L P}{R^2T^2} \qquad A^V = \frac{a^V P}{R^2T^2} \tag{8-12.29a}$$

$$B^L = \frac{b^L P}{RT} \qquad B^V = \frac{b^V P}{RT} \tag{8-12.29b}$$

For benzene (2), Eqs. (8-12.28*a*) and (8-12.28*b*) are used with all subscripts $_1$ and $_2$ interchanged.

For a given value of x_1 and with $T = 323.15$ K, we have four independent equations, Eq. (8-12.1) twice, once for each component using Eqs. (8-12.28), and Eq. (8-12.22) twice, once for each phase. We have four unknowns, y_1, P, V^V, and V^L. When needed, the mole fractions x_2 and y_2 are obtained from

$$x_1 + x_2 = 1 \tag{8-12.30a}$$

$$y_1 + y_2 = 1 \tag{8-12.30b}$$

We substitute Eqs. (8-12.23) to (8-12.26) into Eq. (8-12.22) and into Eqs. (8-12.28) which are used in Eq. (8-12.1). Table 8-40 shows calculated results compared with the experimental results of Kraus and Linek (1971). Note that the values were calculated at the same x_1 as those in the experiments. Figure 8-23 compares calculated and observed results.

In this example, agreement with experiment is excellent because k_{12} is obtained from experimental data at the same temperature, 50°C. In practice, this is often not possible because, if experimental data are available, they are likely to be at a different temperature (e.g., 25°C). For such cases, if the temperature difference is not large, k_{12} values from experimental data at one temperature can be used at the desired temperature. An illustration is given in the next example.

Example 8-24 Using 1-vdW mixing rules and the PRSV EOS, calculate vapor-liquid equilibria *Pxy* for the binary methanol (1)–water (2) at 100°C. Calculate the binary interaction parameter k_{12} by fitting equation-of-state calculations to experimental data for the equilibrium pressure in two ways:
(a) Using experimental data for the same system at 25°C from Butler, et al. (1933).

TABLE 8-40 Vapor-Liquid Equilibria for Acetone (1)–Benzene (2) at 50°C

	Calculated		Experimental	
x_1	y_1	P (mmHg)	y_1	P (mmHg)
0.0417	0.1280	299.10	0.1758	299.42
0.1011	0.2648	330.75	0.2769	335.56
0.1639	0.3725	363.74	0.3689	363.75
0.2700	0.5047	411.64	0.4921	411.55
0.3248	0.5578	433.22	0.5535	432.92
0.3734	0.5993	450.89	0.5946	449.12
0.4629	0.6660	480.38	0.6631	477.91
0.5300	0.7104	500.34	0.7085	500.32
0.5885	0.7466	516.53	0.7481	517.09
0.7319	0.8311	552.35	0.8355	551.16
0.8437	0.8970	577.18	0.8782	578.05
0.9300	0.9520	595.88	0.9238	596.00

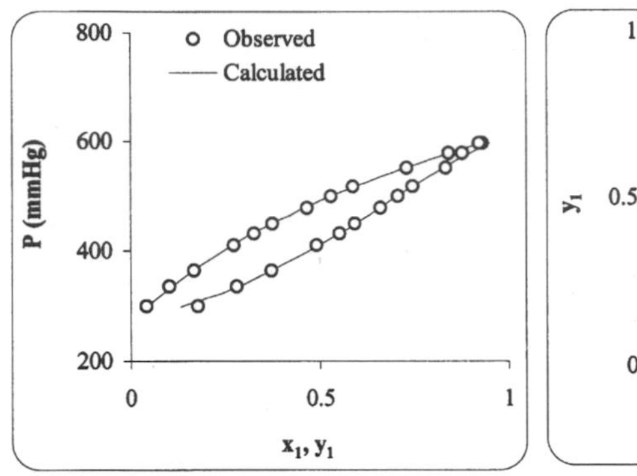

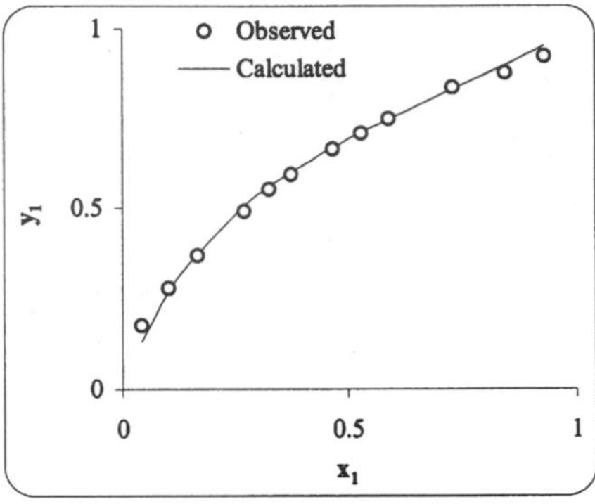

FIGURE 8-23 *Pxy* and *xy* plots for acetone (1)–benzene (2) at 50°C from experiment and from the PRSV EoS.

(b) Using experimental data for the same system at 100°C from Griswold and Wong (1952).

solution The procedure for solving this problem is similar to that shown in Example 8-23. For this binary, parameters $\kappa_i^{(1)}$ $(i = 1,2)$ for the two components are

Component	$\kappa_i^{(1)}$
1	−0.1682
2	−0.0664

(a) Using experimental data at 25°C to find k_{12}: Table 8-41 gives experimental data for methanol (1)–water (2) at 25°C.

TABLE 8-41 Experimental Vapor-Liquid Equilibria for Methanol (1)–Water (2) at 25°C from Butler, et al. (1933)

x_1	y_1	P (mmHg)
0.0202	0.1441	26.7
0.0403	0.2557	30.0
0.0620	0.3463	34.0
0.0791	0.4160	36.3
0.1145	0.5047	42.6
0.2017	0.6474	55.3
0.3973	0.7904	75.4
0.6579	0.8908	96.2
0.8137	0.9521	109.9
1	1	126.6

Fitting the PRSV EOS calculations to experimental total pressures at 25°C, the optimum k_{12} is −0.0965. Using this k_{12}, calculated results are given in Table 8-42. Observed data at 100°C are from Griswold and Wong (1952). Using experimental data at 100°C in Table 8-42, the fitted value of k_{12} is −0.0754. Corresponding calculated Pxy values are shown in Table 8-43. Figure 8-24 compares measured and calculated results by both methods.

Figure 8-24 indicates that for this binary, calculated results at 100°C based on experimental data at 25°C are similar to those calculated based on experimental data at 100°C. For both cases, agreement with experiment is only fair because the PRSV EOS is not truly suitable for strongly polar or hydrogen-bonded fluids like water and methanol. In general, however, results from a k_{12} based on data at the

TABLE 8-42 Vapor-Liquid Equilibria for Methanol (1)–Water (2) at 100°C with k_{12} = −0.0965 from 25°C Data

Calculated, k_{12} = −0.0965			Experimental		
x_1	y_1	P (mmHg)	x_1	y_1	P (mmHg)
0.0	0.000	761.5	0.000	0.000	760.0
0.1	0.353	1076.8	0.011	0.086	827.6
0.2	0.508	1295.1	0.035	0.191	931.0
0.3	0.610	1477.8	0.053	0.245	1003.2
0.4	0.689	1644.7	0.121	0.434	1235.8
0.5	0.756	1810.7	0.281	0.619	1536.0
0.6	0.815	1973.7	0.352	0.662	1624.1
0.7	0.868	2137.7	0.522	0.750	1882.5
0.8	0.916	2303.4	0.667	0.824	2115.1
0.9	0.960	2471.4	0.826	0.911	2337.8
1.0	1.000	2642.1	0.932	0.969	2508.0
			1.000	1.000	2650.9

TABLE 8-43 Calculated VLE for Methanol (1)–Water (2) at 100°C with k_{12} = −0.0754 from 100°C Data

Calculated, k_{12} = −0.0754		
x_1	y_1	P (mmHg)
0.0	0.000	763.9
0.1	0.395	1179.8
0.2	0.527	1401.5
0.3	0.610	1565.8
0.4	0.677	1710.3
0.5	0.737	1847.9
0.6	0.794	1983.3
0.7	0.848	2118.7
0.8	0.901	2254.8
0.9	0.951	2392.2
1.0	1.000	2530.8

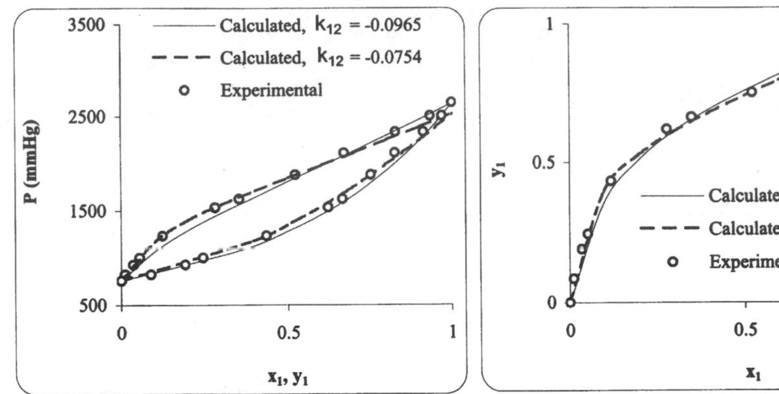

FIGURE 8-24 Vapor-liquid equilibria for methanol(1)–water (2) at 100°C

same temperature of interest are likely to be better than those using a k_{12} based on data at another temperature.

Calculations with Mixing Rules Based on g^E Models. The above examples show the application of 1-*vdW* mixing rules for computing phase equilibria from EoS models. These expressions are adequate for simple and normal fluids (see Sec. 4-2), but for more complex substances, the results can be poor. The principal difficulty is that in complex systems, especially for those with some polar and some nonpolar components, the composition dependence of the fugacity is more complex than that given by simple mixing rules. Activity-coefficient models such as those in Table 8-3 or 8-8 are much more able to describe the experimental behavior. When properly formulated, this procedure can give the quadratic composition dependence of the second virial coefficient.

An important advance in the description of phase equilibria is to combine the strengths of both EoS and activity coefficient approaches by forcing the mixing rule of an EoS to behave with a composition dependence like the g^E model. These are called g^E mixing rules and generally include the direct use of activity coefficient parameters fitted to VLE data. There is a large literature associated with this methodology and the basic techniques are described in Sec. 5-5.

One of the more popular g^E methods is due to Wong and Sandler (1992) with many application details provided by Orbey and Sandler (1995, 1998). The next four examples show how this model can be applied for phase equilibria.

Example 8-25 Use the Wong-Sandler mixing rules and the PRSV EoS to calculate vapor-liquid equilibria for the binary 2-propanol (1)-water (2) at 80°C. To calculate g^E, use the NRTL equation with parameters fitted to data at 30°C.

solution Equation (8-12.1) is used for both components.

$$y_i \phi_i^V = x_i \phi_i^L \qquad (i = 1,2) \tag{8-12.31}$$

where y_i and x_i are the vapor- and liquid-phase mole fractions of component i and ϕ_i^V

and ϕ_i^L are the vapor and liquid fugacity coefficients. The formula and parameters for the PRSV EoS are given in Example 8-23.

The fitted parameters for the EoS are $\kappa_1^{(1)} = 0.2326$ and $\kappa_2^{(1)} = -0.0664$. The mixing rules of Eq. (5-5.10) with a in place of θ give the mixture parameters a^V, a^L, b^V, and b^L as

$$a^V = b^V\left(\frac{g^{EV}}{C} + y_1\frac{a_{11}}{b_1} + y_2\frac{a_{22}}{b_2}\right) \qquad a^L = b^L\left(\frac{g^{EL}}{C} + x_1\frac{a_{11}}{b_1} + x_2\frac{a_{22}}{b_2}\right) \tag{8-12.32a}$$

$$b^V = \frac{\sum_{i=1}^{2}\sum_{j=1}^{2} y_i y_j (b - a/RT)_{ij}}{1 - \frac{1}{RT}\left(\sum_{i=1}^{2} y_i \frac{a_{ii}}{b_i} + \frac{g^{EV}}{C}\right)} \qquad b^L = \frac{\sum_{i=1}^{2}\sum_{j=1}^{2} x_i x_j (b - a/RT)_{ij}}{1 - \frac{1}{RT}\left(\sum_{i=1}^{2} x_i \frac{a_{ii}}{b_i} + \frac{g^{EL}}{C}\right)} \tag{8-12.32b}$$

where g^{EV} is the excess Gibbs energy at the vapor composition and g^{EL} is that at the liquid composition. The constant C for the PRSV EoS is -0.623. It is evident that these are more complex than the 1-vdW rule of Eq. (8-12.23).

With the Wong-Sandler mixing rules, we can use any convenient model for g^E; here we use the NRTL expression which for a binary is

$$\frac{g^E}{RT} = x_1 x_2\left(\frac{\tau_{21}G_{21}}{x_1 + G_{21}x_2} + \frac{\tau_{12}G_{12}}{x_2 + G_{12}x_1}\right) \tag{8-12.33}$$

with $G_{ij} = \exp(-\alpha\tau_{ij})$. There are three parameters: α, τ_{12}, and τ_{21}. Typically α is fixed independently. Here the parameter values are those obtained by Gmehling and Onken (1977a) by fitting the data of Udovenko and Mazanko (1967). The results are $\alpha = 0.2893$, $\tau_{12} = 0.1759$, and $\tau_{21} = 2.1028$.

The cross parameter of Eq. (8-12.32b) is

$$(b - a/RT)_{12} = \frac{1}{2}(b_1 + b_2) - \frac{1}{RT}(a_{11}a_{22})^{1/2}(1 - k_{12}) \tag{8-12.34}$$

Among the two approaches for obtaining a value of the parameter k_{12} described in Sec. 5-5, we choose to match the g^E model by minimizing the objective function F, rather than match second cross virial coefficients as suggested by Kolar and Kojima (1994) and others.

$$F = \sum_{\text{data}}\left|\frac{a^E_{EoS}}{RT} - \frac{g^{EL}}{RT}\right| \tag{8-12.35}$$

The summation of Eq. (8-12.35) is for all data points at the specified temperature of 30°C, and evaluated at the liquid compositions. The molar excess Helmholtz energy from the EoS a^E_{EoS}, is:

$$a^E_{EoS} = \frac{a^L}{b^L} - x_1\frac{a_{11}}{b_1} - x_2\frac{a_{22}}{b_2} \tag{8-12.36}$$

Optimizing Eq. (8-12.35), we obtain $k_{12} = 0.3644$ for use in Eq. (8-12.34). Combining the Wong-Sandler mixing rules and the PRSV EoS, the liquid-phase fugacity coefficient of 2-propanol is

$$\ln \phi_1^L = -\ln \frac{P(V^L - b^L)}{RT} + \frac{1}{b^L}(Z^L - 1)\left(\frac{\partial nb^L}{\partial n_1}\right)_{T,n_2} + \frac{a^L}{2\sqrt{2}b^L RT} \times \left[\frac{1}{na^L}\left(\frac{\partial n^2 a^L}{\partial n_1}\right)_{T,n_2} - \frac{1}{b^L}\left(\frac{\partial nb^L}{\partial n_1}\right)_{T,n_2}\right] \ln \frac{V^L + (1 - \sqrt{2})b^L}{V^L + (1 + \sqrt{2})b^L} \qquad (8\text{-}12.37)$$

Here all properties are those of the liquid phase; $Z^L = PV^L/RT$ is the compressibility factor. The partial derivatives of a^L and b^L are

$$\frac{1}{n}\left(\frac{\partial n^2 a^L}{\partial n_1}\right)_{T,n_2} = RTD^L\left(\frac{\partial nb^L}{\partial n_1}\right) + RTb^L\left(\frac{\partial nD^L}{\partial n_1}\right) \qquad (8\text{-}12.38a)$$

$$\left(\frac{\partial nb^L}{\partial n_1}\right)_{T,n_2} = \frac{1}{1 - D^L}\left[\frac{1}{n}\left(\frac{\partial n^2 Q^L}{\partial n_1}\right)\right] - \frac{Q^L}{(1 - D^L)^2}\left[1 - \left(\frac{\partial nD^L}{\partial n_1}\right)\right] \qquad (8\text{-}12.38b)$$

Q^L, D^L, and their partial derivatives are

$$Q^L = \sum_{i=1}^{2}\sum_{j=1}^{2} x_i x_j (b - a/RT)_{ij}$$

$$D^L = \frac{1}{C}\frac{g^E}{RT} + \sum_{i=1}^{2} x_i \frac{a_i}{RTb_i}$$

$$\frac{1}{n}\left(\frac{\partial n^2 Q^L}{\partial n_1}\right)_{T,n_2} = 2x_2(b - a/RT)_{12} + 2x_1(b - a/RT)_{11} \qquad (8\text{-}12.39)$$

$$\left(\frac{\partial nD^L}{\partial n_1}\right)_{T,n_2} = \frac{a_1}{RTb_1} + \frac{1}{CRT}\left(\frac{\partial ng^E}{\partial n_1}\right)_{T,n_2}$$

In a similar manner, $\ln \phi_2^L$ for water is obtained by interchanging subscripts 1 and 2 in Eqs. (8-12.37) to (8-12.39). For the vapor phase, $\ln \phi_1^V$ and $\ln \phi_2^V$ are computed using Eqs. (8-12.37) to (8-12.39) with the vapor-phase rather than liquid-phase composition and volume.

Table 8-44 shows calculated and experimental results at the x_1 values reported by Wu, et al. (1988).

Figure 8-25 compares calculated and experimental results. The agreement is quite good though the calculated pressures are high. This example illustrates that the Wong-Sandler method can be useful for extrapolating experimental data to higher temperature. In this particular example, data at 30°C were used to predict vapor-liquid equilibria at 80°C.

Example 8-26 Use the Wong-Sandler mixing rules and the PRSV EOS to calculate vapor-liquid equilibria Pxy for the binary methanol (1)-water (2) at 100°C. To calculate g^E, use the UNIFAC correlation at 25°C.

solution The procedure for solving this problem is the same as that of Example 8-25. Here the PRSV parameters $\kappa_i^{(1)}$ for the components are

Component	$\kappa_i^{(1)}$
1	−0.1682
2	−0.0664

For the UNIFAC correlation (Example 8-12), the two components are groups. Group-volume (R_k) and surface-area parameter (Q_k) are

Group	R_k	Q_k
CH_3OH	1.4311	1.432
H_2O	0.9200	1.400

TABLE 8-44 Vapor-Liquid Equilibria for 2-propanol (1)–water (2) at 80°C (Wu, et al., 1988)

	Calculated		Experimental	
x_1	y_1	P (bar)	y_1	P (bar)
0.000	0.000	0.474	0.000	0.475
0.013	0.209	0.592	0.223	0.608
0.098	0.521	0.929	0.504	0.888
0.174	0.555	0.983	0.533	0.922
0.293	0.560	0.990	0.553	0.937
0.380	0.565	0.994	0.569	0.952
0.469	0.580	1.003	0.590	0.985
0.555	0.606	1.012	0.623	0.996
0.695	0.678	1.015	0.700	0.993
0.808	0.766	1.000	0.781	0.990
0.947	0.922	0.953	0.921	0.948
1.000	1.000	0.926	1.000	0.925

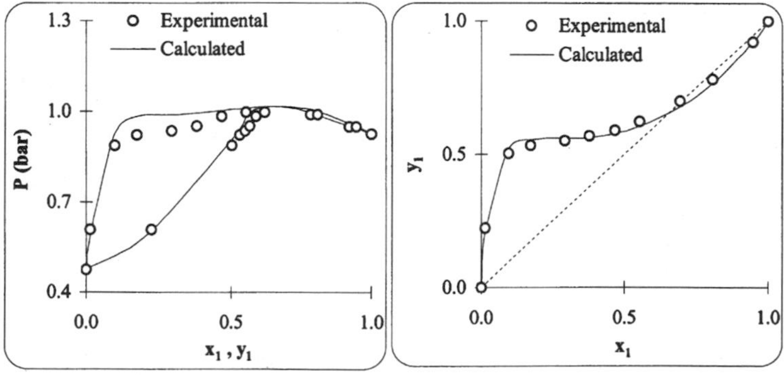

FIGURE 8-25 *Pxy* and *xy* plots for 2-propanol(1)–water (2) at 80°C.

Group-group interaction parameters (K) are

Group	CH_3OH	H_2O
CH_3OH	0.0	−181.0
H_2O	289.6	0.0

Table 8-45 gives calculated results for g^E at 25°C.

Using the results in Table 8-45, we minimize the objective function of Eq. (8-12.35) and obtain an optimum values of 0.0869 for k_{12} at 25°C. Table 8-46 gives VLE calculations at 100°C and the experimental data at the same temperature but different compositions from Griswold and Wong (1952).

TABLE 8-45 Vapor-Liquid Equilibria, Activity Coefficients, and Molar Excess Gibbs Energy for Methanol (1)–water (2) calculated by UNIFAC at 25°C

x_1	y_1	P (mmHg)	γ_1	γ_2	g^E/RT
0.0	0.000	23.7	2.24	1.00	0.000
0.1	0.508	43.9	1.75	1.01	0.065
0.2	0.655	57.3	1.47	1.04	0.108
0.3	0.734	67.8	1.30	1.09	0.139
0.4	0.789	77.0	1.19	1.14	0.148
0.5	0.834	85.5	1.12	1.20	0.148
0.6	0.872	93.7	1.07	1.27	0.136
0.7	0.906	102.0	1.03	1.34	0.108
0.8	0.939	110.3	1.01	1.42	0.078
0.9	0.970	118.9	1.00	1.51	0.041
1.0	1.000	127.7	1.00	1.60	0.000

TABLE 8-46 Vapor-Liquid Equilibria for Methanol (1)–water (2) at 100°C

Calculated			Experimental		
x_1	y_1	P (mmHg)	x_1	y_1	P (mmHg)
0.0	0.000	760.6	0.000	0.000	760.0
0.1	0.401	1161.1	0.011	0.086	827.6
0.2	0.550	1425.9	0.035	0.191	931.0
0.3	0.640	1625.4	0.053	0.245	1003.2
0.4	0.705	1791.9	0.121	0.434	1235.8
0.5	0.759	1941.2	0.281	0.619	1536.0
0.6	0.809	2083.7	0.352	0.662	1624.1
0.7	0.856	2222.4	0.522	0.750	1882.5
0.8	0.903	2360.4	0.667	0.824	2115.1
0.9	0.951	2499.2	0.826	0.911	2337.8
1.0	1.000	2638.7	0.932	0.969	2508.0
			1.000	1.000	2650.9

Figure 8-26 shows calculated and experimental results at 100°C. The agreement is good. The UNIFAC model provides excellent results at 25°C and the system has no azeotrope. As in the previous example, the method of Wong and Sandler (1992) is useful for extrapolation to higher temperatures.

Example 8-27 Use the Wong-Sandler (1992) mixing rules and the PRSV EoS of Stryek and Vera (1986) to calculate vapor-liquid equilibria for the binary CO_2 (1)-propane (2) at 37.8°C. For the molar excess Gibbs energy, g^E, use the van Laar model

$$\frac{g^E}{RT} = \frac{Ax_1x_2}{\left(\frac{A}{B}\right)x_1 + x_2} \tag{8-12.40}$$

where A and B, assumed to be temperature-independent, are calculated from experimental data at 4.44°C reported by Reamer, et al. (1951).

solution Table 8-47 shows experimental data for the mixture at 4.44°C.

The last column in Table 8-47 is obtained using approximations that ignore nonidealities due to pressure

$$\frac{g^E}{RT} = \sum_{i=1}^{2} x_i \ln \gamma_i = \sum_{i=1}^{2} x_i \ln \left(\frac{y_iP}{x_iP_{\text{vp}i}}\right) \tag{8-12.41}$$

where the factor $\mathfrak{F}_i$ of Eq. (8-4.2) is assumed to be unity. Here $P_{\text{vp}i}$ is the pure-component vapor pressure at 4.44°C; 39.06 bar for CO_2 and 5.45 bar for propane. Fitting Eq. (8-12.40) to the last column of Table 8-47, we obtain $A = 1.020$ and $B = 0.924$.

The phase equilibrium calculation procedure is the same as in Examples 8-25 and 8-26. The PRSV parameter $\kappa_i^{(1)}$ fitted to pure-component vapor pressures are $\kappa_1^{(1)} = 0.0429$ and $\kappa_2^{(1)} = 0.0314$. Minimizing the objective function F in Eq. (8-12.35), we obtain the optimum $k_{12} = 0.3572$.

Solving for y_1, P, V^V, V^L at 37.8°C, calculated results are given in Table 8-48. Also shown are experimental data at the same temperature and x_1 values from Reamer, et al. (1951). Note that CO_2 is slightly supercritical at this T.

Figure 8-27 compares calculated and experimental results. Agreement is surprisingly good both because CO_2 is supercritical and because the approximations in Eq. (8-12.41) are unjustified. However, in this case, the Wong-Sandler method

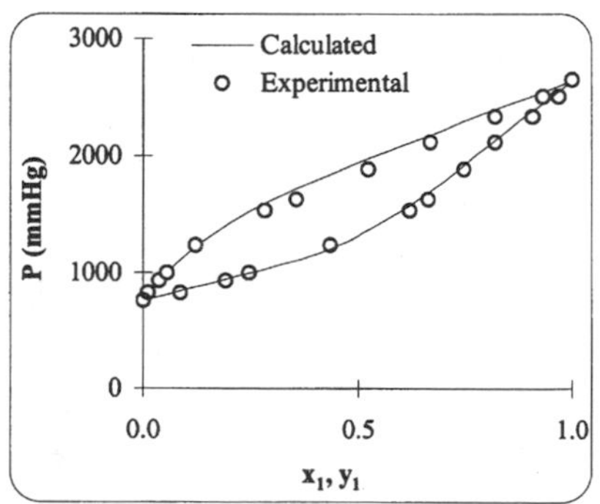

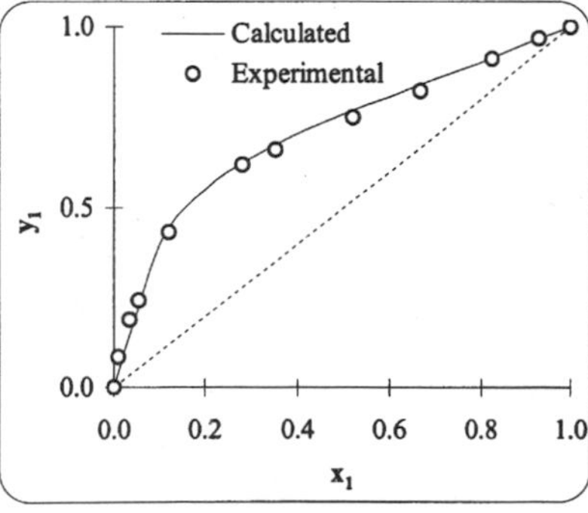

FIGURE 8-26 *Pxy* and *xy* plots for methanol(1)–water (2) at 100°C.

TABLE 8-47 Vapor-Liquid Equilibria for CO_2 (1)–propane (2) at 4.44°C (Reamer, et al., 1951)

x_1	y_1	P (bar)	g^E/RT
0.000	0.000	5.45	0.000
0.025	0.206	6.89	0.019
0.088	0.468	10.34	0.063
0.160	0.604	13.79	0.106
0.240	0.686	17.24	0.144
0.332	0.743	20.68	0.175
0.436	0.788	24.13	0.195
0.553	0.831	27.58	0.196
0.671	0.869	31.03	0.175
0.796	0.910	34.47	0.129
0.940	0.970	37.92	0.045
1.000	1.000	39.06	0.000

TABLE 8-48 Vapor-Liquid Equilibria for CO_2 (1)–propane (2) at 37.8°C

	Calculated		Experimental	
x_1	y_1	P (bar)	y_1	P (bar)
0.000	0.000	13.09	0.000	13.01
0.093	0.355	21.63	0.351	20.68
0.178	0.497	28.52	0.499	27.58
0.271	0.588	35.40	0.588	34.47
0.369	0.653	42.12	0.651	41.37
0.474	0.705	49.02	0.701	48.26
0.581	0.749	55.88	0.750	55.16
0.686	0.790	62.55	0.780	62.05
0.736	0.803	65.70	0.790	65.50

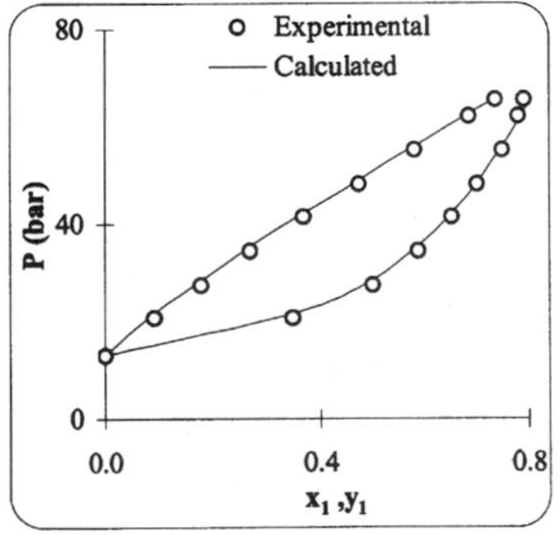

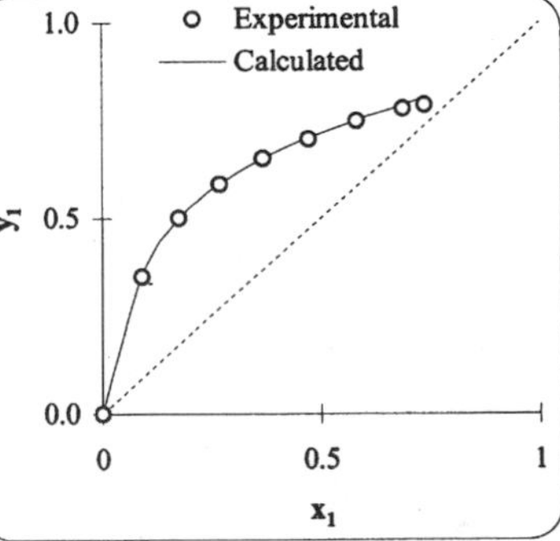

FIGURE 8-27 *Pxy* and *xy* plots for CO_2 (1)–propane(2) at 37.8°C.

is able to extrapolate correctly experimental data over a modest range of temperature. This success is probably because the fluids are normal (see Sec. 4-3) and because the fitting parameter, k_{12}, is able to account for a variety of approximations. For example, when the g^E values of Table 8-47 are obtained with more suitable values for vapor and liquid pressure effects, the results are substantially the same as those from Eq. (8-12.41).

Example 8-28 Use the Wong-Sandler mixing rules and the PRSV EoS to calculate vapor-liquid equilibria for the binary CO_2 (1)-water (2) at 100 and 300°C. For the molar excess Gibbs energy, g^E, use the van Laar model, Eq. (8-12.40). In this case, use the values of $A = 3.12$ and $B = 3.28$ as suggested by Shyu, et al. (1997) from the experimental activity coefficients of water at 200°C. Calculate the results two ways:
(a) A and B are independent of T and equal to those at 200°C from Shyu, et al. (1997).
(b) A and B are inversely proportional to T and obtained from the Shyu, et al. values:

$$A = 3.12(473.15/T) \qquad B = 3.28(473.15/T)$$

In both cases, assume that the binary parameter is $k_{12} = 0.318$ as suggested by Shyu, et al. (1997) from fitting data at 200°C.

solution The procedure is essentially the same as for Examples 8-25 to 8-27. For these two substances, the PRSV fitted parameters are $\kappa_1^{(1)} = 0.0429$ and $\kappa_2^{(1)} = -0.0664$. Table 8-49 compares the calculations with the experimental results of Muller, et al. (1988) at 100°C and of Todheide, et al. (1963) at 300°C.

TABLE 8-49 Comparison of Calculated and Experimental Vapor-Liquid Equilibria for CO_2 (1) and Water (2) at 100°C and 300°C

Experimental#			Calculated*		Calculated+	
x_1	y_1	P, bar	y_1	P, bar	y_1	P, bar
			At 100°C			
0.0005	0.712	3.25	0.714	3.6	0.824	5.93
0.0010	0.845	6.00	0.832	6.21	0.902	11
0.0016	0.893	9.20	0.887	9.39	0.935	17.27
0.0021	0.923	11.91	0.910	12.08	0.949	22.69
0.0026	0.931	14.52	0.926	14.8	0.958	28.3
0.0033	0.946	18.16	0.940	18.68	0.966	36.49
0.0041	0.955	23.07	0.950	23.2	0.971	46.42
			At 300°C			
0.0230	0.352	200	0.376	193	0.337	168
0.0490	0.454	300	0.469	321	0.444	261
0.0790	0.480	400	0.490	476	0.477	367
0.1250	0.460	500	0.472	708	0.481	523
0.2250	0.335	600	0.340	1067	0.394	784
0.2670	0.267	608	0.285	1171	0.292	859

Measurements of Muller, et al. (1988) at 100°C and of Todheide, et al. (1963) at 300°C.
* van Laar parameters in Eq. (8-12.40) $A = 3.12$, $B = 3.28$.
+ van Laar parameters in Eq. (8-12.40) inversely proportional to T:

$$A(100°C) = 3.96,\ B(100°C) = 4.16;\ A(300°C) = 2.58,\ B(300°C) = 2.71$$

Figures 8-28 and 8-29 compare calculated and measured vapor-liquid equilibria at 100°C and 300°C. Figure 8-28 suggests that at 100°C the same parameter values as at 200°C are best while those from *T*-adjustment are too high. However, Figure 8-29 suggests that the lower *T*-adjusted values are better at 300°C and the optimal g^E parameters might be still lower. This example provides a severe test for the Wong-Sandler, or any, method. Failure to achieve good results with temperature independent parameters probably follows because *T* is so much higher than the critical temperature of CO_2 and because of the remote conditions where model parameters $\kappa_i^{(1)}$, *A* and *B* were fitted.

Another possible source of error may be related to ionization effects; at low temperatures, a dilute aqueous solution of CO_2 may have some ionic species, depending upon pH. The amount would change with *T*. Ionization effects are not considered in the calculations described here.

Example 8-29 Use the Wong-Sandler mixing rules (with the NRTL expression for g^E) and the PRSV EoS to calculate vapor-liquid equilibria for the ternary acetone (1)-methanol (2)-water (3) at 100°C. Use the NRTL parameters from fitting the three binaries and the ternary as reported by Gmehling and Onken (1977b).

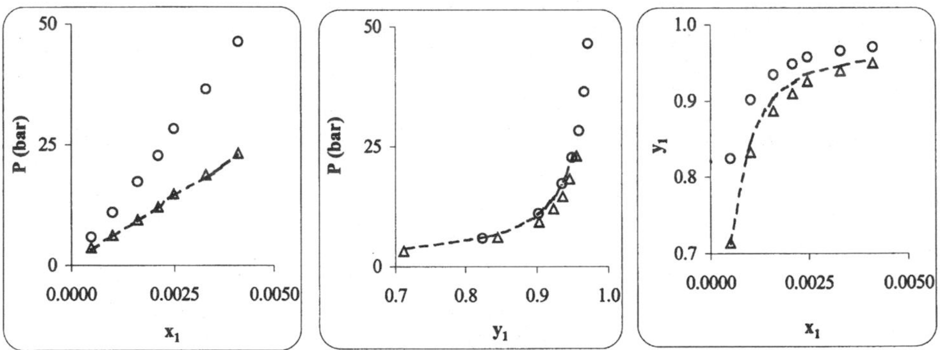

FIGURE 8-28 Vapor-liquid equilibria for CO_2 (1)–water (2) at 100°C. --- denotes experimental data. Δ denotes calculated results with A and B independent of temperature; o denotes calculated results with A and B proportional to (1/T).

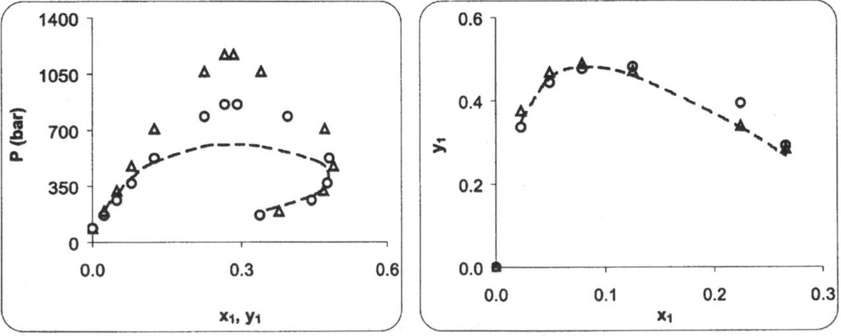

FIGURE 8-29 Vapor-liquid equilibria for CO_2 (1)–water (2) at 300°C. Legends are identical to those in Figure 8-28.

solution The procedure is similar to that in previous Examples 8-25 to 8-28 for binary systems, except that now there are three phase-equilibrium relations

$$y_i \phi_i^V = x_i \phi_i^L \qquad (i = 1, 2, 3) \tag{8-12.42}$$

where all symbols are defined in Eq. (8-12.31) of Example 8-25. Here the PRSV fitted parameters are $\kappa_1^{(1)} = -0.0089$, $\kappa_2^{(1)} = -0.1682$, and $\kappa_3^{(1)} = -0.0664$. Equations (8-12.32) to (8-12.36) are used for the mixture parameters.

The NRTL binary parameters obtained by fitting available binary data are

i	j	α	τ_{ij}	τ_{ji}
1	2	0.3014	1.4400	−0.5783
1	3	0.2862	0.1835	2.0009
2	3	0.3004	−0.5442	1.5011

Minimizing the objective function in Eq. (8-12.35), we obtain the EoS binary parameters, $k_{12} = 0.127$, $k_{13} = 0.189$, and $k_{23} = 0.100$. We use Eqs. (8-12.37) to (8-12.39) for the fugacity coefficients ϕ_i^V and ϕ_i^L. When we specify liquid mole fractions (x_1, x_2) and $T = 100°C$, we can solve five equations (three of the form 8-12.42 and two of the form 8-12.22) for the five unknowns y_1, y_2, P, V^V and V^L. When needed, mole fractions x_3 and y_3 are obtained from

$$y_1 + y_2 + y_3 = 1 \tag{8-12.43}$$

$$x_1 + x_2 + x_3 = 1 \tag{8-12.44}$$

Table 8-50 shows results calculated from the fitting and from the EoS predictions and experimental results at the x_1 and x_2 values of Griswold and Wong (1952). The EoS with the binary parameters is better than the prediction using the NRTL g^E model. For comparison, we note that when ternary data are included in the fitting, the standard deviations are somewhat better than those from the EoS results.

Comparisons of y_1, y_2 and P values are shown in Fig. 8-30. While the agreement with P is good, the vapor mole fraction comparisons must be considered only fair.

A variety of other methods based on g^E-mixing rules have been proposed to calculate vapor-liquid equilibria for mixtures containing one or more polar or hydrogen-bonding components. An issue that has been prominent in the recent literature is the appropriate standard-state pressure to match the EoS to the g^E expression. As noted in Sec. 5-5, several workers, most notably, Twu, et al. (1999b), have discussed this issue at great length, describing the options that various workers have chosen. We illustrate the use of one of these approaches in the next two examples to show its promise and also its complexity.

Example 8-30 Use the Twu, et al. (1997) zero-pressure standard-state g^E-mixing rules and the modified Soave-Redlich-Kwong EoS of Twu, et al. (1991) to calculate vapor-liquid equilibria for the binary ethanol (1)-water (2) at 25°C when $x_1 = 0.536$. Use the NRTL correlation for g^E.

solution The approach is the same as that in the previous Examples 8-25 to 8-29. Here the EoS is that of Soave (1972) as shown in Table 4-6 with the parameterization of Twu, et al. (1991) as shown Table 4-7. The pure-component parameters are (Twu, et al, 1998)

TABLE 8-50 Vapor-Liquid Equilibria for Acetone (1), Methanol (2) and Water (3) at 100°C

Liquid Mole Fractions		Experiment			From NRTL Binary Fitting			Predicted by EoS		
x_1	x_2	y_1	y_2	P	Δy_1	Δy_2	ΔP	Δy_1	Δy_2	ΔP
0.001	0.019	0.110	0.045	1.234	−0.068	−0.007	0.000	0.001	0.006	0.047
0.019	0.029	0.238	0.132	1.545	−0.045	−0.051	0.009	0.005	0.017	−0.058
0.066	0.119	0.341	0.271	2.267	−0.026	−0.049	0.024	0.001	0.006	0.090
0.158	0.088	0.525	0.151	2.668	−0.001	−0.050	0.021	−0.041	0.055	−0.132
0.252	0.243	0.470	0.300	3.103	0.036	−0.038	0.030	0.011	−0.004	0.053
0.385	0.479	0.466	0.470	3.818	0.077	−0.020	0.021	−0.053	0.071	−0.103
0.460	0.171	0.620	0.200	3.398	0.026	−0.024	0.023	−0.064	0.059	−0.256
0.607	0.330	0.622	0.345	4.013	0.128	−0.013	0.015	−0.071	0.074	−0.174
0.770	0.059	0.813	0.081	3.638	0.064	0.010	0.003	−0.076	0.034	−0.151
0.916	0.050	0.902	0.075	3.811	0.100	0.011	−0.006	−0.091	0.076	−0.076
Standard Deviation from Binary Data Parameters					**0.069**	**0.084**	**0.043**	**−0.039**	**0.031**	**0.111**
Standary Deviations Including Ternary Data					**0.034**	**0.016**	**0.062**			

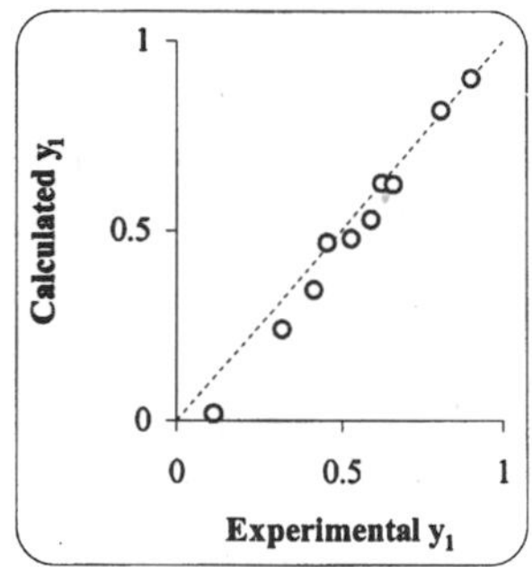

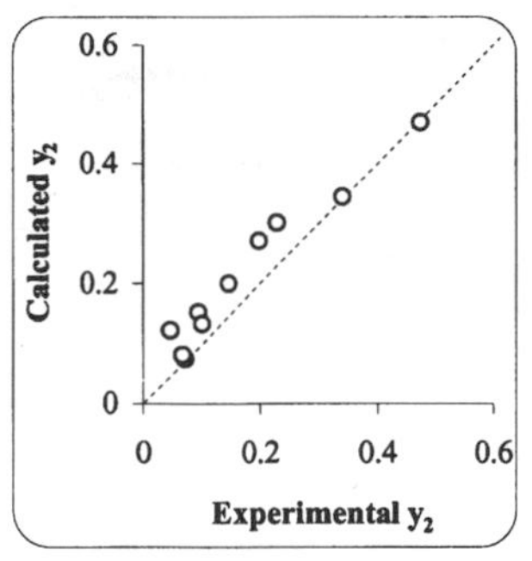

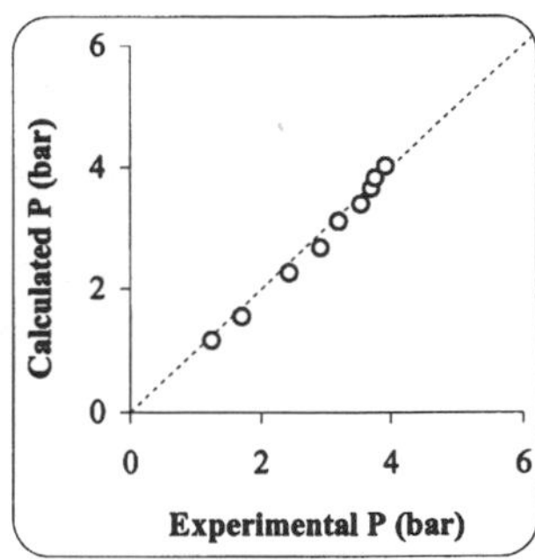

FIGURE 8-30 VLE for acetone (1)–methanol (2)–water (3) at 100°C.

Component	L	M	N	T_c, K	P_c, bar
Ethanol (1)	1.07646	0.96466	1.35369	513.92	61.48
Water (2)	0.41330	0.87499	2.19435	647.3	221.2

As usual, the composition dependence of the model is in the parameters via the mixing rule (Sec. 5-5). The Twu, et al. (1997) rule is complex because it deals explicitly with the problem of matching an EoS expression that has a pressure dependence with a g^E expression that does not. Their approach is to do the match at zero pressure, recognizing that even the 1-*vdW* mixing rules have nonzero g^E at that state. The reader is referred to Chap. 5 and the original references for details.

In the present example, we ignore corrections to the EoS b parameter and set it equal to b_{vdW}, that found from the 1-*vdW* mixing rule, [see Eq. (5-2.3*a*)]. For example in the liquid phase, it would be

$$b^L = b^L_{vdW} = \sum_i x_i b_i \tag{8-12.45}$$

Then the a parameter becomes

$$a = a_{vdW} + \frac{b_{vdW}}{C}(g^E - a^E_{vdW,P\to 0}) \tag{8-12.46}$$

where a_{vdW} is the value of a obtained from the 1-*vdW* mixing rule, [see Eq. (5-2.3*b*)]. Symbol a is an EoS parameter while symbol a^E is the excess Heimholtz energy. For the liquid,

$$a^L_{vdW} = \sum_i \sum_j x_i x_j (a_{ii} a_{jj})^{1/2} \tag{8-12.47}$$

Note that there is no binary parameter, k_{ij} in Eq. (8-12.47). For the vapor phase, a^V and b^V would be computed from Eqs. (8-12.45) to (8-12.47) with vapor mole fractions, y_i. The quantity C is computed from

$$C = -\ln\left(\frac{V_{0,vdW} + 1}{V_{0,vdW}}\right) \tag{8-12.48}$$

where $V_{0,vdW}$ is the smallest positive real volume obtained from the quadratic equation formed when P is set to zero in the EoS. In this case for the liquid, it is

$$V^L_{0,vdW} = \frac{1}{2}\left\{\left(\frac{a^L_{vdW}}{RTb^L_{vdW}} - 1\right) - \left[\left(\frac{a^L_{vdW}}{RTb^L_{vdW}} - 1\right)^2 - 4\frac{a^L_{vdW}}{RTb^L_{vdW}}\right]^{1/2}\right\} \tag{8-12.49}$$

Finally, g_E is obtained from a model such as the NRTL model and $a^E_{vdW,P\to 0}$ is obtained from the Helmholtz departure functions of the EoS with the 1-vdW mixing rules. For example in the liquid,

$$a^{EL}_{vdW,P\to 0} = RT\left[\sum_i x_i \ln\left(\frac{b_i}{b^L_{vdW}}\right) + \frac{1}{C}\left(\frac{a^L_{vdW}}{RTb^L_{vdW}} - \sum_i x_i \frac{a_i}{RTb_i}\right)\right] \tag{8-12.50}$$

where a^L_{vdW} and b^L_{vdW} are the EoS parameters of the mixture, all calculated with the 1-vdW mixing rules at the liquid composition, and a_i and b_i are the pure component parameters for the EoS at the specified T.

The expressions for the fugacity coefficients are also quite complicated for this approach. Here

$$\ln \phi^L_1 = (Z^L - 1)\left[\frac{1}{b^L}\left(\frac{\partial nb^L}{\partial n_1}\right)_{n2}\right] - \ln\left(Z^L - \frac{b^L P}{RT}\right) + \frac{a^L}{RTb^L}\left[\frac{1}{b^L}\left(\frac{\partial nb^L}{\partial n_1}\right)_{n2} - \frac{1}{na^L}\left(\frac{\partial n^2 a^L}{\partial n_1}\right)_{n2}\right]\ln\left(\frac{Z^L + b^L P/RT}{Z^L}\right) \tag{8-12.51}$$

The derivatives here are

$$\frac{1}{na^L}\left(\frac{\partial n^2 a^L}{\partial n_1}\right)_{T,n2} = \frac{1}{b^L}\left(\frac{\partial nb^L}{\partial n_1}\right)_{n2} + \frac{1}{D^L}\left(\frac{\partial nD^L}{\partial n_1}\right)_{T,n2} \tag{8-12.52}$$

$$\frac{1}{b^L}\left(\frac{\partial nb^L}{\partial n_1}\right)_{n2} = \frac{1}{nQ^L}\left(\frac{\partial n^2 Q^L}{\partial n_1}\right)_{n2}\frac{1}{1 - D^L}\left[1 - \left(\frac{\partial nD^L}{\partial n_1}\right)_{T,n2}\right] \tag{8-12.53}$$

$$D^L = \frac{1}{RT}\left[\frac{a^L_{vdW}}{b^L_{vdW}} + \frac{1}{C}[g^E - a^E_{vdW,P\to 0}]\right] \tag{8-12.54}$$

$$Q^L = b^L_{vdW} - a^L_{vdW}/RT \tag{8-12.55}$$

$$\left(\frac{\partial nD^L}{\partial n_1}\right)_{T,n2} = \frac{a^L_{vdW}}{RTb^L_{vdW}}\left[\frac{1}{na^L_{vdW}}\left(\frac{\partial n^2 a^L}{\partial n_1}\right)_{T,n2} - \frac{1}{b^L_{vdW}}\left(\frac{\partial nb^L_{vdW}}{\partial n_1}\right)_{n2}\right] + \frac{1}{C^L}(\ln \gamma_1 - \ln \gamma_{1,vdW}) - \left(D - \frac{a^L_{vdW}}{RTb^L_{vdW}}\right)\left(\frac{n}{C}\right)\left(\frac{\partial C^L}{\partial n_1}\right)_{T,n2} \tag{8-12.56}$$

$$\left(\frac{\partial C^L}{\partial n_1}\right)_{T,n2} = \frac{1}{\left(2V_{0vdW} - \dfrac{a^L_{vdW}}{RTb^L_{vdW} + 1}\right)}\left[\frac{1}{na^L_{vdW}}\left(\frac{\partial n^2 a^L_{vdW}}{\partial n_1}\right)_{T,n2} - \frac{1}{nb^L_{vdW}}\left(\frac{\partial nb^L_{vdW}}{\partial n_1}\right)_{n2} - 1\right] \tag{8-12.57}$$

$$\frac{1}{nQ^L}\left(\frac{\partial n^2 Q^L}{\partial n_1}\right)_{T,n2} = \frac{2\sum_j x_j(b_{1j} - a_{1j}/RT)}{\sum_i\sum_j x_i x_j(b_{1j} - a_{1j}/RT)} \tag{8-12.58}$$

The NRTL activity coefficient, $\ln \gamma_1$, can be found in Table 8-3. The expression for $\ln \gamma_{1,vdW}$ is calculated from the EoS by the ratio of the solution fugacity coefficient, $\ln \phi^L_{1,vdW}(T, P, x_i)$, to the standard state fugacity coefficient, $\ln \phi^L_{\text{pure }1}(T, P)$, both calculated from Eq. (6-7.10) at the specified T and P.

$$\ln \gamma_{1,vdW} = \ln \phi^L_{1,vdW}(T, P, x_1) - \ln \phi^L_{\text{pure }1}(T, P) \tag{8-12.59}$$

Finally the derivatives of a^L_{vdW} and b^L_{vdW} are found from

$$\frac{1}{na^L_{vdW}}\left(\frac{\partial n^2 a^L_{vdW}}{\partial n_1}\right)_{T,n_2} = \frac{2}{a^L_{vdW}}\sum_j x_j a_{1j} \tag{8-12.60a}$$

$$\frac{1}{nb^L_{vdW}}\left(\frac{\partial n b^L_{vdW}}{\partial n_1}\right)_{T,n_2} = \frac{b_1}{b^L_{vdW}} \tag{8-12.60b}$$

Again, for the vapor phase all of the same equations would be used with vapor compositions rather than liquid compositions. For component 2, the same equations would be used, but the subscripts 1 and 2 interchanged.

The NRTL parameters from Twu, et al. (1998) for this system from fitting mixture VLE are $\alpha = 0.2945$, $\tau_{12} = 0.0226$ and $\tau_{21} = 0.7387$.

Solving the phase equilibrium equations and volumetric equations at the specified conditions gives $P = 55.55$ mm Hg, and $y_1 = 0.7059$. The experimental values of Phutela, et al., (1979) are 54.90 mm Hg and 0.6977. This would be considered very good agreement as Twu, et al. (1997, 1998) report for many mixtures.

In addition to implementing the zero-pressure standard state, Twu, et al. (1998) have investigated the infinite-pressure standard state. Example 8-31 shows the differences between the two cases.

Example 8-31 Use the Twu, et al. (1997) infinite-pressure g^E mixing rules and the modified Soave-Redlich-Kwong EoS of Twu, et al. (1991) to calculate vapor-liquid equilibria for the binary ethanol (1)-n-heptane (2) at 70.02°C over all compositions. Use the NRTL correlation for g^E.

solution The EoS is the same as that in the previous Example 8-30. The pure-component parameter values are

Component	T_c, K	P_c, bar	L	M	N
Ethanol (1)	513.92	61.48	1.07646	0.96466	1.35370
n-heptane (2)	540.16	27.36	0.34000	0.84500	2.38300

Most of the relations defined in Eqs. (8-12.45) to (8-12.59) are relevant except that in this case, we do not make the simplifying assumption that $b = b_{vdW}$. The choice of that the standard state makes a difference is in the relations for a and b. The relation for b is now

$$b = \frac{RTb_{vdW} - a_{vdW}}{1 - \left[\dfrac{a_{vdW}}{b_{vdW}} + \dfrac{1}{C'}(g^E - a^E_{vdW,P\to 0})\right]} \tag{8-12.61}$$

where C' is a constant that depends on the EoS. Here $C' = \ln 2$. Instead of (8-12.46) we now have

$$a = b\left[\frac{a_{vdW}}{b_{vdW}} + \frac{1}{C'}(g^E - a^E_{vdW,P\to\infty})\right] \tag{8-12.62}$$

Here the liquid, $a^{EL}_{vdW,P\to\infty}$ is obtained from

$$a^{EL}_{vdW,P\to\infty} = C'b\left[\frac{a^L_{vdW}}{b^L_{vdW}} - \sum_i x_i \frac{a_i}{b_i}\right] \tag{8-12.63}$$

The expression for b_{vdW} is unchanged from that in Example 8-30. However, unlike the last case where no binary parameters were used, two parameters are fitted to obtain the *vdW* expressions for a_{vdW}. Thus, for the liquid,

$$a^L_{vdW} = \sum_i \sum_j x_i x_j (a_{ii}a_{jj})^{1/2}(1 - k_{ij}) + \sum_i x_i \left[\sum_j x_j (a_{ii}a_{jj})^{1/6}(k_{ji} - k_{ij})^{1/3}\right]^3 \tag{8-12.64}$$

Parameters k_{12} and k_{21} are fitted to minimize the objective function

$$F = \sum_{\text{data}} |(\ln \gamma_1)_{EoS} - (\ln \gamma_1)_{g^E}| \tag{8-12.65}$$

With the NRTL parameters of Gmehling, et al. (1988) of $\alpha = 0.4598$, $\tau_{12} = 1.7204$, and $\tau_{21} = 2.3972$ from data at 30.12°C, the optimal values are $k_{12} = 0.4812$ and $k_{21} = 0.1931$.

Using the calculational method as in Example 8-30, we calculate the results shown in Table 8-51 at 70.02° and compare them with the data of Berro, et al (1982). Figure 8-31 also compares the calculated and experimental results. The calculated pressures at low ethanol mole fractions are somewhat high, but the compositions are in very good agreement.

Finally, Twu, et al. (1999a) have developed a way to avoid any explicit reference state pressure for matching the g^E behavior with the EoS parameters.

Many equations of state have been proposed and new ones keep appearing. We cannot here review all developments in this vast field, but we do want to call attention to the application of many of the models that are nonanalytic in density described in Chaps. 4 to 6. The SAFT model has received particular attention because of its theoretical basis and breadth of application to systems with both large and small molecules. Here, "large" refers primarily to chain molecules like normal paraffins and, by extension, to polymers and copolymers. To be successful with such applications, the SAFT EoS is mathematically complex, especially for cases of association and solvation. The literature is rich in publications about SAFT;

TABLE 8-51 Vapor-Liquid Equilibria for Ethanol (1)–*n*-heptane (2) at 70.02°C

	Calculated		Experimental	
x_1	y_1	P (mmHg)	y_1	P (mmHg)
0	0	302.8	0	304.0
0.1013	0.5513	639.3	0.5371	633.3
0.1957	0.6061	707.7	0.5807	682.8
0.3003	0.6215	725.9	0.6018	704.6
0.4248	0.6235	727.8	0.6173	716.5
0.5046	0.6225	727.3	0.6264	720.6
0.6116	0.6243	727.4	0.6381	723.8
0.7116	0.6369	725.8	0.6542	722.1
0.8101	0.672	712.8	0.6833	712.4
0.9259	0.7872	651.7	0.7767	661.9
0.9968	0.9853	551.8	0.9812	551.4

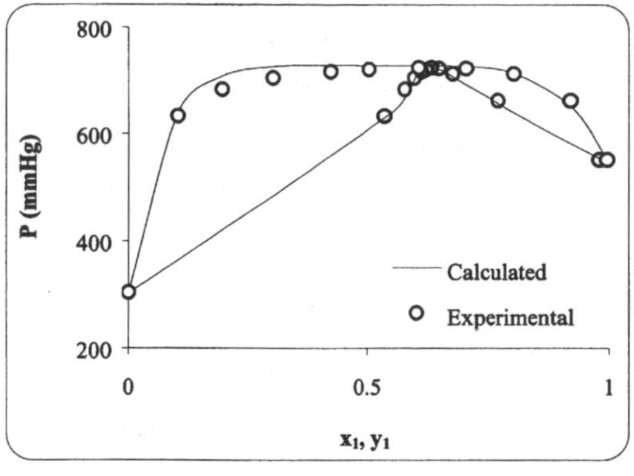

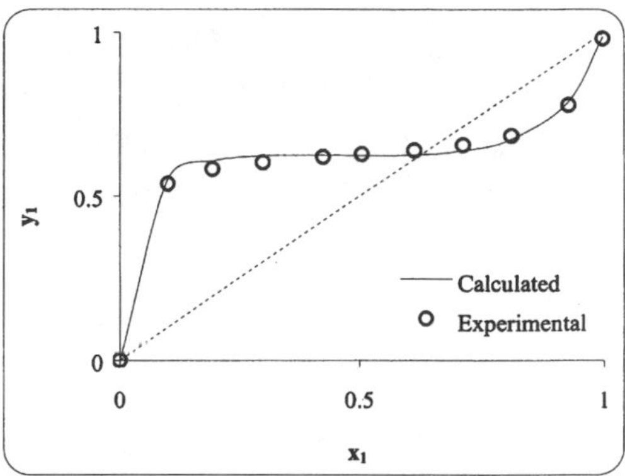

FIGURE 8-31 Vapor-liquid equilibria for ethanol (1)–*n*-heptane (2) at 70.02°C.

some representative basic or recent articles are those by Chapman and coworkers (Jog, et al., 1999; Chapman, et al., 1989), Radosz and coworkers (Adidharma and Radosz, 1998, 1999ab; Kinzl, et al., 2000), and others (Blas and Vega, 1998; Sandler and Fu, 1995; Yu and Chen, 1994).

Some Comparisons Among Different Equations of State. With the recent rapid introduction of EoS models and mixing rules, there has not been adequate time for extensive comparisons among models. However, there have been a few limited studies which we cite here.

One comparison of the Wong-Sandler mixing rule with the Dahl-Michelsen (MHV2) mixing rule for cubic equations of state was done by Huang and Sandler (1993). They used both the Soave-Redlich-Kwong and the Peng-Robinson EoS models. They found that high-pressure predictions of VLE were not sensitive to the data set used to fit the g^E parameters. For bubble points of nine systems of alcohols or water with acetone and hydrocarbons, the W-S rules were slightly more reliable than MHV2 for vapor mole fractions (average deviations ~0.015 vs. ~0.02) but noticeably better for pressures (~3% vs. ~4%).

Knudsen, et al. (1993) investigated the reliability for ternary and higher complex systems, including supercritical components, of five different mixing rules in the Soave-Redlich-Kwong EoS. They examined two to six binary parameters (including T dependence) with a variety of strategies for obtaining the parameters and concluded that binary correlation and ternary prediction improved with three or four parameters, but not with more. They also note the dangers of extrapolating T-dependent parameters.

Fischer and Gmehling (1996) describe comparisons of several mixing rules in the Soave-Redlich-Kwong EoS based on predictive g^E models. Their results for nine systems with a great variety of substances divide the methods into two groups: older data with ~3.5% error in pressure and ~2% error in vapor mole fraction, and new data with ~2.5% error in P and 1% error in y_1.

Finally, Kang, et al. (2000) have compared the Wong-Sandler mixing rule in the Peng-Robinson EoS with the SAFT model and a Nonrandom Lattice Fluid Model (Yoo, et al., 1995; Yeom, et al., 1999). They found that for VLE of systems containing larger molecules (hexane and higher alkanes), SAFT was most reliable, especially for polar/nonpolar mixtures. However, for strongly polar systems of small molecules (e.g., ethanol, methyl acetate) at low pressures, the SAFT model was much worse than the others and when methane was in the system, it did not reproduce the mixture critical region well. For the polar mixtures, the UNIQUAC model was better than any of the EoS models; the Wong-Sandler mixing rules did not reproduce the agreement of the g^E model it was based on, a subject also discussed by Heidemann (1996).

8-13 SOLUBILITIES OF SOLIDS IN HIGH-PRESSURE GASES

Like distillation, the design of the industrially important process supercritical extraction can be based on phase-equilibrium calculations, except that one phase is a gas at elevated pressure and the other phase may be liquid or solid. The book edited by Kiran and Levelt Sengers (Levelt Sengers 1994) provides an extensive treatment of the fundamentals of these systems. Typically, because the condensed phase is from natural products or pharmaceuticals, etc., so it is likely to be poorly characterized. The result is that much of the literature is empirical and the property descriptions of the above sections cannot be easily applied. This has led to two approaches for correlating data: semiempirical and fully empirical. Among the extensive literature of this area, we mention one method of each approach. Insights into the literature can be found in these references and in the Journal of Supercritical Fluids and the Journal of Chemical and Engineering Data.

The first approach is to assume that the properties of the phases can be identified and obtained either from other information or by correlation of specific data. Thus, for the case of solid (1)-gas equilibria, Eq. (8-2.1) becomes

$$f_1^V = f_1^S \tag{8-13.1}$$

where an EoS can be used for the vapor but not for the solid. Thus, we use equations similar to Eq. (8-2.2*a*) for the vapor and similar to Eq. (8-4.1) for the solid

$$f_1^V = y_1 \phi_1^V P = x_1^S \gamma_1^S \phi_1^S P_1^{sub} \exp\left[\int_{P_1^{sub}}^{P} \frac{V_1^S}{RT}\, dP\right] = f_1^S \tag{8-13.2a}$$

where the superscript ^{S} refers to solid phase and P_1^{sub} is the sublimation pressure (the equilibrium two-phase pressure for the pure solid and vapor, equivalent to P_{vp1} for pure liquid and vapor equilibrium). If, as commonly assumed, the solid is pure and incompressible and $P_1^{sub} << P$, Eq. (8-13.2) becomes

$$y_1 \phi_1^V P = P_1^{sub} \exp\left(\frac{V_1^S P}{RT}\right) \tag{8-13.2b}$$

The problem of interest is to obtain y_1, which is usually very small ($y_1 < 10^3$), at a specified T and P. Its behavior is complex when plotted versus P, but considerably simpler when plotted versus the vapor density which is often assumed to be that of the pure gaseous component, ρ_2^o. With this form, correlations or predictions must be used for ϕ_1^V, P_1^{sub}, and V_1^S. Many workers assume that data exist for P_1^{sub} and V_1^S. Then EoS models can be used for ϕ_1^V, typically requiring either binary parameters for cubic equations or cross virial coefficients.

Quiram, et al. (1994) briefly review several such approaches. They claim that the virial EoS of the form of Eq. (5-4.1*a*) is rigorous at lower pressures and effective at the higher pressures typical of separation processing. For a binary solid (1)-gas (2) system, their expression is

$$\ln y_1 = \ln(P_1^{sub}/\rho_2^o RT) + \frac{V_1^S P}{RT} - 2B_{12}\rho_2^o - 1.5C_{122}(\rho_2^o)^2 \qquad (8\text{-}13.3)$$

where B_{12} is the second cross virial coefficient, and C_{122} is the third cross virial coefficient and ρ_2^o is the pure gas density. As discussed by Prausnitz, et al. (1999), usually $B_{12} < 0$ while $C_{122} > 0$ because of attractions between the solid and gas molecule pairs and overlap among the trios. The low solubility is due to the small value of $P_1^{sub}/\rho_2^o RT$ and the dominant term to increase it is B_{12}.

With a correlation for B_{12} such as that given in Sec. 5-4, a single solubility measurement at each T can be used to obtain C_{122} whose T dependence is often not strong. The Quiram method also is reliable for systems containing a gas phase entrainer used to enhance the solid solubility. For 40 binary and ternary systems with CO_2, Quiram, et al. (1994) found about 10% average error in y_1, comparable to the experimental uncertainty.

In many systems, the pure-component properties of Eq. (8-13.2) are not available. Among the fully empirical correlating equations developed, that of Mendez-Santiago and Teja (1999) appears to be the most successful. Their form is

$$\ln y_1 = \ln P + A + \frac{B}{T} + C\frac{\rho_2^o}{T} \qquad (8\text{-}13.4)$$

where A, B, and C are empirical parameters that depend only upon components 1 and 2 and must be fitted to data for the system over a range of P and T. Here ρ_2^o is the molar density of pure component 2. Note that the T, P, and ρ_2^o of Eqs. (8-13.3) and (8-13.4) are not the same, although they are similar. For 41 binary systems of solids in CO_2, agreement with experiment is again comparable to experimental uncertainty.

8-14 LIQUID-LIQUID EQUILIBRIA

Many liquids are only partially miscible, and in some cases, e.g., mercury and hexane at normal temperatures, the mutual solubilities are so small that, for practical purposes, the liquids may be considered immiscible. Partial miscibility is observed not only in binary mixtures but also in ternary (and higher) systems, thereby making extraction a possible separation operation. This section introduces some useful thermodynamic relations which, in conjunction with limited experimental data, can be used to obtain quantitative estimates of phase compositions in liquid-liquid systems.

At ordinary temperatures and pressures, it is (relatively) simple to obtain experimentally the compositions of two coexisting liquid phases and, as a result, the technical literature is rich in experimental results for a variety of binary and ternary systems near 25°C and near atmospheric pressure. However, as temperature and pressure deviate appreciably from those corresponding to normal conditions, the availability of experimental data falls rapidly.

Partial miscibility in liquids is often called *phase splitting*. The thermodynamic criteria which indicate phase splitting are well understood regardless of the number of components (Tester and Modell, 1977), but most thermodynamic texts confine discussion to binary systems. Stability analysis shows that, for a binary system, phase splitting occurs when

$$\left(\frac{\partial^2 g^E}{\partial x_1^2}\right)_{T,P} + RT\left(\frac{1}{x_1} + \frac{1}{x_2}\right) < 0 \tag{8-14.1}$$

where g^E is the molar excess Gibbs energy of the binary mixture (see Sec. 8-5). To illustrate Eq. (8-14.1), consider the simplest nontrivial case. Let

$$g^E = Ax_1x_2 \tag{8-14.2}$$

where A is an empirical coefficient characteristic of the binary mixture. Substituting into Eq. (8-14.1), we find that phase splitting occurs if

$$A > 2RT \tag{8-14.3}$$

In other words, if $A < 2RT$, the two components 1 and 2 are completely miscible; there is only one liquid phase. However, if $A > 2RT$, two liquid phases form because components 1 and 2 are only partially miscible.

The condition when $A = 2RT$ is called *incipient instability,* and the temperature corresponding to that condition is called the *consolute temperature,* designed by T^c. Since Eq. (8-14.2) is symmetric in mole fractions x_1 and x_2, the composition at the consolute or critical point is $x_1^c = x_2^c = 0.5$. In a typical binary mixture, the coefficient A is a function of temperature, and therefore it is possible to have either an upper consolute temperature or a lower consolute temperature, or both, as indicated in Figs. 8-32 and 8-33. Upper consolute temperatures are more common than lower consolute temperatures. Except for those containing polymers, and surfactants systems, both upper and lower consolute temperatures are rare.†

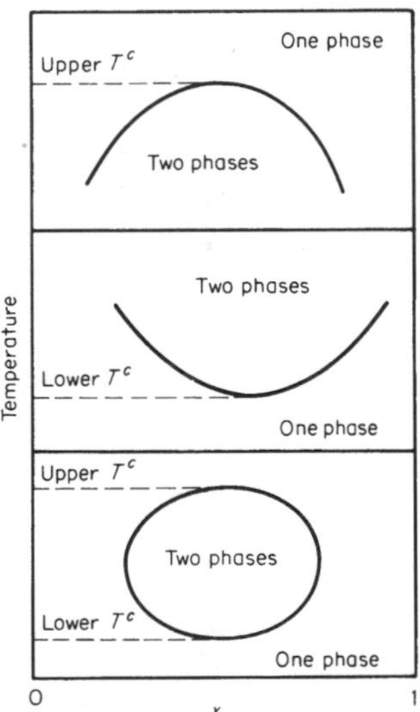

FIGURE 8-32 Phase stability in three binary liquid mixtures. (*Prausnitz, et al., 1999*)

† Although Eq. (8-14.3) is based on the simple two-suffix (one-parameter) Margules equation, similar calculations can be made using other expression for g^E. See, for example, Shain and Prausnitz (1963).

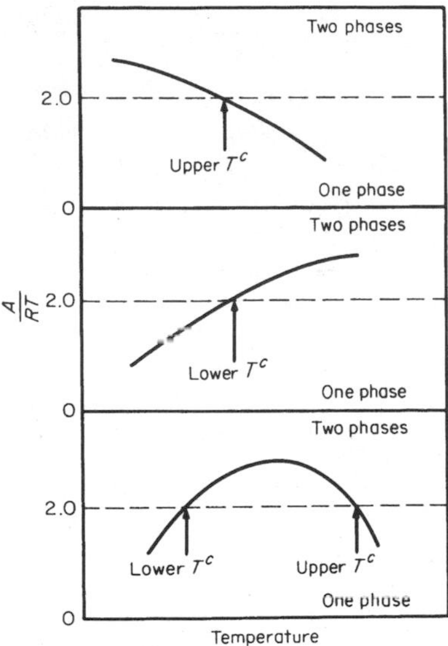

FIGURE 8-33 Phase stability in three binary liquid mixtures whose excess Gibbs energy is given by a two-suffix Margules equation. (*Prausnitz, et al., 1999*)

Stability analysis for ternary (and higher) systems is, in principle, similar to that for binary systems, although the mathematical complexity rises with the number of components. (See, for example, Beegle and Modell, 1974). However, it is important to recognize that stability analysis can tell us only whether a system can or cannot *somewhere* exhibit phase splitting at a given temperature. That is, if we have an expression for g^E at a particular temperature, stability analysis can determine whether or not there is *some* range of composition where two liquids exist. It does *not* tell us what that composition range is. To find the range of compositions in which two liquid phases exist at equilibrium requires a more elaborate calculation. To illustrate, consider again a simple binary mixture whose excess Gibbs energy is given by Eq. (8-14.2). If $A > 2RT$, we can calculate the compositions of the two coexisting equations by solving the two equations of phase equilibrium

$$(\gamma_1 x_1)' = (\gamma_1 x_1)'' \qquad \text{and} \qquad (\gamma_2 x_2)' = (\gamma_2 x_2)'' \tag{8-14.4}$$

where the prime and double prime designate, respectively, the two liquid phases.

From Eq. (8-14.2) we have

$$\ln \gamma_1 = \frac{A}{RT} x_2^2 \tag{8-14.5}$$

and

$$\ln \gamma_2 = \frac{A}{RT} x_1^2 \tag{8-14.6}$$

Substituting into the equation of equilibrium and noting that $x_1' + x_2' = 1$ and $x_1'' + x_2'' = 1$, we obtain

$$x_1' \exp \frac{A(1 - x_1')^2}{RT} = x_1'' \exp \frac{A(1 - x_1'')^2}{RT} \tag{8-14.7}$$

and
$$(1 - x_1') \exp \frac{Ax_1'^2}{RT} = (1 - x_1'') \exp \frac{Ax_1''^2}{RT} \tag{8-14.8}$$

Equations (8-14.7) and (8-14.8) contain two unknowns (x_1' and x_1''), that can be found by iteration. Mathematically, several solutions of these two equations can be obtained. However, to be physically meaningful, it is necessary that $0 < x_1' < 1$ and $0 < x_1'' < 1$.

Similar calculations can be performed for ternary (or higher) mixtures. For a ternary system the three equations of equilibrium are

$$(\gamma_1 x_1)' = (\gamma_1 x_1)''; \qquad (\gamma_2 x_2)' = (\gamma_{22} x_2)''; \qquad (\gamma_3 x_3)' = (\gamma_3 x_3)'' \tag{8-14.9}$$

If we have an equation relating the excess molar Gibbs energy g^E of the mixture to the overall composition (x_1, x_2, x_3), we can obtain corresponding expressions for the activity coefficients γ_1, γ_2, and γ_3, as discussed elsewhere [see Eq. (8-9.4)]. The equations of equilibrium [Eq. (8-14.9)], coupled with material-balance relations (flash calculation), can then be solved to obtain the four unknowns (x_1', x_2' and x_1'', x_2'').

Systems containing four or more components are handled in a similar manner. An expression for g^E for the multicomponent system is used to relate the activity coefficient of each component in each phase to the composition of that phase. From the equations of equilibrium [$(\gamma_i x_i)' = (\gamma_i x_i)''$ for every component i] and from material balances, the phase compositions x_i' and x_i'' are found by iteration.

Considerable skill in numerical analysis is required to construct a computer program that finds the equilibrium compositions of a multicomponent liquid-liquid system from an expression for the excess Gibbs energy for that system. It is difficult to construct a program that always converges to a physically meaningful solution by using only a small number of iterations. This difficulty is especially pronounced in the region near the plait, or critical point, where the compositions of the two equilibrium phases become identical.

King (1980) and Prausnitz, et al. (1980) have given some useful suggestions for constructing efficient programs toward computation of equilibrium compositions in two-phase systems.

Although the thermodynamics of multicomponent liquid-liquid equilibria is, in principle, straightforward, it is difficult to obtain an expression for g^E that is sufficiently accurate to yield reliable results. Liquid-liquid equilibria are much more sensitive to small changes in activity coefficients than vapor-liquid equilibria. In the latter, activity coefficients play a role which is secondary to the all-important pure-component vapor pressures. In liquid-liquid equilibria, however, the activity coefficients are dominant; pure-component vapor pressures play no role at all. Therefore, it has often been observed that good estimates of vapor-liquid equilibria can be made for many systems by using only approximate activity coefficients, provided the pure-component vapor pressures are accurately known. However, in calculating liquid-liquid equilibria, small inaccuracies in activity coefficients can lead to serious errors.

Regardless of which equation is used to represent activity coefficients, much care must be exercised in determining parameters from experimental data. Whenever possible, such parameters should come from mutual solubility data.

When parameters are obtained from reduction of vapor-liquid equilibrium data, there is always some ambiguity. Unless the experimental data are of very high accuracy, it is usually not possible to obtain a truly unique set of parameters; i.e., in a typical case, there is a range of parameter sets such that any set in that range can equally well reproduce the experimental data within the probable experimental error. When multicomponent vapor-liquid equilibria are calculated, results are often not sensitive to which sets of binary parameters are chosen. However, when multicomponent liquid-liquid equilibria are calculated, results are extremely sensitive to the choice of binary parameters. Therefore, it is difficult to establish reliable ternary (or higher) liquid-liquid equilibria by using only binary parameters obtained from binary liquid-liquid and binary vapor-liquid equilibrium data. For reliable results it is usually necessary to utilize at least some multicomponent liquid-liquid equilibrium data.

To illustrate these ideas, we quote some calculations reported by Bender and Block (1975), who considered two ternary systems at 25°C:

System I: Water (1), toluene (2), aniline (3)

System II: Water (1), TCE† (2), acetone (3)

To describe these systems, the NRTL equation was used to relate activity coefficients to composition. The essential problem lies in finding the parameters for the NRTL equation. In system I, components 2 and 3 are completely miscible but components 1 and 2 and components 1 and 3 are only partially miscible. In system II, components 1 and 3 and components 2 and 3 are completely miscible but components 1 and 2 are only partially miscible.

For the completely miscible binaries, Bender and Block set the NRTL parameter $\alpha_{ij} = 0.3$. Parameters τ_{ij} and τ_{ji} were then obtained from vapor-liquid equilibria. Since it is not possible to obtain unique values of these parameters from vapor-liquid equilibria, Bender and Block used a criterion suggested by Abrams and Prausnitz (1975), namely, to choose those sets of parameters for the completely miscible binary pairs which correctly give the limiting liquid-liquid distribution coefficient for the third component at infinite dilution. In other words, NRTL parameters τ_{ij} and τ_{ji} chosen where those which not only represent the *ij* binary vapor-liquid equilibria within experimental accuracy but also give the experimental value of K_k^∞ defined by

$$K_k^\infty = \lim_{\substack{w'_k \to 0 \\ w''_k \to 0}} \frac{w''_k}{w'_k}$$

where w stands for weight fraction, component k is the third component, i.e., the component *not* in the completely miscible *ij* binary, and the prime and double prime designate the two equilibrium liquid phases.

For the partially miscible binary pairs, estimates of τ_{ij} and τ_{ji} are obtained from mutual-solubility data following an arbitrary choice for α_{ij} in the region $0.20 \leq \alpha_{ij} \leq 0.40$. When mutual-solubility data are used, the parameter set τ_{ij} and τ_{ji} depends only on α_{ij}; to find the best α_{ij}, Bender and Block used ternary tie-line data. In other words, since the binary parameters are not unique, the binary parameters chosen where those which gave an optimum representation of the ternary liquid-liquid equilibrium data.

Table 8-52 gives mutual solubility data for the three partially miscible binary systems. Table 8-53 gives NRTL parameters following the procedure outlined

† 1,1,2-Trichloroethane.

TABLE 8-52 Mutual Solubilities in Binary Systems at 25°C (Bender and Block, 1975)

Component		Weight fraction	
i	*j*	*i* in *j*	*j* in *i*
Water	TCE	0.0011	0.00435
Water	Toluene	0.0005	0.000515
Water	Aniline	0.053	0.0368

TABLE 8-53 NRTL Parameters Used by Bender and Block to Calculate Temary Liquid-Liquid Equilibria at 25°C

i	j	τ_{ij}	τ_{ji}	α_{ij}
System 1: water (1), toluene (2), aniline (3)				
1	2	7.77063	4.93035	0.2485
1	3	4.18462	1.27932	0.3412
2	3	1.59806	0.03509	0.3
System II: water (1), TCE (2), acetone (3)				
1	2	5.98775	3.60977	0.2485
1	3	1.38800	0.75701	0.3
2	3	−0.19920	−0.20102	0.3

above. With these parameters, Bender and Block obtained good representation of the ternary phase diagrams, essentially within experimental error. Figure 8-34 and 8-35 compare calculated with observed distribution coefficients for systems I and II.

When the NRTL equation is used to represent ternary liquid-liquid equilibria, there are nine adjustable binary parameters; when the UNIQUAC equation is used, there are six. It is tempting to use the ternary liquid-liquid data alone for obtaining the necessary parameters, but this procedure is unlikely to yield a set of *meaningful* parameters; in this context "meaningful" indicates the parameters which also reproduce equilibrium data for the binary pairs. As shown by Heidemann and others (Heidemann 1973, 1975), unusual and bizarre results can be calculated if the parameter sets are not chosen with care. Experience in this field is not yet plentiful, but all indications are that it is always best to use binary data for calculating binary parameters. Since it often happens that binary parameter sets cannot be determined uniquely, ternary (or higher) data should then be used to fix the best binary sets from the ranges obtained from the binary data. (For a typical range of binary parameter sets, see Fig. 8-4.) It is, of course, always possible to add ternary (or higher) terms to the expression for the excess Gibbs energy and thereby introduce ternary (or higher) constants. This is sometimes justified, but it is meaningful only if the multicomponent data are plentiful and of high accuracy.

In calculating multicomponent equilibria, the general rule is to use binary data first. Then use multicomponent data for fine-tuning.

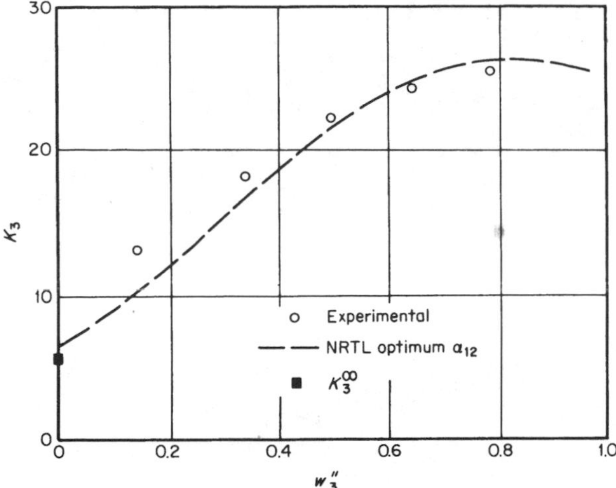

FIGURE 8-34 Distribution coefficient K_3 for the system water (1)–toluene (2)–aniline (3) at 25°C. Concentrations are weight fractions.

$$K_3 = \frac{w_3''}{w_3'} = \frac{\text{weight fraction aniline in toluene-rich phase}}{\text{weight fraction aniline in water-rich phase}}$$

$$K_3^{\infty} = \frac{\gamma_3'^{\infty}}{\gamma_3''^{\infty}} = \frac{\text{activity coefficient of aniline in water-rich phase at infinite dilution}}{\text{activity coefficient of aniline in toluene-rich phase at infinite dilution}}$$

Activity coefficient γ is here defined as the ratio of activity to weight fraction. (*From Bender and Block, 1975*)

Example 8-32 Acetonitrile (1) is used to extract benzene (2) from a mixture of benzene and *n*-heptane (3) at 45°C.

(*a*) 0.5148 mol of acetonitrile is added to a mixture containing 0.0265 mol of benezene and 0.4587 mol of *n*-heptane to form 1 mol of feed.

(*b*) 0.4873 mol of acetonitrile is added to a mixture containing 0.1564 mol of benzene and 0.3563 mol of *n*-heptane to form 1 mol of feed.

For (*a*) and for (*b*), find the composition of the extract phase *E*, the composition of the raffinate phase *R* and α, the fraction of feed in the extract phase.

solution To find the desired quantities, we must solve an isothermal flash problem in which 1 mol of feed separates into α mol of extract and $1 - \alpha$ mol of raffinate.

There are five unknowns: 2 mole fractions in *E*, 2 mole fractions in *R*, and α. To find these five unknowns, we require five independent equations. They are three equations of phase equilibrium

$$(\gamma_i x_i)^E = (\gamma_i x_i)^R \qquad i = 1, 2, 3$$

and two material balances

$$z_i = x_i^E \alpha + x_i^R (1 - \alpha) \qquad \text{for any two components}$$

Here z_i is the mole fraction of component i in the feed; x^E and x^R are, respectively, mole fractions in *E* and in *R*, and γ is the activity coefficient.

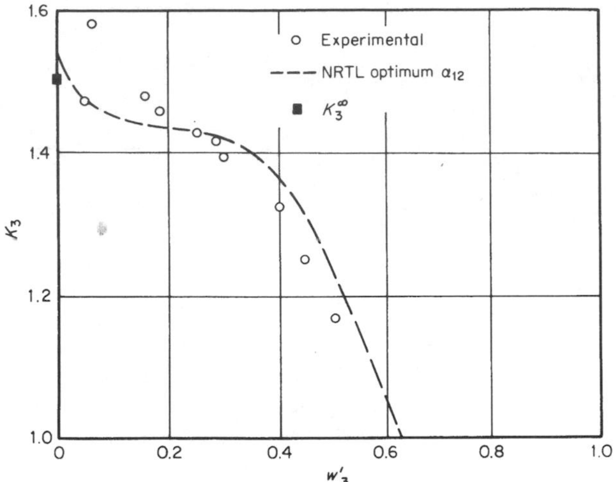

FIGURE 8-35 Distribution coefficient K_3 for the system water (1)–TCE (2)–acetone (3) at 25°C. Concentrations are in weight fractions.

$$K_3 = \frac{w''_3}{w'_3} = \frac{\text{weight fraction acetonein TCE-rich phase}}{\text{weight fraction acetone in water-rich phase}}$$

$$K_3^{\infty} = \frac{\gamma_3'^{\infty}}{\gamma_3''^{\infty}} = \frac{\text{activity coefficient of acetone in water-rich phase at infinite dilution}}{\text{activity coefficient of acetone in TCE-rich phase at infinite dilution}}$$

Activity coefficient γ is here defined as the ratio of activity to weight fraction. (*From Bender and Block, 1975*)

To solve five equations simultaneously, we use an iterative procedure based on the Newton-Raphson method as described, for example, by Prausnitz, et al. (1980). The objective function F is

$$F(x^R, x^E, \alpha) = \sum_{i=1}^{3} \frac{(K_i - 1)z_i}{(K_i - 1)\,\alpha + 1} \rightarrow 0$$

$$\text{where} \qquad K_i = \frac{x_i^E}{x_i^R} = \frac{\gamma_i^R}{\gamma_i^E}$$

For activity coefficients, we use the UNIQUAC equation with the following parameters:

Pure-Component Parameters

Component	r	q
1	1.87	1.72
2	3.19	2.40
3	5.17	4.40

Binary Parameters

$$\tau_{ij} = \exp\left(-\frac{a_{ij}}{T}\right) \qquad \tau_{ji} = \exp\left(-\frac{a_{ji}}{T}\right)$$

Components			
i	j	a_{ij}, K	a_{ji}, K
1	2	60.28	89.57
1	3	23.71	545.8
2	3	−135.9	245.4

In the accompanying table calculated results are compared with experimental data.†

Liquid-Liquid Equilibria in the System Acetonitrile (1)-Benzene (2)-*n*-Heptane (3) at 45°C

			$100x_i^R$			$100x_i^E$	
	i	γ_i^R	Calc.	Exp.	γ_i^E	Calc.	Exp.
(*a*)	1	7.15	13.11	11.67	1.03	91.18	91.29
	2	1.25	3.30	3.41	2.09	1.98	1.88
	3	1.06	83.59	84.92	12.96	6.84	6.83
(*b*)	1	3.38	25.63	27.23	1.17	73.96	70.25
	2	1.01	18.08	17.71	1.41	12.97	13.56
	3	1.35	56.29	55.06	5.80	13.07	16.19

For (*a*), the calculated $\alpha = 0.4915$; for (*b*), it is 0.4781. When experimental data are substituted into the material balance, $\alpha = 0.5$ for both (*a*) and (*b*).

In this case, there is good agreement between calculated and experimental results because the binary parameters were selected by using binary and ternary data.

While activity coefficient models are commonly used for liquid-liquid equilibria, EoS formulations can also be used. For example, Tsonopoulos and Wilson (1983) correlate the solubility of water in hydrocarbons using a variation of the Redlich-Kwong EoS that was proposed by Zudkevitch and Joeffe (1970) with temperature-dependent pure-component a parameter (Table 4-7) and a temperature-independent binary parameter, k_{12}, for Eq. (5-2.4). In this case, because the aqueous phase is essentially pure, the form of Eq. (8-2.1) is

$$f_1^\circ = x_1^\circ \phi_1^\circ P = f_{pure1} \tag{8-14.10}$$

where superscript ° refers to the hydrocarbon phase, and the fugacity coefficient, ϕ_1°, is obtained from the EoS as in Sec. 6-7.

For solubilities in the range of 0 to 200°C, Tsonopoulos and Wilson obtained

† Palmer and Smith (1972).

good agreement with experiment for x_1° with systems such as water with benzene (k_{12} = 0.260), with cyclohexane (k_{12} = 0.519) and with n-hexane (k_{12} = 0.486).

An equivalent to Eq. (8-14.10) can be written for the hydrocarbons. The results are poor if the same values of k_{12} are used. Better agreement is obtained with a different value for the equivalent parameter, k_{21}, but it must be temperature-dependent to be reliable. This experience with solubilities in water is common. The cause is the special properties of aqueous solutions called the "hydrophobic" effect. The nature of water as a small molecule with strong and directional hydrogen bonds makes it unique in properties and uniquely challenging to model. A good review of water-hydrocarbon equilibria is given by Economou and Tsonopoulos (1997). Basic understanding of the theory and phenomena of the hydrophobic effect is given by Ben-Naim (1992), for example.

At present, we do not have any good models for representing liquid-liquid equilibria in the neighborhood of the consolute (or critical or plait) point. Like vapor-liquid critical points (See Chaps. 4 and 5), no current engineering-oriented models take into account the large concentration fluctuations that strongly affect properties under these conditions.

UNIFAC correlations for liquid-liquid equilibria are available (Sorenson and Arlt, 1979) but the accuracy is not good. However, a UNIFAC-type correlation limited to aqueous-organic systems has been established by Hooper, et al. (1988). The Hooper method is illustrated in Example 8-33.

Example 8-33 Use the Modified UNIFAC method suggested by Hooper, et al. (1988) for aqueous systems to calculate liquid-liquid equilibria for the binaries

(a) water (1)–benzene (2) in the range 0 to 70°C

(b) water (1)–aniline (2) in the range 20 to 140°C

solution

(a) water (1)–benzene (2)

The two governing equations at equilibrium are

$$x_i'\gamma_i' = x_i''\gamma_i'' \qquad (i = 1,2) \tag{8-14.11}$$

where subscript i denotes component i; superscript $'$ stands for the water-rich phase and superscript $''$ stands for the benzene-rich phase; x and γ are the liquid-phase mole fraction and activity coefficient, respectively.

At a given temperature, Eq. (8-14.11) can be solved simultaneously for x_2' and x_1''. Mole fractions x_1' and x_2'' are not independent because they are constrained by material balances:

$$x_1' + x_2' = 1$$

$$x_1'' + x_2'' = 1$$

According to Hooper, et al. (1988), the activity coefficient of component i in a water-organic system is

$$\ln\gamma_i = \ln\gamma_i^C + \ln\gamma_i^R \tag{8-14.12}$$

where superscripts C and R denote combinatorial part and residual part of the activity coefficient, respectively.

The combinatorial part is given by:

$$\ln \gamma_i^C = \ln \left(\frac{\phi_i}{x_i}\right) + 1 - \left(\frac{\phi_i}{x_i}\right) \tag{8-14.13}$$

where

$$\phi_i = \frac{x_i r_i^{2/3}}{\sum_j x_j r_j^{2/3}} \tag{8-14.14}$$

r_i is calculated from group contributions as in the original UNIFAC (Example 8-12):

$$r_i = \sum_k \nu_k^{(i)} R_k \tag{8-14.15}$$

where $\nu_k^{(i)}$ denotes the number of times that groups k appears in molecule i; R_k is the volume parameter of group k.

The summation in Eq. (8-14.14) is over all components including component i, whereas the summation in Eq. (8-14.15) is over all distinct groups that appear in molecule i.

The residual part of the activity coefficient is computed in a manner similar to that in original UNIFAC. However, for liquid-liquid equilibria, as in this example, different group-group interaction parameters are required.

The consitutive groups of the two components are

Component	Constitutive Groups
1	H_2O
2	6ACH

Group-volume (R_k) and surface-area (Q_k) parameters are

Group	R_k	Q_k
H_2O	0.9200	1.40
ACH	0.5313	0.40

Group-group interaction parameters (K) are functions of temperature T (K):

$$a_{mn}(T) = a_{mn}^{(0)} + a_{mn}^{(1)}T + a_{mn}^{(2)}T^2$$
$$a_{nm}(T) = a_{nm}^{(0)} + a_{nm}^{(1)}T \tag{8-14.16}$$

where subscript m denotes the H_2O group and subscript n denotes the ACH group.

Coefficients in Eq. (8-14.16) are given by Hooper, et al. (1988):

$a_{mn}^{(0)}$ (K)	$a_{mn}^{(1)}$	$a_{mn}^{(2)}$ $(1/K)$	$a_{nm}^{(0)}$ (K)	$a_{nm}^{(1)}$
−39.04	3.928	−0.00437	2026.5	−3.267

Table 8-54 shows results from solving two Eqs. (8-14.11) simultaneously. Smoothed experimental data at the same temperatures are from Sorensen, et al. (1979).

Figure 8-36 compares calculated and experimental results.

In mole fraction units, the solubility of water in benzene is appreciably larger than the solubility of benzene in water. While overall there is good agreement between calculated and observed results, agreement in the benzene-rich phase is superior to that in the water-rich phase. Neither UNIFAC nor any other currently available engineering-oriented theory is suitable for dilute aqueous solutions of hydrocarbon (or similar) solutes because, as yet, no useful theory has been established for describing the hydrophobic effect. This effect strongly influences the solubility of a nonpolar (or weakly polar) organic solute in water, especially at low or moderate temperatures.

(b) water (1)–aniline (2)

The general procedure here is similar to that shown in part (a). However, different parameters are required.

For UNIFAC, molecules of the two components are broken into groups:

Component	Constitutive Groups
1	H_2O
2	5ACH + $ACNH_2$

TABLE 8-54 Liquid-Liquid Equilibria for Water (1)–Benzene (2) in the range 0 to 70°C

	Calculated		Experimental	
t (°C)	$100x_2'$	$100x_1''$	$100x_2'$	$100x_1''$
0	0.030	0.120	0.040	0.133
10	0.035	0.177	0.040	0.180
20	0.038	0.254	0.040	0.252
30	0.043	0.357	0.041	0.356
40	0.048	0.492	0.044	0.491
50	0.054	0.665	0.047	0.664
60	0.061	0.887	0.053	0.895
70	0.069	1.160	0.062	1.190

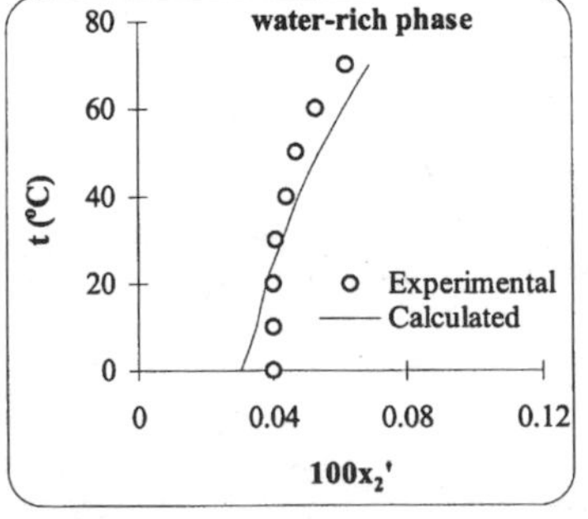

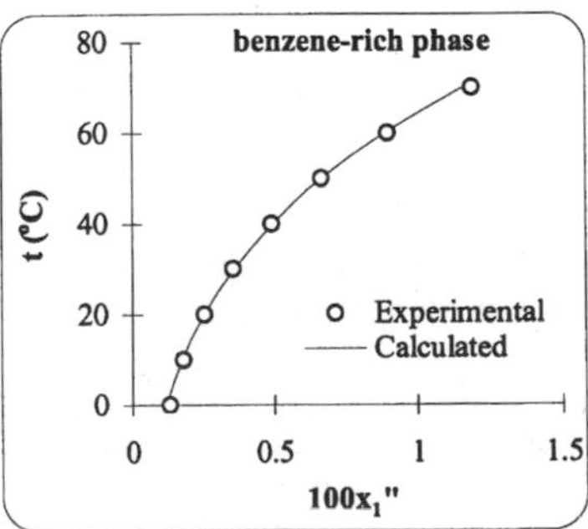

FIGURE 8-36 Temperature-mole fraction plots for water (1)–benzene (2).

Group-volume and surface-area parameters are

Group	R_k	Q_k
H_2O	0.9200	1.400
ACH	0.5313	0.400
$ACNH_2$	1.0600	0.816

According to Hooper, et al. (1988), interaction parameters between organic groups are temperature-independent, whereas those between the water group and an organic group are temperature-dependent. If we denote the H_2O group as m, the ACH group as n, and the $ACNH_2$ group as p, the temperature-independent interaction parameters are a_{np} = 763.6 K and a_{pn} = 9859 K.

The temperature-dependent interaction parameters are

$$a_{mj}(T) = a_{mj}^{(0)} + a_{mj}^{(1)} + a_{mj}^{(2)}T^2$$
$$a_{jm}(T) = a_{jm}^{(0)} + a_{jm}^{(1)}T \tag{8-14.17}$$

where j stands for n or p.

Coefficients in Eq. (8-14.17) are

	$a_{mj}^{(0)}$ (K)	$a_{mj}^{(1)}$	$a_{mj}^{(2)}$ (1/K)	$a_{jm}^{(0)}$ (K)	$a_{jm}^{(1)}$
$j = n$	−39.04	3.928	−0.00437	2026.5	−3.267
$j = p$	−29.31	1.081	−0.00329	−1553.4	6.178

Solving two Eqs. (8-14.11) simultaneously leads to calculated results that are given in Table 8-55. Smoothed experimental data at the same temperatures are from Sorensen, et al. (1979).

Figure 8-37 compares calculated and experimental results. Agreement is satisfactory.

TABLE 8-55 Liquid-Liquid Equilibria for Water (1)–aniline (2) in the range 20 to 140°C

	Calculated		Experimental	
t (°C)	$100x_2'$	$100x_1''$	$100x_2'$	$100x_1''$
20	0.629	24.6	0.674	21.3
25	0.654	23.4	0.679	21.8
40	0.746	21.3	0.721	23.7
60	0.923	22.2	0.847	26.5
80	1.180	27.0	1.110	30.6
100	1.540	34.7	1.520	36.0
120	2.060	43.5	2.110	42.2
140	2.850	52.4	3.020	51.1

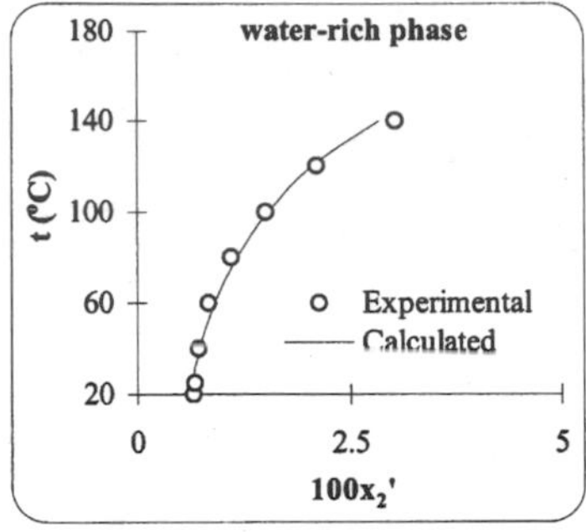

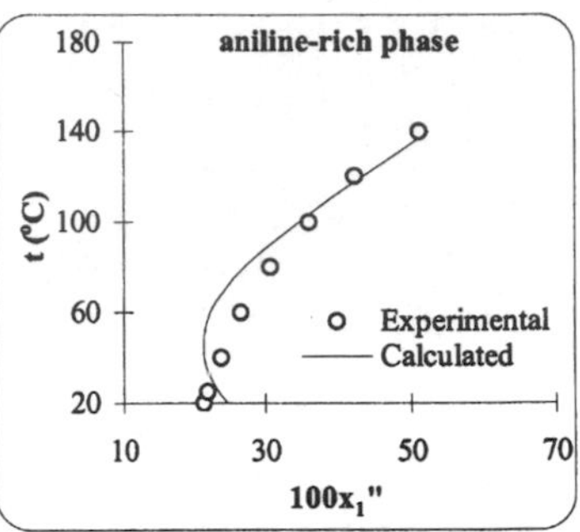

FIGURE 8-37 Temperature-mole fraction plots for water (1)–aniline (2).

Because, unlike benzene, aniline has an amino NH_2 group that can hydrogen-bond with water, mutual solubilities for water-aniline are much larger than those for water-benzene.

Example 8-34 Using the Modified UNIFAC method suggested by Hooper et al. (1988), calculate liquid-liquid equilibria for the ternary water (1)-phenol (2)-toluene (3) at 200°C.

solution The three equations at equilibrium are

$$x_i'\gamma_i' = x_i''\gamma_i'' \qquad (i = 1,2,3) \tag{8-14.18}$$

where all symbols in Eq. (8-14.18) are defined after Eq. (8-14.11).

Here we have Eq. (8-14.18) with four unknowns: x_1', x_2', x_1'', and x_2''. Mole fractions x_3' and x_3'' are not independent because they are constrained by material balances:

$$x_1' + x_2' + x_3' = 1$$

$$x_1'' + x_2'' + x_3'' = 1$$

To calculate liquid-liquid equilibria in a ternary system, we need to perform an isothermal flash calculation where we introduce two additional equations and one additional unknown α:

$$z_i = x_i'\alpha + x_i''(1 - \alpha) \qquad (i = 1,2) \tag{8-14.19}$$

where z_i is the mole fraction of componeent i in the feed; α is the mole fraction of the feed that becomes liquid $'$ in equilibrium with liquid $''$. We do not apply Eq. (8-14.19) to the third component because in addition to the material balance, we also have the overall constraint:

$$z_1 + z_2 + z_3 = 1$$

We solve the isothermal flash for arbitrary positive values of z_1 and z_2 provided $(z_1 + z_2) \leq 1$. For the Modified UNIFAC, molecules of the three components are broken into groups as follows:

Component	Constitutive Groups
1	H_2O
2	5ACH + ACOH
3	5ACH + $ACCH_3$

Group-volume and surface-area parameters are

Group	R_k	Q_k
H_2O	0.9200	1.400
ACH	0.5313	0.400
ACOH	0.8952	0.680
$ACCH_3$	1.2663	0.968

As indicated in Example 8-33, group-group parameters for interaction between organic groups are temperature-independent, whereas those between the H_2O group and an organic group are temperature-dependent. We denote the four groups as follows: m for H_2O, n for ACH, p for ACOH, and q for $ACCH_3$.

The temperature-independent interaction parameters a_{ij} (K) for $i, j = n, p, q$ are

Group	n	p	q
n	0	1208.5	−27.67
p	2717.3	0	7857.3
q	47.31	816.21	0

The temperature-dependent interaction parameters (K) are as in Eq. (8-14.17).

The coefficients are

	$a_{mj}^{(0)}$ (K)	$a_{mj}^{(1)}$	$1000a_{mj}^{(2)}$ (1/K)	$a_{jm}^{(0)}$	$a_{jm}^{(1)}$
$j = n$	−39.04	3.928	−4.370	2026.5	−3.2670
$j = p$	−39.10	2.694	−4.377	−199.25	−0.5287
$j = q$	−50.19	3.673	−5.061	2143.9	−3.0760

At 200°C and a given set (z_1, z_2), three Eqs. (8-14.18) and two Eqs. (8-14.19) can be solved simultaneously for five unknowns: x_1', x_2', x_1'', x_2'', and α. Calculated results are given in Table 8-56. Also shown are experimental data at 200°C from Hooper, et al. (1988a).

TABLE 8-56 Liquid-Liquid Equilibria for Water (1)–Phenol (2)–Toluene (3) at 200°C

Given		Calculated					Observed			
z_1	z_2	α	$100x'_1$	$100x'_2$	$100x''_1$	$100x''_2$	$100x'_1$	$100x'_2$	$100x''_1$	$100x''_2$
0.8	0.000	0.768	99.8	0.000	14.6	0.00	99.05	0.62	17.22	5.92
0.8	0.010	0.767	99.3	0.461	16.7	2.77	98.26	1.12	20.27	9.66
0.8	0.060	0.750	96.5	2.97	30.5	15.1	97.93	1.51	22.32	12.91
0.8	0.100	0.714	93.6	5.44	46.1	21.4	96.96	2.35	28.76	17.69
0.8	0.120	0.675	91.6	7.10	55.9	22.2	96.02	3.27	35.59	21.12
0.8	0.140	0.586	88.3	9.60	68.3	20.2	95.02	4.06	40.16	22.59
0.8	0.150	0.355	83.4	13.0	78.1	16.1	92.76	5.88	52.92	23.14
0.8	0.155	0.762	80.0	15.5	80.0	15.5	91.82	6.57	55.14	23.01

Figure 8-38 compares calculated and experimental results. Agreement is fair. However, we must not conclude from this example that agreement will necessarily be equally good for other aqueous systems.

A frequently encountered problem in liquid-liquid equilibria is to obtain the distribution coefficient (also called the partition coefficient) of a solute between two essentially immiscible liquid solvents when the solute concentration in both solvents is small. In most practical cases, one of the solvents is water and the other solvent is an organic liquid whose solubility in water is negligible.

Because the phases have the solute, s, at infinite dilution in the organic solvent, os, and in water, w, the infinite dilution partition coefficient is a ratio of binary activity coefficients at infinite dilution

$$S^{\infty}_{s,os/w} = \lim_{x_s \to 0} \frac{\rho_{s,os}}{\rho_{s,w}} = \frac{\rho_{os} \lim\limits_{x_w \to 1} \gamma^{\infty}_{s,w}}{\rho_w \lim\limits_{x_{os} \to 1} \gamma^{\infty}_{s,os}} \tag{8-14.20}$$

where $\rho_{s,os}$ and $\rho_{s,w}$ are the molar concentrations (mol L^{-1}) of the solute in the organic and water solvents, respectively and ρ_{os} and ρ_w are the molar densities (mol L^{-1}) of the pure organic and water solvents, respectively. As a result, $S^{\infty}_{s,os/w}$ can be obtained from a linear solvation energy relationship (LSER) similar to the SPACE model of Eq. (8-10.31) for activity coefficients at infinite dilution. Meyer and Maurer (1995) have examined partition coefficients for a wide variety of solutes in such two-phase aqueous-solvent systems. Their correlation is illustrated in Example 8-35.

Example 8-35 Use the Meyer and Maurer correlation to estimate partition coefficients for some organic solutes that distribute between an organic solvent and water at 25°C. Meyer and Maurer (1995) fitted 825 experimental infinite-dilution partition coefficients, $S^{\infty}_{s,os/w}$, for organic solutes in 20 organic solvent/water systems at 25°C and proposed a generalized linear solvation energy relationship (LSER) correlation:

$$\begin{aligned} \log S^{\infty}_{s,os/w}\,(25°C) = {} & K_1 + K_2 V_{os} + K_3 \pi_{os} + K_4 \delta_{os} + K_5 \beta_{os} + K_6 \alpha_{os} \\ & + (M_1 + M_2 V_{os} + M_3 \pi_{os} + M_4 \delta_{os} + M_5 \beta_{os} + M_6 \alpha_{os}) V_s \\ & + (S_1 + S_2 V_{os} + S_3 \pi_{os} + S_4 \delta_{os} + S_5 \beta_{os} + S_6 \alpha_{os}) \pi_s \\ & + (D_1 + D_2 V_{os} + D_3 \pi_{os} + D_4 \delta_{os} + D_5 \beta_{os} + D_6 \alpha_{os}) \delta_s \\ & + (B_1 + B_2 V_{os} + B_3 \pi_{os} + B_4 \delta_{os} + B_5 \beta_{os} + B_6 \alpha_{os}) \beta_s \\ & + (A_1 + A_2 V_{os} + A_3 \pi_{os} + A_4 \delta_{os} + A_5 \beta_{os} + A_6 \alpha_{os}) \alpha_s \end{aligned} \tag{8-14.21}$$

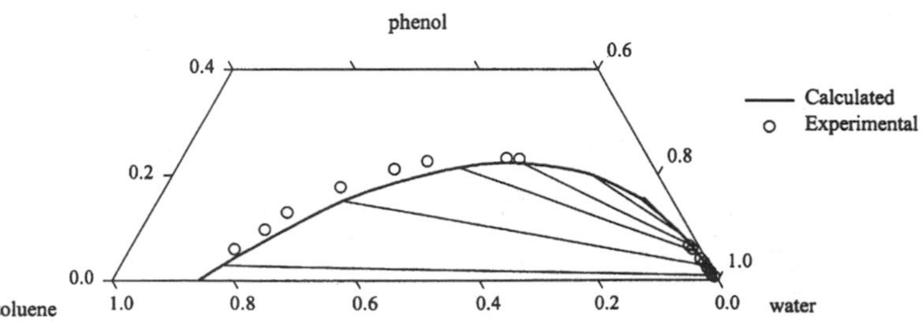

FIGURE 8-38 Liquid-liquid equilibria for water (1)–phenol (2)–toluene (3) at 200°C.

where subscripts *s*, *os*, and *w* denote solute, organic solvent, and water, respectively, K_i, M_i, S_i, D_i, B_i, and A_i (i = 1,2,...,6) are the universal LSER constants:

i	K_i	M_i	S_i	D_i	B_i	A_i
1	0.180	5.049	−0.624	0.464	−2.983	−3.494
2	0.594	0.731	−2.037	0.509	−2.255	0.418
3	1.652	0.090	−1.116	1.202	−1.570	0.134
4	−0.752	0.282	0.531	−0.307	0.625	0.146
5	−1.019	−1.563	2.356	−0.935	1.196	7.125
6	−0.979	0.365	1.761	−1.605	4.015	−0.562

In Eq. (8-14.20) V is a volumetric parameter, π, β, and α are solvachromic parameters; δ is the polarizability parameter. These parameters are given by Meyer and Maurer for various organic compounds.

In this example, we employ the correlation of Meyer and Maurer to calculate infinite-dilution partition coefficients for the three following solutes: benzene, *n*-hexane, 1-heptanol in the 1-octanol/water system at 25°C.

solution For the 1-octanol/water system and the three solutes considered here, Meyer and Maurer give

Compound	V	π	δ	β	α
1-octanol	0.888	0.40	0.18	0.45	0.33
benzene	0.491	0.59	1.00	0.10	0.00
n-hexane	0.648	−0.04	0.00	0.00	0.00
1-heptanol	0.790	0.40	0.08	0.45	0.33

Substituting these parameters and the universal LSER constants into Eq. (8-14.21) leads to the results in Table 8-57. Also shown are experimental data at 25°C taken from Lide and Frederikse (1997). The good results shown here suggest that the correlation of Meyer and Maurer is reliable.

TABLE 8-57 Infinite-Dilution Partition Coefficients in 1-Octanol/Water at 25°C

	$S^{\infty}_{S,OS/W}$		
Solute	Calculated	Observed	% error
benzene	1.35E2	1.35E2	0
n-hexane	1.00E4	1.00E4	0
1-heptanol	3.72E2	4.17E2	10.8

8-15 PHASE EQUILIBRIA IN POLYMER SOLUTIONS

Strong negative deviations from Raoult's law are observed in binary liquid mixtures where one component consists of very large molecules (polymers) and the other consists of molecules of normal size. For mixtures of normal solvents and amorphous polymers, phase-equilibrium relations are usually described by the Flory-Huggins theory, discussed fully in a book by Flory (1953) and in Vol. 15 of the Encylcopedia of Polymer Science and Engineering, Kroschwitz, (1989); a brief introduction is given by Prausnitz et al. (1999). For engineering application, a useful summary is provided by Sheehan and Bisio (1966).

There are several versions of the Flory-Huggins theory, and unfortunately, different authors use different notation. The primary composition variable for the liquid phase is the volume fraction, here designated by Φ and defined by Eqs. (8-10.3) and (8-10.4). In polymer-solvent systems, volume fractions are very different from mole fractions because the molar volume of a polymer is much larger that that of the solvent.

Since the molecular weight of the polymer is often not known accurately, it is difficult to determine the mole fraction. Therefore, an equivalent definition of Φ is frequently useful:

$$\Phi_1 = \frac{w_1/\rho_1}{w_1/\rho_1 + w_2/\rho_2} \qquad \text{and} \qquad \Phi_2 = \frac{w_2/\rho_2}{w_1/\rho_1 + w_2/\rho_2} \tag{8-15.1}$$

where w_i is the weight fraction of component i and ρ_i is the mass density (*not* molar density) of pure component i.

Let subscript 1 stand for solvent and subscript 2 for polymer. The activity a_1 of the solvent, as given by the Flory-Huggins equation, is

$$\ln a_1 = \ln \Phi_1 + \left(1 - \frac{1}{m}\right) \Phi_2 + \chi \Phi_2^2 \tag{8-15.2}$$

where $m = V_2^L/V_1^L$ and the adjustable constant χ is called the *Flory interaction parameter.* In typical polymer solutions $1/m$ is negligibly small compared with unity, and therefore it may be neglected. Parameter χ depends on temperature, but for polymer-solvent systems in which the molecular weight of the polymer is very large, it is nearly independent of polymer molecular weight. In theory, χ is also independent of polymer concentration, but in fact, it often varies with concentration, especially in mixtures containing polar molecules, for which the Flory-Huggins theory provides only a rough approximation.

In a binary mixture of polymer and solvent at ordinary pressures, only the solvent is volatile; the vapor phase mole fraction of the solvent is unity, and therefore, the total pressure is equal to the partial pressure of the solvent.

In a polymer solution, the activity of the solvent is given by

$$a_1 = \frac{P}{P_{\text{vp}1}\mathfrak{F}_1} \tag{8-15.3}$$

where factor $\mathfrak{F}_1$ is defined by Eq. (8-4.2). At low or moderate pressures, $\mathfrak{F}_1$ is equal to unity.

Equation (8-15.2) holds only for temperatures where the polymer in the pure state is amorphous. If the pure polymer has appreciable crystallinity, corrections to Eq. (8-15.2) may be significant, as discussed elsewhere (Flory, 1953).

Equation (8-15.2) is useful for calculating the volatility of a solvent in a polymer solution, provided that the Flory parameter χ is known. Sheehan and Bisio (1966) report Flory parameters for a large number of binary systems†; they, and more recently Barton (1990), present methods for estimating χ from solubility parameters. Similar data are also given in the *Polymer Handbook* (Bandrup and Immergut, 1999). Table 8-58 shows some χ values reported by Sheehan and Bisio.

A particularly convenient and rapid experimental method for obtaining χ is provided by gas-liquid chromatography (Guillet, 1973). Although this experimental technique can be used at finite concentrations of solvent, it is most efficiently used for solutions infinitely dilute with respect to solvent, i.e., at the limit where the volume fraction of polymer approaches unity. Some solvent volatility data obtained from chromatography (Newman and Prausnitz, 1973) are shown in Fig. 8-39. From these data, χ can be found by rewriting Eq. (8-14.2) in terms of a weight fraction activity coefficient Ω

$$\Omega_1 \equiv \frac{a_1}{w_1} = \frac{P}{P_{vp1} w_1 \mathfrak{F}_1} \tag{8-15.4}$$

Combining with Eq. (8-15.2), in the limit as $\Phi_2 \rightarrow 1$, we obtain

$$\chi = \ln \left(\frac{P}{w}\right)_1^\infty - \left(\ln P_{vp1} + \ln \frac{\rho_2}{\rho_1} + 1\right) \tag{8-15.5}$$

where ρ is the mass density (*not* molar density). Equation (8-15.5) also assumes that $\mathfrak{F}_1 = 1$ and that $1/m << 1$. Superscript ∞ denotes that weight fraction w_1 is very small compared with unity. Equation (8-15.5) provides a useful method for finding χ because $(P/w)_1^\infty$ is easily measured by gas-liquid chromatography.

TABLE 8-58 Flory χ Parameters for Some Polymer-Solvent Systems Near Room Temperature (Sheehan and Bisio, 1966)

Polymer	Solvent	χ
Natural rubber	Heptane	0.44
	Toluene	0.39
	Ethyl acetate	0.75
Polydimethyl siloxane	Cyclohexane	0.44
	Nitrobenzene	2.2
Polyisobutylene	Hexadecane	0.47
	Cyclohexane	0.39
	Toluene	0.49
Polystyrene	Benzene	0.22
	Cyclohexane	0.52
Polyvinyl acetate	Acetone	0.37
	Dioxane	0.41
	Propanol	1.2

† Unfortunately, Sheehan and Bisio use completely different notation; v for ϕ, x for m, and μ for χ.

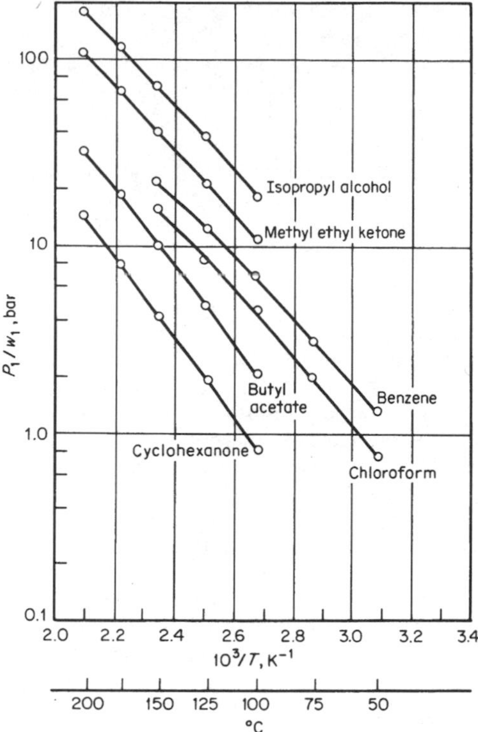

FIGURE 8-39 Volatilities of solvents in Lucite 2044 for a small weight fraction of solute. (*Newman and Prausnitz, 1973*)

Equation (8-15.2) was derived for a binary system, i.e., one in which all polymer molecules have the same molecular weight (monodisperse system). For mixtures containing one solvent and one polymer with a variety of molecular weights (polydisperse system), Eq. (8-15.2) can be used provided m and Φ refer to the polymer whose molecular weight is the number-average molecular weight.

The theory of Flory and Huggins can be extended to multicomponent mixtures containing any number of polymers and any number of solvents. No ternary (or higher) constants are required.

Solubility relations (liquid-liquid equilibria) can also be calculated with the Flory-Huggins theory. Limited solubility is often observed in solvent-polymer systems, and it is common in polymer-polymer systems (incompatibility). The Flory-Huggins theory indicates that, for a solvent-polymer system, limited miscibility occurs when

$$\chi > \frac{1}{2}\left(1 + \frac{1}{m^{1/2}}\right)^2 \tag{8-15.6}$$

For large m, the value of χ may not exceed ½ for miscibility in all proportions.

Liquid-liquid phase equilibria in polymer-containing systems are described in numerous articles published in journals devoted to polymer science and engineering.

The thermodynmamics of such equilibria is discussed in Flory's book and in numerous articles. A comprehensive review of polymer compatibilty and incompatibility is given by Krause (1972) and, more recently, by Danner and High (1993).

For semiquantitative calculations the three-dimensional solubility parameter concept (Hansen et al., 1967, 1971) is often useful, especially for formulations of paints, coatings, inks, etc.

The Flory-Huggins equation contains only one adjustable binary parameter. For simple nonpolar systems one parameter is often sufficient, but for complex systems, much better representation is obtained by empirical extension of the Flory-Huggins theory using at least two adjustable parameters, as shown, for example by Heil and Prausnitz (1966) and by Bae, et al. (1993). Heil's extension is a generalization of Wilson's equation. The UNIQUAC equation with two adjustable parameters is also applicable to polymer solutions.

The theory of Flory and Huggins is based on a lattice model that ignores free-volume differences; in general, polymer molecules in the pure state pack more densely than molecules of normal liquids. Therefore, when polymer molecules are mixed with molecules of normal size, the polymer molecules gain freedom to exercise their rotational and vibrational motions; at the same time, the smaller solvent molecules partially lose such freedom. To account for these effects, an *equation-of-state theory* of polymer solutions has been developed by Flory (1970) and Patterson (1969) based on ideas suggested by Prigogine (1957). The newer theory is necessarily more complicated, but, unlike the older one, it can at least semiquantitatively describe some forms of phase behavior commonly observed in polymer solutions and polymer blends. In particular, it can explain the observation that some polymer-solvent systems exhibit lower consolute temperatures as well as upper consolute temperatures similar to those shown in Figs. 8-32 and 8-33.†

A group-contribution method (UNIFAC) for estimating activities of solvents in polymer-solvent systems was presented by Oishi (1978). Variations on Oishi's work and other correlations for polymer-solvent phase equilibria are reviewed by Danner and High (1993).

8-16 SOLUBILITIES OF SOLIDS IN LIQUIDS

The solubility of a solid in a liquid is determined not only by the intermolecular forces between solute and solvent but also by the melting point and the enthalpy of fusion of the solute. For example, at 25°C, the solid aromatic hydrocarbon phenanthrene is highly soluble in benzene; its solubility is 20.7 mole percent. By contrast, the solid aromatic hydrocarbon anthracene, an isomer of phenanthrene, is only slightly soluble in benzene at 25°C; its solubility is 0.81 mole percent. For both solutes, intermolecular forces between solute and benzene are essentially identical. However, the melting points of the solutes are significantly different: phenanthrene melts at 100°C and anthracene at 217°C. In general, it can be shown that, when other factors are held constant, the solute with the higher melting point has the lower solubility. Also, when other factors are held constant, the solute with the higher enthalpy of fusion has the lower solubility.

† However, in polymer-solvent systems, the upper consolute temperature usually is below the lower consolute temperature.

These qualitative conclusions follow from a quantitative thermodynamic analysis given in numerous texts. (See, for example, Prigogine and Defay, 1954 and Prausnitz, et al., 1999).

In a binary system, let subscript 1 stand for solvent and subscript 2 for solute. Assume that the solid phase is pure. At temperature T, the solubility (mole fraction) x_2 is given by

$$\ln \gamma_2 x_2 = -\frac{\Delta H_m}{RT}\left(1 - \frac{T}{T_t}\right) + \frac{\Delta C_p}{R}\left(\frac{T_t - T}{T}\right) - \frac{\Delta C_p}{R}\ln\frac{T_t}{T} \tag{8-16.1}$$

where ΔH_m is the enthalpy change for melting the solute at the triple-point temperature T_t and ΔC_p is given by the molar heat capacity of the pure solute:

$$\Delta C_p = C_p\ \text{(subcooled liquid solute)} - C_p\ \text{(solid solute)} \tag{8-16.2}$$

The standard state for activity coefficient γ_2 is pure (subcooled) liquid 2 at system temperature T.

To a good approximation, we can substitute normal melting temperature T_{fp} for triple-point temperature T_t, and we can assume that ΔH_m is essentially the same at the two temperatures. In Eq. (8-16.1) the first term on the right-hand side is much more important than the remaining two terms, and therefore a simplified form of that equation is

$$\ln \gamma_2 x_2 = \frac{-\Delta H_m}{RT}\left(1 - \frac{T}{T_{fp}}\right) \tag{8-16.3}$$

If we substitute

$$\Delta S_m = \frac{\Delta H_m}{T_{fp}} \tag{8-16.4}$$

we obtain an alternative simplified form

$$\ln \gamma_2 x_2 = -\frac{\Delta S_m}{R}\left(\frac{T_{fp}}{T} - 1\right) \tag{8-16.5}$$

where ΔS_m is the entropy of fusion. A plot of Eq. (8-16.5) is shown in Fig. 8-40.

If we let $\gamma_2 = 1$, we can readily calculate the ideal solubility at temperature T, knowing only the solute's melting temperature and its enthalpy (or entropy) of fusion. This ideal solubility depends only on properties of the solute; it is independent of the solvent's properties. The effect of intermolecular forces between molten solute and solvent are reflected in activity coefficient γ_2.

To describe γ_2, we can use any of the expressions for the excess Gibbs energy, as discussed in Sec. 8-5. However, since γ_2 depends on the mole fraction x_2, solution of Eq. (8-16.5) requires iteration. For example, suppose that γ_2 is given by a simple one-parameter Margules equation

$$\ln \gamma_2 = \frac{A}{RT}(1 - x_2)^2 \tag{8-16.6}$$

where A is an empirical constant. Substitution into Eq. (8-15.5) gives

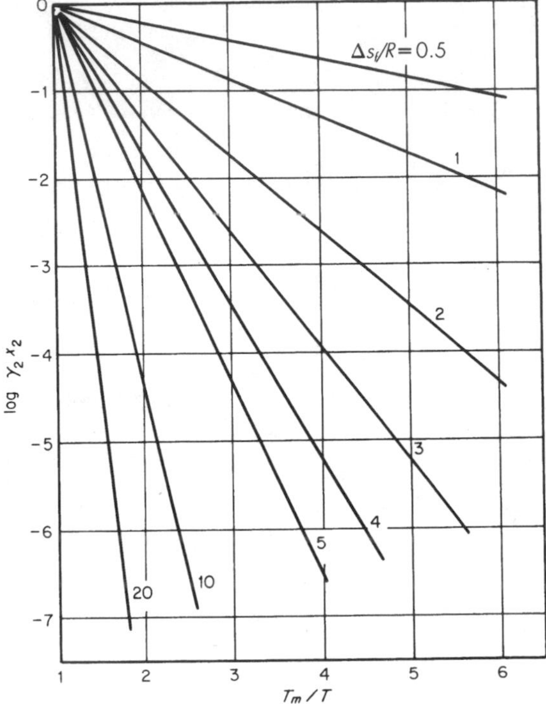

FIGURE 8-40 Activities of solid solutes referred to their pure subcooled liquids. (*From Preston amd Prausnitz, 1970*)

$$\ln x_2 + \frac{A}{RT}(1 - x_2)^2 = -\frac{\Delta S_m}{R}\left(\frac{T_{fp}}{T} - 1\right) \tag{8-16.7}$$

and x_2 must be found by a trial-and-error calculation.

In nonpolar systems, activity coefficient γ_2 can often be estimated by using the Scatchard-Hildebrand equation, as discussed in Sec. 8-10. In that event, since $\gamma_2 \geq 1$, the ideal solubility ($\gamma_2 = 1$) is larger than that obtained from regular solution theory. As shown by Preston and Prausnitz (1970) and as illustrated in Fig. 8-41, regular solution theory is useful for calculating solubilities in nonpolar systems, especially when the geometric-mean assumption is relaxed through introduction of an empirical correction l_{12} (see Sec. 8-10). Figure 8-41 shows three lines: the top line is calculated by using the geometric-mean assumption ($l_{12} = 0$) in the Scatchard-Hildebrand equation. The bottom line is calculated with $l_{12} = 0.11$, estimated from gas-phase *PVTy* data. The middle line is calculated with $l_{12} = 0.08$, the optimum value obtained from solubility data. Figure 8-41 suggests that even an approximte estimate of l_{12} usually produces better results than assuming that l_{12} is zero. Unfortunately, *some* mixture data point is needed to estimate l_{12}. In a few fortunate cases one freezing point datum, e.g., the eutectic point, may be available to fix l_{12}.

In some cases it is possible to use UNIFAC for estimating solubilities of solids, as discussed by Gmehling, el al. (1978).

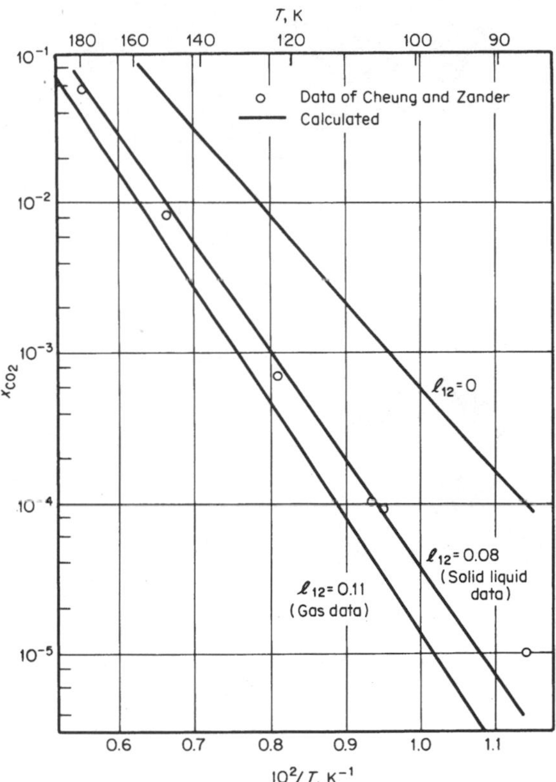

FIGURE 8-41 Solubility of carbon dioxide in propane. (*From Preston and Prausnitz, 1970*)

It is important to remember that the calculations outlined above rest on the assumption that the solid phase is pure, i.e., that there is no solubility of the solvent in the solid phase. This assumption is often a good one, especially if the two components differ appreciably in molecular size and shape. However, in many known cases, the two components are at least partially miscible in the solid phase, and in that event it is necessary to correct for solubility and nonideality in the solid phase as well as in the liquid phase. This complicates the thermodynamic description, but, more important, solubility in the solid phase may significantly affect the phase diagram. Figure 8-42 shows results for the solubility of solid argon in liquid nitrogen. The top line presents calculated results assuming that x^S (argon) = 1, where superscript ^{S} denotes the solid phase. The bottom line takes into account the experimentally known solubility of nitrogen in solid argon [x^S (argon) $\neq$ 1]. In this case it is clear that serious error is introduced by neglecting solubility of the solvent in the solid phase.

Variations of the UNIQUAC equation have been used to correlate experimental solubility data for polar solid organic solutes in water. Examples 8-36 and 8-37 illustrate such correlations for solubilities of sugars and amino acids.

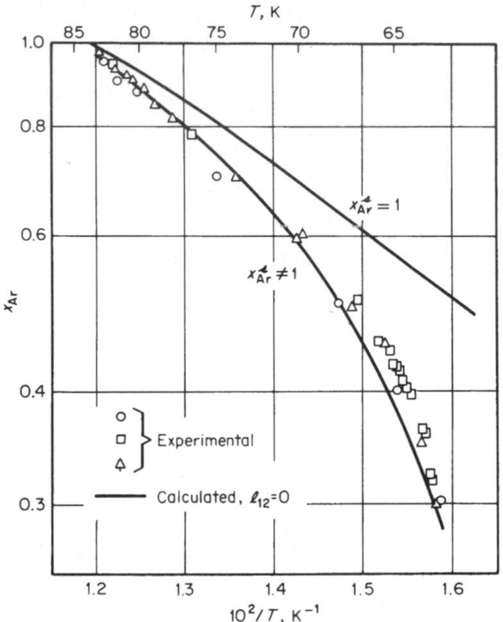

FIGURE 8-42 Solubility of argon in nitrogen: effect of solid-phase composition. (*From Preston and Prausnitz, 1970*)

Example 8-36 Peres and Macedo (1996) proposed a modified UNIQUAC model to describe vapor-liquid equilibria (VLE) and solid-liquid equilibria (SLE) for mixtures containing D-glucose, D-fructose, sucrose, and water.

Use the modified UNIQUAC model to calculate the SLE composition phase diagram for the ternary D-glucose (1)-sucrose (2)-water (3) at 70°C.

solution If each sugar is assumed to exist as a pure solid phase, the two equations at equilibrium are

$$f_i^L = f_{i,p}^S \qquad (i = 1,2) \tag{8-16.8}$$

or equivalently,

$$x_i \gamma_i = f_{i,p}^S / f_{i,p}^L \qquad (i = 1,2) \tag{8-16.9}$$

where subscript i denotes sugar i and subscript p denotes pure. Superscripts L and S stand for liquid and solid; f is fugacity; x is the liquid-phase mole fraction; γ is the liquid-phase activity coefficient.

The standard state for each sugar is chosen as the pure subcooled liquid at system temperature; further, the difference between the heat capacity of pure liquid sugar and that of pure solid sugar $\Delta C_{P,i}$ is assumed to be linearly dependent on temperature:

$$\Delta C_{P,i} = \Delta A_i - \Delta B_i(T - T_o) \qquad (i = 1,2) \tag{8-16.10}$$

where ΔA_i (J mol^{-1} K^{-1}) and ΔB_i (J mol^{-1} K^{-2}) are constants for each sugar; T = 343.15 K; T_o is the reference temperature 298.15 K.

The ratio $f^S_{i,p}/f^L_{i,p}$ [Eq. 8-16.1] is given by Gabas and Laguerie (1993), Raemy and Schweizer (1983), Roos (1993):

$$\ln (f^S_{i,p}/f^L_{i,p}) = \left[-\frac{\Delta H_{m,i}}{R} + \frac{\Delta A_i - \Delta B_i T_o}{R} T_{fp,i} + \frac{\Delta B_i}{2R} T^2_{fp,i} \right] \left(\frac{1}{T} - \frac{1}{T_{fp,i}} \right) + \frac{\Delta A_i - \Delta B_i T_o}{R} \ln \left(\frac{T}{T_{fp,i}} \right) + \frac{\Delta B_i}{2R} (T - T_{fp,i}) \tag{8-16.11}$$

where R is the universal gas constant; $T_{fp,i}$ is the melting temperature of pure sugar i; ΔH_m is the enthalpy change of melting at $T_{fp,i}$; ΔA_i, ΔB_i, T, and T_o are those in Eq. (8-16.10).

Substituting Eq. (8-16.11) into Eq. (8-16.9), we obtain the specific correlation of Eq. (8-16.1) to be used in this case

$$\ln (x_i \gamma_i) = \left[-\frac{\Delta H_{m,i}}{R} + \frac{\Delta A_i - \Delta B_i T_o}{R} T_{fp,i} + \frac{\Delta B_i}{2R} T^2_{fp,i} \right] \left(\frac{1}{T} - \frac{1}{T_{fp,i}} \right) + \frac{\Delta A_i - \Delta B_i T_o}{R} \ln \left(\frac{T}{T_{fp,i}} \right) + \frac{\Delta B_i}{2R} (T - T_{fp,i}) \qquad (i = 1,2) \tag{8-16.12}$$

For the two sugars of interest, physical properites used in Eq. (8-16.12) are given by Peres and Macedo (1996):

	D-glucose	sucrose
$T_{fp,i}$ (K)	423.15	459.15
$\Delta H_{m,i}$ (J mol^{-1})	32432	46187
$\Delta A'_i$ (J mol^{-1} K^{-1})	139.58	316.12
$\Delta B'_i$ (J mol^{-1} K^{-2})	0	−1.15

At fixed T, the right side of Eq. (8-16.12) can be solved simultaneously for $i = 1$ and $i = 2$ to yield x_1 and x_2 (x_3 is given by mass balance $x_1 + x_2 + x_3 = 1$). Activity coefficients are given by a modified form of UNIQUAC. (See Sec. 8-5) where the liquid-phase activity coefficient is divided into a combinatorial part and a residual part:

$$\ln \gamma_i = \ln \gamma_i^C + \ln \gamma_i^R \qquad (i = 1,2,3) \tag{8-16.13}$$

For the combinatorial part, Peres and Macedo (1996) used the expression for liquid-liquid systems suggested by Larsen, et al. (1987):

$$\ln \gamma_i^C = \ln \left(\frac{\varphi_i}{x_i} \right) + 1 - \frac{\varphi_i}{x_i} \tag{8-16.14}$$

where x_i is mole fraction of component i, and the volume fraction φ_i of component i is defined by:

$$\varphi_i = \frac{x_i r_i^{2/3}}{\sum_{j=1}^{3} x_j r_j^{2/3}} \tag{8-16.15}$$

r_i is the UNIQUAC volume parameter for component i.

The residual activity coefficient given by the original UNIQUAC (Abrams and Prausnitz, 1975) is:

$$\ln \gamma_i^R = q_i \left[1 - \ln \left(\sum_{j=1}^{3} \theta_j \tau_{ji} \right) - \sum_{j=1}^{3} \left(\frac{\theta_j \tau_{ij}}{\sum_{k=1}^{3} \theta_k \tau_{kj}} \right) \right] \tag{8-16.16}$$

where q_i is the UNIQUAC surface parameter for component i.

Molecular surface fraction θ_j is given by:

$$\theta_j = \frac{x_j q_j}{\sum_{k=1}^{3} x_k q_k} \tag{8-16.17}$$

and τ_{ij} is given by:

$$\tau_{ij} = \exp\left(-\frac{a_{ij}}{T} \right) \tag{8-16.18}$$

where the a_{ij} (K) are the interaction parameters between components i and j. In general, $a_{ij} \neq a_{ji}$.

For D-glucose, sucrose, and water, volume and surface parameters are

Component	r_i	q_i
D-glucose	8.1528	7.920
sucrose	14.5496	13.764
water	0.9200	1.400

Peres and Macedo (1996) set interaction parameters between the two sugars to zero ($a_{12} = a_{21} = 0$) while they assume that interaction parameters between water and sugars (a_{i3} and a_{3i}, $i = 1,2$) are linearly dependent on temperature:

$$a_{i3} = a_{i3}^{(0)} + a_{i3}^{(1)}(T - T_o)$$
$$a_{3i} = a_{3i}^{(0)} + a_{3i}^{(1)}(T - T_o) \tag{8-16.19}$$

Using experimental binary water-sugar data from 0 to 100°C and from very dilute to saturated concentration, Peres and Macedo (1996) give

	$a_{i3}^{(0)}$ (K)	$a_{3i}^{(0)}$ (K)	$a_{i3}^{(1)}$	$a_{3i}^{(1)}$
i = 1(D-glucose)	−68.6157	96.5267	−0.0690	0.2770
i = 2(sucrose)	−89.3391	118.9952	0.3280	−0.3410

Substituting Eqs. (8-16.13) and (8-16.14) into Eq. (8.16.12) leads to the results in Table 8.59 which shows x_1 and x_2 from solving simultaneously two Eqs. (8-16.2) once for $i = 1$ and once for $i = 2$. Corresponding experimental data are from Abed, et al. (1992).

TABLE 8-59 Solid-Liquid Equilibria for the Ternary D-Glucose (1)–Sucrose (2)–Water (3) at 70°C

Calculated		Experimental	
x_1	x_2	x_1	x_2
0.0000	0.1442	0.0000	0.1462
0.0538	0.1333	0.0545	0.1333
0.1236	0.1193	0.1233	0.1182
0.1777	0.1132	0.1972	0.1060
0.2104	0.0662	0.2131	0.0614
0.2173	0.0304	0.2292	0.0304
0.2475	0.0059	0.2531	0.0059
0.2550	0.0000	0.2589	0.0000

Figure 8-43 suggests that the UNIQUAC equation is useful for describing SLE for this aqueous system where the two solutes (D-glucose and sucrose) are chemically similar enough that the solute-solute parameters a_{12} and a_{21} can be set to zero, yet structurally dissimilar enough that there is no significant mutual solubility in the solid phase.

Example 8-37 Kuramochi, et al., (1996) studied solid-liquid equilibria (SLE) for mixtures containing DL-alanine, DL-serine, DL-valine, and water. Use the modified UNIFAC model of Larsen, et al., (1987) to calculate the solubility diagrams at 25°C for the ternaries

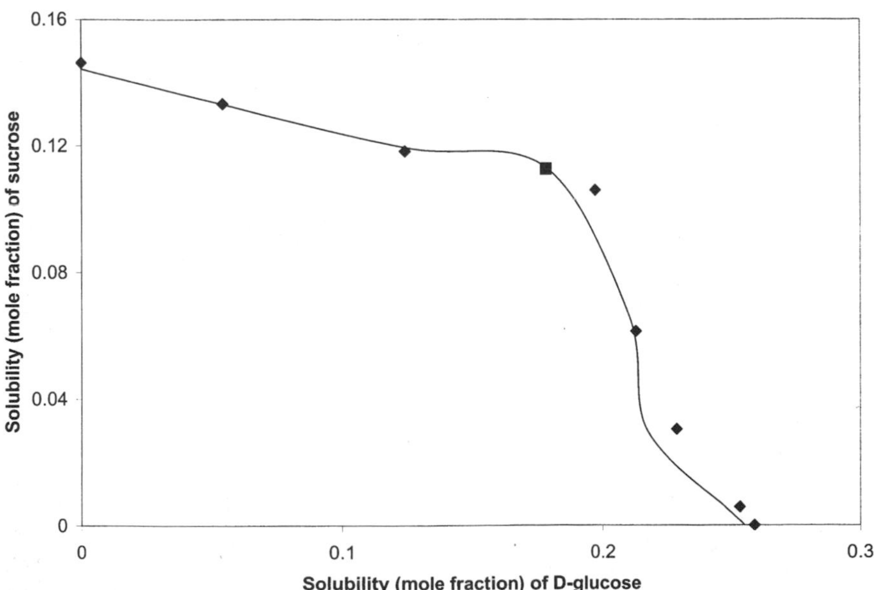

FIGURE 8-43 Solubility plot for D-glucose (1)–sucrose (2)–water (3) at 70°C. ♦ Experimental, —calculated, ■ calculated three-phase point.

(a) DL-alanine (1); DL-valine (2); water (3)

(b) DL-alanine (1); DL-serine (2); water (3)

solution The procedure to solve this problem is similar to that of Example 8-36. The activity-coefficient prediction uses the UNIFAC method established for liquid-liquid systems rather than the method developed for VLE (See Sec. 8-10). If each amino acid is assumed to form a pure solid phase, the two equations at equilibrium are Eqs. (8-16.8) and (8-16.9) of Example 8-36.

In this case, however, since most amino acids decompose before reaching their melting temperatures, the quantities $T_{fp,i}$,$\Delta H_{m,i}$, and $\Delta C_{p,i}$ are not available from experiment. Therefore, Kuramochi, et al. assumed that

$$\ln(f^S_{i,p}/f^L_{i,p}) = A_i - B_i/T + C_i \ln T \qquad (i = 1,2) \qquad (8\text{-}16.20)$$

The final form of Eq. (8-16.1) for this case is then

$$\ln(x_i\gamma_i) = A_i - B_i/T + C_i \ln T \qquad (i = 1.2) \qquad (8\text{-}16.21)$$

Values for γ_i in the amino acid-water binaries and ternaries are calculated from UNIFAC at the same temperatures as those of the experimental data.

To use UNIFAC, Kuramochi, et al., (1996) needed to introduce five new groups: α-CH (α indicates adjacent to an NH_2 group), sc-CH (sc stands for side chain), α-CH_2, sc-CH_2, and CONH. The constitutive groups for the three amino acids and water are as follows:

Compound	Constitutive groups
DL-alanine [$CH_3CH(NH_2)COOH$]	NH_2 + COOH + α−CH + CH_3
DL-valine [$(CH_3)_2CHCH(NH_2)COOH$]	NH_2 + COOH + α−CH + sc−CH + 2CH_3
DL-serine [$OHCH_2CH(NH_2)COOH$]	NH_2 + COOH + α−CH + sc−CH_2 + OH
Water [H_2O]	H_2O

Group-volume (R_i) and surface-area (Q_i) parameters are given as

Group	R_i	Q_i
CH_3	0.9011	0.848
sc-CH_2	0.6744	0.540
sc-CH	0.4469	0.228
α-CH	0.4469	0.228
NH_2	0.6948	1.150
COOH	1.3013	1.224
OH	1.000	1.200
H_2O	0.9200	1.400

For interactions involving the newly assigned groups, Karumochi, et al., (1996) calculated interaction parameters a_{ij} for Eq. 8-12.67 using experimental osmotic coefficients for the three amino acid-water binaries at 25°C. Other interaction a_{ij} values parameters are taken from the LLE UNIFAC table of Larsen, et al., (1987). Pertinent interaction parameters (K) are

Group	CH_3/sc-CH_2/sc-CH	α-CH	NH_2	COOH	OH	H_2O
CH_3/scCH_2/sc−CH	0.0	−896.5	218.6	1554	707.2	49.97
α-CH	−167.3	0.0	−573.2	−960.5	−983.1	−401.4
NH_2	1360	921.8	0.0	867.7	−92.21	86.44
COOH	3085	−603.4	−489.0	0.0	−173.7	−244.5
OH	1674	−1936	61.78	−176.5	0.0	155.6
H_2O	85.70	−1385	8.62	−66.39	−47.15	0.0

To obtain constants A, B, and C, for the amino acids, Kuramochi, et al. minimized an objective function F, defined by

$$F \equiv \sum_{\text{data}} \frac{|x_i^{\text{calc.}} - x_i^{\text{exp.}}|}{x_i^{\text{expt.}}} \qquad (i = 1,2) \tag{8-16.22}$$

where the summation is over all binary experimental data points from 273 to 373 K; $x_i^{\text{exp.}}$ and $x_i^{\text{calc.}}$ are experimental and calculated solubilities of the amino acid (i)-water binary.

Simultaneously solving Eq. (8-16.21) and optimizing Eq. (8-16.22) for the three amino acid-water binaries. Kuramochi, et al., (1996) give constants A, B, and C for each amino acid:

	A	B(K^{-1})	C
DL-alanine	77.052	−2668.6	11.082
DL-valine	−5236,3	−5236.3	17.455
DL-serine	−28.939	−318.35	4.062

For the ternary systems [amino acids (1) and (2) and water (3)] calculated solubilities x_1 and x_2 are found by solving simultaneously two Eqs. (8-16.21) once for each amino acid. Results (in terms of molality m) are shown in Tables 8-60 and 8-61. Experimental ternary data are from Kuramochi, et al., (1996).

Figures 8-44 and 8-45 compare calculated and experimental data.

TABLE 8-60 Solubilities for DL-Alanine (1)–DL-Valine (2)–Water (3) at 25°C

Experimental		Calculated	
m_1 (mol/Kg water)	m_2 (mol/Kg water)	m_1 (mol/Kg water)	m_2(mol/Kg water)
0.0000	0.6099	0.0000	0.6078
0.4545	0.5894	0.4487	0.5862
0.7704	0.5817	0.7693	0.5706
1.1292	0.5623	1.1276	0.5504
1.4962	0.5449	1.4907	0.5331
1.7765	0.5269	1.7862	0.5348
1.7920	0.5010	1.7932	0.5244
1.8350	0.3434	1.8382	0.3402
1.8674	0.1712	1.8634	0.1737
1.8830	0.0000	1.8849	0.0000

TABLE 8-61 Solubilities for DL-Alanine (1)–DL-Serine (2)–Water (3) at 25°C

Experimental		Calculated	
m_1 (mol/Kg water)	m_2 (mol/Kg water)	m_1 (mol/Kg water)	m_2 (mol/Kg water)
0.0000	0.4802	0.0000	0.4853
0.4056	0.4929	0.4021	0.4968
0.7983	0.5053	0.7077	0.5051
1.1703	0.5179	1.1644	0.5188
1.4691	0.5193	1.4690	0.5206
1.9027	0.5234	1.9062	0.5255
1.9012	0.4116	1.9038	0.4187
1.8959	0.3120	1.9014	0.3136
1.8928	0.1993	1.8952	0.2009
1.8887	0.1074	1.8904	0.1123
1.8830	0.0000	1.8849	0.0000

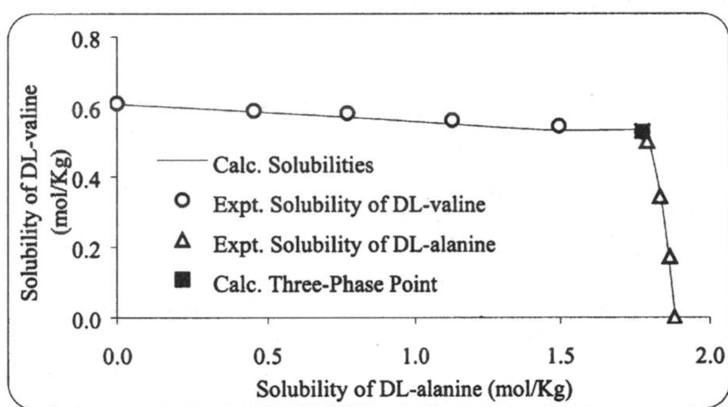

FIGURE 8-44 Solubility plot for DL-alanine (1)–DL-valine (2)–water (3) at 25°C.

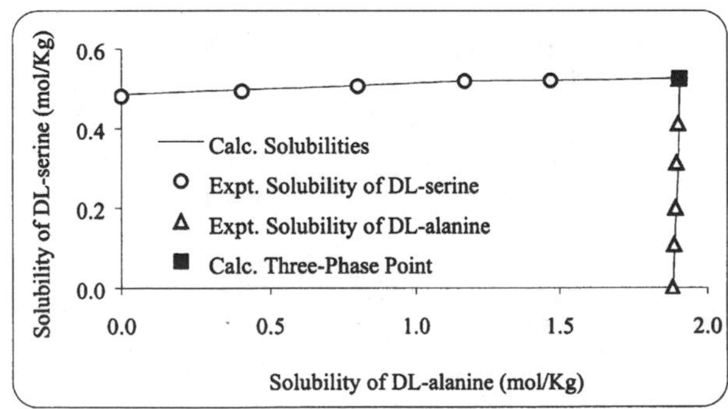

FIGURE 8-45 Solubility plot for DL-alanine (1)–DL-serine (2)–water (3) at 25°C.

8-17 AQUEOUS SOLUTIONS OF ELECTROLYTES

Physical chemists have given much attention to aqueous mixtures containing solutes that ionize either completely (e.g., strong salts like sodium chloride) or partially (e.g., sulfur dioxide and acetic acid). The thermodynamics of such mixtures is discussed in numerous references, but the most useful general discussions are by Pitzer, (1995) and by Robinson and Stokes, (1959). Helpful surveys are given by Pitzer, (1977, 1980), Rafal, (1994), and Zemaitis, (1986). Unfortunately, however, these discussions are primarily concerned with single-solute systems and with non-volatile electrolytes. Further, the discussions are not easily reduced to practice for engineering design, in part because the required parameters are not available, especially at higher temperatures.

Chemical engineers have only recently begun to give careful attention to the thermodynamics of aqueous mixtures of electrolytes. A review for process design is given by Liu and Watanasari (1999). An engineering-oriented introduction is given by Prausnitz et al. (1999).

A monograph edited by Furter (1979) and a review by Ohe (1991) discuss salt effects on vapor-liquid equilibria in solvent mixtures.

Example 8-38 Vapor-liquid equilibria are required for design of a sour-water stripper. An aqueous stream at 120°C has the following composition, expressed in molality (moles per kilogram of water):

CO_2	0.4	NH_3	2.62	CO 0.0016
H_2S	1.22	CH_4	0.003	

Find the total pressure and the composition of the equilibrium vapor phase.

solution This is a bubble-point problem with three volatile weak electrolytes and two nonreacting ("inert") gases. the method for solution follows that outlined by Edwards, et al. (1978).

For the chemical species in the liquid phase, we consider the following equilibria:

$$NH_3 + H_2O \rightleftharpoons NH_4^+ + OH^- \tag{8-17.1}$$

$$CO_2 + H_2O \rightleftharpoons HCO_3^- + H^+ \tag{8-17.2}$$

$$HCO_3^- \rightleftharpoons CO_3^{2-} + H^+ \tag{8-17.3}$$

$$H_2S \rightleftharpoons HS^- + H^+ \tag{8-17.4}$$

$$HS^- \rightleftharpoons S^{2-} + H^+ \tag{8-17.5}$$

$$NH_3 + HCO_3^- \rightleftharpoons NH_2\,COO^- + H_2O \tag{8-17.6}$$

$$H_2O \rightleftharpoons H^+ + OH^- \tag{8-17.7}$$

We assume that "inert" gases CH_4 and CO do not participate in any reactions and that their fugacities are proportional to their molalities (Henry's law). As indicated by Eqs. (8-16.1) to (8-16.7), we must find the concentrations of thirteen species in the liquid phase (not counting water) and six in the vapor phase (the three volatile weak electrolytes, the two inert gases, and water). We have the following unknowns:

m_i molality of noninert chemical species i in solution (11 molalities)
γ_i^* activity coefficient of noninert chemical species i in solution (molality scale, unsymmetric convention) (11 activity coefficients)
a_w activity of liquid water (one activity)
p_i partial pressure of each volatile component i (six partial pressures)
ϕ_i fugacity coefficient of volatile component i in the vapor (six fugacity coefficients)

We have 35 unknowns; we must now find 35 independent relations among these variables.

Reaction Equilibria. The equilibrium constant is known for each of the seven reactions at 120°C. For any reaction of the form

$$A \rightarrow B^+ + C^-$$

the equilibrium constant K is given by

$$K = \frac{a_{B^+}\, a_{C^-}}{a_A} = \frac{\gamma_{B^+}^*\, \gamma_{C^-}^*\, m_{B^+}\, m_{C^-}}{a_A} \tag{8-17.8}$$

where a = activity.

Activity Coefficients. Activity coefficients for the (noninert) chemical species in solution are calculated from an expression based on the theory of Pitzer (1973) as a function of temperature and composition. Binary interaction parameters are given by Edwards, et al., (1978).

Water Activity. By applying the Gibbs-Duhem equation to expressions for the solute activity coefficients, we obtain an expression for the activity of liquid water as a function of temperature and composition.

Vapor-Liquid Equilibria. For each volatile weak electrolyte and for each "inert" gas, we equate fugacities in the two phases:

$$\phi_i p_i = m_i \gamma_i^* H_i \text{ (PC)} \tag{8-17.9}$$

where H_i is Henry's constant for volatile solute i. For the "inert" gases in liquid solution, m_i is given and γ_i^* is taken as unity. PC is the Poynting correction. For water, the equation for phase equilibrium is:

$$\phi_w p_w = a_w P_{\text{vp}w} \phi_w^{\text{sat}} \text{ (PC)} \tag{8-17.10}$$

where the superscript refers to pure, saturated liquid. Henry's constants are available at 120°C. For the Poynting correction, liquid-phase partial molar volumes are estimated.

Vapor Phase Fugacity Coefficients. Since some of the components in the gas phase are polar and the total pressure is likely to exceed 1 bar it is necessary to correct for vapor phase nonideality. The method of Nakamura, et al., (1976) is used to calculate fugacity coefficients as a function of temperature, pressure, and vapor phase composition.

Material Balances. A material balance is written for each weak electrolyte in the liquid phase. For example, for NH_3:

$$\overset{\circ}{m}_{NH_3} = m_{NH_3} + m_{NH_4^+} + m_{NH_2COO^-} \tag{8-17.11}$$

where m° is the nominal concentration of the solute as specified in the problem statement.

Electroneutrality. Since charge is conserved, the condition for electroneutrality is given by

$$\sum_i z_i m_i = 0 \tag{8-17.12}$$

where z_i is the charge on chemical species i.

Total Number of Independent Relations. We have seven reaction equilibria, eleven activity coefficient expressions, one equation for the water activity, six vapor-liquid equilibrium relations, six equations for vapor phase fugacities, three material balances, and one electroneutrality condition, providing a total of 35 independent equations.

Computer program TIDES was designed to perform the tedious trial-and-error calculations. For this bubble-point problem, the only required inputs are the temperature and the molalities of the nominal (stoichiometric) solutes in the liquid phase. The total pressure is 16.4 bar, and the calculated mole fractions in the vapor phase are:

CO_2	0.243	CH_4	0.208
H_2S	0.272	CO	0.127
NH_3	0.026	H_2O	0.122

Table 8-62 shows the program output. Note that activities of solutes have units of molality. However, all activity coefficients and the activity of water are dimensionless.

While this example is based on a correlation prepared in 1978, the essential ideas remain unchanged. More recent data have been presented by Kurz (1995), Lu (1996) and Kuranov (1996) and others. Updated correlations are included in commercial software from companies like ASPEN and OLI SYSTEMS.

8-18 CONCLUDING REMARKS

This chapter on phase equilibria has presented no more than a brief introduction to a very broad subject. The variety of mixtures encountered in the chemical industry is extremely large, and, except for general thermodynamic equations, there are no quantitative relations that apply rigorously to all, or even to a large fraction, of these mixtures. Thermodynamics provides only a coarse but reliable framework; the details must be supplied by physics and chemistry, which ultimately rest on experimental data.

For each mixture it is necessary to construct an appropriate mathematical model for representing the properties of that mixture. Whenever possible, such a model should be based on physical concepts, but since our fundamental understanding of fluids is limited, any useful model is inevitably influenced by empiricism. While at least some empiricism cannot be avoided, the strategy of the process engineer must be to use enlightened rather than blind empiricism. This means foremost that critical and informed judgment must always be exercised. While such judgment is attained only by experience, we conclude this chapter with a few guidelines.

1. Face the facts: you cannot get something from nothing. Do not expect magic from thermodynamics. If you want reliable results, you will need some reliable experimental data. You may not need many, but you do need some. The required data need not necessarily be for the particular system of interest; sometimes they

TABLE 8-62 Program Tides Results

Input specifications:
temperature = 393.15 K

Component	Stoichiometric Concentration
CO_2	0.400
H_2S	1.220
NII_3	2.620
CH_4	0.003
CO	0.0016

	Liquid Phase			Vapor Phase		
Component	Concentration, molality	Activity coef., unitless	Poynt. corr.	Partial pressure, atm.	Fug. coef.	
NH_3	0.97193	1.0147	0.98420	0.41741	0.95997	
NH_4^+	1.4873	0.52287				
CO_2	0.27749×10^{-1}	1.0925	0.98219	3.9433	0.97193	
HCO_3^-	0.20513	0.57394				
CO_3^{2-}	0.63242×10^{-2}	0.15526×10^{-1}				
H_2S	0.11134	1.1784	0.98092	4.4096	0.95444	
HS^-	1.1086	0.56771				
S^{2-}	0.11315×10^{-4}	0.14992×10^{-1}				
NH_2COO^-	0.16080	0.42682				
H^+	0.83354×10^{-7}	0.86926				
OH^-	0.23409×10^{-4}	0.59587				
CH_4	0.30000×10^{-2}		0.97975	3.3627	0.99758	
CO	0.16000×10^{-2}		0.98053	2.0580	1.0082	
	Activity H_2O	0.93393			H_2O vap. press	1.9596 atm.
		Total pressure	16.154 atm.			

Equilibrium Constants at 393.15 K

Reaction	Equil. constant	Units
NH_3 dissociation	0.11777×10^{-4}	molality
CO_2 dissociation	0.30127×10^{-6}	molality
HCO_3 dissociation	0.60438×10^{-10}	molality
H_2S dissociation	0.34755×10^{-6}	molality
HS dissociation	0.19532×10^{-13}	molality
NH_2CO formation	0.55208	1/molality
H_2O dissociation	0.10821×10^{-11}	molality^2

Henry's constants, kg · atm/mol	
CO_2	124.52
H_2S	31.655
NH_3	0.40159
CH_4	1091.3
CO	1263.9

may come from experimental studies on closely related systems, perhaps represented by a suitable correlation. Only in very simple cases can partial thermodynamic properties in a mixture, e.g., activity coefficients, be found from pure-component data alone.

2. Correlations provide the easy route, but they should be used last, not first. The preferred first step should always be to obtain *reliable* experimental data, either from the literature or from the laboratory. Do not at once reject the possibility of obtaining a few crucial data yourself. Laboratory work is more tedious than pushing a computer button, but ultimately, at least in some cases, you may save time by making a few simple measurements instead of a multitude of furious calculations. A small laboratory with a few analytical instruments (especially a chromatograph or a simple boiling-point apparatus) can often save both time and money. If you cannot do the experiment yourself, consider the possibility of having someone else do it for you.

3. It is always better to obtain a few well-chosen and reliable experimental data than to obtain many data of doubtful quality and relevance. Beware of statistics, which may be the last refuge of a poor experimentalist.

4. Always regard published experimental data with skepticism. Many experimental results are of high quality, but many are not. Just because a number is reported by someone and printed by another, do not automatically assume that it must therefore be correct.

5. When choosing a mathematical model for representing mixture properties, give preference if possible to those which have some physical basis.

6. Seek simplicity; beware of models with many adjustable parameters. When such models are extrapolated even mildly into regions other than those for which the constants were determined, highly erroneous results may be obtained.

7. In reducing experimental data, keep in mind the probable experimental uncertainty of the data. Whenever possible, give more weight to those data which you have reason to believe are more reliable.

8. If you do use a correlation, be sure to note its limitations. Extrapolation outside its domain of validity can lead to large error.

9. Never be impressed by calculated results merely because they come from a computer. The virtue of a computer is speed, not intelligence.

10. Maintain perspective. Always ask yourself: Is this result reasonable? Do other similar systems behave this way? If you are inexperienced, get help from someone who has experience. Phase equilibria in fluid mixtures is not a simple subject. Do not hesitate to ask for advice.

NOTATION

a^E	excess molar Helmoltz energy
$a, b, c,$	empirical coefficients
a_i	activity of component i
a_{mn}	group interaction parameter, Eq. (8-10.67)
A, B, C	empirical constants
B_{ij}	second virial coefficient for the ij interaction (Sec. 5-4)
c, d	empirical constants in Eq. (8-7.1)

c_{ij}	cohesive energy density for the *ij* interaction in Sec. 8-10.
c_{12}	empirical constant in Eq. (8-5.15).
C	empirical constant, third virial coefficient
C_p	molar specific heat at constant pressure
D	empirical constant; number of data points, Eq. (8-8.17)
f_i	fugacity of component *i*
f	a function
$\mathcal{F}_i$	nonideality factor defined by Eq. (8-4.2)
g_{ij}	empirical constant (Table 8-3)
g^E	molar excess Gibbs energy
G^E	total excess Gibbs energy
G_{ij}	empirical constant (Table 8-3)
h^E	molar excess enthalpy
ΔH_m	molar enthalpy of fusion
$\bar{h}_i$	partial molar enthalpy of component *i*
H	Henry's constant
ΔH_v	enthalpy of vaporization
I	defined by Eq. (8-8.16)
k_{12}	binary parameter in Sec. 8-12
K	y/x in Sec. 8-12; distribution coefficient in Sec. 8-14; chemical equilibrium constant in Secs. 8-12 and 8-17.
L	fraction liquid
L^{16}	hexadecane-air partition coefficient, Eq. (8-10.29)
l_i	constant defined in Table 8-3
l_{12}	empirical constant in Sec. 8-10
m	defined after Eq. (8-15.2)
m_i	molality in Sec. 8-17
n_i	number of moles of component *i*
n_T	total number of moles
N	number of components; parameter in Tables 8-17 and 8-18
p	partial pressure
P	total pressure
P_{vp}	vapor pressure
q	molecular surface parameter, an empirical constant (Table 8-3)
Q_k	group surface parameter, Eq. (8-10.63)
r	molecular-size parameter, an empirical constant (Table 8-3); number of rings (Table 8-18)
R	gas constant
R_k	group size parameter, Eq. (8-10.63)
$\mathcal{R}$	defined by Eq. (8-10.55)
RI	index of refraction (Table 8-19); also n_D in Eq. (8-10.32)
s^E	molar excess entropy
ΔS_m	molar entropy of fusion
s_j	number of size groups in molecule *j*, Eq. (8-10.55)
$S^{\infty}_{s,os/w}$	infinite dilution partition coefficient of solute *s* between an organic solvent and water, Eq. (8-14.20)
t	temperature, °C
T	absolute temperature, K
T_{fp}	melting point temperature
T_t	triple-point temperature
u_{ij}	empirical constant (Table 8-3)

ΔU change in internal energy
V molar volume
w_k weight fraction of component k
x_i liquid phase mole fraction of component i
X_k group mole fraction for group k
y_i vapor phase mole fraction of component i
z coordination number (Table 8-3)
z_i overall mole fraction in Sec. 8-12; charge on chemical species i in Eq. (8-17.12)
Z compressibility factor

Greek

α parameter in Tables 8-17 and 8-18; acidity parameter (See Sec. 8-10)
α_{ij} empirical constant
β proportionality factor in Eq. (8-5.12); basicity parameter (See Sec. 8-10)
γ_i activity coefficient of component i
Γ_k activity coefficient of group k in Eq. (8-10.57)
δ solubility parameter defined by Eq. (8-10.10); also parameter in Eqs. (8-12.7) and (8-12.12); polarizability parameter in Eqs. (8-10.46) and (8-14.21)
$\bar{\delta}$ average solubility parameter defined by Eq. (8-10.15)
ε parameter in Tables 8-17 and 8-18
ξ parameter in Tables 8-17 and 8-18
η empirical constant in Table 8-14 and Eq. (8-9.23); empirical constant in Tables 8-17 and 8-18
θ parameter in Tables 8-17 and 8-18
Θ_i surface fraction of component i (Table 8-3)
κ isothermal compressibility, $-1/V\,(\partial V/\partial P)_T$, parameter in Eq. (8-12.26)
λ nonpolar solubility parameter or dispersion parameter in Sec. 8-10
λ_{ij} empirical constant in Eq. (8-5.11) and Table 8-14
Λ_{ij} empirical constant in Table 8-3
$\nu_k^{(i)}$ number of groups of type k in molecule i
ν_{kj} number of interaction groups k in molecule j [Eq. (8-10.56)]
π^* dipolarity/polarized scale, see Sec. 8-10
ρ density, g/cm^3
σ^2 variance, Eqs. (8-8.16) and (8-8.17)
τ polar solubility or polar parameter in Eqs. (8-10.17) and (8-10.18)
τ_{ij} empirical constant in Table 8-3
ϕ_i fugacity coefficient of component i
Φ_i site fraction (or volume fraction) of component i
χ Flory interaction parameter, Eq. (8-15.2)
Ψ_{mn} group interaction parameter, Eq. (8-10.67)
ψ_{12} binary (induction) parameter in Eq. (8-10.19)
ω acentric factor, Sec. 2-3
Ω weight fraction activity coefficient in Eq. (8-15.4)

Superscripts

c consolute (Sec. 8-14); calculated quantity, Eq. (8-12.4)
C configurational

e	experimental, Eq. (8-12.4)
E	excess
G	group, Eq. (8-10.53)
H	solute parameter, Table 8-19
KT	Kamlet-Taft solvent parameter, Table 8-19
L	liquid phase
M	measured value, Eq. (8-8.16)
°	standard state as in f_i°, estimated true value, Eq. (8-8.16)
R	residual
s	saturation, susceptibility parameter
S	size (Sec. 8-10), solid phase Eq. (8-10.18)
sub	sublimation
V	vapor phase
∞	infinite dilution

References

Abderafi, S. and T. Bounahmidi: *Fluid Phase Equil.,* **93:** 337 (1994).

Abed, Y., N. Gabas, M. L. Delia and T. Bounahmidi: *Fluid Phase Equil.,* **73:** 175 (1992).

Abbott, M. M., J. K. Floess, G. E. Walsh, and H. C. Van Ness: *AIChE J.,* **21:** 72 (1975).

Abbott, M. M., and H. C. Van Ness: *AIChE J.,* **21:** 62 (1975).

Abraham, M. H.: *Chem. Soc. Rev.,* **22:** 73 (1993).

Abrams, D. S., and J. M. Prausnitz: *AIChE J.,* **21:** 116 (1975).

Adidharma, H., and M. Radosz: *Fluid Phase Equil.,* **158/160:** 165 (1999a).

Adidharma, H., and M. Radosz: *Fluid Phase Equil.,* **161:** 1 (1999b).

Adidharma, H., and M. Radosz: *Ind. Eng. Chem. Res.,* **37:** 4453 (1998).

Adler, S. B.: *Hydrocarbon Process. Intern. Ed.,* **62**(5): 109, **62**(6): 93 (1983).

Anderko, A.: *J. Chem. Soc. Farad. Trans.,* **86:** 2823 (1990).

Anderko, A.: *Fluid Phase Equil.,* **65:** 89 (1991).

Anderson, T. F., D. S. Abrams, and E. A. Grens: *AIChE J.,* **24:** 20 (1978).

Babich, S. V., R. A. Ivanchikova, and L. A. Serafimov: *Zh. Prikl. Khim.,* **42:** 1354 (1969).

Bae, Y. C., J. J. Shim, D. S. Soane, and J. M. Prausnitz: *J. Appl. Polym. Sci.,* **47:** 1193 (1993).

Bandrup, J., E. H. Immergut, and E. A. Grulke: *Polymer Handbook,* 4th ed., Wiley, New York, 1999.

Barton, A. F. M.: *CRC Handbook of Polymer-Liquid Interaction Parameters and Solubility Parameters,* CRC Press, Boca Raton, Fla., 1990. See also the 1983 edition for solvent solubility parameters.

Battino, R.: *Fluid Phase Equil.,* **15:** 231 (1984).

Battino, R., and E. Wilhelm: *J. Chem. Thermodyn.,* **3:** 379 (1971); L. R. Field, E. Wilhelm, and R. Battino: *ibid.,* **6:** 237 (1974).

Bazúa E. R., and J. M. Prausnitz: *Cryogenics,* **11:** 114 (1971).

Beegle, B. L., M. Modell, and R. C. Reid: *AIChE J.,* **20:** 1200 (1974).

Ben Naim, A.: *"Statistical Thermodynamics for Chemists and Biochemists",* New York, Plenum Press, 1982.

Bender, E., and U. Block: *Verfahrenstechnic,* **9:** 106 (1975).

Berro, C., M. Rogalski, and A. Peneloux: *Fluid Phase Equil.,* **8:** 55 (1982).

Blas, F. J., and L. F. Vega: *Ind. Eng. Chem. Res.,* **37:** 660 (1998).

Bondi, A.: *Physical Properties of Molecular Liquids, Crystals and Glasses,* Wiley, New York, 1968.

Brelvi, S. W., and J. P. O'Connell: *AIChE J.,* **18:** 1239 (1972); **21:** 157 (1975).

Brian, P. L. T.: *Ind. Eng. Chem. Fundam.,* **4:** 101 (1965).

Bruin, S., and J. M. Prausnitz: *Ind. Eng. Chem. Process Design Develop.,* **10:** 562 (1971).

Bruin, S.: *Ind. Eng. Chem. Fundam.,* **9:** 305 (1970).

Buck, E.: *CHEMTECH.,* September 1984: 570.

Burrell, H.: *J. Paint Technol.,* **40:** 197 (1968); J. L. Gardon: *ibid.,* **38:** 43 (1966); R. C. Nelson, R. W. Hemwall, and G. D. Edwards: *ibid.,* **42:** 636 (1970); *Encyclopedia of Chemical Technology* (Kirk-Othmer), 2d ed., vol. 18, pp. 564–588.

Bush, D. M., and C. A. Eckert: to be published, 2000.

Butler, A. V., D. W. Thomson, and W. N. McLennon: *J. Chem. Soc. London,* 674 (1933).

Campanella, E. A., P. M. Mathias, and J. P. O'Connell: *AIChE J.,* **33:** 2057 (1987).

Carr, P. W.: *Microchem. J.,* **48:** 1 (1993).

Catte, M., C. G. Dussap, C. Archard, and J. B. Gros: *Fluid Phase Equil.,* **105:** 1 (1995).

Carroll, J. J.: *Chem. Eng. Prog.,* **95:** 49 (1999).

Chapman, W. G., K. E. Gubbins, and G. Jackson: *Fluid Phase Equil.,* **52:** 31 (1989).

Cheung, H., and E. H. Zander: *Chem. Eng. Progr. Symp. Ser.,* **64**(88): **34** (1968).

Clarke, H. A., and R. W. Missen: *J. Chem. Eng. Data,* **19:** 343 (1974).

Coward, I., S. E. Gayle, and D. R. Webb: *Trans. Inst. Chem. Engrs.* (*London*), **56:** 19 (1978).

Danner, R. P., and M. S. High: *Handbook of Polymer Solution Thermodynamics,* DIPPR, American Institute of Chemical Engineers, New York, NY, 1993.

Daubert, T. E., R. P. Danner, H. M. Sibel, and C. C. Stebbins: *Data Physical and Thermodynamics Properties of Pure Chemicals: Data Compilation,* Taylor & Francis, Washington, D. C. (1997).

Derr, E. L., and C. H. Deal: *Inst. Chem. Eng. Symp. Ser. London,* **32**(3): 40 (1969).

Dohrn, R., and G. Brunner: *Fluid Phase Equil.,* **106:** 213 (1995).

Economou, I. G., and C. Tsonopoulos: *Chem. Eng. Sci.,* **52:** 511 (1997).

Edwards, T. J., G. Maurer, J. Newman, and J. M. Prausnitz: *AIChE J.,* **24:** 966 (1978).

Flory, P. J.: *Principles of Polymer Chemistry,* Cornell University Press, Ithaca, N.Y., 1953.

Flory, P. J.: *Discussions Faraday Soc.,* **49:** 7 (1970).

Fredenslund, A., R. L. Jones, and J. M. Prausnitz: *AIChE J.,* **21:** 1086 (1975); A. Fredenslund, J. Gmehling, and P. Rasmussen: *Vapor-Liquid Equilibria Using UNIFAC,* Elsevier, Amsterdam, 1977.

Friend, L., and S. B. Adler: *Chem. Eng. Progr.,* **53:** 452 (1957).

Funk, E. W., and J. M. Prausnitz: *Ind. Eng. Chem.,* **62**(9): 8 (1970).

Furnas, C. C., and W. B. Leighton: *Ind. Eng. Chem.,* **29:** 709 (1937).

Furter, W. F. (ed.): *Advan. Chem. Ser.* **177:** (1979).

Gabas, N., and C. Laguerie: *J. Crystal Growth,* **128:** 1245 (1993).

Gmehling, J., U. Onken, W. Arlt, and J. R. Rarey-Nies: *Dechema Chemistry Series,* **1:** Part 1b, 108 (1974).

Gmehling, J., and U. Onken: *Dechema Chemistry Series,* **1:** Part 1, 408 (1977).

Gmehling, J., and U. Onken: *Dechema Chemistry Series,* **1:** Part 1, 325 (1977a).

Gmehling, J., and U. Onken: *Dechema Chemistry Data Series,* **1:** Part 1: 49 and Part 2a: 77 (1977b).

Gmehling, J., T. F. Anderson, and J. M. Prausnitz: *Ind. Eng. Chem. Fundam.,* **17:** 269 (1978).

Gmehling, J., D. D. Liu, and J. M. Prausnitz: *Chem. Eng. Sci.,* **34:** 951 (1979).

Gmehling, J., U. Onken, and W. Arlt: *Dechema Chemistry Data Series,* **1:** Parts 3 & 4: 389 (1979a).

Gmehling, J., U. Onken, and W. Arlt: *Dechema Chemistry Data Series,* **1:** Part 1a: 344 (1981).

Gmehling, J., Un. Onken, and W. Arlt: *Dechema Chemistry Data Series,* **1:** Part 1a: 194 and 498 (1981a).

Gmehling, J., U. Onken, and J. R. Rarey-Nies: *Dechema Chemistry Data Series,* **1:** Part 2e: 377, 547 (1988).

Gmehling, J. J. Li, and M. Schiller: *Ind. Eng. Chem. Res.,* **32:** 178 (1993).

Gmehling, J., et al.: Azeotropic Data, Wiley-VCH, Weinheim and New York, (1994).

Grenzheuser, P., and J. Gmehling: *Fluid Phase Equil.,* **25:** 1 (1986).

Griswold, J., P. L. Chu, and W. O. Winsauer: *Ind. Eng. Chem.,* **41:** 2352 (1949).

Griswold, J., and S. Y. Wong: *Chem. Eng. Progr., Symp. Ser.* **48**(3): 18 (1952).

Guillet, J. E.: *Advan. Anal. Chem. Instrum.,* **11:** 187 (1973).

Gundersen, T.: *Computers Chem. Eng.,* **6:** 245 (1982).

Hait, M. J., C. L. Liotta, C. A. Eckert, D. L. Bergmann, A. M. Karachewski, A. J. Dallas, D. I. Eikens, J. J. Li, P. W. Carr, R. B. Poe, and S. C. Rutan: *Ind. Eng. Chem. Res.,* **32:** 2905 (1993).

Hansen, C. M.: *J. Paint Technol.,* **39:** 104, 505 (1967); C. M. Hansen and K. Skaarup: *ibid.,* **39:** 511 (1967); C. M. Hansen and A. Beerbower: "Solubility Parameters," in H. F. Mark, J. J. McKetta, and D. F. Othmer (eds.): *Encyclopedia of Chemical Technology,* 2d ed., suppl. vol., Interscience, New York, 1971.

Hansen, H. K., P. Rasmussen, A. Fredenslund, M. Schiller, and J. Gmehling: *Ind. Eng. Chem. Res.,* **30:** 2352 (1991).

Hayden, J. G., and J. P. O'Connell: *Ind. Eng. Chem. Proc. Des. Dev.,* **14:** 3 (1975).

Hayduk, W., and S. C. Cheng: *Can. J. Chem. Eng.,* **48:** 93 (1970); W. Hayduk and W. D. Buckley: *ibid.* **49:** 667 (1971); W. Hayduk and H. Laudie: *AIChE J.,* **19:** 1233 (1973).

Heidemann, R. A.: *Fluid Phase Equil.,* **116:** 454 (1996).

Heidemann, R. A.: *Fluid Phase Equil.,* **14:** 55 (1983).

Heidemann, R. A., and J. M. Prausnitz: *Proc. Natl. Acad. Sci. USA,* **73:** 1773 (1976).

Heidemann, R. A., and J. M. Mandhane: *Chem. Eng. Sci.,* **28:** 1213 (1973); T. Katayama, M. Kato, and M. Yasuda: *J. Chem. Eng. Japan,* **6:** 357 (1973); A. C. Mattelin and L. A. J. Verhoeye: *Chem. Eng. Sci.,* **30:** 193 (1975).

Heil, J. F., and J. M. Prausnitz: *AIChE J.,* **12:** 678 (1966).

Hildebrand, J. H., and R. L. Scott: *Solubility of Nonelectrolytes,* 3rd ed., Reinhold, New York, 1950.

Hildebrand, J. H., and R. L. Scott: *Regular Solutions,* Prentice-Hall, Englewood Cliffs, N.J., 1962.

Hooper, H. H., S. Michel, and J. M. Prausnitz: *Ind. Eng. Chem. Res.,* **27**(11): 2182 (1988).

Hooper, H. H., S. Michel, and J. M. Prausnitz: *J. Chem. Eng. Data,* **33**(4): 502 (1988a).

Horsley, L. H.: *"Azeotropic Data," Advan. Chem. Ser.,* **6:** (1952), **35:** (1962), **116:** (1973).

Howell, W. J., A. M. Karachewski, K. M. Stephenson, C. A. Eckert, J. H. Park, P. W. Carr, and S. C. Rutan: *Fluid Phase Equil.,* **52:** 151 (1989).

Hu, Y., E. Azevedo, D. Lüdecke, and J. M. Prausnitz: *Fluid Phase Equil.,* **17:** 303 (1984).

Huang, H., and S. I. Sandler: *IEC Res.,* **32:** 1498 (1993).

Hwang, Y.-L., G. E. Keller, and J. D. Olson: *Ind. Eng. Chem. Res.,* **31:** 1753 and 1759 (1992).

Jog, P. K., A. Garcia-Cuellar, and W. G. Chapman: *Fluid Phase Equil.,* **158–160:** 321 (1999).

Jones, C. A., A. P. Colburn, and E. M. Schoenborn: *Ind. Eng. Chem.,* **35:** 666 (1943).

Kamlet, M. J., J.-L. M. Aboud, M. H. Abraham, and R. W. Taft: *J. Org. Chem.,* **48:** 2877 (1983).

Kamlet, M J., M. H. Abraham, P. W. Carr, R. M. Doherty, and R. W. Taft: *J. Chem. Soc., Perki Trans.,* **2:** 2087 (1988).

Kang, J. W., J. H. Lee, K-P. Yoo, and C. S. Lee: Korea University, personal communication, 2000.

King, C. J.: *Separation Processes,* 2d ed., McGraw-Hill, New York, 1980, chap. 2.

Kinzl, M., G. Luft, and M. Radosz: *IEC Res.,* **39:** 541 (2000).

Knapp, H., R. Döring, L. Oellrich, U. Plöcker, and J. M. Prausnitz: *Chemistry Data Series,* Vol. VI: *VLE for Mixtures of Low Boiling Substances,* D. Behrens and R. Eckerman (eds.), DECHEMA, Frankfurt a. M., 1982.

Knudsen, K., E. H. Stenby, and Aa. Fredenslund: *Fluid Phase Equil.,* **82:** 361 (1993).

Kojima, K., and K. Tochigi: "Prediction of Vapor-Liquid Equilibria by the ASOG Method," *Physical Sciences Data 3,* Kodansha Ltd., Tokyo, Elsevier Scientific Publishing Company (1979).

Kolar, P., and K. Kojima: *J. Chem. Eng. Japan,* **27:** 460 (1994).

Kraus, J., and J. Linek: *Collect. Czech. Commun.,* **36:** 2547 (1971).

Krause, S.: *J. Macromol. Sci. Rev. Macromol. Chem.,* **C7:** 251 (1972).

Krishnan, T. R., H. Kabra, and D. B. Robinson: *J. Chem. Eng. Data,* **22:** 282 (1977).

Kroschwitz, J.: *Encyclopedia of Polymer Science and Engineering,* 2nd ed., Wiley-Interscience **15:** 419, 1989.

Kuramochi, H., H. Noritomi, D. Hoshino, and K. Nagahama: *Biotech. Progr.,* **12**(3): 371 (1996).

Kuranov, G., B. Rumpf, N. A. Smirnova, and G. Maurer: *Ind. Eng. Chem. Res.,* **35:** 1959 (1996).

Kurz, F., B. Rumpf, and G. Maurer: *Fluid Phase Equil.,* **104:** 261 (1995).

Ladurelli, A. J., C. H. Eon, and G. Guiochon: *Ind. Eng. Chem. Fundam.,* **14:** 191 (1975).

Larsen, B. L., P. Rasmussen, and Aa. Fredenslund: *Ind. Eng. Chem. Res.,* **26:** 2274 (1987).

Laurence, D. A., and G. W. Swift: *J. Chem. Eng. Data,* **17:** 333 (1972).

Lee, B. I., and Kesler, M. G.: *AIChE J.,* **21:** 510 (1975); errata, **21:** 1040 (1975).

Lencka, M., and A. Anderko: *AIChE J.,* **39:** 533 (1993).

Levelt Sengers, J. M. H.: in Kiran, E., and J. M. H. Levelt Sengers (eds.): "Supercritical Fluids: Fundamentals for Applications" NATO ASI Series E: **273:** Klewer Academic Publishers, Dordrecht, Holland, 1994, p. 1.

Li, J., A. J. Dallas, J. I. Eikens, P. W. Carr, D. L. Bergmann, M. J. Hait, and C. A. Eckert: *Anal. Chem.,* **65:** 3312 (1993).

Lide, R. D., and H. P. R. Frederikse: *CRC Handbook of Chemistry and Physics,* CRC Press, Inc., 16–40, 1997.

Liu, Y., and S. Watanasari: *Chem. Eng. Prog.,* **95:** 25 (October 1999).

Lo, T. C., H. H. Bierber, and A. E. Karr: *J. Chem. Eng. Data,* **7:** 327 (1962).

Lu, X., L. Zhang, Y. Wang, J. Shi, and G. Maurer: *Ind. Eng. Chem. Res.,* **35:** 1777 (1996).

Lyckman, E. W., C. A. Eckert, and J. M. Prausnitz: *Chem. Eng. Sci.,* **20:** 685 (1965).

Marinichev, A. N., and M. P. Susarev: *Zh. Prikl. Khim.,* **38:** 1054 (1965).

Mathias, P. M., and J. P. O'Connell: "Equations of State in Engineering and Research", K. C. Chao and R. L. Robinson (eds.), *Adv. Chem. Ser.,* **182:** 97 (1979).

Mathias, P. M., and J. P. O'Connell: *Chem. Eng. Sci.,* **36:** 1123 (1981).

Mathias, P. M., J. F. Boston, and S. Watanasiri: *AIChE J.,* **30:** 182 (1984).

McGlashan, M. L., and A. G. Williamson: *J. Chem. Eng. Data,* **21**(2): 196 (1976).

Mendez-Santiago, J., and A. S. Teja: *Fluid Phase Equil.,* **158–160:** 501 (1999).

Mertl, I.: *Coll. Czech. Chem. Commun.,* **37:** 366 (1972).

Meyer, P., and G. Maurer: *Ind. Eng. Chem. Res.,* **34:** 373 (1995).

Michelsen, M. L.: *Fluid Phase Equil.,* **4:** 1 (1980).

Michelsen, M. L.: *Fluid Phase Equil.,* **9:** 21 (1982).

Mollerup, J.: *Advan. Cryog. Eng.,* **20:** 172 (1975); *Fluid Phase Equil.,* **4:** 99 (1980).

Muller, G., E. Bender, and G. Maurer: *Ber. Buns. Ges. Phys. Chem.,* **92:** 152 (1988).

Murti, P. S., and M. van Winkle: *Chem. Eng. Data Ser.,* **3:** 72 (1978).

Nakamura, R., G. J. F. Breedveld, and J. M. Prausnitz: *Ind. Eng. Chem. Process Design,* **15:** 557 (1976).

Newman, R. D., and J. M. Prausnitz: *J. Paint Technol.,* **43:** 33 (1973).

Null, J. R., and D. A. Palmer: *Chem. Eng. Progr.,* **65:** 47 (1969); J. R. Null: *Phase Equilibrium in Process Design,* Wiley, New York, (1970).

O'Connell, J. P.: in Kiran, E., and J. M. H. Levelt Sengers (eds.) "Supercritical Fluids: Fundamentals for Applications" NATO ASI Series E: **273:** Klewer Academic Publishers, Dordrecht, Holland, 1994, p. 191.

Ohe, S.: *Vapor-Liquid Equilibrium-Salt Effect,* Elsevier, Amsterdam, 1991.

Oishi, T., and J. M. Prausnitz: *Ind. Eng. Chem. Fundam,* **17:** 109 (1978).

Orbey, H., and S. I. Sandler: *AIChE J.,* **41** (3): 683 (1995).

Orbey, H., and S. I. Sandler: *Modeling Vapor-Liquid Equilibria. Cubic Equations of State and Their Mixing Rules,* Cambridge University Press, 1998.

Orentlicher, M., and J. M. Prausnitz: *Chem. Eng. Sci.,* **19:** 775 (1964).

Palmer, D. A.: *Chem. Eng.,* June 9, 1975, p. 80.

Palmer, D. A., and B. D. Smith: *J. Chem. Eng. Data,* **17:** 71 (1972).

Patterson, D.: *Macromolecules,* **2:** 672 (1969).

Peng, D. Y., and D. B. Robinson: *Ind. Eng. Chem. Fundam.,* **15:** 59 (1976).

Peres, A. M., and E. A. Macedo: *Fluid Phase Equil.,* **123:** 71 (1996).

Peres, A. M., and E. A. Macedo: *Fluid Phase Equil.,* **139:** 47 (1997).

Phutela, R. C., Z. S. Kooner, and D. V. Fenby: *Australian J. Chem.,* **32:** 2353 (1979).

Pierotti, G. J., C. H. Deal, and E. L. Derr: *Ind. Eng. Chem.,* **51:** 95 (1959).

Pitzer, K. S.: *J. Phys. Chem.,* **77:** 268 (1973).

Pitzer, K. S.: *Accounts Chem. Res.,* **10:** 371 (1977).

Pitzer, K. S.: *J. Am. Chem. Soc.,* **102:** 2902 (1980).

Pitzer, K. S.: *Thermodynamics,* 3rd ed., McGraw-Hill, New York, 1995.

Plöcker, U., H. Knapp, and J. M. Prausnitz: *Ind. Eng. Chem. Process Design Develop.,* **17:** 324 (1978).

Plyasunov, A., J. P. O'Connell, and R. H. Wood: *Geochimica et Cosmochimica Acta,* **64:** 495 (2000a).

Plyasunov, A., J. P. O'Connell, E. Schock, and R. H. Wood: *Geochimica et Cosmochimicia Acta,* in press, (2000b).

Poling, B. E., E. A. Grens, and J. M. Prausnitz: *Ind. Eng. Chem. Process Design Develop.,* **20:** 127 (1981).

Prausnitz, J. M., R. N. Lichtenthaler, and E. G. Azevedo: *Molecular Thermodynamics of Fluid-Phase Equilibria,* 3rd ed., Prentice Hall, Englewood Cliffs, N.J., 1999.

Prausnitz, J. M., T. A. Anderson, E. A. Grens, C. A. Eckert, R. Hsieh, and J. P. O'Connell: *Computer Calculations for Multicomponent Vapor-Liquid and Liquid-Liquid Equilibria,* Prentice-Hall, Englewood Cliffs, N.J., 1980.

Prausnitz, J. M., and F. H. Shair: AIChE J., **7:** 682 (1961).

Preston, G. T., and J. M. Prausnitz: *Ind. Eng. Chem. Process Design Develop.,* **9:** 264 (1970).

Prigogine, I.: *The Molecular Theory of Solutions,* North-Holland Publishing, Amsterdam, 1957.

Prigogine, I., and R. Defay: *Chemical Thermodynamics,* Longmans, London, 1954.

Quiram, D. J., J. P. O'Connell, and H. D. Cochran: *J. Supercrit. Fluids,* **7:** 159 (1994).

Raemy, A., and T. F. Schweizer: *J. Thermal Anal.,* **28:** 95 (1983).

Rafal, M., J. W. Berthold, N. C. Scrivner, and S. L. Grise: *Models for Electrolyte Solutions,* in *Models for Thermodynamic and Phase-Equilibria Calculations* (S. I. Sandler, ed.), Marcel Dekker, New York, 1994.

Reamer, H. H., B. H. Sage, and W. N. Lacey: *Ind. Eng. Chem.,* **43**(11): 2515 (1951).

Redlich, O., E. L. Derr, and G. Pierotti: *J. Am. Chem. Soc.,* **81:** 2283 (1959); E. L. Derr and M. Papadopoulous: *ibid.,* **81:** 2285 (1959).

Renon, H., and J. M. Prausnitz: *Ind. Eng. Chem. Process Design Develop.,* **8:** 413 (1969).

Riddick, J. A., W. B. Bunger, and T. K. Sakano: *Organic Solvents: Physical Properties and Method of Purification,* 4th ed., John Wiley and Sons, New York, 1986.

Robinson, R. A., and R. H. Stokes: *Electrolyte Solutions,* 2d ed., Academic, New York, 1959.

Roos, Y.: *Carbohydr. Res.,* **238:** 39 (1993).

Sandler, S. I., and Y.-H. Fu: *Ind. Eng. Chem. Res.,* **34:** 1897 (1995).

Schreiber, L. B., and C. A. Eckert: *Ind. Eng. Chem. Process Design Develop.,* **10:** 572 (1971).

Sedlbauer, J., J. P. O'Connell, and R. H. Wood: *Chem. Geology,* **163:** 43 (2000).

Shain, S. A., and J. M. Prausnitz: *Chem. Eng. Sci.,* **18:** 244 (1963).

Sharygin, A. V., J. P. O'Connell, and R. H. Wood: *Ind. Eng. Chem. Res.,* **35:** 2808 (1996).

Sheehan, C. J., and A. L. Bisio: *Rubber Chem. Technol.,* **39:** 149 (1966).

Sherman, S. R., D. Suleiman, M. J. Hait, M. Schiller, C. L. Liotta, C. A. Eckert, J. Li, P. W. Carr, R. B. Poe, and S. C. Rutan: *J. Phys. Chem.,* **99:** 11239 (1995).

Sherman, S. R., D. B. Trampe, D. M. Bush, M. Schiller, C. A. Eckert, A. J. Dallas, J. Li, and P. W. Carr: *Ind. Eng. Chem. Res.,* **35:** 1044 (1996).

Shyu, G. S., N. S. Hanif, K. R. Hall, and P. T. Eubank: *Fluid Phase Equil.,* **130:** 73 (1997).

Sinor, J. E., and J. H. Weber: *J. Chem. Eng. Data,* **5:** 243 (1960).

Soave, G.: *Chem. Eng. Sci.,* **27:** 1197 (1972).

Sorenson, J. M., and W. Arlt: *Dechema Chemistry Data Series,* **5:** Part 1: 341 and 373 (1979).

Starling, K. E.: *Fluid Thermodynamic Properties of Light Petroleum System,* Gulf Publishing Co., 1973.

Steele, K., B. E. Poling, and D. B. Manley: paper presented at *AIChE Mtg., Washington, D.C., December 1974.*

Stryjek, R., and J. H. Vera: *Can. J. Chem. Eng.,* **64:** 323 (1986).

Taft, R. W., J. M. Abboud, M. J. Kamlet, and M. H. Abraham: *J. Soln. Chem.,* **3:** 153 (1985).

Tester, J. W., and M. Modell: *Thermodynamics and its Applications,* 3rd ed., Prentice Hall, Englewood Cliffs, N.J., 1997.

Thomas, E. R., B. A. Newman, G. L. Nicolaides, and C. A. Eckert: *J. Chem. Eng. Data,* **27:** 233 (1982).

Thomas, E. R., and C. A. Eckert: *Ind. Eng. Chem. Process Design Develop.,* **23:** 194 (1984).

Tiegs, D., J. Gmehling, P. Rasmussen, Aa. Fredenslund: *Ind. Eng. Chem. Res.,* **26:** 159 (1986).

Tiepel, E. W., and K. E. Gubbins: *Ind. Eng. Chem. Fundam.,* **12:** 18 (1973); *Can J. Chem. Eng.,* **50:** 361 (1972).

Tochigi, K., D. Tiegs, J. Gmehling, and K. Kojima: *J. Chem. Eng. Japan,* **23:** 453 (1990).

Todheide, K., and E. U. Franck: *Zeit. Phys. Chem.,* **37:** 387 (1963).

Topliss, R. J.: Ph.D. dissertation, University of California, Berkeley, 1985.

Treybal, R. E.: *Liquid Extraction,* 2d ed., McGraw-Hill, New York, 1963.

Tsonopoulos, C., and G. M. Wilson: *AIChE J.,* **29:** 990 (1983).

Turek, E. A., R. S. Metcalfe, L. Yarborough, and R. L. Robinson: *Soc. Petrol. Engrs. J.,* **24:** 308 (1984).

Twu, C. H., D. Bluck, J. R. Cunningham, and J. E. Coon: *Fluid Phase Equil.,* **69:** 33 (1991).

Twu, C. H., and J. E. Coon: *AIChE J.,* **42**(11): 3212 (1996).

Twu, C. H., J. E. Coon, and D. Bluck: *Fluid Phase Equil.,* **139:** 33 (1997); **150–151:** 181 (1998); **153** (1): 29 (1988); *Ind. Eng. Chem. Res.,* **37:** 1580 (1998).

Twu, C. H., J. E. Coon, D. Bluck, and B. Tilton: *Fluid Phase Equil.,* **158–60:** 271 (1999a).

Twu, C. H., B. Tilton, and D. Bluck: "The Strengths and Limitations of Equation of State Models and Mixing Rules", personal communication (1999b).

Udovenko, V. V., and T. B. Frid: *Zh. Fiz. Khim.,* **22:** 1263 (1948).

Udovenko, V. V., and T. F. Mazanko: *Zh. Fiz. Khim.,* **41:** 1615 (1967).

Vidal, J.: *Chem. Eng. Sci.,* **31:** 1077 (1978); *Fluid Phase Equil.,* **13:** 15 (1983).

Villamanan, M. A., A. J. Allawi, and H. C. Van Ness: *J. Chem. Eng. Data,* **29:** 431 (1984).

Weimer, R. F., and J. M. Prausnitz: *Hydrocarbon Process. Petrol. Refiner,* **44:** 237 (1965).

Willock, J. M., and M. van Winkle: *J. Chem. Eng. Data,* **15**(2): 281 (1970).

Wilson, G. M.: *J. Am. Chem. Soc.,* **86:** 127,133 (1964).

Wilson, G. M., and C. H. Deal: *Ind. Eng. Chem. Fundam.,* **1:** 20 (1962).

Wohl, K.: *Chem. Eng. Progr.,* **49:** 218 (1953).

Wohl, K.: *Trans. AIChE,* **42:** 215 (1946).

Wong, D. S. H., and S. I. Sandler: *AIChE J.,* **38**(5): 671 (1992).

Wong, K. F., and C. A. Eckert: *Ind. Eng. Chem. Fundam.,* **10:** 20 (1971).

Wu, H. S., D. Hagewiesche, and S. I. Sandler: *Fluid Phase Equil.,* **43:** 77 (1988).

Yarborough, L.: "Equations of State in Engineering and Research", K. C. Chao and R. L. Robinson (eds.), *Adv. Chem. Ser.,* **182:** 385 (1979).

Yen, L., and J. J. McKetta: *AIChE J.,* **8:** 501 (1962).

Yu, M. L., and Y.-P. Chen: *Fluid Phase Equil.,* **94:** 149 (1994).

Zemaitis, J. F., D. M. Clark, M. Rafal, and N. C. Scrivner: *Handbook of Aqueous Electrolyte Thermodynamics,* AIChE, New York, 1986.

Ziervogel, R. G., and B. E. Poling: *Fluid Phase Equil.,* **11:** 127 (1983).

Zudkevitch, D., and J. Joffe: *AIChE J.,* **16:** 112 (1970).

CHAPTER 9
VISCOSITY

9-1 Scope

The first part of this chapter deals with the viscosity of gases and the second with the viscosity of liquids. In each part, methods are recommended for: (1) correlating viscosities with temperature; (2) estimating viscosities when no experimental data are available; (3) estimating the effect of pressure on viscosity; and (4) estimating the viscosities of mixtures. The molecular theory of viscosity is considered briefly.

9-2 Definitions of Units of Viscosity

If a shearing stress is applied to any portion of a confined fluid, the fluid will move with a velocity gradient with its maximum velocity at the point where the stress is applied. If the local shear stress per unit area at any point is divided by the velocity gradient, the ratio obtained is defined as the viscosity of the medium. Thus, viscosity is a measure of the internal fluid friction, which tends to oppose any dynamic change in the fluid motion. An applied shearing force will result in a large velocity gradient at low viscosity. Increased viscosity causes each fluid layer to exert a larger frictional drag on adjacent layers which in turn decreases the velocity gradient.

It is to be noted that viscosity differs in one important respect from the properties discussed previously in this book; namely, viscosity can only be measured in a nonequilibrium experiment. This is unlike density which can be found in a static apparatus and so is an equilibrium property. On the microscale, however, both properties reflect the effects of molecular motion and interaction. Thus, even though viscosity is ordinarily referred to as a nonequilibrium property; it is, like density, a function of the thermodynamic state of the fluid; in fact, it may even be used to define the state of the material. Brulé and Starling (1984) have emphasized the desirability of using both viscosity and thermodynamic data to characterize complex fluids and to develop correlations. This discussion is limited to Newtonian fluids, i.e., fluids in which the viscosity, as defined, is independent of either the magnitude of the shearing stress or velocity gradient (rate of shear). For polymer solutions which are non-Newtonian, the reader is referred to Ferry (1980) or Larson (1999).

The mechanisms and molecular theory of gas viscosity have been reasonably well clarified by nonequlibrium statistical mechanics and the kinetic theory of gases (Millat, et al., 1996), but the theory of liquid viscosity is less well developed. Brief summaries of both theories will be presented.

Since viscosity is defined as a shearing stress per unit area divided by a velocity gradient, it should have the dimensions of (force) (time)/(length)^2 or mass/(length) (time). Both dimensional groups are used, although for most scientific work, viscosities are expressed in poises, centipoises, micropoises, etc. A poise (P) denotes a viscosity of 0.1 N · s/m^2 and 1.0 cP = 0.01 P. The following conversion factors apply to viscosity units:

$$1 \text{ P} = 100 \text{ cP} = 1.000 \times 10^6 \ \mu\text{P} = 0.1 \text{ N s/m}^2 = 1 \text{ g/(cm} \cdot \text{s)} = 0.1 \text{ Pa} \cdot \text{s}$$
$$= 6.72 \times 10^{-2} \text{ lb-mass/(ft} \cdot \text{s)} = 242 \text{ lb-mass/(ft} \cdot \text{h)}$$

$$1 \text{ cP} = 1 \text{ mPa} \cdot \text{s}$$

The *kinematic viscosity* is the ratio of the viscosity to the density. With viscosity in poises and the density of grams per cubic centimeter, the unit of kinematic viscosity is the *stoke,* with the units square centimeters per second. In the SI system of units, viscosities are expressed in N · s/m^2 (or Pa · s) and kinematic viscosities in either m^2/s or cm^2/s.

9-3 Theory of Gas Transport Properties

The theory of gas transport properties is simply stated, but it is quite complex to express in equations that can be used directly to calculate viscosities. In simple terms, when a gas undergoes a shearing stress so that there is some bulk motion, the molecules at any one point have the bulk velocity vector added to their own random velocity vector. Molecular collisions cause an interchange of momentum throughout the fluid, and this bulk motion velocity (or momentum) becomes distributed. Near the source of the applied stress, the bulk velocity vector is high, but as the molecules move away from the source, they are "slowed down" due to random molecular collisions. This random, molecular momentum interchange is the predominant cause of gaseous viscosity.

Elementary Kinetic Theory

If the gas is modeled in the simplest manner, it is possible to show the general relations among viscosity, temperature, pressure, and molecular size. More rigorous treatments yield similar relations but with important correction factors. The elementary gas model assumes all molecules to be nonattracting rigid spheres of diameter σ (with mass m) moving randomly at a mean velocity v. The molar density is n molecules in a unit volume while the mass density is the mass in a unit volume. Molecules move in the gas and collide transferring momentum in a velocity gradient and energy in a temperature gradient. The motion also transfers molecular species in a concentration gradient. The net flux of momentum, energy, or component mass between two layers is assumed proportional to the momentum, energy, or mass density gradient, i.e.,

$$\text{Flux} \propto -\frac{d\rho'}{dz} \tag{9-3.1}$$

where the density ρ' decreases in the $+z$ direction and ρ' may be ρ_i (mass density),

nmv_y, (momentum density), or $C_v nT$ (energy density). The coefficient of proportionality for all these fluxes is given by elementary kinetic theory as $vL/3$, where v is the average molecular speed and L is the mean free path.

Equation (9-3.1) is also used to define the transport coefficients of diffusivity D, viscosity η, and thermal conductivity λ; that is,

$$\text{Mass flux} = -Dm\frac{dn_i}{dz} = -\frac{vL}{3}\frac{d\rho_i}{dz} \tag{9-3.2}$$

$$\text{Momentum flux} = -\eta\frac{dv_y}{dz} = -\frac{vL}{3}mn\frac{dv_y}{dz} \tag{9-3.3}$$

$$\text{Energy flux} = -\lambda\frac{dT}{dz} = -\frac{vL}{3}C_v n\frac{dT}{dz} \tag{9-3.4}$$

Equations (9-3.2) to (9-3.4) define the transport coefficients D, η, and λ. If the average speed is proportional to $(RT/M)^{1/2}$ and the mean free path to $(n\sigma^2)^{-1}$,

$$D = \frac{vL}{3} = (\text{const})\frac{T^{3/2}}{M^{1/2}P\sigma^2} \tag{9-3.5}$$

$$\eta = \frac{m\rho vL}{3} = (\text{const})\frac{T^{1/2}M^{1/2}}{\sigma^2} \tag{9-3.6}$$

$$\lambda = \frac{vLc_v n}{3} = (\text{const})\frac{T^{1/2}}{M^{1/2}\sigma^2} \tag{9-3.7}$$

The constant multipliers in Eqs. (9-3.5) to (9.3-7) are different in each case; the interesting fact to note from these results is the dependency of the various transfer coefficients on T, P, M, and σ. A similar treatment for rigid, nonattracting spheres having a Maxwellian velocity distribution yields the same final equations but with slightly different numerical constants.

The viscosity relation [Eq. (9-3.6)] for a rigid, non-attracting sphere model is (see page 14 of Hirschfelder, et al., 1954)

$$\eta = 26.69\frac{(MT)^{1/2}}{\sigma^2} \tag{9-3.8}$$

where η = viscosity, μP
M = molecular weight, g/mol
T = temperature, K
σ = hard-sphere diameter, Å

Analogous equations for λ and D are given in Chaps. 10 and 11.

Effect of Intermolecular Forces

If the molecules attract or repel one another by virtue of intermolecular forces, the theory of Chapman and Enskog is normally employed (Chapman and Cowling, 1939; Hirschfelder, et al., 1954). There are four important assumptions in this development: (1) the gas is sufficiently dilute for only binary collisions to occur (ideal gas); (2) the motion of the molecules during a collision can be described by classical mechanics; (3) only elastic collisions occur, and (4) the intermolecular forces act only between fixed centers of the molecules; i.e., the intermolecular potential function is spherically symmetric. With these restrictions, it would appear that the resulting theory should be applicable only to low-pressure, high-temperature monatomic gases. The pressure and temperature restrictions are valid for polyatomic gases and, except for thermal conductivity (See Chap. 10) are adequate for most modeling purposes.

The Chapman-Enskog treatment develops integral relations for the transport properties when the interactions between colliding molecules are described by a potential energy function $\psi(r)$. The equations require complex numerical solution for each choice of intermolecular potential model. In general terms, the first-order solution for viscosity can be written

$$\eta = \frac{(26.69)(MT)^{1/2}}{\sigma^2 \Omega_v} \tag{9-3.9}$$

where the temperature dependence of the collision integral, Ω_v is different for each $\psi(r)$ and all symbols and units are as defined in Eq. (9-3.8). Ω_v is unity if the molecules do not attract each other. Corrections can be found in Chapman and Cowling (1939) and Hirschfelder, et al. (1954). The use of Ω_v from the Lennard-Jones (12-6) potential function is illustrated in Sec. 9-4.

9-4 ESTIMATION OF LOW-PRESSURE GAS VISCOSITY

Essentially all gas viscosity estimation techniques are based on either the Chapman-Enskog theory or the law of corresponding states. Both approaches are discussed below, and recommendations are presented at the end of the section. Experimental values of low-pressure gas viscosities are compiled in Landolt-Bornstein (1955), Stephan and Lucas (1979), and Vargaftik, et al. (1996). Literature references for a number of substances along with equations with which to calculate gas viscosities based on critically evaluated data may be found in Daubert, et al. (1997). Gas phase viscosity information can also be found in Dean (1999), Lide (1999), Perry and Green (1997), and Yaws (1995, 1995a). This information should be used with caution in those cases where constants in equations have been determined from estimated rather than experimental viscosities.

Theoretical Approach

The first-order Chapman-Enskog viscosity equation was given as Eq. (9-3.9). To use this relation to estimate viscosities, the collision diameter σ and the collision integral Ω_v must be found. In the derivation of Eq. (9-3.9), Ω_v is obtained as a function of a dimensionless temperature T^* which depends upon the intermolecular potential chosen. For any potential curve, the dimensionless temperature T^* is related to ε by

$$T^* = \frac{kT}{\varepsilon} \tag{9-4.1}$$

where k is Boltzmann's constant and ε is the minimum of the pair-potential energy. The working equation for η must have as many parameters as were used to define the original $\psi(r)$ relation. While many potential models have been proposed (Hirschfelder, et al., 1954), the Lennard-Jones 12-6 was the first and has most often been applied for ideal gas viscosity.

$$\psi(r) = 4\varepsilon \left[\left(\frac{\sigma}{r}\right)^{12} - \left(\frac{\sigma}{r}\right)^{6} \right] \tag{9-4.2}$$

In Eq. (9-4.2), σ is like a molecular diameter and is the value of r that causes $\psi(r)$ to be zero. With this potential, the collision integral has been determined by a number of investigators (Barker, et al., 1964; Hirschfelder, et al., 1954; Itean, et al., 1961; Klein and Smith, 1968; Monchick and Mason, 1961; and O'Connell and Prausnitz, 1965). Neufeld, et al. (1972) proposed an empirical equation which is convenient for computer application:

$$\Omega_v = [A(T^*)^{-B}] + C[\exp(-DT^*)] + E[\exp(-FT^*)] \tag{9-4.3}$$

where $T^* = kT/\varepsilon$, $A = 1.16145$, $B = 0.14874$, $C = 0.52487$, $D = 0.77320$, $E = 2.16178$, and $F = 2.43787$. Equation (9-4.3) is applicable from $0.3 \leq T^* \leq 100$ with an average deviation of only 0.064%. A graph of log Ω_v as a function of log T^* is shown in Fig. 9-1.

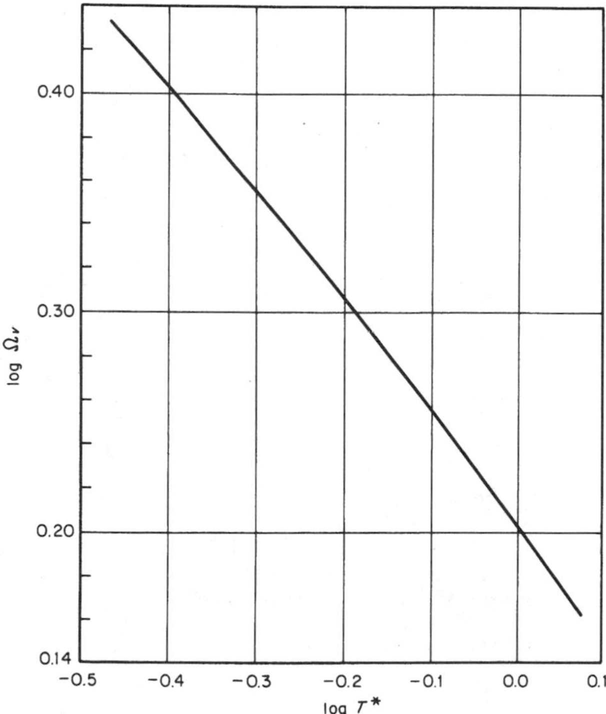

FIGURE 9-1 Effect of temperature on the Lennard-Jones viscosity collision integral.

With values of Ω_v as a function of T^*, a number of investigators have used Eq. (9-3.9) and regressed experimental viscosity-temperature data to find the best values of ε/k and σ for many substances. Appendix B lists a number of such sets as reported by Svehla (1962). It should be noted, however, that there appears also to be a number of other quite satisfactory *sets* of ε/k and σ for any given compound. For example, with *n*-butane, Svehla suggested ε/k = 513.4 K, σ = 4.730 Å, whereas Flynn and Thodos (1962) recommend ε/k = 208 K and σ = 5.869 Å.

Both sets, when used to calculate viscosities, yield almost exactly the same values of viscosity as shown in Fig. 9-2. This interesting paradox has been resolved by Reichenberg (1971), who suggested that log Ω_v is essentially a linear function of log T^* (see Fig. 9-1).

$$\Omega_v = a\,(T^*)^n \tag{9-4.4}$$

Kim and Ross (1967) do, in fact, propose that:

$$\Omega_v = 1.604\,(T^*)^{-0.5} \tag{9-4.5}$$

where $0.4 \leq T^* \leq 1.4$. They note a maximum error of only 0.7%. Substitution of Eq. (9-4.5) into Eq. (9-3.9) leads to

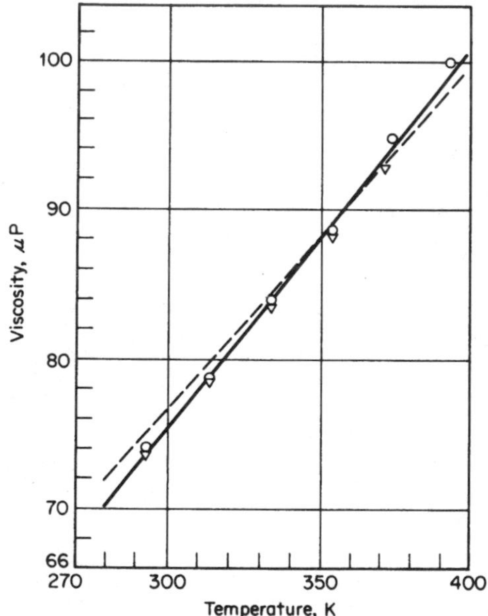

FIGURE 9-2 Comparison of calculated and experimental low-pressure gas viscosity of *n*-butane with Eq. (9-3.9) and the Lennard-Jones Potential: --- Flynn and Thodos (1962) with σ = 5.869 Å and ε/k = 208 K. ——— Svehla (1962) with σ = 4.730 Å and ε/k = 513.4 K. o (*Titani, 1929*); ∇ (*Wobster and Mueller, 1941*).

$$\eta = \frac{16.64(M)^{1/2}T}{(\varepsilon/k)^{1/2}\sigma^2} \tag{9-4.6}$$

where the units are the same as in Eq. (9-3.9). Here the parameters σ and ε/k are combined as a *single* term $(\varepsilon/k)^{1/2}\sigma^2$. There is then no way of delineating individual values of ε/k and σ by using experimental viscosity data, at least over the range where Eq. (9-4.5) applies. Equation (9-4.6) suggests that when $T_r \leq 1.4$, low-pressure gas viscosities are essentially proportional to the absolute temperature.

The conclusion to be drawn from this discussion is that Eq. (9-3.9) can be used to calculate gas viscosity, although the chosen set of ε/k and σ may have little relation to molecular properties. There will be an infinite number of acceptable sets as long as the temperature range is not too broad, e.g., if one limits the estimation to the range of reduced temperatures from about 0.3 to 1.2. In using published values of ε/k and σ for a fluid of interest, the two values from the same set must be used—never ε/k from one set and σ from another.

The difficulty in obtaining a priori meaningful values of ε/k and σ has led most authors to specify rules which relate ε/k and σ to macroscopic parameters such as the critical constants. One such method is shown below.

Method of Chung, et al. (1984, 1988)

These authors have employed Eq. (9-3.9) with

$$\frac{\varepsilon}{k} = \frac{T_c}{1.2593} \tag{9-4.7}$$

$$\sigma = 0.809V_c^{1/3} \tag{9-4.8}$$

where ε/k and T_c are in kelvins, σ is in angstroms, and V_c is in cm^3/mol. Then, using Eqs. (9-4.1) and (9-4.7),

$$T^* = 1.2593T_r \tag{9-4.9}$$

Ω_v in Eq. (9-3.9) is found from Eq. (9-4.3) with T^* defined by Eq. (9-4.9). Chung, et al. also multiply the right-hand side of Eq. (9-3.9) by a factor F_c to account for molecular shapes and polarities of dilute gases. Their final result may be expressed as:

$$\eta = 40.785 \frac{F_c(MT)^{1/2}}{V_c^{2/3}\Omega_v} \tag{9-4.10}$$

where η = viscosity, μP
M = molecular weight, g/mol
T = temperature, K
V_c = critical volume, cm^3/mol
Ω_v = viscosity collision integral from Eq. (9-4.3) and $T^* = 1.2593T_r$
$F_c = 1 - 0.2756\omega + 0.059035\,\mu_r^4 + \kappa$ (9-4.11)

In Eq. (9-4.11), ω is the acentric factor (See Chap. 2) and κ is a special correction for highly polar substances such as alcohols and acids. Values of κ for a few such materials are shown in Table 9-1. Chung, et al. (1984) suggest that for other al-

TABLE 9-1 The Association Factor κ in Eq. (9-4.11) (Chung, et al., 1988)

Compound	κ	Compound	κ
Methanol	0.215	*n*-Pentanol	0.122
Ethanol	0.175	*n*-Hexanol	0.114
n-Propanol	0.143	*n*-Heptanol	0.109
i-Propanol	0.143	Acetic Acid	0.0916
n-Butanol	0.132	Water	0.076
i-Butanol	0.132		

cohols not shown in Table 9-1, $\kappa = 0.0682 + 4.704$[(number of –OH groups)/molecular weight]. The term μ_r is a dimensionless dipole moment. (See discussion in Chap. 2 and also under Eq. (9-4.17) for techniques to nondimensionalize a dipole moment.) When V_c is in cm^3/mole, T_c is in kelvins, and μ is in debyes,

$$\mu_r = 131.3 \frac{\mu}{(V_c T_c)^{1/2}} \tag{9-4.12}$$

Example 9-1 Estimate the viscosity of sulfur dioxide gas at atmospheric pressure and 300°C by using the Chung, et al. method. The experimental viscosity is 246 μP (Landolt-Bornstein, 1955).

solution From Appendix A, $T_c = 430.8$ K, $V_c = 122$ cm^3/mole, $M = 64.065$, and the dipole moment is 1.6 debyes. From Eq. (2-3.3), $\omega = 0.257$. Assume κ is negligible. From Eq. (9-4.12),

$$\mu_r = \frac{(131.3)(1.6)}{[(122)(430.8)]^{1/2}} = 0.916$$

and with Eq. (9-4.11),

$$F_c = 1 - (0.2756)(0.257) + (0.059035)(0.916)^4 = 0.971$$

$$T^* = 1.2593 \frac{300 + 273}{430.8} = 1.675$$

Then, with Eq. (9-4.3), $\Omega_v = 1.256$. The viscosity is determined from Eq. (9-4.10).

$$\eta = (40.785)(0.971) \frac{[(64.065)(300 + 273)]^{1/2}}{(122)^{2/3}(1.256)} = 245.5 \ \mu\text{P}$$

$$\text{Error} = \frac{245.5 - 246}{246} \times 100 = -0.2\%$$

Experimental viscosities and those estimated by the Chung, et al. method are shown in Table 9-2. The critical properties used to prepare Table 9-2 differed slightly from those in Appendix A. The average absolute error was about 1.9%. This agrees well with the more extensive comparison by Chung (1980), who found an average absolute error of about 1.5%.

Corresponding States Methods

From an equation such as (9-3.9), if one associates σ^3 with V_c [as in Eq. (9-4.8)] and assumes V_c is proportional to RT_c/P_c, a dimensionless viscosity can be defined:

$$\eta_r = \xi\eta = f(T_r) \tag{9-4.13}$$

$$\xi = \left[\frac{(RT_c)(N_o)^2}{M^3\,P_c^4}\right]^{1/6} \tag{9-4.14}$$

In SI units, if $R = 8314$ J/(kmol · K) and N_o (Avogadro's number) $= 6.023 \times 10^{26}$ (kmol)$^{-1}$ and with T_c in kelvins, M in kg/kmol, and P_c in N/m^2, ξ has the units of m^2/(N · s) or inverse viscosity. In more convenient units,

$$\xi = 0.176\left(\frac{T_c}{M^3\,P_c^4}\right)^{1/6} \tag{9-4.15}$$

where ξ = reduced, inverse viscosity, $(\mu\text{P})^{-1}$, T_c is kelvins, M is in g/mol, and P_c is in bars.

Equation (9-4.13) has been recommended by several authors (Flynn and Thodos, 1961; Golubev, 1959; Malek and Stiel, 1972; Mathur and Thodos, 1963; Trautz, 1931; and Yoon and Thodos, 1970). The specific form suggested by Lucas (Lucas, 1980; Lucas, 1983; and Lucas, 1984a) is illustrated below.

$$\eta\xi = [0.807\,T_r^{0.618} - 0.357\exp(-0.449T_r) + 0.340\exp(-4.058\,T_r) + 0.018]\,F_P^o\,F_Q^o \tag{9-4.16}$$

ξ is defined by Eq. (9-4.15), η is in μP, T_r is the reduced temperature, and F_P^o and F_Q^o are correction factors to account for polarity or quantum effects. To obtain F_P^o, a reduced dipole moment is required. Lucas defines this quantity somewhat differently than did Chung, et al. in Eq. (9-4.12), i.e.,

$$\mu_r = 52.46\,\frac{\mu^2 P_c}{T_c^2} \tag{9-4.17}$$

where μ is in debyes, P_c is in bars, and T_c is in kelvins. Then F_P^o values are found as:

$$F_P^o = 1 \qquad 0 \le \mu_r < 0.022$$

$$F_P^o = 1 + 30.55(0.292 - Z_c)^{1.72} \qquad 0.022 \le \mu_r < 0.075$$

$$F_P^o = 1 + 30.55(0.292 - Z_c)^{1.72}\,|0.96 + 0.1(T_r - 0.7)| \qquad 0.075 \le \mu_r \tag{9-4.18}$$

The factor F_Q^o is used only for the quantum gases He, H_2, and D_2.

$$F_Q^o = 1.22Q^{0.15}\,\{1 + 0.00385[(T_r - 12)^2]^{1/M}\,\text{sign}\,(T_r - 12)\} \tag{9-4.19}$$

where $Q = 1.38$ (He), $Q = 0.76$ (H_2), $Q = 0.52$ (D_2). Sign( ) indicates that one should use +1 or −1 depending on whether the value of the argument ( ) is greater than or less than zero.

TABLE 9-2 Comparison between Calculated and Experimental Low-Pressure Gas Viscosities

			Percent error†		
Compound	T, °C	Experimental value, μP‡	Chung et al., Eq. (9-4.10)	Lucas, Eq. (9-4.16)	Reichenberg, Eq. (9-4.21)
Acetic acid	150	118	3.4	—	1.7
	250	151	0.2	—	2.4
Acetylene	30	102	0.6	0.6	2.4
	101	126	−0.6	−0.8	0.7
	200	155	−0.5	−0.6	0.7
Ammonia	37	106	2.2	2.3	—
	147	146	0.4	0.2	—
	267	189	−2.1	−1.9	—
Benzene	28	73.2	1.0	3.2	4.4
	100	92.5	0.2	1.3	2.1
	200	117	0.5	1.6	1.8
Bromotrifluoromethane	17	145	7.6	8.3	−0.3
	97	183	8.2	8.6	−0.5
Isobutane	37	79	−1.6	1.6	1.6
	155	105	2.0	4.7	4.0
	287	132	3.9	6.7	5.6
n-butane	7	74	−5.8	−4.1	−5.6
	127	101	−1.0	0.1	−2.3
	267	132	0.6	1.5	−1.4
1-Butene	20	76.1	−1.0	1.1	1.8
	60	86.3	−0.5	1.4	1.7
	120	102	−0.6	0.9	0.9
Carbon dioxide	37	154	−0.6	1.6	—
	127	194	−0.1	2.2	—
	327	272	0.5	3.0	—
Carbon disulfide	30	100	0	5.9	—
	98.2	125	−1.3	3.5	—
	200	161	−1.9	2.2	—
Carbon tetrachloride	125	133	0.8	1.9	−2.8
	200	156	2.4	3.1	−2.1
	300	186	3.8	4.1	−1.4
Chlorine	20	133	2.2	3.5	—
	100	168	3.3	4.1	—
	200	209	4.5	5.1	—
Chloroform	20	100	−3.8	3.4	−1.0
	100	125	−1.3	5.4	0.6
	300	191	−0.9	5.1	−0.6
Cyclohexane	35	72.3	−2.4	−0.1	−4.2
	100	87.3	−1.4	0.3	−4.1
	300	129	2.6	3.6	−1.9
Dimethyl ether	20	90.9	−5.2	−1.0	2.0
	100	117	−6.1	−2.4	0.2

TABLE 9-2 Comparison between Calculated and Experimental Low-Pressure Gas Viscosities (*Continued*)

			Percent error†		
Compound	T, °C	Experimental value, μP‡	Chung et al., Eq. (9-4.10)	Lucas, Eq. (9-4.16)	Reichenberg, Eq. (9-4.21)
Ethane	47	100	0.1	0.7	−3.2
	117	120	0.2	0.8	−3.3
	247	156	−1.0	−0.3	−4.7
Ethyl acetate	125	101	−2.6	9.0	−1.5
	200	120	−2.4	8.8	−2.0
	300	146	−3.3	7.4	−3.5
Ethanol	110	111	−0.5	−2.5	0.6
	197	137	−0.8	1.6	−0.4
	267	156	−0.3	2.0	−0.1
Diethyl ether	125	99.1	−0.4	0.2	0.7
	200	118	−0.8	−0.5	−0.3
	300	141	−0.6	−0.4	−0.5
n-Hexane	107	81	−1.1	0.9	0.3
	267	116	−2.1	−0.7	−2.1
Methane	−13	98	−0.7	−0.5	—
	147	147	0	−0.9	—
Methyl acetate	125	108	0	10	1.7
	200	157	−1.6	7.6	−1.0
Methanol	67	112	−0.4	−0.9	1.1
	127	132	−0.3	−1.1	1.2
	277	181	−0.3	−1.7	0.9
Methyl chloride	50	119	3.9	0.8	−0.6
	130	147	5.0	1.4	1.9
Nitrogen	27	178	0.3	0.3	—
	227	258	0.3	0.4	—
Isopropanol	157	113	0.3	5.9	3.5
	257	139	0.1	5.4	2.7
Propylene	17	83	2.4	2.4	3.0
	127	115	1.5	0.9	0.8
	307	160	1.7	1.2	0.7
Sulfur dioxide	10	120	2.8	5.5	—
	100	163	0.3	2.2	—
	300	246	−0.2	1.5	—
	700	376	1.5	3.4	—
Toluene	60	78.9	−5.2	−3.7	−2.3
	250	123	−3.5	−3.1	−2.5
Average absolute error			1.9	3.0	1.9

† Percent error = [(calc. − exp.)/exp.] × 100.

‡ All experimental viscosity values were obtained from Landolt-Bornstein (1955), Lucas (1984), Stephan and Lucas (1979).

Equation (9-4.16) is similar to an equation proposed by Thodos and coworkers (e.g., Yoon and Thodos, 1970). It is interesting to note that, if $T_r \leq 1$, the $f(T_r)$ in brackets in Eq. (9-4.16) is closely approximated by $0.606T_r$, that is,

$$\eta\xi \approx (0.606T_r)\, F_P^o F_Q^o \qquad T_r \leq 1 \tag{9-4.20}$$

The method of Lucas is illustrated in Example 9-2.

Example 9-2 Estimate the viscosity of methanol vapor at a temperature of 550 K and 1 bar by using Lucas' method. The experimental value is 181 μP (Stephan and Lucas, 1979).

solution From Appendix A, $T_c = 512.64$ K, $P_c = 80.97$ bar, $Z_c = 0.224$, $M = 32.042$, and $\mu = 1.7$ debyes. $T_r = 550/512.64 = 1.07$, and $\mu_r = 52.46[(1.7)^2(80.97)/(512.64)^2] = 4.67 \times 10^{-2}$. From Eq. (9-4.18),

$$F_P^0 = 1 + (30.55)(0.292 - 0.224)^{1.72} = 1.30$$

With Eq. (9-4.15),

$$\xi = 0.176 \left[\frac{512.64}{(32.042)^2\,(80.97)^4}\right]^{1/6} = 4.70 \times 10^{-3}(\mu\text{P})^{-1}$$

Then, with Eq. (9-4.16)

$$\begin{aligned}\eta\xi &= \{(0.807)(1.07)^{0.618} - 0.357 \exp[-(0.449)(1.07)] \\ &\quad +0.34 \exp[-(4.058)(1.07)] + 0.018\}(1.30) \\ &= 0.836\end{aligned}$$

$$\eta = \frac{0.836}{(4.70 \times 10^{-3})} = 178\ \mu\text{ P}$$

$$\text{Error} = \frac{178 - 181}{181} \times 100 = -1.7\%$$

In Table 9-2, experimental viscosities are compared with those computed by Lucas's method. The average absolute error is 3.0%. Even with the correction factor F_P^o, higher errors are noted for polar compounds compared to nonpolar.

Reichenberg (1971, 1979) has suggested an alternate corresponding states relation for low-pressure gas viscosity of organic compounds.

$$\eta = \frac{M^{1/2}T}{a^*[1 + (4/T_c)][1 + 0.36T_r(T_r - 1)]^{1/6}}\,\frac{T_r(1 + 270\,\mu_r^4)}{T_r + 270\,\mu_r^4} \tag{9-4.21}$$

η is in μP; M is the molecular weight; T is the temperature; T_c is the critical temperature, in kelvins; T_r is the reduced temperature; and μ_r, is the reduced dipole moment defined earlier in Eq. (9-4.17). The parameter a^* is defined as

$$a^* = \Sigma n_i C_i \tag{9-4.22}$$

where n_i represents the number of groups of the ith type and C_i is the group contribution shown in Table 9-3.

The term $(1 + 4/T_c)$ in the denominator of Eq. (9-4.21) may be neglected except for treating quantum gases with low values of T_c.

TABLE 9-3 Values of the Group Contributions C_i for the Estimation of a^* in Eq. (9-4-22) (Reichenberg, 1971)

Group	Contribution C_i
—CH_3	9.04
>CH_2 (nonring)	6.47
>CH— (nonring)	2.67
>C< (nonring)	−1.53
=CH_2	7.68
=CH—(nonring)	5.53
>C—(nonring)	1.78
≡CH	7.41
≡C—(nonring)	5.24
>CH_2(ring)	6.91
>CH—(ring)	1.16
>C< (ring)	0.23
=CH—(ring)	5.90
>C=(ring)	3.59
—F	4.46
—Cl	10.06
—Br	12.83
—OH (alcohols)	7.96
>O (nonring)	3.59
>C=O (nonring)	12.02
—CHO (aldehydes)	14.02
—COOH (acids)	18.65
—COO—(esters) or HCOO (formates)	13.41
—NH_2	9.71
>NH (nonring)	3.68

TABLE 9-3 Values of the Group Contributions C_i for the Estimation of a^* in Eq. (9-4-22) (Reichenberg, 1971) (*Continued*)

Group	Contribution C_i
=N—(ring)	4.97
—CN	18.15
\S/ (ring)	8.86

A comparison between calculated and experimental low-pressure gas viscosity values is given in Table 9-2, and the method is illustrated in Example 9-3.

Example 9-3 Use Reichenberg's method to estimate the viscosity of ethyl acetate vapor at 125°C and low pressure. The experimental value is reported to be 101 μP (Landolt-Bornstein Tabellen, 1955).

solution From Appendix A, T_c = 523.2 K, M = 88.106, P_c = 38.3 bar, and μ = 1.9 debyes. With Eq. (9-4.17),

$$\mu_r = \frac{(52.46)(1.9)^2(38.3)}{(523.2)^2} = 0.0265$$

T_r = (125 + 273)/523.2 = 0.761. With Eq. (9-4.22) and Table 9-3

$$a^* = 2(\text{—CH}_3) + (\text{—CH}_2) + (\text{—COO—})$$

$$= (2)(9.04) + 6.47 + 13.41 = 37.96$$

With Eq. (9-4.21),

$$\eta = \frac{(88.106)^{1/2}(125 + 273)}{37.96[1 + (0.36)(0.761)(0.761 - 1)]^{1/6}}$$

$$\frac{(0.761)[1 + (270)(0.0265)^4]}{0.761 + (270)(0.0265)^4} = 99.4\ \mu\text{P}$$

$$\text{Error} = \frac{99.4 - 101}{101} \times 100 = -1.5\%$$

Recommendations for Estimating Low-pressure Viscosities of Pure Gases

Any of the three estimation methods described in this section may be used with the expectation of errors of 0.5 to 1.5% for nonpolar compounds and 2 to 4% for polar compounds. Lucas's method requires as input data T_c, P_c, and M as well as μ and Z_c for polar compounds and is easy to apply. At present, it is not suitable for highly associated gases like acetic acid, but it could probably be extended by multiplication of an appropriate factor as in the Chung, et al. technique. The Chung, et al. method requires somewhat more input (T_c, V_c, and M and μ, ω, and κ for the polar correction). The critical volume is less readily available than the critical pressure (See Chap. 2), and the association factor κ is an empirical constant that must be determined from viscosity data. The method is not suited for quantum gases. Reichenberg's technique requires M, T_c, and structural groups as well as μ for the polar correction. This method is not suitable for inorganic gases and cannot be applied to organic gases for which necessary group contributions have not been determined.

9-5 VISCOSITIES OF GAS MIXTURES AT LOW PRESSURES

The rigorous kinetic theory of Chapman and Enskog can be extended to determine the viscosity of low-pressure multicomponent mixtures (Brokaw, 1964, 1965, 1968, 1965; Chapman and Cowling, 1939; Hirschfelder, et al., 1954; Kestin, et al., 1976). The final expressions are quite complicated and are rarely used to estimate mixture viscosities. Three simplifications of the rigorous theoretical expressions are described below. Reichenberg's equations are the most complex, but, as shown later, the most consistently accurate. Wilke's method is simpler, and that of Herning and Zipperer is even more so. All these methods are essentially interpolative; i.e., the viscosity values for the pure components must be available. The methods then lead to estimations showing how the mixture viscosity varies with composition. Later in this section, two corresponding states methods are described; they do not require pure component values as inputs. A compilation of references dealing with gas mixture viscosities (low and high pressure) has been prepared by Sutton (1976).

Method of Reichenberg (1974, 1977, 1979)

In this technique, Reichenberg has incorporated elements of the kinetic theory approach of Hirschfelder, et al. (1954) with corresponding states methodology to obtain desired parameters. In addition, a polar correction has been included. The general, multicomponent mixture viscosity equation is:

$$\eta_m = \sum_{i=1}^{n} K_i \left(1 + 2\sum_{j=1}^{i-1} H_{ij} K_j + \sum_{j=1\neq i}^{n} \sum_{k=1\neq i}^{n} H_{ij} H_{ik} K_j K_k\right) \tag{9-5.1}$$

where η_m is the mixture viscosity and n is the number of components. With η_i the viscosity of pure i, M_i the molecular weight of i, and y_i the mole fraction of i in the mixture,

$$K_i = \frac{y_i \eta_i}{y_i + \eta_i \sum_{k=1\neq i}^{n} y_k H_{ik}[3 + (2M_k/M_i)]} \tag{9-5.2}$$

Two other component properties used are:

$$U_i = \frac{[1 + 0.36T_{ri}(T_{ri} - 1)]^{1/6} F_{Ri}}{(T_{ri})^{1/2}} \tag{9-5.3}$$

$$C_i = \frac{M_i^{1/4}}{(\eta_i U_i)^{1/2}} \tag{9-5.4}$$

where $T_{ri} = T/T_{ci}$ and F_{Ri} is a polar correction.

$$F_{Ri} = \frac{T_{ri}^{3.5} + (10\,\mu_{ri})^7}{T_{ri}^{3.5}[1 + (10\,\mu_{ri})^7]} \tag{9-5.5}$$

Here μ_{ri} is the reduced dipole moment of i and is calculated as shown earlier in Eq. (9-4.17). For the term $H_{ij} = H_{ji}$,

$$H_{ij} = \left[\frac{M_i M_j}{32(M_i + M_j)^3}\right]^{1/2} (C_i + C_j)^2 \times \frac{[1 + 0.36T_{rij}(T_{rij} - 1)]^{1/6} F_{Rij}}{(T_{rij})^{1/2}} \tag{9-5.6}$$

with

$$T_{rij} = \frac{T}{(T_{ci}\,T_{cj})^{1/2}} \tag{9-5.7}$$

F_{Rij} is found from Eq. (9-5.5) with T_{ri} replaced by T_{rij} and μ_{ri} by $\mu_{rij} = (\mu_{ri}\,\mu_{rj})^{1/2}$.

For a binary gas mixture of 1 and 2, these equations may be written as:

$$\eta_m = K_1(1 + H_{12}^2\,K_2^2) + K_2(1 + 2\,H_{12}\,K_1 + H_{12}^2\,K_1^2) \tag{9-5.8}$$

$$K_1 = \frac{y_1\eta_1}{y_1 + \eta_1\{y_2\,H_{12}[3 + (2\,M_2/M_1)]\}} \tag{9-5.9}$$

$$K_2 = \frac{y_2\eta_2}{y_2 + \eta_2\{y_1\,H_{12}[3 + (2\,M_1/M_2)]\}} \tag{9-5.10}$$

$$U_1 = \frac{[1 + 0.36T_{r1}(T_{r1} - 1)]^{1/6}}{T_{r1}^{1/2}} \; \frac{T_{r1}^{3.5} + (10\,\mu_{r1})^7}{T_{r1}^{3.5}[1 + (10\,\mu_{r1})^7]} \tag{9-5.11}$$

and a comparable expression for U_2. The meaning of C_1 and C_2 is clear from Eq. (9-5.4). Finally, with

$$T_{r12} = \frac{T}{(T_{c1}T_{c2})^{1/2}} \quad \text{and} \quad \mu_{r12} = (\mu_{r1}\,\mu_{r2})^{1/2}$$

$$H_{12} = \frac{(M_1M_2/32)^{1/2}}{(M_1 + M_2)^{3/2}} \; \frac{[1 + 0.36T_{r12}(T_{r12} - 1)]^{1/6}}{(T_{r12})^{1/2}}$$

$$\times\,(C_1 + C_2)^2\,\frac{T_{r12}^{3.5} + (10\,\mu_{r12})^7}{T_{r12}^{3.5}[1 + (10\,\mu_{r12})^7]} \tag{9-5.12}$$

To employ Reichenberg's method, for each component one needs the pure gas viscosity at the system temperature as well as the molecular weight, dipole moment, critical temperature, and critical pressure. The temperature and composition are state variables.

The method is illustrated in Example 9-4. A comparison of experimental and calculated gas-mixture viscosities is shown in Table 9-4.

TABLE 9-4 Comparison of Calculated and Experimental Low-Pressure Gas Mixture Viscosities

System	T, K	Mole fraction first component	Viscosity (exp.) μP	Ref.*	Percent deviation† calculated by method of: Reichenberg, Eq. (9-5.8)	Wilke, Eq. (9-5.16)	Herning and Zipperer, Eq. (9-5.17)	Lucas, Eq. (9-4.16) with Eqs. (9-5.18) through (9-5.23)	Chung, et al., Eq. (9-5.24)
Nitrogen-hydrogen	373	0.0	104.2	6, 11	—	—	—	0.8	−11
		0.2	152.3		4.3	12	2.0	2.1	−23
		0.51	190.3		1.8	5.6	−1.0	−2.0	−11
		0.80	205.8		0.1	1.4	−1.2	3.6	−3.3
		1.0	210.1		—	—	—	0.4	0
Methane-propane	298	0.0	81.0	1	—	—	—	3.5	1.3
		0.2	85.0		0.2	−0.3	−0.2	4.6	1.7
		0.4	89.9		0.1	−0.8	−0.6	5.0	1.7
		0.6	95.0		0.6	−0.4	−0.2	5.4	1.9
		0.8	102.0		0.2	−0.6	−0.5	3.7	1.0
		1.0	110.0		—	—	—	0.3	1.0
	498	0.0	131.0	1	—	—	—	4.0	2.3
		0.2	136.0		0.4	0.0	−0.2	5.2	2.7
		0.4	142.0		0.6	0.0	−0.5	5.6	2.6
		0.6	149.0		0.7	0.0	−0.6	5.2	2.0
		0.8	157.0		0.7	0.0	−0.3	3.7	1.1
		1.0	167.0		—	—	—	−0.2	0.2
Carbon tetrafluoride-sulfur hexafluoride	303	0.0	159.0	8	—	—	—	6.6	0.7
		0.257	159.9		2.0	2.0	1.8	9.3	3.8
		0.491	161.5		3.4	3.4	3.1	11.0	6.2
		0.754	164.3		4.6	4.6	4.3	13.0	8.4
		1.0	176.7		—	—	—	7.4	3.6

TABLE 9-4 Comparison of Calculated and Experimental Low-Pressure Gas Mixture Viscosities (*Continued*)

System	T, K	Mole fraction first component	Viscosity (exp.) μP	Ref.*	Percent deviation† calculated by method of: Reichenberg, Eq. (9-5.8)	Wilke, Eq. (9-5.16)	Herning and Zipperer, Eq. (9-5.17)	Lucas, Eq. (9-4.16) with Eqs. (9-5.18) through (9-5.23)	Chung, et al., Eq. (9-5.24)
Nitrogen-carbon dioxide	293	0.0	146.6	5	—	—	—	1.6	−1.2
		0.213	153.5		0.5	−1.3	−1.0	0.4	−0.3
		0.495	161.8		0.4	−1.8	−1.5	−0.2	0.7
		0.767	172.1		−2.0	−2.8	−2.5	−1.7	−0.7
		1.0	175.8		—	—	—	0.4	−0.2
Ammonia-hydrogen	306	0.0	90.6	6	—	—	—	2.4	−9.9
		0.195	118.4		−4.0	−11	−18	−2.7	2.1
		0.399	123.8		−4.6	−12	−19	−3.0	10.0
		0.536	122.4		−4.5	−11	−16	−2.7	10.0
		0.677	120.0		−4.8	−9.7	−14	−3.1	7.1
		1.0	105.9		—	—	—	1.3	0.9
Hydrogen sulfide-ethyl ether	331	0.0	84.5	7	—	—	—	−1.7	−4.0
		0.204	87		−2.9	−3.2	0.2	2.3	0.4
		0.500	97		−2.2	−2.8	3.2	3.4	1.7
		0.802	116		0.0	−0.4	4.2	0.6	−0.7
		1.0	137		—	—	—	−3.0	−4.1
Ammonia-methylamine	423	0.0	130.0	2	—	—	—	−2.1	−8.0
		0.25	134.5		−0.8	−0.3	−0.6	−1.5	−7.5
		0.75	142.2		−1.0	−0.3	−0.7	0.1	−3.4
		1.0	146.0		—	—	—	1.1	1.1
	673	0.0	204.8	2	—	—	—	−4.6	−11
		0.25	212.8		−2.6	−0.7	−0.9	−4.9	−11
		0.75	228.3		−3.1	−0.7	−0.9	−4.7	−9.3
		1.0	235.0		—	—	—	−4.3	−5.4

Nitrogen-monochlorodi-fluoromethane	323	0.0	134	10	—	—	—	11.0	6.0
		0.286	145		0.8	−0.8	−0.7	11.0	7.3
		0.463	153		1.3	−1.0	−0.3	11.0	7.4
		0.644	164		0.7	−1.8	−1.5	8.9	5.6
		0.824	177		−0.3	−2.2	−2.1	5.0	2.4
		1.0	188		—	—		0.8	0.4
Nitrous oxide-sulfur dioxide	353	0.0	152.3	4	—	—	—	3.9	−1.5
		0.325	161.7		−1.7	−2.2	−2.2	0.6	−1.6
		0.625	167.8		−1.6	−2.2	−2.1	−0.6	−2.3
		0.817	170.7		−0.9	−1.3	−1.2	−0.7	−1.9
		1.0	173.0		—	—	—	−0.6	−0.7
Nitrogen-*n*-heptane	344	0.0	69.4	3, 9	—	—	—	0.8	−3.5
		0.515	104.0		0.7	−6.2	11	−0.2	−4.8
		0.853	154.6		0.9	−4.7	7.4	−0.5	−3.2
		1.0	197.5		—	—	—	0.6	0.2

†Percent deviation = [(calc. − exp.)/(exp.)] × 100.

*References: 1, Bircher, 1943; 2, Burch and Raw, 1967; 3, Carmichael and Sage, 1966; 4, Chakraborti and Gray, 1965; 5, Kestin and Leidenfrost, 1959; 6, Pal and Baruna, 1967; 7, Pal and Bhattacharyya, 1969; 8, Raw and Tang, 1963; 9, Stephan and Lucas, 1979; 10, Tanka, et al., 1977; 11, Trautz and Baumann, 1929.

Example 9-4 Use Reichenberg's method to estimate the viscosity of a nitrogen-monochlorodifluoromethane (R-22) mixture at 50°C and atmospheric pressure. The mole fraction nitrogen is 0.286. The experimental viscosity is 145 μP (Tanaka, et al., 1977).

solution The following pure component properties are used:

	N_2	$CHClF_2$
T_c, K	126.2	369.28
P_c, bar	33.98	49.86
M, g/mol	28.014	86.468
μ, debyes	0	1.4
η, 50°C, μP	188	134

With $T = 50°C$, $T_r\,(N_2) = 2.56$, and $T_r\,(CHClF_2) = 0.875$,

$$T_{r12} = \frac{50 + 273.2}{[(126.2)(369.3)]^{1/2}} = 1.497$$

$\mu_r\,(N_2) = 0$, and from Eq. (9-4.17),

$$\mu_r\,(CHClF_2) = \frac{(52.46)(1.4)^2(49.86)}{(369.28)^2} = 0.0376$$

Since $\mu_{r12} = (\mu_{r1}\,\mu_{r2})^{1/2}$, then for this mixture, $\mu_{r12} = 0$. With Eq. (9-5.11), for CHC1F$_2$,

$$U\,(CHClF_2) = \frac{[1 + (0.36)(0.875)(0.875 - 1)]^{1/6}}{(0.875)^{1/2}} \times \frac{(0.875)^{3.5} + (10)^7\,(0.0376)^7}{(0.875)^{3.5}[1 + (10)^7(0.0376)^7]}$$

$$= 1.063$$

and $U\,(N_2) = 0.725$

Then, $C\,(N_2) = \dfrac{(28.014)^{1/4}}{[(188)(0.725)]^{1/2}} = 0.197$

and $C\,(CHClF_2) = 0.256$
Next,

$$H(N_2\text{—}CHClF_2) = (0.197 + 0.256)^2\,\frac{[(28.014)(86.468)]^{1/2}}{[32(28.014 + 86.468)^3]^{1/2}}$$

$$\times\,\frac{[1 + (0.36)(1.497)(1.497 - 1)]^{1/6}}{(1.497)^{1/2}} \times 1.0$$

$$= 1.237 \times 10^{-3}$$

$$K\,(N_2) = \frac{(0.286)(188)}{0.286 + (188)(0.714)(1.237 \times 10^{-3})\{3 + [(2)(86.469)/28.014]\}} = 29.71$$

and $K\,(CHClF_2) = 107.9$. Substituting into Eq. (9-5.8),

$$\eta_m = (29.71)[1 + (1.237 \times 10^{-3})^2(107.9)^2] + (107.9)[1 + (2)(1.237 \times 10^{-3})(29.71) + (1.237 \times 10^{-3})^2(29.71)^2] = 146.2\ \mu\text{P}$$

$$\text{Error} = \frac{146.2 - 145}{145} \times 100 = 0.8\%$$

Method of Wilke (1950)

In a further simplification of the kinetic theory approach, Wilke (1950) neglected second-order effects and proposed:

$$\eta_m = \sum_{i=1}^{n} \frac{y_i \eta_i}{\sum_{j=1}^{n} y_j \phi_{ij}} \tag{9-5.13}$$

where

$$\phi_{ij} = \frac{[1 + (\eta_i/\eta_j)^{1/2}(M_j/M_i)^{1/4}]^2}{[8(1 + M_i/M_j)]^{1/2}} \tag{9-5.14}$$

ϕ_{ji} is found by interchanging subscripts or by

$$\phi_{ji} = \frac{\eta_j}{\eta_i}\frac{M_i}{M_j}\phi_{ij} \tag{9-5.15}$$

For a binary system of 1 and 2, with Eqs. (9-5.13) to (9-5.15),

$$\eta_m = \frac{y_1 \eta_1}{y_1 + y_2\phi_{12}} + \frac{y_2 \eta_2}{y_2 + y_1\phi_{21}} \tag{9-5.16}$$

where η_m = viscosity of the mixture
$\eta_1, \eta_2,$ = pure component viscosities
y_1, y_2 = mole fractions

and

$$\phi_{12} = \frac{[1 + (\eta_1/\eta_2)^{1/2}(M_2/M_1)^{1/4}]^2}{\{8[1 + (M_1/M_2)]\}^{1/2}}$$

$$\phi_{21} = \phi_{12}\frac{\eta_2 M_1}{\eta_1 M_2}$$

Equation (9-5.13), with ϕ_{ij} from Eq. (9-5.14), has been extensively tested. Wilke (1950) compared values with data on 17 binary systems and reported an average deviation of less than 1%; several cases in which η_m passed through a maximum were included. Many other investigators have tested this method (Amdur and Mason, 1958; Bromley and Wilke, 1951; Cheung, 1958; Dahler, 1959; Gandhi and Saxena, 1964; Ranz and Brodowsky, 1962; Saxena and Gambhir, 1963, 1963a;

Strunk, et al., 1964; Vanderslice, et al. 1962; Wright and Gray, 1962). In most cases, only nonpolar mixtures were compared, and very good results obtained. For some systems containing hydrogen as one component, less satisfactory agreement was noted. In Table 9-4, Wilke's method predicted mixture viscosities that were larger than experimental for the H_2—N_2 system, but for H_2—NH_3, it underestimated the viscosities. Gururaja, et al. (1967) found that this method also overpredicted in the H_2—O_2 case but was quite accurate for the H_2—CO_2 system. Wilke's approximation has proved reliable even for polar-polar gas mixtures of aliphatic alcohols (Reid and Belenyessy, 1960). The principal reservation appears to lie in those cases where $M_i >> M_j$ and $\eta_i >> \eta_j$.

Example 9-5 Kestin and Yata (1968) report that the viscosity of a mixture of methane and *n*-butane is 93.35 μP at 293 K when the mole fraction of *n*-butane is 0.303. Compare this result with the value estimated by Wilke's method. For pure methane and *n*-butane, these same authors report viscosities of 109.4 and 72.74 μP.

solution Let 1 refer to methane and 2 to *n*-butane. $M_1 = 16.043$ and $M_2 = 58.123$.

$$\phi_{12} = \frac{[1 + (109.4/72.74)^{1/2}(58.123/16.043)^{1/4}]^2}{\{8[1 + (16.043/58.123)]\}^{1/2}} = 2.268$$

$$\phi_{21} = 2.268 \frac{72.74}{109.4} \frac{16.043}{58.123} = 0.416$$

$$\eta_m = \frac{(0.697)(109.4)}{0.697 + (0.303)(2.268)} + \frac{(0.303)(72.74)}{0.303 + (0.697)(0.416)}$$

$$= 92.26\ \mu\text{P}$$

$$\text{Error} = \frac{92.26 - 93.35}{93.35} \times 100 = -1.2\%$$

Herning and Zipperer (1936) Approximation of ϕ_{ij}

As an approximate expression for ϕ_{ij} of Eq. (9-5.14) the following is proposed (Herning and Zipperer, 1936):

$$\phi_{ij} = \left(\frac{M_j}{M_i}\right)^{1/2} = \phi_{ji}^{-1} \tag{9-5.17}$$

When Eq. (9-5.17) is used with Eq. (9-5.16) to estimate low-pressure binary gas mixture viscosities, quite reasonable predictions are obtained (Table 9-4) except for systems such as H_2—NH_3. The technique is illustrated in Example 9-6. Note that Examples 9-5 and 9-6 treat the same problem; each provides a viscosity estimate close to the experimental value. But the ϕ_{12} and ϕ_{21} values employed in the two cases are quite different. Apparently, multiple sets of ϕ_{ij} and ϕ_{ji} work satisfactorily in Eq. (9-5.13).

Example 9-6 Repeat Example 9-5 by using the Herning and Zipperer approximation for ϕ_{ij}.

solution As before, with 1 as methane and 2 as *n*-butane

$$\phi_{12} = \left(\frac{58.123}{16.043}\right)^{1/2} = 1.903$$

$$\phi_{21} = \phi_{12}^{-1} = 0.525$$

$$\eta_m = \frac{(0.697)(109.4)}{0.697 + (0.303)(1.903)} + \frac{(0.303)(72.74)}{0.303 + (0.697)(0.525)} = 92.82\ \mu\text{P}$$

$$\text{Error} = \frac{92.82 - 93.35}{93.35} \times 100 = -0.6\%$$

Corresponding States Methods

In this approach, one estimates pseudocritical and other mixture properties (See Sec. 5-3) from pure component properties, the composition of the mixture, and appropriate combining and mixing rules.

Lucas (1980, 1983, 1984a) Rules

Lucas (1980, 1983, 1984a) defined mixture properties as shown below for use in Eqs. (9-4.15) through (9-4.19).

$$T_{cm} = \sum_i y_i T_{ci} \tag{9-5.18}$$

$$P_{cm} = RT_{cm} \frac{\sum_i y_i Z_{ci}}{\sum_i y_i V_{ci}} \tag{9-5.19}$$

$$M_m = \sum_i y_i M_i \tag{9-5.20}$$

$$F^o_{Pm} = \sum_i y_i F^o_{Pi} \tag{9-5.21}$$

$$F^o_{Qm} = \left(\sum_i y_i F^o_{Qi}\right) A \tag{9-5.22}$$

and, letting the subscript H denote the mixture component of highest molecular weight and L the component of lowest molecular weight,

$$A = 1 - 0.01 \left(\frac{M_H}{M_L}\right)^{0.87} \quad \text{for } \frac{M_H}{M_L} > 9 \text{ and } 0.05 < y_H < 0.7; \tag{9-5.23}$$

otherwise, $A = 1$

The method of Lucas does not necessarily lead to the pure component viscosity η_i when all $y_j = 0$ except $y_i = 1$. Thus the method is not interpolative in the same way as are the techniques of Reichenberg, Wilke, and Herning and Zipperer. Nevertheless, as seen in Table 9-4, the method provides reasonable estimates of η_m in most test cases.

Example 9-7 Estimate the viscosity of a binary mixture of ammonia and hydrogen at 33°C and low pressure by using the Lucas corresponding states method.

solution Let us illustrate the method for a mixture containing 67.7 mole percent ammonia. We use the following pure-component values:

	Ammonia	Hydrogen
T_c, K	405.50	33.2
P_c, bar	113.5	13.0
V_c, cm³/mol	72.5	64.3
Z_c	0.244	0.306
M	17.031	2.016
μ, debyes	1.47	0
T_r	0.755	9.223

Using Eqs. (9-5.18) to (9-5.20), $T_{cm} = 285.2$ K, $P_{cm} = 89.6$ bar, and $M_m = 12.18$. From these values and Eq. (9-4.15), $\xi_m = 6.46 \times 10^{-3}$ $(\mu\text{P})^{-1}$. With Eq. (9-4.17), μ_r (NH_3) $= 7.825 \times 10^{-2}$ and μ_r (H_2) $= 0$. Then, with Eq. (9-4.18),

$$F_P^o\,(NH_3) = 1 + 30.55(0.292 - 0.244)^{1.72}\,|0.96 + 0.1(0.755 - 0.7)| = 1.159$$

$$F_P^o\,(H_2) = 1.0$$

$$F_{Pm}^o = (1.159)(0.677) + (1)(0.323) = 1.107$$

For the quantum correction, with Eq. (9-5.23), since $M_H/M_L = 17.031/2.016 = 8.4 < 9$, then $A = 1$. F_Q^o (NH_3) $= 1.0$, and with Eq. (9-4.19),

$$F_Q^o\,(H_2) = (1.22)(0.76)^{0.15}\{1 + 0.00385[(9.209 - 12)^2]^{1/2.016}$$
$$\times \text{sign}\,(9.209 - 12)\}$$
$$= (1.171)[1 + (0.01061)(-1)] = 1.158$$

$$F_{Qm}^o = (1.158)(0.323) + (1)(0.677) = 1.051$$

Next, from Eq. (9-4.16) with $T_{rm} = (33 + 273.2)/285.3 = 1.073$

$$\eta_m \xi_m = (0.645)(1.107)(1.051) = 0.750$$

$$\eta_m = \frac{0.750}{6.46 \times 10^{-3}} = 116.1\ \mu\text{P}$$

The experimental value is 120.0 μP; thus

$$\text{Error} = \frac{116.1 - 120.0}{120.0} \times 100 = -3.2\%$$

The viscosity of the ammonia-hydrogen mixture at 33°C is line 3 in Fig. 9-3.

Chung, et al. rules (1984, 1988)

In this case, Eq. (9-3.9) is employed to estimate the mixture viscosity with, however, a factor F_{cm} as used in Eq. (9-4.10) to correct for shape and polarity.

$$\eta_m = \frac{26.69\, F_{cm}(M_m T)^{1/2}}{\sigma_m^2 \Omega_v} \tag{9-5.24}$$

where $\Omega_v = f(T_m^*)$. In the Chung, et al. approach, the mixing rules are:

$$\sigma_m^3 = \sum_i \sum_j y_i y_j \sigma_{ij}^3 \tag{9-5.25}$$

$$T_m^* = \frac{T}{(\varepsilon/k)_m} \tag{9-5.26}$$

$$\left(\frac{\varepsilon}{k}\right)_m = \frac{\sum_i \sum_j y_i y_j (\varepsilon_{ij}/k)\sigma_{ij}^3}{\sigma_m^3} \tag{9-5.27}$$

$$M_m = \left[\frac{\sum_i \sum_j y_i y_j (\varepsilon_{ij}/k)\sigma_{ij}^2 M_{ij}^{1/2}}{(\varepsilon/k)_m \sigma_m^2}\right]^2 \tag{9-5.28}$$

$$\omega_m = \frac{\sum_i \sum_j y_i y_j \omega_{ij} \sigma_{ij}^3}{\sigma_m^3} \tag{9-5.29}$$

$$\mu_m^4 = \sigma_m^3 \sum_i \sum_j \left(\frac{y_1 y_j \mu_i^2 \mu_j^2}{\sigma_{ij}^3}\right) \tag{9-5.30}$$

$$\kappa_m = \sum_i \sum_j y_i y_j \kappa_{ij} \tag{9-5.31}$$

and the combining rules are:

$$\sigma_{ii} = \sigma_i = 0.809 V_{ci}^{1/3} \tag{9-5.32}$$

$$\sigma_{ij} = \xi_{ij}(\sigma_i \sigma_j)^{1/2} \tag{9-5.33}$$

$$\frac{\varepsilon_{ii}}{k} = \frac{\varepsilon_i}{k} = \frac{T_{ci}}{1.2593} \tag{9-5.34}$$

$$\frac{\varepsilon_{ij}}{k} = \zeta_{ij}\left(\frac{\varepsilon_i}{k}\frac{\varepsilon_j}{k}\right)^{1/2} \tag{9-5.35}$$

$$\omega_{ii} = \omega_i \tag{9-5.36}$$

$$\omega_{ij} = \frac{\omega_i + \omega_j}{2} \tag{9-5.36}$$

$$\kappa_{ii} = \kappa_i \tag{9-5.37}$$

$$\kappa_{ij} = (\kappa_i \kappa_j)^{1/2} \tag{9-5.38}$$

$$M_{ij} = \frac{2M_i M_j}{M_i + M_j} \tag{9-5.39}$$

(9-5.40)

ξ_{ij} and ζ_{ij} are binary interaction parameters which are normally set equal to unity. The F_{cm} term in Eq. (9-5.24) is defined as in Eq. (9-4.11).

$$F_{cm} = 1 - 0.275\omega_m + 0.059035\ \mu_{rm}^4 + \kappa_m \tag{9-5.41}$$

where μ_{rm} is as in Eq. (9-4.12)

$$\mu_{rm} = \frac{131.3\ \mu_m}{(V_{cm}\ T_{cm})^{1/2}} \tag{9-5.42}$$

$$V_{cm} = (\sigma_m/0.809)^3 \tag{9-5.43}$$

$$T_{cm} = 1.2593 \left(\frac{\varepsilon}{k}\right)_m \tag{9-5.44}$$

In these equations, T_c is in kelvins, V_c is in cm^3/mol and μ is in debyes.

The rules suggested by Chung, et al. are illustrated for a binary gas mixture in Example 9-8. As with the Lucas approach, the technique is not interpolative between pure component viscosities. Some calculated binary gas mixture viscosities are compared with experimental values in Table 9-4. Errors vary, but they are usually less than about ±5%.

Example 9-8 Use the Chung, et al. method to estimate the low-pressure gas viscosity of a binary of hydrogen sulfide and ethyl ether containing 20.4 mole percent H_2S. The temperature is 331 K.

solution The properties listed below are from Appendix A, the problem statement, and Table 9-1:

	Hydrogen sulfide	Ethyl ether
T_c, K	373.4	466.70
V_c, cm^3/mol	98	280
ω	0.090	0.281
μ, debyes	0.9	1.3
κ	0	0
M, g/mol	34.082	74.123
y	0.204	0.796

From Eqs. (9-5.32) and (9-5.33),

$$\sigma(\mathrm{H_2S}) = (0.809)(98)^{1/3} = 3.730\ \text{Å}$$

$$\sigma(\mathrm{EE}) = 5.293\ \text{Å}$$

$$\sigma(\mathrm{H_2S\text{-}EE}) = 4.443\ \text{Å}$$

Then, with Eq. (9-5.25),

$$\sigma_m^3 = (0.204)^2(3.730)^3 + (0.796)^2(5.293)^3 + (2)(0.204)(0.796)(4.443)^3 = 124.58\ \text{Å}^3$$

From Eqs. (9-5.34) and (9-5.35)

$$\frac{\varepsilon}{k} = (\mathrm{H_2S}) = \frac{373.4}{1.2593} = 296.5\ \mathrm{K}$$

$$\frac{\varepsilon}{k}(\mathrm{EE}) = 370.6\ \mathrm{K}$$

$$\frac{\varepsilon}{k}(\mathrm{H_2S—EE}) = 331.5\ \mathrm{K}$$

Then, with Eq. (9-5.27),

$$\left(\frac{\varepsilon}{k}\right)_m = [(0.204)^2(296.5)(3.730)^3 + (0.796)^2(370.6)(5.293)^3 + (2)(0.204)(0.796)(331.5)(4.443)^3]/124.58 = 360.4\ \mathrm{K}$$

With Eqs. (9-5.28) and (9-5.40),

$$M_m = (\{0.204)^2(296.5)(3.730)^2(34.082)^{1/2} + (0.796)^2(370.6)(5.293)^2(74.123)^{1/2} + (2)(0.204)(0.796)(331.5)(4.443)^2[(2)(34.082)(74.123)/(34.082 + 74.123)]^{1/2}\}/(360.4)(124.58)^{2/3})^2 = 64.44\ \mathrm{g/mol}$$

With Eq. (9-5.29),

$$\omega_m = \{(0.204)^2(0.090)(3.730)^3 + (0.796)^2(0.281)(5.293)^3 + (2)(0.204)(0.796)[(0.090 + 0.281)/2](4.443)^3\}/124.58 = 0.256$$

and with Eq. (9-5.30),

$$\mu_m^4 = \{[(0.204)^2(0.9)^4/(3.730)^3] + [(0.796)^2(1.3)^4/(5.293)^3] + [(2)(0.204)(0.796)(0.9)^2(1.3)^2/(4.443)^3]\}(124.58) = 2.218$$

$$\mu_m = 1.22\ \text{debyes}$$

so, with Eqs. (9-5.42) to (9-5.44),

$$V_{cm} = \frac{(124.58)}{(0.809)^3} = 235.3\ \mathrm{cm^3/mol}$$

$$T_{cm} = (1.2593)(360.4) = 453.9\ \mathrm{K}$$

$$\mu_{rm} = \frac{(131.3)(1.22)}{[(235.3)(453.9)]^{1/2}} = 0.490$$

Since κ_m, = 0, with Eq. (9-5.41),

$$F_{cm} = 1 - (0.275)(0.256) + (0.059035)(0.490)^4 = 0.933$$

Using T_m^* from Eq. (9-5.26) [= 331/360.4 = 0.918] and Eq. (9-4.3), $\Omega_v = 1.664$. Finally, with Eq. (9-5.24),

$$\eta_m = \frac{(26.69)(0.933)[(64.44)(331)]^{1/2}}{(124.58)^{2/3}(1.664)} = 87.6\ \mu\text{P}$$

The experimental value is 87 μP (Table 9-4).

$$\text{Error} = \frac{87.6 - 87}{87} \times 100 = 0.4\%$$

Discussion and Recommendations to Estimate the Low-pressure Viscosity of Gas Mixtures

As is obvious from the estimation methods discussed in this section, the viscosity of a gas mixture can be a complex function of composition. This is evident from Fig. 9-3. There may be a maximum in mixture viscosity in some cases, e.g., system 3, ammonia-hydrogen. However, cases of a viscosity minimum have been reported.

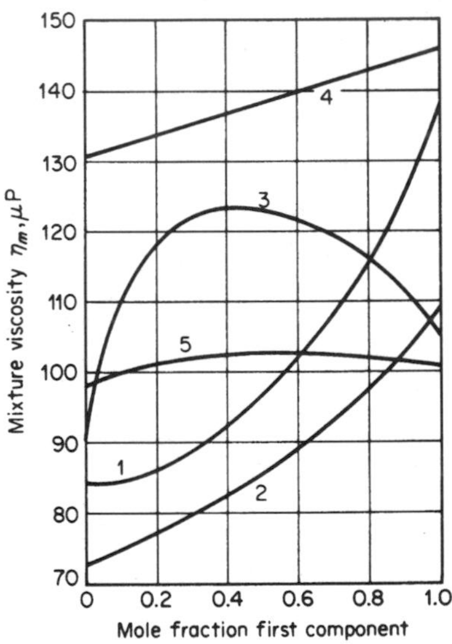

FIGURE 9-3 Gas mixture viscosities.

No.	System	Reference
1	Hydrogen sulfide-ethyl ether	Pal and Bhattacharyya (1969)
2	Methane-*n*-butane	Kestin and Yata (1968)
3	Ammonia-hydrogen	Pal and Barua (1967)
4	Ammonia-methyl amine	Burch and Raw (1967)
5	Ethylene-ammonia	Trautz and Heberling (1931)

Behavior similar to that of the ammonia-hydrogen case occurs most often in polar-nonpolar mixtures in which the pure component viscosities are not greatly different (Hirschfelder, et al., 1960; Rutherford, et al., 1960). Maxima are more pronounced as the molecular weight ratio differs from unity.

Of the five estimation methods described in this section, three (Herning and Zipperer, Wilke, and Reichenberg) use the kinetic theory approach and yield interpolative equations between the pure component viscosities. Reichenberg's method is most consistently accurate, but it is the most complex. To use Reichenberg's procedure, one needs, in addition to temperature and composition, the viscosity, critical temperature, critical pressure, molecular weight, and dipole moment of each constituent. Wilke's and Herning and Zipperer's methods require only the pure component viscosities and molecular weights; these latter two yield reasonably accurate predictions of the mixture viscosity.

Arguing that it is rare to have available the pure gas viscosities at the temperature of interest, both Lucas and Chung, et al. provide estimation methods to cover the entire range of composition. At the end points where only pure components exist, their methods reduce to those described earlier in Sec. 9-3. Although the errors from these two methods are, on the average, slightly higher than those of the interpolative techniques, they are usually less than $\pm 5\%$ as seen from Table 9-4. Such errors could be reduced even further if pure component viscosity data were available and were employed in a simple linear correction scheme. For example, if the pure component viscosity predictions are too high, the mixture prediction would be improved if it were lowered by composition-averaged error of the pure component predictions.

An estimation method recently proposed by Davidson (1993) was judged as effective as those discussed in this section.

It is recommended that Reichenberg's method [Eq. (9-5.8)] be used to calculate η_m if pure component viscosity values are available. Otherwise, either the Lucas method [Eq. (9-4.16)] or the Chung, et al. method [Eq. (9-5.24)] can be employed if critical properties are available for all components.

9-6 EFFECT OF PRESSURE ON THE VISCOSITY OF PURE GASES

Figure 9-4 shows the viscosity of carbon dioxide (T_c = 304.1 K and P_c = 73.8 bar) as a function of temperature and pressure. In some ranges ($T_r > 1.5$ and $P_r < 2$), pressure has little effect on viscosity. But when $1 < T_r < 1.5$ and when $P > P_c$, pressure has a strong effect on viscosity as can be seen by the nearly vertical isobars in this region of Fig. 9-4. Figure 9-4 shows isobars as a function of temperature, while Fig. 9-5 shows isotherms as a function of pressure for nitrogen (T_c = 77.4 K, P_c = 33.9 bar). Lucas (1981, 1983) has generalized the viscosity phase diagrams (for nonpolar gases) as shown in Fig. 9-6. In this case, the ordinate is $\eta\xi$ and the temperatures and pressures are reduced values. ξ is the inverse reduced viscosity defined earlier in Eq. (9-4.15).

At the critical point, the viscosity diverges so that its value is larger than would otherwise be expected. However, this effect is much smaller for viscosity than for thermal conductivity (see Fig. 10-5). Whereas the thermal conductivity can increase by a factor of two near the critical point, the increase in viscosity is on the order

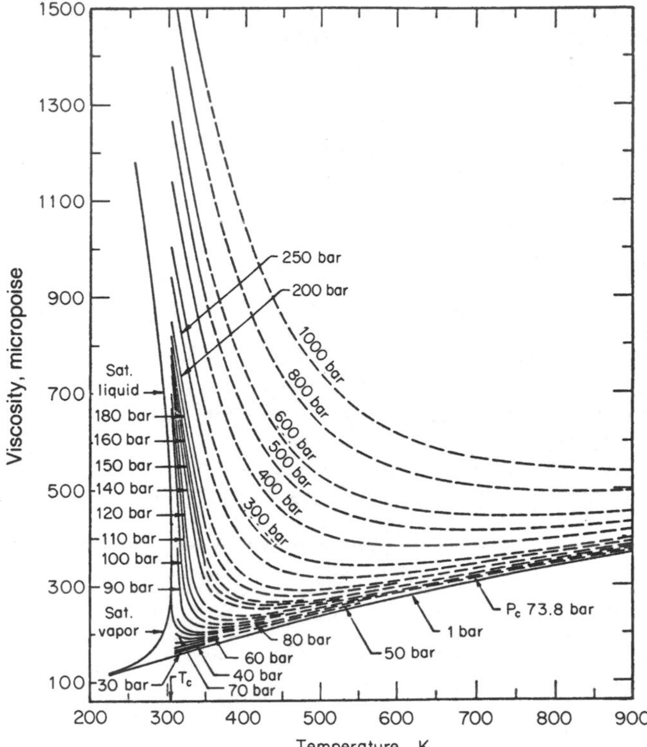

FIGURE 9-4 Viscosity of carbon dioxide. (*Stephan and Lucas, 1979*)

of 1%. In fact Vesovic, et al. (1990) state that for carbon dioxide, "the viscosity enhancement is less than 1% at densities and temperatures outside the range bounded approximately by 300K $< T <$ 310K and 300 kg m^{-3} $< \rho <$ 600 kg m^{-3}."

In Fig. 9-6, the lower limit of the P_r curves would be indicative of the dilute-gas state, as described in Sec. 9-4. In such a state, η increases with temperature. At high reduced pressures, we see there is a wide range of temperatures where η decreases with temperature. In this region the viscosity behavior more closely simulates a liquid state, and, as will be shown in Sec. 9-10, an increase in temperature results in a decrease in viscosity. Finally, at very high-reduced temperatures, a condition again results in which pressure has little effect and viscosities increase with temperature.

The temperature–pressure region in Fig. 9-4 where viscosity changes rapidly with pressure is the very region where density also changes rapidly with pressure. Figure 9-7 shows a plot of the residual viscosity as a function of density for n-butane. A smooth curve results even though values over a range of temperatures are shown. This suggests that density is an important variable when describing viscosity behavior at high pressures, and several of the correlations presented in this section take advantage of this importance.

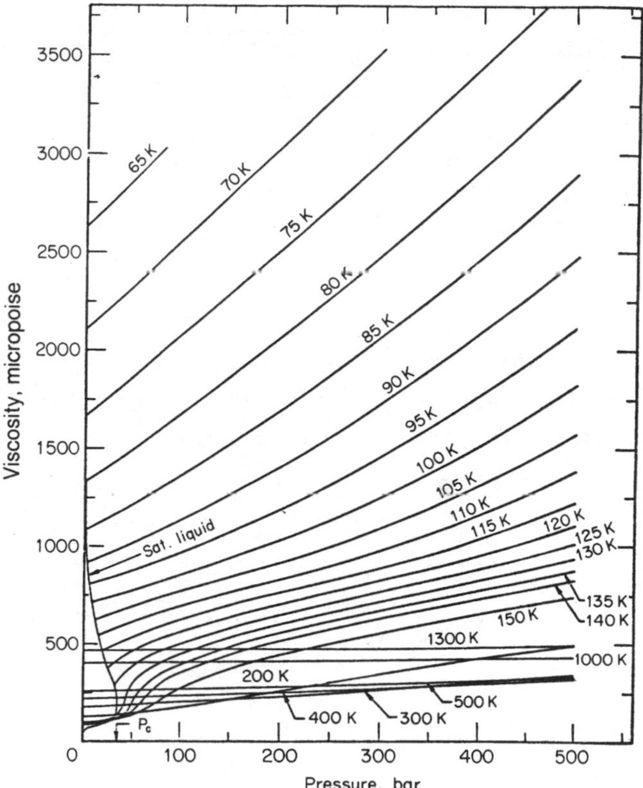

FIGURE 9-5 Viscosity of nitrogen. (*Stephan and Lucas, 1979*)

Enskog Dense-gas Theory

One of the very few theoretical efforts to predict the effect of pressure on the viscosity of gases is due to Enskog and is treated in detail by Chapman and Cowling (1939). The theory has also been applied to dense gas diffusion coefficients, bulk viscosities, and, for monatomic gases, thermal conductivities. The assumption is made that the gas consists of dense, hard spheres and behaves like a low-density hard-sphere system except that all events occur at a faster rate due to the higher rates of collision (Alder, 1966; Alder and Dymond, 1966) The increase in collision rate is proportional to the radial distribution function Ψ. The Enskog equation for shear viscosity is

$$\frac{\eta}{\eta^o} = \Psi^{-1} + 0.8\ b_o\rho + 0.761\ \Psi(b_o\rho)^2 \tag{9-6.1}$$

where η = viscosity, μP
η^o = low-pressure viscosity, μP
b_o = excluded volume = $2/3\ \pi N_o\sigma^3$, cm^3/mol

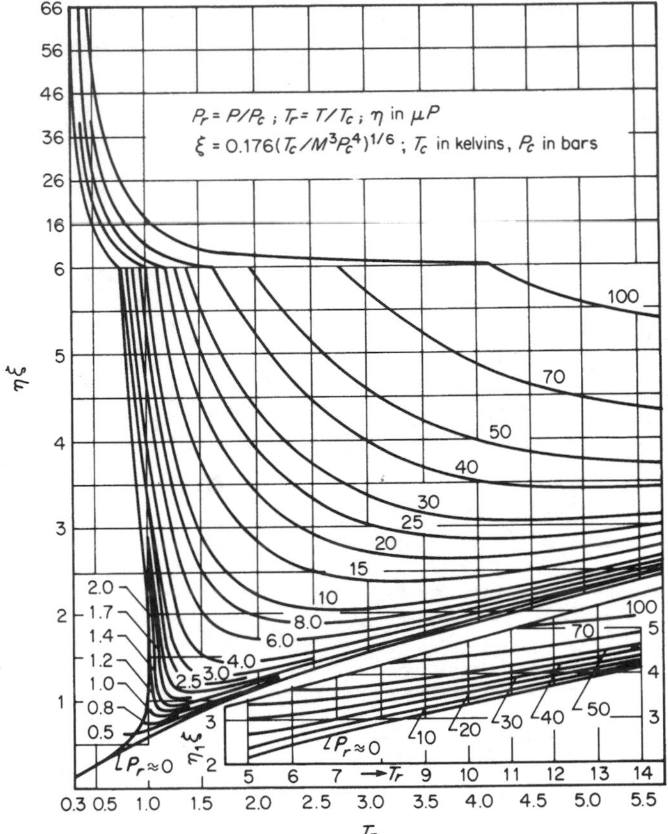

FIGURE 9-6 Generalized phase diagram for gas viscosity. (*Lucas, 1981, 1983*)

N_o = Avogadro's number
σ = hard-sphere diameter, Å
ρ = molar density, mole/cm^3

Ψ is the radial distribution function at contact and can be related to an equation of state by

$$\Psi = \frac{Z - 1}{\rho b_o} \tag{9-6.2}$$

where Z is the compressibility factor.

Dymond among others (Assael, et al., 1996; Dymond and Assael, 1996) has continued efforts to modify the hard sphere approach in order to predict transport properties and has shown that viscosities of dense fluids can be correlated by the universal equation

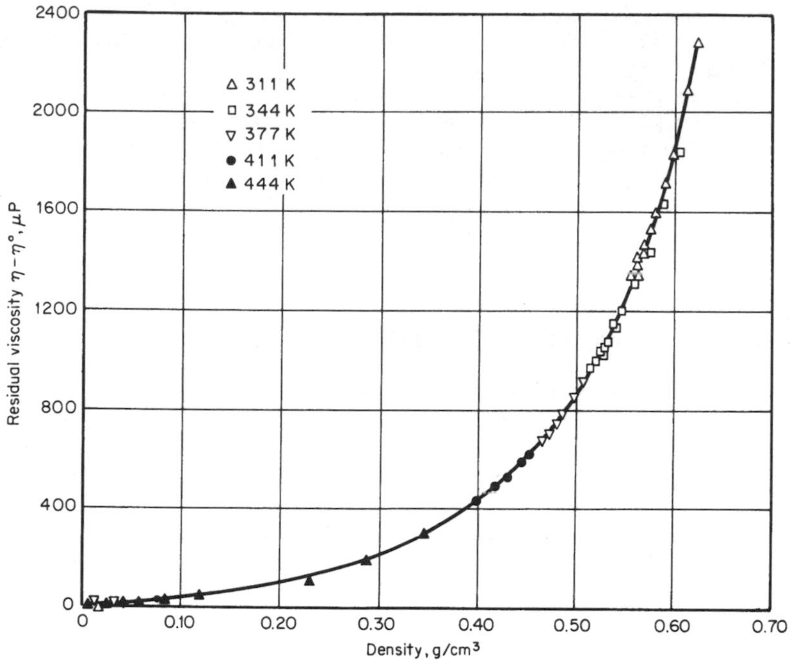

FIGURE 9-7 Residual *n*-butane viscosity as a function of density. (*Dolan et al., 1963*)

$$\log_{10} \eta_r = 1.0945 - \frac{9.26324}{V_r} + \frac{71.038}{V_r^2} - \frac{301.9012}{V_r^3} + \frac{797.69}{V_r^4} - \frac{1221.977}{V_r^5} + \frac{987.5574}{V_r^6} - \frac{319.4636}{V_r^7} \tag{9-6.3}$$

where $V_r = V/V_o$ and V_o is a close-packed volume. η_r is a reduced viscosity defined by

$$\eta_r = 6.619 \times 10^5 \frac{\eta V^{2/3}}{R_\eta (MT)^{1/2}} \tag{9-6.4}$$

where η is in Pa·s, V is in cm^3/mol, M is g/mol and T is in K. R_η is a parameter that accounts for deviations from smooth hard spheres. The two parameters, V_o and R_η are compound specific and are not functions of density. V_o is a function of temperature as is R_η for *n*-alcohols (Assael, et al., 1994). For *n*-alkanes (Assael, et al., 1992a, 1992b), aromatic hydrocarbons (Assael, et al., 1992c), refrigerants (Assael, et al., 1995), and a number of other compounds (Assael, et al. 1992; Bleazard and Teja, 1996), R_η has been found to be independent of temperature. In theory, the two parameters V_o and R_η could be set with two experimental viscosity-density data, but in practice, Eq. (9-6.3) has been used only for systems for which extensive data are available. It has been applied to densities above the critical density and applicability to temperatures down to $T_r \approx 0.6$ has been claimed. Values of V_o and

R_η at 298 K and 350 K (as well as R_λ which is discussed in Chap. 10) for 16 fluids have been calculated with equations given in the above references and are shown in Table 9-5.

Xiang, et al. (1999) have recently extended Eq. (9-6.1) to cover the entire fluid range by introducing a crossover function between the low pressure limit and the high pressure limit. With a single equation, they fit low pressure viscosities to within 4% and liquid and high pressure viscosities generally to within 10% for 18 pure fluids. Their equation requires the density, critical properties, the acentric factor and values for σ and ε/k.

Reichenberg Method (1971, 1975, 1979)

In this case, the viscosity ratio η/η^o is given by Eq. (9-6.5)

$$\frac{\eta}{\eta^o} = 1 + Q \frac{AP_r^{3/2}}{BP_r + (1 + CP_r^D)^{-1}} \tag{9-6.5}$$

The constants, A, B, C, and D are functions of the reduced temperature T_r as shown below, and η^o is the viscosity of the gas at the same T and low pressure.

$$A = \frac{\alpha_1}{T_r} \exp \alpha_2 T_r^a \qquad B = A(\beta_1 T_r - \beta_2)$$

$$C = \frac{\gamma_1}{T_r} \exp \gamma_2 T_r^c \qquad D = \frac{\delta_1}{T_r} \exp \delta_2 T_r^d$$

TABLE 9-5 Typical values of V_o, R_η, and R_λ.

	V_o, cm³/mol		R_η		R_λ	
	298 K	350 K	298 K	350 K	298 K	350 K
Methane	17.9	17.3	1.00	1.00	1.16	1.16
n-Butane	84.1	84.3	1.08	1.08	1.67	1.67
n-Decane	134.3	130.1	1.53	1.53	3.13	3.13
Cyclohexane	77.4	75.7	0.93	0.93	1.35	1.35
CCl_4	65.4	63.8	1.07	1.07	1.57	1.57
R134a	43.7	41.7	1.10	1.10	1.61	1.82
Ethanol	33.8	32.7	4.62	2.80	1.42	1.42
n-Hexanol	91.9	84.4	3.00	3.00	1.95	1.95
Acetic acid	41.3	41.2	0.76	0.76	0.92	0.98
Butyl ethanoate	89.8	88.3	1.10	1.10	2.20	2.25
2-Ethoxyethanol	70.7	68.0	1.33	1.33	1.65	1.78
1,3-Propanediol	62.2	59.1	1.04	1.04	1.16	1.32
Diethylene glycol	78.1	73.6	1.83	1.83	1.65	1.94
Diethanolamine	87.9	80.6	2.38	2.38	1.37	1.72
Triethylamine	73.8	66.4	2.48	2.48	3.03	3.47
Dimethyl disulfide	58.3	57.0	1.12	1.12	1.79	1.79

$\alpha_1 = 1.9824 \times 10^{-3}$ $\alpha_2 = 5.2683$ $a = -0.5767$ $\beta_1 = 1.6552$

$\beta_2 = 1.2760$ $\gamma_1 = 0.1319$ $\gamma_2 = 3.7035$ $c = -79.8678$

$\delta_1 = 2.9496$ $\delta_2 = 2.9190$ $d = -16.6169$

and $Q = (1 - 5.655\ \mu_r)$, where μ_r is defined in Eq. (9-4.17). For nonpolar materials, $Q = 1.0$. Example 9-9 illustrates the application of Eq. (9-6.5), and, in Table 9-7, experimental dense gas viscosities are compared to the viscosities estimated with this method. Errors are generally only a few percent; the poor results for ammonia at 420 K seem to be an anomaly.

Example 9-9 Use Reichenberg's method to estimate the viscosity of *n*-pentane vapor at 500 K and 101 bar. The experimental value is 546 μP (Stephan and Lucas, 1979).

solution Whereas one could estimate the low-pressure viscosity of *n*-pentane at 500 K by using the methods described in Sec. 9-4, the experimental value is available (114 μP) (Stephan and Lucas, 1979) and will be used. The dipole moment of *n*-pentane is zero, so $Q = 1.0$. From Appendix A, $T_c = 469.7$ K and $P_c = 33.7$ bar. Thus $T_r = (500/469.7) = 1.065$ and $P_r = (101/33.7) = 3.00$. From the definitions of A, B, C, and D given under Eq. (9-6.5), $A = 0.2999$, $B = 0.1458$, $C = 1.271$, and $D = 7.785$. With Eq. (9-6.5),

$$\frac{\eta}{\eta^*} = 1 = + \frac{(0.2999)(3.00)^{3/2}}{(0.1458)(3.00) + [1 + (1.271)(3.00)^{7.785}]^{-1}} = 4.56$$

$$\eta = (4.56)(114) = 520\ \mu\text{P}$$

$$\text{Error} = \frac{520 - 546}{546} \times 100 = -4.7\%$$

If one refers back to Fig. 9-6, at $T_r = 1.065$ and $P_r = 3.00$, the viscosity is changing rapidly with both temperature and pressure. Thus, an error of only 5% is quite remarkable.

Lucas (1980, 1981, 1983) Method

In a technique which, in some aspects, is similar to Reichenberg's, Lucas (1980, 1981, 1983) recommends the following procedure. For the reduced temperature of interest, first calculate a parameter Z_1 with Eq. (9-6.6).

$$Z_1 = \eta^o \xi = [0.807 T_r^{0.618} - 0.357 \exp(-0.449\ T_r) + 0.340 \exp(-4.058\ T_r) + 0.018] F_P^o F_Q^o \quad (9\text{-}6.6)$$

where η^o refers to the low-pressure viscosity. Next calculate Z_2. If $T_r \leq 1.0$ and $P_r < (P_{vp}/P_c)$, then

$$Z_2 = 0.600 + 0.760\ P_r^\alpha + (6.990\ P_r^\beta - 0.6)(1 - T_r) \quad (9\text{-}6.7)$$

with $\alpha = 3.262 + 14.98\ P_r^{5.508}$
$\beta = 1.390 + 5.746\ P_r$

If $(1 < T_r < 40)$ and $(0 < P_r \leq 100)$, then

$$Z_2 = \eta^o \xi \left[1 + \frac{aP_r^e}{bP_r^f + (1 + cP_r^d)^{-1}} \right] \tag{9-6.8}$$

where $\eta^o \xi$ is found from Eq. (9-6.6). The term multiplying this group is identical to the pressure correction term in Reichenberg's method, Eq. (9-6.5), but the values of the constants are different.

$$a = \frac{a_1}{T_r} \exp a_2 T_r^\gamma$$

$$b = a(b_1 T_r - b_2)$$

$$c = \frac{c_1}{T_r} \exp c_2 T_r^\delta$$

$$d = \frac{d_1}{T_r} \exp d_2 T_r^\varepsilon$$

$$e = 1.3088$$

$$f = f_1 \exp f_2 T_r^\zeta$$

and

$a_1 = 1.245 \times 10^{-3}$	$a_2 = 5.1726$	$\gamma = -0.3286$
$b_1 = 1.6553$	$b_2 = 1.2723$	
$c_1 = 0.4489$	$c_2 = 3.0578$	$\delta = -37.7332$
$d_1 = 1.7368$	$d_2 = 2.2310$	$\varepsilon = -7.6351$
$f_1 = 0.9425$	$f_2 = -0.1853$	$\zeta = 0.4489$

After computing Z_1 and Z_2, we define

$$Y = \frac{Z_2}{Z_1} \tag{9-6.9}$$

and the correction factors F_P and F_Q,

$$F_P = \frac{1 + (F_P^o - 1)Y^{-3}}{F_P^o} \tag{9-6.10}$$

$$F_Q = \frac{1 + (F_Q^o - 1)[Y^{-1} - (0.007)(\ln Y)^4]}{F_Q^o} \tag{9-6.11}$$

where F_P^o and F_Q^o are low-pressure polarity and quantum factors determined as shown in Eqs. (9-4.18) and (9-4.19). Finally, the dense gas viscosity is calculated as

$$\eta = \frac{Z_2 F_P F_Q}{\xi} \tag{9-6.12}$$

where ξ is defined in Eq. (9-4.15). At low pressures, Y is essentially unity, and $F_P = 1$, $F_Q = 1$. Also Z_2 then equals $\eta^o \xi$ so $\eta \rightarrow \eta^o$, as expected.

The Lucas method is illustrated in Example 9-10, and calculated dense gas viscosities are compared with experimental data in Table 9-7. In Example 9-10 and Table 9-7, the low-pressure viscosity η^o was not obtained from experimental data, but was estimated by the Lucas method in Sec. 9-4. Except in a few cases, the error was found to be less than 5%. The critical temperature, critical pressure, critical compressibility factor, and dipole moment are required, as well as the system temperature and pressure.

Example 9-10 Estimate the viscosity of ammonia gas at 420 K and 300 bar by using Lucas's method. The experimental values of η and η^o are 571 and 146 μP (Stephan and Lucas, 1979).

solution For ammonia we use $M = 17.031$, $Z_c = 0.244$, $T_c = 405.50$ K, $P_c = 113.53$ bar, and $\mu = 1.47$ debyes. Thus, $T_r = (420/405.50) = 1.036$ and $P_r = (300/113.53) = 2.643$. From Eq. (9-4.15),

$$\xi = (0.176)\left[\frac{405.50}{(17.031)^3(113.53)^4}\right]^{1/6} = 4.95 \times 10^{-3}(\mu\text{P})^{-1}$$

with Eq. (9-4.17),

$$\mu_r = (52.46)\left[\frac{(1.47)^2(113.53)}{(405.50)^2}\right] = 7.827 \times 10^{-2}$$

$$F_Q^\circ = 1.0$$

and with Eq. (9-4.18),

$$F_P^\circ = 1 + 30.55(0.292 - 0.244)^{1.72}|0.96 + (0.1)(1.036 - 0.7)| = 1.164$$

From Eq. (9-6.6), $Z_1 = \eta^\circ\xi = 0.7259$

$$\eta^\circ = \frac{0.7258}{4.96 \times 10^{-3}} = 147\ \mu\text{P}$$

$$\text{Error} = \frac{147 - 146}{146} \times 100 = 0.7\%$$

The estimation of the low-pressure viscosity of ammonia agrees very well with the experimental value.

Since $T_r > 1.0$, we use Eq. (9-6.8) to determine Z_2. The values of the coefficients are $a = 0.1998$, $b = 8.834 \times 10^{-2}$, $c = 0.9764$, $d = 9.235$, $e = 1.3088$, and $f = 0.7808$. Then,

$$Z_2 = \left\{1 + \frac{(0.1998)(2.643)^{1.3088}}{(8.834 \times 10^{-2})(2.643)^{0.7808} + [1 + (0.9764)(2.643)^{9.235}]^{-1}}\right\}(0.7259)$$

$$= (4.776)(0.7258)$$

$$= 3.466$$

with Eqs. (9-6.9) to (9-6.11),

$$Y = \frac{3.466}{0.7258} = 4.775$$

$$F_P = \frac{1 + (1.164 - 1)(4.775)^{-3}}{1.164} = 0.860$$

$$F_Q = 1.0$$

and, with Eq. (9-6.12),

$$\eta = \frac{(3.466)(0.860)(1.0)}{4.96 \times 10^{-3}} = 602 \ \mu\text{P}$$

$$\text{Error} = \frac{602 - 571}{571} \times 100 = 5.4\%$$

The Reichenberg and Lucas methods employ temperature and pressure as the state variables. In most other-dense gas viscosity correlations, however, the temperature and density (or specific volume) are used. In those cases, one must have accurate volumetric data or an applicable equation of state to determine the dense gas viscosity. Three different methods are illustrated below.

Method of Jossi, Stiel, and Thodos (Jossi, et al., 1962; Stiel and Thodos, 1964)

In this case, the residual viscosity $\eta - \eta^o$ is correlated with fluid density. All temperature effects are incorporated in the η^o term. To illustrate the behavior of the $\eta - \eta^o$ function, consider Fig. 9-7, which shows $\eta - \eta^o$ for *n*-butane graphed as a function of density (Dolan, et al., 1963). Note that there does not appear to be any specific effect of temperature over the range shown. At the highest density, 0.6 g/cm^3, the reduced density ρ/ρ_c is 2.63. Similar plots for many other substances are available, for example, He, air, O_2, N_2, CH_4 (Kestin and Leidenfrost, 1959); ammonia (Carmichael, et al., 1963; Shimotake and Thodos, 1963); rare gases (Shimotake and Thodos, 1958); diatomic gases (Brebach and Thodos, 1958); sulfur dioxide (Shimotake and Thodos, 1963a); CO_2 (Kennedy and Thodos, 1961; Vesovic, et al., 1990); steam (Kestin and Moszynski, 1959); and various hydrocarbons (Carmichael and Sage, 1963; Eakin and Ellington, 1963; Giddings, 1963; Starling, et al., 1960; Starling and Ellington, 1964). Other authors have also shown the applicability of a residual viscosity-density correlation (Golubev, 1959; Hanley, et al., 1969; Kestin and Moszynski, 1959; Rogers and Brickwedde, 1965; Starling, 1960, 1962).

In the Jossi, Stiel, and Thodos method, separate residual viscosity expressions are given for nonpolar and polar gases, but no quantitative criterion is presented to distinguish these classes.

Nonpolar Gases (Jossi, et al., 1962)

The basic relation is

$$[(\eta - \eta^o)\xi_T + 1]^{1/4} = 1.0230 + 0.23364\,\rho_r + 0.58533\,\rho_r^2 - 0.40758\,\rho_r^3 + 0.093324\,\rho_r^4 \qquad (9\text{-}6.13)$$

where η = dense gas viscosity, μP
η^o = low-pressure gas viscosity, μP
ρ_r = reduced gas density, $\rho/\rho_c = V_c/V$
ξ_T = the group $(T_c/M^3 P_c^4)^{1/6}$, where T_c is in kelvins and P_c is in atmospheres, $(\mu P)^{-1}$
M = molecular weight, g/mol

This relation is reported by Jossi, et al. to be applicable in the range $0.1 \le \rho_r < 3$.

Polar Gases (Stiel and Thodos, 1964)

The relation to be used depends on the reduced density:

$$(\eta - \eta^o)\xi_T = 1.656\rho_r^{1.111} \qquad \rho_r \le 0.1 \quad (9\text{-}6.14)$$

$$(\eta - \eta^o)\xi_T = 0.0607(9.045\rho_r + 0.63)^{1.739} \quad 0.1 \le \rho_r \le 0.9 \quad (9\text{-}6.15)$$

$$\log\{4 - \log[(\eta - \eta^o)\xi_T]\} = 0.6439 - 0.1005\rho_r - \Delta \qquad 0.9 \le \rho_r < 2.6 \quad (9\text{-}6.16)$$

where $\Delta = 0$ when $0.9 \le \rho_r \le 2.2$ and

$$\Delta = (4.75 \times 10^{-4})(\rho_r^3 - 10.65)^2 \qquad \text{when } 2.2 < \rho_r < 2.6 \quad (9\text{-}6.17)$$

and $(\eta - \eta^o)\xi_T = 90.0$ and 250 at $\rho_r = 2.8$ and 3.0, respectively. The notation used in Eqs. (9-6.14) to (9-6.17) is defined under Eq. (9-6.13). Note that the parameter ξ_T is *not* the same as ξ defined earlier in Eq. (9-4.15).

An example of the Jossi, et al. method is shown below, and calculated dense gas viscosities are compared with experimental values in Table 9-7.

Example 9-11 Use the Jossi, Stiel, and Thodos method to estimate the viscosity of isobutane at 500 K and 100 bar. The experimental viscosity is 261 μP (Stephan and Lucas, 1979) and the specific volume is 243.8 cm^3/mol (Waxman and Gallagher, 1983). At low pressure and 500 K, $\eta^o = 120$ μP.

solution Since isobutane is nonpolar, Eq. (9-6.13) is used. From Appendix A, T_c = 407.85 K, P_c = 36.4 bar = 35.9 atm, V_c = 262.7 cm^3/mol, and M = 58.123. Then

$$\xi_T = \left[\frac{(407.85)}{(58.123)^3(35.9)^4}\right]^{1/6} = 3.282 \times 10^{-2}(\mu P)^{-1}$$

The reduced density = $\rho_r = V_c/V = 262.7/243.8 = 1.078$. With Eq. (9-6.13)

$$\begin{aligned}[(\eta - 120)(3.282 \times 10^{-2}) + 1]^{1/4} &= 1.0230 + (0.23364)(1.078) \\ &\quad + (0.58533)(1.078)^2 \\ &\quad -(0.40758)(1.078)^3 \\ &\quad +(0.093324)(1.078)^4 \\ &= 1.571 \\ \eta &= 275\ \mu P\end{aligned}$$

$$\text{Error} = \frac{275 - 261}{261} \times 100 = 5.4\%$$

Chung, et al. Method (1988)

In an extension of the Chung, et al. technique to estimate low-pressure gas viscosities, the authors began with Eq. (9-3.9) and employed empirical correction factors to account for the fact that the fluid has a high density. Their relations are shown below.

$$\eta = \eta^* \frac{36.344(MT_c)^{1/2}}{V_c^{2/3}} \tag{9-6.18}$$

where η = viscosity, μP
M = molecular weight, g/mol
T_c = critical temperature, K
V_c = critical volume, cm^3/mol

and

$$\eta^* = \frac{(T^*)^{1/2}}{\Omega_v} \{F_c[(G_2)^{-1} + E_6 y]\} + \eta^{**} \tag{9-6.19}$$

T^* and F_c are defined as in Eqs. (9-4.9) and (9-4.11). Ω_v is found with Eq. (9-4.3) as a function of T^*, and, with ρ in mol/cm^3,

$$y = \frac{\rho V_c}{6} \tag{9-6.20}$$

$$G_1 = \frac{1 - 0.5\,y}{(1 - y)^3} \tag{9-6.21}$$

$$G_2 = \frac{E_1\{[1 - \exp(-E_4 y)]/y\} + E_2 G_1 \exp(E_5 y) + E_3 G_1}{E_1 E_4 + E_2 + E_3} \tag{9-6.22}$$

$$\eta^{**} = E_7 y^2 G_2 \exp[E_8 + E_9 (T^*)^{-1} + E_{10}(T^*)^{-2}] \tag{9-6.23}$$

and the parameters E_1 to E_{10} are given in Table 9-6 as linear functions of ω (the acentric factor), μ_r^4 [as defined in Eq. (9-4.12)], and the association factor κ (see Table 9-1). One might note that, at very low densities, y approaches zero, G_1 and G_2 approach unity, and η^{**} is negligible. At these limiting conditions, combining

TABLE 9-6 Chung, et al. Coefficients to Calculate $E_i = a_i + b_i\omega + c_i\mu_r^4 + d_i\kappa$

i	a_i	b_i	c_i	d_i
1	6.324	50.412	−51.680	1189.0
2	1.210×10^{-3}	-1.154×10^{-3}	-6.257×10^{-3}	0.03728
3	5.283	254.209	−168.48	3898.0
4	6.623	38.096	−8.464	31.42
5	19.745	7.630	−14.354	31.53
6	−1.900	−12.537	4.985	−18.15
7	24.275	3.450	−11.291	69.35
8	0.7972	1.117	0.01235	−4.117
9	−0.2382	0.06770	−0.8163	4.025
10	0.06863	0.3479	0.5926	−0.727

Eqs. (9-6.18), (9-6.19) and (9-4.9) leads to Eq. (9-4.10), which then applies for estimating η^o.

The application of the Chung, et al. method is shown in Example 9-12. Some calculated values of η are compared with experimental results in Table 9-7. The agreement is quite good and errors usually are below 5%.

Example 9-12 With the Chung, et al. method, estimate the viscosity of ammonia at 520 K and 600 bar. The experimental value of η is 466 μP (Stephan and Lucas, 1979). At this temperature, η^o = 182 μP. The specific volume of ammonia at 520 K and 600 bar is 48.2 cm^3/mol (Haar and Gallagher, 1978).

solution We use T_c = 405.50 K, V_c = 72.4 cm^3/mol, ω = 0.256, M = 17.031, and μ = 1.47 debyes. Thus T_r = 520/405.50 = 1.282 and ρ = 1/48.2 = 2.07×10^{-2} mol/cm^3.

With Eq. (9-4.12),

$$\mu_r = \frac{(131.3)(1.47)}{[(72.4)(405.50)]^{1/2}} = 1.13$$

and with Eq. (9-4.11),

$$F_c = 1 - (0.2756)(0.256) + (0.059035)(1.13)^4 = 1.026$$

$$T^* = (1.2593)(1.282) = 1.1615$$

and with Eq. (9-4.3), Ω_v = 1.275. Using Eqs. (9-6.20) and (9-6.21),

$$y = \frac{(2.075 \times 10^{-2})(72.4)}{6} = 0.250 \text{ and } G_1 = 2.074$$

From Table 9-6, the following coefficients were computed: $E_1 = -65.03$, $E_2 = -9.287 \times 10^{-3}$, $E_3 = -204.3$, $E_4 = 2.575$, $E_5 = -1.706$, $E_6 = 3.018$, $E_7 = 6.749$, $E_8 = 1.103$, $E_9 = -1.552$, and $E_{10} = 1.124$. Then, with Eq. (9-6.22), $G_2 = 1.472$ and, from Eq. (9-6.23), $\eta^{**} = 1.101$. Finally, using Eqs. (9-6.19) and (9-6.18),

$$\eta^* = \frac{(1.615)^{1/2}}{1.275}(1.026)[(1.472)^{-1} + (3.018)(0.250)] + 1.101 = 2.567$$

$$\eta = \frac{(2.567)(36.344)[(17.031)(405.50)]^{1/2}}{(72.4)^{2/3}} = 446\ \mu\text{P}$$

$$\text{Error} = \frac{446 - 466}{466} \times 100 = -4.2\%$$

TRAPP Method (Huber, 1996)

The TRAPP (transport property prediction) method is a corresponding states method to calculate viscosities and thermal conductivities of pure fluids and mixtures. In its original version (Ely, 1981; Ely and Hanely, 1981), it was also used to estimate low pressure values of λ and η and employed methane as a reference fluid. In the most recent version presented below for pure fluids and later in Sec. 9-7 for mixtures, low pressure values are estimated by one of the methods presented earlier in the chapter, propane is the reference fluid, and shape factors are no longer

TABLE 9-7 Comparison of Experimental and Calculated Dense Gas Viscosities

Compound	T, K	P, bar	V, cm^3/mole	Ref.*	η, μP	η°, μP	Reichenberg, Eq. (9-6.5)	Lucas, Eq. (9-6.12)	Jossi, et al., Eq. (9-6.13)	Chung, et al., Eq. (9-6.18) Table 9-6	Brulé and Starling, Eq. (9-6.18) Table 9-8	Trapp
Oxygen	300	30.4	806.1	6	212.8	207.2	−1.0	−1.6	0.6	−1.5	0.2	−0.2
		81.0	295.3		225.7		−1.2	−1.1	−0.6	−1.9	0.8	−0.4
		152.0	155.3		250.3		−0.3	−0.2	−0.8	−0.2	1.6	−0.3
		304.0	81.4		319.3		3.6	0.8	2.8	3.9	4.6	0.6
Methane	200	40.0	282.0	3	90	78.0	7.0	0.6	5.9	3.5	1.7	6.8
		100.0	60.2		296		10.0	8.2	5.1	3.1	14	7.8
		200.0	51.1		415		3.8	5.0	−0.5	−2.2	16	8.1
	500	40	1039.0	3	180	177	−0.4	−5.6	0.9	−5.3	−3.8	0.1
		100	417.7		187		−0.6	−5.1	0.3	−7.2	−2.3	0.0
		200	213.7		204		−1.0	−5.0	−0.3	−2.9	−1.0	−1.0
		500	98.9		263		1.5	−3.3	3.3	2.5	5.3	−1.1
Isobutane	500	20	2396.0	7	127	120	0.9	6.3	0.2	0.8	2.2	−0.3
		50	620.0		146		5.7	12.0	4.5	9.3	12	5.9
		100	244.0		261		−5.2	3.8	5.4	5.5	8.6	2.3
		200	159.0		506		−11	2.3	−9.0	−7.2	−5.0	−6.1
		400	130.0		794		−19	−8.2	−16	−10	−9.9	−11
Ammonia	420	50	588.1	4	149	146	3.0	−2.4	1.7	3.1	−5.2	4.9
		150	61.9		349		−17	−6.5	−15	−13	5.1	14
		300	39.8		571		−21	5.2	3.6	−4.0	22	60
		600	34.3		752		−24	7.8	11	−1.3	31	84
	520	50	807.6	4	185	182	0.7	−5.8	0.5	−0.1	−9.2	2.0
		150	229.6		196		4.5	0.9	2.3	4.0	1.6	9.4
		300	90.7		296		−1.4	5.3	−2.3	0.7	13	5.0
		600	48.2		466		−13	5.8	−3.2	−3.4	12	36

Carbon dioxide	360	50	514.6	1	190	177	3.0	3.1	1.9	1.1	2.4	2.6
		100	211.2		230		2.1	3.3	0.8	3.6	6.1	1.2
		400	55.0		730		1.3	7.7	−3.5	−0.8	1.2	6.3
		800	45.8		1104		−7.0	1.1	−9.6	−2.2	−1.3	4.0
	500	50	802.8	1	243	235	0.0	3.0	1.3	0.7	1.6	0.7
		100	389.2		254		1.7	5.1	1.4	3.3	5.3	2.0
		400	97.1		411		9.8	7.4	2.6	3.6	9.4	2.1
		800	62.9		636		10	9.6	0.9	−3.2	2.8	3.0
n-Pentane	600	20.3	2240	2	143	134	0.0	1.4	0.0	1.2	2.1	0.2
		81.1	418.3		242		−7.5	−5.3	−11	−4.6	−1.0	−8.5
		152	237.5		383		0.9	2.3	−7.9	−7.0	−3.7	−1.3

†Percent error = [(calc. − exp.)/exp.] × 100.

*References: 1, Angus, et al. (1976); 2, Das, et al. (1977); 3, Goodwin (1973); 4, Haar and Gallaghee (1978); 5, Stephan and Lucas (1979); 6, Stewart (1966). Ideal gas values from Stephan and Lucas (1979).

functions of density. Other reference fluids could be chosen and in fact, Huber and Ely (1992) use R134a as the reference fluid to describe the viscosity behavior of refrigerants. The TRAPP method was originally developed only for nonpolar compounds, but there have been efforts to extend the method to polar compounds as well (Hwang and Whiting, 1987).

In the TRAPP method, the residual viscosity of a pure fluid is related to the residual viscosity of the reference fluid, propane:

$$\eta - \eta^o = F_\eta[\eta^R - \eta^{Ro}] \tag{9-6.24}$$

The reference fluid values are evaluated at T_o and density ρ_o, not T and ρ. In Eq. (9-6.24), η^o is the viscosity at low pressure. η^R is the true viscosity of the reference fluid, propane, at temperature T_o and density ρ_o. η^{R0} is the low pressure value for propane at temperature T_o. For propane Younglove and Ely (Zaytsev and Aseyev, 1992) give

$$\eta^R - \eta^{Ro} = G_1 \exp[\rho_o^{0.1} G_2 + \rho_o^{0.5}(\rho_r^R - 1)G_3] - G_1 \tag{9-6.25}$$

where $\rho_r^R = \rho_o/\rho_c^R$, $\eta^R - \eta^{Ro}$ is in μPa·s, and

$$G_1 = \exp(E_1 + E_2/T) \tag{9-6.26}$$

$$G_2 = E_3 + E_4/T^{1.5} \tag{9-6.27}$$

$$G_3 = E_5 + \frac{E_6}{T} + \frac{E_7}{T^2} \tag{9-6.28}$$

$$E_1 = -14.113294896$$

$$E_2 = 968.22940153$$

$$E_3 = 13.686545032$$

$$E_4 = -12511.628378$$

$$E_5 = 0.0168910864$$

$$E_6 = 43.527109444$$

$$E_7 = 7659.4543472$$

T_o, ρ_o, and F_η are calculated by

$$T_o = T/f \tag{9-6.29}$$

$$\rho_o = \rho h \tag{9-6.30}$$

$$F_\eta = \left(\frac{M}{44.094} f\right)^{1/2} h^{-2/3} \tag{9-6.31}$$

where f and h are equivalent substance reducing ratios and are determined as described below.

If vapor pressure and liquid density information are available for the substance of interest, and if $T < T_c$, it is recommended that f be obtained from the equation

$$\frac{P_{vp}}{\rho^S} = f\,\frac{P_{vp}^R(T_o)}{\rho^{SR}(T_o)} \tag{9-6.32}$$

where P_{vp} and ρ^S are the vapor pressure and saturated liquid density at temperature, T. $P_{vp}^R(T_o)$ and $\rho^{SR}(T_o)$ are for the reference fluid, propane. Because the density and vapor pressure of the reference fluid are evaluated at $T_o = T/f$, Eq. (9-6.32) must be solved iteratively. Once f is found from Eq. (9-6.32), h is determined from

$$h = \rho^{SR}(T_o)/\rho^S \tag{9-6.33}$$

If $T > T_c$, or if vapor pressure and saturated liquid density information are not available, h and f can be calculated by

$$f = \frac{T_c}{T_c^R}\,[1 + (\omega - \omega^R)(0.05203 - 0.7498 \ln T_r] \tag{9-6.34}$$

$$h = \frac{\rho_c^R}{\rho_c}\frac{Z_c^R}{Z_c}\,[1 - (\omega - \omega^R)(0.1436 - 0.2822 \ln T_r] \tag{9-6.35}$$

The application of the TRAPP method is shown in Example 9-13. Some calculated values of η are compared with experimental results in Table 9-7. Huber (1996) gives results of additional comparisons and also suggests methods to improve predictions if some experimental data are available.

Example 9-13 Repeat Example 9-11 with the TRAPP method.

solution From Appendix A, for the reference fluid, propane, T_c = 369.83 K, V_c = 200 cm^3/mol, Z_c = 0.276 and ω = 0.152. For isobutane, T_c = 407.85 K, V_c = 259 cm^3/mol, Z_c = 0.278 and ω = 0.186. With Eqs. (9-6.34) and (9-6.35) followed by Eqs. (9-6.29) to (9-6.31)

$$f = \frac{407.85}{369.83}\,[1 + (0.186 - 0.152)(0.05203 - 0.7498 \ln(1.226))] = 1.099$$

$$h = \frac{259 \times 0.276}{200 \times 0.278}\,[1 - (0.186 - 0.152)(0.1436 - 0.2882 \ln(1.226))] = 1.282$$

$$T_o = 500/1.099 = 454.9 \text{ K}, \quad \rho_o = 1.282/243.8 = 0.005258 \text{ mol/cm}^3$$

$$F_\eta = \left(\frac{58.124 \times 1.099}{44.094}\right)^{1/2} (1.282)^{-2/3} .= 1.020$$

For the T_o and ρ_o above, Eqs. (9-6.25) to (9-6.28) give $\eta^R - \eta^{Ro}$ = 14.41 μPa s = 144.1 μP Eq. (9-6.24) gives

$$\eta = 120 + 1.020 \times 144.1 = 267.0\ \mu\text{P}$$

$$\text{Error} = \frac{267 - 261}{261} \times 100 = 2.3\%$$

Other Corresponding States Methods

In a manner identical in form with that of Chung, et al., Brulé and Starling (1984) proposed a different set of coefficients for E_1 to E_{10} to be used instead of those in

Table 9-6. These are shown in Table 9-8. Note that no polarity terms are included and the *orientation* parameter γ has replaced the acentric factor ω. If values of γ are not available, the acentric factor may be substituted.

The Brulé and Starling technique was developed to be more applicable for heavy hydrocarbons rather than for simple molecules as tested in Table 9-7.

Okeson and Rowley (1991) have developed a four-parameter corresponding-states method for polar compounds at high pressures, but did not test their method for mixtures.

Discussion and Recommendations for Estimating Dense Gas Viscosities

Six estimation techniques were discussed in this section. Two (Reichenberg and Lucas) were developed to use temperature and pressure as the input variables to estimate the viscosity. The other four require temperature and density; thus, an equation of state would normally be required to obtain the necessary volumetric data if not directly available. In systems developed to estimate many types of properties, it would not be difficult to couple the *PVT* and viscosity programs to provide densities when needed. In fact, the Brulé and Starling method (Brulé and Starling, 1984) is predicated on combining thermodynamic and transport analyses to obtain the characterization parameters most suitable for both types of estimations.

Another difference to be recognized among the methods noted in this section is that Reichenberg's, Jossi, et al.'s and the TRAPP methods require a low-pressure viscosity at the same temperature. The other techniques bypass this requirement and have imbedded into the methods a low-pressure estimation method; i.e., at low densities they reduce to techniques as described in Sec. 9-4. If the Lucas, Chung, et al., or Brulé-Starling method were selected, no special low-pressure estimation method would have to be included in a property estimation package.

With these few remarks, along with the testing in Table 9-7 as well as evaluations by authors of the methods, we recommend that either the Lucas or Chung, et al. procedure be used to estimate dense (and dilute) gas viscosities of both polar and nonpolar compounds. The Brulé-Starling method is, however, preferable when com-

TABLE 9-8 Brulé and Starling Coefficients to Calculate $E_i = a_i + b_i\gamma$

i	a_i	b_i
1	17.450	34.063
2	-9.611×10^{-4}	7.235×10^{-3}
3	51.043	169.46
4	−0.6059	71.174
5	21.382	−2.110
6	4.668	−39.941
7	3.762	56.623
8	1.004	3.140
9	-7.774×10^{-2}	−3.584
10	0.3175	1.1600

Note: If γ values are not available, use ω, the acentric factor, or, preferably, obtain from multiproperty analysis by using vapor pressure data. (Brulé and Starling, 1984)

plex hydrocarbons are of interest, but even for those materials, the Chung, et al. procedure should be used at low reduced temperatures ($T_r < 0.5$). For nonpolar compounds, we recommend the TRAPP method as well as the Lucas or Chung, et al. methods.

Except when one is working in temperature and pressure ranges in which viscosities are strong functions of these variables (See Fig. 9-6), errors for the recommended methods are usually only a few percent. Near the critical point and in regions where the fluid density is approaching that of a liquid, higher errors may be encountered.

9-7 VISCOSITY OF GAS MIXTURES AT HIGH PRESSURES

The most convenient method to estimate the viscosity of dense gas mixtures is to combine, where possible, techniques given previously in Secs. 9-5 and 9-6.

Lucas Approach (Lucas 1980, 1981, 1983)

In the (pure) dense gas viscosity approach suggested by Lucas, Eqs. (9-6.6) to (9-6.12) were used. To apply this technique to mixtures, rules must be chosen to obtain T_c, P_c, M, and μ as functions of composition. For T_c, P_c, and M of the mixture, Eqs. (9-5.18) to (9-5.20) should be used. The polarity (and quantum) corrections are introduced by using Eqs. (9-6.10) and (9-6.11), where F_P^o and F_Q^o refer to mixture values from Eqs. (9-5.21) and (9-5.22). The parameter Y in Eqs. (9-6.10) and (9-6.11) must be based on T_{cm} and P_{cm}. F_P^o and F_Q^o, for the pure components, were defined in Eqs. (9-4.18) and (9-4.19).

Chung, et al. (1988) Approach

To use this method for dense gas mixtures, Eqs. (9-6.18) to (9-6.23) are used. The parameters T_c, V_c, ω, M, μ and κ in these equations are given as functions of composition in Sec. 9-5. That is,

Parameter	Equations to use
T_{cm}	(9-5.44), (9-5.27)
V_{cm}	(9-5.43), (9-5.25)
ω_m	(9-5.29), (9-5.25)
V_{cm}	(9-5.32), (9-5.33), (9-5.25) and (9-5.43)
μ_m	(9-5.30), (9-5.25)
κ_m	(9-5.31)

TRAPP Method (Huber, 1996)

For gas mixtures at high pressure, the viscosity is determined by a combination of the techniques introduced for high-pressure gases (Sec. 9-6) with appropriate mixing rules. The viscosity of the mixture is given by:

$$\eta_m - \eta_m^o = F_{\eta m}[\eta^R - \eta^{Ro}] + \Delta\eta^{\text{ENSKOG}} \tag{9-7.1}$$

The quantity $\eta^R - \eta^{Ro}$ that appears in Eq. (9-7.1) is for the reference fluid propane and is evaluated with Eq. (9-6.25) at T_o and ρ_o. The following mixing rules are used to determine F_{nm}, T_o and ρ_o.

$$h_m = \sum_i \sum_j y_i y_j h_{ij} \tag{9-7.2}$$

$$f_m h_m = \sum_i \sum_j y_i y_j f_{ij} h_{ij} \tag{9-7.3}$$

$$h_{ij} = \frac{[(h_i)^{1/3} + (h_j)^{1/3}]^3}{8} \tag{9-7.4}$$

$$f_{ij} = (f_i f_j)^{1/2} \tag{9-7.5}$$

f_i and h_i are determined as in Sec. 9-6. T_o and ρ_o are calculated by equations similar to Eqs. (9-6.29) and (9-6.30):

$$T_o = T/f_m \tag{9-7.6}$$

$$\rho_o = \rho h_m = h_m/V \tag{9-7.7}$$

Finally,

$$F_{\eta m} = (44.094)^{-1/2}(h_m)^{-2} \sum_i \sum_j y_i y_j (f_{ij} M_{ij})^{1/2} (h_{ij})^{4/3} \tag{9-7.8}$$

where

$$M_{ij} = \frac{2\, M_i M_j}{M_i + M_j} \tag{9-7.9}$$

The term, $\Delta\eta^{\text{ENSKOG}}$, accounts for size differences (Ely, 1981) and is calculated by

$$\Delta\eta^{\text{ENSKOG}} = \eta_m^{\text{ENSKOG}} - \eta_x^{\text{ENSKOG}} \tag{9-7.10}$$

where

$$\eta_m^{\text{ENSKOG}} = \sum_i \beta_i Y_i + \alpha\rho^2 \sum_i \sum_j y_i y_j \sigma_{ij}^6 \eta_{ij}^o g_{ij} \tag{9-7.11}$$

ρ = density in mols/L, σ is in Å, and η^o and η^{ENSKOG} are in μP

$$\alpha = \frac{48}{25\pi}\left[\frac{2\pi}{3}(6.023 \times 10^{-4})\right]^2 = 9.725 \times 10^{-7}$$

$$\sigma_i = 4.771\, h_i^{1/3} \tag{9-7.12}$$

$$\sigma_{ij} = \frac{\sigma_i + \sigma_j}{2} \tag{9-7.13}$$

Because $\Delta\eta^{\text{ENSKOG}}$ is a correction based on a hard sphere assumption, Eq. (9-3.8) is used to calculate η^o. The radial distribution function, g_{ij}, is calculated (Tham and Gubbins, 1971) by

$$g_{ij} = (1 - \xi)^{-1} + \frac{3\xi}{(1 - \xi)^2} \Theta_{ij} + \frac{2\xi^2}{(1 - \xi)^3} \Theta_{ij}^2 \tag{9-7.14}$$

$$\Theta_{ij} = \frac{\sigma_i \sigma_j}{2\sigma_{ij}} \frac{\sum_k y_k \sigma_k^2}{\sum_k y_k \sigma_k^3} \tag{9-7.15}$$

$$\xi = (6.023 \times 10^{-4}) \frac{\pi}{6} \rho \sum_i y_i \sigma_i^3 \tag{9-7.16}$$

$$Y_i = y_i \left[1 + \frac{8\pi}{15} (6.023 \times 10^{-4}) \rho \sum_j y_j \left(\frac{M_j}{M_i + M_j} \right) \sigma_{ij}^3 g_{ij} \right] \tag{9-7.17}$$

The n values of β_i are obtained by solving the n linear equations of the form

$$\sum_j B_{ij} \beta_j = Y_i \tag{9-7.18}$$

where

$$B_{ij} = 2 \sum_k y_i y_k \frac{g_{ik}}{\eta_{ik}^o} \left(\frac{M_k}{M_i + M_k} \right)^2 \left[\left(1 + \frac{5}{3} \frac{M_i}{M_k} \right) \delta_{ij} - \frac{2}{3} \frac{M_i}{M_k} \delta_{jk} \right] \tag{9-7.19}$$

In Eq. (9-7.19), δ_{ij} is the Kronecker delta function, 1 if $i = j$, and 0 if $i \neq j$. The quantity η_x^{ENSKOG} that appears in Eq. (9-7.10) is for a pure hypothethical fluid with the same density as the mixture and is determined with Eq. (9-7.11) with σ_x defined by

$$\sigma_x = \left(\sum_i \sum_j y_i y_j \sigma_{ij}^3 \right)^{1/3} \tag{9-7.20}$$

$$M_x = \left[\sum_i \sum_j y_i y_j M_{ij}^{1/2} \sigma_{ij}^4 \right]^2 \sigma_x^{-8} \tag{9-7.21}$$

M_{ij} and σ_{ij} are defined in Eqs. (9-7.9) and (9-7.13). Huber (1996) tested the TRAPP method on a number of binary hydrocarbon mixtures over a wide range of densities and reports an average absolute error of about 5%, although, in some cases, significantly larger deviations were found. The method is illustrated in Example 9-14.

Example 9-14 Use the TRAPP method to estimate the viscosity of a mixture of 80 mol % methane (1) and 20 mol % *n*-decane at 377.6 K and 413.7 bar. Lee, et al. (1966) report at these conditions, ρ = 0.4484 g/cm^3 and η_{exp} = 126 μPa · s, although this value is considerably higher than values reported by Knapstad, et al. (1990) at similar conditions.

solution From Appendix A

	M	T_c, K	V_c, cm^3/mol	Z_c	ω
CH_4	16.043	190.56	98.6	0.286	0.011
$C_{10}H_{22}$	142.285	617.7	624	0.256	0.490

Using Eqs. (9-7.2) to (9-7.9) and the procedure illustrated in Example 10-8 leads to $f_m = 0.9819$, $h_m = 0.8664$, $T_o = 384.6$ K, $\rho_o = 9.408$ mol/L, and $\eta^R - \eta^{Ro} = 51.72$ μPa · s. Equation (9-7.8) gives $F_{\eta m} = 1.260$. Calculation of $\Delta\eta^{\text{ENSKOG}}$ requires the application of the method described in Eqs. (9-7.10) through (9-7.21). Intermediate results include $\rho = 10.86$ mol/L, $\xi = 0.3668$ with other values shown below:

ij	σ_{ij}, Å	η^o_{ij}, μ Pa · s	g_{ij}	$B_{ij} \times 10^4$
11	3.716	15.0	2.696	210.7
12	5.314	9.86	3.085	−6.076
22	6.913	13.0	3.873	28.18

When $Y_1 = 2.014$ and $Y_2 = 0.5627$ are used in Eq. (9-7.18), this equation is written for each of the two components and solved to give $\beta_1 = 102.0$ and $\beta_2 = 221.6$. Finally, Eq. (9-7.11) gives $\eta_m^{\text{ENSKOG}} = 911$ μP. From Eqs. (9-7.20) and (9-7.21), $\sigma_x = 4.548$ Å and $M_x = 47.50$. Then, for the hypothetical pure fluid, Eq. (9-7.16) gives

$$\xi = 6.023 \times 10^{-4} \frac{\pi}{6} (10.86)(4.548)^3 = 0.3222$$

Eq. (9-7.14), with $\Theta_{xx} = ½$, gives

$$g_{xx} = \frac{1}{1 - 0.3222} + \frac{1}{2} \frac{3 \times 0.3222}{(1 - 0.3222)^2} + \frac{2(0.3222)^2}{(1 - 0.3222)^3} \frac{1}{4} = 2.694$$

Eq. (9-7.17) gives

$$Y_x = 1 + \frac{8\pi}{15} (6.023 \times 10^{-4})(10.86) \frac{(4.548)^3(2.694)}{2} = 2.389$$

From Eq. (9-3.8)

$$\eta^o_x = 26.69 \frac{(47.50 \times 377.6)^{1/2}}{(4.548)^2} = 172.8 \ \mu\text{P}$$

For a pure component, Eq. (9-7.19) reduces to $B_{xx} = g_{xx}/\eta^o_x$, so that $B_{xx} = 2.694/172.8 = 0.01559$. Then Eq. (9-7.18) gives $\beta_x = 2.389/0.01559 = 153.2$. Applying Eq. (9-7.11) to a pure fluid gives

$$\eta_x^{\text{ENSKOG}} = 153.2 \times 2.389 + 9.725 \times 10^{-7} (10.86)^2(4.548)^6(172.8)(2.694) = 839 \ \mu\text{P}$$

With Eq. (9-7.10)

$$\Delta\eta^{\text{ENSKOG}} = 911 - 839 = 72 \ \mu\text{P} = 7.2 \ \mu\text{Pa} \cdot \text{s}.$$

Then using Eq. (9-7.1)

$$\eta_m = 10.2 + 1.260 \times 51.72 + 7.2 = 82.6 \ \mu\text{Pa} \cdot \text{s}.$$

$$\text{Error} = \frac{82.6 - 126}{126} \times 100 = -34\%$$

In the above example, for the low pressure contribution of 10.2 μPa·s, Eq. (9-3.9) was used for methane, (9-4.16) was used for decane, and (9-5.16) was used for the mixture. Although the TRAPP prediction in Example 9-14 was lower than the experimental value reported by Lee, et al. (1966), TRAPP predictions are considerably higher than the experimental values reported in (Knapstad, et al., 1990) for the same system at similar conditions. High-quality data for mixtures with different size molecules at high pressures are limited. This in turn limits the ability to evaluate models for this case.

Discussion

Both the Lucas and Chung, et al. methods use the relations for the estimation of dense gas viscosity and apply a one-fluid approximation to relate the component parameters to composition. The TRAPP method uses the term $\Delta\eta^{ENSKOG}$ to improve the one-fluid approximation. In the Lucas method, the state variables are T, P, and composition, whereas in the TRAPP and Chung, et al. procedures, T, ρ, and composition are used.

The accuracy of the Lucas and Chung, et al. forms is somewhat less than when applied to pure, dense gases. Also, as noted at the end of Sec. 9-6, the accuracy is often poor when working in the critical region or at densities approaching those of a liquid at the same temperature. The TRAPP procedure can be extended into the liquid region. The paucity of accurate high-pressure gas mixture viscosity data has limited the testing that could be done, but Chung, et al. (1988) report absolute average deviations of 8 to 9% for both polar and nonpolar dense gas mixtures. A comparable error would be expected from the Lucas form. The TRAPP method gives similar deviations for nonpolar mixtures, but has not been tested for polar mixtures. Tilly, et al. (1994) recommended a variation of the TRAPP method to correlate viscosities of supercritical fluid mixtures in which various solutes were dissolved in supercritical carbon dioxide.

As a final comment to the first half of this chapter, it should be noted that, if one were planning a property estimation system for use on a computer, it is recommended that the Lucas, Chung, et al., or Brulé and Starling method be used in the dense gas mixture viscosity correlations. Then, at low pressures or for pure components, the relations simplify directly to those described in Secs. 9-4 to 9-6. In other words, it is not necessary, when using these particular methods, to program separate relations for low-pressure pure gases, low-pressure gas mixtures, and high-pressure pure gases. One program is sufficient to cover all those cases as well as high-pressure gas mixtures.

9-8 LIQUID VISCOSITY

Most gas and gas mixture estimation techniques for viscosity are modifications of theoretical expressions described briefly in Secs. 9-3 and 9-5. There is no comparable theoretical basis for the estimation of liquid viscosities. Thus, it is particularly desirable to determine liquid viscosities from experimental data when such data exist. Viswanath and Natarajan (1989) have published a compilation of liquid viscosity data for over 900 compounds and list constants that correlate these data. Liquid viscosity data can also be found in Gammon, et al. (1993–1998), Riddick, et al. (1986), Stephan and Lucas (1979), Stephen and Hildwein (1987), Stephan and Heckenberger (1988), Timmermans (1965), and Vargaftik, et al. (1996). Data for aqueous electrolyte solutions may be found in Kestin and Shankland (1981), Lobo (1990), and Zaytsev and Aseyev (1992). Tabulations of constants have been published in Daubert, et al. (1997), Duhne (1979), van Velzen et al. (1972), Yaws, et al. (1976), and Yaws (1995, 1995a) that allow estimations of liquid viscosities. When these constants are derived from experimental data they can be used with confidence, but sometimes (Yaws, 1995, 1995a) they are based on estimated viscosities, and in such instances, they should be used only with caution. Liquid phase viscosity values can also be found in Dean (1999), Lide (1999), and Perry and Green (1997).

The viscosities of liquids are larger than those of gases at the same temperature. As an example, in Fig. 9-8, the viscosities of liquid and vapor benzene are plotted as functions of temperature. Near the normal boiling point (353.4 K), the liquid viscosity is about 36 times the vapor viscosity, and at lower temperatures, this ratio increases even further. Two vapor viscosities are shown in Fig. 9-8. The low-pressure gas line would correspond to vapor at about 1 bar. As noted earlier in Eq. (9-4.20), below T_c, low-pressure gas viscosities vary in a nearly linear manner with temperature. The curve noted as saturated vapor reflects the effect of the increase in vapor pressure at higher temperatures. The viscosity of the saturated vapor should equal that of the saturated liquid at the critical temperature (for benzene, T_c = 562.0 K).

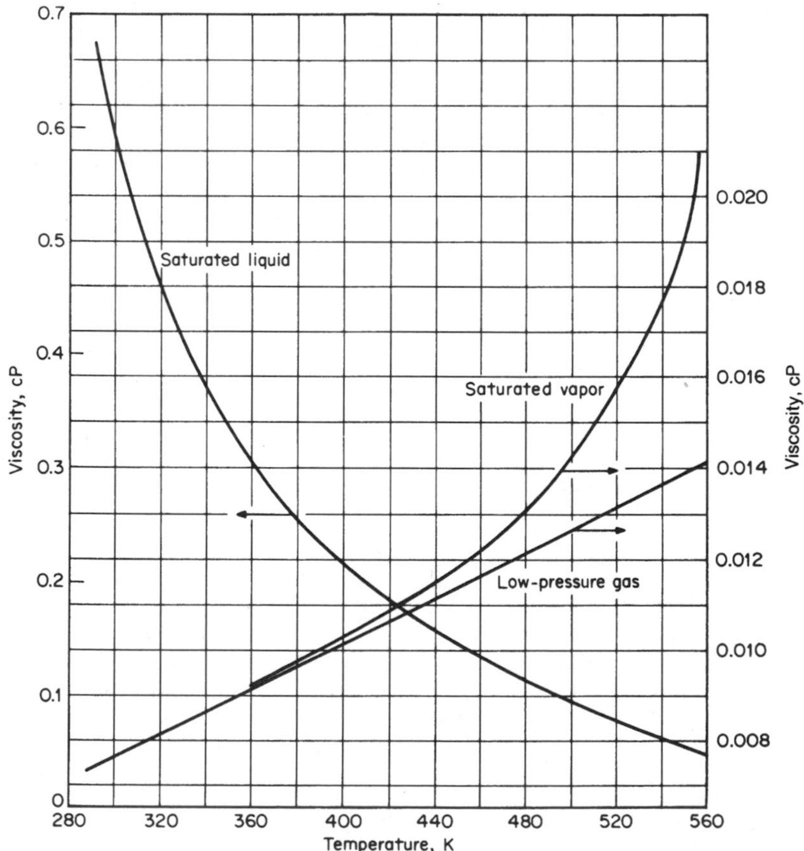

FIGURE 9-8 Viscosities of liquid and vapor benzene. (T_b = 353.2 K; T_c = 562.0 K).

Much of the curvature in the liquid viscosity-temperature curve may be eliminated if the logarithm of the viscosity is plotted as a function of reciprocal (absolute) temperature. This change is illustrated in Fig. 9-9 for four saturated liquids: ethanol, benzene, *n*-heptane, and nitrogen. (To allow for variations in the temperature range, the reciprocal of the reduced temperature is employed.) Typically, the normal boiling point would be at a value of $T_r^{-1} \approx 1.5$. For temperatures below the normal boiling point ($T_r^{-1} > 1.5$), the logarithm of the viscosity varies linearly with T_r^{-1}. Above the normal boiling point, this no longer holds. In the nonlinear region, several corresponding states estimation methods have been suggested, and they are covered in Sec. 9-12. In the linear region, most corresponding states methods have not been found to be accurate, and many estimation techniques employ a group contribution approach to emphasize the effects of the chemical structure on viscosity. The curves in Fig. 9-9 suggest that, at comparable reduced temperatures, viscosities of polar fluids are higher than those of nonpolar liquids such as hydrocarbons, which themselves are larger than those of simple molecules such as nitrogen. If one attempts to replot Fig. 9-9 by using a nondimensional viscosity such as $\eta\xi$ [see, for example, Eqs. (9-4.13) to (9-4.15)] as a function of T_r, the separation between curves diminishes, especially at $T_r > 0.7$. However, at lower values of T_r, there are still significant differences between the example compounds.

In the use of viscosity in engineering calculations, one is often interested not in the dynamic viscosity, but, rather, in the ratio of the dynamic viscosity to the density. This quantity, called the *kinematic viscosity,* would normally be expressed

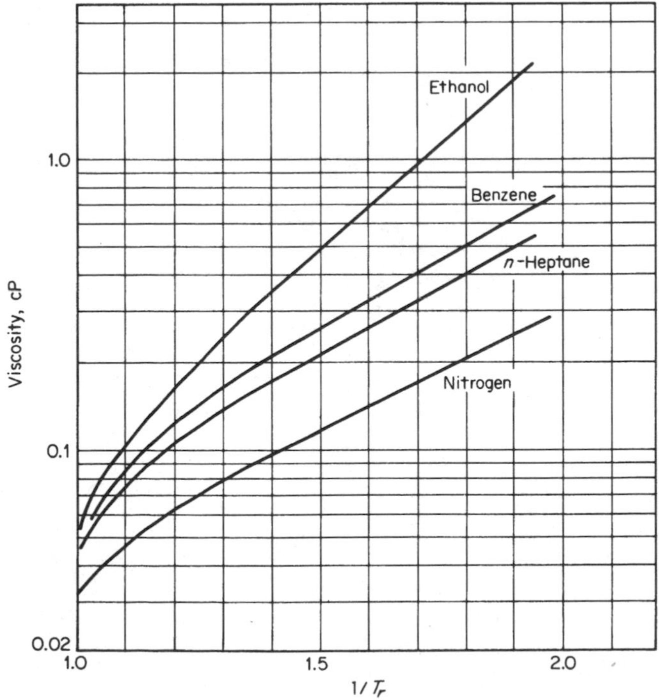

FIGURE 9-9 Viscosities of various liquids as functions of temperature. (*Stephan and Lucas, 1979*)

in m^2/s or in stokes. One stoke (St) is equivalent to 10^{-4} m^2/s. The kinematic viscosity ν, decreases with increasing temperature in a manner such that ln ν is nearly linear in temperature for both the saturated liquid and vapor as illustrated in Fig. 9-10 for benzene. As with the dynamic viscosity, the kinematic viscosities of the saturated vapor and liquid become equal at the critical point.

The behavior of the kinematic viscosity with temperature has led to several correlation schemes to estimate ν rather than η. However, in most instances, ln ν is related to T^{-1} rather than T. If Fig. 9-10 is replotted by using T^{-1}, again there is a nearly linear correlation with some curvature near the critical point (as there is in Fig. 9-9).

In summary, pure liquid viscosities at high reduced temperatures are usually correlated with some variation of the law of corresponding states (Sec. 9-12). At lower temperatures, most methods are empirical and involve a group contribution approach (Sec. 9-11). Current liquid *mixture* correlations are essentially mixing rules relating pure component viscosities to composition (Sec. 9-13). Little theory has been shown to be applicable to estimating liquid viscosities (Andrade, 1954; Brokaw, et al., 1965; Brush, 1962; Gemant, 1941; Hirschfelder, et al., 1954).

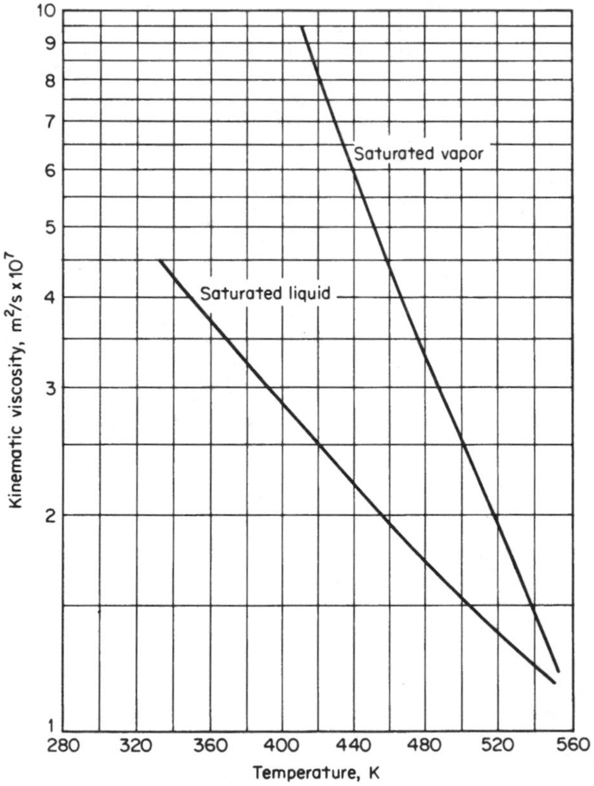

FIGURE 9-10 Kinematic viscosities of saturated liquid and vapor benzene (T_b = 353.2 K; T_c = 562.0 K).

9-9 EFFECT OF HIGH PRESSURE ON LIQUID VISCOSITY

Increasing the pressure over a liquid results in an increase in viscosity. Lucas (1981) has suggested that the change may be estimated from Eq. (9-9.1)

$$\frac{\eta}{\eta_{SL}} = \frac{1 + D(\Delta P_r/2.118)^A}{1 + C\omega\Delta P_r} \tag{9-9.1}$$

where η = viscosity of the liquid at pressure P
η_{SL} = viscosity of the saturated liquid at P_{vp}
$\Delta P_r = (P - P_{vp})/P_c$
ω = acentric factor
$A = 0.9991 - [4.674 \times 10^{-4}/(1.0523T_r^{-0.03877} - 1.0513)]$
$D = [0.3257/(1.0039 - T_r^{2.573})^{0.2906}] - 0.2086$
$C = -0.07921 + 2.1616T_r - 13.4040T_r^2 + 44.1706T_r^3 - 84.8291T_r^4 + 96.1209T_r^5 - 59.8127T_r^6 + 15.6719T_r^7$

In a test with 55 liquids, polar and nonpolar, Lucas found errors in the calculated viscosities of less than 10%. To illustrate the predicted values of Eq. (9.9.1), Figs. 9-11 and 9-12 were prepared. In both, η/η_{SL} was plotted as a function of ΔP_r for various reduced temperatures. In Fig. 9-11, $\omega = 0$, and in Fig. 9-12, $\omega = 0.2$. Except at high values of T_r, η/η_{SL} is approximately proportional to ΔP_r. The effect of pressure is more important at the high reduced temperatures. As the acentric factor increases, there is a somewhat smaller effect of pressure. The method is illustrated in Example 9-15.

Example 9-15 Estimate the viscosity of liquid methylcyclohexane at 300 K and 500 bar. The viscosity of the saturated liquid at 300 K is 0.68 cP, and the vapor pressure is less than 1 bar.

solution From Appendix A, $T_c = 572.19$ K, $P_c = 34.71$ bar, and $\omega = 0.235$. Thus $T_r = 300/572.19 = 0.524$ and $\Delta P_r = 500/34.71 = 14.4$. ($P_{vp}$ was neglected.) Then

$$A = 0.9991 - \frac{4.674 \times 10^{-4}}{(1.0523)(0.524)^{-0.03877} - 1.0513} = 0.9822$$

$$D = \frac{0.3257}{[1.0039 - (0.524)^{2.573}]^{0.2906}} - 0.2086 = 0.1371$$

$$\begin{aligned} C = &-0.07921 + (2.1616)(0.524) - (13.4040)(0.524)^2 + (44.1706)(0.524)^3 \\ &- (84.8291)(0.524)^4 + (96.1209)(0.524)^5 - (59.8127)(0.524)^6 \\ &+ (15.6719)(0.524)^7 = 0.0619 \end{aligned}$$

With Eq. (9-9.1),

$$\frac{\eta}{\eta_{SL}} = \frac{1 + (0.137)(14.4/2.118)^{0.9822}}{1 + (0.235)(14.4)(0.0619)} = 1.57$$

$$\eta = (1.57)(0.68) = 1.07 \text{ cP}$$

The experimental value of η at 300 K and 500 bar is 1.09 cP (Titani, 1929).

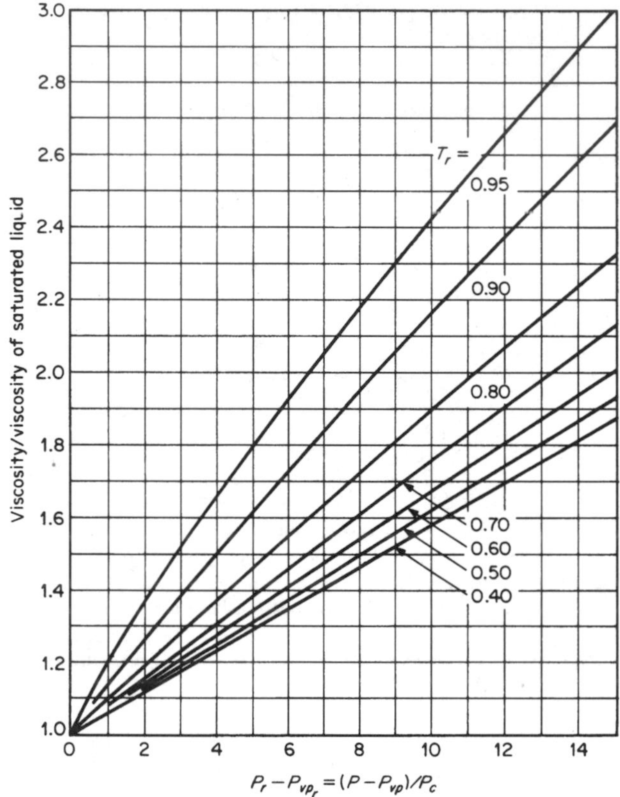

FIGURE 9-11 Effect of pressure on the viscosity of liquids $\omega = 0$.

$$\text{Error} = \frac{1.07 - 1.09}{1.09} \times 100 = -1.8\%$$

Whereas the correlation by Lucas would encompass most pressure ranges, at pressures over several thousand bar the data of Bridgman suggest that the logarithm of the viscosity is proportional to pressure and that the structural complexity of the molecule becomes important. Those who are interested in such high-pressure regions should consult the original publications of Bridgman (1926) or the work of Dymond and Assael (See Sec. 9.6, Assael, et al., 1996, or Dymond and Assael, 1996).

9-10 EFFECT OF TEMPERATURE ON LIQUID VISCOSITY

The viscosities of liquids decrease with increasing temperature either under isobaric conditions or as saturated liquids. This behavior can be seen in Fig. 9-9, where, for

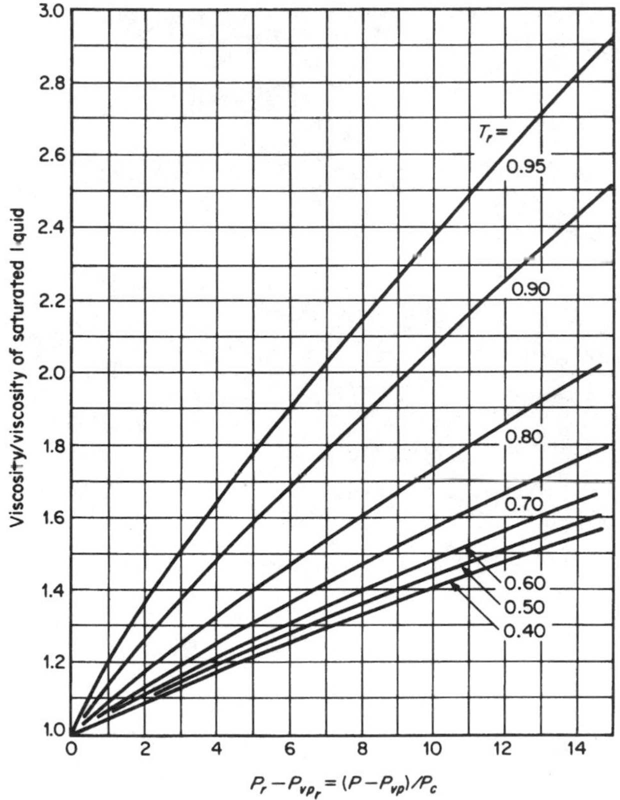

FIGURE 9-12 Effect of pressure on the viscosity of liquids; $\omega = 0.2$.

example, the viscosity of saturated liquid benzene is graphed as a function of temperature. Also, as noted in Sec. 9-8 and as illustrated in Fig. 9-10, for a temperature range from the freezing point to somewhere around the normal boiling temperature, it is often a good approximation to assume ln η_L is linear in reciprocal absolute temperature; i.e.,

$$\ln \eta_L = A + \frac{B}{T} \tag{9-10.1}$$

This simple form was apparently first proposed by de Guzman (1913) (O'Loane, 1979), but it is more commonly referred to as the Andrade equation (1930, 1934). Variations of Eq. (9-10.1) have been proposed to improve upon its correlation accuracy; many include some function of the liquid molar volume in either the A or B parameter (Bingham and Stookey, 1939; Cornelissen and Waterman, 1955; Eversteijn, et al., 1960; Girifalco, 1955; Gutman and Simmons, 1952; Innes, 1956; Marschalko and Barna, 1957; Medani and Hasan, 1977; Miller, 1963, 1963a; Telang, 1945; and van Wyk, et al., 1940). Another variation involves the use of a third constant to obtain the Vogel equation (1921),

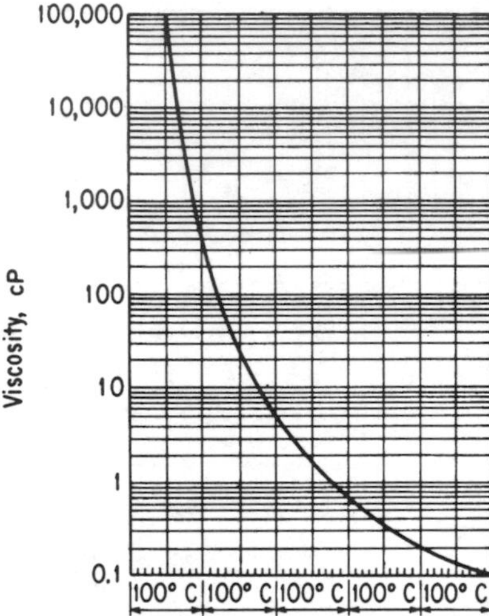

FIGURE 9-13 Lewis and Squires liquid viscosity-temperature correlation. (*Lewis and Squires, 1934 as adapted in Gambill, 1959*)

$$\ln \eta_L = A + \frac{B}{T + C} \tag{9-10.2}$$

Goletz and Tassios (1977) have used this form (for the kinematic viscosity) and report values of A, B, and C for many pure liquids.

Equation (9-10.1) requires at least two viscosity-temperature datum points to determine the two constants. If only one datum point is available, one of the few ways to extrapolate this value is to employ the approximate Lewis-Squires chart (1934), which is based on the empirical fact that the sensitivity of viscosity to temperature variations appears to depend primarily upon the value of the viscosity. This chart, shown in Fig. 9-13, can be used by locating the known value of viscosity on the ordinate and then extending the abscissa by the required number of degrees to find the new viscosity. Figure 9-13 can be expressed in an equation form as

$$\eta_L^{-0.2661} = \eta_K^{-0.2661} + \frac{T - T_K}{233} \tag{9-10.3}$$

where η_L = liquid viscosity at T, cP
η_K = known value of liquid viscosity at T_K, cP

T and T_K may be expressed in either °C or K. Thus, given a value of η_L at T_K, one can estimate values of η_L at other temperatures. Equation (9-10.3) or Fig. 9-13 is only approximate, and errors of 5 to 15% (or greater) may be expected. This method should not be used if the temperature is much above the normal boiling point.

Example 9-16 The viscosity of acetone at 30°C is 0.292 cP; estimate the viscosities at −90°C, −60°C, 0°C, and 60°C.

solution At −90°C, with Eq. (9-10.3),

$$\eta_L^{-0.2661} = (0.292)^{-0.2661} + \frac{-90 - 30}{233}$$

$$\eta_L = 1.7 \text{ cP}$$

For the other cases,

T, °C	η_L, cP Eq. (9-10.3)	η_l, cP Experimental	Precent error
−90	1.7	2.1	−19
−60	0.99	0.98	1
0	0.42	0.39	8
60	0.21	0.23	−9

In summary, from the freezing point to near the normal boiling point, Eq. (9-10.1) is a satisfactory temperature-liquid viscosity function. Two datum points are required. If only one datum point is known, a rough approximation of the viscosity at other temperatures can be obtained from Eq. (9-10.3) or Fig. 9-14.

Liquid viscosities above the normal boiling point are treated in Sec. 9-12.

9-11 ESTIMATION OF LOW-TEMPERATURE LIQUID VISCOSITY

Estimation methods for low-temperature liquid viscosity often employ structural-sensitive parameters which are valid only for certain homologous series or are found from group contributions. These methods usually use some variation of Eq. (9-10.1) and are limited to reduced temperatures less than about 0.75. We present two such methods in this section. We also describe a technique that employs corresponding states concepts. None of the three methods considered is particularly reliable.

Orrick and Erbar (1974) Method

This method employs a group contribution technique to estimate A and B in Eq. (9-11.1).

$$\ln \frac{\eta_L}{\rho_L M} = A + \frac{B}{T} \tag{9-11.1}$$

where η_L = liquid viscosity, cP
ρ_L = liquid density at 20°C, g/cm³
M = molecular weight
T = temperature, K

The group contributions for obtaining A and B are given in Table 9-9. For liquids that have a normal boiling point below 20°C, use the value of ρ_L at 20°C; for liquids

TABLE 9-9 Orrick and Erbar (1974) Group Contributions for A and B in Eq. (9-11.1)

Group	A	B
Carbon atoms†	$-(6.95 + 0.21N)$	$275 + 99N$
R—C(—C)(—R)—R (carbon with C, R, R, R attached)	−0.15	35
R—C(R)(R)—R	−1.20	400
Double bond	0.24	−90
Five-membered ring	0.10	32
Six-membered ring	−0.45	250
Aromatic ring	0	20
Ortho substitution	−0.12	100
Meta substitution	0.05	−34
Para substitution	−0.01	−5
Chlorine	−0.61	220
Bromine	−1.25	365
Iodine	−1.75	400
—OH	−3.00	1600
—COO—	−1.00	420
—O—	−0.38	140
—C(—)=O	−0.50	350
—COOH	−0.90	770

†N = number of carbon atoms not including those in other groups shown above.

whose freezing point is above 20°C, ρ_L at the melting point should be employed. Compounds containing nitrogen or sulfur cannot be treated. Orrick and Erbar tested this method for 188 organic liquids. The errors varied widely, but they reported an average deviation of 15%. This is close to the average value of 16% shown in Table 9-11 for a more limited test. Since ρ_L in Eq. (9-11.1) is at 20°C and not T, the temperature of the liquid, Eq. (9-11.1) is the same form as the Andrade equation, Eq. (9-10.1).

Example 9-17 Estimate the viscosity of liquid n-butyl alcohol at 120°C with the Orrick-Erbar method. The experimental value is 0.394 cP.

solution From Table 9-9

$$A = -6.95 - (0.21)(4) - 3.00 = -10.79$$

$$B = 275 + (99)(4) + 1600 = 2271$$

From Vargaftik, et al. (1996), at 20°C, $\rho_L = 0.8096$ g/cm^3 and $M = 74.123$. Then, with Eq. (9-11.1),

$$\ln \frac{\eta_L}{(0.8096)(74.123)} = -10.79 + \frac{2271}{T}$$

$$\text{At } T = 120°\text{C} = 393 \text{ K}, \eta_L = 0.400 \text{ cP}$$

$$\text{Error} = \frac{0.400 - 0.394}{0.394} \times 100 = 1.5\%$$

Sastri-Rao Method (1992)

In this method, the pure liquid viscosity is calculated with the equation

$$\eta = \eta_B P_{vp}^{-N} \tag{9-11.2}$$

P_{vp} is the vapor pressure in atmospheres and η_B is the viscosity at the normal boiling point, T_b in mPa·s. Below T_b, Sastri and Rao determine P_{vp} with the equation

$$\ln P_{vp} = (4.5398 + 1.0309 \ln T_b) \times \left(1 - \frac{(3 - 2T/T_b)^{0.19}}{T/T_b} - 0.38(3 - 2T/T_b)^{0.19} \ln(T/T_b)\right) \tag{9-11.3}$$

Equation (9-11.3) should be used only when $T < T_b$. Equation (9-11.3) is not necessarily the most accurate equation for vapor pressure predictions but should be used with Eq. (9-11.2) because the group contributions used to estimate η_B and N have been determined when P_{vp} was calculated with Eq. (9-11.3). η_B is determined with the equation

$$\eta_B = \sum \Delta\eta_B + \sum \Delta\eta_{Bcor} \tag{9-11.4}$$

N is determined from

$$N = 0.2 + \sum \Delta N + \sum \Delta N_{cor} \tag{9-11.5}$$

Values for group contributions to determine the summations in Eqs. (9-11.4) and (9-11.5) are given in Table 9-10. The contributions of the functional groups to η_B and N are generaly cumulative. However, if the compound contains more than one identical functional group, its contributions for N should be taken only once unless otherwise mentioned. Thus for branched hydrocarbons with multiple >CH- groups, N is 0.25. In Table 9-10, the term alicyclic means cycloparaffins and cycloolefins and excludes aromatics and heterocyclics. In the contributions of halogen groups, "others" means aromatics, alicyclics, and heterocyclics while the carbon groups listed are meant for aliphatic compounds. Also for halogens, the values of ΔN for aliphatic, alicyclics and aromatics are not used if other non-hydrocarbon groups are present in the cyclic compound (See footnote b in the halogen section of Table 9-10). For example, the corrections for halogenated pyridines and anilines are given in footnote b and are not to be used in conjunction with the corrections listed under "aliphatic, alicyclics and aromatics." Calculation of η_B and N is illustrated in Example 9-18 and typical deviations are shown in Table 9-11.

Example 9-18 Determine the values of η_B and N to be used in Eq. (9-11.2) for o-xylene, ethanol, ethylbenzene, 2-3-dimethylbutane, and o-chlorophenol.

TABLE 9-10 Sastri and Rao (1992) Group Contributions for η_B and N in Eq. (9-11.2)

Hydrocarbon groups				
Group		$\Delta\eta_B$	ΔN	Remarks and examples
		Non-ring		
—CH_3		0.105	0.000	For *n*-alkanes, *n*-alkenes or *n*-alkynes with C > 8 $\Delta N_{cor} = 0.050$
>CH_2		0.000	0.000	
>CH—		−0.110	0.050	(i) if both >CH— and >C< groups are present $\Delta N_{cor} = 0.050$ only
>C<		−0.180	0.100	(ii) ΔN values applicable only for aliphatic hydrocarbons and halogenated derivatives of aliphatic compounds (e.g. 2,2,4 trimethyl pentane, chloroform, bromal) in other cases $\Delta N = 0.000$
=CH_2		0.085	0.000	
=CH—		−0.005	0.000	
=C<		−0.100	0.000	
HC≡C—		−0.115	0.075	
		Ring		Examples of ΔN values
>CH_2		0.060	0.000	
>CH—		−0.085	0.000	ΔN for 2-methyl propane 0.050
>C<		−0.180	0.000	2,3-dimethylbutane 0.050 (see text)
—CH—	Alicylic	0.040	0.000	2,2-dimethylpropane 0.100
=CH—	Others	0.050	0.000	2,2,4,4-tetramethyl pentane 0.100
=C<	Alicylic	−0.100	0.000	2,2,4-trimethylpentane 0.050 (both >CH— and >C< present)
=C<	Others	−0.120	0.000	>CH— in chloroform 0.050 but in isopropylamine 0.000 and
=C<	Fused	−0.040	0.000	isopropylbenzene 0.000
=CH—	Fused	−0.065	0.000	

Contributions of ring structure and hydrocarbon chains to ΔN_{cor}

Structure	ΔN_{cor}	Remarks and examples
All monocyclic and saturated polycyclic hydrocarbon rings (unsubstituted)	0.100	cyclopentane, benzene, or cis-decahydronaphthalene
Methyl substituted compounds of the above	0.050	ethylcyclopentane, toluene
Monocyclic monoalkyl alicyclic hydrocarbons		
$1 < C_{br} \leq 5$	0.025	ethylcyclopentane, *n*-pentylcyclohexane
$C_{br} > 5$	0.050	*n*-hexylcyclopentane
Monocyclic multisubstituted alkyl alicyclic hydrocarbons	0.025	1,3,5-trimethylcyclohexane
Monoalkyl benzenes with $C_{br} > 1$	0.025	ethylbenzene
Bicyclic hydrocarbons partly or fully unsaturated	0.050	tetralin, diphenyl, diphenylmethane
Unsaturated tricyclic hydrocarbons	0.100	*p*-terphenyl, triphenylmethane
Correction for multiple substituition in aromatics by hydrocarbon groups		
ortho	0.050	*o*-xylene, *o*-nitrotoluene
meta and para	0.000	*p*-xylene
1,3,5	0.100	1,3,5-trimethylbenzene
1,2,4	0.050	1,2,4-trimethylbenzene
1,2,6	0.000	
$\Delta \eta_{Bcor}$ for multiple substitution in aromatics by hydrocarbon groups $=0.070$		

TABLE 9-10 Sastri and Rao (1992) Group Contributions for η_B and N in Eq. (9-11.2) (*Continued*)

Contribution of halogen groups

Group	$\Delta\eta_B$ for halogen attached to carbon in: Aliphatic compounds					Others	ΔN^b in halogenated hydrocarbons with no other functional groups		
	—CH_3 or >CH_2^a	>CH—	>C<[a]	=CH—	=C<		Alicyclics	Aromatics	Others
—F[c]	0.185	0.155	0.115	n.d.	n.d.	0.185	0.075	0.025	0.00
—Cl[a]	0.185	0.170	0.170	0.180	0.150	0.170	0.075	0.025	0.00
—Br	0.240	0.235	0.235	0.240	0.210	0.210	0.075	0.025	0.00
—I	0.260	0.260	n.d.	n.d.	n.d.	0.260	0.075	0.025	0.00

[a]Special configurations/function group structure combination	ΔN_{cor}	Remarks and examples
(1) X—$(CH_2)_n$—X where X is halogen	0.050	1,3-dichloropropane
(2) Cl—C—Cl where the C is in a ring	0.050	For each group, one correction in hexachlorocyclopentadiene
[b]Case of non-hydrocarbon group present in cyclic compounds		
(1) Halogen attached to ring carbons in compounds containing		
(A) —NH_2 or phenolic —OH	−0.075	2-chloro-6-methyl aniline
(B) oxygen-containing groups other than OH	0.050	2-chlorophenylmethyl ether
(C) other non-oxygen functional groups	−0.050	2-chloropyridine
(2) Halogen attached to non-hydrocarbon functional group	−0.050	benzoylbromide

[c]Fluorine groups in perfluorocompounds

Group	$\Delta\eta_B$	
Non-ring		
—CF_3	0.210	ΔN=0.150 for all perfluoro *n*-compounds
>CF_2	0.000	
>CF—	−0.080	ΔN=0.200 for all isocompounds
Ring		
—CF_2—	0.145	ΔN=0.200 for all cyclic compounds
>CF—	−0.170	

Contribution of oxygen groups

Group	Structure	$\Delta\eta_B$	ΔN	Remarks
—O—	Non-ring attached to ring carbon	0.020	0.050	For multiple occurrence $\Delta\eta_{Bcor} = 0.050$ (1,3-dimethoxybenzene) In compounds containing $—NH_2$ or phenolic OH group attached to ring carbon $\Delta N_{cor} = -0.050$ (*o*-anisidine, 2-methoxyphenol)
—O—	Ring (Single)	0.120	0.050	
	(Multiple) occurrence)	0.200	0.150	Combined value (dioxane, paraldehyde)
—O—	Others	0.000	0.050	(i) In aliphatic compounds containing —OH, special value for the combination $\Delta N = 0.100$ (2-methoxyethanol) (ii) $\Delta\eta_{Bcor}$ for multiple occurrence 0.05 (dimethoxymethane)
>CO	Non-ring attached to ring carbon	0.030	0.050	(i) In the cyclic compounds containing NH_2 group (with or without other functional groups) special value ΔN −0.100 for the combination (ethylanthranilate) (ii) In cyclic compounds containing >NH group $\Delta\eta_{Bcor} = 0.080$ (acetanilide)
>CO	Ring	0.055	0.100	(i) In cyclic compounds containing >NH group $\Delta\eta_{Bcor} = 0.100$ (ii) In compounds containing —O— group special value $\Delta N =$ 0.125 for the combination
>CO	Others	0.030	0.025	(i) For aliphatic compounds containing $—NH_2$ or >N— groups (acetamide) $\Delta\eta_{Bcor} = 0.080$ (ii) For cyclic compounds containing >NH group (acetanilide) $\Delta\eta_{Bcor} = 0.080$ (iii) In aliphatic compounds containing —OH special value for the combination, $\Delta N = 0.125$ (diacetonealcohol)
$—C(O)_3C—$	Anhydride	0.060	0.050	
—CHO	Aldehyde	0.140	0.050	In compounds containing —OH (phenolic) special value for the combination, $\Delta N = 0.075$ (salicylaldehyde)

TABLE 9-10 Sastri and Rao (1992) Group Contributions for η_B and N in Eq. (9-11.2) (*Continued*)

Contribution of oxygen groups

Group	Structure	$\Delta\eta_B$	ΔN	Remarks
—COO—	Ester	0.040	0.050	(i) For multiple occurance, $\Delta N = 0.100$ (dibutylphthalate) (ii) For —H in formates $\Delta\eta_B = 0.165$ (iii) In cyclic compounds containing NH_2 group, special value for the combination, $\Delta N = 0.100$ (ethylanthanilate) (iv) For aliphatic compounds containing $—NH_2$ or >N—groups, $\Delta\eta_{Bcor} = 0.080$ (methylcarbamate)
—COOH	In aliphatics			
	saturated	0.220	0.100	For C = 3 or 4 $\Delta N = 0.050$
	unsaturated	0.250	0.100	
	In aromatics	0.195	0.175	

Contribution of hydroxyl groups

Structure	$\Delta\eta_B$	ΔN	Remarks
—OH in aliphatics saturated primary	$0.615 - 0.092C + 0.004C^2 - 10^{-0.58C}$ for $C \le 10$ 0.095 for $C > 10$	0.3 for $2 < C < 12$ 0.15 for others	(i) In compounds containing —C— group special value for the combination, $\Delta N = 0.100$ (2 methoxy ethanol) (ii) In compounds containing >NH group, special value for the combination, $\Delta N = 0.300$ (aminoethyl ethanolamine)
Primary branched	$0.615 - 0.092C + 0.004C^2 - 10^{-0.58C}$	0.375	
Secondary straight chain	$0.615 - 0.092C + 0.004C^2 - 10^{-0.58C}$	0.450 for $C \le 5$ 0.300 for $C > 5$	
Secondary branched	$0.615 - 0.092C + 0.004C^2 - 10^{-0.58C}$	0.450 for $C \le 8$ 0.300 for $C > 8$	
Tertiary saturated	$0.615 - 0.092C + 0.004C^2 - 10^{-0.58C}$	0.650 for $C \le 5$ 0.300 for $C > 5$	In compounds containing >CO/—O— groups special value for the combination $\Delta N = 0.125$ (diacetonealcohol)
Unsaturated primary	$0.615 - 0.092C + 0.004C^2 - 10^{-0.58C}$	0.175	
Unsaturated tertiary	$0.615 - 0.092C + 0.004C^2 - 10^{-0.58C}$	0.425	

Structure	$\Delta\eta_B$	ΔN	Remarks
In cyclic alcohols	0.270	0.150	
Phenolic	0.270	0.200	(i) In compounds containing —NH_2 or —CHO groups in ortho position, special value for the combination, $\Delta N = 0.075$ (2-nitrophenol, salicylaldehyde) (ii) In compounds containing —O— $\Delta N_{cor} = 0.050$ (4-methoxphenol)

Contribution of nitrogen groups

Group	Structure	$\Delta\eta_B^a$	ΔN	Remarks
—NH_2	In aliphatic *n*-amines	0.170	0.100	(i) $\Delta\eta_{Bcor} = 0.100$ in $NH_2 - (CH_2)_n - NH_2$ (ethylenediamine) (ii) in compounds containing >COO $\Delta\eta_{Bcor} = 0.080$ (acetamide) (iii) In compounds containing —COO, $\Delta N_{cor} = 0.100$ (ethyl carbamate)
—NH_2	Aliphatic isoamines attached to >CH	0.200	0.100	(isopropylamine)
—NH_2	In monocyclic compounds, attached to side chain	0.170	0.100	(benzylamine)
—NH_2	In monocyclic compounds, attached to ring carbon	0.205	0.150	(i) For compounds containing —O— $\Delta N_{cor} = -0.050$ (2-methoxyaniline) (ii) In cyclic compounds containing —COO—group, special value for the combination, $\Delta N = 0.100$ (ethylanthranilate)
—NH_2	In other aromatics	0.150	0.100	(1-naphthylamine)
>NH	In aliphatics	0.020	0.075	In compounds containing —OH special value for the combination, $\Delta N = 0.300$ (aminoethyl etanolamine)
>NH	In aromatic compounds, attached to side chain	0.020	0.075	(dibenzylamine)
>NH	In aromatic compounds, attached to ring carbon	0.020	0.100	
>NH	Ring	0.160	0.100	

TABLE 9-10 Sastri and Rao (1992) Group Contributions for η_B and N in Eq. (9-11.2) (*Continued*)

Group	Structure	$\Delta\eta_B^a$	ΔN	Remarks
				Contribution of nitrogen groups (continued)
>N—	In aliphatics	−0.115	0.050	For aliphatic compounds containing >CO $\Delta\eta_{Bcor}$ = 0.080 (dimethyl acetamide)
>N—	In aromatic compounds, attached to side chain	−0.115	0.050	(tribenzylamine)
>N—	In aromatic compounds, attached to ring carbon	−0.060	0.050	
>N—	Ring	0.100	0.050	In compounds containing —CN or halogen, ΔN_{cor} = −0.050
—NO_2	In aliphatics	0.180	0.050	For multiple occurrences ΔN_{cor} = 0.050
	In aromatics	0.160	0.050	(i) for multiple occurrences ΔN_{cor} = 0.050 and $\Delta\eta_{Bcor}$ = 0.070 (*m*-dinitrobenzene) (ii) In compounds containing —OH (phenclic) in ortho position, special value for the combination, ΔN = 0.075 (2-nitrophenol)
—CN		0.135	0.025	(i) For multiple occurrence, ΔN_{cor} = 0.075 (ii) With N in ring, ΔN_{cor} = −0.050

[a] $\Delta\eta_B$ = 0.080 for —H in compounds containing hydrocarbon functional groups (e.g. formanilide)

Functional group/structure	$\Delta\eta_B$	ΔN
Contribution of sulphur groups		
—S— Non-ring	0.045	0.000
—S— ring	0.150	0.050
—SH	0.150	0.025

TABLE 9-11 Comparison of Calculated and Experimental Viscosities of Liquids

Compound	T, K	η (exp.) cP**	Percent error* in liquid viscosity calculated by the method of: Orrick and Erbar	Sastri and Rao	Przezdziecki and Sridhar
Acetone	183	2.075	−25	−3.5	−11
	213	0.982	−6.7	1.9	−4.6
	273	0.389	−8.3	3.6	−2.3
	303	0.292	−9.4	1.6	−1.2
	333	0.226	−8.3	1.9	0.2
Acetic acid	283	1.450	−22	−15	8.6
	313	0.901	−15	−15	0
	353	0.561	−9.5	−17	−1.3
	383	0.416	−5.3	−16	0.3
Aniline	263	13.4	—	−24	—
	293	4.38	—	−4.8	—
	333	1.520	—	8.1	−49
	393	0.658	—	−9.6	−33
Benzene	278	0.826	−45	−8.5	1.1
	313	0.492	−35	−6.6	7.3
	353	0.318	−26	−5.5	12
	393	0.219	−46	−5.7	18
	433	0.156	−7.1	−6.4	23
	463	0.121	5.1	−9.5	28
n-Butane	183	0.630	−14	1.6	−9
	213	0.403	−20	−2.8	−8.9
	273	0.210	−23	−0.4	−5.8
1-Butene	163	0.79	−22	0.9	−13
	193	0.45	−20	−2.9	−9.6
	233	0.26	−18	−2.5	−3.3
n-Butyl alcohol	273	5.14	−2.1	0.3	—
	313	1.77	−1.6	−3.4	—
	353	0.762	0.5	−2	—
	393	0.394	−1.4	1.4	—
Carbon tetrachloride	273	1.369	20	−4.4	−24
	303	0.856	22	−2	−15
	343	0.534	20	−0.1	−6.7
	373	0.404	19	−0.7	−2.8
Chlorobenzene	273	1.054	1.4	2.7	−8.3
	313	0.639	−0.6	0.8	−7
	353	0.441	−0.9	−1.2	−5.2
	393	0.326	−5.1	−0.9	−3.8
Chloroform	273	0.700	40	7.4	−11
	303	0.502	34	5.7	−8.1
	333	0.390	27	3.6	−7.9
Cyclohexane	278	1.300	−51	−29.7	−38
	333	0.528	−38	−16.4	−22
Cyclopentane	293	0.439	−32	−5.1	−33
	323	0.323	−28	−7.8	−29

TABLE 9-11 Comparison of Calculated and Experimental Viscosities of Liquids (*Continued*)

			Percent error* in liquid viscosity calculated by the method of		
Compound	*T*, K	η (exp.) cP**	Orrick and Erbar	Sastri and Rao	Przezdziecki and Sridhar
2,2-Dimethylpropane	258	0.431	−3.5	−24.3	20
	283	0.281	−0.8	−15.1	30
Ethane	98	0.985	30	53.6	−24
	153	0.257	−12	26.4	−14
	188	0.162	−22	21.3	−13
Ethylene chloride	273	1.123	−43	−20.7	—
	313	0.644	−35	−15.3	—
	353	0.417	−27	−8.8	—
Ethyl alcohol	273	1.770	27	−14.1	—
	313	0.826	3.5	−6	—
	348	0.465	−5.4	7.7	—
Ethyl acetate	293	0.458	−4.2	−5.5	−16
	353	0.246	0.4	−1.2	−5.3
	413	0.153	7.4	−9.7	−1.8
	463	0.0998	27	−2.2	4.8
Ethylbenzene	253	1.240	−2.9	19.7	−33
	313	0.535	−1.2	7.3	−23
	373	0.308	−1.7	0.1	−16
	413	0.231	−1.2	−1.9	−13
Ethyl bromide	293	0.395	27	0	−23
	333	0.269	32	2.2	−17
	373	0.199	36	4.6	−16
Ethylene	103	0.70	−25	−0.7	25
	133	0.31	−27	−2.3	−17
	173	0.15	−22	4.8	−6.4
Ethyl ether	273	0.289	0	2.7	0
	293	0.236	0	1.6	2.2
	333	0.167	2	−3.5	4
	373	0.118	1	5.5	7.4
Ethyl formate	273	0.507	−18	6.8	−16
	303	0.362	−17	6.6	−11
	328	0.288	−16	7	−9.6
n-Heptane	183	3.77	−21	−33.2	−1.7
	233	0.965	−0.5	−11.3	−27
	293	0.418	−1.9	−7.2	−21
	373	0.209	−3.3	−1	−17
n-Hexacontane	408	7.305	18	37	—
($C_{60}H_{122}$)	466	3.379	−63	19	—
n-Hexane	213	0.888	2.9	−2.3	−8.3
	273	0.381	−2.4	−2.4	−8.2
	343	0.205	−4.9	1.2	−7.1

TABLE 9-11 Comparison of Calculated and Experimental Viscosities of Liquids (*Continued*)

Compound	T, K	η (exp.) cP**	Percent error* in liquid viscosity calculated by the method of: Orrick and Erbar	Sastri and Rao	Przezdziecki and Sridhar
Isobutane	193	0.628	−23	1.1	−37
	233	0.343	−25	12.2	−29
	263	0.239	−24	−17.1	−23
Isopropyl alcohol	283	3.319	−24	−16.2	—
	303	1.811	−15	9.8	—
	323	1.062	−10	4.5	—
Methane	88	0.226	60	1.5	−11
	113	0.115	23	−20.1	−4.3
2-Methylbutane	223	0.550	−13	−10.8	−30
	253	0.353	−12	0.3	−21
	303	0.205	−10	−1.9	−12
n-Pentane	153	2.35	−1	−1.7	11
	193	0.791	3.8	0.1	−7
	233	0.428	−3.3	7.3	−6
	273	0.279	−8.2	2.9	−4.7
	303	0.216	−11	0.6	−4.9
Phenol	323	3.020	0	1.5	−50
	373	0.783	37	12.5	−5.4
Propane	133	0.984	−1.5	45.4	−23
	193	0.327	−22	14	−19
	233	0.205	−25	−1.7	−16
n-Propyl alcohol	283	2.897	−9.1	0.8	—
	313	1.400	−9.8	−6.1	—
	373	0.443	−6.5	−7.4	—
Toluene	253	1.070	−19	0.4	−33
	293	0.587	−13	0.7	−24
	333	0.380	−10	−2.1	−16
	383	0.249	−6.8	−5.1	−10
o-Xylene	273	1.108	3.1	−5.1	−5.5
	313	0.625	5	−4.5	−4.8
	373	0,345	3.7	4.5	−0.3
	413	0.254	3.6	1.8	1.9
				−2.7	
m-Xylene	273	0.808	1.1	−2.6	1.9
	313	0.492	1.4	−0.4	1.8
	353	0.340	0.3	2	2.9
	413	0.218	1.4	3.4	4.6

*[(calc − exp)/exp] × 100.
**Data from Aasen, et al. (1990), Amdur and Mason (1958), and Landolt-Bornstein (1955).

solution o-xylene has 4=CH— (ring, not alicyclic), 2=C< (ring, not alicyclic) and 2 —CH_3 groups. There is a correction to η_{Bcor} of 0.07 for multiple substitution. With values from Table 9-10:

$$\eta_B = 4 \times 0.05 - 2 \times 0.12 + 2 \times 0.105 + 0.07 = 0.24 \text{ mPa·s}$$

$$N = 0.2 + 0.05 = 0.25$$

ethanol has one —CH_3, one >CH_2 (non-ring) and one —OH. With values from Table 9-10:

$$\eta_B = 0.105 + 0.615 - 0.092 \times 2 + 0.004 \times 4 - 10^{-0.58\times 2} = 0.483 \text{ mPa·s}$$

$$N = 0.2 + 0.15 = 0.35$$

ethylbenzene has 5 =CH— (ring, not alicyclic), one =C< (ring not alicyclic), one —CH_3, and one —CH_2— (non-ring). There is a branching correction to ΔN of 0.025. With values from Table 9-10:

$$\eta_B = 5 \times 0.05 - 0.12 + 0.105 = 0.235 \text{ mPa·s}$$

$$N = 0.2 + 0.025 = 0.225$$

2-3 dimethylbutane has 4 —CH_3 and 2 >CH— (non-ring). The value of ΔN of 0.05 is applied only once. With values from Table 9-10:

$$\eta_B = 4 \times 0.105 - 2 \times 0.11 = 0.2 \text{ mPa·s}$$

$$N = 0.2 + 0.05 = 0.25$$

o-chlorophenol has 4=CH— (ring, not alicyclic), 2=C< (ring, not alicyclic), one —Cl attached to an "other", and one —OH (phenolic). Note that the —Cl contribution to ΔN of 0.025 is not used. Footnote b in the halogen section of Table 9-10 applies because of the presence of the non-hydrocarbon —OH group. With values from Table 9-10:

$$\eta_B = 4 \times 0.05 - 2 \times 0.12 + 0.17 + 0.27 = 0.4 \text{ mPa·s}$$

$$N = 0.2 + 0.2 - 0.075 = 0.325$$

Przezdziecki and Sridhar (1985) Method

In this technique, the authors propose using the Hildebrand-modified Batschinski equation (Batschinski, 1913; Hildebrand, 1971; Vogel and Weiss, 1981)

$$\eta_L = \frac{V_o}{E(V - V_o)} \tag{9-11.6}$$

where η_L = liquid viscosity, cP
V = liquid molar volume, cm^3/mol

and the parameters E and V_o are defined below.

$$E = -1.12 + \frac{V_c}{12.94 + 0.10\,M - 0.23\,P_c + 0.0424\,T_{fP} - 11.58(T_{fP}/T_c)} \tag{9-11.7}$$

$$V_o = 0.0085\omega T_c - 2.02 + \frac{V_m}{0.342(T_{fP}/T_c) + 0.894} \tag{9-11.8}$$

where T_c = critical temperature, K
P_c = critical pressure, bar
V_c = critical volume, cm^3 /mol
M = molecular weight, g/mol
T_{fP} = freezing point, K
ω = acentric factor
V_m = liquid molar volume at T_{fP}, cm^3/mol

Thus, to use Eq. (9-11.6), one must have values for T_c, P_c, V_c, T_{fP}, ω, and V_m in addition to the liquid molar volume V at the temperature of interest. The authors recommend that V_m and V be estimated from T_{fP} and T by the Gunn-Yamada (1971) method. In the Gunn-Yamada method, one accurate value of V is required in the temperature range of applicability of Eq. (9-11.6). We define this datum point as V^R at T^R; then at any other temperature T,

$$V(T) = \frac{f(T)}{f(T^R)} V^R \tag{9-11.9}$$

where

$$f(T) = H_1(1 - \omega H_2) \tag{9-11.10}$$

$$H_1 = 0.33593 - 0.33953T_r + 1.51941T_r^2 - 2.02512T_r^3 + 1.11422T_r^4 \tag{9-11.11}$$

$$H_2 = 0.29607 - 0.09045T_r - 0.04842T_r^2 \tag{9-11.12}$$

Equation (9-11.6) was employed with Eqs. (9-11.7) to (9-11.12) to estimate liquid viscosities for the compounds in Table 9-11. The values of T_c, P_c, V_c, T_{fP}, and ω, were obtained from Appendix A. The reference volume for each compound was calculated from the liquid density datum value given in Appendix A. Large errors were noted for alcohols, and those results are not included in the table. For other compounds, the errors varied widely and, except for a few materials, the technique underestimated the liquid viscosity. Larger errors were normally noted at low temperatures, but that might have been expected from the form of Eq. (9-11.6). That is, because V_o is of the order of the volume at the freezing point and $\eta_L \propto (V - V_o)^{-1}$, the estimated value of η_L becomes exceedingly sensitive to the choice of V. This problem was emphasized by Luckas and Lucas (1986), who suggest that Eq. (9-11.6) should not be used below T_r values of about 0.55.

Example 9-19 Use the Przezdziecki and Sridhar correlation to estimate the liquid viscosity of toluene at 383 K. The experimental value is 0.249 cP (Vargaftik, et al., 1996).

solution From Appendix A, for toluene (slightly different values were used to calculate the results shown in Table 9-11)

$$\begin{aligned} T_c &= 591.75 \text{ K} \\ P_c &= 41.08 \text{ bar} \\ V_c &= 316 \text{ cm}^3/\text{mol} \\ T_{fP} &= 178 \text{ K} \\ M &= 92.14 \text{ g/mol} \\ \omega &= 0.264 \\ V_L &= 106.87 \text{ cm}^3/\text{mol at } 298.15 \text{ K} \end{aligned}$$

With $T^R = 298.15$ K, and with Eqs. (9-11.9) to (9-11.12),

$$T_r^R = \frac{298.15}{591.75} = 0.504$$

$$H_1(T_r^R) = 0.33593 - (0.33953)(0.504) + (1.51941)(0.504)^2 - (2.02512)(0.504)^3 + (1.11422)(0.504)^4 = 0.363$$

$$H_2(T_r^R) = 0.29607 - (0.09045)(0.504) - (0.04842)(0.504)^2 = 0.238$$

$$f(T^R) = 0.363[1 - (0.264)(0.238)] = 0.340$$

Similarly,

	T, K	T_r	H_1	H_2	$f(T)$
T_{fP}	178	0.301	0.325	0.264	0.303
T	383	0.647	0.399	0.217	0.376

Then

$$V_m = \frac{0.303}{0.340}(106.87) = 95.2 \text{ cm}^3/\text{mol}$$

$$V = \frac{0.376}{0.340}(106.87) = 118.2 \text{ cm}^3/\text{mol}$$

This value for V agrees with that given in Vargaftik, et al. (1996). With Eqs. (9-11.7) and (9-11.8)

$$E = -1.12 + 316/[12.94 + (0.10)(92.14) - (0.23)(41.08) + (0.0424)(178) - (11.58)(178/591.8)] = 17.72$$

$$V_o = (0.0085)(0.264)(591.75) - 2.02 + \frac{95.2}{[(0.342)(178/591.8) + 0.894]}$$

$$= 94.8 \text{ cm}^3/\text{mol}$$

Then, with Eq. (9-11.6)

$$\eta_L = \frac{94.8}{17.72(118.2 - 94.8)} = 0.229 \text{ cP}$$

$$\text{Error} = \frac{0.229 - 0.249}{0.249} \times 100 = -8\%$$

Other Correlations

Other viscosity-correlating methods have been proposed, and a number of these are summarized in Mehrotra, et al. (1996) and Monnery, et al. (1995). Other recent correlations are given in Mehrotra (1991), and the earlier literature was reviewed in the 4th Edition of this book.

Recommendations for Estimating Low-temperature Liquid Viscosities

Three estimation methods have been discussed. In Table 9-11, calculated liquid viscosities are compared with experimental values for 36 different liquids (usually of simple structure). Large errors may result, as illustrated for all methods. The method of Przezdziecki and Sridhar should not be used for alcohols.

The method of Sastri and Rao assumes that the temperature dependence of η_L is related to the temperature dependence of the vapor pressure, whereas the Orrick and Erbar method is slightly modified to include the liquid density. Neither is reliable for highly branched structures or for inorganic liquids and the Orrick-Erbar method cannot be used for sulfur compounds. Both are limited to a temperature range from somewhat above the freezing point to about $T_r \approx 0.75$. Przezdziecki and Sridhar's method employs the Hildebrand equation, which necessitates knowledge of liquid volumes.

It is recommended that, in general, the method of Sastri and Rao be used to estimate low-temperature liquid viscosities. Errors vary widely, but should be less than 10 to 15% in most instances.

9-12 ESTIMATION OF LIQUID VISCOSITY AT HIGH TEMPERATURES

Low-temperature viscosity correlations as covered in Sec. 9-10 usually assume that ln η_L is a linear function of reciprocal absolute temperature. Above a reduced temperature of about 0.7, this relation is no longer valid, as illustrated in Fig. 9-10. In the region from about $T_r = 0.7$ to near the critical point, many estimation methods are of a corresponding states type that resemble or are identical with those used in the first sections of this chapter to treat gases. For this temperature range, Sastri (1998) recommends

$$\ln \eta = \left[\frac{\ln \eta_B}{\ln(\alpha\eta_B)}\right]^{\phi} \ln(\alpha\eta_B) \tag{9-12.1}$$

where η is in mPa · s

η_B = viscosity at T_b in mPa · s, from contributions in Table 9-10
α = 0.1175 for alcohols and 0.248 for other compounds

$$\phi = \frac{1 - T_r}{1 - T_{br}} \tag{9-12.2}$$

where $T_r = T/T_c$
$T_{br} = T_b/T_c$

Sastri reports average deviations of 10% for $T_r > 0.9$ and 6% for $T_{br} < T_r < 0.9$.

Example 9-20 Estimate the saturated liquid viscosity of *n*-propanol at 433.2 K by using Eq. (9-12.1). The experimental value is 0.188 cP.

solution From Appendix A, T_b = 370.93 K and T_c = 536.78 K. With contributions from Table 9-10, $\eta_B = 0.105 + 0.615 - 0.092 \times 3 + 0.004 \times 9 - 10^{-3\times 0.58} = 0.462$ mPa · s = 462 μPa · s. From Eq. (9-12.2)

$$\phi = \frac{1 - 433.2/536.78}{1 - 370.93/536.78} = 0.624$$

With α = 0.1175, Eq. (9-12.1) gives

$$\ln \eta = \ln(0.1175 \times 462)\left[\frac{\ln(462)}{\ln(0.1175 \times 462)}\right]^{0.624}$$

$$\eta = 185\ \mu\text{Pa} \cdot \text{s} = 0.185\ \text{cP}$$

$$\text{Error} = \frac{0.185 - 0.188}{0.188} \times 100 = -1.6\%$$

A more general estimation method would logically involve the extension of the high-pressure gas viscosity correlations described in Sec. 9-6 into the liquid region. Two techniques have, in fact, been rather widely tested and found reasonably accurate for reduced temperatures above about 0.5. These methods are those of Chung, et al. (1988) and Brulé and Starling (1984). Both methods use Eq. (9-6.16), but they have slightly different coefficients to compute some of the parameters. The Chung, et al. form is preferable for simple molecules and will treat polar as well as nonpolar compounds. The Brulé and Starling relation was developed primarily for complex hydrocarbons, and the authors report their predictions are within 10% of experimental values in the majority of cases. The Chung, et al. method has a similar accuracy for most nonpolar compounds, but significantly higher errors can occur with polar, halogenated, or high-molecular weight compounds. In both cases, one needs accurate liquid density data, and the reliability of the methods decreases significantly for T_r less than about 0.5. The liquids need not be saturated; subcooled compressed liquid states simply reflect a higher liquid density. The Chung, et al. technique was illustrated for dense gas ammonia in Example 9-12. The procedure is identical when applied to high-temperature liquids.

Discussion

The quantity of accurate liquid viscosity data at temperatures much above the normal boiling point is not large. In addition, to test estimation methods such as those of Chung, et al. or Brulé and Starling, one needs accurate liquid density data under the same conditions which apply to the viscosity data. This matching makes it somewhat difficult to test the methods with many compounds. However, Brulé and Starling developed their technique so that they would be coupled to a separate computation program using a modified BWR equation of state to provide densities. They report relatively low errors, and this fact appears to confirm the general approach (See also Brulé and Starling, 1984). Hwang, et al. (1982) have proposed viscosity (as well as density and surface tension) correlations for coal liquids.

Regardless of what high-temperature estimation method is chosen, there is the problem of joining both high- and low-temperature estimated viscosities should that be necessary.

9-13 LIQUID MIXTURE VISCOSITY

Essentially all correlations for liquid mixture viscosity refer to solutions of liquids below or only slightly above their normal boiling points; i.e., they are restricted to reduced temperatures (of the pure components) below about 0.7. The bulk of the discussion below is limited to that temperature range. At the end of the section, however, we suggest approximate methods to treat high-pressure, high-temperature liquid mixture viscosity.

At temperatures below $T_r \approx 0.7$, liquid viscosities are very sensitive to the structure of the constituent molecules (See Sec. 9-11). This generality is also true for liquid mixtures, and even mild association effects between components can often significantly affect the viscosity. For a mixture of liquids, the shape of the curve of viscosity as a function of composition can be nearly linear for so-called ideal mixtures. But systems that contain alcohols and/or water often exhibit a maximum or a minimum and sometimes both (Irving, 1977a).

Almost all methods to estimate or correlate liquid mixture viscosities assume that values of the viscosities of the pure components are available. Thus the methods are interpolative. Nevertheless, there is no agreement on the best way to carry out the interpolation. Irving (1977) surveyed more than 50 equations for binary liquid viscosities and classified them by type. He points out that only very few do not have some adjustable constant that must be determined from experimental mixture data and the few that do not require such a parameter are applicable only to systems of similar components with comparable viscosities. In a companion report from the National Engineering Laboratory, Irving (1977a) has also evaluated 25 of the more promising equations with experimental data from the literature. He recommends the one-constant Grunberg-Nissan (1949) equation [see Eq. (9-13.1)] as being widely applicable yet reasonably accurate except for aqueous solutions. This NEL report is also an excellent source of viscosity data tabulated from the literature. Other data and literature sources for data may be found in Aasen et al. (1990), Aucejo, et al. (1995), supplementary material of Cao, et al. (1993), Franjo, et al. (1995), Kouris and Panayiotou (1989), Krishnan, et al. (1995, 1995a), Kumagai and Takahashi (1995), Petrino, et al. (1995), Stephan and Hildwein (1987), Stephan and Heckenberger (1988), Teja, et al., (1985), and Wu, et al. (1998).

Method of Grunberg and Nissan (1949)

In this procedure, the low-temperature liquid viscosity for mixtures is given as

$$\ln \eta_m = \sum_i x_i \ln \eta_i + \frac{1}{2} \sum_{i=1}^{n} \sum_{j=i}^{n} x_i x_j G_{ij} \tag{9-13.1}$$

or, for a binary of 1 and 2,

$$\ln \eta_m = x_1 \ln \eta_1 + x_2 \ln \eta_2 + x_1 x_2 G_{12} \tag{9-13.2}$$

since $G_{ii} = 0$. In Eqs. (9-13.1) and (9-13.2), x is the liquid mole fraction and G_{ij} is an interaction parameter which is a function of the components i and j as well as the temperature (and, in some cases, the composition). This relation has probably been more extensively examined than any other liquid mixture viscosity correlation. Isdale (1979) presents the results of a very detailed testing using more than 2000 experimental mixture datum points. When the interaction parameter was regressed from experimental data, nonassociated mixtures and many mixtures containing alcohols, carboxylic acids, and ketones were fitted satisfactorily. The overall root mean square deviation for the mixtures tested was 1.6%. More recently, Isdale, et al. (1985) proposed a group contribution method to estimate the binary interaction parameter G_{ij} at 298 K.

The procedure to be followed is:

1. For a binary of i and j, select i by following the priority rules below. (j then becomes the second component.)
a. i = an alcohol, if present
b. i = an acid, if present
c. i = the component with the most carbon atoms
d. i = the component with the most hydrogen atoms
e. i = the component with the most —CH_3 groups
$G_{ij} = 0$ if none of these rules establish a priority.

2. Once the decision has been made which component is i and which is j, calculate $\Sigma\Delta_i$ and $\Sigma\Delta_j$ from the group contributions in Table 9-12.

3. Determine the parameter W. (If either i or j contains atoms other than carbon and hydrogen, set $W = 0$ and go to step 4.) Let the number of carbon atoms in i be N_i and that in j be N_j.

$$W = \frac{(0.3161)(N_i - N_j)^2}{N_i + N_j} - (0.1188)(N_i - N_j) \tag{9-13.3}$$

4. Calculate G_{ij} from

$$G_{ij} = \Sigma\Delta_i - \Sigma\Delta_j + W \tag{9-13.4}$$

G_{ij} is sometimes a function of temperature. However, existing data suggest that, for alkane-alkane solutions or for mixtures of an associated component with an unassociated one, G_{ij} is independent of temperature. However, for mixtures of nonassociated compounds (but not of only alkanes) or for mixtures of associating compounds, G_{ij} is a mild function of temperature. Isdale, et al. (1985) suggest for these latter two cases,

$$G_{ij}(T) = 1 - [1 - G_{ij}(298)] \frac{573 - T}{275} \tag{9-13.5}$$

where T is in kelvins.

Example 9-21 Estimate the viscosity of a mixture of acetic acid and acetone at 323 K (50°C) that contains 70 mole percent acetic acid. Isdale, et al. quote the experimental value to be 0.587 cP, and, at 50°C, the viscosities of pure acetic acid and acetone are 0.798 and 0.241 cP, respectively.

TABLE 9-12 Group Contributions for G_{ij} at 298 K

Group	Notes	Value of Δ_i
$—CH_3$		−0.100
$>CH_2$		0.096
$>CH—$		0.204
$>C<$		0.433
Benzene ring		0.766
Substitutions:		
Ortho		0.174
Meta		—
Para		0.154
Cyclohexane ring		0.887
—OH	Methanol	0.887
	Ethanol	−0.023
	Higher aliphatic alcohols	−0.443
>C=O	Ketones	1.046
—Cl		$0.653–0.161N_{Cl}$
—Br		−0.116
—COOH	Acid with:	
	Nonassociated liquids	$-0.411 + 0.06074N_C$
	Ketones	1.130
	Formic acid with ketones	0.167

N_{Cl} = number of chlorine atoms in the molecule.
N_C = *total* number of carbon atoms in both compounds.

solution First we must estimate G_{ij} at 298 K. Component i is acetic acid (priority rule b). Since the mixture contains atoms other than carbon and hydrogen (i.e., oxygen), $W = 0$. Then, with Table 9-12,

$\Sigma\Delta_i$ (acetic acid) = $—CH_3$ + —COOH = −0.100 + 1.130 = 1.030
$\Sigma\Delta_j$ (acetone) = (2)($—CH_3$)+>C=0 = (2)(−0.100)+ 1.046 = 0.846

With Eq. (9-13.4),

$$G_{ij} = 1.030 - 0.846 = 0.184 \text{ at } 298 \text{ K}$$

At 50°C = 323 K, we need to adjust G_{ij} with Eq. (9-13.5).

$$G_{ij}(323 \text{ K}) = 1 - \frac{(1 - 0.184)(573 - 323)}{275} = 0.258$$

Then, using Eq. (9-13.2),

$$\ln \eta_m = (0.7) \ln (0.798) + (0.3) \ln (0.241) + (0.7)(0.3)(0.258) = -0.531$$

$$\eta_m = 0.588 \text{ cP}$$

This estimated value is essentially identical with the experimental result of 0.587 cP.

To summarize the Isdale modification of the Grunberg-Nissan equation, for each possible binary pair in the mixture, first decide which component is to be labeled i and which j by the use of the priority rules. Determine $\Sigma\Delta_i$ and $\Sigma\Delta_j$ by using

Table 9-12 and W from Eq. (9-13.3), if necessary. Use Eq. (9-13.4) to calculate G_{ij}. Correct for temperatures other than 298 K, if necessary, with Eq. (9-13.5). With the values of G_{ij} so determined, use either Eq. (9-13.1) or (9-13.2) to determine the viscosity of the liquid mixture. This technique yields quite acceptable estimates of low-temperature liquid mixture viscosities for many systems, but Table 9-12 does not allow one to treat many types of compounds. Also, the method does not cover aqueous mixtures.

UNIFAC-VISCO Method (Chevalier, et al., 1988; Gaston-Bonhomme, et al., 1994)

Gaston-Bonhomme, Petrino and Chevalier have modified the UNIFAC activity coefficient method (described in Chap. 8) to predict viscosities. In this method, viscosity is calculated by

$$\ln \eta_m = \sum_i x_i \ln(\eta_i V_i) - \ln V_m + \frac{\Delta^* g^{EC}}{RT} + \frac{\Delta^* g^{ER}}{RT} \tag{9-13.6}$$

The combinatorial term is the same as in the UNIQUAC model (see Table 8-8) and is calculated by

$$\frac{\Delta^* g^{EC}}{RT} = \sum_i x_i \ln \frac{\phi_i}{x_i} + \frac{z}{2} \sum_i q_i x_i \ln \frac{\theta_i}{\phi_i} \tag{9-13.7}$$

where z is the coordination number, equal to 10, θ_i and ϕ_i are the molecular surface area fraction and molecular volume fraction, respectively, given by

$$\theta_i = \frac{x_i q_i}{\sum_j x_j q_j} \tag{9-13.8}$$

and

$$\phi_i = \frac{x_i r_i}{\sum_j x_j r_j} \tag{9-13.9}$$

where q_i, the van der Waals' surface area, and r_i, the van der Waals' volume of component i, are found by summation of the corresponding group contributions. Thus, if $n_k^{(i)}$ is the number of groups of type k in the molecule i,

$$q_i = \sum_k n_k^{(i)} Q_k \tag{9-13.10}$$

$$r_i = \sum_k n_k^{(i)} R_k \tag{9-13.11}$$

where Q_k and R_k are the constants representing the group surface and size and are given in Table 9-13. These values match the UNIFAC values in Table 8-23 in cases where groups are defined the same. The residual term in Eq. (9-13.6) is calculated by

TABLE 9-13 UNIFAC-VISCO, Group Volume and Surface Area Parameters

Group k	R_k	Q_k
CH_2, CH_{2cy}	0.6744	0.540
CH_3	0.9011	0.848
CH_{ar}	0.5313	0.400
Cl	0.7910	0.724
CO	0.7713	0.640
COO	1.0020	0.880
OH	1.0000	1.200
CH_3OH	1.4311	1.432

$$\frac{\Delta^* g^{ER}}{RT} = -\sum x_i \ln \gamma_i^{*R} \tag{9-13.12}$$

where

$$\ln \gamma_i^{*R} = \sum_k n_k^{(i)} [\ln \gamma_k^* - \ln \gamma_k^{*(i)}] \tag{9-13.13}$$

and

$$\ln \gamma_k^* = Q_k \left[(1 - \ln \left(\sum_m \Theta_m \Psi_{mk}^* \right) - \sum_m \frac{\Theta_m \Psi_{km}^*}{\sum_n \Theta_n \Psi_{nm}^*} \right] \tag{9-13.14}$$

$$\Theta_m = \frac{Q_m X_m}{\sum_k X_k Q_k} \tag{9-13.15}$$

In Eq. (9-13.15), Θ_m is the surface area fraction in the mixture of groups and X_m is the mole fraction in the mixture of groups. Except for the minus sign in Eq. (9-13.12), these last four equations are identical to those in the UNIFAC method described in Chap 8. However, the groups are chosen differently and the interaction parameters are different and are calculated by

$$\Psi_{nm}^* = \exp \left(-\frac{\alpha_{nm}}{298} \right) \tag{9-13.16}$$

Values of α_{nm} are given in Table 9-14. γ_k^* is the activity coefficient of group k in a mixture of groups in the actual mixture, and $\gamma_k^{*(i)}$ is the activity coefficient of group k in a mixture of groups formed from the groups in pure component i. Groups in branched hydrocarbons and substituted cyclic and aromatic hydrocarbons are chosen as follows

TABLE 9-14 UNIFAC-VISCO Group Interaction Parameters, α_{nm}

n/m	CH_2	CH_3	CH_{2cy}	CH_{ar}	Cl	CO	COO	OH	CH_3OH
CH_2	0	66.53	224.9	406.7	60.30	859.5	1172.0	498.6	−219.7
CH_3	−709.5	0	−130.7	−119.5	82.41	11.86	−172.4	594.4	−228.7
CH_{2cy}	−538.1	187.3	0	8.958	215.4	−125.4	−165.7	694.4	−381.53
CH_{ar}	−623.7	237.2	50.89	0	177.2	128.4	−49.85	419.3	−88.81
Cl	−710.3	375.3	−163.3	−139.8	0	−404.3	−525.4	960.2	−165.4
CO	586.2	−21.56	740.6	−117.9	−4.145	0	29.20	221.5	55.52
COO	541.6	−44.25	416.2	−36.17	240.5	22.92	0	186.8	69.62
OH	−634.5	1209.0	−138	197.7	195.7	664.1	68.35	0	416.4
CH_3OH	−526.1	653.1	751.3	51.31	−140.9	−22.59	−286.2	−23.91	0

Type of compound	Actual group	Representation
branched cyclic	$>CH—CH_3$	2 CH_2 groups
	$>CH_{cy}—CH_3$	1 CH_{2cy} + 1 CH_2
	$>C_{cy}—(CH_3)_2$	1 CH_{acy} + 2 CH_2
aromatic	$>C_{ar}—CH_3$	1 CH_{ar} + 1 CH_2

Table 9-15 compares results calculated with the UNIFAC-VISCO method to experimental values. Of all the methods evaluated, the UNIFAC-VISCO method was the only one that demonstrated any success in predicting viscosities of mixtures of compounds with large size differences. The method has also been successfully applied to ternary and quaternary alkane systems. The average absolute deviation for 13 ternary alkane systems was 2.6%, while for four quaternary systems it was 3.6%. The method is illustrated in Example 9-22.

Example 9-22 Use the UNIFAC-VISCO method to estimate the viscosity of a mixture of 35.4 mole% *n*-decane (1) and 64.6 mole% *n*-tetratetracontane, $C_{44}H_{90}$ (2) at 397.49 K. The experimental viscosity and density (Aasen, et al., 1990) are 3.278 cP and 0.7447 g/cm³.

solution From Aasen, et al. (1990), $\eta_1 = 0.2938$ cP, $\eta_2 = 4.937$ cP, $V_1 = 220$ g/cm³, and $V_2 = 815.5$ g/cm³.

$$V_m = \frac{\sum_i x_i M_i}{\rho_m} = \frac{0.354 \times 142.28 + 0.646 \times 619.16}{0.7447} = 604.7 \text{ cm}^3/\text{mol}$$

In decane, there are 8 CH_2 groups and 2 CH_3 groups. In tetratetracontane, there are 42 CH_2 groups and 2 CH_3 groups. Equations (9-13.10) and (9-13.11) give

TABLE 9-15 Comparison of Calculated and Experimental Liquid Mixture Viscosities

1st component	2nd component	x_1	T, K	η_{exp} mPa·s	Ref*	η_{calc} mPa·s	% deviation
n-$C_{10}H_{22}$	n-$C_{60}H_{122}$	0.749	384.1	3.075	1	2.309	−25
		0.749	446.4	1.423	1	1.275	−10
n-$C_{10}H_{22}$	n-$C_{44}H_{99}$	0.354	368.8	5.286	1	5.256	−0.6
		0.354	464.1	1.465	1	1.654	13
		0.695	374.1	2.318	1	1.960	−15
butane	squalane	0.839	293.1	1.060	2	0.8812	−17
ethanol	benzene	0.5113	298.1	0.681	3	0.6403	−6.0
acetone	benzene	0.3321	298.1	0.4599	4	0.4553	−1.0
acetone	ethanol	0.3472	298.1	0.5133	5	0.4860	−5.3

*References: 1, Aasen, et al. (1990); 2, Kumagai and Takahashi (1995); 3, Kouris and Panayiotou (1989); 4, Petrino, et al. (1995); 5, Wei, et al. (1985)

$$r_1 = 8 \times 0.6744 + 2 \times 0.9011 = 7.1974$$

$$r_2 = 42 \times 0.6744 + 2 \times 0.9011 = 30.127$$

$$q_1 = 8 \times 0.54 + 2 \times 0.848 = 6.016$$

$$q_2 = 42 \times 0.54 + 2 \times 0.848 = 24.376$$

Equations (9-13.8) and (9-13.9) give

$$\theta_1 = \frac{0.354 \times 6.016}{0.354 \times 6.016 + 0.646 \times 24.376} = 0.1191, \qquad \theta_2 = 0.8809$$

$$\phi_1 = \frac{0.354 \times 7.1974}{0.354 \times 7.1974 + 0.646 \times 30.127} = 0.1158, \qquad \phi_2 = 0.8842$$

Equation (9-13.7) is used to calculate the combinatorial contribution

$$\frac{\Delta^* g^{EC}}{RT} = 0.354 \ln \frac{0.1158}{0.354} + 0.646 \ln \frac{0.8842}{0.646} + 5 \left[0.354 \times 6.016 \ln \frac{0.1191}{0.1158} + 0.646 \times 24.376 \ln \frac{0.8809}{0.8842} \right]$$
$$= 0.1880$$

In the mixture of groups, with CH_2 designated by subscript 1 and CH_3 by subscript 2:

$$X_1 = \frac{8 \times 0.354 + 42 \times 0.646}{8 \times 0.354 + 42 \times 0.646 + 2 \times 0.354 + 2 \times 0.646} = 0.9374,$$
$$X_2 = 0.0626$$

Equation (9-13.15) gives

$$\Theta_1 = \frac{0.9374 \times 0.54}{0.9374 \times 0.54 + 0.0626 \times 0.848} = 0.9051, \qquad \Theta_2 = 0.0949$$

Equation (9-13.16) gives

$$\Psi^*_{12} = \exp\left(-\frac{66.53}{298}\right) = 0.7999, \qquad \Psi^*_{21} = \exp\left(\frac{709.5}{298}\right) = 10.81$$

Equation (9-13.14) gives

$$\ln \gamma_1^* = 0.54 \left[1 - \ln(0.9051 + 0.0949 \times 10.81) - \frac{0.9051}{0.9051 + 0.0949 \times 10.81} - \frac{0.0949 \times 0.7999}{0.9051 \times 0.7999 + 0.0949} \right]$$
$$= -0.1185$$

Similarly, $\ln \gamma_2^* = -3.3791$

In the mixture of groups from pure component 1, $X_1^{(1)} = 0.8$, and $X_2^{(1)} = 0.2$. Using Eq. (9-13.15) for pure component 1 then gives

$$\Theta_1^{(1)} = \frac{0.8 \times 0.54}{0.8 \times 0.54 + 0.2 \times 0.848} = 0.7181, \qquad \Theta_2^{(1)} = 0.2819$$

$$\ln\gamma_1^{*(1)} = 0.54 \left[1 - \ln(0.7181 + 0.2819 \times 10.81) - \frac{0.7181}{0.7181 + 0.2819 \times 10.81} - \frac{0.2819 \times 0.7999}{0.7181 \times 0.7999 + 0.2819} \right]$$
$$= -0.4212$$

Similarly, $\gamma_2^{*(1)} = -1.0479$. In pure component 2, the results are $\Theta_1^{(2)} = 0.9304$, $\Theta_2^{(2)} = 0.0696$, $\ln\gamma_1^{*(2)} = -0.07655$, and $\ln\gamma_2^{(2)} = -4.1201$. Equation (9-13.13) gives

$$\ln\gamma_1^{*R} = 8 \times (-0.1185 + 0.4212) + 2(-3.3791 + 1.0479) = -2.241$$

$$\ln\gamma_2^{*R} = 42(-0.1185 + 0.07655) + 2(-3.3791 + 4.1201) = -0.2799$$

Finally, the residual contribution is calculated with Eq. (9-13.12)

$$\frac{\Delta^* g^{ER}}{RT} = -(-0.354 \times 2.241 - 0.646 \times 0.2799) = 0.9741$$

Equation (9-13.6) is now used to calculate the mixture viscosity

$$\ln\eta_m = 0.354\ln(0.2938) + 0.646\ln(4.937) + 0.354\ln(220) + 0.646\ln(815.5) - \ln(604.7) - 0.1880 + 0.9741$$

$$\eta_m = 3.385 \text{ cP}$$

$$\text{Error} = \frac{3.385 - 3.278}{3.278} \times 100 = 3.26\%$$

Method of Teja and Rice (1981, 1981a)

Based on a corresponding-states treatment for mixture compressibility factors (Teja, 1980; Teja and Sandler, 1980) (See chap. 5), Teja and Rice proposed an analogous form for liquid mixture viscosity.

$$\ln(\eta_m\varepsilon_m) = \ln(\eta\varepsilon)^{(R1)} + [\ln(\eta\varepsilon)^{(R2)} - \ln(\eta\varepsilon)^{(R1)}]\frac{\omega_m - \omega^{(R1)}}{\omega^{(R2)} - \omega^{(R1)}} \qquad (9\text{-}13.17)$$

where the superscripts (R1) and (R2) refer to two reference fluids. η is the viscosity, ω the acentric factor, and ε is a parameter similar to ξ in Eq. (9-4.15) but defined here as:

$$\varepsilon = \frac{V_c^{2/3}}{(T_c M)^{1/2}} \qquad (9\text{-}13.18)$$

The variable of composition is introduced in four places: the definitions of ω_m, V_{cm},

T_{cm}, and M_m. The rules suggested by the authors to compute these mixture parameters are:

$$V_{cm} = \sum_i \sum_j x_i x_j V_{cij} \tag{9-13.19}$$

$$T_{cm} = \frac{\sum_i \sum_j x_i x_j T_{cij} V_{cij}}{V_{cm}} \tag{9-13.20}$$

$$M_m = \sum_i x_i M_i \tag{9-13.21}$$

$$\omega_m = \sum_i x_i \omega_i \tag{9-13.22}$$

$$V_{cij} = \frac{(V_{ci}^{1/3} + V_{cj}^{1/3})^3}{8} \tag{9-13.23}$$

$$T_{cij} V_{cij} = \psi_{ij} (T_{ci} T_{cj} V_{ci} V_{cj})^{1/2} \tag{9-13.24}$$

ψ_{ij} is an interaction parameter of order unity which must be found from experimental data.

It is important to note that, in the use of Eq. (9-13.17) for a given mixture at a specified temperature, the viscosity values for the two reference fluids $\eta^{(R1)}$ and $\eta^{(R2)}$ are to be obtained *not at T*, but at a temperature equal to $T[(T_c)^{(R1)}/T_{cm}]$ for (R1) and $T[(T_c)^{(R2)}/T_{cm}]$ for (R2). T_{cm} is given by Eq. (9-13.20).

Whereas the reference fluids (R1) and (R2) may be chosen as different from the actual components in the mixture, it is normally advantageous to select them from the principal components in the mixture. In fact, for a binary of 1 and 2, if (R1) is selected as component 1 and (R2) as component 2, then, by virtue of Eq. (9-13.22), Eq. (9-13.17) simplifies to

$$\ln(\eta_m \varepsilon_m) = x_1 \ln(\eta\varepsilon)_1 + x_2 \ln(\eta\varepsilon)_2 \tag{9-13.25}$$

but, as noted above, η_1 is to be evaluated at $T(T_{c1}/T_{cm})$ and η_2 at $T(T_{c2}/T_{cm})$.

Our further discussion of this method will be essentially limited to Eq. (9-13.25), since that is the form most often used for binary liquid mixtures and, by this choice, one is assured that the relation gives correct results when $x_1 = 0$ or 1.0. In addition, the assumption is made that the interaction parameter ψ_{ij} is not a function of temperature or composition.

The authors claim good results for many mixtures ranging from strictly nonpolar to highly polar aqueous-organic systems. For nonpolar mixtures, errors averaged about 1%. For nonpolar-polar and polar-polar mixtures, the average rose to about 2.5%, whereas for systems containing water, an average error of about 9% was reported.

In comparison with the Grunberg-Nissan correlation [Eq. (9-13.1)], with G_{ij} found by regressing data, Teja and Rice show that about the same accuracy is achieved for both methods for nonpolar-nonpolar and nonpolar-polar systems, but their technique was significantly more accurate for polar-polar mixtures, and particularly for aqueous solutions for which Grunberg and Nissan's form should not be used.

Example 9-23 Estimate the viscosity of a liquid mixture of water and 1,4-dioxane at 60°C when the mole fraction water is 0.83. For this very nonideal solution, Teja and Rice suggest an interaction parameter $\psi_{ij} = 1.37$. This value was determined by regressing data at 20°C.

solution From Appendix A, for water, $T_c = 674.14$ K, $V_c = 55.95$ cm^3/mol, and $M = 18.02$; for 1,4-dioxane, $T_c = 587$ K, $V_c = 238$ cm^3/mol, and $M = 88.11$. Let 1 be water and 2 be 1,4-dioxane. With Eq. (9-13.18), $\varepsilon_1 = (55.95)^{2/3}/[(647.14)(18.02)]^{1/2} = 0.135$; $\varepsilon_2 = 0.169$. From Eq. (9-13.19),

$$V_{cm} = (0.830)^2(55.95) + (0.170)^2(238) + (2)(0.830)(0.170) \times \frac{[(55.95)^{1/3} + (238)^{1/3}]^3}{8}$$

$$= 80.93 \text{ cm}^3/\text{mol}$$

and with Eq. (9-13.20),

$$T_{cm} = \{(0.830)^2(647.14)(55.95) + (0.170)^2(587)(238) + (2)(0.830)(0.170)(1.37)$$

$$[(647.14)(55.95)(587)(238)]^{1/2}\}/81.29 = 697.9 \text{ K}$$

$$M_m = (0.830)(18.02) + (0.170)(88.11) = 29.94$$

So, with Eq. (9-13.18),

$$\varepsilon_m = \frac{(80.93)^{2/3}}{[(697.9)(29.94)]^{1/2}} = 0.129$$

Next, we need to know the viscosity of water not at 333.2 K (60°C), but at a temperature of (333.2)(647.14)/697.9 = 309.0 K (35.8°C). This value is 0.712 cP (Irving, 1977a). [Note that, at 60°C, η (water) = 0.468 cP.] For 1,4-dioxane, the reference temperature is (333.2)(587)/697.9 = 280.3 K (7.1°C), and at that temperature, $\eta = 1.63$ cP (Irving, 1977a). Again this value is quite different from the viscosity of 1,4-dioxane at 60°C, which is 0.715 cP. Finally, with Eq. (9-13.25),

$$\ln[(\eta_m)(0.129)] = (0.830) \ln[(0.712)(0.135)] + (0.170) \ln[(1.63)(0.169)]$$

$$= -2.163$$

$$\eta_m = 0.891 \text{ cP}$$

The experimental viscosity is 0.89 cP.

Although the agreement between the experimental and estimated viscosity in Example 9-23 is excellent, in other composition ranges, higher errors occur. In Fig. 9-14, we have plotted the estimated and experimental values of the mixture viscosity over the entire range of composition. From a mole fraction water of about 0.8 (weight fraction = 0.45) to unity, the method provides an excellent fit to experimental results. At smaller concentrations of water, the technique overpredicts η_m. Still, for such a nonideal aqueous mixture, the general fit should be considered good.

Discussion

Three methods have been introduced to estimate the viscosity of liquid mixtures: the Grunberg-Nissan relation [Eq. (9-13.1)], the UNIFAC-VISCO method [Eq. (9-13.6] and the Teja-Rice form [Eq. (9-13.24)]. The Grunberg-Nissan and Teja-Rice forms contain one adjustable parameter per binary pair in the mixture. The

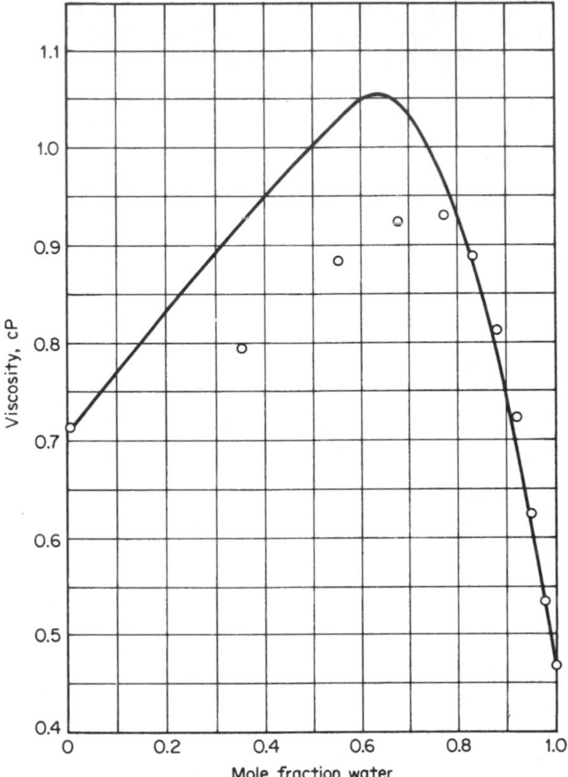

FIGURE 9-14 Viscosity of water and 1,4-dioxane at 333 K. Line is Eq. (9-13.25) with ψ_{ij} = 1.37; o experimental. (*Irving, 1977a*)

UNIFAC-VISCO method is predictive, but limited in the types of compounds to which it can be applied. The method correctly predicts the behavior of the methanol-toluene system which demonstrates both a maximum and minimum in the viscosity vs. concentration curve (Hammond, et al., 1958). An approximate technique is available to estimate the Grunberg-Nissan parameter G_{ij} as a function of temperature [Eq. (9-13.5)] for many types of systems. Teja and Rice suggest that their parameter ψ_{ij} is independent of temperature-at least over reasonable temperature ranges. This latter technique seems better for highly polar systems, especially if water is one of the components, and it has also been applied to undefined mixtures of coal liquids (Teja, et al., 1985; Thurner, 1984) with the introduction of reference components [See Eq. (9-13.16)]. The UNIFAC-VISCO method has been successfully applied to ternary and quaternary alkane mixtures (Chevalier, et al., 1988) but otherwise, evaluation of the above methods for multicomponent mixtures has been limited.

The above three methods are by no means a complete list of available methods. For example, Twu (1985, 1986) presents an equation to estimate the viscosity of petroleum fractions based on the specific gravity and boiling point. This method is particularly useful for cases where the exact chemical composition of a mixture is unknown. Allan and Teja (1991) have also presented a method applicable to petroleum fractions and (Chhabra, 1992) presents a method for mixtures of hydrocarbons. Chhabr and Sridhar (1989) extend Eq. (9-11.6) to mixtures. For the treatment of electrolyte solutions, the reader is referred to Lencka, et al. (1998). Cao, et al. (1993) presented a UNIFAC-based method but our testing did not reproduce their excellent results in a number of cases. Other mixture correlations are reviewed in Monnery, et al. (1995) as well as the 4th edition of this book. For an example of gases dissolved in liquids under pressure, see Tilly, et al. (1994).

An equation developed by McAllister (1960) has been used successfully to correlate data for binary as well as multicomponent mixtures (Aminabhavi, et al., 1982; Aucejo, et al, 1995; Dizechi and Marschall, 1982a; Noda, et al., 1982). For binaries, the McAllister (1960) equation has been written to contain either two or three adjustable parameters. For ternary mixtures, the equation has been used with one (Dizechi and Marschall, 1982a) or three (Noda, et al., 1982) ternary parameters in addition to the binary parameters. Dizechi and Marschall (1982) have extended the equation to mixtures containing alcohols and water and Asfour, et al. (1991) have developed a method to estimate the parameters in the McAllister equation from pure component properties. Because of the variable number of parameters that can be introduced into the McAllister equation, it has had considerable success in the correlation of mixture viscosity behavior.

Lee, et al. (1999) used an equation of state method to successfully correlate the behavior of both binary and multicomponent mixtures. Nonaqueous mixtures required one parameter per binary while aqueous mixtures required two parameters per binary. One drawback of their method is the non-symmetrical mixing rule used for multicomponent aqueous mixtures (Michelsen and Kistenmacher, 1990). The equation of state structure allowed the method to be successfully applied to liquid mixtures at high pressure.

To finish this section, we again reiterate that the methods proposed should be limited to situations in which the reduced temperatures of the components comprising the mixture are less than about 0.7, although the exact temperature range of the Teja-Rice procedure is as yet undefined.

Should one desire the viscosity of liquid mixtures at high pressures and temperatures, it is possible to employ the Chung, et al. (1988) method described in Sec. 9-7 to estimate high-pressure gas mixture viscosities. This recommendation is tempered by the fact that such a procedure has been only slightly tested, and usually with rather simple systems where experimental data exist.

Recommendations to Estimate the Viscosities of Liquid Mixtures

To estimate low-temperature liquid mixture viscosities, either the Grunberg-Nissan equation [Eq. (9-13.1) or (9-13.2)], the UNIFAC-VISCO method [Eq. (9-13.6)] or the Teja-Rice relation [Eq. (9-13.17) or (9-13.25)] may be used. The Grunberg-Nissan and Teja-Rice methods require some experimental data to establish the value of an interaction parameter specific for each binary pair in the mixture. In the absence of experimental data, the UNIFAC-VISCO method is recommended if group interaction parameters are available. The UNIFAC-VISCO method is particularly recommended for mixtures in which the components vary greatly in size. It is possible to estimate the Grunberg-Nissan interaction parameter G_{ij} by a group contribution technique and this technique can be applied to more compounds than can the UNIFAC-VISCO method. All three methods are essentially interpolative in nature, so viscosities of the pure components comprising the mixture must be known (or in the Teja-Rice procedure, one may instead use reference fluids of similar structure rather than the actual mixture components). The errors to be expected range from a few percent for nonpolar or slightly polar mixtures to 5 to 10% for polar mixtures. With aqueous solutions, neither the Grunberg-Nissan form nor the UNIFAC-VISCO method is recommended.

NOTATION

a^*	group contribution sum; Eq. (9-4.22)
b_o	excluded volume, $(2/3)\pi N_o\sigma^3$, Eq. (9-6.1)
C_v	heat capacity at constant volume, J/(mol·K); C_i, structural contribution in Eq. (9-4.22) and Table 9-3 C_{br}, number of carbon atoms in a branch
D	diffusion coefficient, cm²/s or m²/s
F_c	shape and polarity factor in Eq. (9-4.11); F_P^o, low-pressure polar correction factor in Eq. (9-4.18); F_Q^o, low-pressure quantum correction factor in Eq. (9-4.19); F_P, high-pressure polar correction factor in Eq. (9-6.10); F_Q, high-pressure quantum correction factor in Eq. (9-6.11)
g_{ij}	radial distribution function, Eq. (9-7.14)
G_1, G_2	parameters in Eqs. (9-6.21) and (9-6.22); G_{ij} parameter in Eq. (9-13.1)
Δ^*g^{EC}	combinatorial contribution to viscosity in Eq. (9-13.6)
Δ^*g^{ER}	residual contribution to viscosity in Eq. (9-13.6)
k	Boltzmann's constant
L	mean free path
m	mass of molecule
M	molecular weight
n	number density of molecules; number of components in a mixture
N	number of carbon atoms or parameter in Eq. (9-11.2); ΔN, structural contribution in Eq. (9-11.5) and Table 9-10; N_o, Avogadro's number
P	pressure, N/m² or bar (unless otherwise specified); P_c, critical pressure; P_r, reduced pressure, P/P_c; P_{vp}, vapor pressure; $\Delta P_r = (P - P_{vp})/P_c$
q_i	surface area parameter for molecule i
Q	polar parameter in Eq. (9-6.5); Q_k, surface area parameter of group k
r	distance of separation; r_i, volume of molecule i
R	gas constant, usually 8.314 J/(mol · K), R_η, parameter in Eq. (9-6.4); R_k, volume parameter of group k
T^*	kT/ε
T	temperature, K; T_c, critical temperature; T_r, reduced temperature, T/T_c; T_b, boiling point temperature; T_{fp}, melting point temperature
v	molecular velocity
V	volume, cm³/mol; V_c, critical volume; V_r, reduced volume, V/V_c, or in Eq. (9-6.3), V/V_o; V_o, hard packed volume used in Eq. (9-6.3) or parameter in Eq. (9-11.6)
x	mole fraction, liquid
y	mole fraction, vapor; parameter in Eq. (9-6.20)

Y	parameter in Eq. (9-6.9)
Z	compressibility factor; Z_c, critical compressibility factor; Z_1, Z_2, parameters in Eqs. (9-6.6) to (9-6.8)

Greek

γ	orientation factor in the Brulé-Starling method, Table 9-8, or obtain from Brulé-Starling (1984)
Δ	correction term in Eq. (9-6.17)
ε	energy-potential parameter; variable defined in Eq. (9-13.18)
η	viscosity (usually in micropoises for gas and in centipoises for liquids); η^{o}, denotes value at low-pressure (about 1 bar); η_c, at the critical point; η_c^o, at the critical temperature but at about 1 bar; η^*, η^{**}, parameters in Eqs. (9-6.19) and (9-6.23), η_r, reduced viscosity, defined in either Eq. (9-4.13) or (9-6.4), η_b, at the normal boiling point temperature
θ_i, Θ_k	surface area fraction of molecule i or group k
κ	polar correction factor in Eq. (9-4.11), see Table 9-1
λ	thermal conductivity, $W/(m \cdot K)$
μ	dipole moment, debyes; μ_r, dimensionless dipole moment defined in either Eq. (9-4.12) or Eq. (9-4.17)
ν	kinematic viscosity, η/ρ, m^2/s
ξ	inverse viscosity, defined in Eq. (9-4.14) or Eq. (9-4.15); ξ_T, inverse viscosity defined in Eq. (9-6.13)
ϕ_i	volume fraction of molecule i
ρ	density (usually mol/cm^3); ρ_c critical density; ρ_r, reduced density, ρ/ρ_c
σ	molecular diameter, Å
ψ	intermolecular potential energy as a function of r
Ψ	radial distribution function
ψ_{ij}	interaction parameter in Eq. (9-13.24)
ω	acentric factor, Sec. 2-3
Ω_v	collision integral for viscosity

Subscripts

i, j, k	components i, j, k
1, 2	components 1, 2
L	liquid
m	mixture
SL	saturated liquid

REFERENCES

Aasen E., E. Rytter, and H. A. Øye: *Ind. Eng. Chem. Res.*, **29:** 1635 (1990).

Alder, B. J.: "Prediction of Transport Properties of Dense Gases and Liquids," UCRL 14891-T, University of California, Berkeley, Calif., May 1966.

Alder, B. J., and J. H. Dymond: "Van der Waals Theory of Transport in Dense Fluids," UCRL 14870-T, University of California, Berkeley, Calif., April 1966.

Allan, J. M., and A. S. Teja: *Can. J. Chem. Eng.*, **69:** 986 (1991).

Amdur, I., and E. A. Mason: *Phys. Fluids,* **1:** 370 (1958).

American Petroleum Institute, *Selected Values of Physical and Thermodynamic Properties of Hydrocarbons and Related Compounds,* Project 44, Carnegie Press, Pittsburgh, Pa., 1953, and supplements.

Aminabhavi, T. M., R. C. Patel, and K. Bridger: *J. Chem. Eng. Data,* **27:** 125 (1982).

Andrade, E. N. da C.: *Nature,* **125:** 309 (1930).

Andrade, E. N. da C.: *Phil. Mag.,* **17:** 497, 698 (1934).

Andrade, E. N. da C.: *Endeavour,* **13:** 117 (1954).

Angus, S., B. Armstrong, and K. M. deReuck: *International Thermodynamic Tables of the Fluid State-Carbon Dioxide,* Pergamon, New York, 1976.

Asfour, A.-F. A., E. F. Cooper, J. Wu, and R. R. Zahran: *Ind. Eng. Chem. Res.,* **30:** 1669 (1991).

Assael, M. J., J. P. M. Trusler, and T. F. Tsolakis: *Thermophysical Properties of Fluids, An Introduction to their Prediction,* Imperial College Press, London, 1996.

Assael, M. J., J. H. Dymond, M. Papadaki, and P. M Patterson: *Fluid Phase Equil.,* **75:** 245 (1992).

Assael, M. J., J. H. Dymond, M. Papadaki, and P. M Patterson: *Int. J. Thermophys.,* **13:** 269 (1992a).

Assael, M. J., J. H. Dymond, M. Papadaki, and P. M Patterson: *Intern. J. Thermophys.,* **13:** 659 (1992b).

Assael, M. J., J. H. Dymond, M. Papadaki, and P. M Patterson: *Intern. J. Thermophys.,* **13:** 895 (1992c).

Assael, M. J., J. H. Dymond, and S. K. Polimatidou: *Intern. J. Thermophys.,* **15:** 189 (1994).

Assael, M. J., J. H. Dymond, and S. K. Polimatidou: *Intern. J. Thermophys.,* **16:** 761 (1995).

Aucejo, A., M. C. Burget, R. Muñoz, and J. L. Marques: *J. Chem. Eng. Data,* **40:** 141 (1995).

Barker, J. A., W. Fock, and F. Smith: *Phys. Fluids,* **7:** 897 (1964).

Batschinski, A. J.: *Z. Physik. Chim.,* **84:** 643 (1913).

Bingham, E. C., and S. D. Stookey: *J. Am. Chem. Soc.,* **61:** 1625 (1939).

Bircher, L. B.: Ph.D. thesis, University of Michigan, Ann Arbor, Mich., 1943.

Bleazard, J. G., and A. S. Teja: *Ind. Eng. Chem. Res.,* **35,** 2453 (1996).

Brebach, W. J., and G. Thodos: *Ind. Eng. Chem.,* **50:** 1095 (1958).

Bridgman, P. W.: *Proc. Am. Acad. Arts Sci.,* **61:** 57 (1926).

Brokaw, R. S.: *NASA Tech. Note D-2502,* November 1964.

Brokaw, R. S.: *J. Chem. Phys.,* **42:** 1140 (1965).

Brokaw, R. S.: *NASA Tech. Note D-4496,* April 1968

Brokaw, R. S., R. A. Svehla, and C. E. Baker: *NASA Tech. Note D-2580,* January 1965.

Bromley, L. A., and C. R. Wilke: *Ind. Eng. Chem.,* **43:** 1641(1951).

Brulé, M. R., and K. E. Starling: *Ind. Eng. Chem. Process Design Develop.,* **23:** 833 (1984).

Brush, S. G.: *Chem. Rev.,* **62:** 513 (1962).

Burch, L. G., and C. J. G. Raw: *J. Chem. Phys.,* **47:** 2798 (1967).

Cao, W., K. Knudsen, A. Fredenslund, and P. Rasmussen: *Ind. Eng. Chem. Res.,* **32:** 2088 (1993).

Carmichael, L. T., H. H. Reamer, and B. H. Sage: *J. Chem. Eng. Data,* **8:** 400 (1963).

Carmichael, L. T., and B. H. Sage: *J. Chem. Eng. Data,* **8:** 94 (1963).

Carmichael, L. T., and B. H. Sage: *AIChE J.,* **12:** 559 (1966).

Chakraborti, P. K., and P. Gray: *Trans. Faraday Soc.,* **61:** 2422 (1965).

Chapman, S., and T. G. Cowling: *The Mathematical Theory of Nonuniform Gases,* Cambridge, New York, 1939.

Cheung, H.: *UCRL Report 8230,* University of California, Berkeley, Calif., April 1958.

Chevalier, J. L., P. Petrino, and Y. Gaston-Bonhomme: *Chem. Eng. Sci.,* **43:** 1303 (1988).

Chhabra, R. P.: *AIChE J.,* **38:** 1657 (1992).

Chhabra, R. P. and T. Sridhar: *Chem Eng J.,* **40:** 39 (1989).

Chung, T.-H.: Ph.D. thesis, University of Oklahoma, Norman, Okla., 1980.

Chung, T.-H., M. Ajlan, L. L. Lee, and K. E. Starling: *Ind. Eng. Chem. Res.,* **27:** 671 (1988).

Chung, T.-H., L. L. Lee, and K. E. Starling: *Ind. Eng. Chem. Fundam.,* **23:** 8 (1984).

Cornelissen, J., and H. I. Waterman: *Chem. Eng. Sci.,* **4:** 238 (1955).

Dahler, J. S.: *Thermodynamic and Transport Properties of Gases, Liquids, and Solids,* McGraw-Hill, New York, 1959, pp. 14–24.

Das, T. R., C. O. Reed, Jr., and P. T. Eubank: *J. Chem. Eng. Data,* **22:** 3 (1977).

Daubert, T. E., R. P. Danner, H. M. Sibel, and C. C. Stebbins: *Physical and Thermodynamic Properties of Pure Chemicals: Data Compilation,* Taylor & Francis, Washington, D. C., 1997.

Davidson, T. A.: *A Simple and Accurate Method for Calculating Viscosity of Gaseous Mixtures.* U.S. Bureau of Mines, RI9456, 1993.

de Guzman, J.: *Anales Soc. Espan. Fiz. y Quim.,* **11:** 353 (1913).

Dean, J. A.: *Lange's Handbook of Chemistry,* 15th ed, McGraw-Hill, New York, 1999.

Dizechi, M., and E. Marschall: *Ind. Eng. Chem. Process Design Develop.;* **21:** 282 (1982).

Dizechi, M., and E. Marschall: *J. Chem. Eng. Data,* **27:** 358 (1982a).

Dolan, J. P., K. E. Starling, A. L. Lee, B. E. Eakin, and R. T. Ellington: *J. Chem Eng. Data,* **8:** 396 (1963).

Duhne, C. R.: *Chem. Eng.,* **86**(15): 83 (1979).

Dymond, J. H., and M. J. Assael: Chap 10 in *Transport Properties of Fluids, Their Correlation, Prediction and Estimation,* ed. by J. Millat, J. H. Dymond, and C. A. Nieto de Castro (eds.), IUPAC, Cambridge Univ. Press, Cambridge, 1996.

Eakin, B. E., and R. T. Ellington: *J. Petrol. Technol.,* **14:** 210 (1963).

Ely, J. F.: *J. Res. Natl. Bur. Stand.,* **86:** 597 (1981).

Ely, J. F., and H. J. M. Hanley: *Ind. Eng. Chem. Fundam.,* **20:** 323 (1981).

Eversteijn, F. C., J. M. Stevens, and H. I. Waterman: *Chem. Eng. Sci.,* **11:** 267 (1960).

Ferry, J. D.: *Viscoelastic Properties of Polymers,* Wiley, New York (1980).

Flynn, L. W., and G. Thodos: *J. Chem. Eng. Data,* **6:** 457 (1961).

Flynn, L. W., and G. Thodos: *AIChE J.,* **8:** 362 (1962).

Franjo, C., E. Jiménez, T. P. Iglesias, J. L. Legido, and M. I. Paz Andrade: *J. Chem. Eng. Data,* **40:** 68 (1995).

Gambill, W. R.: *Chem. Eng.,* **66**(3): 123 (1959).

Gammon, B. E., K. N. Marsh, and A. K. R. Dewan: *Transport Properties and Related Thermodynamic Data of Binary Mixtures, Parts 1–5,* Design Institute of Physical Property Data (DIPPR), New York, 1993–1998.

Gandhi, J. M., and S. C. Saxena: *Indian J. Pure Appl. Phys.,* **2:** 83 (1964).

Gaston-Bonhomme, Y., P. Petrino, and J. L. Chevalier: *Chem. Eng. Sci.,* **49:** 1799 (1994).

Gemant, A. J.: *Appl. Phys.,* **12:** 827 (1941).

Giddings, J. D.: Ph.D. thesis, Rice University, Houston, Texas, 1963.

Girifalco, L. A.: *J. Chem. Phys.,* **23:** 2446 (1955).

Goletz, E., and D. Tassios: *Ind. Eng. Chem. Process Design Develop.,* **16:** 75 (1977).

Golubev, I. F.: "Viscosity of Gases and Gas Mixtures: A Handbook," *Natl. Tech. Inf. Serv., TT 70 50022,* 1959.

Goodwin, R. D.: "The Thermophysical Properties of Methane from 90 to 500 K at Pressures up to 700 bar," *NBSIR 93-342, Natl. Bur. Stand.,* October 1973.

Grunberg, L., and A. H. Nissan: *Nature,* **164:** 799 (1949).

Gunn, R. D., and T. Yamada: *AIChE J.,* **17:** 1341 (1971).

Gururaja, G. J., M. A. Tirunarayanan, and A. Ramachandran: *J. Chem. Eng. Data,* **12:** 562 (1967).

Gutman, F., and L. M. Simmons: *J. Appl. Phys.,* **23:** 977 (1952).

Haar, L., and J. S. Gallagher: *J. Phys. Chem. Ref. Data,* **7:** 635 (1978).

Hammond, L. W., K. S. Howard, and R. A. McAllister: *J. Phys. Chem.,* **62:** 637 (1958).

Hanley, H. J. M., R. D. McCarty, and J. V. Sengers: *J. Chem. Phys.,* **50:** 857 (1969).

Herning, F., and L. Zipperer: *Gas Wasserfach,* **79:** 49 (1936).

Hildebrand, J. H.: *Science,* **174:** 490 (1971).

Hirschfelder, J. O., C. F. Curtiss, and R. B. Bird: *Molecular Theory of Gases and Liquids,* Wiley, New York, 1954.

Hirschfelder, J. O., M. H. Taylor, and T. Kihara: *Univ. Wisconsin Theoret. Chem, Lab., WIS-OOR-29,* Madison, Wis., July 8, 1960.

Huber, M. L., and J. F. Ely: *Fluid Phase Equil.,* **80:** 239 (1992).

Huber, M. L.: Chap 12 in *Transport Properties of Fluids, Their Correlation, Prediction and Estimation,* J. Millat, J. H. Dymond, and C. A. Nieto de Castro (eds.), IUPAC, Cambridge Univ. Press, Cambridge, 1996.

Hwang, M.-J., and W. B. Whiting: *Ind. Eng. Chem. Res.,* **26:** 1758 (1987).

Hwang, S. C., C. Tsonopoulos, J. R. Cunningham, and G. M. Wilson: *Ind. Eng. Chem. Proc. Des. Dev.,* **21:** 127 (1982).

Innes, K. K.: *J. Phys. Chem.,* **60:** 817 (1956).

Irving, J. B.: "Viscosities of Binary Liquid Mixtures: A Survey of Mixture Equations," *Natl. Eng. Lab., Rept. 630,* East Kilbride, Glasgow, Scotland, February 1977.

Irving, J. B.: "Viscosities of Binary Liquid Mixtures: The Effectiveness of Mixture Equations," *Natl. Eng. Lab., Rept. 631,* East Kilbride, Glasgow, Scotland, February 1977a.

Isdale, J. D.: Symp. Transp. Prop. Fluids and Fluid Mixtures, *Natl. Eng. Lab.,* East Kilbride, Glasgow, Scotland, 1979.

Isdale, J. D., J. C. MacGillivray, and G. Cartwright: "Prediction of Viscosity of Organic Liquid Mixtures by a Group Contribution Method," *Natl. Eng. Lab. Rept.,* East Kilbride, Glasgow, Scotland, 1985.

Itean, E. C., A. R. Glueck, and R. A. Svehla: *NASA Lewis Research Center,* TND 481, Cleveland, Ohio, 1961.

Jossi, J. A., L. I. Stiel, and G. Thodos: *AIChE J.,* **8:** 59 (1962).

Kennedy, J. T., and G. Thodos: *AIChE J.,* **7:** 625 (1961).

Kestin, J., H. E. Khalifa, and W. A. Wakeham: *J. Chem. Phys.,* **65:** 5186 (1976).

Kestin, J., and W. Leidenfrost: *Physica,* **25:** 525 (1959).

Kestin, J., and W. Leidenfrost: in Y. S. Touloukian (ed.), *Thermodynamic and Transport Properties of Gases, Liquids, and Solids,* ASME and McGraw-Hill, New York, 1959, pp. 321–338.

Kestin, J., and J. R. Moszynski: in Y. S. Touloukian (ed.), *Thermodynamic and Transport Properties of Gases, Liquids, and Solids,* ASME and McGraw-Hill, New York, 1959, pp. 70–77.

Kestin, J., and I. R. Shankland: in J. V. Sengers (ed.), *Proc. 8th Symp. Thermophys. Prop., II,* ASME, New York, 1981, p. 352.

Kestin, J., and J. Yata: *J. Chem. Phys.,* **49:** 4780 (1968).

Kim, S. K., and J. Ross: *J. Chem. Phys.,* **46:** 818 (1967).

Klein, M., and F. J. Smith: *J. Res. Natl. Bur. Stand.,* **72A:** 359 (1968).

Knapstad, B., P. A. Skølsvik, and H. A. Øye: *Ber. Bunsenges. Phys. Chem.,* **94:** 1156 (1990).

Kouris, S., and C. Panayiotou: *J. Chem. Eng. Data,* **34:** 200 (1989).

Krishnan, K. M., K. Ramababu, D. Ramachandran, P. Venkateswarlu, and G. K. Raman: *Fluid Phase Equil.,* **105:** 109 (1995).

Krishnan, K. M., K. Ramababu, P. Venkateswarlu, and G. K. Raman: *J. Chem. Eng. Data,* **40:** 132 (1995a).

Kumagai, A., and S. Takahashi: *Intern. J. Thermophys.,* **16:** 773 (1995).

Landolt-Bornstein Tabellen, vol. 4, pt. 1, Springer-Verlag, Berlin, 1955.

Larson, R. G.: *The Structure and Rheology of Complex Fluids,* Oxford Univ. Press (1999).

Lee, A. L., M. H. Gonzalez, and B. E. Eakin: *J. Chem. Eng. Data,* **11:** 281 (1966).

Lee, M.-J., J.-Y. Chiu, S.-M. Hwang, and H.-M. Lin: *Ind. Eng. Chem. Res.,* **38:** 2867 (1999).

Lencka, M. M., A. Anderko, and R. D. Young: *Intern. J. Thermophys.,* **19:** 367 (1998).

Lewis, W. K., and L. Squires: *Refiner Nat. Gasoline Manuf.,* **13**(12): 448 (1934).

Lide, D. R., (ed.): *Handbook of Chemistry and Physics,* 80th ed., CRC Press, Boca Raton, 1999.

Lobo, V. M. M.: "Handbook of Electrolyte Solutions," *Phys. Sci. Data Ser.,* Nos. 41a & 41b, Elsevier, 1990.

Lucas, K.: *Phase Equilibria and Fluid Properties in the Chemical Industry,* Dechema, Frankfurt, 1980, p. 573.

Lucas, K.: *Chem. Ing. Tech.,* **53:** 959 (1981).

Lucas, K.: personal communications, August 1983, September 1984.

Lucas, K.: VDI-Warmeatlas, Abschnitt DA, "Berechnungsmethoden für Stoffeigenschaften," *Verein Deutscher Ingenieure,* Düsseldorf, 1984a.

Luckas, M., and K. Lucas: *AIChE J.,* **32:** 139 (1986).

McAllister, R. A.: *AIChE J.,* **6:** 427 (1960).

Malek, K. R., and L. I. Stiel: *Can. J. Chem. Eng.,* **50:** 491 (1972).

Marschalko, B., and J. Barna: *Acta Tech. Acad. Sci. Hung.,* **19:** 85 (1957).

Mathur, G. P., and G. Thodos: *AIChE J.,* **9:** 596 (1963).

Medani, M. S., and M. A. Hasan: *Can. J. Chem. Eng.,* **55:** 203 (1977).

Mehrotra, A. K.: *Ind. Eng. Chem. Res.,* **30:** 420, 1367 (1991).

Mehrotra, A. K., W. D. Monnery, and W. Y. Scrcek: *Fluid Phase Equil.,* **117:** 344 (1996).

Michelsen, M. L., and H. Kistenmacher: *Fluid Phase Equil.,* **58:** 229 (1990).

Millat, J., V. Vesovic, and W. A. Wakeham: Chap. 4 in *Transport Properties of Fluids, Their Correlation, Prediction and Estimation,* (eds.) J. Millat, J. H. Dymond, and C. A. Nieto de Castro, IUPAC, Cambridge Univ. Press, Cambridge, 1996.

Miller, A. A.: *J. Chem. Phys.,* **38:** 1568 (1963).

Miller, A. A.: *J. Phys. Chem.,* **67:** 1031, 2809 (1963a).

Monchick, L., and E. A. Mason: *J. Chem. Phys.,* **35:** 1676 (1961).

Monnery, W. D., W. S. Svrcek, and A. K. Mehrotra: *Can. J. Chem Eng.,* **73:** 3 (1995).

Neufeld, P. D., A. R. Janzen, and R. A. Aziz: *J. Chem. Phys.,* **57:** 1100 (1972).

Noda, K., M. Ohashi, and K. Ishida: *J. Chem. Eng. Data,* **27:** 326 (1982).

O'Connell, J. P., and J. M. Prausnitz: "Applications of the Kihara Potential to Thermodynamic and Transport Properties of Gases" in *Advances in Thermophysical Properties at Extreme Temperatures and Pressures,* S. Gratch (ed.), ASME, NY, 1965, pp. 19–31.

Okeson, K. J., and R. L. Rowley: *Intern. J. Thermophys.,* **12:** 119 (1991).

O'Loane, J. K.: personal communication, June 1979.

Orrick, C., and J. H. Erbar: personal communication, December 1974.

Pal, A. K., and A. K. Barua: *J. Chem. Phys.,* **47:** 216 (1967).

Pal, A. K., and P. K. Bhattacharyya: *J. Chem. Phys.,* **51:** 828 (1969).

Perry, R. H. and D. W. Green (eds.): *Chemical Engineers' Handbook,* 7th ed., McGraw-Hill, New York, 1997.

Petrino, P. J., Y. H. Gaston-Bonhomme, and J. L. E. Chevalier: *J. Chem. Eng. Data,* **40:** 136 (1995).

Przezdziecki, J. W., and T. Sridhar: *AIChE J.,* **31:** 333 (1985).

Ranz, W. E., and H. A. Brodowsky: Univ. Minn. *OOR Proj. 2340 Tech. Rept. 1,* Minneapolis, Minn., March 15, 1962.

Raw, C. J. G., and H. Tang: *J. Chem. Phys.,* **39:** 2616 (1963).

Reichenberg, D.: "The Viscosities of Gas Mixtures at Moderate Pressures," *NPL Rept. Chem. 29,* National Physical Laboratory, Teddington, England, May, 1974.

Reichenberg, D.: (a) *DCS report 11,* National Physical Laboratory, Teddington, England, August 1971; (b) *AIChE J.,* **19:** 854 (1973); (c) *ibid.,* **21:** 181 (1975).

Reichenberg, D.: "The Viscosities of Pure Gases at High Pressures," *Natl. Eng. Lab., Rept. Chem. 38,* East Kilbride, Glasgow, Scotland, August 1975.

Reichenberg, D.: "New Simplified Methods for the Estimation of the Viscosities of Gas Mixtures at Moderate Pressures," *Natl. Eng. Lab. Rept. Chem. 53,* East Kilbride, Glasgow, Scotland, May 1977.

Reichenberg, D.: *Symp. Transp. Prop. Fluids and Fluid Mixtures, Natl. Eng. Lab.,* East Kilbride, Glasgow, Scotland, 1979.

Reid, R. C., and L. I. Belenyessy: *J. Chem. Eng. Data,* **5:** 150 (1960).

Riddick, J. A., W. B. Bunger, and T. K. Sakano: *Organic Solvents Physical Properties and Methods of Purification,* 4th ed., Wiley, New York.

Rogers, J. D., and F. G. Brickwedde: *AIChE J.,* **11:** 304 (1965).

Rutherford, R., M. H. Taylor, and J. O. Hirschfelder: *Univ. Wisconsin Theoret. Chem. Lab., WIS-OOR-29a,* Madison, Wis., August 23, 1960.

Sastri, S. R. S., and K. K. Rao: *Chem. Eng. J.,* **50:** 9 (1992).

Sastri, S. R. S.: personal communication, Regional Research Laboratory, Bhubaneswar, India (1998).

Saxena, S. C., and R. S. Gambhir: *Brit. J. Appl. Phys.,* **14:** 436 (1963).

Saxena, S. C., and R. S. Gambhir: *Proc. Phys. Sec. London,* **81:** 788 (1963a).

Shimotake, H., and G. Thodos: *AIChE J.,* **4:** 257 (1958).

Shimotake, H., and G. Thodos: *AIChE J.,* **9:** 68 (1963).

Shimotake, H., and G. Thodos: *J. Chem. Eng. Data,* **8:** 88 (1963a).

Starling, K. E.: M.S. thesis, Illinois Institute of Technology, Chicago, I11., 1960.

Starling, K. E.: Ph.D. thesis, Illinois Institute of Technology, Chicago, I11., 1962.

Starling, K. E., B. E. Eakin, and R. T. Ellington: *AIChE J.,* **6:** 438 (1960).

Starling, K. E., and R. T. Ellington: *AIChE J.,* **10:** 11 (1964).

Stephan, K., and K. Lucas: *Viscosity of Dense Fluids,* Plenum, New York, 1979.

Stephan, K., and H. Hildwein: Recommended Data of Selected Compounds and Binary Mixtures, *Chemistry Data Series,* vol. IV, parts 1 & 2, DECHEMA, Frankfurt, 1987.

Stephan, K., and T. Heckenberger: Thermal Conductivity and Viscosity Data of Fluid Mixtures, *Chemistry Data Series,* vol. X, part 1, DECHEMA, Frankfurt, 1988.

Stewart, R. B.: Ph.D. thesis, University of Iowa, Iowa City, Iowa, June 1966.

Stiel, L. I., and G. Thodos: *AIChE J.,* **10:** 275 (1964).

Strunk, M. R., W. G. Custead, and G. L. Stevenson: *AIChE J.,* **10:** 483 (1964).

Sutton, J. R.: "References to Experimental Data on Viscosity of Gas Mixtures," *Natl. Eng. Lab. Rept. 613,* East Kilbride, Glasgow, Scotland, May 1976.

Svehla, R. A.: "Estimated Viscosities and Thermal Conductivities at High Temperatures," *NASA- TRR-132,* 1962.

Tanaka, Y., H. Kubota, T. Makita, and H. Okazaki: *J. Chem. Eng. Japan,* **10:** 83 (1977).

Teja, A. S.: *AIChE J.,* **26:** 337 (1980).

Teja, A. S., and P. Rice: *Chem. Eng. Sci.,* **36:** 7 (1981).

Teja, A. S., and P. Rice: *Ind. Eng. Chem. Fundam.,* **20:** 77 (1981a).

Teja, A. S., and S. I. Sandler: *AIChE J.,* **26:** 341 (1980).

Teja, A. S., P. A. Thurner, and B. Pasumarti: paper presented at the *Annual AIChE Mtg., Washington, D.C., 1983; Ind. Eng. Chem. Process Design Develop.*, **24:** 344 (1985).

Telang, M. S.: *J. Phys. Chem.*, **49:** 579 (1945); **50:** 373 (1946).

Tham, M. K. and K. E. Gubbins: *J. Chem. Phys.*, **55:** 268 (1971).

Thurner, P. A.: S.M. thesis, Georgia Institute of Technology, Atlanta, Ga., 1984.

Tilly, K. D., N. R. Foster, S. J. Macnaughton, and D. L. Tomasko: *Ind. Eng. Chem. Res.*, **33:** 681 (1994).

Timmermans, J.: *Physico-Chemical Constants of Pure Organic Compounds,* vol. 2, Elsevier, Amsterdam, 1965.

Titani, T.: *Bull. Inst. Phys. Chem. Res. Tokyo,* **8:** 433 (1929).

Trautz, M., and P. B. Baumann: *Ann. Phys.*, **5:** 733 (1929).

Trautz, M., and R. Heberling: *Ann. Phys.*, **10:** 155 (1931).

Trautz, M.: *Ann. Phys.*, **11:** 190 (1931).

Twu, C. H.: *Ind. Eng. Chem. Process Des. Dev.*, **24:** 1287 (1985).

Twu, C. H.: *AIChE J.*, **32:** 2091 (1986).

van Velzen, D., R. L. Cardozo, and H. Langenkamp: "Liquid Viscosity and Chemical Constitution of Organic Compounds: A New Correlation and a Compilation of Literature Data," *Euratom,* 4735e, Joint Nuclear Research Centre, Ispra Establishment, Italy, 1972.

van Wyk, W. R., J. H. van der Veen, H. C. Brinkman, and W. A. Seeder: *Physica,* **7:** 45 (1940).

Vanderslice, J. T., S. Weissman, E. A. Mason, and R. J. Fallon: *Phys. Fluids,* **5:** 155 (1962).

Vargaftik, N. B., Y. K. Vinogradov, and V. S. Yargin: *Handbook of Physical Properties of Liquids and Gases,* Begell House, New York, 1996.

Vesovic, V., W. A. Wakeham, G. A. Olchowy, J. V. Sengers, J. T. R. Watson, and J. Millat: *J. Phys. Chem. Ref. Data,* **19:** 763 (1990).

Viswanath, D. S., and G. Natarajan: *Data Book on the Viscosity of Liquids,* Hemisphere Pub. Co., New York, 1989.

Vogel, H.: *Physik Z.*, **22:** 645 (1921).

Vogel. E., and A. Weiss: *Ber. Bunsenges. Phys. Chem.*, **85:** 539 (1981).

Waxman, M., and J. S. Gallagher: *J. Chem. Eng. Data,* **28:** 224 (1983).

Wei, I.-Chien, and R. Rowley: *J. Chem. Eng. Data,* **29:** 332, 336 (1984); *Chem. Eng. Sci.*, **40:** 401 (1985).

Wilke, C. R.: *J. Chem. Phys.*, **18:** 517 (1950).

Wobster, R., and F. Mueller: *Kolloid Beih.*, **52:** 165 (1941).

Wright, P. G., and P. Gray: *Trans. Faraday Soc.*, **58:** 1 (1962).

Wu, J., A. Shan, and A.F. A. Asfour: *Fluid Phase Equil.*, **143:** 263 (1998)

Xiang, H. W., W. A. Wakeham, and J. P. M. Trusler: "Correlation and Prediction of Viscosity Consistently from the Dilute Vapor to Highly Compressed Liquid State," submitted to *High Temp.-High Press* (1999).

Yaws, C. L., J. W. Miller, P. N. Shah, G. R. Schorr, and P. M. Patel: *Chem. Eng.*, **83**(25): 153 (1976).

Yaws, C. L.: *Handbook of Viscosity,* Gulf Pub., Houston, 1995.

Yaws, C. L.: *Handbook of Transport Property Data,* Gulf Pub., Houston, 1995a.

Yoon, P., and G. Thodos: *AIChE J.*, **16:** 300 (1970).

Zaytsev, I. D., and G. G. Aseyev: *Properties of Aqueous Electrolyte Solutions,* CRC Press, Boca Raton, 1992.

CHAPTER TEN
THERMAL CONDUCTIVITY

10-1 SCOPE

Thermal conductivities of both gases and liquids are considered in this chapter. Some background relevant to the theory of thermal conductivity is given in Secs. 10-2 and 10-3 (for gases) and in Sec. 10-8 (for liquids). Estimation techniques for pure gases at near ambient pressures are covered in Sec. 10-4; the effects of temperature and pressure are discussed in Secs. 10-4 and 10-5. Similar topics for liquids are covered in Secs. 10-9 to 10-11. Thermal conductivities for gas and for liquid mixtures are discussed in Secs. 10-6, 10-7, and 10-12. Thermal conductivities of reacting gas mixtures are not covered but are reviewed in Curtiss, et al. (1982).

The units used for thermal conductivity are W/(m·K). To convert these to English or cgs units:

$$W/(m \cdot K) \times 0.5778 = Btu/(hr \cdot ft \cdot {}^\circ R)$$

$$W/(m \cdot K) \times 0.8604 = kcal/(cm \cdot hr \cdot K)$$

$$W/(m \cdot K) \times 2.390 \times 10^{-3} = cal/(cm \cdot s \cdot K)$$

or

$$Btu/(hr \cdot ft \cdot {}^\circ R) \times 1.731 = W/(m \cdot K)$$

$$kcal/(cm \cdot hr \cdot K) \times 1.162 = W/(m \cdot K)$$

$$cal/(cm \cdot s \cdot K) \times 418.4 = W/(m \cdot K)$$

10-2 THEORY OF THERMAL CONDUCTIVITY

In Sec. 9-3, through rather elementary arguments, the thermal conductivity of an ideal gas was found to be equal to $vLC_v n/3$ [Eq. (9-3.7)], where v is the average molecular velocity, L is the mean free path, C_v is the constant volume heat capacity per molecule, and n is the number density of molecules. Similar relations were derived for the viscosity and diffusion coefficients of gases. In the case of the last two properties, this elementary approach yields approximate but reasonable values. For thermal conductivity, it is quite inaccurate. A more detailed treatment is necessary to account for the effect of having a wide spectrum of molecular velocities; also, molecules may store energy in forms other than translational. For monatomic gases, which have no rotational or vibrational degrees of freedom, a more rigorous analysis yields

$$\lambda = \frac{25}{32}(\pi mkT)^{1/2}\frac{C_v/m}{\pi\sigma^2\Omega_v} \qquad (10\text{-}2.1)$$

or, written for computational ease, with $C_v = \frac{3}{2}k$,

$$\lambda = 2.63 \times 10^{-23}\frac{(T/M')^{1/2}}{\sigma^2\Omega_v} \qquad (10\text{-}2.2)$$

where λ = thermal conductivity, W/(m · K)
T = temperature, K
k = Boltzmann's constant = 1.3805×10^{-23} J/K
M' = molecular weight, kg/mol
σ = characteristic dimension of molecule, m
Ω_v = collision integral, dimensionless

Values of λ from Eq. (10-2.2) for xenon and helium at 300 K are 0.008 to 0.1 W/(m · K) respectively. For a hard-sphere molecule, Ω_v is unity; normally, however, it is a function of temperature, and the exact dependence is related to the intermolecular force law chosen. If the Lennard-Jones 12-6 potential [Eq. (9-4.2)] is selected, Ω_v is given by Eq. (9-4.3).

If Eq. (10-2.1) is divided by Eq. (9-3.9),

$$\frac{\lambda M'}{\eta C_v} = 2.5 \qquad (10\text{-}2.3)$$

With $\gamma = C_p/C_v$, the Prandtl number N_{Pr}, is

$$N_{Pr} = \frac{C_p\eta}{\lambda M'} = \frac{\gamma}{2.5} \qquad (10\text{-}2.4)$$

Since γ for monatomic gases is close to $\frac{5}{3}$ except at very low temperatures, Eq. (10-2.4) would indicate that $N_{Pr} \approx \frac{2}{3}$, a value close to that found experimentally. To obtain Eq. (10-2.3), the terms σ^2 and Ω_v cancel and the result is essentially independent of the intermolecular potential law chosen.

The dimensionless group $\lambda M'/\eta C_v$, is known as the *Eucken factor;* it is close to 2.5 for monatomic gases, but it is significantly less for polyatomic gases. Our discussion so far has considered only energy associated with translational motion; since heat capacities of polyatomic molecules exceed those for monatomic gases, a substantial fraction of molecular energy resides in modes other than translational. This has a much greater effect on the thermal conductivity than on viscosity or the diffusion coefficient.

10-3 THERMAL CONDUCTIVITIES OF POLYATOMIC GASES

Eucken and Modified Eucken Models

Eucken proposed that Eq. (10-2.3) be modified for polyatomic gases by separating the contributions due to translational and internal degrees of freedom into separate terms:

$$\frac{\lambda M'}{\eta C_v} = f_{tr}\left(\frac{C_{tr}}{C_v}\right) + f_{int}\left(\frac{C_{int}}{C_v}\right) \tag{10-3.1}$$

Thus the contribution due to translational degrees of freedom has been decoupled from that due to internal degrees of freedom (Cottrell and McCoubrey, 1961; Lambert and Bates, 1962; Mason and Monchick, 1962; O'Neal and Brokaw, 1962; Saxena, et al., 1964; Srivastava and Srivastava, 1959; Vines, 1958; Vines and Bennett, 1954), although the validity of this step has been questioned (Hirschfelder, 1957; Saxena and Agrawal, 1961; Svehla, 1962). Invariably, f_{tr}, is set equal to 2.5 to force Eq. (10-3.1) to reduce to Eq. (10-2.3) for a monatomic ideal gas. C_{tr} is set equal to the classical value of 1.5R, and C_{int} is conveniently expressed as $C_v - C_{tr}$. Then

$$\begin{aligned}\frac{\lambda M'}{\eta C_v} &= \frac{15/4}{C_v/R} + f_{int}\left(1 - \frac{3/2}{C_v/R}\right)\\ &= \frac{15/4}{(C_p/R) - 1} + f_{int}\left[1 - \frac{3/2}{(C_p/R) - 1}\right]\end{aligned} \tag{10-3.2}$$

where the ideal-gas relation ($C_p - C_v = R$) has been used.

Eucken chose $f_{int} = 1.0$, whereby Eq. (10-3.2) reduces to

$$\frac{\lambda M'}{\eta C_v} = 1 + \frac{9/4}{C_v/R} = 1 + \frac{9/4}{(C_p/R) - 1} \tag{10-3.3}$$

the well-known Eucken correlation for polyatomic gases.

Many of the assumptions leading to Eq. (10-3.3) are open to question, in particular, the choice of $f_{int} = 1.0$. Ubbelohde (1935), Chapman and Cowling (1961) Hirschfelder (1957), and Schafer (1943) have suggested that molecules with excited internal energy states could be regarded as separate chemical species, and the transfer of internal energy is then analogous to a diffusional process. This concept leads to a result that

$$f_{int} = \frac{M'\rho D}{\eta} \tag{10-3.4}$$

where M' = molecular weight, kg/mol
η = viscosity, N · s/m^2
ρ = molar density, mol/m^3
D = diffusion coefficient, m^2/s

Most early theories selected D to be equivalent to the molecular self-diffusion coefficient, and f_{int} is then the reciprocal of the Schmidt number. With Eqs. (9-3.9) and (11-3.2), it can be shown that $f_{int} \approx 1.32$ and is almost independent of temperature. With this formulation, Eq. (10-3.2) becomes

$$\frac{\lambda M'}{\eta C_v} = 1.32 + \frac{1.77}{C_v/R} = 1.32 + \frac{1.77}{(C_p/R - 1)} \tag{10-3.5}$$

Equation (10-3.5), often referred to as the modified Eucken correlation, was used by Svehla (1962) in his compilation of high-temperature gas properties.

The modified Eucken relation [Eq. (10-3.5)] predicts larger values of λ than the Eucken form [Eq. (10-3.3)], and the difference becomes greater as C_v increases above the monatomic gas value of about 12.6 J/(mol · K). Both yield Eq. (10-2.3) when $C_v = 3R/2$. Usually, experimental values of λ lie between those calculated by the two Eucken forms except for polar gases, when both predict λ values that are too high. For nonpolar gases, Stiel and Thodos (1964) suggested a compromise between Eqs. (10-3.3) and (10-3.5) as

$$\frac{\lambda M'}{\eta C_v} = 1.15 + \frac{2.03}{C_v/R} = 1.15 + \frac{2.03}{(C_p/R) - 1} \tag{10-3.6}$$

Equations (10-3.3), (10-3.5), and (10-3.6) indicate that the Eucken factor $(\lambda M'/\eta C_v)$ should decrease with increasing temperature as the heat capacity rises, but experimental data indicate that the Eucken factor is often remarkably constant or increases slightly with temperature. In Fig. 10-1 we illustrate the case for ethyl chloride, where the data of Vines and Bennett show the Eucken factor increases from only about 1.41 to 1.48 from 40 to 140°C. On this same graph, the predictions of Eqs. (10-3.3), (10-3.5), and (10-3.6) are plotted and, as noted earlier, all predict a small decrease in the Eucken factor as temperature increases. In Fig. 10-2 we have graphed the experimental Eucken factor as a function of reduced temperature

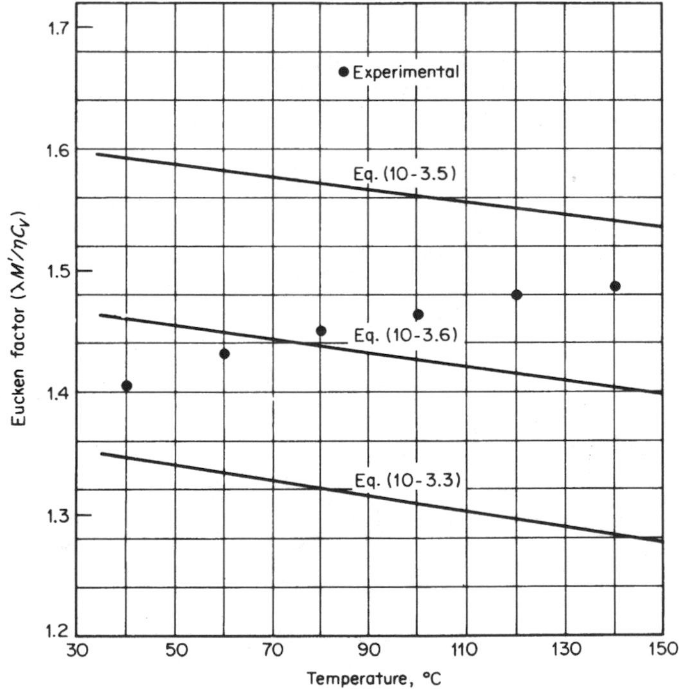

FIGURE 10-1 Eucken factor for ethyl chloride at low pressure. (*Data from Stiel and Thodos, 1964.*)

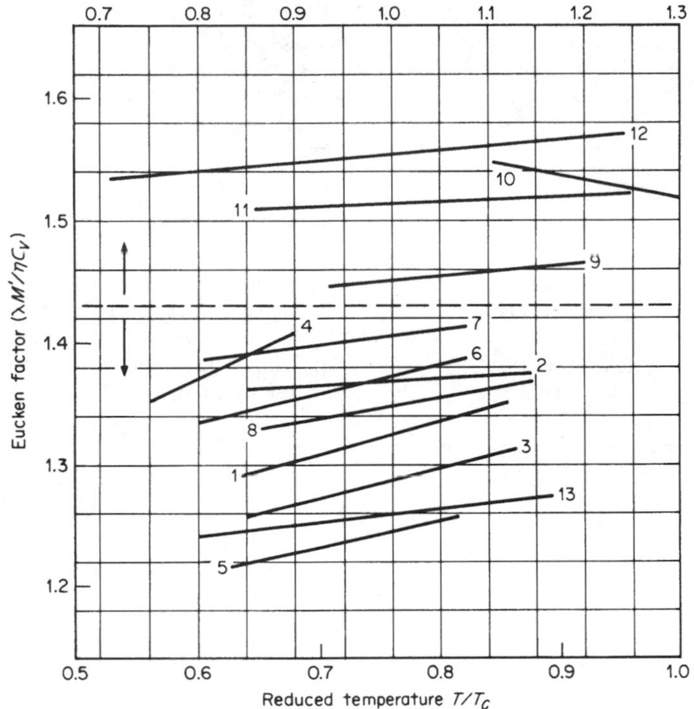

1. Acetone
2. Methyl alcohol
3. Acetaldehyde
4. Water
5. Acetonitrile
6. Cyclohexane
7. Benzene
8. n-Hexane
9. Propane
10. Ethane
11. Acetylene
12. Dichlorodifluoromethane
13. Ethyl acetate

FIGURE 10-2 Variation of the Eucken factor with temperature. (*Data primarily from Stiel and Thodos, 1964.*)

for 13 diverse low-pressure gases. Except for ethane, all show a small rise with an increase in temperature.

Roy and Thodos Estimation Technique

In the same way that the viscosity was nondimensionalized in Eqs. (9-4.13) and (9-4.14), a reduced thermal conductivity may be expressed as

$$\lambda_r = \lambda \Gamma \tag{10-3.7}$$

$$\Gamma = \left[\frac{T_c (M')^3 N_0^2}{R^5 P_c^4} \right]^{1/6} \tag{10-3.8}$$

In SI units, if $R = 8314$ J/(kmol K), N_0 (Avogadro's number) $= 6.023 \times 10^{26}$ $(\text{kmol})^{-1}$, and with T_c in kelvins, M' in kg/kmol, and P_c in N/m^2, Γ has the units of m · K/W or inverse thermal conductivity. In more convenient units,

$$\Gamma = 210 \left(\frac{T_c M^3}{P_c^4} \right)^{1/6} \tag{10-3.9}$$

where Γ is the reduced, inverse thermal conductivity, $[\text{W}/(\text{m} \cdot \text{K})]^{-1}$, T_c is in kelvins, M is in g/mol, and P_c is in bars.

The reduced thermal conductivity was employed by Roy and Thodos (1968, 1970), who, however, separated the $\lambda\Gamma$ product into two parts. The first, attributed only to translational energy, was obtained from a curve fit of the data for the monatomic gases (Roy, 1967); this part varies only with the reduced temperature, $T_r = T/T_c$. In the second, the contribution from rotational, vibrational interchange, etc., was related to the reduced temperature and a specific constant estimated from group contributions. The final equation may be written

$$\lambda_r = \lambda\Gamma = (\lambda\Gamma)_{tr} + (\lambda\Gamma)_{int} \tag{10-3.10}$$

where λ = low-pressure gas thermal conductivity, W/(m · K) and Γ is defined in Eq. (10-3.9).

$$(\lambda\Gamma)_{tr} = 8.757[\exp(0.0464T_r) - \exp(-0.2412T_r)] \tag{10-3.11}$$

$$(\lambda\Gamma)_{int} = Cf(T_r) \tag{10-3.12}$$

Relations for $f(T_r)$ are shown in Table 10-1. The constant C is specific for each material, and it is estimated by a group contribution technique as shown below.

Estimation of Roy-Thodos Constant C. In the discussion to follow, one identifies carbon types as shown:

TABLE 10-1 Recommended $f(T_r)$ Equations for the Roy-Thodos Method

Saturated hydrocarbons†	$-0.152T_r + 1.191T_r^2 - 0.039T_r^3$
Olefins	$-0.255T_r + 1.065T_r^2 + 0.190T_r^3$
Acetylenes	$-0.068T_r + 1.251T_r^2 - 0.183T_r^3$
Naphthalenes and aromatics	$-0.354T_r + 1.501T_r^2 - 0.147T_r^3$
Alcohols	$1.000T_r^2$
Aldehydes, ketones, ethers, esters	$-0.082T_r + 1.045T_r^2 + 0.037T_r^3$
Amines and nitriles	$0.633T_r^2 + 0.367T_r^3$
Halides	$-0.107T_r + 1.330T_r^2 - 0.223T_r^3$
Cyclic compounds‡	$-0.354T_r + 1.501T_r^2 - 0.147T_r^3$

† Not recommended for methane.
‡ For example, pyridine, thiophene, ethylene oxide, dioxane, piperidine.

H H—C— H Type: 1	H —C— H 2	H —C— | 3	| —C— | 4

Parraffinic Hydrocarbons	ΔC
Base group, methane	0.73
First methyl substitution	2.00
Second methyl substitution	3.18
Third methyl substitution	3.68
Fourth and successive methyl substitutions	4.56

For example, C for *n*-octane is equal to $C = \Sigma\Delta C = [0.73 + 2.00 + 3.18 + 3.68 + 4(4.56)] = 27.8$

Isoparaffins are formed by determining the C for the paraffin with the longest possible straight-chain carbon backbone and then making successive substitutions of hydrogen atoms by methyl groups. Values of ΔC attributable to such substitutions are shown below:

Type of substitution	ΔC
$1 \leftarrow 2 \rightarrow 1$	3.64
$1 \leftarrow 2 \rightarrow 2$	4.71
$1 \leftarrow 2 \rightarrow 3$	5.79
$2 \leftarrow 2 \rightarrow 2$	5.79
$1 \leftarrow 3 \rightarrow 1$, $\downarrow$ 1	3.39
$1 \leftarrow 3 \rightarrow 1$, $\downarrow$ 2	4.50
$1 \leftarrow 3 \rightarrow 1$, $\downarrow$ 3	5.61

The type of carbon atom from which the arrow points away is the one involved in the methyl substitution. The arrows point toward the types of adjacent atoms. To calculate C for an isoparaffin, begin with the longest chain, introduce side chains beginning with the left end and proceed in a clockwise direction. To illustrate with 2,2,4-trimethylpentane,

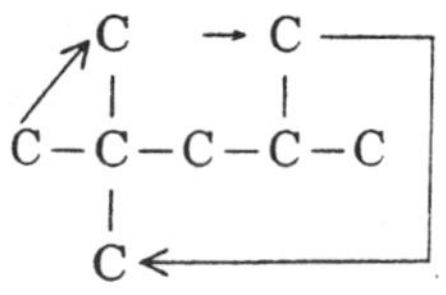

```
  ↗C  → C ——┐
 ╱  |     |   |
C—C—C—C—C     |
    |         |
    C ←———————┘
```

n-Pentane = 0.73 + 2.00 + 3.18 + 3.68 + 4.56 = 14.15. For the methyl substitutions

$$1 \leftarrow 2 \rightarrow 2 = 4.71;\ 2 \leftarrow 2 \rightarrow 1 = 4.71;\ \text{and}\ 1 \leftarrow \underset{\overset{\downarrow}{1}}{3} \rightarrow 2 = 4.50$$

Thus, $C = 14.15 + 4.71 + 4.71 + 4.50 = 28.07$.

Olefinic and Acetylenic Hydrocarbons. First determine C for the corresponding saturated hydrocarbon, as described above; then insert the unsaturated bond(s) and employ the following ΔC contributions:

		ΔC
First double bond	$1 \leftrightarrow 1$	−1.19
	$1 \leftrightarrow 2$	−0.65
	$2 \leftrightarrow 2$	−0.29
Second double bond	$2 \leftrightarrow 1$	−0.17
Any acetylenic bond		−0.83

Naphthenes. Form the paraffinic hydrocarbon with the same number of carbon atoms as in the naphthene ring. Remove two terminal hydrogens and close the ring. $\Delta C = -1.0$.

Aromatics. Benzene has a C value of 13.2. Methyl-substituted benzenes have C values of $13.2 - (5.28)$ (number of methyl substitutions).

Before discussing the estimation of C for nonhydrocarbons, it is important to note that the simple rules shown above are incomplete and do not cover many types of hydrocarbons. There are, however, no experimental data that can be used to obtain additional ΔC contributions. In fact, some of the ΔC values quoted above are based on so few data that they should be considered only approximate. For the 27 *hydrocarbons* studied by Roy and Thodos, one can obtain a rough, but often satisfactory, correlation by using only molecular weight as the correlating parameter, i.e.,

$$C \approx 5.21 \times 10^{-2} M + 1.82 \times 10^{-3} M^2 \qquad M < 120 \qquad (10\text{-}3.13)$$

with M in g/mol and C dimensionless.

For nonhydrocarbons, C is again estimated by a group contribution method wherein one mentally synthesizes the final compound by a particular set of rules and employs ΔC values for each step.

Alcohols. Synthesize the corresponding hydrocarbon with the same carbon structure and calculate C as noted above. Replace the appropriate hydrogen atom by a hydroxyl group and correct C as noted:

Type of —OH substitution	ΔC
On methane	3.79
$1 \leftarrow 1$	4.62
$2 \leftarrow 1$	4.11
$3 \leftarrow 1$	3.55
$4 \leftarrow 1$	3.03
$1 \leftarrow 2 \rightarrow 1$	4.12

The notation is the same as that used earlier; for example, $3 \leftarrow 1$ indicates that the —OH group is replacing a hydrogen atom on a type 1 carbon which is adjacent to a type 3 carbon:

$$\begin{array}{c} \mathrm{C} \\ | \\ \mathrm{C—C—C—C—H} \end{array} \rightarrow \begin{array}{c} \mathrm{C} \\ | \\ \mathrm{C—C—C—C—OH} \end{array}$$

These rules apply only to aliphatic alcohols, and they are incomplete even for them.

Amines. Estimation of C for amines is similar to that described above for alcohols. First, synthesize the corresponding hydrocarbon segment (with the most complex structure) that is finally to be attached to a nitrogen. For *primary* amines, replace the appropriate terminal hydrogen by a $—NH_2$ group with the following ΔC contributions:

Type of substitution	ΔC
On methane	2.60
$1 \leftarrow 1$	3.91
$1 \leftarrow 2 \rightarrow 1$	5.08
$2 \leftarrow 2 \rightarrow 1$	7.85
$1 \leftarrow 3 \rightarrow 1$ $\downarrow$ 1	6.50

For *secondary* amines, there are additional ΔC values:

	ΔC
$CH_3—NH_2 \rightarrow —CH_3—\overset{\displaystyle H}{\overset{\vert}{N}}—CH_3$	3.31
$—CH_2—NH_2 \rightarrow CH_2—\underset{\displaystyle H}{\underset{\vert}{N}}—CH_3$	4.40

Finally, for *tertiary* amines, Roy and Thodos show three types of corrections applicable to the secondary amines:

	ΔC
$CH_3—NH—CH_3 \rightarrow (CH_3)_3{\equiv}N$	2.59
$—CH_2—N(H)—CH_2— \rightarrow —CH_2—N(CH_3)—CH_2—$	3.27
$—CH_2—N(H)—CH_3 \rightarrow —CH_2—N—(CH_3)_2$	2.94

After calculating C for an amine as noted above, any methyl substitutions for a hydrogen on a side chain increase C by 4.56 (the same as shown for fourth and successive methyl substitutions in paraffinic hydrocarbons).

For example,

$$—\overset{|}{N}—CH_3 \rightarrow —\overset{|}{N}—CH_2—CH_3 \qquad \Delta C = 4.56$$

Nitriles. Only three ΔC contributions are shown; they were based on thermal conductivity data for acetonitrile, propionitrile, and acrylic nitrile.

Type of —CN addition	ΔC
On methane	5.43
$CH_3—CH_3 \rightarrow CH_3—CH_2—CN$	7.12
$—CH{=}CH_2 \rightarrow —CH{=}CH—CN$	6.29

Halides. Suggested contributions are shown below; the order of substitution should be F, C1, Br, I.

	ΔC
First halogen substitution on methane:	
Fluorine	0.26
Chlorine	1.38
Bromine	1.56
Iodine	2.70
Second and successive substitutions on methane	
Fluorine	0.38
Chlorine	2.05
Bromine	2.81
Substitution on ethane and higher hydrocarbons	
Fluorine	0.58
Chlorine	2.93

Aldehydes and Ketones. Synthesize the hydrocarbon analog with the same number of carbon atoms and calculate C as noted above. Then form the desired aldehyde or ketone by substituting oxygen for two hydrogen atoms:

	ΔC
$—CH_2—CH_3 \rightarrow —CH_2—CHO$	1.93
$—CH_2—CH_2—CH_2— \rightarrow —CH_2—CO\quad CH_2$	2.80

Ethers. Synthesize the primary alcohol with the longest carbon chain on one side of the ether oxygen. Convert this alcohol to a methyl ether.

$$—CH_2OH \rightarrow —CH_2—O—CH_3 \qquad \Delta C = 2.46$$

Extend the methyl chain, if desired, to an ethyl.

$$—CH_2—O—CH_3 \rightarrow —CH_2—O—CH_2—CH_3 \qquad \Delta C = 4.18$$

Although Roy and Thodos do not propose extensions beyond the ethyl group, presumably more complex chains could be synthesized by using ΔC values obtained from paraffinic and isoparaffinic contributions.

Acids and Esters. Synthesize the appropriate ether so as to allow the following substitutions:

	ΔC
$—CH_2—O—CH_3 \rightarrow —CH_2—O—\underset{\displaystyle \mathrm{H}}{\underset{\vert}{C}}{=}O$	0.75
$—CH_2—O—CH_2— \rightarrow —CH_2—O—\underset{\vert}{C}{=}O$	0.31

Cyclics. Synthesize the ring, if possible, with the following contributions (not substitutions):

Group	ΔC
$—CH_2—$	4.25
$—CH{=}$	3.50
$—NH—$	4.82
$—N{=}$	3.50
$—O—$	3.61
$=S=$	7.01

and determine C as:

$$C = \Sigma\Delta C - 7.83$$

The Roy and Thodos group contributions were obtained from limited data and are averaged values. Calculations cannot be made for many compounds by using the rules given above, but an intelligent guess for a missing increment can often be made. The Roy-Thodos method can also be used in a different way. If a single value of λ is available at a known temperature, Eq. (10-3.10) can be used with Table 10-1 to yield a value of C that can then be employed to determine λ at other temperatures.

Method of Chung, et al. (1984, 1988)

Chung, et al. employed an approach similar to that of Mason and Monchick (1962) to obtain a relation for λ. By using their form and a similar one for low-pressure viscosity [Eq. (9-4.10)], one obtains

$$\frac{\lambda M'}{\eta C_v} = \frac{3.75\Psi}{C_v/R} \tag{10-3.14}$$

where λ = thermal conductivity, W/(m · K)
M' = molecular weight, kg/mol
η = low-pressure gas viscosity, N · s/m^2
C_v = heat capacity at constant volume, J/(mol · K)
R = gas constant, 8.314 J/(mol · K)
$\Psi = 1 + \alpha\{[0.215 + 0.28288\alpha - 1.061\beta + 0.26665Z]/[0.6366 + \beta Z + 1.061\alpha\beta]\}$
$\alpha = (C_v/R) - 3/2$
$\beta = 0.7862 - 0.7109\omega + 1.3168\omega^2$
$Z = 2.0 + 10.5T_r^2$

The β term is an empirical correlation for $(f_{\text{int}})^{-1}$ [Eq. (10-3.4)] and is said to apply only for nonpolar materials. For polar materials, β is specific for each compound; Chung, et al. (1984) list values for a few materials. If the compound is polar and β is not available, use a default value of $(1.32)^{-1} = 0.758$.

Z represents the number of collisions required to interchange a quantum of rotational energy with translational energy. For large values of Z, Ψ reduces to

$$\Psi = 1 + 0.2665\frac{\alpha}{\beta} \qquad \text{for large } Z \tag{10-3.15}$$

If Eq. (10-3.15) is used in Eq. (10-3.14), the Eucken correlation [Eq. (10-3.3)] is obtained when β is set equal to unity. If $\beta = (1.32)^{-1}$, the modified Eucken relation [Eq. (10-3.5)] is recovered. The method is illustrated in Example 10-1.

Example 10-1 Use the above methods to estimate the thermal conductivity of 2-methylbutane (isopentane) vapor at 1 bar and 100°C. The value tabulated in Bretsznajder (1971) is 0.022 W/(m · K).

solution From Appendix A, T_c = 460.39 K, P_c = 33.81 bar, V_c = 308.3 cm^3/mol, Z_c = 0.272, ω = 0.229, and M = 72.151 g/mol = 72.151 × 10^{-3} kg/mol.

First we need to estimate the viscosity of 2-methylbutane. Using the Chung, et al. correlation, Eq. (9-4.10),

$$\eta = \frac{40.785 F_c (MT)^{1/2}}{V_c^{2/3}\Omega_v}$$

where $F_c = 1 - 0.275\omega$, since μ_r and $\kappa = 0$, $T^* = 1.2593T_r = 1.2593[(100 + 273)/460.39] = 1.020$, and $\Omega_v = 1.576$ from Eq. (9-4.3). Thus,

$$\eta = \frac{(40.785)[1 - (0.275)(0.229)][72.151)(373)]^{1/2}}{(308.3)^{2/3}(1.576)} = 87.2\mu\text{P} = 8.72 \times 10^{-5}\text{P}$$

$$= 8.72 \times 10^{-6}\text{N}\cdot\text{s/m}^2$$

The ideal-gas value of C_v is estimated from $(C_p - R)$, where C_p is determined from the polynomial constants in Appendix A,

$$C_v = 144.1 - 8.3 = 135.8 \text{ J/(mol}\cdot\text{K)}$$

$$M' = 72.151 \times 10^{-3} \text{ kg/mol}$$

EUCKEN METHOD, Eq. (10-3.3)

$$\lambda = \frac{\eta C_v}{M'}\left(1 + \frac{9/4}{C_v/R}\right) = \frac{(8.72 \times 10^{-6})(135.8)}{72.151 \times 10^{-3}}\left[1 + \frac{9/4}{135.8/8.314}\right]$$

$$= 1.87 \times 10^{-2} \text{ W/(m}\cdot\text{K)}$$

$$\text{Error} = \frac{1.87 - 2.2}{2.2} \times 100 = -15\%$$

MODIFIED EUCKEN METHOD, Eq. (10-3.5)

$$\lambda = \frac{\eta C_v}{M'}\left(1.32 + \frac{1.77}{C_v/R}\right) = \frac{(8.72 \times 10^{-6})(135.8)}{72.151 \times 10^{-3}}\left(1.32 + \frac{1.77}{135.8/8.314}\right)$$

$$= 2.35 \times 10^{-2}\text{W/(m}\cdot\text{K)}$$

$$\text{Error} = \frac{2.35 - 2.2}{2.2} \times 100 = 6.8\%$$

STIEL AND THODOS METHOD, Eq. (10-3.6)

$$\lambda = \frac{\eta C_v}{M'}\left(1.15 + \frac{2.03}{C_v/R}\right) = \frac{(8.72 \times 10^{-6})(135.8)}{72.151 \times 10^{-3}}\left(1.15 + \frac{2.03}{135.8/8.314}\right)$$

$$= 2.09 \times 10^{-2}\text{W/(m}\cdot\text{K)}$$

$$\text{Error} = \frac{2.09 - 2.2}{2.2} \times 100 = -5.0\%$$

ROY-THODOS METHOD, Eq. (10-3.10)

$$\lambda\Gamma = (\lambda\Gamma)_{tr} + (\lambda\Gamma)_{int}$$

Γ is defined in Eq. (10-3.9),

$$\Gamma = 210 \left(\frac{T_c M^3}{P_c^4}\right)^{1/6} = 210 \left[\frac{(460.39)(72.151)^3}{(33.81)^4}\right]^{1/6} = 474$$

With a reduced temperature, $T_r = (100 + 273)/460.39 = 0.810$, $(\lambda\Gamma)_{tr}$ is found from Eq. (10-3.11),

$$(\lambda\Gamma)_{tr} = 8.757\{\exp[(0.0464)(0.810)] - \exp[(-0.2412)(0.810)]\} = 1.89$$

To find $(\lambda\Gamma)_{int}$, C must first be determined by synthesizing n-butane with the recommended ΔC increments,

$$n\text{-butane} = (0.73 + 2.00 + 3.18 + 3.68) = 9.59$$

Next, to form 2-methylbutane, a $1 \leftarrow 2 \rightarrow 2$ methyl substitution is required. Thus,

$$C(\text{2-methylbutane}) = 9.59 + 4.71 = 14.30$$

The appropriate $f(T_r)$ is given in Table 10-1 for saturated hydrocarbons,

$$f(T_r) = -0.152T_r + 1.191T_r^2 - 0.039T_r^3$$
$$= (-0.152)(0.810) + (1.191)(0.810)^2 - (0.039)(0.810)^3 = 0.638$$

Then,

$$(\lambda\Gamma)_{int} = Cf(T_r) = (14.30)(0.638) = 9.12$$

$$\lambda = \frac{1.89 + 9.12}{474} = 2.32 \times 10^{-2}\ \text{W/(m} \cdot \text{K)}$$

$$\text{Error} = \frac{2.32 - 2.2}{2.2} \times 100 = 5.5\%$$

CHUNG, ET AL. METHOD, Eq. (10-3.14)

$$\frac{\lambda M'}{\eta C_v} = \frac{3.75\Psi}{C_v/R}$$

As defined under Eq. (10-3.14),

$$\alpha = \frac{C_v}{R} - 1.5 = \frac{135.8}{8.314} - 1.5 = 14.83$$

$$\beta = 0.7862 - (0.7109)(0.229) + (1.3168)(0.229)^2 = 0.692$$

$$T_r = \frac{373}{460.39} = 0.810 \text{ and } Z = 2.0 + (10.5)(0.810)^2 = 8.90$$

$$\Psi = 1 + 14.83\,\frac{0.215 + (0.28288)(14.83) - (1.061)(0.692) + (0.26665)(8.90)}{0.6366 + (0.692)(8.90) + (1.061)(14.83)(0.692)}$$
$$= 6.071$$

$$\lambda = \frac{(3.75)(6.071)}{135.8/8.314}\,\frac{(8.72 \times 10^{-6}(135.8)}{72.151 \times 10^{-3}} = 2.29 \times 10^{-2}\ \text{W/(m} \cdot \text{K)}$$

$$\text{Error} = \frac{2.29 - 2.2}{2.2} \times 100 = 4.1\%$$

Example 10-2 Use the Roy-Thodos method to estimate the thermal conductivity of ethyl acetate at 184°C and 1 bar. The experimental value is 2.38×10^{-2} W/(m·K) (Bromley, 1952).

solution From Appendix A, $T_c = 523.2$ K, $P_c = 38.3$ bar, and $M = 88.106$ g/mol. From Eq. (10-3.9),

$$\Gamma = 210(T_c M^3 / P_c^4)^{1/6} = 210 \left[\frac{(523.2)(88.106)^3}{(38.3)^4}\right]^{1/6} = 493$$

With $T = 184 + 273 = 457$ K, with $T_r = 457/523.2 = 0.873$, and using Eq. (10-3.11),

$$(\lambda\Gamma)_{tr} = 8.757\{\exp[(0.0464)(0.873)] - \exp[(-0.2412)(0.873)]\} = 2.02$$

To determine C, the synthesis plan is as follows: ethane → ethanol → methyl ethyl ether → diethyl ether → ethyl acetate. For ethane, $C = 0.73 + 2.00 = 2.73$. Converting to ethanol, $\Delta C = 4.62$. Next, to make methyl ethyl ether, $\Delta C = 2.46$ and then on to diethyl ether, $\Delta C = 4.18$. Finally, we form the ester, ethyl acetate, $\Delta C = 0.31$. Summing, $C = 2.73 + 4.62 + 2.46 + 4.18 + 0.31 = 14.30$. With Table 10-1 for esters,

$$f(T_r) = -0.082T_r + 1.045T_r^2 + 0.037T_r^3$$

$$= (-0.082)(0.873) + (1.045)(0.873)^2 + (0.037)(0.873)^3 = 0.750$$

Then, with Eq. (10-3.10),

$$\lambda_r = \lambda\Gamma = (\lambda\Gamma)_{tr} + (\lambda\Gamma)_{int} = 493\lambda = 2.02 + (14.3)(0.750)$$

$$\lambda = 2.58 \times 10^{-2} \text{ W/(m·K)}$$

$$\text{Error} = \frac{2.58 - 2.38}{2.38} \times 100 = 8.4\%$$

Discussion

Except for the Roy-Thodos method, all methods described in this section for estimating the thermal conductivity of a pure gas at ambient pressure correlate the Eucken factor $\lambda M'/\eta C_v$ as a function of other variables such as C_v, T_r, and ω. To use them, independent values of the gas viscosity and heat capacity are necessary. The Roy-Thodos correlation requires only the critical temperature and pressure and employs a group contribution method to account for the effect of internal degrees of freedom. In Table 10-2, we show the percent errors found when applying all of these techniques to estimate λ for a variety of compounds. As noted earlier, the Eucken equation (10-3.3) tends to underestimate λ, whereas the modified Eucken equation overestimates λ. The Stiel and Thodos equation yields λ values between the two Eucken forms. All three of these relations predict that the Eucken factor should decrease with temperature, whereas, in actuality, the factor appears, in most cases, to increase slightly (see Fig. 10-2). The Chung, et al. modification does predict the correct trend of the Eucken factor with temperature and, except for polar compounds, yields λ values quite close to those reported experimentally. The Roy-Thodos method generally yields the smallest errors, but it is not applicable to inorganic compounds and, even for many types of organic compounds, group contributions are lacking.

TABLE 10-2 Comparison between Calculated and Experimental* Values of the Thermal Conductivity of a Pure Gas at 1 bar

Compound	T, K	λ, exp. W/(m·K) × 10^3	η N·s/m^2 × 10^7	C_v J/(mol·K)	Eucken factor	Percent error† Eucken	Mod. Eucken	Stiel and Thodos	Roy and Thodos	Chung, et al.
Acetaldehyde‡	313	12.6	91.0	48.2	1.27	8.6	27	17	7.5	8.1
$M' = 44.05 \times 10^{-3}$	353	15.9	103	52.3	1.30	4.4	23	13	3.9	6.4
	393	19.4	115	56.5	1.32	1.2	20	10	1.6	5.3
Acetone‡	353	15.7	90.0	77.9	1.30	−4.7	16	5.1	2.9	0
$M' = 58.08 \times 10^{-3}$	393	19.4	100	84.2	1.34	−8.7	12	0.9	1.2	−2.3
	457	24.7	114.5	96.1	1.29	−8.4	13	1.7	4.5	0.8
Acetonitrile‡	353	12.4	85	49.0	1.22	13	33	22	2.7	12
$M' = 41.05 \times 10^{-3}$	393	15.0	95	52.6	1.23	10	30	19	2.9	11
Acetylene	198	11.8	70.1	26.8	1.64	3.8	14	8.8	10	3.0
$M' = 26.04 \times 10^{-3}$	273	18.7	95.5	33.7	1.51	2.8	16	9.1	5.7	8.3
	373	29.8	126.1	40.4	1.52	−4.0	10	2.9	1.9	7.0
Ammonia‡	213	16.5	73.2	25.4	1.51	15	26	20		7.3
$M' = 17.03 \times 10^{-3}$	273	21.9	90.6	26.7	1.54	10	21	15		6.6
Benzene	353	14.6	90	90.4	1.40	−14	5.8	−4.6	1.0	−1.7
$M' = 78.11 \times 10^{-3}$	433	22.6	109.5	114	1.41	−18	2.5	−3.2	0	−1.6
n-Butane	273	13.5	68.8	84.6	1.35	−9.4	11	0.1	−2.0	2.3
$M' = 58.12 \times 10^{-3}$	373	24.6	94.5	110	1.38	−15	5.7	−5.2	−3.7	1.5
Carbon dioxide	200	9.51	101.5	24.7	1.67	5.3	15	9.8		4.9
$M' = 44.01 \times 10^{-3}$	300	16.7	149.5	28.9	1.70	−3.2	7.5	1.9		2.8
	473	28.4	225.0	35.6	1.56	−2.2	11	4.1		11
	598	37.9	272.8	39.5	1.55	−4.8	9.3	1.9		11
	1273	81.7	465.1	48.8	1.58	−13	2.4	−5.6		7.3

Carbon tetrachloride	273	5.95	91.0	72.4	1.39	−9.4	9.7	−0.4	−9.6	−2.4
$M' = 153.82 \times 10^{-3}$	373	8.58	124.3	81.7	1.30	−5.4	15	4.4	5.6	6.3
	457	10.9	151.2	86.3	1.28	−5.3	16	4.7	17	9.4
Cyclohexane	353	16.3	83.0	123	1.34	−14	7.1	−4.2	−0.6	2.6
$M' = 84.16 \times 10^{-3}$	433	25.6	100.5	155	1.38	−19	2.3	−9.0	−1.0	0.5
Dichlorodifluoromethane	273	8.29	114.2	57.8	1.52	−12	4.3	−4.5	0.7	−5.6
$M' = 120.21 \times 10^{-3}$	373	13.8	155.1	69.1	1.56	−18	−1.0	−9.9	−0.4	−6.4
	473	19.4	193.6	77.1	1.57	−20	−3.3	−12	3.8	−5.8
Ethyl acetate‡	319	12.1	81.1	105	1.24	−5.9	17	4.7	6.9	1.3
$M' = 88.11 \times 10^{-3}$	373	16.2	95.5	121	1.24	−6.5	17	4.4	7.3	3.4
	457	23.8	116	142	1.27	−11	12	−0.3	8.4	1.3
Ethyl alcohol‡	293	15.0	84.8	56.5	1.44	−7.7	9.6	0.4	−2.9	−8.8
$M' = 46.07 \times 10^{-3}$	401	24.9	116.6	72.4	1.36	−7.4	12	1.8	−1.5	−2.5
Ethylene	273	17.4	94.2	32.8	1.58	−0.6	12	5.4	−0.3	3.3
$M' = 28.05 \times 10^{-3}$	373	27.8	124.5	43.1	1.46	−1.3	14	6.1	3.0	8.3
Ethyl ether	273	13.0	68.4	104	1.35	−13	7.9	−3.1	−3.2	1.8
$M' = 74.12 \times 10^{-3}$	373	22.2	93.7	121	1.45	−20	−0.7	−11	2.5	−3.6
	486	35.1	120.8	148	1.45	−23	−2.5	−13	8.2	−2.6
n-Hexane	373	20.1	79.0	163	1.34	−17	4.8	−6.8	−0.8	3.4
$M' = 86.18 \times 10^{-3}$	433	27.2	92.0	186	1.37	−20	2.1	−9.4	−1.2	2.1
Isopropyl alcohol‡	304	15.2	78.5	75.4	1.54	−19	−1.8	−11	−6.6	−17
$M' = 60.10 \times 10^{-3}$	400	25.0	105.6	101	1.41	−16	4.0	−6.5	−8.8	8.1
Sulfur dioxide	273	8.29	117	30.6	1.48	8.6	21	15		9.0
$M' = 64.06 \times 10^{-3}$										

† Percent error = [(calc. − exp.)/exp.] × 100.
‡ Compounds for which β was set equal to $(1.32)^{-1}$ as a default value in the Chung, et al. method.
* Data from Bromley (1952), Ehya, et al. (1972), and Vines and Bennett (1954).

It is recommended that, for nonpolar compounds, the Chung et al., or the Roy-Thodos method be used to estimate λ for pure gases at ambient pressure. Errors vary, but, generally, they do not exceed 5 to 7%. For polar compounds, the Roy-Thodos form is recommended.

Thermal conductivity data for pure gases are compiled in Tsederberg (1965) and Vargaftik, et al. (1994, 1996). Constants that may be used to calculate pure gas thermal conductivities at different temperatures are tabulated in Daubert, et al. (1997), Miller, et al. (1976a), and Yaws (1995, 1995a). The constants in Daubert, et al. (1997) are based on critically evaluated data while those in the other references are not. Thermal conductivity values for common gases are tabulated at different temperatures in Dean (1999), Lide (1999), and Perry and Green (1997).

10-4 EFFECT OF TEMPERATURE ON THE LOW-PRESSURE THERMAL CONDUCTIVITIES OF GASES

Thermal conductivities of low-pressure gases increase with temperature. The exact dependence of λ on T is difficult to judge from the λ-estimation methods in Sec. 10-3 because other temperature-dependent parameters (e.g., heat capacities and viscosities) are incorporated in the correlations. Generally, $d\lambda/dT$ ranges from 4×10^{-5} to 1.2×10^{-4} W/(m · K^2), with the more complex and polar molecules having the larger values. Several power laws relating λ with T have been proposed (Correla, et al.,1968; Missenard, 1972), but they are not particularly accurate. To illustrate the trends, Fig. 10-3 has been drawn to show λ as a function of temperature for a few selected gases.

10-5 EFFECT OF PRESSURE ON THE THERMAL CONDUCTIVITIES OF GASES

The thermal conductivities of all gases increase with pressure, although the effect is relatively small at low and moderate pressures. Three pressure regions in which the effect of pressure is distinctly different are discussed below.

Very Low Pressure

Below pressures of about 10^{-3} bar, the mean free path of the molecules is large compared to typical dimensions of a measuring cell, and there λ is almost proportional to pressure. This region is called the *Knudsen domain.* In reported thermal conductivity data, the term *zero-pressure value* is often used; however, it refers to values extrapolated from higher pressures (above 10^{-3} bar) and not to measured values in the very low pressure domain.

Low Pressure

This region extends from approximately 10^{-3} to 10 bar and includes the domain discussed in Secs. 10-3 and 10-4. The thermal conductivity increases about 1% or

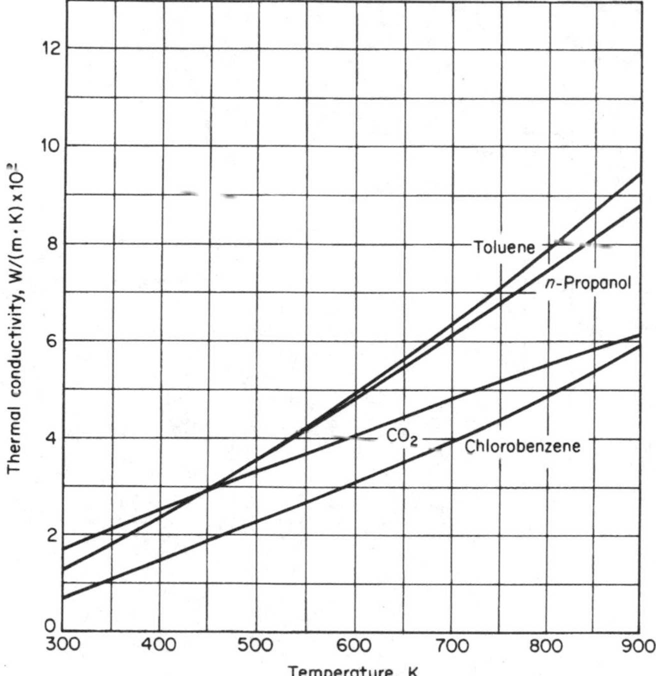

FIGURE 10-3 Effect of temperature on the thermal conductivity of some low-pressure gases.

less per bar (Kannuliuk and Donald, 1950; Vines, 1953, 1958; Vines and Bennett, 1954). Such increases are often ignored in the literature, and either the 1-bar value or the "zero-pressure" extrapolated value may be referred to as the low-pressure conductivity.

High Pressure

In Fig. 10-4 we show the thermal conductivity of propane over a wide range of pressures and temperatures (Holland, et al., 1979). The high-pressure gas domain would be represented by the curves on the right-hand side of the graph above the critical temperature (369.8 K). Increasing pressure raises the thermal conductivity, with the region around the critical point being particularly sensitive. Increasing temperature at low pressures results in a larger thermal conductivity, but at high pressure the opposite effect is noted. Similar behavior is shown for the region below T_c, where λ for liquids decreases with temperature, whereas for gases (see Sec. 10-4), there is an increase of λ with T. Pressure effects (except at very high pressures) are small below T_c. Not shown in Fig. 10-4 is the unusual behavior of λ near the critical point. In this region, the thermal conductivity is quite sensitive to both temperature and pressure (Basu and Sengers, 1983). Figure 10-5 shows a plot of λ for CO_2 near the critical point (Guildner, 1958). It can be seen from Fig. 10-5 that near the critical point, λ can increase by a factor of six due to the enhancement that occurs in the critical region. This enhancement is much bigger for the

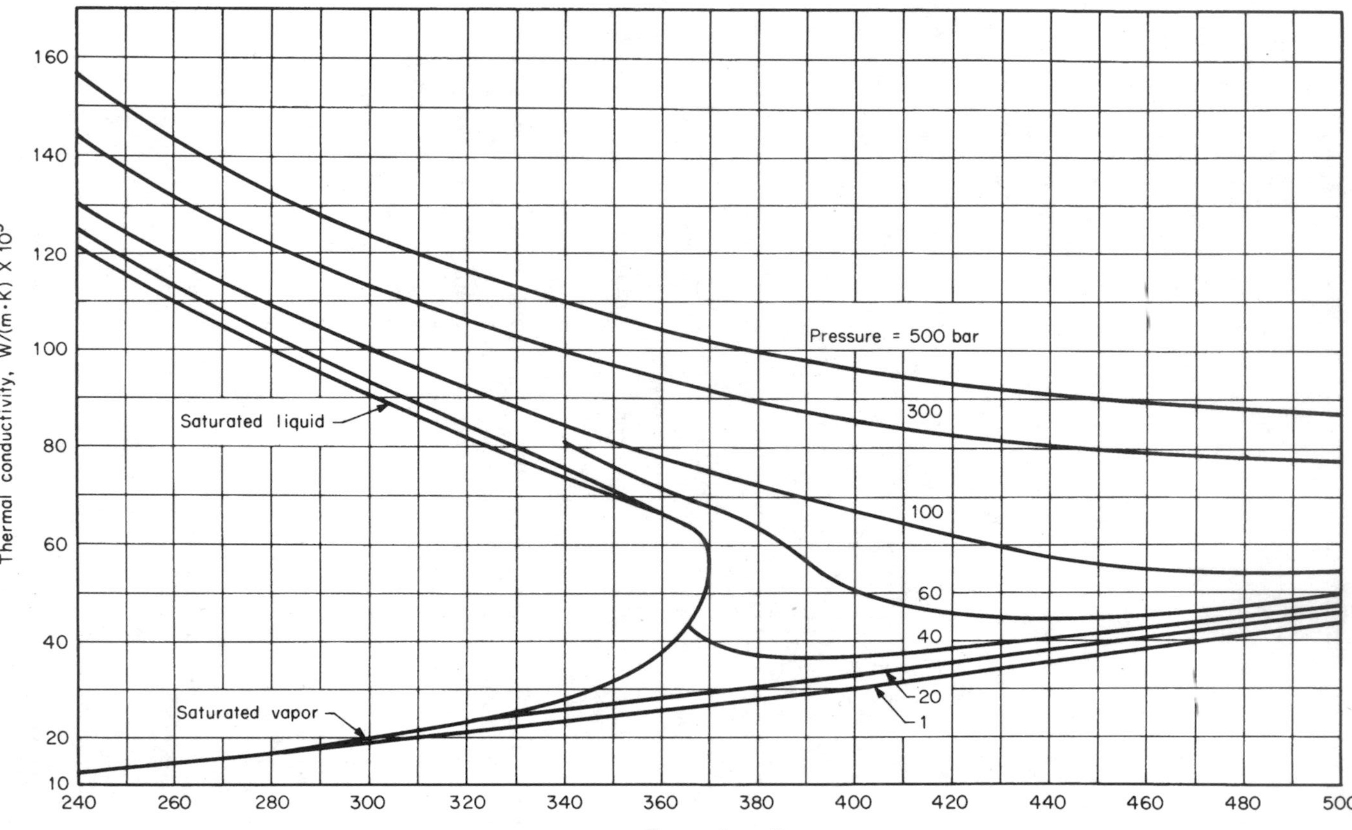

FIGURE 10-4 Thermal conductivity of propane. (*Data from Holland, et al; 1979.*)

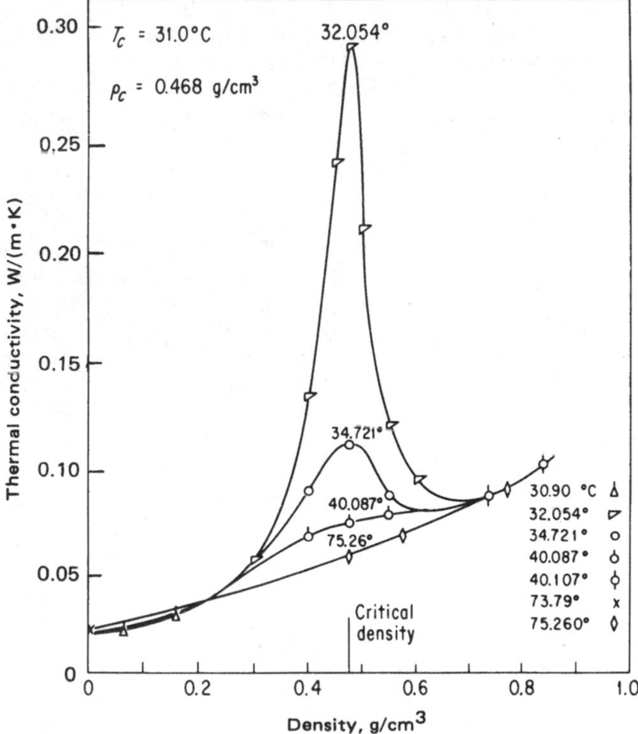

FIGURE 10-5 Thermal conductivity of carbon dioxide near the critical point. *(Data from Guildner, 1958.)*

thermal conductivity than for viscosity. The region for which this enhancement increases λ for CO_2 by at least 1% is bounded approximately by $240 < T < 450$ K ($0.79 < T_r < 1.48$) and 25 kg m^{-3} $< \rho <$ 1000 kg m^{-3} ($0.53 < \rho_r < 2.1$). See Krauss, et al. (1993), Sengers and Luettmer-Strathmann (1996) and Vesocic, et al. (1990) for additional discussion of the critical region enhancement and examples of how to describe the enhancement mathematically. When generalized charts of the effect of pressure on λ are drawn, the enhancement around T_c and P_c is usually smoothed out and not shown.

Excess Thermal Conductivity Correlations

Many investigators have adopted the suggestion of Vargaftik (1951, 1958) that the excess thermal conductivity, $\lambda - \lambda°$, be correlated as a function of the *PVT* properties of the system in a corresponding states manner. (Here $\lambda°$ is the low-pressure thermal conductivity of the gas at the same temperature.) In its simplest form,

$$\lambda - \lambda° = f(\rho) \tag{10-5.1}$$

where ρ is the fluid density. The correlation has been shown to be applicable to ammonia (Groenier and Thodos, 1961; Richter and Sage, 1964), ethane (Carmichael, et al., 1963), n-butane (Carmichael and Sage, 1964; Kramer and Comings, 1960), nitrous oxide (Richter and Sage, 1963), ethylene (Owens and Thodos, 1960), methane (Carmichael, et al., 1966; Mani and Venart, 1973; Owens and Thodos, 1957a), diatomic gases (Misic and Thodos, 1965; Schafer and Thodos, 1959), hydrogen (Schafer and Thodos, 1958), inert gases (Owens and Thodos, 1957), and carbon dioxide (Kennedy and Thodos, 1961). Temperature and pressure do not enter explicitly, but their effects are included in the parameters $\lambda°$ (temperature only) and ρ.

Stiel and Thodos (1964) have generalized Eq. (10-5.1) by assuming that $f(\rho)$ depends only on T_c, P_c, V_c, M, and ρ. By dimensional analysis they obtain a correlation between $\lambda - \lambda°$, Z_c, Γ, and ρ, where Γ was defined in Eq. (10-3.9). From data on 20 nonpolar substances, including inert gases, diatomic gases, CO_2, and hydrocarbons, they established the approximate analytical expressions:

$$(\lambda - \lambda°)\Gamma Z_c^5 = 1.22 \times 10^{-2}[\exp(0.535\rho_r) - 1] \qquad \rho_r < 0.5 \qquad (10\text{-}5.2)$$

$$(\lambda - \lambda°)\Gamma Z_c^5 = 1.14 \times 10^{-2}[\exp(0.67\rho_r) - 1.069] \qquad 0.5 < \rho_r < 2.0 \qquad (10\text{-}5.3)$$

$$(\lambda - \lambda°)\Gamma Z_c^5 = 2.60 \times 10^{-3}[\exp(1.155\rho_r) + 2.016)] \qquad 2.0 < \rho_r < 2.8 \qquad (10\text{-}5.4)$$

where λ is in W/(m·K), Z_c is the critical compressibility, and ρ_r is the reduced density $\rho/\rho_c = V_c/V$.

Equations (10-5.2) to (10-5.4) should not be used for polar substances or for hydrogen or helium. The general accuracy is in doubt, and errors of ±10 to 20% are possible. The method is illustrated in Example 10-3.

Example 10-3 Estimate the thermal conductivity of nitrous oxide at 105°C and 138 bar. At this temperature and pressure, the experimental value is 3.90×10^{-2} W/(m·K) (Richter and Sage, 1963). At 1 bar and 105°C, $\lambda° = 2.34 \times 10^{-2}$ W/(m·K) (Richter and Sage, 1963). From Appendix A, T_c = 309.6 K, P_c = 72.55 bar, V_c = 97.0 cm³/mol, Z_c = 0.273, and M = 44.013 g/mol. At 105°C and 138 bar, Z for N_2O is 0.63 (Couch and Dobe, 1961).

solution With Eq. (10-3.9),

$$\Gamma = 210\left(\frac{T_c M^3}{P_c^4}\right)^{1/6} = 210\left[\frac{(309.6)(44.013)^3}{(72.55)^4}\right]^{1/6} = 208$$

$$V = \frac{ZRT}{P} = \frac{(0.63)(8.314)(378)}{138 \times 10^5} \times 10^6 = 144 \text{ cm}^3/\text{mol}$$

$$\rho_r = \frac{V_c}{V} = \frac{97.0}{144} = 0.674$$

Then, with Eq. (10-5.3),

$$(\lambda - \lambda°)(208)(0.273)^5 = (1.14 \times 10^{-2})\{\exp[(0.67)(0.674)] - 1.069\}$$

$$\lambda - \lambda° = 1.81 \times 10^{-2} \text{ W/(m·K)}$$

$$\lambda = (2.34 + 1.81) \times 10^{-2} = 4.15 \times 10^{-2} \text{ W/(m·K)}$$

$$\text{Error} = \frac{4.15 - 3.90}{3.90} \times 100 = 6.4\%$$

Method of Chung, et al. (1984, 1988)

The low-pressure estimation procedure for pure component thermal conductivities developed by these authors, and given in Eq. (10-3.14), is modified to treat materials at high pressures (or densities).

$$\lambda = \frac{31.2\eta^\circ\Psi}{M'}(G_2^{-1} + B_6 y) + qB_7 y^2 T_r^{1/2} G_2 \tag{10-5.5}$$

where λ = thermal conductivity, W/(m·K)
η° = *low-pressure* gas viscosity, N·s/m²
M' = molecular weight, kg/mol
Ψ = $f(C_v, \omega, T_r)$ [as defined under Eq. (10-3.14)]
q = $3.586 \times 10^{-3} (T_c/M')^{1/2}/V_c^{2/3}$
T = temperature, K
T_c = critical temperature, K
T_r = reduced temperature, T/T_c
V_c = critical volume, cm³/mol

$$y = \frac{V_c}{6V} \tag{10-5.6}$$

$$G_1 = \frac{1 - 0.5y}{(1 - y)^3} \tag{10-5.7}$$

$$G_2 = \frac{(B_1/y)[1 - \exp(-B_4 y)] + B_2 G_1 \exp(B_5 y) + B_3 G_1}{B_1 B_4 + B_2 + B_3} \tag{10-5.8}$$

The coefficients B_1 to B_7 are functions of the acentric factor ω, the reduced dipole moment μ_r [as defined in Eq. (9-4.12)], and the association factor κ. Some values of κ are shown in Table 9-1.

$$B_i = a_i + b_i\omega + c_i\mu_r^4 + d_i\kappa \tag{10-5.9}$$

with a_i b_i, c_i, and d_i given in Table 10-3.

The relation for high-pressure thermal conductivities is quite similar to the Chung, et al. form for high-pressure viscosities [Eqs. (9-6.18) through (9-6.23)].

In Eq. (10-5.5), if V becomes large, y then approaches zero. In such a case, both G_1 and G_2 are essentially unity and Eq. (10-5.5) reduces to Eq. (10-3.14), the

TABLE 10-3 Values of B_i in Eq. (10-5.9)
$B_i = a_i + b_i\omega + c_i\mu_r^4 + d_i\kappa$

i	a_i	b_i	c_i	d_i
1	2.4166 E + 0	7.4824 E − 1	−9.1858 E − 1	1.2172 E + 2
2	−5.0924 E − 1	−1.5094 E + 0	−4.9991 E + 1	6.9983 E + 1
3	6.6107 E + 0	5.6207 E + 0	6.4760 E + 1	2.7039 E + 1
4	1.4543 E + 1	−8.9139 E + 0	−5.6379 E + 0	7.4344 E + 1
5	7.9274 E − 1	8.2019 E − 1	−6.9369 E − 1	6.3173 E + 0
6	−5.8634 E + 0	1.2801 E + 1	9.5893 E + 0	6.5529 E + 1
7	9.1089 E + 1	1.2811 E + 2	−5.4217 E + 1	5.2381 E + 2

relation for λ at low pressures. To use Eq. (10-5.5), it should be noted that the viscosity $\eta°$ is for the *low-pressure*, pure gas. Experimental values may be employed or $\eta°$ can be estimated by the techniques given in Sec. 9-4. The dimensions of $\eta°$ are $N \cdot s/m^2$. The conversion from other viscosity units is $1\ N \cdot s/m^2 = 10\ P = 10^7\ \mu P$.

Chung, et al. tested Eq. (10-5.5) with data from a large range of hydrocarbon types and from data for simple gases. Deviations over a wide pressure range were usually less than 5 to 8%. For highly polar materials, the correlation for β as given under Eq. (10-3.14) is not accurate and, at present no predictive technique to apply to such compounds is available. (See the discussion dealing with polar materials under Eq. (10-3.14).) The high-pressure Chung, et al. method is illustrated in Example 10-4.

TRAPP Method (Huber, 1996)

The TRAPP (transport property prediction) method is a corresponding states method to calculate viscosities and thermal conductivities of pure fluids and mixtures. In its original version (Ely and Hanley, 1983; Hanley, 1976), it was also used to estimate low-pressure values of λ and η and employed methane as a reference fluid. In the most recent version, presented below for pure fluids and later in Sec. 10-7 for mixtures, low-pressure values are estimated by one of the methods presented earlier in the chapter, propane is the reference fluid, and shape factors are no longer functions of density.

In this method, the excess thermal conductivity of a pure fluid is related to the excess thermal conductivity of the reference fluid, propane:

$$\lambda - \lambda° = F_\lambda X_\lambda[\lambda^R - \lambda^{Ro}] \tag{10-5.10}$$

The reference fluid values are evaluated at T_0 and density ρ_0, not T and ρ. In Eq. (10-5.10), $\lambda°$ is the thermal conductivity at low pressure. Equation (10-3.5) leads to

$$\lambda° = (1.32C_p° + 3.741)\frac{\eta°}{M'} \tag{10-5.11}$$

where $\eta°$ = low pressure viscosity in $N \cdot s/m^2$
M' = molecular mass in kg/mol
$\lambda°$ = low pressure thermal conductivity in $W/(m \cdot K)$
$C_p°$ in J/mol K

λ^R is the true thermal conductivity of the reference fluid, propane, at temperature T_0 and density ρ_0. λ^{R_0} is the low pressure value for propane at temperature T_0. Ely (Huber, 1998) has found for propane that

$$\lambda^R - \lambda^{Ro} = C_1\rho_r^R + C_2(\rho_r^R)^3 + (C_3 + C_4/T_r^R)(\rho_r^R)^4 + (C_5 + C_6/T_r^R)(\rho_r^R)^5 \tag{10-5.12}$$

where $T_r^R = T_0/T_c^R$, $\rho_r^R = \rho_0/\rho_c^R$, $\lambda^R - \lambda^{Ro}$ is in $mW/(m \cdot K)$, and

$$C_1 = 15.2583985944$$
$$C_2 = 5.29917319127$$
$$C_3 = -3.05330414748$$
$$C_4 = 0.450477583739$$
$$C_5 = 1.03144050679$$
$$C_6 = -0.185480417707$$

X_λ T_0, ρ_0, and F_λ are calculated by

$$T_0 = T/f \tag{10-5.13}$$

$$\rho_0 = \rho h \tag{10-5.14}$$

$$F_\lambda = \left(\frac{0.044094}{M'} f\right)^{1/2} h^{-2/3} \tag{10-5.15}$$

$$X_\lambda = \left[1 + \frac{2.1866(\omega - \omega^R)}{1 - 0.505(\omega - \omega^R)}\right]^{1/2} \tag{10-5.16}$$

where f and h are equivalent substance reducing ratios and are determined as described below.

If vapor pressure and liquid density information are available, and if $T < T_c$, it is recommended that f be obtained from the equation

$$\frac{P_{\text{vp}}}{\rho^S} = f\,\frac{P^R_{\text{vp}}(T_0)}{\rho^{SR}(T_0)} \tag{10-5.17}$$

where P_{vp} and ρ^S are the vapor pressure and saturated liquid density at temperature, T. $P^R_{\text{vp}}(T_0)$ and $\rho^{SR}(T_0)$ are for the reference fluid, propane. Because the density and vapor pressure of the reference fluid are evaluated at $T_0 = T/f$, Eq. (10-5.17) must be solved iteratively. Once f is found from Eq. (10-5.17), h is determined from

$$h = \rho^{SR}(T_0)/\rho^S \tag{10-5.18}$$

If $T > T_c$, or if vapor pressure and saturated liquid density information are not available, h and f can be calculated by

$$f = \frac{T_c}{T^R_c}[1 + (\omega - \omega^R)(0.05203 - 0.7498 \ln T_r] \tag{10-5.19}$$

$$h = \frac{\rho^R_c}{\rho_c}\frac{Z^R_c}{Z_c}[1 - (\omega - \omega^R)(0.1436 - 0.2822 \ln T_r] \tag{10-5.20}$$

Example 10-4 Estimate the thermal conductivity of propylene at 473 K and 150 bar by using the (*a*) Chung, et al. and (*b*) TRAPP methods. Under these conditions, Vargaftik, et al. (1996) report $\lambda = 6.64 \times 10^{-2}$ W/(m·K) and $V = 172.1$ cm^3/mol. Also, these same authors list the low-pressure viscosity and thermal conductivity of propylene at 473 K as $\eta° = 134 \times 10^{-7}$ N·s/m^2 and $\lambda° = 3.89 \times 10^{-2}$ W/(m·K).

solution For both estimation techniques, we need certain parameters for propylene. From Appendix A:

$$T_c = 364.9 \text{ K}$$

$$P_c = 46.0 \text{ bar}$$

$$V_c = 184.6 \text{ cm}^3/\text{mo}$$

$$Z_c = 0.2798$$

$$\omega = 0.142$$

$$M = 42.081 \text{ g/mol}$$

$$M' = 0.042081 \text{ kg/mol}$$

$$\mu = 0.4 \text{ debye}$$

Also, since propylene is nonpolar, the association factor in Chung, et al.'s method $\kappa = 0$.

The low-pressure heat capacity at constant pressure at 473 K is found from the equation and polynomial constants shown in Appendix A as 90.98 J/(mol · K). Thus

$$C_v = C_p - R = 90.98 - 8.31 = 82.67 \text{ J/(mol} \cdot \text{K)}.$$

METHOD OF CHUNG ET AL. With the definition of Ψ given under Eq. (10-3.14), where

$$\alpha = \frac{C_v}{R} - \frac{3}{2} = \frac{82.67}{8.314} - \frac{3}{2} = 8.443$$

$$\begin{aligned}\beta &= 0.7862 - 0.7109\omega + 1.3168\omega^2 \\ &= 0.7862 - (0.7109)(0.142) + (1.3168)(0.142)^2 = 0.7118\end{aligned}$$

$$T_r = \frac{T}{T_c} = \frac{473}{364.9} = 1.296$$

$$Z = 2.0 + 10.5T_r^2 = 2.0 + (10.5)(1.296)^2 = 19.64$$

then

$$\begin{aligned}\Psi &= 1 + 8.443\frac{0.215 + (0.28288)(8.443) - (1.061)(0.7118) + (0.26665)(19.64)}{0.6366 + (0.7118)(19.64) + (1.061)(8.443)(0.7118)} \\ &= 3.850\end{aligned}$$

From Eq. (10-5.6)

$$y = \frac{\rho V_c}{6} = \frac{V_c}{6V} = \frac{184.6}{(6)(172.1)} = 0.1788$$

The values of B_i are found from Table 10-3, where $\omega = 0.142$ and $\kappa = 0$, and with Eq. (9-4.12),

$$\mu_r^4 = \left\{\frac{(131.3)(0.4)}{[(184.6)(364.9)]^{1/2}}\right\}^4 = 1.68 \times 10^{-3}$$

As an example,

$$B_1 = 2.4166 + (7.4824 \times 10^{-1})(0.142) - (9.1858 \times 10^{-1})(1.68 \times 10^{-3}) = 2.5213$$

$$B_2 = -0.80756 \qquad B_3 = 7.5176 \qquad B_4 = 13.268$$

$$B_5 = 0.90804 \qquad B_6 = -4.0295 \qquad B_7 = 109.19$$

With Eqs. (10-5.8) and (10-5.9),

$$G_1 = \frac{1 - (0.5)(0.1788)}{(1 - 0.1788)^3} = 1.644$$

For G_2

$$\frac{B_1}{y}[1 - \exp(-B_4 y)] = \frac{2.5213}{0.1788}\{1 - \exp[-(13.268)(0.1788)]\} = 12.79$$

$$B_2 G_1 \exp(B_5 y) = (-0.80756)(1.644)\exp[(0.90804)(0.1788)] = -1.562$$

$$B_3 G_1 = (7.5176)(1.644) = 12.36$$

$$B_1 B_4 + B_2 + B_3 = (2.5213)(13.268) - 0.80756 + 7.5176 = 40.16$$

Thus

$$G_2 = \frac{(12.79 - 1.562 + 12.36)}{40.16} = 0.5874$$

and

$$q = \frac{3.586 \times 10^{-3}(T_c/M')^{1/2}}{V_c^{2/3}}$$

$$= \frac{3.586 \times 10^{-3}(364.9/0.042081)^{1/2}}{(184.6)^{2/3}} = 1.030 \times 10^{-2}$$

With Eq. (10-5.5),

$$\lambda = \frac{(31.2)(134 \times 10^{-7})(3.850)}{0.042081}[(0.5874)^{-1} - (4.0295)(0.1788)]$$

$$+ (1.030 \times 10^{-2})(109.19)(0.1788)^2(1.296)^{1/2}(0.5874)$$

$$= 3.825 \times 10^{-2}(0.9819) + 2.404 \times 10^{-2} = 6.16 \times 10^{-2} \text{ W/(m}\cdot\text{K)}$$

$$\text{Error} = \frac{6.16 - 6.64}{6.65} \times 100 = -7.2\%$$

Note that the first term in the final result (3.825×10^{-2}) would represent the estimated value of λ at low pressure. This result is in good agreement with the reported value of 3.89×10^{-2} W/(m · K) in Vargaftik, et al. (1996).

TRAPP METHOD. Properties for the reference fluid propane are

$$T_c^R = 369.83 \text{ K}$$

$$V_c^R = 200.0 \text{ cm}^3/\text{mol}$$

$$Z_c^R = 0.276$$

$$\omega^R = 0.152$$

Equations (10-5.19) and (10-5.20), with $T_r = 473/364.9 = 1.296$, give

$$f = \frac{364.9}{369.83} [1 + (0.142 - 0.152)(0.05203 - 0.7498 \ln(1.296))] = 0.9881$$

$$h = \frac{184.6}{200.0} \frac{0.276}{0.2798} [1 - (0.142 - 0.152)(0.1436 - 0.2822 \ln(1.296))] = 0.9111$$

Equations (10-5.13) and (10-5.14) give

$$T_0 = T/f = 473/0.9881 = 478.7 \text{ K}$$

$$\rho_0 = \rho h = 0.9111/172.1 = 5.294 \times 10^{-3} \text{ mol/cm}^3$$

Equation (10-5.12) with $T_r^R = 480.4/369.83 = 1.299$ and $\rho_r^R = 5.242 \times 10^{-3} \times 200.0 = 1.048$ gives

$$\lambda^R - \lambda^{Ro} = 20.23 \text{ mW/(m} \cdot \text{K)} = 0.02023 \text{ W/(m} \cdot \text{K)}$$

Equations (10-5.15) and (10-5.16) give

$$F_\lambda = \left(\frac{0.044094}{0.042081} 0.9811\right)^{1/2} (0.9111)^{-2/3} = 1.083$$

$$X_\lambda = \left[1 + \frac{2.1866(0.142 - 0.152)}{1 - 0.505(0.142 - 0.152)}\right]^{1/2} = 0.9891$$

Equation (10-5.11), with C_p° and η° from above, gives

$$\lambda^\circ = (1.32 \times 90.98 + 3.741) \frac{134 \times 10^{-7}}{0.042081} = 0.0394 \text{ W/(m} \cdot \text{K)}$$

Equation (10-5.10) gives

$$\lambda = 0.0394 + 1.083 \times 0.02023 \times 0.9891 = 0.0611 \text{ W/(m} \cdot \text{K)}$$

$$\text{Error} = \frac{0.0611 - 0.0664}{0.0664} \times 100 = -8.0\%$$

Discussion

Three methods for estimating the thermal conductivity of pure materials in the dense gas region were presented. All use the fluid density rather than pressure as a system variable. The low-density thermal conductivity is required in the Stiel and Thodos [Eqs. (10.5.2) to (10-5.4)] and TRAPP [Eqs. (10-5.10) to (10-5.19)] methods, but it is calculated as a part of the procedure in the Chung, et al. [Eq. (10-5.5)] method. None of the techniques are applicable for polar gases, and even for nonpolar materials, errors can be large. The Chung, et al. and TRAPP procedures are reported to be applicable over a wide density domain even into the liquid phase. No one of the methods appears to have a clear superiority over the others.

Other methods for dense-fluid thermal conductivites have been presented. Assael, et al. (1996) and Dymond and Assael (1996) have correlated thermal conductivities of dense fluids by the universal equation

$$\log_{10} \lambda_r = 1.0655 - \frac{3.538}{V_r} + \frac{12.12}{V_r^2} - \frac{12.469}{V_r^3} + \frac{4.562}{V_r^4} \quad (10\text{-}5.21)$$

where $V_r = V/V_0$ and V_0 is the same close-packed volume that appears in Eq. (9-6.3). λ_r is a reduced thermal conductivity defined by

$$\lambda_r = 21.23 \frac{\lambda V^{2/3}}{R_\lambda} \left(\frac{M}{T}\right)^{1/2} \quad (10\text{-}5.22)$$

where λ is in W/(m·K), V is in cm^3/mol, M is g/mol and T is in K. R_λ is a parameter that accounts for deviations from smooth hard spheres. The two parameters, V_0 and R_λ are compound specific and are not functions of density. V_0 is a function of temperature as is R_λ for n-alcohols, refrigerants, and a number of other polar compounds (Assael, et al., 1994, 1995; Bleazard and Teja, 1996). For n-alkanes (Assael et al, 1992, 1992a), aromatic hydrocarbons (Assael, et al., 1992b), and other simple molecular fluids (Assael, et al., 1992), R_λ has been found to be independent of temperature. In theory, the two parameters V_0 and R_λ could be set with two experimental viscosity-density data, but in fact Eq. (10-5.21) has been used only for systems for which extensive data are available. It has been applied to densities above the critical density and applicability to temperatures down to $T_r \approx 0.6$ has been claimed. Values of V_0 and R_λ at 298 K and 350 K for 16 fluids are shown in Table 9-5.

Thermal conductivity data for selected fluids appear in the references above as well as in Fleeter, et al. (1980), Le Neindre (1987), Prasad and Venart (1981), Tufeu and Neindre (1981), Yorizane, et al. (1983), and Zheng, et al. (1984).

10-6 THERMAL CONDUCTIVITIES OF LOW-PRESSURE GAS MIXTURES

The thermal conductivity of a gas mixture is not usually a linear function of mole fraction. Generally, if the constituent molecules differ greatly in polarity, the mixture thermal conductivity is larger than would be predicted from a mole fraction average; for nonpolar molecules, the opposite trend is noted and is more pronounced the greater the difference in molecular weights or sizes of the constituents (Gray, et al., 1970; Misic and Thodos, 1961). Some of these trends are evident in Fig. 10-6, which shows experimental thermal conductivities for four systems. The argon-benzene system typifies a nonpolar case with different molecular sizes, and the methanol-n-hexane system is a case representing a significant difference in polarity. The linear systems benzene-n-hexane and ether-chloroform represent a balance between the effects of size and polarity.

Many theoretical papers discussing the problems, approximations, and limitations of the various methods also have appeared. The theory for calculating the conductivity for rare-gas mixtures has been worked out in detail (Brokaw, 1958; Hirschfelder, et al., 1954; Mason, 1958; Mason and Saxena, 1959; Mason and von Ubisch, 1960, Muckenfuss, 1958). The more difficult problem, however, is to modify monatomic mixture correlations to apply to polyatomic molecules.

Many techniques have been proposed; all are essentially empirical, and most reduce to some form of the Wassiljewa equation. Corresponding states methods for

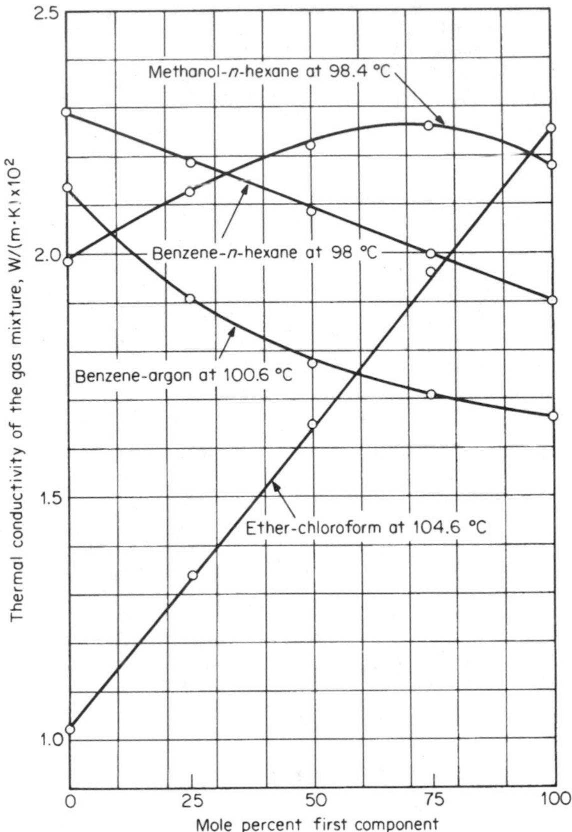

FIGURE 10-6 Typical gas-mixture thermal conductivities. *(From Bennett and Vines, 1955.)*

low-pressure thermal conductivities have also been adapted for mixtures, but the results obtained in testing several were not encouraging.

Wassiljewa Equation

In a form analogous to the theoretical relation for mixture viscosity, Eq. (9-5.13),

$$\lambda_m = \sum_{i=1}^{n} \frac{y_i \lambda_i}{\sum_{j=1}^{n} y_j A_{ij}} \tag{10-6.1}$$

where λ_m = thermal conductivity of the gas mixture
λ_i = thermal conductivity of pure i
y_i, y_j = mole fraction of components i and j
A_{ij} = a function, as yet unspecified
A_{ii} = 1.0

This empirical relation was proposed by Wassiljewa (1904).

Mason and Saxena Modification

Mason and Saxena (1958) suggested that A_{ij} in Eq. (10-6.1) could be expressed as

$$A_{ij} = \frac{\varepsilon[1 + (\lambda_{\text{tr}_i}/\lambda_{\text{tr}_j})^{1/2}(M_i/M_j)^{1/4}]^2}{[8(1 + M_i/M_j)]^{1/2}} \tag{10-6.2}$$

where M = molecular weight, g/mol
λ_{tr} = monatomic value of the thermal conductivity
ε = numerical constant near unity

Mason and Saxena proposed a value of 1.065 for ε, and Tandon and Saxena (1965) later suggested 0.85. As used here, $\varepsilon = 1.0$.

From Eq. (10-2.3), noting for monatomic gases that $C_v = C_{\text{tr}} = 3R/2$,

$$\frac{\lambda_{\text{tr}_i}}{\lambda_{\text{tr}_j}} = \frac{\eta_i}{\eta_j}\frac{M_j}{M_i} \tag{10-6.3}$$

Substituting Eq. (10-6.3) into Eq. (10-6.2) and comparing with Eq. (9-5.15) gives

$$A_{ij} = \phi_{ij} \tag{10-6.4}$$

where ϕ_{ij} is the interaction parameter for gas-mixture viscosity. Thus the relation for estimating mixture viscosities is also applicable to thermal conductivities by simply substituting λ for η. In this approximation, to determine λ_m one needs data giving the pure component thermal conductivities and viscosities. An alternative way to proceed is to use Eqs. (10-6.1) and (10-6.2) but obtain the ratio of translational thermal conductivities from Eq. (10-3.11).

$$\frac{\lambda_{\text{tr}_i}}{\lambda_{\text{tr}_j}} = \frac{\Gamma_j[\exp(0.0464T_{ri}) - \exp(-0.2412T_{ri})]}{\Gamma_i[\exp(0.0464T_{rj}) - \exp(-0.2412T_{rj})]} \tag{10-6.5}$$

where Γ is defined by Eq. (10-3.9). With Eq. (10-6.5), values of A_{ij} become functions of the reduced temperatures of both i and j. However, with this latter approach, pure gas viscosities are not required. Both techniques are illustrated in Example 10-5.

Lindsay and Bromley (1950) have also proposed a technique to estimate A_{ij}. It is slightly more complex than Eq. (10-6.2), and the results obtained do not differ significantly from the Mason-Saxena approach.

The Wassiljewa equation is capable of representing low-pressure mixture thermal conductivities with either a maximum or minimum as composition is varied. As Gray, et al. (1970) have shown, if $\lambda_1 < \lambda_2$,

$$\frac{\lambda_1}{\lambda_2} < A_{12}A_{21} < \frac{\lambda_2}{\lambda_1} \qquad \lambda_m \text{ varies monotonically with composition}$$

$$A_{12}A_{21} \geq \frac{\lambda_2}{\lambda_1} \qquad \lambda_m \text{ has a minimum value below } \lambda_1$$

$$\frac{\lambda_1}{\lambda_2} \geq A_{12}A_{21} \qquad \lambda_m \text{ has a maximum value above } \lambda_2$$

Corresponding States Methods

The Chung, et al. (1984, 1988) method for estimating low-pressure thermal conductivities [Eqs. (10-3.14) and (10-3.15)] has been adapted to handle mixtures. The emphasis of these authors, however, was to treat systems at high pressure and, if possible, as liquids. When their method is used for low-pressure gas mixtures, the accuracy away from the pure components is often not particularly high. However, in their favor is the fact that pure component thermal conductivities are not required as input; the method generates its own values of pure component conductivities.

To use the Chung, et al. form for mixtures, we need to have rules to obtain M', η, C_v, ω, and T_c for the mixture. η_m is found from Eq. (9-5.24), and in using this relation, one also obtains M_m, ω_m, and T_{cm}. [Eqs. (9-5.28), (9-5.29), and (9-5.44)]; $M'_m = M/10^3$. For C_{vm} a mole fraction average rule is used, i.e.,

$$C_{vm} = \sum_i y_i C_{vi} \tag{10-6.6}$$

With these mixture values, the procedure to compute λ_m is identical with that used for the pure component conductivity (see Example 10-1). The method is also illustrated for a mixture in Example 10-5.

Discussion

Three techniques were suggested to estimate the thermal conductivity of a gas mixture at low pressure. Two employ the Wassiljewa formulation [Eq. (10-6.1)] and differ only in the manner $\lambda_{tr_i}/\lambda_{tr_j}$ is calculated. The third method (Chung, et al.) uses a corresponding states approach. It is the least accurate, but it has the advantage that pure component thermal conductivities do not have to be known. The other two methods require either experimental or estimated values of λ for all pure components. All three methods are illustrated in Example 10-5.

For nonpolar gas mixtures, we recommend the Wassiljewa equation with the Mason-Saxena relation for A_{ij}, where $\lambda_{tr_i}/\lambda_{tr_j}$ is calculated from Eq. (10-6.5). Errors will generally be less than 3 to 4%. For nonpolar-polar and polar-polar gas mixtures, none of the techniques examined were found to be particularly accurate. As an example, in Fig. 10-6, none predicted the maximum in λ_m for the methanol-*n*-hexane system. Thus, in such cases, errors greater than 5 to 8% may be expected when one employs the procedures recommended for nonpolar gas mixtures. For mixtures in which the sizes and polarities of the constituent molecules are not greatly different, λ_m can be estimated satisfactorily by a mole fraction average of the pure component conductivities (e.g., the benzene-*n*-hexane and ether-chloroform cases in Fig. 10-6).

Example 10-5 Estimate the thermal conductivity of a gas mixture containing 25 mole % benzene and 75 mole % argon at 100.6°C and about 1 bar. The experimental value is 0.0192 W/(m · K) (Bennett and Vines, 1955).

From Appendix A, the following pure component constants are given:

	Benzene (1)	Argon (2)
T_c, K	562.05	150.86
P_c, bar	48.95	48.98
V_c, cm^3/mol	256.	74.57
ω	0.210	−0.002
Z_c	0.268	0.291
M, g/mol	78.114	39.948
M', kg/mol	0.078114	0.039948

Using the ideal-gas heat capacity constants for the two pure gases (Appendix A), at 373.8 K,

Benzene $\quad C_p = 104.5 \text{ J/(mol·K)}$:

$$C_v = C_p - R = 104.5 - 8.3 = 96.2 \text{ J/(mol·K)}$$

Argon $\quad C_p = 20.8 \text{ J/(mol·K)}$

$$C_v = C_p - R = 20.8 - 8.3 = 12.5 \text{ J/(mol·K)}$$

The pure component viscosities and thermal conductivities at 373.8 K and 1 bar are:

	Benzene (1)	Argon (2)
η, N·s/m^2 × 10^7	92.5	271.
λ, W/(m·K) × 10^2	1.66	2.14

solution Mason and Saxena. Equation (10-6.1) is used with $A_{12} = \phi_{12}$ and $A_{21} = \phi_{21}$ from Eqs. (9-5.14) and (9-5.15).

$$A_{12} = \frac{[1 + (92.5/271)^{1/2}(39.948/78.114)^{1/4}]^2}{\{[8(1 + (78.114/39.948)]\}^{1/2}} = 0.459$$

$$A_{21} = A_{12}\frac{\eta_2}{\eta_1}\frac{M_1}{M_2} = 0.459\,\frac{271}{92.5}\,\frac{78.114}{39.948} = 2.630$$

With Eq. (10-6.1)

$$\begin{aligned}\lambda_m &= \frac{y_1\lambda_1}{y_1 + y_2A_{12}} + \frac{y_2\lambda_2}{y_1A_{21} + y_2}\\ &= \left[\frac{(0.25)(1.66)}{0.25 + (0.75)(0.459)} + \frac{(0.75)(2.14)}{(0.25)(2.630) + 0.75}\right] \times 10^{-2}\\ &= 1.84 \times 10^{-2} \text{ W/(m·K)}\end{aligned}$$

$$\text{Error} = \frac{1.84 - 1.92}{1.92} \times 100 = -4.2\%$$

Mason and Saxena Form with Eq. (10-6.5). In this case, $\lambda_{\text{tr}_1}/\lambda_{\text{tr}_2}$ is obtained from Eq. (10-6.5). Γ is determined with Eq. (10-3.9).

$$\Gamma_1 = 210 \left[\frac{(562.05) \times (78.114)^3}{(48.95)^4} \right]^{1/6} = 398.5$$

$$\Gamma_2 = 210 \left[\frac{(150.86) \times (39.948)^3}{(48.7)^4} \right]^{1/6} = 229.6$$

At 373.8 K, $T_{r_1} = 373.8/562.05 = 0.665$; $T_{r_2} = 373.8/150.86 = 2.478$. Then,

$$\frac{\lambda_{tr_1}}{\lambda_{tr_2}} = \frac{229.6\{\exp[(0.0464)(0.665)] - \exp(-0.2412)(0.665)]\}}{398.5\{\exp[(0.0464)(2.478)] - \exp[(-0.2412)(2.478)]\}}$$

$$= 0.1809$$

$$\frac{\lambda_{tr_2}}{\lambda_{tr_1}} = (0.1809)^{-1} = 5.528$$

Inserting these values into Eq. (10-6.2) with $\varepsilon = 1.0$ gives

$$A_{12} = \frac{[1 + (0.1809)^{1/2}(78.114/39.948)^{1/4}]^2}{\{8[1 + (78.114/39.948)]\}^{1/2}} = 0.4646$$

$$A_{21} = \frac{[1 + (5.528)^{1/2}(39.948/78.114)^{1/4}]^2}{\{8[1 + (39.948/78.114)]\}^{1/2}} = 2.568$$

Then, using Eq. (10-6.1),

$$\lambda_m = \left[\frac{(0.25)(1.66)}{0.25 + (0.75)(0.4646)} + \frac{(0.75)(2.14)}{(0.25)(2.568) + 0.75} \right] \times 10^{-2}$$

$$= 1.85 \times 10^{-2} \text{ W/(m} \cdot \text{K)}$$

$$\text{Error} = \frac{1.85 - 1.92}{1.92} \times 100 = -3.6\%$$

CHUNG ET AL. With this method, we use the relations described in Chap. 9 to determine the mixture properties η_m, M'_m, ω_m, and T_{cm}. (See Example 9-8.) In this case, at 25 mole % benzene, $\eta_m = 182.2\ \mu P = 182.2 \times 10^{-7}$ N · s/m², $M'_m = 0.04631$ kg/mol, $\omega_m = 0.0788$, and $T_{cm} = 276.9$ K. From Eq. (10-6.6),

$$C_{vm} = (0.25)(96.2) + (0.75)(12.5) = 33.4 \text{ J/(mol} \cdot \text{K)}$$

The mixture thermal conductivity is then found from Eq. (10-3.14).

$$\lambda_m = \frac{(\eta_m C_{vm}/M'_m)(3.75\psi_m)}{C_{vm}/R} = \frac{\eta_m R}{M'_m} (3.75\psi_m)$$

$$\psi_m = 1 + \frac{\alpha_m[0.215 + 0.28288\alpha_m - 1.061\beta_m + 0.26665Z_m]}{0.6366 + \beta_m Z_m + 1.061\alpha_m\beta_m}$$

$$\alpha_m = (C_{vm}/R) - \frac{3}{2} = (33.4/8.31) - \frac{3}{2} = 2.52$$

$$\beta_m = 0.7862 - 0.7109\omega_m + 1.3168\omega_m^2 = 0.7384$$

$$T_{rm} = 373.8/276.9 = 1.350$$

$$Z_m = 2.0 + 10.5T_{rm}^2 = 2.0 + (10.5)(1.348)^2 = 21.13$$

$$\psi_m = 1.800$$

$$\lambda_m = \left[(182.2 \times 10^{-7})\frac{8.314}{0.04631}\right][(3.75)(1.800)] = 2.21 \times 10^{-2}\ \text{W/(m}\cdot\text{K)}$$

$$\text{Error} = \frac{2.21 - 1.92}{1.92} \times 100 = 15\%$$

10-7 THERMAL CONDUCTIVITIES OF GAS MIXTURES AT HIGH PRESSURES

There are few experimental data for gas mixtures at high pressures and, even here, most studies are limited to simple gases and light hydrocarbons. The nitrogen-carbon dioxide system was studied by Keyes (1951), and Comings and his colleagues reported on ethylene mixtures with nitrogen and carbon dioxide (Junk and Comings, 1953), rare gases (Peterson, et al., 1971) and binaries containing carbon dioxide, nitrogen, and ethane (Gilmore and Comings, 1966). Rosenbaum and Thodos (1967, 1969) investigated methane-carbon dioxide and methane-carbon tetrafluoride binaries. Binaries containing methane, ethane, nitrogen, and carbon dioxide were also reported by Christensen and Fredenslund (1979), and data for systems containing nitrogen, oxygen, argon, methane, ethylene, and carbon dioxide were published by Zheng, et al. (1984) and Yorizane, et al. (1983). Data for selected systems are summarized in Stephan and Hildwein (1987), Stephan and Heckenberger (1988), and Sutton (1976).

We present below three estimation methods. All are modifications of procedures developed earlier for low- and high-pressure pure gas thermal conductivities. For the extension of Eqs. (10-5.21) and (10-5.22) to mixtures, see Assael, et al. (1992a, 1996).

Stiel and Thodos Modification

Equations (10-5.2) to (10-5.4) were suggested as a way to estimate the high-pressure thermal conductivity of a pure gas. This procedure may be adapted for mixtures if mixing and combining rules are available to determine T_{cm}, P_{cm}, V_{cm}, Z_{cm}, and M_m. Yorizane, et al. (1983a) have studied this approach and recommend the following

$$T_{cm} = \frac{\sum_i \sum_j y_i y_j V_{cij} T_{cij}}{V_{cm}} \tag{10-7.1}$$

$$V_{cm} = \sum_i \sum_j y_i y_j V_{cij} \tag{10-7.2}$$

$$\omega_m = \sum_i y_i \omega_i \tag{10-7.3}$$

$$Z_{cm} = 0.291 - 0.08\omega_m \tag{10-7.4}$$

$$P_{cm} = Z_{cm} R T_{cm} / V_{cm} \tag{10-7.5}$$

$$M_m = \sum_i y_i M_i \tag{10-7.6}$$

$$T_{cii} = T_{ci} \tag{10-7.7}$$

$$T_{cij} = (T_{ci}T_{cj})^{1/2} \tag{10-7.8}$$

$$V_{cii} = V_{ci} \tag{10-7.9}$$

$$V_{cij} = \frac{[(V_{ci})^{1/3} + (V_{cj})^{1/3}]^3}{8} \tag{10-7.10}$$

Using these simple rules, they found they could correlate their high-pressure thermal conductivity data for CO_2—CH_4 and CO_2—Ar systems quite well. In Fig. 10-7, we show a plot of λ_m for the CO_2—Ar system at 298 K. This case is interesting because the temperature is slightly below the critical temperature of CO_2 and, at high pres-

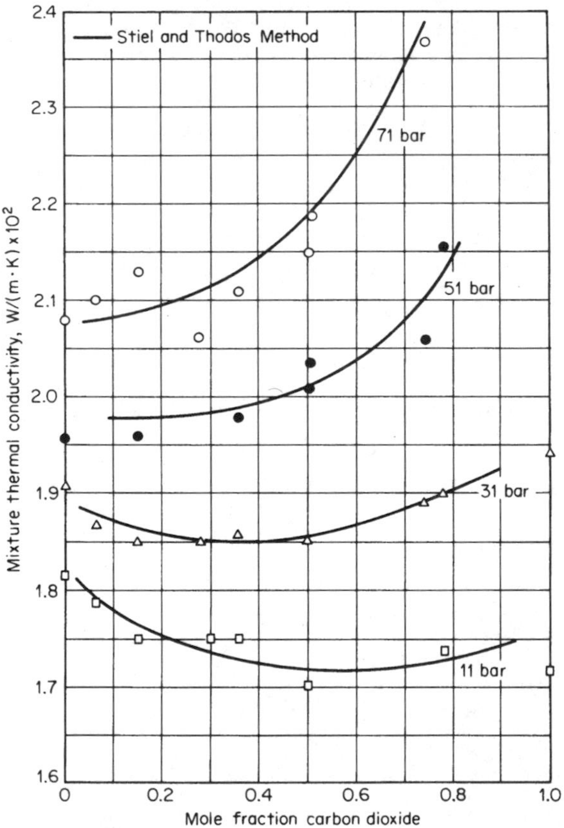

FIGURE 10-7 High-pressure thermal conductivities of the argon–carbon dioxide system. *(Data from Yorizane, et al., 1983a.)*

sure, λ for carbon dioxide increases more rapidly than that for argon. The net result is that the λ_m composition curves are quite nonlinear. Still, the Stiel-Thodos method, with Eqs. (10-7.1) through (10-7.10), appears to give a quite satisfactory fit to the data. We illustrate the approach in Example 10-6.

Example 10-6 Estimate the thermal conductivity of a methane (1)-carbon dioxide (2) mixture containing 75.5 mole % methane at 370.8 K and 174.8 bar. Rosenbaum and Thodos (1969) show an experimental value of 5.08×10^{-2} W/(m · K); these same investigators report experimental values for the mixture of $V = 159$ cm^3/mol and, at 1 bar, $\lambda_m^\circ = 3.77 \times 10^{-2}$ W/(m · K).

solution From Appendix A, we list pure component constants for methane and carbon dioxide that will be used in this example and in Examples 10-7 and 10-8.

	CH_4 (1)	CO_2 (2)
T_c, K	190.56	304.12
P_c, bar	45.99	73.74
V_c, cm^3/mol	98.6	94.07
Z_c	0.286	0.274
ω	0.011	0.225
μ, debye	0	0
C_p J/(mol · K)	39.09	40.04
C_v, J/(mol · K)	30.78	31.73
M, g/mol	16.043	44.010
M', kg/mol	0.01604	0.04401

The heat capacities were calculated from the equation and polynomial constants in Appendix A at $T = 370.8$ K.

With Eqs. (10-7.1) through (10-7.10)

$$T_{c12} = [(190.56)(304.12)]^{1/2} = 240.73 \text{ K}$$

$$V_{c12} = \tfrac{1}{8}[(98.6)^{1/3} + (94.07)^{1/3}]^3 = 96.3 \text{ cm}^3/\text{mol}$$

$$V_{cm} = (0.755)^2(98.6) + (0.245)^2(94.07) + (2)(0.755)(0.245)(96.3) = 97.9 \text{ cm}^3/\text{mol}$$

$$T_{cm} = [(0.755)^2(190.56)(98.6) + (0.245)^2(304.12)(94.07) + (2)(0.755)(0.245)(240.73)(96.3)]/97.5 = 215.4 \text{ K}$$

$$\omega_m = (0.755)(0.011) + (0.245)(0.225) = 0.063$$

$$Z_{cm} = 0.291 - (0.08)(0.063) = 0.286$$

$$P_{cm} = \frac{(0.286)(8.314)(215.4)}{97.5 \times 10^{-6}} = 5.24 \times 10^6 \text{ Pa} = 52.4 \text{ bar}$$

$$M_m = (0.755)(16.04) + (0.245)(44.01) = 22.9 \text{ g/mol}$$

With Eq. (10-3.9),

$$\Gamma = (210)\left[\frac{(215.4)(22.9)^3}{(52.4)^4}\right]^{1/6} = 176$$

and

$$\rho_{rm} = \frac{V_{cm}}{V_m} = \frac{97.5}{159} = 0.613$$

Using Eq. (10-5.3)

$$(\lambda_m - \lambda_m^\circ)[(176)(0.285)^5] = (1.14 \times 10^{-2})\{\exp[(0.67)(0.613)] - 1.069\}$$

$$\lambda_m - \lambda_m^\circ = 1.51 \times 10^{-2} \text{ W/(m}\cdot\text{K)}$$

$$\lambda_m = (1.51 + 3.77)(10^{-2}) = 5.28 \times 10^{-2} \text{ W/(m}\cdot\text{K)}$$

$$\text{Error} = \frac{5.28 - 5.08}{5.08} \times 100 = 4\%$$

Chung, et al. Method (1984, 1988)

To apply this method to estimate the thermal conductivities of high-pressure gas mixtures, one must combine the high-pressure pure component relations with the mixing rules given in Secs. 10-6 and in 9-5. To be specific, Eq. (10-5.5) is employed with all variables subscripted with m to denote them as mixture properties. Example 10-7 illustrates the procedure in detail.

Example 10-7 Repeat Example 10-6 by using the Chung, et al. approach.

solution For the methane (1)-carbon dioxide (2) system, the required pure component properties were given in Example 10-6.

To use Eq. (10-5.5), let us first estimate η_m° with the procedures in Chap. 9. From Eqs. (9-4.7) and (9-4.8),

$$\left(\frac{\varepsilon}{k}\right)_1 = \frac{190.56}{1.2593} = 151.3 \text{ K}$$

$$\left(\frac{\varepsilon}{k}\right)_2 = \frac{304.13}{1.2593} = 241.5 \text{ K}$$

$$\sigma_1 = (0.809)(98.60)^{1/3} = 3.737 \text{ Å}$$

$$\sigma_2 = (0.809)(94.07)^{1/3} = 3.679 \text{ Å}$$

Interaction values are then found from Eqs. (9-5.33), (9-5.35), (9-5.37), and (9-5.40).

$$\sigma_{12} = [(3.737)(3.769)]^{1/2} = 3.711 \text{ Å}$$

$$(\varepsilon/k)_{12} = [(151.3)(241.5)]^{1/2} = 191.1 \text{ K}$$

$$\omega_{12} = \frac{0.011 + 0.225}{2} = 0.118$$

$$M_{12} = \frac{(2)(16.04)(44.01)}{(16.04 + 44.01)} = 23.51$$

With $y_1 = 0.755$ and $y_2 = 0.245$, using Eqs. (9-5.25) to (9-5.29) and Eq. (9-5.41),

$$\sigma_m = 3.723\ \text{Å} \qquad (\varepsilon/k)_m = 171.0\ \text{K} \qquad T^* = 2.168$$

$$M_m = 20.91 \qquad \omega_m = 0.063 \qquad F_{cm} = 0.983$$

So, with Eq. (9-4.3), $\Omega_v = 1.144$. Then, with Eq. (9-5.24),

$$\eta_m^\circ = (26.69)(0.983)\frac{[(20.91)(370.8)]^{1/2}}{(3.723)^2(1.144)} = 145.7\ \mu\text{P} = 145.7\ \text{N}\cdot\text{s/m}^2$$

With Eqs. (9-5.43) and (9-5.44),

$$V_{cm} = \left(\frac{\sigma_m}{0.809}\right)^3 = 97.46\ \text{cm}^3/\text{mol}$$

$$T_{cm} = (1.2593)(\varepsilon/k)_m = 215.3\ \text{K}$$

$$T_r = \frac{T}{T_{cm}} = \frac{370.8}{215.3} = 1.722$$

C_v for the mixture is found with Eq. (10-6.6) as 31.47 J/(mol · K), and Ψ is determined as indicated under Eq. (10-3.14) with

$$\alpha_m = \frac{31.47}{8.314} - 1.5 = 2.285$$

$$\beta_m = 0.7862 - (0.7109)(0.063) + (1.3168)(0.063)^2 = 0.747$$

$$Z = 2.0 + (10.5)(1.722)^2 = 33.14$$

$$\Psi = 1.748$$

and

$$M' = M/10^3 = 20.91 \times 10^{-3}\ \text{kg/mol}$$

With Eqs. (10-5.6) and (10-5.8) and Table 10-3,

$$y_m = 0.1022$$

$$G_1 = 1.311$$

$$G_2 = 0.6519$$

$$B_1 = 2.464 \quad B_2 = -0.6043 \quad B_3 = 6.965 \quad B_4 = 13.98$$

$$B_5 = 0.8444 \quad B_6 = -5.057 \quad B_7 = 99.16$$

$$q = \frac{(3.586 \times 10^{-3})(215.3/20.91 \times 10^{-3})^{1/2}}{(97.46)^{2/3}} = 0.01718$$

Finally, substituting these values into Eq. (10-5.5),

$$\lambda_m = \frac{(31.2)(145.7 \times 10^{-7})(1.748)}{20.91 \times 10^{-3}}[(0.6519)^{-1} - (5.057)(0.1022)]$$
$$+ (0.01718)(99.16)(0.1022)^2(1.722)^{1/2}(0.6519)$$
$$= 0.0539\ \text{W/(m}\cdot\text{K)}$$

$$\text{Error} = \frac{5.39 - 5.08}{5.08} \times 100 = 6\%$$

If the pressure were reduced to 1 bar, y_m would become quite small and G_2 would be essentially unity. In that case, $\lambda_m^\circ = 3.80 \times 10^{-2}$ W/(m·K), a value very close to the value reported experimentally (3.77×10^{-2} W/(m·K).

Whereas the procedure appears tedious, it is readily programmed for computer use. The error found in Example 10-7 is typical for this method when simple gas mixtures are treated. As noted before, the Chung, et al. method should not be used for polar gases. Its accuracy for nonpolar gas mixtures containing other than simple gases or light hydrocarbons is in doubt.

TRAPP Method (Huber, 1996)

For gas mixtures at high pressure, the thermal conductivity is determined by a combination of the techniques introduced for high-pressure gases (Sec. 10-5) with appropriate mixing rules. The thermal conductivity of the mixture is given by:

$$\lambda_m = \lambda_m^\circ + F_{\lambda m} X_{\lambda m} [\lambda^R - \lambda^{Ro}] \tag{10-7.11}$$

where λ_m° is the mixture value at low pressure and may be determined by methods described earlier in the chapter. The quantity $\lambda^R - \lambda^{Ro}$ that appears in Eq. (10-7.11) is for the reference fluid propane and is evaluated with Eq. (10-5.12) at T_0 and ρ_0. The following mixing rules are used to determine $F_{\lambda m}$, T_0 and ρ_0.

$$h_m = \sum_i \sum_j y_i y_j h_{ij} \tag{10-7.12}$$

$$f_m h_m = \sum_i \sum_j y_i y_j f_{ij} h_{ij} \tag{10-7.13}$$

$$h_{ij} = \frac{[(h_i)^{1/3} + (h_j)^{1/3}]^3}{8} \tag{10-7.14}$$

$$f_{ij} = (f_i f_j)^{1/2} \tag{10-7.15}$$

f_i and h_i are determined by the method described in Sec. 10-5. T_0 and ρ_0 are calculated by equations similar to Eqs. (10-5.13) and (10-5.14):

$$T_0 = T/f_m \tag{10-7.16}$$

$$\rho_0 = \rho h_m = h_m/V \tag{10-7.17}$$

Finally,

$$F_{\lambda m} = (44.094)^{1/2}(h_m)^{-2} \sum_i \sum_j y_i y_j (f_{ij}/M_{ij})^{1/2}(h_{ij})^{4/3} \tag{10-7.18}$$

where

$$M_{ij} = \left(\frac{1}{2M_i} + \frac{1}{2M_j}\right)^{-1} \tag{10-7.19}$$

$$X_{\lambda m} = \left[1 + \frac{2.1866(\omega_m - \omega^R)}{1 - 0.505(\omega_m - \omega^R)}\right]^{1/2} \tag{10-7.20}$$

$$\omega_m = \sum_i y_i \omega_i \tag{10-7.21}$$

Huber (1996) tested the TRAPP method on several binary hydrocarbon mixtures and one ternary hydrocarbon mixture over a wide range of densities and reports an average absolute error of about 5%, although, in some cases, significantly larger deviations were found. The technique is illustrated in Example 10-8.

Example 10-8 Repeat Example 10-6 by using the TRAPP procedure.

solution The pure component properties for both components of the methane (1), carbon dioxide (2) binary are given at the beginning of the solution of Example 10-6. Pure component values of h_i and f_i are determined from Eqs. (10-5.19) and (10-5.20) using $T_{r1} = 370.8/190.56 = 1.946$ and $T_{r2} = 370.8/304.12 = 1.219$.

$$f_1 = \frac{190.564}{369.83}[1 + (0.011 - 0.152)(0.05203 - 0.7498 \ln(1.946))] = 0.5478$$

$$h_1 = \left(\frac{98.60}{200.0}\right)\left(\frac{0.276}{0.286}\right)[1 - (0.011 - 0.152)(0.1436 - 0.2822 \ln(1.946))] = 0.4728$$

Similarly, f_2 and h_2 are 0.8165 and 0.4708 respectively. Mixture values along with T_0 and ρ_0 are determined with Eqs. (10-7.12) to (10-7.19).

$$f_{12} = [(0.5478)(0.8165)]^{1/2} = 0.6688$$

$$h_{12} = \frac{[(0.4728)^{1/3} + (0.4708)^{1/3}]^3}{8} = 0.4718$$

$$h_m = (0.755)^2(0.4728) + 2(0.755)(0.245)(0.4718) + (0.245)^2(0.4708) = 0.4723$$

$$f_m h_m = (0.755)^2(0.5478)(0.4728) + 2(0.755)(0.245)(0.6688)(0.4718) + (0.245)^2(0.8165)(0.4708) = 0.2874$$

$$f_m = \frac{0.2873}{0.4721} = 0.6086$$

$$T_0 = 370.8/0.6086 = 609.3 \qquad \rho_0 = 0.4723/159 = 0.002970$$

$$M_{12} = \left(\frac{1}{2(16.043)} + \frac{1}{2(44.010)}\right)^{-1} = 23.5143$$

$$F_{\lambda m} = \frac{(44.094)^{1/2}}{(0.4723)^2}\left[(0.755)^2\left(\frac{0.5478}{16.043}\right)^{1/2}(0.4728)^{4/3} + 2(0.755)(0.245)\left(\frac{0.6688}{23.5143}\right)^{1/2}(0.4718)^{4/3} + (0.245)^2\left(\frac{0.8165}{44.010}\right)^{1/2}(0.4708)^{4/3}\right]$$

$$F_{\lambda m} = 1.926$$

Equations (10-7.20) and (10-7.21) give

$$\omega_m = 0.755 \times 0.011 + 0.245 \times 0.2276 = 0.063$$

$$X_{\lambda m} = \left[1 + \frac{2.1866(0.063 - 0.152)}{1 - 0.505(0.063 - 0.153)}\right]^{1/2} = 0.9026$$

With $T_r^R = 609.3/369.83 = 1.648$ and $\rho_r^R = 0.002970 \times 200 = 0.5941$, Eq. (10-5.12) gives

$$\lambda^R - \lambda^{R_0} = 9.898 \text{ mW/(m} \cdot \text{K)} = 0.009898 \text{ W/(m} \cdot \text{K)}$$

Using Eq. (10-7.11)

$$\lambda_m = 3.77 \times 10^{-2} + (1.926)(0.9026)(0.009898) = 0.0549 \text{ W/(m} \cdot \text{K)}$$

$$\text{Error} = \frac{0.0549 - 0.0508}{0.0508} \times 100 = 8.1\%$$

Discussion

Of the three methods presented to estimate the thermal conductivity of high-pressure (or high-density) gas mixtures, all have been tested on available data and shown to be reasonably reliable with errors averaging about 5 to 7%. However, the database used for testing is small and primarily comprises permanent gases and light hydrocarbons. None are believed applicable to polar fluid mixtures. The Chung, et al. and TRAPP methods have also been tested on more complex (hydrocarbon) systems at densities which are in the liquid range with quite encouraging results.

For simple hand calculation of one or a few values of λ_m, the Stiel and Thodos and TRAPP methods are certainly the simplest. If many values are to be determined, the somewhat more complex, but probably more accurate, method of Chung, et al. or the TRAPP method should be programmed and used.

10-8 THERMAL CONDUCTIVITIES OF LIQUIDS

For many simple organic liquids, the thermal conductivities are between 10 and 100 times larger than those of the low-pressure gases at the same temperature. There is little effect of pressure, and raising the temperature usually decreases the thermal conductivities. These characteristics are similar to those noted for liquid viscosities, although the temperature dependence of the latter is pronounced and nearly exponential, whereas that for thermal conductivities is weak and nearly linear.

Values of λ_L for most common organic liquids range between 0.10 and 0.17 W/(m · K) at temperatures below the normal boiling point, but water, ammonia, and other highly polar molecules have values several times as large. Also, in many cases the dimensionless ratio $M\lambda/R\eta$ is nearly constant (for nonpolar liquids) between values of 2 and 3, so that viscous liquids have a correspondingly larger thermal conductivity. Liquid metals and some organosilicon compounds have large values of λ_L; the former often are 100 times larger than those for normal organic liquids. The solid thermal conductivity at the melting point is approximately 20 to 40% larger than that of the liquid.

The difference between transport property values in the gas phase and the values in the liquid phase indicates a distinct change in mechanism of energy (or momentum or mass) transfer, i.e.,

$$\frac{\lambda_L}{\lambda_G} \cong 10 \text{ to } 100 \qquad \frac{\eta_L}{\eta_G} \cong 10 \text{ to } 100 \qquad \frac{D_L}{D_G} \cong 10^{-4}$$

In the gas phase, the molecules are relatively free to move about and transfer momentum and energy by a collisional mechanism. The intermolecular force fields, though not insignificant, do not drastically affect the value of λ, η, or D. That is, the attractive intermolecular forces are reflected solely in the collision integral terms Ω_v, and Ω_D, which are really ratios of collision integrals for a real force field and an artificial case in which the molecules are rigid, non-attracting spheres. The variation of Ω_v, or Ω_D from unity then yields a rough quantitative measure of the importance of attractive intermolecular forces in affecting gas phase transport coefficients. Reference to Eq. (9-4.3) (for Ω_v) or Eq. (11-3.6) (for Ω_D) shows that Ω values are often near unity. One then concludes that a rigid, non-attracting spherical molecular model yields a low-pressure transport coefficient λ, η, or D not greatly different from that computed when intermolecular forces are included.

In the liquid, however, this hypothesis is not even roughly true. The close proximity of molecules to one another emphasizes strongly the intermolecular forces of attraction. There is little wandering of the individual molecules, as evidenced by the low value of liquid diffusion coefficients, and often a liquid is modeled as a lattice with each molecule caged by its nearest neighbors. Energy and momentum are primarily exchanged by oscillations of molecules in the shared force fields surrounding each molecule. McLaughlin (1964) discusses in more detail the differences in transport mechanisms between a dense gas or liquid and a low-pressure gas.

To date, theory has not been successful in formulating useful and accurate expressions to calculate liquid thermal conductivities, although Eqs. (10-5.21) and (10-5.22) are theory-based and have been used for a number of liquids. Generally however, approximate techniques must be employed for engineering applications.

Only relatively simple organic liquids are considered in the sections to follow. Ho, et al. (1972) have presented a comprehensive review covering the thermal conductivity of the elements, and Ewing, et al. (1957) and Gambill (1959) consider, respectively, molten metals and molten salt mixtures. Cryogenic liquids are discussed by Preston, et al. (1967) and Mo and Gubbins (1974).

Liquid thermal conductivity data have been compiled in Jamieson, et al. (1975), Le Neindre (1987), Liley, et al. (1988), Nieto de Castro, et al. (1986), Stephan and Hildwein (1987), and Vargaftik, et al. (1994). Data for mixtures and electrolyte solution are also given in Jamieson, et al. (1975) and Zaytsev and Aseyev (1992). Other sources include Jamieson and Tudhope (1964), Tsederberg (1965), and Vargaftik, et al. (1996). New liquid thermal conductivity data were reported for alcohols (Cai, et al., 1993; Jamieson, 1979; Jamieson and Cartwright, 1980; Ogiwara, et al., 1982), alkyl amines (Jamieson, 1979; Jamieson and Cartwright, 1980), hydrocarbons (Nieto de Castro et al., 1981; Ogiwara, et al., 1980), nitroalkanes (Jamieson and Cartwright, 1981), ethanolamines (DiGuilio, et al, 1992), glycols (DiGuilio and Teja, 1990), and other organic compounds (Cai, et al. 1993; Qun-Fung, et al., 1997; Venart and Prasad, 1980).

Constants that may be used to calculate thermal conductivities for pure liquids at different temperatures are tabulated in Daubert, et al. (1997), Miller, et al. (1976), and Yaws (1995, 1995a). The constants in Daubert, et al. (1997) are based on critically evaluated data while those in the other references are not. Values are tabulated at various temperatures for common fluids in Dean (1999), Lide (1999), and Perry and Green (1997).

10-9 ESTIMATION OF THE THERMAL CONDUCTIVITIES OF PURE LIQUIDS

All estimation techniques for the thermal conductivity of pure liquids are empirical; and with only limited examination, they often appear rather accurate. As noted earlier, however, below the normal boiling point, the thermal conductivities of most organic, nonpolar liquids lie between 0.10 and 0.17 W/(m · K). With this fact in mind, it is not too difficult to devise various schemes for estimating λ_L within this limited domain.

Two of the estimation methods that were tested are described below. Others that were considered are noted briefly at the end of the section.

Latini, et al. Method

In an examination of the thermal conductivities of many diverse liquids, Latini and his coworkers (Baroncini, et al., 1980, 1981, 1981a, 1983, 1983a, 1984); (Latini and Pacetti, 1977) suggest a correlation of the form:

$$\lambda_L = \frac{A(1 - T_r)^{0.38}}{T_r^{1/6}} \tag{10-9.1}$$

where λ_L = thermal conductivity of the liquid, W/(m · K)
T_b = normal boiling temperature (at 1 atm), K
T_c = critical temperature, K
M = molecular weight, g/mol
T_r = T/T_c

$$A = \frac{A^* T_b^{\alpha}}{M^{\beta} T_c^{\gamma}} \tag{10-9.2}$$

and the parameters A^*, α, β, and γ, are shown in Table 10-4 for various classes of organic compounds. Specific values of A are given for many compounds by (Baroncini, et al., (1981). Equation (10-9.2) is only an approximation of the regressed value of A, and this simplification introduces significant error unless $50 < M < 250$. More recently Latini, et al. (1989, 1996) have suggested a different form than Eq. (10-9.1) for alkanes, aromatics, and refrigerants.

Some estimated values of λ_L found from Eqs. (10-9.1) and (10-9.2) are compared with experimental results in Table 10-6. Errors vary, but they are usually less than 10%. Many types of compounds (e.g., nitrogen or sulfur-containing materials and aldehydes) cannot be treated, and problems arise if the compound may be fitted into two families. *m*-Cresol (Table 10-6) is an example. It could be considered an aromatic compound or an alcohol. In this case, we chose it to be an aromatic

TABLE 10-4 Latini et al. Correlation Parameters for Eq. (10-9.2)

Family	A^*	α	β	γ
Saturated hydrocarbons	0.00350	1.2	0.5	0.167
Olefins	0.0361	1.2	1.0	0.167
Cycloparaffins	0.0310	1.2	1.0	0.167
Aromatics	0.0346	1.2	1.0	0.167
Alcohols	0.00339	1.2	0.5	0.167
Acids (organic)	0.00319	1.2	0.5	0.167
Ketones	0.00383	1.2	0.5	0.167
Esters	0.0415	1.2	1.0	0.167
Ethers	0.0385	1.2	1.0	0.167
Refrigerants				
R20, R21, R22, R23	0.562	0.0	0.5	−0.167
Others	0.494	0.0	0.5	−0.167

material, but the error would not have been very different if it had been considered an alcohol.

Sastri Method

Sastri (1998) recommends

$$\lambda_L = \lambda_b a^m \tag{10-9.3}$$

where

$$m = 1 - \left(\frac{1 - T_r}{1 - T_{br}}\right)^n \tag{10-9.4}$$

For alcohols and phenols, $a = 0.856$ and $n = 1.23$. For other compounds, $a = 0.16$ and $n = 0.2$. The thermal conductivity at the normal boiling point, λ_b, is determined with the group contribution values and corrections in Table 10-5. Sastri reports an average deviation of 8% for 186 points that were tested. Some results are shown in Table 10-6 and the method is illustrated in Example 10-9.

Other Liquid Thermal Conductivity Estimation Techniques

For nonpolar materials, the estimation procedures in Sec. 10-5 may be employed to obtain λ_L when temperatures are well above the normal boiling point and accurate fluid densities are available. In particular, the Chung, et al. and TRAPP methods were specifically devised to treat liquid systems at high reduced temperatures as well as high-pressure gases.

Teja and Rice (1981, 1982) have suggested that, in some cases, values of λ_L are available for compounds similar to the one of interest, and these data could be employed in an interpolative scheme as follows. Two liquids, similar chemically and with acentric factors bracketing the liquid of interest, are selected. The liquid thermal conductivities of these reference liquids should be known over the range

TABLE 10-5 Sastri Group Contributions to Calculate λ_b, $W/(m \cdot K)$

Hydrocarbon groups	$\Delta\lambda_b$	Non-hydrocarbon groups (cont.)	
—CH_3	0.0545	—COOH (acid)	0.0650
—CH_2—	−0.0008	—NH_2	0.0880
>CH—	−0.0600	—NH—	0.0065
>C<	−0.1230	—NH— (ring)	0.0450
=CH_2	0.0545	>N—	−0.0605
=CH—	0.0020	N (ring)	0.0135
=C<	−0.0630	—CN	0.0645
=C=	0.1200	—NO_2	0.0700
Ring[1]	0.1130	S	0.0100
Non-hydrocarbon groups		—F[4]	0.0568
		—F[5]	0.0510
—O—	0.0100	—Cl	0.0550
—OH[2]	0.0830	—Br	0.0415
—OH[3]	0.0680	—I	0.0245
>CO (ketone)	0.0175	—H[6]	0.0675
>CHO (aldehyde)	0.0730	−3 member ring	0.1500
—COO— (ester)	0.0070	Ring (other)[7]	0.1100

[1] In polycyclic compounds, all rings are treated as separate rings.
[2] In aliphatic primary alcohols and phenols with no branch chains.
[3] In all alcohols except as described in [2] above.
[4] In perfluoro carbons.
[5] In all cases except as described in [4] above.
[6] This contribution is used for methane, formic acid, and formates.
[7] In polycyclic non-hydrocarbon compounds, all rings are considered as non-hydrocarbon rings.

Corrections in $W/m \cdot K$, to be added to values from above when calculating λ_b.

	Correction
Hydrocarbons when number of carbon atoms, C, is less than 5	0.0150(5-C)
Compounds with single CH_3 group, no other hydrocarbon groups, and non-hydrocarbon groups other than COOH, Br, or I (e.g., CH_3Cl)[1]	0.0600
Compounds with 2 hydrocarbon groups (2 CH_3, CH_3CH_2, or CH_2=CH) and non-hydrocarbon groups other than COOH, Br, or I.[1]	0.0285
Unsaturated aliphatic compounds with 3 hydrocarbon groups (e.g., allyl amine or vinyl acetate)	0.0285
Special groups $Cl(CH_2)_nCl$	0.0350
Compounds with more than one non-hydrocarbon group and at least one hydrocarbon group (e.g., propyl formate or furfural)[1]	0.0095
Compounds with non-hydrocarbon groups only (e.g., formic acid)	0.1165

[1] Non-hydrocarbons with more than one type of non-hydrocarbon group and either (i) one or two methyl groups, or (ii) one methyl group require both correction factors. (e.g., the correction for methylformate, with one methyl group and two non-hydrocarbon groups is 0.0600 + 0.0095 = 0.0695).

of reduced temperatures of interest. We denote the properties of one reference fluid by a prime and the other by a double prime. Defining

$$\phi = \frac{V_c^{2/3} M^{1/2}}{T_c^{1/2}} \tag{10-9.5}$$

then $\lambda_L \phi$ is found by an interpolation based on the acentric factor ω as shown in Fig. 10-8.

TABLE 10-6 Comparison between Calculated and Experimental Values of Liquid Thermal Conductivity All values of λ_L are in W/(m · K)

Compound	T, K	λ_L, exp	Ref**	λ_b from Table 10-5	Percent error* by method of	
					Sastri	Latini
acetone	273	0.171	2	0.1550	0.5	−9.8
acetonitrile	253	0.2307	5	0.1790	−8.6	−14
	273	0.2168			−5.5	−12
	293	0.2034			−2.2	−10
acrylonitrile	253	0.1837	5	0.1495	−4.4	−6.3
	283	0.1699			−1.1	−4.7
	303	0.1592			2.3	−2.5
n-butane	150	0.1808	6	0.1224	−15	−22
	250	0.1243			3.7	−12
	350	0.084			14	−11
n-butanol	293	0.1535	2	0.1351	−1.3	−3.1
n-butyl acetate	293	0.1369	2	0.1136	−0.8	2.3
carbon tetrachloride	293	0.103	2	0.0970	3.2	−6.9
chloroform	270	0.121	7	0.1050	−3.7	−8.3
	300	0.116			−4.0	−10
	340	0.109			−4.7	−13
m-cresol	300	0.149	7	0.1295	1.6	9.5
	350	0.145			−0.3	4.3
n-decane	240	0.146	7	0.1026	−3.6	1.2
	340	0.120			3.2	3.3
	430	0.100			6.3	2.7
n-decyl alcohol	298	0.162	7	0.1303	2.3	−9.0
	398	0.147			−0.1	−15
dichloromethane	200	0.173	7	0.1422	−0.8	−15
	300	0.139			6.2	−15
diisopropyl ether	293	0.1097	3	0.1080	8.9	4.4
n-eicosane	330	0.147	7	0.0946	−0.6	−3.1
	460	0.118			6.2	−0.1
ethanol	200	0.196	7	0.1652	2.3	−1.0
	300	0.166			5.9	−6
	400	0.143			9.3	−18
ethylbenzene	273	0.1369	6	0.1137	0.9	1.8
	400	0.1044			11	5.2
heptane	200	0.1523	4	0.1050	−9.1	−3.6
	250	0.1373			−5.6	−3.0
	300	0.1223			−1.8	−1.7
	350	0.1073			2.3	0.0
	400	0.0958	6		2.5	−2.5
	500	0.0715			−7.1	−22
methyl propionate	280	0.144	7	0.1152	−8.9	9.4

TABLE 10-6 Comparison between Calculated and Experimental Values of Liquid Thermal Conductivity (*Continued*)
All values of λ_L are in W/(m · K)

Compound	T, K	λ_L, exp	Ref**	λ_b from Table 10-5	Percent error* by method of	
					Sastri	Latini
phenol	301.5	0.179	8	0.1400	−11	−6.2
	353	0.154			−1.0	0.6
	471	0.139			−0.5	−9.6
n-propyl amine	223	0.1878	1	0.1409	−11	−12
	273	0.1789			−14	−17
	298	0.1724			−14	−19
tetrahydrofuran	253	0.168	5	0.1168	−21	7.4
	283	0.1578			−19	7.6
	303	0.1509			−18	7.9
toluene	240	0.1485	4	0.1145	−5.5	4.8
	270	0.1395			−3.0	5.7
	360	0.1125			5.9	10
tributyl amine	273	0.1232	1	0.0958	9.7	−19
	423	0.1028			6.0	−26

*Percent error = [(calc. − exp.)/exp.] × 100.
**Refs: 1, Jamieson and Cartwright, 1980; 2, Jamieson, et al., 1975; 3, Jamieson and Tudhope, 1964; 4, Nieto de Castro, et al., 1986; 5, Qun-Fung, et al., 1997; 6, Stephan and Hildwein, 1987; 7, Vargaftik, 1994; 8, Venart and Prasad, 1980.

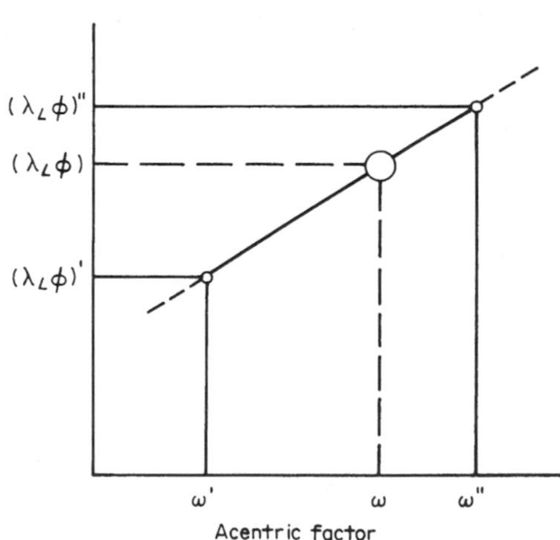

FIGURE 10-8 Schematic representation of the Teja and Rice interpolation procedure. At the circle, $\lambda_L\phi = (\lambda_L\phi)' + [(\omega - \omega')/(\omega'' - \omega')][(\lambda_L\phi)'' - (\lambda_L\phi)']$.

$$\lambda_L \phi = (\lambda_L \phi)' + \frac{\omega - \omega'}{\omega'' - \omega'} [(\lambda_L \psi)'' - (\lambda_L \psi)'] \tag{10-9.6}$$

In Eq. (10-9.10) when one selects λ_L' and λ_L'', they should be evaluated at the same reduced temperature as for the compound of interest. The procedure is illustrated in Example 10-9.

Arikol and Gürbüz (1992) developed a correlation for the thermal conductivity of aliphatic and aromatic hydrocarbons and for ethers, aldehydes, ketones and esters with more than four carbon atoms. Their equation requires the temperature, critical temperature, critical pressure, molecular weight, and normal boiling point and gives average deviations of 2%. Ogiwara, et al. (1982) suggested a general estimation relation for λ_L for aliphatic alcohols. Jamieson (1979) and Jamieson and Cartwright (1980) proposed a general equation to correlate λ_L over a wide temperature range (see Sec. 10-10), and they discuss how the constants in their equation vary with structure and molecular size. Other methods can be found in Assael, et al. (1989), Bleazard and Teja (1996), Dymond and Assael (1996), Klass and Viswanath (1998), Lakshmi and Prasad (1992), Sastri and Rao (1993), and Teja and Tardieu (1988). If critical properties are not available, the methods in Lakshmi and Prasad (1992) or Sastri and Rao (1993) may be used, or the critical properties may be estimated by the methods of Chap. 2. The methods in Klass and Viswanath (1998) and Teja and Tardieu (1988) require an experimental value of the thermal conductivity.

Discussion and Recommendations

The brief comparison shown in Table 10-6 between experimental and estimated values of liquid thermal conductivity would indicate that the Latini, et al. and Sastri methods are reasonably accurate. In many instances, the experimental data are not believed to be particularly reliable and the estimation errors are in the same range as the experimental uncertainty. This is clearly evident from the careful survey of liquid thermal conductivity data provided by the National Engineering Laboratory (Jamieson and Tudhope, 1964; Jamieson, 1979; Jamieson and Cartwright, 1980).

For organic liquids in the temperature region below the normal boiling point, we recommend either the Latini, et al. or Sastri methods. Errors can vary widely, but they are usually less than 15%. There are very few reliable data for liquid thermal conductivities at reduced temperatures exceeding $T_r = 0.65$. If the liquid is nonpolar and is at a reduced temperature greater than about 0.8, one should use the high-pressure fluid correlations given in Sec. 10-5. The Latini, et al. procedure has, however, been applied successfully for refrigerants up to $T_r = 0.9$ (Baroncini, 1981). None of the procedures predict the large increase in λ near the critical point. The estimation techniques described in this section are illustrated in Examples 10-9 through 10-11.

Example 10-9 Using the Teja and Rice scheme, estimate the thermal conductivity of liquid *t*-butyl alcohol at 318 K given the thermal conductivities of *n*-propanol and *n*-hexanol reported by Ogiwara, et al. (1982) as shown below.

n-Propanol $\quad \lambda_L = 0.202 - 1.76 \times 10^{-4} T \quad$ W/(m·K)

n-Hexanol $\quad \lambda_L = 0.190 - 1.36 \times 10^{-4} T \quad$ W/(m·K)

solution From Appendix A:

	V_c, cm^3/mol	T_c, K	M, g/mol	ω
n-Propanol	219	536.78	60.10	0.629
n-Hexanol	381	611.4	102.18	0.573
t-Butyl alcohol	275	506.21	74.12	0.613

Thus, with Eq. (10-9.5),

$$\phi(n\text{-propanol}) - 12.16$$

$$\phi(n\text{-hexanol}) = 21.49$$

$$\phi(t\text{-butyl alcohol}) = 16.18$$

At 318 K, for *t*-butyl alcohol, $T_r = 318/506.21 = 0.629$. At this reduced temperature, the appropriate temperature to use for *n*-propanol is (0.629)(536.78) = 337.6 K, and for *n*-hexanol it is (0.629)(611.4) = 384.6 K. With these, using the Ogiwara, et al. correlations,

$$\lambda_L(n\text{-propanol}) = 0.202 - (1.76 \times 10^{-4})(337.6) = 0.143 \text{ W/(m}\cdot\text{K)}$$

$$\lambda_L(n\text{-hexanol}) = 0.190 - (1.36 \times 10^{-4})(384.6) = 0.138 \text{ W/(m}\cdot\text{K)}$$

Then, using Eq. (10-9.6) with *n*-propanol as the ′ reference and *n*-hexanol as ″ reference,

$$\lambda_L(16.18) = [(0.143)(12.16)] + \frac{0.613 - 0.629}{0.573 - 0.629} \times [(0.138)(21.49) - (0.143)(12.16)] = 2.09$$

$$\lambda_L = \frac{2.07}{16.18} = 0.129 \text{ W/(m}\cdot\text{K)}$$

Ogiwara, et al. (1982) report the experimental value of *t*-butyl alcohol at 318 K to be 0.128 W/(m · K).

Example 10-10 Use Table 10-5 to determine the value of λ_b to be used in Eq. (10-9.3) for tetrahydrofuran, carbon tetrachloride, phenol, and acetonitrile.

solution

(i) Tetrahydrofuran is a non-hydrocarbon with four —CH_2— groups, one —O— and one non-hydrocarbon ring correction. Thus, $\lambda_b = 4\,(-0.0008) + 0.0100 + 0.1100 = 0.1168$ W/(m · K).

(ii) Carbon tetrachloride has one >C< and four —Cl groups. Thus, $\lambda_b = -0.1230 + 4\,(0.0550) = 0.097$ W/(m · K).

(iii) Phenol has five =CH—, one =C<, one —OH and one non-hydrocarbon ring correction. Thus, $\lambda_b = 5\,(0.0020) - 0.0630 + 0.0830 + 0.1100 = 0.14$ W/(m · K).

(iv) Acetonitrile has one —CH_3, one —CN, and one correction for a compound with one hydrocarbon group that is —CH_3 and one non-hydrocarbon group other than COOH, Br, or I. Thus, $\lambda_b = 0.0545 + 0.0645 + 0.0600 = 0.179$ W/(m · K).

Example 10-11 Estimate the thermal conductivity of carbon tetrachloride at 293 K. At this temperature, Jamieson and Tudhope (1964) list 11 values. Six are given a ranking of A and are considered reliable. They range from 0.102 to 0.107 W/ (m · K). Most, however, are close to 0.103 W/(m · K).

solution From Appendix A, $T_c = 556.3$ K, $T_b = 349.79$ K, and $M = 153.822$.

LATINI ET AL. Assuming CCl_4 to be a refrigerant, by Eq. (10-9.2) and Table 10-4,

$$A = \frac{0.494(556.3)^{1/6}}{(153.822)^{1/2}} = 0.114$$

Then, with $T_r = 293/556.3 = 0.527$ and Eq. (10-9.1),

$$\lambda_L = \frac{(0.114)(0.527)^{0.38}}{(0.527)^{1/6}} = 0.0954 \text{ W/(m} \cdot \text{K)}$$

$$\text{Error} = \frac{0.0954 - 0.103}{0.103} \times 100 = -7.4\%$$

SASTRI Using Eqs. (10-9.3) and (10-9.4) and the value of λ_b from Example 10-10,

$$m = 1 - \left(\frac{1 - 293/556.3}{1 - 349.79/556.3}\right)^{0.2} = -0.0498$$

$$\lambda_L = (0.097)(0.16)^{-0.0498} = 0.0106$$

$$\text{Error} = \frac{0.0106 - 0.0103}{0.0106} \times 100 = 3\%$$

10-10 EFFECT OF TEMPERATURE ON THE THERMAL CONDUCTIVITIES OF LIQUIDS

Except for aqueous solutions, water, and some multihydroxy and multiamine molecules, the thermal conductivities of most liquids decrease with temperature. Below or near the normal boiling point, the decrease is nearly linear and is often represented over small temperature ranges by

$$\lambda_L = A - BT \tag{10-10.1}$$

where A and B are constants and B generally is in the range of 1 to 3×10^{-4} W/(m·K^2). In Fig. 10-9, we show the temperature effect on λ_L for a few liquids. For wider temperature ranges, the equations in Sec. 10-9 may be used, or the following correlation suggested by Riedel (1951) may be used.

$$\lambda_L = B[3 + 20(1 - T_r)^{2/3}] \tag{10-10.2}$$

Although not suited for water, glycerol, glycols, hydrogen, or helium, Jamieson (1971) indicates that Eq. (10-10.2) represented well the variation of λ_L with temperature for a wide range of compounds. Although, as noted earlier, few data for λ_L exist over the temperature range from near the melting point to near the critical point, for those that are available, Jamieson (1979) has found that neither Eq. (10-10.1) nor Eq. (10-10.2) is suitable, and he recommends

$$\lambda_L = A(1 + B\tau^{1/3} + C\tau^{2/3} + D\tau) \tag{10-10.3}$$

where A, B, C, and D are constants and $\tau = 1 - T_r$. For nonassociating liquids, $C = 1 - 3B$ and $D = 3B$. With these simplifications, Eq. (10-10.3) becomes

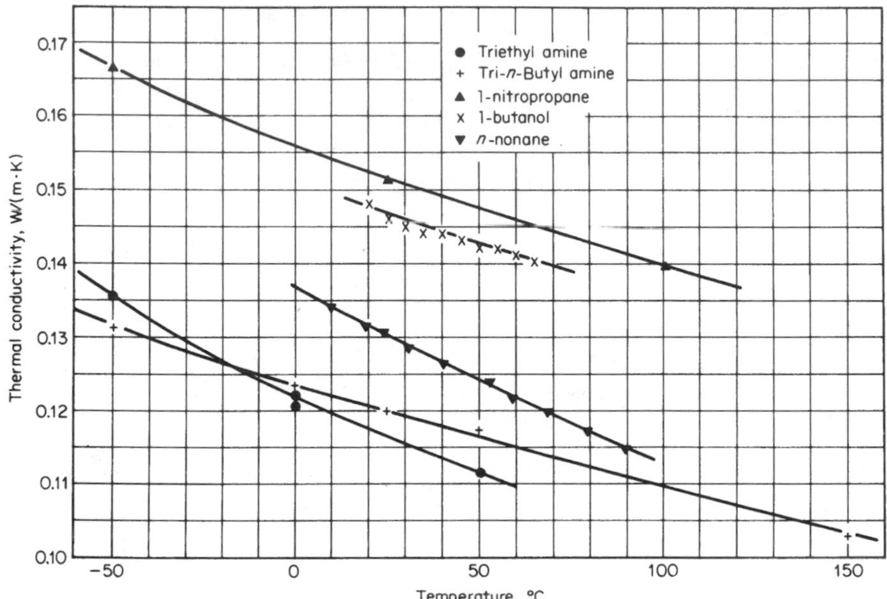

FIGURE 10-9 Thermal conductivity of a few organic liquids as functions of temperature. *(Data from Jamieson and Cartwright, 1980, 1981; Nagata, 1973; Ogiwara, et al., 1982.)*

$$\lambda_L = A[1 + \tau^{2/3} + B(\tau^{1/3} - 3\tau^{2/3} + 3\tau)] \qquad (10\text{-}10.4)$$

As an example, in Fig. 10-9, if one fits the data for tributyl amine (a polar, but nonassociating, liquid), to Eq. (10-10.4), approximate values of A and B are $A = 0.0590$ W/(m·K) and $B = 0.875$. Using them, one can show by differentiating Eq. (10-10.4) that $d\lambda_L/dT$ decreases with increasing temperature, although, as is obvious from Fig. 10-9, the change in slope is not large in the temperature region shown. For other materials for which data are available over a quite wide temperature range, Eq. (10-10.4) is clearly preferable to Eq. (10-10.1) or (10-10.2) (Jamieson, 1984).

For associated liquids, $C = 1 - 2.6B$ and $D \approx 6.5$ for alcohols and 6.0 for alkyd and dialkyd amines. Correlations for C and D for other types of associated molecules are not available. The constants A and B have been correlated, approximately, with carbon number for several homologous series (Jamieson, 1979; Jamieson and Cartwright, 1980, 1981).

For saturated liquids at high pressure, variations of λ_L with temperature should probably be determined by using the high-pressure correlations in Sec. 10-5.

10-11 EFFECT OF PRESSURE ON THE THERMAL CONDUCTIVITIES OF LIQUIDS

At moderate pressures, up to 50 to 60 bar, the effect of pressure on the thermal conductivity of liquids is usually neglected, except near the critical point, where the liquid behaves more like a dense gas than a liquid (see Sec. 10-5). At lower temperatures, λ_L increases with pressure. Data showing the effect of pressure on a number of organic liquids are available in Bridgman (1923) and Jamieson, et al. (1975).

A convenient way of estimating the effect of pressure on λ_L is by Eq. (10-11.1)

$$\frac{\lambda_2}{\lambda_1} = \frac{L_2}{L_1} \tag{10-11.1}$$

where λ_2 and λ_1 refer to liquid thermal conductivities at T and pressures P_2 and P_1 and L_2 and L_1 are functions of the reduced temperature and pressure, as shown in Fig. 10-10. This correlation was devised by Lenoir (1957). Testing with data for 12 liquids, both polar and nonpolar, showed errors of only 2 to 4%. The use of Eq. (10-11.1) and Fig. 10-10 is illustrated in Example 10-12 with liquid NO_2, a material *not* used in developing the correlation.

Example 10-12 Estimate the thermal conductivity of nitrogen dioxide at 311 K and 276 bar. The experimental value quoted is 0.134 W/(m · K) (Richter and Sage, 1957). The experimental value of λ_L for the saturated liquid at 311 K and 2.1 bar is 0.124 W/(m · K) (Richter and Sage, 1957).

solution From Daubert, et al. (1997), T_c = 431.35 K, P_c = 101.33 bar; thus T_r = 311/431.35 = 0.721, P_{r1} = 2.1/101.33 = 0.021, and P_{r2} = 276/101.33 = 2.72. From Fig. 10-10, L_2 = 11.75 and L_1 = 11.17. With Eq. (10-11.1),

$$\lambda_L \text{ (276 bar)} = (0.124)\,\frac{11.75}{11.17} = 0.130 \text{ W(m} \cdot \text{K)}$$

$$\text{Error} = \frac{0.130 - 0.134}{0.134} \times 100 = -3\%$$

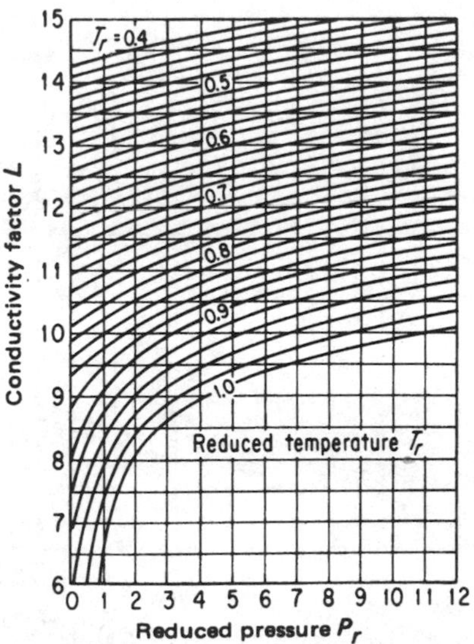

FIGURE 10-10 Effect of pressure on liquid thermal conductivities. (*From Le-Neindre, 1987.*)

Missenard (1970) has proposed a simple correlation for λ_L that extends to much higher pressures. In analytical form

$$\frac{\lambda_L(P_r)}{\lambda_L(\text{low pressure})} = 1 + QP_r^{0.7} \tag{10-11.2}$$

λ_L (P_r) and λ_L (low pressure) refer to liquid thermal conductivities at high and low, i.e., near saturation, pressure, both at the same temperature. Q is a parameter given in Table 10-7. The correlation is shown in Fig. 10-11.

The correlations of Missenard and Lenoir agree up to a reduced pressure of 12, the maximum value shown for the Lenoir form.

Example 10-13 Estimate the thermal conductivity of liquid toluene at 6330 bar and 304 K. The experimental value at this high pressure is 0.228 W/(m · K) (Kandiyoti, et al., 1973). At 1 bar and 304 K. λ_L = 0.129 W/(m · K) (Kandiyoti, et al., 1973).

solution From Appendix A, T_c = 591.75 K and P_c = 41.08 bar. Therefore, T_r = 304/591.75 = 0.514 and P_r = 6330/41.08 = 154. From Table 10-7, Q = 0.0205. Then, using Eq. (10-11.2),

TABLE 10-7 Values of Q in Eq. (10-11.2)

	Reduced pressure					
T_r	1	5	10	50	100	200
0.8	0.036	0.038	0.038	(0.038)	(0.038)	(0.038)
0.7	0.018	0.025	0.027	0.031	0.032	0.032
0.6	0.015	0.020	0.022	0.024	(0.025	0.025
0.5	0.012	0.0165	0.017	0.019	0.020	0.020

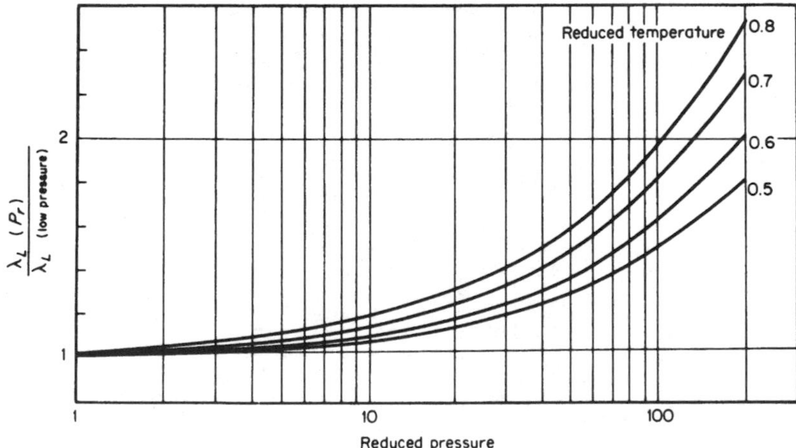

FIGURE 10-11 Missenard (1970) correlation for liquid thermal conductivities at high pressures.

$$\lambda_L(P_r) = (0.129)[1 + (0.0205)(154)^{0.7}] = 0.219 \text{ W/(m}\cdot\text{K)}$$

$$\text{Error} = \frac{0.219 - 0.228}{0.228} \times 100 = -4\%$$

Latini and Baroncini (1983) correlated the effect of pressure on liquid thermal conductivity by using Eq. (10-9.1), but they expressed the A parameter as

$$A = A_0 + A_1 P_r \tag{10-11.3}$$

Thus, A_0 would represent the appropriate A parameter at low pressures, as described in Sec. 10-9 and given by Eq. (10-9.2). Values of A_1 were found to range from 6×10^{-3} to 6×10^{-4} W/(m·K); thus the term $A_1 P_r$ is negligibly small except at quite high values of P_r. The authors have generalized the parameter A_1 for hydrocarbons as

$$A_1 = \frac{0.0673}{M^{0.84}} \quad \text{saturated hydrocarbons} \tag{10-11.4}$$

$$A_1 = \frac{102.50}{M^{2.4}} \quad \text{aromatics} \tag{10-11.5}$$

For hydrocarbons the authors found average errors usually less than 6% with maximum errors of 10 to 15%. The method should not be used for reduced pressures exceeding 50. More recently, Latini, et al. (1989) have extended Eq. (10-9.1) to higher pressures by replacing the exponent of 0.38 with a pressure-dependent expression.

Example 10-14 Rastorguev, et al. (1968), as quoted in Jamieson, et al. (1975), show the liquid thermal conductivity of *n*-heptane at 313 K to be 0.115 W/(m·K) at 1 bar. Estimate the thermal conductivity of the compressed liquid at the same temperature and 490 bar. The experimental value is 0.136 W/(m·K) Rastorguev, et al. (1968).

solution From Appendix A, T_c = 540.2 K, P_c = 27.4 bar, and M = 100.204 g/mol. Since we know the low-pressure value of λ_L, with Eq. (10-9.1), we can estimate A. With T_r = 313/540.2 = 0.580,

$$0.115 = \frac{A(1 - 0.580)^{0.38}}{(0.580)^{1/6}}$$

$$A = 0.146 \text{ W/(m}\cdot\text{K)}$$

This value of A then becomes A_0 in Eq. (10-11.3). Using Eq. (10-11.4),

$$A_1 = \frac{0.0673}{(100.204)^{0.84}} = 1.40 \times 10^{-3}$$

Then, using Eqs. (10-9.1) and (10-11.3) with P_r = 490/27.4 = 17.9,

$$\frac{\lambda_L(P_r = 17.9)}{\lambda_L(\text{low pressure})} = \frac{A_0 + A_1 P_r}{A_0} = \frac{[0.146 + (1.40 \times 10^{-3})(17.9)]}{(0.146)} = 1.17$$

$$\lambda_L(P_r = 17.9) = (0.115)(1.17) = 0.135 \text{ W/(m}\cdot\text{K)}$$

$$\text{Error} = \frac{0.135 - 0.136}{0.136} \times 100 = -1\%$$

10-12 THERMAL CONDUCTIVITIES OF LIQUID MIXTURES

The thermal conductivities of most mixtures of organic liquids are usually less than those predicted by either a mole or weight fraction average, although the deviations are often small. We show data for several binaries in Fig. 10-12 to illustrate this point.

Many correlation methods for λ_m have been proposed (Arikol and Gürbüz, 1992; Assael, et al., 1992a, 1996; Bleazard and Teja, 1996; Fareleira, et al., 1990). Five were selected for presentation in this section. They are described separately and evaluated later when examples are presented to illustrate the methodology in using each of the methods.

There is a surprisingly large amount of experimental mixture data (Baroncini, et al., 1984; Cai, et al., 1993; DiGuilio, 1990; Gaitonde, et al., 1978; Jamieson, et al., 1969, 1973; Jamieson and Irving, 1973; Ogiwara, et al.,1980, 1982; Qun-Fung,

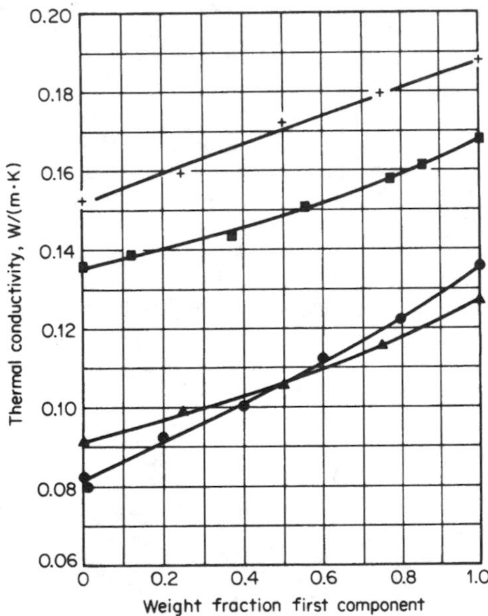

FIGURE 10-12 Thermal conductivities of liquid mixtures. (*Data from Filippov and Novoselova, 1955, and Jamieson and Irving, 1973.*)

et al., 1997; Rabenovish, 1971; Shroff, 1968; Stephan and Hildwein, 1987; Teja and Tardieu, 1988; Usmanov and Salikov, 1977; Vesovic and Wakeham, 1991), although most are for temperatures near ambient.

Filippov Equation

The Filippov equation (Filippov, 1955; Filippov and Novoselova, 1955) is

$$\lambda_m = w_1\lambda_1 + w_2\lambda_2 - 0.72w_1w_2(\lambda_2 - \lambda_1) \tag{10-12.1}$$

where w_1 and w_2 are the weight fractions of components 1 and 2. λ_1 and λ_2 are the pure component thermal conductivities. The components were so chosen that $\lambda_2 \geq \lambda_1$. The constant 0.72 may be replaced by an adjustable parameter if binary mixture data are available. The technique is not suitable for multicomponent mixtures but has been extensively tested for binary mixtures.

Jamieson, et al. Correlation (1975)

Research and data evaluation at the National Engineering Laboratory has suggested, for binary mixtures,

$$\lambda_m = w_1\lambda_1 + w_2\lambda_2 - \alpha(\lambda_2 - \lambda_1)[1 - (w_2)^{1/2}]w_2 \tag{10-12.2}$$

where w_1 and w_2 are weight fractions and, as in the Filippov method, the components are so selected that $\lambda_2 \geq \lambda_1$. α is an adjustable parameter that is set equal to unity if mixture data are unavailable for regression purposes. The authors indicate that Eq. (10-12.2) enables one to estimate λ_m within about 7% (with a 95% confidence limit) for all types of binary mixtures with or without water. It cannot, however, be extended to multicomponent mixtures.

Baroncini, et al. (1981a, 1983, 1984) Correlation

The Latini, et al. method to estimate pure liquid thermal conductivities [Eq. (10-9.1)] has been adapted to treat binary liquid mixtures as shown in Eq. (10-12.3)

$$\lambda_m = \left[x_1^2A_1 + x_2^2A_2 + 2.2\left(\frac{A_1^3}{A_2}\right)^{1/2} x_1x_2\right]\frac{(1 - T_{rm})^{0.38}}{T_{rm}^{1/6}} \tag{10-12.3}$$

where x_1 and x_2 are the mole fractions of components 1 and 2. The A parameters, introduced in Eq. (10-9.1), can be estimated from Eq. (10-9.2) and Table 10-4, or they can be calculated from pure component thermal conductivities (see Example 10-14). The reduced temperature of the mixture is $T_{rm} = T/T_{cm}$, where

$$T_{cm} = x_1T_{c1} + x_2T_{c2} \tag{10-12.4}$$

with T_{c1} and T_{c2} the pure component critical temperatures. The choice of which component is number 1 is made with criterion $A_1 \leq A_2$.

This correlation was tested (Baroncini, et al., 1984) with over 600 datum points on 50 binary systems including those with highly polar components. The average error found was about 3%. The method is not suitable for multicomponent mixtures.

Method of Rowley (1988)

In this procedure, the liquid phase is modeled by using a two-liquid theory wherein the energetics of the mixture are assumed to favor local variations in composition. The basic relation assumed by Rowley is

$$\lambda_m = \sum_{i=1}^{n} w_i \sum_{j=1}^{n} w_{ji} \lambda_{ji} \tag{10-12.5}$$

where λ_m = liquid mixture thermal conductivity, W/(m · K)
w_i = weight fraction of component i
w_{ji} = local weight fraction of component j relative to a central molecule of component i
λ_{ji} = characteristic parameter for the thermal conductivity that expresses the interactions between j and i, W/(m · K)

Mass fractions were selected instead of mole fractions in Eq. (10-12.5) because it was found that the excess mixture thermal conductivity

$$\lambda_m^{ex} = \lambda_m - \sum_{i=1}^{n} w_i \lambda_i \tag{10-12.6}$$

was more symmetrical when weight fractions were employed.

The two-liquid (or local composition) theory was developed in Chap. 8 to derive several of the liquid activity coefficient-composition models. Rowley develops expressions for w_{ij} and relates this quantity to parameters in the NRTL equation (see Chap. 8). In his treatment, he was able to show that Eq. (10-12.5) could be expressed as

$$\lambda_m = \sum_{i=1}^{n} w_i \frac{\sum_{j=1}^{n} w_j G_{ji} \lambda_{ji}}{\sum_{k=1}^{n} w_k G_{ki}} \tag{10-12.7}$$

where G_{ji} and G_{ij} (or G_{ki} and G_{ik}) are the same NRTL parameters as used in activity coefficient correlations for the system of interest.

To obtain $\lambda_{ji}(= \lambda_{ij})$, Rowley makes the important assumption that for any binary, say 1 and 2, $\lambda_m = \lambda_{12} = \lambda_{21}$ when the local *mole fractions* are equal, that is, $x_{12} = x_{21}$. Then, after some algebra, the final correlation is obtained.

$$\lambda_m = \sum_{i=1}^{n} w_i \lambda_i + \sum_{i=1}^{n} w_i \frac{\sum_{j=1}^{n} w_j G_{ji} (\lambda_{ji} - \lambda_i)}{\sum_{k=1}^{n} w_k G_{ki}} \tag{10-12.8}$$

In Eq. (10-12.8), $\lambda_{ii} = \lambda_i$, $\lambda_{ij} = \lambda_{ji}$. In the original formulation (Rowley, 1982), λ_{ij} was given by

$$\lambda_{ij} = \frac{(w_i^*)^2(w_j^* + w_i^* G_{ij})\lambda_i + (w_j^*)^2(w_i^* + w_j^* G_{ji})\lambda_j}{(w_i^*)^2(w_j^* + w_i^* G_{ij}) + (w_j^*)^2(w_i^* + w_j^* G_{ji})} \tag{10-12.9}$$

$$w_i^* = \frac{M_i(G_{ji})^{1/2}}{M_i(G_{ji})^{1/2} + M_j(G_{ij})^{1/2}} \tag{10-12.10}$$

$$w_j^* = 1 - w_i^* \tag{10-12.11}$$

Note that $G_{ii} = G_{jj} = 1$. w_i^* is that weight fraction in a binary *i-j* mixture such that $x_{12} = x_{21}$. More recently, Rowley (1988) recommended for systems not containing water, Eq. (10-12.9) be replaced by

$$\lambda_{ij} = \frac{M_i(w_i^*)^2(w_j^* + w_i^* G_{ij})\lambda_i + M_j(w_j^*)^2(w_i^* + w_j^* G_{ji})\lambda_j}{M_i(w_i^*)^2(w_j^* + w_i^* G_{ij}) + M_j(w_j^*)^2(w_i^* + w_j^* G_{ji})} \tag{10-12.12}$$

For a binary system of 1 and 2, Eq. (10-12.8) becomes

$$\lambda_m = w_1\lambda_1 + w_2\lambda_2 + w_1 w_2 \left[\frac{G_{21}(\lambda_{12} - \lambda_1)}{w_1 + w_2 G_{21}} + \frac{G_{12}(\lambda_{12} - \lambda_2)}{w_1 G_{12} + w_2}\right] \tag{10-12.13}$$

λ_{12} is found from Eq. (10-12.9) if water is one of the components, or Eq. (10-12.12) if water is not present. Using Eq. (10-12.9), with $i = 1$ and $j = 2$ gives

$$\lambda_{12} = \frac{(w_1^*)^2(w_2^* + w_1^* G_{12})\lambda_1 + (w_2^*)^2(w_1^* + w_2^* G_{21})\lambda_2}{(w_1^*)^2(w_2^* + w_1^* G_{12}) + (w_2^*)^2(w_1^* + w_2^* G_{21})} \tag{10-12.14}$$

Equations (10-12.10) and (10-12.11) give

$$w_1^* = \frac{M_1 G_{21}^{1/2}}{M_1 G_{21}^{1/2} + M_2 G_{12}^{1/2}} \tag{10-12.15}$$

$$w_2^* = 1 - w_1^* \tag{10-12.16}$$

With some algebra, it can be shown that the quantity in brackets in Eq. (10-12.13) is equal to $(\lambda_2 - \lambda_1)\, R$ where

$$R = \frac{G_{21}}{(w_1 + w_2 G_{21})(1 + Y)} - \frac{G_{12}}{(w_2 + w_1 G_{12})(1 + Y^{-1})} \tag{10-12.17}$$

$$Y = \left(\frac{w_1^*}{w_2^*}\right)^2 \frac{w_2^* + w_1^* G_{12}}{w_1^* + w_2^* G_{21}} \tag{10-12.18}$$

Thus, it is clear that the entire nonideal effect is included in the R parameter, and the form is quite similar to the Filippov and Jamieson, et al. relations described earlier.

To employ this technique, values for the liquid thermal conductivities of all pure components are required. In addition, from data sources or from regressing vapor-liquid equilibrium data, the NRTL parameters, G_{ij} and G_{ji} must be found. When tested with data on 18 ternary mixtures, Rowley (1988) found an average absolute deviation of 1.86%. The concept of relating transport and thermodynamic properties is an interesting one and bears further study; Brulé and Starling (1984) also have advocated such an approach.

Power Law Method

Following Vredeveld (1973), the following equation may be used for nonaqueous systems in which the ratio of component thermal conductivities does not exceed two:

$$\lambda_m = \left(\sum_i w_i \lambda_i^{-2} \right)^{-1/2} \tag{10-12.19}$$

where w_i is the weight fraction of component i and λ_i is the thermal conductivity of pure i. Eq. (10-12.19) has been used successfully for both binary (Carmichael, et al., 1963) and ternary systems (Rowley, 1988). Attempts to use Eq. (10-12.19) for aqueous binaries have been unsuccessful and typically led to deviations as high as 10% (Rowley, 1988).

Discussion

All five methods for estimating λ_m described in this section have been extensively tested by using binary mixture data. All require the thermal conductivities of the pure components making up the system (or an estimate of the values), and thus they are interpolative in nature. The Filippov, Jamieson, et al., and power-law procedures require no additional information other than the weight fractions and pure component values of λ_L. The Baroncini, et al. method also needs pure component critical properties. Rowley's correlation requires the NRTL parameters G_{ij} and G_{ji} from phase equilibrium data. Only the power-law and Rowley's methods will treat multicomponent mixtures. The power-law method should not be used if water is present or if the ratio of pure component λ values exceeds two. In Fig. 10-13, we show some recent measurements of Usmanov and Salikov (1977) for very polar systems and illustrate how well Filippov's relation (10-12.1) fits these data. Methods described in this section other than the power-law method would have been equally satisfactory. Gaitonde, et al. (1978) measured λ_m for liquid mixtures of alkanes and silicone oils to study systems with large differences in the molecular sizes of the components. They found the Filippov and Jamieson, et al. correlations provide a good fit to the data, but they recommended the general form of McLaughlin (1964) with the inclusion of an adjustable binary parameter. Teja and Tardieu (1988) have used a method based on effective carbon number to estimate thermal conductivities of crude oil fractions.

In summary, one can use any of the relations described in this section to estimate λ_m with the expectation that errors will rarely exceed about 5%.

In the case of aqueous (*dilute*) solutions containing electrolytes, the mixture thermal conductivity usually decreases with an increase in the concentration of the dissolved salts. To estimate the thermal conductivity of such mixtures, Jamieson and Tudhope (1964) recommend the use of an equation proposed originally by Riedel (1951) and tested by Vargaftik and Os'minin (1956). At 293 K:

$$\lambda_m = \lambda(H_2O) + \Sigma \sigma_i C_i \tag{10-12.20}$$

where λ_m = thermal conductivity of the ionic solution at 293 K, W/(m·K)
$\lambda(H_2O)$ = thermal conductivity of water at 293 K, W/(m·K)
C_i = concentration of the electrolyte, mol/L
σ_i = coefficient that is characteristic for each ion

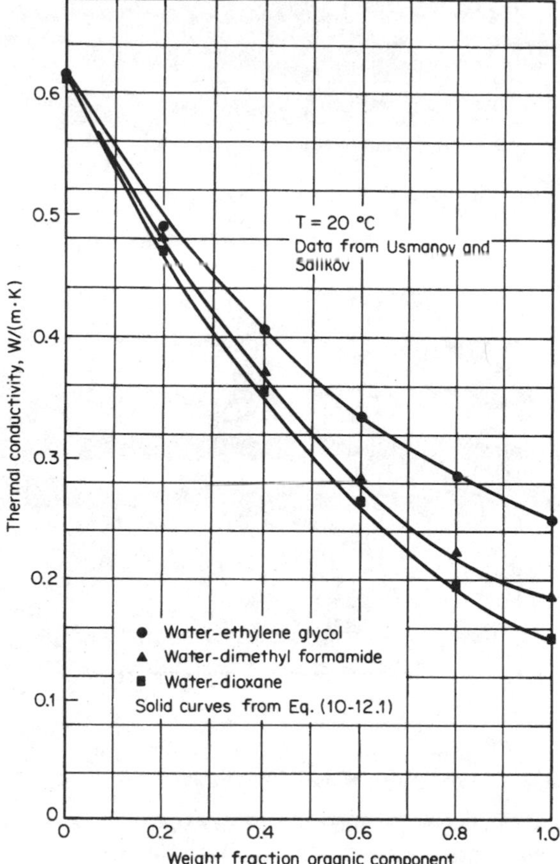

FIGURE 10-13 Filippov correlation of liquid mixture thermal conductivity. (*Data from Osmanov and Salikov, 1977.*)

Values of σ_i are shown in Table 10-8. To obtain λ_m at other temperatures,

$$\lambda_m(T) = \lambda_m(293)\,\frac{\lambda(H_2O \text{ at } T)}{\lambda(H_2O \text{ at 293 K})} \tag{10-12.21}$$

Except for strong acids and bases at high concentrations, Eqs. (10-12.20) and (10-12.21) are usually accurate to within ±5%.

Example 10-15 Using Filippov's and Jamieson, et al.'s methods, estimate the thermal conductivity of a liquid mixture of methanol and benzene at 273 K. The weight fraction methanol is 0.4. At this temperature, the thermal conductivities of pure benzene and methanol are 0.152 and 0.210 W/(m · K) (Jamieson, et al., 1969) respectively. The experimental mixture value (Jamieson, et al., 1969) is 0.170 W/(m · K).

TABLE 10-8 Values of σ_i for Anions and Cations in Eq. (10-12.21) (*Jamieson and Tudhope, 1964*)

Anion	$\sigma_i \times 10^5$	Cation	$\sigma_i \times 10^5$
OH^-	20.934	H^+	−9.071
F^-	2.0934	Li^+	−3.489
Cl^-	−5.466	Na^+	0.000
Br^-	−17.445	K^+	−7.560
I^-	−27.447	NH_4^+	−11.63
NO_2^-	−4.652	Mg^{2+}	−9.304
NO_3^-	−6.978	Ca^{2+}	−0.5815
ClO_3^-	−14.189	Sr^{2+}	−3.954
ClO_4^-	−17.445	Ba^{2+}	−7.676
BrO_3^-	−14.189	Ag^+	−10.47
CO_3^{2-}	−7.560	Cu^{2+}	−16.28
SiO_3^{2-}	−9.300	Zn^{2+}	−16.28
SO_3^{2-}	−2.326	Pb^{2+}	−9.304
SO_4^{2-}	1.163	Co^{2+}	−11.63
$S_2O_3^{2-}$	8.141	Al^{3+}	−32.56
CrO_4^{2-}	−1.163	Th^{4+}	−43.61
$Cr_2O_7^{2-}$	15.93		
PO_4^{3-}	−20.93		
$Fe(CN)_6^{4-}$	18.61		
$Acetate^-$	−22.91		
$Oxalate^{2-}$	−3.489		

solution FILPPOV'S METHOD. We use Eq. (10-12.1). Here methanol is component 2, since λ (methanol) > λ (benzene). Thus,

$$\lambda_m = (0.6)(0.152) + (0.4)(0.210) - (0.72)(0.6)(0.4)(0.210 - 0.152)$$

$$= 0.165 \text{ W/(m} \cdot \text{K)}$$

$$\text{Error} = \frac{0.165 - 0.170}{0.170} \times 100 = -3\%$$

JAMESON ET AL. METHOD. Again methanol is chosen as component 2. With Eq. (10-12.2) and $\alpha = 1$.

$$\lambda_m = (0.6)(0.152) + (0.4)(0.210) - (0.210 - 0.152)[1 - (0.4)^{1/2}](0.4)$$

$$= 0.167 \text{ W/(m} \cdot \text{K)}$$

$$\text{Error} = \frac{0.167 - 0.170}{0.170} \times 100 = -2\%$$

Example 10-16 Estimate the liquid thermal conductivity of a mixture of benzene (1) and methyl formate (2) at 323 K by using the method of Baroncini, et al. At this temperature, the values of λ_L for the pure components are $\lambda_1 = 0.138$ and $\lambda_2 = 0.179$ W/(m · K) (Baroncini, et al., 1984).

solution We will estimate the values of λ_m at 0.25, 0.50, and 0.75 weight fraction benzene. First, however, we need to determine A_1 and A_2. Although Eq. (10-9.2) and Table 10-4 could be used, it is more convenient to employ the pure component values of λ_L with Eq. (10-9.1). From Appendix A, $T_{c1} = 562.05$ K and $T_{c2} = 487.2$ K, so $T_{r1} = 323/562.05 = 0.575$ and $T_{r2} = 323/487.2 = 0.663$. Then, with Eq. (10-9.1), for benzene,

$$0.138 = \frac{A_1(1 - 0.575)^{0.38}}{(0.575)^{1/6}} \qquad A_1 = 0.174$$

Similarly $A_2 = 0.252$. [Note that, if Eq. (10-9.2) and Table 10-4 had been used, we would have $A_1 = 0.176$ and $A_2 = 0.236$.] We have selected components 1 and 2 to agree with the criterion $A_1 \leq A_2$.

Consider first a mixture containing 0.25 weight fraction benzene, i.e., $w_1 = 0.25$ and $w_2 = 0.75$. Then, the mole fractions are $x_1 = 0.204$ and $x_2 = 0.796$. Thus,

$$T_{cm} = (0.204)(562.05) + (0.796)(487.2) = 502.5 \text{ K}$$

$$T_{rm} = \frac{323}{502.5} = 0.643$$

With Eq. (10-12.3),

$$\lambda_m = \left\{(0.204)^2(0.174) + (0.796)^2(0.252) + (2.2)\left[\frac{(0.174)^3}{0.252}\right]^{1/2}(0.204)(0.769)\right\}$$

$$\times \left\{\frac{(1 - 0.643)^{0.38}}{(0.643)^{1/6}}\right\}$$

$$= 0.159 \text{ W/(m} \cdot \text{K)}$$

Calculated results for this and other compositions are shown below with the experimental values (Baroncini, et al., 1984) and percent errors.

Benzene-Methyl Formate Mixtures; $T = 323$ K

Weight fraction benzene	Mole fraction benzene	T_{cm}, K	λ_m, calc, W/(m · K)	λ_m, exp., W/(m · K)	Percent error
0.25	0.204	502.5	0.159	0.158	0.6
0.50	0.435	519.8	0.143	0.151	−5.3
0.75	0.698	539.6	0.135	0.140	−3.6

Example 10-17 Use Rowley's method to estimate the thermal conductivity of a liquid mixture of acetone (1) and chloroform (2) that contains 66.1 weight % of the former. The temperature is 298 K. As quoted by Jamieson, et al. (1975), Rodriguez (1962) reports $\lambda_1 = 0.161$ W/(m · K), $\lambda_2 = 0.119$ W/(m · K), and for the mixture, $\lambda_m = 0.143$ W/(m · K).

solution First, we need the NRTL parameters for this binary at 298 K. Nagata (1973) suggests $G_{12} = 1.360$ and $G_{21} = 0.910$. From Appendix A, $M_1 = 58.08$ and $M_2 = 119.38$ g/mol. Using Eqs. (10-12.15) and (10-12.16)

$$w_1^* = \frac{(58.08)(0.910)^{1/2}}{(58.08)(0.910)^{1/2} + (119.38)(1.360)^{1/2}} = 0.285$$

$$w_2^* = 1 - 0.285 = 0.715$$

With Eq. (10-12.14),

$$\lambda_{12} = \frac{(0.285)^2(0.285 + 0.715 \times 1.360)(0.161) + (0.715)^2(0.285 + 0.0715 \times 0.910)(0.119)}{(0.285)^2(0.285 + 0.715 \times 1.360) + (0.715)^2(0.285 + 0.0715 \times 0.910)}$$
$$= 0.126$$

Then, with Eq. (10-12.13),

$$\lambda_m = (0.661)(0.161) + (0.339)(0.119) + (0.661)(0.339)\left[\frac{(0.910)(0.126 - 0.161)}{0.661 + 0.339 \times 0.910} + \frac{1.360(0.126 - 0.119)}{0.661 \times 1.360 + 0.339}\right]$$
$$= 0.141$$

$$\text{Error} = \frac{0.141 - 0.143}{0.143} \times 100 = -1\%$$

Using Eq. (10-12.12) instead of Eq. (10-12.14) for λ_{12} gives $\lambda_m = 0.140$ for an error of -2%.

NOTATION

- A parameter in Eq. (10-9.1), W/(m·K)
- A_{ij} Wassiljewa coefficient, Eq. (10-6.1)
- B_i parameter in Eq. (10-5.9)
- C_i electrolyte concentration, mol/L, Eq. (10-12.20)
- C heat capacity, J/(mol·K); C_v, at constant volume; C_p, at constant pressure; C_{int}, due to internal degrees of freedom; C_{tr}, due to translational motion
- C group contribution constant in Eq. (10-3.12)
- D diffusion coefficient, m²/s
- f scaling parameter in Eqs. (10-5.13) and (10-7.13)
- f_{tr} translational factor in Eq. (10-3.1)
- f_{int} internal energy factor in Eq. (10-3.1)
- F_λ parameter defined in Eq. (10-5.15)
- G_1 parameter in Eq. (10-5.7)
- G_2 parameter in Eq. (10-5.8)
- G_{ij} NRTL parameter, Eq. (10-12.7)
- h scaling parameter in Eqs. (10-5.14) and (10-7.12)
- k Boltzmann's constant, 1.3804×10^{-23} J/K
- L mean free path, m
- L parameter shown in Fig. 10-10
- m molecular mass, g or kg, exponent in Eq. (10-9.3)
- M molecular weight, g/mol
- M' molecular weight, kg/mol unless otherwise noted
- n number of components in a mixture, exponent in Eq. (10-9.4)
- N_{Pr} Prandtl number, $C_p\eta/\lambda M'$
- N_0 Avogadro's number, 6.022142×10^{23} molecules/mol
- P pressure, N/m² or bar; P_c, critical pressure; P_r, reduced pressure, P/P_c; P_{vp}, vapor pressure
- q parameter defined in Eq. (10-5.5)

Q parameter in Table 10-7
R gas constant, 8.31447 J/(mol · K)
R parameter in Eq. (10-12.17)
T temperature, K; T_c, critical temperature; T_r, reduced temperature, T/T_c; T_0, equivalent temperature in Eqs. (10-5.13) and (10-7.16); T_b, normal boiling point (at 1 atm); $T_{br} = T_b/T_c$
v molecular velocity, m/s
V molar volume, cm^3/mol or m^3/mol; V_c, critical volume
w_i weight fraction of component i
w_i^* weight fraction in a binary where local mole fractions are equal, see Eq. (10-12.10)
x_i mole fraction of component i in a liquid mixture
y_i mole fraction of component i in a vapor mixture
X_λ parameter defined in Eq. (10-5.16)
Y parameter defined in Eq. (10-12.18)
Z compressibility factor PV/RT; Z_c, compressibility factor at the critical point
Z parameter defined under Eq. (10-3.14)

Greek

α parameter defined under Eq. (10-3.14); parameter in Eq. (10-12.2)
β parameter defined under Eq. (10-3.14)
γ C_p/C_v
Γ reduced, inverse thermal conductivity defined in Eq. (10-3.8)
ε interaction energy parameter, J; parameter in Eq. (10-6.2)
η viscosity, N · s/m^2; $\eta°$ low-pressure gas viscosity
κ association constant, see Table 9-1
λ thermal conductivity, W/(m · K); λ_r reduced thermal conductivity, $\Gamma\lambda$ or as in Eq. (10-5.22); $\lambda°$ low-pressure gas thermal conductivity; λ_{tr}, monatomic value of thermal conductivity; λ_L, liquid thermal conductivity; λ_b, liquid thermal conductivity at the normal boiling point
μ dipole moment, debye; μ_r, reduced dipole moment defined in Eq. (9-4.12)
ρ molar density, mol/cm^3 or mol/m^3, ρ_c, critical density; ρ_r, reduced density, ρ/ρ_c; ρ_0, density parameter in Eq. (10-5.14); ρ^S, saturated liquid density, mol cm^{-3}
σ characteristic dimension of the molecule, m or Å; σ_i, ion coefficient in Table 10-8
τ $1 - T_r$, Eq. (10-10.3)
ϕ parameter in Eq. (10-9.5)
Ψ parameter defined under Eq. (10-3.14)
Ω_v collision integral for viscosity and thermal conductivity; Ω_D, collision integral for diffusion coefficients
ω acentric factor

Subscripts

m mixture
L liquid
G gas

Superscripts

R property of reference fluid, propane
o ideal gas value
′,″ reference properties

REFERENCES

Arikol, M., and H. Gürbüz: *Can. J. Chem. Eng.,* **70:** 1157 (1992).

Assael, M. J., J. P. M. Trusler, and T. F. Tsolakis: *Thermophysical Properties of Fluids, An Introduction to Their Prediction,* Imperial College Press, London, 1996.

Assael, M. J., J. H. Dymond, M. Papadaki, and P. M Patterson: *Fluid Phase Equil.,* **75:** 245 (1992).

Assael, M. J., E. Charitidou, and W. A. Wakeham: *Intern. J. Thermophys.,* **10:** 779 (1989).

Assael, M. J., J. H. Dymond, M. Papadaki, and P. M Patterson: *Intern. J. Thermophys.,* **13:** 269 (1992).

Assael, M. J., J. H. Dymond, M. Papadaki, and P. M Patterson: *Intern. J. Thermophys.,* **13:** 659 (1992a).

Assael, M. J., J. H. Dymond, M. Papadaki, and P. M Patterson: *Intern. J. Thermophys.,* **13:** 895 (1992b).

Assael, M. J., J. H. Dymond, and S. K. Polimatidou: *Intern. J. Thermophys.,* **15:** 189 (1994).

Assael, M. J., J. H. Dymond, and S. K. Polimatidou: *Intern. J. Thermophys.,* **16:** 761 (1995).

Baroncini, C., P. Di Filippo, G. Latini, and M. Pacetti: *Intern. J. Thermophys.,* **1**(2)**:** 159 (1980).

Baroncini, C., P. Di Filippo, G. Latini, and M. Pacetti: *Intern. J. Thermophys.,* **2**(1)**:** 21 (1981).

Baroncini, C., P. Di Filippo, G. Latini, and M. Pacetti: *Thermal Cond.,* 1981a (pub. 1983), 17th, Plenum Pub. Co., p. 285.

Baroncini, C., P. Di Filippo, and G. Latini: "Comparison Between Predicted and Experimental Thermal Conductivity Values for the Liquid Substances and the Liquid Mixtures at Different Temperatures and Pressures," paper presented at the *Workshop on Thermal Conductivity Measurement, IMEKO, Budapest,* March 14–16, 1983.

Baroncini, C., P. Di Filippo, and G. Latini: *Intern. J. Refrig.,* **6**(1)**: 60** (1983a).

Baroncini, C., G. Latini, and P. Pierpaoli: *Intern. J. Thermophys.,* **5**(4)**:** 387 (1984).

Basu, R. S., and J. V. Sengers: "Thermal Conductivity of Fluids in the Critical Region," in *Thermal Conductivity,* D. C. Larson (ed.), Plenum, New York, 1983, p. 591.

Bennett, L. A., and R. G. Vines: *J. Chem. Phys.,* **23:** 1587 (1955).

Bleazard, J. G. and A. S. Teja: *Ind. Eng. Chem. Res.,* **35:** 2453 (1996).

Bretsznajder, S.: *Prediction of Transport and Other Physical Properties of Fluids,* trans. by J. Bandrowski, Pergamon, New York, 1971, p. 251.

Bridgman, P. W.: *Proc. Am. Acad. Art Sci.,* **59:** 154 (1923).

Brokaw, R. S.: *J. Chem. Phys.,* **29:** 391 (1958).

Bromley, L. A.: "Thermal Conductivity of Gases at Moderate Pressures," *Univ. California Rad. Lab. UCRL-1852,* Berkeley, Calif., June 1952.

Brulé, M. R., and K. E. Starling: *Ind. Eng. Chem. Proc. Des. Dev.,* **23:** 833 (1984).

Cai, G., H. Zong, Q. Yu, and R. Lin: *J. Chem. Eng. Data,* **38:** 332 (1993).

Carmichael, L. T., V. Berry, and B. H. Sage: *J. Chem. Eng. Data,* **8:** 281 (1963).

Carmichael, L. T., H. H. Reamer, and B. H. Sage: *J. Chem. Eng. Data,* **11:** 52 (1966).

Carmichael, L. T., and B. H. Sage: *J. Chem. Eng. Data,* **9:** 511 (1964).

Chapman, S., and T. G. Cowling: *The Mathematical Theory of Non-uniform Gases,* Cambridge, New York, 1961.

Christensen, P. L., and A. Fredenslund: *J. Chem. Eng. Data,* **24:** 281 (1979).

Chung, T.-H., M. Ajlan, L. L. Lee, and K. E. Starling: *Ind. Eng. Chem. Res.* **27:** 671 (1988).

Chung, T.-H., L. L. Lee, and K. E. Starling: *Ind. Eng. Chem. Fundam.,* **23:** 8 (1984).

Cohen, Y., and S. I. Sandler: *Ind. Eng. Chem. Fundam.,* **19:** 186 (1980).

Correla, F. von, B. Schramm, and K. Schaefer: *Ber. Bunsenges. Phys. Chem.,* **72**(3)**:** 393 (1968).

Cottrell, T. L., and J. C. McCoubrey: *Molecular Energy Transfer in Gases,* Butterworth, London, 1961.

Couch, E. J., and K. A. Dobe: *J. Chem. Eng. Data,* **6:** 229 (1961).

Curtiss, L. A., D. J. Frurip, and M. Blander: *J. Phys. Chem.,* **86:** 1120 (1982).

Daubert, T. E., R. P. Danner, H. M. Sibel, and C. C. Stebbins: *Physical and Thermodynamic Properties of Pure Chemicals: Data Compilation,* Taylor & Francis, Washington, D. C., 1997.

Dean, J. A.: Lange's *Handbook of Chemistry,* 15th ed., McGraw-Hill, New York, 1999.

DiGuilio, R. M., W. L. McGregor, and A. S. Teja: *J. Chem. Eng. Data,* **37:** 342 (1992).

DiGuilio, R. M., and A. S. Teja: *J. Chem. Eng. Data,* **35:** 117 (1990)

Donaldson, A. B.: *Ind. Eng. Chem. Fundam.,* **14:** 325 (1975).

Dymond, J. H., and M. J. Assael: Chap 10 in *Transport Properties of Fluids, Their Correlation, Prediction and Estimation,* J. Millat, J. H. Dymond, and C. A. Nieto de Castro (eds.), IUPAC, Cambridge Univ. Press, Cambridge, 1996.

Ehya, H., F. M. Faubert, and G. S. Springer: *J. Heat Transfer,* **94:** 262 (1972).

Ely, J. F. and H. J. M. Hanley: *Ind. Eng. Chem. Fundam.,* **22:** 90 (1983).

Ewing, C. T., B. E. Walker, J. A. Grand, and R. R. Miller: *Chem. Eng. Progr. Symp. Ser.,* **53**(20)**:** 19 (1957).

Fareleira, J. M. N., C. A. Nieto de Castro, and A. A. H. Pádua: *Ber. Bunsenges. Phys. Chem.,* **94:** 553 (1990).

Filippov, L. P.: *Vest. Mosk. Univ., Ser. Fiz. Mat. Estestv. Nauk,* (8)**10**(5)**:** 67–69 (1955); *Chem. Abstr.,* **50:** 8276 (1956).

Filippov, L. P., and N. S. Novoselova: *Vestn. Mosk. Univ., Ser. Fiz. Mat. Estestv Nauk,* (3)**10**(2)**:** 37–40 (1955); *Chem. Abstr.,* **49:** 11366 (1955).

Fleeter, R. J., Kestin, and W. A. Wakeham: *Physica A (Amsterdam),* **103A:** 521 (1980).

Gaitonde, U. N., D. D. Deshpande, and S. P. Sukhatme: *Ind. Eng. Chem. Fundam.,* **17:** 321 (1978).

Gambill, W. R.: *Chem. Eng.,* **66**(16)**:** 129 (1959).

Gilmore, T. F., and E. W. Comings: *AIChE J.,* **12:** 1172 (1966).

Gray, P., S. Holland, and A. O. S. Maczek: *Trans. Faraday Soc.,* **66:** 107 (1970).

Groenier, W. S., and G. Thodos: *J. Chem. Eng. Data,* **6:** 240 (1961).

Guildner, L. A.: *Proc. Natl. Acad. Sci.,* **44:** 1149 (1958).

Hanley, H. J. M.: *Cryogenics,* **16**(11)**:** 643 (1976).

Hirschfelder, J. O.: *J. Chem. Phys.,* **26:** 282 (1957).

Hirschfelder, J. O., C. F. Curtiss, and R. B. Bird: *Molecular Theory of Gases and Liquids,* Wiley, New York, 1954.

Ho, C. Y., R. W. Powell, and P. E. Liley: *J. Phys. Chem. Ref. Data,* **1:** 279 (1972).

Holland, P. M., H. J. M. Hanley, K. E. Gubbins, and J. M. Haile: *J. Phys. Chem. Ref. Data,* **8:** 559 (1979).

Huber, M. L.: Chap 12 in *Transport Properties of Fluids, Their Correlation, Prediction and Estimation,* J. Millat, J. H. Dymond, and C. A. Nieto de Castro (eds.), IUPAC, Cambridge Univ. Press, Cambridge, 1996.

Huber, M. L.: personal communication, NIST, Boulder, 1998.

Jamieson, D. T.: personal communication, National Engineering Laboratory, East Kilbride, Glasgow, March 1971.

Jamieson, D. T.: *J. Chem. Eng. Data,* **24:** 244 (1979).

Jamieson, D. T.: personal communication, National Engineering Laboratory, East Kilbride, Scotland, October 1984.

Jamieson, D. T., and G. Cartwright: *J. Chem. Eng. Data,* **25:** 199 (1980).

Jamieson, D. T., and G. Cartwright: *Proc. 8th Symp. Thermophys. Prop.,* Vol. 1, *Thermophysical Properties of Fluids,* J. V. Sengers (ed.), ASME, New York, 1981, p. 260.

Jamieson, D. T., and E. H. Hastings: in C. Y. Ho and R. E. Taylor (eds.), *Proc. 8th Conf. Thermal Conductivity,* Plenum, New York, 1969, p. 631.

Jamieson, D. T., and J. B. Irving: paper presented in *13th Intern. Thermal Conductivity Conf., Univ. Missouri,* 1973.

Jamieson, D. T., and J. S. Tudhope: *Natl. Eng. Lab. Glasgow Rep.* 137, March 1964.

Jamieson, D. T., J. B. Irving, and J. S. Tudhope: *Liquid Thermal Conductivity. A Data Survey to 1973,* H. M. Stationary Office, Edinburgh, 1975.

Junk, W. A., and E. W. Comings: *Chem. Eng. Progr.,* **49:** 263 (1953).

Kandiyoti, R., E. McLaughlin, and J. F. T. Pittman: *Chem. Soc. (London), Faraday Trans.,* **69:** 1953 (1973).

Kannuliuk, W. G., and H. B. Donald: *Aust. J. Sci. Res.,* **3A:** 417 (1950).

Kennedy, J. T., and G. Thodos: *AIChE J.,* **7:** 625 (1961).

Keyes, F. G.: *Trans. ASME,* **73:** 597 (1951).

Klass, D. M., and D. S. Viswanath: *Ind. Eng. Chem. Res.,* **37:** 2064 (1998).

Kramer, F. R., and E. W. Comings: *J. Chem. Eng. Data,* **5:** 462 (1960).

Krauss, R., J. Luettmer-Strathmann, J. V. Sengers, and K. Stephan: *Intern. J. Thermophys.,* **14:** 951 (1993).

Lakshmi, D. S. and D. H. L. Prasad: *Chem. Eng. J.,* **48:** 211 (1992).

Lambert, J. D.: in D. R. Bates (ed.), *Atomic and Molecular Processes,* Academic Press, New York, 1962.

Latini, G., and C. Baroncini: *High Temp.-High Press.,* **15:** 407 (1983).

Latini, G., and M. Pacetti: *Therm. Conduct.,* **15:** 245 (1977); pub. 1978.

Latini, G., G. Passerini, and F. Polonara: *Intern. J. Thermophys,* **17:** 85 (1996).

Latini, G., F. Marcotullio, P. Pierpaoli, and A. Ponticiello: *Thermal Conductivity,* vol. 20, p. 205, Plenum Pr., New York, 1989.

Le Neindre: *Recommended Reference Materials for the Realization of Physicochemical Properties,* K. N. Marsh (ed.), IUPAC, Blackwell Sci. Pub., Oxford, 1987.

Lenoir, J. M.: *Petrol. Refiner,* **36**(8)**:** 162 (1957).

Lide, D. R. (ed.): *Handbook of Chemistry and Physics,* 80th ed., CRC Press, Boca Raton, 1999.

Liley, P. E., T. Makita, and Y. Tanaka: *in Cindas Data Series on Material Properties, Properties of Inorganic and Organic Fluids,* C. Y. Ho (ed.), Hemisphere, New York, 1988.

Lindsay, A. L., and L. A. Bromley: *Ind. Eng. Chem.,* **42:** 1508 (1950).

McLaughlin, E.: *Chem. Rev.,* **64:** 389 (1964).

Mani, N., and J. E. S. Venart: *Advan. Cryog. Eng.,* **18:** 280 (1973).

Mason, E. A.: *J. Chem. Phys.,* **28:** 1000 (1958).

Mason, E. A., and L. Monchick: *J. Chem. Phys.,* **36:** 1622 (1962).

Mason, E. A., and S. C. Saxena: *Phys. Fluids,* **1:** 361 (1958).

Mason, E. A., and S. C. Saxena: *J. Chem. Phys.,* **31:** 511 (1959).

Mason, E. A., and H. von Ubisch: *Phys. Fluids,* **3:** 355 (1960).

Miller, J. W., J. J. McGinley, and C. L. Yaws: *Chem. Eng.,* **83**(23)**:** 133 (1976).

Miller, J. W., P. N. Shah, and C. L. Yaws: *Chem. Eng.,* **83**(25)**:** 153 (1976a).

Misic, D., and G. Thodos: *AIChE J.,* **7:** 264 (1961).

Misic, D., and G. Thodos: *AIChE J.,* **11:** 650 (1965).

Missenard, A.: *Rev. Gen. Thermodyn.,* **101**(5)**:** 649 (1970).

Missenard, A.: *Rev. Gen. Thermodyn.,* **11:** 9 (1972).

Mo, K. C., and K. E. Gubbins: *Chem. Eng. Comm.,* **1:** 281 (1974).

Muckenfuss, C., and C. F. Curtiss; *J. Chem. Phys.,* **29:** 1273 (1958).

Nagata, I.: *J. Chem. Eng. Japan,* **6:** 18 (1973).

Nieto de Castro, C. A., J. M. N. A. Fareleira, J. C. G. Calado, and W. A. Wakeham: *Proc. 8th Symp. Thermophys. Prop.,* Vol. 1, *Thermophysical Prop. of Fluids,* J. V. Sengers (ed.), ASME, New York, 1981, p. 247.

Nieto de Castro, C. A., S. F. Y. Li, A. Nagashima, R. D. Trengove, and W. A. Wakeham: *J. Phys. Chem. Ref. Data,* **15:** 1073 (1986).

O'Neal, C. Jr., and R. S. Brokaw: *Phys. Fluids,* **5:** 567 (1962).

Ogiwara, K., Y. Arai, and S. Saito: *Ind. Eng. Chem. Fundam.,* **19:** 295 (1980).

Ogiwara, K., Y. Arai, and S. Saito: *J. Chem. Eng. Japan,* **15:** 335 (1982).

Owens, E. J., and G. Thodos: *AIChE J.,* **3:** 454 (1957).

Owens, E. J., and G. Thodos: *Proc. Joint Conf. Thermodyn. Transport Prop. Fluids, London,* July 1957a, pp. 163–68, Inst. Mech. Engrs., London, 1958.

Owens, E. J., and G. Thodos: *AIChE J.,* **6:** 676 (1960).

Perry, R. H., and D. W. Green (eds.): *Chemical Engineers' Handbook,* 7th ed., McGraw-Hill, New York, 1997.

Peterson, J. N., T. F. Hahn, and E. W. Comings: *AIChE J.,* **17:** 289 (1971).

Prasad, R. C., and J. E. S. Venart: *Proc. 8th Symp. Thermophys. Prop.,* Vol. 1, *Themophysical Prop. of Fluids,* J. V. Sengers (ed.), ASME, New York, 1981, p. 263.

Preston, G. T., T. W. Chapman, and J. M. Prausnitz: *Cryogenics,* **7**(5)**:** 274 (1967).

Qun-Fung, L., L. Rui-Sen, and N. Dan-Yan: *J. Chem. Eng. Data,* **42:** 971, (1997).

Rabenovish, B. A.: *Thermophysical Properties of Substances and Materials,* 3d ed., Standards, Moscow, 1971.

Rastorguev, Yu. L., G. F. Bogatov, and B. A. Grigor'ev: *Isv. vyssh. ucheb. Zaved., Neft'i Gaz.,* **11**(12)**:** 59 (1968).

Richter, G. N., and B. H. Sage: *J. Chem. Eng. Data,* **2:** 61 (1957).

Richter, G. N., and B. H. Sage: *J. Chem. Eng. Data,* **8:** 221 (1963).

Richter, G. N., and B. H. Sage: *J. Chem. Eng. Data,* **9:** 75 (1964).

Riedel, L.: *Chem. Ing. Tech.,* **21:** 349 (1949); **23:** 59, 321, 465 (1951).

Rodriguez, H. V.: Ph.D. thesis, Louisiana State Univ., Baton Rouge, La., 1962.

Rosenbaum, B. M., and G. Thodos: *Physica,* **37:** 442 (1967).

Rosenbaum, B. M., and G. Thodos: *J. Chem. Phys.,* **51:** 1361 (1969).

Rowley, R. L.: *Chem. Eng. Sci.,* **37:** 897 (1982).

Rowley, R. L.: *Chem. Eng. Sci.,* **43:** 361 (1988).

Roy, D.: M.S. thesis, Northwestern University, Evanston, Ill., 1967.

Roy, D., and G. Thodos: *Ind. Eng. Chem. Fundam.,* **7:** 529 (1968).

Roy, D., and G. Thodos: *Ind. Eng. Chem. Fundam.,* **9:** 71 (1970).

Sastri, S. R. S.: personal communication, Regional Research Laboratory, Bhubaneswar (1998).

Sastri, S. R. S., and K. K. Rao: *Chem. Eng.,* **106,** Aug., 1993.

Saxena, S. C., and J. P. Agrawal: *J. Chem. Phys.,* **35:** 2107 (1961).

Saxena, S. C., M. P. Saksena, and R. S. Gambhir: *Brit. J. Appl. Phys.,* **15:** 843 (1964).

Schaefer, C. A., and G. Thodos: *Ind. Eng. Chem,* **50:** 1585 (1958).

Schaefer, C. A., and G. Thodos: *AIChE J.,* **5:** 367 (1959).

Schafer, K.: *Z. Phys. Chem.,* **B53:** 149 (1943).

Sengers, J. V., and J. Luettmer-Strathmann: Chap. 6 in *Transport Properties of Fluids, Their Correlation, Prediction and Estimation,* J. Millat, J. H. Dymond, and C. A. Nieto de Castro (eds.), *IUPAC,* Cambridge Univ. Press, Cambridge, 1996.

Shroff, G. H.: Ph.D. thesis, University of New Brunswick, Fredericton, 1968.

Srivastava, B. N., and R. C. Srivastava: *J. Chem. Phys.,* **30:** 1200 (1959).

Stephan, K. and T. Heckenberger: Thermal Conductivity and Viscosity Data of Fluid Mixtures, *Chem. Data Ser.,* vol. X, Part 1, DECHEMA, Frankfurt, 1988.

Stephan, K., and H. Hildwein: Recommended Data of Selected Compounds and Binary Mixtures, *Chem. Data Ser.,* vol. IV, Parts 1 + 2, DECHEMA, Frankfurt, 1987.

Stiel, L. I., and G. Thodos: *AIChE J.,* **10:** 26 (1964).

Sutton, J. R.: "References to Experimental Data on Thermal Conductivity of Gas Mixtures," *Natl. Eng. Lab., Rept. 612,* East Kilbride, Glasgow, Scotland, May 1976.

Svehla, R. A.: "Estimated Viscosities and Thermal Conductivities of Gases at High Temperatures," *NASA Tech. Rept. R-132,* Lewis Research Center, Cleveland, Ohio, 1962.

Tandon, P. K., and S. C. Saxena: *Appl. Sci. Res.,* **19:** 163 (1965).

Teja, A. S., and P. Rice: *Chem. Eng. Sci.,* **36:** 417 (1981).

Teja, A. S., and P. Rice: *Chem. Eng. Sci.,* **37:** 790 (1982).

Teja, A. S., and G. Tardieu: *Can. J. Chem. Eng.,* **66:** 980 (1988).

Tsederberg, N. V.: *Thermal Conductivity of Gases and Liquids,* The M.I.T. Press Cambridge, Mass., 1965.

Tufeu, R., and B. L. Neindre: *High Temp.-High Press.,* **13:** 31 (1981).

Ubbelohde, A. R.: *J. Chem. Phys.,* **3:** 219 (1935).

Usmanov, I. U., and A. S. Salikov: *Russ. J. Phys. Chem.,* **51**(10), 1488 (1977).

Vargaftik, N. B.: "Thermal Conductivities of Compressed Gases and Steam at High Pressures," *Izu. Vses. Telpotelzh. Inst.,* Nov. 7, 1951; personal communication, Prof. N. V. Tsederberg, Moscow Energetics Institute.

Vargaftik, N. B.: *Proc. Joint Conf. Thermodyn. Transport Prop. Fluids, London,* July 1957, p. 142, Inst. Mech. Engrs., London, 1958.

Vargaftik, N. B., and Y. P. Os'minin: *Teploenergetika,* **3**(7)**:** 11 (1956).

Vargaftik, N. B., L. P. Filippov, A. A. Tarzimanov, and E. E. Totskii: *Handbook of Thermal Conductivity of Liquids and Gases,* CRC Press, Boca Raton, 1994.

Vargaftik, N. B., Y. K. Vinogradov, and V. S. Yargin: *Handbook of Physical Properties of Liquids and Gases,* Begell House, New York, 1996.

Venart, J. E. S., and R. C. Prasad: *J. Chem. Eng. Data,* **25:** 198 (1980).

Vesovic, V., W. A. Wakeham, G. A. Olchowy, J. V. Sengers, J. T. R. Watson, and J. Millat: *J. Phys. Chem. Ref. Data,* **19:** 763 (1990).

Vesovic, V., and W. A. Wakeham: *High Temp.-High Press.,* **23:** 179 (1991).

Vines, R. G.: *Aust. J. Chem.,* **6:** 1 (1953).

Vines, R. G.: *Proc. Joint Conf. Thermodyn. Transport Prop. Fluids, London,* July 1957, Inst. Mech. Engrs., London, 1958, pp. 120–123.

Vines, R. G., and L. A. Bennett: *J. Chem. Phys.,* **22:** 360 (1954).

Vredeveld, D.: personal communication, 1973.

Wassiljewa, A.: *Physik. Z.,* **5:** 737 (1904).

Yaws, C. L.: *Handbook of Thermal Conductivity,* Gulf Pub., Houston, 1995.

Yaws, C. L.: *Handbook of Transport Property Data: Viscosity, Thermal Conductivity, and Diffusion Coefficients of Liquids and Gases,* Gulf Pub., Houston, 1995a.

Yorizane, M., S. Yoshimura, H. Masuoka, and H. Yoshida: *Ind. Eng. Chem. Fundam.,* **22:** 454 (1983).

Yorizane, M., S. Yoshimura, H. Masuoka, and H. Yoshida: *Ind. Eng. Chem. Fundam.,* **22:** 458 (1983a).

Zaytsev, I. D., and G. G. Aseyev: *Properties of Aqueous Solutions of Electrolytes,* CRC Press, Boca Raton, 1992.

Zheng, X.-Y., S. Yamamoto, H. Yoshida, H. Masuoka, and M. Yorizane: *J. Chem. Eng. Japan,* **17:** 237 (1984).

CHAPTER ELEVEN
DIFFUSION COEFFICIENTS

11-1 SCOPE

In Sec. 11-2 we discuss briefly several frames of reference from which diffusion can be related and define the diffusion coefficient. Low-pressure binary gas diffusion coefficients are treated in Secs. 11-3 and 11-4. The pressure and temperature effects on gas-phase diffusion coefficients are covered in Secs. 11-5 and 11-6, respectively. The theory for liquid diffusion coefficients is introduced in Sec. 11-8, and estimation methods for binary liquid diffusion coefficients at infinite dilution are described in Sec. 11-9. Concentration effects are considered in Sec. 11-10 and temperature and pressure effects in Sec. 11-11. Brief comments on diffusion in multicomponent mixtures are made in Secs. 11-7 (gases) and 11-12 (liquids); ionic solutions are covered in Sec. 11-13.

11-2 BASIC CONCEPTS AND DEFINITIONS

The extensive use of the term "diffusion" in the chemical engineering literature is based on an intuitive feel for the concept; i.e., diffusion refers to the net transport of material within a single phase in the absence of mixing (by mechanical means or by convection). Both experiment and theory have shown that diffusion can result from pressure gradients (pressure diffusion), temperature gradients (thermal diffusion), external force fields (forced diffusion), and concentration gradients. Only the last type is considered in this chapter; i.e., the discussion is limited to diffusion in isothermal, isobaric systems with no external force field gradients.

Even with this limitation, confusion can easily arise unless care is taken to define diffusion fluxes and diffusion potentials, e.g., driving forces, clearly. The proportionality constant between the flux and potential is the *diffusion coefficient,* or *diffusivity.*

Diffusion Fluxes

A detailed discussion of diffusion fluxes has been given by Bird, et al. (1960) and Cussler (1997). Various types originate because different reference frames are employed. The most obvious reference plane is fixed on the equipment in which diffusion is occurring. This plane is designated by RR' in Fig. 11-1. Suppose, in a binary mixture of A and B, that A is diffusing to the left and B to the right. If the diffusion rates of these species are not identical, there will be a net depletion on one side and an accumulation of molecules on the other side of RR'. To maintain the requirements of an isobaric, isothermal system, bulk motion of the mixture occurs. Net movement of A (as measured in the fixed reference frame RR') then results from both diffusion and bulk flow.

Although many reference planes can be delineated, a plane of *no net mole flow* is normally used to define a diffusion coefficient in binary mixtures. If J_A^M represents a mole flux in a mixture of A and B, J_A^M is then the net mole flow of A across the boundaries of a hypothetical (moving) plane such that the total moles of A and B are invariant on both sides of the plane. J_A^M can be related to fluxes across RR' by

$$J_A^M = N_A - x_A (N_A + N_B) \tag{11-2.1}$$

where N_A and N_B are the fluxes of A and B across RR' (relative to the fixed plane) and x_A is the mole fraction of A at RR'. Note that J_A^M, N_A, and N_B are vectorial quantities and a sign convention must be assigned to denote flow directions. Equation (11-2.1) shows that the net flow of A across RR' is due to a diffusion contribution J_A^M and a bulk flow contribution $x_A (N_A + N_B)$. For equimolar counterdiffusion, $N_A + N_B = 0$ and $J_A^M = N_A$.

One other flux is extensively used, i.e., one relative to the plane of *no net volume flow*. This plane is less readily visualized. By definition,

$$J_A^M + J_B^M = 0 \tag{11-2.2}$$

and if J_A^V and J_B^V are vectorial molar fluxes of A and B relative to the plane of no net volume flow, then, by definition,

$$J_A^V \overline{V}_A + J_B^V \overline{V}_B = 0 \tag{11-2.3}$$

where $\overline{V}_A$ and $\overline{V}_B$ are the partial molar volumes of A and B in the mixture. It can be shown that

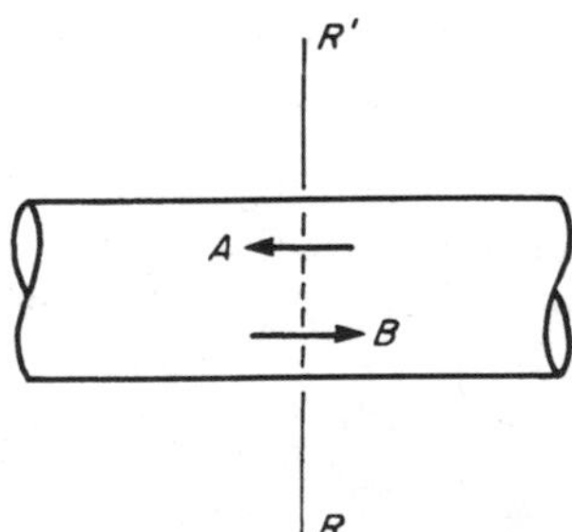

FIGURE 11-1 Diffusion across plane RR'.

$$J_A^V = \frac{\overline{V}_B}{V} J_A^M \quad \text{and} \quad J_B^V = \frac{\overline{V}_A}{V} J_B^M \tag{11-2.4}$$

where V is the volume per mole of mixture. Obviously, if $\overline{V}_A = \overline{V}_B = V$, as in an ideal mixture, then $J_A^V = J_A^M$.

Diffusion Coefficients

Diffusion coefficients for a binary mixture of A and B are commonly defined by

$$J_A^M = -cD_{AB} \frac{dx_A}{dz} \tag{11-2.5}$$

$$J_B^M = -cD_{BA} \frac{dx_B}{dz} \tag{11-2.6}$$

where c is the total molar concentration ($= V^{-1}$) and diffusion is in the z direction. With Eq. (11-2.2), since $(dx_A/dz) + (dx_B/dz) = 0$, we have $D_{AB} = D_{BA}$. The diffusion coefficient then represents the proportionality between the flux of A relative to a plane of no net molar flow and the gradient $c(dx_A/dz)$. From Eqs. (11-2.4) to (11-2.6) and the definition of a partial molar volume it can be shown that, for an isothermal, isobaric binary system,

$$J_A^V = -D_{AB} \frac{dc_A}{dz} \quad \text{and} \quad J_B^V = -D_{AB} \frac{dc_B}{dz} \tag{11-2.7}$$

When fluxes are expressed in relation to a plane of no net volume flow, the potential is the concentration gradient. D_{AB} in Eq. (11-2.7) is identical with that defined in Eq. (11-2.5). When $\overline{V}_A \approx \overline{V}_B \approx V$ as in ideal gases, $J_A^V \approx J_A^M$, $J_B^V \approx J_B^M$.

Mutual, Self-, and Tracer Diffusion Coefficients

The diffusion coefficient D_{AB} introduced above is termed the *mutual diffusion coefficient,* and it refers to the diffusion of one constituent in a binary system. A similar coefficient D_{1m}, would imply the diffusivity of component 1 in a mixture (see Secs. 11-7 and 11-12).

Tracer diffusion coefficients (sometimes referred to as *intradiffusion coefficients*) relate to the diffusion of a labeled component within a *homogeneous* mixture. Like mutual diffusion coefficients, tracer diffusion coefficients can be a function of composition. If D_A^* is the tracer diffusivity of A in a mixture of A and B, then as $x_A \rightarrow 1.0$, $D_A^* \rightarrow D_{AA}$, where D_{AA} is the *self-diffusion coefficient* of A in pure A.

In Fig. 11-2, the various diffusion coefficients noted above are shown for a binary liquid mixture of n-octane and n-dodecane at 60°C (Van Geet and Adamson, 1964). In this case, the mutual diffusion of these two hydrocarbons increases as the mixture becomes richer in n-octane. With A as n-octane and B as n-dodecane, as $x_A \rightarrow 1.0$, $D_{AB} = D_{BA} \rightarrow D_{BA}^{\circ}$, where this notation signifies that this limiting diffusivity represents the diffusion of B in a medium consisting essentially of A, that is, n-dodecane molecules diffusing through almost pure n-octane. Similarly, D_{AB}° is the diffusivity of A in essentially pure B. Except in the case of infinite dilution,

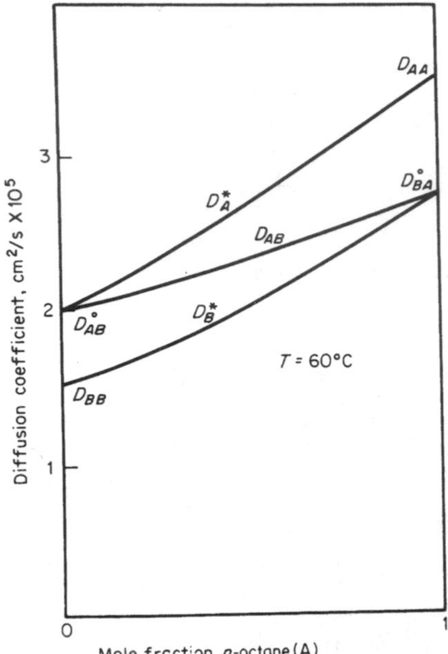

FIGURE 11-2 Mutual, self-, and tracer diffusion coefficients in a binary mixture of *n*-octane and *n*-dodecane. (*Van Geet and Adamson, 1964.*)

tracer diffusion coefficients differ from binary-diffusion coefficients, and there is no way to relate the two coefficients (Cussler, 1997). Similarly, there is no relation between quantities such as D_{BB} and D°_{AB} or D_{AA} and D°_{BA}. In this chapter, only correlation techniques for D_{ij} (or D°_{ij}) are considered; corresponding states methods for D_{ii} have, however, been developed (Lee and Thodos, 1983; Murad, 1981).

Chemical Potential Driving Force

The mutual diffusion coefficient D_{AB} in Eq. (11-2.7) indicates that the flux of a diffusing component is proportional to the concentration gradient. Diffusion is, however, affected by more than just the gradient in concentration, e.g., the intermolecular interactions of the molecules (Dullien, 1974; Turner, 1975) which can give complex composition dependence of the behavior in addition to that from temperature and pressure. Thus, fluxes may not be linear in the concentration gradient; it is even possible that species can diffuse opposite to their concentration gradient.

Modern theories of diffusion (Ghai, et al., 1973) have adopted the premise that if one perturbs the equilibrium composition of a binary system, the subsequent diffusive flow required to attain a new equilibrium state is proportional to the gradient in chemical potential ($d\mu_A/dz$). Since the diffusion coefficient was defined in Eqs. (11-2.5) and (11-2.6) in terms of a mole fraction gradient instead of a chemical potential gradient, it is argued that one should include a thermodynamic correction in any correlation for D_{AB} based on ideal solution considerations. This correction is

$$\alpha = \left[\frac{(\partial \ln a_A)}{(\partial \ln x_A)}\right]_{T,P} \tag{11-2.8}$$

where the activity $a_A = x_A \gamma_A$ and γ_A, the activity coefficient is described in Chap. 8. By virtue of the Gibbs-Duhem equation, α is the same regardless of whether activities and mole fractions of either A or B are used in Eq. (11-2.8). For gases, α is almost always close to unity (except at high pressures), and this correction is seldom used. For liquid mixtures, however, it is widely adopted, as will be illustrated in Sec. 11-10.

11-3 DIFFUSION COEFFICIENTS FOR BINARY GAS SYSTEMS AT LOW PRESSURES: PREDICTION FROM THEORY

The theory describing diffusion in binary gas mixtures at low to moderate pressures has been well developed. As noted earlier in Chaps. 9 (Viscosity) and 10 (Thermal Conductivity), the theory results from solving the Boltzmann equation, and the results are usually credited to both Chapman and Enskog, who independently derived the working equation

$$D_{AB} = \frac{3}{16} \frac{(4\pi kT/M_{AB})^{1/2}}{n\pi\sigma_{AB}^2\Omega_D} f_D \tag{11-3.1}$$

where M_A, M_B = molecular weights of A and B
$M_{AB} = 2[(1/M_A) + (1/M_B)]^{-1}$
n = number density of molecules in the mixture
k = Boltzmann's constant
T = absolute temperature

Ω_D the collision integral for diffusion, is a function of temperature; it depends upon the choice of the intermolecular force law between colliding molecules. σ_{AB} is the characteristic length of the intermolecular force law. Finally, f_D is a correction term, which is of the order of unity. If M_A is of the same order as M_B; f_D lies between 1.0 and 1.02 regardless of composition or intermolecular forces. Only if the molecular masses are very unequal and the light component is present in trace amounts is the value of f_D significantly different from unity, and even in such cases, f_D is usually between 1.0 and 1.1 (Marrero and Mason, 1972).

If f_D is chosen as unity and n is expressed by the ideal-gas law, Eq. (11-3.1) may be written as

$$D_{AB} = \frac{0.00266T^{3/2}}{PM_{AB}^{1/2}\sigma_{AB}^2\Omega_D} \tag{11-3.2}$$

where D_{AB} = diffusion coefficient, cm^2/s
T = temperature, K
P = pressure, bar
σ_{AB} = characteristic length, Å
Ω_D = diffusion collision integral, dimensionless

and M_{AB} is defined under Eq. (11-3.1). The key to the use of Eq. (11-3.2) is the selection of an intermolecular force law and the evaluation of σ_{AB} and Ω_D.

Lennard-Jones 12-6 Potential

As noted earlier [Eq. (9-4.2)], a popular correlation relating the intermolecular energy ψ, between two molecules to the distance of separation r, is given by

$$\psi = 4\varepsilon \left[\left(\frac{\sigma}{r}\right)^{12} - \left(\frac{\sigma}{r}\right)^{6} \right] \tag{11-3.3}$$

with ε and σ as the characteristic Lennard-Jones energy and length, respectively. Application of the Chapman-Enskog theory to the viscosity of pure gases has led to the determination of many values of ε and σ; some of them are given in Appendix B.

To use Eq. (11-3.2), some rule must be chosen to obtain the interaction value σ_{AB} from σ_A and σ_B. Also, it can be shown that Ω_D is a function only of kT/ε_{AB}, where again some rule must be selected to relate ε_{AB} to ε_A and ε_B. The simple rules shown below are usually employed:

$$\varepsilon_{AB} = (\varepsilon_A \varepsilon_B)^{1/2} \tag{11-3.4}$$

$$\sigma_{AB} = \frac{\sigma_A + \sigma_B}{2} \tag{11-3.5}$$

Ω_D is tabulated as a function of kT/ε for the 12-6 Lennard-Jones potential (Hirschfelder, et al., 1954), and various analytical approximations also are available (Hattikudur and Thodos, 1970; Johnson and Colver, 1969; Kestin, et al., 1977; Neufeld, et al., 1972). The accurate relation of Neufield, et al. (1972) is

$$\Omega_D = \frac{A}{(T^*)^B} + \frac{C}{\exp(DT^*)} + \frac{E}{\exp(FT^*)} + \frac{G}{\exp(HT^*)} \tag{11-3.6}$$

where $T^* = kT/\varepsilon_{AB}$ $\quad A = 1.06036$ $\quad B = 0.15610$

$C = 0.19300$ $\quad D = 0.47635$ $\quad E = 1.03587$

$F = 1.52996$ $\quad G = 1.76474$ $\quad H = 3.89411$

Example 11-1 Estimate the diffusion coefficient for the system N_2-CO_2 at 590 K and 1 bar. The experimental value reported by Ellis and Holsen (1969) is 0.583 cm^2/s.

solution To use Eq. (11-3.2), values of σ (CO_2), σ (N_2), ε (CO_2), and ε (N_2) must be obtained. Using the values in Appendix B with Eqs. (11-3.4) and (11-3.5) gives σ (CO_2) = 3.941 Å, σ (N_2) = 3.798 Å; σ (CO_2-N_2) = (3.941 + 3.798)/2 = 3.8695 Å; ε (CO_2)/k = 195.2 K, ε (N_2)/k = 71.4 K; ε (CO_2-N_2)/k = $[(195.2)(71.4)]^{1/2}$ = 118 K. Then $T^* = kT/\varepsilon$ (CO_2-N_2) = 590/118 = 5.0. With Eq. (11-3.6), $\Omega_D = 0.842$. Since M (CO_2) = 44.0 and M (N_2) = 28.0, $M_{AB} = (2)[(1/44.0) + (1/28.0)]^{-1} = 34.22$. With Eq. (11-3.2),

$$D(CO_2\text{-}N_2) = \frac{(0.00266)(590)^{3/2}}{(1)(34.22)^{1/2}(3.8695)^2(0.842)} = 0.52 \text{ cm}^2/\text{s}$$

The error is 11%. Ellis and Holsen recommend values of ε (CO_2-N_2) = 134 K and σ

(CO_2-N_2) = 3.660 Å. With these parameters, they predicted D to be 0.56 cm²/s, a value closer to that found experimentally.

Equation (11-3.2) is derived for dilute gases consisting of nonpolar, spherical, monatomic molecules; and the potential function (11-3.3) is essentially empirical, as are the combining rules [Eqs. (11-3.4) and (11-3.5)]. Yet Eq. (11-3.2) gives good results over a wide range of temperatures and provides useful approximate values of D_{AB} (Gotoh, et al., 1973; Gotch, et al., 1974). The general nature of the errors to be expected from this estimation procedure is indicated by the comparison of calculated and experimental values discussed below in Table 11-2.

The calculated value of D_{AB} is relatively insensitive to the value of ε_{AB} employed and even to the form of the assumed potential function, especially if values of ε and σ are obtained from viscosity measurements.

No effect of composition is predicted. A more detailed treatment does indicate that there may be a small effect for cases in which M_A and M_B differ significantly. In a specific study of this effect (Yabsley, et al., 1973), the low-pressure binary diffusion coefficient for the system He-$CClF_3$ did vary from about 0.416 to 0.430 cm²/s over the extremes of composition. In another study (Mrazek, et al., 1968), no effect of concentration was noted for the methyl alcohol-air system, but a small change was observed with chloroform-air.

Low-pressure Diffusion Coefficients from Viscosity Data

Since the equations for low-pressure gas viscosity [Eq. (9-3.9)] and diffusion [Eq. (11-3.2)] have a common basis in the Chapman-Enskog theory, they can be combined to relate the two gas properties. Experimental data on viscosity as a function of composition at constant temperature are required as a basis for calculating the *binary* diffusion coefficient D_{AB} (DiPippo, et al., 1967; Gupta and Saxena, 1968; Hirschfelder, et al., 1954; Kestin, et al., 1977; Kestin and Wakeham, 1983). Weissman and Mason (1964, 1962) compared the method with a large collection of experimental viscosity and diffusion data and find excellent agreement.

Polar Gases

If one or both components of a gas mixture are polar, a modified Lennard-Jones relation such as the Stockmayer potential is often used. A different collision integral relation [rather than Eq. (11-3.6)] is then necessary and Lennard-Jones σ and ε values are not sufficient.

Brokaw (1969) has suggested an alternative method for estimating diffusion coefficients for binary mixtures containing polar components. Equation (11-3.1) is still used, but the collision integral Ω_D, is now given as

$$\Omega_D = \Omega_D \text{ [Eq. (11-3.6)]} + \frac{0.19\delta_{AB}^2}{T^*} \tag{11-3.7}$$

where $T^* = \dfrac{kT}{\varepsilon_{AB}}$

$$\delta = \frac{1.94 \times 10^3\, \mu_p^2}{V_b T_b} \tag{11-3.8}$$

μ_p = dipole moment, debyes
V_b = liquid molar volume at the normal boiling point, cm^3/mol
T_b = normal boiling point (1 atm), K

$$\frac{\varepsilon}{k} = 1.18(1 + 1.3\delta^2)T_b \tag{11-3.9}$$

$$\sigma = \left(\frac{1.585V_b}{1 + 1.3\delta^2}\right)^{1/3} \tag{11-3.10}$$

$$\delta_{AB} = (\delta_A\delta_B)^{1/2} \tag{11-3.11}$$

$$\frac{\varepsilon_{AB}}{k} = \left(\frac{\varepsilon_A}{k}\frac{\varepsilon_B}{k}\right)^{1/2} \tag{11-3.12}$$

$$\sigma_{AB} = (\sigma_A\sigma_B)^{1/2} \tag{11-3.13}$$

Note that the polarity effect is related exclusively to the dipole moment; this may not always be a satisfactory assumption (Byrne, et al., 1967).

Example 11-2 Estimate the diffusion coefficient for a mixture of methyl chloride (MC) and sulfur dioxide (SD) at 1 bar and 323 K. The data required to use Brokaw's relation from Appendix A and Daubert, et al. (1997) are shown below:

	Methyl chloride (MC)	Sulfur dioxide (SD)
Dipole moment, debyes	1.9	1.6
Liquid molar volume at T_b, cm^3/mol	50.1	44.03
Normal boiling temperature, K	248.95	263.13

solution With Eqs. (11-3.8) and (11-3.11),

$$\delta(\text{MC}) = \frac{(1.94 \times 10^3)(1.9)^2}{(50.1)(248.95)} = 0.56$$

$$\delta(\text{SD}) = \frac{(1.94 \times 10^3)(1.6)^2}{(44.03)(263.1)} = 0.43$$

$$\delta(\text{MC-SD}) = [(0.55)(0.43)]^{1/2} = 0.49$$

Also, with Eqs. (11-3.9) and (11-3.12),

$$\frac{\varepsilon(\text{MC})}{k} = 1.18[1 + 1.3(0.56)^2](248.95) = 414 \text{ K}$$

$$\frac{\varepsilon(\text{SD})}{k} = 1.18[1 + 1.3(0.43)^2](263.1) = 385 \text{ K}$$

$$\frac{\varepsilon(\text{MC} - \text{SD})}{k} = [(414)(385)]^{1/2} = 399 \text{ K}$$

Then, with Eqs. (11-3.10) and (11-3.13),

$$\sigma(\text{MC}) = \left[\frac{(1.585)(50.1)}{1 + (1.3)(0.56)^2}\right]^{1/3} = 3.84\ \text{Å}$$

$$\sigma(\text{SD}) = \left[\frac{(1.585)(44.03)}{1 + (1.3)(0.43)^2}\right]^{1/3} = 3.83\ \text{Å}$$

$$\sigma(\text{MC-SD}) = [(3.84)(3.83)]^{1/2} = 3.84\ \text{Å}$$

To determine Ω_D, $T^* = kT/\varepsilon$ (MC-SD) = 323/399 = 0.810. With Eq. (11-3.6), Ω_D = 1.60. Then with Eq. (11-3.7),

$$\Omega_D = 1.6 + \frac{(0.19)(0.49)^2}{(0.810)} = 1.66$$

With Eq. (11-3.2) and M (MC) = 50.49, M (SD) = 64.06, and M_{AB} = (2)[1/50.49) + (1/64.06)]$^{-1}$ = 56.47

$$D_{\text{MC-SD}} = \frac{(0.00266)(323)^{3/2}}{(1)(56.47)^{1/2}(3.84)^2(1.66)} = 0.084\ \text{cm}^2/\text{s}$$

The experimental value is 0.078 cm^2/s (Brokaw, 1969) and the error is 8%.

Discussion

A comprehensive review of the theory and experimental data for gas diffusion coefficients is available (Marrero and Mason, 1972). There have been many studies covering wide temperature ranges, and the applicability of Eq. (11-3.1) is well verified. Most investigators select the Lennard-Jones potential for its convenience and simplicity. The difficult task is to locate appropriate values of σ and ε. Some values are shown in Appendix B. Brokaw suggests other relations, e.g., Eq. (11-3.9) and (11-3.10). Even after the pure component values of σ and ε have been selected, a combination rule is necessary to obtain σ_{AB} and ε_{AB}. Most studies have employed Eqs. (11-3.4) and (11-3.5) because they are simple and theory suggests no particularly better alternatives. Ravindran, et al. (1979) have used Eq. (11-3.2) to correlate diffusivities of low-volatile organics in light gases.

It is important to employ values of σ and ε obtained from the same source. Published values of these parameters differ considerably, but σ and ε from a single source often lead to the same result as the use of a quite different pair from another source.

The estimation equations described in this section were used to calculate diffusion coefficients for a number of different gases, and the results are shown in Table 11-2. The accuracy of the theoretical relations is discussed in Sec. 11-4 after some empirical correlations for the diffusion coefficient have been described.

11-4 DIFFUSION COEFFICIENTS FOR BINARY GAS SYSTEMS AT LOW PRESSURES: EMPIRICAL CORRELATIONS

Several proposed methods for estimating D_{AB} in low-pressure binary gas systems retain the general form of Eq. (11-3.2), with empirical constants based on experimental data. These include the equations proposed by Arnold (1930), Gilliland

(1934), Wilke and Lee (1955), Slattery and Bird (1958), Bailey (1975), Chen and Othmer (1962), Othmer and Chen (1962), and Fuller, et al. (1965, 1966, 1969). Values of D_{AB} estimated by these equations generally agree with experimental values to within 5 to 10%, although discrepancies of more than 20% are possible. We illustrate two methods which have been shown to be quite general and reliable.

Wilke and Lee (1955)

Equation (11-3.2) is rewritten as

$$D_{AB} = \frac{[3.03 - (0.98/M_{AB}^{1/2})](10^{-3})T^{3/2}}{PM_{AB}^{1/2}\sigma_{AB}^{2}\Omega_D} \tag{11-4.1}$$

where D_{AB} = binary diffusion coefficient, cm^2/s
T = temperature, K
M_A, M_B = molecular weights of A and B, g/mol
M_{AB} = $2[(1/M_A) + (1/M_B)]^{-1}$
P = pressure, bar

The scale parameter σ_{AB} is given by Eq. (11-3.5) where, for each component,

$$\sigma = 1.18V_b^{1/3} \tag{11-4.2}$$

and V_b is the liquid molar volume at the normal boiling temperature, cm^3/mol, found from experimental data or estimated by the methods in Chap. 4. Ω_D is determined from Eq. (11-3.6) with $(\varepsilon/k)_{AB}$ from Eq. (11-3.4) and, for each component,

$$\frac{\varepsilon}{k} = 1.15T_b \tag{11-4.3}$$

with T_b as the normal boiling point (at 1 atm) in kelvins. Note, for systems in which one component is air, σ (air) = 3.62 Å and ε/k (air) = 97.0 K. Eqs. (11-4.2) and (11-4.3) should not be used for hydrogen or helium. We illustrate this method in Example 11-3 below.

Fuller, et al. (1965, 1966, 1969)

These authors modified Eq. (11-3.2) to

$$D_{AB} = \frac{0.00143T^{1.75}}{PM_{AB}^{1/2}[(\Sigma_v)_A^{1/3} + (\Sigma_v)_B^{1/3}]^2} \tag{11-4.4}$$

where the terms have been defined under Eq. (11-4.1) and Σ_v is found for each component by summing atomic diffusion volumes in Table 11-1 (Fuller et al. 1969). These atomic parameters were determined by a regression analysis of many experimental data, and the authors report an average absolute error of about 4% when using Eq. (11-4.4). The technique is illustrated in Example 11-3.

Example 11-3 Estimate the diffusion coefficient of allyl chloride (AC) in air at 298 K and 1 bar. The experimental value reported by Lugg (1968) is 0.098 cm^2/s.

TABLE 11-1 Atomic Diffusion Volumes

Atomic and Structural Diffusion Volume Increments			
C	15.9	F	14.7
H	2.31	Cl	21.0
O	6.11	Br	21.9
N	4.54	I	29.8
Aromatic Ring	−18.3	S	22.9
Heterocyclic ring	−18.3		
Diffusion Volumes of Simple Molecules			
He	2.67	CO	18.0
Ne	5.98	CO_2	26.9
Ar	16.2	N_2O	35.9
Kr	24.5	NH_3	20.7
Xe	32.7	H_2O	13.1
H_2	6.12	SF_6	71.3
D_2	6.84	Cl_2	38.4
N_2	18.5	Br_2	69.0
O_2	16.3	SO_2	41.8
Air	19.7		

solution WILKE AND LEE METHOD. As suggested in the text, for air $\sigma = 3.62$ Å and $\varepsilon/k = 97.0$ K. For allyl chloride, from Daubert, et al. (1997), $T_b = 318.3$ K and $V_b = 84.7$ cm^3/mol. Thus, using Eqs. (11-4.2) and (11-4.3),

$$\sigma(\text{AC}) = (1.18)(84.7)^{1/3} = 5.18 \text{ Å}$$

$$\varepsilon(\text{AC})/k = (1.15)(318.3) = 366 \text{ K}$$

Then, with Eqs. (11-3.4) and (11-3.5)

$$\varepsilon(\text{AC-air})/k = [(366)(97.0)]^{1/2} = 188 \text{ K}$$

$$\sigma(\text{AC-air}) = (5.18 + 3.62)/2 = 4.40 \text{ Å}$$

$$T^* = \frac{T}{\varepsilon(\text{AC-air})/k} = \frac{298}{188} = 1.59$$

and, with Eq. (11-3.6), $\Omega_D = 1.17$. With M (AC) = 76.5 and M (air) = 29.0, $M_{AB} = (2)[(1/76.5) + (1/29.0)]^{-1} = 42.0$. Finally, with Eq. (11-4.1) when $P = 1$ bar,

$$D = \frac{\{3.03 - [0.98/(42.0)^{1/2}]\}(10^{-3})(298)^{3/2}}{(1)(42.0)^{1/2}(4.40)^2(1.17)} = 0.10 \text{ cm}^2/\text{s}$$

$$\text{Error} = \frac{0.10 - 0.098}{0.098} \times 100 = 2\%$$

FULLER ET AL. METHOD. Equation (11-4.4) is used. $P = 1$ bar; M_{AB} was shown above to be equal to 42.0; and $T = 298$ K. For air $(\Sigma_v) = 19.7$, and for allyl chloride, C_3H_5Cl, with Table 11-1, $(\Sigma_v) = (3)(15.9) + (5)(2.31) + 21 = 80.25$. Thus,

$$D = \frac{(0.00143)(298)^{1.75}}{(1)(42.0)^{1/2}[(19.7)^{1/3} + (80.25)^{1/3}]^2} = 0.096 \text{ cm}^2/\text{s}$$

$$\text{Error} = \frac{0.096 - 0.098}{0.098} \times 100 = -2\%$$

Discussion

In Table 11-2 we show experimental diffusion coefficients for a number of binary systems and note the errors found when estimating D_{AB} for (a) the basic theoretical equation (11-3.2), (b) Brokaw's method [Eqs. (11-3.2) and (11-3.7)], (c) Wilke and Lee's method [Eq. (11-4.1)], and (d) Fuller, et al.'s method [Eq. (11-4.4)]. For (a), no calculations were made if σ and ε/k were not available in Appendix B. For (b), calculations were done for systems in which at least one of the species had a nonzero dipole moment. For hydrogen and helium, σ and ε/k were used from Appendix B. For all other compounds, Eqs. (11-3.9) and (11-3.10) were used. For systems in which at least one of the components was polar, Brokaw's method usually, but not always, gave a more accurate prediction than did Eq. (11-3.2). For the 26 cases in Table 11-2 for which predictions are given for both methods, the average absolute percent deviation for Brokaw's method was about 1% less than the predictions of Eq. (11-3.2).

For all methods, there were always a few systems for which large errors were found. These differences may be due to inadequacies of the method or to inaccurate data. In general, however, the Fuller, et al. procedure [Eq. (11-4.4) and Table 11-1] yielded the smallest average error, and it is the method recommended for use. Other evaluations (Elliott and Watts, 1972; Gotoh, et al., 1973; Gotoh, et al., 1974; Lugg, 1968; Pathak, et al., 1981) have shown both the Fuller, et al. and the Wilke-Lee forms to be reliable.

Reviews of experimental data of binary diffusion coefficients are available (Gordon, 1977; Marrero and Mason, 1972) and Massman (1998) presents a review of diffusivities of components commonly found in air.

11-5 THE EFFECT OF PRESSURE ON THE BINARY DIFFUSION COEFFICIENTS OF GASES

At low to moderate pressures, binary diffusion coefficients vary inversely with pressure or density as suggested by Eqs. (11-3.1) and (11-3.2). At high pressures, the product DP or $D\rho$ is no longer constant but decreases with an increase in either P or ρ. Note that it is possible to have a different behavior in the products DP and $D\rho$ as the pressure is raised, since ρ is proportional to pressure only at low pressures, and gas nonidealities with their concomitant effect on the system density may become important. Also, as indicated earlier, at low pressures, the binary diffusion coefficient is essentially independent of composition. At high pressures, where the gas phase may deviate significantly from an ideal gas, small, but finite effects of composition have been noted, e.g., Takahaski and Hongo (1982).

With the paucity of reliable data, it is not surprising that few estimation methods have been proposed. Takahashi (1974) has suggested a very simple corresponding-

TABLE 11-2 Comparison of Methods for Estimating Gas Diffusion Coefficients at Low Pressures

System	T, K	$D_{AB}P$ (exp.), (cm^2/s) bar	Ref.*	Errors as percent of experimental values: Theoretical	Brokaw	Wilke-Lee	Fuller et al.
Air-carbon dioxide	276	0.144	9	−6		2	−3
	317	0.179		−2		6	−1
Air ethanol	313	0.147	13	−10	−16	−11	−8
Air-helium	276	0.632	9	0		1	−5
	346	0.914		0		2	−2
Air-n-hexane	294	0.081	5	−6		−4	−7
	328	0.094		−1		1	−2
Air-2-methylfuran	334	0.107	1		2	9	8
Air-naphthalene	303	0.087	4			−18	−20
Air-water	313	0.292	5	−18	−15	−16	−5
Ammonia-diethyl ether	288	0.101	20	−24	−12	−15	2
	337	0.139		−24	−12	−15	−2
Argon-ammonia	255	0.152	19	3	5	4	13
	333	0.256		3	5	2	7
Argon-benzene	323	0.085	12	9		14	15
	373	0.112		9		13	13
Argon-helium	276	0.655	9	−2		−5	−1
	418	1.417	6	−9		−12	−6
Argon-hexafluorobenzene	323	0.082	12			−5	−18
	373	0.095				8	−9
Argon-hydrogen	295	0.84	22	−9		−16	−4
	628	3.25		−15		−22	−7
	1068	8.21		−19		−25	−7
Argon-krypton	273	0.121	18	−1		3	0
Argon-methane	298	0.205	6	5		4	5
Argon-sulfur dioxide	263	0.078	13	18		24	25
Argon-xenon	195	0.052	6	−2		5	9
	378	0.180		−3		3	0
Carbon dioxide-helium	298	0.620	17	−3		0	−5
	498	1.433		−1		2	1
Carbon dioxide-nitrogen	298	0.169	21	−7		−3	−3
Cargon dioxide-nitrous oxide	313	0.130	3	−6	−4	3	−3
Carbon dioxide-sulfur dioxide	473	0.198	13	7	14	18	15
Carbon dioxide-tetrafluoromethane	298	0.087	11	0		11	−12
	673	0.385		−3		9	−17
Carbon dioxide-water	307	0.201	7	−21	−12	−13	11
Carbon monoxide-nitrogen	373	0.322	2	−6	−13	−8	−4
Ethylene-water	328	0.236	15	−25	−16	−20	−3
Helium-benzene	423	0.618	17	9		8	−4
Helium-bromobenzene	427	0.55	8		28	8	−2
Helium-2-chlorobutane	429	0.568	8		33	12	−2
Helium-n-butanol	423	0.595	17		28	10	−2
Helium-1-iodobutane	428	0.524	8			11	1
Helium-methanol	432	1.046	17	11	8	−3	2
Helium-nitrogen	298	0.696	17	1		−3	2

TABLE 11-2 Comparison of Methods for Estimating Gas Diffusion Coefficients at Low Pressures (*Continued*)

System	T, K	$D_{AB}P$ (exp.), (cm²/s) bar	Ref.*	Errors as percent of experimental values: Theoretical	Brokaw	Wilke-Lee	Fuller et al.
Helium-water	352	1.136	15	1	8	−11	0
Hydrogen-acetone	296	0.430	15	−1	11	−9	2
Hydrogen-ammonia	263	0.58	13	3	8	−7	7
	358	1.11	13	−4	0	−15	−4
	473	1.89		−6	−3	−19	−9
Hydrogen-cyclohexane	289	0.323	10	−4	14	−5	−2
Hydrogen-naphthalene	303	0.305	4			−7	−1
Hydrogen-nitrobenzene	493	0.831	14	16		−10	4
Hydrogen-nitrogen	294	0.773	16	−5		−14	−1
	573	2.449		−8		−15	1
Hydrogen-pyridine	318	0.443	10		9	−8	5
Hydrogen-water	307	0.927	7	−12	−7	−21	4
Methane-water	352	0.361	15	−19	−13	−18	−2
Nitrogen-ammonia	298	0.233	13	−5	−4	−7	−2
	358	0.332		−6	−5	−9	−5
Nitrogen-aniline	473	0.182	14		4	7	9
Nitrogen-sulfur dioxide	263	0.105	13	−3	−4	0	0
Nitrogen-water	308	0.259	15	−11	−8	−12	7
	352	0.364		−18	−15	−20	−4
Nitrogen-sulfur hexafluoride	378	0.146	11	6		12	3
Oxygen-benzene	311	0.102	10	−9		−5	−3
Oxygen-carbon tetrachloride	296	0.076	13	−6		6	6
Oxygen-cyclohexane	289	0.076	10	−7		2	−1
Oxygen-water	352	0.357	15	−17	−12	−16	0
Average absolute error				7.9		9.6	5.4

*References: 1, Alvarez, et al. (1983); 2, Amdur and Shuler (1963); 3, Amdur, et al. (1952); 4, Caldwell (1984); 5, Carmichael, et al. (1955); 6, Carswell and Stryland (1963); 7, Crider (1956); 8, Fuller, et al. (1969); 9, Holson and Strunk (1964); 10, Hudson, et al. (1960); 11, Kestin, et al. (1977); 12, Maczek and Edwards (1979); 13, Mason and Monschick (1962); 14, Pathak, et al. (1981); 15, Schwartz and Brow (1951); 16, Scott and Cox (1960); 17, Seager, et al. (1963); 18, Srivastava and Srivastava (1959); 19, Srivastava and Srivastava (1962); 20, Srivastava and Srivastava (1963); 21, Walker and Westenberg (1958); 22, Westenberg and Frazier (1962).

states method that is satisfactory for the limited database available. His correlation is

$$\frac{D_{AB}P}{(D_{AB}P)^{+}} = f(T_r, P_r) \tag{11-5.1}$$

where D_{AB} = diffusion coefficient, cm²/s
P = pressure, bar

The superscript $^{+}$ indicates that low-pressure values are to be used. The function $f(T_r, P_r)$ is shown in Fig. 11-3, and to obtain pseudocritical properties from which to calculate the reduced temperatures and pressures, Eqs. (11-5.2) to (11-5.5) are used.

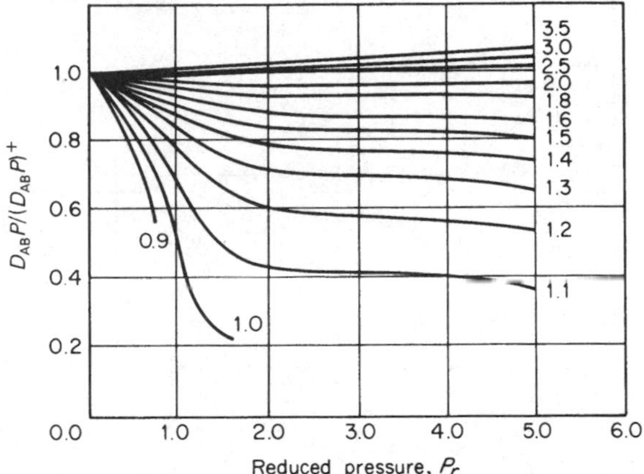

FIGURE 11-3 Takahashi correlation for the effect of pressure and temperature on the binary diffusion coefficient. Lines are at constant reduced temperature.

$$T_r = \frac{T}{T_c} \tag{11-5.2}$$

$$T_c = y_A T_{cA} + y_B T_{cB} \tag{11-5.3}$$

$$P_r = \frac{P}{P_c} \tag{11-5.4}$$

$$P_c = y_A P_{cA} + y_B P_{cB} \tag{11-5.5}$$

As an illustration of this technique, in Fig. 11-4, we have plotted the data of Takahashi and Hongo (1982) for the system carbon dioxide-ethylene. Two cases are considered: one with a very low concentration of ethylene and the other with a very low concentration of carbon dioxide. Up to about 80 bar, the two limiting diffusion coefficients are essentially identical. Above that pressure, D_{AB} for the trace CO_2 system is significantly higher. Plotted as solid curves on this graph are the predicted values of D_{AB} from Fig. 11-3 and Eq. (11-5.1) using the $(D_{AB}P)^+$ product at low pressure to be 0.149 (cm^2/s)bar as found by Takahashi and Hongo. Also, the dashed curve has been drawn to indicate the estimated value of D_{AB} if one had assumed that $D_{AB}P$ was a constant. Clearly this assumption is in error above a pressure of about 10 to 15 bar.

Riazi and Whitson (1993) propose Eq. (11-5.6)

$$\frac{\rho D_{AB}}{(\rho D_{AB})^+} = 1.07 \left(\frac{\mu}{\mu^0}\right)^{b+cP_r} \tag{11-5.6}$$

where $b = -0.27 - 0.38\ \omega$

$c = -0.05 + 0.1\ \omega$

μ^0 is the viscosity at low pressure

ω is the acentric factor

$P_r = P/P_c$

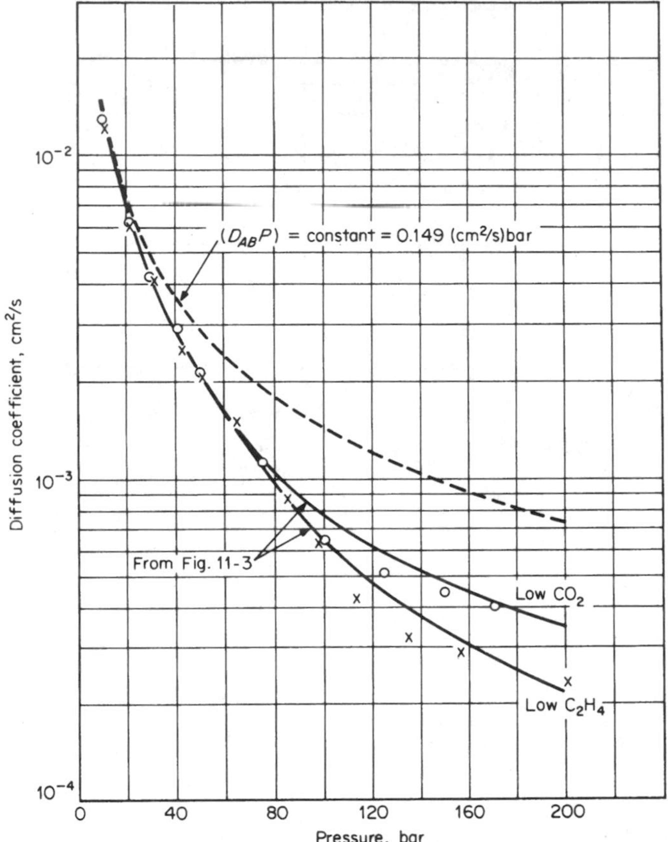

FIGURE 11-4 Effect of pressure and composition on the binary diffusion coefficient in the CO_2-C_2H_4 system at 323.2 K.

$$P_c = y_A P_{cA} + y_B P_{cB} \tag{11-5.7}$$

$$\omega = y_A \omega_A + y_B \omega_B \tag{11-5.8}$$

As in Eq. (11-5.1) the superscript $^+$ represents low-pressure values. Riazi and Whitson (1997) claim that Eq. (11-5.6) is capable of representing high-pressure liquid behavior as well as high-pressure gas behavior. Equation (11-5.6) gives a slightly worse description of the systems shown in Fig. 11-4 than does Eq. (11-5.1). When Eq. (11-5.6) is compared to the data in Fig. 11-4, the average absolute deviation is 14% while the maximum deviation is 30%. Neither Eq. (11-5.1) nor Eq. (11-5.6) is entirely satisfactory. The former requires that values be read from Fig. 11-3. The latter requires viscosity information and does not reproduce the correct value in the limit of low pressure.

Many of the more recent data for diffusion coefficients at high pressure involve a trace solute in a supercritical fluid. To illustrate some data for the diffusion coefficient of complex solutes in supercritical fluids, we show Fig. 11-5 (Debenedetti and Reid, 1986). There the diffusion coefficient is given as a function of reduced pressure from the ideal-gas range to reduced pressures up to about 6. The solutes are relatively complex molecules, and the solvent gases are CO_2 and ethylene. No temperature dependence is shown, since the temperatures studied (see legend) were such that all the reduced temperatures were similar and were, in most cases, in the range of 1 to 1.05. Since the concentrations of the solutes were quite low, the pressure and temperature were reduced by P_c and T_c of the pure solvent. Up to about half the critical pressure, $D_{AB}P$ is essentially constant. Above that pressure, the data show the product $D_{AB}P$ decreasing, and at reduced pressures of about 2, it would appear that $D_{AB} \propto P_r^{1/2}$. As supercritical extractions are often carried out in a reduced temperature range of about 1.1 to 1.2 and in a reduced pressure range of 2 to 4, this plot would indicate that $D_{AB} \cong 10^{-4}$ cm^2/s, a value much less than for a low-pressure gas but still significantly higher than for a typical liquid (see Sec. 11-9).

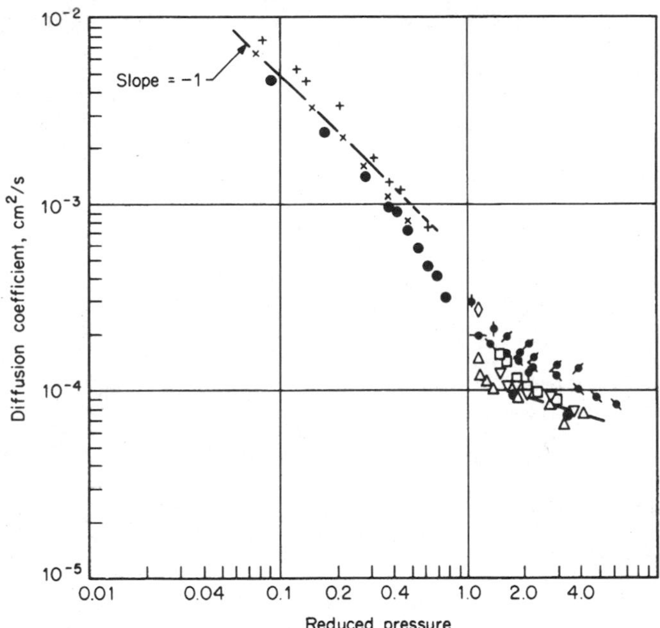

FIGURE 11-5 Diffusion coefficients in supercritical fluids.

Key	System	Ref.*
●, ×, △, +, ▽	CO_2-naphthalene at 20, 30, 35, 40, and 55°C	1, 2, 3, 4
●, ●	Ethylene-naphthalene at 12 and 35°C	3, 4
●	CO_2-benzene at 40°C	5
●	CO_2-propylbenzene at 40°C	6
◇	CO_2-1,2,3-trimethylbenzene at 40°C	6

*References: 1, Morozov and Vinkler (1975); 2, Vinkler and Morozov (1975); 3, Iomtev and Tsekhanskaya (1964); 4, Tsekhanskaya (1971); 5, Schneider (1978); 6, Swaid and Schneider (1979)

Several correlations have been presented for diffusion coefficients of solutes in supercritical fluids. One of the simplest is that of He and Yu (1998) as shown in Eq. (11-5.9)

$$D_{AB} = \alpha \times 10^{-5} \left(\frac{T}{M_A}\right)^{1/2} \exp\left(-\frac{0.3887}{V_{rB} - 0.23}\right) \tag{11-5.9}$$

$$\alpha = 14.882 + 0.005908 \frac{T_{cB}V_{cB}}{M_B} + 2.0821 \times 10^{-6} \left(\frac{T_{cB}V_{cB}}{M_B}\right)^2 \tag{11-5.10}$$

where D_{AB} = diffusion coefficient of solute A in solvent B, cm^2/s
$V_{rB} = V_B/V_{cB}$
V_{cB} = critical volume of the solvent in cm^3/mol
M = molecular weight in g/mol
T_{cB} = critical temperature of the solvent in K

When Eqs. (11-5.9) and (11-5.10) were tested on 1300 data points involving 11 different solvents, the authors found an average error of 8%, which is remarkable for such a simple equation. The cases examined included solvents that were high-temperature liquids as well as supercritical fluids and covered temperature and density ranges of $0.66 < T_r < 1.78$ and $0.22 < \rho_r < 2.62$. Because of its simplicity, Eq. (11-5.9) should likely not be used for conditions outside the range of fit. Furthermore, Eq. (11-5.9) does not include any effect of solvent viscosity or solute density and for solutes or solvents for which these properties are dramatically different than those tested, Eq. (11-5.9) would likely give higher errors. For example, the solutes for the 1300 data points tested were most often organic compounds for which the pure-component liquid density is typically 0.8 to 0.9 g/cm^3. The errors for chloroform and iodine (densities of 1.5 and 4.9 g/cm^3 respectively) in carbon dioxide were −20% and −38% respectively.

There are methods to estimate diffusion coefficients of solutes in supercritical fluids other than Eq. (11-5.9) for which slightly improved accuracy is claimed but which are also more involved. Funazukuri, et al. (1992) propose a method that uses the ratio of the Schmidt number to its value at low pressure. Liu and Ruckenstein (1997) have developed a correlation that uses the Peng-Robinson equation of state to calculate a thermodynamic factor (see Sec. 11-2). References to many of the data on solutes in supercritical fluids are summarized in these latter two references as well as (Catchpole and King, 1994; He and Ye, 1998).

Example 11-4 Estimate the diffusion coefficient of vitamin K_1 in supercritical carbon dioxide at 313 K and 160 bar. The experimental value is reported to be 5.43×10^{-5} cm^2/s (Funazukuri, et al., 1992).

solution For vitamin K_1, M = 450.7 g/mol. From Appendix A, for CO_2, T_c = 304.12 K, P_c = 73.74 bar, V_c = 94.07 cm^3/mol, and ω = 0.225. With these values, the Lee-Kesler equation of state gives V = 54.97 cm^3/mol for pure CO_2 at T = 313 K and P = 160 bar. Thus, V_{rB} = 54.97/94.07 = 0.584. From Eqs. (11-5.9) and (11-5.10)

$$\frac{T_{cB}V_{cB}}{M_B} = \frac{(304.12)(94.07)}{44.01} = 650.$$

$$\alpha = 14.882 + (0.005908)(650) + 2.0821 \times 10^{-6}(650)^2 = 19.60$$

$$D_{AB} = 19.60 \left(\frac{313}{450.7}\right)^{1/2} \exp\left(-\frac{0.3887}{0.584 - 0.23}\right) = 5.45 \times 10^{-5} \text{ cm}^2/\text{s}$$

$$\text{Error} = \frac{5.45 - 5.43}{5.43} \times 100 = 0.4\%$$

11-6 THE EFFECT OF TEMPERATURE ON DIFFUSION IN GASES

At low pressures, where the ideal-gas law approximation is valid, it is seen from Eq. (11-3.2) that

$$D_{AB} \propto \frac{T^{3/2}}{\Omega_D(T)} \tag{11-6.1}$$

or

$$\left(\frac{\partial \ln D_{AB}}{\partial \ln T}\right)_P = \frac{3}{2} - \frac{d \ln \Omega_D}{d \ln T} \tag{11-6.2}$$

Marrero and Mason (1972) indicate that, in most cases, the term $d \ln \Omega_D / d \ln T$ varies from 0 to $-\frac{1}{2}$. Thus D_{AB} varies as $T^{3/2}$ to T^2. This result agrees with the empirical estimation methods referred to in Sec. 11-4, e.g., in the Fuller, et al. method, $D \propto T^{1.75}$. Over wide temperature ranges, however, the exponent on temperature changes. Figure 11-6 shows the approximate variation of this exponent with reduced temperature. The very fact that the temperature exponent increases and then decreases indicates that empirical estimation techniques with a constant exponent will be limited in their range of applicability. The theoretical and the Wilke-Lee methods are therefore preferable if wide temperature ranges are to be covered. Dunlop and Bignell (1997) relate the temperature dependence to the thermal diffusion factor.

11-7 DIFFUSION IN MULTICOMPONENT GAS MIXTURES

A few general concepts of diffusion in multicomponent liquid mixtures presented later (in Sec. 11-12) are applicable for gas mixtures also. One of the problems with

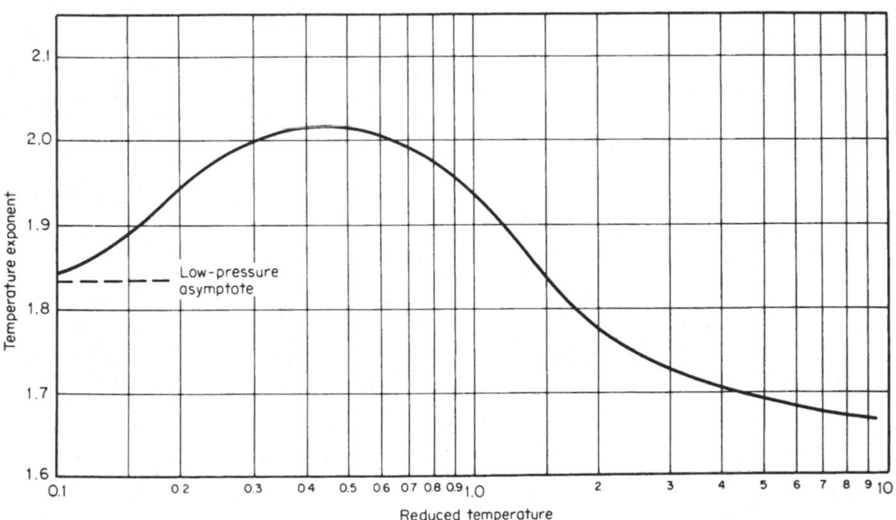

FIGURE 11-6 Exponent of temperature for diffusion in gases. (*Adapted from. Marrero and Mason (1972) with the approximation that* $\varepsilon/k \approx 0.75T_c$.)

diffusion in liquids is that even the binary diffusion coefficients are often very composition-dependent. For multicomponent liquid mixtures, therefore, it is difficult to obtain numerical values of the diffusion coefficients relating fluxes to concentration gradients.

In gases, D_{AB} is normally assumed independent of composition. With this approximation, multicomponent diffusion in gases can be described by the Stefan-Maxwell equation

$$\frac{dx_i}{dz} = \sum_{j=1}^{n} \frac{c_i c_j}{c^2 D_{ij}} \left(\frac{J_j}{c_j} - \frac{J_i}{c_i} \right) \tag{11-7.1}$$

where c_i = concentration of i
c = mixture concentration.
J_i, J_j = flux of i, j
D_{ij} = Binary diffusion coefficient of the ij system
(dx_i/dz) = gradient in mole fraction of i in the z direction

This relation is different from the basic binary diffusion relation (11-2.5), but the employment of common binary diffusion coefficients is particularly desirable. Marrero and Mason (1972) discuss many of the assumptions behind Eq. (11-7.1).

Few attempts have been made by engineers to calculate fluxes in multicomponent systems. However, one important and simple limiting case is often cited. If a dilute component i diffuses into a *homogeneous* mixture, then $J_j \approx 0$. With $c_j/c = x_j$, Eq. (11-7.1) reduces to

$$\frac{dx_i}{dz} = -J_i \sum_{\substack{j=1 \\ j \neq i}}^{n} \frac{x_j}{cD_{ij}} \tag{11-7.2}$$

Defining

$$D_{im} = \frac{-J_i}{dx_i/dz} \tag{11-7.3}$$

gives

$$D_{im} = \left(\sum_{\substack{j=1 \\ j \neq i}}^{n} \frac{x_j}{D_{ij}} \right)^{-1} \tag{11-7.4}$$

This simple relation is sometimes called Blanc's law (Blanc, 1908; Marrero and Mason, 1972). It was shown to apply to several ternary cases in which i was a trace component (Mather and Saxena, 1966). Deviations from Blanc's law are discussed by Sandler and Mason (1968).

The general theory of diffusion in multicomponent gas systems is covered by Cussler (1997) and by Hirschfelder, et al. (1954). The problem of diffusion in three-component gas systems has been generalized by Toor (1957) and verified by Fairbanks and Wilke (1950), Walker, et al. (1960), and Duncan and Toor (1962).

11-8 DIFFUSION IN LIQUIDS: THEORY

Binary liquid diffusion coefficients are defined by Eq. (11-2.5) or (11-2.7). Since molecules in liquids are densely packed and strongly affected by force fields of neighboring molecules, values of D_{AB} for liquids are much smaller than for low-pressure gases. That does not mean that diffusion rates are necessarily low, since concentration gradients can be large.

Liquid state theories for calculating diffusion coefficients are quite idealized, and none is satisfactory in providing relations for calculating D_{AB}. In several cases, however, the form of a theoretical equation has provided the framework for useful prediction methods. A case in point involves the analysis of large spherical molecules diffusing in a dilute solution. Hydrodynamic theory (Bird, et al., 1960; Gainer and Metzner, 1965) then indicates that

$$D_{AB} = \frac{RT}{6\pi\eta_B r_A} \tag{11-8.1}$$

where η_B is the viscosity of the solvent and r_A is the radius of the "spherical" solute. Equation (11-8.1) is the Stokes-Einstein equation which strictly applies to macroscopic systems. However, many authors have used the form as a starting point in developing correlations for molecular diffusion.

Other theories for modeling diffusion in liquids have been based on kinetic theory (Anderson, 1973; Bearman, 1961; Carman, 1973; Carman and Miller, 1959; Darken, 1948; Dullien, 1961; Hartley and Crank, 1949; Kett and Anderson, 1969; Miller and Carman, 1961), absolute-rate theory (Cullinan and Cusick, 1967; Eyring and Ree, 1961; Gainer and Metzner, 1965; Glasstone, et al., 1941; Leffler and Cullinan, 1970; Li and Chang, 1955; Olander, 1963; Ree, et al., 1958), statistical mechanics (Bearman, 1960; Bearman, 1961; Kamal and Canjar, 1962), and other concepts (Albright, et al., 1983; Brunet and Doan, 1970; Horrocks and Mclaughlin, 1962; Kuznetsova and Rashidova, 1980, 1980a; Raina, 1980). Several reviews are available for further consideration (Dullien, 1963; Ghai, et al., 1973; Ghai, et al., 1974; Himmelblau, 1964; Loflin and McLaughlin, 1969).

Diffusion in liquid metals is not treated, although estimation techniques are available (Pasternak and Olander, 1967).

11-9 ESTIMATION OF BINARY LIQUID DIFFUSION COEFFICIENTS AT INFINITE DILUTION

For a binary mixture of solute A in solvent B, the diffusion coefficient D°_{AB} of A diffusing in an infinitely dilute solution of A in B implies that each A molecule is in an environment of essentially pure B. In engineering work, however, D°_{AB} is assumed to be a representative diffusion coefficient even for concentrations of A of 5 to 10 mole %. In this section, several estimation methods for D°_{AB} are introduced; the effect of concentration for mutual diffusion coefficients is covered in Sec. 11-10.

Wilke-Chang Estimation Method (Wilke and Chang, 1955)

An older but still widely used correlation for D°_{AB}, the Wilke-Chang technique is, in essence, an empirical modification of the Stokes-Einstein relation (11-8.1):

$$D^\circ_{AB} = \frac{7.4 \times 10^{-8}(\phi M_B)^{1/2}T}{\eta_B V_A^{0.6}} \tag{11-9.1}$$

where D°_{AB} = mutual diffusion coefficient of *solute* A at very low concentrations in *solvent* B, cm^2/s
M_B = molecular weight of *solvent* B, g/mol
T = temperature, K
η_B = viscosity of *solvent* B, cP
V_A = molar volume of *solute* A at its normal boiling temperature, cm^2/mol
ϕ = association factor of *solvent* B, dimensionless

If experimental data to obtain V_A at T_{Ab} do not exist, estimation methods from Chap. 4 may be used, in particular the Le Bas additive volume table (4-8) is convenient.

Wilke and Chang recommend that ϕ be chosen as 2.6 if the solvent is water, 1.9 if it is methanol, 1.5 if it is ethanol, and 1.0 if it is unassociated. When 251 solute-solvent systems were tested by these authors, an average error of about 10% was noted. Figure 11-7 is a graphical representation of Eq. (11-9.1) with the dashed line representing Eq. (11-8.1); the latter is assumed to represent the maximum value of the ordinate for any value of V_A.

A number of authors have suggested modifications of Eq. (11-9.1) particularly to improve its accuracy for systems where water is the solute and the solvent is an organic liquid (Amourdam and Laddha, 1967; Caldwell and Babb, 1956; Hayduk and Buckley, 1972; Hayduk, et al., 1973; Hayduk and Laudie, 1974; Lees and Sarram, 1971; Lusis, 1971; Lusis and Ratsliff, 1971; Olander, 1961; Scheibel, 1954; Shrier, 1967; Wise and Houghton, 1966; Witherspoon and Bonoli, 1969). However, none of these suggestions have been widely accepted. In Table 11-5, we show a comparison of estimated and experimental values of D°_{AB}. The errors vary so greatly that the concept of an *average* error is meaningless. The method should not be used when water is the *solute*.

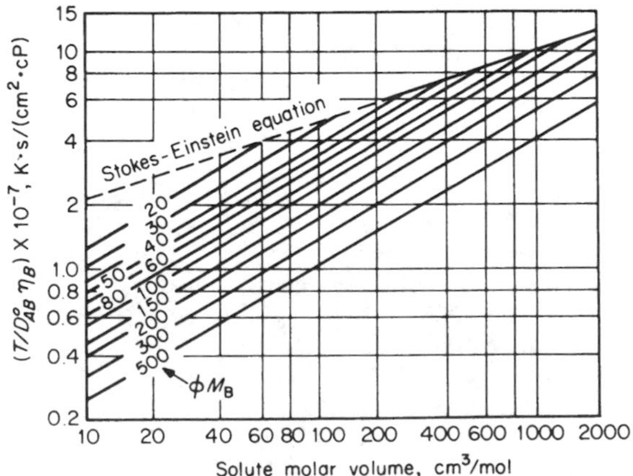

FIGURE 11-7 Graphical representation of Wilke-Chang correlation of diffusion coefficients in dilute solutions. (*Wilke and Chang, 1955.*)

Example 11-5 Use the Wilke-Chang correlation to estimate D°_{AB} for ethylbenzene diffusing into water at 293 K. The viscosity of water at this temperature is essentially 1.0 cP. The experimental value of D°_{AB} is 0.81×10^{-5} cm²/s (Witherspoon and Bonoli, 1969).

solution The normal boiling point of ethylbenzene is 409.36 K (Appendix A). At that temperature, the density is 0.761 g/cm³ (Vargaftik, et al., 1996), so with $M_A = 106.17$, $V_A = 106.17/0.761 = 139.5$ cm³/mol. Then, using Eq. (11-9.1) with $\phi = 2.6$ and $M_B = 18.0$ for water,

$$D^{\circ}_{AB} = 7.4 \times 10^{-8} \frac{[(2.6)(18.0)]^{1/2}(293)}{(1.0)(139.5)^{0.6}} = 0.77 \times 10^{-5} \text{ cm}^2/\text{s}$$

$$\text{Error} = \frac{0.77 - 0.81}{0.81} \times 100 = -5\%$$

Tyn and Calus Method (Tyn and Calus, 1975)

These authors have proposed that D°_{AB} be estimated by the relation

$$D^{\circ}_{AB} = 8.93 \times 10^{-8} \left(\frac{V_A}{V_B^2}\right)^{1/6} \left(\frac{\mathbf{P}_B}{\mathbf{P}_A}\right)^{0.6} \frac{T}{\eta_B} \tag{11-9.2}$$

where V_B = molar volume of the solvent at the normal boiling temperature, cm³/mol, $\mathbf{P}_A$ and $\mathbf{P}_B$ are parachors for the solute and solvent, and the other terms are defined under Eq. (11-9.1).

The parachor is related to the liquid surface tension (see Chap. 12) as

$$\mathbf{P} = V\sigma^{1/4} \tag{11-9.3}$$

where σ is the surface tension in dyn/cm = g/s² = 10^{-3} N/m² and V is the molar volume, cm³/mol, both measured at the same temperature. Thus the units of **P** are cm³ · g$^{1/4}$/(s$^{1/2}$ · mol). Quayle (1953) has tabulated values of **P** for a large number of chemicals; alternatively, **P** may be estimated from additive group contributions as shown in Table 11-3. Over moderate temperature ranges, **P** is essentially a constant.

When using the correlation shown in Eq. (11-9.2), the authors note several restrictions:

1. The method should not be used for diffusion in viscous solvents. Values of η_B above about 20 to 30 cP would classify the solvent as viscous.
2. If the solute is water, a dimer value of V_A and $\mathbf{P}_A$ should be used. In the calculations for Table 11-5, we used $V_A = V_w = 37.4$ cm³/mol and $\mathbf{P}_A = \mathbf{P}_w = 105.2$ cm³ · g$^{1/4}$/(s$^{1/2}$ · mol).
3. If the solute is an organic acid and the solvent is other than water, methanol, or butanol, the acid should be considered a dimer with twice the expected values of V_A and $\mathbf{P}_A$.
4. For nonpolar solutes diffusing into monohydroxy alcohols, the values of V_B and $\mathbf{P}_B$ should be multiplied by a factor equal to $8\eta_B$, where η_B is the solvent viscosity in cP.

TABLE 11-3 Structural Contributions for Calculating the Parachor†

Carbon-hydrogen:		R—[—CO—]—R′ (for the	
C	9.0	—CO— in ketones)	
H	15.5	$R + R' = 2$	51.3
CH_3	55.5	$R + R' = 3$	49.0
CH_2 in $—(CH_2)_n—$		$R + R' = 4$	47.5
$n < 12$	40.0	$R + R' = 5$	46.3
$n > 12$	40.3	$R + R' = 6$	45.3
Alkyl groups		$R + R' = 7$	44.1
1-Methylethyl	133.3	—CHO	66
1-Methylpropyl	171.9	O (if not noted above)	20
1-Methylbutyl	211.7	N (if not noted above)	17.5
2-Methylpropyl	173.3	S	49.1
1-Ethylpropyl	209.5	P	40.5
1,1-Dimethylethyl	170.4	F	26.1
1,1-Dimethylpropyl	207.5	Cl	55.2
1,2-Dimethylpropyl	207.9	Br	68.0
1,1,2-Trimethylpropyl	243.5	I	90.3
C_6H_5	189.6	Ethylenic bonds (=C<)	19.1
Special groups:		Terminal	
—COO— (esters)	63.8	2,3-position	17.7
—COOH (acids)	73.8	3,4-position	16.3
—OH	29.8	Triple bond	40.6
$—NH_2$	42.5	Ring closure:	
—O—	20.0	Three-membered	12
$—NO_2$	74	Four-membered	6.0
$—NO_3$ (nitrate)	93	Five-membered	3.0
$—CO(NH_2)$	91.7	Six-membered	0.8

† As modified from Quayle (1953).

By using Eq. (11-9.2) with the restrictions noted above, values of D°_{AB} were estimated for a number of systems. The results are shown in Table 11-5, along with experimentally reported results. In the majority of cases, quite reasonable estimates of D°_{AB} were found and errors normally were less than 10%.

To use the Tyn-Calus form, however, the parachors of both the solute and the solvent must be known. Although the compilation of Quayle (1953) is of value, it is still incomplete. The structural contributions given in Table 11-3 also are incomplete, and many functional groups are not represented.

A modified form of Eq. (11-9.2) may be developed by combining Eqs. (11-9.2) and (11-9.3) to give

$$D^\circ_{AB} = 8.93 \times 10^{-8} \frac{V_B^{0.267}}{V_A^{0.433}} \frac{T}{\eta_B} \left(\frac{\sigma_B}{\sigma_A}\right)^{0.15} \qquad (11\text{-}9.4)$$

The definitions of the terms are the same as before except, when substituting Eq. (11-9.3), we must define V and σ at T_b. Thus σ_B and σ_A in Eq. (11-9.4) refer to surface tensions at T_b. Note also the very low exponent on this ratio of surface tensions. Since most organic liquids at T_b have similar surface tensions, one might choose to approximate this ratio as equal to unity. (For example, $0.80^{0.15} = 0.97$ and $1.2^{0.15} = 1.03$.) Then,

$$D^\circ_{AB} = 8.93 \times 10^{-8} \frac{V_B^{0.267}}{V_A^{0.433}} \frac{T}{\eta_B} \tag{11-9.5}$$

Alternatively, an approximation to the σ_B/σ_A ratio may be developed by using one of the correlations shown in Chap. 12. For example, if the Brock and Bird corresponding states method [Eqs. (12-3.3) and (12-3.4)] were used, then

$$\sigma = P_c^{2/3} T_c^{1/3} (0.132\alpha_c - 0.279)(1 - T_{br})^{11/9} \tag{11-9.6}$$

with P_c in bars and T_b and T_c in kelvins, $T_{br} = T_b/T_c$, and

$$\alpha_c = 0.9076 \left[1 + \frac{T_{br} \ln (P_c/1.013)}{1 - T_{br}}\right] \tag{11-9.7}$$

Equation (11-9.6) is only approximate, but it may be satisfactory when used to develop the *ratio* (σ_A/σ_B). Also, considering the low power (0.15) to which the ratio is raised, estimates of $(\sigma_B/\sigma_A)^{0.15}$ should be quite reasonable.

When Eq. (11-9.5) was employed to estimate D°_{AB} for the systems shown in Table 11-5, the results, as expected, were very similar to those found from the original Tyn and Calus form Eq. (11-9.2), except when σ_B differed appreciably from σ_A, for example, in the case of water and an organic liquid. In such situations, however, Eq. (11-9.4) with Eqs. (11-9.6) and (11-9.7) still led to results not significantly different from those with Eq. (11-9.2). The various forms of the Tyn-Calus correlation are illustrated in Example 11-6.

Hayduk and Minhas Correlation (Hayduk and Minhas, 1982)

These authors considered many correlations for the infinite dilution binary diffusion coefficient. By regression analysis, they proposed several depending on the type of solute-solvent system.

For *normal paraffin solutions:*

$$D^\circ_{AB} = 13.3 \times 10^{-8} \frac{T^{1.47} \eta_B^{\varepsilon}}{V_A^{0.71}} \tag{11-9.8}$$

where $\varepsilon = (10.2/V_A) - 0.791$ and the other notation is the same as in Eq. (11-9.1). Equation (11-9.8) was developed from data on solutes ranging from C_5 to C_{32} in normal paraffin solvents encompassing C_5 to C_{16}. An average error of only 3.4% was reported.

For *solutes in aqueous solutions:*

$$D^\circ_{AB} = 1.25 \times 10^{-8} (V_A^{-0.19} - 0.292) T^{1.52} \eta_w^{\varepsilon^*} \tag{11-9.9}$$

with $\varepsilon^* = (9.58/V_A) - 1.12$. The rest of the terms are defined in the same manner as under Eq. (11-9.1) except that the subscript w refers to the solvent, water. The authors report that this relation predicted D°_{AB} values with an average deviation of slightly less than 10%.

For *nonaqueous* (*nonelectrolyte*) *solutions:*

$$D^{\circ}_{AB} = 1.55 \times 10^{-8} \frac{T^{1.29}}{\eta_B^{0.92} V_B^{0.23}} \frac{\mathbf{P}_B^{0.5}}{\mathbf{P}_A^{0.42}} \tag{11-9.10}$$

The notation is the same as in Eq. (11-9.2).

The appropriate equation in the set of (11-9.8) to (11-9.10) was used in computing the errors shown in Table 11-5.

It is important to note that, when using the Hayduk-Minhas correlations, the same restrictions apply as in the Tyn-Calus equations.

If Eq. (11-9.3) is used in Eq. (11-9.10) to eliminate the parachors, one obtains

$$D^{\circ}_{AB} = 1.55 \times 10^{-8} \frac{V_B^{0.27}}{V_A^{0.42}} \frac{T^{1.29}}{\eta_B^{0.92}} \frac{\sigma_B^{0.125}}{\sigma_A^{0.105}} \tag{11-9.11}$$

This relation is remarkably similar to the modified Tyn-Calus equation (11-9.4) except for the larger exponent on temperature. As before, when σ_A and σ_B are not greatly different, the surface tension ratio may be set equal to unity as was done to obtain Eq. (11-9.5), or if σ_A and σ_B differ appreciably, Eqs. (11-9.6) and (11-9.7) may be employed.

Example 11-6 Estimate the infinitely dilute diffusion coefficient of acetic acid into acetone at 313 K. The experimental value is 4.04×10^{-5} cm²/s (Wilke and Chang, 1955).

solution The data, from Appendix A, Quayle (1953) and Vargaftik, et al. (1996), are:

	Acetic Acid (solute) A	Acetone (solvent) B
T_b, K	391.0	329.2
T_c, K	594.45	508.1
P_c, bar	57.9	47.0
ρ (at T_b), g/cm³	0.939	0.749
M, g/mol	60.05	58.08
$\mathbf{P}$, cm g$^{1/4}$/(s$^{1/2}$ mol)	129	162
η_B, cP		0.270

TYN-CALUS, Eq. (11-9.2). By rule 3, acetic acid should be treated as a dimer; thus, $V = (2)(64.0) = 128$ cm³/mol and $\mathbf{P} = (2)(129) = 258$ cm³g$^{1/4}$/(s$^{1/2}$ mol).

$$D^{\circ}_{AB} = 8.93 \times 10^{-8} \left(\frac{128}{(77.5)^2}\right)^{1/6} \left(\frac{162}{258}\right)^{0.6} \frac{313}{0.270} = 4.12 \times 10^{-5} \text{ cm}^2/\text{s}$$

$$\text{Error} = \frac{4.12 - 4.04}{4.04} \times 100 = 2\%$$

MODIFIED TYN-CALUS, Eq. (11-9.5)

$$D_{AB}^{\circ} = 8.93 \times 10^{-8} \frac{(77.5)^{0.267}}{(128)^{0.433}} \frac{313}{0.270} = 4.04 \times 10^{-5} \text{ cm}^2/\text{s}$$

Error = 0%

MODIFIED TYN-CALUS, Eqs. (11-9.4), (11-9.6), and (11-9.7). For acetic acid, T_{br} = 391.1/594.45 = 0.658. With Eq. (11-9.7),

$$\alpha_c = 0.9076 \left\{ 1 + (0.658) \left[\frac{\ln(57.9/1.013)}{1 - 0.658} \right] \right\} = 7.972$$

Similarly, α_c for acetone = 7.316. Then, with Eq. (11-9.6)

$$\sigma_A = (57.9)^{2/3}(594.45)^{1/3}[(0.132)(7.972) - 0.278](1 - 0.658)^{11/9} = 26.3 \text{ erg/cm}^2$$

For acetone, σ_B = 19.9 erg/cm² and $(\sigma_B/\sigma_A)^{0.15}$ = 0.959; thus,

$$D_{AB}^{\circ} = (4.04 \times 10^{-5})(0.959) = 3.87 \times 10^{-5} \text{ cm}^2/\text{s}$$

$$\text{Error} = \frac{3.87 - 4.04}{4.04} \times 100 = -4\%$$

In this particular case, the use of the $(\sigma_B/\sigma_A)^{0.15}$ factor actually increased the error; in most other cases, however, errors were less when it was employed.

HAYDUK-MINHAS, Eq. (11-9.10)

$$D_{AB}^{\circ} = 1.55 \times 10^{-8}(313)^{1.29} \frac{(162)^{0.5}/(258)^{0.42}}{(0.270)^{0.92}(77.5)^{0.23}} = 3.89 \times 10^{-5} \text{ cm}^2/\text{s}$$

$$\text{Error} = \frac{3.89 - 4.04}{4.04} \times 100 = -4\%$$

Nakanishi Correlation (Nakanishi, 1978)

In this method, empirical parameters were introduced to account for specific interactions between the solvent and the (infinitely dilute) solute. As originally proposed, the scheme was applicable only at 298.2 K. We have scaled the equation assuming $D_{AB}^{\circ} \eta_B/T$ to be constant.

$$D_{AB}^{\circ} = \left[\frac{9.97 \times 10^{-8}}{(I_A V_A)^{1/3}} + \frac{2.40 \times 10^{-8} A_B S_B V_B}{I_A S_A V_A} \right] \frac{T}{\eta_B} \quad (11\text{-}9.12)$$

where D_{AB}° is the diffusion coefficient of solute A in solvent B at low concentrations, cm³/s. V_A and V_B are the liquid molar volumes of A and B at 298 K, cm³/mol, and the factors I_A, S_A, S_B, and A_B are given in Table 11-4. η_B is the solvent viscosity, in cP, at the system temperature T.

Should the solute (pure) not be a liquid at 298 K, it is recommended that the liquid molar volume at the boiling point be obtained either from data or from correlations in Chap. 4. Then,

TABLE 11-4 Nakanishi Parameter Values for Liquid Diffusion Coefficients

Compound(s)	As solutes (A)†		As solvents (B)	
	I_A	S_A	A_B	S_B
Water	2.8 (1.8)‡	1	2.8	1
Methanol	2.2 (1.5)	1	2.0	1
Ethanol	2.5 (1.5)	1	2.0	1
Other monohydric alcohols	1.5	1	1.8	1
Glycols, organic acids, and other associated compounds	2.0	1	2.0	1
Highly polar materials	1.5	1	1.0	1
Paraffins ($5 \leq n \leq 12$)	1.0	0.7	1.0	0.7
Other substances	1.0	1	1.0	1

† If the solute is He, H_2, D_2, or Ne, the values of V_A should be multiplied by $[1 + (0.85)A^2]$, where A = 3.08 for He^3, 2.67 for He^4, 1.73 for H_2, 1.22 for D_2, and 0.59 for Ne.

‡ The values in parentheses are for cases in which these solutes are dissolved in a solvent which is more polar.

$$V_A(298\text{ K}) = \beta V_A(T_B) \tag{11-9.13}$$

where $\beta = 0.894$ for compounds that are solid at 298 K and $\beta = 1.065$ for compounds that are normally gases at 298 K (and 1 bar). For example, if oxygen is the solute, then, at the normal boiling point of 90.2 K, the molar liquid volume is 27.9 cm^3/mol (Appendix A). With Eq. (11-9.13), $V_A = (1.065)(27.9) = 29.7$ cm^3/mol.

Values of D°_{AB} were estimated for a number of solute-solvent systems and the results were compared with experimental values in Table 11-5. In this tabulation, V_A for water was set equal to the dimer value of 37.4 cm^3/mol to obtain more reasonable results. The poorest estimates were obtained with dissolved gases and with solutes in the more viscous solvents such as *n*-butanol. The use of definite values of I_A to account for solute polarity may cause problems, since it is often difficult to decide whether a compound should be counted as polar ($I_A = 1.5$) or not ($I_A = 1.0$). It might be better to select an average $I_A \cong 1.25$ if there is doubt about the molecular polarity. In Table 11-5, I_A was set equal to 1.5 for pyridine, aniline, nitrobenzene, iodine, and ketones.

Example 11-7 Estimate the value of D°_{AB} for CCl_4 diffusing into ethanol at 298 K. At this temperature, the viscosity of ethanol is 1.08 cP (Riddick, et al., 1986). The experimental value of D°_{AB} is 1.50×10^{-5} cm^2/s (Lusis and Ratcliff, 1971).

solution For this system with CCl_4 as solute A and ethanol as solvent B, from Table 11-4, $I_A = 1$, $S_A = 1$, $A_B = 2$, and $S_B = 1$. From Appendix A, for CCl_4 at 298 K, $V_A = 97.07$ cm^3/mol and for ethanol at 298 K, $V_B = 58.68$ cm^3/mol. Then, with Eq. (11-9.12),

$$D^\circ_{AB} = \left[\frac{9.97 \times 10^{-8}}{(97.07)^{1/3}} + \frac{(2.40 \times 10^{-8})(2)(58.68)}{97.07}\right]\frac{298}{1.08} = 1.40 \times 10^{-5}\text{ cm}^2/\text{s}$$

$$\text{Error} = \frac{1.40 - 1.50}{1.50} \times 100 = -6.7\%$$

Other infinite dilution correlations for diffusion coefficients have been proposed,

TABLE 11-5 Diffusion Coefficients in Liquids at Infinite Dilution

Solute A	Solvent B	T, K	Experimental $D^{\circ}_{AB} \times 10^5$, cm^2/s	Ref.**	Percent error*			
					Wilke and Chang	Tyn and Calus	Hayduk and Minhas	Nakanishi
Acetone	Chloroform	298	2.35	5	42	8.2	5.8	−8.8
		313	2.90		38	5.1	3.2	−11
Benzene		288	2.51	10	1.3	−21	−22	−15
		328	4.25	20	−0.7	−23	−23	−17
Ethanol		288	2.20	10	47	12	8.3	−29
Ethyl ether		298	2.13	20	29	7.4	4.5	4.4
Ethyl acetate		298	2.02	18	36	12	9.4	15
Methyl ethyl ketone		298	2.13	18	37	9.3	6.9	−11
Acetic acid	Benzene	298	2.09	4	28	−9.4	−10	−7.0
Aniline		298	1.96	17	0.4	0.1	0.1	−11
Benzoic acid		298	1.38	4	28	0.6	0.8	−10
Bromobenzene		281	1.45	22	−8.8	−6.4	−6.0	2.4
Cyclohexane		298	2.09	20	−11	−6.9	−7.5	−3.5
		333	3.45		−8.1	−4.2	−5.0	−0.7
Ethanol		288	2.25	10	−1.8	−5.2	−7.0	−39
Formic acid		298	2.28	4	53	−4.2	−4.2	13
n-Heptane		298	2.10	3	−27	−16	−17	−6.2
		353	4.25		−20	−7.1	−8.9	3.6
Methyl ethyl ketone		303	2.09	1	8.7	10	9.0	−8.5
Naphthalene		281	1.19	22	−2.1	5.2	5.6	18
Toluene		298	1.85	20	0.1	4.1	3.5	10
1,2,4-Trichlorobenzene		281	1.34	22	−13	−8.5	−7.7	−0.8
Vinyl chloride		281	1.77	22	8.7	0.1	−0.2	12
Acetic Acid	Acetone	288	2.92	2	35	3.2	−3.2	4.2
		313	4.04	25	33	2.1	−3.7	3.1
Benzoic acid		298	2.62	4	13	−3.5	−8.2	−13
Formic acid		298	3.77	4	56	5.5	0.1	21

TABLE 11-5 Diffusion Coefficients in Liquids at Infinite Dilution (*Continued*)

Solute A	Solvent B	T, K	Experimental $D^{\circ}_{AB} \times 10^5$, cm^2/s	Ref.**	Percent error*			
					Wilke and Chang	Tyn and Calus	Hayduk and Minhas	Nakanishi
Nitrobenzene	Acetone	293	2.94	19	−2.4	8.7	3.0	−6.2
Water		298	4.56	16		5.6	3.7	−19
Bromobenzene	*n*-Hexane	281	2.60	25	16	17	12	26
Carbon tetrachloride		298	3.70	6	12	8.6	4.6	18
Dodecane		298	2.73	23	−17	8.1	−6.7	7.9
n-Hexane		298	4.21	15	−18	−9.0	−4.0	1.3
Methyl ethyl ketone		303	3.74	1	23	23	17	−0.2
Propane		298	4.87	7	3.4	0.6	20	1.3
Toluene		298	4.21	4	−9.0	−6.3	−11	−3.1
Allyl alcohol	Ethanol	293	0.98	9	22	9.7	14	30
Isoamyl alcohol		293	0.81	9	9.8	10	14	15
Benzene		298	1.81	14	−39	−3.0	2.8	−18
Iodine		298	1.32	4	2.4			12
Oxygen		303	2.64	11	−3.2	32	34	47
Pyridine		293	1.10	9	−1.0	−14	−9.1	3.3
Water		298	1.24	12		1.7	11	6.9
Carbon tetrachloride		298	1.50	14	−30	11	18	−6.7
Adipic acid	*n*-Butanol	303	0.40	1	1.1	16	29	8.0
Benzene		298	1.00	14	−52	5.3	20	−23
Butyric acid		303	0.51	1	1.5	7.4	19	2.5
p-Dichlorobenzene		298	0.82	14	−52	13	29	−22
Methanol		303	0.59	14	51	37	50	51
Oleic acid		303	0.25	1	−8.4	26	41	−4.4
Propane		298	1.57	2	−65	−23	−13	−47
Water		298	0.56	14		7.2	26	22
Benzene	*n*-Heptane	298	3.40	3	8.4	1.2	−1.2	13
		372	8.40		1.6	−5.0	−5.9	6.0

Acetic acid	Ethyl acetate	293	2.18	21	69	12	8.5	21
Acetone		293	3.18	21	3.2	−6.8	−9.6	−15
Ethyl benzoate		293	1.85	21	9.0	16	13	−3.9
Methyl ethyl ketone		303	2.93	1	14	8.[illegible]	4.8	−5.4
Nitrobenzene		293	2.25	21	10	6.4	4.2	−2.8
Water		298	3.20	12		16	17	−4.2
Methane	Water	275	0.85	26	10	−3.6	0.0	14
		333	3.55		15	0.7	−2.6	19
Carbon dioxide		298	2.00	24	1.6	−22	−13	−19
Propylene		298	1.44	24	−7.7	−13	−13	−6.2
Methanol		288	1.26	10	5.4	−8.7	−5.4	−9.6
Ethanol		288	1.00	9	5.3	−1.6	−2.7	−8.7
Allyl alcohol		288	0.90	9	5.5	0.5	−2.0	−7.4
Acetic acid		293	1.19	13	2.6	−5.0	−4.7	−24
Ethyl acetate		293	1.00	13	−10	−9.4	−16	−0.9
Aniline		293	0.92	13	−2.5	−5.9	−8.9	−10
Diethylamine		293	0.97	13	−8.6	−7.3	−15	−21
Pyridine		288	0.58	9	49	37	38	31
Ethylbenzene		293	0.81	26	−8.9	−0.2	−18	8.0
Methylcylopentane		275	0.48	26	−2.5	0.3	−14	7.0
		293	0.85		−1.7	1.1	−9.0	7.8
		333	1.92		6.3	9.4	8.5	17
Vinyl chloride		298	1.34	8	3.6	−7.6	−3.3	4.3
		348	3.67		4.2	−7.1	3.0	4.9
ave. abs. % dev.					17	9	11	13

*Percent error = [(calc. - exp.)/exp.] × 100

**References: 1, Amourdam and Laddha (1967); 2, Bidlack and Anderson (1964); 3, Calus and Tyn (1973); 4, Chang and Wilke (1955); 5, Haluska and Colver (1971); 6, Hammond and Stokes (1955); 7, Hayduk, et al. (1973); 8, Hayduk and Laudie (1974); 9, Int. Critical Tables (1926); 10, Johnson and Babb (1956); 11, Krieger, et al. (1967); 12, Lees and Sarram (1971); 13, Lewis (1955); 14, Lusis and Ratcliff (1971); 15, McCall and Douglas (1959); 16, Olander (1961) 17, Rao and Bennett (1971); 18, Ratcliff and Lusis (1971); 19, Reddy and Doraiswamy (1967); 20, Sanni and Hutchinson (1973); 21, Sitaraman, et al. (1963); 22, Stearn, et al. (1940); 23, Vadovic and Colver (1973); 24, Vivian and King (1964); 25, Wilke and Chang (1955); 26, Witherspoon and Bonoli (1969)

but after evaluation they were judged either less accurate or less general than the ones noted above (Akgerman, 1976; Akgerman and Gainer, 1972; Albright, et al., 1983; Brunet and Doan, 1970; Chen, 1984; Faghri and Riazi, 1983; Fedors, 1979; Gainer, 1966; Hayduk and Laudie, 1974; King, et al., 1965; Kuznetsova and Rashidova, 1980; Lusis and Ratcliff, 1968; Othmer and Thakar, 1953; Raina, 1980; Reddy and Doraiswamy, 1967; Siddiqi and Lucas, 1986; Sridhar and Potter, 1977; Teja, 1982; Umesi and Danner, 1981; Vadovic and Colver, 1973).

Effect of Solvent Viscosity

Most of the estimation techniques introduced in this section have assumed that D°_{AB} varies inversely with the viscosity of the solvent. This inverse dependence originated from the Stokes-Einstein relation for a large (spherical) molecule diffusing through a continuum solvent (small molecules). If, however, the solvent is viscous, one may question whether this simple relation is applicable. Davies, et al. (1967) found for CO_2 that in various solvents, $D^\circ_{AB}\eta_B^{0.45} \cong$ constant for solvents ranging in viscosity from 1 to 27 cP. These authors also noted that Arnold (1930a) had proposed an empirical estimation scheme by which $D^\circ_{AB} \propto \eta_B^{-0.5}$. Oosting, et al. (1985) noted that, for the diffusion of 1-hexanol and 2-butanone in malto-dextrin solutions, the viscosity exponent was close to -0.5 over a range of temperatures and concentrations.

Hayduk and Cheng (1971) investigated the effect of solvent viscosity more extensively and proposed that, for nonaqueous systems,

$$D^\circ_{AB} = Q\eta_B^q \tag{11-9.14}$$

where the constants Q and q are particular for a given solute; some values are listed by these authors. In Fig. 11-8, CO_2 diffusion coefficients in various solvents are shown. The solvent viscosity range is reasonably large, and the correlation for organic solvents is satisfactory. In contrast, the data for water as a solvent also are

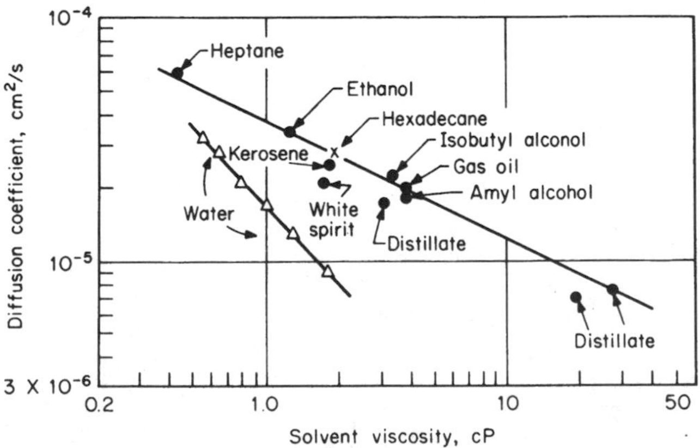

FIGURE 11-8 Diffusion coefficients of carbon dioxide in various solvents. ● Davies et al. (1967); × Hayduk and Cheng (1971); △ Himmelblau (1964)

shown (Himmilblau, 1964). These data fall well below the organic solvent curve and have a slope close to -1. Hiss and Cussler (1973) measured diffusion coefficients of n-hexane and naphthalene in hydrocarbons with viscosities ranging from 0.5 to 5000 cP and report that $D^{\circ}_{AB} \propto \eta_B^{-2/3}$, whereas Hayduk, et al. (1973) found that, for methane, ethane, and propane, D°_{AB} was proportional to $\eta_B^{-0.545}$.

These studies and others (Gainer and Metzner, 1965; Lusis, 1974; Way, 1971) show clearly that, over wide temperature or solvent viscosity ranges, simple empirical correlations, as presented earlier, are inadequate. The diffusion coefficient does not decrease in proportion to an increase in solvent viscosity, but $D^{\circ}_{AB} \propto \eta_B^q$, where q varies, usually from -0.5 to 1.

Discussion

Four estimation techniques were described to estimate the infinite dilution diffusion coefficient of a solute A in a solvent B. In Table 11-5, we show comparisons between calculated and experimental values of D°_{AB} for a number of binary systems. Several comments are pertinent when analyzing the results. First, the temperature range covered is small; thus, any conclusions based upon this sample may not hold at much higher (or lower) temperatures. Second, while D°_{AB} (exp.) is reported to three significant figures, the true accuracy is probably much less because diffusion coefficients are difficult to measure with high precision. Third, all estimation schemes tested showed wide fluctuations in the percent errors. These "failures" may be due to inadequacies in the correlation or to poor data. However, with such wide error ranges, the value of a single average percent error is in doubt.

With these caveats, it is clearly seen that, in general, the Tyn-Calus and the Hayduk-Minhas correlations usually yield the lowest errors; they are, therefore, recommended for calculating D°_{AB}. Both require values of the solute and solvent parachors, but this is obviated with modifications such as Eq. (11-9.5) when $\sigma_A \approx \sigma_B$, or Eq. (11-9.4) [or (11-9.10)] with, say, Eqs. (11-9.6) and (11-9.7) when σ_A differs much from σ_B.

In special situations such as diffusion in n-paraffin solutions, Eq. (11-9.8) is recommended. We did not find a clear advantage for Eq. (11-9.9) over (11-9.2) for solutes diffusing into water, but the former would be more convenient to use.

New experimental data include the systems H_2S-H_2O (Halmour and Sandall, 1984), SO_2-H_2O (Leaist, 1984), CO_2 in binary mixtures (Takahashi, et al., 1982), normal paraffin solutions (Hayduk and Ioakimidis, 1976), hydrocarbons in n-hexane (Dymond, 1981; Dymond and Woolf, 1982), and rare gases in water (Verhallen, et al., 1984). Baldauf and Knapp (1983) studied a wide variety of polar and nonpolar systems at different temperatures and compositions. Mohan and Srinivasan (1984) and McKeigue and Gulari (1989) discuss the reduction of D°_{AB} due to association of alcohols in nonpolar solvents (benzene, carbon tetrachloride).

11-10 CONCENTRATION DEPENDENCE OF BINARY LIQUID DIFFUSION COEFFICIENTS

The concentration dependence of binary diffusion coefficients is not simple. In some cases it varies linearly between the two limiting diffusion coefficients, while in others strong positive or negative deviations from linearity are observed. In Sec. 11-2, it was suggested that the diffusion coefficient D_{AB} in a binary mixture may be proportional to a thermodynamic correction $\alpha = [(\partial \ln a_A / \partial \ln x_A)]_{T,P}$; a_A and x_A are the activity and mole fraction of species A respectively. From the Gibbs-Duhem equation, the derivative $(\partial \ln a_A / \partial \ln x_A)$ is the same whether written for A or B.

Several liquid models purport to relate D_{AB} to composition; e.g., the Darken equation (Darken, 1948; Ghai and Dullien, 1976) predicts that

$$D_{AB} = (D_A^* x_A + D_B^* x_B)\alpha \tag{11-10.1}$$

where D_A^* and D_B^* are tracer diffusion coefficients at x_A, x_B and α is evaluated at the same composition. Equation (11-10.1) was originally proposed to describe diffusion in metals, but it has been used for organic liquid mixtures by a number of investigators (Carman, 1967; Carman and Miller, 1959; Ghai and Dullien 1976; McCall and Douglas, 1967; Miller and Carman, 1961; Tyn and Calus, 1975; Vignes, 1966) with reasonable success except for mixtures in which the components may solvate (Hardt, et al., 1959). The unavailability of tracer diffusion coefficients in most instances has led to a modification of Eq. (11-10.1) as

$$D_{AB} = (D_{BA}^\circ x_A + D_{AB}^\circ x_B)\alpha = [x_A(D_{BA}^\circ - D_{AB}^\circ) + D_{AB}^\circ]\alpha \tag{11-10.2}$$

That is, D_{AB} is a linear function of composition (see Fig. 11-2) corrected by the thermodynamic factor α. Equation (11-10.2) is easier to use because the infinitely dilute diffusion coefficients D_{BA}° and D_{AB}° may be estimated by techniques shown in Sec. 11-9. The thermodynamic term in Eq. (11-10.2) often overcorrects D_{AB}. Rathbun and Babb (1966) suggest α be raised to a fractional power; for associated systems, the exponent chosen was 0.6 unless there were *negative* deviations from Raoult's law when an exponent of 0.3 was recommended. Siddiqi and Lucas (1986) also recommend an exponent of 0.6 when one component is polar and the other nonpolar. When both A and B are polar, Siddiqi and Lucas (1986) recommend using Eq. (11-10.2) with the mole fraction replaced with the volume fraction. It is interesting to note (Sanchez and Clifton, 1977) that curves showing α and D_{AB} as a function of x_A tend to have the same curvature, thus providing some credence to the use of α as a correction factor.

Sanchez and Clifton (1977) found they could correlate D_{AB} with composition for a wide variety of binary systems by using a modification of Eq. (11-10.2):

$$D_{AB} = (D_{BA}^\circ x_A + D_{AB}^\circ x_B)(1 - m + m\alpha) \tag{11-10.3}$$

where the parameter m is to be found from one mixture datum point, preferably in the mid-compositional range. m varies from system to system and may be either greater or less than unity. When $m = 1$, Eq. (11-10.3) reduces to Eq. (11-10.2). Interestingly, for a number of highly associated systems, m was found to be between 0.8 and 0.9. The temperature dependence of m is not known.

Another theory predicts that the group $D_{AB}\eta/\alpha$ should be a linear function of mole fraction (Anderson and Babb, 1961; Bidlack and Anderson, 1964; Byers and King, 1966). Vignes (1966) shows graphs indicating this is not even approximately true for the systems acetone-water and acetone-chloroform. Rao and Bennett (1971) studied several very nonideal mixtures and found that, while the group $D_{AB}\eta/\alpha$ did not vary appreciably with composition, no definite trends could be discerned. One of the systems studied (aniline-carbon tetrachloride) is shown in Fig. 11-9. In this case, D_{AB}, η, α, and $D_{AB}\eta$ varied widely; the group $D_{AB}\eta/\alpha$ also showed an unusual

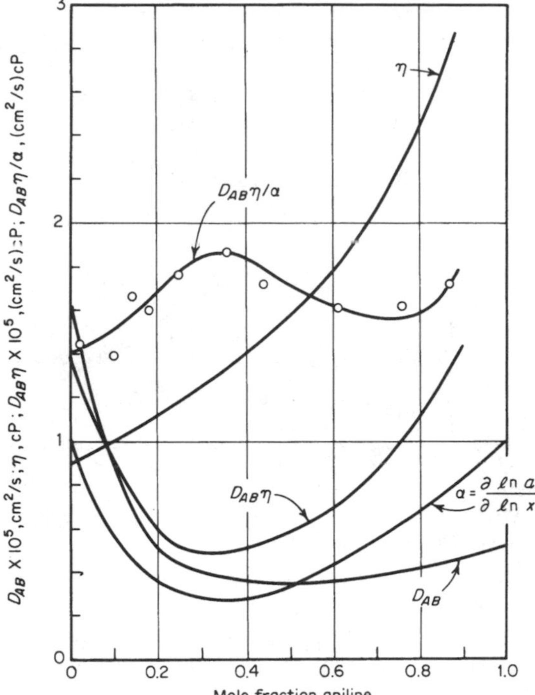

FIGURE 11-9 Diffusion coefficients for the system aniline-carbon tetrachloride at 298 K. (*Rao and Bennett, 1971.*)

variation with composition. Carman and Stein (1956) stated that $D_{AB}\eta/\alpha$ is a linear function of x_A for the nearly ideal system benzene-carbon tetrachloride and for the nonideal system acetone-chloroform but not for ethyl alcohol-water. Vignes (1966) suggested a convenient way of correlating the composition effect on the liquid diffusion coefficient:

$$D_{AB} = [(D^{\circ}_{AB})^{x_B}(D^{\circ}_{BA})^{x_A}]\alpha \tag{11-10.4}$$

and, therefore, a plot of log (D_{AB}/α) vs. mole fraction should be linear. He illustrated this relation with many systems, and, with the exception of strongly associated mixtures, excellent results were obtained. Figure 11-10 shows the same aniline-carbon tetrachloride system plotted earlier in Fig. 11-9. Although not perfect, there is a good agreement with Eq. (11-10.4).

Dullien (1971) carried out a statistical test of the Vignes correlation. It was found to fit experimental data extremely well for ideal or nearly ideal mixtures, but there were several instances when it was not particularly accurate for nonideal, nonassociating solutions. Other authors report that the Vignes correlation is satisfactory for benzene and *n*-heptane (Calus and Tyn, 1973) and toluene and methylcyclohexane (Haluska and Colver, 1970), but not for benzene and cyclohexane (Loflin and McLaughlin, 1969).

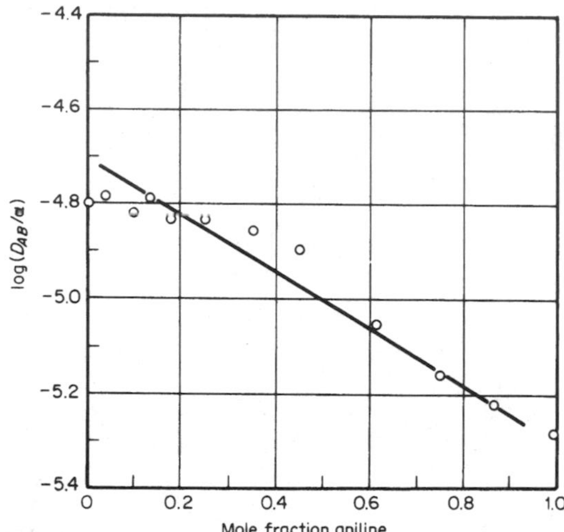

FIGURE 11-10 Vignes plot for the system aniline-carbon tetrachloride at 298 K.

The Vignes relation can be derived from absolute rate theory, and a logical modification of this equation is found to be (Leffler and Cullinan, 1970)

$$D_{AB}\eta = [(D^{\circ}_{AB}\eta_B)^{x_B}(D^{\circ}_{BA}\eta_A)^{x_A}]\alpha \tag{11-10.5}$$

A test of 11 systems showed that this latter form was marginally better in fitting experimental data. In Fig. 11-11, we have plotted both log ($D_{AB}\eta/\alpha$) and log (D_{AB}/α) as a function of composition for the aniline-benzene system. The original Vignes equation fits the data well, but so does Eq. (11-10.5); in fact, for the latter $D_{AB}\eta/\alpha$ is essentially constant.

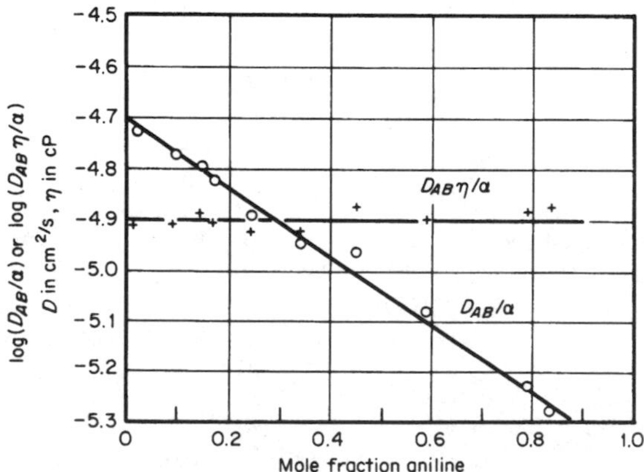

FIGURE 11-11 Vignes plot for the system aniline-benzene at 298 K. (*Rao and Bennett, 1971.*)

Tyn and Calus (1975a) measured the binary diffusion coefficient for several associating systems (ethanol-water, acetone-water, and acetone-chloroform) and found that Eq. (11-10.4) was, generally, preferable to Eq. (11-10.5), although the mean deviation for the Vignes relation was about 14% for the three systems studied.

Other correlation methods have been proposed (Anderson, et al., 1958; Cram and Adamson, 1960; Cullinan, 1971; Gainer, 1970; Haase and Jansen, 1980; Haluska and Colver, 1971; Ratcliff and Holdcroft, 1963; Teja, 1982), but they are either less accurate or less general than those discussed above.

Baldauf and Knapp (1983) present an exceptionally complete data set for eleven binary liquid mixtures giving D_{AB}, η_m, ρ_m, and the refractive index as a function of composition.

In summary, no single correlation is always satisfactory for estimating the concentration effect on liquid diffusion coefficients. The Vignes method [Eq. (11-10.4)] is recommended here as a well-tested and generally accurate correlation. It is also the easiest to apply, and no mixture viscosities are necessary. The thermodynamic correction factor α must, however, be known. If constants are available for a particular activity coefficient model, these constants along with the appropriate equations can be used to calculate α. For example, for the NRTL equation, α is given by the equation

$$\alpha = 1 - 2x_A x_B \left[\frac{\tau_A G_{BA}^2}{(x_A + x_B G_{BA})^3} + \frac{\tau_B G_{AB}^2}{(x_B + x_A G_{AB})^3} \right] \tag{11-10.6}$$

where the symbols are the same as those defined in Chap. 8.

In an approach that does not involve the thermodynamic correction factor α, Hsu and Chen used absolute reaction rate theory (Hsu and Chen, 1998) to develop Eq. (11-10.7). This equation contains two adjustable parameters and has been effective for the correlation of binary diffusion coefficients with composition.

$$\begin{aligned} \ln D_{AB} = {} & x_B \ln D_{AB}^{\circ} + x_A D_{BA}^{\circ} + 2\left(x_A \ln \frac{x_A}{\phi_A} + x_B \ln \frac{x_B}{\phi_B} \right) \\ & + 2x_A x_B \left\{ \frac{\phi_A}{x_A}\left(1 - \frac{\lambda_A}{\lambda_B}\right) + \frac{\phi_B}{x_B}\left(1 - \frac{\lambda_B}{\lambda_A}\right) \right\} \\ & + x_B q_A[(1 - \theta_{BA}^2) \ln \tau_{BA} + (1 - \theta_{BB}^2)\tau_{AB} \ln \tau_{AB}] \\ & + x_A q_B[(1 - \theta_{AB}^2) \ln \tau_{AB} + (1 - \theta_{AA}^2)\tau_{BA} \ln \tau_{BA}] \end{aligned} \tag{11-10.7}$$

$$\theta_{ji} = \frac{\theta_j \tau_{ji}}{\sum_l \theta_l \tau_{li}} \tag{11-10.8}$$

$$\theta_j = \frac{x_j q_j}{\sum_l x_l q_l} \tag{11-10.9}$$

$$\tau_{ji} = \exp\left(-\frac{a_{ji}}{T}\right) \tag{11-10.10}$$

$$\phi_i = \frac{x_i \lambda_i}{\sum_l x_l \lambda_l} \tag{11-10.11}$$

where a_{12}, a_{21} = adjustable parameters
$\lambda_i = (r_i)^{1/3}$
q_i, r_i = UNIFAC volume and surface parameters respectively (see Chap. 8).

Equation (11-10.7) expresses the excess part of the diffusion coefficient in a form similar to that of a UNIQUAC equation. Hsu and Chen give values of a_{12} and a_{21} for 49 systems. For n-alkane systems Hsu and Chen took a_{12} and a_{21} to be zero. For 13 n-alkane systems, they found an average error of 1.6%, while for all systems tested they found an average error of 2.3%.

Example 11-8 Calculate the diffusion coefficient for methanol (A)– water (B) at 313.13 K when $x_A = 0.25$ with the Hsu and Chen method. The experimental value is 1.33×10^{-5} cm²/s (Lee and Li, 1991).

solution From (Hsu and Chen, 1998), $a_{BA} = 194.5302$ and $a_{AB} = -10.7575$. From Chap. 8, $r_A = 1.4311$, $r_B = 0.92$, $q_A = 1.432$, and $q_B = 1.4$. λ_A and λ_B are 1.127 and 0.973 respectively. From (Lee and Li, 1991), $D^\circ_{AB} = 2.1 \times 10^{-5}$ cm²/s and 2.67×10^{-5} cm²/s. Substitution into Eqs. (11-10.8) to (11-10.11) leads to $\theta_A = 0.254$, $\theta_B = 0.721$, $\phi_A = 0.279$, $\phi_B = 0.746$, $\theta_{BA} = 0.612$, $\theta_{AB} = 0.261$, $\theta_{AA} = 0.388$, $\theta_{BB} = 0.739$, $\tau_{AB} = 1.035$, and $\tau_{BA} = 0.5373$. Eq. (11-10.7) then gives

$$\ln D_{AB} = 0.75 \ln (2.10 \times 10^{-5} + 0.25 \ln (2.67 \times 10^{-5})$$

$$+ 2(0.25)(0.75)\left[\frac{0.279}{0.025}\left(1 - \frac{1.127}{0.973}\right) + \frac{0.721}{0.75}\left(1 - \frac{0.973}{1.127}\right)\right]$$

$$+ 2\left(0.25 \ln \frac{0.25}{0.279} + 0.75 \ln \frac{0.75}{0.721}\right)$$

$$+ (0.75)(1.432)[(1 - (0.612)^2)\ln(0.5373) + (1 - (0.739)^2)(1.035)\ln(1.035)]$$

$$+ (0.25)(1.4)[(1 - (0.261)^2)\ln(1.035) + (1 - (0.388)^2)(0.5373)\ln(0.5373)]$$

$$D_{AB} = 1.351 \times 10^{-5} \text{ cm}^2/\text{s}$$

$$\text{Error} = \frac{1.351 - 1.33}{1.33} \times 100 = 1.6\%$$

When the predictive equations, Eqs. (11-10.2), (11-10.4), and (11-10.5) were used to work the problem in Example 11-8, the errors were 21%, 21%, and –32% respectively. Furthermore, these equations required a value for α which itself could not be determined with confidence. For example, Gmehling and Onken, 1977 list three sets of activity coefficient data for the methanol-water system at 313 K. The three different sets of NRTL constants along with Eq. (11-10.6) gave α values that ranged from 0.72 to 0.84.

11-11 THE EFFECTS OF TEMPERATURE AND PRESSURE ON DIFFUSION IN LIQUIDS

For the Wilke-Chang and Tyn-Calus correlations for D°_{AB} in Sec. 11-9, the effect of temperature was accounted for by assuming

$$\frac{D^\circ_{AB}\eta_B}{T} = \text{constant} \tag{11-11.1}$$

In the Hayduk-Minhas method, the (absolute) temperature was raised to a power > 1, and the viscosity parameter was a function of solute volume. While these approximations may be valid over small temperature ranges, it is usually preferable (Sanchez, 1977) to assume that

$$D_{AB} \text{ (or } D^\circ_{AB}) = A \exp \frac{-B}{T} \tag{11-11.2}$$

Equation (11-11.2) has been employed by a number of investigators (Innes and Albright, 1957; McCall, et al., 1959; Robinson, et al., 1966; Tyn, 1975). We illustrate its applicability in Fig. 11-12 with the system ethanol-water from about 298 K to 453 K for both infinitely dilute diffusion coefficients and D_{AB} for a 20 mole % solution (Kircher, et al., 1981). Note that we have not included the thermodynamic correction factor α because it is assumed to be contained in the A and B parameters. Actually, since the viscosity of liquids is an exponentially decreasing function of temperature, below reduced temperatures of about 0.7, the product $D_{AB}\eta$

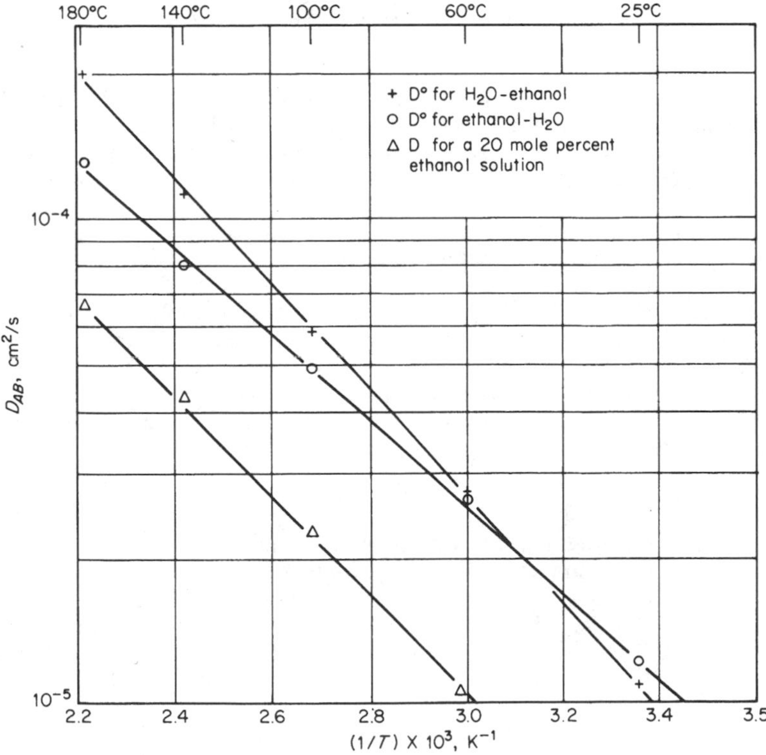

FIGURE 11-12 Variation of diffusion coefficients with temperature. Data from Kircher, et al. (1981).

shows much less variation with temperature than do D_{AB} or η individually. The energies of activation for diffusion and viscosity are opposite in sign and often of the same magnitude numerically (Robinson, et al., 1966).

Tyn (1976, 1981) reviewed the various proposed techniques to correlate infinitely dilute binary (and also self-) diffusion coefficients with temperature. He suggested that

$$\frac{D^{\circ}_{AB}(T_2)}{D^{\circ}_{AB}(T_1)} = \left(\frac{T_c - T_1}{T_c - T_2}\right)^n \tag{11-11.3}$$

where T_c is the critical temperature of the solvent B. T_c, T_1, and T_2 are in kelvins. The parameter n was related to the heat of vaporization of the solvent at T_b (solvent) as follows:

n	$\Delta H_v(T_b)$, J/mol
3	7,900 to 30,000
4	30,000 to 39,700
6	39,700 to 46,000
8	46,000 to 50,000
10	> 50,000

Typical compounds falling into these categories would be $n = 3$, n-pentane, acetone, cyclohexane, chloroform; $n = 4$, benzene, toluene, chlorobenzene, n-octane, carbon tetrachloride; $n = 6$, cyclohexane, propanol, butanol, water; $n = 8$, heptanol; and $n = 10$, ethylene and propylene glycols. See Chap. 7 for more information about $\Delta H_v(T_b)$.

Equation (11-11.3), which does not require mixture viscosity data, was tested with a large number of binary systems, and an error of about 9% was found. When Eq. (11-11.1) also was examined, Tyn reported an error of about 10%. The temperature ranges for Eq. (11-11.3) are about 10 K above the freezing point to about 10 K below the normal boiling point. Larger errors were noted if these ranges were exceeded.

The effect of pressure on liquid diffusion coefficients has received little attention. Easteal (1984) attempted to correlate tracer or self-diffusion coefficients with pressure and suggested

$$\ln D_j^* = a + bP^{0.75} \tag{11-11.4}$$

where D_j^* is a tracer or self-diffusion coefficient and a and b are constants for a given solute, but they do vary significantly with temperature. b is a negative number, and thus D_j^* decreases with an increase in pressure. As an example, the self-diffusion coefficient for n-hexane decreases from about 4.2×10^{-5} cm²/s at 1 bar to about 0.7×10^{-5} cm²/s at 3500 bar at a temperature of 298 K.

From Eq. (11-11.1), at a given temperature, it can be inferred that

$$D^{\circ}_{AB}\eta_B = \text{constant} \tag{11-11.5}$$

If solvent-liquid viscosity data are available at high pressures, or are estimated with the methods in Chap. 9, it should then be possible to employ Eq. (11-11.5) to estimate D°_{AB} at the elevated pressure from low-pressure diffusion coefficient data. Dymond and Woolf (1982) show, however, that this proportionality is only approximate for tracer-diffusion coefficients, but they indicate it may be satisfactory for binaries with large solute molecules.

11-12 DIFFUSION IN MULTICOMPONENT LIQUID MIXTURES

In a binary liquid mixture, as indicated in Secs. 11-2 and 11-8, a single diffusion coefficient was sufficient to express the proportionality between the flux and concentration gradient. In multicomponent systems, the situation is considerably more complex, and the flux of a given component depends upon the gradient of $n - 1$ components in the mixture. For example, in a ternary system of A, B, and C, the flux of A can be expressed as

$$J_A = \mathbf{D}_{AA} \frac{dc_A}{dz} + \mathbf{D}_{AB} \frac{dc_B}{dz} \tag{11-12.1}$$

Similar relations can be written for J_B and J_C. The coefficients $\mathbf{D}_{AA}$ and $\mathbf{D}_{BB}$ are called main coefficients; they are not self-diffusion coefficients. $\mathbf{D}_{AB}$, $\mathbf{D}_{BA}$, etc., are cross-coefficients, because they relate the flux of a component i to a gradient in j. $\mathbf{D}_{ij}$ is normally not equal to $\mathbf{D}_{ji}$ for multicomponent systems.

One important case of multicomponent diffusion results when a solute diffuses through a homogeneous solution of mixed solvents. When the solute is dilute, there are no concentration gradients for the solvent species and one can speak of a single solute diffusivity with respect to the mixture D°_{Am}. This problem has been discussed by several authors (Cullinan and Cusick, 1967a; Holmes, et al., 1962; Perkins and Geankoplis, 1969; Tang and Himmelblau, 1965) and empirical relations for D°_{Am} have been proposed. Perkins and Geankoplis (1969) evaluated several methods and suggested

$$D^{\circ}_{Am}\eta_m^{0.8} = \sum_{\substack{j=1 \\ j \neq A}}^{n} x_j D^{\circ}_{Aj}\eta_j^{0.8} \tag{11-12.2}$$

where D°_{Am} = effective diffusion coefficient for a dilute solute A into the mixture, cm^2/s
D°_{Aj} = infinite dilution binary diffusion coefficient of solute A into solvent j, cm^2/s
x_j = mole fraction of j
η_m = mixture viscosity, cP
η_j = pure component viscosity, cP

When tested with data for eight ternary systems, errors were normally less than 20%, except for cases involving CO_2. These same authors also suggested that the Wilke-Chang equation (11-9.1) might be modified to include the mixed solvent case, i.e.,

$$D^{\circ}_{Am} = 7.4 \times 10^{-8} \frac{(\phi M)^{1/2} T}{\eta_m V_A^{0.6}} \tag{11-12.3}$$

$$\phi M = \sum_{\substack{j=1 \\ j \neq A}}^{n} x_j \phi_j M_j \tag{11-12.4}$$

Although not extensively tested, Eq. (11-12.3) provides a rapid, reasonably accurate estimation method.

For CO_2 as a solute diffusing into mixed solvents, Takahashi, et al. (1982) recommend

$$D^{\circ}_{CO_2 m} \left(\frac{\eta_m}{V_m}\right)^{1/3} = \sum_{\substack{j=1 \\ j \neq CO_2}}^{n} x_j D^{\circ}_{CO_2 j} \left(\frac{\eta_j}{V_j}\right)^{1/3} \tag{11-12.5}$$

where V_m is the molar volume, cm^3/mol, for the mixture at T and V_j applies to the pure component. Tests with a number of ternary systems involving CO_2 led to deviations from experimental values usually less than 4%.

Example 11-9 Estimate the diffusion coefficient of acetic acid diffusing into a mixed solvent containing 20.7 mole % ethyl alcohol in water. The acetic acid concentration is small. Assume the temperature is 298 K. The experimental value reported by Perkins and Geankoplis (1969) is 0.571×10^{-5} cm^2/s.

data Let E = ethyl alcohol, W = water, and A = acetic acid. At 298 K, Perkins and Geankoplis (1969) give $\eta_E = 1.10$ cP, $\eta_W = 0.894$ cP, $D_{AE} = 1.03 \times 10^{-5}$ cm^2/s, $D_{AW} = 1.30 \times 10^{-5}$ cm^2/s, and for the solvent mixture under consideration, $\eta_m = 2.35$ cP.

solution From Eq. (11-12.2)

$$D^{\circ}_{Am} = (2.35)^{-0.8}[(0.207)(1.03 \times 10^{-5})(1.10)^{0.8} + (0.793)(1.30 \times 10^{-5})(0.894)^{0.8}]$$

$$= 0.59 \times 10^{-5} \text{ cm}^2/\text{s}$$

$$\text{Error} = \frac{0.59 - 0.571}{0.571} \times 100 = 3.3\%$$

Note that this diffusion coefficient is significantly below the two limiting binary values. The decrease in the mixture diffusivity appears to be closely related to the increase in solvent mixture viscosity relative to the pure components. Had the modified Wilke-Chang equation been used, with $V_A = 64.1$ cm^3/mol (Chap. 4), $\phi_E = 1.5$, and $\phi_W = 2.6$, and with $M_E = 46$ and $M_W = 18$, using Eqs. (11-12.3) and (11-12.4),

$$\phi M = (0.207)(1.5)(46) + (0.793)(2.6)(18) = 51.39$$

$$D^{\circ}_{Am} = \frac{(7.4 \times 10^{-8})(51.39)^{1/2}(298)}{(2.35)(64.1)^{0.6}} = 0.55 \times 10^{-5} \text{ cm}^2/\text{s}$$

$$\text{Error} = \frac{0.55 - 0.571}{0.571} \times 100 = -3.7\%$$

Example 11-10 Estimate the diffusion coefficient of CO_2 (D) into a mixed solvent of *n*-octanol (L) and carbon tetrachloride (C) containing 60 mole % *n*-octanol. The temperature is 298 K. The experimental value (Takanishi, et al., 1982) is 1.96×10^{-5} cm²/s.

data From Takanishi, et al. (1982), $D^\circ_{DL} = 1.53 \times 10^{-5}$ cm²/s, $D^\circ_{DC} = 3.17 \times 10^{-5}$ cm²/s, $\eta_m = 3.55$ cP, $\eta_L = 7.35$ cP, and $\eta_C = 0.88$ cP. From Appendix A, $V_L = 158.37$ cm³/mol and $V_C = 97.07$ cm³/mol at 298 K. The mixture volume is not known. If we assume that the mole fraction of CO_2 in the liquid mixture is small and that *n*-octanol and carbon tetrachloride form ideal solutions,

$$V_m \approx (0.6)(158.37) + (0.4)(97.07) = 133.8 \text{ cm}^3/\text{mol}$$

solution With Eq. (11-12.5)

$$D^\circ(CO_2\text{-m}) = (3.55/133.8)^{-1/3}[(0.6)(1.53 \times 10^{-5})(7.35/158.37)^{1/3} + (0.4)(3.17 \times 10^{-5})(0.88/97.07)^{1/3}] = 1.99 \times 10^{-5} \text{ cm}^2/\text{s}.$$

$$\text{Error} = \frac{1.99 - 1.96}{1.96} \times 100 = 1.5\%$$

When dealing with the general case of multicomponent diffusion coefficients, there are no convenient and simple estimation methods. Kooijman and Taylor (1991) have discussed extension of the Vignes correlation [Eq. (11-10.4)] to ternary systems; the reader is also referred to page 570 of Bird, et al. (1960) or to Curtiss and Bird (1999) for discussion of this problem. Kett and Anderson (1969, 1969a) apply hydrodynamic theory to estimate ternary diffusion coefficients. Although some success was achieved, the method requires extensive data on activities, pure component and mixture volumes, and viscosities, as well as tracer and binary diffusion coefficients. Bandrowski and Kubaczka (1982) suggest using the mixture critical volume as a correlating parameter to estimate D_{Am} for multicomponent mixtures.

11-13 DIFFUSION IN ELECTROLYTE SOLUTIONS

When a salt dissociates in solution, ions rather than molecules diffuse. In the absence of an electric potential, however, the diffusion of a single salt may be treated as molecular diffusion.

The theory of diffusion of salts at low concentrations is well developed. At concentrations encountered in most industrial processes, one normally resorts to empirical correlations, with a concomitant loss in generality and accuracy. A comprehensive discussion of this subject is available (Newman and Tobias, 1967).

For dilute solutions of a single salt, the diffusion coefficient is given by the Nernst-Haskell equation

$$D^\circ_{AB} = \frac{RT[(1/z_+) + (1/z_-)]}{F^2[(1/\lambda^\circ_+) + (1/\lambda^\circ_-)]} \tag{11-13.1}$$

where D°_{AB} = diffusion coefficient at infinite dilution, based on molecular concentration, cm^2/s
T = temperature, K
R = gas constant, 8.314 J/(mol · K)
λ°_+, λ°_- = limiting (zero concentration) ionic conductances, (A/cm^2) (V/cm) $(g\text{-equiv}/cm^3)$
z_+, z_- = valences of cation and anion, respectively
F = faraday = 96,500 C/g-equiv

Values of λ°_+ and λ°_- can be obtained for many ionic species at 298 K from Table 11-6 or from alternative sources (Moelwyn-Hughes, 1957; Robinson and Stokes, 1959). If values of λ°_+ and λ°_- at other temperatures are needed, an approximate correction factor is $T/334\ \eta_W$, where η_W is the viscosity of water at T in centipoises.

As the salt concentration becomes finite and increases, the diffusion coefficient decreases rapidly and then usually rises, often becoming greater than D°_{AB} at high normalities. Figure 11-13 illustrates the typical trend for three simple salts. The initial decrease at low concentrations is proportional to the square root of the concentration, but deviations from this trend are usually significant above 0.1 N. Figure 11-14 shows the behavior for the rather atypical system HCl in water at 298 K. This system illustrates not only the minimum illustrated in Fig. 11-13, but shows a maximum as well.

Also shown in Fig. 11-14 is the behavior of D°_{AB}/α, where the thermodynamic factor α is given by

$$\alpha = 1 + m\frac{\partial \ln \gamma_\pm}{\partial m} \tag{11-13.2}$$

TABLE 11-6 Limiting Ionic Conductances in Water at 298 K (*Harned and Owen, 1950*) $(A/cm^2)(V/cm)(g\text{-equiv}/cm^2)$

Anion	λ°_-	Cation	λ°_+
OH^-	197.6	H^+	349.8
Cl^-	76.3	Li^+	38.7
Br^-	78.3	Na^+	50.1
I^-	76.8	K^+	73.5
NO_3^-	71.4	NH_4^+	73.4
ClO_4^-	68.0	Ag^+	61.9
HCO_3^-	44.5	Tl^+	74.7
HCO_2^-	54.6	$(1/2)Mg^{2+}$	53.1
$CH_3CO_2^-$	40.9	$(1/2)Ca^{2+}$	59.5
$ClCH_2CO_2^-$	39.8	$(1/2)Sr^{2+}$	50.5
$CNCH_2CO_2^-$	41.8	$(1/2)Ba^{2+}$	63.6
$CH_3CH_2CO_2^-$	35.8	$(1/2)Cu^{2+}$	54
$CH_3(CH_2)_2CO_2^-$	32.6	$(1/2)Zn^{2+}$	53
$C_6H_5CO_2^-$	32.3	$(1/3)La^{3+}$	69.5
$HC_2O_4^-$	40.2	$(1/3)Co(NH_3)_6^{3+}$	102
$(1/2)C_2O_4^{2-}$	74.2		
$(1/2)SO_4^{2-}$	80		
$(1/3)Fe(CN)_6^{3-}$	101		
$(1/4)Fe(CN)_6^{4-}$	111		

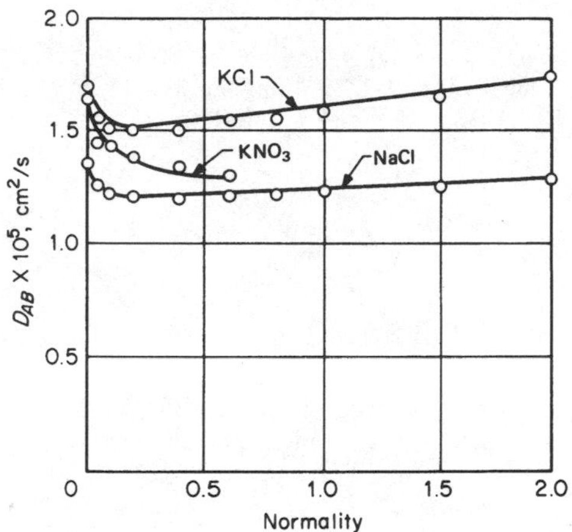

FIGURE 11-13 Effect of concentration on diffusivity of electrolytes in aqueous solution at 18.5°C. Solid lines calculated by using Eq. (11-13.3). (*Data from Gordon* (*1937*).)

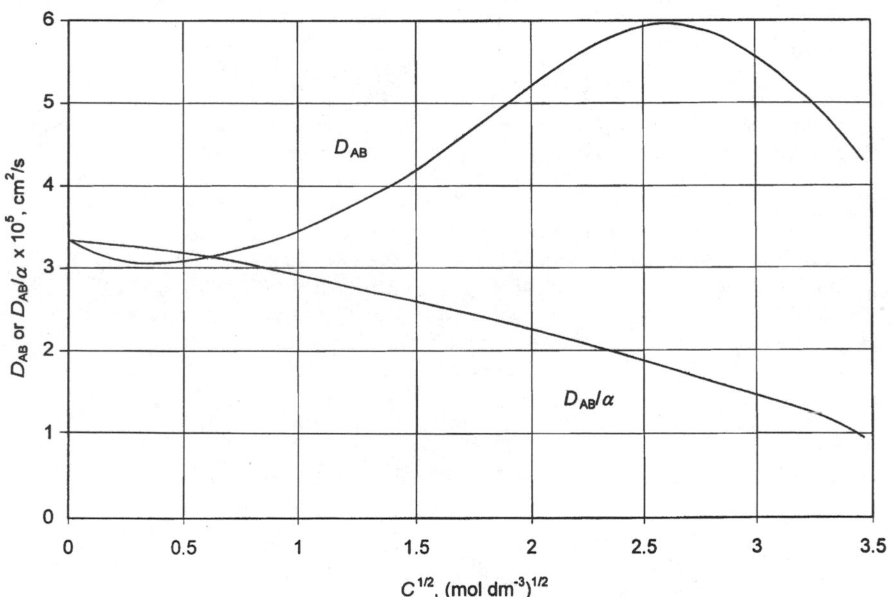

FIGURE 11-14 Effect of concentration on diffusivity of HCl in aqueous solution at 25°C. (*Curves are through data from Rizzo, et al.* (*1997*) *and Stokes* (*1950*).)

m = molality of the solute, mol/kg solvent
$\gamma_{\pm}$ = mean ionic activity coefficient of the solute

From Fig. 11-14, it can be seen that the quantity D°_{AB}/α varies more smoothly with concentration than does D°_{AB}. While D°_{AB}/α for HCl drops smoothly over the whole concentration range, Rizzo, et al. (1997) observe that for most salts, D°_{AB}/α shows a maximum at low concentrations.

The behavior shown in Fig. 11-14 is consistent with an empirical equation proposed by Gordon (1937) that has been applied to systems at concentrations up to 2 N:

$$D_{AB} = D^{\circ}_{AB} \frac{\eta_s}{\eta} (\rho_s \overline{V}_s)^{-1} \alpha \tag{11-13.3}$$

where α is defined in Eq. (11-13.2) and

D°_{AB} = diffusion coefficient at infinite dilution, [Eq. (11-13.1)], cm^2/s
ρ_s = molar density of the solvent, mol/cm^3
$\overline{V}_s$ = partial molar volume of the solvent, cm^3/mol
η_s = viscosity of the solvent, cP
η = viscosity of the solution, cP

In many cases, the product $\rho_s \overline{V}_s$ is close to unity, as is the viscosity ratio η_s/η, so that Gordon's relation provides an activity correction to the diffusion coefficient at infinite dilution. Though Harned and Owen (1950) tabulate $\gamma_{\pm}$ as a function of m for many aqueous solutions, there now exist several semiempirical correlation techniques to relate $\gamma_{\pm}$ to concentration. These correlations are discussed in detail in Prausnitz, et al. (1999).

Data on the diffusion of CO_2 into electrolyte solutions have been reported by Ratcliff and Holdcroft (1963). The diffusion coefficient was found to decrease linearly with an increase in salt concentration.

In summary, for very dilute solutions of electrolytes, employ Eq. (11-13.1). When values of the limiting ionic conductances in water are not available at the desired temperature, use those in Table 11-6 for 298 K and multiply D°_{AB} at 298 K by $T/334\ \eta_w$, where η_w is the viscosity of water at T in centipoises.

For concentrated solutions, use Eq. (11-13.3). If values of $\gamma_{\pm}$ and λ° are not available at T, calculate D°_{AB} at 298 K and multiply it by $(T/298)[(\eta \text{ at } 298)/(\eta \text{ at } T)]$. If necessary, the ratio of the solution viscosity at 298 K to that at T may be assumed to be the same as the corresponding ratio for water.

Example 11-11 Estimate the diffusion coefficient of NaOH in a 2 *N* aqueous solution at 288 K.

solution From data on densities of aqueous solutions of NaOH, it is evident that, up to 12 weight % NaOH (about 3 *N*), the density increases almost exactly in inverse proportion to the weight fraction of water; i.e., the ratio of moles of water per liter is essentially constant at 55.5. Thus both V/n and $\overline{V}_1$ are very nearly 55.5 and cancel in Eq. (11-13.3). In this case, the molality m is essentially identical with the normality. Values of $\gamma_{\pm}$ for NaOH at 298 K are given in Harned and Owen (1950). The value at $m = 2$ is 0.698. When the values are plotted vs. molality m, the slope at 2 m is approximately 0.047. Hence

$$m \frac{\partial \ln \gamma_{\pm}}{\partial m} = \frac{m}{\gamma_{\pm}} \frac{\partial \gamma_{\pm}}{\partial m} = \frac{2}{0.698} (0.047) = 0.135$$

The viscosities of water and 2 *N* NaOH solution at 298 K are 0.894 and 1.42 cP, respectively. Substituting in Eqs. (11-13.1) and (11-13.3) gives

$$D^{\circ}_{AB} = \frac{(2)(8.314)(298)}{[(1/50) + (1/198)](96{,}500)^2}$$

$$= 2.12 \times 10^{-5}\ \text{cm}^2/\text{s}$$

$$D_{AB} = (2.12 \times 10^{-5}) \frac{0.894}{1.42} \frac{55.5}{55.5} [1 + (2)(0.135)]$$

$$= 1.70 \times 10^{-5}\ \text{cm}^2/\text{s}$$

At 288 K, the viscosity of water is 1.144 cP, and so the estimated value of D_{AB} at 288 K is

$$1.70 \times 10^{-5} \frac{(288)}{(334)(1.144)} = 1.28 \times 10^{-5}\ \text{cm}^2/\text{s}$$

which may be compared with the International Critical Tables (1926) value of 1.36 × 10^{-5} cm²/s for an error of −5.9%.

In a system of mixed electrolytes, such as in the simultaneous diffusion of HCl and NaCl in water, the faster-moving H^+ ion may move ahead of its Cl^- partner, the electric current being maintained at zero by the lagging behind of the slower-moving Na^+ ions. In such systems, the unidirectional diffusion of each ion species results from a combination of electric and concentration gradients:

$$N_+ = \frac{\lambda_+}{F^2}\left(-RT\frac{\partial c_+}{\partial z} - Fc_+\frac{\partial E}{\partial z}\right) \tag{11-13.3}$$

$$N_- = \frac{\lambda_-}{F^2}\left(-RT\frac{\partial c_-}{\partial z} - Fc_-\frac{\partial E}{\partial z}\right) \tag{11-13.4}$$

where N_+, N_- = diffusion flux densities of the cation and anion, respectively, g-equiv/cm² · s
c_+, c_- = corresponding ion concentrations, g-equiv/cm³
$\partial E/\partial z$ = gradient in electric potential
λ_+, λ_- = ionic equivalent conductances

Collision effects, ion complexes, and activity corrections are neglected. The electric field gradient may be imposed externally but is present in the ionic solution even if, owing to the small separation of charges that results from diffusion itself, there is no external electrostatic field.

One equation for each cation and one for each anion can be combined with the requirement of zero current at any z to give $\Sigma N_+ = \Sigma N_-$. Solving for the unidirectional flux densities (Ratcliff and Lusis, 1971),

$$n_+N_+ = -\frac{RT\lambda_+}{F^2 n_+}(G_+ - n_+c_+Y) \tag{11-13.5}$$

$$n_-N_- = -\frac{RT\lambda_-}{F^2 n_-}(G_- - n_-c_-Y) \tag{11-13.6}$$

$$Y = \frac{(\Sigma\gamma_+G_+/n_+) - (\Sigma\gamma_-G_-/n_-)}{\Sigma\gamma_+c_+ + \Sigma\gamma_-c_-} \tag{11-13.7}$$

where G_+ and G_- are the concentration gradients $\partial c/\partial z$ in the direction of diffusion.

Vinograd and McBain (1941) have used these relations to represent their data on diffusion in multi-ion solutions. D_{AB} for the hydrogen ion was found to decrease from 12.2 to 4.9×10^{-5} cm^2/s in a solution of HCl and $BaCl_2$ when the ratio of H^+ to Ba^{2+} was increased from zero to 1.3; D_{AB} at the same temperature is 9.03 $\times 10^{-5}$ for the free H^+ ion, and 3.3×10^{-5} for HC1 in water. The presence of the slow-moving Ba^{2+} accelerates the H^+ ion, the potential existing with zero current causing it to move in dilute solution even faster than it would as a free ion with no other cation present. That is, electrical neutrality is maintained by the hydrogen ions moving ahead of the chlorine faster than they would as free ions, while the barium diffuses more slowly than as a free ion.

The interaction of ions in a multi-ion system is important when the several ion conductances differ greatly, as they do when H^+ or OH^- is diffusing. When the diffusion of one of these two ions is not involved, no great error is introduced by the use of "molecular" diffusion coefficients for the salt species present.

The relations proposed by Vinograd and McBain, Eqs. (11-13.5) to (11-13.7), are not adequate to represent a ternary system, in which four independent diffusion coefficients must be known to predict fluxes. The subject of ion diffusion in multicomponent systems is covered in detail by Cussler (1976) and in the papers by Wendt (1965) and Miller (1966, 1967) in which it is demonstrated how one can obtain multicomponent ion diffusion coefficients, although the data required are usually not available. Anderko and Lencka (1998) have used self-diffusion coefficients to model diffusion in multicomponent electrolyte systems; Mills and Lobo (1989) present a compilation of self-diffusion coefficient values.

NOTATION

a_j	activity of component $j = x_j \gamma_j$
A_B	parameter in Table 11-4
c	concentration, mol/cm^3; c_j, for component j; c_+, c_-, ion concentrations
D_{AB}	binary diffusion coefficient of A diffusing into B, cm^2/s; D°_{AB}, at infinite dilution of A in B; D_{AB}, cross-coefficient in multicomponent mixtures; D_{im} of i into a homogeneous mixture; D^+_{AB}, at a low pressure
D^*_A	tracer-diffusion coefficient of A, cm^2/s
D_{AA}	self-diffusion coefficient of A, cm^2/s; D_{AA}, main coefficient for A in multicomponent diffusion
E	electric potential
f_D	correction term in Eq. (11-3.1)
F	faraday = 96,500 C/g-equiv
G_+, G_-	$\partial c_+/\partial z$ and $\partial c_-/\partial z$, Eqs. (11-13.3) to (11-13.7)
I_A	parameter in Table 11-4
J_A	flux of A, mol/(cm^2 · s); J^M_A, flux relative to a plane of no net mole flow; J^V_A, flux relative to a plane of no net volume flow
k	Boltzmann's constant, 1.3805×10^{-23} J/K
m	molality of solute, mol/kg solvent; parameter in Eq. (11-10.3)
M_A	molecular weight of A, g/mol; M_{AB}, $2[(1/M_A) + (1/M_B)]^{-1}$
n	number density of molecules; parameter in Eq. (11-11.3)
N_A	flux of A relative to a fixed coordinate plane, mol/(cm^2 · s), Eq. (11-2.1)

N_+, N_- diffusion flux of cation and anion, respectively, Eqs. (11-13.3) and (11-13.4)
P pressure, bar; P_c, critical pressure; P_r, reduced pressure, P/P_c
$\mathbf{P}_j$ parachor of component j
q, Q parameters in Eq. (11-9.14); q_i, UNIFAC surface parameter
r distance of separation between molecules, Å
r_A molecular radius in the Stokes-Einstein equation, Eq. (11-8.1); r_i, UNIFAC volume parameter
R gas constant, 8.31447 J/(mol · K)
S_A, S_B parameters in Table 11-4
T temperature, kelvins; T_b, at the normal boiling point (at 1 atm); T_c critical temperature; T_r, reduced temperature, T/T_c
V volume, cm^3/mol; V_b, at T_b; $\overline{V}_A$, partial molar volume of A
V_j molar volume of component j at either T_b or T, cm^3/mol
x_j mole fraction of j, usually liquid
y_j mole fraction of j, usually vapor
z direction coordinate for diffusion
z_+, z_- valences of cation and anion, respectively

Greek
α $\partial \ln a_A/\partial \ln x_A$, $\partial \ln a/\partial \ln m$; parameter in Eq. (11-5.9), α_c, parameter in Eq. (11-9.7)
β parameter in Eq. (11-9.13)
γ activity coefficient; $\gamma_\pm$, mean ionic activity coefficient
δ polar parameter defined in Eq. (11-3.8)
ε characteristic energy parameter; ε_A, for pure A; ε_{AB}, for an A-B interaction
ε, ε^* parameters in Eqs. (11-9.8) and (11-9.9)
η viscosity, cP; η_A, for pure A; η_m for a mixture
λ_+°, λ_-° limiting (zero concentration) ionic conductances, (A/cm^2)(V/cm)(g-equiv/cm^3)
μ_A chemical potential of A, J/mol
μ_p dipole moment, debyes, Section 2-6
ρ density, g/cm^3
σ characteristic length parameter, Å; σ_A, for pure A; σ_{AB}, for an A-B interaction; surface tension
Σ_v Fuller et al. volume parameter, Table 11-1
ϕ association parameter for the solvent, Eq. (11-9.1)
ψ intermolecular potential energy of interaction
ω acentric factor
Ω_D collision integral for diffusion, Eq. (11-3.1)

Superscripts
° infinite dilution
* tracer value
+ low pressure

Subscripts
A, B components A and B; usually B is the solvent
m mixture
w water
s solvent

REFERENCES

Akgerman, A.: *Ind. Eng. Chem. Fundam.,* **15:** 78 (1976).

Akgerman, A., and J. L. Gainer: *J. Chem. Eng. Data,* **17:** 372 (1972).

Albright, J. G., A. Vernon, J. Edge, and R. Mills: *J. Chem. Soc., Faraday Trans.* 1, **79:** 1327 (1983).

Alvarez, R., I. Medlina, J. L. Bueno, and J. Coca: *J. Chem. Eng. Data,* **28:** 155 (1983).

Amdur, I., and L. M. Shuler: *J. Chem. Phys.,* **38:** 188 (1963).

Amdur, I., J. Ross, and E. A. Mason: *J. Chem. Phys.,* **20:** 1620 (1952).

Amourdam, M. J., and G. S. Laddha: *J. Chem. Eng. Data,* **12:** 389 (1967).

Anderko, A., and M. M. Lencka: *Ind. Eng. Chem. Res.,* **37:** 2878 (1998).

Anderson, D. K., and A. L. Babb: *J. Phys. Chem.,* **65:** 1281 (1961).

Anderson, D. K., J. R. Hall, and A. L. Babb: *J. Phys. Chem.,* **62:** 404 (1958).

Anderson, J. L.: *Ind. Eng. Chem. Fundam.,* **12:** 490 (1973).

Arnold, J. H.: *Ind. Eng. Chem.,* **22:** 1091 (1930).

Arnold, J. H.: *J. Am. Chem. Soc.,* **52:** 3937 (1930a).

Bailey, R. G.: *Chem. Eng.,* **82**(6)**:** 86 (1975).

Baldauf, W., and H. Knapp: *Ber. Bunsenges. Phys. Chem.,* **87:** 304 (1983).

Bandrowski, J., and A. Kubaczka: *Chem. Eng. Sci.,* **37:** 1309 (1982).

Bearman, R. J.: *J. Chem. Phys.,* **32:** 1308 (1960).

Bearman, R. J.: *J. Phys. Chem.,* **65:** 1961 (1961).

Bidlack, D. L., and D. K. Anderson: *J. Phys. Chem.,* **68:** 3790 (1964).

Bird, R. B., W. E. Stewart, and E. N. Lightfoot: *Transport Phenomena,* Wiley, New York, 1960, chap. 16.

Blanc, A.: *J. Phys.,* **7:** 825 (1908).

Brokaw, R. S.: *Ind. Eng. Chem. Process Design Develop.,* **8:** 240 (1969).

Brunet, J., and M. H. Doan: *Can. J. Chem. Eng.,* **48:** 441 (1970).

Byers, C. H., and C. J. King: *J. Phys. Chem.,* **70:** 2499 (1966).

Byrne, J. J., D. Maguire, and J. K. A. Clarke: *J. Phys. Chem.,* **71:** 3051 (1967).

Caldwell, L.: *J. Chem. Eng. Data,* **29:** 60 (1984).

Caldwell, C. S., and A. L. Babb: *J. Phys. Chem.,* **60:** 14, 56 (1956).

Calus, W. F., and M. T. Tyn: *J. Chem. Eng. Data,* **18:** 377 (1973).

Carman, P. C.: *J. Phys. Chem.,* **71:** 2565 (1967).

Carman, P. C.: *Ind. Eng. Chem. Fundam.,* **12:** 484 (1973).

Carman, P. C., and L. Miller: *Trans. Faraday Soc.,* **55:** 1838 (1959).

Carman, P. C., and L. H. Stein: *Trans. Faraday Soc.,* **52:** 619 (1956).

Carmichael, L. T., B. H. Sage, and W. N. Lacey: *AIChE J.,* **1:** 385 (1955).

Carswell, A. J., and J. C. Stryland: *Can. J. Phys.,* **41:** 708 (1963).

Catchpole, O. J., and M. B. King: *Ind. Eng. Chem. Res.,* **33:** 1828 (1994).

Chang, P., and C. R. Wilke: *J. Phys. Chem.,* **59:** 592 (1955).

Chen, N. H., and D. P. Othmer: *J. Chem. Eng. Data,* **7:** 37 (1962).

Chen, S.-H.: *AIChE J.,* **30:** 481 (1984).

Cram, R. R., and A. W. Adamson: *J. Phys. Chem.,* **64:** 199 (1960).

Crider, W. L.: *J. Am. Chem. Soc.* **78:** 924 (1956).

Cullinan, H. T., Jr.: *Can. J. Chem. Eng.,* **49:** 130 (1971).

Cullinan, H. T., Jr., and M. R. Cusick: *Ind. Eng. Chem. Fundam.,* **6:** 72 (1967).

Cullinan, H. T., Jr., and M. R. Cusick: *AIChE J.,* **13:** 1171 (1967a).

Curtiss, C. F., and R. B. Bird: *Ind. Eng. Chem. Res.,* **38:** 2515 (1999).

Cussler, E. L.: *Multicomponent Diffusion,* Elsevier, New York, 1976.

Cussler, E. L.: *Diffusion: Mass Transfer in Fluid Systems,* 2d ed. Cambridge, 1997, Chaps. 3, 7.

Darken, L. S.: *Trans. Am. Inst. Mining Metall. Eng.,* **175:** 184 (1948).

Daubert, T. E., R. P. Danner, H. M. Sibel, and C. C. Stebbins: *Physical and Thermodynamic Properties of Pure Chemicals: Data Compilation,* Taylor & Francis, Washington, D.C., 1997.

Davies, G. A., A. B. Ponter, and K. Craine: *Can. J. Chem. Eng.,* **45:** 372 (1967).

Debenedetti, P., and R. C. Reid: *AIChE J.,* **12:** 2034 (1986).

Di Pippo, R., J. Kestin, and K. Oguchi: *J. Chem. Phys.,* **46:** 4986 (1967).

Dullien, F. A. L.: *Nature,* **190:** 526 (1961).

Dullien, F. A. L.: *Trans. Faraday Soc.,* **59:** 856 (1963).

Dullien, F. A. L.: *Ind. Eng. Chem. Fundam.,* **10:** 41(1971) .

Dullien, F. A. L.: personal communication, January, 1974.

Duncan, J. B., and H. L. Toor: *AIChE J.,* **8:** 38 (1962).

Dunlop, P. J., and C. M. Bignell: *Intern. J. Thermophys.,* **18:** 939 (1997).

Dymond, J. H.: *J. Phys. Chem,* **85:** 3291 (1981).

Dymond, J. H., and L. A. Woolf: *J. Chem. Soc., Faraday Trans.* 1, **78:** 991 (1982).

Easteal, A. J.: *AIChE J.,* **30:** 641 (1984).

Elliott, R. W., and H. Watts: *Can. J. Chem.,* **50:** 31 (1972).

Ellis, C. S., and J. N. Holsen: *Ind. Eng. Chem. Fundam.,* **8:** 787 (1969).

Eyring, H., and T. Ree: *Proc. Natl. Acad. Sci.,* **47:** 526 (1961).

Faghri, A., and M.-R. Riazi: *Intern. Comm. Heat Mass Transfer,* **10:** 385 (1983).

Fairbanks, D. F., and C. R. Wilke: *Ind. Eng. Chem.,* **42:** 471 (1950).

Fedors, R. F.: *AIChE J.,* **25:** 200, 716 (1979).

Fuller, E. N., and J. C. Giddings: *J. Gas Chromatogr.,* **3:** 222 (1965).

Fuller, E. N., P. D. Schettler, and J. C. Giddings: *Ind. Eng. Chem.,* **58**(5): 18 (1966).

Fuller, E. N., K. Ensley, and J. C. Giddings: *J. Phys. Chem.,* **73:** 3679 (1969).

Funazukuri, T., Y. Ishiwata, and N. Wakao: *AIChE J.,* **38:** 1761 (1992).

Gainer, J. L.: *Ind. Eng. Chem. Fundam.,* **5:** 436 (1966).

Gainer, J. L.: *Ind. Eng. Chem. Fundam.,* **9:** 381 (1970).

Gainer, J. L., and A. B. Metzner: *AIChE-Chem E Symp. Ser.,* no. 6, 1965, p. 74.

Ghai, R. K., and F. A. L. Dullien: personal communication, 1976.

Ghai, R. K., H. Ertl, and F. A. L. Dullien: *AIChE J.,* **19:** 881 (1973).

Ghai, R. K., H. Ertl, and F. A. L. Dullien: *AIChE J.,* **20:** 1 (1974).

Gilliland, E. R.: *Ind. Eng. Chem.,* **26:** 681 (1934).

Glasstone, S., K. J. Laidler, and H. Eyring: *The Theory of Rate Processes,* McGraw-Hill, New York, 1941, chap. 9.

Gmehling, J., and U. Onken: *Vapor-Liquid Equilibrium Data Collection,* Chem. Data Ser., Vol. I, Parts 1 and 1b, DECHEMA, Frankfurt, 1977.

Gordon, A. R.: *J. Chem. Phys.,* **5:** 522 (1937).

Gordon, M.: "References to Experimental Data on Diffusion Coefficients of Binary Gas Mixtures," *Natl. Eng. Lab. Rept.,* Glasgow, Scotland, 1977.

Gotoh, S., M. Manner, J. P. Sørensen, and W. E. Stewart: *Ind. Eng. Chem.,* **12:** 119 (1973).

Gotoh, S., M. Manner, J. P. Sørensen, and W. E. Stewart: *J. Chem. Eng. Data,* **19:** 169, 172 (1974).

Gupta, G. P., and S. C. Saxena: *AIChE J.,* **14:** 519 (1968).

Haase, R., and H.-J. Jansen: *Z. Naturforsch.,* **35A:** 1116 (1980).

Halmour, N., and O. C. Sandall: *J. Chem. Eng. Data,* **29:** 20 (1984).
Haluska, J. L., and C. P. Colver: *AIChE J.,* **16:** 691 (1970).
Haluska, J. L., and C. P. Colver: *Ind. Eng. Chem. Fundam.,* **10:** 610 (1971).
Hammond, B. R., and R. H. Stokes: *Trans. Faraday Soc.,* **51:** 1641 (1955).
Hardt, A. P., D. K. Anderson, R. Rathbun, B. W. Mar, and A. L. Babb: *J. Phys. Chem.,* **63:** 2059 (1959).
Harned, H. S., and B. B. Owen: "The Physical Chemistry of Electrolytic Solutions," *ACS Monogr.* **95,** 1950.
Hartley, G. S., and J. Crank: *Trans. Faraday Soc.,* **45:** 801 (1949).
Hattikudur, U. R., and G. Thodos: *J. Chem. Phys.,* **52:** 4313 (1970).
Hayduk, W., and W. D. Buckley: *Chem. Eng. Sci.,* **27:** 1997 (1972).
Hayduk, W., R. Casteñeda, H. Bromfield, and R. R. Perras: *AIChE J.,* **19:** 859 (1973).
Hayduk, W., and S. C. Cheng: *Chem. Eng. Sci.,* **26:** 635 (1971).
Hayduk, W., and S. Ioakimidis: *J. Chem. Eng. Data,* **21:** 255 (1976).
Hayduk, W., and H. Laudie: *AIChE J.,* **20:** 611 (1974).
Hayduk, W., and B. S. Minhas: *Can. J. Chem. Eng.,* **60:** 295 (1982).
He, C.-H., and Y.-S. Yu: *Ind. Eng. Chem. Res.,* **37:** 3793 (1998).
Himmelblau, D. M.: *Chem. Rev.,* **64:** 527 (1964).
Hirschfelder, J. O., C. F. Curtiss, and R. B. Bird: *Molecular Theory of Gases and Liquids,* Wiley, New York, 1954.
Hiss, T. G., and E. L. Cussler: *AIChE J.,* **19:** 698 (1973).
Holmes, J. T., D. R. Olander, and C. R. Wilke: *AIChE J.,* **8:** 646 (1962).
Holson, J. N., and M. R. Strunk: *Ind. Eng. Chem. Fundam.,* **3:** 163 (1964).
Horrocks, J. K., and E. McLaughlin: *Trans. Faraday Soc.,* **58:** 1367 (1962).
Hsu, Y.-D., and Y.-P. Chen: *Fluid Phase Equil.,* **152:** 149 (1998).
Hudson, G. H., J. C. McCoubrey, and A. R. Ubbelohde: *Trans. Faraday Soc.,* **56:** 1144 (1960).
Innes, K. K., and L. F. Albright: *Ind. Eng. Chem.,* **49:** 1793 (1957).
International Critical Tables, McGraw-Hill, New York, 1926–1930.
Iomtev, M. B., and Y. V. Tsekhanskaya: *Russ. J. Phys. Chem.,* **38:** 485 (1964).
Johnson, D. W., and C. P. Colver: *Hydrocarbon Process. Petrol. Refiner,* **48**(3)**:** 113 (1969).
Johnson, P. A., and A. L. Babb: *Chem. Rev.,* **56:** 387 (1956).
Kamal, M. R., and L. N. Canjar: *AIChE J.,* **8:** 329 (1962).
Kestin, J., H. E. Khalifa, S. T. Ro, and W. A. Wakeham: *Physica,* **88A:** 242 (1977).
Kestin, J., and W. A. Wakeham: *Ber. Bunsenges. Phys. Chem.,* **87:** 309 (1983).
Kett, T. K., and D. K. Anderson: *J. Phys. Chem.,* **73:** 1262 (1969).
Kett, T. K., and D. K. Anderson: *J. Phys. Chem.,* **73:** 1268 (1969a).
King, C. J., L. Hsueh, and K. W. Mao: *J. Chem. Eng. Data,* **10:** 348 (1965).
Kircher, K., A. Schaber, and E. Obermeier: *Proc. 8th Symp. Thermophys. Prop.,* Vol. 1, ASME, 1981, p. 297.
Kooijman, H. A., and R. Taylor: *Ind. Eng. Chem. Res.,* **30:** 1217 (1991).
Krieger, I. M., G. W. Mulholland, and C. S. Dickey: *J. Phys. Chem.,* **71:** 1123 (1967).
Kuznetsova, E. M., and D. Sh. Rashidova: *Russ. J. Phys. Chem.,* **54:** 1332, 1339 (1980).
Leaist, D. G.: *J. Chem. Eng. Data,* **29:** 281 (1984).
Lee, H., and G. Thodos: *Ind. Eng. Chem. Fundam.,* **22:** 17 (1983).
Lee, Y. E., and F. Y Li: *J. Chem. Eng. Data,* **36:** 240 (1991).
Lees, F. P., and P. Sarram: *J. Chem. Eng. Data,* **16:** 41 (1971).
Leffler, J., and H. T. Cullinan, Jr.: *Ind. Eng. Chem. Fundam.,* **9:** 84, 88 (1970).
Lewis, J. B.: *J. Appl. Chem. London,* **5:** 228 (1955).

Li, J. C. M., and P. Chang: *J. Chem. Phys.,* **23:** 518 (1955).
Liu, H., and E. Ruckenstein: *Ind. Eng. Chem. Res.,* **36:** 888 (1997).
Loflin, T., and E. McLaughlin: *J. Phys. Chem.,* **73:** 186 (1969).
Lugg, G. A.: *Anal. Chem.,* **40:** 1072 (1968).
Lusis, M. A.: *Chem. Proc. Eng.,* **5:** 27 (May 1971).
Lusis, M. A.: *Chem. Ind. Devel. Bombay,* January 1972, **48;** *AIChE J.,* **20:** 207 (1974).
Lusis, M. A., and G. A. Ratcliff: *Can. J. Chem. Eng.,* **46:** 385 (1968).
Lusis, M. A., and G. A. Ratcliff: *AIChE J.,* **17:** 1492 (1971).
Maczek, A. O. S., and C. J. C. Edwards: "The Viscosity and Binary Diffusion Coefficients of Some Gaseous Hydrocarbons, Fluorocarbons and Siloxanes," *Symp. Transport Prop. Fluids and Fluid Mixtures, Natl. Eng. Lab.,* East Kilbride, Glasgow, Scotland, April 1979.
McCall, D. W., and D. C. Douglas: *Phys. Fluids,* **2:** 87 (1959).
McCall, D. W., and D. C. Douglas: *J. Phys. Chem.,* **71:** 987 (1967).
McCall, D. W., D. C. Douglas, and E. W. Anderson: *Phys. Fluids,* **2:** 87 (1959); **4:** 162 (1961).
Marrero, T. R., and E. A. Mason: *J. Phys. Chem. Ref. Data,* **1:** 3 (1972).
Mason, E. A., and L. Monchick: *J. Chem. Phys.,* **36:** 2746 (1962).
Massman, W. J.: *Atmos. Environ.,* **32:** 1111 (1998).
Mather, G. P., and S. C. Saxena: *Ind. J. Pure Appl. Phys.,* **4:** 266 (1966).
McKeigue, K., and E. Gulari: *AIChE J.,* **35:** 300 (1989).
Miller, D. G.: *J. Phys. Chem.,* **70:** 2639 (1966); **71:** 616 (1967).
Miller, L., and P. C. Carman: *Trans. Faraday Soc.* **57:** 2143 (1961).
Mills, R., and V. M. M. Lobo: *Self-diffusion in Electrolyte Solutions, A Critical Examination of Data Compiled from the Literature,* Phys. Sci. Data 36, Elsevier, Amsterdam, 1989.
Moelwyn-Hughes, E. A.: *Physical Chemistry,* Pergamon, London, 1957.
Mohan, V., and D. Srinivasan: *Chem. Eng. Comm.,* **29:** 27 (1984).
Morozov, V. S., and E. G. Vinkler: *Russ. J. Phys. Chem.,* **49:** 1404 (1975).
Mrazek, R. V., C. E. Wicks, and K. N. S. Prabhu: *J. Chem. Eng. Data,* **13:** 508 (1968).
Murad, S.: *Chem. Eng. Sci.,* **36:** 1867 (1981).
Nakanishi, K.: *Ind. Eng. Chem. Fundam.,* **17:** 253 (1978).
Neufeld, P. D., A. R. Janzen, and R. A. Aziz: *J. Chem. Phys.,* **57:** 1100 (1972).
Newman, J. S.: in C. W. Tobias (ed.), *Advances in Electrochemistry and Electrochemical Engineering,* Vol. 5, Interscience, New York, 1967.
Olander, D. R.: *AIChE J.,* **7:** 175 (1961); **9:** 207 (1963).
Oosting, E. M., J. I. Gray, and E. A. Grulke: *AIChE J.,* **31:** 773 (1985).
Othmer, D. F., and T. T. Chen: *Ind. Eng. Chem. Process Design Develop.,* **1:** 249 (1962).
Othmer, D. F., and M. S. Thakar: *Ind. Eng. Chem.,* **45:** 589 (1953).
Pasternak, A. D., and D. R. Olander: *AIChE J.,* **13:** 1052 (1967).
Pathak, B. K., V. N. Singh, and P. C. Singh: *Can. J. Chem. Eng.,* **59:** 362 (1981).
Perkins, L. R., and C. J. Geankoplis: *Chem. Eng. Sci.,* **24:** 1035 (1969).
Prausnitz, J. M., R. N. Lichtenthaler, and E. G. de Azevedo: *Molecular Thermodynamics of Fluid-Phase Equilibria,* 3d ed., Prentice Hall, Upper Saddle River, NJ, 1999.
Quayle, O. R.: *Chem. Rev.,* **53:** 439 (1953).
Raina, G. K.: *AIChE J.,* **26:** 1046 (1980).
Rao, S. S., and C. O. Bennett: *AIChE J.,* **17:** 75 (1971).
Ratcliff, G. A., and J. G. Holdcroft: *Trans. Inst. Chem. Eng. London,* **41:** 315 (1963).
Ratcliff, G. A., and M. A. Lusis: *Ind. Eng. Chem. Fundam.,* **10:** 474 (1971).
Rathbun, R. E., and A. L. Babb: *Ind. Eng. Chem. Process Design Develop.,* **5:** 273 (1966).
Ravindran, P., E. J. Davis, and A. K. Ray: *AIChE J.,* **25:** 966 (1979).

Reddy, K. A., and L. K. Doraiswamy: *Ind. Eng. Chem. Fundam.,* **6:** 77 (1967).

Ree, F. H., T. Ree, and H. Eyring: *Ind. Eng. Chem.,* **50:** 1036 (1958).

Riazi, M. R., and C. H. Whitson: *Ind. Eng. Chem. Res.,* **32:** 3081 (1993).

Riddick, J. A., W. B. Bunger, and T. K. Sakano: *Organic Solvents, Physical Properties and Methods of Purification,* 4th ed., Wiley-Interscience, New York, 1986.

Rizzo, R., J. G. Albright, and D. G. Miller: *J. Chem. Eng. Data,* **42:** 623 (1997).

Robinson, R. A., and R. H. Stokes: *Electrolyte Solutions,* 2d ed., Academic, New York, 1959.

Robinson, R. L., Jr., W. C. Edmister, and F. A. L. Dullien: *Ind. Eng. Chem. Fundam.,* **5:** 74 (1966).

Sanchez, V., and M. Clifton: *Ind. Eng. Chem. Fundam.,* **16:** 318 (1977).

Sanchez, V., H. Oftadeh, C. Durou, and J.-P. Hot: *J. Chem. Eng. Data,* **22:** 123 (1977).

Sandler, S., and E. A. Mason: *J. Chem. Phys.,* **48:** 2873 (1968).

Sanni, S. A., and P. Hutchinson: *J. Chem. Eng. Data,* **18:** 317 (1973).

Scheibel, E. G.: *Ind. Eng. Chem.,* **46:** 2007 (1954).

Schneider, G. M.: *Angew. Chem. Intern. Ed. English,* **17:** 716 (1978).

Schwartz, F. A., and J. E. Brow: *J. Chem. Phys.,* **19:** 640 (1951).

Scott, D. S., and K. E. Cox: *Can. J. Chem. Eng.,* **38:** 201 (1960).

Seager, S. L., L. R. Geertson, and J. C. Giddings: *J. Chem. Eng. Data,* **8:** 168 (1963).

Shrier, A. L.: *Chem. Eng. Sci.,* **22:** 1391 (1967).

Siddiqi, M. A., and K. Lucas: *Can. J. Chem. Eng.,* **64:** 839 (1986).

Sitaraman, R., S. H. Ibrahim, and N. R. Kuloor: *J. Chem. Eng. Data,* **8:** 198 (1963).

Slattery, J. C., and R. B. Bird: *AIChE J.,* **4:** 137 (1958).

Sridhar, T., and O. E. Potter: *AIChE J.,* **23:** 590, 946 (1977).

Srivastava, B. N., and K. P. Srivastava: *J. Chem. Phys.,* **30:** 984 (1959).

Srivastava, B. N., and I. B. Srivastava: *J. Chem. Phys.,* **36:** 2616 (1962).

Srivastava, B. N., and I. B. Srivastava: *J. Chem. Phys.,* **38:** 1183 (1963).

Stearn, A. E., E. M. Irish, and H. Eyring: *J. Phys. Chem.,* **44:** 981 (1940).

Stokes, R. H.: *J. Am. Chem. Soc.,* **72:** 2243 (1950).

Swaid, I., and G. M. Schneider: *Ber. Bunsenges. Phys. Chem.,* **83:** 969 (1979).

Takahashi, S.: *J. Chem. Eng. Japan,* **7:** 417 (1974).

Takahashi, S., and M. Hongo: *J. Chem. Eng. Japan,* **15:** 57 (1982).

Takahashi, M., Y. Kobayashi, and H. Takeuchi: *J. Chem. Eng. Data,* **27:** 328 (1982).

Tang, Y. P., and D. M. Himmelblau: *AIChE J.,* **11:** 54 (1965).

Teja, A. S.: personal communication, 1982.

Toor, H. L.: *AIChE J.,* **3:** 198 (1957).

Tsekhanskaya, Y. V.: *Russ. J. Phys. Chem.,* **45:** 744 (1971).

Turner, J. C. R.: *Chem. Eng. Sci.,* **30:** 151 (1975).

Tyn, M. T.: *Chem. Eng.,* **82**(12): 106 (1975).

Tyn, M. T.: *Chem. Eng. J.,* **12:** 149 (1976).

Tyn, M. T.: *Trans. Inst. Chem. Engrs.,* **59**(2), 112 (1981).

Tyn, M. T., and W. F. Calus: *J. Chem. Eng. Data,* **20:** 106 (1975).

Tyn, M. T., and W. F. Calus: *J. Chem. Eng. Data,* **20:** 310 (1975a).

Umesi, N. O., and R. P. Danner: *Ind. Eng. Chem. Process Design Develop.,* **20:** 662 (1981).

Vadovic, C. J., and C. P. Colver: *AIChE J.,* **19:** 546 (1973).

Van Geet, A. L., and A. W. Adamson: *J. Phys. Chem.,* **68:** 238 (1964).

Vargaftik, N. B., Y. K. Vinogradov, and V. S. Yargin: *Handbook of Physical Properties of Liquids and Gases,* Begell House, New York, 1996.

Verhallen, P. T. H. M., L. J. O. Oomen, A. J. J. M. v. d. Elsen, and A. J. Kruger: *Chem. Eng. Sci.,* **39:** 1535 (1984).
Vignes, A.: *Ind. Eng. Chem. Fundam.,* **5:** 189 (1966).
Vinkler, E. G., and V. S. Morozov: *Russ. J. Phys. Chem.,* **49:** 1405 (1975).
Vinograd, J. R., and J. W. McBain: *J. Am. Chem. Soc.,* **63:** 2008 (1941).
Vivian, J. E., and C. J. King: *AIChE J.,* **10:** 220 (1964).
Walker, R. E., N. de Haas, and A. A. Westenberg: *J. Chem. Phys.,* **32:** 1314 (1960).
Walker, R. E., and A. A. Westenberg: *J. Chem. Phys.,* **29:** 1139 (1958).
Way, P.: Ph.D. thesis, Massachusetts Institute of Technology, Cambridge, Mass. 1971
Weissman, S.: *J. Chem. Phys.,* **40:** 3397 (1964).
Weissman, S., and E. A. Mason: *J. Chem. Phys.,* **37:** 1289 (1962).
Wendt, R. P.: *J. Phys. Chem.,* **69:** 1227 (1965).
Westenberg, A. A., and G. Frazier: *J. Chem. Phys.,* **36:** 3499 (1962).
Wilke, C. R., and P. Chang: *AIChE J.,* **1:** 264 (1955).
Wilke, C. R., and C. Y. Lee: *Ind. Eng. Chem.,* **47:** 1253 (1955).
Wise, D. L., and G. Houghton: *Chem. Eng. Sci.,* **21:** 999 (1966).
Witherspoon, P. A., and L. Bonoli: *Ind. Eng. Chem. Fundam.,* **8:** 589 (1969).
Yabsley, M. A., P. J. Carlson, and P. J. Dunlop: *J. Phys. Chem.,* **77:** 703 (1973).

CHAPTER TWELVE
SURFACE TENSION

12-1 SCOPE

The surface tensions of both pure liquids and liquid mixtures are considered in this chapter. For the former, methods based on the law of corresponding states and upon the parachor are judged most accurate when estimated values are compared with experimental determinations. For mixtures, extensions of the pure component methods are presented, as is a method based upon a thermodynamic analysis of the system. Interfacial tensions for liquid-solid systems are not included.

12-2 INTRODUCTION

The boundary between a liquid phase and a gas phase can be considered a third phase with properties distinct from those of the liquid and gas. A qualitative picture of the microscopic surface layer shows that there are unequal forces acting upon the molecules; i.e., at low gas densities, the surface molecules are attracted sidewise and toward the bulk liquid but experience little attraction in the direction of the bulk gas. Thus the surface layer is in tension and tends to contract to the smallest area compatible with the mass of material, container restraints, and external forces, e.g., gravity.

This tension can be presented in various quantitative ways; the most common is the surface tension σ, defined as the force exerted in the plane of the surface per unit length. We can consider a reversible isothermal process whereby surface area A is increased by pulling the surface apart and allowing molecules from the bulk to enter at constant temperature and pressure. The differential reversible work is σdA. Since it is also the differential Gibbs energy change, σ is the surface Gibbs energy per unit of area. At equilibrium systems tend to a state of minimum Gibbs energy at fixed T and P, the product σA also tends to a minimum. For a fixed σ, equilibrium is a state of minimum area consistent with the system conditions.

Analogously, the boundary between two liquid phases may also be considered a third phase which is characterized by the interfacial tension.

Surface tension and interfacial tension are usually expressed in units of dynes per centimeter; which is the same as ergs per square centimeter. With relation to SI units, 1 dyn/cm = 1 erg/cm^2 = 1 mJ/m^2 = 1 mN/m.

The thermodynamics of surface layers furnishes a fascinating subject for study. Guggenheim (1959), Gibbs (1957), and Tester and Modell (1997) have formulated treatments that differ considerably, but reduce to similar equations relating macroscopically measurable quantities. In addition to the thermodynamic aspect, treatments of the physics and chemistry of surfaces have been published (Adamson, 1982; Aveyard and Haden, 1973; Barbulescu, 1974; Brown, 1974; Chattoraj and Birdi, 1984; Evans and Wennerström, 1999; Everett, 1988; Ross, 1965; Rowlinson and Widom, 1982). These subjects are not covered here; instead, the emphasis is placed upon the few reliable methods available to estimate σ from either semitheoretical or empirical equations.

12-3 ESTIMATION OF PURE-LIQUID SURFACE TENSION

As the temperature is raised, the surface tension of a liquid in equilibrium with its own vapor decreases and becomes zero at the critical point (Rowlinson and Widom, 1982). In the reduced-temperature range 0.45 to 0.65, σ for most organic liquids ranges from 20 to 40 dyn/cm, but for some low-molecular-weight dense liquids such as formamide, $\sigma > 50$ dyn/cm. For water $\sigma = 72.8$ dyn/cm at 293 K, and for liquid metals σ is between 300 and 600 dyn/cm; e.g., mercury at 293 K has a value of about 476.

A recent, thorough critical evaluation of experimental surface tensions has been prepared by Jasper (1972). Additional data and tabulations of data are given in Körösi and Kováts (1981), Riddick, et al. (1986), Timmermans (1965), and Vargaftik (1996). Daubert, et al. (1997) list references to original data and Gray, et al. (1983) have correlated surface tensions of coal liquid fractions.

Essentially all useful estimation techniques for the surface tension of a liquid are empirical. Several are discussed below and others are briefly noted at the end of this section.

Macleod-Sugden Correlation

Macleod (1923) suggested a relation between σ and the liquid and vapor densities:

$$\sigma^{1/4} = [P](\rho_L - \rho_V) \tag{12-3.1}$$

Sugden (1924) has called the temperature-independent parameter $[P]$ the parachor and indicated how it might be estimated from the structure of the molecule. Quayle (1953) employed experimental surface tension and density data for many compounds and calculated parachor values. From these, he suggested an additive scheme to correlate $[P]$ with structure, and a modified list of his values is shown in Table 11-3. When $[P]$ values determined in this manner are used, the surface tension is given in dynes per centimeter and the densities are expressed in moles per cubic centimeter. The method is illustrated in Example 12-1, and calculated values of σ are compared with experimental surface tensions in Table 12-1. In Table 12-1, liquid densities were calculated from correlations in Daubert, et al. (1997) or International Critical Tables (1928). Table 11-3 leads to values of $[P]$ that are often as accurate as those from Quayle (1953) (which have been fitted to surface tension data) except for compounds such as benzonitrile, carbon disulfide, carbon tetrachloride, and pyridine. For these four compounds, Table 11-3 leads to errors that are unacceptably high.

Example 12.1 Use the Macleod-Sugden correlation to estimate the surface tension of isobutyric acid at 333 K. The experimental value quoted by Jasper (1972) is 21.36 dyn/cm.

solution At 333 K, the liquid density is 0.9128 g/cm³ (International Critical Tables, 1928), and, with M = 88.107, ρ_L = 0.9128/88.107 = 1.036 × 10⁻² mol/cm³. At 333 K, isobutyric acid is well below the boiling point, and at this low pressure $\rho_V << \rho_L$ and the vapor density term is neglected.

To determine the parachor from Table 11-3,

$$[P] = \mathrm{CH_3{-}CH(CH_3){-}} + \mathrm{{-}COOH} = 133.3 + 73.8 = 207.1$$

Then, with Eq. (12-3.1),

$$\sigma = [(207.1)(1.036 \times 10^{-2})]^4 = 21.19 \text{ dyn/cm}$$

$$\text{Error} = \frac{21.19 - 21.36}{21.36} \times 100 = -0.8\%$$

Since σ is proportional to $([P]\rho_L)^4$, Eq. (12-3.1) is very sensitive to the values of the parachor and liquid density chosen. It is remarkable that the estimated values are as accurate as shown in Table 12-1.

Corresponding States Correlation

Instead of correlating surface tensions with densities, a number of correlations have been developed that relate σ to T. A technique given by Goldhammer (1910) and discussed by Gambill (1959) leads to:

$$\rho_L - \rho_V = \rho_{Lb}\left(\frac{1 - T_r}{1 - T_{rb}}\right)^n \tag{12-3.2}$$

ρ_{Lb}, is the molal liquid density at the normal boiling point in moles per cubic centimeter. Furthermore, the group $\sigma/P_c^{2/3}T_c^{1/3}$ is dimensionless except for a numerical constant which depends upon the units of σ, P_c and T_c. Van der Waals (1894) suggested that this group could be correlated with $1 - T_r$. Brock and Bird (1955) developed this idea for nonpolar liquids and proposed that

$$\frac{\sigma}{P_c^{2/3}T_c^{1/3}} = (0.132\alpha_c - 0.279)(1 - T_r)^{11/9} \tag{12-3.3}$$

where α_c is the Riedel (1954) parameter at the critical point and α is defined as $d \ln P_{vpr}/d \ln T_r$ (see Eq. (7-5.2). Using a suggestion by Miller (1963) to relate α_c to T_{br} and P_c,

$$\alpha_c = 0.9076\left[1 + \frac{T_{br}\ln(P_c/1.01325)}{1 - T_{br}}\right] \tag{12-3.4}$$

it can be shown that

TABLE 12-1 Comparison of Calculated and Experimental Values of Surface Tension of Pure Liquids

			Percent error in method**					
			Eq. 12-3.1, [P] from					
Compound	T, K	σ (exp)* dyn/cm	Table 11-3	Quayle, 1953	Brock & Bird, eq. (12-3.5)	Pitzer, eq. (12-3.7)	Sastri & Rao, eq. (12-3.12)	Zuo & Stenby, eq. (12-3.9)
acetic acid	293	27.59	−5.3	−0.3	53	60	−2.3	53
	333	23.62	−7.2	−2.2	50	57	−4.1	50
acetone	298	24.02	−3.8	−4.5	0.7	4.7	−6.8	1.1
	308	22.34	−2.8	−3.5	2.0	6.1	−5.6	2.4
	318	21.22	−3.7	−4.4	0.9	4.9	−6.7	1.3
aniline	293	42.67	−1.5	−0.8	11	15	−4.9	10
	313	40.5	−3.1	−2.4	10	14	−5.8	9.4
	333	38.33	−4.5	−3.8	9.1	13	−6.7	8.3
	353	36.15	−5.6	−5.0	8.0	11	−7.7	7.3
benzene	293	28.88	−1.9	−0.1	−2.9	1.5	−4.0	−1.6
	313	26.25	−2.1	−0.4	−2.7	1.6	−3.9	−1.7
	333	23.67	−2.5	−0.8	−2.6	1.7	−3.7	−1.7
	353	21.2	−2.4	−0.7	−2.8	1.5	−3.9	−2.0
benzonitrile	293	39.37	−50	−0.4	1.5	5.2	−6.8	1.3
	323	35.89	−51	−1.7	1.4	5.1	−7.0	1.2
	363	31.26	−51	−2.5	1.5	5.1	−6.9	1.3
bromobenzene	293	35.82	1.2	1.8	−0.2	4.0	−2.4	1.0
	323	32.34	0.4	1.1	−0.1	4.1	−2.4	0.9
	373	26.54	1.3	1.9	0.6	4.9	−1.6	1.5
n-butane	203	23.31	2.6	1.1	1.4	6.7	7.1	3.7
	233	19.69	1.8	0.3	0.6	5.8	6.2	2.5
	293	12.46	1.8	0.4	0.6	5.8	6.2	1.7
carbon disulfide	293	32.32	−69	−2.3	2.7	10	2.6	6.8
	313	29.35	−69	−3.0	2.5	10	2.4	6.3

carbon tetrachloride	288	27.65	19	0.1	−4.8	−0.5	−2.6	−3.5
	308	25.21	19	−0.3	−5.0	−0.7	−2.9	−3.9
	328	22.76	19	0.1	−5.1	−0.8	−2.9	−4.1
	348	20.31	19	−0.1	−4.9	−0.6	−2.7	−4.1
	368	17.86	21	1.4	−4.4	−0.1	−2.2	−3.9
chlorobenzene	293	33.59	−0.2	−1.5	−1.6	2.8	−3.8	−0.3
	323	30.01	−0.6	−1.9	−1.7	2.7	−3.8	−0.5
	373	24.06	1.1	−0.2	−1.1	3.3	−3.3	−0.1
p-cresol	313	34.88	2.2	3.5	40	46	12	39
	373	29.32	1.5	2.7	36	42	8.7	35
cyclohexane	293	25.24	−2.1	−3.4	−4.3	0.2	−1.3	−2.9
	313	22.87	−1.5	−2.8	−4.2	0.3	−1.2	−3.0
	333	20.49	0.1	−1.3	−3.9	0.7	−0.8	−2.8
cyclopentane	293	22.61	1.0	5.0	−2.7	2.0	−0.7	−1.3
	313	19.68	2.8	6.9	−0.6	4.2	1.5	0.6
diethyl ether	288	17.56	−0.5	−0.5	0.4	4.5	0.5	0.9
	303	16.2	−2.3	−2.3	−2.3	1.7	−2.1	−1.8
2,3-dimethylbutane	293	17.38	1.2	0.6	−0.5	4.3	6.5	0.9
	313	15.38	1.8	1.2	−0.6	4.1	6.3	0.5
ethyl acetate	293	23.97	−2.8	−0.6	0.8	4.6	−5.5	0.8
	313	21.65	−3.1	−1.0	−0.1	3.6	−6.4	−0.1
	333	19.32	−2.6	−0.4	−0.9	2.7	−7.2	−0.9
	353	17	−2.5	−0.3	−1.7	1.9	−7.9	−1.7
	373	14.68	−0.6	1.7	−2.3	1.3	−8.5	−2.3
ethyl benzoate	293	35.04	−0.1	−1.1	2.5	8.0	−4.5	2.8
	313	32.92	−0.8	−1.8	2.6	8.0	−4.5	2.9
	333	30.81	−1.3	−2.3	2.7	8.2	−4.4	3.1
ethyl bromide	283	25.36	−4.0	−0.4	15	22	5.7	18
	303	23.04	−6.5	−3.0	13	20	3.6	16

TABLE 12-1 Comparison of Calculated and Experimental Values of Surface Tension of Pure Liquids (*Continued*)

			Percent error in method**					
			Eq. 12-3.1, [*P*] from					
Compound	*T*, *K*	σ (exp)* dyn/cm	Table 11-3	Quayle, 1953	Brock & Bird, eq. (12-3.5)	Pitzer, eq. (12-3.7)	Sastri & Rao, eq. (12-3.12)	Zuo & Stenby, eq. (12-3.9)
ethyl mercaptan	288	23.87	−5.4		2.6	7.7	0.5	4.1
	303	22.68	−8.7		−1.4	3.6	−3.3	0.0
formamide	298	57.02	−7.7	−7.7				
	338	53.66	−13	−13				
	373	50.71	−17	−17				
n-heptane	293	20.14	0.6	0.6	0.3	4.5	3.9	0.8
	313	18.18	0.8	0.8	0.2	4.5	3.8	0.8
	333	16.22	1.9	1.9	0.4	4.6	4.0	0.9
	353	14.26	3.0	3.0	0.9	5.1	4.5	1.4
isobutyric acid	293	25.04	−1.3	0.0	60	68	1.1	58
	313	23.2	−2.1	−0.7	59	68	0.7	57
	333	21.36	−2.4	−1.0	59	67	0.4	57
	363	18.6	−1.7	−0.4	59	67	0.2	57
methyl formate	293	24.62	−7.5	3.4	4.6	8.0	−5.4	4.4
	323	20.05	−6.1	5.0	4.6	8.4	−5.3	4.6
	373	12.9	−8.4	2.4	4.3	8.1	−5.6	3.8
	423	6.3	−6.7	4.2	5.6	9.5	−4.4	4.5
	473	0.87	−16	−5.6	21	25	9.5	18
methyl alcohol	293	22.56	−13	−1.0	101	109	0.7	98
	313	20.96	−14	−3.1	92	100	0.5	90
	333	19.41	−17	−5.5	82	90	−0.3	81
phenol	313	39.27	−5.3	−4.3	30	35	3.6	29
	333	37.13	−5.9	−4.9	29	34	2.6	28
	373	32.96	−6.4	−5.4	26	31	0.4	25

n-propyl alcohol	293	23.71	0.8	0.0	62	70	−1.2	60
	313	22.15	−0.7	−1.4	56	64	−1.2	54
	333	20.6	−1.8	−2.5	50	57	−1.4	48
	363	18.27	−3.5	−4.2	39	46	3.5	38
n-propyl benzene	293	29.98	−1.5	−4.1	−3.3	0.6	−3.5	−3.0
	313	26.83	2.0	−0.7	0.4	4.5	0.2	0.8
	333	24.68	2.6	−0.1	1.0	5.1	0.8	1.4
	353	22.53	3.8	1.2	1.9	6.0	1.6	2.3
	373	20.38	5.9	3.2	3.1	7.2	2.8	3.4
pyridine	293	37.21	−29	−1.5	0.2	4.0	−6.7	0.9
	313	34.6	−29	−2.4	−0.2	3.5	−7.1	0.4
	333	31.98	−30	−2.9	−0.6	3.2	−7.4	−0.1
AAPD			8.6	2.4	13.8	17.0	4.0	13.4

*Experimental values for methyl formate from Macleod, 1923, all others from Jasper, 1972

**Error = [(calc.-exp./exp.] × 100

$$\sigma = P_c^{2/3} T_c^{1/3} Q(1 - T_r)^{11/9} \tag{12-3.5}$$

$$Q = 0.1196 \left[1 + \frac{T_{br} \ln(P_c/1.01325)}{1 - T_{br}}\right] - 0.279 \tag{12-3.6}$$

where temperature and pressure are in kelvins and bars respectively.

Pitzer (Curl and Pitzer, 1958; Pitzer, 1995) gives a series of relations for σ in terms of P_c, T_c, and ω that together lead to the following corresponding-states relation for σ

$$\sigma = P_c^{2/3} T_c^{1/3} \frac{1.86 + 1.18\omega}{19.05} \left[\frac{3.75 + 0.91\omega}{0.291 - 0.08\omega}\right]^{2/3} (1 - T_r)^{11/9} \tag{12-3.7}$$

ω is the acentric factor and the units are the same as in Eq. (12-3.5). It is stated that a deviation of more than 5% from this relation for a substance "appears to indicate significant *abnormality*." This means that Eq. (12-3.7) can be used as a test for whether a fluid can be considered a *normal* fluid (See also Chaps. 2 and 4).

In yet another corresponding-states method, Zuo and Stenby (1997) have used a two-reference fluid corresponding-states approach patterned after work done by Rice and Teja (1982) to estimate surface tensions. Unlike Rice and Teja who used T_c and V_c as reducing parameters, Zuo and Stenby used T_c and P_c according to:

$$\sigma_r = \ln\left(1 + \frac{\sigma}{T_c^{1/3} P_c^{2/3}}\right) \tag{12-3.8}$$

The units in Eq. (12-3.8) are bar, kelvins, and dyn/cm. To use this method, σ_r for the fluid of interest is related to σ_r for two reference fluids, methane (1) and *n*-octane (2) by

$$\sigma_r = \sigma_r^{(1)} + \frac{\omega - \omega^{(1)}}{\omega^{(2)} - \omega^{(1)}} (\sigma_r^{(2)} - \sigma_r^{(1)}) \tag{12-3.9}$$

For methane,

$$\sigma^{(1)} = 40.520(1 - T_r)^{1.287} \tag{12-3.10}$$

and for *n*-octane,

$$\sigma^{(2)} = 52.095(1 - T_r)^{1.21548} \tag{12-3.11}$$

The procedure to calculate surface tensions with Eqs. (12-3.8) to (12-3.11) is illustrated in Example 12-2.

While the three corresponding-states methods described above are satisfactory for nonpolar liquids, they are not satisfactory for compounds that exhibit strong hydrogen-bonding (alcohols, acids). To deal with these compounds, Sastri and Rao (1995) present the following modification of the above equations

$$\sigma = K P_c^x T_b^y T_c^z \left[\frac{1 - T_r}{1 - T_{br}}\right]^m \tag{12-3.12}$$

The units in Eq. (12-3.12) are kelvins and bars. Values for the constants are given in Table 12-2.

TABLE 12-2 Values of Constants for Sastri-Rao Method, Eq. (12-3.12)

	K	x	y	z	m
Alcohols	2.28	0.25	0.175	0	0.8
Acids	0.125	0.50	−1.5	1.85	11/9
All others	0.158	0.50	−1.5	1.85	11/9

Sastri's method is illustrated in Example 12-2 and results for a number of compounds are presented in Table 12-1. In Table 12-1, constants for the "all others" category listed in Table 12-2 were used for phenol and *p*-cresol as per Sastri and Rao (1995).

Example 12.2 Estimate the surface tension of liquid ethyl mercaptan at 303 K. The experimental value is 22.68 dyn/cm (Jasper, 1972).

solution From Appendix A, for ethyl mercaptan, T_c = 499 K, T_b = 308.15 K, P_c = 54.9 bar, and from Eq. (2-3.3) ω = 0.192. Thus T_{br} = 308.15/499 = 0.6175 and T_r = 303/499 = 0.6072.

BROCK and BIRD, Eq. (12-3.5).

With Eq. (12-3.6)

$$Q = 0.1196\left[1 + \frac{(0.6175)\ln(54.9/1.01325)}{1 - 0.6175)}\right] - 0.279 = 0.6114$$

$$\sigma = (54.9)^{2/3}\,(499)^{1/3}\,(0.6114)(1 - 0.6072)^{11/9} = 22.36 \text{ dyn/cm}$$

$$\text{Error} = \frac{22.36 - 22.68}{22.68} \times 100 = -1.4\%$$

PITZER, Eq. (12-3.7)

$$\sigma = (54.9)^{2/3}\,(499)^{1/3}\,\frac{1.86 + 1.18 \times 0.192}{19.05}\left[\frac{3.74 + 0.91 \times 0.192}{0.291 - 0.08 \times 0.192}\right]^{2/3}$$

$$(1 - 0.6072)^{11/9}$$

$$\sigma = 23.49 \text{ dyn/cm}$$

$$\text{Error} = \frac{23.49 - 22.68}{22.68} \times 100 = 3.6\%$$

ZUO-STENBY, Eqs. (12-3.8) to (12-3.11)

For methane, from App. A, T_c = 190.56 K, P_c = 45.99 bar, and ω = 0.011. For *n*-octane, T_c = 568.70 K, P_c = 24.90 bar, and ω = 0.399. With Eqs. (12-3.8) and (12-3.10):

$$\sigma^{(1)} = 40.520(1 - 0.607)^{1.287} = 12.180 \text{ dyn/cm}$$

$$\sigma_r^{(1)} = \ln\left(1 + \frac{12.180}{(190.56)^{1/3}\,(45.99)^{2/3}}\right) = 0.1526$$

Similarily for reference fluid 2, *n*-octane, $\sigma^{(2)}$ = 16.74 dyn/cm, and $\sigma_r^{(2)}$ = 0.2127. With Eq. (12-3.9):

$$\sigma_r = 0.1526 + \left(\frac{0.192 - 0.011}{0.399 - 0.011}\right)(0.2127 - 0.1526) = 0.1806$$

Finally, Eq. (12-3.8) may be solved for the desired value of σ to give:

$$\sigma = (499)^{1/3}\,(54.9)^{2/3}[-1 + \exp(0.1806)] = 22.68 \text{ dyn/cm}$$

$$\text{Error} = \frac{22.68 - 22.68}{22.68} \times 100 = 0\%$$

SASTRI-RAO, Eq. (12-3.12)

From Table 12-2, $K = 0.158$, $x = 0.50$, $y = -1.5$, $z = 1.85$, and $m = 11/9$.

$$\sigma = (0.158)(54.9)^{0.50}\,(308.2)^{-1.5}\,(499)^{1.85}\left[\frac{1 - 0.6072}{1 - 0.6176}\right]^{11/9} = 21.92$$

$$\text{Error} = \frac{21.92 - 22.68}{22.68} \times 100 = -3.3\%$$

Discussion

Table 12-1 presents a comparison of results calculated by the methods in this section to experimental surface tension values. The lowest average absolute percent deviation (AAAD) is obtained when parachor values from Quayle (1953) are used. This is to be expected because these values are based on experimental data. Table 11-3 often, but not always, leads to reliable parachor values. The behavior of the three corresponding states methods is nearly identical. All give unacceptable results for alcohols, acids, and phenols. For other polar compounds, results are mixed, and for nonpolar compounds, all three methods predict surface tensions to within 5%. Of the predictive methods, the Sastri-Rao method is the only one that does not give large deviations in at least some cases.

Other approaches to surface tension estimation include a method by Escobedo and Mansoori (1996) in which the parachor is related to the refractive index, and a method by Hugill and van Welsenes (1986) which employs a corresponding-states type expression for the parachor. Both of these methods work well for nonpolar fluids but can lead to large errors for polar compounds.

Another approach to the prediction of surface tensions of pure fluids is to change the reference fluids used in Eq. (12-3.9). For example, Rice and Teja (1982) predicted surface tensions of six alcohols generally to within 5% when ethanol and pentanol were chosen as the reference fluids. This is much better than the results shown in Table 12-1 for methanol and propanol when the reference fluids were methane and octane.

Recommendations

For surface tensions of organic liquids, use the data collection of Jasper (1972). For non-hydrogen-bonded liquids, any of the methods presented in this section may be used [Eqs. (12-3.1) or (12-3.5) to (12-3.12)]. Errors are normally less than 5%.

For hydrogen-bonded liquids, use the Macleod-Sugden form [Eq. (12-3.1)] with the parachor determined from group contributions in Table 11-3 or the Sastri-Rao method, Eq. (12-3.12). Errors are normally less than 5 to 10%.

12-4 VARIATION OF PURE-LIQUID SURFACE TENSION WITH TEMPERATURE

Equations in the previous section indicate that

$$\sigma \propto (1 - T_r)^n \tag{12-4.1}$$

where n varies from 0.8 for alcohols in the Sastri-Rao method to 1.22, or 11/9 for other compounds. In Fig. 12-1, log σ is plotted against log $(1 - T_r)$ with experimental data for acetic acid, diethyl ether, and ethyl acetate. For the latter two, the slope is close to 1.25; for acetic acid, the slope is 1.16. For most organic liquids except alcohols, this encompasses the range normally found for n.

For values of T_r between 0.4 and 0.7, Eq. (12-4.1) indicates that $d\sigma/dT$ is almost constant, and often the surface tension-temperature relation is represented by a linear equation

$$\sigma = a + bT \tag{12-4.2}$$

As an example, for nitrobenzene, between 313 and 473 K, data from Jasper (1972) are plotted in Fig. 12-2. The linear approximation is satisfactory. Jasper lists values of a and b for many materials.

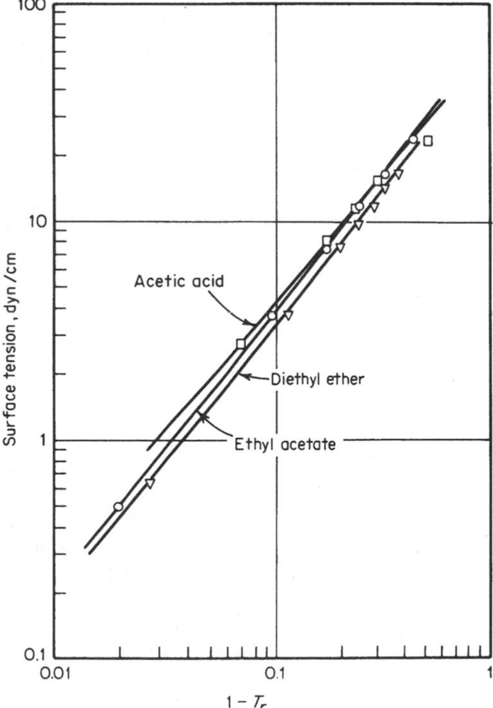

FIGURE 12-1 Variation of surface tension with temperature. *(Data from Macleod, 1923)*

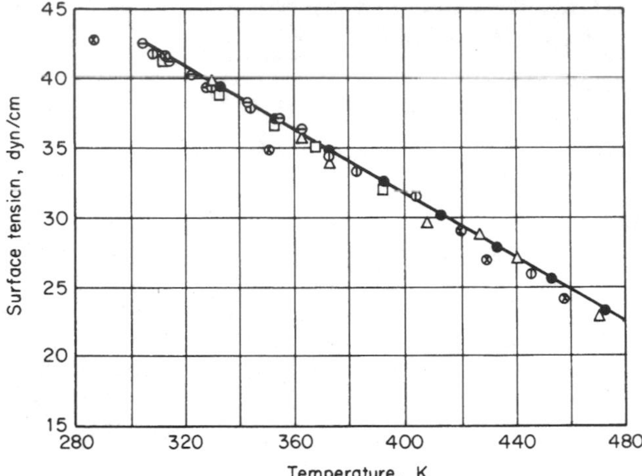

FIGURE 12-2 Surface tension of nitrobenzene, from Jasper (1972) which presents similar graphs for 56 compounds along with references to the original data.

12-5 SURFACE TENSIONS OF MIXTURES

The surface tension of a liquid mixture is not a simple function of the surface tensions of the pure components because, in a mixture, the composition of the surface is not the same as that of the bulk. In a typical situation, we know the bulk composition but not the surface composition. The derivative $d\sigma_m/dx$ usually increases as the concentration of the component with the largest pure component surface tension increases. Usually the component with the lowest surface tension concentrates in the surface phase so that the surface tension of a mixture σ_m is usually, but not always (Agarwal, et al., 1979; Zihao and Jufu, 1982) less than that calculated from a mole fraction average of the surface tensions of the pure components (the excess surface tension is usually negative). This behavior is illustrated in Fig. 12-3 which shows mixture surface tensions for several systems. All illustrate the usual trend with nonlinearity of the σ_m vs. x relation to different degrees. The surface tension of the acetophenone-benzene system is almost linear in composition, whereas the nitromethane-benzene and nitrobenzene-carbon tetrachloride systems are decidedly nonlinear and the diethyl ether-benzene case is intermediate. Systems become more nonlinear as the difference in pure component surface tensions increase, or as the system becomes more nonideal. For the systems shown in Fig. 12-3, system 1 deviates the most from ideal solution behavior while system 4 has the greatest difference between pure component surface tensions. These are also the two systems with the greatest deviation from linearity.

The techniques suggested for estimating σ_m can be divided into two categories: those based on empirical relations suggested earlier for pure liquids and those derived from thermodynamics. The empirical relations can be used when the pure component surface tensions do not differ greatly and when deviations from ideal solution behavior are not large. In reality these relations have been used most often

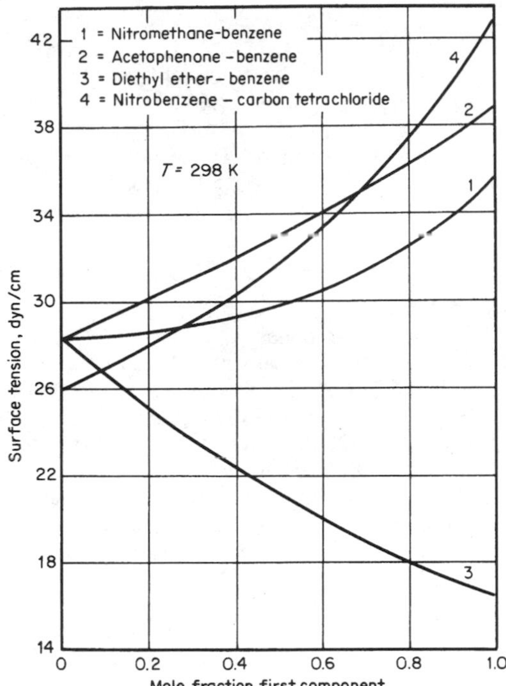

FIGURE 12-3 Mixture surface tensions. (*Hammick and Andrew, 1929*)

to correlate existing data, and in this role, they have been successful at both low and high pressures. Thermodynamic-based methods require more involved calculations, but lead to more reliable results.

Macleod-Sugden Correlation

Applying Eq. (12-3.1) to mixtures gives

$$\sigma_m = [[P_{Lm}]\rho_{Lm} - [P_{Vm}]\rho_{Vm}]^n \tag{12-5.1}$$

where σ_m = surface tension of mixture, dyn/cm
$[P_{Lm}]$ = parachor of the liquid mixture
$[P_{Vm}]$ = parachor of the vapor mixture
ρ_{Lm} = liquid mixture density, mol/cm^3
ρ_{Vm} = vapor mixture density, mol/cm^3

Hugill and van Welsenes (1986) recommend

$$[P_{Lm}] = \sum_i \sum_j x_i x_j [P_{ij}] \tag{12-5.2}$$

$$[P_{Vm}] = \sum_i \sum_j y_i y_j [P_{ij}] \tag{12-5.3}$$

where x_i is the mole fraction of component i in the liquid and y_i is the mole fraction of component i in the vapor. In Eqs. (12-5.2) and (12-5.3)

$$[P_{ij}] = \lambda_{ij} \frac{[P_i] + [P_j]}{2} \tag{12-5.4}$$

where $[P_i]$ = parachor of pure component i. In Eq. (12-5.4), λ_{ij} is a binary interaction coefficient determined from experimental data. In the absence of experimental data, λ_{ij} may be set equal to one, and if n in Eq. (12-5.1) is set equal to 4, Eq. (12-5.1) reduces to the Weinaug-Katz (1943) equation. Recent studies (Gasem, et al., 1989; Zuo and Stenby, 1997) in which n has been fit to experimental data recommend that a value of 3.6 be used for n.

At low pressures, the term involving the vapor density may be neglected; when this simplification is possible, Eq. (12-5.1) has been employed to correlate mixture surface tensions for a wide variety of organic liquids with reasonably good results (Bowden and Butler, 1939; Gambill, 1958; Hammick and Andrew, 1929; Meissner and Michaels, 1949; Riedel, 1955). Many authors, however, do not obtain $[P_i]$ from general group contribution methods or from pure component density and surface tension behavior; instead, they regress mixture data to obtain the best value of $[P_i]$ for each component in the mixture. This procedure leads to an improved description of the mixture data but may not reproduce the pure component behavior. Application of Eq. (12-5.1) is illustrated in Example 12-3.

For gas-liquid systems under high pressure, the vapor term in Eq. (12-5.1) becomes significant. Weinaug and Katz (1943) showed that Eqs. (12-5.1) and (12-5.4) with $n = 4$ and all $\lambda_{ij} = 1.0$ correlate methane-propane surface tensions from 258 to 363 K and from 2.7 to 103 bar. Deam and Maddox (1970) also employed these same equations for the methane-nonane mixture from 239 to 297 K and 1 to 101 bar. Some smoothed data are shown in Fig. 12-4. At any temperature, σ_m decreases with increasing pressure as more methane dissolves in the liquid phase. The effect of temperature is more unusual; instead of decreasing with rising temperature, σ_m increases, except at the lowest pressures. This phenomenon illustrates the fact that at the lower temperatures methane is more soluble in nonane and the effect of liquid composition is more important than the effect of temperature in determining σ_m.

Gasem, et al. (1989) have used Eqs. (12-5.1) to (12-5.4) to correlate the behavior of mixtures of carbon dioxide and ethane in various hydrocarbon solvents including butane, decane, tetradecane, cyclohexane, benzene, and trans-decalin. The measurements range from about 10 bar to the critical point of each system. They recommended a value of $n = 3.6$. When values of $[P_i]$ were regressed and λ_{ij} was set to unity, the average absolute deviations for the ethane and CO_2 systems were 5% and 9%, respectively. When λ_{ij} was also regressed, there was only marginal improvement in the description of the ethane systems while the average deviation in the CO_2 systems decreased to about 5%. Other systems for which Eq. (12-5.1) has been used to correlate high-pressure surface tension data include methane-pentane and methane-decane (Stegemeier, 1959), nitrogen-butane and nitrogen-heptane

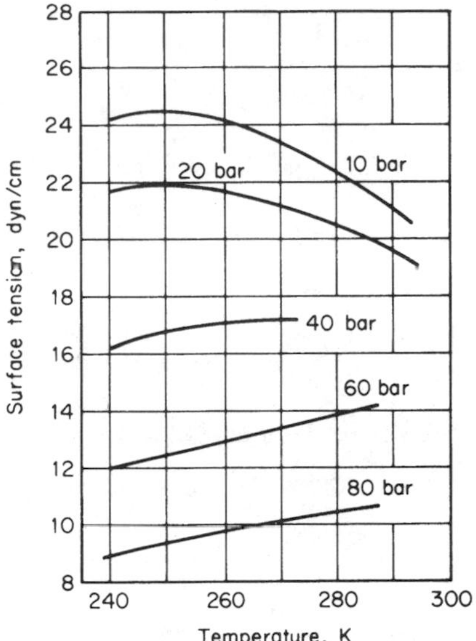

FIGURE 12-4 Surface tension for the system methane-nonane.

(Reno and Katz, 1943), and the effect of pressure of N_2 and H_2 on the surface tension of liquid ammonia (Lefrançois and Bourgeois, 1972).

When the Macleod-Sugden correlation is used, errors at low pressures rarely exceed 5 to 10% and can be much less if $[P_i]$ values are obtained from experimental data. It is desirable that mixture liquid and vapor densities and compositions be known accurately. However, Zuo and Stenby (1997) have correlated the behavior of a number of systems including petroleum fractions by calculating densities with the Soave equation of state, even though this equation does not predict accurate liquid densities. They then fit $[P]$ to surface and interfacial tension data so the error in liquid density is compensated for in the correlation for the parachor. This emphasizes the fact that the parachor is a calculated quantity. Parachor values calculated by Eq. (12-5.1) with an exponent of 4 should obviously not be used in a mixture equation in which the exponent is some other value.

Example 12.3 Use Eq. (12-5.1) to estimate the interfacial tension of a carbon dioxide (1) − *n*-decane (2) mixture at 344.3 K, 11380 kPa, and with $x_1 = 0.775$. At these conditions, Nagarajan and Robinson (1986) report $y_1 = 0.986$, $\rho_{Lm} = 0.7120$ g/cm^3, $\rho_{Vm} = 0.3429$ g/cm^3, and $\sigma_m = 1.29$ mN/m.

solution Use $[P_1] = 73.5$ and $[P_2] = 446.2$. These are the values recommended by Gasem, et al. (1989) when $n = 3.6$ and $\lambda_{ij} = 1$. From Appendix A, $M_1 = 44.010$ and $M_2 = 142.285$.

With Eqs. (12-5.2) to (12-5.4)

$$[P_{Lm}] = (0.775)^2(73.5) + (2(0.775)(0.225)\left(\frac{73.5 + 446.2}{2}\right) + (0.225)^2(446.2)$$

$$= 157.4$$

$$[P_{Vm}] = (0.986)^2(73.5) + 2(0.986)(0.014)\left(\frac{73.5 + 446.2}{2}\right) + (0.014)(446.2)$$

$$= 78.7$$

Converting density to a molar density,

$$\rho_{Lm} = \frac{0.7120}{(0.775)(44.01) + (0.225)(142.285)} = 0.01077 \text{ mol/cm}^3$$

Similarly, $\rho_{Vm} = 0.00756 \text{ mol/cm}^3$. With Eq. (12-5.1)

$$\sigma_m = [(157.4)(0.001077) - (78.7)(0.00756)]^{3.6} = 1.41 \text{ mN/m}$$

$$\text{Error} = \frac{1.41 - 1.29}{1.29} \times 100 = 9.3\%$$

In Example 12-3, the value used for $[P_2]$ of 446.2 was determined by a fit to the data set of Nagarajan and Robinson for which the carbon dioxide liquid phase mole fractions ranged from 0.5 to 0.9. Using the surface tension and density of pure decane to determine $[P_2]$ leads to a value of 465. Using this value in Example 12-3 leads to an error of 25%. In other words, Eq. 12-5.1 does not describe the behavior of the CO_2–decane system over the entire composition range for the temperature in Example 12-3.

Discussion

Often, when only approximate estimates of σ_m are necessary, one may choose the general form

$$\sigma_m^r = \sum_i^n x_i \sigma_i^r \tag{12-5.5}$$

Hadden (1966) recommends $r = 1$ for most hydrocarbon mixtures, which would predict linear behavior in surface tension vs. composition. For the nonlinear behavior as shown in Fig. 12-3, closer agreement is found if $r = -1$ to -3.

Zuo and Stenby (1997) have extended Eqs. (12-3.8) to (12-3.11) to mixtures with success at low to moderate pressures by using a pseudocritical temperature and pressure calculated from the Soave equation of state by applying Eqs. (4-6.5*b*) and (4-6.5*c*) to the mixture EoS. For mixtures containing only hydrocarbons, no interaction parameter was required, but for mixtures containing CO_2 or methane, an interaction parameter was fit to experimental data. Because the pseudocritical point differs from the true critical point, this method breaks down as the true critical point of the mixture is approached. For this case, Eq. (12-5.1) has led to better results because the equation necessarily predicts that σ_m goes to zero as the true critical point is approached. In addition to the work of Gasem, et al. already described, both Hugill and van Welsenes (1986) and Zuo and Stenby (1997) have developed equations for $[P_i]$ in terms of T_{ci}, P_{ci}, and ω_i. These two sets of investigators used different values for n, calculated densities with different equations of state, and ended up with two different equations for $[P_i]$, one predicting that $[P_i]$ goes up with ω_i, while the other predicts that $[P_i]$ goes down with ω_i. This illustrates the importance of documenting how one obtains phase densities and compositions, and illustrates the empirical nature of the parachor approach. When these equations were used to calculate the pure component surface tensions in Table 12-1, deviations were much higher than for the other methods shown in Table 12-1.

Aqueous Systems

Whereas for nonaqueous solutions the mixture surface tension in some cases can be approximated by a linear dependence on mole fraction, aqueous solutions show pronounced nonlinear characteristics. A typical case is shown in Fig. 12-5 for acetone-water at 353 K. The surface tension of the mixture is represented by an approximately straight line on semilogarithmic coordinates. This behavior is typical of organic-aqueous systems, in which small concentrations of the organic material may significantly affect the mixture surface tension. The hydrocarbon portion of the organic molecule behaves like a hydrophobic material and tends to be rejected from the water phase by preferentially concentrating at the surface. In such a case, the bulk concentration is very different from the surface concentration. Unfortunately, the latter is not easily measured. Meissner and Michaels (1949) show graphs similar to Fig. 12-5 for a variety of dilute solutions of organic materials in water and suggest that the general behavior is approximated by the Szyszkowski equation, which they modify to the form

$$\frac{\sigma_m}{\sigma_W} = 1 - (0.411) \log \left(1 + \frac{x}{a}\right) \tag{12-5.6}$$

where σ_W = surface tension of pure water
x = mole fraction of organic material
a = constant characteristic of organic material

Values of a are listed in Table 12-3 for a few compounds. This equation should not be used if the mole fraction of the organic solute exceeds 0.01. For some substances this is well below the solubility limit.

The method of Tamura, et al. (1955) may be used to estimate surface tensions of aqueous binary mixtures over wide concentration ranges of the dissolved organic material and for both low- and high-molecular weight organic-aqueous systems.

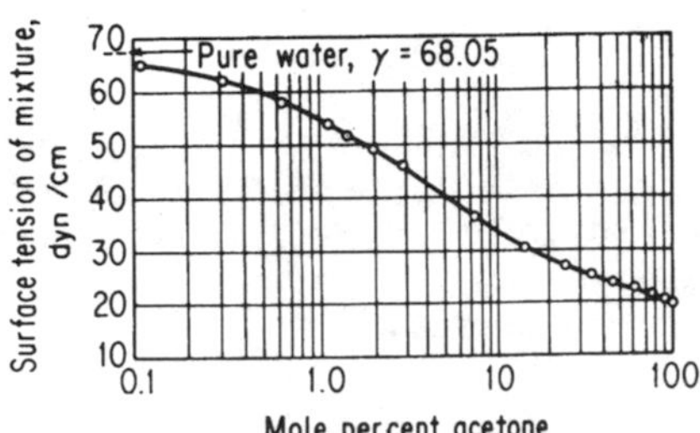

FIGURE 12-5 Surface tensions of water-acetone solutions at 353 K. *(McAllister and Howard, 1957)*

TABLE 12-3 Constants for the Szyszkowski Equation (12-5.6) (Meissner and Michaels, 1949)

Compound	$a \times 10^4$	Compound	$a \times 10^4$
Propionic acid	26	Ethyl propionate	3.1
n-Propyl alcohol	26	Propyl acetate	3.1
Isopropyl alcohol	26	*n*-Valeric acid	1.7
Methyl acetate	26	Isovaleric acid	1.7
n-Propyl amine	19	*n*-Amyl alcohol	1.7
Methyl ethyl ketone	19	Isoamyl alcohol	1.7
n-Butyric acid	7.0	Propyl propionate	1.0
Isobutyric acid	7.0	*n*-Caproic acid	0.75
n-Butyl alcohol	7.0	*n*-Heptanoic acid	0.17
Isobutyl alcohol	7.0	*n*-Octanoic acid	0.034
Propyl formate	8.5	*n*-Decanoic acid	0.0025
Ethyl acetate	8.5		
Methyl propionate	8.5		
Diethyl ketone	8.5		

Equation (12-5.1) is assumed as a starting point, but the significant densities and concentrations are taken to be those characteristic of the surface layer, that is, $(V^\sigma)^{-1}$ replaces ρ_{Lm}, where V^σ is a hypothetical molal volume of the surface layer. V^σ is estimated with

$$V^\sigma = \sum_j x_j^\sigma V_j \tag{12-5.7}$$

where x_j^σ is the mole fraction of j in the surface layer. V_j, however, is chosen as the pure liquid molal volume of j. Then, with Eq. (12-5.1), assuming $\rho_L >> \rho_v$,

$$V^\sigma \sigma_m^{1/4} = x_W^\sigma [P_W] + x_O^\sigma [P_O] \tag{12-5.8}$$

where the subscripts W and O represent water and the organic component. To eliminate the parachor, however, Tamura, et al. introduce Eq. (12-3.1); the result is

$$\sigma_m^{1/4} = \psi_W^\sigma \sigma_W^{1/4} + \psi_O^\sigma \sigma_O^{1/4} \tag{12-5.9}$$

In Eq. (12-5.9), ψ_W^σ is the superficial volume fraction water in the surface layer

$$\psi_W^\sigma = \frac{x_W^\sigma V_W}{V^\sigma} \tag{12-5.10}$$

and similarly for ψ_O^σ.

Equation (12-5.9) is the final correlation. To obtain values of the superficial surface volume fractions ψ_W^σ and ψ_O^σ, equilibrium is assumed between the surface and bulk phases. Tamura's equation is complex, and after rearrangement it can be written in the following set of equations:

$$B = \log \frac{\psi_W^q}{\psi_O} \tag{12-5.11}$$

$$W = 0.441 \frac{q}{T} \left(\frac{\sigma_O V_O^{2/3}}{q} - \sigma_W V_W^{2/3} \right) \tag{12-5.12}$$

$$C = B + W \tag{12-5.13}$$

$$\frac{(\psi_W^\sigma)^q}{\psi_O^\sigma} = 10^C \tag{12-5.14}$$

where ψ_W^σ is defined by Eq. (12-5.10) and ψ_W, ψ_O are the superficial bulk volume fractions of water and organic material, i.e.,

$$\psi_W = \frac{x_W V_W}{x_W V_W + x_O V_O} \qquad \psi_O = \frac{x_O V_O}{x_W V_W + x_O V_O} \tag{12-5.15}$$

where x_W, x_O = bulk mole fraction of pure water and pure organic component
V_W, V_O = molal volume of pure water and pure organic component
σ_W, σ_O = surface tension of pure water and pure organic component
T = temperature, K
q = constant depending upon type and size of organic constituent

Equation (12-5.14) along with the equation, $\psi_W^\sigma + \psi_O^\sigma = 1$ allows values of ψ_W^σ and ψ_O^σ to be determined so that σ_m can be found from Eq. (12-5.9).

The method is illustrated in Example 12-4. Tamura, et al. (1955) tested the method with some 14 aqueous systems and 2 alcohol-alcohol systems; the percentage errors are less than 10% when q is less than 5 and within 20% for q greater than 5. The method cannot be applied to multicomponent mixtures. For nonaqueous mixtures comprising polar molecules, the method is unchanged except that q = ratio of molal volumes of the solute to solvent.

Materials	q	Example
Fatty acids, alcohols	Number of carbon atoms	Acetic acid, $q = 2$
Ketones	One less than the number of carbon atoms	Acetone, $q = 2$
Halogen derivatives of fatty acids	Number of carbons times ratio of molal volume of halogen derivative to parent fattty acid	$q = 2 \dfrac{V_b\ (\textit{chloroacetic acids})}{V_b(\textit{acetic acid})}$

Example 12.4 Estimate the surface tension of a mixture of methyl alcohol and water at 303K when the mole fraction alcohol is 0.122. The experimental value reported is 46.1 dyn/cm (Tamura, et al., 1955).

solution At 303 K (O represents methyl alcohol, W water), σ_W = 71.18 dyn/cm, σ_O = 21.75 dyn/cm, V_W = 18 cm^3/mol, V_O = 41 cm^3/mol, and q = number of carbon atoms = 1. From Eqs. (12-5.15),

$$\frac{\psi_W}{\psi_O} = \frac{(0.878)(18)}{(0.122)(41)} = 3.16$$

and from Eq. (12-5.11),

$$B = \log 3.16 = 0.50$$

$$W \text{ [from Eq. (12-5.12)]} = (0.144)\left(\frac{1}{303}\right)[(21.75)(41)^{2/3} - (71.18)(18)^{2/3}] = -0.34$$

Hence

$$C \text{ [from Eq. (12-5.13)]} = B + W = 0.50 - 0.34 = 0.16$$

From Eq. (12-5.14), with $q = 1$,

$$\frac{\psi_W^\sigma}{\psi_O^\sigma} = 10^{0.16} = 1.45$$

Using $\psi_W^\sigma + \psi_O^\sigma = 1$, we have

$$\frac{\psi_W^\sigma}{1 - \psi_W^\sigma} = 1.45$$

$$\psi_W^\sigma = 0.59 \qquad \psi_O^\sigma = 0.41$$

Finally, from Eq. (12-5.9)

$$\sigma_m = [(0.59)(71.18)^{1/4} + (0.41)(21.75)^{1/4}]^4 = 46 \text{ dyn/cm}$$

$$\text{Error} = \frac{46 - 46.1}{46.1} \times 100 = -0.2\%$$

Thermodynamic-Based Relations

The estimation procedures introduced earlier in this section are empirical; all except the Tamura, et al. method employ the bulk liquid (and sometimes vapor) composition to characterize a mixture. However, the "surface phase" usually differs in composition from that of the bulk phases, and it is reasonable to suppose that, in mixture surface tension relations, surface compositions are more important than bulk compositions. The fact that σ_m is almost always less than the bulk mole fraction average is interpreted as indicating that the component or components with the lower pure component values of σ preferentially concentrate in the surface phase.

The assumptions that the bulk and surface phases are in equilibrium and the partial molar area of component i is the same as the molar area of i leads to the following equations (Sprow and Prausnitz, 1966a):

$$\sigma_m = \sigma_i + \frac{RT}{A_i} \ln \frac{x_i^\sigma \gamma_i^\sigma}{x_i \gamma_i} \quad (i = 1, 2, \ldots N) \tag{12-5.16}$$

$$\sum_i x_i^\sigma = 1. \tag{12-5.17}$$

where σ_m = mixture surface tension, dyn/cm
σ_i = surface tension of pure component i, dyn/cm
R = 8.314×10^7 dyn·cm/(mol·K)
A_i = surface area of component i, cm^2/mol, see Table 12-4
T = temperature, K
x_i = mole fraction of component i in the bulk phase
x_i^σ = mole fraction of component i in the surface phase
γ_i = activity coefficient of component i in the bulk phase
γ_i^σ = activity coefficient of component i in the surface phase

When γ_i^σ is related to the surface composition and γ_i to the bulk liquid composition, Eqs. (12-5.4) and (12-5.5) represent $N + 1$ equations in the $N + 1$ unknowns, σ_m and the N values of x_i^σ. Hildebrand and Scott (1964) have examined the case where $\gamma_i^\sigma = 1$, and Eckert and Prausnitz (1964) and Sprow and Prausnitz (1966, 1966a) have used regular solution theory for γ_i^σ. None of these versions, however, was particularly successful for aqueous mixtures. Suarez, et al. (1989) have used a version of the UNIFAC model by Larsen, et al. (1987) to determine the surface and bulk phase activity coefficients. Larsen's UNIFAC model differs from the one presented in Chap. 8 in that ln γ^c is determined from Eq. (12-5.18) and the a_{mn} parameters of Eq. (8-10.67) are functions of temperature.

$$\ln \gamma_i^c = \ln \frac{\phi_i}{x_i} + 1 - \frac{\phi_i}{x_i} \tag{12-5.18}$$

$$\phi_i = \frac{x_i r_i^{2/3}}{\sum_j x_j r_j^{2/3}} \tag{12-5.19}$$

where x_i = mole fraction of component i in the bulk phase
r_i = UNIFAC volume parameter of molecule i determined by method in Chap. 8

For nonaqueous mixtures with a difference between pure-component surface tensions not exceeding around 20 dyn/cm, Suarez, et al. (1989) claim that σ_m values are predicted with an average error of 3.5% when pure component areas A_i are calculated by

$$A_i = 1.021 \times 10^8 V_c^{6/15} V_b^{4/15} \tag{12-5.20}$$

where V_c and V_b are in cm^3/mol and A_i is in cm^2/mol. Suarez, et al. claim improved results when pure component areas shown in Table 12-4 are used. Table 12-4 values should be used if values of A_i are tabulated for all components. Otherwise, Eq. (12-5.20) should be used for all values. Values from Eq. (12-5.20) and Table 12-4 should not be mixed. Suarez, et al. report average deviations for binary systems including aqueous systems of 3% and deviations of 4% for ternary systems. Zhibao, et al. (1990) have also used UNIFAC to predict surface tensions and report results similar to those of Suarez, et al. The Suarez method is illustrated in Example 12-5.

Example 12.5 Use the Suarez method, Eqs. (12-5.6) and (12-5.7) along with Larsen's (1987) UNIFAC method to estimate σ_m for a mixture of 5 weight % n-propanol(1) and 95 weight % H_2O(2) at 298 K. Vázquez, et al. (1995) report an experimental valve of 41.83 dyn/cm. They also give σ_1 = 23.28 dyn/cm and σ_2 = 72.01 dyn/cm.

TABLE 12-4 Values of A_i to be used in Eq. (12-5.16)

Component	$A_i \times 10^8$ cm^{2}mol^{-1}
Water	0.7225
Methanol	3.987
Ethanol	8.052
1-Propanol	17.41
2-Propanol	20.68
Ethylene glycol	4.123
Glycerol	3.580
1,2-Propanediol	6.969
1,3-Propanediol	8.829
1,3-Butanediol	9.314
1,4-Butanediol	8.736
Acetonitrile	6.058
Acetic acid	6.433
1,4-Dioxane	12.27
Acetone	8.917
Methyl ethyl ketone	12.52
n-Hexane	11.99
Benzene	9.867
Toluene	9.552
Pyridine	10.35

solution There are 4 UNIFAC groups, —CH_3, —CH_2, —OH, and H_2O. R_i and Q_i for these groups are

	—CH_3	—CH_2	—OH	H_2O
R_i	0.9011	0.6744	1.0	0.92
Q_i	0.848	0.54	1.2	1.4

$$r_i = 0.9011 + (2)(0.6744) + 1.0 = 3.2499$$

Similarly, $r_2 = 0.92$, $q_1 = 3.128$, and $q_2 = 1.4$

a_{mn} values at 298 K from Larsen, et al. (1987) are

	—CH_3	—CH_2	—OH	H_2O
—CH_3	0	0	972.8	1857
—CH_2	0	0	972.8	1857
—OH	637.5	637.5	0	155.6
H_2O	410.7	410.7	−47.15	0

From Table 12-4, $A_1 = 17.41 \times 10^8$ cm²/mol and $A_2 = 0.7225 \times 10^8$ cm²/mol. Equations. (12-5.18) and (12-5.19) are used for the combinatorial contribution to γ and Eqs. (8-10.61) and (8-10.64) to (8-10.67) are used for the residual contribution. The bulk composition of 5 weight % *n*-propanol corresponds to $x_1 = 0.01553$ and $x_2 = 0.98447$. At this composition, $\gamma_1 = 10.015$ and $\gamma_2 = 1.002$.

Using Eq. (12-5.16) for component 1

$$\sigma_m - \sigma_1 + \frac{(8.314 \times 10^7)(298)}{8.052 \times 10^8} \ln \frac{x_1^\sigma \gamma_1^\sigma}{(0.01553)(10.015)}$$

$$= 23.28 + 30.77 \ln(6.4295 x_1^\sigma \gamma_1^\sigma)$$

similarly for component 2

$$\sigma_m = 72.01 + 342.9 \ln(1.0136\, x_2^\sigma \gamma_2^\sigma)$$

These two equations plus the condition, $x_1^\sigma + x_2^\sigma = 1.0$, along with the UNIFAC relations for γ_1^σ and γ_2^σ must be solved iteratively. The solution is

$x_1^\sigma = 0.269$, $x_2^\sigma = 0.731$, $\gamma_1^\sigma = 2.05$, $\gamma_2^\sigma = 1.234$ and $\sigma_m = 41.29$ dyn/cm. Thus,

$$\text{Error} = \frac{41.29 - 41.83}{41.83} \times 100 = -1.3\%$$

Note that this model predicts that the component with the lower surface tension, *n*-propanol, is 17 times more concentrated in the surface than the bulk. Using the UNIFAC method in Chap. 8 to calculate activity coefficients at both the bulk and surface concentrations would have predicted a value of σ_m of 43.40 dyn/cm for an error of 3.8%. If one just takes both bulk and surface activity coefficients equal to 1 in Example 12-5, the error is 13%. The Suarez method gives an error of –0.4% for the case of Example 12-4.

Recommendations

For estimating the surface tensions of mixtures, the Suarez method [Eqs. (12-5.18), (12-5.19), and Example 12-5] is generally recommended. However, in certain circumstances, other methods might be preferred. Near mixture critical points, the Macleod-Sugden correlation [Eq. (12-5.1) and Example 12-3] should be used because the form of the equation necessarily gives the correct limit that σ goes to zero at the critical point. For nonpolar mixtures, extension of the corresponding-states method of Zuo and Stenby [Eqs. (12-3.8) to (12-3.11)] to mixtures gives results as reliable as the Suarez method and the calculational procedure is simpler.

For estimating the surface tensions of binary organic-aqueous mixtures, use either the Suarez method or the method of Tamura, et al. as given by Eqs. (12-5.6) to (12-5.14) and illustrated in Example 12-4. For multicomponent mixtures with water as one component, the Suarez method should be used. If the solubility of the organic compound in water is low, the Szyszkowski equation (12-5.6), as developed by Meissner and Michaels, may be used. Of these three methods, the Suarez approach is most broadly applicable but the most complex. The Szyszkowski method is the simplest to use but should be used only when the solute mole fraction is less than 0.01. Furthermore, a value for constant a must be available in Table 12-3.

Interfacial Tensions in Liquid-Liquid Binary Systems

Li and Fu (1991) have presented a UNIQUAC-based equation to predict interfacial tensions in systems with two liquid phases and two components. Unlike the empirical methods presented earlier for interfacial tensions at high pressure (Sec. 12-5), the Li-Fu equation is for highly nonideal systems at low (near atmospheric) pressure. Li and Fu propose

$$\sigma = 3.14 \times 10^{-9}\,(1 - k_{12})W_{12}(\phi_1^I - \phi_1^{II})^2 \tag{12-5.21}$$

where ϕ_{ij} the volume fraction of i, is calculated by

$$\phi_1^I = \frac{x_1^I r_1}{x_1^I + x_2^I r_2} \tag{12.5.22}$$

x_1^I is the mole fraction of component 1 in the phase rich in component 1
x_1^{II} is the mole fraction of component 1 in the phase rich in component 2
r_i is the UNIQUAC volume parameter for component i (See Chap. 8)

$$W_{12} = \frac{R(\Delta U_{12} + \Delta U_{21})}{z} \tag{12-5.23}$$

z is the coordination number, taken as 10
R is the ideal gas constant, here taken as 8.314×10^7 dyn·cm/(mol·K)
ΔU_{12} and ΔU_{21} are UNIQUAC parameters.

UNIQUAC parameters, along with solubility data required in Eq. (12-5.22) have been tabulated for many binary systems in Sørensen and Arlt (1979). In Eq. (12-5.21), the constant 3.14×10^{-9} has units mol/cm^2 and σ is in dyn/cm when the value of R given above is used.

Li and Fu suggest that the parameter k_{12} accounts for orientation effects of molecules at the interface and recommend the empirical equation

$$k_{12} = 0.467 - 0.185\,X + 0.016X^2 \tag{12-5.24}$$

where

$$X = -\ln(x_1^{II} + x_2^I) \tag{12-5.25}$$

For 48 binary systems Li and Fu claim an average absolute percent deviation of 8.8% with Eq. (12-5.21). Other methods (Hecht, 1979; Li and Fu, 1989) give slightly lower deviations, but require either numerical integration or a numerical solution of a set of non-linear equations.

Example 12.6 Use Eq. (12-5.21) to estimate the interfacial tension of the benzene (1)–water (2) system at 20°C. The experimental value (Fu, et al., 1986) is 33.9 dyn/cm. Also, from page 341 of Sørensen and Arlt (1979), $x_1^{II} = 2.52 \times 10^{-3}$, $x_2^I = 4.00 \times 10^{-4}$, $\Delta U_{12} = 882.10$ K and $\Delta U_{21} = 362.50$ K.

solution From Eqs. (12-5.24) and (12-5.25)

$$X = -\ln\,(2.52 \times 10^{-3} + 4.00 \times 10^{-4}) = 5.836$$

$$k_{12} = 0.467 - (0.185)(5.836) + (0.016)(5.836)^2 = -0.0677$$

From Eq. (12-5.23)

$$W_{12} = \frac{8.314 \times 10^7}{10}(882.10 + 362.50) = 1.035 \times 10^{10} \text{ dyn·cm/mol}$$

$$x_1^I = 1 - 4.00 \times 10^{-4} = 0.9996$$

$$x_2^{II} = 1 - 2.52 \times 10^{-3} = 0.99748$$

$$r_1 = 3.1878 \text{ and } r_2 = 0.92$$

$$\phi_1^I = \frac{(0.9996)(3.1878)}{(0.9996)(3.1878) + (4.00 \times 10^{-4})(0.92)} = 0.9999$$

$$\phi_1^{II} = \frac{(2.52 \times 10^{-3})(3.1878)}{(2.52 \times 10^{-3})(3.1878) + (0.99748)(0.92)} = 0.00868$$

$$\sigma = (3.14 \times 10^{-9})(1 + 0.0677)(1.035 \times 10^{10})(0.9999 - 0.00868)^2 = 34.1 \text{ dyn/cm}$$

$$\text{Error} = \frac{34.1 - 33.9}{33.9} \times 100 = 0.6\%$$

NOTATION

- a parameter in Eq. (12-5.6) and obtained from Table 12-3
- A_i area of component i, cm^2/mol
- $[P_i]$ parachor of component i (see Table 11-3)
- P_{vp} vapor pressure, bar; P_{vpr}, reduced vapor pressure, P_{vp}/P_c
- P_c critical pressure, bar
- q parameter in Eqs. (12-5.12) and (12-5.14)
- Q parameter in Eq. (12-3.5)
- R gas constant, 8.314 J/(mol·K) or 8.314 × 10^7 dyne cm/(mol·K)
- T temperature, K; T_c, critical temperature; T_b, normal boiling point, T_r, reduced temperature T/T_c; $T_{br} = T_b/T_c$;
- V liquid molal volume, cm^3/mol; V^σ, for the surface phase; V_c, critical volume; V_b, volume at T_b
- x_i liquid mole fraction; x_i^σ, mole fraction of i in the surface phase
- y_i vapor mole fraction of component i

Greek

- α_c Riedel factor, Eq. (7-5.2)
- γ_i activity coefficient of component i in the bulk liquid; γ_i^σ, in the surface phase; γ_i^c, combinatorial contribution to γ, see Eq. (12-5.18)
- η liquid or vapor viscosity, cP
- ρ liquid or vapor density, mol/cm^3
- σ surface tension, dyn/cm; σ_m for a mixture; σ_O, representing an organic component, σ_r reduced surface tension, see Eq. (12-3.8)
- ϕ_i UNIFAC volume fraction of i, Eqs. (12-5.19) and (12-5.22)
- ψ_i volume fraction of i in the bulk liquid Eq. (12-5.15); ψ_i^σ, in the surface phase, Eq. (12-5.10)
- ω acentric factor, Eq. (2-3.1)

Subscripts

- b normal boiling point
- L liquid

m mixture
O organic component in aqueous solution
r reduced value, i.e., the property divided by its value at the critical point
V vapor
W water

Superscripts

σ surface phase
I, II liquid phase I, liquid phase II

REFERENCES

Adamson, A. W.: *Physical Chemistry of Surfaces,* 4th ed., Wiley, New York, 1982.

Agarwal, D. K., R. Gopal, and S. Agarwal: *J. Chem. Eng. Data,* **24:** 181 (1979).

Aveyard, R., and D. A. Haden: *An Introduction to Principles of Surface Chemistry* Cambridge, London, 1973.

Barbulescu, N.: *Rev. Roum. Chem.,* **19:** 169 (1974).

Bowden, S. T., and E. T. Butler: *J. Chem. Soc.,* **1939:** 79.

Brock, J. R., and R. B. Bird: *AIChE J.,* **1:** 174 (1955).

Brown, R. C.: *Contemp. Phys.,* **15:** 301 (1974).

Chattoraj, D. K., and K. S. Birdi: *Adsorption and the Gibbs Surface Excess,* Plenum Press, New York, 1984.

Curl, R. F., Jr., and K. S. Pitzer: *Ind. Eng. Chem.,* **50:** 265 (1958).

Daubert, T. E., R. P. Danner, H. M. Sibel, and C. C. Stebbins, *Physical and Thermodynamic Properties of Pure Chemicals: Data Compilation,* Taylor & Francis, Washington, D. C. 1997.

Deam, J. R., and R. N. Maddox: *J. Chem. Eng. Data,* **15:** 216 (1970).

Eckert, C. A., and J. M. Prausnitz: *AIChE J.,* **10:** 677 (1964).

Escobedo, J., and G. A. Mansoori: *AIChE J.,* **42:** 1425 (1996).

Evans, D. F., and J. Wennerstöm: The Colloidal Domain, Wiley-VCH, New York, 1999.

Everett, D. H.: *Basic Principles of Colloid Science,* Royal Society of Chemistry, London, 1988.

Fu, J., B. Li, and Z. Wang: *Chem. Eng. Sci.,* **41:** 2673 (1986).

Gambill, W. R.: *Chem. Eng.,* **64**(5)**:** 143 (1958).

Gambill, W. R.: *Chem. Eng.,* **66**(23)**:** 193 (1959).

Gasem, K. A. M., P. B. Dulcamara, Jr., K. B. Dickson, and R. L. Robinson, Jr.: *Fluid Phase Equil.,* **53:** 39 (1989).

Gibbs, J. W.: *The Collected Works of J. Willard Gibbs,* vol. I, *Thermodynamics,* Yale University Press, New Haven, Conn., 1957.

Goldhammer, D. A.: *Z. Phys. Chem.,* **71:** 577 (1910).

Gray, J. A., C. J. Brady, J. R. Cunningham, J. R. Freeman, and G. M. Wilson: *Ind. Eng. Chem. Process Des. Dev.,* **22:** 410 (1983).

Guggenheim, E. A.: *Thermodynamics,* 4th ed., North-Holland, Amsterdam, 1959.

Hadden, S. T.: *Hydrocarbon Process Petrol. Refiner,* **45**(10)**:** 161 (1966).

Hammick, D. L., and L. W. Andrew: *J. Chem. Soc.,* **1929:** 754.

Hecht, G.: *Chem. Technol.,* **31:** 143 (1979).

Hildebrand, J. H., and R. L. Scott: *The Solubility of Nonelectrolytes,* 3d ed., Dover New York, 1964, chap. 21.

Hugill, J. A., and A. J. van Welsenes: *Fluid Phase Equil.,* **29:** 383 (1986).

International Critical Tables, vol. III, McGraw-Hill, New York, 1928, p. 28.
Jasper, J. J.: *J. Phys. Chem. Ref. Data,* **1:** 841(1972).
Körösi, G., and E. sz. Kováts: *J. Chem. Eng. Data,* **26:** 323 (1981).
Larsen, B. L., P. Rasmussen, and A. Fredenslund: *Ind. Eng. Chem. Res.,* **26:** 2274 (1987).
Lefrançois, H., and Y. Bourgeois: *Chim. Ind. Genie Chim.,* **105**(15)**:** 989 (1972).
Li, B., and J. Fu: *Chem Eng Sci.,* **44:** 1519 (1989).
Li, B. and J. Fu: *Fluid Phase Equil.,* **64:** 129 (1991).
Macleod, D. B.: *Trans. Faraday Soc.,* **19:** 38 (1923).
McAllister, R. A., and K. S. Howard: *AIChE J.,* **3:** 325 (1957).
Meissner, H. P., and A. S. Michaels: *Ind. Eng. Chem.,* **41:** 2782 (1949).
Miller, D. G.: *Ind. Eng. Chem. Fundam.,* **2:** 78 (1963).
Nagarajan, N., and R. L. Robinson, Jr.: *J. Chem. Eng. Data,* **31:** 168 (1986).
Pitzer, K. S.: *Thermodynamics,* 3d ed., New York, McGraw-Hill, 1995, p. 521.
Quayle, O. R.: *Chem. Rev.,* **53:** 439 (1953).
Reno, G. J., and D. L. Katz: *Ind. Eng. Chem.,* **35:** 1091 (1943).
Rice, O. K.: *J. Phys. Chem.,* **64:** 976 (1960).
Rice, P., and A. S. Teja: *J. Colloid. Interface Sci.,* **86:** 158 (1982).
Riddick, J. A., W. B. Bunger, and T. K. Sakano: *Organic Solvents, Physical Properties and Methods of Purification,* 4th ed., Techniques of Chemistry, Vol. II, Wiley, New York, 1986.
Riedel, L.: *Chem. Ing. Tech.,* **26:** 83 (1954).
Riedel, L.: *Chem. Ing. Tech.,* **27:** 209 (1955).
Ross, S. (Chairman): *Chemistry and Physics of Interfaces,* American Chemical Society, Washington, D.C., 1965.
Rowlinson, J. S., and G. Widom: *Molecular Theory of Capillarity,* Oxford University Press, New York, 1982.
Sastri, S. R. S., and K. K. Rao: *Chem. Eng. J.,* **59:** 181 (1995).
Sørensen, J. M., and W. Arlt: *Liquid-Liquid Equilibrium Data Collection,* Vol. 1, DECHEMA, 1979.
Sprow, F. B., and J. M. Prausnitz: *Trans. Faraday Soc.,* **62:** 1097 (1966).
Sprow, F. B., and J. M. Prausnitz: *Trans. Faraday Soc.,* **62:** 1105 (1966a); *Can. J. Chem. Eng.,* **45:** 25 (1967).
Stegemeier, G. L.: Ph.D. dissertation, University of Texas, Austin, Tex., 1959.
Suarez, J. T., C. Torres-Marchal, and P. Rasmussen: *Chem Eng. Sci.,* **44:** 782 (1989).
Sugden, S.: *J. Chem. Soc.,* 32, 1177 (1924).
Tamura, M., M. Kurata, and H. Odani: *Bull. Chem. Soc. Japan,* **28:** 83 (1955).
Tester, J. W., and M. Modell: *Thermodynamics and Its Applications,* 3d ed., Prentice Hall, Englewood Cliffs, NJ, 1997.
Timmermans, J.: *Physico-Chemical Constants of Pure Organic Compounds,* Vol. 2, Elsevier, Amsterdam, 1965.
van der Waals, J. D.: *Z. Phys. Chem.,* **13:** 716 (1894).
Vargaftik, N. B.: *Handbook of Physical Properties of Liquids and Gases: Pure Substances and Mixtures,* Begell House, New York, 1996.
Vázquez, G., E. Alvarez, and J. M. Navaza: *J. Chem. Eng. Data,* **40:** 611 (1995).
Weinaug, C. F., and D. L. Katz: *Ind. Eng. Chem.,* **35:** 239 (1943).
Zhibao, L., S. Shiquan, S. Meiren, and S. Jun: *Themochem. Acta,* **169:** 231 (1990).
Zihao, W., and F. Jufu: "An Equation for Estimating Surface Tension of Liquid Mixtures," *Proc. Jt. Mtg. of Chem. Ind. and Eng. Soc. of China and AIChE,* Beijing, China, September 1982.
Zuo, Y.-X., and E. H. Stenby: *Can. J. Chem. Eng.,* **75:** 1130 (1997).

APPENDIX A
PROPERTY DATA BANK

This appendix contains selected property values; many of these have been used to compare the methods of the text. It is limited to those pure substances for which an experimentally validated critical temperature is listed in the Thermodynamics Research Center (TRC) data bank, College Station, TX, USA, or in the other reliable sources listed below. The values have been published with permission of TRC. Not all properties are experimentally available for all substances listed, but some estimates have been made with the methods described in the main body.

The *Formula* listing for the substances has the atoms in alphabetical order except H follows C; the table has the species' formula in alphabetical order. The *Name* used is from the IUPAC as given by TRC, although we have also given common names for some substances.

Our symbols, fonts and equations are given below. The standard-state Gibbs energy and enthalpy of formation are for the species as an ideal gas at 298.15 K and 1.01325 bar (1 atmosphere). The reference states for the elements are as follows:

Ideal gases at 298.15 K and 1.01325 bar: Ar, Cl_2, D_2, F_2, He, H_2, Kr, Ne, O_2, Rn, T_2, Xe.

Crystalline solid at 298.15 K and 1.01325 bar: Al, As, B, C, I_2, P, S, Se, Si, Ti, U

Saturated Liquid at 298.15 K: Br_2, Hg

The table headings are as follows:

Section	Column Heading	Definition
All	No.	Number for this data base
All	Formula	alphabetical listing of atoms in molecule, except H follows C
All	Name	IUPAC name (common name)
All	CAS#	Chemical Abstracts Registry Number
A	Mol. Wt.	molecular weight, g mol^{-1}
A	Tfp	atmospheric (normal) freezing/melting point, K
A	Tb	atmospheric (1.01325 bar) boiling point, K
A	Tc	vapor/liquid critical temperature, K
A	Pc	vapor/liquid critical pressure, bar
A	Vc	vapor/liquid critical volume, cm^3 mol^{-1}
A	Zc	vapor/liquid critical compressibility factor = Pc*Vc/(R*Tc)

Section	Column Heading	Definition
A	Omega	Pitzer acentric factor = $-\log_{10}(Pvp/Pc)_{T/Tc=0.7} - 1$ (see Chap. 2)
B	DelHf0	standard state enthalpy of formation (see Chap. 3), kJ mol^{-1}
B	DelGf0	standard state Gibbs energy of formation (see Chap. 3), kJ mol^{-1}
B	DelHb	enthalpy change of atmospheric boiling, kJ mol^{-1} (see Chap. 7)
B	DelHm	enthalpy change of atmospheric melting, kJ mol^{-1}
B	V liq	liquid molar volume, cm^3 mol^{-1}, at T liq
B	T liq	temperature for V liq, K
B	Dipole	molecular dipole moment, debye, $3.162 \times 10^{-25}(\mathrm{J\ m^3})^{1/2}$ (see Chap. 2)
C	T range	range of temperatures for which ideal gas heat capacity at constant pressure (C_p°) data were fitted by TRC to polynomial $C_p^\circ/R = a0 + a1{*}T + a2{*}T^2 + a3{*}T^3 + a4{*}T^4$ with T in K. The value of C_p° is obtained by multiplying the result of the above equation by a value of the universal gas constant, R, and it will have the same units as the R used (see Table 2-1). Values at any temperature outside of T range are expected to be erroneous.
C	a0; a1; a2; a3; a4	parameters for TRC polynomial equation for C_p°/R.
C	CpIG	TRC tabulated values for C_p° at T = 298.15 K or calculated from the polynomial where R has been taken as 8.3143 J mol^{-1} K^{-1}
C	Cpliq	TRC or other tabulated values for C_p of the liquid at T = 298.15 K; J mol^{-1} K^{-1}
D	Equation #	number of equation for calculating pure vapor pressure, Pvp, bar (see Chap. 7)
	Equation #1:	$\log_{10}(Pvp) = A - B/(T + C - 273.15)$
	Equation #2:	$\log_{10}(Pvp) = A - B/(T + C - 273.15) + 0.43429x^n + E{*}x^8 + F{*}x^{12}$ where $x = (T - t_o - 273.15)/Tc$
	Equation #3:	$\ln(Pvp) = \ln(Pc) + (Tc/T){*}(a{*}tau + b{*}tau^{1.5} + c{*}tau^{2.5} + d{*}tau^5)$ where $tau = (1 - T/Tc)$ Note: for water only the last two terms are $c{*}tau^3 + d{*}tau^6$
D	A/A/Tc; B/B/a; C/C/b; Tc/c; t_o/d; n/Pc; E; F	parameters in Pvp Equations #1, #2, and #3. Usually A, B, C for Equation #1 are the same as those for Equation #2. Note that since Tc is a correlating parameter here, the values may differ from those in Section A.

Section	Column Heading	Definition
D	Pvpmin, bar; Tmin, K	minimum vapor pressure and temperature for an equation to be used. The minima for Equation #3 are at the triple point unless otherwise noted.
D	Pvpmax, bar; Tmax, K	maximum vapor pressure and temperature for an equation to be used. The maxima for Equation #3 are at the critical point unless otherwise noted.

Fonts have been selected to indicate the source and reliability of the values. They are as follows:

Normal (e.g., 123.45) is a value listed in the TRC data base from experiment. The data evaluation and original sources are listed in the TRC data sheets available from the Center or online at STN.

Underlined (123.45) is a value obtained from one of the references given below except a few values in section B that are from handbooks or are from the 4th Edition.

Italicized (*123.45*) is a value computed from other values in the table by the definitions given above. For C_p° the TRC polynomial has been used with R = 8.3145 J mol^{-1} K^{-1}.

Bold (**123.45**) is an estimated value. TRC estimates some Pc values; Ambrose obtains his from the limit of Equation #3 for Pvp; other values have been estimated using methods of the main body. Steele, et al. (1996abc, 1997abcd) also estimate Tc and Pc values by fitting vapor pressure data; when both are fitted, the reliability is questionable so we have included Pc values only if the Tc was experimentally based.

Bold and italicized (***123.45***) indicates that an estimated value was used in a calculation.

Acentric factors were calculated with vapor pressure equations in Section D. Values that required an extrapolation of the Pvp equation of up to 10 K over Tmax are indicated by bold font. Values are not listed when an extrapolation of more than 10 K would have been required. When Pc is in bold font (estimated) the acentric factor is also bold.

REFERENCES

Acree, Jr., W. E.: *Thermochimica Acta,* **189:** 37 (1991).

Ambrose, D., and N. B. Ghiassee: *J. Chem. Thermo.,* **19:** 505 (1987a).

Ambrose, D., and N. B. Ghiassee: *J. Chem. Thermo.,* **19:** 903 (1987b).

Ambrose, D., and N. B. Ghiassee: *J. Chem. Thermo.,* **19:** 911 (1987c).

Ambrose, D., and N. B. Ghiassee: *J. Chem. Thermo.,* **20:** 765 (1988a).

Ambrose, D., and N. B. Ghiassee: *J. Chem. Thermo.,* **20:** 1231 (1988b).

Ambrose, D., N. B. Ghiassee, and R. Tuckerman: *J. Chem. Thermo.,* **20:** 767 (1988c).

Ambrose, D., and J. Walton: *Pure & Appl. Chem.,* **61:** 1395 (1989).

Ambrose, D., and N. B. Ghiassee: *J. Chem. Thermo.,* **22:** 307 (1990).

Ambrose, D., M. B. Ewing, and N. B. Ghiassee: *J. Chem. Thermo.,* **22:** 589 (1990).

Ambrose, D., and C. Tsonopoulos: *J. Chem. Eng. Data,* **41:** 531 (1995).

Brown, J. A.: *J. Chem. Eng. Data,* **18:** 106 (1963).

Daubert, T. E.: *J. Chem. Eng. Data,* **41:** 365 (1996).

Defibaugh, D. R., and M. R. Moldover: *J. Chem. Eng. Data,* **42:** 160 (1997).

Fialho, P. S., and C. A. Nieto de Castro: *Fluid Phase Equil.,* **118:** 103 (1996).

Gude, M. and A. S. Teja: *J. Chem. Eng. Data,* **40:** 1025 (1995).

Lemmon, E. W., M. O. McLinden, and D. G. Friend: "Thermophysical Properties of Fluid Systems" in *NIST Chemistry WebBook, NIST Standard Reference Database Number 69,* W. G. Mallard and P. J. Linstrom (eds.), Nov. 1998, NIST Gaithersburg, MD (http://webbok.nist.gov).

McLinden, M. O., J. S. Gallagher, L. A. Weber, G. Morrison, D. Ward, A. R. H. Goodwin, M. R. Moldover, J. W. Schmidt, H. B. Chae, T. J. Bruno, J. F. Ely, and M. L. Huber: *ASHRAE Trans.,* **95:** 263 (1989).

Schmidt, J. W., D. Carrillo-Nava, and M. R. Moldover: *Fluid Phase Equil.,* **122:** 187 (1996).

Simmrock, K. H., R. Janowsky, and A. Ohnsorge: *Critical Data of Pure Substances,* Chemistry Data Series, V. II, DECHEMA, Frankfurt, 1986.

Steele, W. V., R. D. Chirico, S. E. Knipmeyer, and A. Nguyen: *J. Chem. Eng. Data,* **41:** 1255 (1996a).

Steele, W. V., R. D. Chirico, S. E. Knipmeyer, A. Nguyen, N. K. Smith, and I. R. Tasker: *J. Chem. Eng. Data,* **41:** 1269 (1996b).

Steele, W. V., R. D. Chirico, S. E. Knipmeyer, A. Nguyen, and N. K. Smith: *J. Chem. Eng. Data,* **41:** 1285 (1996c).

Steele, W. V., R. D. Chirico, S. E. Knipmeyer, and A. Nguyen: *J. Chem. Eng. Data,* **42:** 1008 (1997a).

Steele, W. V., R. D. Chirico, S. E. Knipmeyer, and A. Nguyen: *J. Chem. Eng. Data,* **42:** 1021 (1997b).

Steele, W. V., R. D. Chirico, S. E. Knipmeyer, A. Nguyen, and N. K. Smith: *J. Chem. Eng. Data,* **42:** 1037 (1997c).

Steele, W. V., R. D. Chirico, A. B. Crowell, S. E. Knipmeyer, and A. Nguyen: *J. Chem. Eng. Data,* **42:** 1053 (1997d).

Tsonopoulos, C. and D. Ambrose: *J. Chem. Eng. Data,* **40:** 547 (1995).

Tsonopoulos, C. and D. Ambrose: *J. Chem. Eng. Data,* **41:** 645 (1996).

Weber, L. A., and D. E. Defibaugh: *J. Chem. Eng. Data,* **41:** 762 (1996).

Zábranský, M., V. Rùzicka, V. Majer, and E. S. Domalski: Heat Capacity of Liquids: Critical Review and Recommended Values, *Amer. Chem. Soc., and Amer. Inst. Phys. for NIST,* Washington, D.C., 1996.

Section A Basic Constants I.

No.	Formula	Name	CAS #	Mol. Wt.	Tfp, K	Tb, K	Tc, K	Pc, bar	Vc, cm^3/mol	Zc = PcVc/RTc	Omega
1	Ar	argon	7440-37-1	39.948	83.80	*87.27*	150.86	48.98	74.57	*0.291*	−0.002
2	Br_2	bromine	7726-95-6	159.808	265.85	*331.90*	584.10	103.00	135.00	*0.269*	0.119
3	BrD	deuterium bromide	13536-59-9	81.918	185.69	*206.65*	362.00				
4	BrF_3	bromine trifluoride	7787-71-5	136.899	281.92	*398.89*	600.00	49.90	114.70	*0.115*	*0.413*
5	BrF_5	bromine pentafluoride	7789-30-2	174.896	212.65	*314.31*	470.00				
6	BrH	hydrogen bromide	10035-10-6	80.912	186.34	*206.46*	363.20	85.10			0.069
7	$CBrClF_2$	bromochlorodifluoromethane	353-59-3	165.365	113.65	*269.20*	426.90	42.60	246.00	*0.295*	*0.182*
8	$CBrF_3$	bromotrifluoromethane	75-63-8	148.910	105.15	*215.41*	340.15	39.70	200.00	*0.275*	*0.174*
9	CBr_2F_2	dibromodifluoromethane	75-61-6	209.816	163.05	*295.94*	471.30	45.30	250.00	*0.286*	
10	$CClF_3$	chlorotrifluoromethane	75-72-9	104.459	92.00	*191.71*	301.84	38.73	180.30	*0.276*	*0.175*
11	CCl_2F_2	dichlorodifluoromethane (R-12)	75-71-8	120.913	115.19	*243.45*	385.10	41.30	217.00	*0.280*	*0.179*
12	CCl_3F	trichlorofluoromethane (R-11)	75-69-4	137.368	162.69	*296.81*	471.10	44.72	248.00	*0.283*	*0.195*
13	CCl_4	tetrachloromethane	56-23-5	153.822	250.33	*349.79*	556.30	45.57	276.00	*0.271*	
14	CF_4	tetrafluoromethane	75-73-0	88.005	89.55	*145.11*	227.51	37.45	140.70	*0.279*	*0.177*
15	$CHBrF_2$	bromodifluoromethane	1511-62-2	130.920	258.65	*257.68*	412.00	47.90	167.00	*0.234*	*0.172*
16	$CHClF_2$	chlorodifluoromethane (R-22)	75-45-6	86.468	115.73	*232.14*	369.28	49.86	166.00	*0.274*	*0.221*
17	$CHCl_2F$	dichlorofluoromethane	75-43-4	102.923	138.20	*281.97*	451.52	51.87	196.00	*0.271*	*0.207*
18	$CHCl_3$	trichloromethane	67-66-3	119.377	209.74	*334.33*	536.50	55.00	240.00	*0.296*	
19	CHF_3	trifluoromethane (R-23)	75-46-7	70.014	117.96	*191.11*	298.97	48.36	133.00	*0.259*	*0.267*
20	CH_2Cl_2	dichloromethane	75-09-2	84.932	176.00	*312.79*	510.00	61.00			
21	CH_2F_2	difluoromethane	75-10-5	52.024	137.00	*221.43*	351.26	58.05	121.00	*0.245*	*0.278*
22	CH_2O_2	methanoic acid (formic acid)	64-18-6	46.026	281.50	***374.04***	**588.00**	**58.07**			***0.316***
23	CH_3Cl	chloromethane	74-87-3	50.488	175.44	*248.95*	416.20	66.80	143.00	*0.276*	*0.151*
24	CH_3F	fluoromethane	593-53-3	34.033	131.35	*194.88*	315.00	55.48	113.30	*0.240*	*0.204*
25	CH_3NO_2	nitromethane	75-52-5	61.040	244.60	374.35	588.00	58.70	173.00	*0.208*	
26	CH_4	methane	74-82-8	16.043	90.69	*111.66*	190.56	45.99	98.60	*0.286*	*0.011*
27	CH_4O	methanol	67-56-1	32.042	175.49	*337.69*	512.64	80.97	118.00	*0.224*	*0.565*
28	CH_4S	methanethiol (methyl mercaptan)	74-93-1	48.109	150.18	*279.11*	470.00	72.30	145.00	*0.268*	*0.150*
29	CH_5N	methanamine (methyl amine)	74-89-5	31.057	179.69	*266.82*	430.00	74.20	**125.00**	***0.259***	0.283
30	CO	carbon monoxide	630-08-0	28.010	68.15	*81.66*	132.85	34.94	93.10	*0.292*	***0.045***

No.	Formula	Name	CAS #	Mol. Wt.	Tfp, K	Tb, K	Tc, K	Pc, bar	Vc, cm^3/mol	Zc = PcVc/RTc	Omega
31	CO_2	carbon dioxide	124-38-9	44.010	216.58		304.12	73.74	94.07	*0.274*	*0.225*
32	$C_2Br_2ClF_3$	1,2-dibromo-2-chloro-1,1,2-trifluroethane	354-51-8	276.278	323.15	*367.06*	561.20	36.10	357.80	*0.285*	*0.251*
33	C_2ClF_5	1-chloro-1,1,2,2,2-pentafluoroethane	76-15-3	154.467	173.73	*234.08*	353.10	31.29	255.00	*0.273*	*0.251*
34	$C_2Cl_2F_4$	1,1-dichloro-1,2,2,2-tetrafluoroethane	374-07-2	170.921	216.58	*276.59*	418.70	32.13	294.00	*0.271*	*0.244*
35	$C_2Cl_2F_4$	1,2-dichloro-1,1,2,2-tetrafluoroethane	76-14-2	170.921	180.55	*276.58*	418.90	32.37	294.00	*0.273*	*0.246*
36	$C_2Cl_3F_3$	1,1,2-trichloro-1,2,2-trifluoroethane	76-13-1	187.375	236.95	*320.74*	487.40	33.78	325.00	*0.274*	*0.249*
37	C_2F_4	tetrafluoroethene	116-14-3	100.016	142.00	197.51	306.40	39.44	172.00	*0.266*	
38	C_2F_6	hexafluoroethane	76-16-4	138.012	173.05	*195.21*	293.04	30.39	221.90	*0.277*	*0.257*
39	$C_2HBrClF_3$	1-bromo-1-chloro-2,2,2-trifluoroethane	151-67-7	197.382		*323.32*	492.20	38.00	**302.00**	***0.280***	*0.283*
40	$C_2HBrClF_3$	1-bromo-2-chloro-1,1,2-trifluoroethane	354-06-3	197.382		*325.70*	487.30	34.20	**304.00**	***0.257***	*0.320*
41	C_2HClF_4	1-chloro-1,1,2,2-tetrafluoroethane	354-25-6	136.476	156.15	*261.38*	400.00	37.60	244.00	*0.285*	*0.260*
42	C_2HClF_4	1-chloro-1,2,2,2-tetrafluoroethane (R-124)	2837-89-0	136.476	155.00	*261.19*	395.60	36.34	243.80	*0.269*	*0.288*
43	$C_2HCl_2F_3$	1,1-dichloro-2,2,2-trifluoroethane (R-123)	306-83-2	152.931	166.00	*300.81*	456.90	36.74	278.05	*0.269*	*0.282*
44	$C_2HCl_2F_3$	1,2-dichloro-1,2,2-trifluoroethane (R-123a)	354-23-4	152.931	195	301	461.7		278.00		
45	C_2HF_5	pentafluoroethane	354-33-6	120.022	170.15	*225.06*	339.17	36.15	**211.30**	***0.271***	*0.305*
46	C_2HF_5O	pentafluorodimethyl ether (E-125)	3822-68-2	136.022	116.00	*235.00*	354.50	36.31	236.11	*0.237*	*0.326*
47	C_2H_2	ethyne (acetylene)	74-86-2	26.038	192.35	*188.40*	308.30	61.14	112.20	*0.268*	*0.189*
48	$C_2H_2F_2$	1,1-difluoroethene	75-38-7	64.035	129.15	187.50	303.20	44.33	153.50	*0.270*	
49	$C_2H_2F_4$	1,1,1,2-tetrafluoroethane (R-134a)	811-97-2	102.032	172.15	*247.04*	374.26	40.59	**200.80**	***0.262***	*0.326*
50	$C_2H_2F_4$	1,1,2,2-tetrafluoroethane (R-134)	359-35-3	102.032	184.15	*253.10*	391.74	46.40	**190.40**	***0.271***	*0.293*
51	$C_2H_3ClF_2$	1-chloro-1,1-difluoroethane	75-68-3	100.495	142.35	*264.05*	410.30	40.48	231.00	*0.267*	*0.231*
52	$C_2H_3Cl_2F$	1,1-dichloro-1-fluoroethane (R-141b)	1717-00-6	116.950	169.60	*305.20*	477.35	42.50	**254.00**	***0.272***	*0.225*
53	$C_2H_3F_3$	1,1,1-trifluoroethane (R-143a)	420-46-2	84.041	161.85	*225.86*	346.30	37.92	193.60	*0.255*	*0.259*
54	$C_2H_3F_3$	1,1,2-trifluoroethane (R-143)	430-66-0	84.041	189.15	*276.85*	429.80	52.41	**179.20**	***0.263***	*0.315*
55	C_2H_4	ethene (ethylene)	74-85-1	28.054	103.99	*169.42*	282.34	50.41	131.10	*0.282*	***0.087***
56	$C_2H_4Br_2$	1,2-dibromoethane	106-93-4	187.862	283.05	404.50	582.90	71.50	261.70		
57	$C_2H_4Cl_2$	1,1-dichloroethane	75-34-3	98.959	176.19	*330.45*	523.00	51.00	236.00	*0.277*	
58	$C_2H_4Cl_2$	1,2-dichloroethane	107-06-2	98.959	237.46	*356.66*	561.00	54.00	220.00	*0.255*	

59	$C_2H_4F_2$	1,1-difluoroethane (R-152a)	75-37-6	66.051	156.15	*249.10*	386.41	45.16	181.00	*0.252*	*0.276*
60	$C_2H_4O_2$	ethanoic acid (acetic acid)	64-19-7	60.053	289.77	*391.04*	594.45	57.90	171.00	*0.200*	*0.445*
61	$C_2H_4O_2$	methyl methanoate (methyl formate)	107-31-3	60.053	174.15	*304.90*	487.20	60.00	172.00	*0.255*	
62	C_2H_5Br	bromoethane	74-96-4	108.966	154.55	*311.50*	503.80	62.30	214.90		
63	C_2H_5Cl	chloroethane	75-00-3	64.514	134.82	*285.42*	460.30	53.00	199.00	*0.276*	
64	C_2H_5F	fluoroethane	353-36-6	48.060	129.95	*235.43*	375.28	50.27	164.00	*0.267*	*0.217*
65	C_2H_6	ethane	74-84-0	30.070	90.35	*184.55*	305.32	48.72	145.50	*0.279*	*0.099*
66	C_2H_6O	ethanol	64-17-5	46.069	159.05	*351.80*	513.92	61.48	167.00	*0.240*	*0.649*
67	C_2H_6O	dimethyl ether	115-10-6	46.069	131.65	*248.31*	400.10	54.00	170.00	*0.276*	
68	C_2H_6S	ethanethiol (ethyl mercaptan)	75-08-1	62.136	152.15	*308.15*	499.00	54.90	207.00	*0.274*	
69	C_2H_6S	2-thiapropane (dimethylsulfide)	75-18-3	62.136	174.85	*310.48*	503.00	55.30	201.00	*0.266*	
70	C_2H_7N	ethanamine (ethyl amine)	75-04-7	45.084	192.15	*289.75*	456.40	56.30	181.80	*0.267*	*0.276*
71	C_2H_7N	N-methylmethanamine (dimethyl amine)	124-40-3	45.084	180.96	280.00	437.20	53.40	180.00	*0.264*	
72	C_3F_8	octafluoropropane (R-218)	76-19-7	188.020	125.60	236.60	345.10	26.80	299.82	*0.280*	0.325
73	C_3HF_7	1,1,1,2,3,3,3-heptafluoropropane (R-227ea)	431-89-0	170.030		293.00	376.00		293.15		
74	$C_3H_2ClF_5O$	2-chloro-1,1,2-trifluoroethyl difluoromethyl ether (enflurane)	13838-16-9	184.493		*329.98*	**475.03**	**29.80**			***0.430***
75	$C_3H_2ClF_5O$	1-chloro-2,2,2-trifluoroethyl difluoromethyl ether (isoflurane)	26675-46-7	184.493		*322.42*	**467.80**	**30.46**			***0.402***
76	$C_3H_2F_6$	1,1,1,2,3,3-hexafluoropropane (R-236ea)	431-63-0	152.050		279.00	412.50		268.88		0.382
77	$C_3H_2F_6$	1,1,1,3,3,3-hexafluoropropane (R-236fa)	690-39-1	152.050	232.65		398.10		277.21		0.377
78	$C_3H_3F_5$	1,1,1,2,2-pentafluoropropane (R-245cb)	1814-88-6	134.050		255.10	380.40	31.48	268.74	*0.267*	0.297
79	$C_3H_3F_5$	1,1,1,3,3-pentafluoropropane (R-245fa)	460-73-1	134.050			427.00		259.28		0.385
80	$C_3H_3F_5$	1,1,2,2,3-pentafluoropropane (R-245ca)	679-86-7	134.050	191.20	301.20	447.60		256.02		0.353
81	$C_3H_3F_5O$	2-(difluoromethoxy)-1,1,1-trifluoroethane (E-245)	1885-48-9	150.050		302.20	444.00		291.02		
82	C_3H_3NO	1,2-oxazole(isoxazole)	288-14-2	55.058		*368.61*	590.00	**61.00**	**190.94**	***0.237***	***0.258***
83	C_3H_4	1-propyne (methyl acetylene)	74-99-7	40.065	170.50	*250.12*	402.40	56.30	163.50	*0.275*	
84	C_3H_4	1,2-propadiene	463-49-0	40.065	136.85	*238.77*	394.00	52.50	**173.90**	***0.279***	*0.122*
85	C_3H_6	propene (propylene)	115-07-1	42.081	87.89	*225.46*	364.90	46.00	184.60	*0.280*	*0.142*
86	C_3H_6	cyclopropane	75-19-4	42.081	145.73	*240.34*	398.25	55.75	162.80	*0.274*	*0.130*
87	$C_3H_6Cl_2$	1,2-dichloropropane	78-87-5	112.187	172.65	*369.43*	578.00	**46.50**	287.66	***0.278***	***0.255***
88	C_3H_6O	2-propen-1-ol (allyl alcohol)	107-18-6	58.080		370.23	545.10				
89	C_3H_6O	propanone (acetone)	67-64-1	58.080	178.50	*329.22*	508.10	47.00	209.00	*0.233*	*0.307*

Section A Basic Constants I (*Continued*)

No.	Formula	Name	CAS #	Mol. Wt.	Tfp, K	Tb, K	Tc, K	Pc, bar	Vc, cm^3/mol	Zc = PcVc/RTc	Omega
90	$C_3H_6O_2$	propanoic acid	79-09-4	74.079	252.31	*414.31*	604.00	**45.30**	233.00	***0.207***	***0.539***
91	$C_3H_6O_2$	methyl ethanoate (methyl acetate)	79-20-9	74.079	175.15	*330.09*	506.80	46.90	228.00	*0.254*	
92	$C_3H_6O_2$	ethyl methanoate (ethyl formate)	109-94-4	74.079	193.55	*327.47*	508.50	47.40	229.00	*0.257*	***0.282***
93	$C_3H_6O_3$	dimethylcarbonate	616-38-6	90.084		*363.24*	557.00	**48.00**	251.63	***0.261***	***0.336***
94	C_3H_7Cl	1-chloropropane	540-54-5	78.541	150.35	*319.67*	503.10	45.80	254.00	*0.278*	
95	C_3H_8	propane	74-98-6	44.097	91.45	*231.02*	369.83	42.48	200.00	*0.276*	*0.152*
96	C_3H_8O	1-propanol	71-23-8	60.096	147.00	*370.93*	536.78	51.75	219.00	*0.254*	*0.629*
97	C_3H_8O	2-propanol	67-63-0	60.096	183.65	*355.39*	508.30	47.62	220.00	*0.248*	*0.665*
98	C_3H_8O	methyl ethyl ether	540-67-0	60.096	160.00	*280.50*	437.80	44.00	221.00	*0.267*	***0.236***
99	C_3H_8S	2-thiabutane (methyl ethyl sulfide)	624-89-5	76.163	167.24	*339.80*	533.00	42.60	**260.00**	***0.250***	
100	C_3H_9N	1-propanamine (propyl amine)	107-10-8	59.111	188.35	*320.38*	497.00	48.00	**235.00**	***0.273***	*0.283*
101	C_3H_9N	2-propanamine (methyl ethyl amine)	75-31-0	59.111	177.95	*304.93*	471.80	45.40	221.00	*0.256*	***0.277***
102	C_3H_9N	N,N-dimethylmethanamine (trimethyl amine)	75-50-3	59.111	155.85	276.02	433.30	40.75	254.00	*0.266*	
103	C_3H_9NO	2-ethoxymethanamine (2-methylaminoethanol)	109-83-1	75.112		*432.39*	630.00	**53.00**	259.01	***0.262***	***0.604***
104	C_4F_8	octafluorocyclobutane	115-25-3	200.031	232.96	267.17	388.37	27.78	324.80	*0.279*	
105	C_4F_{10}	decafluoro-2-methylpropane	354-92-7	238.028		273.15	395.40	24.20	378.00	*0.278*	
106	C_4H_4O	furan	110-00-9	68.075	187.54	*304.44*	490.15	55.00	218.00	*0.281*	
107	C_4H_4S	thiophene	110-02-1	84.142	233.75	*357.31*	580.00	56.60	219.00	*0.252*	
108	C_4H_5N	pyrrole	109-97-7	67.090	249.74	403.00	639.70	63.30	**200.00**	***0.238***	
109	C_4H_6	1-butyne	107-00-6	54.092	147.29	*281.21*	440.00	46.00	208.00	*0.262*	***0.245***
110	C_4H_6	1,3-butadiene	106-99-0	54.092	164.24	*268.62*	425.00	43.20	221.00	*0.270*	*0.195*
111	$C_4H_6O_3$	acetic anhydride	108-24-7	102.090	199.00	*412.69*	**606.00**	**40.00**			***0.456***
112	C_4H_8	cyclobutane	287-23-0	56.108	182.42	*285.64*	460.00	49.90	218.00	*0.275*	*0.185*
113	C_4H_8	1-butene	106-98-9	56.108	87.79	*266.92*	419.50	40.20	240.80	*0.278*	*0.194*
114	C_4H_8	trans-2-butene	624-64-6	56.108	167.58	*274.03*	428.60	41.00	237.70	*0.276*	*0.218*
115	C_4H_8	cis-2-butene	590-18-1	56.108	134.25	*276.87*	435.50	42.10	233.80	*0.269*	*0.203*
116	C_4H_8	2-methylpropene	115-11-7	56.108	132.81	*266.24*	417.90	40.00	238.80	*0.275*	*0.199*
117	C_4H_8O	butanone (methyl ethyl ketone)	78-93-3	72.107	186.51	*352.71*	536.80	42.10	267.00	*0.252*	*0.322*

118	C_4H_8O	tetrahydrofuran	109-99-9	72.107	164.61	*339.12*	540.20	51.90	224.00	*0.259*	
119	$C_4H_8O_2$	butanoic acid	107-92-6	88.106	267.97	*436.87*	624.00	**40.30**	292.00	***0.227***	***0.600***
120	$C_4H_8O_2$	2-methylpropanoic acid	79-31-2	88.106	227.05	*427.57*	605.00	**37.00**	290.00	***0.213***	***0.618***
121	$C_4H_8O_2$	1,3-dioxane	505-22-6	88.106	228.15	379.20	588.00	45.80	**257.00**	***0.241***	
122	$C_4H_8O_2$	1,4-dioxane	123-91-1	88.106	284.15	374.50	587.00	51.70	238.00	*0.255*	
123	$C_4H_8O_2$	methyl propanoate	554-12-1	88.106	185.65	*352.60*	530.60	40.00	282.00	*0.256*	*0.349*
124	$C_4H_8O_2$	ethyl ethanoate (ethyl acetate)	141-78-6	88.106	189.55	*350.21*	523.20	38.30	286.00	*0.252*	*0.361*
125	$C_4H_8O_2$	propyl methanoate (propyl formate)	110-74-7	88.106	180.25	*354.69*	538.00	40.60	285.00	*0.259*	*0.320*
126	C_4H_9Cl	2-chlorobutane	78-86-4	92.568	141.85	*341.24*	520.60	36.80	312.00	*0.265*	*0.267*
127	C_4H_{10}	butane	106-97-8	58.123	134.79	*272.66*	425.12	37.96	255.00	*0.274*	*0.200*
128	C_4H_{10}	2-methylpropane (isobutane)	75-28-5	58.123	113.54	*261.34*	407.85	36.40	262.70	*0.278*	*0.186*
129	$C_4H_{10}N_2$	piperazine	110-85-0	86.136	384.6	*421.772*	661.00		**267.50**		
130	$C_4H_{10}O$	1-butanol	71-36-3	74.123	183.35	*390.88*	563.05	44.23	275.00	*0.260*	*0.590*
131	$C_4H_{10}O$	2-methyl-1-propanol (isobutanol)	78-83-1	74.123		*381.04*	547.78	43.00	273.00	*0.258*	*0.590*
132	$C_4H_{10}O$	2-methyl-2-propanol (tert-butanol)	75-65-0	74.123	298.55	*355.49*	506.21	39.73	275.00	*0.260*	*0.613*
133	$C_4H_{10}O$	2-butanol (sec-butanol)	78-92-2	74.123	158.50	*372.66*	536.05	41.79	269.00	*0.252*	*0.574*
134	$C_4H_{10}O$	diethyl ether	60-29-7	74.123	156.86	*307.59*	466.70	36.40	280.00	*0.263*	*0.281*
135	$C_4H_{10}O_2$	1,2-dimethoxyethane	110-71-4	90.126		*358.15*	537.00		270.64		
136	$C_4H_{10}O_2$	1,2-butanediol	26171-83-5	90.126		*469.58*	680.00		303.05		
137	$C_4H_{10}O_2$	1,3-butanediol	107-88-0	136.154		*481.38*	676.00		305.00		
138	$C_4H_{10}S$	3-thiapentane (diethyl sulfide)	352-93-2	90.189	169.22	*365.25*	557.00	39.60	318.00	*0.269*	*0.295*
139	$C_4H_{11}N$	1-butanamine (butyl amine)	109-73-9	73.138	224.05	*349.44*	526.80	40.40	**290.00**	***0.267***	*0.338*
140	$C_4H_{11}N$	N-ethylethanamine (diethyl amine)	109-89-7	73.138	223.35	328.60	496.60	37.10	301.00	*0.270*	
141	$C_4H_{11}N$	2-methyl-1-propanamine (isobutyl amine)	78-81-9	73.138	188.55	*340.81*	528.50	40.20	**286.00**	***0.262***	***0.220***
142	C_5F_{12}	dodecafluoropentane	678-26-2	288.036	263.15	302.35	422.00	20.40	383.10	*0.272*	**0.415**
143	C_5H_5N	pyridine	110-86-1	79.101	231.43	*388.37*	620.00	**56.70**	254.00	***0.267***	***0.242***
144	C_5H_6O	2-methylfuran	534-22-5	82.102	178.87	*337.87*	527.85	47.20	246.40	*0.255*	***0.278***
145	C_5H_8	1-pentyne	627-19-0	68.119	167.50	*313.38*	470.00	41.70	**277.00**	***0.296***	*0.394*
146	C_5H_8	cyclopentene	142-29-0	68.119	138.05	317.35	506.50	48.00	245.00	*0.279*	
147	C_5H_8O	cyclopentanone	120-92-3	84.118	222.50	*403.72*	**624.50**	**46.00**	268.00	***0.237***	***0.288***
148	C_5H_{10}	cyclopentane	287-92-3	70.134	179.28	322.38	511.60	45.08	260.00	*0.276*	
149	C_5H_{10}	1-pentene	109-67-1	70.134	106.95	*303.11*	464.80	35.60	298.40	*0.275*	*0.237*
150	C_5H_{10}	cis-2-pentene	627-20-3	70.134	121.78	*310.07*	475.00	36.90	302.10	***0.273***	***0.253***
151	C_5H_{10}	2-methyl-2-butene	513-35-9	70.134	139.40	*311.70*	470.00	38.60	292.00	*0.288*	*0.339*
152	C_5H_{10}	3-methyl-1-butene	563-45-1	70.134	104.65	*293.21*	452.70	35.50	304.90	*0.288*	*0.211*

Section A Basic Constants I (*Continued*)

No.	Formula	Name	CAS #	Mol. Wt.	Tfp, K	Tb, K	Tc, K	Pc, bar	Vc, cm^3/mol	Zc = PcVc/RTc	Omega
153	$C_5H_{10}O$	cyclopentanol	96-41-3	86.134		*413.49*	**619.50**	**49.00**			***0.420***
154	$C_5H_{10}O$	2-pentanone (methyl propyl ketone)	107-87-9	86.134	196.34	*375.39*	561.10	36.90	301.00	*0.238*	*0.346*
155	$C_5H_{10}O$	3-pentanone (diethyl ketone)	96-22-0	86.134	234.20	*375.14*	561.50	37.30	336.00	*0.268*	*0.342*
156	$C_5H_{10}O$	3-methyl-2-butanone (methyl isopropyl ketone)	563-80-4	86.134	181.15	*367.55*	567.70	36.20	310.00	*0.238*	***0.216***
157	$C_5H_{10}O$	2-methyltetrahydrofuran	96-47-9	86.134	137.04	*353.37*	537.00	37.60	267.00	*0.225*	*0.292*
158	$C_5H_{10}O_2$	pentanoic acid	109-52-4	102.133	239.45	*459.31*	643.00	**35.80**	336.20		***0.670***
159	$C_5H_{10}O_2$	3-methylbutanoic acid	503-74-2	102.133	243.85	*449.68*	629.00	**34.00**			***0.651***
160	$C_5H_{10}O_2$	methyl butanoate	623-42-7	102.133	187.35	*375.90*	554.40	34.80	340.00	*0.257*	*0.381*
161	$C_5H_{10}O_2$	ethyl propanoate	105-37-3	102.133	199.25	*372.18*	546.00	33.60	345.00	*0.255*	*0.390*
162	$C_5H_{10}O_2$	methyl 2-methylpropanoate	547-63-7	102.133	188.45	365.45	540.80	34.30	339.00	*0.259*	
163	$C_5H_{10}O_2$	propyl ethanoate (propyl acetate)	109-60-4	102.133	178.15	*374.65*	549.40	33.30	345.00	*0.252*	*0.389*
164	$C_5H_{10}O_2$	2-methylpropyl methanoate (isobutyl formate)	542-55-2	102.133	177.35	*371.22*	551.00	38.80	352.00	*0.298*	*0.400*
165	$C_5H_{11}Cl$	1-chloropentane	543-59-9	106.595	174.15	*381.54*	552.00				
166	C_5H_{12}	pentane	109-66-0	72.150	143.43	*309.22*	469.70	33.70	311.00	*0.268*	*0.252*
167	C_5H_{12}	2-methylbutane	78-78-4	72.150	113.26	*300.99*	460.39	33.81	308.30	*0.272*	*0.229*
168	C_5H_{12}	2,2-dimethylpropane (neopentane)	463-82-1	72.150	256.58	*282.65*	433.75	31.99	303.20	*0.269*	*0.197*
169	$C_5H_{12}O$	1-pentanol	71-41-0	88.150	194.25	*411.16*	588.15	39.09	326.00	*0.262*	*0.579*
170	$C_5H_{12}O$	2-pentanol	6032-29-7	88.150		**392.30**	560.30	36.75	329.00	*0.260*	***0.561***
171	$C_5H_{12}O$	2-methyl-1-butanol	137-32-6	88.150		**403.79**	575.40	39.40			***0.605***
172	$C_5H_{12}O$	2-methyl-2-butanol	75-85-4	88.150	264.40	*375.15*	545.00	37.90	**323.00**	***0.270***	*0.478*
173	$C_5H_{12}O$	3-methyl-1-butanol	123-51-3	88.150	155.95	*403.69*	579.40	39.10	**325.00**	***0.264***	*0.559*
174	$C_5H_{12}O$	3-methyl-2-butanol	598-75-4	88.150		385.20	556.10	38.70			
175	$C_5H_{12}O$	ethyl propyl ether	628-32-0	88.150	146.45	*337.01*	500.60	32.50	339.00	*0.265*	*0.328*
176	$C_5H_{12}S$	3-methyl-1-butanethiol (isopentyl mercaptan)	541-31-1	104.216	139.64	*391.50*	604.00	35.00	**364.00**	***0.254***	***0.191***
177	C_6ClF_5	chloropentafluorobenzene	344-07-0	202.511		*391.11*	**570.81**	**32.37**	376.00	***0.256***	***0.400***
178	C_6F_6	hexafluorobenzene	392-56-3	186.056		*353.40*	**516.73**	**32.75**	335.00	***0.255***	***0.396***
179	C_6F_{12}	dodecafluorocyclohexane	355-68-0	300.047	335.65		457.29	22.37	**459.00**	***0.270***	

180	C_6F_{14}	tetradecafluorohexane	355-42-0	338.044	186.05	*329.75*	448.70	18.70	573.20	*0.274*	**0.513**
181	C_6F_{14}	tetradecafluoro-2-methylpentane	355-04-4	338.044		*330.75*	453.00	18.20	550.00	*0.266*	
182	C_6F_{14}	tetradecafluoro-3-methylpentane	865-71-4	338.044	158.15	*331.55*	450.00	16.90	**531.00**	***0.240***	
183	C_6F_{14}	tetradecafluoro-2,3-dimethyl butane	354-96-1	338.044	258.15	*332.95*	463.00	18.70	528.00	*0.255*	
184	C_6HF_5	pentafluorobenzene	363-72-4	168.066	225.85	*358.89*	**530.97**	**35.37**	324.00	***0.260***	***0.374***
185	$C_6H_2F_4$	1,2,4,5-tetrafluorobenzene	327-54-8	150.076		*363.50*	**543.35**	**37.99**			***0.357***
186	C_6H_5Cl	chlorobenzene	108-90-7	112.558	227.90	*404.91*	632.40	45.20	308.00	*0.265*	*0.251*
187	C_6H_6	benzene	71-43-2	78.114	278.68	*353.24*	562.05	48.95	256.00	*0.268*	*0.210*
188	C_6H_6O	phenol	108-95-2	94.113	314.05	*455.04*	694.25	61.30	229.00	*0.243*	***0.442***
189	C_6H_7N	benzeneamine (aniline)	62-53-3	93.128	266.85	*457.17*	699.00	53.10	273.90	*0.256*	*0.380*
190	C_6H_7N	2-methylpyridine (2-picoline)	109-06-8	93.128	206.45	*402.50*	621.00	46.00	335.00	*0.260*	***0.299***
191	C_6H_7N	3-methylpyridine (3-picoline)	108-99-6	93.128	254.96	*417.28*	645.00	44.80	**288.00**	***0.241***	***0.279***
192	C_6H_7N	4-methylpyridine (4-picoline)	108-89-4	93.128	276.80	*418.49*	645.70	46.60	325.62	*0.253*	***0.305***
193	C_6H_8O	2-cyclohexen-1-one	930-68-7	96.131		*445.49*	685.00	**45.30**	305.18	***0.243***	***0.308***
194	C_6H_{10}	cyclohexene	110-83-8	82.145	169.67	356.12	560.40		296.88		
195	$C_6H_{10}O$	cyclohexanone	108-94-1	98.144		*428.59*	**653.00**	**40.00**			***0.299***
196	$C_6H_{10}O$	4-methyl-3-penten-2-one (mesityl oxide)	141-79-7	98.147		*402.86*	605.00		353.43		
197	C_6H_{12}	cyclohexane	110-82-7	84.161	279.69	*353.93*	553.50	40.73	308.00	*0.273*	*0.211*
198	C_6H_{12}	methylcyclopentane	96-37-7	84.161	130.72	*344.98*	532.79	37.84	319.00	*0.272*	*0.227*
199	C_6H_{12}	1-hexene	592-41-6	84.161	133.34	*336.63*	504.00	31.43	355.10	*0.266*	*0.281*
200	C_6H_{12}	4-methylpent-1-ene	691-37-2	84.161	119.5	*326.82*	495.00	**32.90**	**357.9**	***0.286***	***0.257***
201	$C_6H_{12}O$	cyclohexanol	108-93-0	100.161	297.65	*433.94*	648.00	**40.75**	333.88	***0.253***	***0.366***
202	$C_6H_{12}O$	2-hexanone (methyl butyl ketone)	591-78-6	100.161	217.40	*400.70*	587.00	33.20	378.00	*0.254*	*0.393*
203	$C_6H_{12}O$	3-hexanone (ethyl propyl ketone)	589-38-8	100.161	217.50	*396.65*	582.80	33.20	378.00	*0.256*	*0.380*
204	$C_6H_{12}O$	4-methyl-2-pentanone (methyl isobutyl ketone)	108-10-1	100.161	189.15	*389.15*	574.60	32.70	340.60	*0.256*	*0.351*
205	$C_6H_{12}O$	butylvinylether	111-34-2	100.163		*367.13*	540.50	**32.00**	**383.62**	***0.273***	***0.358***
206	$C_6H_{12}O_2$	hexanoic acid	142-62-1	116.160		*478.38*	**662.00**	**32.00**	377.20		***0.694***
207	$C_6H_{12}O_2$	methyl pentanoate	624-24-8	116.160		400.55	567.00	31.90	416.30	*0.267*	
208	$C_6H_{12}O_2$	ethyl butanoate	105-54-4	116.160	175.15	*394.69*	566.00	30.60	421.00	*0.274*	*0.463*
209	$C_6H_{12}O_2$	propyl propanoate	106-36-5	116.160	197.25	***395.64***	578.00	30.90	**394.00**	***0.253***	*0.373*
210	$C_6H_{12}O_2$	ethyl 2-methylpropanoate	97-62-1	116.160	185.00	383.00	553.00	30.70	421.00	*0.281*	
211	$C_6H_{12}O_2$	butyl ethanoate (butyl acetate)	123-86-4	116.160	199.65	*399.12*	579.00	30.90	412.80	*0.253*	*0.407*
212	$C_6H_{12}O_2$	2-methylpropyl ethanoate (isobutyl acetate)	110-19-0	116.160	174.25	*389.72*	561.00	31.60	413.00	*0.266*	*0.456*

Section A Basic Constants I (*Continued*)

No.	Formula	Name	CAS #	Mol. Wt.	Tfp, K	Tb, K	Tc, K	Pc, bar	Vc, cm^3/mol	Zc = PcVc/RTc	Omega
213	$C_6H_{12}O_2$	pentyl methanoate (pentyl formate)	638-49-3	116.160	199.65	403.55	576.00	30.90	412.00	*0.254*	
214	$C_6H_{12}O_2$	3-methylbutyl methanoate (isopentyl formate)	110-45-2	116.160	179.65	*397.28*	578.00	31.30	411.40	*0.252*	*0.400*
215	$C_6H_{12}O_3$	2-ethoxyethylacetate	111-15-9	132.165		*429.74*	610.60	**31.66**	**443.5**	***0.277***	***0.523***
216	C_6H_{14}	hexane	110-54-3	86.177	177.84	*341.88*	507.60	30.25	368.00	*0.264*	*0.300*
217	C_6H_{14}	2-methylpentane	107-83-5	86.177	119.48	*333.40*	497.50	30.10	366.70	*0.267*	*0.278*
218	C_6H_{14}	3-methylpentane	96-14-0	86.177	110.26	*336.40*	504.40	31.20	366.70	*0.273*	*0.273*
219	C_6H_{14}	2,2-dimethylbutane	75-83-2	86.177	173.33	*322.87*	488.70	30.80	359.10	*0.272*	*0.233*
220	C_6H_{14}	2,3-dimethylbutane	79-29-8	86.177	144.35	*331.12*	499.90	31.30	357.60	*0.269*	*0.248*
221	$C_6H_{14}O$	1-hexanol	111-27-3	102.177	229.20	*430.44*	611.40	35.10	381.00	*0.263*	*0.573*
222	$C_6H_{14}O$	2-hexanol	626-93-7	102.177	223.00	*413.02*	586.20	33.80	**383.00**	***0.266***	*0.562*
223	$C_6H_{14}O$	3-hexanol	623-37-0	102.177		*408.94*	582.40	33.60	383.00	*0.266*	*0.539*
224	$C_6H_{14}O$	2-methyl-1-pentanol	105-30-6	102.177		*420.76*	604.40	34.50			*0.498*
225	$C_6H_{14}O$	2-methyl-2-pentanol	590-36-3	102.177	171.15	*394.51*	559.50	**34.70**	**380.00**	***0.283***	***0.573***
226	$C_6H_{14}O$	2-methyl-3-pentanol	565-67-3	102.177		401.20	576.00	34.60			
227	$C_6H_{14}O$	4-methyl-1-pentanol	626-89-1	102.177		*424.93*	603.50	**36.30**	**380.00**	***0.275***	***0.588***
228	$C_6H_{14}O$	4-methyl-2-pentanol	108-11-2	102.177		*404.86*	574.50	33.10	**380.00**	***0.263***	*0.552*
229	$C_6H_{15}N$	N,N-diethylethanamine (triethyl amine)	121-44-8	101.192	158.45	362.90	535.00	30.00	390.00	*0.263*	
230	$C_6H_{15}N$	N-propyl-1-propanamine (dipropyl amine)	142-84-7	101.192	210.15	382.30	550.00	31.40	**402.00**	***0.276***	
231	C_7F_{14}	tetradecafluoromethylcyclohexane	355-02-2	350.055	228.45	349.45	485.91	20.19	**570.00**	***0.285***	
232	C_7F_{16}	hexadecafluoroheptane	335-57-9	388.052	195.15	*355.59*	475.00	**16.50**	664.30	***0.248***	***0.561***
233	$C_7H_3F_5$	pentafluorotoluene	771-56-2	182.093	243.35	*390.65*	**566.52**	**31.24**	384.00	***0.255***	***0.415***
234	C_7H_8	toluene	108-88-3	92.141	178.16	*383.79*	591.75	41.08	316.00	*0.264*	*0.264*
235	C_7H_8O	benzyl alcohol	100-51-6	108.140	257.80	*478.46*	**715.00**	**43.00**			***0.390***
236	C_7H_8O	2-methylphenol (o-cresol)	95-48-7	108.140	302.95	*464.17*	697.60	50.00	232.00	*0.243*	*0.436*
237	C_7H_8O	3-methylphenol (m-cresol)	108-39-4	108.140	284.95	*475.38*	705.70	45.60	312.00	*0.241*	*0.452*
238	C_7H_8O	4-methylphenol (p-cresol)	106-44-5	108.140	307.89	*475.12*	704.50	51.50	277.00	*0.244*	*0.510*
239	C_7H_9N	2,3-dimethylpyridine (2,3 lutidine)	583-61-9	107.155	249.55	*434.30*	655.40	40.90	**367.00**	***0.275***	*0.345*
240	C_7H_9N	2,4-dimethylpyridine (2,4 lutidine)	108-47-4	107.155	209.23	*431.55*	647.00	38.70	**367.00**	***0.264***	*0.351*

241	C_7H_9N	2,5-dimethylpyridine (2,5 lutidine)	589-93-5	107.155	257.65	*430.14*	644.20	39.80	**367.00**	***0.273***	*0.369*
242	C_7H_9N	2,6-dimethylpyridine (2,6 lutidine)	108-48-5	107.155	267.03	*416.91*	623.80	39.80	**367.00**	***0.282***	*0.373*
243	C_7H_9N	3,4-dimethylpyridine (3,4 lutidine)	583-58-4	107.155	262.15	*452.29*	683.80	40.90	**367.00**	***0.264***	*0.337*
244	C_7H_9N	3,5-dimethylpyridine (3,5 lutidine)	591-22-0	107.155	266.65	*445.06*	667.20	38.70	**367.00**	***0.256***	*0.351*
245	$C_7H_{12}O_2$	butyl-2-propenoate(butylacrylate)	141-32-2	128.175		*419.77*	644.00		427.54		*0.312*
246	C_7H_{14}	cycloheptane	291-64-5	98.188	265.15	*391.95*	604.30	**38.40**	359.00	***0.274***	***0.242***
247	C_7H_{14}	methylcyclohexane	108-87-2	98.188	146.56	*374.09*	572.19	34.71	368.00	*0.268*	*0.235*
248	C_7H_{14}	ethylcyclopentane	1640-89-7	98.188	134.70	*376.59*	569.50	33.97	375.00	*0.269*	*0.270*
249	C_7H_{14}	cis-1,3-dimethylcyclopentane	2532-58-3	98.188	139.45	*364.71*	551.00	**34.00**	363.30	***0.277***	***0.276***
250	C_7H_{14}	trans-1,3-dimethylcyclopentane	1759-58-6	98.188	139.18	*363.90*	553.00	**34.00**	363.30	***0.276***	***0.253***
251	C_7H_{14}	1-heptene	592-76-7	98.188	153.45	*366.79*	537.30	29.20	409.00	*0.267*	*0.343*
252	$C_7H_{14}O_2$	heptanoic acid	111-14-8	130.187		*495.35*	**679.00**	**29.00**	429.70		***0.712***
253	$C_7H_{14}O_2$	ethyl pentanoate	539-82-2	130.187	181.95	419.25	570.00	27.80	**449.00**	***0.263***	
254	$C_7H_{14}O_2$	ethyl 3-methylbutanoate	108-64-5	130.187	173.85	407.45	588.00	27.30	**447.00**	***0.250***	
255	$C_7H_{14}O_2$	propyl butanoate	105-66-8	130.187	177.95	*416.50*	600.00	27.80	**449.00**	***0.250***	*0.399*
256	$C_7H_{14}O_2$	2-methylpropyl propanoate	540-42-1	130.187	201.75	*409.75*	592.00	27.30	**447.00**	***0.248***	*0.375*
257	$C_7H_{14}O_2$	propyl 2-methylpropanoate	644-49-5	130.187		408.55	589.00	27.30	**447.00**	***0.249***	
258	$C_7H_{14}O_2$	3-methylbutyl ethanoate (isopentyl acetate)	123-92-2	130.187	194.65	415.20	599.00	28.10	**442.00**	***0.249***	
259	C_7H_{16}	heptane	142-82-5	100.204	182.59	*371.57*	540.20	27.40	**428.00**	***0.261***	*0.350*
260	C_7H_{16}	2-methylhexane	591-76-4	100.204	154.89	*363.18*	530.10	27.30	421.00	*0.261*	*0.331*
261	C_7H_{16}	3-methylhexane	589-34-4	100.204	149.35	*365.00*	535.20	28.10	404.00	*0.255*	*0.323*
262	C_7H_{16}	3-ethylpentane	617-78-7	100.204	154.57	*366.63*	540.50	28.90	415.80	*0.267*	*0.311*
263	C_7H_{16}	2,2-dimethylpentane	590-35-2	100.204	149.37	*352.32*	520.40	27.70	415.80	*0.266*	*0.287*
264	C_7H_{16}	2,3-dimethylpentane	565-59-3	100.204	82.60	*362.91*	537.30	29.10	393.00	*0.256*	*0.297*
265	C_7H_{16}	2,4-dimethylpentane	108-08-7	100.204	153.94	*353.62*	519.70	27.40	417.50	*0.265*	*0.304*
266	C_7H_{16}	3,3-dimethylpentane	562-49-2	100.204	138.25	*359.19*	536.30	29.50	414.10	*0.274*	*0.269*
267	C_7H_{16}	2,2,3-trimethylbutane	464-06-2	100.204	248.15	*354.01*	531.10	29.50	397.60	*0.266*	*0.250*
268	$C_7H_{16}O$	1-heptanol	111-70-6	116.203	239.20	*449.81*	631.90	31.50	435.00	*0.261*	*0.588*
269	$C_7H_{16}O$	2-heptanol	543-49-7	116.203		434.20	608.30	30.21	442.00		
270	$C_7H_{16}O$	4-heptanol	589-55-9	116.203			602.60		432.00		
271	C_8F_{18}	octadecafluorooctane	307-34-6	438.059		379.05	502.20	16.60	534.40	*0.281*	**0.619**
272	C_8H_8O	methylphenylketone(acetophenone)	98-86-2	120.153		*475.26*	713.00	**40.30**	380.23	***0:258***	***0.361***
273	C_8H_{10}	ethylbenzene	100-41-4	106.167	178.18	*409.36*	617.15	36.09	374.00	*0.263*	*0.304*

No.	Formula	Name	CAS #	Mol. Wt.	Tfp, K	Tb, K	Tc, K	Pc, bar	Vc, cm^3/mol	Zc = PcVc/RTc	Omega
274	C_8H_{10}	1,2-dimethylbenzene (o-xylene)	95-47-6	106.167	247.97	*417.59*	630.30	37.32	370.00	*0.263*	*0.312*
275	C_8H_{10}	1,3-dimethylbenzene (m-xylene)	108-38-3	106.167	225.28	*412.34*	617.00	35.41	375.00	*0.259*	*0.327*
276	C_8H_{10}	1,4-dimethylbenzene (p-xylene)	106-42-3	106.167	286.41	*411.53*	616.20	35.11	378.00	*0.259*	*0.322*
277	$C_8H_{10}O$	2-ethylphenol	90-00-6	122.167	269.84	*477.67*	703.00	**43.00**	**342.00**	***0.252***	***0.475***
278	$C_8H_{10}O$	3-ethylphenol	620-17-7	122.167	269.15	*491.58*	718.80	**41.50**	**342.00**	***0.237***	***0.489***
279	$C_8H_{10}O$	4-ethyl-phenol	123-07-9	122.167	318.18	*491.15*	716.40	**40.50**	**342.00**	***0.233***	***0.491***
280	$C_8H_{10}O$	2,3-dimethylphenol (2,3-xylenol)	526-75-0	122.167	345.95	*490.03*	722.80	**46.30**	470.00	***0.263***	***0.496***
281	$C_8H_{10}O$	2,4-dimethylphenol (2,4-xylenol)	105-67-9	122.167	297.68	*484.09*	707.60	**42.80**	510.00	***0.249***	***0.506***
282	$C_8H_{10}O$	2,5-dimethylphenol (2,5-xylenol)	95-87-4	122.167	347.97	*484.29*	706.90	**42.80**	470.00	***0.249***	***0.514***
283	$C_8H_{10}O$	2,6-dimethylphenol (2,6-xylenol)	576-26-1	122.167	318.75	*474.18*	701.00	**46.30**	520.00	***0.272***	***0.489***
284	$C_8H_{10}O$	3,4-dimethylphenol (3,4-xylenol)	95-65-8	122.167	333.95	*500.11*	729.80	**42.80**	460.00	***0.241***	***0.512***
285	$C_8H_{10}O$	3,5-dimethylphenol (3,5-xylenol)	108-68-9	122.167	336.75	*494.85*	715.60	**41.30**	610.00	***0.237***	***0.544***
286	C_8H_{16}	cyclooctane	292-64-8	112.215	287.95	*424.31*	647.20	35.70	410.00	*0.271*	*0.254*
287	C_8H_{16}	t-1,4-dimethylcyclohexane	2207-04-7	112.215	239.85	*429.75*	587.70				
288	C_8H_{16}	1-octene	111-66-0	112.215	171.46	*394.44*	567.00	26.80	468.00	*0.266*	*0.393*
289	$C_8H_{16}O_2$	octanoic acid	124-07-2	112.215	289.45	***512.01***	**695.00**	**26.40**			***0.734***
290	$C_8H_{16}O_2$	2-ethylhexanoic acid	149-57-5	144.218		*500.66*	675.00		554.68		
291	$C_8H_{16}O_2$	propyl pentanoate	141-06-0	144.214	202.45	440.65	613.00	25.20	**504.00**	***0.249***	
292	$C_8H_{16}O_2$	2-methylpropyl butanoate	539-90-2	144.214		430.05	611.00	24.90	**502.00**	***0.246***	
293	$C_8H_{16}O_2$	propyl 3-methylbutanoate	557-00-6	144.214		429.05	609.00	24.90	**502.00**	***0.247***	
294	$C_8H_{16}O_2$	3-methylbutyl propanoate	105-68-0	144.214		433.35	611.00	25.50	**497.00**	***0.249***	
295	$C_8H_{16}O_2$	2-methylpropyl-2-methylpropanoate	97-85-8	144.214	192.45	421.75	602.00	24.50	**501.00**	***0.245***	
296	C_8H_{18}	octane	111-65-9	114.231	216.39	*398.82*	568.70	24.90	492.00	*0.259*	*0.399*
297	C_8H_{18}	2-methylheptane	592-27-8	114.231	164.13	*390.80*	559.60	24.80	488.20	*0.260*	*0.378*
298	C_8H_{18}	3-methylheptane	589-81-1	114.231	152.63	*392.08*	563.60	25.50	471.10	*0.253*	*0.371*
299	C_8H_{18}	4-methylheptane	589-53-7	114.231	152.21	*390.86*	561.70	25.40	476.00	*0.259*	*0.371*
300	C_8H_{18}	3-ethylhexane	619-99-8	114.231		*391.69*	565.40	26.10	460.50	*0.253*	*0.362*
301	C_8H_{18}	2,2-dimethylhexane	590-73-8	114.231	151.98	*379.99*	549.80	25.30	478.00	*0.265*	*0.339*
302	C_8H_{18}	2,3-dimethylhexane	584-94-1	114.231		*388.76*	563.40	26.30	468.20	*0.263*	*0.347*
303	C_8H_{18}	2,4-dimethylhexane	589-43-5	114.231		*382.58*	553.50	25.60	472.00	*0.263*	*0.344*

304	C_8H_{18}	2,5-dimethylhexane	592-13-2	114.231	181.99	*382.26*	550.00	24.90	482.00	*0.262*	*0.357*
305	C_8H_{18}	3,3-dimethylhexane	563-16-6	114.231	147.06	*385.12*	562.00	26.50	442.80	*0.251*	*0.320*
306	C_8H_{18}	3,4-dimethylhexane	583-48-2	114.231		*390.88*	568.80	26.90	458.80	*0.265*	*0.338*
307	C_8H_{18}	3-ethyl-2-methylpentane	609-26-7	114.231	158.18	*388.81*	567.00	27.00	445.30	*0.254*	*0.331*
308	C_8H_{18}	3-ethyl-3-methylpentane	1067-08-9	114.231	182.31	*391.42*	576.50	28.10	455.10	*0.267*	*0.305*
309	C_8H_{18}	2,2,3-trimethylpentane	564-02-3	114.231	160.90	*382.99*	563.40	27.30	436.00	*0.254*	*0.298*
310	C_8H_{18}	2,2,4-trimethylpentane (isooctane)	540-84-1	114.231	165.80	*372.39*	543.90	25.70	469.70	*0.266*	*0.304*
311	C_8H_{18}	2,3,3-trimethylpentane	560-21-4	114.231	172.48	*387.92*	573.50	28.20	455.10	*0.269*	*0.291*
312	C_8H_{18}	2,3,4-trimethylpentane	565-75-3	114.231	163.97	*386.62*	566.30	27.30	456.20	*0.267*	*0.316*
313	C_8H_{18}	2,2,3,3-tetramethylbutane	594-82-1	114.231	373.94	*379.44*	567.80	28.70	482.00	*0.280*	*0.248*
314	$C_8H_{18}O$	1-octanol	111-87-5	130.230	257.65	*468.33*	652.50	28.60	490.00	*0.258*	*0.594*
315	$C_8H_{18}O$	2-octanol	123-96-6	130.230	241.15	*453.03*	637.00	28.10	**493.00**	***0.262***	*0.531*
316	$C_8H_{18}O$	3-octanol	589-98-0	130.230			628.50		515.00		
317	$C_8H_{18}O$	4-octanol	589-62-8	130.230		*440.72*	625.10		515.00		
318	$C_8H_{18}O$	2-ethyl-1-hexanol	104-76-7	130.230	203.20	*457.77*	**640.50**	**27.99**			***0.558***
319	$C_8H_{19}N$	n-octylamine	111-86-4	129.246		*451.70*	641.00	**26.17**	**517.00**	***0.254***	***0.446***
320	$C_8H_{19}N$	N-butyl-1-butanamine (dibutyl amine)	111-92-2	129.246	211.15	432.80	602.30	25.70	**512.00**	***0.263***	
321	C_9F_{20}	eicosafluorononane	375-96-2	488.067		398.45	523.90	15.60	590.60	*0.282*	**0.635**
322	C_9H_7N	quinoline	91-22-5	129.161	257.55	*510.30*	782.00	**45.00**	**402.00**	***0.278***	***0.315***
323	C_9H_7N	isoquinoline	119-65-3	129.161	299.62	516.35	803.00	**47.00**	**402.00**	***0.283***	
324	C_9H_{10}	indan	496-11-7	118.178	221.74	451.00	684.90	39.50	**393.00**	***0.273***	
325	C_9H_{12}	propylbenzene	103-65-1	120.194	173.65	*432.35*	638.35	32.00	440.00	*0.265*	*0.345*
326	C_9H_{12}	1-methylethylbenzene (cumene)	98-82-8	120.194	177.12	*425.52*	631.00	32.09	434.70	*0.261*	*0.326*
327	C_9H_{12}	1-ethyl-4-methylbenzene	622-96-8	120.194	210.81	*435.13*	640.20	32.30	440.00	*0.259*	*0.364*
328	C_9H_{12}	1,2,3-trimethylbenzene	526-73-8	120.194	247.77	*449.23*	664.50	34.54	435.00	*0.267*	*0.367*
329	C_9H_{12}	1,2,4-trimethylbenzene	95-63-6	120.194	229.35	*442.49*	649.10	32.32	435.00	*0.256*	*0.377*
330	C_9H_{12}	1,3,5-trimethylbenzene (mesitylene)	108-67-8	120.194	228.43	*437.90*	637.30	31.27	430.00	*0.252*	*0.399*
331	C_9H_{18}	1-nonene	124-11-8	126.242	191.80	*420.03*	594.00	**23.30**	526.00	***0.248***	***0.411***
332	$C_9H_{18}O_2$	nonanoic acid	112-05-0	158.241		*527.74*	**711.00**	**24.30**			***0.748***
333	$C_9H_{18}O_2$	3-methylbutyl butanoate	106-27-4	158.241		*452.09*	619.00	23.30	**500.00**	***0.226***	*0.583*
334	C_9H_{20}	nonane	111-84-2	128.258	219.66	*423.97*	594.60	22.90	555.00	*0.257*	*0.445*
335	C_9H_{20}	2-methyloctane	3221-61-2	128.258	192.79	*416.44*	587.00	**23.10**	**529.00**	***0.250***	***0.423***
336	C_9H_{20}	2,2-dimethylheptane	1071-26-7	128.258	160.16	*405.97*	577.80	23.50	**525.00**	***0.257***	0.383
337	C_9H_{20}	2,2,5-trimethylhexane	3522-94-9	128.258	167.37	*397.24*	569.80		519.00		
338	C_9H_{20}	2,2,3,3-tetramethylpentane	7154-79-2	128.258	263.26	*413.44*	607.60	27.40	478.00	*0.269*	*0.304*

No.	Formula	Name	CAS #	Mol. Wt.	Tfp, K	Tb, K	Tc, K	Pc, bar	Vc, cm^3/mol	Zc = PcVc/RTc	Omega
339	C_9H_{20}	2,2,3,4-tetramethylpentane	1186-53-4	128.258	152.06	*406.18*	592.70	25.30	490.00	*0.258*	*0.301*
340	C_9H_{20}	2,2,4,4-tetramethylpentane	1070-87-7	128.258	206.64	*395.44*	574.60	24.90	504.00		*0.314*
341	C_9H_{20}	2,3,3,4-tetramethylpentane	16747-38-9	128.258	171.05	*414.71*	607.10	**26.70**	493.00	***0.261***	***0.309***
342	$C_9H_{20}O$	1-nonanol	143-08-8	144.257	268.15	*486.52*	668.90	**26.30**	544.00	***0.261***	***0.633***
343	$C_9H_{20}O$	2-nonanol	628-99-9	144.257		466.70	649.60	25.30	575.00	*0.269*	
344	$C_{10}F_{22}$	docosafluorodecane	307-45-9	538.075		417.35	542.30	14.50	624.20	*0.279*	
345	$C_{10}H_8$	naphthalene	91-20-3	128.174	351.35	*491.16*	748.40	40.50	407.00	*0.265*	*0.304*
346	$C_{10}H_{12}$	1,2,3,4-tetrahydronaphthalene	119-64-2	132.205	237.35	480.75	720.00	36.50	408.00	*0.249*	
347	$C_{10}H_{12}$	1-methylindan	767-58-8	132.205		463.80	694.10	35.30	**448.00**	***0.274***	
348	$C_{10}H_{12}$	2-methylindan	824-63-5	132.205		464.50	695.30	35.30	**448.00**	***0.274***	
349	$C_{10}H_{12}$	4-methylindan	824-22-6	132.205		478.60	716.40	35.30	**448.00**	***0.266***	
350	$C_{10}H_{12}$	5-methylindan	874-35-1	132.205		475.10	711.20	35.30	**448.00**	***0.267***	
351	$C_{10}H_{14}$	butylbenzene	104-51-8	134.221	185.19	*456.42*	660.50	28.90	497.00	*0.262*	*0.393*
352	$C_{10}H_{14}$	2-methylpropylbenzene (isobutylbenzene)	538-93-2	134.221	221.67	*445.90*	650.00	30.50	**480.00**	***0.271***	*0.383*
353	$C_{10}H_{14}$	1,4-diethylbenzene	105-05-5	134.221	230.30	*456.90*	657.90	28.03	480.50	*0.247*	*0.403*
354	$C_{10}H_{14}$	1-(1-methylethyl)-4-methylbenzene (p-cymene)	99-87-6	134.221	205.22	*450.26*	652.00	28.00	497.00	*0.248*	*0.376*
355	$C_{10}H_{14}$	1,2,4,5-tetramethylbenzene (durene)	95-93-2	134.221	352.45	*470.00*	676.00	29.00	**482.00**	***0.249***	*0.423*
356	$C_{10}H_{18}$	cis-bicyclo[4.4.0]decane (cis-decalin)	493-01-6	138.253	230.14	*468.92*	703.60	32.00	480.00	*0.265*	*0.276*
357	$C_{10}H_{18}$	trans-bicyclo[4.4.0]decane (trans-decalin)	493-02-7	138.253	242.75	*460.42*	687.00	32.00	480.00	*0.272*	*0.303*
358	$C_{10}H_{20}$	1-decene	872-05-9	140.269	206.89	443.75	617.00	22.20	584.00	*0.253*	
359	$C_{10}H_{20}O_2$	decanoic acid	334-48-5	172.268	305.15	*541.92*	**726.00**	**22.30**			***0.749***
360	$C_{10}H_{22}$	decane	124-18-5	142.285	243.49	*447.30*	617.70	21.10	624.00	*0.256*	*0.490*
361	$C_{10}H_{22}$	2,2,5-trimethylheptane	20291-95-6	142.285		*423.90*	598.90	**22.40**	**569.00**	***0.256***	***0.398***
362	$C_{10}H_{22}$	3,3,5-trimethylheptane	7154-80-5	142.285		*428.83*	609.60	23.20	**564.00**	***0.258***	*0.383*
363	$C_{10}H_{22}$	2,2,3,3-tetramethylhexane	13475-81-5	142.285	219.19	*433.46*	623.00	25.10			*0.366*
364	$C_{10}H_{22}$	2,2,5,5-tetramethylhexane	1071-81-4	142.285	260.60	*410.61*	581.40	21.90			*0.377*
365	$C_{10}H_{22}O$	1-decanol	112-30-1	158.284	280.05	*504.25*	684.40	23.70	600.00	*0.252*	*0.661*
366	$C_{10}H_{24}N_4$	octamethylethenetetramine	996-70-3	200.326		*477.44*	680.00		646.21		
367	$C_{11}H_{10}$	1-methylnaphthalene	90-12-0	142.200	242.69	*517.84*	772.00	36.00	**462.00**	***0.259***	*0.348*

368	$C_{11}H_{10}$	2-methylnaphthalene	91-57-6	142.200	307.71	*514.20*	761.00	35.40	**462.00**	***0.258***	*0.374*
369	$C_{11}H_{24}$	undecane	1120-21-4	156.312	247.57	*469.08*	639.00	19.80	689.00	*0.257*	*0.537*
370	$C_{11}H_{24}O$	1-undecanol	112-42-5	172.311		*521.24*	**705.00**	**22.40**			***0.656***
371	$C_{12}H_{10}$	1,1'-biphenyl	92-52-4	154.211	342.35	*528.23*	773.00	33.80	497.00	*0.261*	*0.404*
372	$C_{12}H_{12}$	1,6-dimethylnaphthalene	575-43-9	156.227	257.00	539.50	784.00	31.00	**517.00**	***0.246***	
373	$C_{12}H_{12}$	2,7-dimethylnaphthalene	582-16-1	156.227	368.85	535.00	775.00	32.30			
374	$C_{12}H_{18}$	1,3,5-triethylbenzene	102-25-0	162.276		*488.93*	679.00	**24.35**	624.14	***0.269***	***0.527***
375	$C_{12}H_{20}$	1,3-dimethyltricyclo[3.3.1.1^{3,7}]decane (1,3-dimethyladamantane)	702-79-4	164.292		*476.44*	708.00		571.45		
376	$C_{12}H_{24}$	1-dodecene	112-41-4	168.323	237.95	486.95	658.00	19.30			
377	$C_{12}H_{26}$	dodecane	112-40-3	170.338	263.57	*489.48*	658.00	18.20	754.00	*0.251*	*0.576*
378	$C_{12}H_{26}O$	1-dodecanol	112-53-8	186.338	297.10	***537.79***	**720.00**	**20.80**	718.00	***0.249***	**0.684**
379	$C_{13}H_{12}$	diphenylmethane	101-81-5	168.238	298.39	*537.65*	760.00	27.10	563.00	*0.241*	*0.481*
380	$C_{13}H_{28}$	tridecane	629-50-5	184.365	267.76	*508.63*	675.00	16.80	823.00	*0.246*	*0.618*
381	$C_{13}H_{28}O$	1-tridecanol	112-70-9	200.365		*553.72*	**734.00**	**19.35**			***0.712***
382	$C_{14}H_{10}$	phenanthrene	85-01-8	178.233	372.35	*611.55*	869.00	28.70	554.00	*0.220*	*0.479*
383	$C_{14}H_{10}$	anthracene	120-12-7	178.233	492.65	*614.39*	869.30	28.70	554.00	*0.220*	*0.501*
384	$C_{14}H_{22}$	1,4-di(trimethylmethyl)benzene (p-ditertbutylbenzene)	1012-72-2	190.330	350.80	*510.43*	708.00	**23.00**	**732.0**	***0.286***	***0.506***
385	$C_{14}H_{30}$	tetradecane	629-59-4	198.392	279.01	*526.76*	693.00	15.70	894.00	*0.244*	*0.644*
386	$C_{14}H_{30}O$	1-tetradecanol	112-72-1	214.392		*569.04*	**747.00**	**18.10**			***0.744***
387	$C_{15}H_{32}$	pentadecane	629-62-9	212.419	283.08	*543.83*	708.00	14.80	966.00	*0.243*	*0.685*
388	$C_{15}H_{32}O$	1-pentadecanol	629-76-5	228.419		*583.68*	**759.00**	**17.00**			***0.778***
389	$C_{16}H_{34}$	hexadecane	544-76-3	226.446	291.32	*559.98*	723.00	14.00	1034.00	*0.241*	*0.718*
390	$C_{16}H_{34}$	2,2,4,4,6,8,8-heptamethylnonane	4390-04-9	226.446		*520.25*	693.00	15.70			*0.548*
391	$C_{16}H_{34}O$	1-hexadecanol	4485-13-6	242.446	322.45	*597.53*	**770.00**	**16.10**			***0.818***
392	$C_{17}H_{36}$	heptadecane	629-78-7	240.473	295.13	*574.56*	736.00	13.40	1103.00	*0.242*	*0.753*
393	$C_{17}H_{36}O$	1-heptadecanol	1454-85-9	256.472	327.00	*611.12*	**780.00**	**15.00**			***0.853***
394	$C_{18}H_{14}$	1,2-diphenylbenzene	84-15-1	230.309	329.35	605.15	857.00	29.90			
395	$C_{18}H_{14}$	1,3-diphenylbenzene	92-06-8	230.309	360.15	636.15	883.00	24.80	724.00	*0.245*	
396	$C_{18}H_{14}$	1,4-diphenylbenzene	92-94-4	230.309	483.25	649.15	908.00	29.90	729.00	*0.289*	
397	$C_{18}H_{38}$	octadecane	593-45-3	254.500	301.32	*588.30*	747.00	12.90	1189.00	*0.247*	*0.800*
398	$C_{18}H_{38}O$	1-octadecanol	112-92-5	270.499	331.00	*623.57*	**790.00**	**14.40**			***0.892***
399	$C_{19}H_{40}$	nonadecane	629-92-5	268.527	305.25	*602.34*	755.00	11.60			*0.845*

No.	Formula	Name	CAS #	Mol. Wt.	Tfp, K	Tb, K	Tc, K	Pc, bar	Vc, cm^3/mol	Zc = PcVc/RTc	Omega
400	$C_{19}H_{40}O$	1-nonadecanol	1454-84-8	284.526		*635.41*	**799.00**	**13.80**			***0.934***
401	$C_{20}H_{42}$	eicosane	112-95-8	282.554	309.95	*616.84*	768.00	10.70			*0.865*
402	$C_{20}H_{42}O$	1-eicosanol	629-96-9	298.553	339.00	*647.69*	**809.00**	**13.00**			***0.954***
403	$C_{21}H_{44}$	heneicosane	629-94-7	296.580	313.65	629.65	778.00	10.30			
404	$C_{22}H_{46}$	docosane	629-97-0	310.607	317.55	641.75	786.00	9.80			
405	$C_{23}H_{48}$	tricosane	638-67-5	324.634	320.65	653.35	790.00	9.20			
406	$C_{24}H_{50}$	tetracosane	646-31-1	338.661	324.05	664.45	800.00	8.70			
407	ClD	deuterium chloride	7698-05-7	37.467	158.51	*188.43*	323.50				
408	$ClFO_3$	perchloryl fluoride	7616-94-6	102.449	125.41	*226.49*	368.40	53.70	161.00	*0.282*	
409	ClF_5	chlorine pentafluoride	13637-63-3	130.445	171.15	*259.28*	416.00	53.00	230.40		
410	ClH	hydrogen chloride	7647-01-0	36.461	158.97	*188.15*	324.69	83.10	81.00	*0.249*	
411	ClH_4N	ammonium chloride	12125-02-9	53.491	793.20	*613.16*	882.00		165.00		
412	ClNO	nitrogen oxychloride	2696-92-6	65.459	213.55	*267.77*	440.60	91.20			
413	Cl_2	chlorine	7782-50-5	70.905	172.19	*239.12*	417.00	77.00	124.00	*0.275*	
414	DH	deuterium hydride	13983-20-5	3.022	16.59	*22.13*	35.90	14.80	**62.70**	***0.311***	*−0.176*
415	DI	deuterium iodide	14104-45-1	128.919	221.28	*237.52*	421.80				
416	D_2	deuterium	7782-39-0	4.028	18.63	*22.13*	38.25	16.50	60.20	*0.312*	
417	D_2	deuterium, normal	800000-54-8	4.028	18.72	*23.65*	38.35	16.65	**60.20**	***0.314***	*−0.143*
418	D_2O	deuterium oxide	7789-20-0	20.028	276.96	*374.55*	643.89	216.71	56.26	*0.228*	
419	D_2S	deuterium sulfide	13536-94-2	36.094	187.15		372.30	89.00	**96.00**	***0.276***	
420	D_3N	trideuteroammonia	13550-49-7	20.049	198.82	*242.10*	405.50	113.00	**72.00**	***0.241***	
421	D_3P	trideuterophosphine	13537-03-6	37.016			323.60	65.00			
422	FH	hydrogen fluoride	7664-39-3	20.006	189.58	*292.68*	461.00	65.00	69.00	*0.117*	
423	FNO_2	nitrogen dioxyfluoride	10022-50-1	65.004	107.15	*200.75*	349.30	**82.00**			
424	F_2	fluorine	7782-41-4	37.997	53.48	*84.95*	144.30	52.15	66.20	*0.288*	*0.051*
425	F_2HN	difluoroamine	10405-27-3	53.012	157.15	250.15	403.00	63.00	**97.00**	***0.182***	
426	F_2N_2	cis-difluorodiazene	13812-43-6	66.010		167.40	272.00	**68.00**	113.30		
427	F_2N_2	trans-difluorodiazene	13776-62-0	66.010	101.15	161.70	260.00	**64.00**	113.30		
428	F_2O	oxygen difluoride	7783-41-7	53.996	49.35	*128.38*	215.20	49.50	97.60	*0.270*	
429	F_2Xe	xenon difluoride	13709-36-9	169.287	402.18		631.00	93.00	149.00	*0.264*	
430	F_3N	nitrogen trifluoride	7783-54-2	71.002	66.37	144.11	234.00	45.30	118.75	*0.256*	0.124
431	F_3NO	trifluoroamine oxide	13847-65-9	87.001	112.15	185.65	302.60	**64.00**	150.10		

432	F_4N_2	tetrafluorohydrazine	10036-47-2	104.007	111.65	198.95	309.20	37.00	181.00	*0.222*	
433	F_4S	sulfur tetrafluoride	7783-60-0	108.060	148.15	*233.15*	364.00	43.30	**146.00**	***0.209***	*0.258*
434	F_4Xe	xenon tetrafluoride	13709-61-0	207.284	390.25		612.00	70.40	**189.00**	***0.261***	
435	F_6S	sulfur hexafluoride	2551-62-4	146.056	222.45	*209.25*	318.72	37.60	198.40	*0.282*	*0.208*
436	F_6U	uranium hexafluoride	7783-81-5	352.070	337.00	*324.96*	**503.35**	**45.31**	250.00	***0.271***	**0.277**
437	HI	hydrogen iodide	10034-85-2	127.912	222.38	*237.57*	423.90	90.00	132.70		0.038
438	H_2	hydrogen	1333-74-0	2.016	13.83	*20.27*	32.98	12.93	64.20	*0.303*	*−0.217*
439	H_2	hydrogen, normal	800000-51-5	2.016	13.56	*20.38*	33.25	12.97	65.00	*0.305*	−0.216
440	H_2O	water	7732-18-5	18.015	273.15	373.15	647.14	220.64	55.95	*0.229*	*0.344*
441	H_2S	hydrogen sulfide	7783-06-4	34.082	187.62	*212.84*	373.40	89.63	98.00	*0.283*	*0.090*
442	H_2S_2	dihydrogen disulfide	13465-07-1	66.148	183.35	*344.25*	572.00	59.10	**150.00**	***0.186***	
443	H_2S_3	dihydrogen trisulfide	13845-23-3	98.214		*443.13*	738.00	51.30	**205.00**	***0.171***	
444	H_2S_4	dihydrogen tetrasulfide	13845-25-5	130.280		*513.14*	855.00	43.70	**260.00**	***0.160***	
445	H_2S_5	dihydrogen pentasulfide	13845-24-4	162.346		*558.10*	930.00	38.90	**315.00**	***0.158***	
446	H_2Se	hydrogen selenide	7783-07-5	80.976	207.42	*228.25*	411.00	89.20			
447	H_3N	ammonia	7664-41-7	17.031	195.41	***239.82***	405.40	113.53	72.47	*0.255*	*0.257*
448	H_3P	phosphine	7803-51-2	33.998	139.37	*185.42*	324.50	65.40			
449	H_4N_2	hydrazine	302-01-2	32.045	274.68	*386.65*	653.01	147.00	101.10	*0.282*	
450	He	helium	7440-59-7	4.003	2.15	*4.30*	5.19	2.27	57.30	*0.301*	−0.390
451	He	helium-3	14762-55-1	3.017	1.01	*3.33*	3.31	1.14	72.50	*0.300*	*−0.480*
452	I_2	iodine	7553-56-2	253.809	386.76	*457.56*	819.00		155.00		
453	Kr	krypton	7439-90-9	83.800	115.77	*119.74*	209.40	55.00	91.20	*0.288*	
454	NO	nitrogen monoxide (nitric oxide)	10102-43-9	30.006	109.51	*121.38*	180.00	64.80	58.00	*0.251*	*0.582*
455	N_2	nitrogen	7727-37-9	28.014	63.15	***77.35***	126.20	33.98	90.10	*0.289*	*0.037*
456	N_2O	dinitrogen oxide (nitrous oxide)	10024-97-2	44.013	182.33	*184.67*	309.60	72.55	97.00	*0.273*	
457	N_2O_4	dinitrogen tetroxide (nitrogen dioxide)	10544-72-6	92.011	261.95	*302.22*	431.01	101.00			*1.007*
458	Ne	neon	7440-01-9	20.180	24.56	*27.07*	44.40	27.60	41.70	*0.312*	*−0.016*
459	OT_2	tritium oxide	14940-65-9	22.032	277.64		641.72	214.10	**56.00**	***0.225***	
460	O_2	oxygen	7782-44-7	31.999	54.36	*90.17*	154.58	50.43	73.37	*0.288*	
461	O_2S	sulfur dioxide	7446-09-5	64.065	197.67	*263.13*	430.80	78.84	122.00	*0.269*	
462	O_3	ozone	10028-15-6	47.998	80.65	*161.80*	261.05	55.70	89.40	*0.229*	***0.224***
463	O_3S	sulfur trioxide	7446-11-9	80.064	289.95	*317.90*	490.90	82.10	126.50	*0.254*	
464	Rn	radon	10043-92-2	222.018	202.15	*209.80*	377.00	63.00	140.00	*0.281*	
465	S	sulfur	7704-34-9	32.066	392.75	*717.75*	1313.01	182.00	158.00	*0.263*	
466	Se	selenium	7782-49-2	78.960	*494.00*	*957.95*	1766.00	271.60	62.30	*0.118*	
467	T_2	tritium	10028-17-8	6.032			40.00	18.50	**55.30**	***0.308***	
468	Xe	xenon	7440-63-3	131.290	161.25	*165.01*	289.74	58.40	118.00	*0.286*	

No.	Formula	Name	CAS #	DelHf0, kJ/mol	DelGf0, kJ/mol	DelHb, kJ/mol	DelHm, kJ/mol	V liq, cm³/mol	T liq, K	Dipole, Debye
1	Ar	argon	7440-37-1	0.00	0.00	6.43		29.10	90.00	0.0
2	Br_2	bromine	7726-95-6	30.91	3.13	58.80		51.51	298.15	0.0
3	BrD	deuterium bromide	13536-59-9	−37.12	−53.79					
4	BrF_3	bromine trifluoride	7787-71-5	−255.64	−229.51	47.57		48.84	298.15	1.1
5	BrF_5	bromine pentafluoride	7789-30-2	−428.86	−351.65	30.60		71.02	298.15	1.5
6	BrH	hydrogen bromide	10035-10-6	−36.26	−53.30			35.85	193.15	0.8
7	$CBrClF_2$	bromochlorodifluoromethane	353-59-3	−435.20	−411.80			91.67	298.15	
8	$CBrF_3$	bromotrifluoromethane	75-63-8	−649.80	−622.90			74.42	213.15	0.7
9	CBr_2F_2	dibromodifluoromethane	75-61-6	−386.60	−375.70			93.04	298.15	0.7
10	$CClF_3$	chlorotrifluoromethane	75-72-9	−704.20	−663.60	15.73		68.87	193.15	0.5
11	CCl_2F_2	dichlorodifluoromethane (R-12)	75-71-8	−490.80	−451.70	20.08		82.91	243.15	0.5
12	CCl_3F	trichlorofluoromethane (R-11)	75-69-4	−283.70	−244.40	25.06	6.89	93.82	298.15	0.5
13	CCl_4	tetrachloromethane	56-23-5	−95.81	−53.53	29.82	3.28	97.07	298.15	0.0
14	CF_4	tetrafluoromethane	75-73-0	−933.50	−888.80			56.41	153.15	0.0
15	$CHBrF_2$	bromodifluoromethane	1511-62-2	−429.50	−412.90			64.81	253.15	1.5
16	$CHClF_2$	chlorodifluoromethane (R-22)	75-45-6	−482.80	−451.70	20.22	4.12	59.08	213.15	1.4
17	$CHCl_2F$	dichlorofluoromethane	75-43-4	−284.90	−254.40	25.13		75.96	298.15	1.3
18	$CHCl_3$	trichloromethane	67-66-3	−102.93	−70.09	29.24	8.80	80.68	298.15	1.1
19	CHF_3	trifluoromethane (R-23)	75-46-7	−693.30	−658.80			51.66	213.15	1.6
20	CH_2Cl_2	dichloromethane	75-09-2	−95.40	−68.84	28.06	4.60	64.53	298.15	1.8
21	CH_2F_2	difluoromethane	75-10-5	−452.30	−424.70			42.91	223.15	2.0
22	CH_2O_2	methanoic acid (formic acid)	64-18-6	−378.60	−35.06	22.69	12.72			1.5
23	CH_3Cl	chloromethane	74-87-3	−81.96	−58.42	21.58	6.43	50.59	253.15	1.9
24	CH_3F	fluoromethane	593-53-3	−237.70	−213.70			40.60	213.15	1.8
25	CH_3NO_2	nitromethane	75-52-5	−74.70	−6.90	33.99		53.96	298.15	3.1
26	CH_4	methane	74-82-8	−74.52	−50.45	8.17	0.94	35.54	90.68	0.0
27	CH_4O	methanol	67-56-1	−200.94	−162.24	35.21	3.18	40.73	298.15	1.7
28	CH_4S	methanethiol (methyl mercaptan)	74-93-1	−22.59	−9.52		5.91	55.52	293.15	1.3
29	CH_5N	methanamine (methyl amine)	74-89-5	−22.53	32.73	25.60	6.13	47.34	298.15	1.3

30	CO	carbon monoxide	630-08-0	−110.53	−137.16	6.04	0.84	34.88	81.00	0.1
31	CO_2	carbon dioxide	124-38-9	−393.51	−394.38		9.02			0.0
32	$C_2Br_2ClF_3$	1,2-dibromo-2-chloro-1,1,2-trifluroethane	354-51-8			31.17		123.67	298.15	
33	C_2ClF_5	1-chloro-1,1,2,2,2-pentafluoroethane	76-15-3	−1123.00	−1042.00	19.22	1.88	98.79	233.15	0.3
34	$C_2Cl_2F_4$	1,1-dichloro-1,2,2,2-tetrafluoroethane	374-07-2	−926.80	−845.40	23.08		116.41	298.15	
35	$C_2Cl_2F_4$	1,2-dichloro-1,1,2,2-tetrafluoroethane	76-14-2	−900.40	−818.10	22.94	1.51	117.37	298.15	0.5
36	$C_2Cl_3F_3$	1,1,2-trichloro-1,2,2-trifluoroethane	76-13-1	−705.80	−627.30	28.08	2.47	119.78	293.15	
37	C_2F_4	tetrafluoroethene	116-14-3	−659.00	−624.10			65.84	197.00	0.0
38	C_2F_6	hexafluoroethane	76-16-4	−1343.00	−1257.00	16.15				0.0
39	$C_2HBrClF_3$	1-bromo-1-chloro-2,2,2-trifluoroethane	151-67-7			28.08		106.35	298.15	
40	$C_2HBrClF_3$	1-bromo-2-chloro-1,1,2-trifluoroethane	354-06-3			28.31		106.29	298.15	
41	C_2HClF_4	1-chloro-1,1,2,2-tetrafluoroethane	354-25-6	−903.30	−830.90	22.30		105.06	273.15	
42	C_2HClF_4	1-chloro-1,2,2,2-tetrafluoroethane	2837-89-0	−924.70	−851.80	22.53		99.88	295.00	
43	$C_2HCl_2F_3$	1,1-dichloro-2,2,2-trifluoroethane (R-123)	306-83-2	−743.90	−668.90	25.72		103.90	295.00	
44	$C_2HCl_2F_3$	1,2-dichloro-1,2,2-trifluoroethane (R-123a)	354-23-4					103.72	295	
45	C_2HF_5	pentafluoroethane	354-33-6	−1105.00	−1030.00			98.61	293.48	1.5
46	C_2HF_5O	pentafluorodimethyl ether (E-125)	3822-68-2					105.66	295.00	
47	C_2H_2	ethyne (acetylene)	74-86-2	190.92	201.30		21.28	43.47	203.15	0.0
48	$C_2H_2F_2$	1,1-difluoroethene	75-38-7	−336.81	−313.06			103.78	297.00	1.4
49	$C_2H_2F_4$	1,1,1,2-tetrafluoroethane (R-134a)	811-97-2	−907.10	−838.40			75.38	255.00	
50	$C_2H_2F_4$	1,1,2,2-tetrafluoroethane (R-134)	359-35-3	−892.40	−824.60			74.42	275.00	0.0
51	$C_2H_3ClF_2$	1-chloro-1,1-difluoroethane	75-68-3	−529.70	−465.70	22.38	2.69	90.49	298.15	2.1
52	$C_2H_3Cl_2F$	1,1-dichloro-1-fluoroethane (R-1416)	1717-00-6	−339.70	−276.20	25.94		93.63	290.00	
53	$C_2H_3F_3$	1,1,1-trifluoroethane (R-143a)	420-46-2	−745.60	−678.70	18.99	6.19	75.38	245.00	2.3
54	$C_2H_3F_3$	1,1,2-trifluoroethane (R-143)	430-66-0	−669.40	−609.40			69.22	300.00	
55	C_2H_4	ethene (ethylene)	74-85-1	52.50	68.48	13.53	3.35	51.07	183.15	0.0
56	$C_2H_4Br_2$	1,2-dibromoethane	106-93-4	−38.94	−1.06	34.77		86.62	298.15	1.0
57	$C_2H_4Cl_2$	1,1-dichloroethane	75-34-3	−130.12	−73.23	28.85	7.87	84.73	298.15	2.0
58	$C_2H_4Cl_2$	1,2-dichloroethane	107-06-2	−126.78	−70.20	31.98	8.84	79.45	298.15	1.8
59	$C_2H_4F_2$	1,1-difluoroethane (R-152a)	75-37-6	−500.80	−443.30	21.56		65.76	252.47	2.3
60	$C_2H_4O_2$	ethanoic acid (acetic acid)	64-19-7	−432.25	−374.27	23.70	11.72	57.53	298.15	1.3
61	$C_2H_4O_2$	methyl methanoate (methyl formate)	107-31-3	−352.40	−294.90	27.92		62.14	298.15	1.8
62	C_2H_5Br	bromoethane	74-96-4	−63.60	−25.70	27.04		75.12	298.15	2.0
63	C_2H_5Cl	chloroethane	75-00-3	−112.26	−60.43	24.53	4.45	72.58	298.15	2.0

No.	Formula	Name	CAS #	DelHf0, kJ/mol	DelGf0, kJ/mol	DelHb, kJ/mol	DelHm, kJ/mol	V liq, cm³/mol	T liq, K	Dipole, Debye
64	C_2H_5F	fluoroethane	353-36-6	−264.40	−212.40			56.67	213.15	2.0
65	C_2H_6	ethane	74-84-0	−83.82	−31.86	14.70	2.86	46.15	90.36	0.0
66	C_2H_6O	ethanol	64-17-5	−234.95	−167.73	38.56	5.01	58.68	298.15	1.7
67	C_2H_6O	dimethyl ether	115-10-6	−184.11	−112.92	21.51	4.94	69.07	293.15	1.3
68	C_2H_6S	ethanethiol (ethyl mercaptan)	75-08-1	−46.02	−2.07	26.79	4.98	74.58	298.15	1.6
69	C_2H_6S	2-thiapropane (dimethylsulfide)	75-18-3	−37.20	7.41	27.00	7.98	73.77	298.15	1.5
70	C_2H_7N	ethanamine (ethyl amine)	75-04-7	−47.47	36.28			65.59	298.15	1.3
71	C_2H_7N	*N*-methylmethanamine (dimethyl amine)	124-40-3	−18.80	68.80	26.40	5.94	68.73	293.15	1.0
72	C_3F_8	octafluoropropane (R-218)	76-19-7	−1737.62	−1610.31			140.00	295.00	0.0
73	C_3HF_7	1,1,1,2,3,3,3-heptafluoropropane (R-227ea)	431-89-0					121.40	295.00	
74	$C_3H_2ClF_5O$	2-chloro-1,1,2-trifluoroethyl difluoromethyl ether (enflurane)	13838-16-9			30.09				
78	$C_3H_3F_5$	1,1,1,2,2-pentafluoropropane (R-245cb)	1814-88-6					112.82	295.00	
79	$C_3H_3F_5$	1,1,1,3,3-pentafluoropropane (R-245fa)	460-73-1					99.50	295.00	
81	$C_3H_3F_5O$	2-(difluoromethoxy)-1,1,1-trifluoroethane (E-245)	1885-48-9					107.71	295.00	
82	C_3H_3NO	1,2-oxazole(isoxazole)	288-14-2	82.02		37.17		51.408	298.15	
83	C_3H_4	1-propyne (methyl acetylene)	74-99-7	110.00	150.60	22.15		62.31	273.15	0.7
84	C_3H_4	1,2-propadiene	463-49-0	190.92	201.30	20.60	4.40	60.95	243.15	0.2
85	C_3H_6	propene (propylene)	115-07-1	20.00	62.50	18.42	3.00			0.4
86	C_3H_6	cyclopropane	75-19-4	53.30	104.40	20.05	5.44	71.76	233.15	0.0
87	$C_3H_6Cl_2$	1,2-dichloropropane	78-87-5			36.17		100.815	323.14	
88	C_3H_6O	2-propen-1-ol	107-18-6	−123.60		47.30				
89	C_3H_6O	propanone (acetone)	67-64-1	−217.10	−152.60	29.10	5.69	73.94	298.15	2.9

90	$C_3H_6O_2$	propanoic acid	79-09-4	−452.80	−369.60	31.14	10.66	74.97	298.15	1.5
91	$C_3H_6O_2$	methyl ethanoate (methyl acetate)	79-20-9	−408.80	−321.40	30.32	7.49	79.89	298.15	1.7
92	$C_3H_6O_2$	ethyl methanoate (ethyl formate)	109-94-4	−388.30	−303.00	29.91		80.93	298.15	2.0
93	$C_3H_6O_3$	dimethylcarbonate	616-38-6	−571		37.7		84.82	298.15	
94	C_3H_7Cl	1-chloropropane	540-54-5	−133.18	−52.51	27.18		88.95	298.15	2.0
95	C_3H_8	propane	74-98-6	−104.68	−24.29	19.04	3.53	74.87	233.15	0.0
96	C_3H_8O	1-propanol	71-23-8	−255.20	−159.81	41.44	5.20	75.14	298.15	1.7
97	C_3H_8O	2-propanol	67-63-0	−272.70	−173.32	39.85	5.38	76.92	298.15	1.7
98	C_3H_8O	methyl ethyl ether	540-67-0	−216.50	−117.13	25.69		85.85	293.15	1.2
99	C_3H_8S	2-thiabutane (methyl ethyl sulfide)	624-89-5	−59.29	11.92	29.53	9.76	91.02	298.15	
100	C_3H_9N	1-propanamine (propyl amine)	107-10-8	−70.10	41.90	29.55		83.11	298.15	1.3
101	C_3H_9N	2-propanamine (methyl ethyl amine)	75-31-0	−83.70	32.23	27.83		86.33	298.15	
102	C_3H_9N	*N,N*-dimethylmethanamine (trimethyl amine)	75-50-3	−23.60	99.30	22.94	6.55	94.28	298.15	0.6
103	C_3H_9NO	2-ethoxymethanamine (2-methylaminoethanol)	109-83-1			57		82.21	323.14	
104	C_4F_8	octafluorocyclobutane	115-25-3					120.94	253.15	0.0
105	C_4F_{10}	decafluoro-2-methylpropane	354-92-7					156.91	293.15	
106	C_4H_4O	furan	110-00-9	−34.73	0.94	27.10	3.80	73.09	298.15	0.7
107	C_4H_4S	thiophene	110-02-1	114.90	126.10	31.48	4.97	79.47	298.15	0.5
108	C_4H_5N	pyrrole	109-97-7	108.20	160.30	38.75	7.91	69.38	294.00	1.8
109	C_4H_6	1-butyne	107-00-6	162.26	198.77	24.52	6.03	83.22	298.15	0.8
110	C_4H_6	1,3-butadiene	106-99-0	110.00	150.60	22.47	7.98	88.04	298.15	0.0
111	$C_4H_6O_3$	acetic anhydride	108-24-7	−575.70	−476.80					3.0
112	C_4H_8	cyclobutane	287-23-0	28.40	112.20	24.19	1.09	81.43	298.15	0.0
113	C_4H_8	1-butene	106-98-9	−0.54	70.37	22.07	3.96	95.34	298.15	0.3
114	C_4H_8	trans-2-butene	624-64-6	−11.00	63.34	23.34	9.76	93.65	298.15	0.0
115	C_4H_8	cis-2-butene	590-18-1	−7.40	65.46	22.72	7.58	91.01	298.15	0.3
116	C_4H_8	2-methylpropene	115-11-7	−17.10	58.18	21.53	5.93	95.24	298.15	0.5
117	C_4H_8O	butanone (methyl ethyl ketone)	78-93-3	−238.60	−146.50	31.30	8.44	90.13	298.15	3.3
118	C_4H_8O	tetrahydrofuran	109-99-9	−184.18	−79.57	29.81	8.54	81.71	298.15	1.7
119	$C_4H_8O_2$	butanoic acid	107-92-6	−473.60		40.45	11.08	92.46	298.15	1.5
120	$C_4H_8O_2$	2-methylpropanoic acid	79-31-2	−484.20				93.42	298.15	1.3
121	$C_4H_8O_2$	1,3-dioxane	505-22-6	−337.30	−203.90	34.37				
122	$C_4H_8O_2$	1,4-dioxane	123-91-1	−314.70	−180.20	34.16	12.85	85.29	293.15	0.0

No.	Formula	Name	CAS #	DelHf0, kJ/mol	DelGf0, kJ/mol	DelHb, kJ/mol	DelHm, kJ/mol	V liq, cm^3/mol	T liq, K	Dipole, Debye
123	$C_4H_8O_2$	methyl propanoate	554-12-1	−427.50	−310.90	32.24		95.93	298.15	1.7
124	$C_4H_8O_2$	ethyl ethanoate (ethyl acetate)	141-78-6	−444.50	−328.00	31.94	10.48	98.55	298.15	1.9
125	$C_4H_8O_2$	propyl methanoate (propyl formate)	110-74-7	−407.60	−293.50	33.61		97.94	298.15	1.9
126	C_4H_9Cl	2-chlorobutane	78-86-4	−165.69	−55.06	29.17		105.76	298.15	2.1
127	C_4H_{10}	butane	106-97-8	−125.79	−16.57	22.44	4.66	100.48	298.15	0.0
128	C_4H_{10}	2-methylpropane	75-28-5	−134.99	−21.44	21.30	4.61	104.36	298.15	0.1
129	$C_4H_{10}N_2$	piperazine	110-85-0	31.6		26.7				
130	$C_4H_{10}O$	1-butanol	71-36-3	−274.60	−150.17	43.29	9.28	91.96	298.15	1.8
131	$C_4H_{10}O$	2-methyl-1-propanol (isobutanol)	78-83-1	−282.90	−167.40	41.82		92.91	298.15	1.7
132	$C_4H_{10}O$	2-methyl-2-propanol (tert-butanol)	75-65-0	−325.81	−191.20	39.07	6.79	94.88	298.15	1.7
133	$C_4H_{10}O$	2-butanol (sec-butanol)	78-92-2	−292.75	−167.71	40.75		92.35	298.15	1.7
134	$C_4H_{10}O$	diethyl ether	60-29-7	−250.80	−120.70	26.52	7.27	104.75	298.15	1.3
135	$C_4H_{10}O_2$	1,2-dimethoxyethane	110-71-4	−342.80		36.69		104.56	298.15	
136	$C_4H_{10}O_2$	1,2-butandiol	26171-83-5			71.55		92.06	323.14	
137	$C_4H_{10}O_2$	1,3-butanediol	107-88-0			74.46		138.54	323.14	
138	$C_4H_{10}S$	3-thiapentane (diethyl sulfide)	352-93-2	−83.22	18.24	31.77	11.90	108.51	298.15	1.6
139	$C_4H_{11}N$	1-butanamine (butyl amine)	109-73-9	−91.76	49.03	31.81		98.97	298.15	1.3
140	$C_4H_{11}N$	*N*-ethylethanamine (diethyl amine)	109-89-7	−71.70	73.00	29.06		104.24	298.15	1.1
141	$C_4H_{11}N$	2-methyl-1-propanamine (isobutyl amine)	78-81-9	−98.55		30.61		100.45	298.15	1.2
142	C_5F_{12}	dodecafluoropentane	678-26-2	−2561.86	−2351.67			177.80	293.15	0.0
143	C_5H_5N	pyridine	110-86-1	140.37	190.55	35.09	8.28	80.88	298.15	2.3
144	C_5H_6O	2-methylfuran	534-22-5	−66.32	−2.36	28.35	8.55	89.73	293.15	0.7
145	C_5H_8	1-pentyne	627-19-0	143.10	209.20	26.86		98.91	298.15	0.9
146	C_5H_8	cyclopentene	142-29-0	32.95	110.90	26.76	3.36	91.43	293.15	0.9
147	C_5H_8O	cyclopentanone	120-92-3	−192.80		36.35				3.0
148	C_5H_{10}	cyclopentane	287-92-3	−77.10	38.92	27.30	0.61	94.73	298.15	0.0
149	C_5H_{10}	1-pentene	109-67-1	−21.30	78.60	25.20	5.81	110.40	298.15	0.4
150	C_5H_{10}	cis-2-pentene	627-20-3	−26.30	73.50	26.06	7.11	107.85	298.15	
151	C_5H_{10}	2-methyl-2-butene	513-35-9	−40.80	61.60	26.31	7.60	107.57	298.15	

152	C_5H_{10}	3-methyl-1-butene	563-45-1	−27.60	76.00	23.94	5.36	112.77	298.15	
154	$C_5H_{10}O$	2-pentanone (methyl propyl ketone)	107-87-9	−259.20	−138.20	33.44	10.63	107.33	298.15	2.5
155	$C_5H_{10}O$	3-pentanone (diethyl ketone)	96-22-0	−257.90	−134.30	33.45	11.59	106.41	298.15	2.7
156	$C_5H_{10}O$	3-methyl-2-butanone (methyl isopropyl ketone)	563-80-4	−262.60	−139.30	32.35	9.34	107.91	298.15	2.8
157	$C_5H_{10}O$	2-methyltetrahydrofuran	96-47-9			30.13		100.39	298.15	
158	$C_5H_{10}O_2$	pentanoic acid	109-52-4	−497.00	−351.00	62.40	14.16	109.30	298.15	
159	$C_5H_{10}O_2$	3-methylbutanoic acid	503-74-2					110.79	298.15	1.0
160	$C_5H_{10}O_2$	methyl butanoate	623-42-7	−450.70	−305.10	33.79		114.42	298.15	1.7
161	$C_5H_{10}O_2$	ethyl propanoate	105-37-3	−463.60	−319.10	33.88		115.54	298.15	1.8
162	$C_5H_{10}O_2$	methyl 2-methylpropanoate	547-63-7	−464.00	−317.00	32.61		114.63	293.15	2.0
163	$C_5H_{10}O_2$	propyl ethanoate (propyl acetate)	109-60-4	−464.80	−320.20	33.92		115.72	298.15	1.8
164	$C_5H_{10}O_2$	2-methylpropyl methanoate (isobutyl formate)	542-55-2	−436.30	−293.00	33.60		116.96	298.15	1.9
165	$C_5H_{11}Cl$	1-chloropentane	543-59-9	−175.02	−36.68	33.15		121.53	298.15	2.2
166	C_5H_{12}	pentane	109-66-0	−146.76	−8.65	25.79	8.40	115.22	298.15	0.0
167	C_5H_{12}	2-methylbutane	78-78-4	−153.70	−13.86	24.69	5.16	116.46	298.15	0.1
168	C_5H_{12}	2,2-dimethylpropane (neopentane)	463-82-1	−168.11	−15.24	22.74	3.26	122.16	298.15	0.0
169	$C_5H_{12}O$	1-pentanol	71-41-0	−295.60	−142.20	44.40	9.38	108.63	298.15	1.7
170	$C_5H_{12}O$	2-pentanol	6032-29-7	−313.80		41.40				
171	$C_5H_{12}O$	2-methyl-1-butanol	137-32-6	−302.00		55.18				
172	$C_5H_{12}O$	2-methyl-2-butanol	75-85-4	−329.00	−166.00	40.00	4.46	109.50	298.15	1.9
173	$C_5H_{12}O$	3-methyl-1-butanol	123-51-3	−300.00		44.10		109.22	298.15	1.8
174	$C_5H_{12}O$	3-methyl-2-butanol	598-75-4	−316.40		53.00				
175	$C_5H_{12}O$	ethyl propyl ether	628-32-0	−272.20	−115.00	28.94	8.40	121.35	298.15	1.2
176	$C_5H_{12}S$	3-methyl-1-butanethiol (isopentyl mercaptan)	541-31-1	−114.60	17.70		7.45	125.32	298.15	
177	C_6ClF_5	chloropentafluorobenzene	344-07-0			34.76				
178	C_6F_6	hexafluorobenzene	392-56-3			31.66				0.0
179	C_6F_{12}	dodecafluorocyclohexane	355-68-0							0.0
180	C_6F_{14}	tetradecafluorohexane	355-42-0	−2973.99	−2722.34			198.91	293.15	0.0
181	C_6F_{14}	tetradecafluoro-2-methylpentane	355-04-4					195.06	293.15	

No.	Formula	Name	CAS #	DelHf0, kJ/mol	DelGf0, kJ/mol	DelHb, kJ/mol	DelHm, kJ/mol	V liq, cm^3/mol	T liq, K	Dipole, Debye
184	C_6HF_5	pentafluorobenzene	363-72-4			32.15	10.85			
186	C_6H_5Cl	chlorobenzene	108-90-7	51.09	98.36	35.19	9.61	102.22	298.15	1.6
187	C_6H_6	benzene	71-43-2	82.88	129.75	30.72	9.95	89.41	298.15	0.0
188	C_6H_6O	phenol	108-95-2	−96.40	−32.55	46.18	11.29	87.87	298.15	1.6
189	C_6H_7N	benzeneamine (aniline)	62-53-3	87.45	167.90	42.44	10.56	91.52	298.15	1.6
190	C_6H_7N	2-methylpyridine (2-picoline)	109-06-8	99.16	177.37	36.17	9.72	99.11	298.15	1.9
191	C_6H_7N	3-methylpyridine (3-picoline)	108-99-6	106.36	184.62	37.35	14.18	97.81	298.15	2.4
192	C_6H_7N	4-methylpyridine (4-picoline)	108-89-4	102.13	182.08	37.51	12.58	98.02	298.15	
193	C_6H_8O	2-cyclohexen-1-one	930-68-7			48.36		99.59	323.14	
194	C_6H_{10}	cyclohexene	110-83-8	−4.32	106.90	33.42	3.29	101.89	298.15	0.6
195	$C_6H_{10}O$	cyclohexanone	108-94-1	−230.12	−90.87					
196	$C_6H_{10}O$	4-methyl-3-penten-2-one(mesityl oxide)	141-79-7	−178.28		42.7		115.47	298.15	
197	C_6H_{12}	cyclohexane	110-82-7	−123.10	32.26	29.97	2.63	108.75	298.15	0.3
198	C_6H_{12}	methylcyclopentane	96-37-7	−106.00	36.51	29.08	6.93	113.13	298.15	0.0
199	C_6H_{12}	1-hexene	592-41-6	−41.95	86.90	28.28	7.52	125.90	298.15	0.4
200	C_6H_{12}	4-methylpent-1-ene	691-37-2	−49.44	85.69	28.57	3.60	129.44	298.15	
201	$C_6H_{12}O$	cyclohexanol	108-93-0	−294.55	−118.05		1.76			1.7
202	$C_6H_{12}O$	2-hexanone (methyl butyl ketone)	591-78-6	−279.80	−129.90	36.35	14.90	124.10	298.15	
203	$C_6H_{12}O$	3-hexanone (ethyl propyl ketone)	589-38-8	−278.60	−126.10	35.36	13.49	123.43	298.15	
204	$C_6H_{12}O$	4-methyl-2(methyl isobutyl ketone)	108-10-1	−286.40	−135.10	34.49		125.81	298.15	2.8
205	$C_6H_{12}O$	butylvinylether	111-34-2	−179.20		36.59		130.25	298.15	
206	$C_6H_{12}O_2$	hexanoic acid	142-62-1							
207	$C_6H_{12}O_2$	methyl pentanoate	624-24-8	−471.10	−296.50	35.36		132.75	298.00	
208	$C_6H_{12}O_2$	ethyl butanoate	105-54-4	−485.50	−312.00	35.47		132.95	298.15	1.8
209	$C_6H_{12}O_2$	propyl propanoate	106-36-5	−483.10	−309.70	35.54		132.50	298.15	1.8
210	$C_6H_{12}O_2$	ethyl 2-methylpropanoate	97-62-1	−499.60	−324.00	33.67		133.67	293.15	2.1
211	$C_6H_{12}O_2$	butyl ethanoate (butyl acetate)	123-86-4	−485.30	−312.10	36.28	14.59	132.51	298.15	1.8
212	$C_6H_{12}O_2$	2-methylpropyl ethanoate (isobutyl acetate)	110-19-0	−494.70	−322.00	35.90		133.87	298.15	1.9
213	$C_6H_{12}O_2$	pentyl methanoate (pentyl formate)	638-49-3	−448.20	−276.40			131.72	298.15	1.9

214	$C_6H_{12}O_2$	3-methylbutyl methanoate (isopentyl formate)	110-45-2	−453.80	−279.80			132.85	298.15	
215	$C_6H_{12}O_3$	2-ethoxyethylacetate	111-15-9			52.61		140.62	323.14	
216	C_6H_{14}	hexane	110-54-3	−166.92	0.15	28.85	13.07	131.59	298.15	0.0
217	C_6H_{14}	2-methylpentane	107-83-5	−174.55	−5.14	27.79	6.27	132.89	298.15	0.1
218	C_6H_{14}	3-methylpentane	96-14-0	−172.00	−3.20	28.06	5.30	130.62	298.15	
219	C_6H_{14}	2,2-dimethylbutane	75-83-2	−183.97	−9.63	26.31	0.58	133.73	298.15	
220	C_6H_{14}	2,3-dimethylbutane	79-29-8	−175.90	−2.05	27.28	0.79	131.17	298.15	0.2
221	$C_6H_{14}O$	1-hexanol	111-27-3	−316.50	−134.13	44.50	15.40	125.19	298.15	1.8
222	$C_6H_{14}O$	2-hexanol	626-93-7	−333.50		41.01		126.07	298.15	
223	$C_6H_{14}O$	3-hexanol	623-37-0							
224	$C_6H_{14}O$	2-methyl-1-pentanol	105-30-6							
225	$C_6H_{14}O$	2-methyl-2-pentanol	590-36-3	−349.00		39.59		126.22	298.15	
226	$C_6H_{14}O$	2-methyl-3-pentanol	565-67-3							
227	$C_6H_{14}O$	4-methyl-1-pentanol	626-89-1	−324.90		44.46		126.22	298.15	
228	$C_6H_{14}O$	4-methyl-2-pentanol	108-11-2	−339.20		44.20		127.20	298.15	
229	$C_6H_{15}N$	*N,N*-diethylethanamine (triethyl amine)	121-44-8	−92.70	118.00	31.01		139.96	298.15	0.9
230	$C_6H_{15}N$	*N*-propyl-1-propanamine (dipropyl amine)	142-84-7	−113.20		33.47		138.07	298.15	1.0
231	C_7F_{14}	tetradecafluoromethylcyclohexane	355-02-2	−2898.00				195.67	298.15	
232	C_7F_{16}	hexadecafluoroheptane	335-57-9	−3386.11	−3093.02	36.29		223.92	293.15	0.0
233	$C_7H_3F_5$	pentafluorotoluene	771-56-2			34.75	12.99			
234	C_7H_8	toluene	108-88-3	50.17	122.29	33.18	6.85	106.87	298.15	0.4
235	C_7H_8O	benzyl alcohol	100-51-6	−72.38	18.20		8.97			1.7
236	C_7H_8O	2-methylphenol (*o*-cresol)	95-48-7	−128.57	−34.27	45.30	13.94	103.64	298.15	1.6
237	C_7H_8O	3-methylphenol (*m*-cresol)	108-39-4	−132.30	−40.07	47.45	9.41	104.99	298.15	1.8
238	C_7H_8O	4-methylphenol (*p*-cresol)	106-44-5	−125.35	−31.55	47.45	11.89	105.00	298.15	1.6
239	C_7H_9N	2,3-dimethylpyridine (2,3-lutidine)	583-61-9	68.28	177.59	39.08	13.48	113.73	298.15	2.2
240	C_7H_9N	2,4-dimethylpyridine (2,4-lutidine)	108-47-4	63.89	172.04	38.53	8.83	115.53	298.15	2.3
241	C_7H_9N	2,5-dimethylpyridine (2,5-lutidine)	589-93-5	66.44	174.34	38.68	14.65	115.87	298.15	2.2
242	C_7H_9N	2,6-dimethylpyridine (2,6-lutidine)	108-48-5	58.70	168.40	37.46	13.04	116.52	298.15	1.7
243	C_7H_9N	3,4-dimethylpyridine (3,4-lutidine)	583-58-4	70.04	179.34	39.99	14.70	112.32	298.15	1.9
244	C_7H_9N	3,5-dimethylpyridine (3,5-lutidine)	591-22-0	72.80	182.44	39.46	13.11	114.26	298.15	2.6
245	$C_7H_{12}O_2$	butyl-2-propenoate(butylacrylate)	141-32-2	−375.30		47.31				

No.	Formula	Name	CAS #	DelHf0, kJ/mol	DelGf0, kJ/mol	DelHb, kJ/mol	DelHm, kJ/mol	V liq, cm³/mol	T liq, K	Dipole, Debye
246	C_7H_{14}	cycloheptane	291-64-5	−118.10	64.30	33.18	1.88	121.73	298.15	0.0
247	C_7H_{14}	methylcyclohexane	108-87-2	−154.70	27.64	31.27	6.75	128.35	298.15	0.0
248	C_7H_{14}	ethylcyclopentane	1640-89-7	−126.90	45.07	31.96	6.87	128.83	298.15	0.0
249	C_7H_{14}	cis-1,3-dimethylcyclopentane	2532-58-3	−135.90	39.23	30.40	7.37	131.91	298.15	
250	C_7H_{14}	trans-1,3-dimethylcyclopentane	1759-58-6	−133.50	41.42	30.80	7.27	132.64	298.15	
251	C_7H_{14}	1-heptene	592-76-7	−62.76	95.06	31.09	12.66	141.77	298.15	0.3
252	$C_7H_{14}O_2$	heptanoic acid	111-14-8							
253	$C_7H_{14}O_2$	ethyl pentanoate	539-82-2	−505.90	−303.50	36.96		148.45	293.15	
254	$C_7H_{14}O_2$	ethyl 3-methylbutanoate	108-64-5	−527.90	−324.00	37.00		149.99	293.15	
255	$C_7H_{14}O_2$	propyl butanoate	105-66-8	−505.30	−303.00			149.95	298.15	1.8
256	$C_7H_{14}O_2$	2-methylpropyl propanoate	540-42-1	−512.70	−311.20			150.21	298.15	
257	$C_7H_{14}O_2$	propyl 2-methylpropanoate	644-49-5	−518.10	−314.00			147.27	273.15	
258	$C_7H_{14}O_2$	3-methylbutyl ethanoate (isopentyl acetate)	123-92-2	−511.20	−306.80	37.50		150.26	298.15	1.8
259	C_7H_{16}	heptane	142-82-5	−187.80	8.20	31.77	14.03	147.47	298.15	0.0
260	C_7H_{16}	2-methylhexane	591-76-4	−194.60	3.70	30.62	9.19	148.60	298.15	0.0
261	C_7H_{16}	3-methylhexane	589-34-4	−191.30	5.30	30.89		146.74	298.15	0.0
262	C_7H_{16}	3-ethylpentane	617-78-7	−189.50	11.40	31.12	9.55	144.40	298.15	0.0
263	C_7H_{16}	2,2-dimethylpentane	590-35-2	−205.81	0.80	29.23	5.82	149.67	298.15	0.0
264	C_7H_{16}	2,3-dimethylpentane	565-59-3	−194.10	5.82	30.46		145.05	298.15	0.0
265	C_7H_{16}	2,4-dimethylpentane	108-08-7	−201.67	3.51	29.55	6.84	149.95	298.15	0.0
266	C_7H_{16}	3,3-dimethylpentane	562-49-2	−201.40	3.50	29.62	6.85	145.40	298.15	0.0
267	C_7H_{16}	2,2,3-trimethylbutane	464-06-2	−204.40	5.00	28.90	2.26	146.15	298.15	0.0
268	$C_7H_{16}O$	1-heptanol	111-70-6	−330.90	−119.56	48.10	13.20	141.95	298.15	1.7
269	$C_7H_{16}O$	2-heptanol	543-49-7							
270	$C_7H_{16}O$	4-heptanol	589-55-9							
271	C_8F_{18}	octadecafluorooctane	307-34-6	−3798.24	−3463.70	33.38		253.21	293.15	0.0
272	C_8H_8O	methylphenylketone(acetophenone)	98-86-2			55.40		119.94	323.14	
273	C_8H_{10}	ethylbenzene	100-41-4	29.92	130.73	35.57	9.18	123.08	298.15	0.4
274	C_8H_{10}	1,2-dimethylbenzene (*o*-xylene)	95-47-6	19.08	122.05	36.24	13.60	121.25	298.15	0.5

275	C_8H_{10}	1,3-dimethylbenzene (*m*-xylene)	108-38-3	17.32	118.89	35.66	11.57	123.47	298.15	0.3
276	C_8H_{10}	1,4-dimethylbenzene (*p*-xylene)	106-42-3	18.03	121.48	35.67	16.81	123.93	298.15	0.1
277	$C_8H_{10}O$	2-ethylphenol	90-00-6	−145.23	−23.15	46.20	13.94	120.41	298.15	
278	$C_8H_{10}O$	3-ethylphenol	620-17-7	−146.06	−25.01	48.90	9.41	121.25	298.15	
279	$C_8H_{10}O$	4-ethyl-phenol	123-07-9	−144.05	−21.43	48.90	11.89	120.84	298.15	
280	$C_8H_{10}O$	2,3-dimethylphenol (2,3 xylenol)	526-75-0	−157.19	−33.20	47.60	21.02			
281	$C_8H_{10}O$	2,4-dimethylphenol (2,4 xylenol)	105-67-9	−162.88	−41.07	47.40		119.75	293.15	2.0
282	$C_8H_{10}O$	2,5-dimethylphenol (2,5 xylenol)	95-87-4	−161.63	−39.52	47.10	23.38	119.30	293.15	1.5
283	$C_8H_{10}O$	2,6-dimethylphenol (2,6 xylenol)	576-26-1	−161.74	−38.89	44.70	18.90			
284	$C_8H_{10}O$	3,4-dimethylphenol (3,4 xylenol)	95-65-8	−156.56	−34.13	50.10	18.13	118.04	293.15	1.7
285	$C_8H_{10}O$	3,5-dimethylphenol (3,5 xylenol)	108-68-9	−161.54	−39.26	49.70	18.00	119.54	293.15	1.8
286	C_8H_{16}	cyclooctane	292-64-8	−124.40	91.38	35.90	2.41	134.87	298.15	0.0
287	C_8H_{16}	1-octene	111-66-0	−83.59	103.20	34.07	15.57	157.85	298.15	0.3
288	C_8H_{16}	*t*-1,4-dimethylcyclohexane	2207-04-7	−168.300	62.43	37.88	12.33	147.97	298.15	
289	$C_8H_{16}O_2$	octanoic acid	124-07-2				21.36			
290	$C_8H_{16}O_2$	2-ethylhexanoic acid	149-57-5			76.31		161.81	323.15	
291	$C_8H_{16}O_2$	propyl pentanoate	141-06-0	−525.70	−294.40					
292	$C_8H_{16}O_2$	2-methylpropyl butanoate	539-90-2	−533.00	−302.50			168.24	293.15	
293	$C_8H_{16}O_2$	propyl 3-methylbutanoate	557-00-6	−547.10	−314.00			167.11	293.15	
294	$C_8H_{16}O_2$	3-methylbutyl propanoate	105-68-0	−529.20	−295.80			166.72	298.15	
295	$C_8H_{16}O_2$	2-methylpropyl 2-methylpropanoate	97-85-8	−546.20	−313.00			164.82	273.15	
296	C_8H_{18}	octane	111-65-9	−208.75	16.27	34.41	20.65	163.53	298.15	0.0
297	C_8H_{18}	2-methylheptane	592-27-8	−215.35	11.96	33.26	11.92	164.63	298.15	
298	C_8H_{18}	3-methylheptane	589-81-1	−212.50	13.00	33.66	11.38	162.78	298.15	
299	C_8H_{18}	4-methylheptane	589-53-7	−211.96	15.98	33.35	10.84	163.06	298.15	
300	C_8H_{18}	3-ethylhexane	619-99-8	−210.71	17.11	33.59		161.01	298.15	
301	C_8H_{18}	2,2-dimethylhexane	590-73-8	−224.60	10.68	32.07	6.78	165.29	298.15	
302	C_8H_{18}	2,3-dimethylhexane	584-94-1	−213.80	15.65	33.17		161.31	298.15	
303	C_8H_{18}	2,4-dimethylhexane	589-43-5	−219.24	11.51	32.51		164.08	298.15	
304	C_8H_{18}	2,5-dimethylhexane	592-13-2	−222.51	9.95	32.54	12.95	165.70	298.15	
305	C_8H_{18}	3,3-dimethylhexane	563-16-6	−219.99	13.68	32.31	7.11	161.81	298.15	
306	C_8H_{18}	3,4-dimethylhexane	583-48-2	−212.67	16.99	33.24		159.73	298.15	
307	C_8H_{18}	3-ethyl-2-methylpentane	609-26-7	−212.80	19.23	32.93	11.34	159.72	298.15	
308	C_8H_{18}	3-ethyl-3-methylpentane	1067-08-9	−214.85	22.88	32.78	10.84	157.88	298.15	
309	C_8H_{18}	2,2,3-trimethylpentane	564-02-3	−219.95	17.97	31.94	8.62	160.43	298.15	

No.	Formula	Name	CAS #	DelHf0, kJ/mol	DelGf0, kJ/mol	DelHb, kJ/mol	DelHm, kJ/mol	V liq, cm³/mol	T liq, K	Dipole, Debye
310	C_8H_{18}	2,2,4-trimethylpentane (isooctane)	540-84-1	−224.01	14.21	30.79	9.04	166.07	298.15	
311	C_8H_{18}	2,3,3-trimethylpentane	560-21-4	−218.45	18.56	32.12	0.86	158.15	298.15	
312	C_8H_{18}	2,3,4-trimethylpentane	565-75-3	−217.32	19.28	32.36	9.26	159.74	298.15	
313	C_8H_{18}	2,2,3,3-tetramethylbutane	594-82-1	−225.73	10.31	31.42	7.54	139.02	298.15	
314	$C_8H_{18}O$	1-octanol	111-87-5	−356.90	−116.59	46.90		158.37	298.15	2.0
315	$C_8H_{18}O$	2-octanol	123-96-6			44.40		159.38	298.15	1.6
318	$C_8H_{18}O$	2-ethyl-1-hexanol	104-76-7							1.8
319	$C_8H_{19}N$	*n*-octylamine	111-86-4	−173.46		54.63		162.33	298.15	
320	$C_8H_{19}N$	*N*-butyl-1-butanamine (dibutyl amine)	111-92-2			38.44		170.58	298.15	1.1
321	C_9F_{20}	eicosafluorononane	375-96-2	−4210.36	−3834.38			271.15	298.15	0.0
322	C_9H_7N	quinoline	91-22-5			49.70	10.80	118.50	298.15	2.3
323	C_9H_7N	isoquinoline	119-65-3			49.00		118.39	293.15	2.7
324	C_9H_{10}	indan	496-11-7	60.75	166.61	39.63		123.15	298.15	
325	C_9H_{12}	propylbenzene	103-65-1	7.91	137.58	38.20	9.27	140.20	298.15	
326	C_9H_{12}	1-methylethylbenzene	98-82-8	4.00	139.05	37.50	7.79	140.17	298.15	0.8
327	C_9H_{12}	1-ethyl-4-methylbenzene	622-96-8	−2.05	130.28	38.40	13.36	139.60	293.15	
328	C_9H_{12}	1,3,5-trimethylbenzene (mesitylene)	526-73-8	−9.50	124.96	40.00	8.18	133.22	278.70	
329	C_9H_{12}	1,2,4-trimethylbenzene	95-63-6	−13.81	117.50	39.20	13.19	136.00	278.70	
330	C_9H_{12}	1,3,5-trimethylbenzene	108-67-8	−15.90	118.26	39.00	9.51	142.99	298.15	0.1
331	C_9H_{18}	1-nonene	124-11-8	−104.00	111.80	36.31	18.08	174.05	298.15	
332	$C_9H_{18}O_2$	nonanoic acid	112-05-0				20.28			
333	$C_9H_{18}O_2$	3-methylbutyl butanoate	106-27-4	−551.50	−289.20			184.02	298.15	
334	C_9H_{20}	nonane	111-84-2	−228.86	25.00	36.91	15.50	179.70	298.15	0.0
335	C_9H_{20}	2-methyloctane	3221-61-2	−235.85	20.30	36.10	18.00	180.75	298.15	
336	C_9H_{20}	2,2-dimethylheptane	1071-26-7	−246.10	18.10	34.60	8.90	181.51	298.15	
337	C_9H_{20}	2,2,5-trimethylhexane	3522-94-9	−253.26	14.00	40.30	6.20	182.39	298.15	
338	C_9H_{20}	2,2,3,3-tetramethylpentane	7154-79-2	−237.11	37.60	34.30	2.33	170.34	298.15	
339	C_9H_{20}	2,2,4,4-tetramethylpentane	1070-87-7	−242.25	34.20	38.30	9.75	179.23	298.15	
340	C_9H_{20}	2,2,3,4-tetramethylpentane	1186-53-4	−234.97	35.40	33.70	0.50	174.45	298.15	
341	C_9H_{20}	2,3,3,4-tetramethylpentane	16747-38-9	−236.31	38.30	34.50	9.00	170.76	298.15	

342	$C_9H_{20}O$	1-nonanol	143-08-8	−381.20	−111.92	54.40		174.92	298.15	1.7
343	$C_9H_{20}O$	2-nonanol	628-99-9							
344	$C_{10}F_{22}$	docosafluorodecane	307-45-9	−4622.48	−4205.05					0.0
345	$C_{10}H_8$	naphthalene	91-20-3	150.30	223.50	43.40	19.12	129.13	333.15	0.0
346	$C_{10}H_{12}$	1,2,3,4-tetrahydronaphthalene	119-64-2	26.61	166.89			136.27	293.15	
347	$C_{10}H_{12}$	1-methylindan	767-58-8							
348	$C_{10}H_{12}$	2-methylindan	824-63-5							
349	$C_{10}H_{12}$	4-methylindan	824-22-6							
350	$C_{10}H_{12}$	5-methylindan	874-35-1							
351	$C_{10}H_{14}$	butylbenzene	104-51-8	−13.14	145.39	38.87	11.22	156.78	298.15	0.4
352	$C_{10}H_{14}$	2-methylpropylbenzene	538-93-2	−20.34	140.20	37.80	12.50	158.08	298.15	0.3
353	$C_{10}H_{14}$	1,4-diethylbenzene	105-05-5	−20.37	140.30	39.40	10.60	156.45	298.15	0.1
354	$C_{10}H_{14}$	1-(1-methylethyl)-4-methylbenzene	99-87-6	−27.74	136.70	38.20	9.66	157.49	298.15	0.0
355	$C_{10}H_{14}$	1,2,4,5-tetramethylbenzene	95-93-2	−44.56	120.50	45.52	21.00	152.28	298.15	0.0
356	$C_{10}H_{18}$	cis-bicyclo[4.4.0]decane (cis-decalin)	493-01-6	−169.20	85.60	41.00		154.83	298.15	0.0
357	$C_{10}H_{18}$	trans-bicyclo[4.4.0]decane (trans-decalin)	493-02-7	−182.10	74.20	40.20		159.66	298.15	0.0
358	$C_{10}H_{20}$	1-decene	872-05-9	−124.20	121.10			189.30	293.15	
359	$C_{10}H_{20}O_2$	decanoic acid	334-48-5				28.01			
360	$C_{10}H_{22}$	decane	124-18-5	−249.53	33.30	38.75	28.78	195.95	298.15	0.0
361	$C_{10}H_{22}$	2,2,5-trimethylheptane	20291-95-6	−272.21	22.20	36.20	12.00	196.44	298.15	
362	$C_{10}H_{22}$	3,3,5-trimethylheptane	7154-80-5	−259.87	32.60	36.40	14.00	192.48	298.15	
363	$C_{10}H_{22}$	2,2,3,3-tetramethylhexane	13475-81-5	−257.99	47.20	36.20	12.40	187.02	298.15	
364	$C_{10}H_{22}$	2,2,5,5-tetramethylhexane	1071-81-4	−285.89	19.50	42.40	9.80	199.06	298.15	
365	$C_{10}H_{22}O$	1-decanol	112-30-1	−396.70	−98.45	49.80	38.00	191.51	298.15	1.8
366	$C_{10}H_{24}N_4$	octamethylethenetetramine	996-70-3	132.9		53.85		232.40	298.15	
367	$C_{11}H_{10}$	1-methylnaphthalene	90-12-0	115.20	216.40	46.00	6.94	139.37	293.15	0.5
368	$C_{11}H_{10}$	2-methylnaphthalene	91-57-6	114.90	215.00	46.50	12.13	145.77	333.15	0.4
369	$C_{11}H_{24}$	undecane	1120-21-4	−270.16	41.25	41.20	22.32	212.24	298.15	0.0
370	$C_{11}H_{24}O$	1-undecanol	112-42-5							
371	$C_{12}H_{10}$	1,1′-biphenyl	92-52-4	182.42	281.08	48.30	18.57	155.77	347.00	0.0
372	$C_{12}H_{12}$	1,6-dimethylnaphthalene	575-43-9	79.80	210.90			155.93	293.15	
373	$C_{12}H_{12}$	2,7-dimethylnaphthalene	582-16-1	79.00	210.70	59.50	23.35			
374	$C_{12}H_{18}$	1,3,5-triethylbenzene	102-25-0			59.22		193.39	323.14	
375	$C_{12}H_{20}$	1,3-dimethyltricyclo[3.3.1.1^{3,7}]decane (1,3-dimethyladamantane)	702-79-4			49.37		186.61	323.14	

Section B Basic Constants II (*Continued*)

No.	Formula	Name	CAS #	DelHf0, kJ/mol	DelGf0, kJ/mol	DelHb, kJ/mol	DelHm, kJ/mol	V liq, cm^3/mol	T liq, K	Dipole, Debye
376	$C_{12}H_{24}$	1-dodecene	112-41-4	−165.50	138.00			222.06	293.15	
377	$C_{12}H_{26}$	dodecane	112-40-3	−290.79	49.53	43.40	36.58	228.59	298.15	0.0
378	$C_{12}H_{26}O$	1-dodecanol	112-53-8	−443.10	−87.13					
379	$C_{13}H_{12}$	diphenylmethane	101-81-5	165.20				167.23	293.15	0.4
380	$C_{13}H_{28}$	tridecane	629-50-5	−311.42	57.81	45.65	28.50	244.94	298.15	0.0
381	$C_{13}H_{28}O$	1-tridecanol	112-70-9							
382	$C_{14}H_{10}$	phenanthrene	85-01-8	207.50	308.20		16.47			0.0
383	$C_{14}H_{10}$	anthracene	120-12-7	230.90	333.70		28.83			0.0
384	$C_{14}H_{22}$	1,4-di(trimethylmethyl)benzene (*p*-ditertbutylbenzene)	1012-72-2				22.48	236.17	373.15	
385	$C_{14}H_{30}$	tetradecane	629-59-4	−332.05	66.09	47.61	45.61	261.32	298.15	0.0
386	$C_{14}H_{30}O$	1-tetradecanol	112-72-1							
387	$C_{15}H_{32}$	pentadecane	629-62-9	−352.68	74.37	49.45	34.80	277.71	298.15	0.0
388	$C_{15}H_{32}O$	1-pentadecanol	629-76-5							
389	$C_{16}H_{34}$	hexadecane	544-76-3	−373.31	82.65	51.21	51.84	294.11	298.15	0.0
390	$C_{16}H_{34}$	2,2,4,4,6,8,8-heptamethylnonane	4390-04-9							
391	$C_{16}H_{34}O$	1-hexadecanol	4485-13-6				34.29			
392	$C_{17}H_{36}$	heptadecane	629-78-7	−393.94	90.93	52.89	40.50	310.45	298.15	0.0
393	$C_{17}H_{36}O$	1-heptadecanol	1454-85-9	−546.30	−44.67					
394	$C_{18}H_{14}$	1,2-diphenylbenzene	84-15-1							
395	$C_{18}H_{14}$	1,3-diphenylbenzene	92-06-8							
396	$C_{18}H_{14}$	1,4-diphenylbenzene	92-94-4				35.50			0.7
397	$C_{18}H_{38}$	octadecane	593-45-3	−414.57	99.21	54.46	61.39	326.66	298.15	0.0
398	$C_{18}H_{38}O$	1-octadecanol	112-92-5	−566.90	−36.22					1.7
399	$C_{19}H_{40}$	nonadecane	629-92-5	−435.20	107.49	56.02	45.82	343.25	298.15	0.0
400	$C_{19}H_{40}O$	1-nonadecanol	1454-84-8							
401	$C_{20}H_{42}$	eicosane	112-95-8	−455.83	115.77	57.49	69.87	361.18	298.15	0.0
402	$C_{20}H_{42}O$	1-eicosanol	629-96-9	−608.10	−19.43					
403	$C_{21}H_{44}$	heneicosane	629-94-7				47.70	381.11	313.15	0.0
404	$C_{22}H_{46}$	docosane	629-97-0				49.96	399.14	318.15	0.0

405	$C_{23}H_{48}$	tricosane	638-67-5				41.76			0.0
406	$C_{24}H_{50}$	tetracosane	646-31-1				54.89	434.96	324.25	0.0
407	ClD	deuterium chloride	7698-05-7	−93.33	−95.93			31.02	193.15	
408	$ClFO_3$	perchloryl fluoride	7616-94-6	−21.44	50.62	19.33		62.70	243.15	0.0
409	ClF_5	chlorine pentafluoride	13637-63-3	−238.49	−147.11			73.41	298.15	
410	ClH	hydrogen chloride	7647-01-0	−92.31	−95.19	16.15		30.28	183.15	1.1
411	ClH_4N	ammonium chloride	12125-02-9							
412	ClNO	nitrogen oxychloride	2696-92-6	51.71	65.97	25.78		46.10	261.00	1.8
413	Cl_2	chlorine	7782-50-5	0.00	0.00	20.41		45.36	239.00	0.0
414	DH	deuterium hydride	13983-20-5	0.32	−1.46	1.08	0.16			0.0
415	DI	deuterium iodide	14104-45-1	26.23	1.84			46.04	237.00	
416	D_2	deuterium	7782-39-0	0.00	0.00	1.23	0.20	24.41	22.70	0.0
417	D_2	deuterium, normal	800000-54-8	0.00	0.00	1.23	0.20			0.0
418	D_2O	deuterium oxide	7789-20-0	−249.20	−234.53	41.46	6.38	18.13	298.15	1.9
419	D_2S	deuterium sulfide	13536-94-2	−23.89	−35.39	18.85	2.37			
420	D_3N	trideuteroammonia	13550-49-7	−64.28	−31.69					
421	D_3P	trideuterophosphine	13537-03-6							
422	FH	hydrogen fluoride	7664-39-3	−273.30	−275.40			20.69	293.15	1.8
423	FNO_2	nitrogen dioxyfluoride	10022-50-1	−108.78	−66.55	18.05				0.5
424	F_2	fluorine	7782-41-4	0.00	0.00	6.62		25.16	85.00	0.0
425	F_2HN	difluoroamine	10405-27-3							1.9
426	F_2N_2	cis-difluorodiazene	13812-43-6	74.89	114.89					0.2
427	F_2N_2	trans-difluorodiazene	13776-62-0	81.17	120.35					0.0
428	F_2O	oxygen difluoride	7783-41-7	24.70	42.01	11.09		34.93	123.15	0.2
429	F_2Xe	xenon difluoride	13709-36-9					54.09	298.15	0.0
430	F_3N	nitrogen trifluoride	7783-54-2	−129.70	−88.29	11.56		46.11	144.00	0.2
431	F_3NO	trifluoroamine oxide	13847-65-9	−163.30	−96.46					
432	F_4N_2	tetrafluorohydrazine	10036-47-2	−8.37	79.66	13.27		69.34	163.00	0.3
433	F_4S	sulfur tetrafluoride	7783-60-0	−774.04	−731.01	26.44		55.82	195.00	1.0
434	F_4Xe	xenon tetrafluoride	13709-61-0	−187.60				68.41	298.15	0.0
435	F_6S	sulfur hexafluoride	2551-62-4	−1220.89	−1117.09			79.81	223.15	0.0
436	F_6U	uranium hexafluoride	7783-81-5	−2139.00	−2060.00					0.0
437	HI	hydrogen iodide	10034-85-2	26.50	1.70	19.76		46.06	243.15	0.5
438	H_2	hydrogen	1333-74-0	0.00	0.00	0.89	0.12	28.39	20.00	0.0

Section B Basic Constants II (*Continued*)

No.	Formula	Name	CAS #	DelHf0, kJ/mol	DelGf0, kJ/mol	DelHb, kJ/mol	DelHm, kJ/mol	V liq, cm^3/mol	T liq, K	Dipole, Debye
439	H_2	hydrogen, normal	800000-51-5	0.00	0.00	0.90	0.12			0.0
440	H_2O	water	7732-18-5	−241.81	−228.42	40.66	6.01	18.07	298.15	1.8
441	H_2S	hydrogen sulfide	7783-06-4	−20.63	−33.43	18.68	2.38	34.32	214.00	0.9
442	H_2S_2	dihydrogen disulfide	13465-07-1	15.52	−1.67	35.10	7.53	49.85	298.15	
443	H_2S_3	dihydrogen trisulfide	13845-23-3					66.23	298.15	
444	H_2S_4	dihydrogen tetrasulfide	13845-25-5					82.66	298.15	
445	H_2S_5	dihydrogen pentasulfide	13845-24-4							
446	H_2Se	hydrogen selenide	7783-07-5			19.70		38.20	231.00	
447	H_3N	ammonia	7664-41-7	−45.94	−16.41	23.35	5.66	24.96	239.15	1.5
448	H_3P	phosphine	7803-51-2	5.44	13.44	14.60		45.57	183.15	0.6
449	H_4N_2	hydrazine	302-01-2	95.40	159.38	44.77	12.66	31.79	293.15	3.0
450	He	helium	7440-59-7	0.00	0.00	0.08		32.54	4.30	0.0
451	He	helium-3	14762-55-1	0.00	0.00					0.0
452	I_2	iodine	7553-56-2	62.42	19.33	41.57		67.86	453.00	1.3
453	Kr	krypton	7439-90-9	0.00	0.00	9.08		34.63	120.00	0.0
454	NO	nitrogen monoxide (nitric oxide)	10102-43-9	90.25	86.58	13.78	2.30	23.44	121.00	0.2
455	N_2	nitrogen	7727-37-9	0.00	0.00	5.58	0.72	34.84	78.00	0.0
456	N_2O	dinitrogen oxide (nitrous oxide)	10024-97-2	82.05	104.18	16.55	6.54	35.90	184.00	0.2
457	N_2O_4	dinitrogen tetroxide (nitrogen dioxide)	10544-72-6	9.16	97.85	29.00	14.65	63.59	293.15	
458	Ne	neon	7440-01-9	0.00	0.00	1.71		16.76	27.00	0.0
459	OT_2	tritium oxide	14940-65-9	−249.37	−226.24			18.15	298.15	
460	O_2	oxygen	7782-44-7	0.00	0.00	6.82	0.44	27.85	90.00	0.0
461	O_2S	sulfur dioxide	7446-09-5	−296.81	−300.14	24.94	7.40	44.03	263.15	1.6
462	O_3	ozone	10028-15-6	142.70	163.10	14.20		35.40	161.00	0.6
463	O_3S	sulfur trioxide	7446-11-9	−395.72	−370.93	40.69	7.53	42.10	298.15	
464	Rn	radon	10043-92-2	0.00	0.00			50.46	211.15	0.0
465	S	sulfur	7704-34-9	276.98	236.50	10.46	1.61			0.0
466	Se	selenium	7782-49-2	0.00	0.00	95.48				0.0
467	T_2	tritium	10028-17-8	0.00	0.00					0.0
468	Xe	xenon	7440-63-3	0.00	0.00	12.57		42.91	165.00	0.0

Section C Ideal Gas and Liquid Heat Capacities. $C_p^\circ/R = \text{a0} + \text{a1}T + \text{a2}T^2 + \text{a3}T^3 + \text{a4}T^4$.
CpIG and Cpliq at 298.15 K, J mol^{-1} K^{-1}.

No.	Formula	Name	CAS #	Trange, K	a0	a1 × 10^3	a2 × 10^5	a3 × 10^8	a4 × 10^{11}	CpIG	Cpliq
1	Ar	argon	7440-37-1	—	2.500	0.000	0.000	0.000	0.000	20.79	
2	Br_2	bromine	7726-95-6	50-1000	3.212	7.160	−1.528	1.445	−0.499	36.05	75.67
3	BrD	deuterium bromide	13536-59-9	50-1000	3.716	−2.318	0.738	−0.717	0.250	29.23	
6	BrH	hydrogen bromide	10035-10-6	50-1000	3.842	−3.098	0.917	−1.032	0.426	29.14	
7	$CBrClF_2$	bromochlorodifluoromethane	353-59-3	100-1000	1.968	36.592	−5.489	4.036	−1.170	74.65	
8	$CBrF_3$	bromotrifluoromethane	75-63-8	100-1000	1.959	30.789	−3.782	2.236	−0.515	69.24	
9	CBr_2F_2	dibromodifluoromethane	75-61-6	100-1000	2.476	36.115	−5.666	4.368	−1.324	77.02	
10	$CClF_3$	chlorotrifluoromethane	75-72-9	50-1000	2.369	23.861	−1.579	−0.366	0.528	66.87	
11	CCl_2F_2	dichlorodifluoromethane (R-12)	75-71-8	50-1000	2.185	31.251	−3.724	1.930	−0.323	72.28	119.00
12	CCl_3F	trichlorofluoromethane (R-11)	75-69-4	50-1000	2.090	38.890	−6.079	4.542	−1.316	78.09	121.80
13	CCl_4	tetrachloromethane	56-23-5	200-1000	2.518	41.882	−7.160	5.739	−1.756	83.43	131.60
14	CF_4	tetrafluoromethane	75-73-0	50-1000	2.643	15.383	0.850	−2.940	1.469	61.05	
15	$CHBrF_2$	bromodifluoromethane	1511-62-2	100-1000	3.254	13.871	−0.070	−1.130	0.606	58.76	
16	$CHClF_2$	chlorodifluoromethane (R-22)	75-45-6	50-1000	3.164	10.422	1.179	−2.650	1.222	55.85	110
17	$CHCl_2F$	dichlorofluoromethane	75-43-4	50-1000	2.949	17.130	−0.629	−0.821	0.573	60.94	108
18	$CHCl_3$	trichloromethane	67-66-3	200-1000	2.389	26.218	−3.145	1.857	−0.423	65.40	113.80
19	CHF_3	trifluoromethane (R-23)	75-46-7	50-1000	3.450	3.480	3.012	−4.452	1.834	50.98	
20	CH_2Cl_2	dichloromethane	75-09-2	200-1000	2.710	11.561	0.324	−1.370	0.662	50.88	100.00
21	CH_2F_2	difluoromethane	75-10-5	50-1000	4.150	−5.584	4.384	−5.160	1.920	42.88	
22	CH_2O_2	methanoic acid (formic acid)	64-18-6	50-1000	3.809	1.568	3.587	−4.410	1.672	*53.45*	99.17
23	CH_3Cl	chloromethane	74-87-3	200-1000	3.578	−1.750	3.071	−3.714	1.408	40.74	81.84
24	CH_3F	fluoromethane	593-53-3	50-1000	4.561	−10.437	4.813	−5.069	1.769	37.51	
25	CH_3NO_2	nitromethane	75-52-5	50-1000	4.196	−1.102	5.158	−6.721	2.660	57.22	106.80
26	CH_4	methane	74-82-8	50-1000	4.568	−8.975	3.631	−3.407	1.091	35.69	
27	CH_4O	methanol	67-56-1	50-1000	4.714	−6.986	4.211	−4.443	1.535	44.06	81.08
28	CH_4S	methanethiol (methyl mercaptan)	74-93-1	50-1000	4.119	1.313	2.591	−3.212	1.208	50.26	90.50
29	CH_5N	methanamine (methyl amine)	74-89-5	50-1000	4.193	−2.122	4.039	−4.738	1.751	50.05	102.09
30	CO	carbon monoxide	630-08-0	50-1000	3.912	−3.913	1.182	−1.302	0.515	29.14	
31	CO_2	carbon dioxide	124-38-9	50-1000	3.259	1.356	1.502	−2.374	1.056	37.13	
33	C_2ClF_5	1-chloro-1,1,2,2,2-pentafluoroethane	76-15-3	50-1000	2.355	50.469	−5.156	2.041	−0.139	111.10	

Section C Ideal Gas and Liquid Heat Capacities. $C_p^\circ/R = a0 + a1T + a2T^2 + a3T^3 + a4T^4$.
CpIG and Cpliq at 298.15 K, J mol^{-1} K^{-1} (*Continued*)

No.	Formula	Name	CAS #	Trange, K	a0	a1 × 10^3	a2 × 10^5	a3 × 10^8	a4 × 10^{11}	CpIG	Cpliq
34	$C_2Cl_2F_4$	1,1-dichloro-1,2,2,2-tetrafluoroethane	374-07-2	50-1000	2.268	56.415	−6.908	3.953	−0.861	115.80	
35	$C_2Cl_2F_4$	1,2-dichloro-1,1,2,2-tetrafluoroethane	76-14-2	50-1000	2.525	53.644	−5.771	2.417	−0.199	116.60	
36	$C_2Cl_3F_3$	1,1,2-trichloro-1,2,2-trifluoroethane	76-13-1	50-1000	2.133	63.238	−8.916	6.140	−1.683	121.00	
37	C_2F_4	tetrafluoroethene	116-14-3	200-1000	2.223	36.551	−4.776	3.283	−0.931	80.41	
38	C_2F_6	hexafluoroethane	76-16-4	50-1000	2.525	43.543	−2.948	−0.630	0.967	106.54	
41	C_2HClF_4	1-chloro-1,1,2,2-tetrafluoroethane	354-25-6	50-1000	2.888	38.360	−2.468	−0.397	0.677	100.40	
42	C_2HClF_4	1-chloro-1,2,2,2-tetrafluoroethane	2837-89-0	50-1000	3.022	35.834	−1.744	−1.211	0.994	99.06	
43	$C_2HCl_2F_3$	1,1-dichloro-2,2,2-trifluoroethane (R-123)	306-83-2	50-1000	2.996	39.490	−2.743	−0.122	0.572	102.60	
44	$C_2HCl_2F_3$	1,2-dichloro-1,2,2-trifluoroethane (R-123a)	354-23-4	50-1000	2.699	43.299	−3.663	0.697	0.322	*104.45*	
45	C_2HF_5	pentafluoroethane	354-33-6	50-1000	3.146	29.937	−0.056	−3.019	1.669	94.40	
47	C_2H_2	ethyne (acetylene)	74-86-2	50-1000	2.410	10.926	−0.255	−0.790	0.524	59.03	
48	$C_2H_2F_2$	1,1-difluoroethene	75-38-7	200-1000	0.749	26.756	−1.905	0.245	0.204	59.08	
49	$C_2H_2F_4$	1,1,1,2-tetrafluoroethane (R-134a)	811-97-2	50-1000	3.064	25.420	0.586	−3.339	1.716	86.64	
50	$C_2H_2F_4$	1,1,2,2-tetrafluoroethane (R-134)	359-35-3	50-1000	3.084	32.841	−2.425	0.488	0.162	90.32	
51	$C_2H_3ClF_2$	1-chloro-1,1-difluoroethane	75-68-3	50-1000	2.338	29.791	−1.048	−1.336	0.927	83.26	130
52	$C_2H_3Cl_2F$	1,1-dichloro-1-fluoroethane (R-1416)	1717-00-6	50-1000	2.140	36.934	−3.121	0.927	0.068	88.37	
53	$C_2H_3F_3$	1,1,1-trifluoroethane (R-143a)	420-46-2	50-1000	2.577	23.727	0.480	−2.824	1.439	78.61	
54	$C_2H_3F_3$	1,1,2-trifluoroethane (R-143)	430-66-0	50-1000	3.531	16.450	2.074	−4.217	1.869	77.34	
55	C_2H_4	ethene (ethylene)	74-85-1	50-1000	4.221	−8.782	5.795	−6.729	2.511	42.90	
56	$C_2H_4Br_2$	1,2-dibromoethane	106-93-4	298-1000	3.784	24.587	−0.750	−0.886	0.601	*85.31*	135.6
57	$C_2H_4Cl_2$	1,1-dichloroethane	75-34-3	200-1000	2.610	24.853	−0.675	−1.035	0.643	76.32	126.4
58	$C_2H_4Cl_2$	1,2-dichloroethane	107-06-2	298-1000	2.990	23.197	−0.404	−1.133	0.617	77.32	126.30
59	$C_2H_4F_2$	1,1-difluoroethane (R-152a)	75-37-6	50-1000	3.292	11.749	2.835	−4.645	1.941	*68.49*	118.00
60	$C_2H_4O_2$	ethanoic acid (acetic acid)	64-19-7	50-1000	4.375	−2.397	6.757	−8.764	3.478	63.44	123.10
61	$C_2H_4O_2$	methyl methanoate (methyl formate)	107-31-3	298-1000	2.277	18.013	1.160	−2.921	1.342	66.50	119.70
62	C_2H_5Br	bromoethane	74-96-4	100-1000	3.636	6.861	3.749	−5.446	2.231	64.23	99.8
63	C_2H_5Cl	chloroethane	75-00-3	200-1000	3.029	9.885	2.967	−4.550	1.871	62.64	106
64	C_2H_5F	fluoroethane	353-36-6	50-1000	3.881	1.616	4.799	−6.161	2.364	59.61	

65	C_2H_6	ethane	74-84-0	50-1000	4.178	−4.427	5.660	−6.551	2.487	52.47	231.50
66	C_2H_6O	ethanol	64-17-5	50-1000	4.396	0.628	5.546	−7.024	2.685	65.21	112.25
67	C_2H_6O	dimethyl ether	115-10-6	100-1000	4.361	6.070	2.899	−3.581	1.282	65.57	
68	C_2H_6S	ethanethiol (ethyl mercaptan)	75-08-1	50-1000	3.894	12.951	2.052	−3.287	1.312	73.01	117.80
69	C_2H_6S	2-thiapropane (dimethylsulfide)	75-18-3	273-1000	3.535	17.530	0.596	−1.632	0.696	74.06	118.10
70	C_2H_7N	ethanamine (ethyl amine)	75-04-7	50-1000	4.640	2.069	5.797	−7.659	3.043	71.54	129.70
71	C_2H_7N	*N*-methylmethanamine (dimethyl amine)	124-40-3	273-1000	2.469	15.462	2.642	−4.025	1.564	70.50	
72	C_3F_8	octafluoropropane (R-218)	76-19-7	200-1000	1.605	76.488	−8.707	4.540	−0.856	147.95	
73	C_3HF_7	1,1,1,2,3,3,3-heptafluoropropane (R-227ea)	431-89-0								137.00
82	C_3H_3NO	1,2-oxazole (isoxazole)	288-14-2	50-1000	3.911	−9.705	10.380	−13.472	5.359	*59.01*	96.48
83	C_3H_4	1-propyne (methyl acetylene)	74-99-7	50-1000	3.158	12.210	1.167	−2.316	1.002	60.73	
84	C_3H_4	1,2-propadiene	463-49-0	50-1000	3.403	6.271	3.388	−5.113	2.161	81.82	
85	C_3H_6	propene (propylene)	115-07-1	50-1000	3.834	3.893	4.688	−6.013	2.283	64.32	112.00
86	C_3H_6	cyclopropane	75-19-4	50-1000	4.493	−18.097	12.744	−16.049	6.426	55.57	
87	$C_3H_6Cl_2$	1,2-dichloropropane	78-87-5	298-1000	1.697	40.582	−2.247	−0.038	0.377	*98.27*	
88	C_3H_6O	2-propen-1-ol	107-18-6	298-1000	0.248	34.938	−1.685	−0.192	0.324	*76.01*	138.90
89	C_3H_6O	propanone (acetone)	67-64-1	200-1000	5.126	1.511	5.731	−7.177	2.728	74.52	126.60
90	$C_3H_6O_2$	propanoic acid	79-09-4								152.80
91	$C_3H_6O_2$	methyl ethanoate (methyl acetate)	79-20-9	298-1000	4.242	14.388	3.338	−4.930	1.931	85.30	143.90
92	$C_3H_6O_2$	ethyl methanoate (ethyl formate)	109-94-4							89.00	146.60
94	C_3H_7Cl	1-chloropropane	540-54-5	200-1000	4.365	9.895	5.366	−7.708	3.120	85.30	131
95	C_3H_8	propane	74-98-6	50-1000	3.847	5.131	6.011	−7.893	3.079	73.60	120.00
96	C_3H_8O	1-propanol	71-23-8	50-1000	4.712	6.565	6.310	−8.341	3.216	85.56	143.73
97	C_3H_8O	2-propanol	67-63-0	50-1000	3.334	18.853	3.644	−6.115	2.543	89.32	154.40
98	C_3H_8O	methyl ethyl ether	540-67-0	100-1000	4.008	21.493	1.803	−3.333	1.331	93.30	
99	C_3H_8S	2-thiabutane (methyl ethyl sulfide)	624-89-5	273-1000	2.816	29.186	0.807	−2.888	1.325	95.06	144.60
100	C_3H_9N	1-propanamine (propyl amine)	107-10-8	50-1000	4.142	12.606	5.471	−7.524	2.918	91.80	136.20
101	C_3H_9N	2-propanamine (methyl ethyl amine)	75-31-0	50-1000	3.633	22.221	3.094	−5.375	2.236	97.55	163.88
102	C_3H_9N	*N,N*-dimethylmethanamine (trimethyl amine)	75-50-3	298-1000	1.660	27.899	2.517	−5.097	2.190	91.80	136.20
104	C_4F_8	octafluorocyclobutane	115-25-3	298-1000	0.949	80.942	−7.976	2.970	−0.087	*79.48*	
105	C_4F_{10}	decafluoro-2-methylpropane	355-25-9	200-1000	1.965	99.798	−11.830	6.680	−1.457	78.02	
106	C_4H_4O	furan	110-00-9	50-1000	3.816	−10.453	12.446	−16.907	7.020	65.40	114.64

Section C Ideal Gas and Liquid Heat Capacities. $C_p^\circ/R = a0 + a1T + a2T^2 + a3T^3 + a4T^4$.
CpIG and Cpliq at 298.15 K, J mol^{-1} K^{-1} *(Continued)*

No.	Formula	Name	CAS #	Trange, K	a0	a1 × 10^3	a2 × 10^5	a3 × 10^8	a4 × 10^{11}	CpIG	Cpliq
107	C_4H_4S	thiophene	110-02-1	50-1000	3.063	1.520	9.514	−14.129	6.088	72.78	123.88
108	C_4H_5N	pyrrole	109-97-7	50-1000	3.554	−6.426	12.231	−16.957	7.095	71.60	127.74
109	C_4H_6	1-butyne	107-00-6	50-1000	2.995	20.800	1.560	−3.462	1.524	81.42	122.80
110	C_4H_6	1,3-butadiene	106-99-0	50-1000	3.607	5.085	8.253	−12.371	5.321	79.54	125.20
111	$C_4H_6O_3$	acetic anhydride	108-24-7	298-1000	−1.274	50.172	−1.459	−1.951	1.244	*99.51*	124.10
112	C_4H_8	cyclobutane	287-23-0	50-1000	4.739	−16.423	14.488	−18.041	7.089	70.56	109.3
113	C_4H_8	1-butene	106-98-9	50-1000	4.389	7.984	6.143	−8.197	3.165	85.56	124.90
114	C_4H_8	trans-2-butene	624-64-6	50-1000	5.584	−4.890	9.133	−10.975	4.085	87.67	
115	C_4H_8	cis-2-butene	590-18-1	50-1000	3.689	19.184	2.230	−3.426	1.256	80.15	127.00
116	C_4H_8	2-methylpropene	115-11-7	50-1000	3.231	20.949	2.313	−3.949	1.566	88.09	
117	C_4H_8O	butanone (methyl ethyl ketone)	78-93-3	200-1000	6.349	11.062	4.851	−6.484	2.469	103.26	158.90
118	C_4H_8O	tetrahydrofuran	109-99-9	50-1000	5.171	−19.464	16.460	−20.420	8.000	76.53	124.10
119	$C_4H_8O_2$	butanoic acid	107-92-6								177.70
120	$C_4H_8O_2$	2-methylpropanoic acid	79-31-2								173.00
121	$C_4H_8O_2$	1,3-dioxane	505-22-6	50-1000	3.834	−0.249	11.985	−15.494	6.047	89.47	143.90
122	$C_4H_8O_2$	1,4-dioxane	123-91-1	50-1000	3.730	1.851	11.781	−15.602	6.177	92.18	154.50
124	$C_4H_8O_2$	ethyl ethanoate (ethyl acetate)	141-78-6	298-1000	10.228	−14.948	13.033	−15.736	5.999	113.64	170.60
125	$C_4H_8O_2$	propyl methanoate (propyl formate)	110-74-7							110.10	178.10
126	C_4H_9Cl	2-chlorobutane	78-86-4	200-1000	4.450	22.285	4.350	−7.215	3.015	110.22	
127	C_4H_{10}	butane	106-97-8	200-1000	5.547	5.536	8.057	−10.571	4.134	98.49	142.89
128	C_4H_{10}	2-methylpropane (isobutane)	75-28-5	50-1000	3.351	17.883	5.477	−8.099	3.243	96.65	142.50
130	$C_4H_{10}O$	1-butanol	71-36-3	50-1000	4.467	16.395	6.688	−9.690	3.864	108.03	177.06
131	$C_4H_{10}O$	2-methyl-1-propanol (isobutanol)	78-83-1							113.00	183.00
132	$C_4H_{10}O$	2-methyl-2-propanol (tert-butanol)	75-65-0	50-1000	2.611	36.052	1.517	−4.360	1.947	113.63	220.10
133	$C_4H_{10}O$	2-butanol (sec-butanol)	78-92-2	50-1000	3.860	28.561	2.728	−5.140	2.117	112.74	199.00
134	$C_4H_{10}O$	diethyl ether	60-29-7	100-1000	4.612	37.492	−1.870	1.316	−0.698	119.46	172.60
138	$C_4H_{10}S$	3-thiapentane (diethyl sulfide)	352-93-2	273-1000	4.335	26.082	3.959	−6.881	2.900	116.57	171.50
139	$C_4H_{11}N$	1-butanamine (butyl amine)	109-73-9	298-1000	2.668	38.366	1.150	−3.817	1.712	113.90	294.30
140	$C_4H_{11}N$	n-ethylethanamine (diethyl amine)	109-89-7	298-1000	3.028	32.373	2.828	−5.501	2.300	116.00	178.10
141	$C_4H_{11}N$	2-methyl-1-propanamine (isobutyl amine)	78-81-9	298-1000	0.380	53.027	−2.436	0.207	0.038	*117.09*	194.0

142	C_5F_{12}	dodecafluoropentane	678-26-2	200-1000	2.315	123.238	−14.997	8.875	−2.081	231.95	
143	C_5H_5N	pyridine	110-86-1	298-1000	−3.505	49.389	−1.746	−1.595	1.097	78.23	132.70
144	C_5H_6O	2-methylfuran	534-22-5	50-1000	3.952	5.535	9.252	−13.046	5.353	89.66	143.7
145	C_5H_8	1-pentyne	627-19-0	50-1000	3.382	31.688	0.790	−3.109	1.417	106.69	167.00
146	C_5H_8	cyclopentene	142-29-0	50-1000	4.555	−12.408	15.195	−19.676	7.900	*81.25*	122.4
147	C_5H_8O	cyclopentanone	120-92-3	50-1000	4.294	−1.236	13.080	−17.531	7.071	*95.32*	154.5
148	C_5H_{10}	cyclopentane	287-92-3	50-1000	5.019	−19.734	17.917	−21.696	8.215	82.76	126.80
149	C_5H_{10}	1-pentene	109-67-1	200-1000	5.079	11.919	7.838	−10.952	4.381	108.20	154.00
150	C_5H_{10}	cis-2-pentene	627-20-3	298-1000	2.901	31.785	1.842	−3.953	1.609	*108.87*	151.80
151	C_5H_{10}	2-methyl-2-butene	513-35-9	298-1000	1.240	39.303	0.270	−2.500	1.120	*105.00*	152.80
152	C_5H_{10}	3-methyl-1-butene	563-45-1	298-1000	2.108	41.912	−0.416	−1.937	0.954	115.00	156.10
153	$C_5H_{10}O$	cyclopentanol	96-41-3	50-1000	4.370	5.723	13.357	−18.752	7.733	*113.00*	184.5
154	$C_5H_{10}O$	2-pentanone (methyl propyl ketone)	107-87-9	200-1000	7.836	9.051	8.063	−10.847	4.283	125.90	184.30
155	$C_5H_{10}O$	3-pentanone (diethyl ketone)	96-22-0	200-1000	8.071	13.654	6.120	−8.337	3.253	129.87	191.00
156	$C_5H_{10}O$	3-methyl-2-butanone (methyl isopropyl ketone)	563-80-4							124.10	180.10
158	$C_5H_{10}O_2$	pentanoic acid	109-52-4								210.30
159	$C_5H_{10}O_2$	3-methylbutanoic acid	503-74-2								197.10
160	$C_5H_{10}O_2$	methyl butanoate	623-42-7							133.10	200.80
161	$C_5H_{10}O_2$	ethyl propanoate	105-37-3							131.90	197.60
162	$C_5H_{10}O_2$	methyl 2-methylpropanoate	547-63-7							132.00	191.00
163	$C_5H_{10}O_2$	propyl ethanoate (propyl acetate)	109-60-4							134.90	202.20
164	$C_5H_{10}O_2$	2-methylpropyl methanoate (isobutyl formate)	542-55-2							136.60	217.00
165	$C_5H_{11}Cl$	1-chloropentane	543-59-9	200-1000	7.052	9.759	10.210	−14.077	5.640	130.58	
166	C_5H_{12}	pentane	109-66-0	200-1000	7.554	−0.368	11.846	−14.939	5.753	120.04	167.19
167	C_5H_{12}	2-methylbutane	78-78-4	200-1000	1.959	38.191	2.434	−5.175	2.165	118.97	164.80
168	C_5H_{12}	2,2-dimethylpropane (neopentane)	463-82-1	200-1000	−11.428	156.037	−33.383	40.127	−17.806	120.80	172.00
169	$C_5H_{12}O$	1-pentanol	71-41-0	200-1000	5.530	16.887	9.430	−13.725	5.607	130.70	208.10
172	$C_5H_{12}O$	2-methyl-2-butanol	75-85-4	298-1000	4.282	27.248	6.649	−10.683	4.474	*131.69*	247.9
173	$C_5H_{12}O$	3-methyl-1-butanol	123-51-3								211.00
174	$C_5H_{12}O$	3-methyl-2-butanol	598-75-4								245.90
175	$C_5H_{12}O$	ethyl propyl ether	628-32-0							140.00	197.40

Section C Ideal Gas and Liquid Heat Capacities. $C_p^\circ/R = a0 + a1T + a2T^2 + a3T^3 + a4T^4$.
CpIG and Cpliq at 298.15 K, J mol^{-1} K^{-1} (*Continued*)

No.	Formula	Name	CAS #	Trange, K	a0	a1 × 10³	a2 × 10⁵	a3 × 10⁸	a4 × 10¹¹	CpIG	Cpliq
176	$C_5H_{12}S$	3-methyl-1-butanethiol (isopentyl mercaptan)	541-31-1							142.30	200.30
177	C_6Cl_5F	chloropentafluorobenzene	344-07-0	200-1000	2.994	74.969	−8.448	4.889	−1.177	*158.30*	
178	C_6F_6	hexafluorobenzene	392-56-3	200-1000	2.531	75.268	−8.410	4.845	−1.166	*155.38*	
180	C_6F_{14}	tetradecafluorohexane	355-42-0	200-1000	2.660	146.733	−18.179	11.086	−2.710	273.99	
184	C_6HF_5	pentafluorobenzene	363-72-4	200-1000	1.197	72.572	−7.369	3.676	−0.718	*143.02*	
185	$C_6H_2F_4$	1,2,4,5-tetrafluorobenzene	327-54-8	200-1000	0.007	69.341	−6.192	2.343	−0.202	*131.21*	
186	C_6H_5Cl	chlorobenzene	108-90-7	200-1000	0.104	38.288	1.808	−5.732	2.718	97.99	150.80
187	C_6H_6	benzene	71-43-2	50-1000	3.551	−6.184	14.365	−19.807	8.234	82.43	135.95
188	C_6H_6O	phenol	108-95-2	50-1000	2.582	17.501	8.894	−14.435	6.317	103.22	
189	C_6H_7N	benzeneamine (aniline)	62-53-3	50-1000	2.598	19.936	8.438	−13.368	5.630	107.90	191.90
190	C_6H_7N	2-methylpyridine (2-picoline)	109-06-8	50-1000	4.156	2.699	12.517	−17.424	7.163	100.17	158.40
191	C_6H_7N	3-methylpyridine (3-picoline)	108-99-6	50-1000	4.140	2.780	12.458	−17.328	7.118	99.99	158.70
192	C_6H_7N	4-methylpyridine (4-picoline)	1108-89-4	50-1000	3.904	4.296	12.062	−16.884	6.942	99.70	159.00
194	C_6H_{10}	cyclohexene	110-83-8	50-1000	3.874	−0.909	14.902	−19.907	8.011	*101.49*	
195	$C_6H_{10}O$	cyclohexanone	108-94-1	50-1000	4.416	−1.248	17.367	−23.640	9.595	*116.19*	
197	C_6H_{12}	cyclohexane	110-82-7	100-1000	4.035	−4.433	16.834	−20.775	7.746	106.10	156.20
198	C_6H_{12}	methylcyclopentane	96-37-7	50-1000	5.379	−8.258	17.293	−21.646	8.263	109.50	158.70
199	C_6H_{12}	1-hexene	592-41-6	200-1000	6.303	12.352	10.258	−14.272	5.708	130.83	183.30
200	C_6H_{12}	4-methylpent-1-ene	691-37-2	298-1000	−1.326	65.625	−3.560	0.514	0.176	*126.59*	
201	$C_6H_{12}O$	cyclohexanol	108-93-0	50-1000	3.239	21.585	10.322	−14.762	5.885	*128.06*	
202	$C_6H_{12}O$	2-hexanone (methyl butyl ketone)	591-78-6	200-1000	9.146	8.701	10.736	−14.496	5.768	148.53	213.40
203	$C_6H_{12}O$	3-hexanone (ethyl propyl ketone)	589-38-8	200-1000	9.357	13.505	8.735	−11.918	4.710	152.51	216.80
206	$C_6H_{12}O$	4-methyl-2-pentanone (methyl isobutyl ketone)	108-10-1							147.40	209.60
207	$C_6H_{12}O_2$	methyl pentanoate	624-24-8							155.80	229.30
208	$C_6H_{12}O_2$	ethyl butanoate	105-54-4							154.60	255.70
209	$C_6H_{12}O_2$	propyl propanoate	106-36-5							153.10	229.10
210	$C_6H_{12}O_2$	ethyl 2-methylpropanoate	97-62-1							154.00	222.00
211	$C_6H_{12}O_2$	butyl ethanoate (butyl acetate)	123-86-4							151.50	228.40
212	$C_6H_{12}O_2$	2-methylpropyl ethanoate (isobutyl acetate)	110-19-0							154.30	241.10

213	$C_6H_{12}O_2$	pentyl methanoate (pentyl formate)	638-49-3							156.20	243.00
214	$C_6H_{12}O_2$	3-methylbutyl methanoate (isopentyl formate)	110-45-2							155.80	241.30
216	C_6H_{14}	hexane	110-54-3	200-1000	8.831	−0.166	14.302	−18.314	7.124	142.59	195.43
217	C_6H_{14}	2-methylpentane	107-83-5	200-1000	2.096	46.419	3.124	−6.829	2.902	142.21	193.93
218	C_6H_{14}	3-methylpentane	96-14-0	200-1000	0.433	11.143	0.730	−1.6[illegible]2	0.690	140.10	190.66
219	C_6H_{14}	2,2-dimethylbutane	75-83-2	200-1000	3.007	39.059	4.851	−8.243	3.367	141.50	188.70
220	C_6H_{14}	2,3-dimethylbutane	79-29-8	200-1000	−2.214	74.352	−3.697	0.273	0.308	139.41	188.66
221	$C_6H_{14}O$	1-hexanol	111-27-3	200-1000	6.784	17.060	11.935	−17.147	6.985	153.30	232.50
223	$C_6H_{14}O$	3-hexanol	623-37-0								269.30
224	$C_6H_{14}O$	2-methyl-1-pentanol	105-30-6								249.20
229	$C_6H_{15}N$	*N,N*-diethylethanamine (triethyl amine)	121-44-8	298-1000	4.581	40.089	5.793	−10.032	4.193	161.00	
231	$C_7F_1$6	hexadecafluoroheptane	335-57-9	200-1000	3.002	170.245	−21.365	13.301	−3.341	316.03	
232	$C_7H_3F_5$	pentafluorotoluene	771-56-2	298-1000	6.079	48.333	0.158	−3.861	1.948	164.30	
234	C_7H_8	toluene	108-88-3	50-1000	3.866	3.558	13.356	−18.659	7.690	103.75	157.29
235	C_7H_8O	2-methylphenol (o-cresol)	95-48-7	50-1000	3.123	31.032	6.152	−10.805	4.642	127.30	
236	C_7H_8O	3-methylphenol (m-cresol)	108-39-4	50-1000	2.876	26.142	8.544	−14.238	6.189	124.68	225.02
237	C_7H_8O	4-methylphenol (p-cresol)	106-44-5	50-1000	2.881	27.407	7.943	−13.423	5.843	124.97	
238	C_7H_9N	2,3-dimethylpyridine (2,3 lutidine)	583-61-9							129.74	189.30
239	C_7H_9N	2,3-dimethylpyridine (2,3 lutidine)	108-47-4	50-1000	4.225	13.393	11.364	−16.169	6.585	120.97	186.00
240	C_7H_9N	2,4-dimethylpyridine (2,4 lutidine)	589-93-5	50-1000	4.247	13.487	11.535	−16.564	6.803	122.01	183.40
241	C_7H_9N	2,5-dimethylpyridine (2,5 lutidine)	108-48-5	50-1000	4.183	14.253	11.226	−16.133	6.610	121.86	184.30
242	C_7H_9N	2,6-dimethylpyridine (2,6 lutidine)	583-58-4	50-1000	3.413	31.807	5.284	−8.851	3.610	128.75	189.60
243	C_7H_9N	3,4-dimethylpyridine (3,4 lutidine)	591-22-0	50-1000	4.135	15.216	10.792	−15.509	6.324	121.78	186.30
244	C_7H_{14}	3,5-dimethylpyridine (3,5 lutidine)	291-64-5	50-1000	3.995	5.299	17.971	−24.179	9.665	132.01	200.00
247	C_7H_{14}	methylcyclohexane	108-87-2	50-1000	3.148	18.438	13.624	−18.793	7.364	135.80	184.50
248	C_7H_{14}	ethylcyclopentane	1640-89-7	50-1000	5.847	−0.048	17.507	−22.495	8.656	133.60	186.60
249	C_7H_{14}	cis-1,3-dimethylcyclopentane	2532-58-3	298-1000	−2.522	60.538	2.703	−7.572	3.361	134.50	190.00
250	C_7H_{14}	trans-1,3-dimethylcyclopentane	1759-58-6	298-1000	−2.522	60.538	2.703	−7.572	3.361	134.50	190.90
251	C_7H_{14}	1-heptene	592-76-7	200-1000	7.520	12.824	12.670	−17.578	7.035	153.46	211.8
253	$C_7H_{14}O_2$	ethyl pentanoate	539-82-2							177.30	257.50
254	$C_7H_{14}O_2$	ethyl 3-methylbutanoate	108-64-5							176.00	253.00
255	$C_7H_{14}O_2$	propyl butanoate	105-66-8							175.80	257.20

Section C Ideal Gas and Liquid Heat Capacities. $C_p^\circ/R = a0 + a1T + a2T^2 + a3T^3 + a4T^4$.
CpIG and Cpliq at 298.15 K, J mol^{-1} K^{-1} *(Continued)*

No.	Formula	Name	CAS #	Trange, K	a0	a1 × 10^3	a2 × 10^5	a3 × 10^8	a4 × 10^{11}	CpIG	Cpliq
256	$C_7H_{14}O_2$	2-methylpropyl propanoate	540-42-1							179.50	268.00
257	$C_7H_{14}O_2$	propyl 2-methylpropanoate	644-49-5							175.00	254.00
258	$C_7H_{14}O_2$	3-methylbutyl ethanoate (isopentyl acetate)	123-92-2							173.50	265.40
259	C_7H_{16}	heptane	142-82-5	200-1000	9.634	4.156	15.494	−20.066	7.770	165.20	224.98
260	C_7H_{16}	2-methylhexane	591-76-4	200-1000	3.452	46.373	5.446	−9.875	4.089	164.50	222.97
261	C_7H_{16}	3-methylhexane	589-34-4							163.60	221.25
262	C_7H_{16}	3-ethylpentane	617-78-7	200-1000	7.598	19.547	11.641	−16.107	6.378	166.00	219.58
263	C_7H_{16}	2,2-dimethylpentane	590-35-2	200-1000	1.315	60.462	2.813	−7.620	3.434	166.70	221.20
264	C_7H_{16}	2,3-dimethylpentane	565-59-3	200-1000	−4.314	96.708	−6.454	2.227	−0.313	160.83	218.28
265	C_7H_{16}	2,4-dimethylpentane	108-08-7	200-1000	−3.133	97.923	−6.912	2.424	−0.250	170.75	224.22
266	C_7H_{16}	3,3-dimethylpentane	562-49-2	200-1000	−0.480	73.415	−0.740	−3.562	1.822	165.80	214.76
267	C_7H_{16}	2,2,3-trimethylbutane	464-06-2	200-1000	−0.711	73.562	−1.007	−3.018	1.583	163.30	213.51
268	$C_7H_{16}O$	1-heptanol	111-70-6	200-1000	7.935	18.023	14.223	−20.320	8.262	175.90	270.60
270	$C_7H_{16}O$	4-heptanol	589-55-9								317.60
271	C_8F_{18}	octadecafluorooctane	307-34-6	200-1000	3.352	193.679	−24.528	15.491	−3.962	358.07	429
273	C_8H_{10}	ethylbenzene	100-41-4	50-1000	4.544	10.578	13.644	−19.276	7.885	*127.40*	185.96
274	C_8H_{10}	1,2-dimethylbenzene (o-xylene)	95-47-6	50-1000	3.289	34.144	4.989	−8.335	3.338	132.31	188.07
275	C_8H_{10}	1,3-dimethylbenzene (m-xylene)	108-38-3	50-1000	4.002	17.537	10.590	−15.037	6.008	125.71	188.44
276	C_8H_{10}	1,4-dimethylbenzene (p-xylene)	106-42-3	50-1000	4.113	14.909	11.810	−16.724	6.736	126.02	181.66
277	$C_8H_{10}O$	2-ethylphenol	90-00-6	298-1000	−2.392	82.472	−4.476	−0.434	0.890	151.16	
278	$C_8H_{10}O$	3-ethylphenol	620-17-7	298-1000	−4.029	90.167	−5.877	0.702	0.546	148.54	
279	$C_8H_{10}O$	4-ethyl-phenol	123-07-9	298-1000	−3.628	88.368	−5.660	0.616	0.548	148.83	
280	$C_8H_{10}O$	2,3-dimethylphenol (2,3 xylenol)	526-75-0							164.10	
281	$C_8H_{10}O$	2,4-dimethylphenol (2,4 xylenol)	105-67-9	50-1000	3.752	41.738	5.814	−11.112	4.855	156.10	
282	$C_8H_{10}O$	2,5-dimethylphenol (2,5 xylenol)	95-87-4	50-1000	3.008	50.507	2.960	−7.548	3.351	157.09	
283	$C_8H_{10}O$	2,6-dimethylphenol (2,6 xylenol)	576-26-1	50-1000	2.604	51.990	2.683	−7.380	3.340	155.97	
284	$C_8H_{10}O$	3,4-dimethylphenol (3,4 xylenol)	95-65-8	50-1000	1.407	67.846	−1.345	−3.401	1.952	163.52	
285	$C_8H_{10}O$	3,5-dimethylphenol (3,5 xylenol)	108-68-9	50-1000	2.869	43.266	5.901	−11.468	5.053	152.70	
286	C_8H_{16}	cyclooctane	292-64-8	50-1000	4.236	13.119	16.313	−21.072	7.987	146.19	215.50
287	C_8H_{16}	*t*-1,4-dimethylcyclohexane	2207-04-7	50-1000	3.902	20.058	15.345	−20.707	7.974	*155.19*	210.3
288	C_8H_{16}	1-octene	111-66-0	200-1000	8.745	13.240	15.096	−20.895	8.366	176.10	241.40
289	$C_8H_{16}O_2$	octanoic acid	124-07-02								298

296	C_8H_{18}	octane	111-65-9	200-1000	10.824	4.983	17.751	−23.137	8.980	187.78	254.15
297	C_8H_{18}	2-methylheptane	592-27-8	200-1000	5.257	41.188	9.739	−15.692	6.570	187.23	251.99
298	C_8H_{18}	3-methylheptane	589-81-1							185.80	250.20
299	C_8H_{18}	4-methylheptane	589-53-7	200-1000	2.182	62.828	4.419	−10.021	4.358	187.02	251.08
300	C_8H_{18}	3-ethylhexane	619-99-8	200-1000	7.178	33.720	10.365	−15.312	6.088	189.62	250.29
301	C_8H_{18}	2,2-dimethylhexane	590-73-8	200-1000	3.452	55.222	6.066	−11.271	4.732	188.30	249.20
302	C_8H_{18}	2,3-dimethylhexane	584-94-1	200-1000	−3.594	100.983	−4.880	−0.342	0.789	184.10	248.78
303	C_8H_{18}	2,4-dimethylhexane	589-43-5	200-1000	−3.372	108.645	−7.267	2.175	−0.103	192.30	250.08
304	C_8H_{18}	2,5-dimethylhexane	592-13-2	200-1000	−1.367	87.285	−1.799	−3.343	1.857	185.48	249.20
305	C_8H_{18}	3,3-dimethylhexane	563-16-6	200-1000	−2.093	94.480	−2.808	−2.811	1.816	190.87	246.60
306	C_8H_{18}	3,4-dimethylhexane	583-48-2	200-1000	−6.148	116.522	−8.351	2.961	−0.344	182.34	246.90
307	C_8H_{18}	3-ethyl-2-methylpentane	609-26-7	200-1000	−0.873	95.193	−5.319	1.180	0.000	192.05	248.91
308	C_8H_{18}	3-ethyl-3-methylpentane	1067-08-9	200-1000	−1.350	88.584	−1.961	−3.139	1.789	187.99	245.89
309	C_8H_{18}	2,2,3-trimethylpentane	564-02-3	200-1000	−4.490	108.022	−6.486	1.441	0.142	185.64	245.39
310	C_8H_{18}	2,2,4-trimethylpentane (isooctane)	540-84-1	200-1000	0.384	77.059	0.665	−5.565	2.619	188.41	238.55
311	C_8H_{18}	2,3,3-trimethylpentane	560-21-4	200-1000	−5.726	117.939	−8.940	4.012	−0.816	187.02	245.56
312	C_8H_{18}	2,3,4-trimethylpentane	565-75-3							191.59	248.61
313	C_8H_{18}	2,2,3,3-tetramethylbutane	594-82-1	200-1000	0.768	72.950	1.659	−6.322	2.886	187.19	279.10
314	$C_8H_{18}O$	1-octanol	111-87-5	200-1000	9.193	18.228	16.682	−23.641	9.580	198.60	302.40
316	$C_8H_{18}O$	3-octanol	589-98-0								338.50
317	$C_8H_{18}O$	4-octanol	589-62-8								337.60
321	C_9F_{20}	eicosafluorononane	375-96-2	200-1000	3.697	217.163	−27.706	17.697	−4.589	400.11	
322	C_9H_7N	quinoline	91-22-5								194.90
324	C_9H_{10}	indan	496-11-7								190.30
325	C_9H_{12}	propylbenzene	103-65-1	50-1000	4.759	23.956	11.859	−17.393	7.064	152.30	214.71
326	C_9H_{12}	1-methylethylbenzene	98-82-8	50-1000	2.985	34.196	11.938	−20.152	8.923	159.69	213.30
327	C_9H_{12}	1-ethyl-4-methylbenzene	622-96-8	50-1000	5.097	17.385	13.600	−19.299	7.817	148.25	210.30
328	C_9H_{12}	1,2,3-trimethylbenzene	526-73-8	50-1000	4.042	31.152	10.185	−16.262	6.922	155.07	216.10
329	C_9H_{12}	1,2,4-trimethylbenzene	95-63-6	50-1000	5.319	20.074	12.034	−16.873	6.687	149.71	215.00
330	C_9H_{12}	1,3,5-trimethylbenzene (mesitylene)	108-67-8	50-1000	5.305	20.039	11.606	−16.317	6.503	147.63	209.10
331	C_9H_{18}	1-nonene	124-11-8	200-1000	9.963	13.704	17.511	−24.203	9.694	198.74	271.20
333	$C_9H_{18}O_2$	3-methylbutyl butanoate	106-27-4							221.40	320.40
334	C_9H_{20}	nonane	111-84-2	200-1000	12.152	4.575	20.416	−26.777	10.465	210.41	284.45
335	C_9H_{20}	2-methyloctane	3221-61-2	200-1000	5.914	47.039	10.387	−16.685	6.852	212.05	271.50
336	C_9H_{20}	2,2-dimethylheptane	1071-26-7							211.30	283.40
337	C_9H_{20}	2,2,5-trimethylhexane	3522-94-9	200-1000	−0.880	92.152	−0.423	−5.261	2.601	*208.11*	
338	C_9H_{20}	2,2,3,3-tetramethylpentane	7154-79-2	200-1000	−6.019	131.511	−9.916	4.520	−0.934	209.80	284.20

Section C Ideal Gas and Liquid Heat Capacities. C_p°/R = a0 + a1T + a2T^2 + a3T^3 + a4T^4.
CpIG and Cpliq at 298.15 K, J mol^{-1} K^{-1} *(Continued)*

No.	Formula	Name	CAS #	Trange, K	a0	a1 × 10^3	a2 × 10^5	a3 × 10^8	a4 × 10^{11}	CpIG	Cpliq
339	C_9H_{20}	2,2,3,4-tetramethylpentane	1186-53-4	200-1000	−5.422	123.507	−8.031	2.590	−0.261	207.30	252.70
340	C_9H_{20}	2,2,4,4-tetramethylpentane	1070-87-7	200-1000	3.621	67.875	4.875	−10.109	4.292	*214.94*	
341	C_9H_{20}	2,3,3,4-tetramethylpentane	16747-38-9	200-1000	−9.189	161.921	−17.927	12.689	−3.869	218.30	275.70
342	$C_9H_{20}O$	1-nonanol	143-08-8	200-1000	10.350	19.105	19.007	−26.878	10.891	221.20	334.20
344	$C_{10}F_{22}$	docosafluorodecane	307-45-9	200-1000	4.042	240.657	−30.888	19.907	−5.219	442.15	
345	$C_{10}H_8$	naphthalene	91-20-3	50-1000	2.889	14.306	15.978	−23.930	10.173	<u>132.55</u>	
351	$C_{10}H_{14}$	butylbenzene	104-51-8	200-1000	6.490	19.080	15.665	−22.059	8.887	173.86	243.39
352	$C_{10}H_{14}$	2-methylpropylbenzene (isobutylbenzene)	538-93-2							173.90	241.00
353	$C_{10}H_{14}$	1,4-diethylbenzene	105-05-5	298-1000	−0.359	75.371	0.442	−5.736	2.783	<u>176.15</u>	239.10
354	$C_{10}H_{14}$	1-(1-methylethyl)-4-methylbenzene (*p*-cymene)	99-87-6							173.70	237.70
355	$C_{10}H_{14}$	1,2,4,5-tetramethylbenzene	95-93-2	298-1000	3.352	67.376	0.527	−4.883	2.290	186.50	
356	$C_{10}H_{18}$	cis-bicyclo[4.4.0]decane (cis-decalin)	493-01-6	298-1000	−5.445	80.068	5.065	−11.756	5.088	168.10	232.00
357	$C_{10}H_{18}$	trans-bicyclo[4.4.0]decane (trans-decalin)	493-02-7	298-1000	−2.155	53.852	12.610	−20.981	9.066	168.60	228.50
358	$C_{10}H_{20}$	1-decene	872-05-9	200-1000	11.175	14.222	19.908	−27.488	11.012	*221.97*	<u>300.3</u>
360	$C_{10}H_{22}$	decane	124-18-5	200-1000	13.467	4.139	23.127	−30.477	11.970	233.05	314.54
361	$C_{10}H_{22}$	2,2,5-trimethylheptane	20291-95-6	200-1000	−0.961	100.294	0.237	−6.883	3.357	229.20	306.40
362	$C_{10}H_{22}$	3,3,5-trimethylheptane	7154-80-5	200-1000	−2.999	118.108	−4.182	−2.362	1.717	232.80	295.30
363	$C_{10}H_{22}$	2,2,3,3-tetramethylhexane	13475-81-5	200-1000	−7.678	153.766	−12.616	6.171	−1.323	*236.82*	
364	$C_{10}H_{22}$	2,2,5,5-tetramethylhexane	1071-81-4	200-1000	0.8	87.376	3.168	−9.35	4.141	*228.78*	
365	$C_{10}H_{22}O$	1-decanol	112-30-1	200-1000	11.637	19.130	21.517	−30.271	12.247	243.80	366.00
367	$C_{11}H_{10}$	1-methylnaphthalene	90-12-0	298-1000	−5.637	98.625	−4.956	−1.033	1.281	159.30	224.40
368	$C_{11}H_{10}$	2-methylnaphthalene	91-57-6	298-1000	−4.671	93.882	−4.334	−1.331	1.317	154.60	
369	$C_{11}H_{24}$	undecane	1120-21-4							255.69	345.05
370	$C_{11}H_{24}O$	1-undecanol	112-42-5	200-1000	12.923	18.973	24.124	−33.816	13.675	*267.24*	
371	$C_{12}H_{10}$	1,1′-biphenyl	92-52-4	200-1000	−0.843	61.392	6.352	−13.754	6.169	165.28	250.95
372	$C_{12}H_{12}$	1,6-dimethylnaphthalene	575-43-9	298-1000	−4.332	103.947	−4.556	−1.769	1.571	185.10	
373	$C_{12}H_{12}$	2,7-dimethylnaphthalene	582-16-1	298-1000	−3.288	101.288	−4.748	−1.079	1.219	*187.08*	
374	$C_{12}H_{18}$	1,3,5-triethylbenzene	102-25-0	298-1000	1.319	87.791	1.406	−7.682	3.595	*224.42*	

376	$C_{12}H_{24}$	1-dodecene	112-41-4	200-1000	13.617	15.108	24.747	−34.1[illegible]1	13.669	*267.38*	360.70
377	$C_{12}H_{26}$	dodecane	112-40-3	200-1000	17.229	−7.242	31.922	−42.322	17.022	278.33	375.47
378	$C_{12}H_{26}O$	1-dodecanol	112-53-8	200-1000	14.073	19.938	26.412	−36.989	14.951	*289.95*	
379	$C_{13}H_{12}$	diphenylmethane	101-81-5								266.10
380	$C_{13}H_{28}$	tridecane	629-50-5	200-1000	18.546	−7.636	34.604	−45.978	18.509	300.97	406.89
382	$C_{14}H_{10}$	phenanthrene	85-01-8	50-1000	2.374	38.372	16.471	−26.813	11.640	*185.16*	
383	$C_{14}H_{10}$	anthracene	120-12-7	50-1000	2.577	31.826	18.811	−29.722	12.840	*182.29*	
385	$C_{14}H_{30}$	tetradecane	629-59-4	200-1000	18.375	6.585	32.307	−42.663	16.590	323.61	438.48
387	$C_{15}H_{32}$	pentadecane	629-62-9	200-1000	21.180	−8.424	39.969	−53.290	21.482	346.25	469.95
389	$C_{16}H_{34}$	hexadecane	544-76-3	200-1000	39.747	−206.152	114.814	−155.548	67.534	368.89	501.45
392	$C_{17}H_{36}$	heptadecane	629-78-7	200-1000	23.813	−9.210	45.333	−60.601	24.455	391.53	534.34
397	$C_{18}H_{38}$	octadecane	593-45-3	200-1000	25.130	−9.603	48.015	−64.256	25.942	414.17	564.45
399	$C_{19}H_{40}$	nonadecane	629-92-5	200-1000	26.447	−9.998	50.697	−67.912	27.428	436.81	595.94
401	$C_{20}H_{42}$	eicosane	112-95-8	200-1000	27.764	−10.389	53.379	−71.567	28.914	459.45	627.45
407	ClD	deuterium chloride	7698-05-7	50-1000	3.917	−3.965	1.205	−1.323	0.521	29.17	
408	$ClFO_3$	perchloryl fluoride	7616-94-6	298-1000	0.470	36.338	−4.796	3.147	−0.83	*64.93*	
410	ClH	hydrogen chloride	7647-01-0	50-1000	3.827	−2.936	0.879	−1.031	0.439	29.17	
413	Cl_2	chlorine	7782-50-5	50-1000	3.0560	5.3708	−0.8098	0.5693	−0.15256	29.14	
414	DH	deuterium hydride	13983-20-5	50-1000	3.893	−3.508	1.083	−1.337	0.580	29.20	
415	DI	deuterium iodide	14104-45-1	50-1000	3.741	−2.862	1.000	−1.051	0.382	29.36	
416	D_2	deuterium	7782-39-0	50-1000	3.590	−0.462	0.057	0.036	−0.026	29.19	
417	D_2	deuterium, normal	800000-54-8							29.20	
418	D_2O	deuterium oxide	7789-20-0	50-1000	4.274	−3.465	1.376	−1.482	0.568	34.26	84.35
419	D_2S	deuterium sulfide	13536-94-2	50-1000	4.290	−3.944	1.974	−2.268	0.872	36.13	
420	D_3N	trideuteroammonia	13550-49-7	50-1000	4.090	−3.243	2.367	−0.264	0.961	38.23	
422	FH	hydrogen fluoride	7664-39-3	50-1000	3.901	−3.708	1.165	−1.465	0.639	29.14	52.00
423	FNO_2	nitrogen dioxyfluoride	10022-50-1	298-1000	1.620	20.883	−2.512	1.586	−0.420	*49.89*	
424	F_2	Fluorine	7782-41-4	50-1000	3.347	0.467	0.526	−0.794	0.330	31.30	
428	F_2O	oxygen difluoride	7783-41-7	50-1000	3.437	5.527	0.502	−1.453	0.735	43.31	
433	F_4S	sulfur tetrafluoride	7783-60-0	298-1000	−0.808	51.235	−8.251	6.335	−1.887	*72.03*	
437	HI	hydrogen iodide	10034-85-2	50-1000	3.648	−1.392	0.389	−0.326	0.110	29.16	
438	H_2	hydrogen	1333-74-0	50-1000	2.883	3.681	−0.772	0.692	−0.213	28.84	
439	H_2	hydrogen, normal	800000-51-5							28.83	
440	H_2O	water	7732-18-5	50-1000	4.395	−4.186	1.405	−1.564	0.632	33.58	75.29
441	H_2S	hydrogen sulfide	7783-06-4	50-1000	4.266	−3.438	1.319	−1.331	0.488	34.12	74.68

Section C Ideal Gas and Liquid Heat Capacities. $C_p^\circ/R = a0 + a1T + a2T^2 + a3T^3 + a4T^4$.
CpIG and Cpliq at 298.15 K, J mol^{-1} K^{-1} (*Continued*)

No.	Formula	Name	CAS #	Trange, K	a0	a1 × 10³	a2 × 10⁵	a3 × 10⁸	a4 × 10¹¹	CpIG	Cpliq
442	H_2S_2	dihydrogen disulfide	13465-07-1	50-1000	3.364	8.093	0.636	−1.991	1.048	49.21	92.74
443	H_2S_3	dihydrogen trisulfide	13845-23-3								123.42
444	H_2S_4	dihydrogen tetrasulfide	13845-25-5								154.10
445	H_2S_5	dihydrogen pentasulfide	13845-24-4								184.78
447	H_3N	ammonia	7664-41-7	50-1000	4.238	−4.215	2.041	−2.126	0.761	35.65	83
449	H_4N_2	hydrazine	302-01-2	50-1000	3.627	2.239	2.876	−4.060	1.690	49.12	96.8
450	He	helium	9440-59-7	—	2.500	0.000	0.000	0.000	0.000	*20.79*	
451	He	helium-3	14762-55-1	—	2.500	0.000	0.000	0.000	0.000	*20.79*	
452	I2	iodine	7553-56-2	50-1000	3.508	6.303	−1.461	1.470	−0.531	*36.88*	
453	Kr	krypton	7439-90-9	—	2.500	0.000	0.000	0.000	0.000	*20.79*	
454	NO	nitrogen monoxide (nitric oxide)	10102-43-9	50-1000	4.534	−7.644	2.066	−2.156	0.806	29.87	
455	N_2	nitrogen	7727-37-9	50-1000	3.539	−0.261	0.007	0.157	−0.099	29.12	
456	N_2O	dinitrogen oxide (nitrous oxide)	10024-97-2	50-1000	3.165	3.401	0.989	−1.880	0.890	38.64	
457	N_2O_4	dinitrogen tetroxide (nitrogen dioxide)	10544-72-6	50-1000	3.374	27.257	−1.917	−0.616	0.859	81.07	142.2
458	Ne	neon	7440-01-9	—	2.500	0.000	0.000	0.000	0.000	*20.79*	
459	OT_2	tritium oxide	14940-65-9							34.96	
460	O_2	oxygen	7782-44-7	50-1000	3.630	−1.794	0.658	−0.601	0.179	29.38	
461	O_2S	sulfur dioxide	7446-09-5	50-1000	4.417	−2.234	2.344	−3.271	1.393	40.05	
462	O_3	ozone	10028-15-6	50-1000	4.106	−3.809	3.131	−4.300	1.813	39.60	
463	O_3S	sulfur trioxide	7446-11-9	50-1000	3.426	6.479	1.691	−3.356	1.590	50.86	226.8
464	Rn	radon	10043-92-2	—	2.500	0.000	0.000	0.000	0.000	*20.79*	
465	S	sulfur	7704-34-9	50-1000	2.803	−0.036	0.143	−0.435	0.268	23.67	
467	T_2	tritium	10028-17-8							29.20	
468	Xe	xenon	7440-63-3	—	2.500	0.000	0.000	0.000	0.000	*20.79*	

Section D Vapor Pressure Correlations Parameters

No.	Formula	Name	CAS #	Eq. #	A / A / Tc	B / B / a	C / C / b	Tc / c	to / d	n / Pc	E	F	Pvpmin, bar	Tmin, K	Pvpmax bar	Tmax, K
1	Ar	argon	7440-37-1	1	3.74141	304.2270	267.320						0.60	82.59	2	94.26
2	Br_2	bromine	7726-95-6	1	4.00270	1119.680	221.380						0.06	266.00	2	354.25
3	BrD	deuterium bromide	13536-59-9	1	3.28728	505.680	220.600						0.30	185.20	2	221.89
4	BrF_3	bromine trifluoride	7787-71-5	1	4.85464	1673.950	219.480						0.02	309.09	2	421.28
5	BrF_5	bromine pentafluoride	7789-30-2	1	4.39858	1219.280	236.400						0.02	236.80	2	334.31
6	BrH	hydrogen bromide	10035-10-6	1	3.41243	540.8200	225.440						0.30	185.10	2	221.53
7	$CBrClF_2$	bromochlorodifluoromethane	353-59-3	1	3.95850	933.0400	240.000						0.02	198.00	2	288.26
				2	3.95850	933.0400	240.000	426.90	3	2.26960	−54.789	3324.1	2	288.15	37.06	418.15
8	$CBrF_3$	bromotrifluoromethane	75-63-8	1	3.89640	731.3100	245.700						0.02	158.10	2	230.85
				2	3.89640	731.3100	245.700	340.15	−53	2.39700	4.095	941.51	2.2	233.15	34.50	333.15
9	CBr_2F_2	dibromodifluoromethane	75-61-6	1	4.18780	1127.430	246.800						0.02	217.80	2	316.42
10	$CClF_3$	chlorotrifluoromethane	75-72-9	1	3.90353	654.6560	249.390						0.02	140.61	2	205.48
				2	3.90353	654.6560	249.390	301.84	−76	2.46214	62.986	−2130.8	2.26	208.15	28.39	288.15
11	CCl_2F_2	dichlorodifluoromethane (R-12)	75-71-8	1	4.01171	868.0760	246.390						0.02	178.77	2	260.70
				2	4.01171	868.0760	246.390	385.10	−23	3.27101	104.141	−3216.3	2.19	263.15	39.92	383.15
12	CCl_3F	trichlorofluoromethane (R-11)	75-69-4	1	4.00905	1043.313	236.950						0.02	218.98	2	317.57
				2	4.00905	1043.313	236.950	471.10	40	2.40860	75.083	−1375.6	2.04	318.15	36.89	458.15
13	CCl_4	tetrachloromethane	56-23-5	1	4.10445	1265.632	232.148						0.02	259.00	2	373.76
14	CF_4	tetrafluoromethane	75-73-0	1	3.95894	510.5950	257.200						0.02	106.20	2	155.54
				2	3.95894	510.5950	257.200	227.51	−120	2.41377	−93.740	7425.9	2.33	158.15	28.37	218.15
15	$CHBrF_2$	bromodifluoromethane	1511-62-2	1	3.40030	640.3200	204.100						0.02	194.50	2	275.65
				2	3.40030	640.3200	204.130	412.00	−9	0.98620	189.780	−6582.6	1.88	273.15	44.70	403.15
16	$CHClF_2$	chlorodifluoromethane (R-22)	75-45-6	1	4.13253	835.4620	243.460						0.02	173.13	2	247.74
				2	4.13253	836.4620	243.460	369.28	−30	2.76007	37.609	−369.26	2.01	248.15	48.82	368.15
17	$CHCl_2F$	dichlorofluoromethane	75-43-4	1	4.02473	959.9340	230.030						0.02	210.83	2	300.91
				2	4.02473	959.9340	230.030	451.52	12	2.55869	9.610	574.28	2.15	303.15	45.58	443.15
18	$CHCl_3$	trichloromethane	67-66-3	1	3.96288	1106.904	218.552						0.02	250.10	2	356.89
19	CHF_3	trifluoromethane (R-23)	75-46-7	1	4.22140	707.3960	249.840						0.02	142.79	2	203.75
				2	4.22140	707.3960	249.840	298.97	−70	2.79148	70.243	2833.0	2.48	208.15	36.91	288.15
20	CH_2Cl_2	dichloromethane	75-09-2	1	4.07622	1070.070	223.240						0.02	235.20	2	333.36
21	CH_2F_2	difluoromethane	75-10-5	1	4.29712	833.1370	245.860						0.02	166.23	2	235.78
				2	4.29712	833.1370	245.860	351.36	−40	2.48212	61.006	−747.4	2.22	238.15	43.87	338.15
22	CH_2O_2	methanoic acid (formic acid)	64-18-6	3	588.00	−7.24917	0.44255	−0.35558	−0.96906	58.07					58.07	588.00
23	CH_3Cl	chloromethane	74-87-3	1	4.16533	920.8600	245.580						0.02	184.60	2	265.87
24	CH_3F	fluoromethane	593-53-3	1	4.19421	734.2220	253.570						0.02	144.17	2	208.17
				2	4.19421	734.2220	253.570	317.36	−73	2.60926	57.676	−1868.	2.52	213.15	53.67	313.15
26	CH_4	methane	74-82-8	1	3.76870	395.7440	266.681						0.15	92.64	2	120.59
				3	190.551	−6.02242	1.26652	−0.5707	−1.366	45.992					45.99	190.55

Section D Vapor Pressure Correlations Parameters (*Continued*)

No.	Formula	Name	CAS #	Eq. #	A / A / Tc	B / B / a	C / C / b	Tc / c	to / d	n / Pc	E	F	Pvpmin, bar	Tmin, K	Pvpmax bar	Tmax, K
27	CH_4O	methanol	67-56-1	1	5.20277	1580.080	239.500						0.02	262.59	2	356.00
				3	512.64	−8.63571	1.17982	−2.4790	−1.0240	80.92					80.92	512.64
28	CH_4S	methanethiol (methyl mercaptan)	74-93-1	1	4.15653	1015.547	238.706						0.02	207.80	2	297.85
29	CH_5N	methanamine (methyl amine)	74-89-5	1	4.54420	1050.660	237.830						0.02	203.61	2	282.93
				2	4.21300	899.0300	220.000	430.00	0	2.13900	−151.85	7356.00	2.44	288.15	65.98	423.15
30	CO	carbon monoxide	630-08-0	1	3.81912	291.7430	267.996						0.20	69.73	2	88.08
32	$C_2Br_2ClF_3$	1,2-dibromo-2-chloro-1,1,2-trifluroethane	354-51-8	1	3.84523	1166.348	209.870						0.02	272.65	2	392.37
33	C_2ClF_5	1-chloro-1,1,2,2,2-pentafluoroethane	76-15-3	1	3.93652	795.2120	241.370						0.02	172.89	2	250.52
				2	3.93652	795.2120	241.370	353.10	−33	2.47050	82.646	−1205.4	2.21	253.15	31.29	353.10
34	$C_2Cl_2F_4$	1,1-dichloro-1,2,2,2-tetrafluoroethane	374-07-2	1	3.83243	875.9380	225.460						0.02	206.04	2	295.73
				2	3.83243	875.9380	225.460	418.70	10	2.12840	699.960	−66758	2.17	298.15	31.95	418.15
35	$C_2Cl_2F_4$	1,2-dichloro-1,1,2,2-tetrafluoroethane	76-14-2	1	3.93549	930.7340	233.410						0.02	204.93	2	295.83
				2	3.93549	930.7340	233.410	418.90	12	4.45933	849.560	−57942	2.16	298.15	32.02	418.15
36	$C_2Cl_3F_3$	1,1,2-trichloro-1,2,2-trifluoroethane	76-13-1	1	4.00134	1107.71	229.640						0.02	237.83	2	342.87
				2	4.00134	1107.719	229.640	487.40	55	2.89655	69.650	−2236.1	2.02	343.15	31.79	483.15
38	C_2F_6	hexafluoroethane	76-16-4	1	3.68388	572.7330	233.650						0.30	175.65	2	208.80
				2	3.68388	572.7330	233.650	293.04	−73	1.89050	156.827	−7370.8	2.44	213.15	21.02	278.15
39	$C_2HBrClF_3$	1-bromo-1-chloro-2,2,2-trifluoroethane	151-67-7	1	4.20682	1199.262	235.290						0.02	240.92	2	344.91
40	$C_2HBrClF_3$	1-bromo-2-chloro-1,1,2-trifluoroethane	354-06-3	1	3.48366	841.1410	189.300						0.02	246.15	2	348.14
41	C_2HClF_4	1-chloro-1,1,2,2-tetrafluoroethane	354-25-6	1	4.25710	1006.840	248.600						0.02	193.60	2	279.06
				2	4.25710	1006.840	248.560	399.87	−10	2.77560	−538.004	30952.0	1.93	278.15	36.33	398.15
42	C_2HClF_4	1-chloro-1,2,2,2-tetrafluoroethane	2837-89-0	1	4.0536	900.49	234.389						0.14	222.00	2.59	286.00
				2	3.98581	872.8360	231.260	395.85	−5	2.05345	4.517	−583.96	2.34	283.15	29.00	383.15
43	$C_2HCl_2F_3$	1,1-dichloro-2,2,2-trifluoroethane (R-123)	306-83-2	1	4.21161	1132.447	241.590						0.02	223.16	2	321.15
				2	4.21161	1132.447	241.590	456.83	35	4.59524	179.953	−6961.2	2.13	323.15	29.46	443.15
45	C_2HF_5	pentafluoroethane	354-33-6	1	4.13392	800.8690	242.090						0.02	168.36	2	240.01
				2	4.13392	800.8700	242.090	339.17	−40	2.91989	164.960	−6993.2	2.79	248.15	31.73	333.15
47	C_2H_2	ethyne (acetylene)	74-86-2	1	3.67374	528.6700	228.790						1.20	191.44	2	201.11
				3	308.35	−6.87886	1.30164	−1.22474	−3.59556	61.39					61.39	308.35
49	$C_2H_2F_4$	1,1,1,2-tetrafluoroethane (R-134a)	811-97-2	1	4.11874	850.8810	232.990						0.02	186.41	2	263.04
				2	4.11874	850.8810	232.990	374.26	−20	2.39793	31.124	2784.80	2.44	268.15	39.69	373.15
50	$C_2H_2F_4$	1,1,2,2-tetrafluoroethane (R-134)	359-35-3	1	4.12013	885.5970	235.900						0.02	190.05	2	269.75
				2	4.12013	885.5970	235.290	391.74	−10	1.97108	−196.89	17336.0	2.27	273.15	35.70	378.15

51	$C_2H_3ClF_2$	1-chloro-1,1-difluoroethane	75-68-3	1	4.05053	928.6450	238.690						0.02	195.98	2	282.13
				2	4.05053	928.6450	238.690	410.30	0	2.94747	115.850	−3920.5	2.07	283.15	35.82	403.15
52	$C_2H_3Cl_2F$	1,1-dichloro-1-fluoroethane (R-141b)	1717-00-6	1	4.03117	1062.074	231.799						0.02	226.71	2	326.09
				2	4.03117	1062.074	231.790	477.35	40	4.49103	752.781	−43010	2.46	333.15	40.56	473.15
53	$C_2H_3F_3$	1,1,1-trifluoroethane (R-143a)	420-46-2	1	4.06800	801.3400	244.550						0.02	167.55	2	241.33
				2	4.06800	801.3400	244.550	346.30	−40	2.50293	63.440	−981.56	2.15	243.15	35.53	343.15
54	$C_2H_3F_3$	1,1,2-trifluoroethane (R-143)	430-66-0	1	4.13152	928.1770	221.270						0.02	211.07	2	294.19
				2	4.13152	928.1770	221.270	429.80	10	2.13500	8.924	−587.53	2.31	298.15	43.09	418.15
55	C_2H_4	ethene (ethylene)	74-85-1	1	3.91382	596.5260	256.370						0.02	123.06	2	181.90
				2	3.91382	596.5300	256.370	282.34	−99	2.79132	9.717	52.77	2.71	188.15	40.99	273.15
57	$C_2H_4Cl_2$	1,1-dichloroethane	75-34-3	1	4.16780	1201.050	231.270						0.02	246.60	2	352.49
58	$C_2H_4Cl_2$	1,2-dichloroethane	107-06-2	1	4.28356	1341.370	230.050						0.02	267.40	2	379.91
59	$C_2H_4F_2$	1,1-difluoroethane (R-152a)	75-37-6	1	3.75231	735.1600	220.270						0.02	187.74	2	265.89
				2	3.75231	735.1600	220.270	386.41	−18	1.38810	72.728	−1421.1	2.19	268.15	45.16	386.74
60	$C_2H_4O_2$	ethanoic acid (acetic acid)	64-19-7	1	4.54456	1555.120	224.650						0.02	297.58	2	414.97
				3	592.71	−8.29430	0.97928	−0.21745	−5.72367	57.86					57.86	592.71
61	$C_2H_4O_2$	methyl methanoate (methyl formate)	107-31-3	1	4.29529	1125.200	230.560						0.02	230.30	2	324.29
62	C_2H_5Br	bromoethane	74-96-4	1	4.04485	1090.811	231.710						0.02	231.35	2	332.80
63	C_2H_5C	chloroethane	75-00-3	1	4.09088	1020.630	237.570						0.02	211.86	2	304.89
64	C_2H_5F	fluoroethane	353-36-6	1	4.21998	897.3680	250.660						0.02	174.10	2	251.47
				2	4.21998	897.3680	250.660	375.28	−20	2.97505	352.246	−24619	2.58	258.15	50.27	375.28
65	C_2H_6	ethane	74-84-0	1	3.95405	663.720	256.681						0.02	133.80	2	198.16
				3	305.33	−6.47500	1.41071	−1.1440	−1.8590	48.71					48.71	305.33
66	C_2H_6O	ethanol	64-17-5	1	5.33675	1648.220	230.918						0.02	276.50	2	369.54
				3	513.92	−8.68587	1.17831	−4.8762	−1.5880	61.32					61.32	513.92
67	C_2H_6O	dimethyl ether	115-10-6	1	4.44136	1025.560	256.050						0.02	184.10	2	264.80
68	C_2H_6S	ethanethiol (ethyl mercaptan)	75-08-1	1	4.07696	1084.531	231.385						0.02	229.50	2	328.99
69	C_2H_6S	2-thiapropane (dimethylsulfide)	75-18-3	1	4.07369	1090.755	230.799						0.02	231.30	2	331.47
70	C_2H_7N	ethanamine (ethyl amine)	75-04-7	1	4.43400	1102.880	232.450						0.02	220.53	2	307.55
				2	3.88560	840.4800	200.000	456.35	23	2.09210	90.941	−3179.C	2.04	308.15	45.83	443.15
74	$C_3H_2ClF_5O$	2-chloro-1,1,2-trifluoroethyl difluoromethyl ether (enflurane)	13838-16-9	3	475.03	−8.32915	2.37044	−3.75113	−4.6033	29.80					29.80	475.03
75	$C_3H_2ClF_5O$	1-chloro-2,2,2-trifluoroethyl difluoromethyl ether (isoflurane)	26675-46-7	3	467.80	−8.08994	2.07729	−3.32	−4.2641	30.46					30.46	467.80
78	$C_3H_3F_5$	1,1,1,2,2-pentafluoropropane (R-245cb)	1814-88-6	3	380.38	−7.67509	2.38205	−3.6522	0	31.483			0.74	248.00	9.95	326.00
83	C_3H_4	1-propyne (methyl acetylene)	74-99-7	1	4.24555	935.0900	243.580						0.02	186.87	2	266.63
84	C_3H_4	1,2-propadiene	463-49-0	1	2.83860	458.0600	196.070						0.02	178.00	2	257.00
				2	3.67520	734.5680	234.740	393.00	−20	1.13600	−264.98	16325.0	2.16	258.15	54.60	423.15
85	C_3H_6	propene (propylene)	115-07-1	1	3.95606	789.6240	247.580						0.02	165.20	2	241.61
				2	3.95606	789.6200	247.580	365.57	−41	2.67417	22.130	−199.34	1.74	238.15	44.67	363.15
86	C_3H_6	cyclopropane	75-19-4	1	4.03084	866.1500	248.000						0.02	176.30	2	257.37
				2	4.03084	866.1500	248.000	398.25	−26	2.66720	−2.153	567.17	2.058	258.15	55.75	398.25
87	$C_3H_6Cl_2$	1,2-dichloropropane	78-87-5	3	578.00	−7.70557	2.62197	−2.74104	−3.08934	46.5			0.05333	293.67	2.7	406.5
88	C_3H_6O	2-propen-1-ol (allyl alcohol)	107-18-6	1	8.78252	4510.213	416.797						0.03	294.00	1	370.23

Section D Vapor Pressure Correlations Parameters (*Continued*)

No.	Formula	Name	CAS #	Eq. #	A/A/Tc	B/B/a	C/C/b	Tc/c	to/d	n/Pc	E	F	Pvpmin, bar	Tmin, K	Pvpmax bar	Tmax, K
89	C_3H_6O	propanone (acetone)	67-64-1	1	4.21840	1197.010	228.060						0.02	247.38	2	350.65
				3	508.10	−7.55098	1.60784	−1.9944	−3.2002	47.02					47.02	508.10
90	$C_3H_6O_2$	propanoic acid	79-09-4	1	4.75466	1662.582	209.046						0.02	321.72	2	437.41
				3	604.00	−8.14882	0.79590	−3.1836	−3.81338	45.30					45.30	604.00
91	$C_3H_6O_2$	methyl ethanoate (methyl acetate)	79-20-9	1	4.18621	1156.430	219.690						0.02	249.90	2	351.11
92	$C_3H_6O_2$	ethyl methanoate (ethyl formate)	109-94-4	1	4.07899	1101.000	215.980						0.02	247.80	2	348.60
93	$C_3H_6O_3$	dimethylcarbonate	616-38-6	3	557.00	−8.24279	3.25566	−4.2825	−2.1194	48			0.13322	310.56	2.7	397.5
94	C_3H_7Cl	1-chloropropane	540-54-5	1	4.07655	1125.009	229.860						0.02	238.09	2	341.29
95	C_3H_8	propane	74-98-6	1	3.92828	803.9970	247.040						0.02	168.90	2	247.76
				3	369.85	−6.76368	1.55481	−1.5872	−2.024	69.90					69.90	369.85
96	C_3H_8O	1-propanol	71-23-8	1	4.99991	1512.940	205.807						0.02	293.19	2	389.32
				3	536.78	−8.53706	1.96214	−7.6918	−2.9450	51.68					51.68	536.78
97	C_3H_8O	2-propanol	67-63-0	1	5.24268	1580.920	219.610						0.02	281.28	2	373.46
				3	508.30	−8.73656	2.16240	−8.70785	4.77927	47.63					47.63	508.30
98	C_3H_8O	methyl ethyl ether	540-67-0	1	3.00683	504.4900	160.750						0.06	232.00	2	298.85
99	C_3H_8S	2-thiabutane (methyl ethyl sulfide)	624-89-5	1	4.06339	1182.562	224.784						0.02	253.50	2	362.68
100	C_3H_9N	1-propanamine (propyl amine)	107-10-8	1	4.34440	1186.390	226.210						0.02	258.31	2	340.36
				2	3.50110	759.5000	170.000	497.00	55	2.13340	1429.00	−80295	1.86	338.15	33.44	473.15
101	C_3H_9N	2-propanamine (methyl ethyl amine)	75-31-0	1	4.05530	1005.490	216.510						0.02	231.38	2	324.47
106	C_4H_4O	furan	110-00-9	1	4.11990	1070.200	830						0.02	228.20	2	324.56
107	C_4H_4S	thiophene	110-02-1	1	4.08416	1246.020	221.350						0.02	267.20	2	381.16
109	C_4H_6	1-butyne	107-00-6	1	4.16676	1014.450	235.740						0.02	210.35	2	299.83
110	C_4H_6	1,3-butadiene	106-99-0	1	3.96640	927.2100	238.630						0.02	198.00	2	287.49
				2	3.96640	927.2100	238.630	425.00	2	2.51460	23.653	1970.80	2.04	348.15	41.90	343.15
111	$C_4H_6O_3$	acetic anhydride	108-24-7	3	606.00	−8.35130	1.89050	−2.8357	−5.1156	40.00					40.00	606.00
112	C_4H_8	cyclobutane	287-23-0	1	4.04436	1025.500	241.430						0.02	210.20	2	305.67
				2	4.04436	1025.500	241.430	460.00	20	2.17400	0.000	0.00	2.161	308.15	49.90	460.00
113	C_4H_8	1-butene	106-98-9	1	3.91780	908. 800	238.540						0.02	196.41	2	285.88
				2	3.91780	908.800	238.540	419.95	1	2.10580	−66.740	5100.70	2.16	288.15	36.18	413.15
114	C_4H_8	trans-2-butene	624-64-6	1	4.00827	967.5000	240.840						0.02	201.83	2	293.29
				2	4.00827	967.5000	240.840	428.63	8	2.71670	49.772	−1061.2	1.69	288.15	30.82	413.15
115	C_4H_8	cis-2-butene	590-18-1	1	4.00958	967.3200	237.873						0.02	204.73	2	296.11
				2	4.00958	967.3200	237.870	435.58	11	2.60300	47.148	−1082.1	2.14	298.15	34.81	423.15
116	C_4H_8	2-methylpropene	115-11-7	1	3.80956	866.2500	234.640						0.02	195.77	2	285.41
				2	3.80956	866.2500	234.640	417.90	0	1.59900	−150.96	9633.00	2.2	288.15	37.00	413.15
117	C_4H_8O	butanone (methyl ethyl ketone)	78-93-3	1	4.13860	1232.630	218.690						0.02	265.62	2	375.66
				2	4.13860	1232.630	218.690	536.80	87	2.31490	−4.900	3279.00	2.14	378.15	39.69	533.15
118	C_4H_8O	tetrahydrofuran	109-99-9	1	4.12142	1203.110	226.355						0.02	253.50	2	361.71
119	$C_4H_8O_2$	butanoic acid	107-92-6	1	4.82340	1731.708	195.955						0.02	342.70	2	460.12
				3	624.00	−8.42953	1.34333	−5.37332	−2.74438	40.30					40.30	624.00

120	$C_4H_8O_2$	2-methylpropanoic acid	79-31-2	1	3.71153	1097.830	141.740						0.02	334.00	2	453.31
				3	605.00	−8.53258	1.30605	−5.2242	−2.05813	37.00					37.00	605.00
123	$C_4H_8O_2$	methyl propanoate	554-12-1	1	3.98745	1129.570	204.240						0.02	267.50	2	375.32
124	$C_4H_8O_2$	ethyl ethanoate (ethyl acetate)	141-78-6	1	4.13361	1195.130	212.470						0.02	265.50	2	372.51
125	$C_4H_8O_2$	propyl methanoate (propyl formate)	110-74-7	1	3.97008	1132.300	204.080						0.02	268.90	2	377.68
126	C_4H_9Cl	2-chlorobutane	78-86-4	1	4.12220	1245.200	234.400						0.02	252.60	2	364.62
127	C_4H_{10}	butane	106-97-8	1	3.93266	935.7730	238.789						0.02	200.50	2	292.03
				3	425.25	−7.01763	1.67770	−1.9739	−2.1720	37.92					37.92	425.25
128	C_4H_{10}	2-methylpropane (isobutane)	75-28-5	1	4.00272	947.5400	248.870						0.02	190.40	2	280.25
				2	4.00272	947.5400	248.870	408.14	−5	2.67050	−19.640	2792.0C	1.863	278.15	19.87	373.15
129	$C_4H_{10}N_2$	piperazine	110-85-0	3	661.00	−8.10664	3.36281	−4.52962	−3.8278	58			0.90976	418.00	2.7	460.48
130	$C_4H_{10}O$	1-butanol	71-36-3	1	4.64930	1395.140	182.739						0.02	310.18	2	411.26
				3	563.05	−8.40615	2.23010	−8.2486	−0.7110	44.24					44.24	563.05
131	$C_4H_{10}O$	2-methyl-1-propanol (isobutanol)	78-83-1	1	4.34504	1190.380	166.670						0.02	303.40	2	400.84
				3	547.78	−8.31460	2.13678	−8.4832	−0.79774	43.04					43.04	547.78
132	$C_4H_{10}O$	2-methyl-2-propanol (tert-butanol)	75-65-0	1	4.44484	1154.480	177.650						0.02	283.00	2	374.10
				3	506.20	−8.47927	2.47845	−9.27918	−2.53992	39.73					39.73	506.20
133	$C_4H_{10}O$	2-butanol (sec-butanol)	78-92-2	3	536.01	−8.09820	1.64406	−7.4900	−5.27355	41.98					41.98	536.01
134	$C_4H_{10}O$	diethyl ether	60-29-7	1	4.10962	1090.640	231.200						0.02	229.71	2	328.31
				3	466.74	−7.43301	1.78847	−2.4793	−3.2811	36.50					36.50	466.74
135	$C_4H_{10}O_2$	1,2-dimethoxyethane	110-71-4	3	537.00	−8.0898	2.53555	−3.4809	−3.65036	39.60			0.13332	305.86	2.7	392.29
136	$C_4H_{10}O_2$	1,2-butandiol	26171-83-5	3	680.00	−9.98662	5.09869	−9.38593	−2.85378	52.10			0.02	372.55	2.7	506.40
137	$C_4H_{10}O_2$	1,3-Butanediol	107-88-0	3	676.00	−9.29011	3.03108	−9.27334	−1.05346	40.20			0.006902	364.98	2.32	512.05
138	$C_4H_{10}S$	3-thiapentane (diethyl sulfide)	352-93-2	1	4.05326	1257.833	218.662						0.02	273.10	2	389.71
139	$C_4H_{11}N$	1-butanamine (butyl amine)	109-73-9	1	4.30770	1276.870	220.520						0.02	265.20	2	371.32
				2	3.90120	1041.310	191.000	526.80	84	2.03520	1398.00	−126749.	1.82	368.15	30.36	503.15
141	$C_4H_{11}N$	2-methyl-1-propanamine (isobutyl amine)	78-81-9	1	3.90070	1055.560	203.350						0.02	258.31	2	363.04
143	C_5H_5N	pyridine	110-86-1	1	4.16749	1373.026	214.690						0.02	292.51	2	413.57
				2	4.16750	1373.030	214.690	620.00	127	2.71070	−45.881	3987.7€	1.98	413.15	52.26	613.15
144	C_5H_6O	2-methylfuran	534-22-5	1	3.70410	991.2000	203.290						0.02	253.30	2	361.13
145	C_5H_8	1-pentyne	627-19-0	1	4.00260	1068.100	227.000						0.02	233.00	2	334.00
147	C_5H_8O	cyclopentanone	120-92-3	3	624.50	−7.36589	1.54092	−2.28143	−3.0514	46.00					46.00	624.50
148	C_5H_{10}	cyclopentane	287-92-3	1	4.06783	1152.574	234.510						0.02	238.50	2	344.62
149	C_5H_{10}	1-pentene	109-67-1	1	3.96914	1044.010	233.450						0.02	223.89	2	324.32
				2	3.96914	1044.010	233.450	464.78	38	2.57510	122.880	−4873.4	1.66	318.15	25.17	443.15
150	C_5H_{10}	cis-2-pentene	627-20-3	1	3.96798	1052.440	228.693						0.02	229.40	2	331.46
151	C_5H_{10}	2-methyl-2-butene	513-35-9	1	4.09149	1124.330	236.630						0.02	230.69	2	333.14
152	C_5H_{10}	3-methyl-1-butene	563-45-1	1	3.94945	1012.370	236.647						0.02	215.73	2	221.50
				2	3.94945	1012.370	236.650	453.15	28	2.72220	95.875	−3435.8	1.67	308.15	35.50	453.15
153	$C_5H_{10}O$	cyclopentanol	96-41-3	3	619.50	−7.40984	1.71852	−6.8471	−4.36177	49.00					49.00	619.50
154	$C_5H_{10}O$	2-pentanone (methyl propyl ketone)	107-87-9	1	4.15140	1316.730	215.380						0.02	282.84	2	399.74
				2	4.15140	1316.730	215.380	561.10	120	2.06640	−348.80	52963.C	2.18	403.15	36.08	558.15
155	$C_5H_{10}O$	3-pentanone (diethyl ketone)	96-22-0	1	4.42708	1481.170	233.010						0.02	281.90	2	399.12

Section D Vapor Pressure Correlations Parameters (*Continued*)

No.	Formula	Name	CAS #	Eq. #	A/A/Tc	B/B/a	C/C/b	Tc/c	to/d	n/Pc	E	F	Pvpmin, bar	Tmin, K	Pvpmax bar	Tmax, K
156	$C_5H_{10}O$	3-methyl-2-butanone (methyl isopropyl ketone)	563-80-4	1	3.46583	955.4300	181.730						0.02	276.40	2	393.31
157	$C_5H_{10}O$	2-methyltetrahydrofuran	96-47-9	1	3.95009	1175.510	217.802						0.02	263.44	2	377.49
158	$C_5H_{10}O_2$	pentanoic acid	109-52-4	1	4.16920	1405.800	151.800						0.02	361.00	2	484.78
				3	643.00	−8.76701	1.54990	−6.19961	−4.21927	35.80					35.80	643.00
159	$C_5H_{10}O_2$	3-methylbutanoic acid	503-74-2	1	4.58470	1676.300	189.500						0.02	350.00	2	474.97
				3	629.00	−8.67381	1.62939	−6.51756	−2.08757	34.00					34.00	629.00
160	$C_5H_{10}O_2$	methyl butanoate	623-42-7	1	4.10641	1271.060	207.210						0.02	284.90	2	399.96
				2	4.10641	1271.060	207.210	554.45	120	2.46460	543.870	−34817.6	2.17	403.15	31.80	548.15
161	$C_5H_{10}O_2$	ethyl propanoate	105-37-3	1	4.14400	1274.700	209.000						0.02	282.40	2	395.85
163	$C_5H_{10}O_2$	propyl ethanoate (propyl acetate)	109-60-4	1	4.05548	1233.46	203.080						0.02	284.40	2	398.60
164	$C_5H_{10}O_2$	2-methylpropyl methanoate (isobutyl formate)	542-55-2	1	3.98450	1195.900	202.500						0.03	288.00	2	395.32
165	$C_5H_{11}Cl$	1-chloropentane	543-59-9	1	3.93641	1271.160	215.000						0.02	283.72	2	407.81
166	C_5H_{12}	pentane	109-66-0	1	3.97786	1064.840	232.014						0.02	228.71	2	330.75
				3	469.80	−7.30698	1.75845	−2.1629	−2.9130	33.75					33.75	469.80
167	C_5H_{12}	2-methylbutane	78-78-4	1	3.92023	1022.880	233.460						0.02	221.72	2	322.32
				2	3.92023	1022.880	233.460	460.43	36	2.14912	−227.07	19674.0	1.766	318.15	15.77	413.15
168	C_5H_{12}	2,2-dimethylpropane (neopentane)	463-82-1	1	3.83916	938.2340	235.249						0.40	259.33	2	303.08
				2	3.83916	938.2340	235.249	433.78	17	2.42328	34.505	580.56	1.7142	298.15	16.28	393.15
169	$C_5H_{12}O$	1-pentanol	71-41-0	1	4.39646	1336.010	166.320						0.02	326.01	2	433.05
				3	588.15	−8.98005	3.91624	−9.9081	−2.1910	39.09					39.09	588.15
170	$C_5H_{12}O$	2-pentanol	6032-29-7	1	4.42349	1291.212	173.130						0.008	298.12	0.735	383.34
171	$C_5H_{12}O$	2-methyl-1-butanol	137-32-6	1	4.48266	1360.367	173.220						0.004	298.12	0.711	393.70
172	$C_5H_{12}O$	2-methyl-2-butanol	75-85-4	1	3.64420	863.4000	135.300						0.02	299.00	2	396.11
173	$C_5H_{12}O$	3-methyl-1-butanol	123-51-3	1	4.07851	1128.190	146.470						0.02	321.90	2	425.34
175	$C_5H_{12}O$	ethyl propyl ether	628-32-0	1	3.83648	1052.470	210.880						0.02	252.40	2	359.96
176	$C_5H_{12}S$	3-methyl-1-butanethiol (isopentyl mercaptan)	541-31-1	1	4.03981	1342.509	214.446						0.02	292.60	2	417.78
177	C_6ClF_5	chloropentafluorobenzene	344-07-0	3	570.81	−8.10119	1.95485	−2.79778	−4.1940	32.37					32.37	570.81
178	C_6F_6	hexafluorobenzene	392-56-3	3	516.73	−8.04104	1.93510	−2.9390	−4.5480	32.75					32.75	516.73
184	C_6HF_5	pentafluorobenzene	363-72-4	3	530.97	−7.86799	1.71659	−2.53582	−4.59937	35.37					35.37	530.97
185	$C_6H_2F_4$	1,2,4,5-tetrafluorobenzene	327-54-8	3	543.35	−7.85347	1.94620	−2.8652	−3.80563	37.99					37.99	543.35
186	C_6H_5Cl	chlorobenzene	108-90-7	1	4.02012	1378.790	211.700						0.02	302.50	2	432.18
				2	4.02012	1378.790	211.700	632.43	137	2.20300	18.280	674.77	1.82	428.15	44.07	630.15
187	C_6H_6	benzene	71-43-2	1	3.98523	1184.240	217.572						0.05	279.64	2	377.06
				3	562.16	−7.01433	1.55256	−1.8479	−3.7130	48.98					48.98	562.16
188	C_6H_6O	phenol	108-95-2	1	4.26960	1523.420	175.400						0.02	353.00	2	481.62

189	C_6H_7N	benzeneamine (aniline)	62-53-3	1	4.40870	1692.770	200.440						0.02	349.86	2	484.81
				2	4.40870	1692.770	200.440	699.00	197	4.90600	452.800	−239100	2.2	488.15	40.30	673.15
190	C_6H_7N	2-methylpyridine (2-picoline)	109-06-8	1	4.15550	1415.410	211.730						0.02	303.19	2	428.63
191	C_6H_7N	3-methylpyridine (3-picoline)	108-99-6	1	4.18930	1492.130	212.530						0.02	314.03	2	444.37
192	C_6H_7N	4-methylpyridine (4-picoline)	108-89-4	1	4.16750	1481.571						210.650	0.02	315.05	2	445.68
194	C_6H_{10}	Cyclohexene	110-83-8	3	560.40	−9.08102	5.75488	−5.17505	−1.0489	49.05			0.06417	285.39	1.04	356.99
195	$C_6H_{10}O$	cyclohexanone	108-94-1	3	653.00	−7.49380	1.63094	−2.12212	−3.91327	40.00					40.00	653.00
196	$C_6H_{10}O$	4-methyl-3-penten-2-one (mesityloxide)	141-79-7	3	605.00	−8.68118	3.99203	−4.81662	−1.73164	40			0.02	303.67	1.985	428.56
197	C_6H_{12}	cyclohexane	110-82-7	1	3.93002	1182.774	220.618						0.06	282.11	2	378.46
				2	3.93002	1182.770	220.618	553.50	25	3.40407	10.048	−126.96	1.9871	378.15	40.48	553.15
198	C_6H_{12}	methylcyclopentane	96-37-7	1	4.18199	1295.543	238.390						0.02	255.06	2	368.58
				2	4.18199	1295.543	238.390	532.79	80	2.70504	−741.05	43373.0	2.26	373.15	37.50	532.79
199	C_6H_{12}	1-hexene	592-41-6	1	3.98260	1148.620	225.340						0.02	249.98	2	359.80
				2	3.98260	1148.620	225.340	504.03	72	2.45920	106.260	−3773.6	1.91	358.15	26.86	493.15
200	C_6H_{12}	4-methylpent-1-ene	691-37-2	1	3.96019	1121.302	229.687						0.02	241.60	2	349.90
201	$C_6H_{12}O$	cyclohexanol	108-93-0	3	650.00	−7.12838	1.40189	−5.60756	−9.57158	42.60					42.60	650.00
202	$C_6H_{12}O$	2-hexanone (methyl butyl ketone)	591-78-6	1	4.15330	1395.800	208.980						0.02	302.68	2	426.50
203	$C_6H_{12}O$	3-hexanone (ethyl propyl ketone)	589-38-8	1	4.11658	1359.880	207.300						0.02	299.60	2	422.25
204	$C_6H_{12}O$	4-methyl-2-pentanone (methyl isobutyl ketone)	108-10-1	1	3.82220	1190.6904	195.450						0.02	293.40	2	415.85
				3	574.60	−7.70040	1.69968	−2.80448	−3.81623	32.70					32.70	574.60
205	$C_6H_{12}O$	butylvinylether	111-34-2	3	540.50	−8.04744	2.31158	−2.91499	−4.09565	32.00			0.13332	311.89	2.7	403.37
206	$C_6H_{12}O_2$	hexanoic acid	142-62-1	3	662.00	−8.86570	1.95079	−7.80315	−2.85006	32.00					32.00	662.00
208	$C_6H_{12}O_2$	ethyl butanoate	105-54-4	1	3.27456	921.5600	160.380						0.02	298.00	2	422.69
209	$C_6H_{12}O_2$	propyl propanoate	106-36-5	1	4.44890	1545.300	225.300						0.02	299.20	2	420.40
211	$C_6H_{12}O_2$	butyl ethanoate (butyl acetate)	123-86-4	1	4.50000	1596.700	229.300						0.02	301.00	2	424.11
212	$C_6H_{12}O_2$	2-methylpropyl ethanoate (isobutyl acetate)	110-19-0	1	4.35460	1462.400	219.700						0.02	295.00	2	414.22
214	$C_6H_{12}O_2$	3-methylbutyl methanoate (isopentyl formate)	110-45-2	1	4.24880	1439.400	215.100						0.02	300.00	2	422.66
215	$C_6H_{12}O_3$	2-ethoxyethylacetate	111-15-9	3	610.60	−9.64168	4.58179	−6.25993	−4.12066	31.80			0.02	330.00	2.7	468.70
216	C_6H_{14}	hexane	110-54-3	1	4.00139	1170.875	224.317						0.02	254.24	2	365.25
				3	507.90	−7.53998	1.83759	−2.5438	−3.1630	30.35					30.35	507.90
217	C_6H_{14}	2-methylpentane	107-83-5	1	3.98332	1145.800	227.815						0.02	246.90	2	356.50
				2	3.98332	1145.800	227.820	497.50	69	2.27660	0.000	0.00	1.583	348.15	24.60	483.15
218	C_6H_{14}	3-methylpentane	96-14-0	1	3.99283	1162.370	228.286						0.02	249.00	2	359.72
				2	3.99283	1162.370	228.290	504.40	72	5.74154	690.900	−40238.	1.917	358.15	26.42	493.15
219	C_6H_{14}	2,2-dimethylbutane	75-83-2	1	3.89590	1090.160	230.517						0.02	237.40	2	345.89
				2	3.89590	1090.160	230.520	488.70	59	2.17300	0.000	0.00	1.611	338.15	24.71	473.15
220	C_6H_{14}	2,3-dimethylbutane	79-29-8	1	3.93486	1127.400	228.966						0.02	244.20	2	354.43
				2	3.93486	1127.400	228.970	499.90	67	2.51900	332.500	−24950.	1.682	348.15	24.69	483.15
221	$C_6H_{14}O$	1-hexanol	111-27-3	1	4.18948	1295.590	152.510						0.02	340.80	2	453.83
				3	610.70	−9.49034	5.13288	−10.5817	−5.1540	34.70					34.70	610.70
222	$C_6H_{14}O$	2-hexanol	626-93-7	1	4.93223	1696.190	204.430						0.02	324.50	2	434.97

Section D Vapor Pressure Correlations Parameters (*Continued*)

No.	Formula	Name	CAS #	Eq. #	A/A/Tc	B/B/a	C/C/b	Tc/c	to/d	n/Pc	E	F	Pvpmin, bar	Tmin, K	Pvpmax bar	Tmax, K
223	$C_6H_{14}O$	3-hexanol	623-37-0	1	6.16250	2662.265	296.620						0.008	298.00	1	411.00
224	$C_6H_{14}O$	2-methyl-1-pentanol	105-30-6	1	6.19790	2625.143	276.330						0.003	298.00	1	423.00
225	$C_6H_{14}O$	2-methyl-2-pentanol	590-36-3	1	3.27663	811.0500	126.600						0.02	309.50	2	419.12
227	$C_6H_{14}O$	4-methyl-1-pentanol	626-89-1	1	4.17605	1273.350	153.560						0.02	336.40	2	448.19
228	$C_6H_{14}O$	4-methyl-2-pentanol	108-11-2	1	4.66180	1566.760	204.790						0.02	315.00	2	427.65
232	C_7F_{16}	hexadecylfluoroheptane (perfluoroheptane)	335-57-9	3	475.00	−9.18955	3.15138	−5.41934	−4.11174	16.5			0.1[illegible]335	303.68	2.7	389.70
233	$C_7H_3F_5$	pentafluorotoluene	771-56-2	3	566.52	−8.08717	1.76131	−2.72838	−4.13797	31.24					31.24	566.52
234	C_7H_8	toluene	108-88-3	1	4.05043	1327.62000	217.62500						0.02	286.44	2	409.61
				3	591.80	−7.31600	1.59425	−1.93165	−3.72220	41.06					41.06	591.80
235	C_7H_8O	benzyl alcohol	100-51-6	3	715.00	−7.29099	1.17084	−4.7167	−5.5300	43.00					43.00	715.00
236	C_7H_8O	2-methylphenol (o-cresol)	95-48-7	1	4.18340	1534.540	176.300						0.02	357.80	2	492.11
				2	4.18340	1534.540	176.300	697.60	200	1.70720	463.530	−36925	2.05	493.15	50.00	697.57
237	C_7H_8O	3-methylphenol (m-cresol)	108-39-4	1	4.21530	1556.830	167.600						0.02	368.80	2	503.28
				2	4.21530	1556.830	167.600	705.70	215	2.19340	−549.69	67638.0	1.99	503.15	45.60	705.69
238	C_7H_8O	4-methylphenol (p-cresol)	106-44-5	1	4.18050	1525.320	163400						0.02	369.10	2	502.93
				2	4.18050	1525.320	163400	704.50	215	2.10170	65.801	77063	2.01	503.15	51.50	704.49
239	C_7H_9N	2,3-dimethylpyridine (2,3 lutidine)	583-61-9	1	4.18570	1536.350	206400						0.02	327.90	2	462.24
240	C_7H_9N	2,4-dimethylpyridine (2,4 lutidine)	108-47-4	1	4.20962	1542.940	208.630						0.02	325.65	2	459.28
241	C_7H_9N	2,5-dimethylpyridine (2,5 lutidine)	589-93-5	1	4.20857	1541.780	209.850						0.02	324.20	2	457.87
242	C_7H_9N	2,6-dimethylpyridine (2,6 lutidine)	108-48-5	1	4.08748	1407.250	201.001						0.02	315.34	2	443.79
243	C_7H_9N	3,4-dimethylpyridine (3,4 lutidine)	583-58-4	1	4.18920	1605.140	204.550						0.02	341.10	2	481.43
244	C_7H_9N	3,5-dimethylpyridine (3,5 lutidine)	591-22-0	1	4.21290	1595.150	207.240						0.02	335.73	2	473.68
245	$C_7H_{12}O_2$	Butyl-2-propenoate (Butylacrylate)	141-32-2	3	644.00	−7.59083	1.96932	−3.05837	−4.17604	45.40			0.02	318.51	1.0133	419.77
246	C_7H_{14}	cycloheptane	291-64-5	1	3.96330	1322.21997	215.297						0.02	291.40	2	418.89
				2	3.96330	1322.220	215.297	604.30	129	2.52840	250.300	−13243.	1.5443	408.15	33.70	593.15
247	C_7H_{14}	methylcyclohexane	108-87-2	1	3.98232	1290.968	223.701						0.02	276.68	2	400.13
				2	3.98232	1290.97	223.701	572.19	115	2.79424	53.706	2916.13	1.9059	398.15	34.71	572.19
248	C_7H_{14}	ethylcyclopentane	1640-89-7	1	4.00408	1293.712	220.120						0.02	279.88	2	402.39
				2	4.00408	1293.712	220.120	569.52	110	2.66692	561.915	−45612.	2.30	408.15	33.60	569.52
249	C_7H_{14}	cis-1,3-dimethylcyclopentane	2532-58-3	1	4.00405	1259.821	223.530						0.02	270.52	2	389.83
250	C_7H_{14}	trans-1,3-dimethylcyclopentane	1759-58-6	1	3.95279	1232.161	221.420						0.02	269.74	2	389.15
251	C_7H_{14}	1-heptene	592-76-7	1	4.02677	1258.340	219.300						0.02	273.62	2	391.59
				2	4.02677	1258.340	219.300	537.30	103	2.61660	290.600	−17516.	.83	388.15	23.23	523.15
252	$C_7H_{14}O_2$	heptanoic acid	111-14-8	3	679.00	−8.94240	2.20536	−8.82144	−1.9710	29.00					29.00	679.00
255	$C_7H_{14}O_2$	propyl butanoate	105-66-8	1	3.40455	1019.490	156.600						0.02	316.00	2	445.04
256	$C_7H_{14}O_2$	2-methylpropyl propanoate	540-42-1	1	3.56180	1042.300	156.500						0.03	321.60	2	436.30
259	C_7H_{16}	heptane	142-82-5	1	4.02023	1263.909	216.432						0.02	277.71	2	396.53
				3	540.15	−7.77404	1.85614	−2.8298	−3.5070	27.35					27.35	540.15

260	C_7H_{16}	2-methylhexane	591-76-4	1	3.99739	1235.520	219.497						0.02	270.55	2	387.88
				2	3.99739	1235.520	219.500	530.10	100	2.04000	575.200	−40292	1.548	378.15	21.38	513.15
261	C_7H_{16}	3-methylhexane	589-34-4	1	3.99571	1242.018	219.435						0.02	271.90	2	389.82
				2	3.99571	1242.020	219.440	535.20	100	1.89740	267.300	−9936.0	1.471	378.15	20.51	513.15
262	C_7H_{16}	3-ethylpentane	617-78-7	1	4.00449	1254.055	220.136						0.02	272.90	2	391.62
				2	4.00449	1254.060	220.140	540.50	103	2.38910	565.800	−38997	1.404	378.15	19.64	513.15
263	C_7H_{16}	2,2-dimethylpentane	590-35-2	1	3.94392	1191.959	223.498						0.02	260.90	2	376.84
				2	3.94392	1191.906	223.500	520.40	89	2.20020	515.600	−33215	1.59	368.15	21.68	503.15
264	C_7H_{16}	2,3-dimethylpentane	565-59-3	1	3.98066	1238.986	221.942						0.04	281.56	2	387.89
				2	3.98066	1238.990	221.940	537.30	99	1.97920	282.400	−12835	1.553	378.15	20.74	513.15
265	C_7H_{16}	2,4-dimethylpentane	108-08-7	1	3.95442	1193.612	221.807						0.02	262.40	2	378.01
				2	3.95442	1193.160	221.810	519.70	90	1.92600	224.400	−4163.0	1.538	368.15	21.54	503.15
266	C_7H_{16}	3,3-dimethylpentane	562-49-2	1	3.94912	1227.020	225.121						0.02	265.20	2	384.36
				2	3.94912	1227.020	225.120	536.30	96	2.15280	420.700	−24617	1.707	378.15	21.38	513.15
267	C_7H_{16}	2,2,3-trimethylbutane	464-06-2	1	3.91555	1199.397	225.908						0.02	260.90	2	379.04
				2	3.91555	1199.400	225.910	531.10	91	1.98860	309.700	−16910	1.507	368.15	20.09	503.15
268	$C_7H_{16}O$	1-heptanol	111-70-6	1	4.01991	1274.890	140.940						0.02	355.10	2	475.03
				3	632.50	−9.68778	5.35716	−10.1672	−8.0100	31.35					31.35	632.50
272	C_8H_8O	MethylPhenylKetone (Acetophenone)	98-86-2	3	713.00	−8.9386	4.01161	−4.5941	−2.57768	44.00			0.02	360.46	2.7	520.00
273	C_8H_{10}	ethylbenzene	100-41-4	1	4.06861	1415.770	212.300						0.02	306.32	2	436.63
				3	617.20	−7.53139	1.75439	−2.42012	−3.57146	36.00					36.00	617.20
274	C_8H_{10}	1,2-dimethylbenzene (o-xylene)	95-47-6	1	4.09789	1458.706	212.041						0.02	312.75	2	445.30
				3	630.33	−7.60491	1.75383	−2.27531	−3.73771	37.35					37.35	630.33
275	C_8H_{10}	1,3-dimethylbenzene (m-xylene)	108-38-3	1	4.14051	1468.703	216.120						0.02	308.54	2	439.56
				3	617.05	−7.67717	1.80240	−2.47745	−3.66068	35.38					35.38	617.05
276	C_8H_{10}	1,4-dimethylbenzene (p-xylene)	106-42-3	1	4.10494	1446.832	214.627						0.02	307.81	2	438.88
				3	616.23	−7.71694	1.89119	−2.39695	−3.63026	35.16					35.16	616.23
277	$C_8H_{10}O$	2-ethylphenol	90-00-6	1	4.13365	1550.440	171.074						0.02	367.90	2	506.61
278	$C_8H_{10}O$	3-ethylphenol	620-17-7	1	4.16568	1572.260	159.52399						0.02	381.72	2	520.46
279	$C_8H_{10}O$	4-ethyl-phenol	123-07-9	1	4.13227	1545.23999	156.468						0.02	381.67	2	520.01
280	$C_8H_{10}O$	2,3-dimethylphenol (2,3 xylenol)	526-75-0	1	4.12202	1576.780	166.173						0.02	377.86	2	519.64
281	$C_8H_{10}O$	2,4-dimethylphenol (2,4 xylenol)	105-67-9	1	4.18688	1592.780	170.004						0.02	373.76	2	513.04
282	$C_8H_{10}O$	2,5-dimethylphenol (2,5 xylenol)	95-87-4	1	4.13449	1563.140	167.453						0.02	373.66	2	513.46
283	$C_8H_{10}O$	2,6-dimethylphenol (2,6 xylenol)	576-26-1	1	4.19336	1627.230	187.547						0.02	361.76	2	503.66
284	$C_8H_{10}O$	3,4-dimethylphenol (3,4 xylenol)	95-65-8	1	4.21183	1627.780	160.041						0.02	388.50	2	529.34
285	$C_8H_{10}O$	3,5-dimethylphenol (3,5 xylenol)	108-68-9	1	4.26229	1645.270	164.821						0.02	384.32	2	523.67
286	C_8H_{16}	cyclooctane	292-64-8	1	3.98125	1434.670	209.712						0.02	316.00	2	453.27
				2	3.98125	1434.670	209.712	647.20	162	2.30600	325.500	−31112.	1.787	448.15	31.30	633.15
287	C_8H_{16}	*t*-1,4-dimethylcyclohexane	2207-04-7	1	4.02425	1457.08	205.99						0.02	321.75	2	458.51
288	C_8H_{16}	1-octene	111-66-0	1	4.05985	1355.460	213.050						0.02	295.47	2	420.71
				2	4.05985	1355.460	213.050	566.65	131	2.68960	512.500	−40092.	1.88	418.15	21.34	553.15
289	$C_8H_{16}O_2$	octanoic acid	124-07-2	3	695.00	−9.04015	2.16529	−8.66117	−4.69516	26.40					26.40	695.00

No.	Formula	Name	CAS #	Eq. #	A/A/Tc	B/B/a	C/C/b	Tc/c	to/d	n/Pc	E	F	Pvpmin, bar	Tmin, K	Pvpmax bar	Tmax, K
296	C_8H_{18}	octane	111-65-9	1	4.05075	1356.360	209.635						0.02	299.42	2	425.23
				3	568.95	−8.04937	2.03865	−3.3120	−3.6480	24.90					24.90	568.95
297	C_8H_{18}	2-methylheptane	592-27-8	1	4.03877	1335.220	213.415						0.03	299.81	2	416.95
				2	4.03877	1335.220	213.415	559.60	128	2.47135	255.100	−7424.0	1.6088	408.15	19.73	543.15
298	C_8H_{18}	3-methylheptane	589-81-1	1	4.01533	1326.140	211.813						0.04	306.32	2	418.36
				2	4.01533	1326.140	211.813	563.67	129	2.43555	315.600	−15218.	1.5543	408.15	19.13	543.15
299	C_8H_{18}	4-methylheptane	589-53-7	1	4.02214	1325.704	212.367						0.02	292.50	2	417.05
				2	4.02214	1325.740	212.367	561.74	128	2.41333	240.700	−8481.0	1.6055	408.15	19.63	543.15
300	C_8H_{18}	3-ethylhexane	619-99-8	1	4.01533	1327.930	212.645						0.02	292.90	2	418.01
				2	4.01533	1327.930	212.645	565.49	129	2.39952	227.500	−2817.0	1.5687	408.15	19.17	543.15
301	C_8H_{18}	2,2-dimethylhexane	590-73-8	1	3.95748	1271.180	214.830						0.02	283.00	2	405.96
				2	3.95748	1271.180	214.830	549.87	117	2.40185	301.200	−17401	1.6476	398.15	20.05	533.15
302	C_8H_{18}	2,3-dimethylhexane	584-94-1	1	3.99236	1314.290	214.059						0.02	290.00	2	415.13
				2	3.99236	1314.290	214.059	563.49	126	2.33502	185.000	−3318.0	1.6871	408.15	19.87	543.15
303	C_8H_{18}	2,4-dimethylhexane	589-43-5	1	3.97399	1285.850	214.600						0.02	285.20	2	408.63
				2	3.97399	1285.850	214.600	553.52	119	2.36737	149.300	482.00	1.5404	398.15	19.25	533.15
304	C_8H_{18}	2,5-dimethylhexane	592-13-2	1	3.98112	1285.470	214.248						0.02	285.20	2	408.20
305	C_8H_{18}	3,3-dimethylhexane	563-16-6	1	3.97403	1306.960	217.376						0.02	286.10	2	411.59
				2	3.97403	1306.960	217.376	562.02	122	2.39488	144.900	−2353.0	1.3412	408.15	20.57	543.15
306	C_8H_{18}	3,4-dimethylhexane	583-48-2	1	4.00310	1329.400	214.836						0.02	291.40	2	417.40
				2	4.00310	1329.400	214.836	568.85	128	2.50297	320.500	−18497.	1.596	408.15	18.98	543.15
307	C_8H_{18}	3-ethyl-2-methylpentane	609-26-7	1	3.98610	1317.050	215.229						0.02	289.50	2	415.31
				2	3.98610	1317.050	215.229	567.09	126	2.38973	174.900	−4584.0	1.6811	408.15	19.54	543.15
308	C_8H_{18}	3-ethyl-3-methylpentane	1067-08-9	1	3.98950	1345.920	219.584						0.02	290.10	2	418.46
				2	3.98950	1345.920	219.584	576.58	129	2.43672	182.800	−7717.0	1.5622	408.15	18.05	543.15
309	C_8H_{18}	2,2,3-trimethylpentane	564-02-3	1	3.94826	1293.940	218.355						0.02	283.90	2	409.56
				2	3.94826	1293.940	218.355	563.50	120	2.45345	162.400	−5383.0	1.934	408.15	20.84	543.15
310	C_8H_{18}	2,2,4-trimethylpentane (isooctane)	540-84-1	1	3.93646	1257.850	220.767						0.02	275.50	2	398.38
				2	3.93646	1257.850	220.767	543.90	124	2.13261	134.500	12998.0	1.9889	398.15	25.42	543.15
311	C_8H_{18}	2,3,3-trimethylpentane	560-21-4	1	3.96421	1325.810	220.161						0.02	287.00	2	414.91
				2	3.96421	1325.810	220.161	573.56	125	2.37930	76.300	1851.00	1.7032	408.15	18.93	543.15
312	C_8H_{18}	2,3,4-trimethylpentane	565-75-3	1	3.97700	1314.310	217.481						0.04	300.19	2	413.19
				2	3.97700	1314.310	217.481	566.41	124	2.39574	169.400	−4867.0	1.7713	408.15	19.98	543.15
313	C_8H_{18}	2,2,3,3-tetramethylbutane	594-82-1	1	3.90420	1270.100	219.500						0.90	375.20	2	406.00
314	$C_8H_{18}O$	1-octanol	111-87-5	1	3.90225	1274.800	131.990						0.02	368.80	2	495.15
				3	652.50	−10.01437	5.90629	−10.4026	−9.0480	28.60					28.60	652.50
315	$C_8H_{18}O$	2-octanol	123-96-6	1	3.51370	1060.400	122.500						0.02	354.00	2	480.72
				3	638.00	−9.37352	4.73760	−8.3382	−11.646	28.90					28.90	638.00
317	$C_8H_{18}O$	4-octanol	589-62-8	1	5.08522	1816.393	190.020						0.0001	283.00	0.02	353.00
318	$C_8H_{18}O$	2-ethyl-1-hexanol	104-76-7	3	640.50	−9.61812	5.17861	−9.1144	−11.004	27.99					27.99	640.50

No.	Formula	Name	CAS No.	Eq.												
319	$C_8H_{19}N$	n-octanamine(Octylamine)	111-86-4	3	641.00	−7.99396	1.40573	−2.98188	−6.60435	26.17			0.02	434.49	2.7	494.5
322	C_9H_7N	quinoline	91-22-5	1	4.19490	1812.250	195.450						0.02	385.18	2	543.11
				2	4.19490	1812.250	195.450	782.00	247	1.73760	28.233	−2288.0	2.01	543.15	44.70	773.15
325	C_9H_{12}	propylbenzene	103-65-1	1	4.07664	1491.800	207.250						0.02	324.19	2	461.01
				2	4.07664	1491.800	207.250	638.28	170	2.19580	0.000	0.00	2.097	463.15	31.80	638.28
326	C_9H_{12}	1-methylethylbenzene	98-82-8	1	4.06112	1460.766	207.830						0.02	318.92	2	453.81
327	C_9H_{12}	1-ethyl-4-methylbenzene	622-96-8	1	4.10862	1517.577	207.900						0.02	326.56	2	463.82
328	C_9H_{12}	1,2,3-trimethylbenzene	526-73-8	1	4.17110	1598.241	207.620						0.02	337.80	2	478.50
329	C_9H_{12}	1,2,4-trimethylbenzene	95-63-6	1	4.17692	1579.353	209.290						0.02	332.64	2	471.34
330	C_9H_{12}	1,3,5-trimethylbenzene (mesitylene)	108-67-8	1	4.22541	1581.360	210.010						0.02	330.06	2	466.10
331	C_9H_{18}	1-nonene	124-11-8	1	4.07920	1436.200	205.690						0.02	316.02	2	447.59
				2	4.07920	1436.200	205.690	593.20	157	2.60900	655.800	−55549	1.6	438.15	17.99	573.15
332	$C_9H_{18}O_2$	nonanoic acid	112-05-0	3	711.00	−9.10090	2.49646	−9.98583	−2.13513	24.30					24.30	711.00
333	$C_9H_{18}O_2$	3-methylbutyl butanoate	106-27-4	1	4.50447	1805.080	222.300						0.02	342.00	2	480.28
334	C_9H_{20}	nonane	111-84-2	1	4.07356	1438.030	202.694						0.02	319.57	2	451.64
				3	594.90	−8.32886	2.25707	−3.8257	−3.7320	22.90					22.90	594.90
335	C_9H_{20}	2-methyloctane	3221-61-2	1	4.03660	1399.900	204.000						0.02	313.00	2	444.00
336	C_9H_{20}	2,2-dimethylheptane	1071-26-7	1	3.95530	1346.100	208.000						0.02	303.00	2	433.00
337	C_9H_{20}	2,2,5-trimethylhexane	3522-94-9	1	3.97372	1332.86	211.81						0.02	296.30	2	424.25
338	C_9H_{20}	2,2,3,3-tetramethylpentane	7154-79-2	1	3.95319	1397.690	213.780						0.02	306.60	2	442.00
339	C_9H_{20}	2,2,3,4-tetramethylpentane	1186-53-4	1	3.95552	1373.790	214.780						0.02	301.40	2	434.20
340	C_9H_{20}	2,2,4,4-tetramethylpentane	1070-87-7	1	3.92055	1324.65	216.08						0.02	292.79	2	423.04
341	C_9H_{20}	2,3,3,4-tetramethylpentane	16747-38-9	1	3.99105	1422.030	215.256						0.02	307.81	2	443.27
342	$C_9H_{20}O$	1-nonanol	143-08-8	1	3.83303	1297.750	125.000						0.02	382.10	2	515.58
				3	671.50	−9.91542	5.13670	−8.8075	−12.497	26.30					26.30	671.50
345	$C_{10}H_8$	naphthalene	91-20-3	1	4.13555	1733.710	201.859						0.02	368.44	2	523.40
				3	748.40	−7.61444	1.91553	−2.5075	−3.2300	40.50					40.50	748.40
351	$C_{10}H_{14}$	butylbenzene	104-51-8	1	4.10345	1575.470	201.200						0.02	343.50	2	486.20
352	$C_{10}H_{14}$	2-methylpropylbenzene	538-93-2	1	4.05978	1529.960	204.640						0.02	334.19	2	475.50
353	$C_{10}H_{14}$	1,4-diethylbenzene	105-05-5	1	4.12958	1592.590	202.440						0.02	343.95	2	486.60
354	$C_{10}H_{14}$	1-(1-methylethyl)-4-methylbenzene	99-87-6	1	4.17215	1606.890	208.570						0.02	338.28	2	479.60
355	$C_{10}H_{14}$	1,2,4,5-tetramethylbenzene	95-93-2	1	4.18329	1660.560	200.640						0.02	354.80	2	500.20
356	$C_{10}H_{18}$	cis-bicyclo[4.4.0]decane (cis-decalin)	493-01-6	1	4.00019	1594.460	203.392						0.02	349.53	2	500.79
357	$C_{10}H_{18}$	trans-bicyclo[4.4.0]decane (trans-decalin)	493-02-7	1	3.98171	1564.683	206.259						0.02	342.33	2	492.00
359	$C_{10}H_{20}O_2$	decanoic acid	334-48-5	3	726.00	−9.07060	2.77535	−11.1014	−2.43545	22.30					22.30	726.00
360	$C_{10}H_{22}$	decane	124-18-5	1	4.06853	1495.170	193.858						0.02	338.53	2	476.15
				3	617.65	−8.60643	2.44659	−4.2925	−3.9080	21.05					21.05	617.65
361	$C_{10}H_{22}$	2,2,5-trimethylheptane	20291-95-6	1	4.00345	1417.400	203.800						0.02	318.00	2	452.00
362	$C_{10}H_{22}$	3,3,5-trimethylheptane	7154-80-5	1	3.98014	1435.430	205.490						0.02	320.00	2	458.00
363	$C_{10}H_{22}$	2,2,3,3-tetramethylhexane	13475-81-5	1	3.96928	1464.03	209.06						0.02	322.38	2	463.20
364	$C_{10}H_{22}$	2,2,5,5-tetramethylhexane	1071-81-4	1	4.00614	1377.98	207.00						0.02	307.68	2	438.06

Section D Vapor Pressure Correlations Parameters (*Continued*)

No.	Formula	Name	CAS #	Eq. #	A/A/Tc	B/B/a	C/C/b	Tc/c	to/d	n/Pc	E	F	Pvpmin, bar	Tmin, K	Pvpmax bar	Tmax, K
365	$C_{10}H_{22}O$	1-decanol	112-30-1	1	3.84905	1369.000	125.078						0.02	394.80	2	533.92
				3	689.00	−9.75478	4.18634	−7.0572	−15.980	24.10					24.10	689.00
366	$C_{10}H_{24}N_4$	octamethylethenetetramine	996-70-3	3	680.00	−8.33725	2.87447	−4.08037	−3.54204	24			0.02	357.70	1.208	485.20
367	$C_{11}H_{10}$	1-methylnaphthalene	90-12-0	1	4.16082	1826.948	195.002						0.02	389.93	2	551.40
368	$C_{11}H_{10}$	2-methylnaphthalene	91-57-6	1	4.19340	1840.268	198.395						0.02	387.07	2	547.50
369	$C_{11}H_{24}$	undecane	1120-21-4	1	4.09710	1569.570	187.700						0.02	356.25	2	499.00
				3	638.85	−8.85076	2.60205	−4.7305	−4.0810	19.55					19.55	638.85
370	$C_{11}H_{24}O$	1-undecanol	112-42-5	3	705.00	−9.85733	3.97841	−6.6002	−16.691	22.40					22.40	705.00
371	$C_{12}H_{10}$	1,1′-biphenyl	92-52-4	1	4.18870	1841.480	185.150						0.02	400.77	2	561.60
				2	4.18870	1841.480	185.150	770.00	270	2.75420	0.000	0.00	1.87	623.15	31.90	841.69
374	$C_{12}H_{18}$	1,3,5-triethylbenzene	102-25-0	3	679.00	−9.35738	3.7883	−5.45184	−2.91351	24.35			0.02	371.64	2.7	534.70
375	$C_{12}H_{20}$	1,3-dimethyltricyclo [3.3.1.1^{3,7}]decane (1,3-dimethyladamantane)	702-79-4	3	708.00	−8.17338	3.28872	−3.47324	−2.48597	30.00			0.02	352.17	2.7	526.20
377	$C_{12}H_{26}$	dodecane	112-40-3	1	4.12285	1639.270	181.840						0.02	372.89	2	520.24
				3	658.65	−9.08593	2.77846	−5.1985	−4.1730	18.30					18.30	658.65
378	$C_{12}H_{26}O$	1-dodecanol	112-53-8	3	720.00	−9.91901	3.61884	−5.8537	−18.204	20.80					20.80	720.00
379	$C_{13}H_{12}$	diphenylmethane	101-81-5	1	4.18060	1862.640	181.650						0.02	408.20	2	571.70
				2	4.18060	1862.640	181.650	770.00	270	2.01000	260.720	0.00	1.876	633.15	30.70	813.15
380	$C_{13}H_{28}$	tridecane	629-50-5	1	4.13246	1690.670	174.220						0.02	388.85	2	540.19
				3	676.00	−9.32959	2.89925	−5.5550	−4.4700	17.10					17.10	676.00
381	$C_{13}H_{28}O$	1-tridecanol	112-70-9	3	734.00	−9.99402	3.36986	−5.4865	−18.592	19.35					19.35	734.00
382	$C_{14}H_{10}$	phenanthrene	85-01-8	1	4.37081	2329.540	195.280						0.02	461.60	2	650.00
383	$C_{14}H_{10}$	anthracene	120-12-7	1	4.79891	2819.630	247.020						0.02	460.00	2	653.00
384	$C_{14}H_{22}$	1,4-di(trimethylmethyl)benzene (*p*-ditertbutylbenzene)	1012-72-2	3	708.00	−9.28468	3.89231	−5.55138	−3.34144	23			0.02	387.02	2.7	559.10
385	$C_{14}H_{30}$	tetradecane	629-59-4	1	4.13790	1740.880	167.720						0.02	403.69	2	559.15
				3	693.00	−9.54470	3.06637	−6.0070	−4.5300	16.10					16.10	693.00
386	$C_{14}H_{30}O$	1-tetradecanol	112-72-1	3	747.00	−10.13519	3.27661	−5.3447	−18.711	18.10					18.10	747.00
387	$C_{15}H_{32}$	pentadecane	629-62-9	1	4.14849	1789.950	161.380						0.02	417.80	2	576.90
				3	708.00	−9.80239	3.29217	−6.5317	−4.5840	15.15					15.15	708.00
388	$C_{15}H_{32}O$	1-pentadecanol	629-76-5	3	759.00	−10.32431	3.32013	−5.4784	−18.263	17.00					17.00	759.00
389	$C_{16}H_{34}$	hexadecane	544-76-3	1	4.15357	1830.510	154.450						0.02	431.47	2	593.80
				3	722.00	−10.03664	3.41426	−6.8627	−4.8630	14.35					14.35	722.00
390	$C_{16}H_{34}$	2,2,4,4,6,8,8-heptamethylnonane	4390-04-9	3	693.00	−8.90870	2.27470	−3.6490	−6.6600	15.70					15.70	693.00
391	$C_{16}H_{34}O$	1-hexadecanol	4485-13-6	3	770.00	−10.54087	3.47260	−6.0770	−15.939	16.10					16.10	770.00
392	$C_{17}H_{36}$	heptadecane	629-78-7	1	4.13920	1865.100	149.200						0.02	443.50	2	610.00
				3	735.00	−10.23600	3.54177	−7.1898	−5.0000	13.70					13.70	735.00
393	$C_{17}H_{36}O$	1-heptadecanol	1454-85-9	3	780.00	−10.73125	3.55515	−6.3591	−15.696	15.00					15.00	780.00
397	$C_{18}H_{38}$	octadecane	593-45-3	1	4.12710	1894.300	143.300						0.02	455.00	2	625.00
				3	746.00	−10.47230	3.69655	−7.5779	−5.1090	13.00					13.00	746.00
398	$C_{18}H_{38}O$	1-octadecanol	112-92-5	3	790.00	−10.91637	3.57835	−6.6199	−15.060	14.40					14.40	790.00

399	$C_{19}H_{40}$	nonadecane	629-92-5	1	4.14020	1932.800	137.600				0.02	466.50	2	639.00
				3	758.00	−10.68217	3.98054	−8.3030	−4.9950	12.30			12.30	758.00
400	$C_{19}H_{40}O$	1-nonadecanol	1454-84-8	3	799.00	−11.22657	4.03454	−7.7867	−11.970	13.80			13.80	799.00
401	$C_{20}H_{42}$	eicosane	112-95-8	1	4.27710	2032.700	132.100				0.02	481.10	2	652.00
				3	769.00	−10.97958	4.25588	−8.9573	−5.0430	11.60			11.60	769.00
402	$C_{20}H_{42}O$	1-eicosanol	629-96-9	3	809.00	−11.23154	3.66900	−7.0775	−14.321	13.00			13.00	809.00
407	ClD	deuterium chloride	7698-05-7	1	4.06086	668.2000	249.499				0.15	160.44	2	201.37
408	$ClFO_3$	perchloryl fluoride	7616-94-6	1	4.02009	791.7270	243.880				0.02	167.80	2	242.15
409	ClF_5	chlorine pentafluoride	13637-63-3	1	3.39423	653.0600	206.60				0.02	194.80	2	277.68
410	ClH	hydrogen chloride	7647-01-0	1	4.29490	745.7800	258.88				0.15	159.97	2	201.00
411	ClH_4N	ammonium chloride	12125-02-9	1	6.48060	3703.700	232.00				0.02	494.00	2	640.5.0
412	ClNO	nitrogen oxychloride	2696-92-6	1	4.48644	1094.730	249.70				0.04	209.40	2	285.01
413	Cl_2	chlorine	7782-50-5	1	4.06280	861.3400	246.33				0.02	176.31	2	255.79
414	DH	deuterium hydride	13983-20-5	1	3.14102	77.13490	275.62				0.08	15.73	2	24.69
415	DI	deuterium iodide	14104-45-1	1	2.72964	414.6800	187.87				0.40	217.80	2	256.03
416	D_2	deuterium	7782-39-0	1	3.14102	77.13490	275.62				0.02	15.20	2	18.97
417	D_2	deuterium, normal	800000-54-8	1	3.25315	83.52510	275.22				0.10	17.57	2	26.23
418	D_2O	deuterium oxide	7789-20-0	1	5.04327	1616.760	219.54				0.20	335.17	2	394.54
420	D_3N	trideuteroammonia	13550-49-7	1	4.61234	966.2260	240.80				0.05	195.75	2	256.46
422	FH	hydrogen fluoride	7664-39-3	1	4.80588	1475.600	287.88				0.02	212.10	2	312.83
423	FNO_2	nitrogen dioxyfluoride	10022-50-1	1	3.95830	654.55	238.00				0.02	151.00	2	214.12
424	F_2	fluorine	7782-41-4	1	3.89078	304.3500	266.54				0.02	61.00	2	91.39
428	F_2O	oxygen difluoride	7783-41-7	1	4.36109	545.0500	269.91				0.02	93.10	2	137.40
430	F_3N	nitrogen trifluoride	7783-54-2	1	3.90456	501.9130	216.00				0.02	146.72	2	196.43
433	F_4S	sulfur tetrafluoride	7783-60-0	1	3.96440	823.4000	248.00				0.02	170.00	2	249.92
435	F_6S	sulfur hexafluoride	2551-62-4	1	5.54090	1096.500	262.00				0.02	162.00	2	220.40
436	F_6U	uranium hexafluoride	7783-81-5	3	503.35	−7.37599	1.8001	−2.69686	−3.13299	45.31			45.31	503.35
437	HI	hydrogen iodide	10034-85-2	1	2.69803	405.3300	186.13				0.40	217.90	2	256.12
438	H_2	hydrogen	1333-74-0	1	2.93954	66.79540	275.65				0.05	10.25	2	22.82
439	H_2	hydrogen, normal	800000-51-5	1	2.94928	67.50780	275.70				0.05	13.33	2	22.94
440	H_2O	water	7732-18-5	1	5.11564	1687.537	230.17				0.01	273.20	16	473.20
				3	647.300	−7.77224	1.45684	−2.71942*	−1.41336*		0.01	273.20	221	647.30
441	H_2S	hydrogen sulfide	7783-06-4	1	4.22882	806.9330	251.39				0.20	185.51	2	227.20
442	H_2S_2	dihydrogen disulfide	13465-07-1	1	4.05500	1199.000	225.00				0.02	256.00	2	367.55
443	H_2S_3	dihydrogen trisulfide	13845-23-3	1	3.93200	1488.000	209.00				0.02	328.00	2	473.96
444	H_2S_4	dihydrogen tetrasulfide	13845-25-5	1	4.07000	1772.000	196.00				0.02	384.00	2	547.30
445	H_2S_5	dihydrogen pentasulfide	13845-24-4	1	4.44500	2104.000	189.00				0.02	426.00	2	591.88
446	H_2Se	hydrogen selenide	7783-07-5	1	4.76030	927.6000	240.00				0.02	177.00	2	213.00
447	H_3N	ammonia	7664-41-7	1	4.48540	926.1320	240.17				0.05	193.03	2	254.31
				3	405.500	−7.28322	1.5716	−1.85672	−2.39312	113.530			113.5	405.50
448	H_3P	phosphine	7803-51-2	1	3.84049	645.5120	256.07				0.03	137.44	2	199.46
449	H_4N_2	hydrazine	302-01-2	1	4.92680	1679.07	227.70				0.02	298.90	2	408.43

Section D Vapor Pressure Correlations Parameters (*Continued*)

No.	Formula	Name	CAS #	Eq. #	A/A/Tc	B/B/a	C/C/b	Tc/c	to/d	n/Pc	E	F	Pvpmin, bar	Tmin, K	Pvpmax bar	Tmax, K
450	He	helium	7440-59-7	1	1.68360	8.15480	273.71						0.02	1.85	2	5.34
451	He	helium-3	14762-55-1	1	1.39750	5.59400	273.84						0.02	1.12	2	4.41
452	I2	iodine	7553-56-2	1	4.14310	1611.900	205.18						0.15	392.49	2	487.51
453	Kr	krypton	7439-90-9	1	3.75560	416.3800	264.45						0.50	111.34	2	129.23
454	NO	nitrogen monoxide (nitric oxide)	10102-43-9	1	5.86790	682.9386	268.27						0.15	106.94	2	127.56
455	N_2	nitrogen	7727-37-9	1	3.61947	255.68	266.55						0.08	60.81	2	83.65
				3	126.20	−6.11102	1.2189	−0.69366	−1.89893	34.00					34.00	126.20
456	N_2O	dinitrogen oxide (nitrous oxide)	10024-97-2	1	4.12884	654.2600	247.16						0.80	180.82	2	196.91
457	N_2O_4	dinitrogen tetroxide (nitrogen dioxide)	10544-72-6	1	4.50989	1185.722	234.18						0.10	254.17	2	320.69
458	Ne	neon	7440-01-9	1	3.20934	78.38000	270.55						0.40	24.33	2	29.55
460	O_2	oxygen	7782-44-7	1	3.81634	319.0130	266.70						0.02	64.29	2	97.20
461	O_2S	sulfur dioxide	7446-09-5	1	4.40720	999.9000	237.19						0.02	199.71	2	279.47
462	O_3	ozone	10028-15-6	1	3.96200	552.5000	251.00						0.02	120.00	2	173.07
463	O_3S	sulfur trioxide	7446-11-9	1	6.17575	1735.310	236.50						0.15	284.50	2	332.04
464	Rn	radon	10043-92-2	1	4.62040	884.4100	255.00						0.02	158.00	2	222.90
465	S	sulfur	7704-34-9	1	3.96853	2500.120	186.30						0.02	527.98	2	768.55
466	Se	selenium	7782-49-2	1	4.75650	4213.000	202.00						0.02	724.00	2	1017.00
468	Xe	xenon	7440-63-3	1	3.76779	566.2820	258.66						0.60	156.43	2	177.84

*For water the exponents on the last two terms are 3 and 6.

APPENDIX B
LENNARD-JONES POTENTIALS AS DETERMINED FROM VISCOSITY DATA†

	Substance	b_0,‡ cm^3/g-mol	σ, Å	ϵ/k, K
Ar	Argon	56.08	3.542	93.3
He	Helium	20.95	2.551§	10.22
Kr	Krypton	61.62	3.655	178.9
Ne	Neon	28.30	2.820	32.8
Xe	Xenon	83.66	4.047	231.0
Air	Air	64.50	3.711	78.6
AsH_3	Arsine	89.88	4.145	259.8
BCl_3	Boron chloride	170.1	5.127	337.7
BF_3	Boron fluoride	93.35	4.198	186.3
$B(OCH_3)_3$	Methyl borate	210.3	5.503	396.7
Br_2	Bromine	100.1	4.296	507.9
CCl_4	Carbon tetrachloride	265.5	5.947	322.7
CF_4	Carbon tetrafluoride	127.9	4.662	134.0
$CHCl_3$	Chloroform	197.5	5.389	340.2
CH_2Cl_2	Methylene chloride	148.3	4.898	356.3
CH_3Br	Methyl bromide	88.14	4.118	449.2
CH_3Cl	Methyl chloride	92.31	4.182	350
CH_3OH	Methanol	60.17	3.626	481.8
CH_4	Methane	66.98	3.758	148.6
CO	Carbon monoxide	63.41	3.690	91.7
COS	Carbonyl sulfide	88.91	4.130	336.0
CO_2	Carbon dioxide	77.25	3.941	195.2
CS_2	Carbon disulfide	113.7	4.483	467
C_2H_2	Acetylene	82.79	4.033	231.8
C_2H_4	Ethylene	91.06	4.163	224.7
C_2H_6	Ethane	110.7	4.443	215.7
C_2H_5Cl	Ethyl chloride	148.3	4.898	300
C_2H_5OH	Ethanol	117.3	4.530	362.6
C_2N_2	Cyanogen	104.7	4.361	348.6
CH_3OCH_3	Methyl ether	100.9	4.307	395.0
CH_2CHCH_3	Propylene	129.2	4.678	298.9

Substance		b_0,‡ $cm^3/g\text{-mol}$	σ, Å	ϵ/k, K
CH_3CCH	Methylacetylene	136.2	4.761	251.8
C_3H_6	Cyclopropane	140.2	4.807	248.9
C_3H_8	Propane	169.2	5.118	237.1
$n\text{-}C_3H_7OH$	*n*-Propyl alcohol	118.8	4.549	576.7
CH_3COCH_3	Acetone	122.8	4.600	560.2
CH_3COOCH_3	Methyl acetate	151.8	4.936	469.8
$n\text{-}C_4H_{10}$	*n*-Butane	130.0	4.687	531.4
$iso\text{-}C_4H_{10}$	Isobutane	185.6	5.278	330.1
$C_2H_5OC_2H_5$	Ethyl ether	231.0	5.678	313.8
$CH_3COOC_2H_5$	Ethyl acetate	178.0	5.205	521.3
$n\text{-}C_5H_{12}$	*n*-Pentane	244.2	5.784	341.1
$C(CH_3)_4$	2,2-Dimethylpropane	340.9	6.464	193.4
C_6H_6	Benzene	193.2	5.349	412.3
C_6H_{12}	Cyclohexane	298.2	6.182	297.1
$n\text{-}C_6H_{14}$	*n*-Hexane	265.7	5.949	399.3
Cl_2	Chlorine	94.65	4.217	316.0
F_2	Fluorine	47.75	3.357	112.6
HBr	Hydrogen bromide	47.58	3.353	449
HCN	Hydrogen cyanide	60.37	3.630	569.1
HCl	Hydrogen chloride	46.98	3.339	344.7
HF	Hydrogen fluoride	39.37	3.148	330
HI	Hydrogen iodide	94.24	4.211	288.7
H_2	Hydrogen	28.51	2.827	59.7
H_2O	Water	23.25	2.641	809.1
H_2O_2	Hydrogen peroxide	93.24	4.196	289.3
H_2S	Hydrogen sulfide	60.02	3.623	301.1
Hg	Mercury	33.03	2.969	750
$HgBr_2$	Mercuric bromide	165.5	5.080	686.2
$HgCl_2$	Mercuric chloride	118.9	4.550	750
HgI_2	Mercuric iodide	224.6	5.625	695.6
I_2	Iodine	173.4	5.160	474.2
NH_3	Ammonia	30.78	2.900	558.3
NO	Nitric oxide	53.74	3.492	116.7
NOCl	Nitrosyl chloride	87.75	4.112	395.3
N_2	Nitrogen	69.14	3.798	71.4
N_2O	Nitrous oxide	70.80	3.828	232.4
O_2	Oxygen	52.60	3.467	106.7
PH_3	Phosphine	79.63	3.981	251.5
SF_6	Sulfur hexafluoride	170.2	5.128	222.1
SO_2	Sulfur dioxide	87.75	4.112	335.4
SiF_4	Silicon tetrafluoride	146.7	4.880	171.9
SiH_4	Silicon hydride	85.97	4.084	207.6
$SnBr_4$	Stannic bromide	329.0	6.388	563.7
UF_6	Uranium hexafluoride	268.1	5.967	236.8

†R. A. Svehla, *NASA Tech. Rep.* R-132, Lewis Research Center, Cleveland, Ohio, 1962.
‡$b_0 = \frac{2}{3}\pi N_0 \sigma^3$, where N_0 is Avogadro's number.
§The parameter σ was determined by quantum-mechanical formulas.

APPENDIX C

GROUP CONTRIBUTIONS FOR MULTIPROPERTY METHODS

This appendix contains the group definitions, correlating equations and parameter values of group contribution methods for pure component property estimation that are found in more than one chapter. Specifically, basic equations, tables of group identities, example molecules and values for contributions to properties are given for the Constantinou and Gani (1994; 1995) Nielsen (1998) and Joback (1984, 1987) methods. The Constantinou/Gani table has been assembled with additional assistance of Dr. Jens Abildskov, Department of Chemical Engineering, Technical University of Denmark, Lyngby, DK-2800, including use of the computer software of ProPred in the ICAS suite of the Computer Aided Process Engineering Center (CAPEC) jointly led by Drs. Rafiqul Gani and Sten Bay Jørgensen of the Technical University of Denmark (http://www.capec.kt.dtu.dk). The Joback table was assembled with the assistance of Dr. K. G. Joback, especially using the computer software of CRANIUM provided by Molecular Knowledge Systems, Inc. Bedford, NH, 03110 (http://www.molknow.com). The authors are grateful to all of these individuals and organizations for their help.

The symbol * has been included with the group formula if it is not present in the substances of the data base in Appendix A and therefore has not been directly tested in the methods of Chaps. 2, 3, and 4. Results using such groups are included in summaries of original and later sources. Many small molecules with two or fewer carbon atoms are treated as single groups. It is possible to form some of them from the groups listed; comparisons have been made when this was possible, but the results were often poor. However, since property values for most small molecules are in Appendix A, it is not recommended that they be predicted anyway. If no contribution was assigned for a group in a method, the symbol X is used in this appendix.

C-1 JOBACK PROPERTY FUNCTIONS FROM GROUP CONTRIBUTIONS

For a molecule with N_i groups, a property denoted as in the main text and Appendix A is given by the equation below with values of the individual parameters in Table C-1. T_b is the normal boiling temperature in Kelvins; accuracy is much greater if the experimental value is used instead of an estimation (See Chap. 2). N_{atoms} is the total number of atoms in the molecule.

TABLE C-1 Joback Group Contributions for Various Properties

Property	*tfpk*	*tbk*	*tck*	*pck*	*vck*	*hfk*	*gfk*	*hvk*	*hmk*	*CpAk*	*CpBk*	*CpCk*	*CpDk*
Units	K	K	K	bar	cm^3 mol^{-1}	kJ mol^{-1}	kJ mol^{-1}	kJ mol^{-1}	kJ mol^{-1}	J mol^{-1} K^{-1}	J mol^{-1} K^{-1}	J mol^{-1} K^{-1}	J mol^{-1} K^{-1}
Group k													
CH_3 (1)	−5.10	23.58	0.0141	−0.0012	65	−76.45	−43.96	567	217	19.500	−8.08E-03	1.53E-04	−9.67E-08
CH_2 (2)	11.27	22.88	0.0189	0.0000	56	−20.64	8.42	532	619	−0.909	9.50E-02	−5.44E-05	1.19E-08
CH (3)	12.64	21.74	0.0164	0.0020	41	29.89	58.36	404	179	−23.000	2.04E-01	−2.65E-04	1.20E−07
C (4)	46.43	18.25	0.0067	0.0043	27	82.23	116.02	152	−349	−66.200	4.27E-01	−6.41E-04	3.01E-07
$=CH_2$ (1)	−4.32	18.18	0.0113	−0.0028	56	−9.63	3.77	412	−113	−23.600	−3.81E-02	1.72E-04	−1.03E-07
=CH (2)	8.73	24.96	0.0129	−0.0006	46	37.97	48.53	527	643	−8.000	1.05E-01	−9.63E-05	3.56E-08
=C (3)	11.14	24.14	0.0117	0.0011	38	83.99	92.36	511	732	−28.100	2.08E-01	−3.06E-04	1.46E-07
=C= (2)	17.78	26.15	0.0026	0.0028	36	142.14	136.70	636	1128	27.400	−5.57E-02	1.01E-04	−5.02E-08
≡CH (1)	−11.18	9.20	0.0027	−0.0008	46	79.30	77.71	276	555	24.500	−2.71E-02	1.11E-04	−6.78E-08
≡C (2)	64.32	27.38	0.0020	0.0016	37	115.51	109.82	789	992	7.870	2.01E-02	−8.33E-06	1.39E-09
CH_2(ss) (2)	7.75	27.15	0.0100	0.0025	48	−26.80	−3.68	573	117	−6.030	8.54E-02	−8.00E-06	−1.80E-08
CH(ss) (3)	19.88	21.78	0.0122	0.0004	38	8.67	40.99	464	775	8.670	1.62E-01	−1.60E-04	6.24E-08
C(ss) (4)	60.15	21.32	0.0042	0.0061	27	79.72	87.88	154	−328	−90.900	5.57E-01	−9.00E-04	4.69E-07
=CH(ds) (2)	8.13	26.73	0.0082	0.0011	41	2.09	11.30	608	263	−2.140	5.74E-02	−1.64E-06	−1.59E-08
=C(ds) (3)	37.02	31.01	0.0143	0.0008	32	46.43	54.05	731	572	−8.250	1.01E-01	−1.42E-04	6.78E-08
F (1)	−15.78	−0.03	0.0111	−0.0057	27	−251.92	−247.19	−160	334	26.500	−9.13E-02	1.91E-04	−1.03E-07
Cl (1)	13.55	38.13	0.0105	−0.0049	58	−71.55	−64.31	1083	601	33.300	−9.63E-02	1.87E-04	−9.96E-08
Br (1)	43.43	66.86	0.0133	0.0057	71	−29.48	−38.06	1573	861	28.600	−6.49E-02	1.36E-04	−7.45E-08

I (1)*	41.69	93.84	0.0068	−0.0034	97	21.06	5.74	2275	651	32.100	−6.41E-02	1.26E-04	−6.87E-08
OH (1)	44.45	92.88	0.0741	0.0112	28	−208.04	−189.20	4021	575	25.700	−6.91E-02	1.77E-04	−9.88E-08
ACOH (1)	82.83	76.34	0.0240	0.0184	−25	−221.65	−197.37	2987	1073	−2.810	1.11E-01	−1.16E-04	4.94E-08
O (2)	22.23	22.42	0.0168	0.0015	18	−132.22	−105.00	576	284	25.500	−6.32E-02	1.11E-04	−5.48E-08
O(ss) (2)	23.05	31.22	0.0098	0.0048	13	−138.16	−98.22	1119	1405	12.200	−1.26E-02	6.03E-05	−3.86E-08
C=O (2)	61.20	76.75	0.0380	0.0031	62	−133.22	−120.50	2144	1001	6.450	6.70E-02	−3.57E-05	2.86E-09
C=O (ss) (2)	75.97	94.97	0.0284	0.0028	55	−164.50	−126.27	1588	X	30.400	−8.29E-02	2.36E-04	−1.31E-07
CH=O (1)*	36.90	72.20	0.0379	0.0030	82	−162.03	−143.48	2173	764	30.900	−3.36E-02	1.60E-04	−9.88E-08
COOH (1)	155.50	169.09	0.0791	0.0077	89	−426.72	−387.87	4669	2641	24.100	4.27E-02	8.04E-05	−6.87E-08
COO (2)	53.60	81.10	0.0481	0.0005	82	−337.92	−301.95	2302	1663	24.500	4.02E-02	4.02E-05	−4.52E-08
=O (1)*	2.08	−10.50	0.0143	0.0101	36	−247.61	−250.83	1412	866	6.820	1.96E-02	1.27E-05	−1.78E-08
NH_2 (1)	66.89	73.23	0.0243	0.0109	38	−22.02	14.07	2578	840	26.900	−4.12E-02	1.64E-04	−9.76E-08
NH (2)	52.66	50.17	0.0295	0.0077	35	53.47	89.39	1538	1197	−1.210	7.62E-02	−4.86E-05	1.05E-08
NH (ss) (2)	101.51	52.82	0.0130	0.0114	29	31.65	75.61	1656	1790	11.800	−2.30E-02	1.07E-04	−6.28E-08
N (3)	48.84	11.74	0.0169	0.0074	9	123.34	163.16	453	1124	−31.100	2.27E-01	−3.20E-04	1.46E-07
=N− (2)	X	74.60	0.0255	−0.0099	X	23.61	X	797	X	X	X	X	X
=N− (ds) (2)	68.40	57.55	0.0085	0.0076	34	55.52	79.93	1560	872	8.830	−3.84E-03	4.35E-05	−2.60E-08
=NH (1)*	X	X	X	X	X	93.70	119.66	2908	X	5.690	−4.12E-03	1.28E-04	−8.88E-08
CN (1)*	59.89	125.66	0.0496	−0.0101	91	88.43	89.22	3071	577	36.500	−7.33E-02	1.84E-04	−1.03E-07
NO_2 (1)	127.24	152.54	0.0437	0.0064	91	−66.57	−16.83	4000	2313	25.900	−3.74E-03	1.29E-04	−8.88E-08
SH (1)	20.09	63.56	0.0031	0.0084	63	−17.33	−22.99	1645	564	35.300	−7.58E-02	1.85E-04	−1.03E-07
S (2)	34.40	68.78	0.0119	0.0049	54	41.87	33.12	1629	987	19.600	−5.61E-03	4.02E-05	−2.76E-08
S(ss) (2)	79.93	52.10	0.0019	0.0051	38	39.10	27.76	1430	372	16.700	4.81E-03	2.77E-05	−2.11E-08

*Group could not be tested with substances available in App. A.

Property	Function
T_{fp}	$T_{fp} = 122 + \sum_k N_k(tfpk)$
T_b	$T_b = 198 + \sum_k N_k(tbk)$
T_c	$T_c = T_b \left[0.584 + 0.965 \left\{\sum_k N_k(tck)\right\} - \left\{\sum_k N_k(tck)\right\}^2\right]^{-1}$
P_c	$P_c = \left[0.113 + 0.0032N_{atoms} - \sum_k N_k(pck)\right]^{-2}$
V_c	$V_c = 17.5 + \sum_k N_k(vck)$
ΔH_f°	$\Delta H_f^\circ = 68.29 + \sum_k N_k(hfk)$
ΔG_f°	$\Delta G_f^\circ = 53.88 + \sum_k N_k(gfk)$
ΔH_v	$\Delta H_v = 15.30 + \sum_k N_k(hvk)$
ΔH_m	$\Delta H_m = -0.88 + \sum_k N_k(hmk)$
C_p°	$C_p^\circ = \left[\sum_k N_k(CpAk) - 37.93\right] + \left[\sum_k N_k(CpBk) + 0.21\right] T + \left[\sum_k N_k(CpCk) - 3.91\text{E}{-04}\right] T^2 + \left[\sum_k N_k(CpDk) - 2.06\text{E}{-07}\right] T^3 .$

The headings of Table C-1 are quantities in the summations of the above formulae. The number of other groups that each group is bonded to is given in parenthesis. Thus, $=CH_2$ is bonded to 1 other group (which must be $=CH_2$, $=CH$ or $=C$), $=CH$ is bonded to 2 groups (one of which must be $=CH_2$, $=CH$ or $=C$) , and $=C$ is bonded to 3 groups (one of which must be $=CH_2$, $=CH$ or $=C$). The symbol (ss) indicates a group in a nonaromatic ring; (ds) indicates a group in an aromatic ring. Note that particular units have been included for most properties. ΔH_f° and ΔG_f° are at 298.15 K and 1 atm.

C-2 CONSTANTINOU/GANI PROPERTY FUNCTIONS FROM GROUP CONTRIBUTIONS

First-Order groups are indicated by k, Second-Order groups by j. For a molecule with N_i First-Order groups and M_j Second-Order groups, a property is given by the formula below with values of the individual parameters in Tables C-2 (First Order) and C-3 (Second Order). When doing only a First-Order calculation, set $W = 0$; to include Second-Order groups, set $W = 1$.

Property	Function
T_{fp}	$T_{fp} = 102.425 \ln \left[\sum_k N_k(tfp1k) + W \sum_j M_j(tfp2j) \right]$
T_b	$T_b = 204.359 \ln \left[\sum_k N_k(tb1k) + W \sum_j M_j(tb2j) \right]$
T_c	$T_c = 181.128 \ln \left[\sum_k N_k(tc1k) + W \sum_j M_j(tc2j) \right]$
P_c	$P_c = \left[\sum_k N_k(pc1k) + W \sum_j M_j(pc2j) + 0.10022 \right]^{-2} + 1.3705$
V_c	$V_c = -0.00435 + \left[\sum_k N_k(vc1k) + W \sum_j M_j(vc2j) \right]$
ω	$\omega = 0.4085 \left\{ \ln \left[\sum_k N_k(w1k) + W \sum_j M_j(w2j) + 1.1507 \right] \right\}^{(1/0.5050)}$
ΔG_f°	$\Delta G_f^\circ = -14.83 + \left[\sum_k N_k(gf1k) + W \sum_j M_j(gf2j) \right]$
ΔH_f°	$\Delta H_f^\circ = 10.835 + \left[\sum_k N_k(hf1k) + W \sum_j M_j(hf2j) \right]$
ΔH_{v298}	$\Delta H_{v298} = 6.829 + \left[\sum_k N_k(hv1k) + W \sum_j M_j(hv2j) \right]$
$V_{liq(298)}$	$V_{liq} = -0.00435 + \left[\sum_k N_k(v_{liq}1k) + W \sum_j M_j(v_{liq}2j) \right]$
C_p°	$C_p^\circ = \left[\sum_k N_k(CpA1k) + W \sum_j M_j(CpA2j) - 19.7779 \right]$ $+ \left[\sum_\lambda N_k(CpB1k) + W \sum_j M_j(CpB2j) + 22.5981 \right] \theta$ $+ \left[\sum_k N_k(CpC1k) + W \sum_j M_j(CpC2j) - 10.7983 \right] \theta^2$ $\theta = (T - 298)/700$

The headings of Tables C-2 and C-3 are quantities in the summations of the above formulae. The number of other groups that each group is bonded to is given in parenthesis. Thus, $CH_2{=}CH$ is bonded to 1 other group, $CH{=}CH$ and $CH_2{=}C$ to 2 groups, and $CH{=}C$ to 3 groups. AC is an aromatic carbon. F*Special* is any fluorine not found in another group. Note that particular units have been included

TABLE C-2 First-Order Constantinou/Gani Group Contributions for Various Properties

Property	*tfp1k*	*tb1k*	*tc1k*	*pc1k*	*vc1k*	*w1k*	*hf1k*	*gf1k*	*hv1k*	$v_{liq}1k$	*CpA1k*	*CpB1k*	*CpC1k*
Units	K	K	K	$bar^{1/2}$	m^3 $kmol^{-1}$		kJ mol^{-1}	kJ mol^{-1}	kJ mol^{-1}	cm^3 mol^{-1}	J mol^{-1} K^{-1}	J mol^{-1} K^{-1}	J mol^{-1} K^{-1}
CH_3 (1)	0.4640	0.8894	1.6781	0.0199	0.0750	0.296	−45.947	−8.030	4.116	0.0261	35.1152	39.5923	−9.9232
CH_2 (2)	0.9246	0.9225	3.4920	0.0106	0.0558	0.147	−20.763	8.231	4.650	0.0164	22.6346	45.0933	−15.7033
CH (3)	0.3557	0.6033	4.0330	0.0013	0.0315	−0.071	−3.766	19.848	2.771	0.0071	8.9272	59.9786	−29.5143
C (4)	1.6479	0.2878	4.8823	−0.0104	−0.0003	−0.351	17.119	37.977	1.284	−0.0038	0.3456	74.0368	−45.7878
CH_2=CH (1)	1.6472	1.7827	5.0146	0.0250	0.1165	0.408	53.712	84.926	6.714	0.0373	49.2506	59.3840	−21.7908
CH=CH (2)	1.6322	1.8433	7.3691	0.0179	0.0954	0.252	69.939	92.900	7.370	0.0269	35.2248	62.1924	−24.8156
CH_2=C (2)	1.7899	1.7117	6.5081	0.0223	0.0918	0.223	64.145	88.402	6.797	0.0270	37.6299	62.1285	−26.0637
CH=C (3)	2.0018	1.7957	8.9582	0.0126	0.0733	0.235	82.528	93.745	8.178	0.0161	21.3528	66.3947	−29.3703
C=C (4)	5.1175	1.8881	11.3764	0.0020	0.0762	−0.210	104.293	116.613	9.342	0.0030	10.2797	65.5372	−30.6057
CH_2=C=CH(1)	3.3439	3.1243	9.9318	0.0313	0.1483	0.152	197.322	221.308	12.318	0.0434	66.0574	69.3936	−25.1081
ACH (2)	1.4669	0.9297	3.7337	0.0075	0.0422	0.027	11.189	22.533	4.098	0.0132	16.3794	32.7433	−13.1692
AC (3)	0.2098	1.6254	14.6409	0.0021	0.0398	0.334	27.016	30.485	12.552	0.0044	10.4283	25.3634	−12.7283
$ACCH_3$ (2)	1.8635	1.9669	8.2130	0.0194	0.1036	0.146	−19.243	22.505	9.776	0.0289	42.8569	65.6464	−21.0670
$ACCH_2$ (3)	0.4177	1.9478	10.3239	0.0122	0.1010	−0.088	9.404	41.228	10.185	0.0192	32.8206	70.4153	−28.9361
ACCH (4)	−1.7567	1.7444	10.4664	0.0028	0.0712	1.524	27.671	52.948	8.834	0.0099	19.9504	81.8764	−40.2864
OH (1)	3.5979	3.2152	9.7292	0.0051	0.0390	0.737	−181.422	−158.589	24.529	0.0055	27.2107	2.7609	1.3060
ACOH (3)	13.7349	4.4014	25.9145	−0.0074	0.0316	1.015	−164.609	−132.097	40.246	0.0113	39.7712	35.5676	−15.5875
CH_3CO (1)	4.8776	3.5668	13.2896	0.0251	0.1340	0.633	−182.329	−131.366	18.999	0.0365	59.3032	67.8149	−20.9948
CH_2CO (2)	5.6622	3.8967	14.6273	0.0178	0.1119	0.963	−164.410	−132.386	20.041	0.0282	X	X	X
CHO (1)*	4.2927	2.8526	10.1986	0.0141	0.0863	1.133	−129.2	−107.858	12.909	0.0200	40.7501	19.6990	−5.4360
CH_3COO (1)	4.0823	3.6360	12.5965	0.0290	0.1589	0.756	−389.737	−318.616	22.709	0.0450	66.8423	102.4553	−43.3306
CH_2COO (2)	3.5572	3.3953	13.8116	0.0218	0.1365	0.765	−359.258	−291.188	17.759	0.0357	X	X	X
HCOO (1)	4.2250	3.1459	11.6057	0.0138	0.1056	0.526	−332.822	−288.902	X	0.0267	51.5048	44.4133	−19.6155
CH_3O (1)	2.9248	2.2536	6.4737	0.0204	0.0875	0.442	−163.569	−105.767	10.919	0.0327	50.5604	38.9681	−4.7799
CH_2O (2)	2.0695	1.6249	6.0723	0.0151	0.0729	0.218	−151.143	−101.563	7.478	0.0231	39.5784	41.8177	−11.0837

CH−O (3)	4.0352	1.1557	5.0663	0.0099	0.0587	0.509	−129.488	−92.099	5.708	0.0180	25.6750	24.7281	4.2419
FCH_2O (1)*	4.5047	2.5892	9.5059	0.0090	0.0686	0.800	−140.313	−90.883	11.227	0.0206	X	X	X
CH_2NH_2 (1)	6.7684	3.1656	12.1726	0.0126	0.1313	X	−15.505	58.085	14.599	0.0265	57.6861	64.0768	−21.0480
$CHNH_2$ (2)	4.1187	2.5983	10.2075	0.0107	0.0753	0.953	3.320	63.051	11.876	0.0195	44.1122	77.2155	−33.5086
CH_3NH (2)	4.5341	3.1376	9.8544	0.0126	0.1215	0.550	5.432	82.471	14.452	0.0267	53.7012	71.7948	−22.9685
CH_2NH (3)	6.0609	2.6127	10.4677	0.0104	0.0996	0.386	23.101	95.888	14.481	0.0232	44.6388	68.5041	−26.7106
CHNH (4)*	3.4100	1.5780	7.2121	−0.0005	0.0916	0.384	26.718	85.001	X	0.0181	X	X	X
CH_3N (2)	4.0580	2.1647	7.6924	0.0159	0.1260	0.075	54.929	128.602	6.947	0.0191	41.4064	85.0996	−35.6318
CH_2N (3)	0.9544	1.2171	5.5172	0.0049	0.0670	0.793	69.885	132.756	6.918	0.0168	30.1561	81.6814	−36.1441
$ACNH_2$ (2)	10.1031	5.4736	28.7570	0.0011	0.0636	X	20.079	68.861	28.453	0.0137	47.1311	51.3326	−25.0276
C_5H_4N (1)	X	6.2800	29.1528	0.0296	0.2483	X	134.062	199.958	31.523	0.0608	84.7602	177.2513	−72.3213
C_5H_3N (2)	12.6275	5.9234	27.9464	0.0257	0.1703	X	139.758	199.288	31.005	0.0524	X	X	X
CH_2CN (1)*	4.1859	5.0525	20.3781	0.0361	0.1583	1.670	88.298	121.544	23.340	0.0331	58.2837	49.6388	−15.6291
COOH (1)	11.5630	5.8337	23.7593	0.0115	0.1019	0.570	−396.242	−349.439	43.046	0.0223	46.5577	48.2322	−20.4868
CH_2Cl (1)	3.3376	2.9637	11.0752	0.0198	0.1156	X	−73.568	−33.373	13.780	0.0337	48.4648	37.2370	−13.0635
CHCl (2)	2.9933	2.6948	10.8632	0.0114	0.1035	X	−63.795	−31.502	11.985	0.0266	36.5885	47.6004	−22.8148
CCl (3)	9.8409	2.2073	11.3959	0.0031	0.0792	0.716	−57.795	−25.261	9.818	0.0202	29.1848	52.3817	−30.8526
$CHCl_2$ (1)*	5.1638	3.9300	16.3945	0.0268	0.1695	X	−82.921	−35.814	19.208	0.0468	60.8262	41.9908	−20.4091
CCl_3 (1)	X	3.5600	X	X	X	0.617	X	X	17.574	0.0620	56.1685	46.9337	−31.3325
CCl_2 (2)	10.2337	4.5797	18.5875	0.0349	0.2103	X	−107.188	−53.332	X	X	78.6054	32.1318	−19.4033
ACCl (2)	2.7336	2.6293	14.1565	0.0131	0.1016	0.296	−16.752	−0.596	11.883	0.0241	33.6450	23.2759	−12.2406
CH_2N02 (1)*	5.5424	5.7619	24.7369	0.0210	0.1653	X	−66.138	17.963	30.644	0.0338	63.7851	83.4744	−35.1171
CHN02 (2)*	4.9738	5.0767	23.2050	0.0122	0.1423	X	−59.142	18.088	26.277	0.0262	51.1442	94.2934	−45.2029
$ACNO_2$ (2)*	8.4724	6.0837	34.5870	0.0150	0.1426	X	−7.365	60.161	X	0.0250	X	X	X
CH_2SH (1)	3.0044	3.2914	13.8058	0.0136	0.1025	X	−8.253	16.731	14.931	0.0345	58.2445	46.9958	−10.5106
I (1)*	4.6089	3.6650	17.3947	0.0028	0.1081	0.233	57.546	46.945	14.364	0.0279	29.1815	−9.7846	3.4554
Br (1)	3.7442	2.6495	10.5371	−0.0018	0.0828	0.278	1.834	−1.721	11.423	0.0214	28.0260	−7.1651	2.4332
CH≡C (1)	3.9106	2.3678	7.5433	0.0148	0.0933	0.618	220.803	217.003	7.751	X	45.9768	20.6417	−8.3297
C≡C (2)*	9.5793	2.5645	11.4501	0.0041	0.0763	X	227.368	216.328	11.549	0.0145	26.7371	21.7676	−6.4481
Cl—(C═C) (3)*	1.5598	1.7824	5.4334	0.0160	0.0569	X	−36.097	−28.148	X	0.0153	25.8094	−5.2241	1.4542

TABLE C-2 First-Order Constantinou/Gani Group Contributions for Various Properties (*Continued*)

Property	*tfp1k*	*tb1k*	*tc1k*	*pc1k*	*vc1k*		*hf1k*	*gf1k*	*hv1k*	$v_{liq}1k$	*CpA1k*	*CpB1k*	*CpC1k*
Units	K	K	K	bar$^{1/2}$	m^3 kmol^{-1}	*w1k*	kJ mol^{-1}	kJ mol^{-1}	kJ mol^{-1}	cm^3 mol^{-1}	J mol^{-1} K^{-1}	J mol^{-1} K^{-1}	J mol^{-1} K^{-1}
ACF (2)	2.5015	0.9442	2.8977	0.0130	0.0567	0.263	−161.740	−144.549	4.877	0.0173	30.1696	26.9738	−13.3722
$HCON(CH_2)_2$ (2)*	X	7.2644	X	X	X	0.500	X	X	X	X	X	X	X
CF_3 (1)	3.2411	1.2880	2.4778	0.0442	0.1148	X	−679.195	−626.580	8.901	X	63.2024	51.9366	−28.6308
CF_2 (2)	X	0.6115	1.7399	0.0129	0.0952	X	X	X	1.860	X	44.3567	44.5875	−23.2820
CF (3)	X	1.1739	3.5192	0.0047	X	X	X	X	8.901	X	X	X	X
COO (2)	3.4448	2.6446	12.1084	0.0113	0.0859	X	−313.545	−281.495	X	0.0192	X	X	X
CCl_2F (1)	7.4756	2.8881	9.8408	0.0354	0.1821	0.503	−258.960	−209.337	13.322	0.0538	X	X	X
HCClF (1)	X	2.3086	X	X	X	X	X	X	X	X	X	X	X
$CClF_2$ (1)	2.7523	1.9163	4.8923	0.0390	0.1475	0.547	−446.835	−392.975	8.301	0.0538	X	X	X
F*Special* (1)	1.9623	1.0081	1.5974	0.0144	0.0378	X	−223.398	−212.718	X	X	22.2082	−2.8385	1.2679
$CONH_2$ (1)*	31.2786	10.3428	65.1053	0.0043	0.1443	X	−203.188	−136.742	X	X	X	X	X
$CONHCH_3$ (1)*	X	X	X	X	X	X	−67.778	X	X	X	X	X	X
$CONHCH_2$ (1)*	X	X	X	X	X	X	−182.096	X	51.787	X	X	X	X
$CON(CH_3)_2$ (1)*	11.3770	7.6904	36.1403	0.0401	0.2503	X	−189.888	−65.642	X	0.0548	X	X	X
$CONCH_2CH_2$ (3)*	X	X	X	X	X	X	−46.562	X	X	X	X	X	X
$CON(CH_2)_2$ (3)*	X	6.7822	X	X	X	X	X	X	X	X	X	X	X
C_2H_5O2 (1)*	X	5.5566	17.9668	0.0254	0.1675	0.428	−344.125	−241.373	X	0.0410	X	X	X
C_2H_4O2 (2)	X	5.4248	X	X	X	X	X	X	X	X	X	X	X
CH_3S (1)	5.0506	3.6796	14.3969	0.0160	0.1302	X	−2.084	30.222	16.921	0.0348	57.7670	44.1238	−9.5565
CH_2S (2)	3.1468	3.6763	17.7916	0.0111	0.1165	0.438	18.022	38.346	17.117	0.0273	45.0314	55.1432	−18.7776
CHS (3)*	X	2.6812	X	X	X	0.739	X	X	13.265	X	40.5275	55.0141	−31.7190
C_4H_3S (1)	X	5.7093	X	X	X	X	X	X	27.966	X	80.3010	132.7786	−58.3241
C_4H_2S (2)*	X	5.8260	X	X	X	X	X	X	X	X	X	X	X

*Groups could not be tested with substances available in App. A.

TABLE C-3 Second-Order Constantinou/Gani Group Contributions for Various Properties

Property	$tfp2j$	$tb2j$	$tc2j$	$pc2j$	$vc2j$	$w2j$	$hf2ji$	$gf2j$	$hv2j$	$V1iq$	$CpA2j$	$CpB2j$	$CpC2j$
Units	K	K	K	$bar^{1/2}$	$m^3\ kmol^{-1}$		$kJ\ mol^{-1}$	$kJ\ mol^{-1}$	$kJ\ mol^{-1}$	$cm^3\ mol^{-1}$	$J\ mol^{-1}\ K^{-1}$	$J\ mol^{-1}\ K^{-1}$	$J\ mol^{-1}\ K^{-1}$
Group j													
$(CH_3)_2CH$	0.0381	−0.1157	−0.5334	0.000488	0.00400	0.01740	−0.860	0.297	0.292	0.00133	0.5830	−1.2002	−0.0584
$(CH_3)_3C$	−0.2355	−0.0489	−0.5143	0.001410	0.00572	0.01922	−1.338	−0.399	−0.720	0.00179	0.3226	2.1309	−1.5728
$CH(CH_3)CH(CH_3)$	0.4401	0.1798	1.0699	−0.001850	−0.00398	−0.00475	6.771	6.342	0.868	−0.00203	0.9668	−2.0762	0.3148
$CH(CH_3)C(CH_3)_2$	−0.4923	0.3189	1.9886	−0.005200	−0.01081	−0.02883	7.205	7.466	1.027	−0.00243	−0.3082	1.8969	−1.6454
$C(CH_3)_2C(CH_3)_2$	6.0650	0.7273	5.8254	−0.013230	−0.02300	−0.08632	14.271	16.224	2.426	−0.00744	−0.1201	4.2846	−2.0262
3 membered ring	1.3772	0.4745	−2.3305	0.003714	−0.00014	0.17563	104.800	94.564	X	X	8.5546	−22.9771	10.7278
4 membered ring	X	0.3563	−1.2978	0.001171	−0.00851	0.22216	99.455	92.573	X	X	3.1721	−10.0834	4.9674
5 membered ring	0.6824	0.1919	−0.6785	0.000424	−0.00866	0.16284	13.782	5.733	−0.568	0.00213	−5.9060	−1.8710	4.2945
6 membered ring	1.5656	0.1957	0.8479	0.002257	0.01636	−0.03065	−9.660	−8.180	−0.905	0.00063	−3.9682	17.7889	−3.3639
7 membered ring	6.9709	0.3489	3.6714	−0.009800	−0.02700	−0.02094	15.465	20.597	−0.847	−0.005[illegible]9	−3.2746	32.1670	−17.8246
$CH_n{=}CH_m{-}CH_p{=}CH_k$ $m, p \in (0,1), k, n \in (0,2)$	1.9913	0.1589	0.4402	0.004186	−0.00781	0.01648	−8.392	−5.505	2.057	−0.00188	2.6142	4.4511	−5.9808
$CH_3{-}CH_m{=}CH_n$ $m \in (0,1), n \in (0,2)$	0.2476	0.0668	0.0167	−0.000180	−0.00098	0.00619	0.474	0.950	−0.073	0.00009	−1.3913	−1.5496	2.5899
$CH_2{-}CH_m{=}CH_n$ $m \in (0,1), n \in (0,2)$	−0.5870	−0.1406	−0.5231	0.003538	0.00281	−0.01150	1.472	0.699	−0.369	0.00012	0.2630	−2.3428	0.8975
$CH{-}CH_m{=}CH_n$ or $C{-}CH_m{=}CH_n$* $m \in (0,1), n \in (0,2)$	−0.2361	−0.0900	−0.3850	0.005675	0.00826	0.02778	4.504	1.013	0.345	0.00142	6.5145	−17.5541	10.6977
Alicyclic side-chain $Ccyclic C_m$ $m > 1$	−2.8298	0.0511	2.1160	−0.002550	−0.01755	−0.11024	1.252	1.041	−0.114	−0.00107	4.1707	−3.1964	−1.1997
CH_3CH_3	1.4880	0.6884	2.0427	0.005175	0.00227	−0.11240	−2.792	−1.062	X	X	X	X	X
CHCHO or CCHO*	2.0547	−0.1074	−1.5826	0.003659	−0.00664	X	−2.092	−1.359	0.207	−0.00009	X	X	X
CH_3COCH_2	−0.2951	0.0224	0.2996	0.001474	−0.00510	−0.20789	0.975	0.075	−0.668	−0.00030	3.7978	−7.3251	2.5312
CH_3COCH or CH_3COC	−0.2986	0.0920	0.5018	−0.002300	−0.00122	−0.16571	4.753	X	0.071	−0.00108	X	X	X
$Ccyclic{=}0$	0.7143	0.5580	2.9571	0.003818	−0.01966	X	14.145	23.539	0.744	−0.00111	X	X	X
ACCHO*	−0.6697	0.0735	1.1696	−0.002480	0.00664	X	−3.173	−2.602	−3.410	−0.00035	X	X	X
CHCOOH or CCOOH*	−3.1034	−0.1552	−1.7493	0.004920	0.00559	0.08774	1.279	2.149	X	−0.00050	X	X	X
ACCOOH*	28.4324	0.7801	6.1279	0.000344	−0.00415	X	12.245	10.715	8.502	0.00777	−15.7667	−0.1174	6.1191
CH_3COOCH or CH_3COOC	0.4838	−0.2383	−1.3406	0.000659	−0.00293	−0.26623	−7.807	−6.208	−3.345	0.00083	X	X	X
$COCH_2COO$ or COCHCOO or COCCOO*	0.0127	0.4456	2.5413	0.001067	−0.00591	X	37.462	29.181	X	0.00036	X	X	X
$CO{-}O{-}CO$	−2.3598	−0.1977	−2.7617	−0.004880	−0.00144	0.91939	−16.097	−11.809	1.517	0.00198	−6.4072	15.2583	−8.3149

TABLE C-3 Second-Order Constantinou/Gani Group Contributions for Various Properties (*Continued*)

Property	*tfp2j*	*tb2j*	*tc2j*	*pc2j*	*vc2j*	*w2j*	*hf2ji*	*gf2j*	*hv2j*	*V1iq*	*CpA2j*	*CpB2j*	*CpC2j*
Units	K	K	K	$bar^{1/2}$	$m^3\ kmol^{-1}$		$kJ\ mol^{-1}$	$kJ\ mol^{-1}$	$kJ\ mol^{-1}$	$cm^3\ mol^{-1}$	$J\ mol^{-1}\ K^{-1}$	$J\ mol^{-1}\ K^{-1}$	$J\ mol^{-1}\ K^{-1}$
ACCOO*	−2.0198	0.0835	−3.4235	−0.000540	0.02605	X	−9.874	−7.415	X	0.00001	X	X	X
CHOH	−0.5480	−0.5385	−2.8035	−0.004390	−0.00777	0.03654	−3.887	−6.770	−1.398	−0.00092	2.4484	−0.0765	0.1460
COH	0.3189	−0.6331	−3.5442	0.000178	0.01511	0.21106	−24.125	−20.770	0.320	0.00175	−1.5252	−7.6380	8.1795
$CH_m(OH)CH_n(OH)$* $m, n \in (0,2)$	0.9124	1.4108	5.4941	0.005052	0.00397	X	0.366	3.805	−3.661	0.00235	X	X	X
CH_m cyclic−OH $m \in (0,1)$	9.5209	−0.0690	0.3233	0.006917	−0.02297	X	−16.333	−5.487	4.626	−0.00250	X	X	X
$CH_n(OH)CH_m(NH_p)$* $m \in (0,1), n, p \in (0,2)$	2.7826	1.0682	5.4864	0.001408	0.00433	X	−2.992	−1.600	X	0.00046	X	X	X
$CH_m(NH_2)CH_n(NH_2)$* $m, n \in (0,2)$	2.5114	0.4247	2.0699	0.002148	0.00580	X	2.855	1.858	X	X	X	X	X
CH_m cyclic—NH_p—CH_n cyclic* $m, n, p \in (0,1)$	1.0729	0.2499	2.1345	−0.005950	−0.01380	−0.13106	0.351	8.846	2.311	−0.00179	X	X	X
CH_n−O−CH_m=CH_p* $m \in (0,1), n, p \in (0,2)$	0.2476	0.1134	1.0159	−0.000880	0.00297	X	−8.644	−13.167	X	−0.00206	X	X	X
AC—O—CH_m* $m \in (0,3)$	0.1175	−0.2596	−5.3307	−0.002250	−0.00045	X	1.532	−0.654	X	0.01203	X	X	X
CH_m cyclic—S—CH_n cyclic $m, n \in (0,1)$	−0.2914	0.4408	4.4847	X	X	−0.01509	−0.329	−2.091	0.972	−0.00023	−2.7407	11.1033	−11.0878
CH_n=CH_m—F $m \in (0,1), n \in (0,2)$	−0.0514	−0.1168	−0.4996	0.000319	−0.00596	X	X	X	X	X	X	X	X
CH_n=CH_m—Br* $m \in (0,1), n \in (0,2)$	−1.6425	−0.3201	−1.9334	0	0.00510	X	11.989	12,373	X	0	−1.6978	1.0477	0.2002
CH_n=CH_m—I* $m \in (0,1), n \in (0,2)$	X	−0.4453	X	X	X	X	X	X	X	X	X	X	X
ACBr*	2.5832	−0.6776	−2.2974	0.009027	−0.00832	−0.03078	12.285	14.161	−7.488	0.00178	−2.2923	3.1142	−1.4995
ACI*	−1.5511	−0.3678	2.8907	0.008247	−0.00341	0.00001	11.207	12.530	−4.864	0.00171	−0.3162	2.3711	−1.4825
$CH_m(NH_2)$—COOH* $m \in (0,2)$	X	X	X	X	X	X	11.740	X	X	X	X	X	−0.0584

* Group could not be tested with substances available in App. A.

TABLE C-4 Sample Assignments for Second−Order Groups (Constantinou and Gani, 1994)

Group j	Example Molecule (# of Groups)
$(CH_3)_2CH$	2-Methylpentane (1)
$(CH_3)_3C$	2,2,4,4-Tetramethylpentane (2)
$CH(CH_3)CH(CH_3)$	2,3,4,4-Tetramethylpentane (2)
$CH(CH_3)C(CH_3)_2$	2,2,3,4,4-Pentamethylpentane (2)
$C(CH_3)_2C(CH_3)_2$	2,2,3,3,4,4-Hexamethylpentane (2)
3 membered ring	cyclopropane (1)
4 membered ring	cyclobutane (1)
5 membered ring	cyclopentane (1)
6 membered ring	cyclohexane (1)
7 membered ring	cycloheptane (1)
$CH_n{=}CH_m{-}CH_p{=}CH_k$, $m, p \in (0,1)$, $k, n \in (0,2)$	1,3 butadiene (1)
$CH_3{-}CH_m{=}CH_n$, $m \in (0,1)$, $n \in (0,2)$	2-Methyl−2−Butene (3)
$CH_2{-}CH_m{=}CH_n$, $m \in (0,1)$, $n \in (0,2)$	1,4 Pentadiene (2)
$CH{-}CH_m{=}CH_n$ or $C{-}CH_m{=}CH_n$, $m \in (0,1)$, $n \in (0,2)$*	4-Methyl-2-Pentene (2)
Alicyclic side−chain $C\mathit{cyclic}C_m$ $m > 1$	Propylcycloheptane (1)
CH_3CH_3	Ethane (only)
CHCHO or CCHO*	2-Methylbutylaldehyde (1)
CH_3COCH_2	2-Pentanone (1)
CH_3COCH or CH_3COC	3-Methyl-2-Pentanone (1)
C*cyclic*=O	Cyclohexanone (1)
ACCHO*	Benzaldehyde (1)
CHCOOH or CCOOH	2-MethylButanoic acid (1)
ACCOOH*	Benzoic acid (1)
CH_3COOCH or CH_3COOC	2-Methylethyl Ethanoate (1)
$COCH_2COO$ or COCHCOO or COCCOO*	Ethylacetoethanoate (1)
CO—O—CO	Acetic Anhydride (1)
ACCOO*	Ethyl benzoate (1)
CHOH	2-Butanol (1)
COH	2-Methyl-2-Butanol (1)
$CH_m(OH)CH_n(OH)$, $m, n \in (0,2)$*	1,2,3-Propantriol (1)
$CH_m\mathit{cyclic}{-}OH$, $m \in (0,1)$	Cyclopentanol (1)
$CH_n(OH)CH_m(NH_p)$, $m \in (0,1)$, $n, p \in (0,2)$*	1-Amino-2-Butanol (1)
$CH_m(NH_2)CH_n(NH_2)$, $m, n \in (0,2)$*	1,2-Diaminopropane (1)
CH_mcyclic—NH_p—$CH_n\mathit{cyclic}$, $m, n, p \in (0,1)$*	Pyrrolidine (1)
$CH_n{-}O{-}CH_m{=}CH_p$, $m \in (0,1)$, $n, p \in (0,2)$*	Ethylvinylether (1)
$AC{-}O{-}CH_m$, $m \in (0,3)$*	Ethylphenylether (1)
$CH_m\mathit{cyclic}{-}S{-}CH_n\mathit{cyclic}$, $m, n \in (0,1)$	Tetrahydrothiophene (1)
$CH_n{=}CH_m{-}F$, $m \in (0,1)$, $n \in (0,2)$	1-Fluoro-1-propene (1)
$CH_n{=}CH_m{-}Br$, $m \in (0,1)$, $n \in (0,2)$*	1-Bromo-1-propene (1)
$CH_n{=}CH_m{-}I$, $m \in (0,1)$, $n \in (0,2)$*	1-Iodo−1−propene (1)
ACBr*	Bromobenzene
ACI*	Iodobenzene
$CH_m(NH_2){-}COOH$, $m \in (0,2)$*	2-Aminohexanoic acid

*Group could not be tested with substances available in App. A.

for most properties. ΔH_f° and ΔG_f° are at 298.15 K and 1 atm. ΔH_{v298} and V_{liq} are at 298 K.

For Second-Order groups, the letters k, m, n and p refer to the number of hydrogen atoms that can be attached to a carbon in a group; their range is given in the parenthesis following $\in$. Bonds to other carbons or atoms will complete full coordination. For example, the six forms of the group $CH_3—CH_m{=}CH_n$, $m \in (0,1)$, $n \in (0,2)$ can be developed as:

m	n	$CH_3—CH_m{=}CH_n$	No. Other Groups Bonded
0	0	$CH_3—C(—){=}C<$	3
1	0	$CH_3—CH{=}C<$	2
0	1	$CH_3—C(—){=}CH—$	2
1	1	$CH_3—CH{=}CH—$	1
0	2	$CH_3—C(—){=}CH_2$	1
1	2	$CH_3—CH{=}CH_2$	0

The molecules $CH_3—CH{=}C(CH_3)_2$, $CH_3—C(CH_3){=}CH(CH_3)$, $CH_3—C(NH_2){=}CHCH_2CH_3$ are among those which have this group at the Second Order. The descriptions are written so that each index k, m, n and p can take on all values indicated. C*cyclic* is a carbon atom in a ring compound. The 3-, 4-, 5-, 6-, 7-membered ring contributions are not used when aromatic First-Order carbons have been used. Table C-4 shows at least one molecule that each Second-Order group appears in; determining Second-Order contributions can be challenging.

C-3 REFERENCES

Constantinou, L., and R. Gani: *AIChE J.,* **40:** 1697 (1994).

Constantinou, L., R. Gani, and J. P. O'Connell: *Fluid Phase Equil.,* **103:** 11 (1995).

Joback, K. G.: "A Unified Approach to Physical Property Estimation Using Multivariate Statistical Techniques," S.M. Thesis, Department of Chemical Engineering, Massachusetts Institute of Technology, Cambridge, MA, 1984.

Joback, K. G., and R. C. Reid: *Chem. Eng. Comm.,* **57:** 233 (1987).

Nielsen, T. L.: "Molecular Structure Based Property Prediction," 15-point Project Department of Chemical Engineering, Technical University of Denmark., Lyngby, DK-2800, 1998.

INDEX